CRITICAL REVIEWS™
QUARTERLY JOURNALS

CRITICAL REVIEWS in ANALYTICAL CHEMISTRY
Edited by Louis Meites, Ph.D., Chairman, Department of Chemistry, Clarkson College of Technology. Associate Editor: Gunter Zweig, Ph.D., Director, Life Sciences Division, Syracuse University Research Corp. Assistant Editor: Irving Sunshine, Ph.D., Chief Toxicologist, Cuyahoga County Coroner's Office, Ohio.

CRITICAL REVIEWS in CLINICAL LABORATORY SCIENCES
Edited by John W. King, M.D., Ph.D., Director of Clinical Laboratories, Cleveland Clinic Foundation, and Willard R. Faulkner, Ph.D., Vanderbilt University Medical Center.

CRITICAL REVIEWS in ENVIRONMENTAL CONTROL
Edited by Richard G. Bond, M.P.S., Director of Environmental Health, University of Minnesota, and Conrad P. Straub, Director, Environmental Health Research and Training Center, University of Minnesota.

CRITICAL REVIEWS in FOOD TECHNOLOGY
Edited by Thomas E. Furia, Technical Development Manager, Industrial Chemicals Division, Geigy Chemical Corporation.

CRITICAL REVIEWS in RADIOLOGICAL SCIENCES
Edited by Yen Wang, M.D., D.Sc., Chairman, Department of Radiology, Homestead Hospital, Pittsburgh.

CRITICAL REVIEWS in SOLID STATE SCIENCES
Edited by Donald E. Schuele, Ph.D., and Richard Hoffman, Ph.D., both of the Department of Physics, Case Western Reserve University.

THE CHEMICAL RUBBER CO. (CRC) HANDBOOK SERIES

Handbook of Chemistry and Physics, 50th edition
Standard Mathematical Tables, 17th edition
Handbook of Tables for Mathematics, 3rd edition
Handbook of Tables for Organic Compound Identification, 3rd edition
Handbook of Tables for Probability and Statistics, 2nd edition
Handbook of Clinical Laboratory Data, 2nd edition
Manual for Clinical Laboratory Procedures, 2nd edition
Handbook of Laboratory Safety, 1st edition
Handbook of Food Additives, 1st edition
Handbook of Biochemistry: Selected Data for Molecular Biology, 1st edition
Handbook of Analytical Toxicology, 1st edition
*Handbook of Tables for Applied Engineering Science, 1st edition
*Handbook of Lasers and Laser Applications, 1st edition
*Handbook of Chromatography, 1st edition
*Handbook of Properties of Engineering Materials, 1st edition
*Handbook of Environmental Pollution Control, 1st edition
*CRC-Fenaroli Handbook of Flavors, 1st edition
*Handbook of Laboratory Animal Care, 1st edition

*Currently in preparation

EDITORS FOR THE CHEMICAL RUBBER CO.

Editor-in-Chief
ROBERT C. WEAST, Ph.D.
Vice-President, Research, Consolidated Natural Gas Service Company, Inc.
Formerly Professor of Chemistry at Case Institute of Technology

Coordinating Editor
GEORGE L. TUVE, Sc.D.
Formerly Professor of Engineering at Case Institute of Technology

Editor-in-Chief of Mathematics	**Editor-in-Chief of Biosciences**
SAMUEL M. SELBY, Ph.D.	IRVING SUNSHINE, Ph.D.
Distinguished Professor Emeritus of Mathematics	*Chief Toxicologist Cuyahoga County Coroner's Office Cleveland, Ohio*
Formerly Chairman of Mathematics Department University of Akron	*Associate Professor of Toxicology Case Western Reserve University*
Presently Chairman of Mathematics Department Hiram College	

Editors

Chromatography
    Gunter Zweig, Ph.D.   *Syracuse University Research Corporation*

Engineering Sciences
    Ray E. Bolz, D.Eng.   *Case Western Reserve University*
    Richard G. Bond, M.P.H.   *University of Minnesota*
    Donald F. Gibbons, Ph.D.   *Case Western Reserve University*
    W. Bruce Johnson, Ph.D.   *Case Western Reserve University*
    Conrad P. Straub, Ph.D., D.Eng.   *University of Minnesota*

Laboratory Safety
    Norman V. Steere   *The 3M Company*

Life Sciences
    Willard R. Faulkner, Ph.D.   *Vanderbilt University Medical Center*
    Thomas E. Furia   *Geigy Chemical Corporation*
    John W. King, M.D., Ph.D.   *The Cleveland Clinic Foundation*
    Neal S. Nelson, Ph.D.   *Radionuclide Toxicology Laboratory*
    Herbert A. Sober, Ph.D.   *National Institutes of Health*

Mathematics
    Brian Girling, M.Sc., F.I.M.A.   *The City University, London, England*

Nuclear Sciences
    Yen Wang, M.D., D.Sc.(Med.)   *Homestead Hospital*

Organic Chemistry
    Saul Patai, Ph.D.   *Hebrew University of Jerusalem*
    Zvi Rappoport, Ph.D.   *Hebrew University of Jerusalem*

Statistics
    William H. Beyer, Ph.D.   *University of Akron*

Handbook of Radioactive Nuclides

EDITOR

Yen Wang, M.D., D.Sc.(Med.)

Director, Department of Radiology
Homestead Hospital
Homestead, Pennsylvania

Published by
THE CHEMICAL RUBBER CO.
18901 Cranwood Parkway, Cleveland, Ohio 44128

This book presents data obtained from authentic and highly regarded sources. Reprinted material is quoted with permission, and sources are indicated. A wide variety of references are listed. Every reasonable effort has been made to give reliable data and information, but the editor and the publisher cannot assume responsibility for the validity of the material or for the consequences of its application.

© 1969 by The Chemical Rubber Co.
All Rights Reserved
Library of Congress Card No. 75-81089

PREFACE

In the preparation of this reference volume, the aim has been to provide a convenient single source for basic information on radioactive nuclides, instrumentation, dosimetry, and applications, as well as on general radiation protection. Most of the information is in tabular or graphical form, but narrative presentation and explanation are also included. For convenient reference, the Handbook has been divided into several sections, many of which cover special applications of radionuclides.

A considerable proportion of the material herein has been compiled especially for this book from current scientific journals and from various authoritative collections of basic radionuclide data and applications. The editors are grateful for permissions to reprint these materials and have made every effort to designate the original sources and to give references to further information available on each subject.

Although the editors have tried to include in condensed form the information most frequently used and needed in the fields of radionuclide application, there is always a question of what to include in a one-volume desk-reference book. Suggestions from readers for additions or modifications in future editions are invited.

We would like to take this opportunity to express our deep gratitude to our contributors and to the members of the editorial board as well as to others who have assisted us in the compilation of this volume. The information on radionuclides is so voluminous that the completion of the book would not have been possible without their effort and assistance.

Yen Wang

June 1969

ADVISORY BOARD

EDITOR AND CHAIRMAN

Yen Wang, M.D., D.Sc.(Med.)
 Director, Department of Radiology
 Homestead Hospital
 Homestead, Pennsylvania

MEMBERS

P. S. Baker, Ph.D.
 Director, Isotopes Information Center
 Oak Ridge National Laboratory
 Oak Ridge, Tennessee

William H. Beierwaltes, M.D.
 Professor of Medicine
 University of Michigan Medical Center
 Ann Arbor, Michigan

Nathaniel I. Berlin, M.D., Ph.D.
 Clinical Director
 National Cancer Institute
 Bethesda, Maryland

John R. Cameron, Ph.D.
 Professor of Radiology and Physics
 University of Wisconsin
 Madison, Wisconsin

D. Harold Copp, M.D.
 Professor and Head of the Department
  of Physiology
 University of British Columbia
 Vancouver, B.C., Canada

R. F. Glascock
 National Institute for Research in
  Dairying
 University of Reading
 Shinfield, Reading, England

Paul F. Hahn, Ph.D.
 Division of Radiological Health
 Bureau of State Service
 Department of Health, Education and
  Welfare
 Washington, D.C.

Hirotake Kakehi, M.D.
 Department of Radiology
 Chiba University of Medicine
 Chiba, Japan

John H. Lawrence, M.D.
 Director, Donner Laboratory and
  Donner Pavilion
 University of California
 Berkeley, California

Lloyd J. Roth, M.D., Ph.D.
 Professor and Chairman of the
  Department of Pharmacology
 University of Chicago
 Chicago, Illinois

Edward M. Smith, D.Sc.
 Technical Director, Division of Nuclear
  Medicine
 University of Miami
 Miami, Florida

Ernest J. Sternglass, Ph.D.
 Westinghouse Research Laboratories
 Pittsburgh, Pennsylvania

Henry N. Wagner, Jr., M.D.
 Associate Professor of Medicine and
  Radiological Sciences
 Johns Hopkins Medical Institutions
 Baltimore, Maryland

Niel Wald, M.D.
 Professor of Radiation Health
 University of Pittsburgh
 Pittsburgh, Pennsylvania

CONTRIBUTORS

H. L. Atkins
Medical Research Center
Brookhaven National Laboratory
Upton, New York

R. J. Bailey
The Radiochemical Centre
Amersham, Buckinghamshire, England

P. S. Baker
Isotopes Information Center
Oak Ridge National Laboratory
Oak Ridge, Tennessee

Riad Barmada
Department of Orthopedic Surgery
University of Illinois Medical Center
Chicago, Illinois

Francis J. Bradley
New York State Department of Labor
New York, New York

Allen Brodsky
Graduate School of Public Health
University of Pittsburgh
Pittsburgh, Pennsylvania

D. F. Bunch
Atomics International
Canoga Park, California

Elizabeth T. Bush
Nuclear-Chicago Corporation
Des Plaines, Illinois

J. R. Cameron
Department of Radiology
University of Wisconsin Medical Center
Madison, Wisconsin

E. A. Evans
The Radiochemical Centre
Amersham, Buckinghamshire, England

Robert G. Gallaghar
Applied Health Physics, Inc.
Bethel Park, Pennsylvania

William D. Gibbs
Department of Radiology
University of Wisconsin Medical Center
Madison, Wisconsin

E. C. Gregg
Department of Radiology
Case Western Reserve University
Cleveland, Ohio

D. L. Hansen
Nuclear-Chicago Corporation
Des Plaines, Illinois

Richard A. Holmes
Department of Radiological Science
Johns Hopkins Medical Institutions
Baltimore, Maryland

Charles A. Kelsey
Department of Radiology
University of Wisconsin Medical Center
Madison, Wisconsin

Vincent Lopez-Majano
Department of Radiological Science
Johns Hopkins Medical Institutions
Baltimore, Maryland

Clarence C. Lushbaugh
Department of Applied Radiobiology
Oak Ridge Institute of Nuclear Studies
Oak Ridge, Tennessee

August Miale, Jr.
Division of Nuclear Medicine
Georgetown University Hospital
Washington, D.C.

Margaret Minsky
Radiological Protection Service
Belmont, Surrey, England

Malcolm R. Powell
Department of Radiology
University of California Medical Center
San Francisco, California

Robert Radtke
Department of Radiology
University of Wisconsin Medical Center
Madison, Wisconsin

Robert D. Ray
Department of Orthopedic Surgery
Presbyterian-St. Luke's Hospital
Chicago, Illinois

L. Rosenthall
Division of Nuclear Medicine
Montreal General Hospital
Department of Diagnostic Radiology
McGill University
Montreal, P.Q., Canada

Edward M. Smith
Division of Nuclear Medicine
University of Miami School of Medicine
Miami, Florida

James A. Sorenson
Department of Radiology
University of Wisconsin Medical Center
Madison, Wisconsin

Michael M. Ter-Pergossian
The Edward Mallinckrodt Institute of Radiology
Washington University School of Medicine
St. Louis, Missouri

Donald E. Tow
Department of Radiological Science
Johns Hopkins Medical Institutions
Baltimore, Maryland

J. C. Turner
Organic Department
The Radiochemical Centre
Amersham, Buckinghamshire, England

J. Vennart
Radiological Protection Service
Belmont, Surrey, England

Henry N. Wagner, Jr.
Department of Radiological Science
Johns Hopkins Medical Institutions
Baltimore, Maryland

Niel Wald
Department of Radiation Health
University of Pittsburgh
Pittsburgh, Pennsylvania

Yen Wang
Department of Radiology
Homestead Hospital
Homestead, Pennsylvania

TABLE OF CONTENTS

PART I. NUCLEAR DATA

Isotopes 3
Radioisotope Production and Processing 7
Physical and Nuclear Data 16

PART II. ESSENTIAL PHYSICS DATA

Radiation Interaction with Matter 67
Statistical Aspects of Nuclear Counting 77

PART III. NUCLEAR INSTRUMENTATION

Radiation Detectors and Equipment 89
Radioisotope Counting and Calibration 106
Improvement of Liquid Scintillation Counting Efficiencies by Optimization of Scintillator
    Composition 114
Modulation Transfer Function for Radioisotope Imaging Systems 123
Personnel Monitoring 130
Whole-Body Counter Systems 134
Information Storage and Retrieval in Radioisotope Imaging Systems 141
Radioisotope Cameras 148

PART IV. RADIATION DOSIMETRY

Radiation Absorbed-Dose Calculations for Biologically Distributed Radionuclides 167
Radiation Doses from Administered Radionuclides 201
Dose from Ingestion or Inhalation of Soluble Radionuclides 220

PART V. BIOCHEMISTRY

Standards of Activity 229
Radioactive Isotope Dilution Analysis 237
Improved Solubilization Procedures for Liquid Scintillation Counting of Biological
    Materials 246
Sample Preparation for Liquid Scintillation Counting 256
The Stability of Labeled Organic Compounds 274
Storage and Stability of Compounds Labeled with Radioisotopes. I 285
Storage and Stability of Compounds Labeled with Radioisotopes. II 315
Synthesis of Labeled Compounds 339

PART VI. RADIONUCLIDES FOR MEDICAL APPLICATION

Cerebronervous Applications. I. Measurement of Cerebral Blood Flow 383
Cerebronervous Applications. II. Brain Scanning 388
Radionuclides and the Endocrine System 395
Radionuclide Diagnosis of Cardiovascular Disease 423

Radionuclides in Respiratory-System Studies 434
Nuclear-Medicine Techniques in Gastroenterologic Diagnosis 443
The Use of Radioisotopes in the Osseous and Cartilaginous System 466
Radioisotopes and the Hematopoietic System 477
Radionuclide Techniques Applied to the Genitourinary System 492

PART VII. RADIONUCLIDES FOR INDUSTRIAL APPLICATIONS

Characteristic Effects of Radiation 503
Availability of Isotopes 505
Radioisotope Utilization 507
Applications in the Metals Industries 509
Applications in the Electrical Industry 515
Applications in Transportation-Equipment Industries 517
Applications in Chemical Processing 519
Applications in Consumer-Products Industries 531
Applications in Crude-Petroleum and Natural-Gas Industries 540
Applications in Mining 544
Applications in the Utilities 547
Applications in Agriculture 550
Applications in Aerospace and Other Environmental Uses 557
Glossary 569

PART VIII. RADIATION PROTECTION AND REGULATIONS

Basic Units of Radiation Measurement 573
Radiation Protection Guides and Regulatory Limits of Exposure 609
Data and Methods for Estimating Radiation Exposures from Internal and External
   Radiation Sources 647
Determination of Facilities, Equipment, and Procedures Required for Various Types of
   Operations 664
Personnel Dosimetry 711
Transportation of Radioactive Materials 719
Radioactive-Waste Disposal 781
Administration of a Radiation-Protection Program 795
Emergency Planning and Procedures 799
Appendix 1. Emergency Notification Instructions 816
Appendix 2. Radiation-Incident Evaluation Record 818
Appendix 3. Radiation-Incident Evaluation Record 819
Appendix 4. U.S. Atomic Energy Commission Regional Office Areas of Responsibility
   for Radiological Assistance 820
Appendix 5. Evaluation of Personnel-Monitoring Results 822
Appendix 6. A Suggested Check List on Radiation-Accident Preparedness ... 823
Appendix 7. Evaluation of Incidents or Occurrences Involving Radiation ... 826
Appendix 8. Radiation-Incident Reporting Procedures 828
Appendix 9. U.S. Atomic Energy Commission Regional Compliance Offices ... 830
Appendix 10. Records, Reports, and Notifications pertinent to Radiation Incidents ... 831

PART IX. RADIATION INJURY AND ITS MANAGEMENT

Standard Man 837
Radiation Injury 845
Medical Management of Radiation Emergencies 868

PART X. REFERENCE DATA

Greek Alphabet	887
Signs and Symbols of Particular Interest in Radiation and Radioactivity	888
Signs and Symbols in Mathematics	890
Abbreviations	891
Commonly Used Units	894
Fundamental Constants	895
The Standard Man (Conventional)	897
Densities of Common Metals	899
Relation Between Thicknesses of Ordinary Concrete and of Lead for Radium and ^{60}Co Gamma Rays	900
Nomogram of Absorption of Beta Particles	901
Decay Curve for Radioactive Materials	902
Radioactive Decay	903
Referential Conversion Factors	904
Convenient Conversion Factors	905
Equations	909
Four-Place Mantissas for Common Logarithms of Decimal Fractions	919
Four-Place Mantissas for Common Logarithms	921
Natural Trigonometric Functions, Sine, Cosine, Tangent, Cotangent, for Angles, in Degrees and in Decimals	923
Exponential Functions	928
Squares, Square Roots, Cubes, and Cube Roots	932

INDEX 943

PART I
NUCLEAR DATA

P. S. Baker
Isotopes Information Center
Oak Ridge National Laboratory*

Prepared at the request of Division of Isotopes Development of USAEC for inclusion in the CRC Handbook of Radioactive Nuclides.

* Operated by Union Carbide Corporation for the U.S. Atomic Energy Commission.

ISOTOPES

With the exception of normal hydrogen, all atoms—whether existing naturally or produced artificially—are made up of protons, electrons and neutrons. For each element the number of positively charged protons or negatively charged electrons is defined by the atomic number, Z, with the number of protons and electrons being equal, so that the atom is electrically neutral. The number of neutrons that can occur in the nuclei of atoms can vary greatly—from zero for the normal hydrogen mentioned above to more than 150 for some of the transuranic elements.

The mass of the neutron is very nearly that of the proton, but since the neutron is uncharged, it essentially contributes mass to a nucleus, without having much effect on the chemical properties of the atom. The sum of protons and neutrons gives the atomic weight, A. (The electrons are ignored, for the most part, since an electron weighs only a little more than 1/2000 of a proton or neutron.) However, a more important consideration is the fact that, for a given value of Z, there can be more than one value of A; that is, for a given atomic number there can be several atomic weights. Atoms of an element differing from each other *only* in the number of neutrons are known as *isotopes*.

In nature, there are more than twenty elements that have no isotopes; they are referred to as *anisotopic* or *mononuclidic* elements. These are shown in Table 1. The other elements all have isotopes, varying from two for elements such as antimony, chlorine, boron, bromine, and lithium to ten for tin.

Table 1. ANISOTOPIC (STABLE) ELEMENTS

Aluminum	Gold	Rhodium
Arsenic	Holmium	Scandium
Beryllium	Iodine	Sodium
Bismuth	Manganese	Terbium
Cesium	Niobium	Thallium
Cobalt	Phosphorus	Thulium
Fluorine	Praseodymium	Yttrium

STABLE VS. RADIOACTIVE ISOTOPES

For some reason, the ratio of neutrons to protons is of significance to the stability of a nucleus. If there are either too many or too few neutrons for the number of protons, the nucleus finds itself in a state of distress, and, in an attempt to reorganize itself into a more stable configuration, may undergo various types of rearrangements that involve the release of radiations of one kind or another. Such nuclei are said to be *radioactive*, and the atoms are referred to as *radionuclides*, *radioactive isotopes*, or *radioisotopes*. Eventually, each distressed nucleus will undergo a rearrangement to become more stable. Approximately 1500 such nuclides have been identified, with some elements having as many as 30 isotopes and isomers; they are not all "available", in the usual sense of the word.

HALF-LIFE

A peculiar thing about an atom with an unstable nucleus is the fact that, although it is known that the stabilizing rearrangement will occur sometime, it is not known just when it will occur. But if a statistically significant number of these unstable isotopes can be gathered together, the rearrangements will occur at a definite and predictable rate that is characteristic of the isotope. A convenient method for describing this characteristic "decay" is in terms of *half-life*, which is the time required for one half of any starting amount of a radionuclide to undergo rearrangement. Figure 1 shows a typical decay curve, based on the fact that at the end of each half-life interval one half of the starting material will be left (at the end of two half-lives, one fourth; at the end of three half-lives, one eighth, etc.).

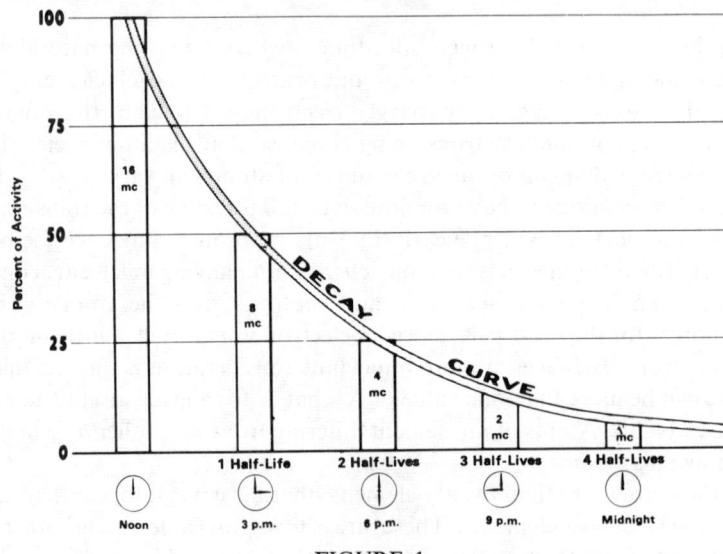

FIGURE 1.
Typical Radioisotope Decay Curve.

Table 2 is included to allow convenient calculation of decay in relation to time; this, of course, is based on the radioactive-decay curve shown in Figure 1.

Table 2. DECAY OF A RADIOELEMENT

Half-Lives	F*	Half-Lives	F*	Half-Lives	F*	Half-Lives	F*
0.00	1.000	0.70	0.616	1.65	0.319	3.20	0.109
0.02	0.986	0.75	0.595	1.70	0.308	3.30	0.102
0.04	0.973	0.80	0.574	1.75	0.297	3.40	0.095
0.06	0.959	0.85	0.555	1.80	0.287	3.50	0.088
0.08	0.946	0.90	0.535	1.85	0.277	3.60	0.083
0.10	0.933	0.95	0.518	1.90	0.268	3.70	0.077
0.12	0.920	1.00	0.500	1.95	0.259	3.80	0.072
0.14	0.908	1.05	0.483	2.00	0.250	3.90	0.067
0.16	0.895	1.10	0.467	2.10	0.233	4.00	0.063
0.18	0.883	1.15	0.451	2.20	0.218	4.10	0.058
0.20	0.871	1.20	0.435	2.30	0.203	4.20	0.054
0.25	0.841	1.25	0.421	2.40	0.189	4.30	0.051
0.30	0.812	1.30	0.406	2.50	0.177	4.40	0.047
0.35	0.785	1.35	0.393	2.60	0.165	4.50	0.044
0.40	0.758	1.40	0.379	2.70	0.154	4.60	0.041
0.45	0.732	1.45	0.367	2.80	0.144	4.70	0.039
0.50	0.707	1.50	0.354	2.90	0.134	4.80	0.036
0.55	0.683	1.55	0.342	3.00	0.125	4.90	0.034
0.60	0.660	1.60	0.330	3.10	0.117	5.00	0.031
0.65	0.638						

* F=fraction remaining.

The values of the half-lives vary from microseconds to megayears. As a matter of interest, there are some who feel that every isotope is radioactive; the half-lives of some are simply so long that, to all purposes, they are called stable. For most considerations, any isotope with a half-life greater than 10^{10} years is essentially stable. On the other hand, if an isotope has a half-life of less than a few seconds, it can be considered nonexistent, as far as any practical uses are concerned. Table 3 lists a few isotopes and their half-lives. Between 175 and 200 radioisotopes can be considered as "commercially available", although perhaps 500 have half-lives long enough to make the isotopes suitable for use.

Table 3. TYPICAL ISOTOPES AND HALF-LIVES

Isotope	Half-Life	Isotope	Half-Life
^{37}K	1.2 secs	^{45}Ca	165 days
^{37}S	5.1 mins	^{110m}Ag	255 days
^{18}F	1.8 hrs	^{55}Fe	2.6 yrs
^{24}Na	15 hrs	^{90}Sr	27.7 yrs
^{131}I	8.1 days	^{63}Ni	92 yrs
^{125}I	60 days	^{14}C	5730 yrs
^{35}S	88 days	^{36}Cl	3×10^5 yrs

RADIATIONS

As mentioned earlier, when radioisotopes undergo decay, the rearrangements are accompanied by radiations, which may be electrons (positive or negative), alpha particles, or electromagnetic radiations such as gamma rays (so called when they come from the nucleus) or X rays (so called when the nuclear rearrangement results in *K-capture*—the "capture" of an electron from an "electron shell"—and a subsequent shifting of electrons remaining in the electron shells, with coincident release of electromagnetic radiation).

Table 4. TYPICAL NUCLIDES, WITH ENERGIES AND HALF-LIVES

Isotope	Type of Radiation	Energy, Mev	Half-Life
^3H	β^-	0.019	12.3 yrs
^{32}P	β^-	1.71	14.3 days
^{51}Cr	γ	0.32	27.8 days
^{64}Cu	β^-	0.57	12.9 hrs
	β^+	0.65	
	γ	1.34	
^{99m}Tc	γ	0.140	6.0 hrs
^{109}Cd	γ	0.09	453 days
^{125}I	X ray	0.027	60 days
	γ	0.035	
	e^-	0.03	
^{198}Au	β^-	0.97	2.7 days
	γ	0.41	
^{204}Tl	β^-	0.77	3.8 yrs
	Hg X rays		

The type of radiation can often be predicted from a consideration of the status of the unstable nucleus itself. If there are too few neutrons for the number of protons, one would expect essentially a "conversion" of a proton to a neutron. This is accomplished by emission of a *positron*, or by electron capture, in which an electron from an orbital shell "falls" into the nucleus. The above-mentioned subsequent rearrangement of the electrons gives rise to a characteristic X ray. On the other hand, if there are too many neutrons for the number of protons, one would expect a "conversion" of a neutron to a proton plus an electron. Such radioisotopes are the *beta emitters*. Alpha particles usually result from decay of heavier elements. As a rule, gamma rays also result from these transformations.

The energies of the particles or rays vary from a few electron volts to several million volts. Table 4 lists a few typical nuclides, with types and energies of radiation as well as half-lives. In general, the shorter the half-life, the more energetic the radiation; but this is only a rule of thumb.

RADIOISOTOPE PRODUCTION AND PROCESSING

Knowing that radioisotopes result when the neutron-to-proton ratio becomes too high or too low, one is not hard pressed to come up with the two principal ways of producing these unstable configurations: by adding or removing neutrons, or by adding or removing protons. Actually, it is usually easier to add than to remove. There are also combinations of these methods, which are often used. A third method is related to the process of *fission*, in which large nuclei, such as ^{235}U, are bombarded with neutrons and split into relatively large "chunks" of matter, which are usually radioactive.

For adding neutrons to produce *neutron-excess* isotopes, a reactor is perhaps the best device, although neutron generators (either isotopic, such as antimony-beryllium and plutonium-beryllium, or machine-activated) are also very useful. For adding protons to form *neutron-deficient* isotopes, a charged-particle accelerator such as a cyclotron is an excellent device.

REACTOR PRODUCTION

The important steps in the reactor production of isotopes are as follows:
1. A suitable target is prepared, and irradiated with neutrons.
2. The irradiated target is processed, to put it into a form amenable to shipment and ultimate use. This may involve simple dissolution, or it may involve more complicated separations, including ion exchange, precipitation and distillation, to remove undesirable impurities or to concentrate the product nuclides.
3. The processed material is placed in inventory.
4. The isotopes are dispensed, and packaged for shipment.

Figure 2 shows a typical irradiation ampule and can for insertion in one of the Oak Ridge reactors.

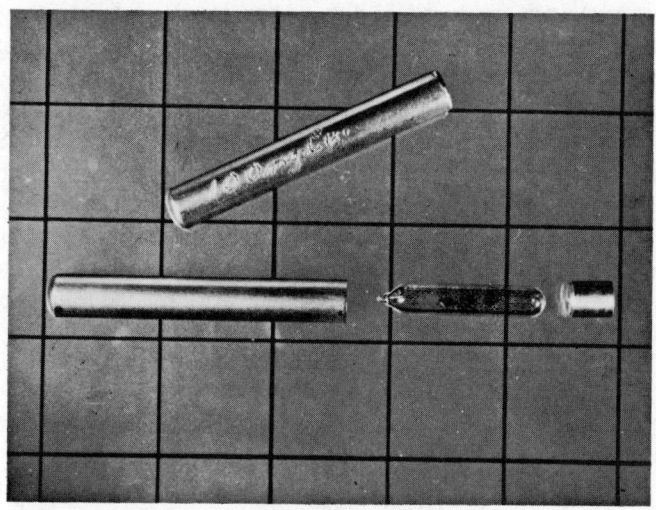

FIGURE 2.
Quartz and Aluminum Capsules for Reactor Irradiations.

Figure 3 shows a "stringer", which holds a number of ampules for irradiation.

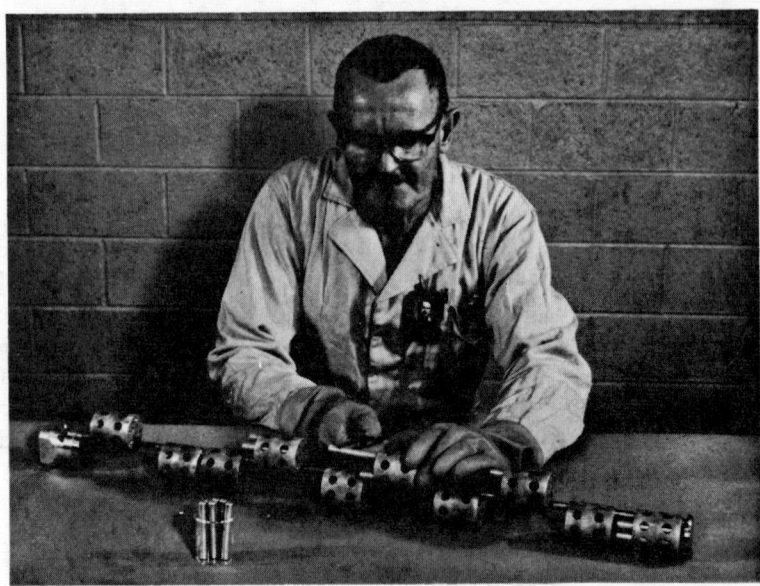

FIGURE 3.
Inserting Target Capsules in Stringer.

Figure 4 is a cutaway drawing of the core of the Oak Ridge Research Reactor (ORR), showing the location of stringers and hydraulic tubes for insertion and removal of targets. The irradiated target capsule is transported to the processing area, where it is processed in suitable hot-cell facilities.

The chemical processing of the irradiated target may be as simple as dissolving a soluble salt in water. At other times it may require evaporation, precipitation, extraction, distillation, and ion exchange. The purified product is generally stored in inventory, and eventually shipped to a customer. When a shipment is to be made, an appropriate amount of solution is transferred to a shipping bottle, the bottle is packaged and checked for surface contamination, and the external radiation level is measured before shipment.

The early isotopes were produced in low-power natural-uranium graphite-moderated reactors, with fluxes ranging from approximately 10^{10} to 10^{13} neutrons/cm^2/second. The newer reactors, for the most part, use various enrichments of ^{235}U and have fluxes up to and above 10^{15} neutrons/cm^2/second. Until recently, the ORR has been the major workhorse for the production of radioisotopes in this country. Amersham in England and Saclay in France have been major producers overseas. Gradually the domestic production of the more profitable isotopes is being taken over by private enterprise. However, the ORR will continue to be used for the production of those isotopes not available commercially and for the production of some of the so-called "commercial" isotopes that have special technical qualities. The reactors will also be used for research and development work related to improving production techniques and providing new (not currently routine) isotopes.

Neutron Reactions

Since reactor activations are neutron-induced, it is important to understand the various possible reactions in order to evaluate the potential production methods.

A. Types of Reactions. Table 5 lists the types of neutron reactions that are commonly used for activation of stable nuclides, along with a number of specific reactions in each category. The most important of these by far is the (n,γ) reaction, involving so-called "thermal"

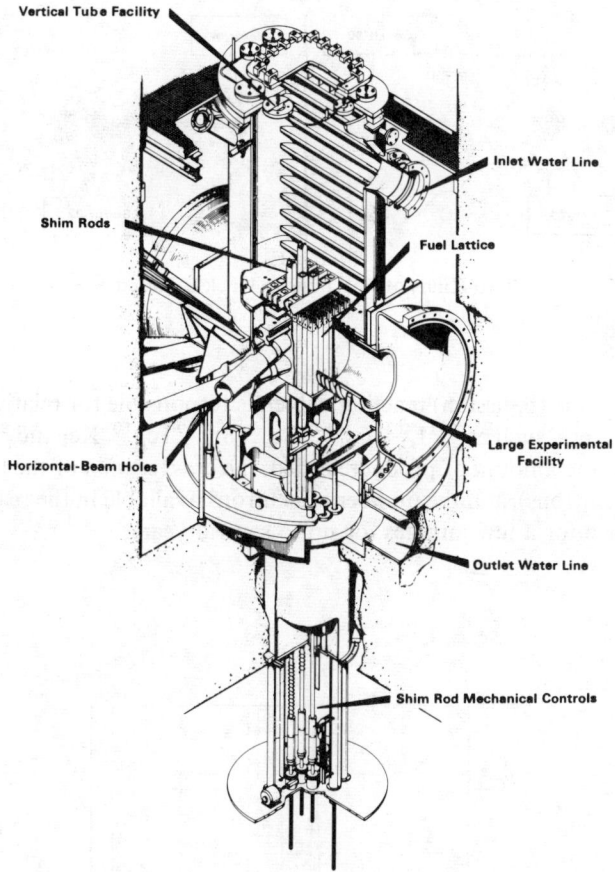

FIGURE 4.
Location of Fuel Elements, Experiments, and Radioisotope Targets in the Oak Ridge Research Reactor.

neutrons, which are captured with the coincident formation of a gamma ray. Also shown are a number of common reactions involving fast neutrons that are sufficiently energetic to remove other particles as they penetrate the nuclei.

Table 5. NEUTRON-INDUCED REACTIONS

Type	Example
(n,γ)	$^{46}Ca(n,\gamma)^{47}Ca$
	$^{197}Au(n,\gamma)^{198}Au$
$(n,fission)$	$^{235}U(n,fission)^{131}I$
	$^{235}U(n,fission)^{90}Sr$
(n,p)	$^{32}S(n,p)^{32}P$
	$^{14}N(n,p)^{14}C$
(n,α)	$^{40}Ca(n,\alpha)^{37}Ar$
	$^{6}Li(n,\alpha)^{3}H$
$(n,\gamma) \xrightarrow{decay}$	$^{124}Xe(n,\gamma)^{125}Xe \xrightarrow{EC} {}^{125}I$
	$^{198}Pt(n,\gamma)^{199}Pt \xrightarrow{\beta} {}^{199}Au$

In general, the more energetic the impinging neutron, the more extensive the "damage" to the nucleus, i.e., the more mass is knocked out. Figure 5 illustrates two of these reactions.

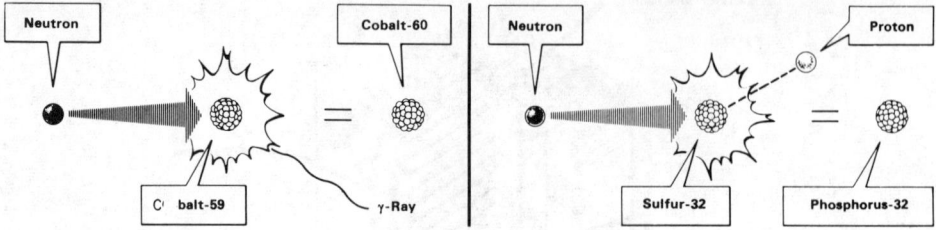

FIGURE 5.
Two Kinds of Reactions in Nuclear Reactors.

Finally, there is the (n, fission) reaction, which is responsible for relatively large quantities of a number of isotopes, such as ^{131}I, ^{137}Cs, ^{144}Ce, ^{90}Sr, ^{99}Tc, ^{133}Xe, and ^{91}Y. The irradiation time depends upon the amount of product needed, the *cross section* of the target atom (i.e., the ability to capture neutrons), and the number of neutrons available in the reactor (*flux*). Irradiations may take place for a few minutes, or up to several years.

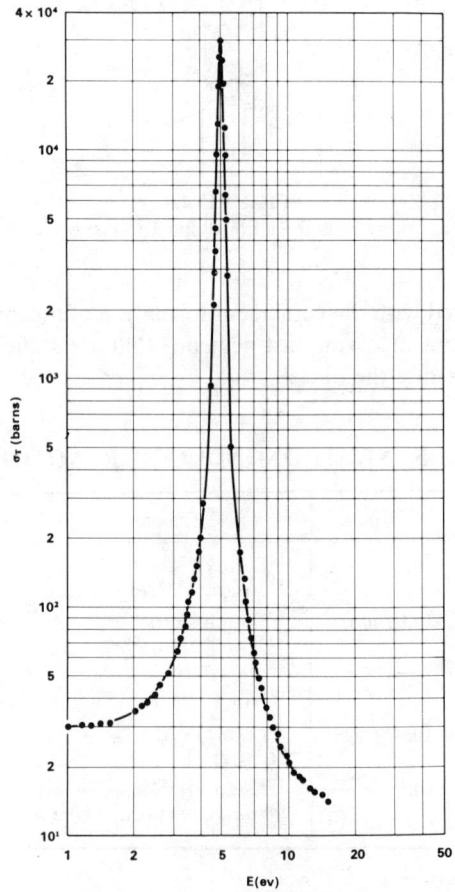

FIGURE 6.
Total Neutron Cross Section for ^{197}Au.

B. Energetics. Nuclei are peculiar entities, and the energetics of neutron capture are not always simple. The cross sections, i.e., the abilities to capture neutrons, vary from target to target as well as with the energy of the bombarding neutron. Figure 6 is a cross-section curve for the target nucleus ^{197}Au. This type of information becomes extremely important from the standpoint of reactor yields of radioactive products, since production is often very sensitive to neutron energy. Hence, if a resonance energy can be attained, the yield can be increased manyfold. In this particular case, gold has a resonance at 4 ev, so cadmium can be wrapped around the gold to absorb thermal neutrons. This allows the high production rate for ^{198}Au and prevents the secondary ^{198}Au(n,γ)^{199}Au reaction; the result is a high-purity ^{198}Au product.

As shown in Table 5, some of the neutron reactions result in transmutations. This in itself is of considerable significance, since any transmutation makes the production of carrier-free material possible. Hence it is sometimes desirable to search for a transmutation reaction, although a more economical high-yield (n,γ) reaction is available.

C. Secondary Capture. A factor that was not of serious concern before the availability of high fluxes is secondary neutron capture. As a product nuclide is formed, it, in turn, serves as a target for subsequent neutron capture; the higher the flux, the more serious the effect. This factor is being evaluated for many isotopes and has been handled satisfactorily for gold. It may often be necessary to optimize the various irradiation parameters so as to get suitable products. For example, in the production of 60-day ^{125}I, the formation of 14-day ^{126}I is serious and introduces into the product a considerable impurity that can be removed only by decay. This, of course, results in wasted ^{125}I, since it is decaying at the same time. For many of the rare earths, cross sections are high, and secondary capture seriously limits the production of the desired isotopes.

Product Yields and Characteristics

The energies and half-lives suitable for radioisotope applications are characteristics defined by the choice of isotope. However, it is possible to control a number of parameters that enhance the value of a particular product.

A. Advantages of Enriched Stable-Isotope Targets. Table 6 lists typical isotopic abundances of nuclides enriched by the electromagnetic process, which are used for radioisotope production.

Table 6. ENRICHED STABLE TARGETS FOR RADIOISOTOPE PRODUCTION

Reaction	Abundance, %	
	Normal	Enriched
^{44}Ca(n,γ)^{45}Ca	2.1	90
^{50}Cr(n,γ)^{51}Cr	4.3	95
^{54}Fe(n,γ)^{55}Fe	5.8	95
^{58}Fe(n,γ)^{59}Fe	0.3	80
^{74}Se(n,γ)^{75}Se	0.9	15
^{84}Sr(n,γ)^{85}Sr	0.6	64
^{124}Sn(n,γ)^{125}Sn $\xrightarrow{\beta}$ ^{125}Sb	6.0	95
^{132}Ba(n,γ)^{133}Ba	0.1	12
^{114}Cd(n,γ)^{115}Cd	28.9	98
^{184}W(n,γ)^{185}W	30.6	96
^{235}U(n,fission)^{131}I	0.7	95

There are a number of advantages to the use of these materials as targets.

1. In general, as one nuclide is enriched, the others are depleted. This permits production of a radioisotope of higher purity than is possible with normal targets. For example, Table 7 compares the normal and enriched abundances of ^{54}Fe and ^{58}Fe used in the production of ^{55}Fe ($T_{1/2} \sim 3$ years) and ^{59}Fe ($T_{1/2} = 44$ days) respectively. Iron-55 decays by electron capture, emitting a characteristic 5.9-kev manganese X ray. The product resulting from the activation of normal iron contains about 10 percent ^{59}Fe with its approximately 1-Mev gammas and several betas, and 90 percent ^{55}Fe radiations. Enriching the ^{54}Fe to 98 percent reduces the ^{58}Fe to almost nothing; so, a nearly pure ^{55}Fe product results. On the other hand, if a target enriched to 85 percent ^{58}Fe is used, the ^{55}Fe is less than 5 percent of the ^{59}Fe, a reduction by a factor of nearly 200 from the ratio resulting from a normal iron target.

Table 7. ENRICHED ^{54}Fe AND ^{58}Fe.

Isotope	Abundance, %		
	Normal	^{54}Fe	^{58}Fe
54	5.8	97.2	0.5
56	91.7	2.8	15.6
57	2.2	< 0.05	1.9
58	0.3	< 0.05	82.0

A similar illustration involves the production of ^{47}Ca by the ^{46}Ca(n,γ)^{47}Ca reaction, as shown in Table 8. In normal calcium the ^{46}Ca abundance is 0.0033 percent, and the ^{44}Ca

Table 8. ENRICHED ^{44}Ca AND ^{46}Ca.

Isotope	Abundance, %		
	Normal	^{44}Ca	^{46}Ca
40	97.0	1.36	48.7
42	0.6	0.06	0.54
43	0.1	0.04	0.13
44	2.1	98.55	2.86
46	0.003	< 0.01	46.50
48	0.2	< 0.01	1.31

abundance is 2.06 percent. When the normal material is irradiated to prepare ^{47}Ca, the ^{45}Ca ($T_{1/2}$ = 165 days, 0.254-Mev β) is present in very large quantities (> 50 percent of the activity) and makes the material essentially unusable as a source of ^{47}Ca. On the other hand, with targets enriched to approximately 40 percent ^{46}Ca, the ^{45}Ca content in the product is less than 1 percent of the activity. Furthermore, the specific activity of the product is increased many thousands of times.

2. Flux, irradiation, and yield can be manipulated in a number of ways. For a given flux, the irradiation time or sample size can be reduced, as compared to a normal target, for a given yield. For a given flux, irradiation time and total target weight, the yield increases as a function of the increased abundance of the target nuclide. The specific activity or activity concentration also increases as a function of enrichment for a fixed flux, irradiation time and target weight.

B. Short-Lived vs. Long-Lived Products. The production of short-lived radionuclides is rather difficult, since the rate of decay soon equals the rate of production. High fluxes and enriched targets are highly advantageous in activating such isotopes, since a higher yield is obtained before equilibrium is reached. Incidentally, reactors located at sites near the ultimate users can be used to great advantage as sources of short-lived isotopes. Long-lived products, of course, are produced at a rate commensurate with flux, time and target size. It is not always recognized, however, that the preparation of short-lived radioisotopes, as compared with long-lived isotopes, is favored by high fluxes, and particularly by flux "bursts" in pulse-type (TRIGA) reactors. This method is especially valuable in examples such as that of ^{47}Ca–^{45}Ca previously discussed, since the ^{47}Ca/^{45}Ca ratio can be improved still further, if necessary.

C. Secondary X-Ray and Bremsstrahlung Sources. More and more effort is being directed to the use of long-lived isotopes, often fission products, to provide characteristic X rays or secondary X rays, or both, as well as bremsstrahlung for a variety of applications. By suitable choice of absorber and isotope, it is possible to fabricate sources with almost any desired energy. Unfortunately the yields are not very high, but the source designs are improving and the future looks promising. These sources result in the availability of an almost continuous spectrum of radiant energies. Table 9 lists a few such sources currently being used. Obviously, the applications cannot conveniently include tracer studies.

Table 9. SECONDARY GAMMA AND X-RAY SOURCES

Isotope	Target	Energy, kev
Bremsstrahlung		
^{147}Pm	Ag	20 to 40
	Al	10 to 40
^{85}Kr	C	30 to 80
^{3}H	Ti	4 to 8
	Zr	5 to 9
^{90}Sr	Al	70 to 150
Characteristic X Ray*		
^{109}Cd	Self	22 (Ag-K)
^{3}H	Zr	2 (Zr-L)
^{55}Fe	Self	5.9 (Mn-K)

* From electron capture or internal conversion.

CHARGED-PARTICLE ACTIVATIONS

Accelerators bombard stable nuclides with charged particles (e.g., protons, deuterons, alpha particles), with the result that the charged particle is accepted by the nucleus; many times a neutron is ejected. Table 10 lists the more common charged-particle activation reactions.

Table 10. IMPORTANT REACTIONS FOR CYCLOTRON-PRODUCED RADIOISOTOPES

1. p,n	7. p,pn
2. p,2n	8. p,2p
3. d,n	9. d,p
4. d,2n	10. d,α
5. p,α	11. p,d
6. p,αn	12. α,n

The procedure for making isotopes with an accelerator differs from the reactor technique primarily in the irradiation techniques. The target is quite different and usually must be water-cooled. The chemical processing, on the other hand, is essentially the same as for reactor-produced radioisotopes. Figure 7 shows a schematic drawing of a cyclotron, and Figures 8 and 9 show, respectively, an open-faced and a capsule-type target for cyclotron irradiations.

There are several advantages to charged-particle activations.

1. The products usually result from transmutations, and are therefore carrier-free.

2. Since the isotopes are neutron-deficient, any K-capture decay gives rise to characteristic X rays, which are extremely useful.

3. Often the only practical way to obtain a useful isotope of an element is in a cyclotron, e.g., ^{7}Be and ^{48}V, which are the only radionuclides of these elements with half-lives long enough for tracer experiments.

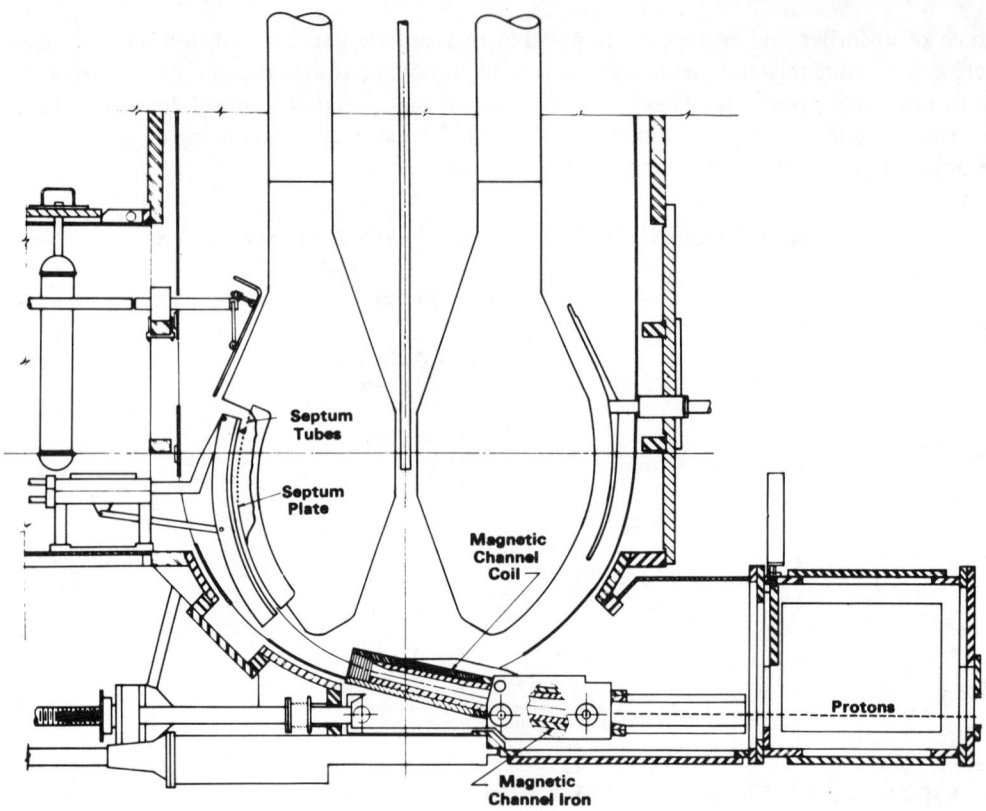

FIGURE 7.
86-inch Cyclotron Beam Deflection System.

FIGURE 8.
Open-Faced Cyclotron Target for Irradiating Metals.

FIGURE 9.
Capsule Cyclotron Target for Irradiating Powders.

FISSION PRODUCTS

In all fission-type nuclear reactors the fissioning of uranium and plutonium results in a large number of fission-product isotopes; some of these are stable, while others are radioactive. It is possible to separate individual fission-product isotopes and to use them. This is currently being done for such nuclides as ^{144}Ce, ^{147}Pm, ^{90}Sr, and ^{137}Cs.

There are two principal differences between fission products and the other two types of artificial radioisotopes: 1) the large quantities that are available and can be processed at any one time, and 2) the fact that they are produced whether they are needed or not. Ordinarily, yields from the reactor and accelerator productions run from a few millicuries to several curies; one notable exception to this is the kilocurie production of ^{60}Co. With fission products, a single batch may easily involve hundreds of thousands of curies. Although the chemical processing techniques are similar (e.g., precipitation, filtration, evaporation, extraction), the facilities are much larger and the shielding requirements much greater.

RADIOISOTOPE GENERATORS (COWS)

Although not exactly production methods in the usual sense of the word, isotope generators are becoming quite important as reservoirs of radioisotopes. If a particular radioisotope happens to decay to a "daughter" that is also a radioisotope, then there exists the possibility of separating "mother" and "daughter". In particular, if the parent is of a long half-life, then it can be stored and "milked" as the "daughter" is required. Examples of commercially available generators are ^{132}Te → ^{132}I, ^{99}Mo → ^{99m}Tc, and ^{90}Sr → ^{90}Y. Brucer[1] has written a very comprehensive report on possible generators.

---

1. Brucer, M., 118 Medical Radioisotope Cows. *Isotopes and Radiation Technology*, 3(1): 1-13, 1965.

PHYSICAL AND NUCLEAR DATA

PHYSICAL DATA

Except for the elements with very low atomic numbers, isotope effects other than mass are very difficult to detect. As a result, the physicochemical properties of isotopes are essentially those of the natural elements. Table 11 lists some of the isotopes for which the isotope differences are sufficient to permit separation. Otherwise, those physicochemical properties that are considered to be of possible interest to radioisotope users are included in Table 16 with the General Data given in Section A.

Table 11.
ISOTOPES WITH SIGNIFICANT PHYSICOCHEMICAL DIFFERENCES

Isotope	Separation Method
^1H, ^2H, ^3H	Chemical exchange; distillation
^6Li, ^7Li	Electrolysis; electromigration
^{10}B, ^{11}B	Exchange distillation
^{12}C, ^{13}C	Chemical exchange
^{14}N, ^{15}N	Chemical exchange
^{16}O, ^{17}O, ^{18}O	Chemical exchange; distillation

NUCLEAR DATA

The choice of nuclear data that are presented in the following tables is based on the anticipated needs of the users of radioisotopes. Much of the basic information concerning actual decay schemes is of no particular importance to a user. On the other hand, half-life, cross section, yields, specific activities and activity concentrations at various fluxes, isotopic impurities expected, types of decay, types and energies of radiation, and availability are all of interest to users.

Tables 12 to 15 consist of those isotopes that 1) are not commercially available, 2) have suitable half-lives and energies for possible use, and 3) have feasible production reactions. The isotopes are tabulated according to half-lives, and the data include types of radiations and approximate energies, with ranges shown where several different energies exist. No attempt is made to identify relative intensities. Also, "trace" is used to show low-intensity β^+ and α radiations. The symbol "ε" is used to denote electron capture (which may or may not result in gamma radiation). The symbol "e^-" is used to designate internal-conversion electrons. The latter can be of considerable significance in connection with dose calculations; the dose calculations are different than those for β^-, since e^- are monoenergetic. Unfortunately, this particular area of nuclear data is sadly lacking for information—primarily because of the complexity of the measurements and calculations. Isomeric transitions are indicated by "IT".

Table 12.
RADIOISOTOPES WITH REASONABLE EXPECTANCY OF ROUTINE PRODUCTION
(Half-Lives > 1 hour, < 1 day)

Nuclide	Half-Life, hours	Principal Radiations and Approximate Energies, Mev	Nuclide	Half-Life, hours	Principal Radiations and Approximate Energies, Mev
^{41}Ar	1.8	γ: 1.3 β$^-$: 1.2; 2.5	^{82m}Rb	6.3	γ: 0.5–1.5 β$^+$: 0.8 ε
^{45}Ti	3.1	γ: some β$^+$: 1.02	^{86}Y	15	γ: 1.1; others to 1.9 β$^+$: 1.2; 0.6–3.1
^{48}Cr	23	γ: 0.3; 0.1 ε	^{87m}Y	14	γ: 0.4 IT (e$^-$: 0.37)
^{55}Co	18	γ: 0.9; 1.4; 0.5 β$^+$: 1.5; 1.0	^{92}Y	3.5	γ: 0.4–1.8 β$^-$: 3.6; 1.3; others
^{58m}Co	9	IT (e$^-$: 0.03)	^{86}Zr	16.5	γ: 0.2; others ε
^{61}Co	1.65	γ: 0.07 β$^-$: 1.2	^{87}Zr	1.6	γ: 1.2; others β$^+$: 2.1
^{62}Zn	9.3	γ: 0.04; 0.6 β$^+$: 0.7 ε	^{89}Nb	1.9	γ: 0.5–3.8 β$^+$: 2.9
^{71m}Zn	3.9	γ: 0.4–1.1 β$^-$: 1.5	^{90}Nb	14.6	γ: 0.1–2.3 β$^+$: 1.5 e$^-$: 0.12
^{68}Ga	1.1	γ: 0.5; 0.8–1.9 β$^+$: 1.9 ε	^{96}Nb	23	γ: 0.4–1.2 β$^-$: 0.7; 0.4
^{73}Ga	4.8	γ: 0.3; 0.7 β$^-$: 1.2	^{97}Nb	1.2	γ: 0.7; others β$^-$: 1.3; others
^{66}Ge	2.4	γ: 0.4; 0.05; others β$^+$: 1.3; 2.0 ε	^{93m}Mo	6.9	γ: 0.26–1.5 IT (e$^-$: 0.25)
^{75}Ge	1.4	γ: 0.3; 0.07–0.63 β$^-$: 1.2; 0.9	^{93}Tc	2.7	γ: 0.5–1.5 β$^+$: 0.8; 0.6 ε
^{78}Ge	1.5	γ: 0.3 β$^-$: 0.7	^{94}Tc	4.8	γ: 0.5–0.9 β$^+$: 0.8 ε
^{78}As	1.5	γ: 0.6; 0.7; 1.3; others β$^-$: 4.1; 1.4	^{95}Tc	20	γ: 0.8; 0.2–1.1 ε
^{73}Se	7.1	γ: 0.4; 0.07; 0.5 β$^+$: 1.3; 1.7 e$^-$: 0.06; 0.35	^{95}Ru	1.7	γ: 0.3; 1.4; others β$^+$: 1.2; 0.7 ε
^{75}Br	1.6	γ: 0.3–0.6 β$^+$: 1.7 ε	^{99m}Rh	4.7	γ: 0.4; 0.6; 0.9–1.4 β$^+$: 0.8; others ε
^{76}Br	16.2	γ: 0.5–3.6 β$^+$: 3.6; 0.9 ε	^{100}Rh	21	γ: 0.5; 0.3–2.4 β$^+$: 2.6 e$^-$: 0.5 ε
^{80m}Br	4.5	γ: 0.04 IT(e$^-$: 0.05)	^{106m}Rh	2.2	γ: 0.5; 0.2–1.6 β$^-$: 0.8–1.6
^{83}Br	2.4	γ: to 0.5 β$^-$: 0.9	^{101}Pd	8.5	γ: 0.02; 0.2–1.3 β$^+$: 0.8 e$^-$: 0.02 ε
^{83m}Kr	1.86	γ: 0.01 IT (e$^-$: to 0.03)			
^{87}Kr	1.3	γ: 0.4–2.6 β$^-$: 3.8; 1.3	^{111m}Pd	5.5	γ: 1.7; 0.5 β$^-$: 2.0 IT (e$^-$: 0.15)
^{81}Rb	4.7	γ: 0.2; 0.5; others β$^+$: 1.03 ε			

Table 12. (Continued)
RADIOISOTOPES WITH REASONABLE EXPECTANCY OF ROUTINE PRODUCTION
(Half-Lives > 1 hour, < 1 day)

Nuclide	Half-Life, hours	Principal Radiations and Approximate Energies, Mev	Nuclide	Half-Life, hours	Principal Radiations and Approximate Energies, Mev
^{103}Ag	1.1	γ: 0.13–1.3 β$^-$: 1.3 β$^+$: 1.6 ε	^{116}Te	2.5	γ: from daughter β$^+$: from daughter e$^-$: 0.6; 0.09 ε
^{104}Ag	1.1	γ: 0.7; 0.9; 0.2–1.8 β$^+$: 1.0 e$^-$: 0.5–0.7 ε	^{117}Te	1.0	γ: 0.5–1.8 β$^+$: 1.8
			^{119}Te	16	γ: 0.7; others to 1.8 β$^+$: 0.6 ε
^{112}Ag	3.2	γ: 0.6; 1.4; 0.7–3.3 β$^-$: 4.1			
^{113}Ag	5.3	γ: 0.3; 0.1–1.2 β$^-$: 2.0	^{125}Xe	17	γ: 0.05–0.24 e$^-$: 0.02–0.21 ε
^{107}Cd	6.5	γ: Ag X rays β$^+$: 0.3 ε	^{135}Xe	9.2	γ: 0.3; 0.6 β$^-$: 0.9; 0.6 e$^-$: 0.2
^{117}Cd	2.5	γ: 0.3–1.6 β$^-$: 0.7; 2.0	^{134m}Cs	2.9	γ: 0.13 β$^-$: 0.6 IT (e$^-$: to 0.12)
^{109}In	4.3	γ: 0.2; 0.6; 0.2–1.2 β$^+$: 0.8 e$^-$: 0.03–0.2 ε	^{139}Ba	1.4	γ: 0.2; 1.4 β$^-$: 2.2 e$^-$: 0.13; 0.16
^{110m}In	4.9	γ: 0.1; 0.6; others e$^-$: 0.1–0.9 ε	^{135}La	19.4	γ: 0.5; 0.1–0.9 e$^-$: 0.2–0.5 ε
^{110}In	1.1	γ: 0.7; others β$^+$: 2.3 e$^-$: 0.6 ε	^{142}La	1.5	γ: 0.6; 2.4; 0.9–3.6 β$^-$: 2.1; 0.9; 4.5
			^{137}Ce	9	γ: 0.01; 0.5 e$^-$: 0.4 ε
^{113m}In	1.7	γ: 0.4 IT (e$^-$: 0.4)			
^{115m}In	4.4	γ: 0.3 β$^-$: 0.8 IT (e$^-$: 0.3)	^{141}Nd	2.5	γ: 1.2; others β$^+$: 0.8 ε
^{117m}In	1.9	γ: 0.3 β$^-$: 1.7 IT (e$^-$: 0.3)	^{150}Pm	2.7	γ: 0.3–2.5 β$^-$: 3.1
^{110}Sn	4.0	γ: 0.3 e$^-$: 0.25 ε	^{152m}Eu	1.6	γ: 0.02; 0.09 IT (e$^-$: 0.01–0.04)
			^{156m}Tb	5.5	γ: Tb X rays IT(e$^-$: 0.04; 0.08) ε
^{116m}Sb	1	γ: 0.1–1.3 β$^+$: 1.2; others e$^-$: 0.07–0.1 ε	^{165}Er	10.4	γ: Ho X rays ε
^{117}Sb	2.8	γ: 0.16 β$^+$: 0.6 ε	^{177}Yb	1.9	γ: 0.1–1.2 β$^-$: 1.4 e$^-$: 0.06–0.14
^{118m}Sb	5.1	γ: 0.04–1.2 e$^-$: 0.01–0.22 ε	^{176m}Lu	3.7	γ: 0.1 β$^-$: 1.3 e$^-$: 0.022–0.09
^{128}Sb	8.6	γ: 0.3–0.9 β$^-$: 1.0	^{180m}Hf	5.5	γ: 0.06–0.5 IT (e$^-$: 0.03–0.27)

Table 12. (*Continued*)
RADIOISOTOPES WITH REASONABLE EXPECTANCY OF ROUTINE PRODUCTION
(Half-Lives > 1 hour, < 1 day)

Nuclide	Half-Life, hours	Principal Radiations and Approximate Energies, Mev	Nuclide	Half-Life, hours	Principal Radiations and Approximate Energies, Mev
^{176}Ta	8.0	γ: 0.1–3.0 e⁻: 0.02–3.0 ε	^{196m}Au	10	γ: 0.15–0.32 IT (e⁻: 0.07–0.16)
^{178}Ta	2.2	γ: 0.1–0.4 e⁻: 0.02–0.26 ε	^{198m}Tl	1.9	γ: 0.28–0.64 IT (e⁻: 0.03–0.25) ε
^{180m}Ta	8.1	γ: 0.1 β⁻: 0.7 e⁻: 0.03–0.09 ε	^{198}Tl	5.3	γ: 0.4; 0.2–2.8 β⁺: 2.4 (trace) e⁻: 0.11–0.33 ε
^{182m}Re	13	γ: 1.2; 0.07–2.0 β⁺: 1.7 (trace) e⁻: 0.01–2.1 ε	^{199}Tl	7.4	γ: 0.16–0.45 e⁻: 0.04–0.19 ε
^{182}Os	22	γ: 0.5; 0.2 e⁻: 0.02–0.44 ε	^{200}Pb	21	γ: 0.11–0.61 e⁻: 0.02–0.18 ε
^{189m}Os	5.7	γ: Os X rays IT (e⁻: 0.02; 0.03)	^{201}Pb	9.4	γ: 0.33–1.4 β⁺: 0.6 (trace) e⁻: 0.24–0.32 ε
^{187}Ir	10.5	γ: 0.03–1.0 e⁻: 0.01–1.1 ε	^{202m}Pb	3.6	γ: 0.39–0.96 IT (e⁻: 0.12–0.77) ε
^{189}Pt	11	γ: 0.1–0.8 e⁻: 0.04–0.8 ε	^{204m}Pb	1.1	γ: 0.3–0.9 IT (e⁻: 0.3–0.9)
^{197m}Pt	1.4	γ: 0.3; 0.1 β⁻: 0.7; others IT (e⁻: 0.04–0.33)	^{209}Pb	3.3	γ: None β⁻: 0.6
^{193}Au	16	γ: 0.2; 0.1; 0.01–0.5 e⁻: 0.03–0.18 ε	^{204}Bi	11.2	γ: 0.2–1.2 e⁻: 0.06–0.90 ε

Table 13.
RADIOISOTOPES WITH REASONABLE EXPECTANCY OF ROUTINE PRODUCTION
(Half-Lives > 1 day, < 1 week)

Nuclide	Half-Life, days	Principal Radiations and Approximate Energies, Mev	Nuclide	Half-Life, days	Principal Radiations and Approximate Energies, Mev
^{48}Sc	1.8	γ: 1.3; 1.0 β⁻: 0.7; 0.5	^{131m}Te	1.2	γ: 0.8; 0.1–2.0 β⁻: 0.4; 0.6–2.5 IT (e⁻: 0.05–0.18)
^{56}Ni	6.1	γ: 0.2–1.6 e⁻: 0.16 ε	^{133m}Xe	2.3	γ: 0.23 IT (e⁻: 0.2)
^{57}Ni	1.5	γ: 1.4; 1.9; 0.1 β⁺: 0.8 ε	^{129}Cs	1.3	γ: 0.04; 0.4; 0.6 e⁻: to 0.38 ε
^{66}Ni	2.3	γ: None β⁻: 0.2	^{133m}Ba	1.6	γ: 0.28 IT (e⁻: to 0.27)
^{69}Ge	1.6	γ: 1.1; 0.6; 0.5–1.3 β⁺: 1.2; 0.6 ε	^{137m}Ce	1.4	γ: to 0.8 IT (e⁻: 0.2) ε
^{71}As	2.6	γ: 0.2; 0.5 β⁺: 0.8 ε	^{148}Pm	5.4	γ: 0.5–1.5 β⁻: 2.5; 1.0; 1.9
^{72}As	1.1	γ: 0.5–3.7 β⁺: 2.5; 3.3 e⁻: 0.7 ε	^{145}Eu	5.9	γ: 0.6–2.0 β⁺: Some e⁻: 0.06–0.85 ε
^{79}Kr	1.4	γ: 0.04–1.3 β⁺: 0.6; 0.3 e⁻: to 0.4 ε	^{146}Eu	4.6	γ: 0.7; 0.1–2.9 β⁺: 1.5 e⁻: 0.6; 0.7 ε
^{83}Sr	1.4	γ: 0.8; 0.4 β⁺: 1.2; 0.8 ε	^{147}Gd	1.0	γ: 0.2; 0.4; 0.9; 0.1–1.3 e⁻: 0.18–0.39 ε
^{89}Zr	3.2	γ: 0.5–1.7 β⁺: 0.9 e⁻: 0.9 ε	^{153}Tb	2.6	γ: 0.1; 0.2; 0.02–1.0 e⁻: 0.01–0.16
			^{156}Tb	5.4	γ: 0.1–2.0 e⁻: 0.04–0.15 ε
^{95m}Nb	3.8	γ: Nb X rays IT(e⁻: 0.2)			
^{96}Tc	4.3	γ: 0.3–1.1 e⁻: 0.3–0.8 ε	^{166}Dy	3.4	γ: 0.1–0.4 β⁻: 0.4; 0.5 e⁻: 0.02–0.05
^{101m}Rh	4.5	γ: 0.3; 0.5 IT (e⁻: 0.2) ε	^{166}Ho	1.1	γ: 0.1; 1.4; 0.7–1.8 β⁻: 1.8; others
			^{160}Er	1.2	γ: Ho X rays ε
^{100}Pd	3.7	γ: 0.1 e⁻: 0.01–0.08 ε	^{169}Lu	1.5	γ: 0.1; 0.07–2.2 β⁺: 1.2 e⁻: 0.01–2.2 ε
^{119}Sb	1.6	γ: 0.02 e⁻: 0.02 ε	^{170}Lu	2	γ: 0.6; 0.1–3.1 β⁺: 2.4 e⁻: 0.02–3.2 ε
^{120m}Sb	5.8	γ: 0.1–1.2 e⁻: 0.06–0.2 ε	^{172}Lu	6.7	γ: 0.08–1.09 e⁻: 0.02–2.1 ε
^{127}Sb	3.9	γ: 0.06–1.3 β⁻: 1.5; 1.1; others			
^{119m}Te	4.7	γ: 1.5; 1.2; 0.2–2.1 e⁻: 0.12–0.27 ε	^{173}Hf	1.0	γ: 0.1–1.2 e⁻: 0.06–1.1 ε

Table 13. (Continued)
RADIOISOTOPES WITH REASONABLE EXPECTANCY OF ROUTINE PRODUCTION
(Half-Lives >1 day, <1 week)

Nuclide	Half-Life, days	Principal Radiations and Approximate Energies, Mev	Nuclide	Half-Life, days	Principal Radiations and Approximate Energies, Mev
^{183}Ta	5.0	γ: 0.2; 0.3; 0.04–0.4 β^-: 0.6 e^-: 0.03–0.4	^{196}Au	6.2	γ: 0.3–1.1 β^-: 0.26 e^-: 0.3 ε
^{182}Re	2.7	γ: 0.02–1.4 e^-: 0.01–1.4 ε	^{195m}Hg	1.7	γ: 0.2–0.6 IT (e^-: 0.001–0.18) ε
^{188}Ir	1.7	γ: 0.1; 0.6; 0.3–2.7 β^+: 1.7 (trace) e^-: 0.08–2.7 ε	^{200}Tl	1.1	γ: 0.37–1.52 β^+: 1.4 (trace) e^-: 0.29; 0.35 ε
^{191}Pt	3.0	γ: 0.1–0.6 e^-: 0.02–0.6 ε	^{201}Tl	3.0	γ: 0.13; 0.17 e^-: 0.02–0.08 ε
^{195m}Pt	4.1	γ: 0.1 IT (e^-: 0.02–0.13)	^{203}Pb	2.2	γ: 0.28–0.68 e^-: 0.19; 0.26 ε
^{194}Au	1.6	γ: 0.3; 1.0–2.3 β^+: 1.2; 1.5 e^-: 0.02–2.4 ε			

Table 14.
RADIOISOTOPES WITH REASONABLE EXPECTANCY OF ROUTINE PRODUCTION
(Half-Lives >1 week, <1 year)

Nuclide	Half-Life, days	Principal Radiations and Approximate Energies, Mev	Nuclide	Half-Life, days	Principal Radiations and Approximate Energies, Mev
^{73}As	80	γ: 0.05 e⁻: 0.05	^{148}Eu	54	γ: 0.4–1.6 β⁺: 0.9 e⁻: 0.02–0.58 ε
^{72}Se	8.4	γ: 0.05 e⁻: 0.04	^{149}Eu	106	γ: 0.3 e⁻: 0.02–0.28 ε
^{82}Sr	25	γ: 0.5; 0.8; Rb X rays β⁺: 3.2			
^{88}Zr	85	γ: 0.4 e⁻: 0.4	^{146}Gd	48	γ: 0.08–0.16 e⁻: 0.07; 0.11 ε
^{91m}Nb	62	γ: 1.2; Nb X rays IT (e⁻: 0.1)			
^{92m}Nb	10.1	γ: 0.9	^{149}Gd	9	γ: 0.1–0.9 α: 3.0 (trace) e⁻: 0.1–0.3 ε
^{95m}Tc	60	γ: 0.2–1.0 β⁺: 0.6 IT (e⁻: 0.02–0.2)			
^{97m}Tc	91	γ: Tc X rays IT (e⁻: 0.08)	^{151}Gd	120	γ: 0.02–0.31 α: 2.6 (trace) e⁻: 0.01–0.17 ε
^{99}Rh	16	γ: 0.1; 0.4; 0.1–2.7 β⁺: 1.0			
^{102}Rh	206	γ: 0.5; 0.4–2.1 β⁺: 1.3 β⁻: 1.2	^{159}Dy	144	γ: 0.06; Tb X rays e⁻: 0.01–0.06 ε
^{106m}Ag	8.3	γ: 0.2–1.8 e⁻: 0.2–0.7	^{167}Tm	9.6	γ: 0.06–0.53 e⁻: 0.05–0.20 ε
^{117m}Sn	14	γ: 0.16 IT (e⁻: 0.14)	^{168}Tm	86	γ: 0.08–1.28 e⁻: 0.02–0.04 ε
^{123}Sn	125	γ: 1.1 β⁻: 1.4			
^{125}Sn	9.4	γ: trace 0.3–2.2 β⁻: 2.3	^{171}Lu	8.2	γ: 0.02–0.84 e⁻: 0.01–0.85 ε
^{126}Sb	12.5	γ: 0.4; 0.7 β⁻: 2.0			
^{118}Te	6	γ: Sb X rays ε	^{177m}Lu	155	γ: 0.1–0.5 β⁻: Some, 0.17 IT (e⁻: to 0.47)
^{121m}Te	150	γ: 0.2; 1.1 IT (e⁻: 0.01–0.18) ε	^{188}W	69	γ: 0.3 (trace) β⁻: 0.35
^{121}Te	17	γ: 0.6 e⁻: 0.01–0.54 ε	^{184}Re	38	γ: 0.9; 0.1–0.8 e⁻: 0.04–0.10 ε
^{123m}Te	117	γ: 0.16 IT (e⁻: 0.06–0.13)	^{189}Ir	13.3	γ: 0.25 e⁻: 0.02–0.27 ε
^{127}Xe	36.4	γ: 0.06–0.38 e⁻: 0.02–0.20 ε	^{190}Ir	12	γ: 0.19–1.7 e⁻: 0.11–1.7 ε
^{129m}Xe	8	γ: 0.04; 0.2 IT (e⁻: to 0.19)	^{188}Pt	10	γ: 0.2; 0.5 α: 3.9 (trace) e⁻: 0.04–0.4 ε
^{143}Pr	13.6	γ: None β⁻: 0.9			
^{148m}Pm	41	γ: 0.1–1.0 β⁻: 0.7 IT (e⁻: 0.03–0.6)	^{194}Hg	146	γ: Au X rays ε
^{145}Sm	340	γ: 0.06 e⁻: 0.02; 0.05 ε	^{202}Tl	12	γ: 0.4; 1.0 e⁻: 0.36 ε
^{147}Eu	24	γ: 0.1; 0.2; 0.6–1.3 α: 2.9 β⁺: Some e⁻: 0.03–0.15 ε	^{205}Bi	15.3	γ: 0.1–1.9 β⁺: 1.0 (trace) e⁻: 0.01–0.02 ε

Table 15.

RADIOISOTOPES WITH REASONABLE EXPECTANCY OF ROUTINE PRODUCTION

(Half-Lives 1 year or more)

Nuclide	Half-Life, years	Principal Radiations and Approximate Energies, Mev	Nuclide	Half-Life, years	Principal Radiations and Approximate Energies, Mev
^{32}Si	~650	γ: None β⁻: 0.2	^{108m}Ag	5	γ: 0.7 IT (e⁻: 0.03) ε
^{39}Ar	270	γ: None β⁻: 0.6	^{113m}Cd	14	γ: Cd X rays β⁻: 0.6
^{42}Ar	33	β⁻: (from ^{42}K)	^{121m}Sn	76	γ: 0.04 β⁻: 0.4
^{41}Ca	8 × 10⁴	γ: K X rays ε	^{135}Cs	2 × 10⁶	γ: None β⁻: 0.2
^{59}Ni	8 × 10⁴	γ: Co X rays ε	^{144}Pm	1	γ: 0.7; 0.5 e⁻: 0.4–0.6 ε
^{79}Se	7 × 10⁴	γ: None β⁻: 0.2	^{145}Pm	18	γ: 0.07 e⁻: 0.02–0.06 ε
^{81}Kr	2 × 10⁵	γ: Br X rays ε	^{146}Pm	4.4	γ: 0.4; 0.8 β⁻: 0.8 ε
^{93}Zr	1.5 × 10⁶	β⁻: 0.06; 0.03			
^{91}Nb	Long	γ: Zr X rays ε	^{151}Sm	90	γ: 0.02 β⁻: 0.08 e⁻: 0.01; 0.02
^{93m}Nb	13.6	γ: Nb X rays IT (e⁻: 0.03)	^{171}Tm	1.9	γ: 0.07 β⁻: 0.1; 0.03 e⁻: 0.06
^{94}Nb	2 × 10⁴	γ: 0.9; 0.7 β⁻: 0.5			
^{93}Mo	10⁴	γ: Nb X rays	^{192m}Ir	5	γ: Ir X rays IT (e⁻: 0.15)
^{101}Rh	3	γ: 0.1; 0.2; 0.3 e⁻: 0.1–0.2 ε	^{205}Pb	3 × 10⁷	γ: Tl X rays ε
^{107}Pd	7 × 10⁶	γ: None β⁻: 0.04			

Table 16 comprises two sections and is a rather comprehensive listing of those radioisotopes that can be considered "commercially available", since they are listed in sales catalogs. For convenience, the isotopes are listed by elements.[2] In Section A, general data, such as average atomic weight and a few chemical and physical properties of the elements, are shown; biological half-lives,[3] as available, are also included. No information on chemical toxicity is included, since it is presumed that the quantities injected, in case of medical or biomedical applications, will be well below the toxic limits.

The nuclear data are tabulated in Section B, which is divided into two parts. Part I covers radioactive isotopes, Part II covers stable isotopes. Column 1 of Part I shows the mass number of the nuclide, and Column 2 the half-life, Column 3 lists the usual production reaction (or reactions, where more than one may be used), and Column 4 the production cross section. For cyclotron reactions the yield information is not always available and, for this reason, has been omitted. Column 5 shows the *theoretical* specific activity; this can sometimes be approached as the result of transmutation reactions, where carrier-free separation can be effected. These values were calculated from the following equations:

---

2. In a few cases, where there are no isotopes of an element commercially available, the element is not included.
3. Taken from *Health Physics*, 3: 191. (The data are expected to be updated during 1969.)

$$\text{Curies/gram or SpA } (T_{1/2} \text{ in hrs}) = \frac{3.14 \times 10^9}{(T_{1/2}) \text{ (atomic weight)}}, \qquad (1)$$

$$\text{Curies/gram or SpA } (T_{1/2} \text{ in days}) = \frac{1.308 \times 10^8}{(T_{1/2}) \text{ (atomic weight)}}, \qquad (2)$$

$$\text{Curies/gram or SpA } (T_{1/2} \text{ in yrs}) = \frac{3.59 \times 10^5}{(T_{1/2}) \text{ (atomic weight)}}. \qquad (3)$$

Production yields are shown in the next three columns, 6 to 8, and are calculated from the relationship

$$A_s = \frac{N \times \sigma \times 16.3}{\text{atomic weight}}, \qquad (4)$$

where A_s is the saturation activity per gram of sample being irradiated at a flux of 10^{12} n/cm^2/second, N the abundance of target atoms in the normal material, σ the flux in n/cm^2/second, and 16.3 a constant that takes into consideration Avogadro's number, the area of a barn, and the definition of a curie. For isotopes with long half-lives the activity may not be listed for 1 or 2 half-lives, since irradiations longer than 1 year are generally impractical.

Columns 9 to 12 deal with the types and energies of the radiations. Energies are approximate, and branching ratios (or percentages) are shown in parentheses. The highest energy is always indicated, although it may be present only to a very slight extent (e.g., "trace" or "others to—"). Energies shown do not include radiations from impurities, nor even from daughter isotopes, unless specifically stated. The data are taken from References 4 and 5. Internal-conversion information is not being included in this first edition, since data are spotty. However, the presence of conversion electrons is indicated where the contribution is felt to be of significance (e.g., no β^- present, or the γ/e^- conversion ratio is high).

In order to keep the table simple, energies are "rounded off", and the percentages of similar energies are combined; e.g. 0.292 γ (15) and 0.313 γ (22) would be combined to give 0.30 γ (37). It should also be mentioned in the presence of β^+ always guarantees twice the β^+ percentages as an accompanying 0.511–Mev annihilation gamma. Thus, although this particular energy may not be listed specifically in the γ-ray column, it will always be present whenever positrons (β^+) are shown.

Part II is a simple table, listing only the natural abundance for each stable isotope.

Additional material on each isotope can be found in other standard references.[4,5,6,7,8,9] The new *Table of Isotopes*[5] is particularly valuable for evaluating decay schemes, and *Reactor Yield Calculations*[8] presents yields of most of the more common radioisotopes in tabular form. The Oak Ridge National Laboratory Catalog[9] lists available enriched stable isotopes, which are often quite advantageous for use as targets.

---

4. Goldman, D. T. and Stehn, R. J., *Chart of the Nuclides*. General Electric Co. (Knolls Atomic Power Laboratory), Schenectady, N.Y., 1966.

5. Lederer, C. M., et al., *Table of Isotopes*, 6th ed. John Wiley and Sons, New York, 1967.

6. Goldberg, M. D., et al., Compilers, *Neutron Cross Sections*. BNL-325 and Supplements 1958–1966. Sigma Center, Brookhaven National Laboratory, Associated Universities, Inc., available from Clearing House for Federal Scientific and Technical Information, National Bureau of Standards, U.S. Department of Commerce, Springfield, Virginia 22151.

7. International Atomic Energy Agency, *International Directory of Isotopes*, 3rd ed. International Atomic Energy Agency, Vienna, Austria, 1964.

8. Whitson, T. C. and Friend, C. W., Reactor Yield Calculations for 81 Radioisotopes Produced by (n.γ) Reactions at Fluxes of 10^7 to 10^{16} n/cm^2/sec for Irradiation Times of 30 Minutes to One Year. *USAEC Report ORNL-IIC-8*, 1966.

9. Oak Ridge National Laboratory, *Radioisotopes, Stable Isotopes, Research Materials* (Catalog). Oak Ridge National Laboratory, Oak Ridge, Tennessee, 1967.

Table 16. ELEMENTS AND THEIR COMMERCIALLY AVAILABLE ISOTOPES
A. General Data: Elements

Element (Symbol)	Atomic No.	Atomic Weight	Physical and Chemical Properties	Soluble Compounds	Biological Half-Life
Hydrogen (H)	1	1.008	Colorless, odorless and tasteless gas; D.: 0.087 g/l; reacts with oxygen to form water.	Acids Bases Water	Total Body: 12 days Body Water: 12 days
Helium (He)	2	4.003	Colorless, odorless and tasteless gas; D.: 0.18 g/l; inert.	None	Insoluble
Lithium (Li)	3	6.939	Shiny soft metal; D.: 0.53 g/ml; M.P.: 181°C; tarnishes rapidly in air, reacts vigorously with water.	Halides Nitrate Sulfate	Total Body: 2 days Ovaries: 2 days Testes: 2 days
Beryllium (Be)	4	9.012	Dark-gray metal; D.: 1.8 g/ml; M.P.: 1283°C; stable in air; soluble in acids; element and its compounds are poisonous.	Chloride Nitrate Sulfate	Total Body: 53.6 days Bone: 53.6 days Kidneys: 53.6 days Liver: 53.6 days Spleen: 53.6 days
Boron (B)	5	10.811	Gray or black amorphous or crystalline solid; D.: 2.5 g/ml; M.P.: 2040°C; relatively unreactive; soluble in HNO_3.	Borates	Total Body: 0.5 days Brain: 0.5 days Liver: 0.5 days Pancreas: 0.5 days
Carbon (C)	6	12.010	Black (as diamond, colorless), odorless and tasteless solid; D.: 2 g/ml (as diamond, 3.5 g/ml); M.P.: > 3500°C; stable in air, oxidized by hot concentrated HNO_3.	Alkali metal carbonates Cyanides Many organic compounds	Total Body: 10 days Fat: 12 days Bone: 40 days
Nitrogen (N)	7	14.007	Colorless, odorless and tasteless gas; D.: 1.25 g/l; relatively inert at ordinary temperatures.	Ammonium salts Cyanides Nitrates	Total Body: 90 days
Oxygen (O)	8	16.000	Colorless, odorless and tasteless gas; D.: 1.43 g/l; reacts with many reducing agents at room temperature, and with nearly all at elevated temperature.	Bases Oxygen acids Water	Total Body: 14 days
Fluorine (F)	9	18.998	Greenish-yellow gas with pungent odor; D.: 1.7 g/l; element and compounds toxic.	Alkali metal fluorides Ammonium salts Hydrofluoric acid	Total Body: 800 days Bone: 1,450 days Teeth: 1,450 days
Neon (Ne)	10	20.183	Colorless, odorless and tasteless gas; D.: 0.89 g/l; inert.	None	Insoluble
Sodium (Na)	11	22.990	Shiny soft metal; D.: 0.97 g/ml; M.P.: 97.5°C; very reactive with water.	Most salts Sodium hydroxide	Total Body: 11 days
Magnesium (Mg)	12	24.312	Shiny white metal, rigid; D.: 1.74 g/ml; M.P.: 650°C; stable in air; soluble in acids.	Chloride Nitrate Sulfate	Total Body: 180 days Bone: 180 days

Table 16. ELEMENTS AND THEIR COMMERCIALLY AVAILABLE ISOTOPES

A. General Data: Elements (*Continued*)

Element (Symbol)	Atomic No.	Atomic Weight	Physical and Chemical Properties	Soluble Compounds	Biological Half-Life
Aluminum (Al)	13	26.982	Shiny white metal, rigid; D.: 2.7 g/ml; M.P.: 660°C; stable in air; soluble in acids.	Chloride Nitrate Sulfate	Total Body: 550 days Lungs: 600 days Spleen: 500 days
Silicon (Si)	14	28.086	Gray or brown crystalline or amorphous solid; D.: 2-2.5 g/ml; M.P.: 1410°C; soluble in $HNO_3 + HF$.	Silicates	Total Body: 60 days Skin 60 days Lungs: 60 days Adrenals: 60 days Ovaries: 60 days Testes: 60 days
Phosphorus (P)	15	30.974	Yellow waxy or red amorphous nonmetallic solid; D.: 1.8-2.2 g/ml; M.P.: 45°C (yellow), 590°C (red); unstable in air; soluble in CS_2; yellow form is extremely toxic.	Alkali metal phosphates Phosphoric acid Phosphorous acid	Total Body: 257 days Bone: 1,155 days Brain: 257 days Liver: 18 days
Sulfur (S)	16	32.064	Yellow crystalline non-metallic solid; D.: 2 g/ml; M.P.: 120°C; stable in air; soluble in CS_2.	Alkali metal sulfates Sulfuric acid	Total Body: 90 days Skin: 1,530 days Bone: 600 days Testes: 623 days
Chlorine (Cl)	17	35.453	Greenish gas with pungent odor; D.: 3.1 g/l; soluble in alkalies; very toxic.	Most metals and non-metals, except Ag, Tl, Hg^+, Pb^{++}	Total Body: 29 days
Argon (Ar)	18	39.948	Colorless, odorless and tasteless gas; D.: 1.8 g/l; inert.	None	Not reported
Potassium (K)	19	39.102	Shiny soft metal; D.: 0.86 g/ml; M.P.: 63.4°C; spontaneously flammable in air; soluble in water (with formation of H_2).	Almost all	Total Body: 58 days Muscle: 58 days Brain: 58 days Liver: 58 days Spleen: 58 days
Calcium (Ca)	20	40.08	Gray soft metal; D.: 1.55 g/ml; M.P.: 850°C; tarnishes in air, reacts with water and acids.	Acetate Chloride Nitrate	Total Body: 16,000-18,000 days Bone: 16,000-18,000 days
Scandium (Sc)	21	44.956	Available primarily in compound form; D.: 3.02 g/ml.	Chloride Nitrate Sulfate	Total Body: 30 days Bone: 33 days Kidneys: 75 days Liver: 36 days
Titanium (Ti)	22	47.90	White light metal, rigid; D.: 4.5 g/ml; M.P.: 1677°C; stable in air; soluble in acids.	Chloride Nitrate Sulfate	Total Body: 320 days Lungs: 320 days Spleen: 320 days Pancreas: 320 days
Vanadium (V)	23	50.942	Steel-gray very hard metal; D.: 5.9 g/ml; M.P.: 1917°C; stable in air; soluble in HNO_3 and H_2SO_4.	Chloride Fluoride	Total Body: 42 days Bone: 135 days Kidneys: 74 days Liver: 70 days Spleen: 90 days

Table 16. ELEMENTS AND THEIR COMMERCIALLY AVAILABLE ISOTOPES

A. General Data: Elements (*Continued*)

Element (Symbol)	Atomic No.	Atomic Weight	Physical and Chemical Properties	Soluble Compounds	Biological Half-Life
Chromium (Cr)	24	51.996	Bright silvery hard metal; D.: 7.1 g/ml; M.P.: 1903°C; stable in air; soluble in HCl and H_2SO_4.	Chloride Chromate Nitrate Sulfate	Total Body: 616 days Lungs: 616 days Kidneys: 616 days Thyroid: 616 days Prostate: 616 days
Manganese (Mn)	25	54.938	Lustrous reddish hard metal, brittle; D.: 7.3 g/ml; M.P.: 1244°C; stable in air; soluble in acids; an essential trace element.	Chloride Nitrate Sulfate	Total Body: 17 days Kidneys: 7 days Liver: 25 days Pancreas: 6 days
Iron (Fe)	26	55.847	Shiny silvery metal; D.: 7.8 g/ml; M.P.: 1535°C; stable in air; soluble in acids; abundantly available.	Acetate Chloride Nitrate Sulfate	Total Body: 800 days Bone: 665 days Lungs: 3,200 days Liver: 550 days Spleen: 600 days
Cobalt (Co)	27	58.933	Bright silvery metal, somewhat magnetic; D.: 8.7 g/ml; M.P.: 1495°C; stable in air; soluble in acids.	Chloride Nitrate Sulfate	Total Body: 9.5 days Liver: 9.5 days Spleen: 9.5 days Pancreas: 9.5 days
Nickel (Ni)	28	58.71	Bright silvery hard metal, malleable, ductile, slightly magnetic; D.: 8.9 g/ml; M.P.: 1455°C; stable in air; soluble in acids.	Chloride Nitrate Sulfate	Total Body: 667 days Bone: 800 days Liver: 500 days
Copper (Cu)	29	63.546	Reddish-brown metal, ductile; D.: 8.9 g/ml; M.P.: 1083°C; stable in air; soluble in HNO_3.	Chloride Nitrate Sulfate	Total Body: 80 days Brain: 800 days Heart: 80 days Kidneys: 16 days Liver: 150 days Spleen: 2 days
Zinc (Zn)	30	65.37	Bluish-white metal, malleable, ductile; D.: 7.1 g/ml; M.P.: 419°C; stable in air; soluble in acids.	Chloride Nitrate Sulfate	Total Body: 930 days Muscle: 1,960 days Bone: 1,300 days Kidneys: 149 days Liver: 90 days Pancreas: 25 days Ovaries: 107 days Testes: 270 days Prostate: 14 days
Gallium (Ga)	31	69.72	Steel-gray metal, liquid above 30°C; D.: 5.9 g/ml; stable in air; soluble in acids and alkalies.	Chloride Nitrate Sulfate	Total Body: 6 days Bone: 12 days Kidneys: 9 days Liver: 5.8 days Spleen: 6 days
Germanium (Ge)	32	72.59	Gray-white metal, brittle; D.: 5.4 g/ml; M.P.: 938°C; stable in air; soluble in hot H_2SO_4	Fluoride Iodide	Total Body: 1 day Kidneys: 12 days Liver: 7.5 days
Arsenic (As)	33	74.92	Steel-gray conductive metal, or gray or brown nonconductive solid; D.: 4.7–5.7 g/ml; M.P.: 817°C; stable in air; soluble in HNO_3.	Chloride Oxide (in acid solution)	Total Body: 280 days Kidneys: 550 days Liver: 550 days

Table 16. ELEMENTS AND THEIR COMMERCIALLY AVAILABLE ISOTOPES

A. General Data: Elements (*Continued*)

Element (Symbol)	Atomic No.	Atomic Weight	Physical and Chemical Properties	Soluble Compounds	Biological Half-Life
Selenium (Se)	34	78.96	Nonmetallic element occurring in several allotropic forms, usually gray or red; D.: 4.8 g/ml; M.P.: 50–217°C, depending on form; stable in air; soluble in CS_2 and H_2SO_4.	Oxide (in acid solution)	Total Body: 11 days Kidneys: 11 days Liver: 24 days Spleen: 18 days
Bromine (Br)	35	79.90	Red dense liquid; D.: 3.1 g/ml; B.P.: 59°C; stable in air, volatile; soluble in alkalies; toxic.	Alkali metal Alkaline earth metal salts	Total Body: 8 days
Krypton (Kr)	36	83.80	Inert gas: D.: 3.7 g/ml.	None	Not reported
Rubidium (Rb)	37	85.47	Silvery soft metal; D.: 1.5 g/ml; M.P.: 39°C; spontaneously flammable in air: explosively soluble in water.	Hydroxide Most salts	Total Body: 45 days Muscle: 80 days Liver: 63 days Spleen: 45 days Pancreas: 60 days
Strontium (Sr)	38	87.62	Silvery-white metal; D.: 2.6 g/ml; M.P.: 770°C; tarnishes in air; reacts vigorously with water.	Acetate Chloride Nitrate	Total Body: 13,000 days Bone: 18,000 days
Yttrium (Y)	39	88.905	Gray rare-earth metal; D.: 4.5 g/ml; M.P.: 1500°C; tarnishes in air; soluble in acids.	Chloride Nitrate	Total Body: 14,000 days Bone: 18,000 days
Zirconium (Zr)	40	91.22	Silvery-white metal, ductile; D.: 6.4 g/ml; M.P.: 1239°C; stable in air; slightly soluble in acids.	Chloride Nitrate Sulfate	Total Body: 450 days Bone: 1,000 days Kidneys: 900 days Liver: 320 days Spleen: 900 days
Niobium (Nb)	41	92.906	Lustrous platinum-white metal; D.: 8.6 g/ml; M.P.: 2500°C; stable in air; soluble in HCl, HNO_3, and hot H_2SO_4.	Chloride (in acid solution)	Total Body: 760 days Bone: 1,000 days Kidneys: 760 days Liver: 845 days Spleen: 950 days
Molybdenum (Mo)	42	95.94	Silvery-white hard metal, tough; D.: 9 g/ml; M.P.: 2620°C; stable in air; soluble in hot, concentrated H_2SO_4.	Bromide Chloride Oxybromide	Total Body: 5 days Kidneys: 3 days Liver: 45 days
Technetium (Tc)	43	(~99)	Shiny dark metal; D.: 11.5 g/ml; M.P.: 2140°C; stable in air; soluble in HNO_3; not occurring in nature.	Ammonium pertechnetate	Total Body: 1 day Skin: 10 days Bone: 25 days Lungs: 5 days Kidneys: 20 days Liver: 30 days
Ruthenium (Ru)	44	101.07	Silvery-white hard metal of the platinum family, brittle; D.: 12.2 g/ml; M.P.: > 1950°C; stable in air; soluble in aqua regia.	Chloride Nitrate	Total Body: 7.3 days Bone: 16 days Kidneys: 2.5 days

Table 16. ELEMENTS AND THEIR COMMERCIALLY AVAILABLE ISOTOPES

A. General Data: Elements (*Continued*)

Element (Symbol)	Atomic No.	Atomic Weight	Physical and Chemical Properties	Soluble Compounds	Biological Half-Life
Rhodium (Rh)	45	102.91	Silvery-white metal of the platinum family; D.: 12.5 g/ml; M.P.: 1960°C; stable in air, resistant to corrosion; soluble in aqua regia.	Chloride Nitrate Sulfate	Total Body: 10.4 days Bone: 16.6 days Kidneys: 28 days Liver: 18.2 days Spleen: 20.8 days
Palladium (Pd)	46	106.4	Silvery-white metal of the platinum family; D.: 12 g/ml; M.P.: 1550°C; stable in air; soluble in aqua regia and hot H_2SO_4.	Chloride Nitrate Sulfate	Total Body: 5 days Kidneys: 30 days Liver: 19 days Spleen: 15 days
Silver (Ag)	47	107.87	White metal, ductile; D.: 10.5 g/ml; M.P.: 961°C; soluble in HNO_3.	Acetate Nitrate	Total Body: 5 days Bone: 30 days Kidneys: 10 days Liver: 15 days
Cadmium (Cd)	48	112.40	Silver-white metal with a bluish tinge, malleable and ductile; D.: 8.65 g/ml; M.P.: 321°C; stable in air; soluble in acids.	Chloride Nitrate Sulfate	Total Body: 200 days Kidneys: 300 days Liver: 200 days
Indium (In)	49	114.82	Silver-white soft metal, malleable, ductile; D.: 7.3 g/ml; M.P.: 156°C; stable in air; soluble in acids.	Chloride Nitrate Sulfate	Total Body: 48 days Skin: 67 days Bone: 57 days Kidneys: 60 days Liver: 58 days Spleen: 48 days Thyroid: 8.4 days
Tin (Sn)	50	118.69	White soft metal with a bluish tinge, malleable; D.: 7.3 g/ml; M.P.: 232°C; stable in air; soluble in acids.	Chloride Nitrate Sulfate (in excess acid)	Total Body: 35 days Bone: 100 days Liver: 70 days Thyroid: 70 days Prostate: 35 days
Antimony (Sb)	51	121.75	Gray metal, brittle; D.: 6.68 g/ml; M.P.: 631°C; stable in air; soluble in aqua regia and hot concentrated H_2SO_4.	Chloride Nitrate Sulfate (in excess acid)	Total Body: 38 days Bone: 100 days Lungs: 100 days Liver: 38 days Thyroid: 4 days
Tellurium (Te)	52	127.60	Shiny tin-white crystalline nonmetallic solid of sulfur family, brittle; D.: 6.2 g/ml; M.P.: 450°C; stable in air; soluble in HNO_3 and H_2SO_4.	Bromide Chloride Nitrate (in excess acid)	Total Body: 15 days Bone: 30 days Kidneys: 30 days Liver: 30 days Spleen: 30 days Thyroid: 9 days Testes: 30 days
Iodine (I)	53	126.90	Purple volatile solid; D.: 4.9 g/ml; poisonous.	Most metal salts, except those of Ag, Hg, Pb, Tl	Total Body: 138 days Bone: 14 days Kidneys: 7 days Liver: 7 days Spleen: 7 days Thyroid: 138 days Testes: 7 days
Xenon (Xe)	54	131.30	Colorless gas; D.: 5.9 g/l; inert.	None	Short

Table 16. ELEMENTS AND THEIR COMMERCIALLY AVAILABLE ISOTOPES

A. General Data: Elements (*Continued*)

Element (Symbol)	Atomic No.	Atomic Weight	Physical and Chemical Properties	Soluble Compounds	Biological Half-Life
Cesium (Cs)	55	132.91	Shiny very soft alkali metal; D.: 1.87 g/ml; M.P.: 29°C; reacts with air; soluble in water (explosively) and alcohol.	Chloride Nitrate Phosphate Sulfate	Total Body: 70 days Muscle: 140 days Bone: 140 days Lungs: 140 days Kidneys: 42 days Liver: 90 days Spleen: 98 days
Barium (Ba)	56	137.34	White soft alkaline earth metal; D.: 3.5 g/ml; M.P.: 704°C; stable in dry air, but reacts with moist air; soluble in acids.	Chloride Nitrate	Total Body: 65 days Muscle: 2,000 days Bone: 65 days Lungs: 6,500 days Kidneys: 8.5 days Liver: 975 days Spleen: 13 days Ovaries: 4.9 days Testes: 6.2 days
Lanthanum (La)	57	138.91	Tin-white metal, malleable; D.: 6.15 g/ml; M.P.: 920°C; oxidizes in air; soluble in acids.	Chloride Nitrate	Total Body: 500 days Bone: 1,000 days Liver: 400 days
Cerium (Ce)	58	140.12	Soft metal with steel-like appearance, malleable, ductile; D.: 6.9 g/ml; M.P.: ~804°C; oxidizes slowly in air; soluble in acids.	Chloride Nitrate Sulfate	Total Body: 563 days Bone: 1,500 days Kidneys: 563 days Liver: 293 days
Praseodymium (Pr)	59	140.91	Yellowish rare-earth metparamagnetic; D.: 6.77 g/ml; M.P.: 932°C; oxidizes slowly in air; soluble in acids.	Chloride Sulfate	Total Body: 750 days Bone: 1,500 days Kidneys: 750 days Liver: 375 days
Neodymium (Nd)	60	144.24	Yellowish rare-earth metal; D.: 6.8–7 g/ml; M.P.: 1024°C; tarnishes quickly in air; soluble in acids.	Chloride Sulfate	Total Body: 656 days Bone: 1,500 days Kidneys: 656 days Liver: 131 days
Promethium (Pm)	61	(~147)	Artificial element recovered from fission products, primarily as ^{147}Pm; D.: 7.3 g/ml; M.P.: 1035°C; tarnishes in air; soluble in HCl and HNO_3; available usually as oxide.	Chloride	Total Body: 656 days Bone: 1,500 days Kidneys: 656 days Liver: 656 days
Samarium (Sm)	62	150.35	Shiny yellowish-gray hard metal, brittle; D.: 7.54 g/ml; M.P.: 1072°C; tarnishes slowly in air; soluble in acids.	Chloride	Total Body: 656 days Bone: 1,500 days Kidneys: 656 days Liver: 187 days
Europium (Eu)	63	151.96	Shiny soft metal, ductile; D.: 5.26 g/ml; M.P.: 826°C; oxidizes in air, reacts slowly with water; soluble in acids.	Chloride	Total Body: 635 days Bone: 1,500 days Kidneys: 1,480 days Liver: 127 days

Table 16. ELEMENTS AND THEIR COMMERCIALLY AVAILABLE ISOTOPES

A. General Data: Elements (*Continued*)

Element (Symbol)	Atomic No.	Atomic Weight	Physical and Chemical Properties	Soluble Compounds	Biological Half-Life
Gadolinium (Gd)	64	157.25	Shiny grayish-white metal, ductile; D.: 7.9 g/ml; M.P.: 1312°C; tarnishes in moist air; soluble in acids.	Chloride Nitrate Sulfate	Total Body: 550 days Bone: 1,000 days Liver: 460 days
Terbium (Tb)	65	158.92	Shiny grayish-white soft metal; D.: 8.27 g/ml; M.P.: 1356°C; stable in air; soluble in acids.	Chloride Nitrate	Total Body: 670 days Bone: 1,000 days Kidneys: 700 days
Dysprosium (Dy)	66	162.50	Shiny grayish-white metal; D.: 8.54 g/ml; M.P.: 1407°C; tarnishes in moist air; soluble in acids.	Chloride Sulfate	Total Body: 700 days Bone: 1,000 days Liver: 500 days
Holmium (Ho)	67	164.93	Shiny soft metal, malleable; D.: 8.8 g/ml; M.P.: 1461°C; tarnishes in moist air; soluble in acids.	Sulfate	Total Body: 750 days Bone: 1,000 days Kidneys: 800 days Liver: 875 days
Erbium (Er)	68	167.26	Shiny dark-gray soft metal; D.: 9.05 g/ml; M.P.: 1497°C; stable in air; soluble in acids.	Chloride Nitrate Sulfate	Total Body: 650 days Bone: 1,000 days Kidneys: 650 days Liver: 433 days
Thulium (Tm)	69	168.93	Shiny grayish-white soft metal; D.: 9.33 g/ml; M.P.: 1545°C; stable in air; soluble in acids.	Chloride	Total Body: 675 days Bone: 1,000 days Kidneys: 335 days
Ytterbium (Yb)	70	173.04	Shiny grayish-white soft metal; D.: 6.98 g/ml; M.P.: 824°C; stable in air; soluble in acids	Chloride Sulfate	Total Body: 685 days Bone; 1,000 days Kidneys: 685 days
Lutetium (Lu)	71	174.97	Shiny grayish-white metal; D.: 9.84 g/ml; M.P.: 1652°C; stable in air; soluble in acids.	Sulfate	Total Body: 750 days Bone: 1,000 days Kidneys: 750 days
Hafnium (Hf)	72	178.49	Silvery-white metal resembling zirconium, ductile; D.: 11.4 g/ml; M.P.: 1975°C; stable in air; soluble in aqua regia.	Oxychloride (in HCl solution)	Total Body: 563 days Bone: 600 days Kidneys: 563 days Liver: 625 days Spleen: 350 days
Tantalum (Ta)	73	180.95	Shiny bluish-gray metal; D.: 16.6 g/ml; M.P.: 2977°C; stable in air; soluble in acids, except HF.	Fluoride Tantalate (in KOH solution)	Total Body: 240 days Bone: 300 days Kidneys: 400 days Liver: 400 days Spleen: 240 days
Tungsten (W)	74	183.85	Shiny steel-gray metal; D.: 19.3 g/ml; M.P.: 3380°C (highest of any metal); stable in air; soluble in hot KOH, insoluble in dilute acids.	Oxychloride Tungstate (in KOH solution)	Total Body: 1 day Bone: 9 days Liver: 4 days
Rhenium (Re)	75	186.2	Platinum-like heavy metal; D.: 20.5 g/ml; M.P.: 3180°C; very stable in air; soluble in HNO_3	Oxide Rhenate (in HNO_3 solution)	Total Body: 7 days Skin: 25 days Bone: 3.5 days Liver: 14 days Thyroid: 3 days

Table 16. ELEMENTS AND THEIR COMMERCIALLY AVAILABLE ISOTOPES

A. General Data: Elements (*Continued*)

Element (Symbol)	Atomic No.	Atomic Weight	Physical and Chemical Properties	Soluble Compounds	Biological Half-Life
Osmium (Os)	76	190.2	Brilliant bluish hard metal, brittle; D.: 22.48 g/ml (highest of any known substance); M.P.: 2725°C; stable in air; soluble in fuming HNO_3; oxide very poisonous.	Alkali metal complexes (in NaOH) Chloride	Total Body: 2 days Kidneys: 5 days Liver: 5.5 days
Iridium (Ir)	77	192.2	Shiny hard metal, brittle and nonductile, D.: 22.4 g/ml; M.P.: 2450°C; unreactive; insoluble in acids.	Chloride	Total Body: 20 days Kidneys: 50 days Liver: 27 days Spleen: 50 days
Platinum (Pt)	78	195.0	Tin-white soft metal, malleable, ductile; D.: 21.45 g/ml; M.P.: 1769°C; stable in air; soluble in acids.	Chloride Sulfate	Total Body: 24 days Kidneys: 60 days Liver: 20 days Spleen: 60 days
Gold (Au)	79	196.97	Bright-yellow soft metal, malleable, ductile; D.: 19.3 g/ml; M.P.: 1063°C; stable in air; insoluble in acids.	Chloride Sulfate	Total Body: 120 days Kidneys: 280 days Liver: 300 days Spleen: 240 days
Mercury (Hg)	80	200.59	Shiny liquid metal; D.: 13.55 g/ml; B.P.; 357°C; stable in air; soluble in HNO_3.	Chloride Nitrate	Total Body: 10 days Kidneys: 14.5 days Liver: 13.5 days Spleen: 10 days
Thallium (Tl)	81	204.37	Gray soft metal, malleable; D.: 11.85 g/ml; M.P.: 304°C; stable in air; soluble in HNO_3 and H_2SO_4.	Chloride Nitrate	Total Body: 5 days Muscle: 5.5 days Bone: 7 days Lungs: 6 days Kidneys: 7 days Liver: 5 days
Lead (Pb)	82	207.19	Dull-gray soft metal, malleable; D.: 11.34 g/ml; M.P.: 327°C; tarnishes in air; soluble in acids.	Acetate Chloride (slightly) Nitrate	Total Body: 1,500 days Bone: 3,700 days Kidneys: 531 days Liver: 1,900 days
Bismuth (Bi)	83	208.98	Lustrous tin-white soft metal, brittle; has lowest heat conductivity of any metal; D.: 9.78 g/ml; M.P.: 271°C; stable in air; soluble in HNO_3.	Chloride Nitrate Sulfate (in acid solution)	Total Body: 5 days Bone: 13 days Kidneys: 6 days Liver: 15 days Spleen: 10 days
Polonium (Po)	84	(210)	Radioactive element; D.: 9.4 g/ml; M.P.: 254°C; soluble in HCl and HNO_3.	Chloride Nitrate	Total Body: 30 days Bone: 24 days Kidneys: 70 days Liver: 41 days Spleen: 60 days
Radon (Rn)	86	(222)	Radioactive gas, decay product of radium; D.: 10 g/l; B.P.: −62°C; chemically inert.	None	Short

Table 16. ELEMENTS AND THEIR COMMERCIALLY AVAILABLE ISOTOPES

A. General Data: Elements (*Continued*)

Element (Symbol)	Atomic No.	Atomic Weight	Physical and Chemical Properties	Soluble Compounds	Biological Half-Life
Radium (Ra)	88	(226)	White metal; D.: ~5 g/ml; reactive; salts are poisonous.	Bromide Chloride	Total Body: 8,100 days Bone: 16,000 days Kidneys: 10 days Liver: 10 days
Actinium (Ac)	89	(227)	Has not been isolated in pure state.	Chloride Nitrate	Total Body: 24,000 days Bone: 73,000 days Kidneys: 24,000 days Liver: 24,000 days
Thorium (Th)	90	232.04	Gray soft metal, malleable, ductile; D.: 11.7 g/ml; M.P.: 1730°C; burns in air; soluble in HCl and H_2SO_4.	Halides Nitrate	Total Body: 57,000 days Bone: 73,000 days Kidneys: 22,000 days Liver: 57,000 days
Protactinium (Pa)	91	(231)	Rare element associated with uranium, < 1 ppm.	Chloride Nitrate	Total Body: 41,000 days Bone: 73,000 days Kidneys: 51,000 days Liver: 58,000 days
Uranium (U)	92	238.03 (highest of naturally occurring elements)	Lustrous white metal, malleable; D.: 18.9 g/ml; reactive.	Chloride Sulfate	Total Body: 100 days Bone: 300 days Kidneys: 15 days
Neptunium (Np)	93	(237)	Silvery metal; D.: 20.4 g/ml; stable in air; synthetic long-lived radioisotope.	Chloride Sulfate	Total Body: 39,000 days Bone: 73,000 days Kidneys: 64,000 days Liver: 54,000 days
Plutonium (Pu)	94	(242)	Metal, very similar to uranium.	Chloride Sulfate	Total Body: 65,000 days Bone: 73,000 days Kidneys: 30,000 days Liver: 30,000 days
Americium (Am)	95	(241)	Silvery metal, artificially prepared; radioactive.	Chloride	Total Body: 20,000 days Bone: 70,000 days Kidneys: 27,000 days Liver: 35,000 days
Curium (Cm)	96	(242)	Synthetic element, available primarily as oxide; radioactive.	Chloride	Total Body: 24,000 days Bone: 73,000 days Kidneys: 24,000 days Liver: 3,000 days

Table 16. ELEMENTS AND THEIR COMMERCIALLY AVAILABLE ISOTOPES

B. Nuclear Data: I. Radioactive Isotopes

Nuclides of (Mass No.)	Half-Life	Production Reaction	Production Cross Section, barns	Specific Activity (Theory), Ci/g	Yield, Ci/g* Half-Lives 1	2	1 year (or saturation,** if shorter)	Gamma Energies, Mev (%)	Beta Energies, Mev (%)	Interfering Isotopes Isotope	Energy and Type	Decay Product
Hydrogen												
(3)	12.3 y	^{6}Li(n,α)^{3}H	950	9.8×10^3			~10		0.018(100)			^{3}He
Beryllium												
(7)	53 d	^{7}Li(p,n)^{7}Be		3.6×10^5	Cyclotron, 30 mc/hr			0.48(10)				^{7}Li
(10)	2.7×10^6 y	^{9}Be(n,γ)^{10}Be	0.01	0.013			≪0.001		0.56(100)			^{10}B
Carbon												
(14)	5.73×10^3 y	^{14}N(n,p)^{14}C	1.8	4.6			<0.001		0.155(100)			^{14}N
Fluorine												
(18)	1.8 h	^{19}F(p,pn)^{18}F		1.0×10^8	Cyclotron, 1500 mc/hr			0.5(O X ray)	β^+0.65(97)			^{18}O
Sodium												
(22)	2.58 y	^{25}Mg(p,α)^{22}Na ^{26}Mg(p,αn)^{22}Na		6.0×10^3	Cyclotron, 0.35 mc/hr			0.511(180) 1.27(100)	β^+0.55(90) Trace 1.8			^{22}Ne
(24)	15.0 h	^{23}Na(n,γ)^{24}Na	0.53	9.0×10^6	0.17	0.24	0.33	1.37(100) 2.75(100)	Trace 4.2 1.4(100)			^{24}Mg
Magnesium												
(28)	21.3 h	^{26}Mg(T,p)^{28}Mg		5.3×10^6	Cyclotron, yield not available			1.35(70) 3 others to 0.95	0.46(100)	^{28}Al	1.78γ 2.87β	^{28}Al → ^{28}Si
Aluminum												
(26)	7.4×10^5 y	^{26}Mg(p,n)^{26}Al		0.02	Cyclotron, yield not available			0.511(170) 1.81(100)	β^+1.17(85)			^{26}Mg
Silicon												
(31)	2.62 h	^{30}Si(n,γ)^{31}Si	0.11	4.0×10^7	<0.001		~0.002	1.26(<0.1)	1.48(99)			^{31}P

Table 16. ELEMENTS AND THEIR COMMERCIALLY AVAILABLE ISOTOPES

B. Nuclear Data: I. Radioactive Isotopes (Continued)

Nuclides of (Mass No.)	Half-Life	Production Reaction	Production Cross Section, barns	Specific Activity (Theory), Ci/g	Yield, Ci/g* Half-Lives 1	2	1 year (or saturation,** if shorter)	Gamma Energies, Mev (%)	Beta Energies, Mev (%)	Interfering Isotopes Isotope	Energy and Type	Decay Product
Phosphorus												
(32)	14.3 d	$^{32}S(n,p)^{32}P$ $^{31}P(n,\gamma)^{32}P$	0.10 0.19	3.0×10^5	0.03 0.05	0.04 0.07	0.05 0.1		1.70(100)	^{33}P	$0.25\beta^-$	^{32}S
(33)	25 d	$^{33}S(n,p)^{33}P$	0.002	1.6×10^5			≪0.001		0.25(100)	^{32}P	$1.70\beta^-$	^{33}S
Sulfur												
(35)	86.7 d	$^{35}Cl(n,p)^{35}S$ $^{34}S(n,\gamma)^{35}S$	0.40 0.27	4.0×10^4	0.07 0.003	0.10 0.004	0.12 0.005		0.17(100)			^{35}Cl
Chlorine												
(36)	3.0×10^5 y	$^{35}Cl(n,\gamma)^{36}Cl$	44	0.033			≪0.001		0.714(98)			^{36}Ar
(38)	37.3 m	$^{37}Cl(n,\gamma)^{38}Cl$	0.4	1.3×10^8	0.02	0.03	0.04	1.60(38) 2.17(47)	4.9(53)	^{36}Cl	$0.7\beta^-$	^{38}Ar
Argon												
(37)	35.1 d	$^{40}Ca(n,\alpha)^{37}Ar$	0.43	1.0×10^5	0.09	0.13	0.18	~3 kev X ray Brehm. to 0.8				^{37}Cl
Potassium												
(42)	12.4 h	$^{41}K(n,\gamma)^{42}K$	1.2	6.0×10^6	0.015	0.023	0.031	Trace to 2.4 1.5(20)	2.0(20) 3.5(80)			^{42}Ca
(43)	22 h	$^{40}Ar(\alpha,p)^{43}K$		3.0×10^6	Cyclotron, yield not available				0.83(87)			^{43}Ca
Calcium												
(45)	165 d	$^{44}Ca(n,\gamma)^{45}Ca$	0.7	1.8×10^4	0.003	0.004	0.004		0.25(100)			^{45}Sc
(47)	4.7 d	$^{46}Ca(n,\gamma)^{47}Ca$	0.3	6.0×10^5			<0.001	1.3(74)	2.0(18) 0.67(82)	^{47}Sc ^{45}Ca	$0.6\beta^-$; 0.16γ, 0.25β	$^{47}Sc \to ^{47}Ti$

Table 16. ELEMENTS AND THEIR COMMERCIALLY AVAILABLE ISOTOPES

B. Nuclear Data: I. Radioactive Isotopes (Continued)

Nuclides of (Mass No.)	Half-Life	Production Reaction	Production Cross Section, barns	Specific Activity (Theory), Ci/g	Yield, Ci/g* Half-Lives 1	Yield, Ci/g* Half-Lives 2	1 year (or saturation,** if shorter)	Gamma Energies, Mev (%)	Beta Energies, Mev (%)	Interfering Isotopes Isotope	Interfering Isotopes Energy and Type	Decay Product
Scandium												
(43)	3.9 h	$^{40}Ca(\alpha,p)^{43}Sc$		1.9×10^7	Cyclotron, yield not available			0.38(22) 0.51(176)	0.8 1.2β^+(88)			^{43}Ca
(44m)	2.4 d	$^{41}K(\alpha,n)^{44m}Sc$		1.2×10^6	Cyclotron, yield not available			0.27(86) 1.14(2.7)	Trace 0.27e^-	^{44}Sc	1.5β^+; 0.5, 1.16γ	^{44}Ca
(44)	4.0 h	Daughter ^{44}Ti		1.8×10^7				0.51(188) 1.16(100)	1.5β^+(94)			^{44}Ca
(46)	84 d	$^{45}Sc(n,\gamma)^{46}Sc$	23	3.3×10^4	4.0	6.0	7.9	0.89(100) 1.12(100)	0.36($\sim$100) 1.5(< 0.01)			^{46}Ti
(47)	3.4 d	Daughter ^{47}Ca		8.0×10^5				0.16(73)	0.60(27) 0.43(73)			^{47}Ti
Titanium												
(44)	48 y	$^{45}Sc(p,2n)^{44}Ti$		1.7×10^2	Cyclotron, yield not available			0.07(90) 0.08(98)	0.07e^-	^{44}Sc	0.5, 1.16γ; 1.5β^+	$^{44}Sc \rightarrow ^{44}Ti$
Vanadium												
(48)	16.1 d	$^{48}Ti(p,n)^{48}V$		1.6×10^5	Cyclotron, 25 mc/hr			0.5(100) 0.98(100) 1.3(97) 2.2(3)	0.70β^+(50)			^{48}Ti
(49)	330 d	$^{52}Cr(p,\alpha)^{49}V$		8.0×10^3	Cyclotron, yield not available			Brems. to 0.6 Ti X rays				^{49}Ti

Table 16. ELEMENTS AND THEIR COMMERCIALLY AVAILABLE ISOTOPES

B. Nuclear Data: I. Radioactive Isotopes (Continued)

Nuclides of (Mass No.)	Half-Life	Production Reaction	Production Cross Section, barns	Specific Activity (Theory), Ci/g	Yield, Ci/g* Half-Lives 1	Yield, Ci/g* Half-Lives 2	1 year (or saturation,** if shorter)	Gamma Energies, Mev (%)	Beta Energies, Mev (%)	Interfering Isotopes Isotope	Interfering Isotopes Energy and Type	Decay Product
Chromium												
(51)	27.8 d	$^{50}Cr(n,\gamma)^{51}Cr$ $^{51}V(p,n)^{51}Cr$	17	9.0×10^4	0.11	0.17	0.23 Cyclotron, 120 mc/hr	0.32(9)	Trace 0.3 e$^-$			^{51}V
Manganese												
(52)	5.7 d	$^{52}Cr(p,n)^{52}Mn$		5.0×10^5	Cyclotron, 65 mc/hr			1.4(100) 0.9(84) + 0.5, 0.7	0.6β^+(34)			^{52}Cr
(53)	2.0×10^6 y	$^{53}Cr(p,n)^{53}Mn$ $^{52}Cr(d,n)^{53}Mn$		0.003	Cyclotron, yield not available			Cr X rays				^{53}Cr
(54)	303 d	$^{54}Cr(p,n)^{54}Mn$		8.0×10^3	Cyclotron, 0.5 mc/hr			0.83(100)	Trace 0.83 e$^-$			^{54}Cr
(56)	2.6 h	$^{55}Mn(n,\gamma)^{56}Mn$	13.3	2.0×10^7	2.1	3.2	4.1	0.8(99) 1.8(29) 2.1(15)	2.85			^{56}Fe
Iron												
(52)	8.3 h	$^{50}Cr(\alpha,2n)^{52}Fe$		7.3×10^6	Cyclotron, yield not available			0.16(100) 0.51(112)	0.8β^+(56)	^{52}Mn	1.6β^+ 1.4γ	$^{52}Mn \to ^{52}Cr$
(55)	2.7 y	$^{54}Fe(n,\gamma)^{55}Fe$ $^{55}Mn(p,n)^{55}Fe$	2.9	2.5×10^3	Cyclotron, 8 mc/hr		0.001	~6 kev Mn X ray		^{59}Fe	0.48β^- γ to 1.3	^{55}Mn
(59)	45 d	$^{58}Fe(n,\gamma)^{59}Fe$	1.1	5.0×10^4	< 0.001	0.001	0.001	1.1(56) 1.3(44)	0.48(99) 1.57(0.3)	^{55}Fe	Mn X ray	^{59}Co

Table 16. ELEMENTS AND THEIR COMMERCIALLY AVAILABLE ISOTOPES

B. Nuclear Data: I. Radioactive Isotopes (*Continued*)

Nuclides of (Mass No.)	Half-Life	Production Reaction	Production Cross Section, barns	Specific Activity (Theory), Ci/g	Yield, Ci/g* Half-Lives 1	2	1 year (or saturation,** if shorter)	Gamma Energies, Mev (%)	Beta Energies, Mev (%)	Interfering Isotopes Isotope	Energy and Type	Decay Product
Cobalt												
(56)	77.3 d	^{56}Fe(p,n)^{56}Co		3.0×10^4	Cyclotron, 125 mc/hr			0.85(100) to 3.3(13)	1.5β+(20)			^{56}Fe
(57)	267 d	^{60}Ni(p,α)^{57}Co		8.0×10^3	Cyclotron, 25 mc/hr			0.12(87) 0.14(11)	4 e− to 0.13			^{57}Fe
(58)	71 d	^{55}Mn(α,n)^{58}Co ^{58}Ni(n,p)^{58}Co	0.1 (fast)	3.0×10^4	Cyclotron, yield not available 0.0009	0.0013	0.0018	0.81(99) 1.7(0.6)	0.47β+(15)			^{58}Fe
(60)	5.26 y	^{59}Co(n,γ)^{60}Co	37	1.1×10^3			1.26	1.17(100) 1.33(100)	0.31(99+) 1.5(0.12)			^{60}Ni
Nickel												
(63)	92 y	^{62}Ni(n,γ)^{63}Ni	15	6.0×10^2			0.001		0.067(100)			^{63}Cu
(65)	2.56 h	^{64}Ni(n,γ)^{65}Ni	1.5	2.0×10^7			≪ 0.001	1.1(16) 1.5(25)	Up to 2.1	^{63}Ni	0.067β−	^{65}Cu
Copper												
(61)	3.3 h	^{60}Ni(d,n)^{61}Cu		1.6×10^7	Cyclotron, yield not available			0.06–1.2	1.2β+(60) Some 0.06 e−			^{61}Ni
(64)	12.9 h	^{63}Cu(n,γ)^{64}Cu	4.5	3.8×10^6	0.4	0.6	0.8	0.51(38) 1.34(0.5)	0.57β(39) 0.66β+(19)			^{64}Zn; ^{64}Ni
(67)	61 h	^{67}Zn(n,p)^{67}Cu	∼0.01	8.0×10^5			≪ 0.001	0.18(40) 0.09(23)	4 to 0.57 maximum			^{67}Zn

Table 16. ELEMENTS AND THEIR COMMERCIALLY AVAILABLE ISOTOPES

B. Nuclear Data: I. Radioactive Isotopes (Continued)

Nuclides of (Mass No.)	Half-Life	Production Reaction	Production Cross Section, barns	Specific Activity (Theory), Ci/g	Yield, Ci/g* Half-Lives			Gamma Energies, Mev (%)	Beta Energies, Mev (%)	Interfering Isotopes		Decay Product
					1	2	1 year (or saturation,** if shorter)			Isotope	Energy and Type	
Zinc												
(65)	245 d	$^{64}Zn(n,\gamma)^{65}Zn$	0.46	8.0×10^3	0.3		0.36	1.12(49)	Trace 1.1e$^-$ 0.33(β^+(2)			^{65}Cu
(69m)	14 h	$^{68}Zn(n,\gamma)^{69m}Zn$	0.1	3.0×10^6	0.02	0.03	0.04	0.44(95)	0.43e$^-$(5)	^{69}Zn	0.90β^-	^{69}Ga
Gallium												
(66)	9.5 h	$^{63}Cu(\alpha,n)^{66}Ga$		4.5×10^6	Cyclotron, yield not available			0.51(114) 1.1(37) 2.7(25)	4.2β^+(57)			^{66}Zn
(67)	78 h	$^{67}Zn(p,n)^{67}Ga$		5.0×10^5	Cyclotron, 120 mc/hr			0.09(40) 0.18(24) 0.30(22) 0.39(7)	0.09e$^-$(15)			^{67}Zn
(72)	14.1 h	$^{71}Ga(n,\gamma)^{72}Ga$	5.0	3.0×10^6	0.23	0.35	0.48	0.6–1.9(142) 2.2–2.5(59)	0.6–1.5(83) 2.5–3.2(17)			^{72}Ge
Germanium												
(68)	282 d	$^{69}Ga(p,2n)^{68}Ge$		7.0×10^3	Cyclotron, 3 mc/hr			Ga X rays		^{68}Ga	1.9β^+; 0.51 + 1.1γ	^{68}Zn
(71)	11 d	$^{70}Ge(n,\gamma)^{71}Ge$	3.5	1.6×10^5	0.080	0.120	0.161	Ga X rays		^{77}Ge (~1% for 1 day)	2.8β^- 2.4γ	^{71}Ga
(77)	11 h	$^{76}Ge(n,\gamma)^{77}Ge$	0.2	4.4×10^6	0.0015	0.0024	0.0033	>10 to 2.4	10 to 2.8	^{71}Ge, ^{77}As	2.8β^- 2.4γ	$^{77}As \rightarrow {}^{77}Se$

Table 16. ELEMENTS AND THEIR COMMERCIALLY AVAILABLE ISOTOPES

B. Nuclear Data: I. Radioactive Isotopes (*Continued*)

Nuclides of (Mass No.)	Half-Life	Production Reaction	Production Cross Section, barns	Specific Activity (Theory), Ci/g	Yield, Ci/g* Half-Lives 1	Yield, Ci/g* Half-Lives 2	1 year (or saturation,** if shorter)	Gamma Energies, Mev (%)	Beta Energies, Mev (%)	Interfering Isotopes Isotope	Interfering Isotopes Energy and Type	Decay Product
Arsenic												
(74)	18 d	^{71}Ga(α,n)^{74}As		1.0×10^5	Cyclotron, yield not available			0.5–0.6(124)	1.36(18) 0.7(14) β+:1.5(3.5) β+:0.9(26)			^{74}Ge ^{74}Se
(76)	26.5 h	^{75}As(n,γ)^{76}As	4.5	1.6×10^6	0.5	0.8	1.0	0.56(43) ~7 to 2.1	2.97(53) 7 to 2.4			^{76}Se
(77)	39 h	^{77}Ge daughter		1.0×10^6	Carrier-free daughter of ^{77}Ge			3 to 0.5(3)	0.6(97)	^{77m}Se	0.16γ 0.16 e$^-$	^{77}Se
Selenium												
(75)	120 d	^{74}Se(n,γ)^{75}Se	30	1.4×10^4	0.03	0.04	0.05	0.14(57) 0.27(60)	~8 to 0.4 e$^-$ to 0.25			^{75}As
Bromine												
(77)	58 h	^{75}As(α,2n)^{77}Br		7.0×10^5	Cyclotron, yield not available			0.24(30) 0.52(24) 5 to 1.0	β+:0.34(1) 3 e$^-$ to 0.5	^{77m}Se	0.16γ 0.16 e$^-$	^{77}Se
(82)	35.3 h	^{81}Br(n,γ)^{82}Br	3.2	1.0×10^6	0.14	0.22	0.30	0.55(66) 0.78(83) 6 to 1.48	0.44(100)			^{82}Kr

Table 16. ELEMENTS AND THEIR COMMERCIALLY AVAILABLE ISOTOPES

B. Nuclear Data: I. Radioactive Isotopes (*Continued*)

Nuclides of (Mass No.)	Half-Life	Production Reaction	Production Cross Section, barns	Specific Activity (Theory), Ci/g	Yield, Ci/g* Half-Lives 1	Yield, Ci/g* Half-Lives 2	1 year (or saturation,** if shorter)	Gamma Energies, Mev (%)	Beta Energies, Mev (%)	Interfering Isotopes Isotope	Interfering Isotopes Energy and Type	Decay Product
Krypton												
(83m)	1.86 h	^{83}Rb daughter		2.0×10^7				0.01(9) Kr X rays	3 e$^-$ to 0.03			^{83}Kr
(85m)	4.4 h	^{84}Kr(n,γ)^{85}Kr	0.10	8.0×10^6	0.107	0.160	0.214	0.15(74) 0.31(13)	0.82(77)	^{85}Kr	0.67β^-	^{85}Kr → ^{85}Rb
(85)	10.76 y	Fission		4.0×10^2				0.5(0.4)	0.67(99)			^{85}Rb
(79)	34.9 h	^{78}Kr(n,γ)^{79}Kr	2	1.0×10^6	< 0.001	< 0.001	< 0.001	0.51(15) 7 to 1.3	0.6β^+(8) Some e$^-$ to 0.4	^{85m}Kr	0.7β^-	^{79}Br
Rubidium												
(83)	83 d	^{81}Br(α,2n)^{83}Rb		2.0×10^4	Cyclotron, yield not available			0.53(93) 0.79(1) Kr X rays	Some e$^-$ to 0.03	^{83m}Kr	0.01γ Kr X rays 0.03e$^-$	^{83}Kr
(84)	33 d	^{84}Kr(p,n)^{84}Rb ^{84}Sr(n,p)^{84}Rb		5.0×10^4	Cyclotron, 1 mc/hr Cyclotron, yield not available			0.9(74) 0.52(42)	β^+0.8(11) 1.6(10) β^-0.9(3)			^{84}Kr; ^{84}Sr
(86)	18.7 d	^{85}Rb(n,γ)^{86}Rb	1.0	9.0×10^4	0.07	0.09	0.14	1.1(9)	0.7(9) 1.8(9)			^{86}Sr
Strontium												
(85)	64 d	^{84}Sr(n,γ)^{85}Sr	1.45	2.4×10^4	< 0.001		0.001	0.51(100)	0.5 e$^-$ (1)			^{85}Rb
(87m)	2.8 h	^{87}Y daughter		1.2×10^7				Sr X rays 0.39(80)	0.4 e$^-$ (22)			^{87}Rb

Table 16. ELEMENTS AND THEIR COMMERCIALLY AVAILABLE ISOTOPES

B. Nuclear Data: I. Radioactive Isotopes (Continued)

Nuclides of (Mass No.)	Half-Life	Production Reaction	Production Cross Section, barns	Specific Activity (Theory), Ci/g	Yield, Ci/g* Half-Lives 1	Yield, Ci/g* Half-Lives 2	1 year (or saturation,** if shorter)	Gamma Energies, Mev (%)	Beta Energies, Mev (%)	Interfering Isotopes Isotope	Interfering Isotopes Energy and Type	Decay Product
Strontium (cont.)												
(89)	50.4 d	Fission		3.0×10^4					1.46(99)	^{90}Sr; ^{89}Y; ^{90}Y	0.6 2.2β− 0.9γ	^{89}Y
(90)	28 y	Fission		1.44×10^2					0.6(100) 2.2(100) from ^{90}Y	^{89}Sr	1.5β−	^{90}Zr
Yttrium												
(87)	80 h	^{87}Sr(p,n)^{87}Y		4.5×10^5	Cyclotron, 8.4 mc/hr			0.48(99)	0.7β+(0.3)	^{87m}Sr	0.39γ	^{87}Sr
(88)	108 d	^{88}Sr(p,n)^{88}Y		1.4×10^4	Cyclotron, 2.3 mc/hr			0.90(91) 1.84(100)	0.76β+(0.2)			^{88}Sr
(90)	64.2 h	^{90}Sr daughter		5.5×10^5					2.27(99)			^{90}Zr
(91)	59 d	Fission		2.5×10^4				1.22(0.3)	1.55(100)			^{91}Zr
Zirconium												
(95)	65 d	Fission		2.1×10^4				0.72(49) 0.76(49)	0.36−0.40(98) 0.89(2)	^{95}Nb ^{95m}Nb	0.16β− 0.77γ	^{95}Nb→^{95}Mo
(97)	17 h	^{96}Zr(n,γ)^{97}Zr Fission	0.05	2.0×10^6			<0.001	0.75(93) from ^{97m}Zr	β−1.9(93)	^{97}Nb	1.9β−(90) 0.66γ(99)	^{97}Nb→^{97}Mo
Niobium												
(95)	35 d	Daughter ^{95}Zr		4.0×10^4				0.77(100)	0.16(100)			^{95}Mo

Table 16. ELEMENTS AND THEIR COMMERCIALLY AVAILABLE ISOTOPES

B. Nuclear Data: I. Radioactive Isotopes (Continued)

Nuclides of (Mass No.)	Half-Life	Production Reaction	Production Cross Section, barns	Specific Activity (Theory), Ci/g	Yield, Ci/g* Half-Lives 1	Yield, Ci/g* Half-Lives 2	Yield, Ci/g* 1 year (or saturation,** if shorter)	Gamma Energies, Mev (%)	Beta Energies, Mev (%)	Interfering Isotopes Isotope	Interfering Isotopes Energy and Type	Decay Product
Molybdenum												
(99)	66 h	^{98}Mo(n,γ)^{99}Mo	0.15	4.8×10^5	0.003	0.004	0.006	5 to 0.78	1.23(82) 0.45(17)	^{99m}Tc ^{99}Tc	0.29β$^-$ 0.14γ	^{99}Tc→^{99}Ru
Technetium												
(99m)	6.0 h	Daughter ^{99}Mo		5.0×10^6				0.14(90)	Trace 0.12 e$^-$	^{99}Tc	0.29β$^-$	^{99}Tc→^{99}Ru
(99)	2.1×10^5 y	Daughter ^{99m}Tc Fission		1.7×10^{-2}					0.29(100)			^{99}Ru
Ruthenium												
(97)	2.9 d	^{96}Ru(n,γ)^{97}Ru	0.2	4.0×10^5	0.0009	0.0014	0.0018	0.22(91) 0.32(8)	Some 0.19 e$^-$			^{97}Tc→^{97}Mo
(103)	40 d	Fission ^{102}Ru(n,γ)^{103}Ru	1.4	3.0×10^4	0.03	0.05	0.07	0.50(88) 0.61(6)	0.70(3) 0.21(89)	^{106}Ru ^{106}Rh	1.5γ 3.5β$^-$	^{103}Rh
(106)	1.0 y	Fission		3.4×10^3					0.04(100)	^{103}Ru ^{106}Rh	3.5β$^-$(80) to 1.5γ	^{106}Pd
Rhodium												
(102m)	2.9 y	^{102}Ru(p,n)^{102}Rh		1.2×10^3	Cyclotron, 0.01 mc/hr			0.48(95) 6 to 1.1				^{102}Ru
(105)	36 h	^{104}Ru(n,γ) ^{105}Ru $\xrightarrow{\beta^-}$ ^{105}Rh	0.48	8.0×10^5	0.008	0.012	0.016	0.31(24)	0.56(100)	^{103m}Rh	negligible after 5 hrs	^{105}Pd

Table 16. ELEMENTS AND THEIR COMMERCIALLY AVAILABLE ISOTOPES

B. Nuclear Data: I. Radioactive Isotopes (Continued)

Nuclides of (Mass No.)	Half-Life	Production Reaction	Production Cross Section, barns	Specific Activity (Theory), Ci/g	Yield, Ci/g* Half-Lives 1	2	1 year (or saturation,** if shorter)	Gamma Energies, Mev (%)	Beta Energies, Mev (%)	Interfering Isotopes Isotope	Energy and Type	Decay Product
Palladium												
(103)	17 d	$^{102}Pd(n,\gamma)^{103}Pd$ $^{103}Rh(p,n)^{103}Pd$	4.8	7.7×10^4	0.004	0.005	0.007	Rh X rays trace to 0.5		^{103m}Rh	0.04γ 0.02 0.04 e⁻	^{103}Rh
					Cyclotron, 8 mc/hr							
(109)	13.5 h	$^{108}Pd(n,\gamma)^{109}Pd$	12	2.0×10^6	0.25	0.37	0.50	Many to 0.77 (< 0.1%)	1.03(100)	^{109m}Ag	Ag X rays	^{109}Ag
Silver												
(105)	40 d	$^{106}Pd(p,2n)^{105}Ag$		3.0×10^4	Cyclotron, 6 mc/hr			0.34(42) many to 1.1	4 e⁻ to 0.32(10)			^{105}Pd
(110m)	260 d	$^{109}Ag(n,\gamma)^{110m}Ag$	3	4.0×10^3	0.11		0.14	0.89(71) many to 1.5	Several to 2.8			^{110}Pd
(111)	7.5 d	$^{110}Pd(n,\gamma)$ $^{111}Pd \xrightarrow{\beta^-} {}^{111}Ag$	0.2	1.2×10^5	0.0015	0.0022	0.0030	0.25(1) 0.34(6)	1.05(93)			^{111}Cd
Cadmium												
(109)	1.3 y	$^{108}Cd(n,\gamma)^{109}Cd$ $^{109}Ag(p,n)^{109}Cd$	10	2.6×10^3	0.005		0.005	0.09(100) (from ^{109m}Ag)	Some 0.08 e⁻	^{115m}Cd	1.6β⁻	^{109}Ag
					Cyclotron, 9 mc/hr							
(115m)	43 d	$^{114}Cd(n,\gamma)^{115m}Cd$	0.14	2.7×10^4	0.0003	0.0004	0.006	Several to 1.3(< 4)	1.61(97)	^{115}Cd	1.1β⁻ 0.5γ	^{115}In
(115)	2.3 d	$^{114}Cd(n,\gamma)^{115}Cd$	1.1	5.0×10^5	0.024	0.035	0.046	0.5(36)	1.12(62) 0.60(37)			^{115}In

Table 16. ELEMENTS AND THEIR COMMERCIALLY AVAILABLE ISOTOPES

B. Nuclear Data: I. Radioactive Isotopes (*Continued*)

Nuclides of (Mass No.)	Half-Life	Production Reaction	Production Cross Section, barns	Specific Activity (Theory), Ci/g	Yield, Ci/g* Half-Lives 1	2	1 year (or saturation,** if shorter)	Gamma Energies, Mev (%)	Beta Energies, Mev (%)	Interfering Isotopes Isotope	Energy and Type	Decay Product
Indium												
(111)	2.8 d	$^{111}Cd(p,n)^{111}In$		4.0×10^5	Cyclotron, 50 mc/hr			0.17(89) 0.25(94)	0.15–0.24 e^- (15)			^{111}Cd
(114m)	50 d	$^{113}In(n,\gamma)^{114}In$	8	2.3×10^4	0.028	0.042	0.056	0.19(17) 0.56(3.5) 0.72(3.5)	Appreciable 0.2 e^-	^{114}In	2.0β^- 0.4β^+ 1.3γ	^{114}Cd; ^{114}Sn
(116m)	54 m	$^{115}In(n,\gamma)^{116m}In$	154	3.0×10^7	10.0	15.0	20.0	1.29(80) 6 to 2.1	1.0(51) 0.87(28) 0.60(21)			^{116}Sn
Tin												
(113)	118 d	$^{112}Sn(n,\gamma)^{113}Sn$	1.3	1.0×10^4	< 0.0010	0.0012	0.0015	In X rays 0.26(2)		^{113m}In	0.39(98) 0.37 e^-	^{113}In
(119m)	250 d	$^{118}Sn(n,\gamma)^{119m}Sn$	0.01	5.0×10^3			< 0.001	Sn X rays 0.024(16)	0.02–0.06 e^- (80)	^{113}Sn ^{113m}In	0.39γ 0.37 e^-	^{119}Sn
(121)	27 h	$^{120}Sn(n,\gamma)^{121}Sn$	0.14	1.0×10^6	0.003	0.0045	0.006		0.38(100)			^{121}Sb
Antimony												
(122)	2.8 d	$^{121}Sb(n,\gamma)^{122}Sb$	6	4.0×10^5	0.23	0.34	0.46	0.56(66) others to 1.26	Trace 0.56β^+ 1.4(63) 2.0(30)	^{124}Sb	2.3β^- 2.1γ	^{122}Sn; ^{122}Te
(124)	60 d	$^{123}Sb(n,\gamma)^{124}Sb$	3.3	2.0×10^4	0.10	0.15	0.19	0.6(97) 1.7(50) 8 to 2.1	5 to 2.3	None if aged		^{124}Te

Table 16. ELEMENTS AND THEIR COMMERCIALLY AVAILABLE ISOTOPES

B. Nuclear Data: I. Radioactive Isotopes (*Continued*)

Nuclides of (Mass No.)	Half-Life	Production Reaction	Production Cross Section, barns	Specific Activity (Theory), Ci/g	Yield, Ci/g* Half-Lives 1	Yield, Ci/g* Half-Lives 2	1 year (or saturation,** if shorter)	Gamma Energies, Mev (%)	Beta Energies, Mev (%)	Interfering Isotopes Isotope	Interfering Isotopes Energy and Type	Decay Product
Antimony (*cont.*)												
(125)	2.7 y	^{124}Sn(n,γ)^{125}Sb $\xrightarrow{\beta^-}$ ^{125}Sb	0.1	1.1×10^3	< 0.001			0.44(40) 4 to 0.66	0.6(50) 2.3(20) trace β$^+$	^{125m}Te	0.11 e$^-$ Te X rays	^{125}Te
Tellurium												
(125m)	58 d	^{124}Te(n,γ)^{125m}Te	5	2.0×10^4	0.015	0.022	0.030	Te X rays 0.035(7) 0.11(0.3)	4 e$^-$ to 0.11	^{127m}Te ^{129m}Te	1.5 β$^-$ 0.7 γ	^{125}Te
(127m)	105 d	^{126}Te(n,γ)^{127m}Te	0.1	4.0×10^4	0.0013	0.002	0.0026	Te X rays trace to 0.7	0.08 e$^-$	^{127}Te	0.7 β$^-$ I X rays	^{127}I
(127)	9.3 h	^{126}Te(n,γ)^{127}Te	0.9	2.5×10^6	0.011	0.016	0.022	I X rays trace to 0.4	0.69(99)			^{127}I
(129m)	34 d	^{128}Te(n,γ)^{129m}Te	0.02	3.0×10^4			< 0.001	0.69(6) Te X rays	1.6(36) 0.07 0.10 e$^-$	^{125m}Te ^{127m}Te ^{129m}Te	1.48 β$^-$ I and Te X rays	^{129}I → ^{129}Xe
(129)	1.1 h	^{128}Te(n,γ)^{129}Te	0.14	2.5×10^7	0.003	0.004	0.005	0.03(19) 0.46(15) others to 1.1	6 to 1.48 0.025 e$^-$			^{129}I → ^{129}Xe
(132)	3.2 d	Fission		3.0×10^5				0.23(90)	0.2 e$^-$ 0.22(100)	^{132}I	Many β$^-$ to 2.1; many γ to 2.2	^{132}Xe

Table 16. ELEMENTS AND THEIR COMMERCIALLY AVAILABLE ISOTOPES

B. Nuclear Data: I. Radioactive Isotopes (*Continued*)

Nuclides of (Mass No.)	Half-Life	Production Reaction	Production Cross Section, barns	Specific Activity (Theory), Ci/g	Yield, Ci/g* Half-Lives 1	2	1 year (or saturation,** if shorter)	Gamma Energies, Mev (%)	Beta Energies, Mev (%)	Interfering Isotopes Isotope	Energy and Type	Decay Product
Iodine												
(123)	13 h	^{123}Te(p,n)^{123}I		1.9×10^6	Cyclotron, > 100 mc/hr			0.16(97); 3% others to 0.53	Some 0.13 e$^-$			^{123}Te
(124)	4.2 d	^{121}Sb(α,n)^{124}I		2.5×10^5	Cyclotron, yield not available			0.61(67) 0.51(50) others to 2.3	2.1β^+(46) 1.5β^+(46)			^{124}Te
(125)	60.2 d	^{124}Xe(n,γ) $\rightarrow$ ^{125}Xe $\xrightarrow{ec}$ ^{125}I	110	1.7×10^4	0.008	0.012	0.016	0.027 X rays 0.035(7)	0.03 e$^-$(90)			^{125}Te
(126)	13.2 d	^{126}Te(p,n)^{126}I		8.0×10^4	Cyclotron, 1 mc/hr			0.67(33) 0.39(34)	Trace 1.1β^+ 0.87(24) 1.25(7)			^{126}Te; ^{126}Xe
(129)	1.6×10^7 y	Fission		1.8×10^{-4}				0.04(9)	0.15(100)	None if sufficiently aged		^{129}Xe
(130)	12.5 h	^{129}I(n,γ)^{130}I ^{130}Te(p,n)^{130}I	28	2.0×10^6	0.9	1.3	1.8 Cyclotron, 40 mc/hr	0.54(99) 0.67(100) 0.74(87) others to 1.15	0.62(52) 1.04(48)	^{129}I	0.15β 0.04γ	^{130}Xe
(131)	8.05 d	Fission ^{130}Te(n,γ) ^{131}Te $\xrightarrow{\beta^-}$ ^{131}I	0.24	1.2×10^5	0.004	0.006	0.008	0.36(80) others to 0.72	0.61(87) others to 0.81	^{131m}Xe	Xe X rays 0.13 and 0.16 e$^-$	^{131}Xe
(132)	2.3 h	Daughter ^{132}Te		1.2×10^7				0.67(144) 0.77(89) others to 1.99	2.16(18) 1.61(21) 1.22(24) others to 2.1			^{132}Xe
(133)	21 h	Fission		1.0×10^6				0.53(90) others to 1.3	1.27(90)	^{133}Xe ^{133m}Xe	0.23γ 0.35β^-	^{133}Xe $\rightarrow$ ^{133}Cs

Table 16. ELEMENTS AND THEIR COMMERCIALLY AVAILABLE ISOTOPES

B. Nuclear Data: I. Radioactive Isotopes (*Continued*)

Nuclides of (Mass No.)	Half-Life	Production Reaction	Production Cross Section, barns	Specific Activity (Theory), Ci/g	Yield, Ci/g* Half-Lives 1	2	1 year (or saturation,** if shorter)	Gamma Energies, Mev (%)	Beta Energies, Mev (%)	Interfering Isotopes Isotope	Energy and Type	Decay Product
Xenon												
(131m)	12 d	$^{130}Xe(n,\gamma)^{131m}Xe$	5	1.0×10^5	0.012	0.018	0.024	0.164(2) Xe X rays	0.13, 0.16 e^-			^{131}Xe
(133)	5.3 d	Fission		2.0×10^5				0.08(37) 0.08 e^-(63)	0.35(100)	^{85}Kr ^{131m}Xe		^{133}Cs
Cesium												
(131)	9.7 d	$^{130}Ba(n,\gamma)$ $^{131}Ba \xrightarrow{ec} {}^{131}Cs$	8.8	1.0×10^5	< 0.001	< 0.001	0.001	Xe X rays				^{131}Xe
(132)	6.6 d	$^{133}Cs(p,pn)^{132}Cs$		1.5×10^5	Cyclotron, yield not available			0.67(97) trace to 1.3	0.4β^+(0.6) 0.7β^-(2)			^{132}Ba; ^{132}Xe
(134)	2.1 y	$^{133}Cs(n,\gamma)^{134}Cs$	28	1.3×10^3			1.05	0.6(98) 0.8(98) others to 1.4	0.09(27) 0.66(71)			^{134}Ba
(137)	30 y	Fission		$\sim 1.0 \times 10^2$				0.66(85) from ^{137m}Ba	1.18(7) 0.51(93)	^{137m}Ba	0.66γ	^{137}Ba
Barium												
(131)	11.6 d	$^{130}Ba(n,\gamma)^{131}Ba$	8.8	8.7×10^4	< 0.001	< 0.001	0.001	Several to 1.05 0.5(48)		^{131}Cs	Xe X rays	^{131}Xe
(133)	7.2 y	$^{132}Ba(n,\gamma)^{133}Ba$	7	4.0×10^2			< 0.001	Several to 0.38 0.36(69)	4 e^- to 0.32			^{133}Cs
(140)	12.8 d	Fission		7.0×10^4				Several to 0.44 0.54(34)	0.48(40) 1.02(60)	^{140}La	β^- to 2.2 γ to 2.5	^{140}Ce

Table 16. ELEMENTS AND THEIR COMMERCIALLY AVAILABLE ISOTOPES

B. Nuclear Data: I. Radioactive Isotopes (Continued)

Nuclides of (Mass No.)	Half-Life	Production Reaction	Production Cross Section, barns	Specific Activity (Theory), Ci/g	Yield, Ci/g* Half-Lives 1	Yield, Ci/g* Half-Lives 2	Yield, Ci/g* 1 year (or saturation,** if shorter)	Gamma Energies, Mev (%)	Beta Energies, Mev (%)	Interfering Isotopes Isotope	Interfering Isotopes Energy and Type	Decay Product
Lanthanum												
(140)	40 h	Daughter ^{140}Ba		5.3×10^5				0.48(40) 1.6(96) others to 2.5	2.2(27) 5 others			^{140}Ce
Cerium												
(139)	140 d	^{139}La(p,n)^{139}Ce		6.5×10^3	Cyclotron, 1.2 mc/hr			0.165(80)	0.13, 0.16 e$^-$			^{139}La
(141)	32.5 d	Fission	0.6	2.7×10^4				0.15(48)	0.44(70) 0.58(30) 0.10, 0.14 e$^-$	^{144}Ce ^{144}Pr	β$^-$ to 3.0 0.13γ	^{141}Pr
(143)	1.4 d	^{142}Ce(n,γ)^{143}Ce	1	6.5×10^5	0.007	0.010	0.013	0.29(46) others to 1.1	1.40(38) 1.1(42) others 3 e$^-$ to 0.25	^{143}Pr	0.93β$^-$	^{143}Nd
(144)	285 d	Fission		3.0×10^3				0.13(11)	0.31(76) 0.18(24)	^{144}Pr	3.0β$^-$ trace γ to 2.2	^{144}Nd
Praseodymium												
(142)	19.2 h	^{141}Pr(n,γ)^{142}Pr	12	1.1×10^6	0.70	1.05	1.40	1.6(4)	2.15(96) 0.64(4)			^{142}Nd
(143)	13.7 d	Fission		7.0×10^4					0.93(100)			^{143}Nd

Table 16. ELEMENTS AND THEIR COMMERCIALLY AVAILABLE ISOTOPES

B. Nuclear Data: I. Radioactive Isotopes (Continued)

Nuclides of (Mass No.)	Half-Life	Production Reaction	Production Cross Section, barns	Specific Activity (Theory), Ci/g	Yield, Ci/g* Half-Lives 1	2	1 year (or saturation,** if shorter)	Gamma Energies, Mev (%)	Beta Energies, Mev (%)	Interfering Isotopes Isotope	Energy and Type	Decay Product
Neodymium												
(147)	11.1 d	Fission		8.0×10^4				0.09(20) 0.53(13) others	0.80(76) 0.36(20)	^{147}Pm	0.22β$^-$	^{147}Pm → ^{147}Sm
(149)	1.8 h	^{148}Nd(n,γ)^{149}Nd	4	1.1×10^7	0.014	0.021	0.028	0.11(18) 0.21–0.27(53) others to 0.65	1.0–1.1(56) 1.4(38) others	^{149}Pm	1.07β$^-$ some γ to 0.85	^{149}Sm
Promethium												
(147)	2.7 y	Fission		9.0×10^2					0.23(100)			^{147}Sm
(149)	2.2 d	Daughter ^{149}Nd		4.1×10^5				< 3% to 0.85	1.1(97)			^{149}Sm
(151)	1.2 d	^{150}Nd(n,γ) ^{151}Nd $\xrightarrow{\beta^-}$ ^{151}Pm	1.5	2.4×10^5	< 0.001	< 0.001	0.001	0.17(18) 0.34(21) others to 0.96	1.2(100) 4 e$^-$ to 0.06			^{151}Sm → ^{151}Eu
Samarium												
(153)	2 d	^{152}Sm(n,γ)^{153}Sm	210	4.3×10^5	3.1	4.6	6.2	0.10(28) others to 0.64	5 e$^-$ to 0.10 0.81(20) 0.71(50) 0.64(30)			^{153}Eu
Europium												
(152m)	9.3 h	^{151}Eu(n,γ)^{152m}Eu	2800.0	2.0×10^6	750	1100	1500	0.12(8) 0.84(13) 0.96(12) others to 1.4	1.88(90) trace β$^+$			^{152}Sm; ^{152}Gd

Table 16. ELEMENTS AND THEIR COMMERCIALLY AVAILABLE ISOTOPES

B. Nuclear Data: I. Radioactive Isotopes (Continued)

Nuclides of (Mass No.)	Half-Life	Production Reaction	Production Cross Section, barns	Specific Activity (Theory), Ci/g	Yield, Ci/g* Half-Lives 1	Yield, Ci/g* Half-Lives 2	1 year (or saturation,** if shorter)	Gamma Energies, Mev (%)	Beta Energies, Mev (%)	Interfering Isotopes Isotope	Interfering Isotopes Energy and Type	Decay Product
Europium (cont.)												
(152)	12.4 y	^{151}Eu(n,γ)^{152}Eu	5900	2.0×10^2			18.65	0.12(37) 0.34(27) 1.4(14) others	8 to 1.48 trace β⁺	^{154}Eu	1.85β⁻ 1.28γ	^{152}Sm; ^{152}Gd
(154)	16 y	^{153}Eu(n,γ)^{154}Eu	320	1.5×10^2			0.738	0.12(38) 0.72(21) 1.0(31) 1.28(3) others	0.27(20) 0.59(45) 0.89(23) 1.86(12)	^{152}Eu	1.41β⁻ 1.9γ	^{154}Gd
(155)	1.8 y	^{154}Sm(n,γ) ^{155}Sm $\xrightarrow{\beta^-}$ ^{155}Eu	5	1.3×10^3	0.14	0.21	0.27	0.09(32) 0.11(20)	0.25(16) 0.15(14)			^{155}Gd
Gadolinium												
(153)	240 d	^{152}Gd(n,γ)^{153}Gd	180	3.3×10^3	0.019	0.022	0.024	0.10(55) 0.07(2) Eu X rays	4 e⁻ to 0.10			^{153}Eu
(159)	18 h	^{158}Gd(n,γ)^{159}Gd	3.4	1.1×10^6	0.044	0.066	0.088	Tb X rays	0.60(13) 0.89(24) 0.95(63)			^{159}Tb
Terbium												
(160)	72 d	^{159}Tb(n,γ)^{160}Tb	46	1.1×10^4	2.28	3.42	4.56	0.30(30) 0.88(31) 1.0(31) others to 1.27	7 to 0.86 trace to 1.7			^{160}Dy

Table 16. ELEMENTS AND THEIR COMMERCIALLY AVAILABLE ISOTOPES

B. Nuclear Data: I. Radioactive Isotopes (*Continued*)

Nuclides of (Mass No.)	Half-Life	Production Reaction	Production Cross Section, barns	Specific Activity (Theory), Ci/g	Yield, Ci/g* Half-Lives 1	2	1 year (or saturation,** if shorter)	Gamma Energies, Mev (%)	Beta Energies, Mev (%)	Interfering Isotopes Isotope	Energy and Type	Decay Product
Terbium (*cont.*)												
(161)	6.9 d	^{160}Gd(n,γ)^{161}Gd $\xrightarrow{\beta^-}$ ^{161}Tb	0.8	1.2×10^5	0.008	0.012	0.016	0.03(21) 0.05(19) 0.06(5) 0.08(10)	0.45(24) 0.51(66) 0.58(10)			^{161}Dy
Dysprosium												
(165)	2.3 h	^{164}Dy(n,γ)^{165}Dy	800	9.0×10^6	12	18	24	Many to 1.08 (total ~20%)	1.21(16) 1.29(83)			^{165}Ho
Holmium												
(166m)	1.2×10^3 y	^{165}Ho(n,γ)^{166}Ho	1	1.8	0.05	0.07	0.10	0.18(90) 0.28(30) 0.7–0.8(118) others to 1.43	0.07(100) 5 e to 0.18			^{166}Er
Erbium												
(169)	9.4 d	^{168}Er(n,γ)^{169}Er	2	8.0×10^4	0.027	0.040	0.054	Tm M X rays	0.34(100)			^{169}Tm
(171)	7.5 h	^{170}Er(n,γ)^{171}Er	9	2.3×10^6	0.06	0.09	0.12	0.11(25) 0.20(91) others to 0.96	1.07(90) 1.5(2)	^{169}Er	0.34 β– Tm X rays	^{171}Tm → ^{171}Yb
Thulium												
(170)	130 d	^{169}Tm(n,γ)^{170}Tm	125	6.0×10^3	5.6	8.4	10.4	0.08(10)	0.97(78) 0.89(22)			^{170}Yb

Table 16. ELEMENTS AND THEIR COMMERCIALLY AVAILABLE ISOTOPES

B. Nuclear Data: I. Radioactive Isotopes (Continued)

Nuclides of (Mass No.)	Half-Life	Production Reaction	Production Cross Section, barns	Specific Activity (Theory), Ci/g	Yield, Ci/g* Half-Lives 1	2	1 year (or saturation,** if shorter)	Gamma Energies, Mev (%)	Beta Energies, Mev (%)	Interfering Isotopes Isotope	Energy and Type	Decay Product
Ytterbium												
(169)	32 d	^{168}Yb(n,γ)^{169}Yb	1.1×10^4	2.5×10^4	0.068	1.05	1.35	Tm X rays 0.06(45) 0.20(35) others to 0.31	Many e⁻ to 0.14	^{175}Yb	0.40γ, 0.46β⁻	^{169}Tm
(175)	4.2 d	^{174}Yb(n,γ)^{175}Yb	55	1.8×10^5	0.80	1.20	1.60	10% to 0.40	0.46(87) 4 e⁻ to 0.33	^{169}Yb	0.31γ, 0.14 e⁻	^{175}Lu
Lutetium												
(177)	6.8 d	^{176}Lu(n,γ)^{177}Lu	2.1×10^3	1.1×10^5	2.5	3.75	4.90	0.21(6) Hf X rays	0.50(86) 4 e⁻ to 0.14	^{176}Lu	0.43β⁻, 0.31γ	^{177}Hf
Hafnium												
(175)	70 d	^{174}Hf(n,γ)^{175}Hf	400	1.1×10^4	0.037	0.054	0.073	0.34(85) 2 to 0.43	4 e⁻ to 0.33	^{181}Hf	0.41β⁻, 0.48γ	^{175}Lu
(181)	45 d	^{180}Hf(n,γ)^{181}Hf	10	1.7×10^4	0.16	0.24	0.32	0.48(81) Ta X rays	0.41(93) 4 e⁻ to 0.42	^{174}Hf	0.43γ, 0.33 e⁻	^{181}Ta
Tantalum												
(182)	115 d	^{181}Ta(n,γ)^{182}Ta	21	6.0×10^3	0.87	1.30	1.55	0.07(42) 1.1(34) 1.2(56) others to 1.6	0.52(65) 1.7(0.3) many e⁻ to 1.6			^{182}W
Tungsten												
(181)	130 d	^{180}W(n,γ)^{181}W ^{181}Ta(p,n)^{181}W	~20	5.5×10^3	0.0012 Cyclotron, 13 mc/hr	0.0018	0.0021	~1% to 0.15 Ta X rays	0.004 0.006 e⁻	^{185}W	0.43β⁻	^{181}Ta

Table 16. ELEMENTS AND THEIR COMMERCIALLY AVAILABLE ISOTOPES

B. Nuclear Data: I. Radioactive Isotopes (Continued)

Nuclides of (Mass No.)	Half-Life	Production Reaction	Production Cross Section, barns	Specific Activity (Theory), Ci/g	Yield, Ci/g* Half-Lives 1	Yield, Ci/g* Half-Lives 2	1 year (or saturation,** if shorter)	Gamma Energies, Mev (%)	Beta Energies, Mev (%)	Interfering Isotopes Isotope	Interfering Isotopes Energy and Type	Decay Product
Tungsten (cont.)												
(185)	74 d	$^{184}W(n,\gamma)^{185}W$	2.1	1.0×10^4	0.028	0.042	0.055		0.43(100)	^{181}W	0.006 e$^-$ Ta X rays	^{185}Re
(187)	24 h	$^{186}W(n,\gamma)^{187}W$	40	6.9×10^5	0.50	0.75	1.01	0.07(11) 0.48(23) 0.69(27) others to 0.77	0.63(70) 1.31(15) many e$^-$ to 0.8			$^{187}Re \rightarrow$ ^{187}Os
Rhenium												
(183)	70 d	$^{184}W(p,2n)^{183}Re$		1.0×10^4	Cyclotron, < 0.1 mc/hr			0.05(50) 0.16(20) many to 0.21	Many e$^-$ to 0.40			^{183}W
(186)	3.8 d	$^{185}Re(n,\gamma)^{186}Re$	110	1.8×10^5	1.78	2.67	3.56	0.14(9) trace to 0.77	1.07(74) 0.93(21)	^{188}Re	2.1β^- 0.16γ	^{186}W; ^{186}Os
(188)	17 h	$^{187}Re(n,\gamma)^{188}Re$	71	9.9×10^5	1.96	2.94	3.92	0.16(10) < 3% to 2.0	1.96(20) 2.12(80)	^{186}Re	1.07β^- 0.14γ	^{188}Os
Osmium												
(185)	94 d	$^{184}Os(n,\gamma)^{185}Os$	~200	7.0×10^3	0.0017	0.0026	< 0.0035	0.65(80) 0.88(14)	4 e$^-$ to 0.63	^{191}Os	0.14β^- 0.13γ	^{185}Re
(191m)	14 h	$^{190}Os(n,\gamma)^{191m}Os$	8.6	1.2×10^6	0.10	0.15	0.20	Os L X rays	0.06, 0.07 e$^-$	^{191}Os	0.14β^- 0.13γ	^{191}Ir
(191)	15 d	$^{190}Os(n,\gamma)^{191}Os$	12.5	4.5×10^4	0.14	0.21	0.28	0.13(25) Ir X rays	0.14(99) 5 e$^-$ to 0.13	^{185}Os ^{193}Os	1.13β^- 0.88γ	^{191}Ir
(193)	32 h	$^{192}Os(n,\gamma)^{193}Os$	1.6	5.0×10^5	0.028	0.042	0.056	Many (12%) to 0.56	1.13(70) 9 others	^{191}Os ^{191m}Os	0.14β^- 0.13γ	^{193}Ir

Table 16. ELEMENTS AND THEIR COMMERCIALLY AVAILABLE ISOTOPES

B. Nuclear Data: I. Radioactive Isotopes (Continued)

Nuclides of (Mass No.)	Half-Life	Production Reaction	Production Cross Section, barns	Specific Activity (Theory), Ci/g	Yield, Ci/g* Half-Lives 1	Yield, Ci/g* Half-Lives 2	1 year (or saturation,** if shorter)	Gamma Energies, Mev (%)	Beta Energies, Mev (%)	Interfering Isotopes Isotope	Interfering Isotopes Energy and Type	Decay Product
Iridium												
(192)	74 d	^{191}Ir(n,γ)^{192}Ir	1.0×10^3	9.0×10^3	14.9	22.3	29.7	0.32(80) 0.47(49) others to 0.61	0.67(46) 0.24(81) 0.54(41) 4 e− to 0.39			^{192}Os ^{192}Pt
(194)	19 h	^{193}Ir(n,γ)^{194}Ir	110	8.0×10^5	2.92	4.38	5.84	Many (13%) to 1.7	2.24(89)	^{192}Ir	0.47γ 0.67β−	^{194}Pt
Platinum												
(193m)	4.4 d	^{192}Pt(n,γ)^{193m}Pt	2	1.5×10^5	<0.001	<0.001	0.0013	Pt X rays	4 e− to 0.13	^{197}Pt	0.67β− 0.80γ	^{193}Ir
(197)	20 h	^{196}Pt(n,γ)^{197}Pt	0.9	7.7×10^5	0.010	0.15	0.020	0.80(20) 0.19(6) Au X rays	0.67(90) 3 e− to 0.11	^{193m}Pt	Pt X rays 0.13 e−	^{197}Au
Gold												
(195)	183 d	^{195}Pt(p,n)^{195}Au		3.6×10^3	Cyclotron, 8 mc/hr			0.10(10) 0.13(1) Pt X rays	3 e− to 0.09		0.47γ 0.67β−	^{195}Pt
(198)	2.7 d	^{197}Au(n,γ)^{198}Au	99	2.5×10^5	4.0	6.0	8.0	0.41(95) trace to 1.1	0.97(99)	^{199}Au	0.46β− 0.21γ	^{198}Au
(199)	3.15 d	^{198}Pt(n,γ) ^{199}Pt $\xrightarrow{\beta^-}$ ^{199}Au	4	2.1×10^5	0.012	0.018	0.023	0.16(37) 0.21(8) Hg X rays	0.46(6) 0.30(72) 0.25(22) 3 e− to 0.15	^{198}Au	0.97β− 0.41γ	^{199}Hg

Table 16. ELEMENTS AND THEIR COMMERCIALLY AVAILABLE ISOTOPES

B. Nuclear Data: I. Radioactive Isotopes (*Continued*)

Nuclides of (Mass No.)	Half-Life	Production Reaction	Production Cross Section, barns	Specific Activity (Theory), Ci/g	Yield, Ci/g* Half-Lives 1	2	1 year (or saturation,** if shorter)	Gamma Energies, Mev (%)	Beta Energies, Mev (%)	Interfering Isotopes Isotope	Energy and Type	Decay Product
Mercury												
(197m)	24 h	^{196}Hg(n,γ)^{197m}Hg	25	6.4×10^5	0.0014	0.0021	0.0028	0.13(42) 0.28(7) Hg X rays	6 e⁻ to 0.16	^{197}Hg	0.08γ 0.07 e⁻	^{197}Au
(197)	65 h	^{196}Hg(n,γ)^{197}Hg	880	2.4×10^5	0.05	0.08	0.11	0.08(18) trace to 0.27 Au X rays	0.06, 0.07 e⁻	^{197m}Au ^{203}Hg†	0.28γ 0.28 e⁻	^{197}Au
(203)	47 d	^{202}Hg(n,γ)^{203}Hg	4	1.4×10^4	0.24	0.36	0.48	0.279(77)	0.214(100) 3 e⁻ to 0.28	None if aged		^{203}Tl
Thallium												
(202)	12 d	^{202}Hg(p,n)^{202}Tl		5.0×10^4	Cyclotron, 0.025 mc/hr			0.44(95) others to 0.96	0.36 e⁻			^{202}Hg
(204)	3.8 y	^{203}Tl(n,γ)^{204}Tl	11	4.9×10^2			0.043	Hg X rays	0.765(98)			^{204}Hg(2) ^{204}Pb(98)
Lead												
(210)	22 y	Descendant ^{226}Ra		0.8×10^2				0.05(4) Bi L X rays	0.02(81) 0.06(19) 0.03, 0.04 e⁻ ‡	^{210}Bi ^{210}Po	5.0α 1.16β⁻ Po X rays	Continues decay chain
Bismuth												
(206)	6.24 d	^{207}Pb(p,2n)^{206}Bi ^{206}Pb(p,n)^{206}Bi		1.0×10^5	Cyclotron, 70 mc/hr Cyclotron, yield not available			0.80(99) 0.88(72) 1.7(36) many others	3 e⁻ to 0.26			^{206}Pb

† If reactor-produced   ‡ Also trace 3.8α

Table 16. ELEMENTS AND THEIR COMMERCIALLY AVAILABLE ISOTOPES

B. Nuclear Data: I. Radioactive Isotopes (Continued)

Nuclides of (Mass No.)	Half-Life	Production Reaction	Production Cross Section, barns	Specific Activity (Theory), Ci/g	Yield, Ci/g* Half-Lives 1	2	1 year (or saturation,** if shorter)	Gamma Energies, Mev (%)	Beta Energies, Mev (%)	Interfering Isotopes Isotope	Energy and Type	Decay Product
Bismuth (cont.)												
(207)	30 y	^{207}Pb(p,n)^{207}Bi		0.6×10^2	Cyclotron, 0.3 mc/hr			0.57(98) 1.06(77) 1.78(9)	3 e$^-$ to 1.05	^{207}Pb	1.1γ 1.0 e$^-$	^{207}Pb
(210m)	2.6×10^6 y	^{209}Bi(n,γ)^{210}Bi	0.02	7.0×10^{-4}			≪0.001	0.26(45) 0.30(23) others to 0.61	4.6α(100)	^{206}Tl	1.5β$^-$	^{206}Pb
Polonium												
(208)	2.9 y	^{209}Bi(p,2n)^{208}Po		5.8×10^2	Cyclotron, 0.3 mc/hr			Trace to 0.285 Bi X rays	5.11α(99)			^{204}Pb
(210)	138 d	^{209}Bi(n,γ) ^{210}Bi $\xrightarrow{\beta^-}$ ^{210}Po	0.015	4.5×10^3	0.06	0.09	0.11	Trace 0.803	5.3α(100)			^{206}Pb
Radon												
(222)	3.8 d	Radium decay		1.6×10^5				Trace 0.51	5.5α(100)	^{218}Po etc.	6.0α etc.	Decay chain
Radium												
(224)	3.64 d	Daughter natural ^{228}Th		1.7×10^5				0.24(4) trace others to 0.65	0.14, 0.23 e$^-$ 5.68α(94)	^{220}Rn ^{210}Po ^{212}Pb etc.	6.8α 0.58β$^-$ 0.24γ	Decay chain
(226)	1.62×10^3 y	Daughter ^{230}Th		1.0				0.19(4) trace others to 0.6 Rn X rays	0.09, 0.17 e$^-$ 4.8α(95)	^{222}Rn ^{218}Po ^{214}Pb etc.	α,β,γ	Decay chain

Table 16. ELEMENTS AND THEIR COMMERCIALLY AVAILABLE ISOTOPES

B. Nuclear Data: I. Radioactive Isotopes (Continued)

Nuclides of (Mass No.)	Half-Life	Production Reaction	Production Cross Section, barns	Specific Activity (Theory), Ci/g	Yield, Ci/g* Half-Lives 1	2	1 year (or saturation,** if shorter)	Gamma Energies, Mev (%)	Beta Energies, Mev (%)	Interfering Isotopes Isotope	Energy and Type	Decay Product
Radium (cont.)												
(228)	5.7 y	Daughter ^{232}Th		2.4×10^2					0.055(100) 0.005 e$^-$	^{228}Ac ^{228}Th ^{224}Ra etc.	α,β,γ	Decay chain
Actinium												
(227)	21.8 y	Daughter ^{231}Pa		0.8×10^2				Trace to 0.19 Th L X rays	0.04(90) 0.02(10) 4.95α(1.4) 0.01 e$^-$	^{227}Rh ^{223}Ra ^{223}Fr etc.		Decay chain
Thorium												
(228)	1.91 y	Daughter ^{228}Ac		8.4×10^2				~2% to 0.21 Ra L X rays	0.07 0.08 e$^-$ 5.4 α (100)	^{224}Ra ^{220}Rn etc.	α,β,γ	Decay chain
(230)	7.6×10^4 y	Daughter ^{234}U		0.02				<1% to 0.25 Ra L X rays	0.05 0.06 e$^-$ 4.6 α (100)	^{226}Ra ^{222}Rn etc.	α,β,γ	Decay chain
Protactinium												
(231)	3.2×10^4 y	Daughter ^{231}Th		0.05				0.29(12) Ac X rays	4 e$^-$ to 0.35 5α between 4.7 and 5.1 (total 90)	^{227}Ac ^{227}Th etc.	α,β,γ	Decay chain
(233)	27.4 d	Daughter ^{233}Th		2.2×10^4				0.31(44) U X rays	0.26(58) 0.15(37) 0.57(5) 8 e$^-$ to 0.29			^{233}U
(234)	6.7 h	Daughter ^{234m}Pa		2.0×10^6				0.10(76) 0.9(70) others to 1.1 U X rays	1.13(13) 0.53(66) others			^{234}U
Uranium												
(232)	72 y	Daughter ^{232}Pa		22.0				Trace to 0.33 Th L X rays	0.04, 0.05 e$^-$ 5.3α(100)	^{228}Th ^{224}Ra etc.	α,β,γ	Decay chain

Table 16. ELEMENTS AND THEIR COMMERCIALLY AVAILABLE ISOTOPES

B. Nuclear Data: I. Radioactive Isotopes (*Continued*)

Nuclides of (Mass No.)	Half-Life	Production Reaction	Production Cross Section, barns	Specific Activity (Theory), Ci/g	Yield, Ci/g* Half-Lives 1	2	1 year (or saturation,** if shorter)	Gamma Energies, Mev (%)	Beta Energies, Mev (%)	Interfering Isotopes Isotope	Energy and Type	Decay Product
Uranium (*cont.*)												
(233)	1.6×10^5 y	Daughter ^{233}Pa		0.01				Many to 0.32 Th X rays	0.02, 0.04 e$^-$ 4.8α(98)	^{229}Th ^{225}Ra ^{225}Ac	α,β,γ	Decay chain
Neptunium												
(237)	2.1×10^6 y	Daughter ^{237}U		7.0×10^{-4}				0.03(14) 0.09(14) 0.145(1) Pa L X rays	6 e$^-$ to 0.08 4.7α(87)	^{233}Pa	α,β,γ	Decay chain
Plutonium												
(237)	45.6 d	^{235}U(α,2n)^{237}Pu		1.2×10^4	Cyclotron, yield not available			0.06(5) Np X rays	5 e$^-$ to 0.06, trace 5.7α			^{237}Np
(239)	2.4×10^4 y	^{238}U(n,γ)^{239}U $\xrightarrow{\beta^-}$ ^{239}Np $\xrightarrow{\beta^-}$ ^{239}Pu	2.7	0.06			$\ll 0.001$	Trace to 0.65 U X rays	4 e$^-$ to 0.05 5.1α(100)			^{235}U
(240)	6.7×10^3 y	^{238}U(2n,2γ)^{240}Pa		0.25	Very slow; requires high fluxes and long irradiations			Trace 0.65 U L X rays	0.03, 0.04 e$^-$ 5.1α(100)			^{236}U
Americium												
(241)	458 y	Daughter ^{241}Pu		3.3				0.06(36) trace to 0.72 Np L X rays	3 e$^-$ to 0.05 5.5α(98)			^{237}Np
Curium												
(242)	163 d	^{241}Am(n,γ)^{242}Am $\xrightarrow{\beta^-}$ ^{242}Cm	700	3.3×10^3	23	35	37	Trace to 0.89 Pu L X rays	0.02, 0.04 e$^-$ 6.1α(100)			^{238}Pu
(244)	18 y	^{242}Am(2n,2γ)^{244}Cm		0.8×10^2	Low, because of double capture and fissioning			Trace to 0.82 Cm L X rays	0.02, 0.04 e$^-$ 5.8α(100)			^{240}Pu

* Normal target at a flux of 10^{12} n/cm^2/sec for indicated irradiation times; with enriched targets, the increased yield is proportional to the extent of enrichment of the target atom; for yields at other fluxes, multiply value given in table by ratio $\frac{\text{new flux}}{10^{12}}$

** Saturation = > ~ 5 half-lives.

Table 16. ELEMENTS AND THEIR COMMERCIALLY AVAILABLE ISOTOPES
B. Nuclear Data: II. Stable Isotopes

Nuclides of (Mass No.)	Natural Abundance, %	Nuclides of (Mass No.)	Natural Abundance, %
Hydrogen		**Potassium**	
(1)	99.985	(39)	93.1
(2)	0.015	(40)[1*]	0.01
Helium		(41)	6.9
(3)	0.00013	**Calcium**	
(4)	~100.0	(40)	96.97
Lithium		(42)	0.64
(6)	7.42	(43)	0.14
(7)	92.58	(44)	2.06
Beryllium		(46)	0.003
(9)	100.0	(48)	0.18
Boron		**Scandium**	
(10)	19.78	(45)	100.0
(11)	80.22	**Titanium**	
Carbon		(46)	7.93
(12)	98.89	(47)	7.28
(13)	1.11	(48)	73.94
Nitrogen		(49)	5.51
(14)	99.63	(50)	5.34
(15)	0.37	**Vanadium**	
Oxygen		(50)[2*]	0.24
(16)	99.76	(51)	99.76
(17)	0.04	**Chromium**	
(18)	0.20	(50)	4.31
Fluorine		(52)	83.76
(19)	100.0	(53)	9.55
Neon		(54)	2.38
(20)	90.92	**Manganese**	
(21)	0.26	(55)	100.0
(22)	8.82	**Iron**	
Sodium		(54)	5.82
(23)	100.0	(56)	91.66
Magnesium		(57)	2.19
(24)	78.70	(58)	0.33
(25)	10.13	**Cobalt**	
(26)	11.17	(59)	100.0
Aluminum		**Nickel**	
(27)	100.0	(58)	67.84
Silicon		(60)	26.23
(28)	92.21	(61)	1.19
(29)	4.70	(62)	3.66
(30)	3.09	(64)	1.08
Phosphorus		**Copper**	
(31)	100.0	(63)	63.09
Sulfur		(65)	30.91
(32)	95.0	**Zinc**	
(33)	0.76	(64)	48.89
(34)	4.22	(66)	27.81
(36)	0.014	(67)	4.11
Chlorine		(68)	18.57
(35)	75.53	(70)	0.62
(37)	24.47	**Gallium**	
Argon		(69)	60.4
(36)	0.34	(71)	39.6
(38)	0.06		
(40)	99.60		

1* Half-life = 1.3×10^9 y   2* Half-life $> 10^{15}$ y

Table 16. ELEMENTS AND THEIR COMMERCIALLY AVAILABLE ISOTOPES
B. Nuclear Data: II. Stable Isotopes (*Continued*)

Nuclides of (Mass No.)	Natural Abundance, %	Nuclides of (Mass No.)	Natural Abundance, %
Germanium		**Rhodium**	
(70)	20.52	(103)	100.0
(72)	27.43	**Palladium**	
(73)	7.76	(102)	0.96
(74)	36.54	(104)	10.97
(76)	7.76	(105)	22.23
Arsenic		(106)	27.33
(75)	100.0	(108)	26.71
Selenium		(110)	11.81
(74)	0.87	**Silver**	
(76)	9.02	(107)	51.82
(77)	7.58	(109)	48.18
(78)	23.52	**Cadmium**	
(80)	49.82	(106)	1.22
(82)	9.19	(108)	0.88
Bromine		(110)	12.39
(79)	50.54	(111)	12.75
(81)	49.46	(112)	24.07
Krypton		(113)	12.26
(78)	0.35	(114)	28.86
(80)	2.27	(116)	7.58
(82)	11.56	**Indium**	
(83)	11.55	(113)	4.28
(84)	56.90	(115)[3*]	95.72
(86)	17.37	**Tin**	
Rubidium		(112)	0.96
(85)	72.15	(114)	0.66
(87)	27.85	(115)	0.35
Strontium		(116)	14.30
(84)	0.56	(117)	7.61
(86)	9.86	(118)	24.03
(87)	7.02	(119)	8.58
(88)	82.56	(120)	32.85
Yttrium		(122)	4.72
(89)	100.0	(124)	5.94
Zirconium		**Antimony**	
(90)	51.46	(121)	57.25
(91)	11.23	(123)	42.75
(92)	17.11	**Tellurium**	
(94)	17.40	(120)	0.09
(96)	2.80	(122)	2.46
Niobium		(123)	0.87
(93)	100.0	(124)	4.61
Molybdenum		(125)	6.99
(92)	15.84	(126)	18.71
(94)	9.04	(128)	31.79
(95)	15.72	(130)	34.48
(96)	16.53	**Iodine**	
(97)	9.46	(127)	100.0
(98)	23.78	**Xenon**	
(100)	9.63	(124)	0.096
Ruthenium		(126)	0.090
(96)	5.51	(128)	1.92
(98)	1.87	(129)	26.44
(99)	12.72	(130)	4.08
(100)	12.62	(131)	21.18
(101)	17.07	(132)	26.89
(102)	31.61	(134)	10.44
(104)	18.60	(136)	8.87

[3*] Half-life = 5×10^{14} y

Table 16. ELEMENTS AND THEIR COMMERCIALLY AVAILABLE ISOTOPES

B. Nuclear Data: II. Stable Isotopes (*Continued*)

Nuclides of (Mass No.)	Natural Abundance, %	Nuclides of (Mass No.)	Natural Abundance, %
Cesium		**Dysprosium**	
(133)	100.0	(156)[9*]	0.052
		(158)	0.090
Barium		(160)	2.29
(130)	0.101	(161)	18.88
(132)	0.097	(162)	25.53
(134)	2.42	(163)	24.97
(135)	6.59	(164)	28.18
(136)	7.81		
(137)	11.30	**Holmium**	
(138)	71.66	(165)	100.0
Lanthanum		**Erbium**	
(138)	0.09	(162)	0.136
(139)	99.91	(164)	1.56
		(166)	33.41
Cerium		(167)	22.94
(136)	0.193	(168)	27.07
(138)	0.250	(170)	14.88
(140)	88.48		
(142)[4*]	11.07	**Thulium**	
		(169)	100.0
Praseodymium			
(141)	100.0	**Ytterbium**	
		(168)	0.135
Neodymium		(170)	3.03
(142)	27.11	(171)	14.31
(143)	12.17	(172)	21.82
(144)	23.85	(173)	16.13
(145)	8.30	(174)	31.84
(146)	17.22	(176)	12.73
(148)	5.73		
(150)	5.62	**Lutetium**	
		(175)	97.40
Samarium		(176)[10*]	2.60
(144)	3.09		
(147)[5*]	14.97	**Hafnium**	
(148)[6*]	11.24	(174)[11*]	0.18
(149)[7*]	13.83	(176)	5.20
(150)	7.44	(177)	18.50
(152)	26.72	(178)	27.14
(154)	22.71	(179)	13.75
		(180)	35.24
Europium			
(151)	47.82	**Tantalum**	
(153)	52.18	(180)	0.012
		(181)	99.988
Gadolinium			
(152)[8*]	0.20	**Tungsten**	
(154)	2.15	(180)	0.14
(155)	14.73	(182)	26.41
(156)	20.47	(183)	14.40
(157)	15.68	(184)	30.64
(158)	24.87	(186)	28.41
(160)	21.90		
		Rhenium	
Terbium		(185)	37.07
(159)	100.0	(187)[12*]	62.93

4* Half-life = 5×10^{15} y
5* Half-life = 1.06×10^{11} y
6* Half-life = 1.2×10^{13} y
7* Half-life = 4×10^{14} y
8* Half-life = 1.1×10^{14} y
9* Half-life = 2×10^{14} y
10* Half-life = 2.2×10^{10} y
11* Half-life = 4.3×10^{15} y
12* Half-life = 4×10^{10} y

Table 16. ELEMENTS AND THEIR COMMERCIALLY AVAILABLE ISOTOPES

B. Nuclear Data: II. Stable Isotopes *(Continued)*

Nuclides of (Mass No.)	Natural Abundance %	Nuclides of (Mass No.)	Natural Abundance %
Osmium		**Mercury**	
(184)	0.018	(196)	0.146
(186)	1.59	(198)	10.02
(187)	1.64	(199)	16.84
(188)	13.3	(200)	23.13
(189)	16.1	(201)	13.22
(190)	26.4	(202)	29.80
(192)	41.0	(204)	6.85
		Thallium	
		(203)	29.50
Iridium		(205)	70.50
(191)	37.3	**Lead**	
(193)	62.7	(204)	1.48
		(206)	23.6
		(207)	22.6
Platinum		(208)	52.3
(190)[13*]	0.013	**Bismuth**	
(192)	0.78	(209)	100.0
(194)	32.9	**Thorium**	
(195)	33.8	(232)[14*†]	100.0
(196)	25.3	**Uranium**	
(198)	7.2	(234)[15*†]	0.0006
Gold		(235)[16*†]	0.72
(197)	100.0	(238)[17*†]	99.27

13* Half-life = 6×10^{11} y   14* Half-life = 1.4×10^{10} y   15* Half-life = 2.5×10^5 y   16* Half-life = 7.1×10^8 y
17* Half-life = 4.5×10^9 y   † Naturally occurring

Data in Table 16 first published in *Health Phys.* 3: 191, 1960. Reprinted by permission of Pergamon Press, Inc., New York.

PART II
ESSENTIAL PHYSICS DATA

Compiled by
J. R. Cameron

RADIATION INTERACTION WITH MATTER

MASS ABSORPTION COEFFICIENTS VS. PHOTON ENERGY

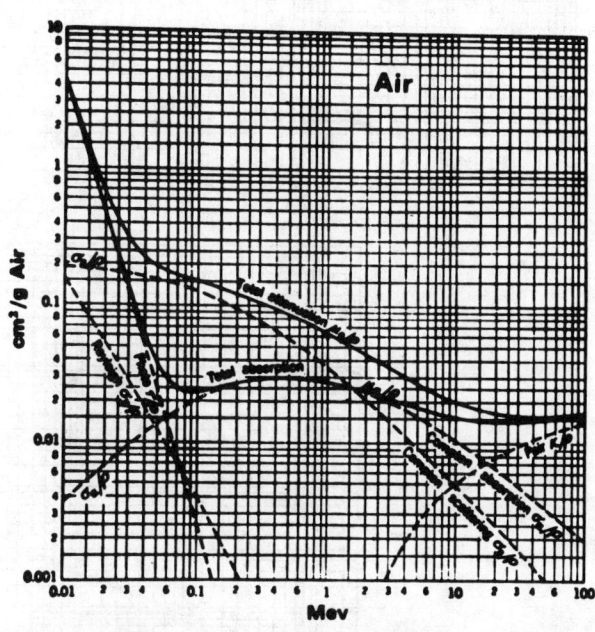

FIGURE 1.
Mass Absorption Coefficients
vs. Photon Energy for Air.

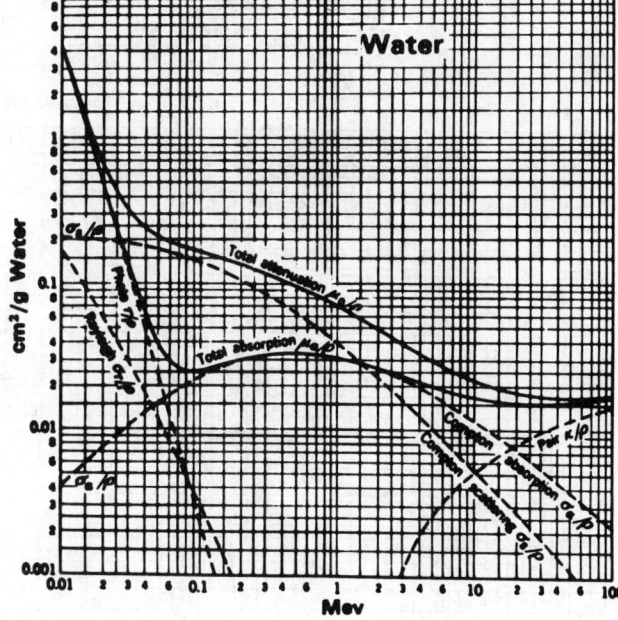

FIGURE 2.
Mass Absorption Coefficients
vs. Photon Energy for Water.

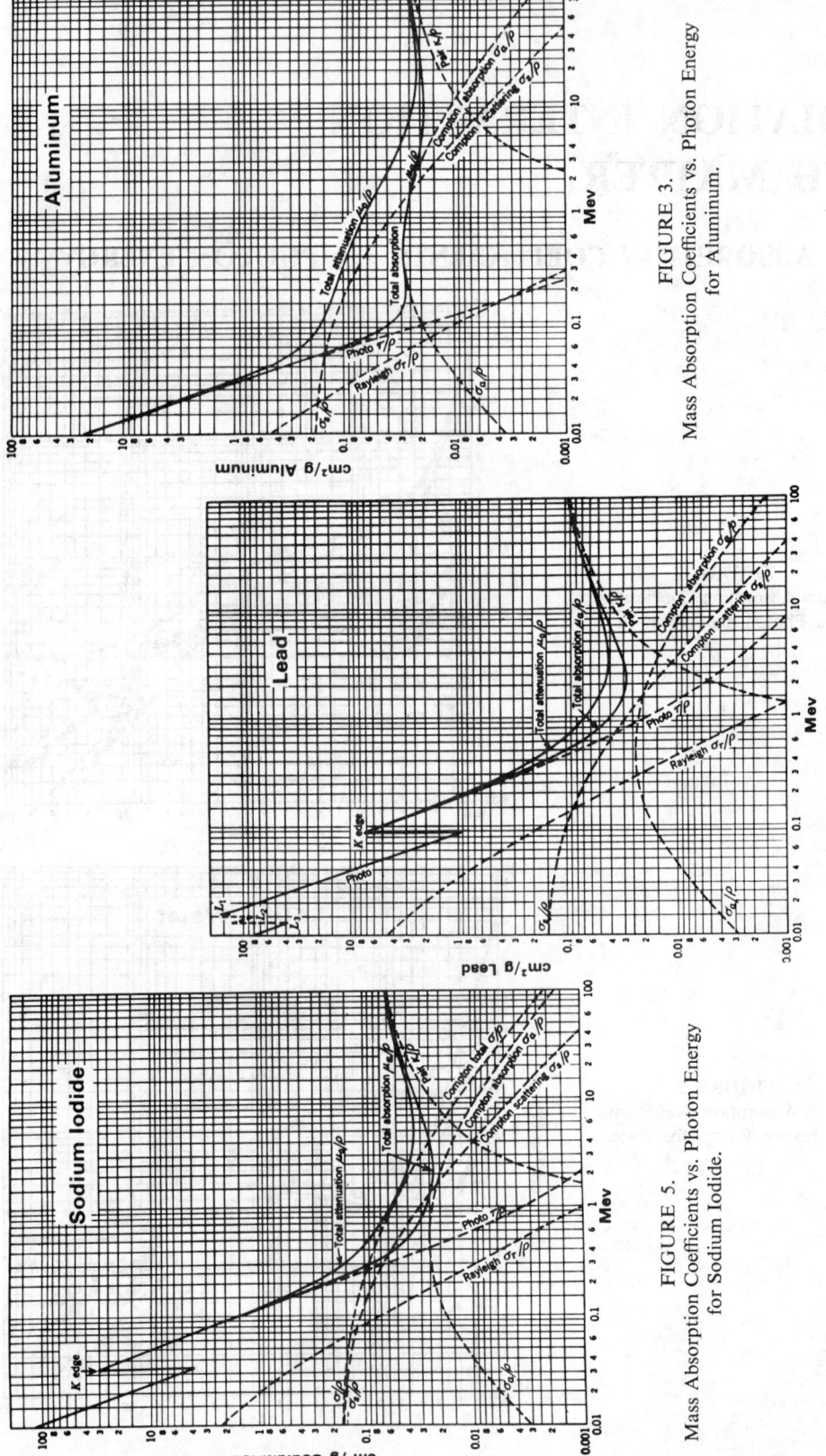

FIGURE 3. Mass Absorption Coefficients vs. Photon Energy for Aluminum.

FIGURE 4. Mass Absorption Coefficients vs. Photon Energy for Lead.

FIGURE 5. Mass Absorption Coefficients vs. Photon Energy for Sodium Iodide.

HALF-THICKNESS VS. PHOTON ENERGY

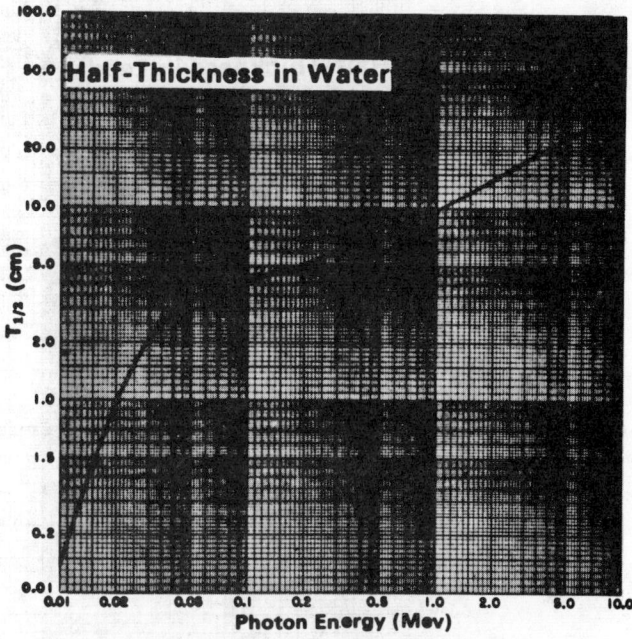

FIGURE 6.
Half-Thickness vs. Photon Energy for Water.
(Centimeters of water necessary to reduce the number
of gamma rays in a broad beam by a factor of 2.)

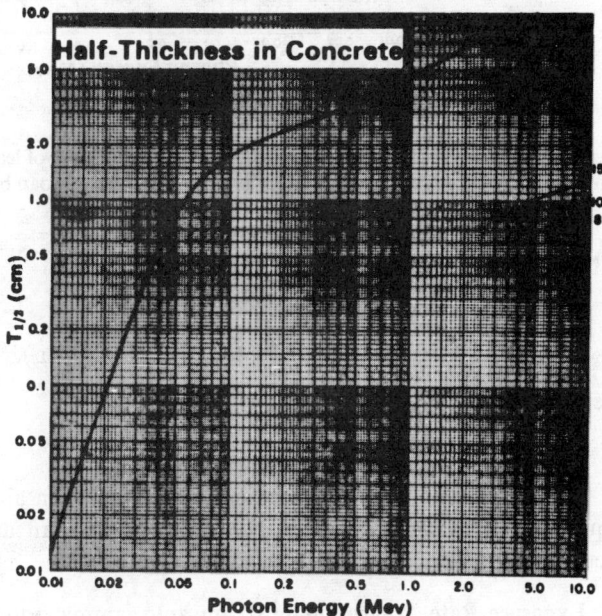

FIGURE 7.
Half-Thickness vs. Photon Energy for Concrete.
(Centimeters of concrete of density 2.35 g/cm^3 necessary
to reduce the number of gamma rays in a broad beam
by a factor of 2.)

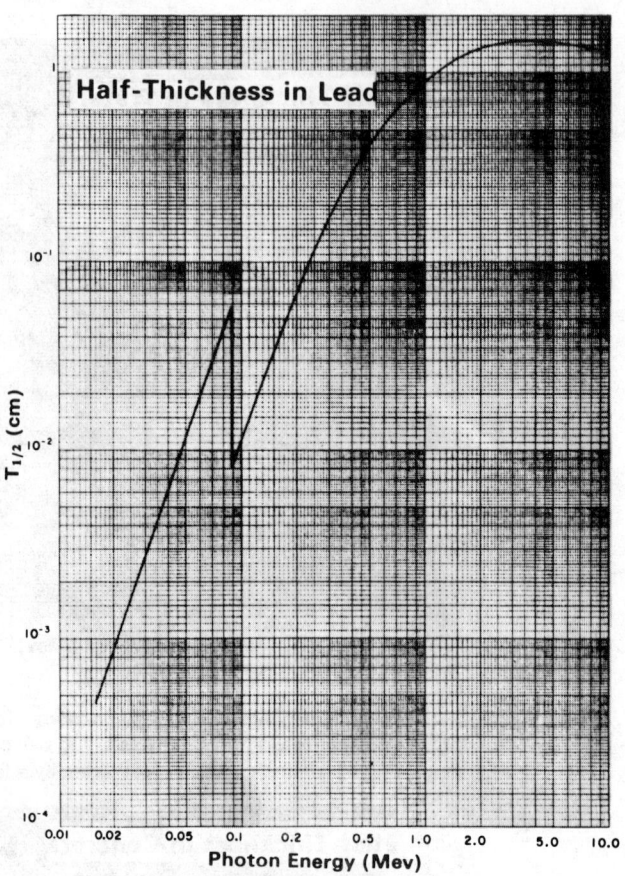

FIGURE 8.
Half-Thickness vs. Photon Energy for Lead. (Centimeters of lead of density 11.29 g/cm³ necessary to reduce the number of gamma rays in a broad beam by a factor of 2.)

GAMMA-EMITTER DOSE

Exposure Rate (from a point source).

$$\Gamma = 0.156nE(10^5 \mu_a),$$

where Γ = mR/hr at 1 meter per mCi,
  n = gamma quanta per disintegration,
  E = energy of gamma quanta in Mev,
  μ_a = energy absorption coefficient for gamma in air (S.T.P.) in cm^{-1}.

(Equation assumes that one ion pair in air causes an average energy expenditure of 32.7 electron volts.)

Exposure Rate, Approximate (from any gamma point source).

$$\text{R/hr at 1 foot} \cong 6CEn, \qquad \text{mR/hr/mCi at 1 meter} \cong 0.5\, nE,$$

where C = number of curies,
  E = gamma-ray energy in Mev,
  n = gamma quanta/disintegration.

Essential Physics Data

Exposure Rate (from any gamma point source).

$$\text{mR/hr} = \frac{n\Gamma}{s^2},$$

where n = number of mCi,
Γ = mR/hr at 1 meter per mCi,
s = distance in meters.

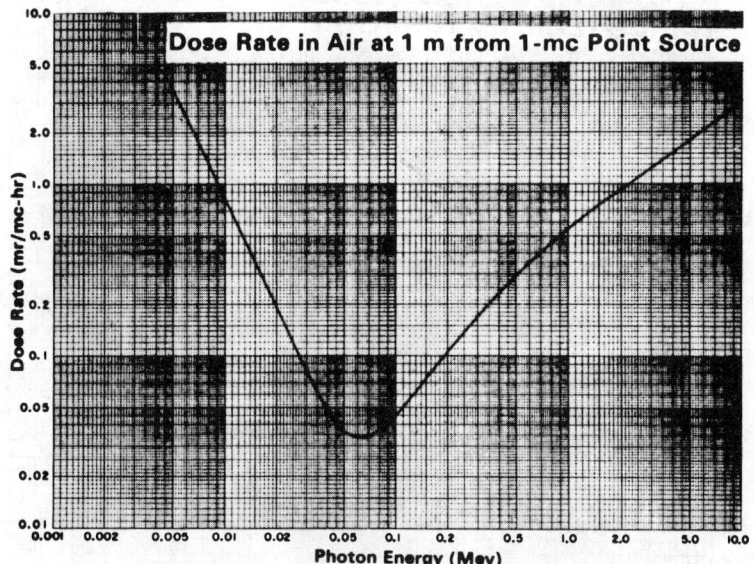

FIGURE 9. Γ Factor vs. Photon Energy.
(Dose rate in milliroentgens per millicurie-hour
at 1 meter in air—without absorption—from a point source.)

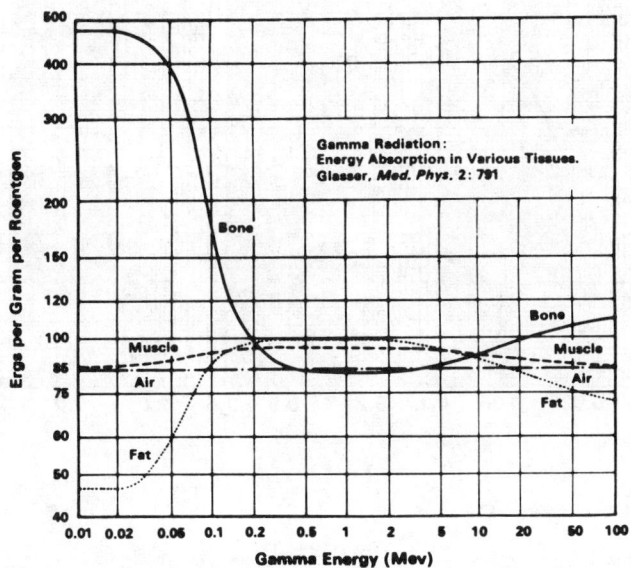

FIGURE 10.
Energy Absorption in Various Tissues vs. Photon Energy Emitted for Each Disintegration.
(From: *Medical Physics*, Vol. 2, by O. Glasser. Copyright © 1950, Year Book Medical Publishers, Inc.
Used by permission of Year Book Medical Publishers.)

RANGE–ENERGY RELATION FOR ELECTRONS

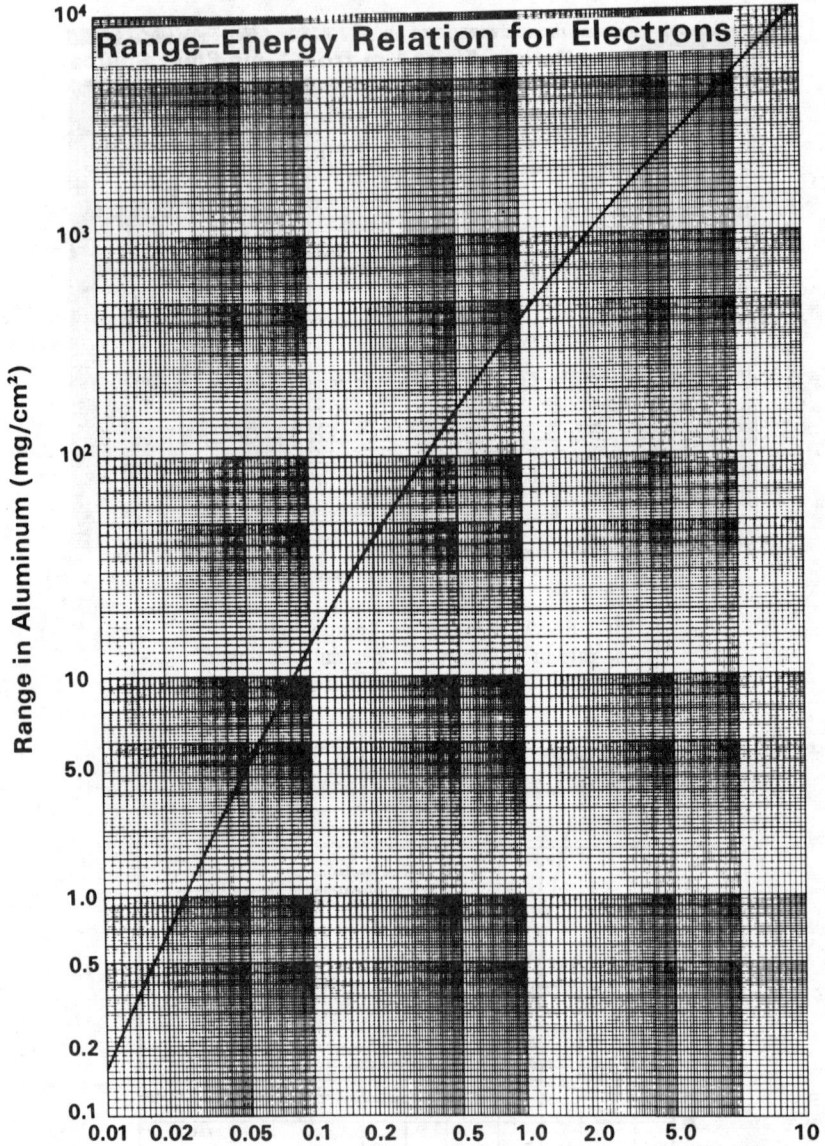

FIGURE 11.
Range in Aluminum vs. Electron Energy.
(From: Katz, L. and Penfold, A. S., Range–Energy Relation for Monoenergetic Electrons; *Rev. Mod. Phys.*, 24:28, 1962.)

PENETRATION ABILITY OF BETA RADIATION

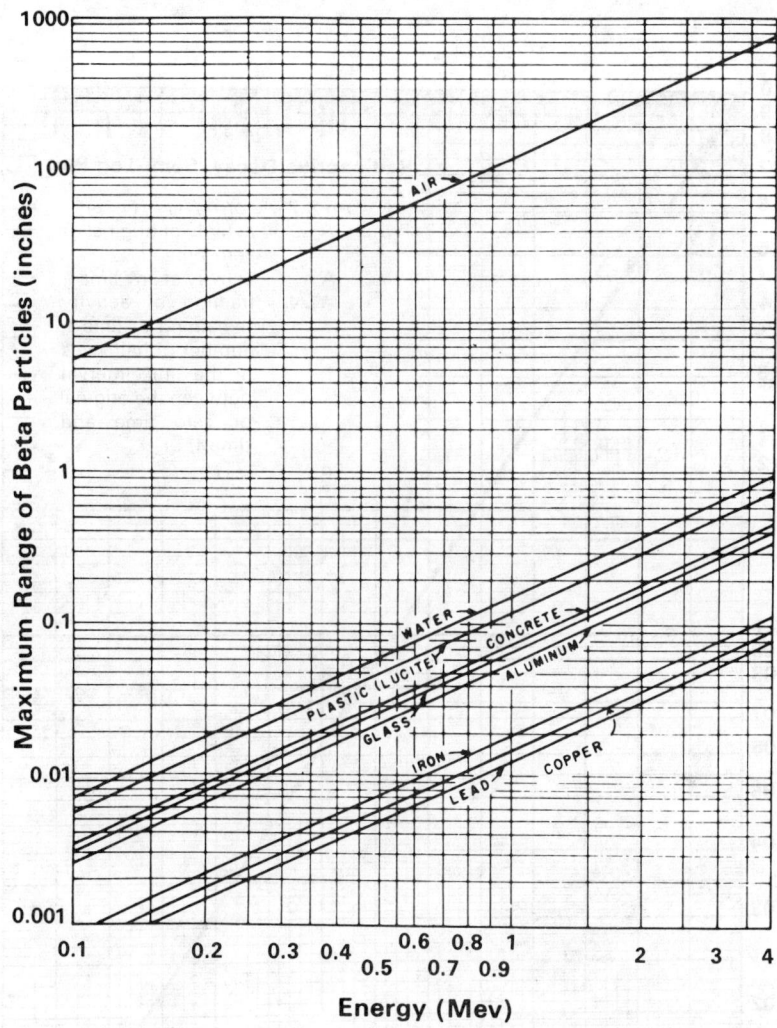

FIGURE 12.
Maximum Range of Beta Particles vs. Energy.
(The maximum range of beta particles as a function of energy in the various materials indicated. From SRI Report No. 361, *The Industrial Uses of Radioactive Fission Products*, with permission of the Stanford Research Institute and the U.S. Atomic Energy Commission.)

RADIOACTIVE DECAY

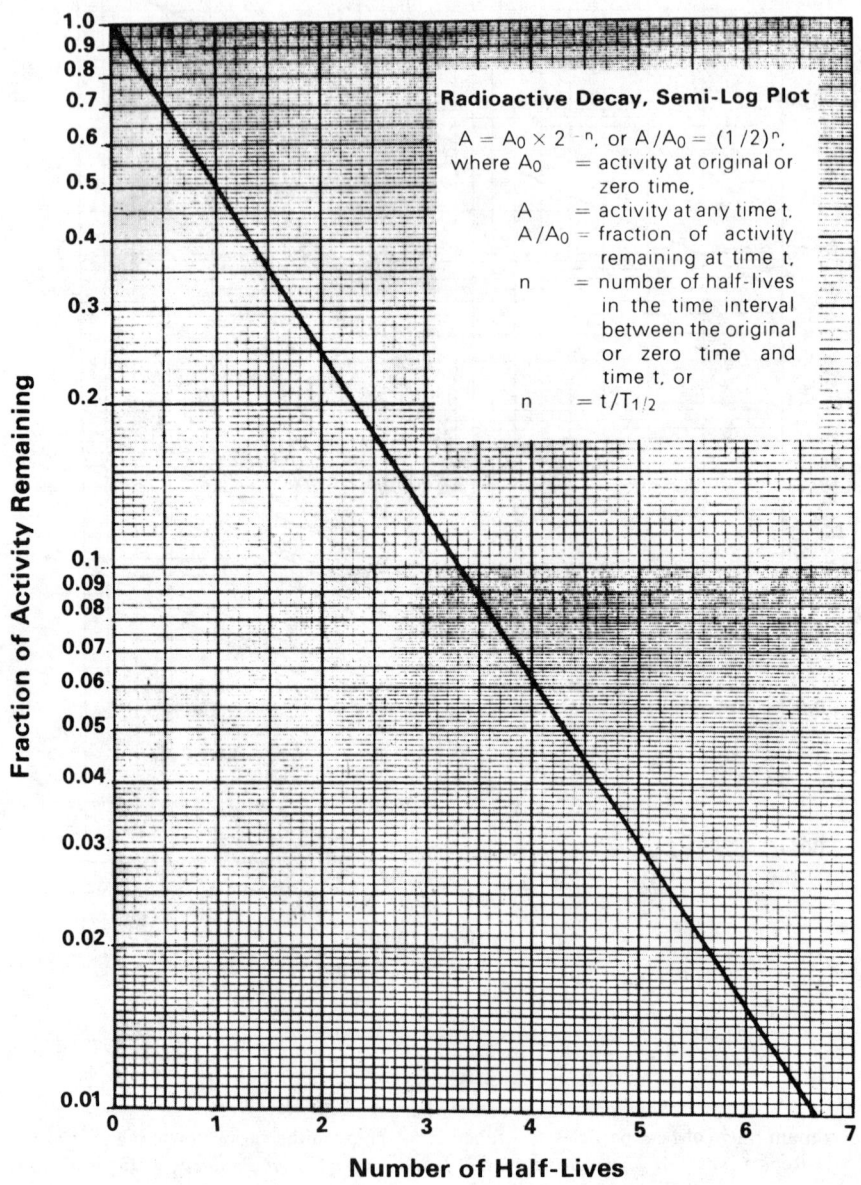

FIGURE 13.
Radioactive Decay, 0–7 Half-Lives.

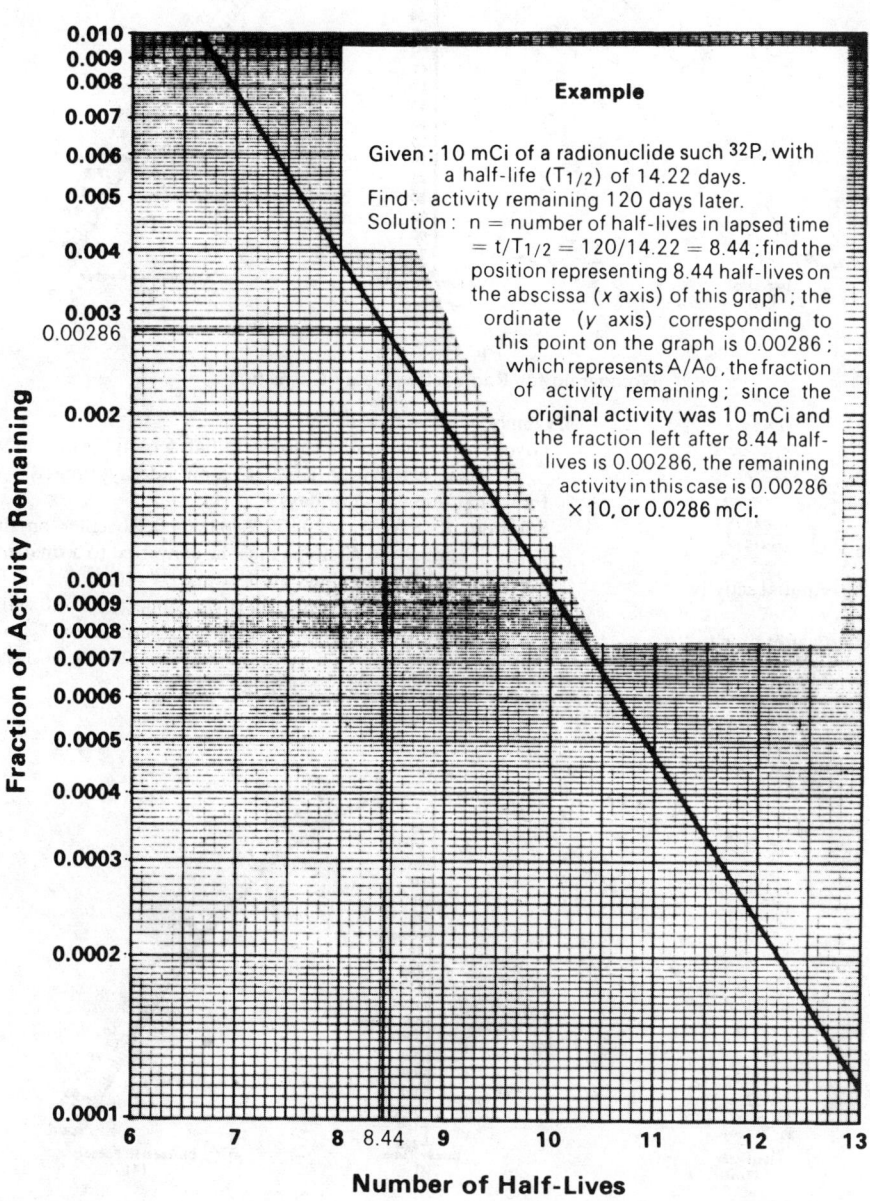

FIGURE 14.
Radioactive Decay, 7–13 Half-Lives.

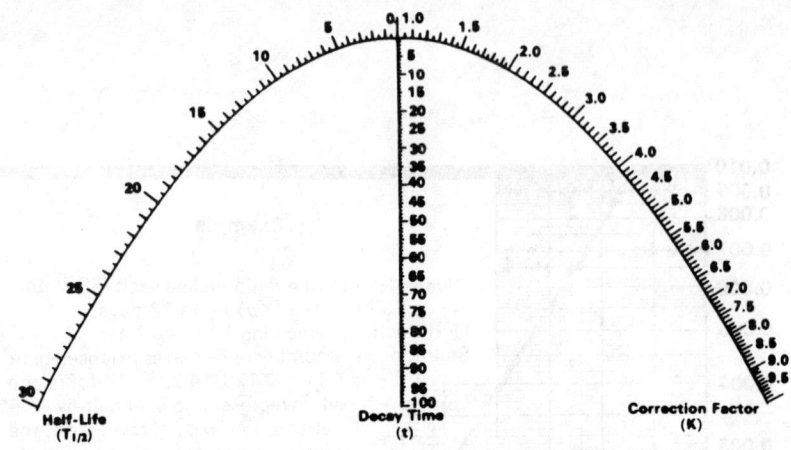

FIGURE 15.
Nomogram for Radioactive-Decay Correction.

Equation

$$K = \frac{A_0}{A_t} = e^{\left(\frac{0.693\,t}{T_{1/2}}\right)}$$

$$A_t = A_0 \times e^{\left(-\frac{0.693\,t}{T_{1/2}}\right)}$$

where A_0 = initial activity
($t = 0$)
A_t = activity after time t.

Example

Given: unknown radioisotope; initial activity (A_0) = 10,972 dpm; final activity (A_t) = 2,743 dpm; decay time (t) = 62 hours.
Find: half-life ($T_{1/2}$) in days.
Solution: $K = A_0/A_t = 10,972/2,743 = 4.0$; to enter on the nomogram, t = 62 hours must be converted to a smaller number, e.g. by dividing by 4; thus t/4 = 62/4 = 15.5 four-hour periods; using straight edge, connect K = 4.0 with t = 15.5 and read $T_{1/2}$ = 7.75 four-hour periods, or $T_{1/2} = 4 \times 7.75 = 31$ hours; $T_{1/2} = 31$ hour = 31/24 = 1.29 days half-life of the unknown isotope.

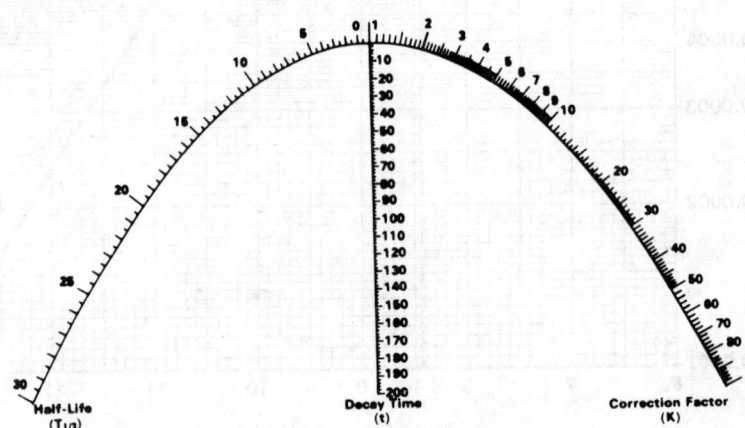

FIGURE 16.
Nomogram for Radioactive-Decay Correction.

Equation

$$K = \frac{A_0}{A_t} = e^{\left(\frac{0.693\,t}{T_{1/2}}\right)}$$

$$A_t = A_0 \times e^{\left(-\frac{0.693\,t}{T_{1/2}}\right)}$$

where A_0 = initial activity
($t = 0$)
A_t = activity after time t.

Example

Given: ^{24}Na; half-life ($T_{1/2}$) = 14.9 hours; decay time (t) = 2 days; final activity (A_t) = 62 dpm.
Find: initial activity (A_0).
Solution: $T_{1/2}$ = 14.9 hours, t = 2 days = $2 \times 24 = 48$ hours; using straight edge, connect $T_{1/2}$ = 14.9 with t = 48 and read K = 9.32; then, $A_0 = K \times A_t = 9.32 \times 62 = 578$ dpm initial activity.

STATISTICAL ASPECTS OF NUCLEAR COUNTING

The statistically random nature of radioactive decay—i.e., of the emission of radiation from the nucleus of an unstable atom—leads to uncertainty about the accuracy of any radiation measurement. The degree of uncertainty can be calculated by means of simple formulas, in which the only variable is the number of emitted photons or particles detected. The larger the number of detected events, the smaller the uncertainty.

If one observed an infinite number of events in infinite time, the frequency of the events would be known with certainty. For observations in finite (but substantial) time, probability figures can be established; deviations inherent in this type of statistical analysis must, however, be taken into account. Many events—i.e., emissions of rays—must be observed, before significant conclusions can be drawn.

STATISTICAL-ERROR CALCULATIONS

If the average counting rate recorded for a radioactive source is n counts per unit time, then the number of counts obtained in any time interval t will not be exactly equal to n times t, but will be subject to fluctuations governed by the laws of probability. From a determined count rate or total number of counts, and knowing the total counting time, the following nomograms and graph can be used for calculating the maximum deviation to be expected a certain percent of the time. The examples below will illustrate the use of the nomograms and graph.

Given:

1128 counts in first 16 minutes	$n_1 = 70.5$ cpm, $t_1 = 16$ minutes
1040 counts in second 16 minutes	$n_2 = 65.0$ cpm, $t_2 = 16$ minutes
2168 counts total in 32 minutes	$n_s = 67.8$ cpm, $t_s = 32$ minutes
3736 counts background in 64 minutes	$n_b = 58.4$ cpm, $t_b = 64$ minutes

Example 1. Consistency of Counts.

Find the permissible difference in cpm at the 95% confidence level.

By Equation.

$$E = z\left[\frac{n_1}{t_1} + \frac{n_2}{t_2}\right]^{1/2},$$

where E = permissible error, cpm;

z = constant associated with a given confidence level, dependent upon a certain percent of area under the normal curve: at 95%, $z = 1.96$; at 90%, $z = 1.645$;

n_1 = counting rate, cpm;

n_2 = counting rate, cpm;

t_1 = total time n_1 counted, in minutes;

t_2 = total time n_2 counted, in minutes.

$$E = 1.96\left[\frac{70.5}{16} + \frac{65.0}{16}\right]^{1/2},$$

$$E = 1.96(4.40 + 4.06)^{1/2}.$$

$$E = 5.7$$

Nomogram Solution. In Figure 17 connect $n_s/t_s = 4.40$, on the left scale, with a straight line to $n_b/t_b = 4.06$, on the right scale. Read 0.95 error on the middle scale, 5.7 cpm.

Graphic Solution. In Figure 18, if the total number of counts is $1128 + 1040 = 2168$ for counting times of 16 minutes, the error is found by interpolation between counting times of 15 and 20 minutes, and is approximately 5.7 cpm.

Actual difference $= n_1 - n_2 = 70.5 - 65.0 = 5.5$. Since 5.5. is less than 5.7, the counts are consistent.

Example 2. Determination of Net Counting Rate of Sample.

Net counting rate of sample $n_n = n_s - n_b = 67.8 - 58.4 = 9.3$ cpm.

Example 3. Precision of Count.

Find the error of the net counting rate at the 95% confidence level.

By Equation.

$$E_{95} = z(n_s/t_s + n_b/t_b)^{1/2} = 1.96(67.8/32 + 58.4/64)^{1/2} = 1.96(2.12 + 0.925)^{1/2} = 3.42 \text{ cpm.}$$

Nomogram Solution. In Figure 19 connect $n_s/t_s = 2.12$, on the left scale, with a straight line to $n_b/t_b = 0.925$, on the right scale. Read the 0.95 error on the middle scale, 3.4 cpm.

Graphic Solution. In Figure 18 this must be done in two steps, since the counting times are different. E_s for 2168 counts total for 32 minutes is 2.8 cpm. E_b for 3736 counts total for 64 minutes is 1.9 cpm. The error of the net count $E_n = (E_s^2 + E_b^2) = (2.8^2 + 1.9^2) = 3.4$ cpm. The E_{95} in percent $= 100(E_{95})/n_n = 100(3.4)/9.3 = 37\%$. A count rate of $9.3 + 3.4$ cpm at the 95% confidence level means that, if the sample were counted a large number of times, 95% of the counting rate determinations for this sample would be between 5.9 and 12.7 cpm.

Essential Physics Data

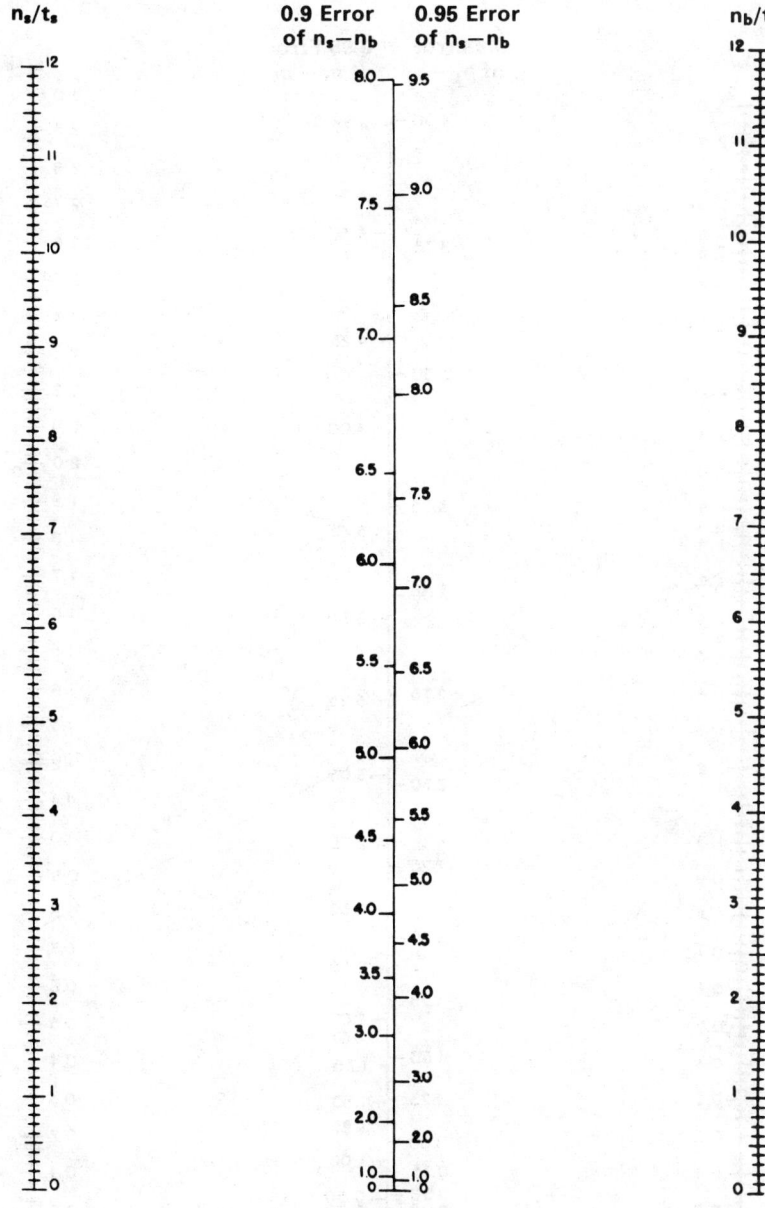

FIGURE 17.
Nomogram for 0.9 and 0.95 Error of Low Counting Rates.
(Jarrett, AECU-262, Mon P-126.)

Explanation of Symbols
n_s = counting rate of the sample, including the background, in counts per minute.
t_s = number of minutes the sample was counted.
n_b = counting rate of the background, in counts per minute.
t_b = number of minutes the background was counted.

Instructions for Use
Draw a straight line from a point on the left scale that corresponds to the quotient n_s/t_s through the point on the right scale that corresponds to the quotient n_b/t_b; the point where this line crosses the center scale will correspond to the 0.9 and the 0.95 error of the determination $n_s - n_b$.

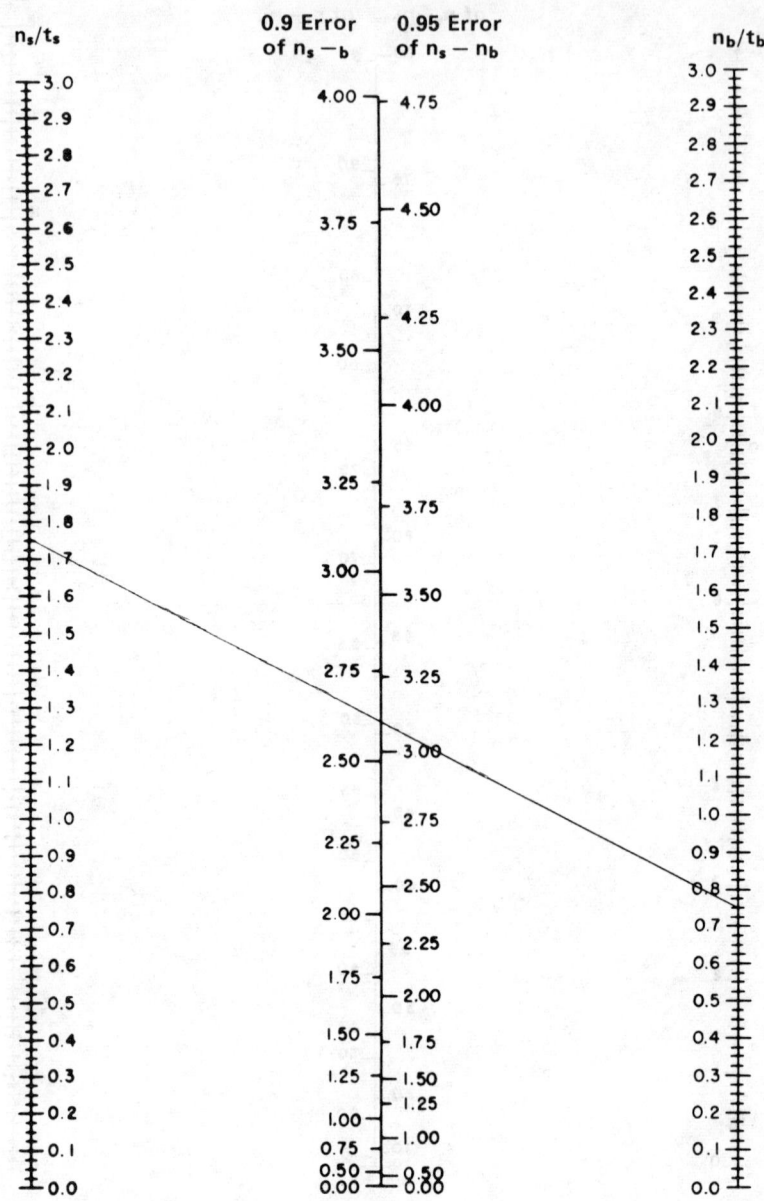

FIGURE 18.
Nomogram for 0.9 and 0.95 Error of Low Counting Rates.
(Jarrett, AECU-262, Mon P-126.)

Explanation of Symbols

n_s = counting rate of the sample, including the background, in counts per minute.

t_s = number of minutes the sample was counted.

n_b = counting rate of the background, in counts per minute.

t_b = number of minutes the background was counted.

Instructions for Use

Draw a straight line from a point on the left scale that corresponds to the quotient n_s/t_s through the point on the right scale that corresponds to the quotient n_b/t_b; the point where this line crosses the center scale will correspond to the 0.9 and the 0.95 error of the determination $n_s - n_b$.

Example

Given: $n_s = 35$ cpm
$n_b = 15$ cpm
$t_s = 20$ min
$t_b = 20$ min

Find: 0.95 error in net count.

Solution: $\frac{n_s}{t_s} = 1.75$ $\frac{n_b}{t_b} = 0.75$

Draw connecting line between given values, and read 3.1 cpm on center line.

Essential Physics Data

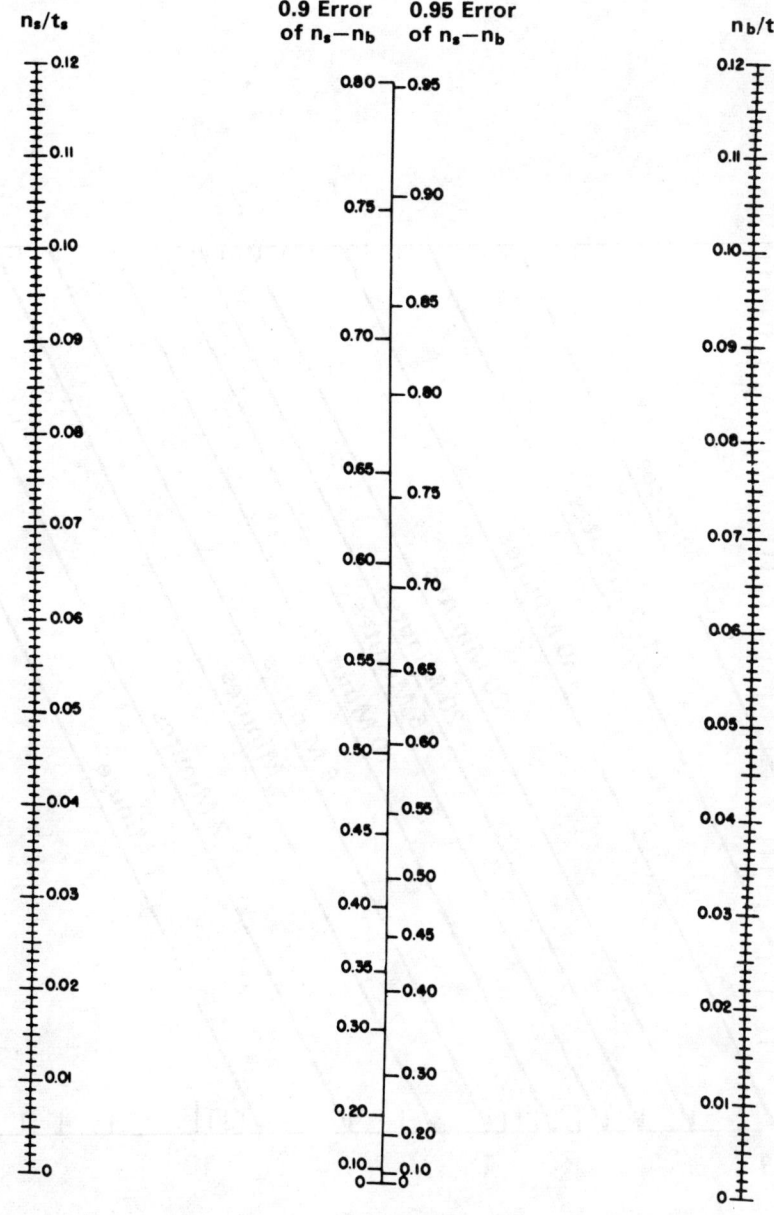

FIGURE 19.
Nomogram for 0.9 and 0.95 Error of Low Counting Rates.
(Jarrett, AECU-262, Mon P-126.)

Explanation of Symbols

n_s = counting rate of the sample, including the background, in counts per minute.

t_s = number of minutes the sample was counted.

n_b = counting rate of the background, in counts per minute.

t_b = number of minutes the background was counted.

Instructions for Use

Draw a straight line from a point on the left scale that corresponds to the quotient n_s/t_s through the point on the right scale that corresponds to the quotient n_b/t_b; the point where this line crosses the center scale will correspond to the 0.9 and the 0.95 error of the determination $n_s - n_b$.

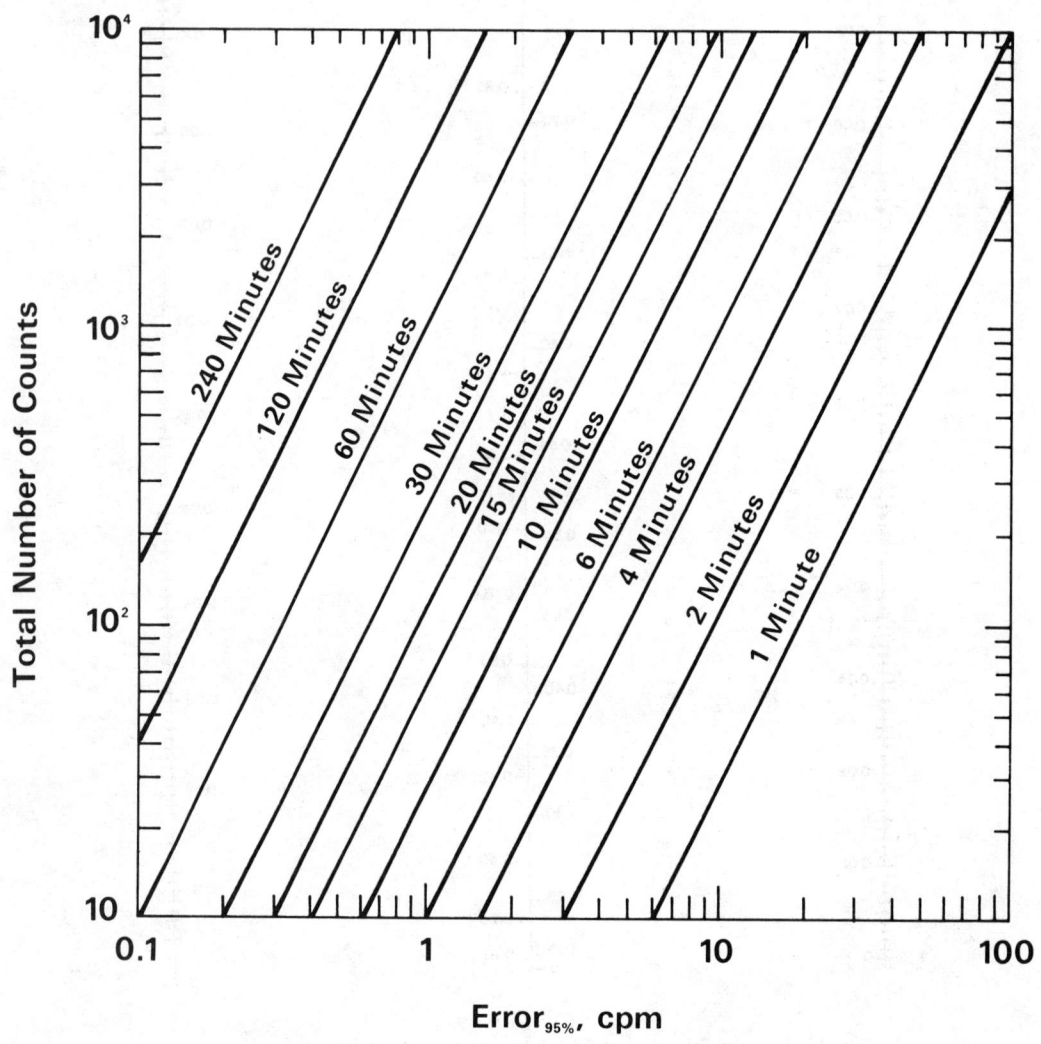

FIGURE 20.
Error in Counts per Minute
as a Function of Total Count and Length of Count.
(95% Confidence Level.)

Essential Physics Data

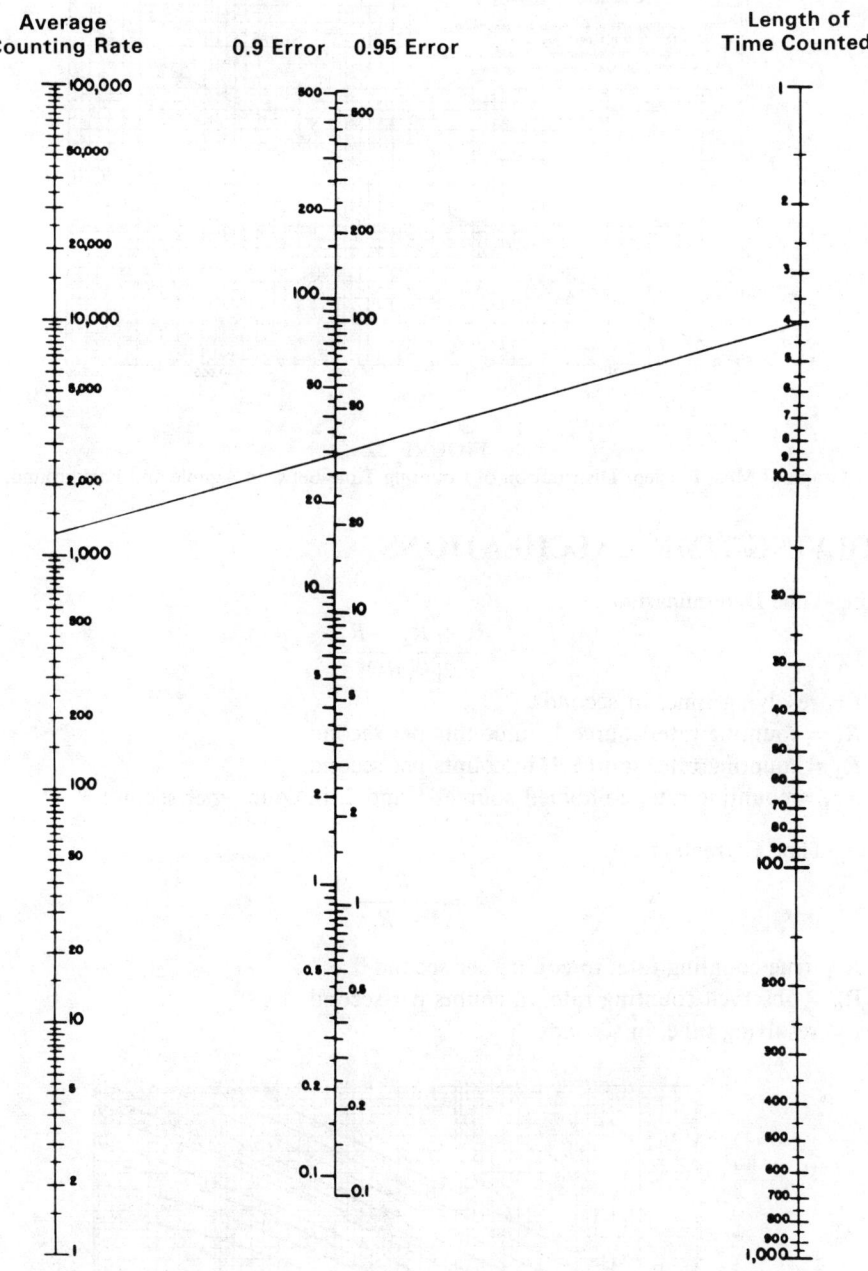

FIGURE 21.
Nomogram for 0.9 and 0.95 Error of Counting Rate Determinations.
(Jarrett, AECU-262, Mon P-126.)

Instructions for Use
Draw a straight line from a point on the left scale corresponding to the counting rate of the sample through the point on the right scale corresponding to the length of time the sample was counted; the point where this line crosses the center scale corresponds to the 0.9 and the 0.95 error of the determination.

Example
The 0.9 error of a sample that averaged 1250 counts per minute during a four-minute determination is 29 counts per minute.

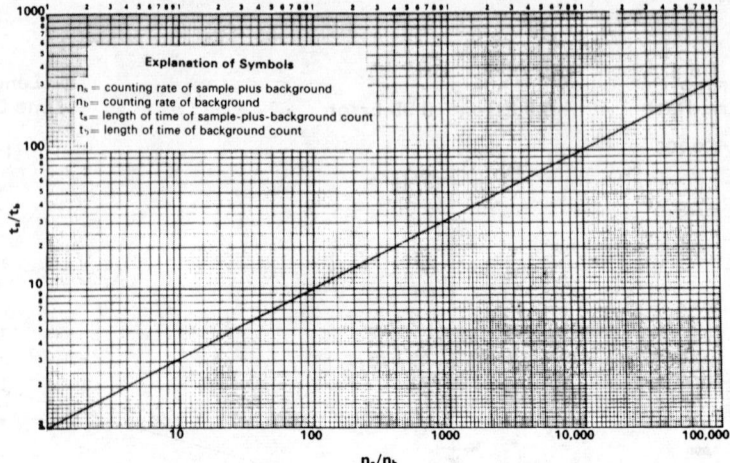

FIGURE 22.
Graph of Most Efficient Distribution of Counting Time Between Sample and Background.

RESOLVING-TIME CALCULATIONS

Resolving-Time Determination

$$\tau = \frac{R_1 + R_2 - R_{12}}{2(R_1 R_2)},$$

where τ = resolving time, in seconds,
R_1 = counting rate, source 1, in counts per second,
R_2 = counting rate, source 2, in counts per second,
R_{12} = counting rate, combined sources 1 and 2, in counts per second.

Resolving-Time Correction

$$R = \frac{R_0}{1 - R_0 \tau},$$

where R = true counting rate, in counts per second,
R_0 = observed counting rate, in counts per second,
τ = resolving time, in seconds.

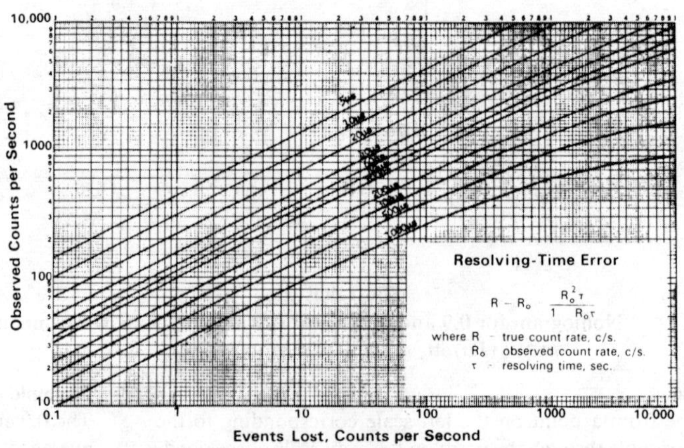

FIGURE 23.
Resolving-Time Error.

STATISTICS OF COUNTING*

- n   number of counts, one observation
- t   counting time, one observation
- $\bar{n}$ mean number of counts, series of observations
- $\bar{t}$ mean counting time, series of observations
- m   number of observations
- σ theoretical standard deviation
- S_t observed standard deviation
- S_n observed mean standard deviation of the number of counts recorded in a preset time
- r   average number of counts per unit time interval.

Theoretical Standard Deviation.

1. *For Single Observation.*

$$\sigma_n = \sqrt{rt} \cong \sqrt{n}$$

2. *For Average Number of Counts per Interval.*

$$\sigma_{\bar{n}} = \sqrt{rt/m} \cong \sqrt{\bar{n}/m}$$

Observed (Experimental) Standard Deviation.

1. *Series of Observations, Preset Time.*

$$S_n \left[= \sum_{i=1}^{m} (m_i - \bar{n})^2/(m-1) \right]^{1/2}$$

2. *Series of Observations, Preset Count.*

$$S_t \left[= \sum_{i=1}^{m} (t_i - \bar{t})^2/(m-1) \right]^{1/2}$$

$$S_n = (n/\bar{t})S_t$$

3. *Reliability Factor.*

$$R.F. = S_n/\sigma_{\bar{n}}$$

* Bleuler, E. and Goldsmith, G. J., *Experimental Nucleonics*. Holt, Rinehart and Winston, New York, 1952.

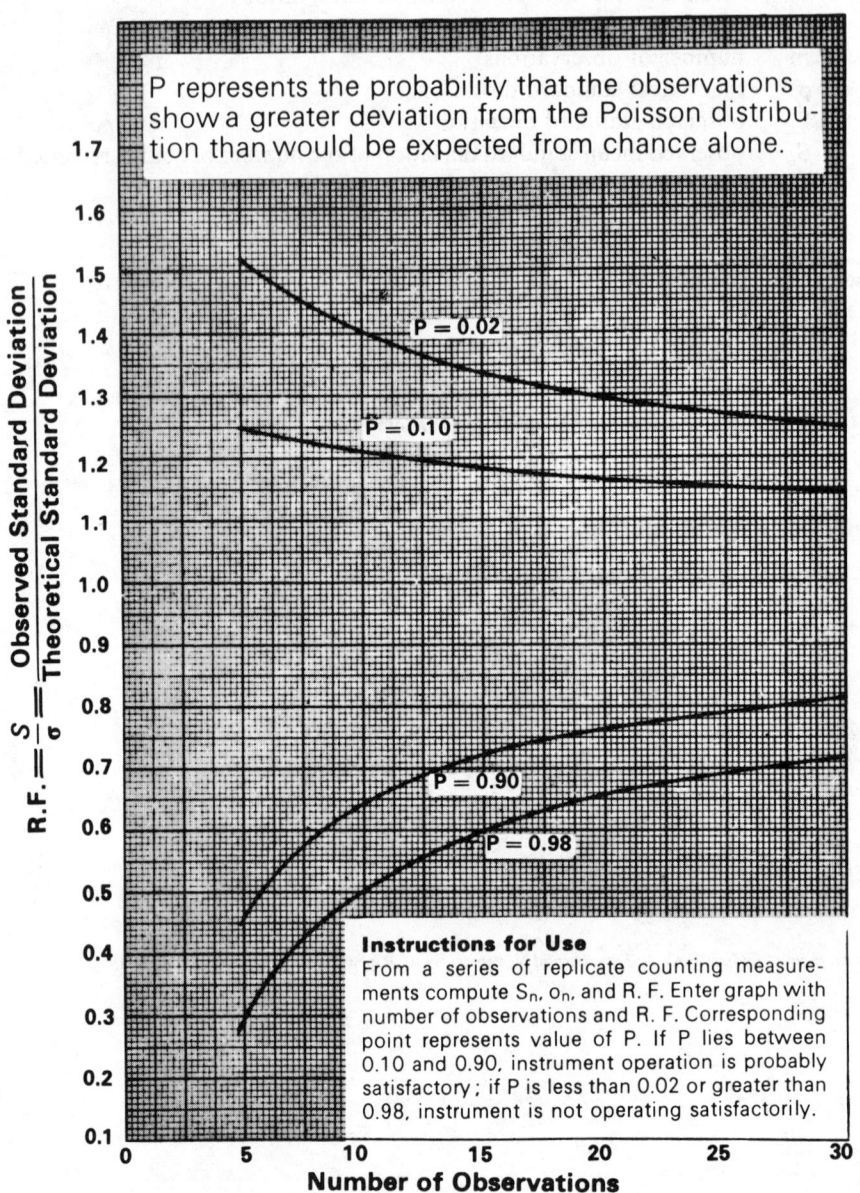

FIGURE 24.
Statistical Limits of Counter Reliability.

PART III
NUCLEAR INSTRUMENTATION

PART III

NUCLEAR INSTRUMENTATION

RADIATION DETECTORS AND EQUIPMENT

C. A. Kelsey

INTRODUCTION

The purpose of any radiation detector is to convert the ionization produced by radiation into some kind of detectable signal, usually electronic. In the ionization process electrons are removed from an atom, resulting in the formation of a positive ion core and free electrons. The ionization is produced by direct collisions in both alpha and beta interactions. In the case of gamma interactions there are three processes which can be important: photoelectric interaction, Compton scattering, and pair production. All gamma-ray detectors utilize at least one of these processes to obtain a measurable output signal. For more complete treatment of radiation detectors, see References 1–4.

GAS-FILLED DETECTORS

The basic configuration of all gas-filled detectors is essentially the same and is shown in Figure 1.5–8

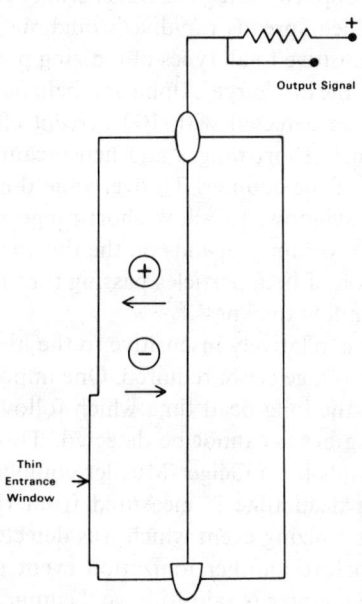

FIGURE 1.
Schematic Diagram of Typical Gas-Filled Detector.

A central electrode is surrounded by detector gas, which is contained by an outer shell. An ionizing particle passing through the chamber forms ions along its path. If a positive voltage is applied to the center wire, electrons will be attracted to the central electrode and positive ions will drift towards the outer shell. The resulting current flow can be measured in an external circuit. Above an applied voltage of a few hundred volts, a maximum current is reached, indicating that essentially all ions are being collected. A gas-filled detector operated in this manner is called an *ionization chamber*. Ionization chambers are normally used as integral detectors, and usually measure the number of ions produced during a given period of time. Ion chambers are easy to operate and relatively insensitive to the applied voltage. However, they are sensitive to physical shock and moisture. Because they are simple, reliable, and can operate over large ranges of radiation levels, ion chambers find their greatest uses in survey meters, personnel dosimeters, and in calibration equipment.

PROPORTIONAL COUNTERS

The basic design of a proportional counter is identical to that of an ion chamber.[9-11] If the applied voltage is increased above that required to collect all ions, it is possible to give the ionization electrons enough energy to produce secondary ionization. Under these conditions a single ionization event can produce many secondary electrons, resulting in amplification factors between 100 and 10,000. Operation of a proportional counter requires an extremely stable high-voltage supply, because the amount of amplification in the gas is dependent on the high voltage applied to the center wire. Individual ionzation events can be detected with only moderate amplification in external electronic circuits. Proportional counters are employed primarily in research applications where the required output is proportional to the ionization produced by the radiation.

GEIGER-MUELLER COUNTERS

The physical design of a Geiger-Mueller tube is the same as that of an ion chamber or of a proportional counter, but the applied voltage is considerably higher.[10-13] A single ionization event produces a discharge which spreads rapidly throughout the entire volume of the tube. A Geiger-Mueller detector is sensitive to all types of ionizing particles, because only one ionization event is required to trigger the discharge. Alpha and beta particles within the active volume of a Geiger-Mueller counter are detected with 100 percent efficiency. However, most alpha and beta particles have extremely short ranges and hence cannot penetrate the counter wall to reach the detecting volume of the counter. To overcome this problem, many counter tubes are provided with a thin foil window, to allow short-range particles to enter the detector. The number reaching the active volume depends on the thickness of the foil material. Figure 2 presents a plot of the absorption of beta particles passing through a thin window as a function of beta-particle energy and window thickness.

Because the output pulse is relatively insensitive to the high voltage applied to the center wire, regulation of the supply voltage is not required. One important undesirable characteristic of Geiger-Mueller counters is the long dead time which follows every ionizing event. During the dead time another ionizing event cannot be detected. This dead time is normally on the order of 100 to 200 microseconds for a Geiger-Mueller counter. It should be understood that in a Geiger-Mueller tube the dead time is measured from the time of passage of the last ionizing event and *not* the last ionizing event which was detected. In a high-radiation field the Geiger tube cannot recover before another ionization event occurs. In such a situation the output drops to zero, and the counter is said to have "jammed".

Although the basic physical design of the ionization chamber, the proportional chamber and the Geiger-Mueller counters is the same, each counting tube must be designed for a specific application, because the insulation, design of the tube, applied high voltage, and gas-fillings are different for the different types of detectors.

SCINTILLATION DETECTORS

Scintillation detectors are widely used as gamma-ray detectors, because the detecting medium is a solid or liquid rather than gas, and the detection efficiency is correspondingly

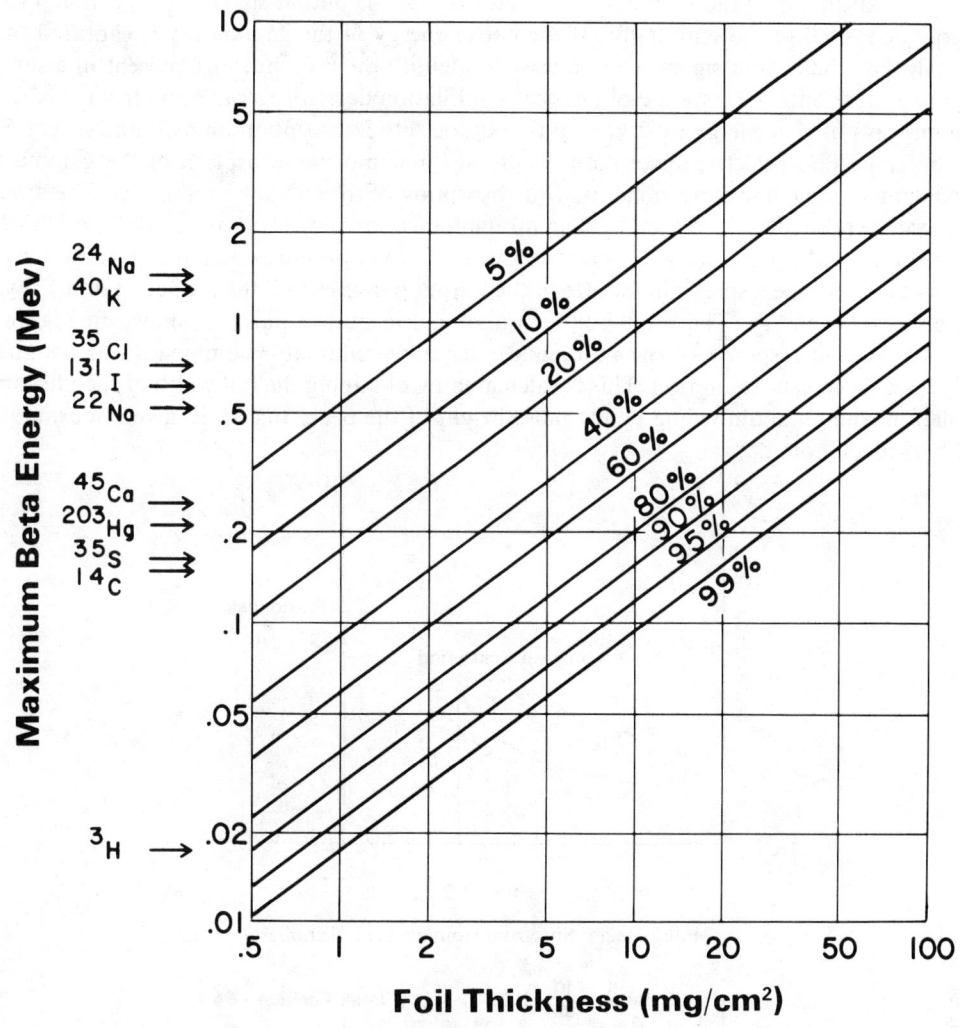

FIGURE 2.
Attenuation of Beta Particles as a Function of Energy and Absorber Thickness.

higher because of the increased density. In general, there are four classes of scintillators: inorganic crystals, organic crystals, plastic phosphors, and liquid phosphors. Table 1 presents some representative characteristics of these four classes of scintillators.[14-16] The most commonly used scintillation material is sodium iodide activated with thallium. The efficiency for gamma detection depends on crystal size, gamma-ray energy, and source–detector geometry.

Table 1. PROPERTIES OF COMMON SCINTILLATING MATERIALS

Material	Form	Density, g/cm^3	Wave Length of Maximum Emission, Å	Decay Time, μ sec	Relative Pulse Height
NaI (Tl)	Crystal	3.7	4100	.25	210
KI (Tl)	Crystal	3.2	4000	5.0	80
CsI (Tl)	Crystal	4.5	White	1.0	50
LiI (Eu)	Crystal	4.1	4400	1.0	75
Anthracene	Crystal	1.2	4400	.03	100
Stilbene	Crystal	1.2	4100	.008	60
Plastic Phosphor	Solid	1.1	3500–4500	.005	35
Liquid Phosphor	Liquid	0.9	3500–4500	.005	35

One advantage of the scintillation detector is that the output signal is proportional to the energy deposited in the scintillator. If the entire energy of the gamma ray is captured in the crystal, then the output signals can be used to identify the radionuclides present in a sample. Figure 3 represents the response of an ideal scintillation detector to radiation from ^{137}Cs. For gamma rays of this energy (662 kev), only photoelectric absorption and Compton scattering are possible. Photoelectric absorption results in the complete absorption of the gamma ray, producing a single line corresponding with absorption of the 662-kev gamma ray. The amount of energy deposited in the crystal by Compton interactions varies with the angle through which the photon has been scattered. In practice the ideal situation illustrated in Figure 3a does not occur. There will be a spread in the size of the output pulses even if the incident gamma rays all have the same energy. The results of such a spread in output pulses is shown in Figure 3b. The spread and response of the photomultiplier tube–scintillator combination is commonly expressed in percent resolution. This is calculated by obtaining the full width of the photopeak at half maximum and dividing by the pulse height of the peak. Figure 3b gives an example of such a calculation.

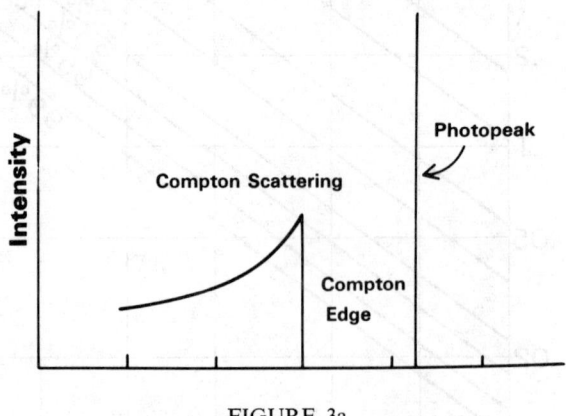

FIGURE 3a.

Pulse-Height Spectrum from an Ideal Scintillator

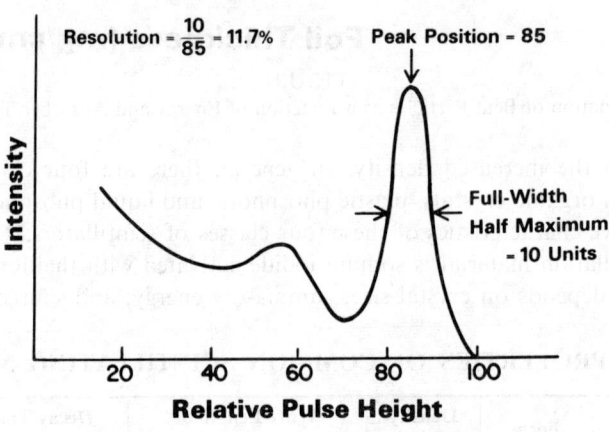

FIGURE 3b.

Typical Pulse-Height Spectrum from a Real Scintillator

The resolution of a scintillation detector is determined by the characteristics of the photomultiplier tube as well as of the scintillator itself. Table 2 presents some representative values for some common photomultiplier tubes. The entry under Cathode Response deals with the wavelength sensitivity of the cathode, and is discussed in detail by Birks.[17]

Table 2. CHARACTERISTICS OF PHOTOMULTIPLIERS

A. RCA Photomultiplier Tubes

Type	4440	4459	4523	4524
Cathode Response	S-11	S-20	Bialkali	Bialkali
Minimum Cathode Diameter	1.24"	1.68"	1.68"	2.59"
Window Material	Lime glass	Borosilicate glass	Lime glass	Lime glass
Window Shape	Flat	Curved	Flat	Flat
Dynode Material	CsSb	Be—O—Cs	Be—O—Cs	Be—O—Cs
Dynode Structure	Cage	Linear	Venetian blind	Venetian blind
Number of Stages	10	12	10	10
Nominal Diameter	1.5"	2.0"	2.0"	3.0"
Maximum Diameter	1.56"	2.06"	2.31"	3.06"
Maximum Overall Length	4.12"	6.31"	5.81"	6.31"
Cathode Sensitivity, $\mu A/1\, m$	45	150	60	60
Typical Anode–Cathode Voltage	1,000	2,300	1,500	1,500
Gain	6.0×10^5	6.6×10^6	4.5×10^5	4.5×10^5
Dark Current, nA	16	30	0.5	1
Anode Pulse Rise Time, ns	2.8	2.3	12	14

Type	4525	5819	6199	6342A
Cathode Response	Bialkali	S-11	S-11	S-11
Minimum Cathode Diameter	4.38"	1.68"	1.24"	1.68"
Window Material	Lime glass	Lime glass	Lime glass	Lime glass
Window Shape	Flat	Curved	Flat	Flat
Dynode Material	Be—O—Cs	CsSb	CsSb	Be—O—Cs
Dynode Structure	Venetian blind	Cage	Cage	Cage
Number of Stages	10	10	10	10
Nominal Diameter	5.0"	2.0"	1.5"	2.0"
Maximum Diameter	5.31"	2.31"	1.56"	2.31"
Maximum Overall Length	7.69"	5.81"	4.57"	5.81"
Cathode Sensitivity, $\mu A/1\, m$	67	50	45	80
Typical Anode–Cathode Voltage	1,500	1,000	1,000	1,250
Gain	4.0×10^5	2.0×10^6	1.0×10^6	4.0×10^5
Dark Current, nA	1.5	6	4.5	4
Anode Pulse Rise Time, ns	18	3.4	2.8	3.2

Table 2. CHARACTERISTICS OF PHOTOMULTIPLIERS (Continued)

Type	6655A	6810A	6903	7746
Cathode Response	S-11	S-11	S-13	S-11
Minimum Cathode Diameter	1.68"	1.68"	1.63"	1.68"
Window Material	Lime glass	Lime glass	UV-transmitting silica	Lime glass
Window Shape	Flat	Flat	Flat	Curved
Dynode Material	CsSb	Be—O—Cs	CsSb	Be—O—Cs
Dynode Structure	Cage	Linear	Cage	Linear
Number of Stages	10	14	10	10
Nominal Diameter	2.0"	2.0"	2.0"	2.0"
Maximum Diameter	2.31"	2.38"	2.31"	2.31"
Maximum Overall Length	5.81"	7.50"	6.56"	6.12"
Cathode Sensitivity, μA/1 m	76	70	60	70
Typical Anode–Cathode Voltage	1,000	2,000	1,000	2,000
Gain	1.6×10^6	5.4×10^7	4.0×10^5	1.7×10^7
Dark Current, nA	6	1,000	10	200
Anode Pulse Rise Time, ns	3.4	3.8	—	1.8

Type	7850	8053	8054	8055
Cathode Response	S-11	S-11	S-11	S-11
Minimum Cathode Diameter	1.68"	1.68"	2.59"	4.38"
Window Material	Lime glass	Lime glass	Lime glass	Lime glass
Window Shape	Curved	Flat	Flat	Flat
Dynode Material	Be—O—Cs	Be—O—Cs	Be—O—Cs	Be—O—Cs
Dynode Structure	Linear	Venetian blind	Venetian blind	Venetian blind
Number of Stages	12	10	10	10
Nominal Diameter	2.0"	2.0"	3.0"	5.0"
Maximum Diameter	2.06"	2.31"	3.06"	5.31"
Maximum Overall Length	6.31"	5.81"	6.31"	7.69"
Cathode Sensitivity, μA/1 m	70	70	80	110
Typical Anode–Cathode Voltage	2,300	1,500	1,500	1,500
Gain	8.6×10^7	6.0×10^5	5.4×10^5	4.0×10^5
Dark Current, nA	2,400	4	4	4
Anode Pulse Rise Time	2.0	12	14	18

Table 2. CHARACTERISTICS OF PHOTOMULTIPLIERS (Continued)

Type	8575
Cathode Response	Bialkali
Minimum Cathode Diameter	1.80"
Window Material	Lime glass
Window Shape	Flat
Dynode Material	Be—O—Cs
Dynode Structure	Linear
Number of Stages	12
Nominal Diameter	2.0"
Maximum Diameter	2.10"
Maximum Overall Length	5.71"
Cathode Sensitivity, μA/1 m	85
Typical Anode–Cathode Voltage	2,000
Gain	1.0×10^7
Dark Current, nA	1
Anode Pulse Rise Time, ns	2.1

B. Dumont Photomultiplier Tubes

Type	K1390	K1391	K2199	K2482
Cathode Response	S-11	S-11	S-11	S-11
Minimum Cathode Diameter	2.50"	4.20"	1.50"	1.60"
Window Material	Glass	Glass	Glass	Glass
Dynode Material	CsSb	CsSb	CsSb	CsSb
Dynode Structure	Box	Box	Box	Box
Number of Stages	10	10	10	10
Nominal Diameter	3.0"	5.0"	2.0"	2.0"
Maximum Diameter	3.0"	5.3"	2.0"	2.0"
Maximum Overall Length	5.70"	6.80"	5.70"	5.70"
Cathode Sensitivity, μA/1 m	60	60	60	60
Typical Anode–Cathode Voltage	1,000	1,000	1,100	1,100
Gain	7.5×10^5	7.5×10^5	3.0×10^6	2.0×10^5
Dark Current, nA	50	50	10	150
Anode Pulse Rise Time, ns	—	—	—	—
Related Tubes	—	—	—	—

Table 2. CHARACTERISTICS OF PHOTOMULTIPLIERS (*Continued*)

Type	6291	6292	6362	6363
Cathode Response	S-11	S-11	S-11	S-11
Minimum Cathode Diameter	1.25"	1.5"	0.5"	2.5"
Window Material	Glass	Glass	Glass	Glass
Dynode Material	AgMg	AgMg	AgMg	AgMg
Dynode Structure	Box	Box	Box	Box
Number of Stages	10	10	10	10
Nominal Diameter	1.5"	2.0"	0.75"	3.0"
Maximum Diameter	1.5"	2.0"	0.75"	3.0"
Maximum Overall Length	4.25"	4.9"	4.9"	5.4"
Cathode Sensitivity, μA/1 m	60	60	60	60
Typical Anode–Cathode Voltage	1,100	1,100	1,100	1,100
Gain	2.0×10^5	2.0×10^5	2.0×10^5	2.0×10^5
Dark Current, nA	50	50	50	50
Anode Pulse Rise Time, ns	—	—	—	—
Related Tubes	—	7664 S-13 response	7860	—

Type	6364	6810	6935	7064
Cathode Response	S-11	S-11	S-11	S-11
Minimum Cathode Diameter	4.2"	1.68"	05."	1.5"
Window Material	Glass	Glass	Glass	Glass
Dynode Material	AgMg	AgMg	CsSb	CsSb
Dynode Structure	Box	Box	Box	Box
Number of Stages	10	14	10	10
Nominal Diameter	5.0"	2.0"	0.75"	2.0"
Maximum Diameter	5.25"	2.0"	0.75"	2.0"
Maximum Overall Length	6.75"	6.69"	4.9"	5.7"
Cathode Sensitivity, μA/1 m	60	70	50	60
Typical Anode–Cathode Voltage	1,100	2,100	1,100	1,000
Gain	2.0×10^5	1.2×10^6	3.0×10^5	7.5×10^5
Dark Current, nA	50	2,000	100	50
Anode Pulse Rise Time, ns	—	3	—	—
Related Tubes	—	—	—	—

Table 2. CHARACTERISTICS OF PHOTOMULTIPLIERS (Continued)

C. EMI Photomultiplier Tubes

Type	6094B	6097B	6097S	6255B
Cathode Response	S-11	S-11	EMI-S	S-13
Minimum Cathode Diameter	0.4"	1.75"	1.75"	1.75"
Window Material (Flat)	Glass	Glass	Glass	Fused silica
Dynode Material	CsSb	CsSb	CsSb	CsSb
Dynode Structure	Venetian blind	Venetian blind	Venetian blind	Venetian blind
Number of Stages	11	11	11	13
Nominal Diameter	2.0"	2.0"	2.0"	2.0"
Maximum Diameter	2.02"	2.02"	2.02"	2.02"
Maximum Overall Length	4.2"	4.92"	4.92"	5.27"
Cathode Sensitivity, µA/1 m	70	70	50	70
Typical Anode–Cathode Voltage	1,400	1,400	1,500	1,500
Gain	4.0×10^6	3.0×10^6	4.0×10^6	4.0×10^7
Dark Current, nA	2	5	1	50
Anode Pulse Rise Time, ns	7	7	7	8
Related Tubes	6256B	—	—	6255S
	Fused-quartz window			Low dark current

Type	9502B	9502S	9514B	9524B
Cathode Response	S-11	EMI-S	S-11	S-11
Minimum Cathode Diameter	0.4"	0.4"	1.75"	0.91"
Window Material (Flat)	Glass	Glass	Glass	Glass
Dynode Material	CsSb	CsSb	CsSb	CsSb
Dynode Structure	Venetian blind	Venetian blind	Venetian blind	Box
Number of Stages	13	13	13	11
Nominal Diameter	2.0"	2.0"	2.0"	1.0"
Maximum Diameter	2.02"	2.02"	2.02"	1.18"
Maximum Overall Length	4.64"	4.64"	5.27"	4.92"
Cathode Sensitivity, µA/1 m	70	50	70	70
Typical Anode–Cathode Voltage	1,500	1,700	1,500	1,000
Gain	3.0×10^7	4.0×10^7	3.0×10^7	3.0×10^6
Dark Current, nA	3	0.5	50	2
Anode Pulse Rise Time, ns	8	8	8	10
Related Tubes	6256S	6256S	9514S	9526B
	Fused-quartz window	Fused-quartz window	Low dark current	Quartz window

Table 2. CHARACTERISTICS OF PHOTOMULTIPLIERS (Continued)

Type	9530B	9531B	9531S	9545B
Cathode Response	S-11	S-11	EMI-S	S-11
Minimum Cathode Diameter	4.4"	3.03"	3.03"	9.8"
Window Material (Flat)	Glass	Glass	Glass	Glass
Dynode Material	CsSb	CsSb	CsSb	CsSb
Dynode Structure	Venetian blind	Venetian blind	Venetian blind	Venetian blind
Number of Stages	11	11	11	11
Nominal Diameter	5.0"	3.0"	3.0"	12.0"
Maximum Diameter	5.12"	3.58"	3.58"	11.81"
Maximum Overall Length	6.21"	6.61"	6.61"	14.72"
Cathode Sensitivity, μA/1 m	70	70	60	60
Typical Anode–Cathode Voltage	1,500	1,400	1,500	1,700
Gain	3.0×10^6	3.0×10^6	4.0×10^6	3.0×10^6
Dark Current, nA	10	100	8	1,000
Anode Pulse Rise Time, ns	16	12	12	25
Related Tubes	9530S Low dark current	9531Q Quartz window	—	—

Type	9558B	9594B	9623B	9634B
Cathode Response	S-20	S-11	S-11	S-11
Minimum Cathode Diameter	1.73"	—	6.8"	1.73"
Window Material (Flat)	Glass	Glass	Glass	Glass
Dynode Material	CsSb	CsSb	CsSb	CsSb
Dynode Structure	Venetian blind	Linear	Venetian blind	Venetian blind
Number of Stages	11	14	11	13
Nominal Diameter	2.0"	2.0"	7.5"	2.0"
Maximum Diameter	2.02"	2.02"	7.48"	2.02"
Maximum Overall Length	6.02"	7.5"	9.33"	5.27"
Cathode Sensitivity, μA/1 m	150	70	60	70
Typical Anode–Cathode Voltage	1,150	1,700	1,700	1,500
Gain	2.0×10^6	1.0×10^8	3.0×10^6	2.0×10^7
Dark Current, nA	2	400	300	50
Anode Pulse Rise Time, ns	8	2	20	8
Related Tubes	9558Q Quartz window	—	—	—

Section A of Table 2 reproduced from Booklet PIT-703 by permission of Radio Corporation of America.
Section B of Table 2 reproduced from Dumont Multiplier Tubes, 4th ed., by permission of Fairchild Camera and Instrument Co.
Section C of Table 2 reproduced from Booklet 20M/6–67/PMT, Issue 1, by permission of Whittaker Corp., Gencom Division.

LIQUID SCINTILLATION COUNTERS

Liquid scintillation counters are primarily designed to provide an efficient method of counting beta particles.[18-20] Table 3 lists some common radioisotopes that may be used with liquid scintillation systems, together with their half-lives and decay energies. The advantage of liquid scintillation counters arises from the fact that the beta sample is in intimate contact with the scintillator.

Table 3. SOME ISOTOPES FOR USE WITH SCINTILLATION COUNTERS

Isotope	Particle	Energy, Mev	Half-Life
^{3}H	β^-	0.018	12.3 years
^{14}C	β^-	0.156	5730 years
^{22}Na	β^+	0.54	2.58 years
	γ	1.28	
^{24}Na	β^-	1.39	15 hours
	γ	1.37, 2.76	
^{32}P	β^-	1.71	14.3 days
^{35}S	β^-	0.168	86.7 days
^{36}Cl	β^-	0.71	3×10^5 years
^{40}K	β^-	1.33	1.3×10^9 years
	γ	1.46	
^{45}Ca	β^-	0.25	165 days
^{51}Cr	γ	0.32	27.8 days
^{54}Mn	γ	0.84	314 days
^{59}Fe	β^-	0.27, 0.46	45 days
	γ	1.1, 1.3	
^{63}Ni	β^-	0.067	92 years
^{64}Cu	β^-	0.57	12.9 hours
	β^+	0.66	
^{65}Zn	β^+	0.33	245 days
^{85}Kr	β^-	0.67	10.8 years
^{90}Sr	β^-	0.54	28 years
^{115}Cd	β^-	1.63	43 days
^{123}Sn	β	1.42	125 days
	γ	1.08	
^{131}I	β	0.61, 0.33, 0.25	8 days
	γ	0.36	
^{137}Cs	β	0.51, 1.17	30 years
^{203}Hg	β^-	0.21	47 days
	γ^-	0.279	

One problem common to all liquid scintillation systems is that of determining the detection efficiency and quenching. Quenching refers to any process in the liquid scintillator that reduces the amount of light reaching the photomultiplier tube. These processes can be either chemical or optical. Optical quenching consists of absorption of the light, for example, by a colored solution. All other processes that reduce the amount of light reaching the photomultiplier tubes are lumped under the category of chemical quenching. Even clear, colorless solutions can have high chemical quenching. Figure 4 illustrates the effect of increasing amounts of a quenching agent (nitromethane) on the pulse-height spectra of both tritium and ^{14}C.

The recorded counts per minute (cpm) observed in a liquid scintillator can be related to the activity of the sample measured in disintegrations per minute (dpm) by

$$\text{dpm} = \text{cpm}/\text{eff}, \tag{1}$$

where eff is the efficiency of detection for the particular sample of interest. Determination of the efficiency is a central problem in liquid scintillation counting, and many methods of determination have been suggested. One approach is to add a standard of the same isotope to the sample after it has been counted and then repeat the count. This is termed *internal*

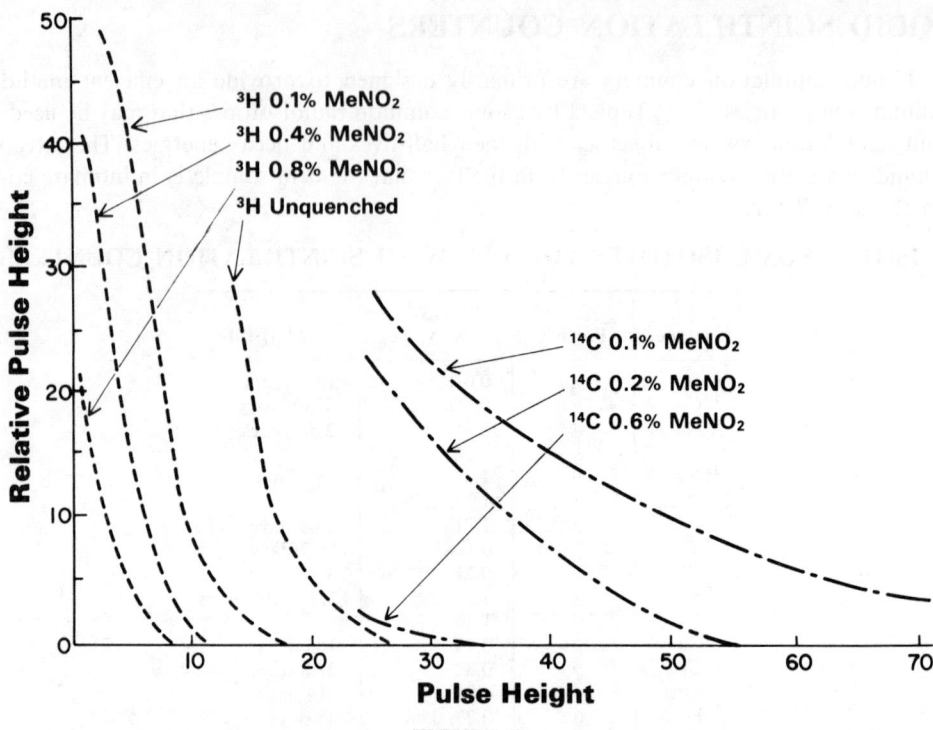

FIGURE 4.
Liquid Scintillation Spectra as a Function of Quenching-Agent Concentration.

standardization. The specific activity (mCi/ml) of the standard must be high enough, so that good counting statistics are obtained without significantly altering the volume or composition of the original sample. The efficiency is then given by

$$\text{eff} = \frac{R_2 - R_1}{A}, \qquad (2)$$

where R_1 is the counting rate observed with the sample alone, R_2 a counting rate after the standard has been added, and A the activity of the standard in dpm. A major disadvantage of this method is that each sample must be counted twice, and the sample count cannot be repeated if necessary.

Another method of determining the efficiency is known as the *Channel's-Ratio Method*. Quenching results in a decreased amount of light reaching the photomultiplier tube and a reduction in the size of the pulses out of the detecting circuits. The net result is a spectrum of lower overall pulse height and lower efficiency for a given control setting. Figure 5 illustrates this point. The ratio of counts recorded in a channel including most of the beta spectrum to the counts recorded in a channel accepting only the higher energy pulses can be used to determine the amount of quenching present, and hence is a measure of counting efficiency.[21,22] The advantages of this technique are that it requires only a single counting, is independent of sample volume, the results are immediately available, and the samples are preserved for later studies, if desired. Long counting times are required for an accurate determination of the channel's ratio for samples with low count rates.

A third method of determining efficiency is known as the *External-Standardization Method*. After a sample has been counted, an external gamma-ray source is brought next to the sample. The gamma rays from the radioactive source produce Compton scattered electrons in the scintillator solution. These Compton scattered electrons undergo quenching processes that are similar to those affecting beta particles from the sample. From the response of the

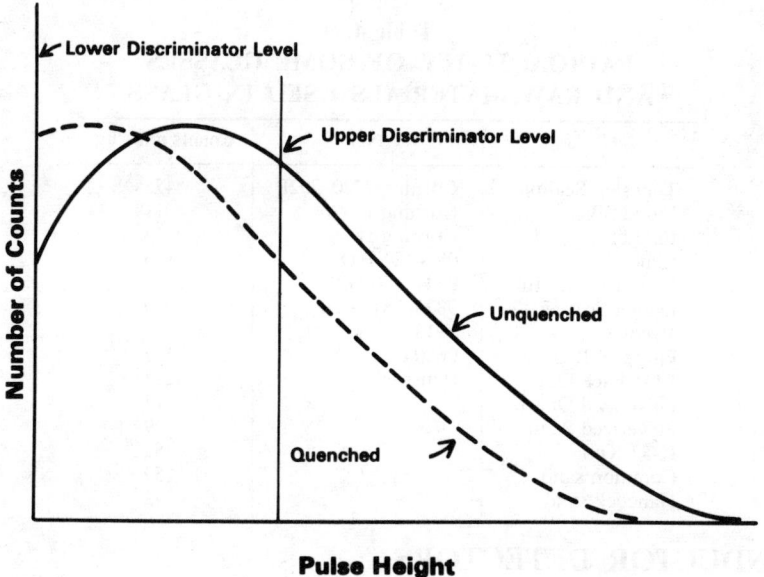

FIGURE 5. Beta Spectrum Shift Produced by Quenching in a Liquid Scintillator.

sample with the external standard in place it is possible to determine the efficiency correction required. Because the external source can have high activity, the second counting can be done quickly. External standardization is rapid, does not involve additional handling of the sample, and preserves the sample for future studies, if necessary; it is, however, affected to some extent by changes in sample volume. Figure 6 presents a typical pulse height spectrum from a liquid scintillator using two sources.

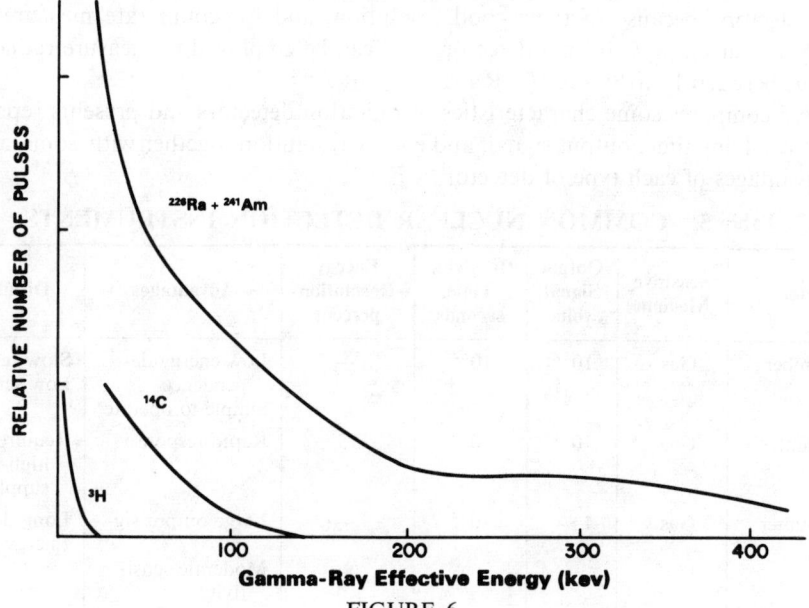

FIGURE 6.
Representative Liquid Scintillation Spectrum
from a Two-Source External Gamma-Ray Standard.

In order to reduce the background as much as possible, it may be necessary at times to use special photomultiplier tubes constructed of low-radioactive materials. Table 4 presents the radioactivity of some glasses and raw materials used in glass.[23]

Table 4.
RADIOACTIVITY OF SOME GLASSES AND RAW MATERIALS USED IN GLASS

Glass	No.	Counts min⁻¹ kg
Tungsten Sealing	Corning 7720–702P	475
Fused Silica	Corning I	11
Fused Silica	Corning II	4
Lime Glass	081–124 HD	175
U.V. Transmitting	9741–970 O.P.	1193
Dumet Sealing	9823–981 A 20	979
Vycor	7913	34
Philips OB Glass	06004	337
EMI Face Plate	(Unspecified)	213
Pulverized Quartz		117
Pulverized Sand		59
K-52 Kona		82
Common sand		59
Hancock sand		42

SEMICONDUCTOR DETECTORS

The principle of operation of a semiconductor detector is similar to that of a gas-filled detector, except that the detecting medium is a solid.[4,24-26] An applied electric field sweeps out free electrons and holes, producing a depletion layer containing practically no free-charged carriers.[27,28] The extent of this carrier-free region is determined by the applied electric field, and can be changed by changing electric field. When an ionizing particle enters the depletion region, it produces electron–hole pairs, which are swept apart by the applied electric field. This produces an output signal. The dead time of semiconductor detectors is very short, and they can be used to measure high count rates. Solid-state semiconductors also can be made extremely small (1 mm³), and therefore may be employed for dosimetry inside patients. They are useful both as pulse detectors, because of their good resolution, and for count rate measurements. By measuring the current flow from a detector, they can be employed to measure radiation fields in the range between 1 mr/hr and 500 R/hr.

Table 5 compares some characteristics of radiation detectors and presents representative figures for resolving time, output signal, and energy resolution together with some advantages and disadvantages of each type of detector.

Table 5. COMMON NUCLEAR DETECTION INSTRUMENTS

Detector	Sensitive Medium	Output Signal, volts	Resolving Time, seconds	Energy Resolution, percent	Advantages	Disadvantages
Ion Chamber	Gas	10^{-6}	10^{-4}	—	Low energy dependence Simple to operate	Slow response Low sensitivity
Prop Counter	Gas	10^{-2}	10^{-6}	15	Rapid response	Requires stable high-voltage supply
G-M Counter	Gas	1	10^{-4}	—	Large output signal Moderate sensitivity	Long dead time Energy dependent
Scintillation Crystal	Solid	1	10^{-7}	10	High sensitivity Rapid response Good energy resolution	Fragile Expensive
Semiconductor Detector	Solid	10^{-3}	10^{-9}	1	Excellent energy response Short dead time	Requires high amplification

DEAD TIME AND COUNTING LOSSES

The dead time or resolving time of a detection system is that time during which the system is unable to respond to an ionizing event. It is also defined as the time which must separate two events in order for the system to recognize them as two distinct events. Dead time can occur both in the detector and in the associated electronic counting equipment. There are two classes of detector systems: "paralyzing" and "nonparalyzing". A "nonparalyzing" system can respond to subsequent events which occur τ seconds later, regardless of how many events occur during the dead time. A "paralyzing" system, however, is insensitive for τ seconds after every event that occurs. For recovery of a paralyzing system, a period of τ seconds is required, during which no events occur. Geiger-Mueller counters and mechanical registers are two common examples of paralyzing nuclear detection equipment. Figure 7 illustrates counting losses in both paralyzing and nonparalyzing equipment.

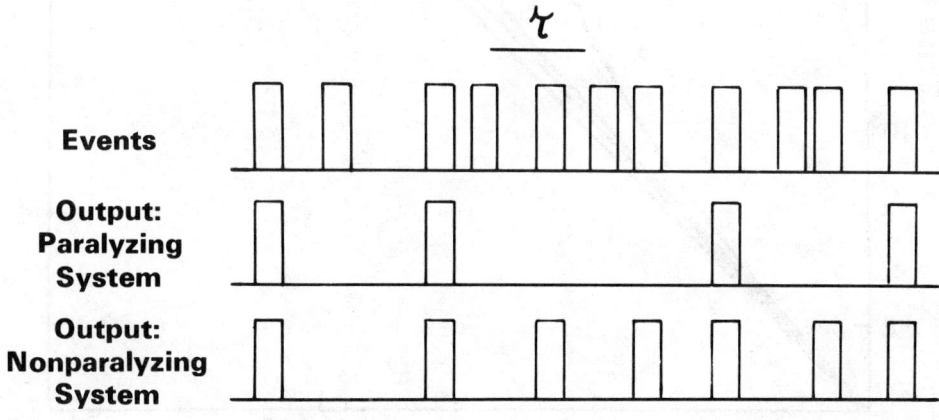

FIGURE 7.
Pulse Loss Due to Dead Time in Paralyzing and Nonparalyzing Detection Equipment.

As an example, consider an observed counting rate of R_o counts per second being detected by a counter having resolving time τ. If the detector is operated for a time T, it is actually "on" only for a time

$$T(1 - R_o \tau), \tag{3}$$

and all events occuring during the dead time are lost. The number of counts which are lost is given by $R_o \tau$. The true count rate is given by

$$R_{true} = \frac{R_o}{1 - R_o}. \tag{4}$$

The relation between the observed and true count rate for several representative resolving times is shown in Figure 8.

An experimental determination of resolving time can be made by counting two sources (A and B) individually and then simultaneously. If source A gives a counting rate R_A, source B gives a counting rate R_B, and the combination gives a rate R_{AB}, then the resolving time τ is given by

$$\tau = \frac{R_A + R_B - R_{AB} - R_b}{R_{12}^2 - R_1^2 - R_2^2}, \tag{5}$$

where R_b is the background counting rate.[29]

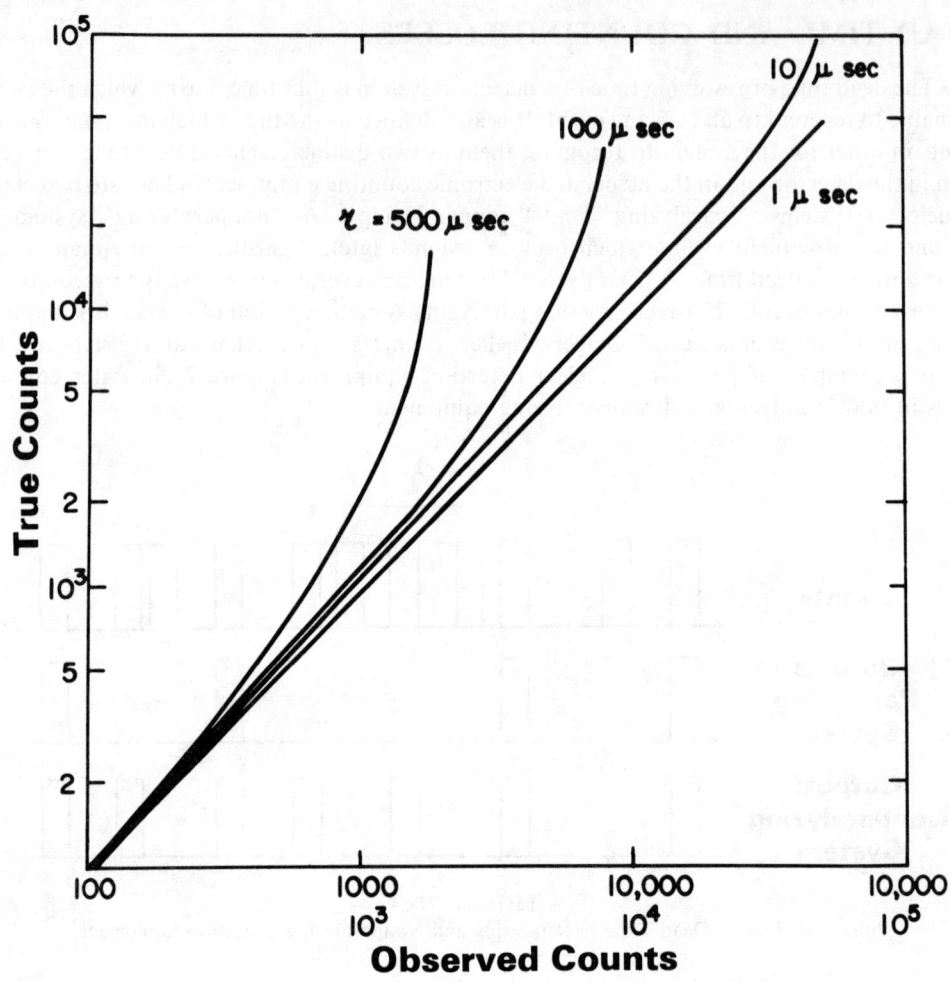

FIGURE 8.
Count Rate Correction
for Various Representative Dead Times.

REFERENCES

1. Attix, F. H. and Roesch, W. C., eds., *Radiation Dosimetry*, Vol. 2, 2nd ed. Academic Press, New York, 1966.
2. Burcham, W. E., *Nuclear Physics—An Introduction*, Ch. 6. McGraw-Hill, New York, 1963.
3. Hine, G., ed., *Instrumentation for Nuclear Medicine*, Vol. 1. Academic Press, New York, 1967.
4. Price, W. J., *Nuclear Radiation Detection*, 2nd ed. McGraw-Hill, New York, 1964.
5. Boag, J. W., Ionization Chambers. *Radiation Dosimetry*, Vol. 2, Ch. 9, p. 2, Attix, F. H. and Roesch, W. C., eds. Academic Press, New York, 1966.
6. Rossi, B. B. and Staub, H. H., *Ionization Chambers and Counters*. McGraw-Hill, New York, 1950.
7. Spiers, F. W., Radiation Units and Theory of Ionization Dosimetry. *Radiation Dosimetry*, Ch. 1, p. 1, Hine, G. J. and Brownell, G. L., eds. Academic Press, New York, 1956.
8. Wilkinson, D. H., *Ionization Chambers and Counters*. Cambridge University Press, New York, 1960.
9. Curran, S. C., The Proportional Counter as a Detector and Spectrometer. *Encyclopedia of Physics*, 45:175, Springer Verlag, Berlin, Germany, 1958.
10. Emery E. W., Geiger-Mueller and Proportional Counters. *Radiation Dosimetry*, Vol. 2, Ch. 10, p. 73, Attix, F. H. and Roesch, W. C., eds. Academic Press, New York, 1966.
11. Sinclair, W. K., Geiger-Mueller Counters and Proportional Counters. *Radiation Dosimetry*, Ch. 5, p. 213, Hine, G. J. and Brownell, G. L., eds. Academic Press, New York, 1956.
12. Korff, S. A., Geiger Counters. *Encyclopedia of Physics*, 45:52, Springer Verlag, Berlin, Germany, 1958.

13. Robinson, C. V., Geiger-Mueller and Proportion Counters. *Instrumentation in Nuclear Medicine*, Vol. 1, Ch. 4, p. 57, Hine, G., ed. Academic Press, New York, 1967.
14. Hine, G. J., Sodium Iodide Scintillators. *Instrumentation in Nuclear Medicine*, Vol. 1, p. 95, Hine, G., ed. Academic Press, New York, 1967.
15. Ramm, W. J., Scintillation Detectors. *Radiation Dosimetry*, Vol. 2, Ch. 11, p. 123, Attix. F. H. and Roesch, W. C., eds. Academic Press, New York, 1966.
16. Swank, R. K., Characteristics of Scintillators. *Ann. Rev. Nucl. Sci.*, 4:11—140, 1954.
17. Birks, J. B., *The Theory and Practice of Scintillation Counting*. Pergamon Press, Macmillan Co., New York, 1964.
18. Bell, C. G., and Hayes, F. N., eds., *Liquid Scintillation Counting*. Pergamon Press, Oxford, England, 1958.
19. Polic, E. F., Liquid Scintillation Counting Equipment. *Instrumentation for Nuclear Medicine*, Vol. 1, Ch. 10, p. 227, Hine, G., ed. Academic Press, New York, 1967.
20. Rapkin, E., Preparation of Samples for liquid Scintillation Counting. *Instrumentation for Nuclear Medicine*, Vol. 1, Ch. 9, p. 181, Hine, G., ed. Academic Press, New York, 1967.
21. Baille, L. A., Determination of Liquid Scintillation Counting Efficiency by Pulse-Height Shift. *Int. J. Appl. Radiat. Isotop.*, 8:1, 1960.
22. Bruno, G. A. and Christian, J. E., Correction for Quenching Associated with Liquid Scintillation Counting. *Anal. Chem.*, 33: 650, 1961.
23. LeVine, H. D., Chariton, L., and Graveson, R. T. AEC Report *HASL* 60, p. 7, 1959.
24. Brown, W. L., Introduction to Semiconductor Particle Detectors. IRE *Trans. Nucl. Sci.*, 8:2, 1961.
25. Friedland, S. S. and Zatzick, M. R., Semiconductor Detectors. *Instrumentation in Nuclear Medicine*, Vol.1, Ch. 5, p. 73, Hine, G., ed. Academic Press, New York, 1967.
26. Miller, G. L., Gibson, W. M., and Donovan, P. F., Semiconductor Particle Detectors. *Ann. Rev. Nucl. Sci.*, 12:189, 1962.
27. Fowler, J. L., Solid-State Dosimetry. *Phys. Med. Biol.*, 8:1, 1963.
28. Fowler, J. L., Solid-State Electrical-Conductivity Dosimeters. *Radiation Dosimetry*, Vol. 2, Ch. 14, p. 291, Attix, F. H. and Roesch, W. C., eds. Academic Press, New York, 1966.
29. Friedlander, G. and Kennedy, J. W., *Nuclear and Radiochemistry*, p. 265. Wiley and Sons, New York, 1955.

RADIOISOTOPE COUNTING AND CALIBRATION

James A. Sorenson

SODIUM IODIDE WELL COUNTERS

A sodium iodide well counter is the most commonly used device for measuring samples with low levels of gamma activity. It is basically a sodium iodide crystal and photomultiplier tube assembly, with a hole in the crystal for insertion of a sample. This allows the sample to be positioned near the center of the crystal. Figure 9 presents a schematic cross-sectional view of a typical well counter.

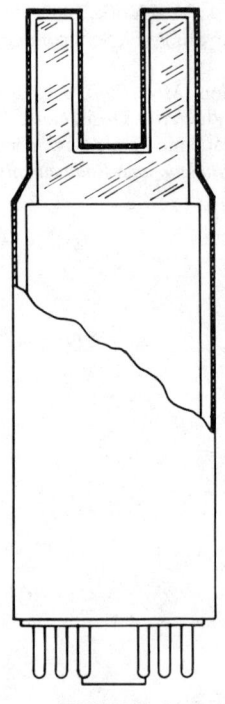

FIGURE 9.
Sodium Iodide Well Counter.

Sodium iodide well counters have the same advantages and disadvantages as ordinary sodium iodide assemblies. They have high detection efficiency and good energy resolution for gamma rays, but are fragile and expensive. Because of the necessity for a lightproof enclosure around the crystal, sodium iodide well counters cannot be used to count beta radiation efficiently. For beta particles of ^{32}P, for example, the counting efficiency including bremsstrahlung is about 0.8 percent per disintegration in a standard well counter.[1] For lower-energy beta emitters the counting efficiency is effectively zero. The components of a complete sodium iodide well counter system include lead shielding around the detector assembly, preamplifiers and amplifiers, a high-voltage supply, a pulse-height analyzer and/or pulse discriminator, and a scaler and timer. Shielding around the detector assembly usually amounts to about three inches of lead.

General Characteristics of Sodium Iodide Well Counters

Various factors can affect the number of radiations detected and counted from a radioactive sample with a well counter system. The number of counts can vary, for example, with the sample volume. Volume effects are caused primarily by increased losses of radiation through the end of the well as the sample volume increases and rises in the well. When, for instance, samples of fixed total activity and increasing volume are placed in progressively larger well detector systems, the number of counts decreases. If, on the other hand, increasing sample volumes of a solution of constant concentration are counted in a standard well, the number of counts detected will rise—although not linearly, as one would expect. For this reason it is recommended that a full test tube of material be counted, if it is available. This will also yield the highest counting rate. Only test tubes of the same diameter and wall thickness should be used for such applications, so that the volume of sample in the well does not change significantly.

Pulse-height spectra obtained for ^{131}I and ^{137}Cs with a standard 2" × 2" well counter are shown in Figure 10, A and B respectively. There are a considerable number of pulses found in the Compton region. For this reason it is usually desirable to count all of the pulses above a certain energy level, using a pulse-height discriminator. The counting rate will vary with the discriminator setting as more or fewer of the pulses in the Compton region of the spectrum are counted. This variation is shown in Figure 11 for ^{131}I. To obtain a minimum of variation in count rate with the setting, thus reducing the possibility of errors, it is desirable to operate with the discriminator level set at a plateau region, i.e., in the 50- to 100-kev range for the spectra shown in Figure 11. Figure 11 also shows how in a typical standard well counter system the background counting rate increases as the discriminator setting is lowered.

For routine operation it is desirable to adjust the gain of the system, so that the lower-level discriminator can be read directly in kev. Thus, for a 10-turn 1000-division dial, as is commonly found on spectrometers, the calibration might well be 1 kev per division. A simple way of making this calibration is to insert a sample of ^{131}I in the well and set the base level and window width to 344 and 40 divisions respectively. By adjusting the high voltage or amplifier gain until a maximum counting rate is observed, a 1-kev-per-division ratio can be obtained. This method is time-consuming, because the direction and magnitude of the required adjustment are not always apparent. A more convenient and accurate procedure is to adjust the gain or high-voltage settings until the counting rate at equal intervals on either side of a desired maximum is the same. Thus, for ^{131}I with a 40-kev window, the count rates at 324 and 364 divisions should be equal. If after initial rough adjustments the count rate at 324 is higher than at 364, the gain or high voltage must be increased. Figure 12 illustrates correct and incorrect adjustments. This procedure has the advantage of indicating both the direction and the approximate magnitude of the necessary adjustments. The final settings can usually be obtained with only three or four counting measurements.

The efficiency with which gamma rays are detected in a well counting system depends strongly on the energy of the gamma ray being measured. This is because the higher-energy gamma rays are more likely to escape the crystal without an interaction. The probability of escape can be reduced by using a larger crystal, thus surrounding the sample with more detecting material. The percentage of gamma rays absorbed is a function of energy for sodium iodide well counters of various sizes. Below 100 kev the detection efficiency is practically 100 percent for all well counter sizes shown, but at 1 Mev it decreases to about 25 percent for the standard 2" × 2" well crystal, and to 45 percent for a 3" × 3" crystal with a 1-inch diameter wall.

The detection efficiency, theoretically, assumes no losses through the open end of the well. It is also assumed that all gamma rays interacting with the crystal will be counted. In fact some of the gamma rays undergoing Compton interactions will not be detected, since these might not produce pulses large enough to pass the pulse-height discriminator. This amounts to a further loss of detection efficiency, which becomes more serious as the gamma-ray energy—

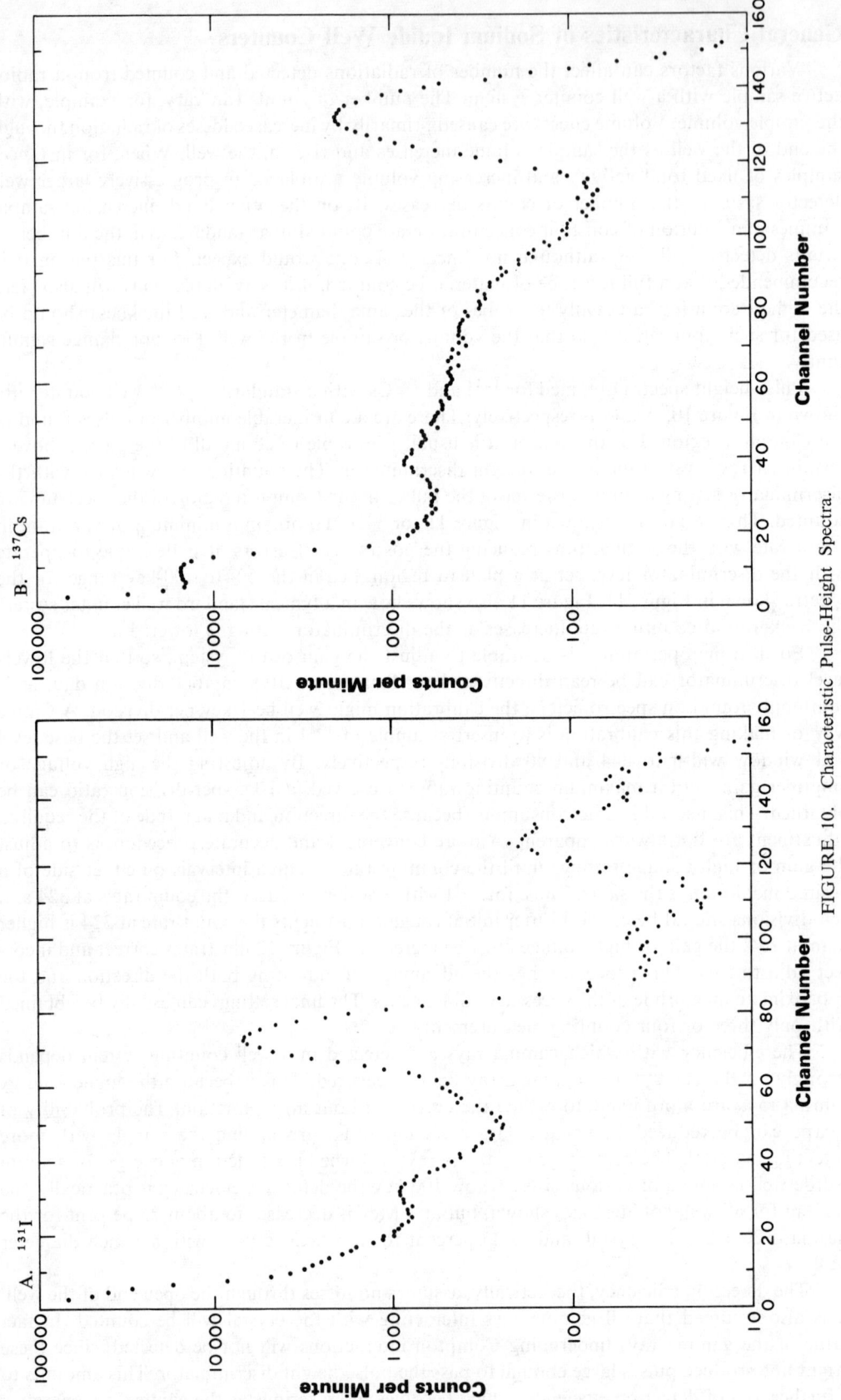

FIGURE 10. Characteristic Pulse-Height Spectra.

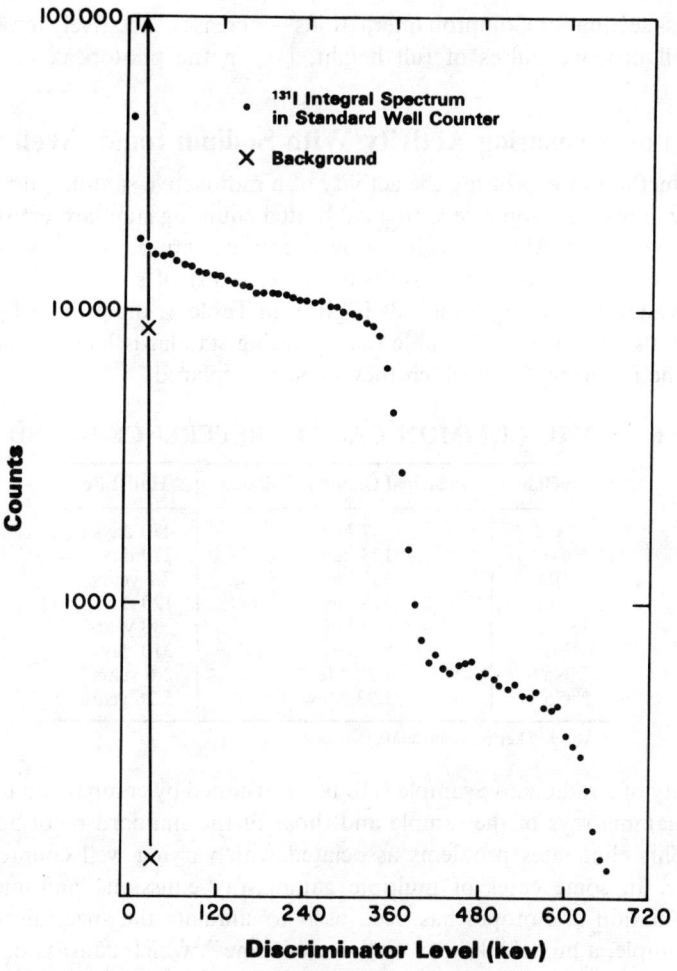

FIGURE 11.
Counting Rates in Relation to Discriminator Settings.

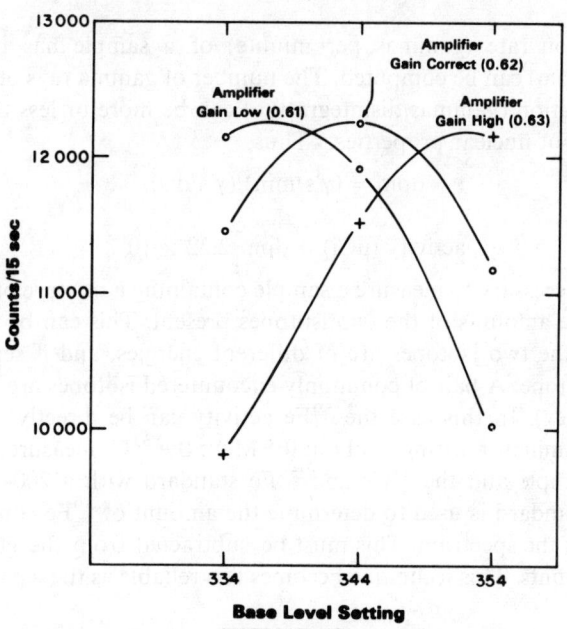

FIGURE 12.
Calibration by Adjustment of Amplifier Gain.

and thus the cross section for Compton interactions—increases. Relatively fewer of the detected gamma rays will produce pulses of full height, i.e., in the photopeak of the pulse-height spectrum.

Basic Method of Measuring Activity With Sodium Iodide Well Counters

The usual method of measuring the activity of a radioactive sample with a well counter is to compare its gamma emission rate with a calibrated counting standard emitting gamma rays of nearly the same energy. Absolute calibration procedures are not used, because of the many factors affecting counting efficiency described above. A list of some of the commonly used, commercially available counting standards is given in Table 6; the principal gamma emission and half-life are also given. It is desirable that counting standards have a relatively long half-life, to reduce the frequency with which they must be replaced.

Table 6. SOME COMMON GAMMA REFERENCE STANDARDS

Nuclide	Principal Gamma Emission	Half-Life
^{109}Cd	87 kev	470 days
^{57}Co	123 kev	270 days
^{133}Ba	357 kev	7.2 years
^{113}Sn	393 kev	120 days
137Cs	662 kev	30 years
^{54}Mn	840 kev	310 days
^{22}Na	1.28 Mev*	2.6 years
^{60}Co	1.33 Mev	5.26 years

* Also 540 kev β^+ annihilation radiation.

If the activity of a radioactive sample is to be determined by comparison to a standard, it is necessary that gamma rays of the sample and those of the standard be of approximately the same energy. This eliminates problems associated with varying well counter efficiencies, as discussed above. In some cases of multiple gamma-ray emissions and more complicated spectra a combination of isotopes has been used to simulate the spectrum of the measured sample. For example, a mixture known as "mock iodine", which consists of ^{133}Ba (355 and 380 kev) and ^{137}Cs (662 kev), has been used as a standard for measuring ^{131}I activities.[2] The gamma-ray ratios are enough like those of ^{131}I to make the life of the standard useful for about ten years.

Once the emission rate (gammas per minute) of a sample has been determined, its disintegration rate (dpm) can be computed. The number of gamma rays of the principal energy emitted per disintegration (gammas/disintegration) can be more or less than one, and can be obtained from tables of nuclear properties.[3] Thus,

$$\text{dpm} = (\gamma\text{'s/min})/(\gamma\text{'s/dis}), \tag{1}$$

and to obtain activity,

$$\text{activity } (\mu\text{Ci}) = \text{dpm}/2.22 \times 10^6. \tag{2}$$

It is sometimes necessary to measure a sample containing a mixture of two isotopes and to determine the relative amounts of the two isotopes present. This can be done if the principal gamma emission of the two isotopes are of different energies, and if separate standards are available for each isotope. A pair of commonly encountered isotopes are ^{51}Cr (0.32 Mev) and ^{59}Fe (1.1 and 1.3 Mev). In this case the ^{59}Fe activity can be directly assessed by the ^{59}Fe standard with a discriminator setting of about 0.5 Mev; the ^{51}Cr measurement can be obtained by measuring the sample and the ^{51}Cr and ^{59}Fe standard with a 200- to 400-kev analyzer window. The ^{59}Fe standard is used to determine the amount of ^{59}Fe contribution to the 200- to 400-kev portion of the spectrum. This must be subtracted from the gross counts in order to get the net ^{51}Cr counts. This technique becomes less reliable as the separation of gamma-ray energies decreases.

There are also special problems in counting low-energy gamma emitters in well counters, e.g. ^{125}I (27 kev). Self-absorption in the sample volume and absorption by test tube walls can be considerable at these lower energies. It is necessary with these low-energy gamma rays that the sample and standard volumes and the containers be as nearly identical as possible, to equalize these effects.

Other Arrangements of Sodium Iodide Crystals For Sample Counting and Calibration

The principal restrictions of sodium iodide well counters for gamma-ray counting are that they are useful only for small sample volumes of low activities. For activities greater than about one microcurie the counting rates in well counters become so high that counter dead-time losses may become excessive. Larger activities can be measured by placing the sample in front of and at some distance from a single crystal. The calibrated standard, also of a larger activity, must be measured at the same distance and under the same geometric conditions. In this type of measurement nearby objects can produce considerable scattered radiation, which must also be measured by the detector. Care must be taken not to rearrange nearby objects between the measurements of the standard and of the sample. Smaller activities can be measured by placing the sample directly on the face of the crystal. Measurements with this arrangement are particularly sensitive to changes in sample volume, as is illustrated in Figure 13.

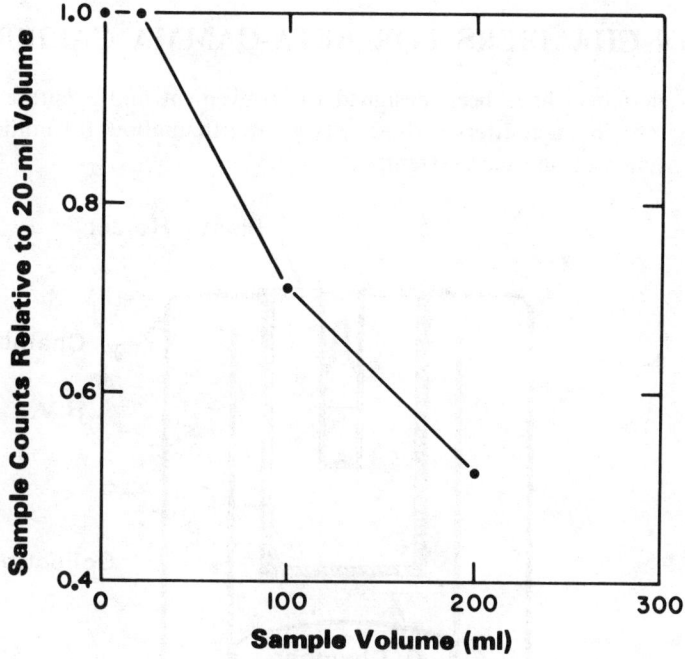

FIGURE 13.
Sensitivity to Volume Changes in Measurements Using a Single Crystal.

Paired, opposed sodium iodide crystals have been used for high-efficiency counting of larger volumes. Such an arrangement can efficiently count volumes of one liter or more. Some radioisotope scanners employ paired, opposed sodium iodide detectors; these scanners can also be used as large-volume counters, with the sample simply placed on the subject table. The counting efficiency of such paired detector systems decreases from a maximum of about 30 percent as the crystal separation increases, but is constant for a fixed separation. The effects of volume changes for a fixed crystal separation are shown in Figure 14.

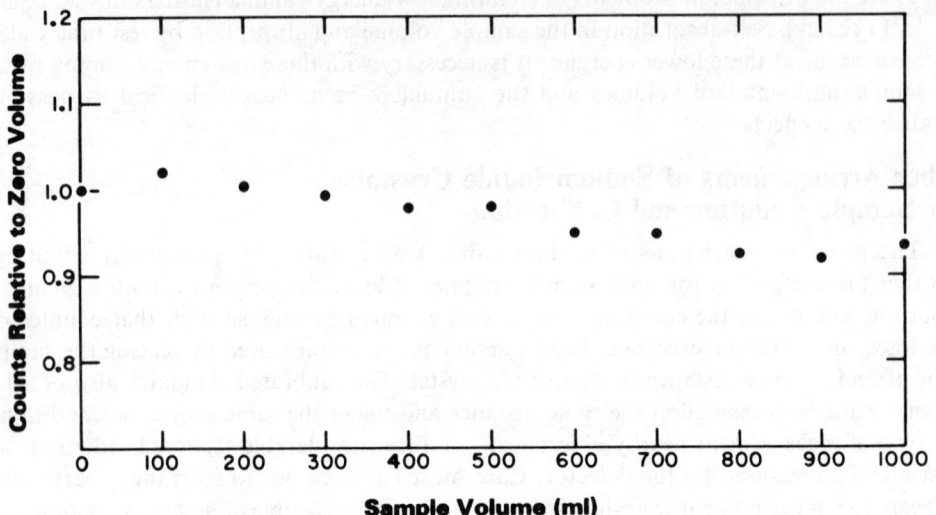

FIGURE 14.
Effect of Volume Changes in Measurements
Using Paired, Opposed Crystals at Fixed Separation.

IONIZATION CHAMBERS FOR BETA-GAMMA CALIBRATION

Ionization chambers have been designed for convenient and accurate calibrations of beta emitters and of gamma emitters with activity greater than about 0.1 millicurie. Figure 15 shows the cross section of one such system.

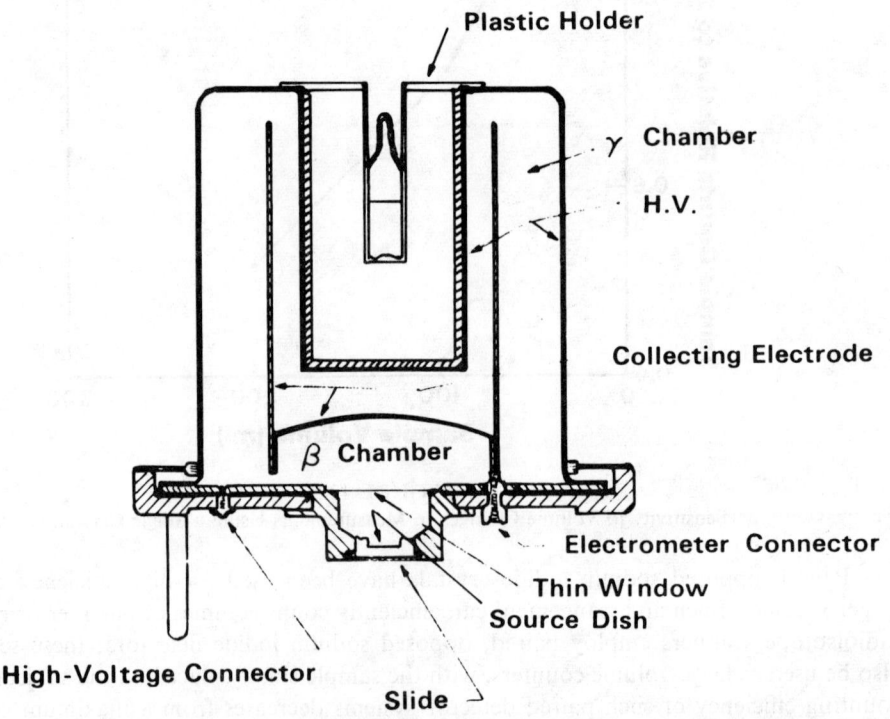

FIGURE 15.
Ionization Chamber.

For gamma-ray counting the sample is placed in a test tube or an ampule in the reentrant cavity. The gamma-counting cavity has brass walls, to absorb beta particles. The ion chamber is not sealed, and contains air at atmospheric pressure. The ionization current per millicurie for a few gamma emitters is given in Table 2. Saturation occurs at an ionization current of about 10^{-10} amps. Background currents are of the order of 10^{-14} amps.[4]

Beta-emitting samples are counted on a source dish beneath the beta chamber, which has a thin duraluminum entrance window (0.68 mg/cm^2). The ionization current for some beta emitters is also given in Table 7.

Table 7.
IONIZATION CURRENT PER MILLICURIE FOR BETA-GAMMA IONIZATION CHAMBER

Radionuclide	Gamma-Ray Chamber, amp × 10^{-12} per mCi	Beta-Ray Chamber, amp × 10^{-12} per mCi/ml
^{24}Na	75.6 (b)	—
^{32}P	0.79* (b)	268 (c)
^{42}K	6.0 (b)	433 (c)
^{59}Fe	27.1 (a)	—
^{60}Co	53.2 (a)	—
^{82}Br	63.0 (a)	103 (c)
^{90}Y	—	335 (c)
^{131}I	10.6 (a)	73 (c)
^{198}Au	10.8 (b)	137 (c)
^{226}Ra + decay products	35.6 (b)	—

* Bremsstrahlung
(a) 1 g of solution in 2-ml ampule
(b) 4 g of solution in 5-ml ampule
(c) 1 ml of solution in polyethylene dish

An ionization chamber is a rugged and easy-to-use instrument that does not require critical high-voltage adjustment. It is useful for measuring millicurie amounts of gamma-ray-emitting isotopes and of beta-emitting isotopes. The ionization current per millicurie varies with temperature and pressure of the gas, and it is necessary to make corrections for these factors. If the ionization chamber is used for calibration purposes with little or no shielding, there is a possibility of scattered radiation from nearby objects scattering back into the chamber. The ionization current can be affected by as much as several percent by varying the arrangement of these objects, and care must be taken to minimize these effects. Accuracies of present-day ionization chambers are claimed to be of the order of ± 3 percent.

REFERENCES

1. Loevinger, R., and Feitelberg, S., Using Bremsstrahlung Detection by a Scintillator for Simplified Beta Counting. *Nucleonics*, 13(4): 42, 1955.
2. Brucer, M., Oddie, T. H., and Eldridge, J. S., Thyroid Uptake Calibration. I. Mock-Iodine, A Radioactive Iodine Gamma-Ray Standard. *ORINS-14*. Oak Ridge Institute of Nuclear Studies, Inc., 1956.
3. Strominger, D., Hollander, J. M., and Seaborg, G. T., Table of Isotopes. *Rev. Mod. Phys.*, 30(2): 585, 1958.
4. Dale, J. W. G., Perry, W. E., and Pulfer, R. F., A Beta-Gamma Ionization Chamber for Substandards of Radioactivity. I. Uses and Calibration. *Int. J. Appl. Radiat. Isotop.*, 10(2): 72–78, 1960.

IMPROVEMENT OF LIQUID SCINTILLATION COUNTING EFFICIENCIES BY OPTIMIZATION OF SCINTILLATOR COMPOSITION

E. T. Bush and D. L. Hansen
Nuclear-Chicago Corporation, Des Plaines, Illinois.
Reprinted with permission of the *International Atomic Energy Agency, Vienna, Austria*, from *Radioisotope Sample Measurement in Medicine and Biology*, pp. 395—408, 1955.

INTRODUCTION

A specific objective of this study was the evaluation of three new liquid scintillation counting solutes by comparison with some of those which have been most commonly used. At the time the work was begun, only one of the new fluors was commercially available, and reports as to its effectiveness were somewhat contradictory. This was the secondary solute "dimethyl-POPOP", *p*-bis[2-(4-methyl-5-phenyl-oxazolyl)] benzene, which was to be compared with another new secondary solute, "BPSB", bis(isopropylstyryl) benzene, and with the widely used POPOP as a standard. A new primary solute, "BBOT", 2,5-bis [5,-tert.butylbenzoxazolyl(2')] thiophene, having an extraordinarily long wavelength emission spectrum peaked at 435 nm, was to be compared with PPO, 2,5-diphenyloxazole, as the most commonly used primary, and with PBD, phenylbiphenyloxadiazole-1,3,4, as the most efficient primary available. Both BPSB and BBOT are now marketed.

A more general objective of the study was the determination of the extent to which the counting efficiency in modern liquid scintillation counters is increased by the use of secondary solutes, the so-called "wavelength shifters". It was realized early in the history of liquid scintillation counting[1-3] that the usefulness of wavelength shifters is strongly dependent on the multiplier phototube being used and on the "optics" of the system, that is, on the reflecting and transmitting media in the path of the emitted light and on the path length. There have been numerous reports that very little, if any, improvement is seen in the newer counters when secondary solutes are added to PPO solutions. Yet almost all the published formulae for preparing liquid scintillation samples include a secondary. The effect of these secondary solutes and of the long-wavelength BBOT on counting efficiency was therefore studied in counters representing three different types of optical systems used in American commercial automatic instruments. Differences in the responses of certain types of multiplier phototubes, as well as differences between individual tubes of the same type, were studied by using numerous counters having the same optical system.

The problem of choosing valid conditions for comparison of the fluors is not limited to the selection of optics. It is necessary to consider whether the relative efficiencies of the different fluors will be the same in quenched as in unquenched samples and in different solvents. Since nearly all counting samples of practical interest are quenched to some degree, the prediction of optimal scintillator composition in quenched samples is of primary importance.

Although the necessity to study the fluors in quenched samples was recognized from the outset, the extent to which their relative efficiencies would depend on the choice of quenching agent and of fluor concentration was not anticipated. Some unexpected results led to a broader study of optimal fluor concentration in solutions of the traditional as well as the new fluors. Although the literature of liquid scintillation counting is not lacking in suggestions that the conventional 4 to 6 gram per liter of PPO and 0.05 to 0.1 gram per liter of POPOP are suboptimal for various kinds of samples, there has been no systematic study of the relationship of optimal fluor concentration to the scintillation solvent or the type of quenching agent. For practical purposes, the most useful comparison of different fluors in a given sample type is considered to be based on counting efficiencies at the optimal concentrations of the respective fluors.

EXPERIMENTAL PROCEDURE

BBOT was obtained from CIBA Ltd., BPSB and PBD from Pilot Chemicals Inc., dimethyl-POPOP from Arapahoe Chemicals Inc., and PPO and POPOP from both Pilot and Arapahoe. Solutions of these fluors were prepared in toluene (Baker reagent grade), in p-dioxane (Eastman Organic Chemicals), m.p. 10.5 to 11°C, or in p-dioxane (Matheson, Coleman & Bell spectroquality reagent grade). Counting samples were labeled with toluene-^{3}H or toluene-^{14}C of known activity, in amounts which constituted less than 10 per cent of the total sample volume. All samples were in equilibrium with air, and were counted in 20-ml screw-cap vials supplied by Wheaton Glass Co.

The counters were standard production models manufactured by Nuclear-Chicago Corporation and Packard Instrument Co. They may be classified according to optical system as follows:
1. Titanium dioxide optics—reflecting surfaces of the counting chamber are titanium dioxide paint.
2. Aluminum optics—reflecting surfaces are formed aluminum.
3. Light guide—a plastic block surrounding the sample vial is separated from the vial by an air interface, but is optically coupled to the phototube faces; the reflecting surfaces consist of an evaporated aluminum coating on the outside of this block; this system is found in the Packard Model 3214.

All multiplier phototubes were manufactured by EMI Ltd. They were of three types:
1. 6097S—a glass-face tube classed as having S response characteristics, which is used in counters operated at ambient temperatures.
2. 6097B—a glass-face tube with S-11 response, used in cooled counters.
3. 6255B—a quartz-face tube with S-13 response, used in cooled counters.

An Aminco-Bowman spectrophotofluorometer was used to study the effects of various solvents and quenching agents on the ultraviolet-excited fluors. A Cary 14 recording spectrophotometer (Applied Physics Corporation) was used to examine the ultraviolet absorption spectra of solvents, quenching agents and fluor solutions.

RESULTS AND DISCUSSION
Secondary Solutes

The optical system in which the efficiency is most dependent on the wavelength of the fluorescent spectrum is the titanium dioxide system. Earlier work[4] has shown that titanium dioxide has a somewhat higher reflectivity than aluminum at wavelengths above 430 nm, but that reflectivity falls off rapidly in the region of most primary-solute fluorescence. As anticipated, the addition of secondary solutes to PPO solutions produced the greatest increase in efficiency when counting in titanium dioxide optics. This is illustrated in Figure 16 for tritium samples counted in titanium dioxide optics and in aluminum optics.

The counting efficiency in each counter in the absence of secondary solute has been assigned the value of unity. The actual efficiency for the sample containing primary solute only was much lower in the titanium dioxide system than in the aluminum system, but the values were not directly comparable, because different pairs of phototubes were used in the two counters. The PPO concentration of 17.5 gram per liter was optimal for these quenched samples. It has been reported[5-7] that very small concentrations of secondary solute are sufficient to suppress the emission spectrum of the primary solute and shift the observed emission almost entirely to that characteristic of the secondary. This appears to be confirmed by the rapid initial rise in efficiency at low secondary concentration in the titanium dioxide system. The convention of using 0.05 gram per liter of POPOP is attributable to this observation, although, as shown here, such an amount may be very suboptimal in quenched samples. The curves were carried out to the saturation limits at 0°C, and precipitates of the secondary solutes were visible in those samples of high concentration that showed decreased efficiency. DimethylPOPOP was subsequently found to be less soluble in toluene at 0°C than is indicated in Figure 16, and it is possible that the most concentrated sample was supersaturated when these data were taken. DimethylPOPOP is the most soluble of the three secondaries, but more of it is required to reach the same efficiency.

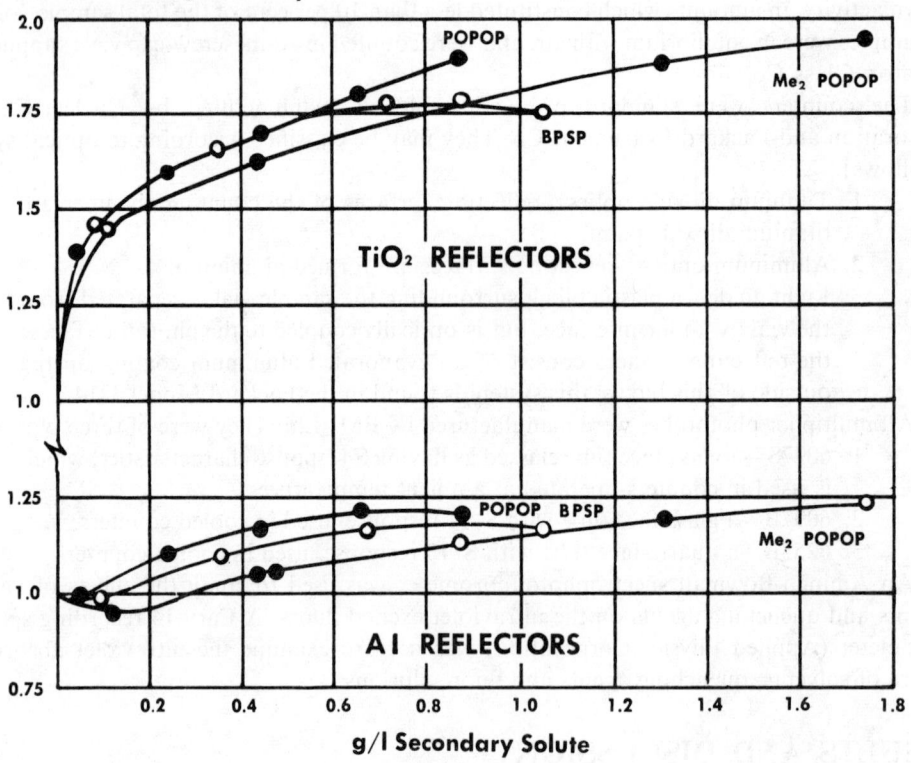

FIGURE 16.
Relative Counting Efficiencies of Three Secondary Solutes in Two Different Optical Systems, in Samples of Tritiated Toluene Quenched with CCl_4 and Containing 17.5 g/l PPO.

From the results in these two counters and in many others it was concluded that the three secondary solutes give nearly equal counting efficiencies at their respective maximal concentrations. DimethylPOPOP appeared slightly inferior, but observed differences between the fluors were not greater than 5 percent in relative counting efficiency. At lower concentrations, dimethylPOPOP was less efficient than the other fluors at equal molarity or at equal weight per unit volume. In all samples, both quenched and unquenched, the highest efficiency for any secondary solute was obtained at its maximal concentration. "Concentration quenching" or self-quenching, if it exists for these secondaries, does not cause a loss in efficiency at the concentrations attainable in toluene.

The pattern of efficiency as secondary-solute concentration increased was not the same in all counters with aluminum optics. When the samples shown in Figure 16 were counted in other counters of the same type, the efficiency did not always drop when going from zero to small concentrations of secondary solute; the curves sometimes had an initial flat portion or rose very slowly. A slow initial rise indicates that better light collection or better spectral matching with the phototube outweighs the loss of light, due to the fact that the quantum efficiency of the secondary solute is less than 100 percent. When the curve dips, it indicates that wavelength shifting produces no very significant increase in photocathode quantum efficiency. Differences in spectral response between the different tube types examined were not larger than the difference between individual tubes of the same type.

The counter with the light pipe showed a continuous increase in efficiency as secondary-solute concentration increased. Only one such counter was available for the study, and therefore the investigators are unable to say whether this is characteristic of these optics or whether such counters would show about the same range of curve shapes as others. Since the plastic is less transparent to ultraviolet wavelengths than air, this optical system might be expected to be more wavelength-dependent than the air-coupled system.

The counter which gave the curves shown at the bottom in Figure 16 evidently had a very flat spectral response over the emission range of the primary and secondary solutes. When unquenched tritium samples containing 4 gram per liter of PPO and various secondaries were counted in it, the count rates with secondary solutes were found to be less than that with PPO alone, except at the maximal concentration of BPSB and POPOP, where they were equal to that with PPO. At the highest concentration of dimethylPOPOP the count rate was about 5 percent lower. The inefficiency of the wavelength-shifting process is responsible for the loss of counts at low secondary-solute concentrations, and the restoration of counts at higher concentration in these unquenched samples can be explained by the measurable efficiencies of the wavelength shifters as primary solutes. When used alone, the tritium efficiency of POPOP or BPSB varied from 1 percent or less at 0.05 gram per liter to 12 or 15 percent at 0.5 gram per liter. In the quenched samples, however, the increase in efficiency obtained by adding a large amount of secondary solute was much greater than could be accounted for on this basis. This was an unexpected result, and it cannot be explained on the basis of spectrum shifting alone; it was observed in every counter, regardless of its wavelength sensitivity. The addition of a secondary solute to a sample quenched with CCl_4 produced a much greater increase in efficiency than did the same amount of secondary solute in an unquenched sample.

A possible explanation of this effect might lie in the reduced transparency of the quenched sample to light of shorter wavelengths, so that more light can escape the counting vial when a wavelength shifter is used. This was ruled out with CCl_4 and at least some of the other quenching agents for which the same phenomenon was demonstrated, after examination of their ultraviolet absorption spectra.

The increase in counting efficiency of the quenched sample when secondary solute is added is, therefore, due to increased light output from the sample rather than to a change in wavelength. Since the quantum efficiency of the secondary solute, or the number of photons emitted per photon absorbed, cannot exceed unity, some secondary-solute molecules must have been excited by a mechanism other than absorption of a photon emitted by the primary, i.e., by nonradiative energy transfer from the primary. Since the secondary solute increases total photon emission only in the quenched samples, it must be concluded that the quenching agent acts to decrease the probability of primary-solute emission. The secondary solute, in effect, competes with the quenching agent for the excitation energy of the primary by nonradiative transfer. Evidently the secondary solute is significantly less subject to direct solute quenching than the primary.

This hypothesis can be tested, if direct solute quenching can be measured independently of solvent quenching; the latter is generally assumed to be by far the larger effect in liquid scintillation counting. Such a measurement can be made by exciting the fluors directly by ultraviolet light rather than through the transfer of energy from the solvent. This is described below. The results were used in a further test of the hypothesis. If a compound that is quenching in liquid scintillation counting is found not to quench the primary fluor, i.e., if it is a solvent quencher only, then addition of a secondary solute to such a quenched solution should produce no greater increase in efficiency than the secondary solute produces in an unquenched sample in the same counter. Quenching agents whose effects on PPO had been examined by spectrofluometry were added to toluene solutions of low PPO concentration (2 gram per liter), in which self-quenching of PPO was minimized. A second set of samples was prepared in identical manner, except for the inclusion of 0.5 gram per liter of POPOP. All were labeled with tritium and counted in various counters. The observed increments in efficiency due to the presence of POPOP followed closely the order of fluor quenching observed by spectrofluorometry. Two possible exceptions were noted, and these compounds—acetone and ethyl acetate—will be studied further.

Spectrofluorometry

Excitation and fluorescent emission spectra of the fluors were obtained in various solvents and in the presence of compounds which act as quenching agents under high-energy excitation. These were selected either because they are often found in liquid scintillation counting samples or because they are representative of classes of chemical compounds.

In general, the shape of the excitation or emission spectrum of any fluor at a given concentration was not dependent on the solvent or added impurity, and the only differences were those of intensity. Exceptions to this were found in the very slight displacement of the emission peaks of the fluors to longer wavelengths (by 5 to 10 nm) in the presence of quaternary ammonium salts and bases (e.g. Hyamine* hydroxide), and in the large displacement of the emission spectra of PPO and POPOP to longer wavelengths in the presence of aqueous hydrochloric acid. The latter appeared to react with these fluors.

Of the five fluors (not including dimethylPOPOP) that were tested, PPO was quenched to a greater degree and by more compounds than any of the others. The order of the quenching agents as to severity of quenching was not the same with the different fluors, and no pattern was established with respect to chemical class. The intensity of the fluors in some common solvents used in liquid scintillation counting samples, namely p-dioxane, anisole, dimethoxyethane, ethanol and aqueous ethanol, was as great or greater than in toluene, presumably because of the reduced transparency of the latter at the exciting wavelengths, as verified by absorption measurements.

Primary-Solute Concentration

Self-quenching in PPO solutions is well known, and it has been shown that the optimal concentration in air-equilibrated toluene solutions is 4 to 5 gram per liter.[5] At higher concentrations, light output is decreased as the probability for further energy transfer from the solvent to the fluor is exceeded by the probability of the quenching of excited fluor molecules by other fluor molecules. The concentration at which maximum light output occurs depends on the probability of transfer from solvent to fluor, and thus on the degree of solvent quenching. As more external quenching agent is added, the peak light output is decreased and is found at higher fluor concentrations. Figure 17 shows tritium counting efficiency as a function of fluor concentration at three different concentrations of a quenching agent, diethylamine. For some other quenching agents a more dramatic increase in optimal concentration is seen, but even here, at counting efficiencies that are quite reasonable, one can see that the usual amount of PPO is suboptimal. Rarely does any recipe require as much as 10 gram per liter of PPO.

If we consider those quenching agents which are solvent quenchers only, i.e., those which act only to reduce the probability of energy transfer from solvent to primary solute, then the same curve of counting efficiency versus fluor concentration should be obtained at an equal quenching concentration of any of these in a given solvent; that is, two such quenchers giving equal efficiencies at one fluor concentration should give equal efficiencies at any other con-

---

* Hyamine 10-X, trademark of Rohm and Haas Co.

centration of the same fluor, and the optimal fluor concentration should be the same in both solutions. This was demonstrated in solutions of two quenching agents that were shown by spectrofluorometry not to quench PPO. It was found that a tritiated toluene solution containing 30 percent ethanol by volume and an arbitrary amount of PPO could be counted with the same efficiency as one containing 26 percent dimethylformamide by volume and the same amount of PPO. When PPO concentration was varied, these two solutions maintained equal efficiencies at equal solute concentrations and showed the same maximal counting rate at 6 gram per liter of PPO.

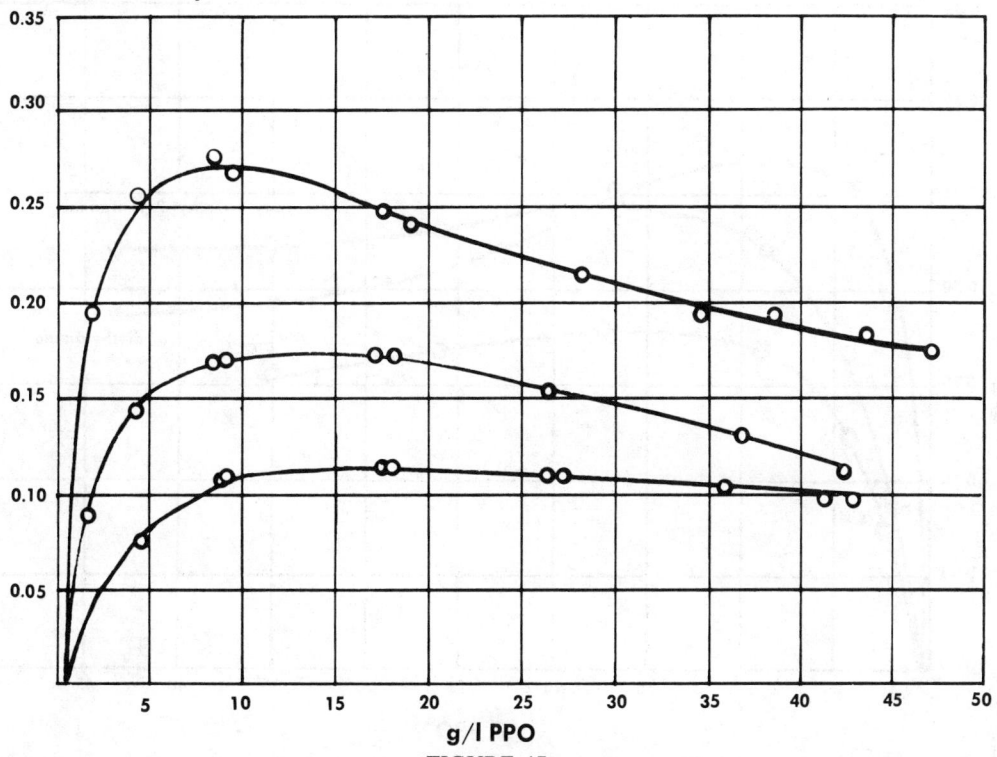

FIGURE 17.
Change in Optimal Primary-Solute Concentration with Increasing Quenching by Diethylamine in Tritiated Toluene.

When the fluor as well as the solvent is quenched, a difference may be expected in the dependence of efficiency on fluor concentration, and adjustment of quencher concentration will not give a curve that can be superimposed on one of the type described above. This is illustrated in Figure 18 for three fluor-quenching agents in toluene. Optimal primary-solute concentration is not uniquely associated with counting efficiency as it is in solutions in which only the solvent is quenched. This is presumed to be due to differences among the quenchers in their relative quenching of solvent and fluor.

When optimal primary-solute concentration occurs at high levels, where self-quenching is considerable, addition of a secondary solute increases efficiency significantly by reducing self-quenching of the primary solute, and also shifts optimal primary-solute concentration to slightly higher values. Addition of a secondary solvent such as naphthalene, if it is less quenched by a solvent quencher than toluene, shifts optimal primary-solute concentration to lower values. For instance, the optimal concentration of PPO in dioxane (Eastman Organic Chemicals) was approximately 7 gram per liter, and in dioxane containing 20 percent water it was about 30 gram per liter. When naphthalene at 50 gram per liter was included in the water solution, optimal PPO concentration dropped to around 5 gram per liter; efficiency was more than doubled. Use of spectroquality-grade dioxane reduced the required amount of PPO somewhat, and increased the relative efficiency for tritium by 10 to 20 percent. Naphthalene is more resistant than toluene to quenching by all compounds, however. In some quenched samples it reduced counting efficiency markedly.

Optimal primary-solute concentrations were found to be the same for all counters, and for both ^{3}H and ^{14}C.

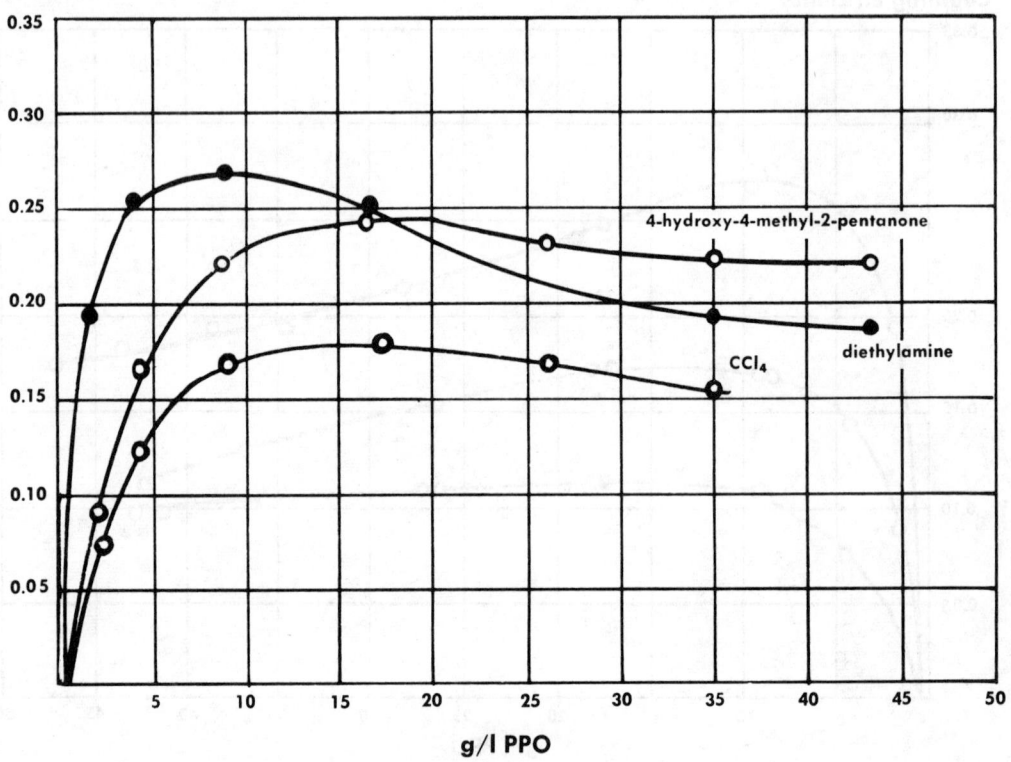

FIGURE 18.
Dependence of Tritium Counting Efficiency on Primary-Solute Concentration in Toluene Solutions of Three Direct Solute-Quenching Agents.

Primary-Solute Efficiencies

Except in counters with titanium dioxide optics, the order of efficiencies among the three primary solutes at their respective optimal concentrations in unquenched samples was PBD > PPO > BBOT. With few exceptions, the same order was observed in quenched samples. PBD is less subject than PPO to quenching by many compounds; it also shows less self-quenching, having an optimal concentration of about 9 gram per liter in unquenched toluene samples. It has, however, limited solubility—about 13 gram per liter in toluene at room temperature. PPO is soluble to 270 gram per liter,[5] and BBOT to 53 gram per liter.[8] When the quenched system is such that optimal primary-solute concentration is very high, PPO has an advantage. Among the samples prepared, the efficiencies in PPO solutions reached or surpassed those in PBD solutions only when concentrations of 40 to 50 gram per liter of PPO were required and the PBD concentration could be brought only to 11 or 12 gram per liter. With quenching agents that affect PPO, but not PBD, the quenching of the PPO may be partly compensated by addition of secondary solute. This is also important at high PPO concentrations, where self-quenching of PPO is reduced by the secondary, as illustrated in Figure 19. This graph also shows the desirability of adding a large amount of secondary solute.

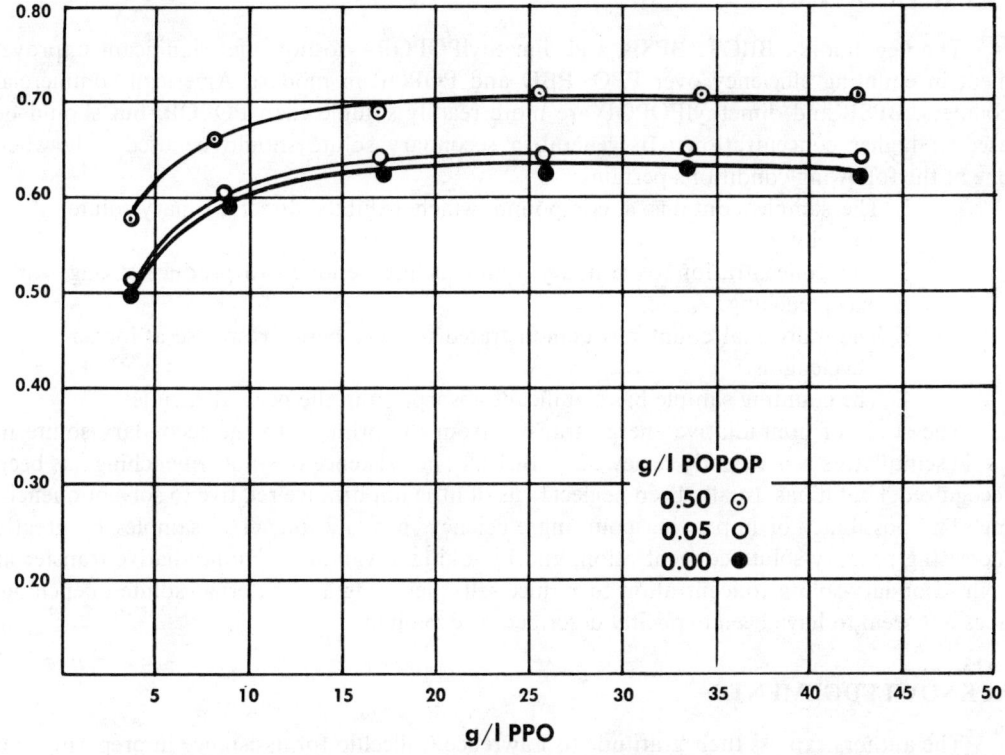

FIGURE 19.
Effect of Secondary-Solute Concentration on ^{14}C Counting Efficiency
in the Presence of CCl$_4$ Direct Solute Quenching and Self-Quenching of PPO in Toluene.

The best showing made by BBOT was in counters with titanium dioxide optics. In unquenched solutions BBOT gave higher efficiency than PPO alone, and approximately the same as PBD. Results in quenched solutions varied with the nature of the quenching agent and the degree of direct solute quenching. In both quenched and unquenched solutions, however, use of a secondary solute with PPO or PBD resulted in higher efficiencies than use of BBOT. (BBOT was not improved by addition of a secondary solute.)

In aluminum optics BBOT was superior to PPO only in samples containing very severe fluor quenchers, and was never superior to optimal amounts of PPO and POPOP together. BBOT showed very limited compatibility with water in dioxane solutions, as shown in Table 8.

Table 8.
FIGURES OF MERIT FOR COUNTING TRITIATED WATER IN SOLUTIONS OF BBOT AND PPO-POPOP

Concentration in Dioxane Solution			Counting Efficiency %	Figure of Merit, % efficiency ×%H$_2$O
H$_2$O (vol.%)	BBOT (g/l)	Naphthalene (g/l)		
2	10*	205	24.5	49
5	10*	119	19	95
10	5**	155	17.5	175
H$_2$O (vol. %)	PPO† (g/l)	POPOP (g/l)	Naphthalene (g/l)	
2	6	0.275	168	28	56
10	6	0.275	155‡	23	230
15	6	0.275	112‡	17.5	262
20	6	0.275	80‡	12.5	250

* Approximate optimal concentration of BBOT. ** Approximate solubility limit of BBOT at this water concentration at 8°C.
† Approximate optimal concentration in these solutions. ‡ Maximal concentration in homogeneous solution at 8°C.

CONCLUSIONS

The new fluors—BBOT, BPSB, and dimethylPOPOP—do not offer significant improvement in counting efficiency over PPO, PBD and POPOP in modern American commercial counters. BPSB and dimethylPOPOP are more readily soluble than POPOP, but should be used at higher concentrations. In general, a secondary solute should be used only when one of the following conditions pertains:

1. The sample contains a compound which exhibits direct primary-solute quenching.
2. The concentration of primary solute is high enough to produce strong self-quenching.
3. The individual counter is demonstrated to have better response at longer wavelengths.
4. The counting sample has significant absorption in the near ultraviolet.

The study of nonradiative energy transfer from the primary to the secondary solute in liquid scintillators was recently reviewed by Birks.[8] The existence of solute quenching has been recognized,[9] but it has usually been neglected as of little importance relative to solvent quenching. The possibility of improving counting efficiency in many quenched samples by greatly increasing primary-solute concentration, and by taking advantage of nonradiative transfer at high secondary-solute concentration to reduce self-quenching and external solute quenching, does not seem to have been explicitly described or exploited.

ACKNOWLEDGEMENTS

The authors express their gratitude to Lawrence Calicchio for assistance in preparing and counting many of the samples, and to the Research Institute of the Illinois Institute of Technology for making the spectrophotometer and spectrophotofluorometer available.

REFERENCES

1. Hayes, F. N., Ott, D. G., and Kerr, V. N., Pulse-Height Comparison of Secondary Solutes. *Nucleonics*, 14(1): 42, 1956.
2. Ott, D. G., Round Table on Chemistry of the Counting Sample Scintillation Solutes. *Liquid Scintillation Counting*, pp. 101–107, Bell, C. G. and Hayes, F. N., eds. Pergamon Press, Oxford, 1958.
3. Swank, R. K., Buck, W. L., Hayes, F. N., and Ott, D. G., Spectral Effects in the Comparison of Scintillators and Photomultipliers. *Rev. Sci. Instrum.*, 29: 279, 1958.
4. Swank, R. N., Limits of Sensitivity of Liquid Scintillation Counters. *Liquid Scintillation Counting*, p. 23, Bell, C. G. and Hayes, F. N., eds. Pergamon Press, Oxford, 1958.
5. Hayes, F. N., Rogers, B. S., Sanders, P., Schuch, R. L., and Williams, D. L., *Liquid Solution Scintillators*. Los Alamos Report LA-1639, 1953.
6. Davidson, J. D. and Feigelson, P., Practical Aspects of Internal-Sample Liquid Scintillation Counting. *Int. J. Appl. Radiat. Isotop.*, 2: 1, 1957.
7. Roser, F. S., Spectral Emission of Composite Liquid Phosphor. *Science*, 121: 806, 1955.
8. Birks, J. B., *The Theory and Practice of Scintillation Counting,* pp. 314–315. Macmillan Co., New York, 1964.
9. Kallman, H. and Furst, M., The Basic Processes Occurring in the Liquid Scintillator. *Liquid Scintillation Counting*, pp. 3–22, Bell, C. G. and Hayes, F. N., eds. Pergamon Press, Oxford, 1958.

MODULATION TRANSFER FUNCTION FOR RADIOISOTOPE IMAGING SYSTEMS

Robert N. Beck
*Argonne Cancer Research Hospital**
The University of Chicago, Chicago, Illinois

The modulation transfer function, MTF(ν), describes the response of a linear image-forming system to a sinusoidal object with spatial frequency ν (cycles/cm). This is analogous to the temporal frequency response of an electronic amplifier or filter. The rationale for this description arises in a natural manner from the fact that any object and its image can be described in terms of the amplitudes and phases of their respective frequency components;[1] the MTF(ν) is simply a measure of the efficiency with which the modulation at each frequency, ν, is transferred by the imaging system from the object to the image, and is unaffected by the average or "DC component" of the object.

The MTF(ν) has been used in optics,[2] photography,[3] and diagnostic radiology[4] as a measure of spatial resolution which is more general than a single number, such as the resolving power for line patterns. The latter may be thought of as indicating the highest spatial frequency to which a system responds adequately. The need for this more general measure of resolution exists because image quality is not necessarily correlated with resolving power, since the high-frequency response does not indicate how well an imaging system responds to the object components of lower frequency.[5] In short, the intuitive notion that the system that best reproduces the small structures (high-frequency components) in an object will certainly reproduce the large structures best is false.

When applied to radioisotope imaging devices,[6,7] the MTF(ν) can be used to describe the response of a collimated radiation detector to a planar source of radioactivity (located at a distance z from the collimator face) in which the concentration varies sinusoidally with spatial frequency ν (cycles/cm) = $1/\lambda$, where λ (cm) is the wave length. Using the notation of Figure 20, the typical object component† is a plane–wave distribution, uniform in the y direction, described by

$$\sigma(x)\left[\frac{\text{photons emitted}}{\text{sec-cm}^2}\right] = \bar{\sigma} + \tilde{\sigma}(\nu)\sin 2\pi\nu x. \qquad (1)$$

The (true mean) *object modulation* or contrast, $m_o(\nu)$, is defined by

$$m_o(\nu) \equiv \frac{\tilde{\sigma}(\nu)}{\bar{\sigma}} = \frac{\sigma_t - \sigma_o}{\sigma_t + \sigma_o}. \qquad (2)$$

When the detector axis is at the point x, this object structure gives rise to a detector count rate with a true mean value $C(x)$, which is also sinusoidal and is described by

$$C(x)\left[\frac{\text{counts}}{\text{sec}}\right] = \bar{C} + \tilde{C}(\nu)\sin 2\pi\nu x. \qquad (3)$$

---

* Operated by the University of Chicago for the United States Atomic Energy Commission.
† More explicit notation would indicate that this source is the differential element of a volume distribution of radioactivity, $\rho(x,y,z)\left[\frac{\text{photons emitted}}{\text{sec-cm}^3}\right]$, at a distance z from the collimator and at a depth z_0 below the tissue surface, with effective source strength given by $d\sigma(x,y,z) = \rho(x,y,z)\,e^{-\mu z_0}\,dz$. In addition, it is assumed here that the frequency spectra for the x and y directions are identical, so that a one-dimensional treatment suffices.

Since this count rate gives rise to an image, the (true mean) *image modulation* or contrast, $m_i(v)$, is defined by

$$m_i(v) \equiv \frac{\tilde{C}(v)}{\bar{C}} = \frac{C_t - C_o}{C_t + C_o}. \tag{4}$$

The modulation transfer function is simply the ratio of image modulation to object modulation, defined by

$$\text{MTF}(v) \equiv \frac{m_i(v)}{m_o(v)}. \tag{5}$$

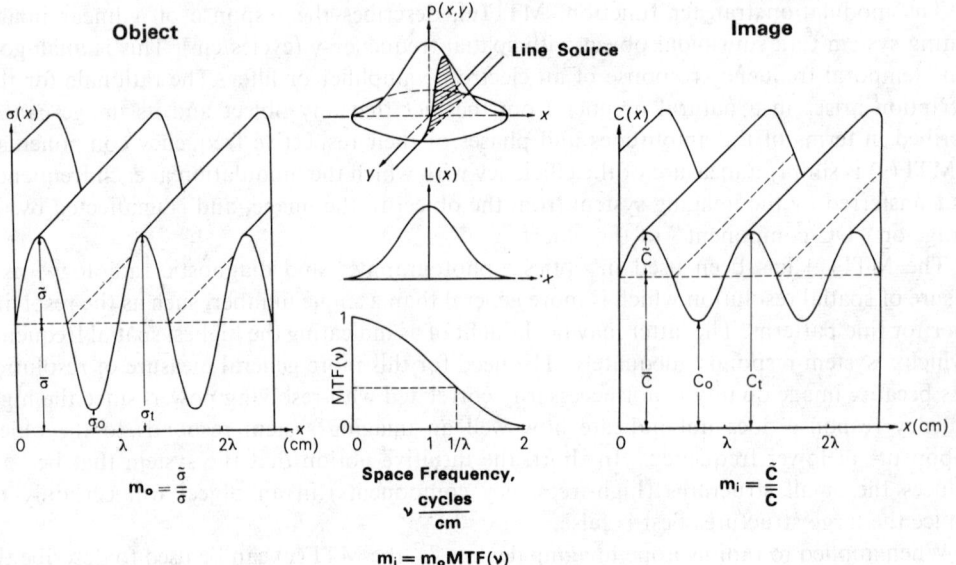

FIGURE 20. Imaging Process.

From equations (2) and (5) it is clear that MTF(v) can be determined directly by measuring $m_i(v)$, described by equation (4), with a sinusoidal source for which $\sigma_o = 0$ and $m_o(v) = 1$ (or with any other sinusoidal source for which $m_o(v)$ is known). For these measurements it is convenient to use as a test object a sinusoidal sunburst pattern of radioactivity that contains the useful range of v.[6,7]

Alternatively, it has been shown[2,3,7] that the MTF(v) can be determined from the line spread function, $L(x)$, which describes the response to a uniform line source of radioactivity, parallel to the y axis, and at a distance z from the collimator face (see Figure 21). In this case MTF(v) is equal to the Fourier transform of $L(x)$, normalized so that MTF(0) = 1; that is,

$$\text{MTF}(v) = \frac{\int_{-\infty}^{\infty} L(x) e^{-2\Pi i v x}\, dx}{\int_{-\infty}^{\infty} L(x)\, dx}. \tag{6}$$

If $L(x)$ is an even function, as is usually the case, equation (6) reduces to

$$\text{MTF}(v) = \frac{\int_{-\infty}^{\infty} L(x) \cos 2\Pi v x\, dx}{\int_{-\infty}^{\infty} L(x)\, dx} = \frac{\sum_{j=1}^{m} L(x_j) \cos 2\Pi v x_j}{\sum_{j=1}^{m} L(x_j)}, \tag{7}$$

where $x_j = (j - 1)\Delta x, j = 1, 2, 3, \ldots, m$.

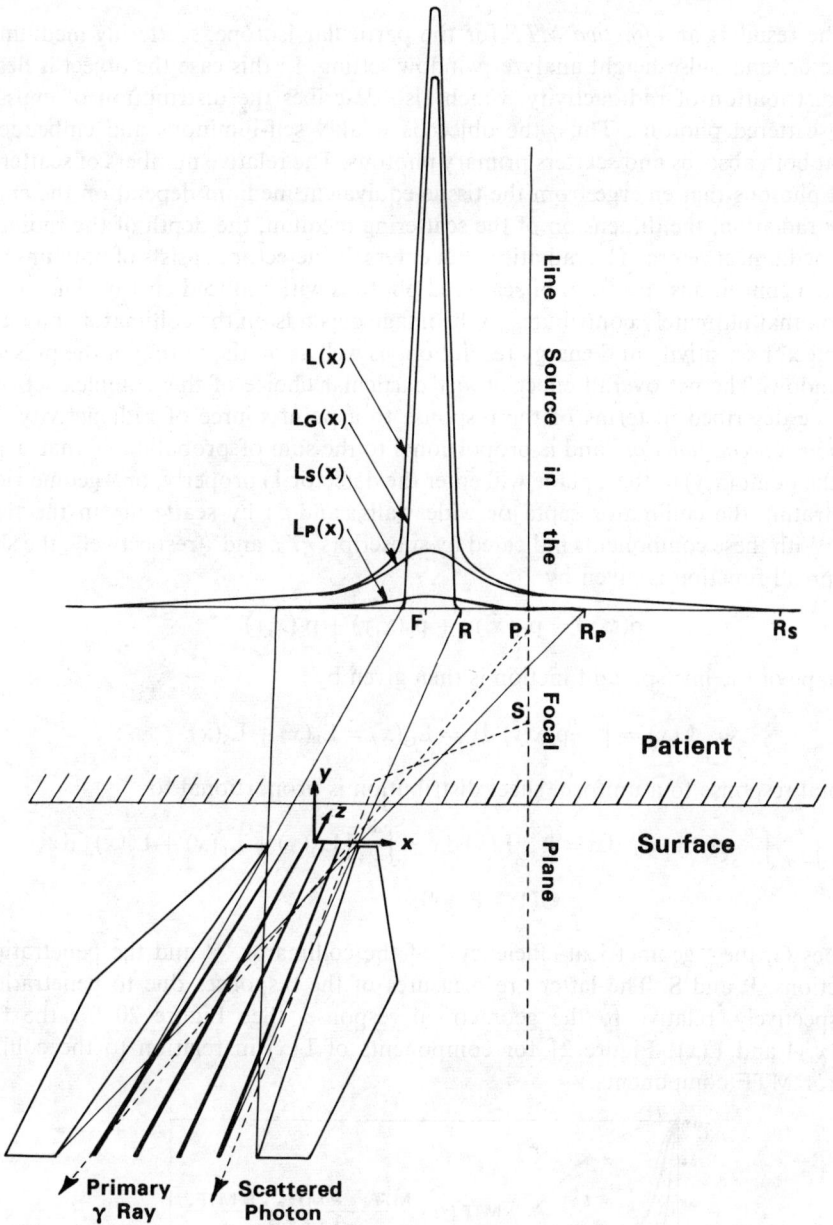

FIGURE 21. Determination of MTF(ν) from the Line Spread Function.
The line spread function, $L(x)$, depicted here for a line source in the focal plane at a distance $z = F$ from the collimator face, consists of three principal components: 1) $L_G(x)$, due to properly collimated γ rays emitted within radius R; 2) $L_P(x)$, due to primary γ rays emitted within the region with radius R_p, which penetrate the collimator septa; 3) $L_S(x)$, due to γ rays emitted within the region with radius R_S, which are scattered within the patient or collimator.

The approximation given in equation (7) provides a useful basis for computing MTF(ν) from measured values of $L(x)$. To measure $L(x)$, a line source in a small-diameter polyethylene tubing 30 cm in length is recommended,[8] $L(x)$ being measured over the range -15 cm $\leq x \leq +15$ cm. Because septal penetration and scattering within the collimator are energy-dependent phenomena, and both affect $L(x)$, the measurements should be made with the isotope to be used in clinical practice. If scattering within the tissue is to be taken into account, the measurements are made with the line source in a suitable tissue-equivalent scattering medium. The result is an *effective MTF* for the particular isotope, scattering medium, collimated detector, and pulse-height analyzer window setting. In this case the object is described simply by distribution of radioactivity, which also describes the distribution of emission of primary unscattered photons. Thus, the object is weakly self-luminous and embedded in a medium that both absorbs and scatters primary photons. The relative numbers of scattered and unscattered photons that emerge from the tissue-equivalent medium depend on the energy of the primary radiation, the dimension of the scattering medium, the depth of the radioactivity within this medium, et cetera. The radiation that enters the detector consists of both unscattered photons and a continuous spectrum of scattered photons with reduced energy. The portion of this radiation that ultimately contributes to the image depends on the collimator solid angle of view, the detector sensitivity and energy resolution, as well as on the setting of the pulse-height analyzer window. The net overall effect of any particular choice of this complex set of parameters can be described in terms of the response to a point source of radioactivity. This is called the *point spread function*, and is proportional to the sum of probabilities that a photon emitted at the point (x,y) in the z plane will enter the detector 1) properly, or "geometrically"; 2) by penetrating the collimator septa or wide walls; and 3) by scattering in the tissue or collimator. With these components indicated by subscripts $_G$, $_P$, and $_S$ respectively, the shape of the point spread function is given by

$$p(x,y) = p_G(x,y) + p_P(x,y) + p_S(x,y). \tag{8}$$

The shape of the line spread function is then given by

$$L(x) = \int_{-\infty}^{\infty} p(x,y)\,dy = L_G(x) = L_P(x) + L_S(x). \tag{9}$$

The total response to a uniform sheet distribution is proportional to

$$\int_{-\infty}^{\infty}\int_{-\infty}^{\infty} p(x,y)\,dy\,dx = \int_{-\infty}^{\infty} L(x)\,dx = \int_{-\infty}^{\infty}[L_G(x) + L_P(x) + L_S(x)]\,dx$$
$$\equiv G(1 + P + S), \tag{10}$$

which defines G, the "geometrical efficiency" of the collimator,[7,9] and the penetration and scatter fractions, P and S. The latter are measures of the responses due to penetration and scatter, respectively, relative to the geometrical response. (See Figure 20 for the relation between $p(x,y)$ and $L(x)$; Figure 21 for components of $L(x)$ in relation to the collimator; Figure 22 for MTF components.)

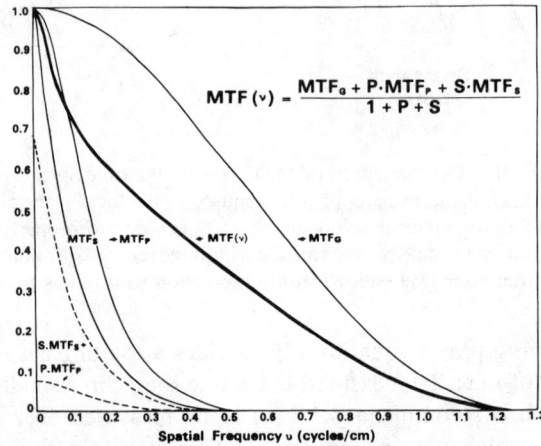

FIGURE 22. MTF Components.
Large structures are partially resolved, even by radiation that enters the detector by septal penetration and scattering processes (MTF$_P$ and MTF$_S$ >0 for ν< 0.5 cycles/cm or for λ >2 cm). This radiation simply reduces MTF(ν) for λ >0.5 cycles/cm.

The MTF due to properly collimated radiation alone is defined by

$$\mathrm{MTF_G}(\nu) \equiv \frac{\int_{-\infty}^{\infty} L_G(x) e^{-2\pi i \nu x} \, dx}{\int_{-\infty}^{\infty} L_G(x) \, dx} = \frac{\int_{-\infty}^{\infty} L_G(x) e^{-2\pi i \nu x} \, dx}{G}. \qquad (11)$$

Similarly, considering penetration and scatter responses separately, we can define

$$\mathrm{MTF_P}(\nu) \equiv \frac{\int_{-\infty}^{\infty} L_P(x) e^{-2\pi i \nu x} \, dx}{GP}, \qquad (12)$$

and

$$\mathrm{MTF_S}(\nu) \equiv \frac{\int_{-\infty}^{\infty} L_S(x) e^{-2\pi i \nu x} \, dx}{GS}. \qquad (13)$$

From equations (6) and (10), the MTF due to the combination of these components of detector response is

$$\mathrm{MTF}(\nu) = \frac{\int_{-\infty}^{\infty} [L_G(x) + L_P(x) + L_S(x)] e^{-2\pi i \nu x} \, dx}{G(1 + P + S)}, \qquad (14)$$

and, using equations (11), (12) and (13), equation (14) reduces to

$$\mathrm{MTF}(\nu) = \frac{\mathrm{MTF_G}(\nu) + P \cdot \mathrm{MTF_P}(\nu) + S \cdot \mathrm{MTF_S}(\nu)}{1 + P + S}. \qquad (15)$$

If the collimator septa and side shielding are sufficiently thick, $P \ll 1$; furthermore, if the detector has good energy resolution, so that most of the scattered radiation can be eliminated by pulse-height selection, $S \ll 1$.

In addition, the magnitude of each component, $\mathrm{MTF_{G, P, S}}$, depends on the shape, and especially the width, of the corresponding line spread component, $L_{G, P, S}$, relative to the wave length $\lambda = \frac{1}{\nu}$. Near $\nu = 0$, λ is very large, compared to the width of all these components, since all of them decrease at least as fast as the inverse of the distance from the detector axis. In this case $\cos 2\pi\nu x \cong 1$ in equation (7), $\mathrm{MTF_{G, P, S}} = 1$ in equation (15), and $\mathrm{MTF}(\nu) = 1$.

In the mid-frequency range that is of primary interest (see Figure 22 for $\nu > 0.5$ cycles/cm), where λ is comparable to the width of $L_G(x)$, but small compared to $L_P(x)$ and $L_S(x)$, we find that $\mathrm{MTF_P}(\nu)$ and $\mathrm{MTF_S}(\nu)$ are small compared to $\mathrm{MTF_G}(\nu)$, and equation (15) reduces to

$$\mathrm{MTF}(\nu) \cong \frac{\mathrm{MTF_G}(\nu)}{(1 + P + S)}. \qquad (16)$$

In this frequency range, penetration and scatter are essentially equivalent to background radiation* in the sense that they reduce the image contrast and contribute nothing to the formation of a structured image. In the high-frequency range, $\mathrm{MTF_G}(\nu) = 0$ also; thus, the detector does not resolve very small structures.

It is clear from equation (7) that, for $\nu = 0$, $\cos 2\pi\nu x = 1$ and $\mathrm{MTF}(0) = 1$ for any line spread function $L(x)$. In addition, the normalization factor $1/\int_{-\infty}^{\infty} L(x) dx$ in equation (7) insures that $\mathrm{MTF}(\nu)$ is independent of the *magnitude* of $L(x)$ (and, therefore, independent of such factors as the counting rate, counting time, concentration of radioactivity in the line source, et cetera). On the contrary, $\mathrm{MTF}(\nu)$ depends only on the *shape* of $L(x)$, which should be determined accurately, i.e., with negligible statistical error. In this way, the two main causes of poor

---

* Here it is assumed that "background radiation" is independent of the source to be imaged, and is due to cosmic rays, radioactivity in the materials comprising the detector, et cetera. If C_b is the count rate due to such sources, it can be seen, by adding C_b to C_t and C_o in equation (4), that image contrast is thereby reduced.

image quality, namely *distortion* and *noise*, are separated experimentally as they are separated conceptually. In the context of scanning, distortion occurs primarily in the form of smoothing due to imperfect spatial resolution,[7] which is associated with decreasing values of MTF(v) as v increases. Noise, in the form of random fluctuation in the number of photons recorded per unit area, is associated with limited detector sensitivity[9] (in addition to limited source strength, scanning time, et cetera). In practical systems, a compromise between resolution and sensitivity is required,[10] and this can be seen as a compromise between distortion and noise.

When the counting rate in equation (3) is written explicitly in terms of parameters used to describe the detector–pulse-height analyzer system,[7] we obtain

$$C(x)\left[\frac{\text{counts}}{\text{sec}}\right] = G(1 + P + S)\eta\psi[\bar{\sigma} + \text{MTF}(v)\tilde{\sigma}(v)\sin 2\pi vx]. \tag{17}$$

Here, η is the photopeak crystal efficiency, and ψ is the fraction of unscattered photopeak pulses passed by the pulse-height analyzer.

In the mid-frequency range, where equation (16) holds, equation (17) reduces to

$$C(x)\left[\frac{\text{counts}}{\text{sec}}\right] = G\eta\psi[\bar{\sigma}(1 + P + S) + \text{MTF}_G(v)\tilde{\sigma}(v)\sin 2\pi vx] \tag{18a}$$

$$= \bar{C} + \tilde{C}(v)\sin 2\pi vx, \tag{18b}$$

where equation (18b) is the same as equation (3). Comparing equations (18a) and (18b), it is clear that, in the mid-frequency range, penetration and scatter contribute only to the average count rate, essentially as background radiation does.

If the scanning time per unit area is $t\left[\dfrac{\text{sec}}{\text{cm}^2}\right]$, the count rate in equation (18) gives rise to a true mean image defined by

$$C(x)t \equiv N(x)\left[\frac{\text{counts}}{\text{cm}^2}\right] = \bar{N} + \tilde{N}(v)\sin 2\pi vx, \tag{19a}$$

$$N(x) = G\eta\psi t[\bar{\sigma}(1 + P + S) + \text{MTF}_G(v)\tilde{\sigma}(v)\sin 2\pi vx]. \tag{19b}$$

Observed images will be distributed about $N(x)$ with a standard deviation $\text{S.D.}[N(x)] = \sqrt{N(x)}$, or with a fractional standard deviation given by

$$\varepsilon = \frac{\text{S.D.}[N(x)]}{N(x)} = \frac{1}{\sqrt{N(x)}}. \tag{20}$$

The quantity ε is a measure of the statistical error or noise in the image due to random fluctuation of $\sigma(x)$. For a noise-free image, it is necessary that $\varepsilon = 0$; however, since $N(x)$ is always finite in practice, image quality is always limited (to the extent that $\varepsilon > 0$) by noise.

Comparing equation (17) with equation (1), it is clear that $C(x)$ can yield an undistorted image of $\sigma(x)$ only if $\text{MTF}(v) = 1$. From equation (16), this condition is satisfied in the mid-frequency range only if $\text{MTF}_G(v) = 1$, and $P = 0$ and $S = 0$. In practice, these conditions are never met, and to the extent that $\text{MTF}_G(v) < 1$, image quality is limited by spatial resolution.

Equation (19b) provides a useful description of the true mean image of a sine-wave distribution of radioactivity in terms that relate the noise and distortion to parameters which describe the situation. To summarize:

1. Noise in the image results from the fact that the scanning time, concentration of radioactivity, and sensitivity to properly collimated radiation are necessarily finite. ($G\eta\psi t\bar{\sigma} < \infty$ implies that $\varepsilon > 0$.)
2. Distortion in the image results from the limited spatial resolution for properly collimated radiation. [$\text{MTF}(v) < 1$ implies that $N(x) \neq$ constant $X\sigma(x)$.]
3. Image contrast is further reduced by the response to essentially uncollimated radiation from the source. ($P, S \neq 0$ implies that $m_i = \dfrac{\text{MTF}_G m_o}{1 + P + S} < \text{MTF}_G m_o$.)

The total count rate due to a more complex source in the z plane is found by summing the count rates due to the average concentration of activity $\bar{\sigma}(z)$ and all frequency components $\tilde{\sigma}(v,z)$. For a volume distribution of radioactivity, the total count rate is found by summing count rates from all z planes, taking attenuation into account.

The above analysis is based on the assumption that the radiation detector is linear, i.e., that the counting rate is proportional to the concentration of radioactivity. This assumption is valid for scintillation detectors over the range of concentrations usually encountered in scanning. Within the linear range, the same type of analysis can be carried out for a camera-type detector.[11,12]

If the recording device is also linear (and this may be the case even if film is used, provided that the object contrast is sufficiently low), the MTF of the scanning system is simply given by the product of the MTF's of the detector and recorder.

This analysis may be extended to include the MTF of the human visual system[10,13] under conditions where the latter is linear.

REFERENCES

1. Elias, P., Grey, D. S., and Robinson, D. Z., Fourier Treatment of Optical Processes. *J. Opt. Soc. Am.*, 42(2): 127–134, 1952.
2. Lamberts, R. L., Relationship Between Sine-Wave Response and the Distribution of Energy in the Optical Image of a Line. *J. Opt. Soc. Am.*, 48(7): 490–495, 1958.
3. Perrin, F. H., Methods of Appraising Photographic Systems. *Soc. Motion Pict. Telev. Eng.*, 69: 151, 239, 1960.
4. Moseley, R. D., ed., *Proceedings of the Second Colloquium on Radiologic Instrumentation, University of Chicago, 1964*. Charles C Thomas, Springfield, Ilinois, 1965.
5. Blaschke, W. S., A Scientific Approach to Assessing Image Quality. *J. Photogr. Sci.*, 7: 162–172, 1959.
6. Beck, R. N., A Theory of Radioisotope Scanning Systems. *Medical Radioisotope Scanning*, 1: 35. International Atomic Energy Agency, Vienna, Austria, 1694.
7. Beck, R. N., The Scanning System as a Whole: General Considerations. *Fundamental Problems in Scanning*, Gottschalk, A. and Beck, R. N., eds. Charles C Thomas, Springfield, Illinois, 1968.
8. Hine, G. J., Personal Communication. IAEA Consultants' Meeting, Vienna, Austria, 1965.
9. Beck, R. N., Collimation of γ Rays. *Fundamental Problems in Scanning*, Gottschalk, A. and Beck, R. N., eds. Charles C Thomas, Springfield, Illinois, 1968.
10. Beck, R. N., Criteria for Evaluating Radioisotope Imaging Systems. *Fundamental Problems in Scanning*, Gottschalk, A. and Beck, R. N., eds. Charles C Thomas, Springfield, Illinois, 1968.
11. Cradduck, T. D., Fedoruk, S. O., and Reid, W. B., A New Method for Assessing the Performance of Scintillation Cameras and Scanners. *Phys. Med. Biol.*, 11(3): 423–435, 1966.
12. Gottschalk, A., Modulation Transfer Function Studies with a Gamma Scintillation Camera. *Fundamental Problems in Scanning*, Gottschalk, A. and Beck, R. N., eds. Charles C Thomas, Springfield, Illinois, 1968.
13. Morgan, R. H., Threshold Visual Perception and Its Relationship to Photon Fluctuation and Sine-Wave Response. *Am. J. Roentgenol. Radiat. Therapy Nucl. Med.*, 93(4): 982–997, 1965.

PERSONNEL MONITORING
Robert Radtke

Personnel monitoring is the determination of the amount of ionizing radiation to which an individual has been exposed. The determination is commonly made by the use of photographic films, pocket ionization chambers, thermoluminescent materials, or radiophotoluminescent materials.

When the use of radioactive material is under AEC or state regulations, personnel monitoring is required for:

1. Each individual who receives, or is likely to receive, a dose in any calendar quarter in excess of 25 percent of the applicable value in Table 9.
2. Each individual under 18 years of age who receives, or is likely to receive, a dose in any calendar quarter in excess of 5 percent of the applicable value in Table 9.
3. Each individual who enters an area where a major portion of his body could receive in any one hour a dose in excess of 100 millirem.*†

Table 9. REMS PER CALENDAR QUARTER‡

1. Whole body, head and trunk, active blood-forming organs, lenses of eyes, or gonads	1¼
2. Hands and forearms, feet and ankles	18¾
3. Skin of whole body	7½

The most widely used method of monitoring personnel for radiation exposure is the photographic film, or "film badge". Basically the method is a comparison of the film exposure due to known and unknown amounts of radiation. The physical principles of the photographic-film response and the practical problems encountered in the use of photographic film for personnel monitoring are extensively covered in the literature.[3,4,9]

The pocket ionization chamber is a device used for specific applications in personnel monitoring. It is a small pencil-size instrument with an ionization chamber that discharges a capacitor when in the presence of ionizing radiation. The rate of discharge depends primarily on the intensity of the incident radiation. The ionization chamber may be either a direct- or indirect-reading instrument, but either type can be read at short intervals.

Thermoluminescent materials will release an amount of light that is proportional to the ionizing-radiation exposure, when heated. This property forms the basis for their use in dosimetry and personnel monitoring. Thermoluminescent dosimetry (TLD) systems are available to individuals providing their own personnel-monitoring programs, or the TLD monitoring may be purchased as a service.

Radiophotoluminescent (RPL) materials used in personnel monitoring are commonly called glasses.[13,14] Measurement of either the intensity of luminescence or changes in optical absorption of these RPL materials are proportional to the ionizing-radiation exposure. Table 10 gives a comparison of the various personnel-monitoring techniques, and Table 11 of portable radiation-detecting devices.

---

* Title 10, Code of Federal Regulations, Part 20.202, April, 1967.
† The Council of State Governments, Suggested State Regulations for Control of Radiation, Sec. C. 202, Chicago, 1964.
‡ Title 10, Code of Federal Regulations, Part 20.101(a) April, 1967.

Table 10. PERSONNEL-MONITORING DETECTORS

Detector	Radiation Detected	Range	Minimum Energy Detected	Advantages	Possible Disadvantages
Film	gamma beta thermal neutron fast neutron	0.01 to 10,000 rem	20 kev for gamma rays, 200 kev for beta rays	1. Inexpensive. 2. Gives estimate of integrated dose. 3. Provides permanent record.	1. A moderate directional dependence. 2. Strong energy dependence for low-energy X rays. 3. False readings produced by heat, pressure, and certain vapors.
Pocket Ionization Chambers	gamma beta minus gamma thermal neutron fast neutron minus gamma	0.001 to 2000 R	30 kev for gamma rays, 20 kev for fast neutron	1. Yield fairly accurate information quickly. 2. Small size, low directional dependence. 3. Reasonably uniform in response to radiation in the energy range of 50 kev to 2 Mev. 4. Economical for long-term use. 5. Require little maintenance. 6. Reusable.	1. There is no permanent record. 2. Frequent reading, tabulation, and recharging may be required. 3. Subject to accidental discharge (through shock and, sometimes, electrical leakage). 4. Range of measurement is limited; full scale ranges from 0.2 to 2000 R available.
TLD	gamma beta thermal neutron fast neutron	10^5 Rad	20 kev	1. Indefinite shelf life within the useful range. 2. Small size and low directional dependence. 3. Small energy dependence. 4. Reusable. 5. Inexpensive. 6. Give estimate of integrated dose over long periods.	1. Limited TLD systems supplied as commercial service. 2. Cancellation of dose upon reading. 3. Dose range depends on the sensitivity of the reader. 4. Radiations detected depend on type of TL material.
RPL	gamma beta thermal neutron fast neutron	0.01 to 10^6 Rad	40 kev	1. Indefinite shelf life within the useful range. 2. Small size, low directional dependence. 3. Reusable. 4. Give estimate of integrated dose over long periods, with unlimited number of interval measurements. 5. Provide a permanent record.	1. Primary use today is in civil defense and military dosimetry. 2. Commercial service limited. 3. Number of RPL systems available is limited. 4. Luminescent contaminations on the glass surface are possible. 5. Build-up of the RPL immediately after a short-time exposure.

Table 11. PORTABLE RADIATION-DETECTING DEVICES

Detector	Radiation Detected	Ranges	Use	Comments
Geiger–Mueller (G–M) Tube	alpha beta X gamma	0.04 mr/hr to 500 mr/hr	1. Low-dose-rate surveys 2. Area monitors 3. Personnel radiation monitors	1. Radiation detected depends on the type of G–M tube. 2. Energy dependent. 3. Some models saturate—do not use in high radiation fields. 4. Sensitive to microwave fields. 5. Ratemeter and audible pulse. 6. Rapid response. 7. Rugged, dependable.
Ion Chamber	alpha beta X gamma	3 mr/hr to 10,000 R/hr	1. Medium-and high-dose-rate surveys 2. Area monitors	1. Wide dose-rate range on a single instrument. 2. Low energy dependence. 3. Some models can be used in RF fields. 4. Some models slow to respond.
Scintillation	alpha beta X gamma neutrons	0.025 mr/hr to 200 mr/hr or to 800,000 c/m	1. Low-level-contamination surveys	1. High sensitivity. 2. Rapid response. 3. Fragile. 4. Audible signal and ratemeter. 5. Radiation detected depends on instrument and crystal. 6. Fast neutron detector where dose rate is not required.
Proportional Counter	alpha beta X gamma neutrons	to 500,000 c/m to 20,000 n_{th}/cm^2-sec to 100 mrads/hr n_f	1. Low-level-contamination surveys 2. Neutron survey	1. Primary use is for alpha detection or neutron surveys. 2. Alpha detector can discriminate between alpha and beta-gamma. 3. Neutron detector can discriminate against gamma radiation. 4. Maintenance may be a problem.
BF_3 Counter	Neutrons	to 100,000 c/m	1. Survey	1. Rather low sensitivity. 2. Bulky. 3. Used with various moderators.

For further information on rate-measuring instruments (their detectors, selection, maintenance, and calibration) see *Radiation Accidents and Emergencies* (pp. 209–216) by Lanzl, Pingle, and Rust; Charles C Thomas, Springfield, Illinois, 1965.

REFERENCES

1. Attix, F. H. and Roesch, W. C., eds., *Radiation Dosimetry*, Vol. 2. Academic Press, New York and London, 1966.
2. Spiers, F. W. and Reed, G. W., eds., *Radiation Dosimetry*. Proceedings of the International School of Physics "Enrico Fermi" Course 30, Academic Press, New York and London, 1964.
3. Brodsky, A., Spritzer, A. A., Feagin, F. E., Bradley, F. J., Karches, G. J., and Mandelberg, H. I., Accuracy and Sensitivity of Film Measurements of Gamma Radiation—Part IV. *Health Physics*, 11: 1071–1082, 1965.
4. Storm, E. and Shaler, S., Development of Energy-Independent Film Badges with Multielement Filters. *Health Physics*, 11: 1127–1144, 1965.
5. Eastman Kodak Company, Kodak Personnel-Monitoring Films, *Kodak Pamphlet No. P-31*, Second Edition, 1965.
6. Blatz, H., *Radiation Hygiene Handbook*. McGraw-Hill, New York, 1959.
7. Barber, D. B., *Standards of Performance for Film Badge Services*. U.S. Dept. of Health, Education, and Welfare, Public Health Service, Div. of Radiological Health, 1966.
8. Langmead, W. A. and Adams, N., Investigations of the Accuracy Attained in Routine Film Badge Dosimetry. Paper presented at Health Physics Society Meeting, Los Angeles, California, June 16, 1965.
9. Gorson, R. O., Suntharalingam, N., and Thomas, J. W., Results of a Film-Badge Reliability Study. *Radiology*, 84: 333–346, 1965.

10. Lin, F. M. and Cameron, J. R., A Bibliography of Thermoluminescent Dosimetry. *USAEC Report COO-1105-124*, 1967.
11. Cameron, J. R., Suntharalingam, N., and Kenney, G. N., *Thermoluminescent Dosimetry*. University of Wisconsin Press, Madison, 1967.
12. Symposium, Personnel Dosimetry Techniques for External Radiation. European Nuclear Energy Agency of the Organization for Economic Cooperation and Development, Madrid, 1963.
13. National Bureau of Standards, Report of the ICRU, 1959. *Handbook 78*. Government Printing Office, Washington, D.C., 1961.
14. Becker, K., Photographic, Glass, or Thermoluminescent Dosimetry. *Health Physics*, 12: 955–964, 1966.
15. Becker, K., Radiophotoluminescence Dosimetry—A Bibliography. *Health Physics*, 12: 1367–1374, 1966.
16. National Bureau of Standards, Radiological Monitoring Methods and Instruments. *Handbook 51*. Government Printing Office, Washington, D.C., 1952.
17. NCRP Report No. 32, *Radiation Protection in Educational Institutions*. NCRP Publications, Washington, D.C., 1966.
18. Saenger, E. L., *Medical Aspects of Radiation Accidents*. Prepared under AEC Contract AT (30-1)-2106, Government Printing Office, Washington, D.C., 1963.

WHOLE-BODY COUNTER SYSTEMS
William D. Gibbs and *C. C. Lushbaugh*

The growth of whole-body counting systems has been directed toward ultrasensitive, low-background, heavily shielded types of whole-body counters for detecting minute amounts of radioactivity. While such instruments can be of great value in medical diagnosis and research, their sensitivity restricts their use to studies where other forms of radioassay (i.e., blood samples) are not required. Although it is often desirable to correlate whole-body retention of therapeutic and large test doses of radionuclides in patients with the radioassay of blood, urine and feces, the ultrasensitive systems cannot tolerate the amounts of radioisotope that are required for these purposes.

Ideally, a whole-body counter for clinical diagnostic and research applications should have a range of operating sensitivity that will allow measurement of the largest clinically acceptable dose (about 200 millicuries of ^{131}I) down to natural body background (about 1 nanocurie) in a reasonable counting time (maximum 30 minutes). A single instrument with such a wide (200,000,000 to 1) operating range would be difficult and tremendously expensive to design and construct. However, it is quite feasible to design and construct a "family" of intercalibrated whole-body counters with sensitivity ranges that overlap within this span. General requirements for each member of such a "family" of instruments are shown in Table 12.

Table 12.
REQUIREMENTS FOR INDIVIDUAL UNITS IN A FULL-RANGE WHOLE-BODY COUNTING SYSTEM

Instrument	Operating Range	Shielding	Detector(s)	Electronic Instrumentation	Cost
High-Range Counter	200–0.1 mc	Minimal	Single small (2″ × 2″) NaI	Single-channel pulse-height analyzer with scaler and timer	$5,000
Medium-Range Counter	1–0.01 mc	Medium	Several (2–6) medium-size (3″ × 3″) NaI	Single- or multi-channel pulse-height analyzer	$8,000 to $25,000
Low-Range Counter	0.03–0.0001 mc	Massive	Single large (8″ × 4″) or several (4–10) moderate (5″ × 4″) NaI, large plastic scintillators, or large-volume liquid scintillator	Multichannel pulse height analyzer	$100,000 and up

The subject–detector geometry of all the whole-body counters must be arranged so that changes in distribution of the radioisotope within the subject during the course of time will produce minimal changes in detection efficiency. For example, Morris et al.[1] have described a high-range (low-sensitivity) counter design that satisfied this requirement. This instrument consists of a single 2″ × 2″ NaI detector suspended near the ceiling, looking at a stretcher cot on a cement floor. The midpoint distance from detector to subject is 250 cm. The detector views all portions of the subject from head to foot with approximately equal efficiency (Figure 23). When the subject contains more than about 7 millicuries, a curved lead attenuator placed over the front of the detector allows assay of up to 200 millicuries of most radioisotopes.

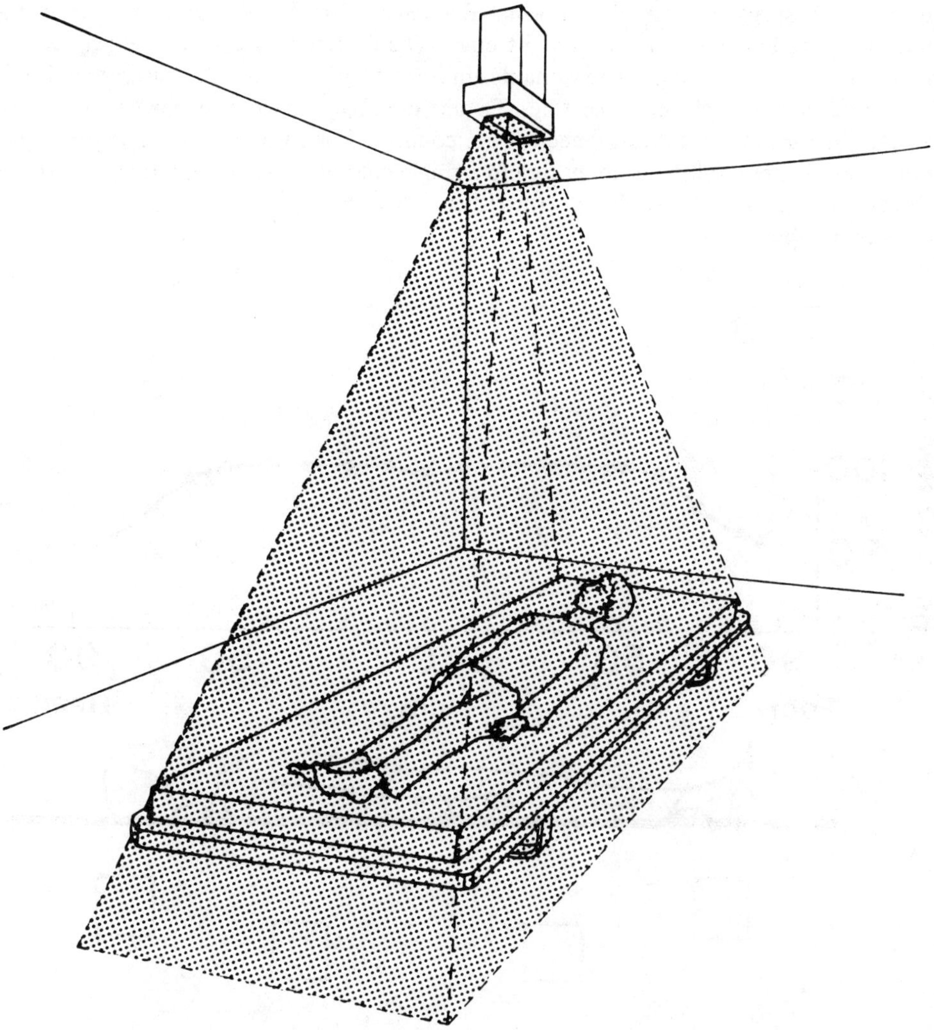

Nuclear Instrumentation

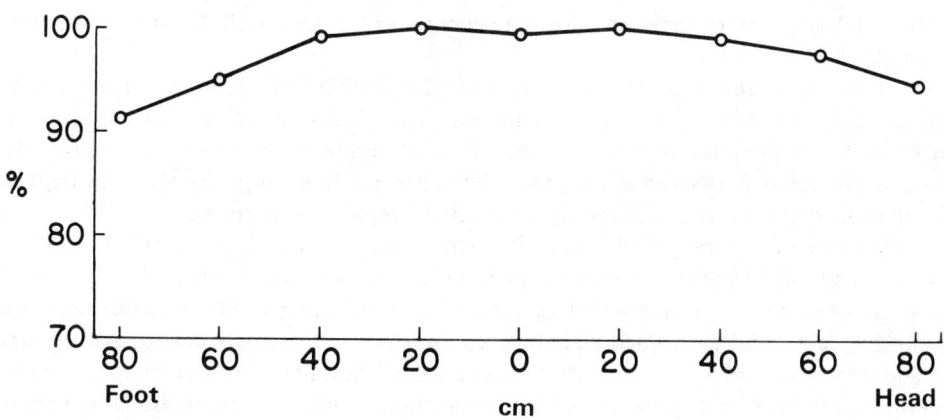

FIGURE 23a.

Subject-Detector Geometry of Single-Crystal High-Range Whole-Body Counter.

FIGURE 23b. Measured Response Curve of High-Range Whole-Body Counter.

The medium-range whole-body counter in this system, also designed and constructed by Morris,[2] consists of an array of four 3-inch diameter × 3-inch thick NaI detectors arranged in a line along the bottom of a lead shield trough. The detectors at each end are raised above the middle detectors, to obtain a reasonably uniform response over the entire length of the hospital stretcher on which the subject lies. Details of subject-detector geometry are shown in Figure 24. The outputs of all four detectors are connected in parallel to a single preamplifier. The output of the preamplifier is connected to a single-channel pulse-height analyzer, although a multichannel pulse-height analyzer with either a perforated-tape or magnetic-tape output would be preferable.

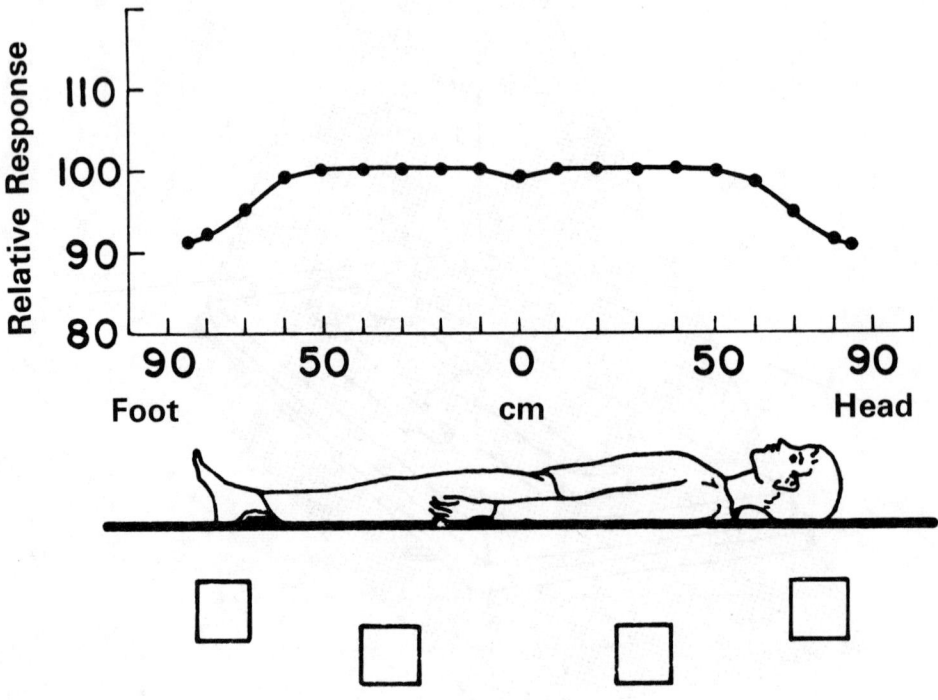

FIGURE. 24.
Subject–Detector Geometry and Response Curve of a Four-Detector Medium-Range Whole-Body Counter.

Such a system could prepare data easily and economically for computer analysis of complex spectra in retention studies involving two or more radioisotopes. Multichannel analyzers also allow more rapid and accurate energy calibration of all detectors in the system than single-channel analyzers.

These high-range and medium-range whole-body counters are presented here merely as examples, and do not exhaust by any means the possible designs of such instruments. They should not be accepted as optimal instruments for all applications. However, as examples of existing working whole-body counters, they define the relative simplicity that can be used in assaying usual clinical tracer and therapeutic radioisotope body burdens.

A directory of low-range, ultrasensitive whole-body counters, published in 1964,[3] lists 134 instruments throughout the world that were in use or under construction at that time. Detailed information is given for 111 of these. The type and number of detectors, subject–detector geometry, and associated electronic equipment are extremely varied. All the whole-body counters described have a few factors in common. They all incorporate massive shielding, to minimize radiation background. All use multichannel pulse-height analysis, varying from 2 to 512 channels. In general, the liquid and plastic scintillator systems use 2 to 6 channels, while the crystal detector systems tend to use 200 or more channels. Other general characteristics of existing whole-body counting systems using liquid, plastic, and crystal detectors are summarized in Table 13.

Table 13.
GENERAL CHARACTERISTICS OF EXISTING WHOLE-BODY COUNTING SYSTEMS

Detector	Energy Resolution	Geometric Efficiency, %	Background 0.1–2.0 Mev, c/m	Actual Efficiency 0.662 Gamma Ray, %
Liquid	Poor	50–90	100,000	5–10
Plastic	Poor	10–75	50,000	1–8
NaI Crystal	Good	Usually 5	Usually 5000	Usually 1

Examples of subject–detector geometry of several types of whole-body counters, along with some advantages and disadvantages of each, are shown in Figures 25 to 29. These illustrations do not include details of detector size, subject–detector distance, or shield characteristics, because of the wide variety found among these factors in existing whole-body counting systems.

Any system requires extremely stable electronics. Drifts in the output voltage must be minimized. Minute electronic drifts in any part of multidetection systems cause large reduction in energy resolution of the system, which cannot be corrected in retrospect. To avoid line-voltage variations, a separate transformer should supply the entire electronic system for any whole-body counter. Also, provision must be made in multidetector systems for individual gain adjustment of each detector assembly from the operating console, so that energy calibration of each detector can be made individually and as frequently as necessary.

In selecting site and construction materials for a low-range whole-body counter, extensive testing for radioactivity should be done, to insure that both site and all construction materials will result in a background as low as can be obtained with available funds.

The whole-body counter systems described here are only examples of existing types of systems. The optimal all-purpose geometry-free whole-body counter system is yet to be built.

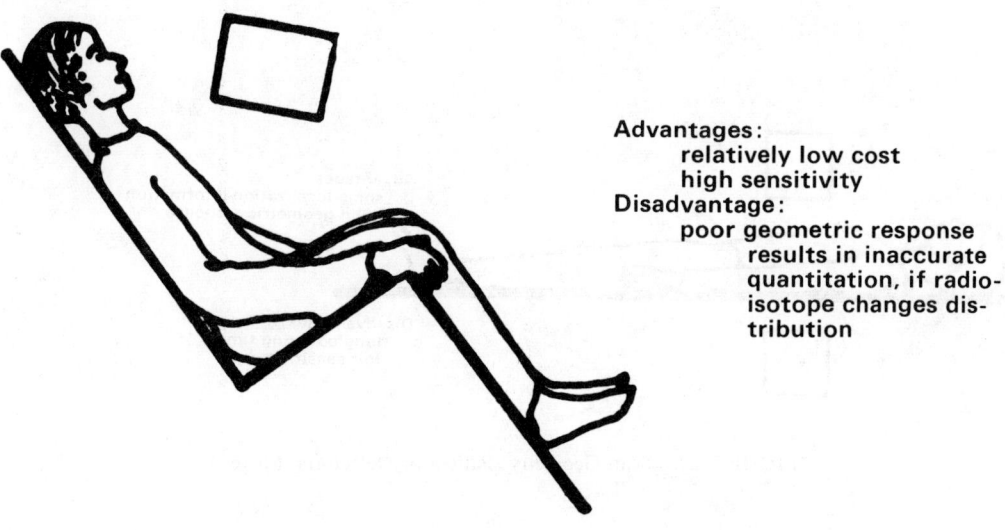

Advantages:
  relatively low cost
  high sensitivity
Disadvantage:
  poor geometric response results in inaccurate quantitation, if radio-isotope changes distribution

FIGURE 25. Tilting-Chair Geometry.

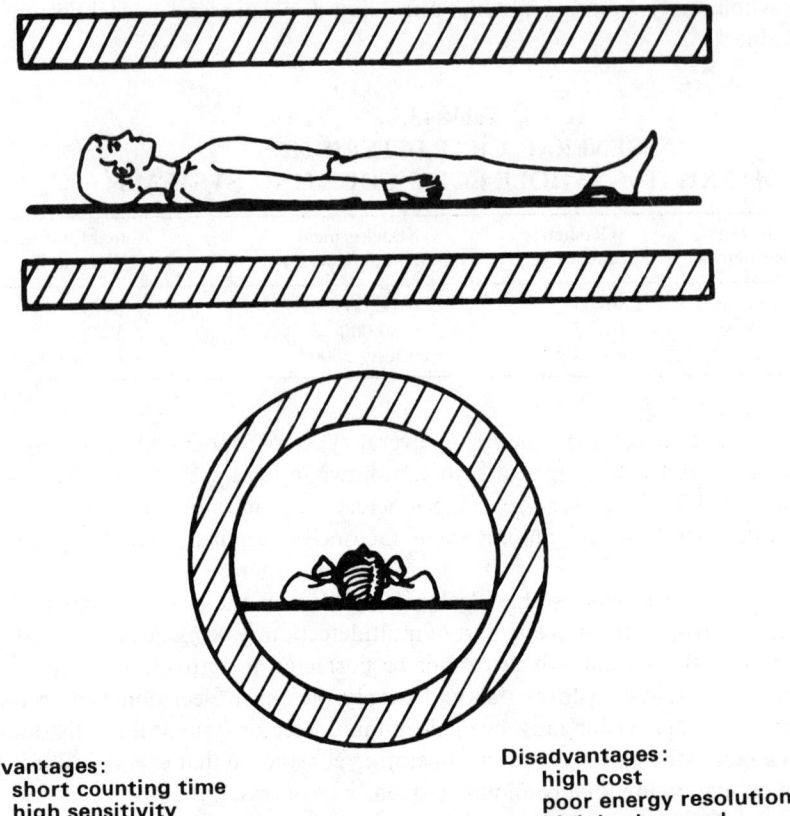

FIGURE 26. Large-Volume Liquid or Plastic Scintillator Geometry.

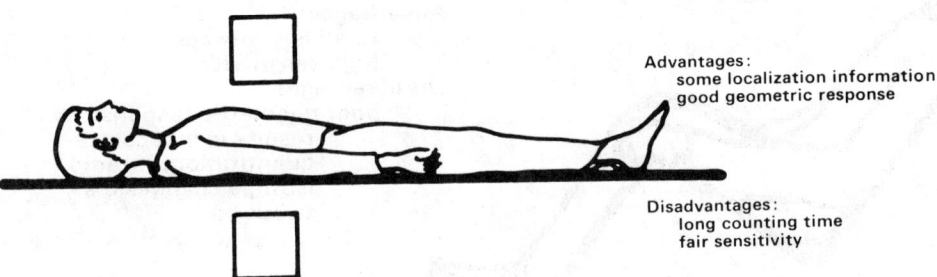

FIGURE 27. Scan Geometry; Patient or Detector(s) Move(s).

Nuclear Instrumentation

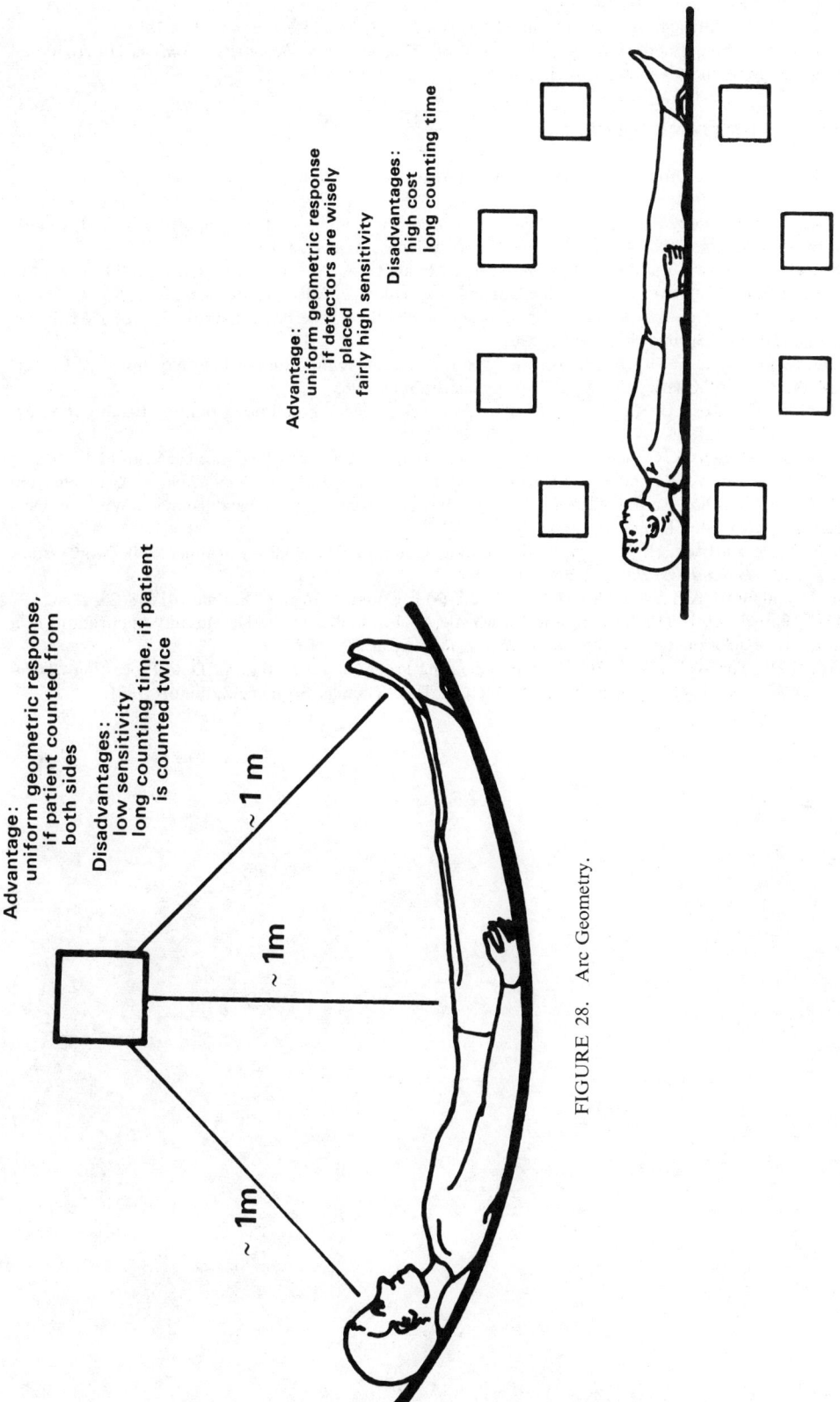

FIGURE 28. Arc Geometry.

FIGURE 29. Multidetector Array.

REFERENCES

1. Morris, A. C., Jr., Ross, D. A., and Travis, J. C., A High-Level Whole-Body Counter. *Int. J. Appl. Radiat.*, 15: 391–396, 1964.
2. Morris, A. C., Jr., A Diagnostic-Level Whole-Body Counter. *J. Nucl. Med.*, 6: 481–488, 1965.
3. International Atomic Energy Agency, *Directory of Whole-Body Radioactivity Monitors.* International Atomic Energy Agency, Vienna, Austria, 1964.

SUGGESTED READING MATERIAL

Anderson, E. C. and Van Dilla, M.A., Low-Level Gamma-Ray Detection in Humans. *IRE Trans. Nucl. Sci.*, NS–5(3): 194–198, 1958.

International Atomic Energy Agency, *Whole-Body Counting*, Proceedings of the Symposium held at Vienna, June 16, 1961. International Atomic Energy Agency, Vienna, Austria, 1962.

Katsunuma, H., Yoshizawa, Y., Maeda, K., Takeuchi, Y., Imahori, A., Anzai, I., Kusama, T., and Kaneko, M., The Whole-Body (Human) Counter of the University of Tokyo. *J. Nucl. Sci. Technol.*, 3: 114–117, 1966.

Meneely, G. R., ed., *Radioactivity in Man*, a Symposium, held at the Vanderbilt University School of Medicine. Charles C Thomas, Springfield, Illinois, 1961.

Meneely, G. R. and Linde, S. M., eds., *Radioactivity in Man*, Second Symposium, held at Northwestern University Medical School. Charles C Thomas, Springfield, Illinois, 1965.

Morris, A. C., Jr. and Ross, D. A., The Design of a Low-Background Whole-Body Counting Facility. *USAEC Report, ORINS–51*, 1966.

Palmer, H. E. and Roesch, W. C., A Shadow Shield Whole-Body Counter. *Health Phys.*, 11: 1213–1219, 1965.

Peabody, C. O., Fraser, V. M., and Speight, R. G., *The A.E.E. Winfrith Whole-Body Monitor*. Available from H.M. Stationery Office, Radiological and Safety Division, Atomic Energy Establishment, Winfrith, Dorchester, Dorset, United Kingdom.

Reizenstein, P. and Karlsson, H. A., Clinical Whole-Body Counting: Whole-Body Scanner with Two Crystals. *Acta Radiol. Therap. Phys. Biol.*, 4: 209–220, 1966.

Tauxe, W. N. and Orvis, A. L., The Mayo Clinic Whole-Body Counter. *Mayo Clin. Proc.*, 41: 18–23, 1966.

Trott, N. G., Parnell, C. J., Hodt, H. J., and Entwhistle, R. F., Studies in the Design and Applications of a Clinical Low-Background Counting Room. *Brit. J. Radiol.*, 36: 592–607, 1963.

Woodward, K. T., The Walter Reed Whole-Body Counting Program and Method of Data Processing. *Radioactivity in Man*, pp. 117–128, Meneely, G. R., ed. Charles C Thomas, Springfield, Illinois, 1961.

INFORMATION STORAGE AND RETRIEVAL IN RADIOISOTOPE IMAGING SYSTEMS

E. C. Gregg
Department of Radiology, Case-Western Reserve University

INTRODUCTION

Although the original two-dimensional record of distributed activity as seen by a scanning collimator or two-dimensional device and presented in a currently conventional manner is adequate for a fair number of diagnoses, experience has shown that in many cases further operations are desirable.[1] The most prominent of these is background erase, where only signals above a certain level are presented, so that it becomes easier to detect defects in some broad distribution. Other operations being used or attempted are profiling (either closed contour lines of constant count rate or a graph of count rate vs. distance along a line through the object), contrast enhancement, contour or edge enhancement, color discrimination of intensity levels, and, finally, analysis of original data in order to assess statistical certainty and perhaps local uptakes. Some of these operations—particularly background erase and color discrimination—have been performed while the data are being collected, but since these operations require preset conditions and some of the data may be irretrievably lost, experience again has shown that more permanent storage of all the original unmodified data and subsequent operation based thereon are preferred.[2,3] This has been termed "total information storage", which may be interpreted as very-long-time memory (perhaps years), and is characterized by the use of either magnetic tape, magnetic drum, or photographic film as the storage element. On the other hand, because of the relatively low speed of handling input data, most computer operations require a temporary storage and some modification of the data (i.e., dividing it into "bins") before presenting them to the computer in the form of punched tape or equivalent.[4,5] In these cases, either the magnetic tape, if used, or the punched tape may act as the permanent data storage. However, the division of the data into bins does introduce a data modification, which may or may not be important, and—although little used in computer operations today—the original data on magnetic tape or drum are preferable.

It should be pointed out that most data-handling operations today are one-dimensional (on-line or sequential), so that storage is quite necessary for those devices that are truly two-dimensional. Examples of these are the autofluoroscope and the image-amplifier–scintillator combination, where data are collected by a large number of individual detectors acting simultaneously or in parallel. This initial storage is necessary in order to rescan the data one-dimensionally, perform the desired operations on the data, and then finally present it again in two-dimensional form. This same technique is also used on scintillation cameras, although they detect only one event and its position at a time. The difficulty is that the position coordinates occur at random in both space and time, and must somehow be stored and rescanned for the more regular operations later. While film is the preferred method today, it is possible to use either temporary storage devices, such as electronic storage tubes designed for coordinate transformations,[6] or a combination of pulse-height analyzers with magnetic-core storage that can be in turn connected to a computer.[7] Another device that has been used is the cathode-ray storage tube, or storage videcon, where the input position and count data may be converted to a continuously stored visual image, so that one may observe the actual building up of the final image as data are collected. While such tubes generally have very limited gray scales, they have proved of value in instantaneously assessing data collection, particularly for the scintillation camera, autofluoroscope, and the multicrystal scanner. The latter is a combination of serial and parallel detectors, and uses multiple parallel channels on magnetic tape for permanent storage.[8]

THEORETIC CONSIDERATIONS

The final presentation of any system will be affected both in linearity of output signal relative to the original signal and in resolution. Regarding the latter, it must be remembered that one is dealing with two-dimensional distributions, and any comments for one direction will also hold in the orthogonal direction. Since at the moment little has been accomplished with true two-dimensional image handling, and since most operations are one-dimensional (i.e., scanning the original object and/or rescanning the stored data with subsequent signal manipulation), discussion shall be confined to the one-dimensional case. If a detector with polar symmetry or a system with isotropic response is caused to scan or look at an infinitely narrow line of infinite extent, a bell-shaped curve of intensity vs. distance perpendicular to the line is obtained. This is called the *line spread function*, and the reciprocal of the width at half-maximum (WHM) is usually called the resolution, although in some cases the width itself has been listed. A collimator with a WHM of $1/2$ inch will have a (defined) resolution of two lines per inch. Since the ability to resolve detail depends upon the signal intensity, the WHM is only a rough measure, and recourse is generally made to the modulation transfer function (MTF) for a more complete description of the geometric characteristics of the system. Now, the MTF of any system is simply the Fourier Transform of the line spread function,[9] and may be represented by

$$\tau(K) = \int_{-\infty}^{\infty} A(x)\cos(2\pi Kx)\,dx, \tag{1}$$

where K is the spatial frequency coordinate (lines per unit distance), and $A(x)$ the line spread function along the x axis. Normalization requires that

$$\int_{-\infty}^{\infty} A(x)\,dx = \tau(0) = 1. \tag{2}$$

Graphically, the plot of $\tau(K)$ vs. K shows the relative "gain" of the system as a function of the spatial frequency K, and since it generally falls off with increasing K in scintiscanning systems, it gives a measure of the difficulty of observing small objects relative to large objects. This results from the fact that Fourier analysis of an image produces a spectrum of amplitudes as a function of K, with smaller objects possessing much higher frequencies than the larger objects. The most interesting property of the MTF is that, if a system consists of elements in series, each with its own MTF, then the MTF of the overall system (assuming linearity) is simply the product of all those in the series:

$$\tau(K) = \tau_1(K)\tau_2(K)\tau_3(K)\ldots\tau_n(K). \tag{3}$$

For example, Figure 30 shows the MTF of a detector with a WHM of $1/2$ inch, a storage recording spot of WHM $1/8$ inch, a subsequent replay spot with a WHM of $3/8$ inch, and finally the product of all three, which would be the MTF of the final presentation. A Gaussian distribution (line spread function) was assumed for all three, which is a reasonable approximation to most practical cases.[12] It is apparent that each device or operation in series degenerates the response, so that ultimately only very gross details are observable with the overall system, in spite of the resolution of each element. For devices or operations with Gaussian line spread functions it can be shown that

$$\tau_n(K) = \exp(-\pi^2 K^2 d_n^2 / 2.78), \tag{4}$$

where d_n is the WHM of the nth operation. For series elements the overall MTF then becomes

$$\tau(K) = \exp\left(-\frac{\pi^2 K^2}{2.78}\right)(d_1^2 + d_2^2 + \cdots + d_n^2),\tag{5}$$

or the overall WHM becomes

$$d^2 = d_1^2 + d_2^2 + \cdots + d_n^2.\tag{6}$$

In the above example $d^2 = (^1/_2)^2 + (^3/_8)^2 + (^1/_8)^2$, or $d = 0.64''$, compared to $d = 0.5''$ for the detector alone, which again illustrates the broadening of the line spread function, or loss of detail, caused by elements in series. It is to be noted that $\tau(K) = 0.0287$ for $K = 1/d$, which means that the (system) amplitude is down by a factor of about 35 for image spatial frequencies on the order of the resolution as defined above.

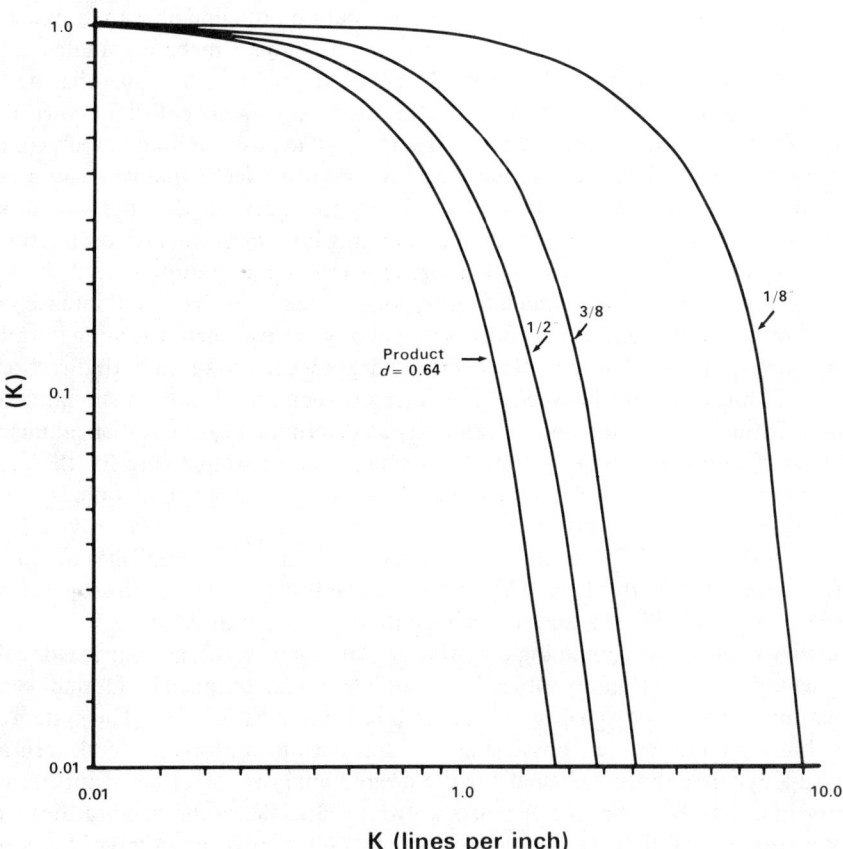

FIGURE 30.
MTF of Three Different Devices with WHM's as Shown, and of Their Product, if Used in Series in a Linear System.

Electronic systems may also affect the overall resolution and presentation, depending both on their location and on the type of scanning used. In mechanical scanners moving continuously in one direction and stepwise in the other, any electronic system with time delay (i.e., a ratemeter) will always present the data at a later time, determined by the actual count rate. Thus the image is shifted in the direction of scan by a changing distance, which not only distorts the image, but, for scans in alternate directions, will produce scalloping. This effect is not observed in the stepwise direction, since the circuit time constants are generally shorter than the time between steps. In scintillation cameras, the positioning circuits are important factors in determining the resolution in both directions.[7,10]

Regarding the distance between scan steps for either a rectangular or spiral raster, it is obvious that one is approximating the MTF of the system orthogonal to the continuous scan direction by a series of discrete steps. Although such steps may produce artifacts in the presentation, these have not been shown to be of importance, and steps on the order of or less than $1/3$ the WHM are generally acceptable. The same holds true in both directions when preparing data for a digital computer, where it is necessary to subdivide the scanned area into square "bins" for analysis. The lower limitation is either the time required for the total scan or the statistical uncertainty due to the limited number of events in each bin.

PRACTICAL CONSIDERATIONS

In the case of two-dimensional presentations (or storage) with a one-dimensional scanning device, it is obvious that, if one records the detected data on the line of scan, a pattern will be obtained that is structured into lines. As in television,[11] this can be eliminated by using a recording spot shaped according to a $\cos^2\theta$ distribution and of such a size that half-overlap occurs between scan lines. Another technique is to use a very small spot that is turned on for a short time only during each detected event, with its possible position modulated perpendicular to the direction of scan by the $\cos^2\theta$ function.[12] Still another technique is to use a large spot consisting of a bell-shaped distribution of small intense spots, so that the whole pattern is recorded for each event. These techniques of eliminating line structure have been termed "data blending" in scintiscanning.[13] In cases of records for visual presentation (i.e., "photoscans"), there has been some argument for much larger spots, on the basis that small intense spots may produce higher spatial frequency components that annoy the eye when searching for more gross detail.[14] However, the use of such spots obviously degrades the MTF in both directions, and a choice generally must be made between a pleasing picture and resolution. A similar choice must also be made between resolution and detectability as determined by the statistical uncertainties in the resolvable areas. For presentations with small spots, a simple cure for the problem of visualization is minification to the point where the spatial frequency components of the sharp dots lie outside the range (resolution) of the eye—a technique that has proved of value in radiology.[15] In the case of TV scanning and replay, defocusing accomplishes about the same results. While the bandwidth of the TV channel could also be lowered, this operation would only affect the direction of scan, and not change that in the step direction.

Regarding linearity and obtaining quantitative data for analysis, as contrasted with strictly visual presentations, the primary objective is to obtain an output signal that bears some known relation to the input signal, with as high a resolution as possible. The system need not be linear, but if nonlinear (i.e., possessing low end cut-off and/or high end saturation), it should have a dynamic range sufficient for the desired analysis. A further—but not necessary —requirement is that the technique be reasonably reproducible, so that calibrations will not be necessary for each operation. In the case of magnetic-tape or drum storage, the advent and degree of saturation is determined by the overall system time constant, which, like the MTF, is determined by all the elements in the system. Both recording-pulse duration and tape speed are important parameters here. Such a nonlinearity is really pulse-coincidence loss, and is generally negligible over several decades in well-designed systems.

With two-dimensional photographic storage, one is concerned not only with spatial coincidence, but also with the nonlinearities resulting from film characteristics. The latter can be particularly troublesome in producing a preset erase level, due to the relative insensitivity of most film at low light levels. Film nonlinearities are important for superposed large spots, while for very small dots spatial coincidence becomes important. Such records are usually made by photographing a cathode-ray oscilloscope whose beam position is electrically slaved to the detector position. Since conventional tubes produce a spot about 0.020 inch in diameter, it is obvious that for the usual 5-inch tube with a 2-inch by 3-inch recording area one is almost always bothered by spatial-coincidence loss, unless density addition of larger superposed spots is used. Of course, for this latter technique, the final relationship between optical-transmission density and event density depends critically on film characteristics, and preset erase may be produced. However, because of practicality and availability, this is till the method of choice for most systems. If the saturated-microdot technique is preferred, since it is more independent of film characteristics and high count rates are encountered, recourse may be made to a small (1-inch diameter) CRO tube, with a small spot (~0.005 inch) modulated perpendicular to the scan direction and moved mechanically in spatial synchrony with the detector, to produce a 1:1 record.[12] This larger record and much smaller intense spot with modulation materially reduce the spatial-coincidence loss due to a finite spot size as well as eliminate any preset erase conditions. In general, for well-designed systems a range of at least two decades can be obtained for linearity of transmission density vs. events per unit area. This has proved adequate for most diagnoses and data analyses. Some examples of the above are shown in Figure 31.

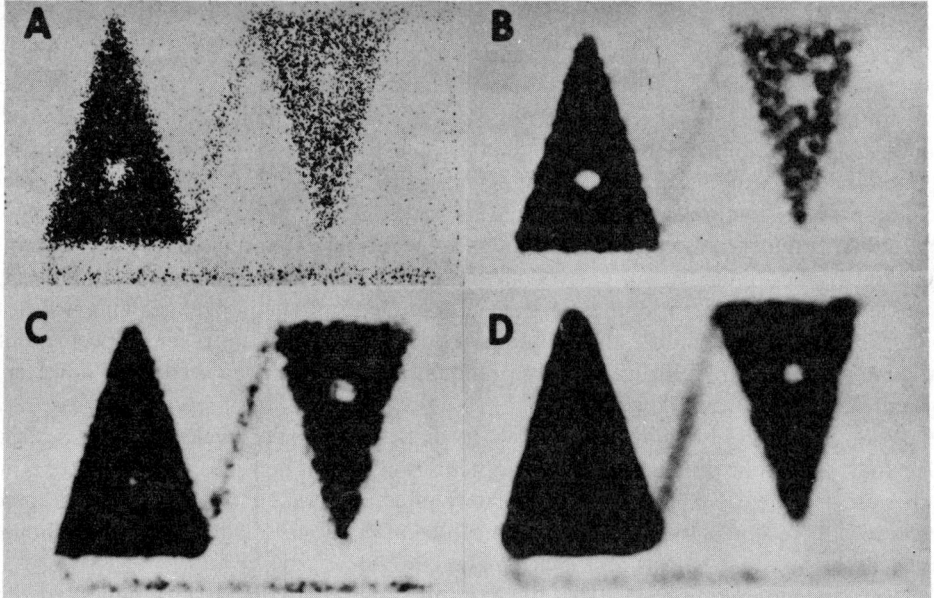

FIGURE 31. Effects of Different Recording Spot WHM's.
Scans of two hollow triangular prisms of equal viewed size (short base 3″), but with a depth ratio of 3:1, filled with a radioactive liquid in series through a 1/16″-diameter tube; each contains a void (plug) 0.5″ in diameter; the collimator had a WHM of 0.33″; the scan steps were 0.08″; the recording spot WHM's were (A) 0.005″ (microdot), (B) 1/16″, (C) 1/4″, and (D) 3/8″; identical scan times were used, producing about 25,000 events per picture; note the erase present in the last three, due to film and paper characteristics.

While digital computers produce a spatial array of numbers representing the detected events in a given "bin", this is very difficult to scan visually, and recourse is generally made to producing patterns pleasing to the eye for the rapid detection of any abnormalities in the scan.[16] While some computers use more dense printed symbols for higher detected count densities and/or a photographic gray scale selected by the digital count density, programming to produce isodensity lines of accurately preset steps (i.e., 2 or 3 standard deviations) is more popular.[17]

QUANTITATION AND IMAGE MANIPULATION

It must be stressed that the fundamental limitation in scintiscanning is the statistics of the source itself. As the resolution becomes greater or the bins become smaller, the smaller the number of counts in each resolvable area will become, and hence the larger the statistical variations. For example, in a typical scintiscan which collects 100,000 counts over a scan area of 20 × 25 cm, the average density would be about 200 counts per cm². Thus, the uncertainty of measuring the average of any 1-cm² area (the fractional standard deviation) would be about 7 percent. If the area of interest were 0.25 cm square, this would increase to about 28 percent. Thus, for a given average signal difference required for detection of two adjacent areas of fixed size, it is obvious that the ability to resolve or the apparent resolution will depend critically both on the total signal detected (signal-to-noise ratio) and on the MTF of the system. While the philosophical implications of this have been discussed elsewhere,[12] of primary interest here is the fact that these same variations (small signal-to-noise ratio) exist in and limit all the subsequent operations on the stored data. Although digital computers can be programmed to compare the events in adjacent bins and arrive at an average,[17] and although TV rescanning or equivalent can be done with a larger spot for such data smoothing,[18] all techniques devised today still result in a reduction of resolution, as mentioned previously. This statistical limitation and its influence on resolution and/or detectability exist particularly in contour enhancement and profiling, which in essence require differentiation that, in turn, only makes the signal-to-noise ratio worse.

Contrast enhancement is the production of larger differences in output signal for a given change in input signal. It is obvious that this is only amplification, which does not change the signal-to-noise ratio and, hence, should not change detectability for the poor signals in scintiscanning. Signal amplification is of value only when it is necessary to overcome subsequent system noise or some limiting threshold of detectability at the output. However, in film recording there is some compression of the data, and enhancement can be used to bring the data back to their original true gray scale. Although some further enhancement may be used (usually unnecessarily), too much enhancement will cause the signal to saturate while amplifying the somewhat smaller noise, since most systems are limited in dynamic range. This will in turn produce a much poorer picture, and should be used with caution. Contrast enhancement of the final analogue image by utilizing digital techniques has also been reported;[17,19] however, the same considerations would apply.

Background erase is simply a variable bias diode circuit or equivalent that passes signals higher than a given preset level. In digital systems, only count densities above a certain preset number are presented. Systems with immediately adjustable erase and contrast enhancement to overcome film compression have been found to materially aid diagnoses,[20] even though in retrospect the information was on the original record.

Regarding quantitation of data, there is no doubt that digital techniques are highly accurate, and particularly valuable when performing simple uptake measurements in selected areas of the scan. However, with the large statistical uncertainties in the original data, analogue film and rescanning techniques are also quite acceptable to many investigators. Although flying spot scanners are superior in accuracy and stability,[2] videcon viewing with appropriate circuitry that presents any selectable scan line as a "profile slice" through the area of interest is reasonably accurate after calibration.[3] In this case, the observed voltage may be related to the average count density as modified by the videcon optics and as seen by the scanning spot.

It should be emphasized that the terms "digital" and "analogue" refer only to the method of storage, read-out, and manipulation of the data. Both film and magnetic tape can be used for either procedure, depending on the recording and read-out method used. Likewise, analogue computers can perform the same operations as their digital equivalents, and many systems have combinations of both. The method of choice depends mostly on ultimate accuracy, speed, convenience, and cost.

Regarding color presentations, while these are aesthetically pleasing, there is doubt that color discrimination will produce as many observable steps of intensity as a gray scale.[19] This is particularly true of the high-count-density pictures and certainty dependent on the type of diagnosis desired. Directly produced color presentations are usually plagued by long time constants and the need to preset levels before scanning. Methods to overcome these shortcomings are to rescan the original data either mechanically[21] or electronically (color TV) with adjustable conditions of color separation.

Lastly, it is possible to perform true two-dimensional operations optically on a given image by using special optical filters and systems.[22] Unfortunately the state of the art is such that only simple changes in the two-dimensional MTF are possible, and no filters have been constructed for the more complex operations described above. Two-dimensional flying spot scanning and image manipulation has also been reported,[23] but has not yet been applied to scintiscan records with their low signal-to-noise ratio.

REFERENCES

1. Mallard, J. R., Medical Radioisotope Visualization. *J. Appl. Radiat. Isotop.*, 17: 205–249, 1966.
2. Gregg, E. C., Voelker, W. H., Storaasli, J. P., and Friedell, H. L., Information Storage and Recall in Scintiscanning. *Radiology*, 78: 114, 1962.
3. Gregg, E. C., Voelker, W. H., Storaasli, J. P., and Friedell, H. L., Basic Concepts and Design of a Total Information Storage and Data Extraction System for Radioisotope Scanning. *Am. J. Roentgenol. Radium Therapy Nucl. Med.*, 93: 733–746, 1965.
4. Kawin, B., Huston, R. S., and Cope, C. B., Digital Processing/Display System for Radioisotope Scanning. *J. Nucl. Med.*, 5: 500–514, 1964.
5. Kuhl, D. E. and Edwards, R. Q., Perforated Tape Recorder for Digital Scan Data Store with Gray Shade and Numeric Readout. *J. Nucl. Med.*, 7: 261–280, 1966.
6. Knoll, M. and Kazan, B. *Storage Tubes*. John Wiley and Sons, New York, 1952.
7. Meyers, M. J., Kenny, P. J., Laughlin, J. S., and Lundy, P., Quantitative Analysis of Data from Scintillation Cameras. *Nucleonics*, 24(2): 58–61, 1966.
8. Hindel, R. and Gilson, A. J., Multicrystal Scanner is Rapid and Versatile. *Nucleonics*, 26(2): 52–57, 1967.
9. Gregg, E. C., Pedagogical Note on Modulation Transfer Function. *Invest. Radiol.*, 1: 418–421, 1966.
10. Cradduck, T., Fedoruk, S., and Reid, W., A New Method of Assessing the Performance of Scintillation Cameras and Scanners. *Phys. Med. Biol.*, 11: 423–435, 1966.
11. Zworkin, V. K. and Morton, G. A., *Television*. John Wiley and Sons, New York, 1954.
12. Gregg, E. C., Modulation Transfer Function, Information Capacity and Performance Criteria of Scintiscans *J. Nucl. Med.*, 9(3): 116–127, 1968.
13. MacIntyre, W. J. and Christie, J. H., The Use of Data Blending to Reduce Statistical Fluctuations in Scintiscanning. *Radiology*, 86: 141, 1966.
14. Beck, R. N., Charleston, D. B., Eidelberg, P., and Harper, P. V., The ACRH Brain Scanning System. *J. Nucl. Med.*, 8: 1–24, 1967.
15. Tuddenham, W. J., The Visual Physiology of Roentgen Diagnosis. *Radiology*, 78: 116–123, 1957.
16. Brown, D. W., Digital Computer Analysis and Display of the Radionuclide Scan. *J. Nucl. Med.*, 7: 740–753, 1966.
17. Kuhl, D. E., Digital Approach to Smoothing Area Count Integration, Contrast Control and Isocount Generation. *Fundamental Problems in Radioisotope Scanning*, Gottschalk, A. and Beck, R. N., eds. Charles C Thomas, Springfield, Illinois, 1968.
18. Charleston, D. G., Beck, R. N., Eidelberg, P., and Schuh, M., Techniques which Aid in Quantitative Interpretation of Scan Data. *Medical Radioisotope Scanning*. International Atomic Energy Agency, Vienna, Austria, 1964.
19. Tauxe, W. N., Chaapel, D. W., and Sprau, A. C., Contrast Enhancement of Scanning Procedures by High-Speed Digital Computers. *J. Nucl. Med.*, 7: 647–656, 1966.
20. Rejali, A. M., Friedell, H. L., and Gregg, E. C., Radioisotope Scanning with a System for Total Information Storage and Controlled Retrieval. *Am. J. Roentgenol. Radium Therapy Nucl. Med.*, 97: 837–849, 1966.
21. Harris, C. C., Bell, P. R., Francis, J. E., Ross, D. A., Jordan, J. C., and Satterfield, M. M., The Rescanner. *Medical Radioisotope Scanning*, p. 529. International Atomic Energy Agency, Vienna, Austria, 1964.
22. O'Neill, E. L., Spatial Filtering in Optics. *IRE Trans. Inform. Theor.*, June 1965.
23. Kovanasznay, L. S. G. and Joseph, H. M., Image Processing. *Proc. I.R.E.*, 54: 560, 1955.

RADIOISOTOPE CAMERAS

Michel M. Ter-Pogossian, Ph.D.
The Edward Mallinckrodt Institute of Radiology,
Washington University School of Medicine, St. Louis, Missouri.

INTRODUCTION

A large fraction of the examinations carried out in nuclear medicine consists of the visualization of tumors, organs, or other anatomical structures by the detection of the electromagnetic radiation emitted by radioactive substances concentrated in these structures. These electromagnetic radiations may be either emitted by the radioactive nuclide (gamma rays), or they may result from a process secondary to radioactive decay (X rays and annihilation radiation). For the sake of simplicity, these different electromagnetic radiations will, in this text, mostly be called gamma rays.

The instruments used for this purpose can be broadly divided into three categories: 1) scanners, 2) stationary devices, often called radioisotope cameras, and 3) hybrid devices. Scanners utilize radiation detectors collimated to accept photons from a single resolution element of the structure to be visualized, and the image of the structure is formed by a scanning motion of the detector over the area of interest. Radioisotope cameras, on the other hand, are stationary devices; their radiation detector continuously observes the whole field to be visualized, and the imaging of the distribution of the radioactive isotope within the field may be accomplished in a number of ways. The hybrid devices combine in their design some of the principles of both scanners and radioisotope cameras. It is the purpose of this chapter to describe four types of radioisotope cameras, which are based on different designs in regard to the method used in processing the image supplied to their detector by the collimator.

GENERAL CONSIDERATIONS

The evaluation of the usefulness of radioisotope cameras in nuclear medicine, and particularly the intercomparison of different cameras and the comparison of these devices with scanners, is based on the evaluation of a number of parameters governing the visualization of objects by means of gamma rays.

Any device designed for the imaging of structures by means of gamma rays must include three components: 1) a device capable of channeling the gamma-ray photons with the purpose of forming an image, usually called the collimator, 2) a detector sensitive to the gamma-ray photons, and 3) a system capable of establishing a spatial relationship between the signal supplied by the detector and the position of the gamma-ray photons in the image supplied by the collimator. A parallel situation exists in optical imaging, where the lens channels light photons which, in turn, impinge upon a detector (either a photographic emulsion or another type of detector, such as the input screen of a television pick-up tube). In photographic recording the emulsion itself serves as the imaging device, while in television recording the scanning electron beam establishes the spatial relationship between the intensity of the video signal and the optical image.

Extrinsic and Intrinsic Resolution of Radioisotope Cameras

The resolution that can be achieved by means of a nuclear imaging device, be it a scanner or a radioisotope camera, is affected by the collimator used and by the detector–imaging system of the device. The contribution of the collimator can be referred to as extrinsic resolution of the device, because it is independent of the detector-imaging system used, while the contribution of the detector-imaging system is called the intrinsic resolution. In scanners the intrinsic resolution is limited by the interspace between scanning lines, which can be improved by altering the design of the mechanical drive; this improvement in intrinsic resolution is, of course, achieved at the cost of the time involved in a given examination. The intrinsic resolution of radioisotope cameras depends on the detector-imaging system used. It varies considerably from one design to another and, in general, cannot be easily improved. The intrinsic resolution of various radioisotope cameras will be discussed in some detail in this text.

The intrinsic resolution of a nuclear imaging device may be either superior or inferior to the extrinsic resolution of the collimating system used. It is interesting to note that a similar situation exists in optical imaging devices. In photographic recording the resolution of the film used is generally superior to the resolution of the lens; thus, an improvement of the quality of the lens used improves also the quality of photographic recording. On the other hand, in television recording, in most instances, the resolution of the lens is superior to the resolution of the television pick-up tube; an improvement of the lens quality does not improve appreciably the quality of the television image.

Since the overall resolution of a nuclear imaging system is limited by the combination of the resolutions of the collimator and of the detector-imaging system, it is always desirable to design the detector-imaging system with resolution capabilities superior to that of the collimator to be used. Under such circumstances the overall resolution of the system will be close to that of the selected collimator.

Collimators for Radioisotope Cameras

A number of different methods of collimation have been used with the purpose of providing the detector of a radioisotope camera with an image formed by gamma-ray photons. These collimators can be classified in three categories: 1) multichannel collimators, 2) pinhole collimators, and 3) collimators for the annihilation radiation.[1] Other collimating devices, such as converging collimators for tomography,[2] have also been used with radioisotope cameras; however, these designs have not yet found broad acceptance.

Multichannel (or parallel-hole) collimators (Figure 32) are the most widely used image-forming devices in radioisotope cameras. They consist of a series of holes or channels with parallel axes in a plate made of a material as opaque as possible to the image-forming gamma rays; the material generally used for this purpose is lead. The cross section of the holes may assume a number of shapes, such as circles or polygons; the thickness of the septa between the holes is dictated by the energy of the radiation used. The channels may be cylindrical or conical; the diameter and height are dictated by the desired resolution.[3] The gamma-ray image provided by such collimators is equal in size to the object observed. The best resolution is obtained by means of a multichannel collimator in contact with the object observed, and this resolution decreases with distance.

The principle of operation of pinhole collimators for gamma rays is identical to that of the pinhole optical camera (Figure 33). The collimator consists of a hole that is provided in a plate opaque to the gamma-ray photons. The photons, originating at the object to be visualized and traveling towards the hole, form an inverted image of the object at the detector. If the distance between the hole and the detector is equal to the distance between the object and the detector, the image and the object are of equal size. A relative variation of these distances alters the size of the image relative to that of the object. Pinhole collimators may thus provide smaller or larger images of the examined object. The efficiency of pinhole collimators in utilizing the photons emitted by the object is inversely proportional to the square of the distance from the hole to the object. Thus, such collimators are best suited for imaging small objects, because they can be positioned close to the hole. Larger objects are not efficiently imaged, because they must be observed at a larger distance. By and large, pinhole collimators are incorporated in isotope cameras for two purposes: 1) for the magnification of small objects, such as the thyroid gland, if the resolution of the camera is deemed inadequate for the observation of the object without magnification, and 2) for the minification of large objects, such as the lungs, if the detector of the camera is too small to encompass the whole organ. The latter examinations are very inefficient, and in clinical practice, in most instances, multichannel collimators are preferred to pinhole collimators.

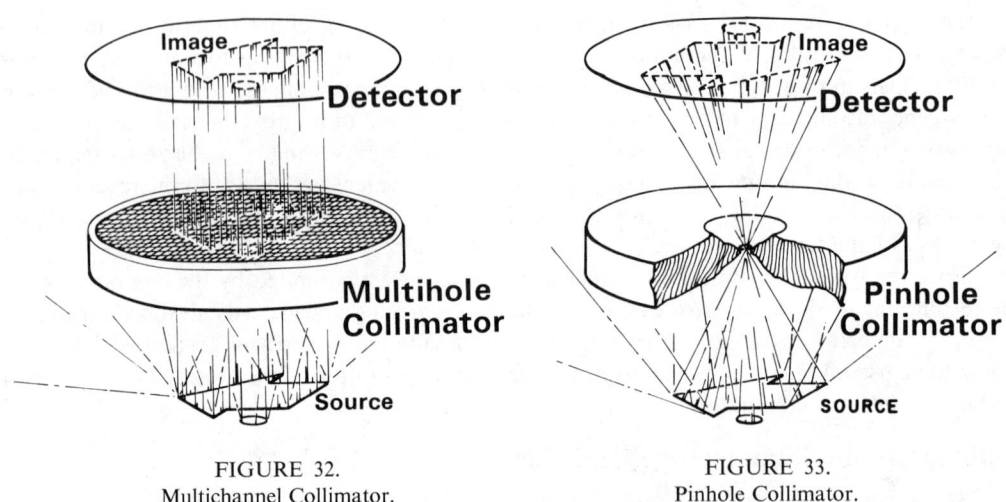

FIGURE 32.
Multichannel Collimator.

FIGURE 33.
Pinhole Collimator.

Imaging of radioactive isotopes emitting positrons can also be accomplished with certain isotope cameras, by utilizing the property exhibited by positrons to undergo annihilation in matter, with the resulting emission of two 511-kev photons traveling in opposite directions.[1] This method is based on the use of an isotope camera and a focal detector (Figure 34), both connected to a coincidence circuit, which allows the registration of a signal only if annihilation photons are simultaneously sensed by the image and by the focal detectors. Suitable electronic circuitry, the description of which is outside the scope of this text, permits the imaging of the positron-emitting object without the use of any collimation as such. It should be noted that, with suitable electronic circuitry, this method of imaging exhibits a plane of best focus, which allows a laminographic study of the object observed.[1] Positron collimation exhibits a number of disadvantages. This method is, of course, limited to the use of positron emitters; it requires complex electronic circuitry, and high counting rates are limited by the resolving time of the electronic circuitry, which must process a large number of single events for a few coincidences. The latter factor limits the activity that can be placed in the field of the image detector (about 50 microcuries at the present time[1]). Furthermore, the relatively high energy of the annihilation radiation drastically reduces the efficiency of detection in certain isotope cameras.

Detector Efficiency and Figure of Merit

One of the useful parameters in the evaluation of a radioisotope camera is the detector efficiency, which can be defined as the ratio of the number of photons sensed by the detector to the number of photons impinging upon it. The efficiency is generally expressed as a percentage. In general the detector efficiency varies with the energy of the gamma-ray photons detected.

Another useful parameter in the evaluation of the imaging efficiency of a radioisotope camera is its figure of merit, which is expressed as the product of the useful area of the detector multiplied by the detector efficiency. This parameter is particularly useful in the comparison of radioisotope cameras with scanners.

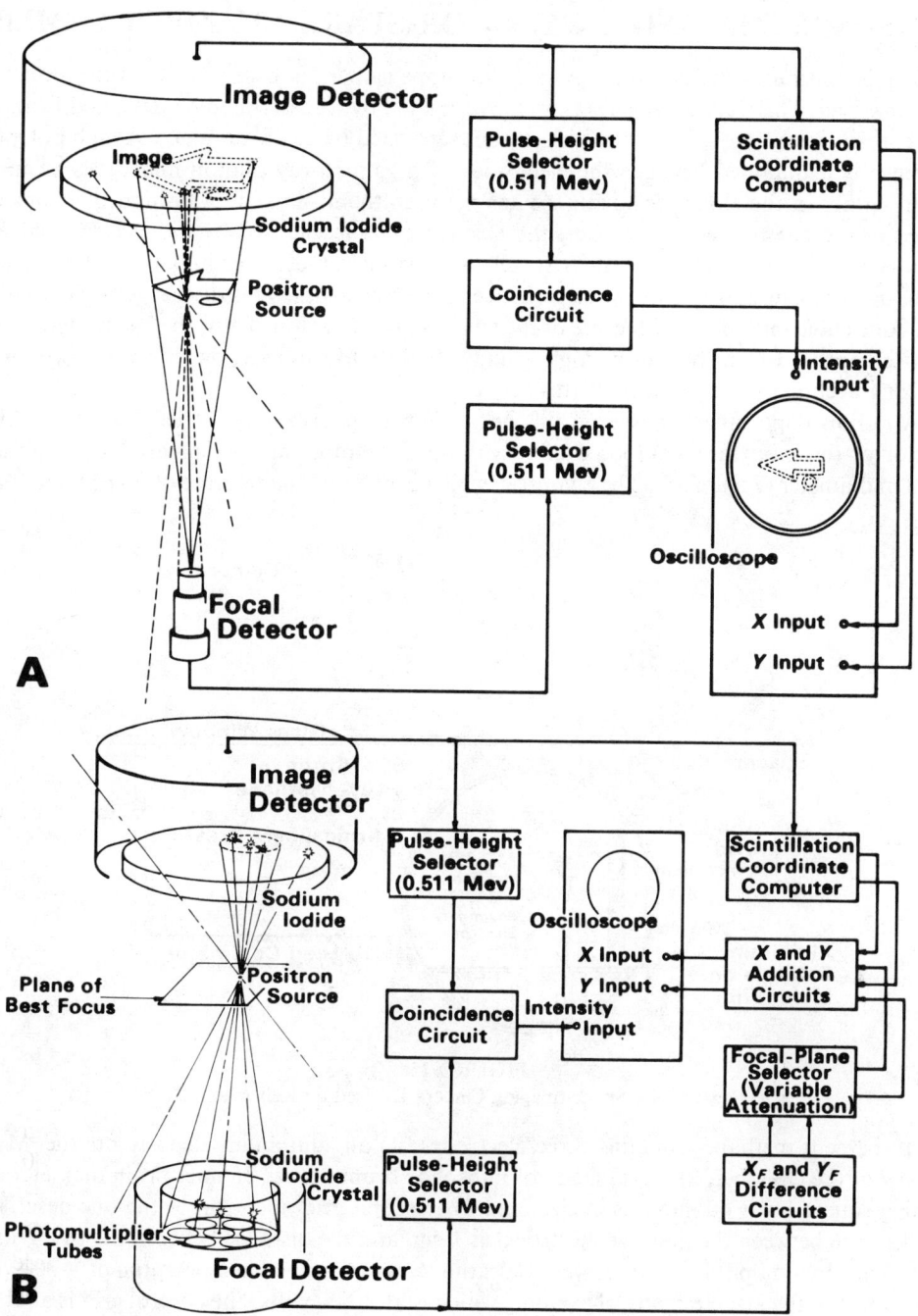

FIGURE 34.
Two Positron Collimating Systems
Devised by H. O. Anger for the Scintillation Camera.
(Photograph by courtesy of Mr. O. H. Anger.)

There are a number of other criteria that can be—and have been—used in the evaluation and in the intercomparison of radioisotope cameras. The number of these criteria is so great and, in many instances, so specific to a given application that a thorough discussion of these criteria is outside the scope of this review, and only the most significant aspects of the features and of the limitations of the design of isotope cameras will be discussed.

THE SPARK-IMAGING CAMERA OR SPARK CHAMBER CAMERA

The operation of the spark-imaging radioisotope camera is based on the utilization of a spark chamber. This device, which was first investigated by Keussel[4] and by Bella and Franzinetti,[5] consists essentially of two plane electrodes separated by gas-filled space. A high potential difference is maintained between the electrodes. If a gamma-ray photon undergoes an interaction—either in the electrodes or in the gas between them—in such a fashion that ions are formed in the gas between electrodes, the electrons are accelerated by the electric field and collide with the gas molecules in their travel towards the anode, thus initiating an avalanche which, in turn, causes a spark discharge between the electrodes. This discharge is localized to within one cubic millimeter of the site of the triggering interaction. Thus, the spark chamber is a device capable of converting a single gamma-ray photon interaction into a spark and of localizing accurately the position of this event.

A radioisotope camera based on this principle was described by Kellershohn et al.[6] The detector of this camera is composed of a cylindrical chamber approximately 6 cm high and 20 cm in diameter (Figure 35). The chamber, which is filled with a mixture of 90 percent argon

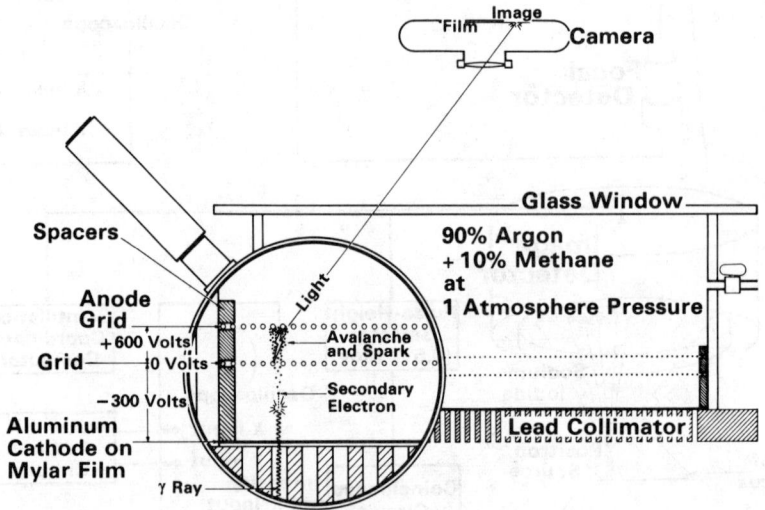

FIGURE 35.
Diagram of a Spark-Imaging Camera Devised by Kellershohn.[3]

and 10 percent methane, contains three electrodes: 1) an aluminum deposit on the Mylar window of the chamber, 2) a grid made of phosphor bronze wires 35 microns in diameter and forming square holes 40 microns in size, and 3) a second grid, identical to the one described. The distance between the first two electrodes is 1 cm, and the distance between the two grids is 5 mm. The aluminum foil, which serves as a cathode, is maintained at a potential of −300 volts with respect to the last electrode, the anode, which is at +600 volts; the control grid is at about 10 volts. The electrons formed subsequently to the interaction of gamma-ray photons either in the cathode or in the gas are accelerated by the potential difference between the cathode and the grid. They pass through the holes of the grid and generate a Townsend avalanche under the influence of the highly electric field that exists between the anode and the grid. It should be noted that this avalanche is insufficient to generate a spark, but it produces a negative pulse on the anode. This pulse is highly amplified, and is used to apply a 3-kv pulse on the grid; this, in turn, triggers a spark in the grid anode space. This spark coincides well with the ions' trajectory created by the avalanche. The sparks created in the detector chamber can be photographed by means of a relatively conventional camera through the glass backing of the chamber. Improved versions of this device have also been described by the original authors.[6-9]

FIGURE 36.
A 5″ Spark-Imaging Camera Devised by Horwitz.
(Photograph by courtesy of Dr. Norman H. Horwitz.)

A spark-imaging camera of somewhat different design (Figure 36) was described by Horwitz et al.[10-12] The detector of this unit is a spark chamber containing only two electrodes. The cathode consists of a gold or silver film deposited on the surface of a flat substrate; the anode is an optically transparent stannic oxide film deposited on a flat sheet of NESA glass. The spark discharge can thus be viewed through the glass substrate and the conductive anode. The filling gas most successfully used is neon, with a quenching gas composed of methyl

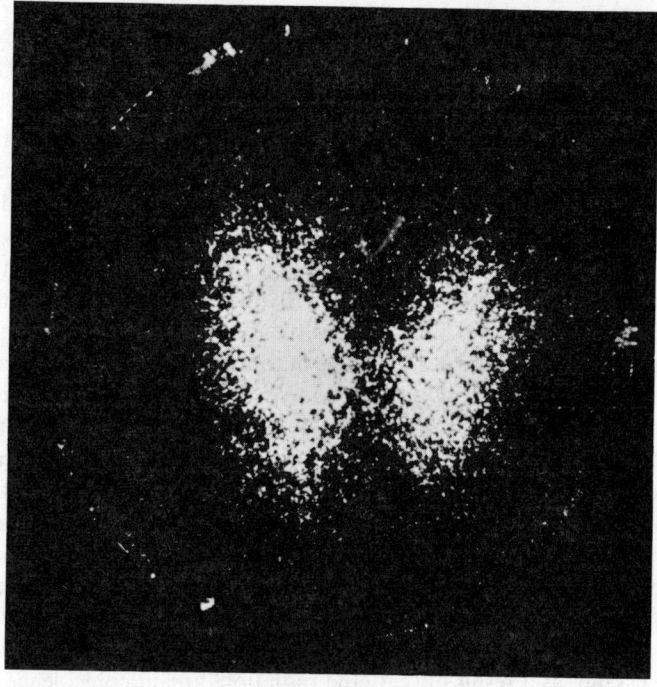

FIGURE 37.
Image of a Thyroid Gland Containing Iodine-125.
Obtained with a Spark-Imaging Camera.
(Photograph by courtesy of Dr. Norman H. Horwitz.)
Activity in the gland is about 40 μCi of ^{125}I; image consists of 15,000 image points accumulated in 7 minutes; the image shows a pyramidal lobe well above the thyroid structure.

alcohol and iodine vapor. Under these circumstances the sparking potential is about 1500 volts. The voltage depends upon the spacing of the electrodes, on the electrode material and structure, and on the quantity and type of quenching and filling gas. After sparking, there is a dead time of about 15 milliseconds, following which the chamber returns to the sparking potential. The dead time can be reduced by operating the chamber well above the sparking potential; thus a dead time of about 2 milliseconds has been obtained. The sparks are recorded by means of a Polaroid camera optically coupled to the chamber. Polaroid ASA-3000 film is sufficiently sensitive to record individual events.

The diameter of the chamber is 5 inches, and it is mostly designed for thyroid imaging. A battery-operated converter energizes the chamber for about 20 hours on a pair of flashlight cells.[13] Figure 37 shows a typical image of a thyroid, which was obtained with this camera.

Spark-imaging cameras exhibit two principal disadvantages: 1) their efficiency decreases very rapidly with gamma-ray photon energy, and 2) their counting rate, because of the dead time, is limited. The Horwitz camera with the gold cathode has an efficiency of the order of 1.5 percent for ^{99m}Tc, 140 kev photons, and it is about 1 percent at 30 kilovolts. More recently, Lanisart and Kellershohn[14] reported a detection efficiency for their spark chamber of about 7 percent for ^{197}Hg radiation. It is probable that these disadvantages may be partly overcome by operating the cameras at higher gas pressures in an attempt to increase the efficiency. Other improvements can be made to reduce the dead time. Anger[14] suggests the use of a spark chamber optically coupled to an image intensifier tube, to overcome this deficiency. Indeed, under these circumstances the chamber could be operated under Townsend avalanche conditions without sparking and, consequently, with a much shorter dead time. Pullan and Perry[8] are developing a spark chamber with cross-wired electrodes for direct read-out of spark positions without photographic recording. The principal advantages of these devices are their great simplicity and low cost.

THE DIGITAL AUTOFLUOROSCOPE

The digital autofluoroscope is a radioisotope camera designed for the purpose of imaging the distribution of radioisotopes within the body, as are other cameras, and particularly for the purpose of obtaining quantitative distribution information on a regional basis. In general terms, the detector of the autofluoroscope consists of a mosaic of activated sodium iodide crystals optically coupled to a series of photomultiplier tubes. The gamma-ray image supplied to the detector by the collimator is converted into light flashes in the crystals. These light flashes are converted by the photomultiplier tubes into electrical pulses, which are, in turn, processed by electronic circuitry for imaging and recording.

Since its original inception, the design of the autofluoroscope underwent a series of modifications.[16-20] The present detector of the autofluoroscope is a mosaic of 294 activated sodium iodide crystals 1 cm square and $1^1/_2''$ thick. The crystals are arranged in a 14×21 rectangular array which is 6×9 inches in size. Each of the crystals in the array is optically coupled by plastic light guides to one of fourteen rank photomultiplier tubes and one of twenty-one file photomultiplier tubes (Figures 38, 39 and 40). Thus, a pulse occurring simultaneously in one rank and one file photomultiplier tube uniquely identifies the crystal in which the scintillation occurred. A pulse passed by a single-channel pulse-height analyzer is stored in a corresponding position in a magnetic-core memory that has one location for each of the 294 crystals. Each core position has a storage capacity of 1,000 counts. Should scintillations occur simultaneously in more than one crystal, as a result of scattered radiation, an anticoincidence system prevents the event from being recorded. At the end of the study the data can be extracted from the memory and displayed on a cathode-ray tube. The cathode-ray tube image consists of 294 squares, and their brightness is proportional to the number of counts stored in the corresponding position. Variable contrast enhancement, background subtract, or isocount display can be varied without destroying the digital data stored in the memory. The stored information can

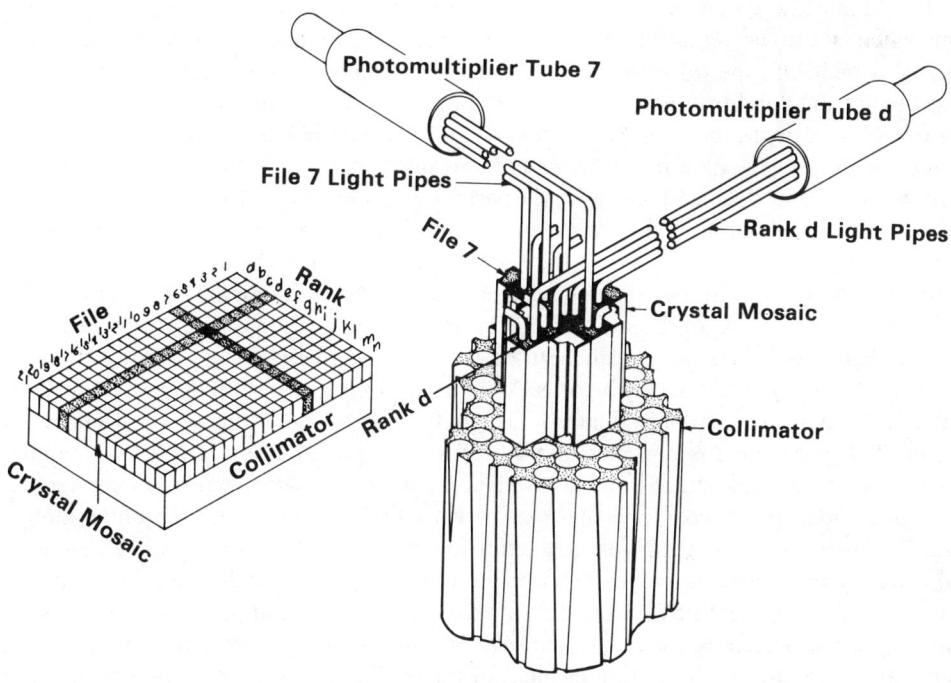

FIGURE 38.
Diagram Showing a Portion of the Imaging System Used in the Digital Autofluoroscope.
(Photograph by courtesy of Baird-Atomic, Inc.)

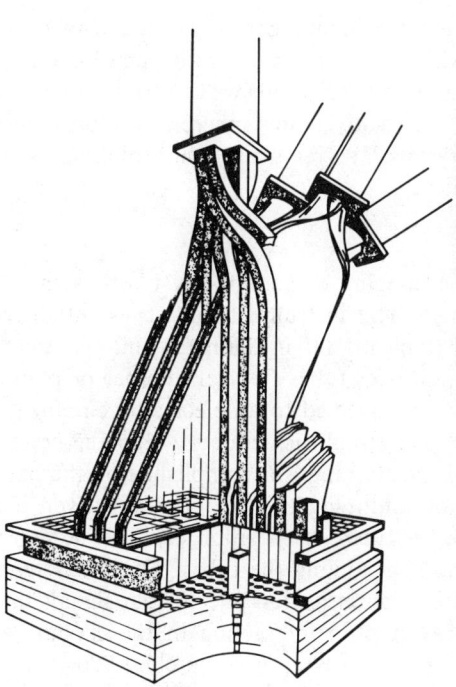

FIGURE 39.
Drawing of the Light Guides
Used in the Imaging System
of the Digital Autofluoroscope.
(Photograph by courtesy of Baird-Atomic, Inc.)

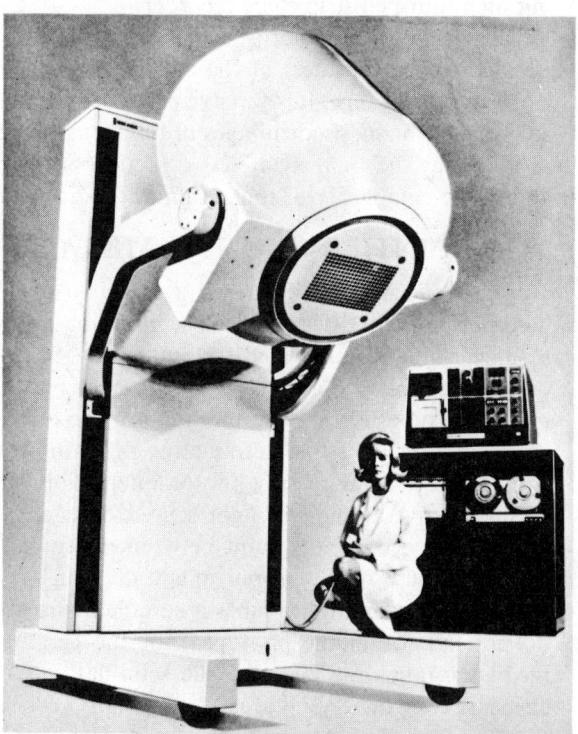

FIGURE 40.
Photograph of the Digital Autofluoroscope.
(Courtesy of Baird-Atomic, Inc.)

also be "dumped" destructively or nondestructively on incremental digital magnetic tape, from which it can be replayed for future reference and analysis. A second magnetic-core memory is built into the instrument. To correct for differences in crystal and light pipe deficiency, data from a uniform plane source of radioactivity can be stored in memory-2, while the data from the patient are stored in memory-1, and the final image can be normalized by displaying the ratio of the data in the two memories on a point-by-point system. This technique is particularly useful when correcting for the peripheral decrease in efficiency associated with the use of a pinhole collimator. The detector of the autofluoroscope can be utilized with a number of image-forming collimators. Two kinds are routinely used with the instrument. One is a multihole collimator with 294 tapered square holes, each corresponding to one crystal; the other is a pinhole collimator, used for minification or magnification of the object seen.

The digital autofluoroscope can be used for most routine clinical scanning studies, and has been extensively employed for a variety of clinical investigations[5]; it decreases the examination time by a factor of approximately 10, compared to that required by a conventional scanner. This instrument is particularly useful for the regional quantitation of dynamic processes. To perform these studies, the instrument is operated in the dynamic mode, and counts are accumulated for a preset time that can be varied from 0.03 seconds to one minute. The stored information is "dumped" on magnetic tape, the memory is cleared, and a new accumulation cycle commences. Upon completion of the study the tape is replayed, and specific anatomical sites are identified on the display. To retrieve quantitative data from the study, a memory-flagging system is used. This consists of a light-sensitive element that emits a signal when played over a given resolution element of the CRT and flags the corresponding memory position. The sum of the counts stored in all the flagged memory positions is then converted to an analog signal, which is read by a strip chart recorder. Three different areas of any size or shape can be flagged. If the tape is replayed, the accumulated counts within these areas are read out on a four-pen strip chart recorder.

The digital autofluoroscope is a device that exhibits a number of attractive features. Because of the thickness of the crystals used, it exhibits an excellent photon efficiency over a broad energy range. It is capable of processing data rapidly; thus, it is well suited for dynamic studies. The main disadvantages of this design are: 1) a relatively low inherent resolution that is limited by the 1-cm square cross section of the crystals, 2) a relatively small total-field size, and 3) the complexity of the design of this instrument.

THE SCINTILLATION CAMERA

In 1957, Anger[21] described a principle for the imaging of gamma-ray photons that is incorporated in the design of the scintillation camera.[22] This instrument consists essentially of a single cylindrical activated sodium iodide crystal (typically 11" in diameter and $1/2$" thick) optically coupled to an hexagonal array of photomultiplier tubes (typical number of photomultiplier tubes is 19). The photomultiplier tubes are connected to an electronic circuit, the purpose of which is to determine the position of a flash of light generated within the crystal subsequent to a gamma-ray photon interaction. This positioning is accomplished on the basis of the relative amount of light sensed by each photomultiplier tube (Figure 41). Even if a scintillation occurs at a point between photomultiplier tubes, its position can still be determined, because of the proportionate division of the light among the tubes. The total signal from the photomultiplier tubes is summed, and pulse-height selection rejects light signals that do not fall within the photopeak of the gamma-ray photon interacting in the crystal. At medium gamma-ray energies, the scintillation camera resolves approximately 1,000 picture elements over the crystal area.[23]

The scintillation camera can be fitted with a number of different collimators. Pinhole as well as multichannel collimators are widely used with this camera for clinical purposes.[24]

The inherent resolution of the scintillation camera detector is limited by two factors: 1) the statistical fluctuations of the light photons generated in the crystal and divided among the photomultiplier tubes, and 2) multiple scattering of high-energy photons in the scintillator. The first factor establishes a lower-light gamma-ray photon energy usable without an excessive loss of resolution; with the present scintillation camera, this limit is about 70 kev. The upper limit of the usefulness is dictated mostly by X-ray penetration of the collimator septa and by the low efficiency of the relatively thin (0.5″) crystal. The inherent resolution of the image

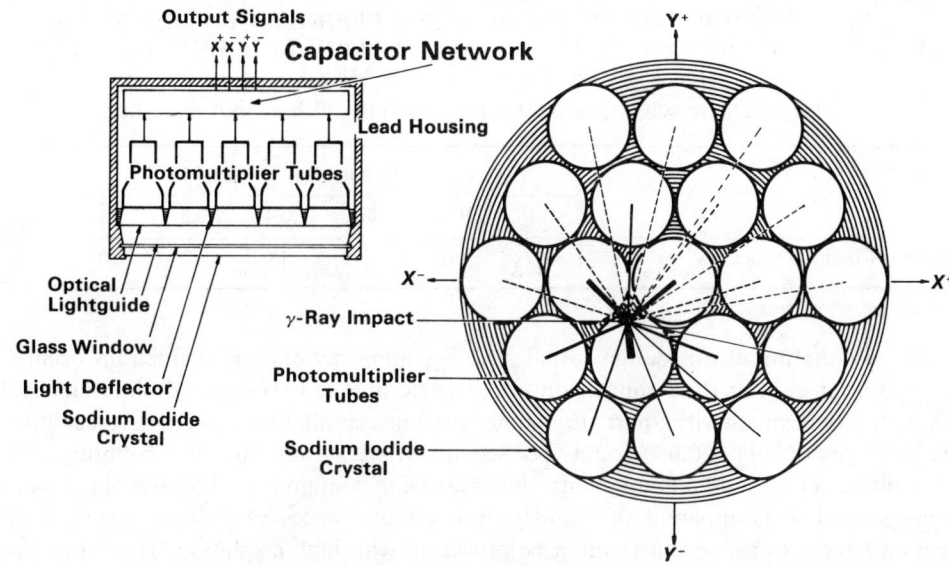

FIGURE 41.
Diagram of a Portion of the Anger Scintillation Camera.

detector was measured by Anger by moving a thin beam of gamma rays from a slit collimator across the crystal and by taking exposures of the image thus obtained; the inherent-resolution distance of the image detector is defined as the distance between two barely distinguishable double lines. The results of these studies are shown in Table 14. It is apparent that the resolution improves with photon energy and reaches an optimal value at about 360 kilo-electron volts.

Table 14.
INHERENT RESOLUTION OF THE 11″ ANGER SCINTILLATION DETECTOR CAMERA VERSUS GAMMA-RAY ENERGY*
Pulse-height selector window
set to 1.5× full width at half maximum.

	Resolution Distance, cm				
Gamma- Ray Energy	0.1 Mev	0.2 Mev	0.3 Mev	0.4 Mev	0.5 Mev
S-11 Photocathodes (Myers, Kenny & Laughlin)		1.1	0.86	0.76	0.76
Bialkali Photocathodes (Anger)	1.12	0.86	0.68	0.61	0.61

* Data adapted from Reference 1.

The scintillation camera is a very ingenious device, which has found broad applications in nuclear medicine. This design exhibits several drawbacks. The inherent resolution of the image detector is relatively poor, particularly for lower-energy gamma-ray photons; indeed, between

100 and 200 kev it is of the order of 1 cm, and it reaches an optimal value of about 6 mm for photons with an energy greater than 350 kev. While the efficiency of the relatively thin sodium iodide detector is quite adequate for lower-energy gamma-ray photons, it drops rapidly with photon energy (Table 15); thus, at 350 kev this efficiency is only 20 percent, and it drops down to about 12 percent at 500 kev.

TABLE 15.
MEASURED PHOTOPEAK COUNTING EFFICIENCY VERSUS GAMMA-RAY ENERGY FOR AN 11 × ½-INCH SODIUM IODIDE CRYSTAL IN A SCINTILLATION CAMERA.*

Pulse-height selector window set to 1.5 × full width at half maximum.

	Gamma-Ray Energy, Mev						
	0.15	0.2	0.3	0.4	0.5	0.6	0.7
Photopeak Counting Efficiency, %	77	57	29	18	12	10	9 —

* Data adapted from Reference 1.

A third limitation of this design is that all the gamma-ray photon interactions occurring in the crystal must, after their conversion into electric pulses, be processed simultaneously. This calls for electronics with short dead time. Present scintillation cameras are capable of processing approximately 50,000 counts per second. If we assume that this counting rate is distributed over a thousand picture points, this means a maximum rate of 50 events per picture point per second. It is apparent that under these circumstances very fast dynamic studies (more than 5 pictures per second) cannot be processed with high resolution. This latter situation may be remedied by the use of very fast electronics; furthermore, if the examined object is relatively small, this restriction becomes less serious, and scintillation cameras, even in the present state of the art, appear to be adequate for studies as fast as several images per second.

The scintillation camera can also be used for the imaging of positron-emitting radioactive isotopes through the use of positron collimation.[1] This is achieved by placing the object to be imaged between the scintillation camera (without a lead collimator) and a focal detector consisting either of an hexagonal array of 19 scintillation counters or of a solid sodium iodide crystal viewed by an array of seven or more photomultiplier tubes. This method of collimation is based on the simultaneous detection of the annihilation radiation photons generated when positrons are absorbed in matter. The recording of simultaneous events by means of a coincidence circuit between an event localized in the image detector crystal and an event in the focal detector permits the imaging of the source of positrons with a high sensitivity and resolution. This method of collimation permits tomographic representation of the object examined, with simultaneous read-out of several planes;[25] also, deep-lying structures are just as visible as those at the surface, because total travel of the two annihilation photons through tissues is the same in both cases. On the other hand, the scintillation camera used as a positron camera overloads easily, because a high number of events must be processed in a short period of time by the image detector circuits; thus, long exposure times are needed. Furthermore, tissue absorption and scattering are relatively great in thick areas of the body, because both annihilation photons must be detected, to be recorded. To date, positron collimation has not been extensively used clinically.

THE IMAGE INTENSIFIER RADIOISOTOPE CAMERA (THE MAGNACAMERA)

The possibility of using an image intensifier tube for the imaging of gamma-ray photons was suggested by a number of authors.[26-29] The proposed designs, however, did not result in the construction of useful radioactive isotope cameras, mostly because the signal-to-noise ratio in the image intensifiers used was so unfavorable that the imaging of objects containing radioactive isotopes could be achieved only with quantities of radioactivity prohibitively high for applications in nuclear medicine. In 1962, Ter-Pogossian et al.[30] described an isotope camera that was based on the use of a high-gain, low-noise image intensifier tube and was able to visualize a thyroid phantom containing about one microcurie of ^{125}I per cubic centimeter in approximately 3 minutes. This instrument was capable of visualizing single gamma-ray photons. After a series of improvements,[31] a clinically useful camera based on this principle was developed under the name of Magnacamera.*

The Magnacamera detector consists essentially of a high-gain (approximately 50,000), low-noise image intensifier tube containing a cesium iodide input crystal (Figure 42). The output screen of the image intensifier is optically coupled to an intensifier orthicon television pick-up tube. The video signal from the television tube is displayed on two television monitors for visual observation and for recording by means of a camera, and it is also simultaneously recorded on video tape (Figure 43).

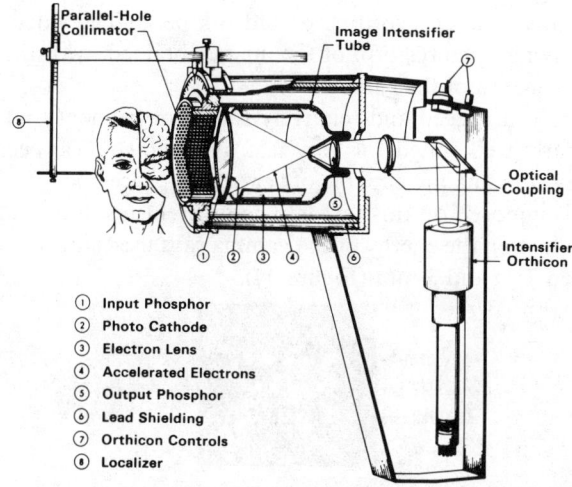

FIGURE 42.
Diagram of the Detector Head of the Magnacamera.
(Courtesy of Picker X-Ray Company)

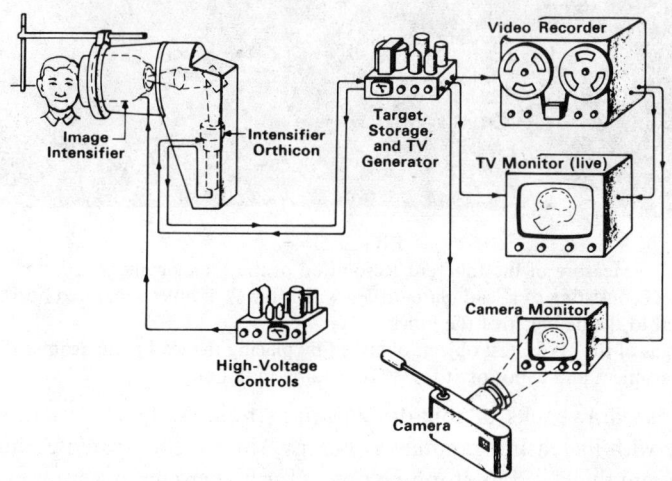

FIGURE 43.
Block Diagram of the Components of the Magnacamera.
(Courtesy of Picker X-Ray Company)

* Picker X-Ray Corporation.

The image-detecting system functions as follows. The gamma-ray image formed by a collimator (either multichannel or pinhole) impinges upon the cesium iodide input crystal contained in the vacuum envelope of the image intensifier tube, and is thus converted into a light image. The cesium iodide crystal is in intimate contact with a photoelectric surface which converts the light image into electrons. These electrons are accelerated in the image intensifier tube and directed by electron optics to form a minified image on the output fluorescent screen, which converts the electron image into a light image of greater energy density than the original gamma-ray image. This light image is further intensified in brightness and converted into a video signal in the television pick-up tube. From then on it is handled as a video signal.

The coupling of the image intensifier tube with a television intensifier tube accomplishes a number of purposes. 1) The intensifier television tube provides additional image intensification; thus the image intensifier acts as a low-noise, high-signal-to-noise-ratio preamplifier coupled to the high-gain intensifier orthicon tube, which functions best at the higher light levels provided by the image intensifier preamplification; the orthicon target acts as a light storage device that integrates information between readouts. 2) The conversion of the image into a video signal allows its easy recording on video tape. 3) The video signal thus obtained can be used for the regional quantization of the image obtained, with the purpose of carrying out dynamic studies by means of the camera.[32]

The Magnacamera is a rugged and relatively simple instrument that requires little maintenance. It is particularly well suited for very fast dynamic studies, because both the image intensifier and the television link are capable of very rapid processing of information. Thus, no electronic restriction is imposed on this factor. The inherent resolution of this system is quite high, and is not dependent on the energy of the gamma rays used; indeed, the system is capable of a resolution between $2^1/_2$ and 3 mm (Figure 44).

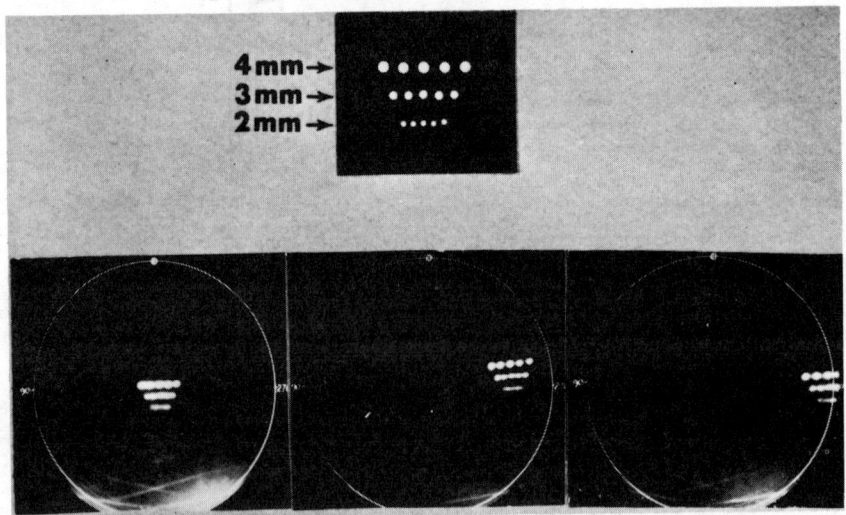

FIGURE 44.
Measure of the Inherent Resolution of the Magnacamera.
Top: test object consisting of a lead plate drilled with a series of holes separated horizontally by a distance equal to the diameter of the holes.
Bottom: images of the above test object, obtained by placing the lead plate against the input face of the Magnacamera and exposing it to ^{99m}Tc gamma radiation.

The most serious drawbacks of this design are as follows. 1) The usefulness of this unit decreases rapidly with increasing gamma-ray energy, since the activated cesium iodide crystal is only about 1.5 mm thick and its stopping power for higher-energy gamma rays is low; thus, the energy of the detector for 300-kev gamma rays is only 10 percent (Figure 45). 2) The unit is incapable of pulse-height discrimination. 3) The field size is limited to a little over 20 cm by the fact that larger image intensifier tubes are difficult to build.

Another approach to the use of image intensifiers for isotope imaging consists in the use of a light intensifier tube optically coupled to an external crystal. This approach has been studied by a number of authors. It is appealing in a number of respects, particularly because it allows the use of larger and thicker crystals, giving a broader field of acceptance and a higher efficiency for gamma-ray collection. Also, the light intensifier tubes presumably would be cheaper than the special nuclear tubes required in the Magnacamera. Such a design would also facilitate pulse-height analysis. The most serious weakness of this approach lies in the optical coupling between the crystal and the input screen of the light intensifier tube. Indeed, an inadequately efficient coupling at this stage brings the quantum sink of the detecting system to such a low value that single events in the crystal can no longer be recorded. To date, this approach has not yielded a clinically useful radio-isotope camera; however, efficient coupling between a crystal and an image intensifier tube may render this design practical.

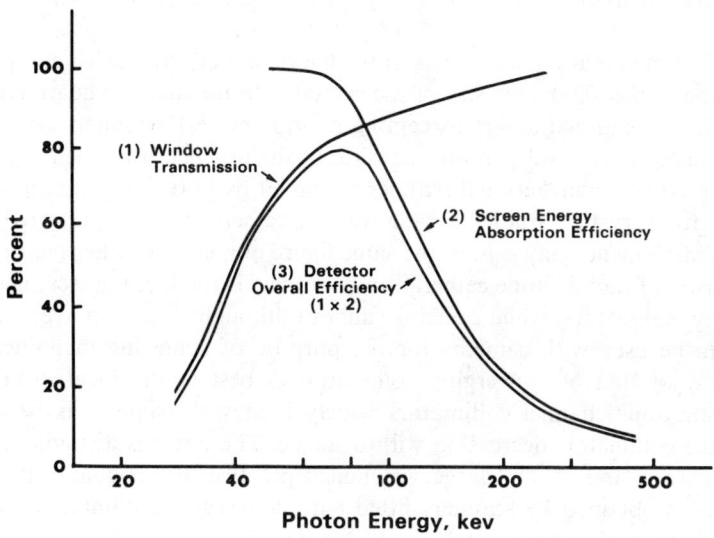

FIGURE 45.
Detector Efficiency of the Magnacamera vs. Gamma-Ray Energy.

COMPARISON OF RADIOISOTOPE CAMERAS

It is perhaps wise to emphasize that a close comparison between the different radioisotope cameras described in this chapter is particularly difficult. Indeed, the differences that exist in the desirable features and in the shortcomings of these devices, with respect to their usefulness in nuclear medicine, are so profound that these cameras should be considered as different instruments rather than as different designs of the same instrument. Of course, in many areas the applications of these cameras overlap; nevertheless, their specific applications are as numerous as their common ones. It is noteworthy that all the cameras use fairly similar image-forming collimators; with the exception of spark chamber imaging devices, the detection of gamma-ray photons is achieved through the use of scintillating crystals.

The possibility of using solid-state semiconductor detectors in radioisotope cameras is presently under study.[15] Such a detector may consist of a mosaic of semiconductor detectors connected to a suitable read-out system. At present the detection efficiency of the largest available germanium solid-state detectors is approximately 30 percent for 300-kev gamma rays, and the photopeak detection efficiency is only 5 percent for the same radiation.[15] Another disadvantage of these detectors is that, for best pulse-height resolution, they must be operated at liquid-nitrogen temperatures. While the incorporation of such detectors in radioisotope cameras appears to be promising, practical models of these cameras have not yet been developed.

RADIOISOTOPE CAMERAS AND SCANNERS

The main reason for using radioisotope cameras in nuclear medicine is to reduce the time of examinations normally carried out by scanners without affecting the clinical value of the examination. The degree of success achieved in pursuing this goal is difficult to assess, because a large number of parameters must enter into the comparison of cameras and scanners. The obvious advantage of cameras over scanners is that a camera detector observes the field under examination continuously, while the detector of a scanner, at a given time, observes only a small portion of the field, thus wasting the photons emitted outside its field of view. In fact, the multichannel converging collimators that are widely used with scintiscanners subtend a considerably larger solid angle of photon acceptance than each channel of a camera collimator of similar resolution. Taking this factor into account, the best method for the comparison of a camera and a scanner is supplied by their relative figures of merit, obtained by multiplying the total detector area by its efficiency in detecting the gamma radiation for which the comparison is carried out.

It should be noted that the usefulness of the figure of merit for such a comparison exhibits a number of limitations. Thus, the size of the crystal detector that can be practically used in a scintiscanner of conventional design (excepting hybrid devices) is limited, because a too-large solid angle of acceptance results in an undesirably shallow depth of focusing. Furthermore, the scanning speed of a conventional scanner is limited by purely mechanical considerations. Consequently, for very fast studies a radioisotope camera may be preferred to a scanner, although both instruments may exhibit the same figure of merit. Another factor of importance in the comparison of radioisotope cameras and scanners is the fact that scanners can be fitted with converging collimators, while cameras cannot (although slightly converging or divergent collimators can be used with cameras for the purpose of achieving magnification or minification). The resolution of converging collimators is best at the focal distance, while the resolution of the multichannel collimators widely used with cameras is best in immediate contact with the collimator, decreasing with distance. The latter is a fundamental deficiency of these collimators, and by and large, in clinical practice, better-quality imaging of deep-lying structures is obtained by scanners fitted with converging collimators than by cameras fitted with multichannel collimators.

While any comparison between scanners and radioisotope cameras is valid only for a given instrument, for a given gamma-ray photon energy, and for a given clinical examination, a certain number of general comments can be made. Most cameras are faster for low-energy gamma-ray photons (under 350 kev), while scanners—and particularly hybrid devices—are at least as fast, if not faster, than most cameras for higher energies. Because of the possible use of converging collimators, scanners, at any gamma-ray photon energy, are capable of better resolution than cameras for deep-seated lesions. On the other hand, scanners cannot be practically used for very fast studies, no matter how large and efficient their detector, because of the difficulties encountered in rapidly moving the heavy detector. Thus, cameras are best suited for dynamic studies that must be carried out in short periods of time.

In conclusion, the development of radioisotope cameras has added a very useful instrument to the nuclear-medicine armamentarium. However, at this stage of development of these instruments it does not appear that cameras are replacing scanners; they complement their use by allowing fast dynamic studies at the cost of the resolution obtained.

ACKNOWLEDGEMENTS

Much of the material in this section was contributed by Mr. Hal O. Anger, Dr. Merrill A. Bender and Dr. Norman H. Horwitz, and many of the sentences in this text are paraphrased or directly transposed from their work. The author is very grateful to these investigators for their willingness in contributing to this section most of the descriptive material about the instruments they invented, creating thus the field under discussion.

REFERENCES

1. Anger, H. O., Sensitivity, Resolution, and Linearity of the Scintillation Camera. *IEEE Trans. Nucl. Sci.*, NS-13(1-3): 380-392, 1966.
2. Anger, H. O., Tomographic Gamma-Ray Scanner with Simultaneous Readout of Several Planes. *UCRL-16899 Rev.*, University of California.
3. International Atomic Energy Agency, *Medical Radioisotope Scanning*, Vol. 1. International Atomic Energy Agency, Vienna, Austria, 1964.
4. Keuffel, J. W. *Phys. Rev.* 73: 531, 1948.
5. Bella, F. and Franzinetti, C. *Nuov. Cimento*, 10: 1461, 1953.
6. Kellershohn, C., Lansiart, A., and Desgrez, A. *Medical Radioisotope Scanning*, 1:333. International Atomic Energy Agency, Vienna, Austria, 1964.
7. Lansiart, A. and Kellershohn, C. *1st International Symposium on Medical Physics, Harrogate, England, 1965*.
8. Pullan, B. R. and Perry, B. T. *1st International Conference on Medical Physics, Harrogate, England, 1965*.
9. Pullan, B. R. and Perry, B. T. *Nucleonics*, 24 (3): 59, 1966.
10. Horwitz, N. H., Lofstrom, J. E., and Forsaith, A. L. *Radiology*, 84: 125, 1965.
11. Horwitz, N. H., Lofstrom, J. E., and Forsaith, A. L. *J. Nucl. Med.*, 6: 724, 1965.
12. Forsaith, A. L., Horwitz, N. H., and Lofstrom, J. E. *Phys. Med. Biol.*, 11: 411, 1966.
13. Widman, J. C. *Med. Biol.* (England), in press.
14. Lansiart, A. and Kellershohn, C., Spark Chambers in Nuclear Medicine. *Nucleonics*, 24(3): 56, 1966.
15. Anger, H. O. *I.S.A. Trans.*, 5: 311, 1966.
16. Bender, M. A. and Blau, M., Autofluoroscopy: The Use of a Non-Scanning Device for Tumor Localization with Radioisotopes. *J. Nucl. Med.*, 1: 105, 1960.
17. Bender, M. A. and Blau, M., The Autofluorscope. *Progress in Medical Radioisotope Scanning*, p. 151, Kniseley, R. M. et al., eds. Oak Ridge Institute of Nuclear Studies, Oak Ridge, Tennessee, 1962.
18. Bender, M. A. and Blau, M., The Clinical Use of the Autofluoroscope. *J. Nucl. Med.*, 3: 202, 1962.
19. Bender, M. A. and Blau M., The Autofluoroscope. *Nucleonics*, 21: 52, 1963.
20. Bender, M. A., The Digital Autofluoroscope. *Medical Radioisotope Scanning*, 1: 391. Proceedings of the Symposium on Medical Radioisotope Scanning, Athens, April 20-24, 1964. International Atomic Energy Agency, Vienna, Austria, 1964.
21. Anger, H. O., A New Instrument for Mapping Gamma-Ray Emitters. *Biology and Medicine Quarterly Report, UCRL-3653*, 1957.
22. Anger, H. O., Scintillation Camera. *Rev. Sci. Instr.*, 29: 27, 1958.
23. Anger, H. O., Scintillation Camera with 11-Inch Crystal. *Donner Laboratory Semi-Annual Report, UCRL-11184*: 69, 1963.
24. Anger, H. O. and VanDyke, D. C., Human Bone Marrow Distribution Shown *In Vivo* by Iron-52 and the Positron Scintillation Camera. *Science*, 144: 1587, 1964.
25. Anger, H. O., Gamma-Ray and Positron Scintillation Camera. *Nucleonics*, 21(10): 56, 1963.
26. Mortimer, R. K., Anger, H. O., and Tobias, C. A., The Gamma-Ray Pinhole Camera with Image Amplifier. *University of California Radiation Laboratory Report UCRL-2624*, 1954.
27. Mortimer, R. K., Anger, H. O., and Tobias, C. A., The Gamma-Ray Pinhole Camera with Image Amplifier. *Convention Record of the Institute of Radio Engineers*. National Convention, The Institute of Radio Engineers, Part 9, pp. 2-5, New York, 1954.
28. Kellershohn, C. and Pellerin, P., Sur la Possibilité d'Utiliser un Amplification d'Image pour Mettre en Evidencede la Localization et la Distribution d'un Corps Radioactif. *Comptes Rendues des Séances de la Société de Biologie*, 149: 533, 1955.
29. Aronow, S., Burnham, C., and Chomichi, O., A Positron Camera Using Image Intensification. *Meeting of the Society of Nuclear Medicine, California, 1964*.
30. Ter-Pogossian, M., Kastner, J., and Vest, T. B., Autofluorography of the Thyroid Gland by Means of Image Amplification. (Presented at the Meeting of the RSNA, 1962.) *Radiology*, 81 (6): 984, 1963.
31. Ter-Pogossian, M., Niklas, W. F., Ball, J., and Eichling, J. O., An Image Tube Scintillation Camera for Use with Radioactive Isotopes Emitting Low-Energy Photons. *Radiology*, 86(3): 463, 1966.
32. Ashburn, W. L., Harbert, J. C., Whitehouse, W. C., and Murphy, R. E., Image Recording and Integration System for Dynamic Radioisotope Studies Based on Advanced Television Circuitry. *Annual Meeting of the Society of Nuclear Medicine, St. Louis, Missouri, 1968*.

PART IV
RADIATION DOSIMETRY

RADIATION ABSORBED-DOSE CALCULATIONS FOR BIOLOGICALLY DISTRIBUTED RADIONUCLIDES

Edward M. Smith, D.Sc.
University of Miami School of Medicine, Division of Nuclear Medicine

INTRODUCTION

Absorbed-dose calculations are made to estimate the quantity of energy deposited in tissues of the body from intentionally or unintentionally administered radionuclides. These calculations are necessary for evaluating the radiation risk from chronic exposure to radioactive debris and from accidental ingestion of radioactive materials, as well as for medical uses of radiopharmaceuticals. The symbols, units, and methods of calculating the absorbed dose will follow the recommendations of the Medical Internal Radiation Dose (MIRD) Committee of the Society of Nuclear Medicine.[1]

Factors that affect the absorbed dose include: the activity of the administered radionuclide (radiopharmaceutical), the route of administration, the chemical and physical state of the radionuclide, the energy released per disintegration, and the metabolic fate of the radionuclide. When radioactive materials accidentally enter the body, the above factors are usually known with very little accuracy, and must be approximated, or else evaluated by reconstructing the incident. However, for the radiopharmaceuticals used in nuclear medicine all but the last factor—the metabolic fate of the administered radiopharmaceutical—are known with a moderate degree of accuracy. The metabolic fate of the radiopharmaceutical is a biological factor, and consequently is associated with variability that usually can be evaluated only from limited data on experimental animals and a small number of human beings under various physiological and pathological conditions. These data must be obtained by the individual investigator, or he must utilize data existing in the literature; if so, he must satisfy himself that these data are compatible with the situation at hand.

The multiplicity of factors that must be taken into account when calculating the absorbed dose decreases the precision and accuracy of the calculation. For most situations the calculated absorbed dose is only an estimate, and it is advisable to determine an upper and lower value for this estimate, based on the best available data.

The equations for calculating the absorbed dose given in this chapter are based on a paper by Loevinger and Berman,[2] and form the basic framework for the formalism recommended by the MIRD Committee.[3] This scheme for absorbed-dose calculation of biologically distributed radionuclides was developed because of a need for a flexible method applicable to any type of radiation, eliminating the dichotomy of particulate and electromagnetic radiations, as well as for any specified biological model, whether it be for a uniform distribution in the total body or for a nonhomogeneous distribution of the radionuclide at the cellular level.

A detailed discussion, with examples of medical internal radiation absorbed-dose calculations, is given by Smith et al.[4]

UNITS

In Report 10a, Radiation Quantities and Units,[5] the following definitions are recommended:

rad = unit of absorbed dose
  The absorbed dose (D) is the energy deposited by ionizing radiation matter.
  One rad equals 100 ergs per gram of matter.

curie = unit of activity
  One curie (Ci) is equal to 3.7×10^{10} disintegrations per second.

The biological effects of ionizing radiation depend on the energy absorbed and on various modifying factors. The *dose equivalent* unit is the "rem". The dose in rems is equal to the absorbed dose in rads multiplied by the modifying factors related to tissue sensitivity, biological endpoint and quality of the radiation. The use of this unit is described in the report of the Relative Biological Effectiveness (RBE) Committee.[6] The dose equivalent in rems and the absorbed dose in rads are numerically identical for most clinically useful radionuclides, but this is not necessarily so for radionuclides that may be introduced into the body as the result of chronic environmental exposure or by accidental ingestion.

MAXIMUM PERMISSIBLE DOSE

The maximum permissible dose (MPD), originally developed as a guideline for occupational exposure, may be used to evaluate the significance of the absorbed dose in diagnostic procedures. The dose limitations recommended[7,8] are concerned entirely with exposures other than those received by an individual in the course of medical procedures. In designating the MPD, organs in the body are divided into four groups. All organs within each group have the same maximum permissible dose.

Gonads, red bone marrow (and, in the case of uniform radiation, the whole body)	5 rems in a year
Skin, thyroid, bone	30 rems in a year
Hands and forearms, feet and ankles	75 rems in a year
All other organs	15 rems in a year

One-half the annual limit is permitted within a period of three months.

QUANTITIES, SYMBOLS, AND NOTATION

Table 1 give the various quantities, along with their symbols and units used in this formalism. The equations as presented yield the absorbed dose in rads, if the source activity is given in microcuries and time is given in hours. Other units may be substituted, if the appropriate change is made in the numerical value and in the units of the equilibrium of absorbed-dose constant.

A bar over a symbol denotes a mean value; for example, $\overline{D}$ is the mean absorbed dose. A *tilde* over a symbol or the word "cumulated" denotes a time integral; for example, $\tilde{A} = \int A dt$ is the cumulated activity. Quantities that are specifically related to a certain type of radiation are so specified by a right subscript, for example, ϕ_i is the absorbed fraction due to *i*-type radiation.

Certain quantities depend on source and target regions, and since, in general, it is necessary to distinguish between source and target, the dependence is shown in the form $r_1 \leftarrow r_2$, where r_1 is the target and r_2 is the source region; for example, $\Phi_i(r_1 \leftarrow r_2)$. This is read "the specific absorbed fraction in r_1 from r_2". When a quantity is a symmetrical function of source and target regions, this is shown by a double arrow, for example, $\Phi_i(r_1 \leftrightarrow r_2)$, which can be read "the specific absorbed fraction in r_1 from r_2 or in r_2 from r_1". When the quantity is a function of only one variable, in this case distance, it is shown in the conventional manner, $\Phi_i(x)$.

Table 1.

PRINCIPAL QUANTITIES AND SUBSCRIPTS WITH SYMBOLS AND UNITS

Functions of Radiation Type Only
Δ	(g-rad/μCi-hr)	Equilibrium dose constant
n	(particles or photons per disintegration)	Mean number of particles or photons per disintegration
$\bar{E}$	(Mev/particle or photon)	Mean energy per particle or photon
λ	(hr^{-1})	Decay constant
T	(hours)	Half-life or half-time

Functions of Radiation Type and Absorbing Material
$\phi(x)$	(g^{-1})	Point isotropic specific absorbed fraction
$B_{en}(\mu x)$	(dimensionless)	Point isotropic energy-transfer build-up factor
μ_{en}	(cm^{-1})	Linear energy-absorption coefficient
μ	(cm^{-1})	Linear attenuation coefficient

Functions of Radiation Type, Absorbing Material, and Source and Target
D	(rad)	Absorbed dose
R	(rad/hr)	Absorbed-dose rate
ϕ	(dimensionless)	Absorbed fraction
Φ	(g^{-1})	Specific absorbed fraction
g	(cm)	Geometrical factor

Subscripts Specifying Radiation Type
i	Symbol for all types, used with summation sign
n–p	Nonpenetrating
ph	Photon
$\alpha, \beta, \gamma,$ X, etc.	Symbols for specific radiation types

Quantities Characterizing Source and Target
A	(μCi)	Activity
C	(μCi/g)	Concentration
q	(μCi)	Distribution function (activity corrected for physical decay)
m	(g)	Mass
v	(cm^3)	Volume
r	(cm^i, i = 0, 1, 2, or 3)	Region, i.e., point, line, surface, or volume
v, r, and p	(for point)	Subscripts

Other Quantities
x	(cm)	Distance
ρ	(g/cm^3)	Mass density
t	(hours)	Time

From: Loevinger, R. and Berman, M., A Schema for Absorbed-Dose Calculations for Biologically Distributed Radionuclides. *J. Nucl. Med.*, Suppl. 1: 10, 1968. (Courtesy of the Society of Nuclear Medicine.)

EQUATIONS FOR ABSORBED-DOSE CALCULATIONS

The unifying feature of this formalism is the use of the quantity *absorbed fraction*, first introduced by Ellett, Calahan and Brownell,[9,10] and an extension of this concept, *specific absorbed fraction*.[2] The absorbed fraction is simply the fraction of the emitted energy that is absorbed in the region of interest.

$$\phi = \frac{\text{energy absorbed by target}}{\text{energy emitted by source}}. \tag{1}$$

Defining ϕ symbolically,

$$\phi_i(v \leftarrow r) = \frac{m_v \bar{D}_i(v \leftarrow r)}{\tilde{A}_r \Delta_i}, \tag{2}$$

where

$\phi_i(v \leftarrow r) = \dfrac{\text{energy from the } i^{\text{th}} \text{ type of radiation in source } r, \text{ absorbed in target volume } v}{\text{energy of the } i^{\text{th}} \text{ type of radiation emitted by source region } r}$,

m_v = mass of volume v, in grams,

$\bar{D}_i(v \leftarrow r)$ = mean absorbed dose to the volume v, from the source, r, emitting i^{th} type radiations,

$\tilde{A}_r$ = cumulative activity of the region r, µCi-hr,

Δ_i = equilibrium absorbed-dose constant for radiations of type $i = 1, 2, 3, \ldots$, with a fractional frequency n_i per disintegration and a mean energy $n_i \bar{E}_i$ Mev per disintegration,

Δ_i = 2.13 $n_i \bar{E}_i$ g-rad/µCi-hr.

Equation (2) has the limitation that the target region must be a volume. This limitation is circumvented by defining the specific absorbed fraction, Φ, that is, the absorbed fraction per unit mass of target.

$$\Phi_i(v \leftarrow r) = \dfrac{\phi_i(v \leftarrow r)}{m_v} \, \text{g}^{-1}. \qquad (3)$$

Generally, the specific absorbed fraction in any target region r_1 from any source region r_2 is defined by a limiting process,

$$\Phi_i(r_1 \leftarrow r_2) = \lim_{v \to r_1} \Phi(v \leftarrow r_2) = \dfrac{\bar{D}_i(r_1 \leftarrow r_2)}{\tilde{A}_2 \Delta_i}, \qquad (4)$$

where v is a volume containing the region r_1. This equation is subject to the limitation that the two regions must have no points in common unless at least one is a volume.

The general equation for the calculation of the mean absorbed dose is:

$$\bar{D}(r_1 \leftarrow r_2) = \tilde{A}_2 \sum_i \Delta_i \Phi_i(r_1 \leftarrow r_2) \text{rad}. \qquad (5)$$

If the dose-reciprocity theorem[2,3] is valid for the situation in question, then equation (5) may be written in the form:

$$\bar{D}(r_1 \leftarrow r_2) = \tilde{A}_2 \sum_i \Delta_i \Phi_i(r_1 \leftrightarrow r_2) \text{rad}. \qquad (5a)$$

A very important practical case occurs if the target region r_1 and the source region r_2 are the same volume v. The basic absorbed-dose equation then takes on a simplified form, giving the mean self-dose to the volume v:

$$\bar{D}(v \leftarrow v) = \tilde{C}_v \sum_i \Delta_i \phi_i(v \leftarrow v) \text{rad}, \qquad (6)$$

where $\tilde{C}_v = \tilde{A}_v/m_v$, and $\phi_i = m_v \Phi_i$.

Another important practical case has to do with the absorbed dose to a volume from a point, which can be given in the form

$$\bar{D}(v \leftarrow p) = \dfrac{\tilde{A}_p}{m_v} \sum_i \Delta_i \phi_i(v \leftarrow p) \text{rad}. \qquad (7)$$

If the dose-reciprocity theorem is valid, the dose to a point from a volume is given by the equation

$$\bar{D}(p \leftarrow v) = \tilde{C}_v \sum_i \Delta_i \phi_i(v \leftarrow p) \text{rad}. \qquad (8)$$

THE DOSE-RECIPROCITY THEOREM AND THE UNIFORM ISOTROPIC MODEL

The name *uniform isotropic model*[2,3] is used here to mean that the source and target are regions in a homogeneous absorbing material sufficiently large so that edge effects are negligible and the activity is uniformly distributed in the source region. For such a model, there exist certain reciprocal relationships between any pair of regions.

1. The specific absorbed fraction is independent of which region is designated as source and which as target, or in symbols,

$$\Phi_i(r_1 \leftarrow r_2) = \Phi_i(r_2 \leftarrow r_1) \equiv \Phi_i(r_1 \leftrightarrow r_2) g^{-1} \tag{9a}$$

and

$$\frac{\phi_i(v_1 \leftarrow v_2)}{m_1} = \frac{\phi_i(v_2 \leftarrow v_1)}{m_2} g^{-1}. \tag{9b}$$

2. The mean absorbed dose per µCi-hr is independent of which region is designated as source and which as target, or in symbols,

$$\frac{\bar{D}(r_1 \leftarrow r_2)}{\tilde{A}_2} = \frac{\bar{D}(r_2 \leftarrow r_1)}{\tilde{A}_1} \text{ rad/µCi-hr}. \tag{9c}$$

Dose may be changed to dose-rate, provided cumulated activity is changed to activity (see Table 2).

Table 2.
CONVERSION BETWEEN DOSE AND DOSE-RATE SYSTEMS OF EQUATIONS

Dose-Rate Equations			Dose Equations		
Quantity	Units	Symbol	Symbol	Units	Quantity
Absorbed dose rate	rad/hr	R	D	rad	Absorbed dose
Activity	µCi	A	Ã	µCi-hr	Cumulated activity
Concentration	µCi/g	C	Č	µCi-hr/g	Cumulated concentration

From: Loevinger, R. and Berman, M., A Schema for Absorbed-Dose Calculations for Biologically Distributed Radionuclides. *J. Nucl. Med.*, Suppl. 1: 14, 1968. (Courtesy of the Society of Nuclear Medicine.)

The reciprocity equations and the dose equations, which include the validity of the dose-reciprocity theorem, do not require exponential absorption of the radiation, but are valid in the presence of scattered radiation, provided the conditions of the uniform isotropic model are met. Only limited experimental or theoretical information is available about the validity of the dose-reciprocity theorem in practice.[10] It does, however, seem probable that under most circumstances of interest in medical internal absorbed-dose computations the dose-reciprocity theorem can be used without substantial error.

EQUILIBRIUM ABSORBED-DOSE CONSTANT

The equilibrium absorbed-dose constant, Δ, is the total energy emitted by a radionuclide per disintegration.

$$\Delta = \sum_i \Delta_i = \sum_i n_i \bar{E}_i \text{ Mev/disintegration}, \tag{10a}$$

$$\Delta_i = 2.13 \, n_i \bar{E}_i \text{ g-rad/µCi-hr}, \tag{10b}$$

$$\Delta_i = 51.12 \, n_i \bar{E}_i \text{ g-rad/µCi-d}. \tag{10c}$$

The equilibrium absorbed-dose constant represents the amount of energy that would be deposited in an absorber if absorption were complete. It is necessary to compute Δ_i separately for each type and energy of radiation emitted by the radionuclide, since the absorbed fraction depends upon both these factors. Dillman[11] has developed a computer code to compute the parameters needed to calculate the equilibrium absorbed-dose constant directly from the Nuclear Data Sheets.

Beta-Minus and Beta-Plus Particles

If values of $\bar{E}_{\beta-}$ or $\bar{E}_{\beta+}$ are not available, and only the end point energy of the beta spectra is available, the mean energy may be obtained graphically, utilizing the work of Loevinger[12] or Dillman.[13] If these graphs are not available, the following two equations may be used, to approximate the mean beta-particle energy:[14]

$$n_{\beta-} \bar{E}_{\beta-} = 0.33 \, n_{\beta-} E_o \left(1 - \frac{Z^{1/2}}{50}\right)\left(1 + \frac{E_o^{1/2}}{4}\right) \text{ Mev/disintegration} \quad (11a)$$

and

$$n_{\beta+} \bar{E}_{\beta+} = 0.33 \, n_{\beta+} E_o \left(1 + \frac{E_o^{1/2}}{4}\right) \text{ Mev/disintegration}, \quad (11b)$$

where

Z = atomic number,

E_o = maximum beta-particle energy (Mev)

Although equations (7a) and (7b) are usually accurate to within five percent, in some cases the beta-particle spectrum has an atypical shape that results in a significant error in the calculated value.[15]

Electron Capture

When a radionuclide decays by electron capture, X rays and Auger electrons are generated. Smith et al.[16] have developed equations for the calculation of the mean energy per disintegration associated with nonpenetrating types of radiation resulting from electron capture and internal conversion, which is an extension of earlier work by Loevinger.[15] The computer code developed by Dillman[11,13] extends the work of Smith et al. for electron capture by computing theoretical K/L/M capture ratios when experimental values are not available from the Nuclear Data Sheets. In addition, it removes the limitation that photons of an energy equal to or less than 0.0113 Mev are considered as nonpenetrating radiations. This division at 0.0113 Mev is based on the fact that photons of this energy lose 95 percent of their energy in water within 10 mm of the site of their emission. This is approximately the equivalent of the range of beta particles from ^{32}P in water.

The following equations are taken from Smith et al.[4, 16]

$$n_\varepsilon \bar{E}_\varepsilon = n_{\varepsilon_K} \omega_K \left[\left(\frac{K_\alpha}{K_\alpha + K_\beta}\right)E_{L_{II-III}} + \left(\frac{K_\beta}{K_\alpha + K_\beta}\right)E_{M_{II-III}}\right]$$
$$+ n_{\varepsilon_K}(1 - \omega_K)E_K + n_{\varepsilon_{L_I}} E_{L_I} + n_{\varepsilon_{L_{II}}} E_{L_{II}} + n_{\varepsilon_{M_I}} E_{M_I} \text{ Mev/disintegration}, \quad (12)$$

where

$n_\varepsilon \bar{E}_\varepsilon$ = mean local energy deposited per disintegration due to electron capture and subsequent products (Mev per disintegration),

$n_{\varepsilon_K}, n_{\varepsilon_{L_I}}, n_{\varepsilon_{L_{II}}}, n_{\varepsilon_{M_I}}$ = fractional frequency of K, L_I, L_{II}, and M_I electron capture per disintegration respectively,

$E_{L_{II-III}}$ = average binding energy of L_{II} and L_{III} electrons (Mev),

$E_{M_{II-III}}$ = average binding energy of M_{II} and M_{III} electrons (Mev),

$E_{L_I}, E_{L_{II}}, E_{M_I}$ = binding energies of the L_I, L_{II}, and M_I electrons (Mev) respectively,

$\dfrac{K_\alpha}{K_\alpha + K_\beta}$ = relative abundance of K_α X rays,

$\dfrac{K_\beta}{K_\alpha + K_\beta}$ = relative abundance of K_β X rays,

ω_K = K-shell fluorescent yield.

The values of 0.75 and 0.25 may be used for the relative abundance of K_α and K_β X rays respectively with reasonable accuracy for elements with atomic numbers between 35 and 82. More accurate values may be found in nuclear spectroscopy tables.[17] Table 3 gives values for the K-shell fluorescent yield, and Table 4 gives values for electron-binding energies and energies of characteristic X rays for various elements. Some radionuclides decay by electron capture to more than one energy state, which is indicated in the capture-branching ratios; each transition should be calculated separately. Then,

$$n_\varepsilon \bar{E}_\varepsilon = n_{\varepsilon_1} \bar{E}_{\varepsilon_1} + n_{\varepsilon_2} \bar{E}_{\varepsilon_2} + \cdots \text{ Mev/disintegration.} \quad (12a)$$

Table 3. K-SHELL FLUORESCENT YIELD (ω_K)

Element	Atomic No.	ω_K	Element	Atomic No.	ω_K
O	8	0.005	Zr	40	0.735
Ne	10	0.02	Mo	42	0.775
Mg	12	0.035	Rh	45	0.81
Si	14	0.06	Ag	47	0.835
S	16	0.09	Sb	51	0.865
Ar	18	0.13	Ba	56	0.90
Ca	20	0.17	Nd	60	0.915
Sc	21	0.195	Tb	65	0.93
V	23	0.24	Tm	69	0.94
Mn	25	0.29	Ta	73	0.95
Co	27	0.345	Re	75	0.95
Cu	29	0.40	Au	79	0.96
Ga	31	0.46	Bi	83	0.965
As	33	0.535	Fr	87	0.965
Kr	36	0.64	Th	90	0.965
Sr	38	0.695			

From: Fink, R. W., Jopson, R. C., Mark, H., and Swift, C. D., Atomic Fluorescence Yields. *Rev. Mod. Phys.*, 38: 513, 1966.

Internal Conversion

When internal conversion occurs, monoenergetic conversion electrons, X rays and Auger electrons are generated. One may use the results of Smith et al.[4,16] or the computer code of Dillman[11,13] to calculate the energy per disintegration emitted as a result of nonpenetrating type of radiations. From Smith et al.,

$$n_e \bar{E}_e = f_\gamma N_{e_K} \left[E_\gamma - \omega_K E_K + \omega_K \left\{ \left(\frac{K_\alpha}{K_\alpha + K_\beta} \right) E_{L_{II-III}} + \frac{K_\beta}{K_\alpha + K_\beta} E_{M_{II-III}} \right\} \right]$$
$$+ f_\gamma E_\gamma (N_{e_L} + N_{e_M} + \cdots) \text{ Mev/disintegration} \quad (13)$$

$$\alpha = \frac{N_e}{N_\gamma}; \quad (13a)$$

$$\alpha_{total} = \alpha_t = \alpha_K + \alpha_L + \alpha_M + \cdots; \quad (13b)$$

$$N_{e_K} = \frac{\alpha_K}{1 + \alpha_t}; \quad N_{e_L} = \frac{\alpha_L}{1 + \alpha_t}; \quad N_{e_M} = \frac{\alpha_M}{1 + \alpha_t}, \quad (13c)$$

where

$n_e \bar{E}_e$ = mean energy per disintegration due to conversion electrons and subsequent products (Mev per disintegration),

α_t = total-internal-conversion coefficient, which is the ratio of the number of conversion electrons, N_e, emitted to the number of gamma rays emitted, N_γ,

$\alpha_K, \alpha_L, \alpha_M \ldots$ = the internal-conversion coefficient for the K, L, M, ... electron shells respectively,

$N_{e_K}, N_{e_L}, N_{e_M} \ldots$ = the number of K, L, M, ... conversion electrons arising from photons of energy E_γ,

f_γ = the fractional frequency of the disintegrations that give rise to an energy state that may give rise to a photon of energy, E_γ.

The effects of internal conversion of gamma rays of different energies in the same or competitive decay process are some, then

$$n_e \bar{E}_e = n_{e_1} \bar{E}_{e_1} + n_{e_2} \bar{E}_{e_2} + \cdots \text{ Mev/disintegration.} \tag{13d}$$

Characteristic X Rays

Characteristic X rays are produced as a result of electron capture and internal conversion of photons. The fractional frequency of these X rays may be calculated from the work of Smith et al.[16] or from the computer code developed by Dillman.[11,13] Dillman's code allows for the computation of the fractional frequency of K-, L-, and M-characteristic X rays, while Smith's equations are limited to K_α and K_β X rays for elements with an atomic number less than or equal to 82. This is because of the 0.0113-Mev photon energy cut-off inherent in these equations. The following equations are taken from Smith et al.[16]

$$n_{K\alpha} = \omega_K \left(\frac{K_\alpha}{K_\alpha + K_\beta}\right)\left[\sum_i n_{\varepsilon K_i} + \sum_j f_{\gamma_j} N_{eK_j}\right] \text{ number/disintegration} \tag{14a}$$

$$n_{K\beta} = \omega_K \left(\frac{K_\beta}{K_\alpha + K_\beta}\right)\left[\sum_i n_{\varepsilon K_i} + \sum_j f_{\gamma_j} N_{eK_j}\right] \text{ number/disintegration} \tag{14b}$$

where $n_{K\alpha}$ = fractional frequency of K_α X rays emitted per disintegration,

$n_{K\beta}$ = fractional frequency of K_β X rays emitted per disintegration,

$n_{\varepsilon K_i}$ = fractional frequency of K electron capture per disintegration to the ith energy level,

f_{γ_j} = fractional frequency of the disintegrations that give rise to an energy state E_j, which may give rise to a photon of energy E_γ (this is the fractional frequency before the photon yield has been corrected for internal conversion),

N_{eK_j} = number of K conversion electrons arising from the conversion of a photon of energy E_{γ_j} per disintegration.

Gamma Rays

The fractional frequency of unconverted gamma rays, that is, the proportion of the gamma rays that are not internally converted, may be calculated from the following equation.[4,16]

$$n_{\gamma_i} = f_{\gamma_i}\left(\frac{1}{1 + \alpha_{t_i}}\right) \text{ number/disintegration.} \tag{15}$$

Table 4. ELECTRON-BINDING ENERGY AND CHARACTERISTIC X-RAY ENERGY

Element (Atomic No.)	Electron Binding Energy, kev				Characteristic X-Ray Energy, kev							
	E_K	E_{L_I}	$E_{L_{II}}$	$E_{L_{III}}$	K_{β_2}	K_{β_1}	K_{α_1}	K_{α_2}	L_{γ_1}	L_{β_2}	L_{β_1}	L_{α_1}	L_{α_2}
H (1)	0.0136												
He (2)	0.0246												
Li (3)	0.055												
Be (4)	0.116												
B (5)	0.192						0.052						
							0.110						
							0.185						
C (6)	0.283						0.282						
N (7)	0.399						0.392						
O (8)	0.531						0.523						
F (9)	0.687			0.022			0.677						
Ne (10)	0.874	0.048	0.022				0.851						
Na (11)	1.08	0.055	0.034	0.034		1.067	1.041						
Mg (12)	1.303	0.063	0.050	0.049		1.297	1.254						
Al (13)	1.559	0.087	0.073	0.072		1.553	1.487	1.486					
Si (14)	1.838	0.118	0.099	0.098		1.832	1.740	1.739					
P (15)	2.142	0.153	0.129	0.128		2.136	2.015	2.014					
S (16)	2.470	0.193	0.164	0.163		2.464	2.308	2.306					
Cl (17)	2.826	0.238	0.203	0.202		2.815	2.622	2.621					
Ar (18)	2.203	0.287	0.247	0.245		3.192	2.957	2.955					
K (19)	3.607	0.341	0.297	0.294		3.589	3.313	3.310					
Ca (20)	4.038	0.399	0.352	0.349		4.012	3.691	3.688			0.344		0.341
Sc (21)	4.496	0.462	0.411	0.406		4.460	4.090	4.085			0.399		0.395
Ti (22)	4.964	0.530	0.460	0.454		4.931	4.510	4.504			0.458		0.452
V (23)	5.463	0.604	0.519	0.512		5.427	4.952	4.944			0.519		0.510
Cr (24)	5.988	0.679	0.583	0.574		5.946	5.414	5.405			0.581		0.571
Mn (25)	6.537	0.762	0.650	0.639		6.490	5.898	5.887			0.647		0.636
Fe (26)	7.111	0.849	0.721	0.708		7.057	6.403	6.390			0.717		0.704
Co (27)	7.709	0.929	0.794	0.779		7.649	6.930	6.915			0.790		0.775
Ni (28)	8.331	1.015	0.871	0.853	8.328	8.264	7.477	7.460			0.866		0.849
Cu (29)	8.980	1.100	0.953	0.933	8.976	8.904	8.047	8.027			0.948		0.928
Zn (30)	9.660	1.200	1.045	1.022	9.657	9.571	8.638	8.615			1.032		1.009

Table 4. ELECTRON-BINDING ENERGY AND CHARACTERISTIC X-RAY ENERGY (Continued)

Element (Atomic No.)	Electron-Binding Energy, kev					Characteristic X-Ray Energy, kev							
	E_K	E_{L_I}	$E_{L_{II}}$	$E_{L_{III}}$		K_{β_2}	K_{β_1}	K_{α_1}	K_{α_2}	L_{γ_1}	L_{β_2}	L_{β_1}	L_{α_1}	L_{α_2}
Ga (31)	10.368	1.30	1.134	1.117		10.365	10.263	9.251	9.234			1.122		1.096
Ge (32)	11.103	1.42	1.248	1.217		11.100	10.981	9.885	9.854			1.216		1.186
As (33)	11.863	1.529	1.359	1.323		11.863	11.725	10.543	10.507			1.317		1.282
Se (34)	12.652	1.652	1.473	1.434		12.651	12.495	11.221	11.181			1.419		1.379
Br (35)	13.475	1.794	1.599	1.552		13.465	13.290	11.923	11.877			1.526		1.480
Kr (36)	14.323	1.931	1.727	1.675		14.313	14.112	12.648	12.597			1.638		1.587
Rb (37)	15.201	2.067	1.866	1.806		15.184	14.960	13.394	13.335			1.752	1.694	1.692
Sr (38)	16.106	2.221	2.008	1.941		16.083	15.834	14.164	14.097			1.872	1.806	1.805
Y (39)	17.037	2.369	2.154	2.079		17.011	16.736	14.957	14.882			1.996	1.922	1.920
Zr (40)	17.998	2.547	2.305	2.220		17.969	17.666	15.774	15.690	2.302	2.219	2.124	2.042	2.040
Nb (41)	18.987	2.706	2.467	2.374		18.951	18.621	16.614	16.520	2.462	2.367	2.257	2.166	2.163
Mo (42)	20.002	2.884	2.627	2.523		19.964	19.607	17.478	17.373	2.623	2.518	2.395	2.293	2.290
Tc (43)	21.054	3.054	2.795	2.677		21.012	20.612	18.370	18.250	2.792	2.674	2.538	2.424	2.420
Ru (44)	22.118	3.236	2.966	2.837		22.072	21.655	19.278	19.149	2.964	2.836	2.683	2.558	2.554
Rh (45)	23.224	3.419	3.145	3.002		23.169	22.721	20.214	20.072	3.144	3.001	2.834	2.696	2.692
Pd (46)	24.347	3.617	3.329	3.172		24.297	23.816	21.175	21.018	3.328	3.172	2.990	2.838	2.833
Ag (47)	25.517	3.810	3.528	3.352		25.454	24.942	22.162	21.988	3.519	3.348	3.151	2.984	2.978
Cd (48)	26.712	4.019	3.727	3.538		26.641	26.093	23.172	22.982	3.716	3.528	3.316	3.133	3.127
In (49)	27.928	4.237	3.939	3.729		27.859	27.274	24.207	24.000	3.920	3.713	3.487	3.287	3.279
Sn (50)	29.190	4.464	4.157	3.928		29.106	28.483	25.270	25.042	4.131	3.904	3.662	3.444	3.435
Sb (51)	30.486	4.697	4.381	4.132		30.387	29.723	26.357	26.109	4.347	4.100	3.843	3.605	3.595
Te (52)	31.809	4.938	4.613	4.341		31.698	30.993	27.471	27.200	4.570	4.301	4.029	3.769	3.758
I (53)	33.164	5.190	4.856	4.559		33.016	32.292	28.610	28.315	4.800	4.507	4.220	3.937	3.926
Xe (54)	34.579	5.452	5.104	4.782		34.398	33.644	29.779	29.463	5.036	4.720	4.422	4.111	4.098
Cs (55)	35.959	5.720	5.358	5.011		35.819	34.984	30.970	30.623	5.280	4.936	4.620	4.286	4.272
Ba (56)	37.410	5.995	5.623	5.247		37.255	36.376	32.191	31.815	5.531	5.156	4.828	4.467	4.451
La (57)	38.931	6.283	5.894	5.489		38.728	37.799	33.440	33.033	5.789	5.384	5.043	4.651	4.635
Ce (58)	40.449	6.561	6.165	5.729		40.231	39.255	34.717	34.276	6.052	5.613	5.262	4.840	4.823
Pr (59)	41.998	6.846	6.443	5.968		41.772	40.746	36.023	35.548	6.322	5.850	5.489	5.034	5.014
Nd (60)	43.571	7.144	6.727	6.215		43.349	42.269	37.359	36.845	6.602	6.090	5.722	5.230	5.208

Table 4. ELECTRON-BINDING ENERGY AND CHARACTERISTIC X-RAY ENERGY (Continued)

Element (Atomic No.)	Electron-Binding Energy, kev					Characteristic X-Ray Energy, kev.								
	E_K	E_{L_I}	$E_{L_{II}}$	$E_{L_{III}}$		K_{β_2}	K_{β_1}	K_{α_1}	K_{α_2}	L_{γ_1}	L_{β_2}	L_{β_1}	L_{α_1}	L_{α_2}
Pm (61)	45.207	7.448	7.018	6.466		44.955	43.811	38.726	38.180	6.891	6.336	5.956	5.431	5.408
Sm (62)	46.846	7.754	7.312	6.721		46.581	45.400	40.124	39.523	7.180	6.587	6.206	5.636	5.609
Eu (63)	48.515	8.069	7.624	6.983		48.241	47.027	41.529	40.877	7.478	6.842	6.456	5.846	5.816
Gd (64)	50.229	8.393	7.940	7.252		49.961	48.718	42.983	42.280	7.788	7.102	6.714	6.059	6.027
Tb (65)	51.998	8.724	8.258	7.519		51.737	50.391	44.470	43.737	8.104	7.368	6.979	6.275	6.241
Dy (66)	53.789	9.083	8.591	7.790		53.491	52.187	45.985	45.193	8.418	7.638	7.249	6.495	6.457
Ho (67)	55.615	9.411	8.920	8.074		55.292	53.934	47.528	46.686	8.748	7.912	7.528	6.720	6.680
Er (68)	57.483	9.776	9.263	8.364		57.088	55.690	49.099	48.205	9.089	8.188	7.810	6.948	6.904
Tm (69)	59.382	10.144	9.628	8.652		58.969	57.487	50.730	49.762	9.424	8.472	8.103	7.181	7.135
Yb (70)	61.303	10.486	9.977	8.943		60.959	59.352	52.360	51.326	9.779	8.758	8.401	7.414	7.367
Lu (71)	63.304	10.867	10.345	9.241		62.946	61.282	54.063	52.959	10.142	9.048	8.708	7.654	7.604
Hf (72)	65.313	11.264	10.734	9.556		64.936	63.209	55.757	54.759	10.514	9.346	9.021	7.898	7.843
Ta (73)	67.400	11.676	11.130	9.876		66.999	65.210	57.524	56.270	10.892	9.649	9.341	8.145	8.087
W (74)	69.508	12.090	11.535	10.198		69.090	67.233	59.310	57.973	11.283	9.959	9.670	8.396	8.333
Re (75)	71.662	12.522	11.955	10.531		71.220	69.298	61.131	59.707	11.684	10.273	10.008	8.651	8.584
Os (76)	73.860	12.965	12.383	10.869		73.393	71.404	62.991	61.477	12.094	10.596	10.354	8.910	8.840
Ir (77)	76.097	13.413	12.819	11.211		75.605	73.549	64.886	63.278	12.509	10.918	10.706	9.173	9.098
Pt (78)	78.379	13.873	13.268	11.559		77.866	75.736	66.820	65.111	12.939	11.249	11.069	9.441	9.360
Au (79)	80.713	14.353	13.733	11.919		80.165	77.968	68.794	66.980	13.379	11.582	11.439	9.711	9.625
Hg (80)	83.106	14.841	14.212	12.285		82.526	80.258	70.821	68.894	13.828	11.923	11.823	9.987	9.896
Tl (81)	85.517	15.346	14.697	12.657		84.904	82.558	72.860	70.820	14.288	12.268	12.210	10.266	10.170
Pb (82)	88.001	15.870	15.207	13.044		87.343	84.922	74.957	72.794	14.762	12.620	12.611	10.549	10.448
Bi (83)	90.521	16.393	15.716	13.424		89.833	87.335	77.097	74.805	15.244	12.977	13.021	10.836	10.729
Po (84)	93.112	16.935	16.244	13.817		92.386	89.809	79.296	76.868	15.740	13.338	13.441	11.128	11.014
At (85)	95.740	17.490	16.784	14.215		94.976	92.319	81.525	78.956	16.248	13.705	13.873	11.424	11.304
Rn (86)	98.418	18.058	17.337	14.618		97.616	94.877	83.800	81.080	16.768	14.077	14.316	11.724	11.597
Fr (87)	101.147	18.638	17.904	15.028		100.305	97.483	86.119	83.243	17.301	14.459	14.770	12.029	11.894
Ra (88)	103.927	19.233	18.481	15.442		103.048	100.136	88.485	85.446	17.845	14.839	15.233	12.338	12.194
Ac (89)	106.759	19.842	19.078	15.865		105.838	102.846	90.894	87.681	18.405	15.227	15.712	12.650	12.499
Th (90)	109.630	20.460	19.688	16.296		108.671	105.592	93.334	89.942	18.977	15.620	16.200	12.966	12.808

Table 4. ELECTRON-BINDING ENERGY AND CHARACTERISTIC X-RAY ENERGY (Continued)

Element (Atomic No.)	Electron-Binding Energy, kev				Characteristic X-Ray Energy, kev								
	E_K	E_{L_I}	$E_{L_{II}}$	$E_{L_{III}}$	K_{β_2}	K_{β_1}	K_{α_1}	K_{α_2}	L_{γ_1}	L_{β_2}	L_{β_1}	L_{α_1}	L_{α_2}
Pa (91)	112.581	21.102	20.311	16.731	111.575	108.408	95.851	92.271	19.559	16.022	16.700	13.291	13.120
U (92)	115.591	21.753	20.943	17.163	114.549	111.289	98.428	94.648	20.163	16.425	17.218	13.613	13.438
Np (93)	118.619	22.417	21.596	17.614	117.533	114.181	101.005	97.023	20.774	16.837	17.740	13.945	13.758
Pu (94)	121.720	23.097	22.262	18.066	120.592	117.146	103.653	99.457	21.401	17.254	18.278	14.279	14.082
Am (95)	124.876	23.793	22.944	18.525	123.706	120.163	106.351	101.932	22.042	17.677	18.829	14.618	14.411
Cm (96)	128.088	24.503	23.640	18.990	126.875	123.235	109.098	104.448	22.699	18.106	19.393	14.961	14.743
Bk (97)	131.357	25.230	24.352	19.461	130.101	126.362	111.896	107.023	23.370	18.540	19.970	15.309	15.079
Cf (98)	134.683	25.971	25.080	19.938	133.383	129.544	114.745	109.603	24.056	18.980	20.562	15.661	15.420
Es (99)	138.067	26.729	25.824	20.422	136.724	132.781	117.646	112.244	24.758	19.426	21.166	16.018	15.764
Fm (100)	141.510	27.503	26.584	20.912	140.122	136.075	120.598	114.926	25.475	19.879	21.785	16.379	16.113

Reprinted by special permission from Nucleonics, 13: 36, 1955; copyright © 1955 by McGraw-Hill, New York, N.Y. 10036.

NONPENETRATING RADIATION

Nonpenetrating radiation includes both charged particles and photons in which 95 percent or more of the energy of the particle or photon is transferred to the volume in which it originated; that is, the source and target are identical. For nonpenetrating (n-p) radiation distributed in a volume v_1, the absorbed fraction is by definition unity or zero.

$$\phi_{n-p}(v_1 \leftrightarrow v_1) = 1 \quad \text{and} \quad \Phi_{n-p}(v_1 \leftarrow v_1) = m_1^{-1}, \tag{16a}$$

if $v_1 = v_1$;

$$\phi_{n-p}(v_2 \leftarrow v_1) = 0 \quad \text{and} \quad \Phi_{n-p}(v_2 \leftarrow v_1) = 0, \tag{16b}$$

if $v_2 \neq v_1$.

In many instances the biological target—for example, the organ—will be one centimeter in radius, or larger. For this situation, all charged particles from radionuclides and all photons with an energy less than or equal to 0.0113 Mev are considered nonpenetrating.

In some instances the biological target will be less than one centimeter in radius, and the distance required for each type of radiation to deposit 95 per cent of its energy must be determined, before the radiation can be classified as nonpenetrating.

PENETRATING PHOTON RADIATIONS

Absorbed-dose calculations for penetrating photon radiations have, in the past, been carried out by several methods. In order that these results may be used with the recommendations of the MIRD Committee,[1] tables of absorbed fractions, of specific absorbed fractions, and of energy absorption build-up factors, as well as equations for converting the previously used parameters to absorbed fractions or specific fractions, will be presented.

Absorbed Fraction

The absorbed fraction was originally developed by Ellett et al.[9] for calculating the penetrating photon component of the absorbed dose, since the previous methods using the geometrical factor and specific gamma-ray constant had serious limitations, especially at low photon energies, as pointed out by Quimby.[18] The absorbed-fraction method and the geometrical-factor–specific-gamma-ray-constant method for calculating the penetrating component of the absorbed dose have been discussed and quantitatively compared by Ellett et al.[9,10] and by Smith.[19]

Ellett et al.,[9,10,20] Reddy,[21] and Brownell[22] have calculated values for the absorbed fractions for various geometrical shapes, radionuclides distributions, and photon energies. Table 5 gives the tissue composition used in the absorbed fraction calculation.

Table 5.

TISSUE COMPOSITION USED IN
ABSORBED-FRACTION CALCULATIONS

Element	% Mass
Oxygen	71.39
Carbon	14.89
Hydrogen	10.00
Nitrogen	3.47
Chlorine	0.10
Sodium	0.15

From: Brownell, G. L., Ellett, W. H., and Reddy, A. R., Absorbed Fractions for Photon Dosimetry. *J. Nucl. Med.*, Suppl. 1: 33, 1968. (Courtesy of the Society of Nuclear Medicine.)

Tables 6 to 9 give values of absorbed fraction for a central point source distribution as a function of photon energy for a variety of geometrical shapes.

Table 10 gives values of absorbed fraction for point sources at distances one-half, one-sixth, and one-twelfth the height of an elliptical cylinder.

Tables 11 to 13 give values of the absorbed fraction for a uniform distribution as a function of photon energy for a variety of geometrical shapes.

Tables 6 to 13 give the absorbed fractions for targets of finite volume, and these values do not include the contribution due to photons coming from outside the target volume and interacting within it. For the situation where organs are located within the trunk of the body, the average absorbed dose will contain a contribution due to the back-scatter photons.[10,22] Table 14 may be used to obtain the fractional increase in the absorbed dose due to backscatter.

Berger[23] has calculated absorbed-fraction values, using the moments method. He has calculated the fraction $\phi_{ph}(x)$ of energy emitted by a point isotropic source that is deposited in a sphere of radius x around the source in an unbounded homogeneous medium. The absorbed fraction for water has been calculated (Table 15) for a large set of sphere radii, and by interpolation those radii for which the absorbed fraction assumed values of 0.05, 0.10, . . ., 0.95 have been determined.

Fisher and Snyder[24] have developed a Monte Carlo code, which computes the absorbed fraction of photon energy from photon sources uniformly distributed in various organs of the human body. The homogeneous phantom used in this code has twenty-two regions, which have been defined mathematically and approximate the size and shape of the organ they represent. In this paper, it graphically presents the absorbed fraction for the various organs in their phantom for four photon energies between 0.2 and 4 Mev for uniform source distribution in the total body and in the skeleton. Ford and Snyder,[25] using the same computer code, have developed a phantom that is nonhomogeneous and uses the appropriate tissue compositions for soft tissue, lung, and bone. In this paper, it graphically presents the absorbed fraction for the various organs in their phantom for photon energies from 0.010 Mev to 4 Mev for a source uniformly distributed in the skeleton and in the lungs.

The absorbed-fraction data computed by Ellett et al.[20] and by Snyder and Ford,[26] using different Monte Carlo codes, and the absorbed-fraction data computed by Berger,[23] using the moments method, have compared quite favorably.[22,27]

Energy Absorption Build-Up Factor

Berger[23] has calculated *energy absorption build-up factors*, B_{en}, for point isotropic sources in an infinite water phantom for photon energies from 15 kev to 3 Mev. Equation (17a) gives the absorbed-dose rate, $R(x)$, in rads per second for a point isotropic monoenergetic source of radiation in an unbounded homogeneous medium at a distance x from the source.

$$R(x) = A\Delta\Phi(x) \text{ rad/sec,} \qquad (17a)$$

where $\Delta = knE$ g-rad/sec.

Values for the point isotropic specific absorbed fraction $\Phi(x)$, i.e., the fraction of the emitted energy absorbed per gram at a distance x from a point isotropic photon source, are given in Table 16 or may be calculated from equation (17b).

$$\Phi(x) = \Phi_{ph}(x) = \left[\frac{\mu_{en}}{\rho} \cdot \frac{1}{4\pi x^2} \cdot e^{-\mu x}\right] \cdot B_{en}(\mu x). \qquad (17b)$$

The quantity $B_{en}(\mu x)$ is the energy absorption build-up factor, and takes into account the contribution of scattered photons.

Table 6.
ABSORBED FRACTIONS FOR CENTRAL POINT SOURCES IN ELLIPTICAL CYLINDERS OR ELLIPSOIDS*

φ

Mass, kg	0.020 Mev	0.030 Mev	0.040 Mev	0.060 Mev	0.080 Mev	0.100 Mev	0.140 Mev	0.160 Mev	0.279 Mev	0.364 Mev	0.662 Mev	1.460 Mev	2.750 Mev	Ra-dium†
2	0.924	0.605	0.367	0.203	0.166	0.159	0.160	0.161	0.171	0.176	0.176	0.157	0.125	0.163
4	0.956	0.697	0.453	0.263	0.212	0.202	0.200	0.201	0.214	0.218	0.216	0.193	0.154	0.201
6	0.971	0.749	0.509	0.305	0.247	0.233	0.229	0.230	0.242	0.245	0.243	0.217	0.174	0.227
8	0.980	0.785	0.551	0.340	0.276	0.258	0.252	0.253	0.265	0.267	0.263	0.236	0.190	0.247
10	0.985	0.811	0.585	0.368	0.301	0.280	0.272	0.272	0.283	0.285	0.281	0.252	0.204	0.263
20	0.995	0.881	0.689	0.468	0.388	0.358	0.342	0.340	0.348	0.348	0.340	0.306	0.251	0.320
30	0.998	0.914	0.749	0.534	0.447	0.413	0.391	0.387	0.391	0.389	0.379	0.341	0.282	0.357
40	0.998	0.934	0.788	0.581	0.491	0.456	0.429	0.423	0.424	0.421	0.409	0.368	0.306	0.385
50	0.999	0.946	0.817	0.619	0.527	0.491	0.460	0.453	0.450	0.447	0.433	0.390	0.325	0.408
60	0.999	0.956	0.838	0.649	0.557	0.520	0.487	0.478	0.473	0.469	0.453	0.409	0.342	0.427
70	0.999	0.963	0.855	0.674	0.583	0.545	0.511	0.500	0.493	0.488	0.471	0.425	0.356	0.444
80	0.999	0.968	0.869	0.696	0.606	0.567	0.532	0.520	0.511	0.505	0.486	0.439	0.369	0.458
90	0.999	0.972	0.881	0.715	0.626	0.587	0.551	0.538	0.527	0.520	0.500	0.452	0.380	0.471
100	0.999	0.976	0.890	0.731	0.643	0.604	0.568	0.554	0.542	0.534	0.513	0.463	0.391	0.483
120	0.999	0.981	0.906	0.759	0.674	0.634	0.597	0.583	0.567	0.558	0.535	0.483	0.409	0.505
140	0.999	0.985	0.918	0.782	0.700	0.660	0.622	0.607	0.589	0.579	0.555	0.501	0.425	0.523
160	0.999	0.987	0.928	0.801	0.722	0.681	0.644	0.629	0.609	0.598	0.572	0.516	0.439	0.539
180	0.999	0.989	0.936	0.818	0.740	0.700	0.662	0.648	0.627	0.615	0.587	0.530	0.452	0.554
200	0.999	0.990	0.944	0.833	0.756	0.717	0.678	0.665	0.643	0.630	0.601	0.542	0.463	0.568

* The principal axes of the elliptical cylinders and the ellipsoids are in the ratios of 1/1.8/6.19 and 1/1.8/9.27 respectively.
† Weighted value for ^{226}Ra decay product spectrum given by H. E. Johns, 1953.

From: Brownell, G. L., Ellett, W. H., and Reddy, A. R., Absorbed Fractions for Photon Dosimetry. *J. Nucl. Med.*, Suppl. 1: 34, 1968. (Courtesy of the Society of Nuclear Medicine.)

Table 7.

ABSORBED FRACTIONS FOR CENTRAL POINT SOURCES IN SPHERES

Mass, kg	ϕ									
	0.020 Mev	0.030 Mev	0.040 Mev	0.060 Mev	0.100 Mev	0.140 Mev	0.160 Mev	0.279 Mev	0.662 Mev	2.750 Mev
2	0.989	0.794	0.537	0.322	0.243	0.233	0.234	0.241	0.235	0.168
4	0.996	0.878	0.658	0.421	0.317	0.301	0.297	0.302	0.293	0.209
6	0.999	0.916	0.727	0.488	0.370	0.348	0.342	0.344	0.330	0.238
8	0.999	0.938	0.772	0.540	0.413	0.386	0.379	0.377	0.359	0.259
10	0.999	0.952	0.806	0.581	0.448	0.418	0.409	0.405	0.382	0.277
20	0.999	0.982	0.894	0.709	0.569	0.529	0.517	0.500	0.461	0.339
30	0.999	0.991	0.932	0.780	0.644	0.600	0.587	0.562	0.514	0.380
40	0.999	0.995	0.954	0.826	0.698	0.652	0.639	0.608	0.554	0.411
50	0.999	0.996	0.966	0.859	0.738	0.692	0.679	0.644	0.586	0.436
60	0.999	0.997	0.974	0.882	0.770	0.725	0.712	0.675	0.613	0.457
70	0.999	0.998	0.980	0.900	0.796	0.752	0.739	0.700	0.637	0.476
80	0.999	0.998	0.983	0.915	0.818	0.775	0.762	0.722	0.657	0.492
90	0.999	0.999	0.986	0.926	0.836	0.794	0.781	0.741	0.675	0.507
100	0.999	0.999	0.988	0.935	0.851	0.811	0.799	0.758	0.691	0.520
120	0.999	0.999	0.991	0.948	0.876	0.839	0.827	0.786	0.719	0.544
140	0.999	0.999	0.993	0.958	0.895	0.860	0.849	0.809	0.742	0.564
160	0.999	0.999	0.995	0.965	0.910	0.878	0.867	0.829	0.761	0.582
180	0.999	0.999	0.996	0.971	0.923	0.892	0.882	0.845	0.778	0.598
200	0.999	0.999	0.998	0.976	0.933	0.904	0.894	0.858	0.792	0.612

From: Brownell, G. L., Ellett, W. H., and Reddy, A. R., Absorbed Fractions for Photon Dosimetry. *J. Nucl. Med.*, Suppl. 1: 34, 1968. (Courtesy of the Society of Nuclear Medicine.)

Table 8.

ABSORBED FRACTIONS FOR CENTRAL POINT SOURCES IN RIGHT CIRCULAR CYLINDERS*

Mass kg	ϕ					
	0.040 Mev	0.080 Mev	0.160 Mev	0.364 Mev	0.662 Mev	1.460 Mev
2	0.528	0.258	0.224	0.240	0.229	0.200
4	0.645	0.336	0.290	0.295	0.288	0.253
6	0.712	0.391	0.335	0.333	0.326	0.286
8	0.757	0.435	0.370	0.363	0.354	0.311
10	0.789	0.471	0.399	0.387	0.376	0.332
20	0.878	0.593	0.501	0.472	0.453	0.401
30	0.917	0.668	0.568	0.528	0.504	0.446
40	0.940	0.721	0.618	0.571	0.543	0.480
50	0.954	0.761	0.658	0.605	0.575	0.509
60	0.964	0.792	0.691	0.633	0.602	0.533
70	0.971	0.818	0.719	0.658	0.625	0.553
80	0.977	0.838	0.743	0.679	0.646	0.572
90	0.981	0.856	0.763	0.698	0.664	0.588
100	0.984	0.871	0.781	0.714	0.680	0.603
120	0.989	0.894	0.811	0.742	0.708	0.629
140	0.992	0.911	0.834	0.765	0.730	0.651
160	0.994	0.924	0.852	0.784	0.749	0.669
180	0.994	0.933	0.866	0.800	0.765	0.685
200	0.994	0.939	0.877	0.813	0.777	0.698

* The principal axes of the right circular cylinders are in the ratio of 1/1/0.75.

From: Brownell, G. L., Ellett, W. H., and Reddy, A. R., Absorbed Fractions for Photon Dosimetry *J. Nucl. Med.*, Suppl. 1: 35, 1968. (Courtesy of the Society of Nuclear Medicine.)

Table 9.

ABSORBED FRACTIONS FOR CENTRAL POINT SOURCES IN RIGHT CIRCULAR CYLINDERS*

Height, cm	Radius, cm		
	3	5	10
0.040 Mev			
5	0.219	0.299	0.379
10	0.266	0.403	0.582
15	0.279	0.442	0.682
20	0.283	0.458	0.731
0.080 Mev			
5	0.092	0.128	0.174
10	0.113	0.179	0.292
15	0.119	0.200	0.365
20	0.122	0.210	0.409
0.160 Mev			
5	0.095	0.128	0.168
10	0.115	0.170	0.261
15	0.122	0.189	0.318
20	0.129	0.198	0.353
0.364 Mev			
5	0.103	0.137	0.183
10	0.124	0.180	0.268
15	0.131	0.199	0.318
20	0.134	0.208	0.348
0.662 Mev			
5	0.104	0.140	0.185
10	0.124	0.181	0.265
15	0.131	0.199	0.311
20	0.134	0.208	0.339
1.460 Mev			
5	0.093	0.124	0.165
10	0.111	0.160	0.234
15	0.117	0.175	0.273
20	0.120	0.184	0.298

* These data were obtained for Dr. G. K. Bahr of Cincinnati General Hospital, and are presented here through his courtesy. The mass density of the material is unity.

From: Brownell, G. L., Ellett, W. H., and Reddy, A. R., Absorbed Fractions for Photon Dosimetry. *J. Nucl. Med.*, Suppl. 1: 35, 1968. (Courtesy of the Society of Nuclear Medicine.)

Table 10.

ABSORBED FRACTIONS FOR POINT SOURCES AT DISTANCES 1/2, 1/6, AND 1/12 HEIGHT OF ELLIPTICAL CYLINDERS*

Mass, kg	φ								
	0.080 Mev			0.662 Mev			2.750 Mev		
	1/2	1/6	1/12	1/2	1/6	1/12	1/2	1/6	1/12
2	0.166	0.158	0.149	0.176	0.165	0.151	0.125	0.116	0.105
4	0.212	0.203	0.187	0.215	0.204	0.186	0.154	0.145	0.133
6	0.247	0.238	0.217	0.242	0.230	0.211	0.174	0.164	0.152
8	0.276	0.266	0.242	0.263	0.250	0.230	0.190	0.179	0.166
10	0.301	0.291	0.264	0.281	0.266	0.245	0.204	0.192	0.178
20	0.388	0.378	0.343	0.340	0.323	0.299	0.251	0.238	0.220
30	0.447	0.437	0.399	0.379	0.361	0.335	0.282	0.268	0.249
40	0.491	0.482	0.442	0.409	0.390	0.362	0.306	0.292	0.271
50	0.527	0.518	0.477	0.433	0.413	0.384	0.325	0.312	0.289
60	0.557	0.548	0.507	0.453	0.434	0.403	0.342	0.328	0.304
70	0.583	0.575	0.533	0.470	0.452	0.419	0.356	0.343	0.318
80	0.606	0.598	0.556	0.486	0.467	0.434	0.369	0.356	0.330
90	0.626	0.618	0.576	0.500	0.482	0.448	0.380	0.368	0.341
100	0.644	0.636	0.595	0.513	0.495	0.460	0.391	0.378	0.351
120	0.674	0.668	0.626	0.535	0.518	0.482	0.409	0.397	0.368
140	0.700	0.694	0.653	0.555	0.537	0.501	0.425	0.413	0.384
160	0.722	0.717	0.676	0.572	0.555	0.518	0.439	0.428	0.397
180	0.740	0.736	0.696	0.587	0.571	0.533	0.452	0.440	0.410
200	0.756	0.753	0.713	0.601	0.585	0.547	0.463	0.451	0.421

* The principal axes of the elliptical cylinders are in the ratio of 1/1.8/6.19.

From: Brownell, G. L., Ellett, W. H., and Reddy, A. R., Absorbed Fractions for Photon Dosimetry. *J. Nucl. Med.*, Suppl. 1:36, 1968. (Courtesy of the Society of Nuclear Medicine.)

Table 11.

ABSORBED FRACTIONS FOR UNIFORM DISTRIBUTION OF ACTIVITY IN ELLIPSOIDS*

Mass, kg	φ										
	0.020 Mev	0.030 Mev	0.040 Mev	0.060 Mev	0.080 Mev	0.100 Mev	0.160 Mev	0.364 Mev	0.662 Mev	1.460 Mev	2.750 Mev
2	0.702	0.407	0.317	0.131	0.072	0.099	0.113	0.112	0.134	0.099	0.096
4	0.762	0.485	0.325	0.176	0.127	0.133	0.144	0.148	0.155	0.133	0.120
6	0.795	0.529	0.345	0.206	0.157	0.155	0.163	0.170	0.173	0.155	0.134
8	0.815	0.560	0.366	0.228	0.179	0.172	0.178	0.187	0.189	0.171	0.147
10	0.830	0.583	0.385	0.247	0.196	0.185	0.190	0.200	0.202	0.183	0.156
20	0.868	0.649	0.460	0.308	0.250	0.233	0.234	0.245	0.250	0.223	0.187
30	0.884	0.685	0.508	0.346	0.284	0.265	0.264	0.273	0.280	0.248	0.207
40	0.893	0.709	0.541	0.374	0.310	0.290	0.287	0.294	0.301	0.267	0.222
50	0.900	0.727	0.567	0.397	0.332	0.312	0.305	0.312	0.317	0.282	0.235
60	0.905	0.741	0.585	0.416	0.351	0.330	0.321	0.327	0.330	0.294	0.247
70	0.909	0.753	0.600	0.432	0.368	0.346	0.335	0.340	0.341	0.306	0.257
80	0.912	0.763	0.613	0.446	0.383	0.361	0.348	0.351	0.351	0.316	0.265
90	0.916	0.772	0.624	0.459	0.397	0.374	0.359	0.362	0.360	0.325	0.274
100	0.918	0.780	0.634	0.471	0.409	0.386	0.369	0.371	0.368	0.334	0.283
120	0.924	0.793	0.652	0.492	0.431	0.407	0.388	0.389	0.384	0.350	0.298
140	0.929	0.804	0.670	0.511	0.450	0.425	0.405	0.405	0.399	0.364	0.310
160	0.933	0.814	0.688	0.528	0.466	0.440	0.421	0.420	0.415	0.378	0.321
180	0.937	0.821	0.708	0.544	0.480	0.454	0.436	0.433	0.432	0.391	0.331
200	0.940	0.828	0.729	0.559	0.491	0.466	0.451	0.446	0.449	0.403	0.340

* The principal axes of the ellipsoids are in the ratio of 1/1.8/9.27.

From: Brownell, G. L., Ellett, W. H., and Reddy, A. R., Absorbed Fractions for Photon Dosimetry. *J. Nucl. Med.*, Suppl. 1: 37, 1968. (Courtesy of the Society of Nuclear Medicine.)

Table 12.

ABSORBED FRACTIONS FOR UNIFORM DISTRIBUTION OF ACTIVITY IN SMALL SPHERES AND THICK ELLIPSOIDS*

Mass, kg	φ										
	0.020 Mev	0.030 Mev	0.040 Mev	0.060 Mev	0.080 Mev	0.100 Mev	0.160 Mev	0.364 Mev	0.662 Mev	1.460 Mev	2.750 Mev
0.3	0.684	0.357	0.191	0.109	0.086	0.085	0.087	0.099	0.096	0.092	0.077
0.4	0.712	0.388	0.212	0.121	0.096	0.093	0.097	0.108	0.108	0.099	0.083
0.5	0.731	0.412	0.229	0.131	0.104	0.099	0.104	0.116	0.117	0.104	0.089
0.6	0.745	0.431	0.244	0.140	0.111	0.105	0.111	0.122	0.124	0.109	0.093
1.0	0.780	0.486	0.289	0.167	0.135	0.125	0.130	0.142	0.144	0.125	0.106
2.0	0.818	0.559	0.360	0.212	0.173	0.160	0.162	0.174	0.173	0.153	0.127
3.0	0.840	0.600	0.405	0.245	0.201	0.188	0.186	0.197	0.195	0.174	0.143
4.0	0.856	0.629	0.438	0.271	0.222	0.209	0.205	0.216	0.213	0.190	0.156
5.0	0.868	0.652	0.464	0.294	0.241	0.227	0.222	0.231	0.228	0.204	0.167
6.0	0.876	0.671	0.485	0.312	0.258	0.241	0.236	0.245	0.240	0.216	0.177

* The principal axes of the small spheres and thick ellipsoids are in the ratios of 1/1/1 and 1/0.667/1.333.

From: Brownell, G. L., Ellett, W. H., and Reddy, A. R., Absorbed Fractions for Photon Dosimetry. *J. Nucl. Med.*, Suppl. 1:37, 1968. (Courtesy of the Society of Nuclear Medicine.)

Table 13.

ABSORBED FRACTIONS FOR UNIFORM DISTRIBUTION OF ACTIVITY IN FLAT ELLIPSOIDS*

Mass, kg	φ								
	0.020 Mev	0.030 Mev	0.040 Mev	0.060 Mev	0.080 Mev	0.100 Mev	0.160 Mev	0.662 Mev	2.75 Mev
0.3	0.627	0.306	0.164	0.090	0.075	0.072	0.078	0.084	0.062
0.4	0.654	0.334	0.179	0.098	0.081	0.079	0.085	0.095	0.069
0.5	0.674	0.356	0.192	0.106	0.087	0.085	0.090	0.103	0.074
0.6	0.690	0.374	0.204	0.112	0.092	0.090	0.095	0.109	0.079
1.0	0.731	0.423	0.243	0.134	0.109	0.106	0.112	0.128	0.093
2.0	0.779	0.492	0.305	0.173	0.140	0.133	0.140	0.154	0.112
3.0	0.803	0.533	0.344	0.200	0.162	0.154	0.159	0.171	0.125
4.0	0.820	0.564	0.372	0.221	0.181	0.171	0.174	0.185	0.136
5.0	0.833	0.588	0.394	0.238	0.197	0.185	0.187	0.197	0.146
6.0	0.844	0.608	0.414	0.254	0.211	0.198	0.198	0.209	0.156

* The principal axes of the flat ellipsoids are in the ratio of 1/0.5/2.0.

From: Brownell, G. L., Ellett, W. H., and Reddy, A. R., Absorbed Fractions for Photon Dosimetry. *J. Nucl. Med.*, Suppl. 1: 38, 1968. (Courtesy of the Society of Nuclear Medicine.)

Table 14.

FRACTIONAL INCREASE OF ABSORBED FRACTION IN CENTRAL ORGAN DUE TO BACKSCATTERED RADIATION FROM 70-kg PHANTOM SURROUNDING ORGAN

Energy, Mev	0.020	0.030	0.040	0.060	0.080	0.100	0.160	0.364	0.662	1.46	2.75
Fractional Increase of φ	1.01	1.08	1.19	1.24	1.28	1.26	1.17	1.05	1.04	1.02	1.01

From: Brownell, G. L., Ellett, W. H., and Reddy, A. R., Absorbed Fractions for Photon Dosimetry. *J. Nucl. Med.*, Suppl. 1: 38, 1968. (Courtesy of the Society of Nuclear Medicine.)

Table 15. RADIUS OF SPHERE FOR $\phi_{ph}(x)$*

$\phi_{ph}(x)$	Photon Energy (E), Mev								
	0.015	0.02	0.03	0.04	0.05	0.06	0.08	0.10	0.15
	Radius of Sphere, cm								
0.05	4.00E − 02	9.84E − 02	3.26E − 01	6.74E − 01	1.04E 00	1.32E 00	1.63E 00	1.72E 00	1.68E 00
0.10	8.18E − 02	2.00E − 01	6.41E − 01	1.27E 00	1.91E 00	2.40E 00	2.96E 00	3.16E 00	3.20E 00
0.15	1.26E − 01	3.04E − 01	9.52E − 01	1.84E 00	2.70E 00	3.36E 00	4.14E 00	4.47E 00	4.62E 00
0.20	1.72E − 01	4.13E − 01	1.27E 00	2.39E 00	3.45E 00	4.27E 00	5.25E 00	5.69E 00	5.98E 00
0.25	2.20E − 01	5.27E − 01	1.58E 00	2.94E 00	4.19E 00	5.15E 00	6.32E 00	6.88E 00	7.32E 00
0.30	2.72E − 01	6.47E − 01	1.91E 00	3.49E 00	4.93E 00	6.03E 00	7.38E 00	8.05E 00	8.64E 00
0.35	3.27E − 01	7.73E − 01	2.25E 00	4.06E 00	5.68E 00	6.91E 00	8.44E 00	9.22E 00	9.98E 00
0.40	3.86E − 01	9.08E − 01	2.61E 00	4.65E 00	6.44E 00	7.81E 00	9.51E 00	1.04E 01	1.13E 01
0.45	4.50E − 01	1.05E 00	2.99E 00	5.26E 00	7.24E 00	8.73E 00	1.06E 01	1.16E 01	1.27E 01
0.50	5.19E − 01	1.21E 00	3.39E 00	5.91E 00	8.07E 00	9.71E 00	1.18E 01	1.29E 01	1.42E 01
0.55	5.96E − 01	1.38E 00	3.83E 00	6.60E 00	8.96E 00	1.07E 01	1.30E 01	1.42E 01	1.57E 01
0.60	6.81E − 01	1.57E 00	4.31E 00	7.35E 00	9.91E 00	1.18E 01	1.43E 01	1.57E 01	1.74E 01
0.65	7.77E − 01	1.79E 00	4.84E 00	8.18E 00	1.10E 01	1.30E 01	1.57E 01	1.72E 01	1.92E 01
0.70	8.87E − 01	2.03E 00	5.45E 00	9.11E 00	1.21E 01	1.44E 01	1.73E 01	1.90E 01	2.11E 01
0.75	1.02E 00	2.32E 00	6.15E 00	1.02E 01	1.35E 01	1.59E 01	1.91E 01	2.09E 01	2.34E 01
0.80	1.18E 00	2.67E 00	6.99E 00	1.14E 01	1.50E 01	1.77E 01	2.12E 01	2.33E 01	2.60E 01
0.85	1.38E 00	3.11E 00	8.05E 00	1.30E 01	1.70E 01	2.00E 01	2.38E 01	2.61E 01	2.93E 01
0.90	1.67E 00	3.74E 00	9.51E 00	1.52E 01	1.97E 01	2.30E 01	2.73E 01	3.00E 01	3.37E 01
0.95	2.15E 00	4.79E 00	1.20E 01	1.88E 01	2.41E 01	2.80E 01	3.31E 01	3.62E 01	4.08E 01

Table 15. RADIUS OF SPHERE FOR $\phi_{ph}(x)$* (Continued)

$\phi_{ph}(x)$	Photon Energy (E), Mev									
	0.2	0.3	0.4	0.5	0.6	0.8	1.0	1.5	2.0	
	Radius of Sphere, cm									
0.05	1.62E 00	1.53E 00	1.51E 00	1.51E 00	1.52E 00	1.57E 00	1.62E 00	1.78E 00	1.93E 00	
0.10	3.13E 00	3.02E 00	3.01E 00	3.03E 00	3.07E 00	3.17E 00	3.29E 00	3.61E 00	3.92E 00	
0.15	4.59E 00	4.49E 00	4.51E 00	4.56E 00	4.63E 00	4.81E 00	5.00E 00	5.50E 00	5.98E 00	
0.20	6.01E 00	5.96E 00	6.02E 00	6.11E 00	6.23E 00	6.49E 00	6.76E 00	7.46E 00	8.12E 00	
0.25	7.41E 00	7.43E 00	7.55E 00	7.69E 00	7.86E 00	8.22E 00	8.58E 00	9.49E 00	1.03E 01	
0.30	8.82E 00	8.92E 00	9.11E 00	9.32E 00	9.54E 00	1.00E 01	1.05E 01	1.16E 01	1.27E 01	
0.35	1.02E 01	1.05E 01	1.07E 01	1.10E 01	1.13E 01	1.19E 01	1.24E 01	1.38E 01	1.51E 01	
0.40	1.17E 01	1.20E 01	1.24E 01	1.27E 01	1.31E 01	1.38E 01	1.45E 01	1.62E 01	1.77E 01	
0.45	1.32E 01	1.37E 01	1.41E 01	1.46E 01	1.50E 01	1.59E 01	1.67E 01	1.87E 01	2.05E 01	
0.50	1.48E 01	1.54E 01	1.60E 01	1.65E 01	1.70E 01	1.80E 01	1.90E 01	2.13E 01	2.35E 01	
0.55	1.64E 01	1.72E 01	1.79E 01	1.86E 01	1.92E 01	2.04E 01	2.15E 01	2.42E 01	2.67E 01	
0.60	1.82E 01	1.92E 01	2.00E 01	2.08E 01	2.15E 01	2.29E 01	2.42E 01	2.73E 01	3.02E 01	
0.65	2.02E 01	2.13E 01	2.23E 01	2.32E 01	2.41E 01	2.57E 01	2.72E 01	3.08E 01	3.41E 01	
0.70	2.23E 01	2.37E 01	2.49E 01	2.59E 01	2.69E 01	2.88E 01	3.06E 01	3.47E 01	3.85E 01	
0.75	2.47E 01	2.64E 01	2.78E 01	2.90E 01	3.02E 01	3.24E 01	3.45E 01	3.93E 01	4.36E 01	
0.80	2.76E 01	2.96E 01	3.12E 01	3.27E 01	3.41E 01	3.67E 01	3.91E 01	4.47E 01	4.97E 01	
0.85	3.12E 01	3.35E 01	3.55E 01	3.73E 01	3.90E 01	4.21E 01	4.49E 01	5.15E 01	5.74E 01	
0.90	3.59E 01	3.88E 01	4.13E 01	4.35E 01	4.55E 01	4.93E 01	5.28E 01	6.08E 01	6.80E 01	
0.95	4.36E 01	4.75E 01	5.07E 01	5.36E 01	5.62E 01	6.12E 01	6.58E 01	7.62E 01	8.56E 01	

* The values of radii x (cm) of spherical volumes centered around point isotropic sources in which indicated fraction $\phi_{ph}(x)$ of energy E(Mev) is deposited. The digits following the symbol E indicate the powers of ten by which each number is to be multiplied.

From: Berger, M. J., Energy Deposition in Water by Photons from Point Isotropic Sources, *J. Nucl. Med.*, Suppl. 1: 25, 1968. (Courtesy of the Society of Nuclear Medicine.)

Table 16.
SPECIFIC ABSORBED FRACTION $\phi_{ph}(x)$ FOR POINT ISOTROPIC SOURCES IN WATER*

x, cm	0.015	0.02	0.03	0.04	0.05	0.06	0.08	0.10	0.15
0.1	9.16E+00	3.97E+00	1.21E+00	5.57E-01	3.44E-01	2.63E-01	2.13E-01	2.07E-01	2.22E-01
0.2	2.04E+00	9.63E-01	3.07E-01	1.43E-01	8.87E-02	6.77E-02	5.45E-02	5.26E-02	5.61E-02
0.4	4.00E-01	2.24E-01	7.84E-02	3.76E-02	2.35E-02	1.79E-02	1.42E-02	1.36E-02	1.43E-02
0.6	1.37E-01	9.14E-02	3.53E-02	1.74E-02	1.10E-02	8.33E-03	6.56E-03	6.21E-03	6.45E-03
0.8	5.90E-02	4.68E-02	1.99E-02	1.01E-02	6.44E-03	4.90E-03	3.83E-03	3.59E-03	3.68E-03
1.0	2.87E-02	2.71E-02	1.27E-02	6.68E-03	4.29E-03	3.26E-03	2.54E-03	2.36E-03	2.39E-03
1.5	6.36E-03	9.13E-03	5.50E-03	3.12E-03	2.06E-03	1.58E-03	1.22E-03	1.12E-03	1.10E-03
2.0	1.77E-03	3.83E-03	2.94E-03	1.80E-03	1.23E-03	9.53E-04	7.31E-04	6.63E-04	6.40E-04
2.5	5.58E-04	1.80E-03	1.76E-03	1.17E-03	8.21E-04	6.44E-04	4.95E-04	4.45E-04	4.22E-04
3.0	1.90E-04	9.16E-04	1.13E-03	8.05E-04	5.87E-04	4.67E-04	3.60E-04	3.22E-04	3.00E-04
4.0	2.53E-05	2.73E-04	5.25E-04	4.35E-04	3.39E-04	2.78E-04	2.18E-04	1.92E-04	1.76E-04
5.0	3.79E-06	9.15E-05	2.72E-04	2.60E-04	2.16E-04	1.83E-04	1.46E-04	1.30E-04	1.16E-04
6.0	6.16E-07	3.30E-05	1.50E-04	1.64E-04	1.46E-04	1.27E-04	1.04E-04	9.29E-05	8.23E-05
8.0	1.90E-08	4.88E-06	5.15E-05	7.31E-05	7.32E-05	6.83E-05	5.91E-05	5.33E-05	4.70E-05
10.0	6.31E-10	8.01E-07	1.95E-05	3.54E-05	3.98E-05	3.95E-05	3.62E-05	3.34E-05	2.96E-05
12.0	2.38E-11	1.42E-07	7.84E-06	1.81E-05	2.27E-05	2.39E-05	2.33E-05	2.19E-05	1.98E-05
14.0		2.67E-08	3.28E-06	9.51E-06	1.33E-05	1.49E-05	1.53E-05	1.48E-05	1.37E-05
16.0		5.19E-09	1.41E-06	5.13E-06	7.94E-06	9.41E-06	1.03E-05	1.02E-05	9.68E-06
18.0		1.02E-09	6.18E-07	2.81E-06	4.81E-06	6.03E-06	7.00E-06	7.16E-06	6.96E-06
20.0		2.03E-10	2.76E-07	1.56E-06	2.95E-06	3.91E-06	4.80E-06	5.05E-06	5.07E-06
25.0		4.05E-12	3.88E-08	3.76E-07	8.99E-07	1.36E-06	1.92E-06	2.17E-06	2.36E-06
30.0			5.80E-09	9.42E-08	2.83E-07	4.84E-07	7.82E-07	9.50E-07	1.13E-06
35.0			8.99E-10	2.43E-08	9.09E-08	1.76E-07	3.24E-07	4.22E-07	5.49E-07
40.0			1.43E-10	6.40E-09	2.97E-08	6.46E-08	1.35E-07	1.88E-07	2.69E-07

Table 16. (Continued)
SPECIFIC ABSORBED FRACTION $\phi_{ph}(x)$ FOR POINT ISOTROPIC SOURCES IN WATER*

x, cm	\multicolumn{8}{c}{Photon Energy (E), Mev}
	0.20	0.30	0.40	0.50	0.60	0.80	1.00	1.50	2.00
0.1	2.38E − 01	2.54E − 01	2.61E − 01	2.63E − 01	2.62E − 01	2.55E − 01	2.47E − 01	2.26E − 01	2.09E − 01
0.2	5.97E − 02	6.38E − 02	6.54E − 02	6.58E − 02	6.54E − 02	6.37E − 02	6.17E − 02	5.65E − 02	5.22E − 02
0.4	1.51E − 02	1.60E − 02	1.64E − 02	1.65E − 02	1.63E − 02	1.59E − 02	1.54E − 02	1.41E − 02	1.30E − 02
0.6	6.78E − 03	7.14E − 03	7.28E − 03	7.31E − 03	7.25E − 03	7.04E − 03	6.81E − 03	6.24E − 03	5.76E − 03
0.8	3.85E − 03	4.03E − 03	4.10E − 03	4.11E − 03	4.07E − 03	3.95E − 03	3.82E − 03	3.50E − 03	3.23E − 03
1.0	2.49E − 03	2.59E − 03	2.63E − 03	2.63E − 03	2.60E − 03	2.52E − 03	2.44E − 03	2.23E − 03	2.06E − 03
1.5	1.13E − 03	1.16E − 03	1.17E − 03	1.17E − 03	1.15E − 03	1.11E − 03	1.08E − 03	9.84E − 04	9.08E − 04
2.0	6.49E − 04	6.58E − 04	6.60E − 04	6.55E − 04	6.45E − 04	6.23E − 04	6.01E − 04	5.49E − 04	5.06E − 04
2.5	4.23E − 04	4.24E − 04	4.23E − 04	4.19E − 04	4.11E − 04	3.96E − 04	3.82E − 04	3.48E − 04	3.21E − 04
3.0	2.98E − 04	2.96E − 04	2.94E − 04	2.90E − 04	2.84E − 04	2.73E − 04	2.63E − 04	2.40E − 04	2.21E − 04
4.0	1.72E − 04	1.68E − 04	1.65E − 04	1.62E − 04	1.58E − 04	1.52E − 04	1.46E − 04	1.33E − 04	1.22E − 04
5.0	1.12E − 04	1.08E − 04	1.05E − 04	1.03E − 04	1.00E − 04	9.56E − 05	9.17E − 05	8.34E − 05	7.70E − 05
6.0	7.86E − 05	7.46E − 05	7.24E − 05	7.05E − 05	6.86E − 05	6.53E − 05	6.26E − 05	5.69E − 05	5.25E − 05
8.0	4.43E − 05	4.13E − 05	3.98E − 05	3.85E − 05	3.74E − 05	3.55E − 05	3.39E − 05	3.08E − 05	2.85E − 05
10.0	2.78E − 05	2.57E − 05	2.46E − 05	2.38E − 05	2.30E − 05	2.18E − 05	2.08E − 05	1.89E − 05	1.75E − 05
12.0	1.86E − 05	1.71E − 05	1.63E − 05	1.58E − 05	1.53E − 05	1.45E − 05	1.38E − 05	1.26E − 05	1.17E − 05
14.0	1.29E − 05	1.19E − 05	1.14E − 05	1.10E − 05	1.06E − 05	1.01E − 05	9.68E − 06	8.85E − 06	8.24E − 06
16.0	9.23E − 06	8.53E − 06	8.18E − 06	7.92E − 06	7.69E − 06	7.32E − 06	7.03E − 06	6.46E − 06	6.03E − 06
18.0	6.71E − 06	6.25E − 06	6.03E − 06	5.86E − 06	5.70E − 06	5.45E − 06	5.25E − 06	4.85E − 06	4.55E − 06
20.0	4.94E − 06	4.65E − 06	4.52E − 06	4.41E − 06	4.30E − 06	4.14E − 06	4.00E − 06	3.73E − 06	3.52E − 06
25.0	2.39E − 06	2.33E − 06	2.31E − 06	2.29E − 06	2.26E − 06	2.22E − 06	2.17E − 06	2.07E − 06	1.98E − 06
30.0	1.20E − 06	1.22E − 06	1.24E − 06	1.25E − 06	1.26E − 06	1.26E − 06	1.26E − 06	1.23E − 06	1.20E − 06
35.0	6.08E − 07	6.51E − 07	6.85E − 07	7.09E − 07	7.24E − 07	7.44E − 07	7.56E − 07	7.67E − 07	7.63E − 07
40.0	3.13E − 07	3.54E − 07	3.85E − 07	4.09E − 07	4.26E − 07	4.51E − 07	4.68E − 07	4.92E − 07	5.02E − 07

* The corresponding absorbed-dose rate (rad/sec) for a source emitting 1 photon per second is $k\Phi_{ph}(x)$, where $k = 1.60 \times 10^{-8}$ g-rad/Mev and E is the source energy in Mev. The digits following the symbol E indicate the powers of ten by which each number is to be multiplied.

From: Berger, M. J., Energy Deposition in Water by Photons from Point Isotropic Sources. *J. Nucl. Med.*, Suppl. 1: 23, 1968 (Courtesy of the Society of Nuclear Medicine.)

$R(x)$ = absorbed dose rate, rad/sec;

x = distance from source, cm;

$\Phi(x)$ = point isotropic specific absorbed fraction, g^{-1};

$k = 1.60 \times 10^{-8}$ g-rad/Mev;

n = number of photons of energy E emitted; (corrected for internal conversion) per disintegration;

E = energy of photons emitted by the source, Mev per photon;

A = source activity, number of disintegrations/sec;

ρ = density of the medium, g/cm^3;

μ = linear photon-attenuation coefficient at the source energy, cm^{-1};

μ_{en} = linear photon-energy-absorption coefficient at the source energy, cm^{-1}.

Values for $B_{en}(\mu x)$ are tabulated in Table 17 for various mean-free paths (mfp) versus photon energy in an infinite water medium.

Values for the *linear photon-attenuation coefficient*, μ, and for the linear *photon-energy-absorption coefficient* at the source energy, μ_{en}, may be found in Table 18.

Geometrical Factors

The *geometry factor*, g_p or $\bar{g}$, has been used in calculation of radiation exposure for many years. These values may be calculated from equations 18a and 18b.

$$g_p = g(p \leftarrow v) = \int_v \frac{e^{-\mu_{eff} x}}{x^2} \, dv \text{ cm,} \tag{18a}$$

$$\bar{g} = g(v \leftrightarrow v) = \frac{1}{v} \int_v g_p \, dv \text{ cm,} \tag{18b}$$

where

$e^{-\mu_{eff} x}$ = the tissue attenuation of the photons originating from the elemental volume (dv) and incident on p;

$\frac{1}{x^2}$ = the inverse-square law diminution of the radiation field emanating from dv and incident on p.

The "effective" tissue absorption coefficient, $e^{-\mu_{eff} x}$, which is a function of the distance from the point source, is usually taken to be 0.028 cm^{-1}. These calculations are valid only to the extent that this value is an adequate representation of the distribution of absorbed energy around a point source, and do not take into account degradation of photon energies. Focht[29] has made some recent calculations for the average geometrical factor, which are given in Table 19.

The average geometrical factor, $\bar{g}$, can be approximated for spheres of unit density with a radius (R) less than 10 cm by equation (19).

$$\bar{g} = 3\pi R \text{ cm.} \tag{19}$$

The geometrical factors may be converted into a "pseudo" absorbed fraction[2] by the use of equation (20).

$$\phi_{ph}(v \leftarrow r) = \frac{\mu_{en}}{4\pi} g(r \leftarrow v),$$

$$r = p \quad \text{or} \quad v,$$

$$g(p \leftarrow v) = g_p = g(v \leftrightarrow v) = \bar{g}, \tag{20}$$

Table 17.
ENERGY ABSORPTION BUILD-UP FACTOR $B_{en}(\mu x)$ FOR POINT ISOTROPIC SOURCES IN WATER*

μx (mfp)	0.015	0.02	0.03	0.04	0.05	0.06	0.08	0.10	0.15
0.05	1.02E 00	1.03E 00	1.08E 00	1.12E 00	1.13E 00	1.13E 00	1.12E 00	1.10E 00	1.08E 00
0.10	1.03E 00	1.06E 00	1.16E 00	1.23E 00	1.27E 00	1.27E 00	1.24E 00	1.21E 00	1.17E 00
0.20	1.06E 00	1.12E 00	1.31E 00	1.47E 00	1.55E 00	1.57E 00	1.51E 00	1.45E 00	1.36E 00
0.30	1.08E 00	1.18E 00	1.45E 00	1.71E 00	1.86E 00	1.89E 00	1.82E 00	1.72E 00	1.57E 00
0.40	1.10E 00	1.22E 00	1.59E 00	1.96E 00	2.18E 00	2.23E 00	2.15E 00	2.02E 00	1.80E 00
0.50	1.12E 00	1.27E 00	1.72E 00	2.21E 00	2.51E 00	2.61E 00	2.52E 00	2.34E 00	2.05E 00
0.60	1.14E 00	1.31E 00	1.85E 00	2.46E 00	2.86E 00	3.01E 00	2.92E 00	2.70E 00	2.33E 00
0.80	1.17E 00	1.38E 00	2.10E 00	2.97E 00	3.62E 00	3.89E 00	3.81E 00	3.51E 00	2.96E 00
1.00	1.19E 00	1.45E 00	2.34E 00	3.50E 00	4.43E 00	4.87E 00	4.85E 00	4.46E 00	3.69E 00
1.20	1.21E 00	1.50E 00	2.57E 00	4.05E 00	5.30E 00	5.95E 00	6.03E 00	5.55E 00	4.53E 00
1.40	1.23E 00	1.55E 00	2.79E 00	4.60E 00	6.23E 00	7.14E 00	7.37E 00	6.79E 00	5.49E 00
1.60	1.24E 00	1.60E 00	3.00E 00	5.17E 00	7.22E 00	8.43E 00	8.85E 00	8.19E 00	6.58E 00
1.80	1.26E 00	1.64E 00	3.21E 00	5.76E 00	8.27E 00	9.83E 00	1.05E 01	9.76E 00	7.80E 00
2.00	1.28E 00	1.68E 00	3.42E 00	6.36E 00	9.37E 00	1.13E 01	1.23E 01	1.15E 01	9.16E 00
2.50	1.31E 00	1.78E 00	3.93E 00	7.91E 00	1.24E 01	1.55E 01	1.76E 01	1.67E 01	1.32E 01
3.00	1.34E 00	1.87E 00	4.42E 00	9.53E 00	1.57E 01	2.04E 01	2.40E 01	2.31E 01	1.83E 01
3.50	1.37E 00	1.96E 00	4.91E 00	1.12E 01	1.93E 01	2.60E 01	3.17E 01	3.09E 01	2.46E 01
4.00	1.40E 00	2.04E 00	5.39E 00	1.30E 01	2.33E 01	3.22E 01	4.06E 01	4.02E 01	3.21E 01
4.50	1.43E 00	2.11E 00	5.86E 00	1.48E 01	2.76E 01	3.92E 01	5.10E 01	5.11E 01	4.09E 01
5.00	1.45E 00	2.18E 00	6.32E 00	1.67E 01	3.22E 01	4.69E 01	6.28E 01	6.37E 01	5.13E 01
6.00	1.48E 00	2.30E 00	7.22E 00	2.06E 01	4.23E 01	6.46E 01	9.13E 01	9.49E 01	7.69E 01
7.00	1.51E 00	2.40E 00	8.09E 00	2.48E 01	5.38E 01	8.56E 01	1.27E 02	1.35E 02	1.10E 02
8.00	1.54E 00	2.49E 00	8.96E 00	2.93E 01	6.67E 01	1.10E 02	1.70E 02	1.85E 02	1.51E 02
9.00	1.57E 00	2.60E 00	9.84E 00	3.40E 01	8.10E 01	1.38E 02	2.23E 02	2.46E 02	2.03E 02
10.00	1.60E 00	2.71E 00	1.07E 01	3.90E 01	9.68E 01	1.70E 02	2.86E 02	3.20E 02	2.65E 02
11.00	1.64E 00	2.81E 00	1.16E 01	4.42E 01	1.14E 02	2.07E 02	3.59E 02	4.09E 02	3.40E 02
12.00	1.66E 00	2.89E 00	1.25E 01	4.98E 01	1.33E 02	2.48E 02	4.45E 02	5.15E 02	4.28E 02
13.00	1.67E 00	2.95E 00	1.33E 01	5.56E 01	1.54E 02	2.95E 02	5.45E 02	6.39E 02	5.33E 02
14.00	1.66E 00	2.98E 00	1.42E 01	6.17E 01	1.77E 02	3.46E 02	6.60E 02	7.83E 02	6.54E 02
15.00	1.66E 00	3.00E 00	1.50E 01	6.80E 01	2.01E 02	4.04E 02	7.91E 02	9.51E 02	7.95E 02
16.00	1.66E 00	3.04E 00	1.59E 01	7.47E 01	2.28E 02	4.68E 02	9.41E 02	1.14E 03	9.57E 02
17.00	1.69E 00	3.14E 00	1.68E 01	8.18E 01	2.57E 02	5.38E 02	1.11E 03	1.37E 03	1.14E 03
18.00	1.76E 00	3.30E 00	1.78E 01	8.91E 01	2.87E 02	6.16E 02	1.30E 03	1.62E 03	1.35E 03
19.00	1.88E 00	3.55E 00	1.89E 01	9.69E 01	3.21E 02	7.01E 02	1.52E 03	1.90E 03	1.59E 03
20.00	2.02E 00	3.83E 00	2.01E 01	1.05E 02	3.56E 02	7.94E 02	1.76E 03	2.23E 03	1.86E 03

Photon Energy (E), Mev

Table 17. (Continued)
ENERGY ABSORPTION BUILD-UP $B_{en}(\mu x)$ FOR POINT ISOTROPIC SOURCES IN WATER*

μx (mfp)	Photon Energy (E), Mev							
	0.20	0.30	0.40	0.50	0.60	0.80	1.00	1.50	2.00	3.00
0.05	1.07E 00	1.06E 00	1.06E 00	1.05E 00	1.05E 00	1.04E 00	1.04E 00	1.04E 00	1.03E 00	1.03E 00
0.10	1.15E 00	1.12E 00	1.11E 00	1.10E 00	1.10E 00	1.09E 00	1.08E 00	1.07E 00	1.07E 00	1.06E 00
0.20	1.31E 00	1.26E 00	1.23E 00	1.22E 00	1.20E 00	1.18E 00	1.17E 00	1.15E 00	1.14E 00	1.13E 00
0.30	1.49E 00	1.41E 00	1.37E 00	1.34E 00	1.31E 00	1.29E 00	1.27E 00	1.24E 00	1.22E 00	1.19E 00
0.40	1.69E 00	1.57E 00	1.51E 00	1.47E 00	1.44E 00	1.39E 00	1.37E 00	1.32E 00	1.30E 00	1.26E 00
0.50	1.91E 00	1.74E 00	1.66E 00	1.60E 00	1.56E 00	1.51E 00	1.47E 00	1.41E 00	1.38E 00	1.33E 00
0.60	2.14E 00	1.93E 00	1.82E 00	1.75E 00	1.70E 00	1.63E 00	1.58E 00	1.51E 00	1.46E 00	1.39E 00
0.80	2.66E 00	2.34E 00	2.18E 00	2.07E 00	2.00E 00	1.89E 00	1.82E 00	1.71E 00	1.63E 00	1.53E 00
1.00	3.27E 00	2.82E 00	2.59E 00	2.44E 00	2.33E 00	2.18E 00	2.07E 00	1.91E 00	1.81E 00	1.68E 00
1.20	3.97E 00	3.35E 00	3.04E 00	2.83E 00	2.69E 00	2.49E 00	2.35E 00	2.13E 00	2.00E 00	1.82E 00
1.40	4.75E 00	3.95E 00	3.54E 00	3.27E 00	3.08E 00	2.82E 00	2.64E 00	2.36E 00	2.19E 00	1.97E 00
1.60	5.64E 00	4.62E 00	4.09E 00	3.75E 00	3.50E 00	3.17E 00	2.94E 00	2.59E 00	2.38E 00	2.12E 00
1.80	6.64E 00	5.36E 00	4.70E 00	4.26E 00	3.96E 00	3.54E 00	3.26E 00	2.84E 00	2.58E 00	2.27E 00
2.00	7.74E 00	6.17E 00	5.35E 00	4.81E 00	4.44E 00	3.93E 00	3.59E 00	3.08E 00	2.78E 00	2.41E 00
2.50	1.10E 01	8.54E 00	7.21E 00	6.37E 00	5.79E 00	5.00E 00	4.48E 00	3.73E 00	3.30E 00	2.79E 00
3.00	1.51E 01	1.14E 01	9.42E 00	8.16E 00	7.31E 00	6.18E 00	5.45E 00	4.40E 00	3.82E 00	3.16E 00
3.50	2.00E 01	1.48E 01	1.20E 01	1.02E 01	9.02E 00	7.47E 00	6.48E 00	5.11E 00	4.37E 00	3.54E 00
4.00	2.59E 01	1.88E 01	1.49E 01	1.25E 01	1.09E 01	8.86E 00	7.59E 00	5.84E 00	4.92E 00	3.93E 00
4.50	3.28E 01	2.34E 01	1.81E 01	1.50E 01	1.30E 01	1.03E 01	8.75E 00	6.60E 00	5.49E 00	4.31E 00
5.00	4.08E 01	2.86E 01	2.18E 01	1.78E 01	1.52E 01	1.19E 01	9.98E 00	7.38E 00	6.07E 00	4.71E 00
6.00	6.04E 01	4.11E 01	3.02E 01	2.40E 01	2.02E 01	1.54E 01	1.26E 01	9.04E 00	7.28E 00	5.51E 00
7.00	8.53E 01	5.65E 01	4.02E 01	3.13E 01	2.59E 01	1.93E 01	1.55E 01	1.08E 01	8.53E 00	6.32E 00
8.00	1.16E 02	7.51E 01	5.19E 01	3.96E 01	3.23E 01	2.35E 01	1.87E 01	1.26E 01	9.82E 00	7.14E 00
9.00	1.54E 02	9.73E 01	6.54E 01	4.90E 01	3.95E 01	2.82E 01	2.20E 01	1.45E 01	1.11E 01	7.97E 00
10.00	1.99E 02	1.23E 02	8.07E 01	5.96E 01	4.74E 01	3.33E 01	2.56E 01	1.65E 01	1.25E 01	8.80E 00
11.00	2.53E 02	1.54E 02	9.81E 01	7.13E 01	5.61E 01	3.87E 01	2.94E 01	1.86E 01	1.39E 01	9.63E 00
12.00	3.17E 02	1.88E 02	1.17E 02	8.41E 01	6.56E 01	4.45E 01	3.35E 01	2.07E 01	1.53E 01	1.05E 01
13.00	3.90E 02	2.28E 02	1.39E 02	9.82E 01	7.59E 01	5.06E 01	3.77E 01	2.29E 01	1.67E 01	1.13E 01
14.00	4.75E 02	2.73E 02	1.63E 02	1.14E 02	8.69E 01	5.71E 01	4.21E 01	2.52E 01	1.82E 01	1.22E 01
15.00	5.72E 02	3.24E 02	1.89E 02	1.30E 02	9.87E 01	6.40E 01	4.67E 01	2.75E 01	1.97E 01	1.31E 01
16.00	6.83E 02	3.80E 02	2.17E 02	1.48E 02	1.11E 02	7.12E 01	5.14E 01	2.99E 01	2.12E 01	1.40E 01
17.00	8.09E 02	4.43E 02	2.48E 02	1.67E 02	1.25E 02	7.87E 01	5.64E 01	3.23E 01	2.27E 01	1.49E 01
18.00	9.51E 02	5.13E 02	2.81E 02	1.87E 02	1.39E 02	8.66E 01	6.15E 01	3.47E 01	2.42E 01	1.57E 01
19.00	1.11E 03	5.90E 02	3.17E 02	2.09E 02	1.54E 02	9.49E 01	6.68E 01	3.72E 01	2.57E 01	1.65E 01
20.00	1.29E 03	6.74E 02	3.56E 02	2.32E 02	1.70E 02	1.03E 02	7.23E 01	3.96E 01	2.72E 01	1.73E 01

* The digits following the symbol E indicate the powers of ten by which each number is to be multiplied.

From: Berger, M. J., Energy Deposition in Water by Photons from Point Isotropic Sources. *J. Nucl. Med.*, Suppl. 1: 20-21, 1968. (Courtesy of the Society of Nuclear Medicine.)

Table 18.

MASS-ATTENUATION COEFFICIENT μ/ρ AND ENERGY-ABSORPTION COEFFICIENT μ_{en}/ρ FOR WATER AND MUSCLE[28]*

E, Mev	Water		Muscle	
	μ/ρ, cm²/g	μ_{en}/ρ, cm²/g	μ/ρ, cm²/g	μ_{en}/ρ, cm²/g
0.01	4.99	4.79	5.09	4.87
0.015	1.48	1.28	1.53	1.32
0.02	0.711	0.512	0.730	0.533
0.03	0.337	0.149	0.342	0.154
0.04	0.248	0.0677	0.249	0.0701
0.05	0.214	0.0418	0.214	0.0431
0.06	0.197	0.0320	0.196	0.0328
0.08	0.179	0.0262	0.178	0.0264
0.1	0.168	0.0256	0.167	0.0256
0.15	0.149	0.0277	0.147	0.0275
0.2	0.136	0.0297	0.135	0.0294
0.3	0.118	0.0319	0.117	0.0317
0.4	0.106	0.0328	0.105	0.0325
0.5	0.0966	0.0330	0.0958	0.0328
0.6	0.0894	0.0329	0.0886	0.0325
0.8	0.0785	0.0321	0.0778	0.0318
1.0	0.0706	0.0311	0.0699	0.0308
1.5	0.0575	0.0284	0.0570	0.0282
2.0	0.0493	0.0263	0.0489	0.0259
3.0	0.0396	0.0233	0.0392	0.0227

* Assumed composition of muscle (fraction by weight): 0.1020 H, 0.1230 C, 0.0350 N, 0.7290 O, 0.0008 Na, 0.0002 Mg, 0.0020 P, 0.0050 S, and 0.0030 K.

From: Berger, M. J., Energy Deposition in Water by Photons from Point Isotropic Sources. *J. Nucl. Med.*, Suppl. 1: 19, 1968. (Courtesy of the Society of Nuclear Medicine.)

Table 19.

AVERAGE GEOMETRICAL FACTOR, $\bar{g}$, FOR CYLINDERS CONTAINING A UNIFORMLY DISTRIBUTED γ-RAY EMITTER ($\mu = 0.028$)

Length of Cylinder, cm	Radius of Cylinder, cm										
	1	2	3	5	7	10	15	20	25	30	35
1	3.8	7.5	10.2	13.0	13.5	13.8	15.1	16.0	17.5	18.0	19.0
2	6.5	11.7	15.7	21.6	23.2	25.2	28.1	30.5	32.8	35.4	37.3
3	8.4	14.7	19.8	27.7	31.0	34.5	39.2	42.9	46.5	49.5	52.5
5	10.6	18.8	25.6	36.0	42.4	48.5	56.1	62.6	68.2	73.0	77.2
7	11.6	21.4	29.3	41.4	50.0	59.0	68.7	77.8	84.7	90.2	93.8
10	12.7	23.6	33.0	47.1	57.8	70.2	83.2	94.0	103	109	113
15	13.7	25.6	36.4	53.2	66.1	81.4	99.7	113	123	130	135
20	14.2	26.7	38.0	56.3	72.2	89.6	111	127	139	147	152
30	14.5	27.6	39.7	59.9	76.8	98.8	124	144	159	172	179
40	14.8	28.2	40.7	62.4	80.0	103	133	156	175	187	197
50	14.8	28.4	41.3	64.1	82.2	106	139	165	185	199	208
60	14.8	28.7	41.7	65.5	84.0	109	143	171	193	206	216
70	14.8	28.8	41.9	65.6	85.3	111	146	174	196	212	222
80	14.8	28.8	42.1	65.8	86.0	112	148	176	198	214	226
90	14.8	28.9	42.3	66.0	86.5	113	149	177	199	216	228
100	14.8	29.2	42.5	66.2	86.8	114	150	179	201	218	230

From: Focht, E. F., Quimby, E. H., and Gershowitz, M., Revised Average Geometric Factors for Cylinders in Isotope Dosage, Part I *Radiology*, 85: 151, 1965. (Courtesy of *Radiology*.)

where μ_{en} is the linear photon-energy-absorption coefficient at the source energy (Table 18). The values for the absorbed fraction calculated by equation (20) carry only the degree of validity attached to the original g_p or $\bar{g}$ values. Quimby[18] has stated that the use of g_p or $\bar{g}$ is valid only with photon energies between 0.1 and 2.0 Mev.

CONCENTRATION OF ACTIVITY AS A FUNCTION OF TIME

Tissue Distribution Data

The concentration of an administered radionuclide must be determined as a function of time in various organs, tissues and body compartments. These data are obtained from sampling blood, urine, and feces, from external counting, and from tissues removed at operation or autopsy. Methods of obtaining these data are given by Ross et al.[30]

Biological, Physical, and Effective Half-Life

The quantity of a radionuclide remaining in an organ can often be represented by a single exponential function of time, or as a sum of several independent exponentials. The sum of several independent exponentials is also an exponential. Thus the combined rate of elimination due to both biological and physical processes can be described by an exponential function. If the biological elimination of the radionuclide from an organ can be represented by a single biological clearance rate, λ_1, and the physical decay of the radionuclide is given by its physical decay constant, λ, then the effective decay constant, λ_{eff1}, is:

$$\lambda_{eff1} = \lambda_1 + \lambda. \tag{21a}$$

Each biological clearance rate will have its own effective clearance rate, which is the sum of the biological clearance-rate constant and the physical-decay constant. The effective biological clearance rate of a radionuclide in an organ is related to the effective biological half-time of the radionuclide in that organ in the same manner as the physical-decay constant is related to the half-life of a radionuclide. This relationship is given in equation (21b).

$$\lambda = \frac{0.693}{T}. \tag{21b}$$

Substituting into equations (21a) and (21b), the effective half-life is given by equation (22):

$$T_{eff1} = \frac{T \times T_1}{T + T_1}. \tag{22}$$

The effective half-life is always less than both the physical half-life and biological half-life.

Equations for Cumulated Activity and Cumulated Concentration

The absorbed dose is determined from the cumulated activity, $\tilde{A}$, and the cumulated concentration, $\tilde{C}$, by using the equations that follow.[4] A set of equations for calculation of absorbed-dose rate from the activity, A, and the concentration, C, may be derived by making the appropriate changes in the units of these equations.

General Equations.

$$\tilde{A} = \int_{t_1}^{t_2} A(t)\, dt\, \mu\text{ Ci-hr}, \tag{23a}$$

$$\tilde{C} = \int_{t_1}^{t_2} C(t)\, dt\, \frac{\mu\text{Ci-hr}}{g}, \tag{23b}$$

where A(t) = the activity, in µCi, irradiating the tissue at any time t, in hours;

C(t) = the concentration, in µCi/g, of the radionuclide in the tissue at any time t, in hours, for which the absorbed dose is being calculated;

t_1 and t_2 = the time interval, in hours, for which the absorbed dose is being calculated; $0 \leq t_1 < t_2$ and $t_1 < t_2 \leq$ infinity.

The expressions $A(t)$ and $C(t)$ often consist of one or more exponentials; therefore,

$$A(t) = \sum_j A_j(0) e^{-\lambda_{effj} t},$$

and

$$\lambda_{eff} = \frac{0.693}{T_{eff}};$$

then

$$A(t) = \sum_j A_j(0) e^{-0.693 t/T_{effj}} \; \mu Ci, \tag{24a}$$

$$C(t) = \sum_j C_j(0) e^{-0.693 t/T_{effj}} \frac{\mu Ci}{g}, \tag{24b}$$

where $A_j(0)$ = the initial activity, in μCi, in the tissue for the jth component of the disappearance curve;

T_{effj} = the effective half-life, in hours, of the jth component of the disappearance curve for a given organ;

$C_j(0)$ = the initial concentration, in $\mu Ci/g$ in the tissue for the disappearance curve.

If $A(t)$ and $C(t)$ cannot be fitted by a series of exponentials or by a power function, they may be determined by graphical integration. The equation for cumulated activity may be obtained directly from the equation for cumulated concentration by substituting $\tilde{A}$ for $\tilde{C}$ and $A(t)$ for $C(t)$. The unit should then be changed from $\mu Ci\text{-}hr/g$ to $\mu Ci\text{-}hr$ for cumulated activity.

Integrated Equations for Any Time Interval.

The integrated form of equation (24b) is:

$$\tilde{C} = 1.44 \sum_j C_j(0) T_{effj} (e^{-0.693 t_1/T_{effj}} - e^{-0.693 t_2/T_{effj}}) \frac{\mu Ci\text{-}hr}{g}. \tag{25}$$

The $\tilde{C}$ may be calculated for any time interval from t_1 to t_2. When using these equations, it is assumed that the time during which the radioactivity is accumulated by the tissue is negligible compared to the time during which it is eliminated.

Equation for Complete Elimination of the Radionuclide.

To determine the absorbed dose over the entire period during which the radionuclide is present, it is necessary to determine $\tilde{C}$ from time $t_1 = 0$ to $t_2 =$ infinity. When integrated between these limits, equation (25) becomes:

$$\tilde{C} = \int_0^\infty C(t)\, dt = 1.44 \sum_j C_j(0) T_{effj} \frac{\mu Ci\text{-}hr}{g}. \tag{26}$$

It is sometimes assumed that the slope of the last component of the concentration curve describes the rate of elimination after the last measurement is made. It is more reliable to use the physical half-life of the radionuclide, rather than the extrapolated effective half-life, for the time after the last datum point was obtained. If this is done, equation (26) becomes:

$$\tilde{C} = 1.44 \sum_{j=1}^{j=n-1} C_j(0) T_{effj} + 1.44 C_n(0) T_{effn} (1 - e^{-0.693 t^*/T_{effn}})$$

$$+ 1.44 C_n(t^*) T (e^{-0.693 t^*/T}) \frac{\mu Ci\text{-}hr}{g}, \tag{27}$$

where $C_n(0)$ = The initial concentration, in µCi/g, in the tissue for the longest-lived component of the disappearance curve;

$C_n(t^*)$ = the concentration, in µCi/g, in the tissue for the longest-lived component of the disappearance curve at time t^*, when the last datum point was determined;

T_{effn} = the effective half-life, in hours, of the longest-lived component of the disappearance curve for a given organ;

t^* = the time, in hours, after the activity was administered to the time the last datum point on the disappearance curve was determined;

T = the physical half-life, in hours, of the radionuclide.

Equations to be Used When the Amount of Time Required for Uptake is Significant.

An error may be introduced into the calculated value of the cumulated concentration, if one assumes instantaneous uptake of the radionuclide by the tissue; this is especially true when the physical half-life of the radionuclide (radiopharmaceutical) is comparable to the biological half-time for uptake by the tissue in question. If a short-lived radionuclide is used, the absorbed dose will be overestimated, unless the physical decay of the radionuclide is taken into account during the uptake period. The concentration of the radionuclide at any time during the period of uptake is given by:

$$C(t) = C(0)(e^{-0.693t/T_{eff}} - e^{-0.693t/T_{up}}) \frac{\mu Ci}{g}, \qquad (28a)$$

where $C(t)$ = concentration, in µCi/g, in the tissue at time t;

$C(0)$ = maximum concentration, in µCi/g, in the tissue, if there were instantaneous uptake; i.e., $T_{up} = 0$ or $C(0)$ would be approached asymptotically, if there were no elimination from the tissue; i.e., $T_{eff} \to$ infinity;

T_{up} = effective half-time, in hours, for uptake in the tissue, or

$$T_{up} = \frac{T \times T_{bu}}{T + T_{bu}};$$

T_{bu} = biological half-time, in hours, for uptake in the tissue.

To evaluate $C(0)$, one extrapolates back to time $t = 0$ on the curve describing the exponential elimination of the radionuclide.

The cumulated concentration may be calculated from equation (28b):[15]

$$\tilde{C} = 1.44 C(0) T_{eff} \left(1 - \frac{T_{up}}{T_{eff}}\right) \frac{\mu Ci\text{-}hr}{g}. \qquad (28b)$$

As a rule of thumb, if the effective half-life of the radionuclide is at least 20 times greater than the effective biological uptake, then the uptake can be neglected in absorbed-dose calculations, since the error would be 5 percent or less.

Equations to be Used When Only Limited Tissue Distribution Data Are Available.

The cumulated concentration must at times be evaluated from limited distribution data. In such cases, $\tilde{C}$ can be approximated by graphical integration of the concentration-versus-time curve. The activity remaining in the tissue after the last observation is assumed to be eliminated at a rate equal to the decay constant of the radionuclide. Equation (29) is used to make this calculation:

$$\tilde{C} = \sum_{j=1}^{j=n+1} (C_j - C_{j+1}) \left[T_j + \left(\frac{T_{j+1} - T_j}{2}\right) \right] + C_n T_n + 1.44 C_n T \frac{\mu Ci\text{-}hr}{g}, \qquad (29)$$

where C_j = concentration, in µCi/g, in the tissue at the jth observation made at time T_j, in hours;

$C_j + 1$ = concentration, in µCi/g, in the tissue at the $j + 1^{th}$ observation made at time $T_j + 1$, in hours;

C_n = concentration, in µCi/g, in the tissue when the last observation was made, $j = n$, at time T_n, in hours.

ANATOMICAL AND METABOLIC DATA

The mass of an organ must be known, to calculate the concentration of the radionuclide in the organ, if external counting techniques are used. The mass of an organ may be estimated on the basis of scanning, radiography, physical examination, or other data. Values of organ mass for a 70-kg "standard" man are given in Table 20.[14]

Table 20.

ORGANS OF STANDARD MAN.
MASS AND EFFECTIVE RADIUS OF ORGANS
OF THE ADULT HUMAN BODY.

Organ	Mass (m), cm	Percent of Total Body	Effective Radius (X) cm
Total body†	70,000	100	30
Muscle	30,000	43	30
Skin and subcutaneous tissue‡	6,100	8.7	0.1
Fat	10,000	14	20
Skeleton			
without bone marrow	7,000	10	5
red marrow	1,500	2.1	
yellow marrow	1,500	2.1	
Blood	5,400	7.7	
Gastrointestinal tract‡	2,000	2.9	30
Contents of G.I. tract			
lower large intestine	150		5
stomach	250		10
small intestine	1,100		30
upper large intestine	135		5
Liver	1,700	2.4	10
Brain	1,500	2.1	15
Lungs (2)	1,000	1.4	10
Lymphoid tissue	700	1.0	
Kidneys (2)	300	0.43	7
Heart	300	0.43	7
Spleen	150	0.21	7
Urinary bladder	150	0.21	
Pancreas	70	0.10	5
Salivary glands (6)	50	0.071	
Testes (2)	40	0.057	3
Spinal cord	30	0.043	1
Eyes (2)	30	0.043	0.25
Thyroid gland	20	0.029	3
Teeth	20	0.029	
Prostate gland	20	0.029	3
Andrenal glands or suprarenal (2)	20	0.029	3
Thymus	10	0.014	
Ovaries (2)	8	0.011	3
Hypophysis (pituitary)	0.6	8.6×10^{-6}	0.5
Pineal gland	0.2	2.9×10^{-6}	0.04
Parathyroids (4)	0.15	2.1×10^{-6}	0.06
Miscellaneous (blood vessels, cartilage, nerves, etc.)	390	0.56	

† The mass of the skin alone is taken to be 2000g.
‡ Does not include contents of the gastrointestinal tract.

From: International Commission on Radiation Protection, Committee II, 1959, Permissible Dose for Internal Radiation. *Health Phys.*, 3: 1, 1960. (Courtesy of *Health Physics*.)

Recently Cook and Snyder[31] have pointed out the limitations in the use of a standard man, and the ICRP has a task group revising the data characterizing standard man. These data may be a very strong function of age, especially in the preadult. Seltzer et al.[32] have studied the radiation exposure from the use of radionuclides in pediatric cases (Tables 21 and 22).

Table 21.
BODY WEIGHTS, ORGAN WEIGHTS, AND CALCULATED GEOMETRICAL FACTORS FOR VARIOUS AGES

Age	Whole Body		Kidney		Liver		Spleen	
	Weight, g	g, cm	Weight, g	g, cm	Weight, g	g, cm	Weight, g	g, cm
Newborn	3,540	64	23	16	136	27	9.4	12
1 yr.	12,100	89	72	22	333	36	31.0	17
5 yr.	20,300	94	112	25	591	44	54.0	20
10 yr.	33,500	102	187	30	918	50	101.0	25
15 yr.	55,000	112	247	33	1289	55	138.0	28
Standard man	70,000	126	300	35	1700	59	150.0	29

From: Seltzer, R. A., Kereiakes, J. G., and Saenger, E. L., Radiation Exposure from Radioisotopes in Pediatrics, *New Engl. J. Med.*, 271: 84, 1964. (Courtesy of *New England Journal of Medicine*.)

Table 22.
ABSORBED DOSE FROM THE ORAL ADMINISTRATION OF RADIOACTIVE IODINE FOR DIAGNOSTIC TESTS AT VARIOUS AGES

Organ	Radionuclide	Newborn	1 Year	5 Years	10 Years	15 Years	Standard Man
		Absorbed Dose in mrads/μCi Administered					
Whole body	$Na^{131}I$	10.	2.0	1.3	0.81	0.53	0.45
	$Na^{125}I$	9.4	1.6	1.0	0.67	0.44	0.39
	$Na^{132}I$	1.1	0.44	0.27	0.17	0.11	0.09
		Absorbed Dose in mrads/μCi Administered					
Thyroid	$Na^{131}I$	32.	10.0	4.3	3.1	1.7	1.3
	$Na^{125}I$	19.	6.2	2.6	1.8	1.0	0.82
	$Na^{132}I$	1.2	0.4	0.17	0.12	0.07	0.05

From: Kereiakes, J. G., Seltzer, R. A., Blackburn, B., and Saenger, E. L., Radionuclide Doses to Infants and Children: A Plea for a Standard Child *Health Phys.*, 11: 11: 999, 1965. (Courtesy of *Health Physics*.)

In many situations where the radionuclide is excreted in the feces, additional information is needed in regard to the gastrointestinal tract of man.[33,34] Table 23 gives the essential parameters for the gastrointestinal tract of the standard man.

Table 23.
GASTROINTESTINAL TRACT OF THE STANDARD MAN

Portion of G.I. Tract That is the Critical Tissue	Mass of Contents, g	Time Food Remains (τ), day	Fraction from Lung to G.I. Tract, f_a	
			Soluble	Insoluble
Stomach (S)	230	1/24	0.50	0.625
Small intestine (SI)	1,100	4/24	0.50	0.625
Upper large intestine (ULI)	135	8/24	0.50	0.625
Lower large intestine (LLI)	150	18/24	0.50	0.625

From: International Commission on Radiological Protection, Committee 2, 1959. *Health Phys.*, 3: 1, 1960. (Courtesy of *Health Physics*.)

In many instances one must estimate the water balance and the air balance for man. These data, as applied to the standard man, are given in Table 24.

Table 24.
INTAKE AND EXCRETION OF THE STANDARD MAN

Water Balance

	Intake, cm³/day		Excretion, cm³/day
Food	1,000	Urine	1,400
Fluids	1,200	Sweat	600
Oxidation	300	From Lungs	300
		Feces	200
Total	2,500	Total	2,500

Air Balance

	O_2, vol %	CO_2, vol %	N_2 + Others, vol %
Inspired air	20.94	0.03	79.03
Expired air	16	4.0	80
Alveolar air (inspired)	15	5.6	—
Alveolar air (expired)	14	6.0	—

Vital Capacity of Lungs	
Men	3 to 4 liters
Women	2 to 3 liters
Air Inhaled	
During 8-hour work day	10^7 cm³/day
During 16 hours not at work	10^7 cm³/day
Total	2×10^7 cm³/day
Interchange area of lungs	50 m²
Area of upper respiratory tract, trachea, bronchi	20 m²
Total surface area of respiratory tract	70 m²

From: International Commission on Radiological Protection. Committee II, 1959. Permissible Dose for Internal Radiation. *Health Phys.* 3: 1, 1960. (Courtesy of *Health Physics*.)

The ICRP task group on lung dynamics has prepared a detailed report on the deposition and retention of particulate matter in the human respiratory tract.[35] Spiers[36] has considered in depth the anatomical considerations for absorbed-dose calculations for bone and lung tissues.

REFERENCES

1. Smith, E. M., Activities of the Medical Internal Radiation Dose Committee. *J. Nucl. Med.*, Suppl. 1: 5–6, 1968.
2. Loevinger, R. and Berman, M., A Formalism for Calculation of Absorbed Dose from Radionuclides. *Phys. Med. Biol.*, 13: 205, 1968.
3. Loevinger, R. and Berman, M., MIRD Pamphlet No. 1—A Schema for Absorbed-Dose Calculations for Biologically Distributed Radionuclides. *J. Nucl. Med.*, Suppl. 1: 7, 1968.
4. Smith, E. M., Radiation Dosimetry. *Principles of Nuclear Medicine*, pp. 742–784, Wagner, H. N., Jr., ed. W. B. Saunders, Philadelphia, 1968.
5. International Commission on Radiological Units and Measurements, Report 10a, Radiation Quantities and Units. *NBS Handbook 84*, U.S. Government Printing Office, Washington, D.C., 1962.
6. International Commission on Radiological Protection and International Commission on Radiological Units and Measurements, Report of the RBE Committee. *Health Phys.*, 9: 357, 1963.
7. International Commission of Radiological Protection, *Recommendations of the International Commission on Radiological Protection, Publication, 6.* Pergamon Press, New York, 1964.

8. International Commission on Radiological Protection, *Recommendations of the International Commission on Radiological Protection, Publication 9.* Pergamon Press, New York, 1966.
9. Ellett, W. H., Callahan, A. B., and Brownell, G. L., Gamma-Ray Dosimetry of Internal Emitters. I. Monte Carlo Calculations of Absorbed Dose from Point Sources. *Brit. J. Radiol.*, 37: 45, 1964.
10. Ellett, W. H., Callahan, A. B., and Brownell, G. L., Gamma-Ray Dosimetry of Internal Emitters. II. Monte Carlo Calculations of Absorbed Dose from Uniform Sources. *Brit. J. Radiol.*, 38: 541, 1965.
11. Dillman, L. T., Calculating the Effect of Energy per Radioactive Decay for Use in Internal Dose Calculations. *ORNL-4168*, pp. 233–245, 1967.
12. Loevinger, R., Average Energy of Allowed Beta-Particle Spectra. *Phys. Biol. Med.*, 1: 330, 1957.
13. Dillman, L. T. and Snyder, W. S., *Calculations of Physical Parameters for Use in Dose Estimation.* To be published as a pamphlet in the Report of the Medical Internal Radiation Dose Committee, Society of Nuclear Medicine, 1967.
14. International Commission on Radiological Protection, Committee II, 1959, Permissible Dose for Internal Radiation. *Health Phys.*, 3: 1, 1960.
15. Loevinger, R., Holt, J. G., and Hine, G. J., Internally Administered Radioisotopes. *Radiation Dosimetry*, Chapter 17, Hine, G. J. and Brownell, G. L., eds. Academic Press, New York, 1956.
16. Smith, E. M., Harris, C. C., and Rohrer, R. H., Calculation of Local Energy Deposition Due to Electron Capture and Internal Conversion. *J. Nucl. Med.*, 7: 23, 1965.
17. Wapstra, A. H., Nijh, G. J., and Van Lieshout, R., *Nuclear Spectroscopy Tables.* Interscience Publishers, New York, 1959.
18. Quimby, E. H., Dosage Calculations for Radioactive Isotopes. *Radioactive Isotopes in Medicine and Biology —Basic Physics and Instrumentation*, Quimby, E. H. and Feitelberg, S., eds., Lea and Febiger, Philadelphia, 1963.
19. Smith, E. M., Calculating Absorbed Doses from Radiopharmaceuticals. *Nucleonics*, 24: 33–39, 1966.
20. Ellett, W. H., Brownell, G. L., and Reddy, A. B., Assessment of Monte Carlo Calculations for Determining the Gamma-Ray Dose from Internal Emitters. *Phys. Med. Biol.*, 13: 219, 1968.
21. Reddy, A. R., Ellett, W. H., and Brownell, G. L., Gamma-Ray Dosimetry of Internal Emitters. III. Monte Carlo Calculations of Absorbed Dose for Low-Energy Gamma Rays. *Brit. J. Radiol.*, 42: 512, 1967.
22. Brownell, G. L., Ellett, W. H., and Reddy, R., MIRD Pamphlet No. 3—Absorbed Fractions for Photon Dosimetry. *J. Nucl. Med.*, Suppl. 1: 27, 1968.
23. Berger, M. J., MIRD Pamphlet No. 2—Energy Deposition in Water by Photons from Point Isotropic Sources. *J. Nucl. Med.*, Suppl. 1: 15, 1968.
24. Fisher, H. L. and Snyder, W. S., Distribution of Dose in the Body from a Source of Gamma Rays Distributed Uniformly in an Organ. *ORNL-4168*, pp. 245–257, 1967.
25. Ford, M. R. and Snyder, W. S., *A Monte Carlo Estimation of Observed Fractions in Body Organs of Realistic Size, Shape, Composition and Density from a Source of Photons Distributed Uniformly in an Organ.* Presented at the meeting of the Health Physics Society, and to be published as a pamphlet in the report of the Medical Internal Radiation Dose Committee, Society of Nuclear Medicine, 1968.
26. Snyder, W. S. and Ford, M. R., A Monte Carlo Code for Estimation of Dose from Gamma-Ray Sources. *Health Phys.*, 11: 838, 1965 and *ORNL-3849*.
27. Snyder, W. S., Personal Communications and Unpublished Data, 1965.
28. Hubbell, J. H. and Berger, M. J., *NBS Report 8681*, Sept. 1966 (unpublished). To appear in *Engineering Compendium on Radiation Shielding*, Jaeger, R. G., ed. International Atomic Energy Agency, Vienna, Austria. Data for muscle obtained in private communication from J. H. Hubbell.
29. Focht, E. F., Quimby, E. H., and Gershowitz, M., Revised Average Geometric Factors for Cylinders in Isotope Dosage. *Radiology*, 85: 151, 1965.
30. Ross, D. A., Harris, C. C., Kuhl, D. D., Reba, R. C., Wagner, H. N., Jr., and Karmen, A., Measurement of Radioactivity. Physical Principles of Radionuclide Scanning. Whole-Body Counting. Liquid Scintillation Counting. *Principles of Nuclear Medicine.* pp. 129–258, Wagner, H. N., Jr., ed. W. B. Saunders, Philadelphia, 1968.
31. Cook, M. J. and Snyder, W. S., Estimation of Population Exposure. *Health Phys.*, 11: 810, 1965.
32. Seltzer, R. A., Kereiakes, J. G., and Saenger, E. L., Radiation Exposure from Radioisotopes in Pediatrics. *New Engl. J. Med.*, 271: 84, 1964.
33. Dolphin, G. W. and Eve, I. S., Dosimetry of the Gastrointestinal Tract. *Health Phys.*, 12: 163, 1966.
34. Eve, I. S., A Review of the Physiology of the Gastrointestinal Tract in Relation to Radiation Doses from Radioactive Materials. *Health Phys.*, 12: 131, 1966.
35. International Commission on Radiological Protection, Committee II, Task Group on Lung Dynamics, Deposition and Retention Models for Internal Dosimetry of the Human Respiratory Tract. *Health Phys.*, 12: 173, 1966.
36. Spiers, F. W., *Radioisotopes in the Human Body: Physical and Biological Aspects.* Academic Press, New York, 1968.

RADIATION DOSES FROM ADMINISTERED RADIONUCLIDES

J. Vennart, B.Sc., F.Inst.P., and *Margaret Minski*, B.Sc.
Radiological Protection Service, Belmont, Surrey, England
By Courtesy of the Editors of the *British Journal of Radiology*, 35: 414, 1962,
With permission of the *Ministry of Health and Medical Research Council Radiological Protection Service, Belmont, Surrey, England.*

The increasing use of radionuclides for medical purposes and the importance of reducing to a minimum the exposure of the persons concerned to ionizing radiations, which has been emphasized in a recent official report,[1] stress the need for information concerning the radiation doses received by body tissues after administration of radioactive materials. What is needed, therefore, in addition to a knowledge of the dose received by the "target" organ or tissue, is information concerning the doses received by other organs and tissues, as well as the whole body, which are also irradiated as a result of the investigation. The average doses received by several important organs and tissues and by the whole body of adult males following the injection or ingestion of 1 µCi of most of those radionuclides that are at present being used in hospitals are shown in Table 25. These values are intended as an aid in planning the administration of radionuclides to patients, and they should also be of value to those responsible for assessing the risks accompanying the accidental intake of radioactive materials. However, it is important to emphasize that the corresponding doses for children and small individuals will usually be higher than those shown in Table 25, because the radionuclide will be distributed throughout a smaller mass. On the other hand, as a result of simplifying the calculation of dose in order to reduce labor, the doses shown for adult males are in some cases overestimated. The values are not intended, therefore, to be used for prescribing therapeutic administrations of radionuclides without paying attention to the approximations employed. Details of these and of the method of calculation are discussed in detail below.

CALCULATION OF DOSE

The dose, D rems, received by any organ or tissue of the body during a time, t days, following the administration of 1 µCi of a radionuclide is expressed by the equation

$$D(\text{in rems}) = (3.2 \times 10^9) I_t E (1.6 \times 10^{-6})/100m, \qquad (1)$$

where 3.2×10^9 is the number of disintegrations per day per µCi, I_t (in µCi-days) the integrated activity in the organ or tissue considered during a time (t days) after the administration of 1 µCi, E (in Mev) the effective energy absorbed in the organ per disintegration and suitably modified for the relative biological effectiveness (RBE) of each radiation concerned, 1.6×10^{-6} the number of ergs per Mev, 100 the number of ergs per gram-rad, and m (in grams) the mass of the organ. Simplifying equation (1), the following results:

$$D(\text{in rems}) = 51.21 I_t E/m. \qquad (2)$$

When prescribing a therapeutic dose, it is essential that the parameters I_t, E and m should be accurately determined for each patient. The integrated activity, I_t, would normally be determined by the administration of a tracer dose. Both m and E are dependent upon the size of the organ, and this might be determined either by a scanning technique following the administration of the tracer dose or by palpation. The purpose of this paper, however, is to predict the approximate magnitude of the dose (D) for a variety of organs and for different radionuclides as an aid to planning, and it is necessary, therefore, to choose average values for the parameters I_t, E and m. The most comprehensive data extant concerning these parameters are contained in the Report of Committee 2 of the International Commission on Radiological Protection,[2] and accordingly this information is used in the present calculations. The values for m, the mass of the organ, that are shown in Table 25 have been taken directly from ICRP-2, and they are average values for adult male Caucasians. Typical values for children and females, as well as for some of the Asiatic races, will in general be less, and the value of D, the dose to the organ, will be correspondingly greater. These values are not found in ICRP-2, but the information required may be obtained from standard textbooks and other relevant publications, such as *Tabulae Biologicae*.[3]

The values of E, the effective energy, used in the calculations are shown in Table 25. Many of the approximations used to calculate this parameter, and also to calculate I_t, are identical with those adopted by ICRP-2. These are detailed in the sections below. A further section deals with the calculation of the dose to the gastrointestinal tract that is irradiated following the administration of radionuclides by mouth.

The Integrated Activity, I_t

After any material enters into the body, it will be offered to the various organs and tissues, to be retained there for different periods of time before being excreted. The general appearance of the curve depicting these processes is shown as the full line in Figure 1. In the large majority of cases, the amount of material in the organ rises to a maximum within a few hours following its administration, and thereafter decreases due to biological elimination. The pattern of biological elimination can usually be described by a series of exponential terms with negative exponents, which are represented by straight lines of negative slope in the semi-logarithmic type of plot shown in Figure 1. Each of these terms probably represents the mode of excretion from a particular compartment of the organ concerned, and for some organs, like bone, where many different metabolic processes are involved, the excretion pattern will be very complex. The integrated activity, I_t, from 0 to t days is given by the area under the retention curve between these limits of time, and to compute I_t accurately, the shape of the curve must be known precisely. Fortunately, in a large number of instances a close approximation to I_t can be made quite simply, because one particular component of the retention curve—often it is the longest-lived component—dominates the situation. For example, in Figure 1 the area under the dotted line *ab* is approximately equal to that under the full line depicting the actual retention. Therefore, if it is assumed that a fraction (f) of the administered material, given by the intercept of *ab* on the ordinate, reaches the organ instantaneously and thereafter is biologically eliminated in an exponential manner, with a half-period (T_b) determined by the slope of *ab*, the integrated activity so calculated will be approximately the same as that for the true circumstances. This simplification is adopted by ICRP-2, and the values of f and T_b that are recommended in that report have been chosen, wherever possible, from actual observations on humans with administered radionuclides or from a knowledge of the intake of elements in food and of their concentrations in the different parts of the body. In a great many cases, however, human data are not available, and the values of f and T_b were then extrapolated from experiments on animals. Sometimes even these data are lacking, and it is then necessary to make estimates from a comparison with elements having similar chemical behavior. It is also important to emphasize that the values of f and T_b given in ICRP-2 are average values for adult males. There may be considerable variations between individuals of the same sex, and also between males and females. Furthermore, both f and T_b may vary throughout life, reflecting the different pattern of growth and remodeling at different ages. Finally, although in ICRP-2 different compounds of the same element are only broadly classified as "soluble" and "insoluble", in fact both f and T_b will also vary quite markedly with their chemical composition. Much remains to be done, therefore, by direct observations on humans to whom the material concerned was administered.

In preparing Table 25, values of f (which are shown as f'_2 for injection and f_w for ingestion) and T_b, with two exceptions, were taken directly from ICRP-2. The two exceptions are for the radioisotopes of cobalt and for ^{32}P. In the former case, ICRP-2 gives no information about the uptake of cobalt by the kidney; values of f_w, f'_2 and T_b were supplied by Ford.[4] The value of f'_2 for bone following the administration of ^{32}P, namely 0.2, that is given in ICRP-2 is incorrect,[4] and the value 0.5 was used when preparing Table 25.

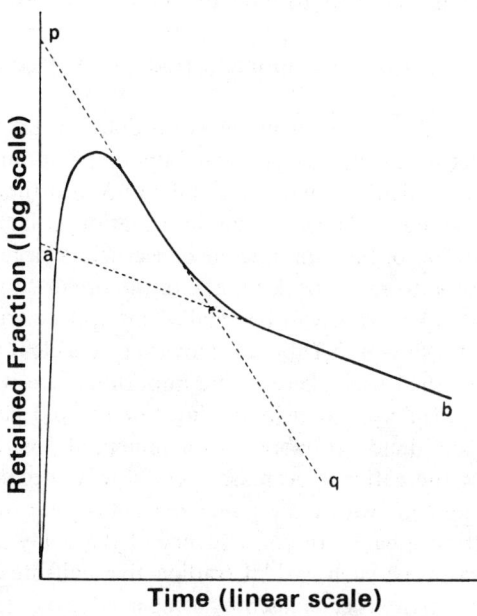

FIGURE 1. Biological Retention Curve.

When the administered material is radioactive, its activity in the organ decreases due to radioactive decay as well as by biological elimination. The process of radioactive decay follows a simple exponential law, and the amount of the radionuclide present in the organ at any time according to the simplified calculation is, therefore, proportional to the product of two exponential terms—one representing biological elimination, and the other radioactive decay. The resultant is an exponential whose exponent is the sum of those representing the two types of elimination. It is, therefore, possible to speak in terms of an effective decay constant (λ) or, more commonly, of an effective half-period (T), which is related to T_r, the radioactive half-period, and T_b in the following manner:

$$T = T_b T_r / (T_b + T_r), \qquad (3)$$

where $T = \log_e 2/\lambda = 0.693/\lambda$.

Values of T and T_r were taken, when available, from ICRP-2. The report does not list values for ^{15}O, ^{43}K, ^{52}Fe, ^{56}Co, ^{123}I, ^{124}I, ^{125}I, ^{126}I, and ^{130}I. Values for T_r in days were obtained from the literature. T was then calculated from equation (3), using values of T_b given in ICRP-2 for an appropriate isotope.

The integrated activity, I_t, following the administration of 1 µCi of a radionuclide is then given by the following equation:

$$I_t \text{ (in µCi-days)} = f \int_0^t e^{(-0.693t/r)} \, dt = fT[1 - e^{(-0.693t/r)}]/0.693 \qquad (4)$$

The total integrated activity throughout infinite time, I_∞, is given by

$$I_\infty \text{ (in µCi-days)} = fT/0.693. \qquad (5)$$

The total dose resulting from the administration of 1 μCi of any radionuclide is proportional to I_∞, and it follows from equation (2), therefore, that the total dose, D, is given by

$$D \text{ (in rems)} = 51.2 f T E / 0.693 m = 74 f T E / m. \tag{6}$$

Table 25 shows the total dose resulting from the administration of 1 μCi of the various radionuclides in different organs, and since it is often of interest to know the rate at which the dose is delivered, Table 26 was compiled, to show the fraction of these total doses delivered one week, four weeks, thirteen weeks, and one year after administration. These fractions were obtained from the ratio of I_t for the appropriate period, given by equation (5), to I_∞, given by equation (6).

It is worth noting that the simplified calculation of I_t does not take into account the loss of material by radioactive decay during the period of uptake. The amount lost in this way is dependent upon the rate at which the material is taken up into the organ, and for the large majority of radionuclides listed in Table 25 this information is lacking. However, provided that the radioactive half-period of the radionuclide concerned is large compared with the time required for its stable isotope to reach peak uptake in the organ considered, the error caused by adopting the simplified calculation will be small. This will be true for a large number of the organs and radionuclides shown in Table 25. However, it is clear that neglect of this factor leads to an overestimate of the dose. There is one important case where information on this matter is available, namely, the uptake of iodine into the thyroid gland. In normal thryroids the amount of iodine in the gland probably rises exponentially with a half-life of about six hours to a flat maximum; thereafter it decreases very slowly, because the rate of biological elimination is small. For persons with toxic goiter the uptake is usually more rapid, and the rate of biological elimination greater. In consequence of the delay in entering the gland, the integrated activity for an isotope such as ^{132}I (radioactive half-life 2.33 hours) will be overestimated if the integrated activity is calculated by using a value of f (the fraction of ingested or injected material which reaches the gland) that is strictly applicable only to stable iodine or one of its long-lived radioisotopes. For either f'_2 or f_w ICRP-2 recommends the value 0.3, irrespective of the radioisotope that is being considered. Neglecting the small loss due to biological elimination during the period of uptake, it may be shown that the value of I_∞, and consequently the total dose, calculated according to equation (6), should be multiplied by a factor $T_r/(T_u + T_r)$, to allow for radioactive decay during the period of uptake. It is assumed that uptake into the gland is exponential with a half-period T_u. For normal thyroids, where T_u is about six hours, the multiplying factors for ^{123}I, ^{124}I, ^{125}I, ^{126}I, ^{130}I, ^{131}I, and ^{132}I are 0.68, 0.94, 1.00, 0.98, 0.68, 0.97, and 0.28 respectively. The values shown in Table 25 for the total dose to the thyroid following injection or ingestion of 1 μCi of these radioisotopes of iodine have been modified in this way. No correction factor has been used for any other organ or radionuclide, since the information was not available. The factor is also usually neglected in ICRP-2, except insofar as the values of f'_2 and f_w given in the recommendations may have been obtained from actual experimental data on humans to whom the isotope concerned was administered. There are, however, two elements for which ICRP-2 gives different values of f for different radioisotopes of the element. These are the elements strontium and radium, when bone is the organ considered. For stable strontium and for the very long-lived strontium-90, f'_2 (the fraction of the material going from blood to bone) is given as 0.3, but for the shorter-lived strontium isotopes, ^{85}Sr, ^{89}Sr, ^{91}Sr, and ^{92}Sr, the value recommended for f'_2 is 0.7. This reflects the fact that, whereas in Figure 1 the line ab might best represent the logarithm of the fractional retention against time for stable strontium and the long-lived strongium-90, the line pq might be more representative for those strontium isotopes where radioactive decay dominates the situation. Similarly, the value of f'_2 shown for ^{224}Ra in Table 25, 0.5, is greater than that recommended for ^{226}Ra, 0.1, which has a longer radioactive half-period.

Table 25.
AVERAGE RADIATION DOSES RECEIVED BY DIFFERENT ORGANS OF ADULT MALES FOLLOWING THE ADMINISTRATION OF 1 μCi OF VARIOUS RADIONUCLIDES

Radio-nuclide	Organ or Tissue of Reference	Fraction Reaching Organ After Ingestion (f_w)*	Fraction Reaching Organ After Injection (f'_2)*	Radio-active Half-Period (T_r), days*	Biological Half-Period (T_b), days*	Effective Half-Period (T), days*	Mass of Organ (m), grams*	Effective Energy (E), Mev	Dose per μCi Administered, rems By Ingestion	Dose per μCi Administered, rems By Injection
^{3}H	Body tissue	1.0	1.0	4.5×10^3	12	12	4.3×10^{-4}	0.01	2.1×10^{-4}	2.1×10^{-4}
	Total body	1.0	1.0		12	12	7×10^4	0.01	1.3×10^{-4}	1.3×10^{-4}
^{14}C	Fat	0.5	0.5	2.10^6	12	12	10^4	0.054	2.4×10^{-3}	2.4×10^{-3}
	Total body	1.0	1.0		10	10	7×10^4	0.054	5.7×10^{-4}	5.7×10^{-4}
	Bone	0.025	0.025		40	40	7,000	0.054	5.7×10^{-4}	5.7×10^{-4}
^{15}O	Total body	1.0	1.0	1.50×10^{-3}	14	1.50×10^{-3}	7×10^4	1.36	2.2×10^{-6}	2.2×10^{-6}
^{18}F	Bone and teeth	0.53	0.53	0.078	1,450	0.078	7,000	0.41	1.8×10^{-4}	1.8×10^{-4}
	Total body	1.0	1.0		808	0.078	7×10^4	0.89	7.3×10^{-5}	7.3×10^{-5}
	G.I. (S) sol.	1.0					250	0.54	1.9×10^{-3}	
	G.I. (ULI) insol.	1.0					135	0.41	1.3×10^{-3}	
^{22}Na	Total body	1.0	1.0	950	11	11	7×10^4	1.6	1.9×10^{-2}	1.9×10^{-2}
	G.I. (LLI) sol.	0.05					150	0.53	3.4×10^{-3}	
	G.I. (LLI) insol.	1.0					150	0.53	6.8×10^{-2}	
^{24}Na	Total body	1.0	1.0	0.63	11	0.6	7×10^4	2.7	1.7×10^{-3}	1.7×10^{-3}
	G.I. (S) sol.	1.0					250	1.5	6.2×10^{-3}	
	G.I. (LLI) insol.	1.0					150	1.0	4.8×10^{-2}	
^{32}P	Bone	0.375	0.5	14.3	1,155	14.1	7,000	0.69	3.8×10^{-2}	5.1×10^{-2}
	Liver	0.05	0.07		18	8	1,700	0.69	1.2×10^{-2}	1.7×10^{-2}
	Total body	0.75	1.0		257	13.5	7×10^4	0.69	7.4×10^{-3}	9.8×10^{-3}
	Brain	5.3×10^{-3}	7×10^{-3}		257	13.5	1,500	0.69	2.4×10^{-3}	3.2×10^{-3}
	G.I. (LLI) sol.	0.25					150	0.69	2.1×10^{-2}	
	G.I. (LLI) insol.	1.0					150	0.69	8.4×10^{-2}	
^{35}S	Testes	1.3×10^{-3}	1.3×10^{-3}	87.1	623	76.4	40	0.056	1.0×10^{-2}	1.0×10^{-2}
	Total body	1.0	1.0		90	44.3	7×10^4	0.056	2.6×10^{-3}	2.6×10^{-3}
	Skin	0.01	0.01		1,530	82.4	2,000	0.056	1.7×10^{-3}	1.7×10^{-3}
	Bone	0.03	0.03		600	76.1	7,000	0.056	1.4×10^{-3}	1.4×10^{-3}
	G.I. (LLI) sol.	0.05					150	0.056	3.6×10^{-4}	
	G.I. (LLI) insol.	1.0					150	0.056	7.1×10^{-3}	

* Values taken from the Report of Committee 2 of the International Commission on Radiological Protection (ICRP—2).

Table 25. (Continued)

AVERAGE RADIATION DOSES RECEIVED BY DIFFERENT ORGANS OF ADULT MALES FOLLOWING THE ADMINISTRATION OF 1 μCi OF VARIOUS RADIONUCLIDES

Radio-nuclide	Organ or Tissue of Reference	Fraction Reaching Organ After Ingestion (f_w)*	Fraction Reaching Organ After Injection (f'_2)*	Radio-active Half-Period (T_r), days*	Biological Half-Period (T_b), days*	Effective Half-Period (T), days*	Mass of Organ (m), grams*	Effective Energy (E), Mev	Dose per μCi Administered, rems By Ingestion	By Injection
^{36}Cl	Total body	1.0	1.0	1.2×10^8	29	29	7×10^4	0.26	8.0×10^{-3}	8.0×10^{-3}
	G.I. (LLI) sol.	0.05					150	0.26	1.7×10^{-3}	
	G.I. (LLI) insol.	1.0					150	0.26	3.3×10^{-2}	
^{38}Cl	Total body	1.0	1.0	0.026	29	0.026	7×10^4	2.3	6.3×10^{-5}	6.3×10^{-5}
	G.I. (S) sol.	1.0					250	1.9	4.9×10^{-3}	
	G.I. (S) insol.	1.0					250	1.9	4.9×10^{-3}	
^{42}K	Brain	0.04	0.04	0.52	58	0.52	1,500	1.5	1.5×10^{-3}	1.5×10^{-3}
	Spleen	4×10^{-3}	4×10^{-3}		58	0.52	150	1.5	1.5×10^{-3}	1.5×10^{-3}
	Muscle	0.65	0.65		58	0.52	3×10^4	1.6	1.3×10^{-3}	1.3×10^{-3}
	Total body	1.0	1.0		58	0.52	7×10^4	1.6	8.8×10^{-4}	8.8×10^{-4}
	Liver	0.02	0.02		58	0.52	1,700	1.5	6.8×10^{-4}	6.8×10^{-4}
	G.I. (S) sol.	1.0					250	1.5	6.2×10^{-3}	
	G.I. (LLI) insol.	1.0					150	1.5	5.9×10^{-2}	
^{43}K	Muscle	0.65	0.65	0.93	58	0.92	3×10^4	0.94	1.4×10^{-3}	1.4×10^{-3}
	Brain	0.04	0.04		58	0.92	1,500	0.71	1.3×10^{-3}	1.3×10^{-3}
	Spleen	4×10^{-3}	4×10^{-3}		58	0.92	150	0.52	9.4×10^{-4}	9.4×10^{-4}
	Total body	1.0	1.0		58	0.92	7×10^4	0.94	9.1×10^{-4}	9.1×10^{-4}
	Liver	0.02	0.02		58	0.92	1,700	0.60	4.8×10^{-4}	4.8×10^{-4}
	G.I. (LLI) sol.	0.05					150	0.48	1.6×10^{-3}	
	G.I. (LLI) insol.	1.0					150	0.48	3.1×10^{-2}	
^{45}Ca	Bone	0.54	0.9	164	1.8×10^4	162	7,000	0.086	7.9×10^{-2}	1.3×10^{-1}
	Total body	0.6	1.0		1.64×10^4	162	7×10^4	0.086	8.8×10^{-3}	1.5×10^{-2}
	G.I. (LLI) sol.	0.4					150	0.086	4.4×10^{-3}	
	G.I. (LLI) insol.	1.0					150	0.086	1.1×10^{-2}	
^{47}Ca + ^{47}Sc	Bone	0.54	0.9	4.9	1.8×10^4	4.9	7,000	0.67	1.9×10^{-2}	3.1×10^{-2}
	Total body	0.6	1.0		1.64×10^4	4.9	7×10^4	1.4	4.3×10^{-3}	7.2×10^{-3}
	G.I. (LLI) sol.	0.4					150	0.67	3.0×10^{-2}	
	G.I. (LLI) insol.	1.0					150	0.67	7.6×10^{-3}	

Table 25. (Continued)
AVERAGE RADIATION DOSES RECEIVED BY DIFFERENT ORGANS OF ADULT MALES FOLLOWING THE ADMINISTRATION OF 1 μCi OF VARIOUS RADIONUCLIDES

Radio-nuclide	Organ or Tissue of Reference	Fraction Reaching Organ After Ingestion (f_w)*	Fraction Reaching Organ After Injection (f'_2)*	Radio-active Half-Period (T_r), days*	Biological Half-Period (T_b), days*	Effective Half-Period (T), days*	Mass of Organ (m), grams*	Effective Energy (E), Mev	Dose per μCi Administered, rems By Ingestion	Dose per μCi Administered, rems By Injection
^{51}Cr	Prostate	4.5×10^{-6}	9×10^{-4}	27.8	616	26.6	20	8.4×10^{-3}	3.7×10^{-6}	7.4×10^{-4}
	Thyroid	4.5×10^{-6}	9×10^{-4}		616	26.6	20	8.4×10^{-3}	3.7×10^{-6}	7.4×10^{-4}
	Total body	<0.005	1.0		616	26.6	7×10^4	0.025	3.5×10^{-6}	7.0×10^{-6}
	Kidney	1.3×10^{-5}	2.7×10^{-3}		616	26.6	300	0.012	1.0×10^{-6}	2.1×10^{-4}
	G.I. (LLI) sol.	1.0					150	0.01	1.2×10^{-3}	
	G.I. (LLI) insol.	1.0					150	0.01	1.2×10^{-3}	
^{52}Fe + ^{52m}Mn	Liver	0.013	0.13	0.325	554	0.325	1,700	3.22	5.9×10^{-4}	5.9×10^{-3}
	Total body	0.1	1.0		800	0.324	7×10^4	2.64	9.0×10^{-5}	9.0×10^{-4}
	G.I. (ULI) sol.	0.9					135	2.46	6.3×10^{-2}	
	G.I. (ULI) insol.	1.0					135	2.46	7.0×10^{-2}	
^{55}Fe	Spleen	2×10^{-3}	0.02	1.1×10^3	600	388	150	6.5×10^{-3}	2.5×10^{-3}	2.5×10^{-2}
	Liver	0.013	0.13		554	368	1,700	6.5×10^{-3}	1.4×10^{-3}	1.4×10^{-2}
	Bone	0.01	0.1		1,680	665	7,000	6.5×10^{-3}	4.6×10^{-4}	4.6×10^{-3}
	Total body	0.1	1.0		800	463	7×10^4	6.5×10^{-3}	3.2×10^{-4}	3.2×10^{-3}
	G.I. (LLI) sol.	0.9					150	6.5×10^{-3}	7.5×10^{-4}	
	G.I. (LLI) insol.	1.0					150	6.5×10^{-3}	8.3×10^{-4}	
^{59}Fe	Spleen	2×10^{-3}	0.02	45.1	600	41.9	150	0.34	1.4×10^{-2}	1.4×10^{-1}
	Liver	0.013	0.13		554	41.7	1,700	0.42	9.9×10^{-3}	9.9×10^{-2}
	Total body	0.1	1.0		800	42.7	7×10^4	0.81	3.6×10^{-3}	3.6×10^{-2}
	Bone	0.01	0.1		1,680	43.9	7,000	0.29	1.3×10^{-3}	1.3×10^{-2}
	G.I. (LLI) sol.	0.9					150	0.29	3.3×10^{-3}	
	G.I. (LLI) insol.	1.0					150	0.29	3.7×10^{-2}	
^{56}Co	Total body	0.3	1.0	72	9.5	8.39	7×10^4	0.824	2.2×10^{-3}	7.3×10^{-3}
	Pancreas	6×10^{-4}	2×10^{-3}		9.5	8.39	70	0.332	1.8×10^{-3}	5.9×10^{-3}
	Liver	7×10^{-3}	0.04		9.5	8.39	1,700	0.460	1.2×10^{-3}	6.7×10^{-3}
	Spleen	4.2×10^{-4}	1.4×10^{-3}		9.5	8.39	150	0.384	6.7×10^{-4}	2.2×10^{-3}
	Kidney	6×10^{-4}	2×10^{-3}		9.5	8.39	300	0.384	4.8×10^{-4}	1.6×10^{-3}
	G.I. (LLI) sol.	0.7					150	0.332	3.0×10^{-2}	
	G.I. (LLI) insol.	1.0					150	0.332	4.2×10^{-2}	

Table 25. (Continued)

AVERAGE RADIATION DOSES RECEIVED BY DIFFERENT ORGANS OF ADULT MALES FOLLOWING THE ADMINISTRATION OF 1 µCi OF VARIOUS RADIONUCLIDES

Radio-nuclide	Organ or Tissue of Reference	Fraction Reaching Organ After Ingestion (f_w)*	Fraction Reaching Organ After Injection (f'_2)*	Radioactive Half-Period (T_r), days*	Biological Half-Period (T_b), days*	Effective Half-Period (T), days*	Mass of Organ (m), grams*	Effective Energy (E), Mev	Dose per µCi Administered, rems By Ingestion	By Injection
^{57}Co	Total body	0.3	1.0	270	9.5	9.2	7×10^4	0.09	2.6×10^{-4}	8.7×10^{-4}
	Pancreas	6×10^{-4}	2×10^{-3}		9.5	9.2	70	0.04	2.3×10^{-4}	7.8×10^{-4}
	Liver	7×10^{-3}	0.04		9.5	9.2	1,700	0.053	1.5×10^{-4}	8.5×10^{-4}
	Spleen	4.2×10^{-4}	1.4×10^{-3}		9.5	9.2	150	0.045	8.6×10^{-5}	2.9×10^{-4}
	Kidney	6×10^{-4}	2×10^{-3}		9.5	9.2	300	0.045	6.1×10^{-5}	2.0×10^{-4}
	G.I. (LLI) sol.	0.7					150	0.04	3.6×10^{-3}	
	G.I. (LLI) insol.	1.0					150	0.04	5.1×10^{-3}	
^{58}Co	Total body	0.3	1.0	72	9.5	8.4	7×10^4	0.61	1.6×10^{-3}	5.4×10^{-3}
	Pancreas	6×10^{-4}	2×10^{-3}		9.5	8.4	70	0.17	9.0×10^{-4}	3.0×10^{-3}
	Liver	7×10^{-3}	0.04		9.5	8.4	1,700	0.29	7.4×10^{-4}	4.2×10^{-3}
	Spleen	4.2×10^{-4}	1.4×10^{-3}		9.5	8.4	150	0.22	3.8×10^{-4}	1.3×10^{-3}
	Kidney	6×10^{-4}	2×10^{-3}		9.5	8.4	300	0.22	2.7×10^{-4}	9.1×10^{-4}
	G.I. (LLI) sol.	0.7					150	0.17	1.5×10^{-2}	
	G.I. (LLI) insol.	1.0					150	0.17	2.2×10^{-2}	
^{60}Co	Total body	0.3	1.0	1.9×10^3	9.5	9.5	7×10^4	1.5	4.5×10^{-3}	1.5×10^{-2}
	Pancreas	6×10^{-4}	2×10^{-3}		9.5	9.5	70	0.44	2.6×10^{-3}	8.8×10^{-3}
	Liver	7×10^{-3}	0.04		9.5	9.5	1,700	0.72	2.1×10^{-3}	1.2×10^{-2}
	Spleen	4.2×10^{-4}	1.4×10^{-3}		9.5	9.5	150	0.56	1.1×10^{-3}	3.7×10^{-3}
	Kidney	6×10^{-4}	2×10^{-3}		9.5	9.5	300	0.56	7.9×10^{-4}	2.6×10^{-3}
	G.I. (LLI) sol.	0.7					150	0.44	3.9×10^{-2}	
	G.I. (LLI) insol.	1.0					150	0.44	5.6×10^{-2}	
^{64}Cu	Spleen	0.02	0.07	0.53	2	0.42	150	0.17	7.0×10^{-4}	2.5×10^{-3}
	Kidney	0.01	0.05		16	0.51	300	0.17	2.1×10^{-4}	1.1×10^{-3}
	Liver	0.02	0.08		150	0.53	1,700	0.19	8.8×10^{-5}	3.5×10^{-4}
	Heart	3×10^{-3}	0.01		80	0.53	300	0.17	6.6×10^{-5}	2.2×10^{-4}
	Total body	0.28	1.0		80	0.53	7×10^4	0.25	3.9×10^{-5}	1.4×10^{-4}
	Brain	3×10^{-3}	0.01		800	0.53	1,500	0.21	1.6×10^{-5}	5.6×10^{-5}
	G.I. (LLI) sol.	0.72					150	0.16	4.6×10^{-3}	
	G.I. (LLI) insol.	1.0					150	0.16	6.4×10^{-3}	

Table 25. (Continued)

AVERAGE RADIATION DOSES RECEIVED BY DIFFERENT ORGANS OF ADULT MALES FOLLOWING THE ADMINISTRATION OF 1 μCi OF VARIOUS RADIONUCLIDES

Radio-nuclide	Organ or Tissue of Reference	Fraction Reaching Organ After Ingestion (f_w)*	Fraction Reaching Organ After Injection (f'_2)*	Radio-active Half-Period (T_r), days*	Biological Half-Period (T_b), days*	Effective Half-Period (T), days*	Mass of Organ (m), grams*	Effective Energy (E), Mev	Dose per μCi Administered, rems By Ingestion	Dose per μCi Administered, rems By Injection
^{65}Ni	Bone	0.15	0.5	0.11	800	0.11	7,000	1.1	1.9×10^{-4}	6.4×10^{-4}
	Liver	0.02	0.07		500	0.11	1,700	1.2	1.2×10^{-4}	4.0×10^{-4}
	Total body	0.3	1.0		667	0.11	7×10^4	1.4	4.9×10^{-5}	1.6×10^{-4}
	G.I. (ULI) sol.	0.7					135	1.1	5.5×10^{-3}	
	G.I. (ULI) insol.	1.0					135	1.1	7.8×10^{-3}	
^{65}Zn	Prostate	6×10^{-3}	0.06	245	14	13	20	0.056	1.6×10^{-2}	1.6×10^{-1}
	Liver	0.035	0.35		91	66	1,700	0.15	1.5×10^{-2}	1.5×10^{-1}
	Kidney	4×10^{-3}	0.04		149	93	300	0.11	1.0×10^{-2}	1.0×10^{-1}
	Total body	0.1	1.0		933	194	7×10^4	0.32	6.5×10^{-3}	6.5×10^{-2}
	Pancreas	3×10^{-3}	0.03		25	23	70	0.084	6.1×10^{-3}	6.1×10^{-2}
	Muscle	0.03	0.3		1,959	218	3×10^4	0.32	5.1×10^{-3}	5.1×10^{-2}
	Bone	0.015	0.15		1,300	206	7,000	0.084	2.7×10^{-3}	2.7×10^{-2}
	Ovary	4×10^{-5}	4×10^{-4}		107	74	8	0.056	1.5×10^{-3}	1.5×10^{-2}
	Testes	9×10^{-5}	9×10^{-4}		270	128	40	0.056	1.2×10^{-3}	1.2×10^{-2}
	G.I. (LLI) sol.	0.9					150	0.084	9.6×10^{-3}	
	G.I. (LLI) insol.	1.0					150	0.084	1.1×10^{-2}	
^{71}Ge	Kidney	3×10^{-4}	0.03	12	12	6	300	0.01	4.4×10^{-4}	4.4×10^{-2}
	Liver	2×10^{-4}	0.02		7.5	4.6	1,700	0.01	4.0×10^{-7}	4.0×10^{-5}
	Total body	0.01	1.0		1	0.92	7×10^4	0.01	9.7×10^{-8}	9.7×10^{-6}
	G.I. (LLI) sol.	1.0					150	0.01	1.2×10^{-3}	
	G.I. (LLI) insol.	1.0					150	0.01	1.2×10^{-3}	
^{74}As	Kidney	3×10^{-4}	0.01	17.5	550	17	300	0.34	4.3×10^{-4}	1.4×10^{-2}
	Total body	0.03	1.0		280	16.5	7×10^4	0.56	2.9×10^{-4}	9.8×10^{-3}
	Liver	9×10^{-4}	0.03		550	17	1,700	0.38	2.5×10^{-4}	8.4×10^{-3}
	G.I. (LLI) sol.	0.97					150	0.32	3.8×10^{-2}	
	G.I. (LLI) insol.	1.0					150	0.32	3.9×10^{-2}	
^{76}As	Kidney	3×10^{-4}	0.01	1.1	550	1.1	300	1.1	8.9×10^{-5}	3.0×10^{-3}
	Liver	9×10^{-4}	0.03		550	1.1	1,700	1.1	4.7×10^{-5}	1.6×10^{-3}
	Total body	0.03	1.0		280	1.1	7×10^4	1.3	4.5×10^{-5}	1.5×10^{-3}
	G.I. (LLI) sol.	0.97					150	1.1	7.7×10^{-2}	
	G.I. (LLI) insol.	1.0					150	1.1	8.0×10^{-2}	

Table 25. (Continued)

AVERAGE RADIATION DOSES RECEIVED BY DIFFERENT ORGANS OF ADULT MALES FOLLOWING THE ADMINISTRATION OF 1 μCi OF VARIOUS RADIONUCLIDES

Radionuclide	Organ or Tissue of Reference	Fraction Reaching Organ After Ingestion (f_w)*	Fraction Reaching Organ After Injection (f'_2)	Radioactive Half-Period (T_r), days*	Biological Half-Period (T_b), days*	Effective Half-Period (T), days*	Mass of Organ (m), grams*	Effective Energy (E), Mev	Dose per μCi Administered, rems By Ingestion	Dose per μCi Administered, rems By Injection
^{82}Br	Total body	1.0	1.0	1.5	8	1.3	7×10^4	1.8	2.5×10^{-3}	
	G.I. (S) sol.	1.0					250	0.85	3.6×10^{-3}	
	G.I. (LLI) insol.	~1.0					150	0.53	4.5×10^{-2}	
^{85}Sr	Bone	0.21	0.7	65	1.8×10^4	64.8	7,000	0.091	1.3×10^{-2}	4.4×10^{-2}
	Total body	0.3	1.0		1.3×10^4	64.7	7×10^4	0.33	6.8×10^{-3}	2.3×10^{-2}
	G.I. (LLI) sol.	0.7					150	0.091	8.1×10^{-3}	
	G.I. (LLI) insol.	1.0					150	0.091	1.2×10^{-2}	
^{86}Rb	Pancreas	3×10^{-3}	3×10^{-3}	18.6	60	14.3	70	0.65	2.9×10^{-2}	2.9×10^{-2}
	Liver	0.05	0.05		63	14.4	1,700	0.66	2.1×10^{-2}	2.1×10^{-2}
	Spleen	4×10^{-3}	4×10^{-3}		45	13.2	150	0.66	1.7×10^{-2}	1.7×10^{-2}
	Muscle	0.45	0.45		80	15.1	3×10^4	0.70	1.2×10^{-2}	1.2×10^{-2}
	Total body	1.0	1.0		45	13.2	7×10^4	0.70	9.8×10^{-3}	9.8×10^{-3}
	G.I. (LLI) sol.	0.05					150	0.65	4.0×10^{-2}	
	G.I. (LLI) insol.	1.0					150	0.65	8.0×10^{-2}	
^{90}Y	Bone	7.5×10^{-5}	0.75	2.68	1.8×10^4	2.68	7,000	0.89	1.9×10^{-6}	1.9×10^{-2}
	Total body	10^{-4}	1.0		1.4×10^4	2.68	7×10^4	0.89	2.5×10^{-7}	2.5×10^{-3}
	G.I. (LLI) sol.	1.0					150	0.89	9.0×10^{-2}	
	G.I. (LLI) insol.	1.0					150	0.89	9.0×10^{-2}	
^{95}Nb	Liver	9×10^{-6}	0.09	35	845	33.6	1,700	0.26	3.4×10^{-6}	3.4×10^{-2}
	Kidney	2×10^{-6}	0.02		760	33.5	300	0.20	3.3×10^{-6}	3.3×10^{-2}
	Spleen	8×10^{-7}	8×10^{-3}		950	33.8	150	0.20	2.7×10^{-6}	2.7×10^{-2}
	Bone	3.8×10^{-5}	0.38		1,000	33.8	7,000	0.16	2.2×10^{-6}	2.2×10^{-2}
	Total body	10^{-4}	1.0		760	33.5	7×10^4	0.51	1.8×10^{-6}	1.8×10^{-2}
	G.I. (LLI) sol.	~1.0					150	0.16	2.0×10^{-2}	
	G.I. (LLI) insol.	1.0					150	0.16	2.0×10^{-2}	
^{110m}Ag + ^{110}Ag	Kidney	2×10^{-4}	0.02	270	10	10	300	0.65	3.2×10^{-4}	3.2×10^{-2}
	Liver	3×10^{-4}	0.03		15	14.2	1,700	0.84	1.6×10^{-4}	1.6×10^{-2}
	Total body	0.01	1.0		5	4.9	7×10^4	1.7	8.8×10^{-5}	8.8×10^{-3}
	Bone	5×10^{-4}	0.05		30	27	7,000	0.51	7.3×10^{-5}	7.3×10^{-3}
	G.I. (LLI) sol.	0.99					150	0.51	6.4×10^{-2}	
	G.I. (LLI) insol.	1.0					150	0.51	6.5×10^{-2}	

Table 25. (Continued)

AVERAGE RADIATION DOSES RECEIVED BY DIFFERENT ORGANS OF ADULT MALES FOLLOWING THE ADMINISTRATION OF 1 µCi OF VARIOUS RADIONUCLIDES

Radionuclide	Organ or Tissue of Reference	Fraction Reaching Organ After Ingestion (f_w)*	Fraction Reaching Organ After Injection (f'_2)*	Radioactive Half-Period (T_r), days*	Biological Half-Period (T_b), days*	Effective Half-Period (T), days*	Mass of Organ (m), grams*	Effective Energy (E), Mev	Dose per µCi Administered, rems By Ingestion	Dose per µCi Administered, rems By Injection
^{111}Ag + ^{111m}Cd	Kidney	2×10^{-4}	0.02	7.5	10	4	300	0.37	7.3×10^{-5}	7.3×10^{-3}
	Liver	3×10^{-4}	0.03		15	5	1,700	0.38	2.5×10^{-5}	2.5×10^{-2}
	Total body	0.01	1.0		5	3	7×10^4	0.40	1.3×10^{-5}	1.3×10^{-3}
	Bone	5×10^{-4}	0.05		30	6	7,000	0.37	1.2×10^{-5}	1.2×10^{-3}
	G.I. (LLI) sol.	0.99					150	0.37	4.3×10^{-2}	
	G.I. (LLI) insol.	1.0					150	0.37	4.3×10^{-2}	
^{123}I	Thyroid	0.3	0.3	0.54	138	0.54	20	0.034	1.4×10^{-2}	1.4×10^{-2}†
	Total body	1.0	1.0		138	0.54	7×10^4	0.105	6.0×10^{-5}	6.0×10^{-5}
	G.I. (LLI) sol.	0.05					150	0.04	8.4×10^{-5}	
	G.I. (LLI) insol.	1.0					150	0.04	1.7×10^{-3}	
^{124}I	Thyroid	0.3	0.3	4	138	3.9	20	0.33	1.3	1.3†
	Total body	1.0	1.0		138	3.9	7×10^4	1.01	4.1×10^{-3}	4.1×10^{-3}
	G.I. (LLI) sol.	0.05					150	0.41	2.2×10^{-3}	
	G.I. (LLI) insol.	1.0					150	0.41	4.4×10^{-2}	
^{125}I	Thyroid	0.3	0.3	60	138	42	20	0.012	0.6	0.6†
	Total body	1.0	1.0		138	42	7×10^4	0.030	1.3×10^{-3}	1.3×10^{-3}
	G.I. (LLI) sol.	0.05					150	0.017	1.1×10^{-4}	
	G.I. (LLI) insol.	1.0					150	0.017	5.4×10^{-3}	
^{126}I	Thyroid	0.3	0.3	13	138	12	20	0.18	2.4	2.4†
	Total body	1.0	1.0		138	12	7×10^4	0.41	5.2×10^{-3}	5.2×10^{-3}
	G.I. (LLI) sol.	0.05					150	0.21	1.3×10^{-3}	
	G.I. (LLI) insol.	1.0					150	0.21	2.5×10^{-2}	
^{130}I	Thyroid	0.3	0.3	0.52	138	0.52	20	0.49	0.19	0.19†
	Total body	1.0	1.0		138	0.52	7×10^4	1.55	8.5×10^{-4}	8.5×10^{-4}
	G.I. (LLI) sol.	0.05					150	0.57	1.1×10^{-3}	
	G.I. (LLI) insol.	1.0					150	0.57	2.2×10^{-2}	
^{131}I + ^{131m}Xe	Thyroid	0.3	0.3	8.05	138	7.6	20	0.23	1.9	1.9†
	Total body	1.0	1.0		138	7.6	7×10^4	0.44	3.5×10^{-3}	3.5×10^{-3}
	G.I. (LLI) sol.	0.05					150	0.25	1.5×10^{-3}	
	G.I. (LLI) insol.	1.0					150	0.25	2.9×10^{-2}	

Table 25. (Continued)

AVERAGE RADIATION DOSES RECEIVED BY DIFFERENT ORGANS OF ADULT MALES FOLLOWING THE ADMINISTRATION OF 1 µCi OF VARIOUS RADIONUCLIDES

Radio-nuclide	Organ or Tissue of Reference	Fraction Reaching Organ After Ingestion (f_w)*	Fraction Reaching Organ After Injection (f'_2)*	Radio-active Half-Period (T_r), days*	Biological Half-Period (T_b), days*	Effective Half-Period (T), days*	Mass of Organ (m), grams*	Effective Energy (E), Mev	Dose per µCi Administered, rems By Ingestion	Dose per µCi Administered, rems By Injection
^{132}I	Thyroid	0.3	0.3	0.097	138	0.097	20	0.65	2.0×10^{-2}	2.0×10^{-2}†
	Total body	1.0	1.0		138	0.097	7×10^4	1.7	1.7×10^{-4}	1.7×10^{-4}
	G.I. (S) sol.	1.0					250	1.0	3.7×10^{-3}	
	G.I. (ULI) insol.	1.0					135	0.76	4.1×10^{-3}	
^{132}Te + ^{132}I	Spleen	2.5×10^{-3}	0.01	3.2	30	2.9	150	0.96	3.4×10^{-3}	1.4×10^{-2}
	Testes	7.5×10^{-4}	3×10^{-3}		30	2.9	40	0.73	2.9×10^{-3}	1.2×10^{-2}
	Thyroid	2.5×10^{-4}	10^{-3}		9	2.4	20	0.74	1.6×10^{-3}	6.6×10^{-3}
	Liver	0.01	0.05		30	2.9	1,700	1.1	1.4×10^{-3}	6.9×10^{-3}
	Kidney	0.02	0.07		30	2.9	300	0.96	1.4×10^{-3}	4.8×10^{-3}
	Total body	0.25	1.0		15	2.6	7×10^4	1.9	1.3×10^{-3}	5.2×10^{-3}
	Bone	0.023	0.09		30	2.9	7,000	0.86	6.1×10^{-4}	2.3×10^{-3}
	G.I. (LLI) sol.	0.75					150	0.86	6.8×10^{-2}	
	G.I. (LLI) insol.	1.0					150	0.86	9.0×10^{-2}	
^{140}Ba + ^{140}La	Bone	0.035	0.7	12.8	65	10.7	7,000	1.12	4.4×10^{-3}	8.8×10^{-2}
	Total body	0.05	1.0		65	10.7	7×10^4	2.4	1.3×10^{-3}	2.6×10^{-2}
	Liver	3×10^{-5}	6×10^{-4}		975	12.6	1,700	1.4	2.3×10^{-5}	4.6×10^{-4}
	Muscle	1.5×10^{-4}	3×10^{-3}		2,000	12.7	3×10^4	2.3	1.1×10^{-4}	2.2×10^{-4}
	Spleen	2.5×10^{-6}	5×10^{-5}		13	6.4	150	1.2	9.5×10^{-6}	1.9×10^{-6}
	Kidney	5×10^{-6}	10^{-4}		8.5	5.1	300	1.2	7.5×10^{-6}	1.5×10^{-4}
	G.I. (LLI) sol.	0.95					150	1.12	1.3×10^{-1}	
	G.I. (LLI) insol.	1.0					150	1.12	1.4×10^{-1}	
^{140}La	Liver	1.5×10^{-5}	0.15	1.68	400	1.68	1,700	1.1	1.2×10^{-6}	1.2×10^{-2}
	Bone	4×10^{-5}	0.4		1,000	1.68	7,000	0.80	5.7×10^{-7}	5.7×10^{-3}
	Total body	10^{-4}	1.0		500	1.68	7×10^4	1.9	3.4×10^{-7}	3.4×10^{-3}
	G.I. (LLI) sol.	1.0					150	0.80	7.0×10^{-2}	
	G.I. (LLI) insol.	1.0					150	0.80	7.0×10^{-2}	

Table 25. (Continued)

AVERAGE RADIATION DOSES RECEIVED BY DIFFERENT ORGANS OF ADULT MALES FOLLOWING THE ADMINISTRATION OF 1 μCi VARIOUS RADIONUCLIDES

Radio-nuclide	Organ or Tissue of Reference	Fraction Reaching Organ After Ingestion (f_w)*	Fraction Reaching Organ After Injection (f'_2)	Radioactive Half-Period (T_r), days*	Biological Half-Period (T_b), days*	Effective Half-Period (T), days*	Mass of Organ (m), grams*	Effective Energy (E), Mev	Dose per μCi Administered, rems	
									By Ingestion	By Injection
^{147}Pm	Bone	3.5×10^{-5}	0.35	920	1,500	570	7,000	0.069	1.5×10^{-5}	1.5×10^{-1}
	Kidney	2×10^{-6}	0.02		656	383	300	0.069	1.3×10^{-5}	1.3×10^{-1}
	Liver	6×10^{-6}	0.06		656	383	1,700	0.069	6.9×10^{-6}	6.9×10^{-2}
	Total body	10^{-4}	1.0		656	383	7×10^4	0.069	2.8×10^{-6}	2.8×10^{-2}
	G.I. (LLI) sol.	$\sim$1.0					150	0.069	8.8×10^{-3}	
	G.I. (LLI) insol.	1.0					150	0.069	8.8×10^{-3}	
^{177}Lu	Bone	6.8×10^{-5}	0.68	6.8	1,000	6.75	7,000	0.16	7.7×10^{-7}	7.7×10^{-3}
	Kidney	10^{-6}	0.01		750	6.7	300	0.16	2.6×10^{-7}	2.6×10^{-3}
	Total body	10^{-4}	1.0		750	6.7	7×10^4	0.17	1.2×10^{-7}	1.2×10^{-3}
	G.I. (LLI) sol.	$\sim$1.0					150	0.16	1.9×10^{-2}	
	G.I. (LLI) insol.	1.0					150	0.16	1.9×10^{-2}	
^{182}Ta +	Liver	3×10^{-5}	0.3	112	400	88	1,700	0.56	6.4×10^{-5}	6.4×10^{-1}
	Kidney	3×10^{-6}	0.03		400	88	300	0.45	2.9×10^{-5}	2.9×10^{-1}
^{182m}W	Spleen	10^{-6}	0.01		240	76	150	0.45	1.7×10^{-5}	1.7×10^{-1}
	Total body	10^{-4}	1.0		240	76	7×10^4	1.1	8.8×10^{-6}	8.8×10^{-2}
	Bone	2×10^{-5}	0.2		300	82	7,000	0.38	6.6×10^{-6}	6.6×10^{-2}
	G.I. (LLI) sol.	$\sim$1.0					150	0.38	4.8×10^{-2}	
	G.I. (LLI) insol.	1.0					150	0.38	4.8×10^{-2}	
^{192}Ir	Liver	0.023	0.23	74.5	27	20	1,700	0.60	2.0×10^{-2}	2.0×10^{-1}
	Kidney	4.5×10^{-3}	0.045		50	30	300	0.50	1.7×10^{-2}	1.7×10^{-1}

* Values taken from Report of Committee 2 of the International Commission on Radiological Protection (ICRP–2).
† Assuming that the half-period of uptake into the thyroid is six hours.

Table 26.

FRACTIONS OF THE TOTAL DOSES, FOR ORGANS OTHER THAN THE GASTROINTESTINAL TRACT SHOWN IN TABLE 25, WHICH ARE DELIVERED IN PERIODS OF 1 WEEK, 4 WEEKS, 13 WEEKS, AND 1 YEAR FOLLOWING THE ADMINISTRATION OF RADIONUCLIDES

Effective Half-Period (T) in the Organ, days	1 Week	4 Weeks	13 Weeks	1 Year
1	0.99	1.00	1.00	1.00
2	0.91	1.00	1.00	1.00
3	0.80	1.00	1.00	1.00
4	0.70	0.99	1.00	1.00
5	0.62	0.98	1.00	1.00
10	0.39	0.86	1.00	1.00
15	0.27	0.72	0.99	1.00
20	0.21	0.62	0.96	1.00
30	0.15	0.48	0.88	1.00
40	0.11	0.39	0.79	1.00
50	0.10	0.32	0.72	0.99
60	0.08	0.27	0.65	0.99
80	0.06	0.21	0.55	0.96
100	0.05	0.17	0.47	0.91
200	0.02	0.095	0.27	0.72
300	0.02	0.068	0.19	0.57
400	0.01	0.05	0.15	0.47
500	0.01	0.04	0.11	0.40
600	0.01	0.03	0.10	0.34
700	0.01	0.03	0.09	0.30

The Effective Energy, E

The effective energy, E, per disintegration of the radionuclide concerned will depend upon the following factors.

1. The decay scheme of the radionuclide.
2. The energy and spectra of the radiations emitted by the radionuclide and its daughters (making an appropriate allowance for the different rates of excretion of the parent and daughters).
3. The dimensions of the organ and of the radioactive deposit within it.
4. The relative biological effectiveness (RBE) of the radiations concerned.

Wherever appropriate, the values of E shown in Table 25 were taken directly from ICRP-2. They were not available in the case of ^{56}Co, ^{52}Fe, ^{15}O, ^{43}K, ^{123}I, ^{124}I, ^{125}I, and ^{130}I, and for these E was calculated, using the latest information on their decay schemes in the same manner as described in ICRP-2, modified where necessary as described below. The factors 1 and 4 are, for the most part, well defined. In factor 4, RBE is taken as 1 for X rays, γ, β radiation, and electrons (1.7, if the maximum energy for β radiation and electrons is less than 0.03 Mev), and as 10 for α particles. The factor 2 is also fairly well defined with respect to the average energy emitted by the different radiations concerned, although an approximation has to be made to allow for the shape of β-ray spectra. This approximation is described in detail in ICRP-2. The contribution to E from daughter elements varies in a complex manner with the radioactive half-period of the parent in relation to those of its daughters and with their relative rates of biological elimination from the organ. When the radioactive half-period of the daughter is very long, compared with that of its parent, as in the case of ^{147}Pm and its daughter ^{147}Sm, the daughter does not contribute significantly to the effective energy per disintegration of its parent. In the ICRP-2 recommendations, a fraction, F, of the effective energy contributed by each daughter nuclide is added to that of the parent. This factor F is defined as the ratio (number of disintegrations per unit time of daughter atoms to the number of disintegrations per unit time of parent atoms) that exists in the organ after 50 years of continuous intake of the parents alone.* For all the radionuclides with daughters listed in Table 25, excepting ^{147}Pm, for which F is negligibly small, the factor F is very nearly equal to unity (in no case is it less than 0.9). Since in most practical circumstances the administered radionuclide will be accompanied by an equilibrium amount of its daughters, it follows that for the present purpose the value of F for the radionuclides concerned will be very near to unity at all times following their administration. The values of E shown in Table 25 for these radionuclides include the contributions from their daughters, calculated from the values of F given in ICRP-2; when these were not available, F was taken to be 1.

If the calculation of E is not to be too complex, some approximation is required with respect to factor 3, namely, the dimensions of the organ and of the radioactive deposit within it. In the case of γ rays, because of their high penetrating power, part of their energy may be absorbed some distance from their point of origin. In fact, some γ rays may escape from the body or from the organ where they originated without any appreciable loss of energy. It is clear, therefore, that, even though the distribution of radioactive material in the organ is uniform, the distribution of absorbed energy within it will depend markedly upon its shape and size. These matters are discussed in detail by Loevinger, Holt and Hine.[5] For simplicity, in ICRP-2 the whole body and each organ are assumed to be spherical and are assigned average radii.

In an organ of radius x cm containing a radionuclide emitting a photon of energy E, the effective energy is given by

$$\text{Effective energy} = E[1 - e^{(-\delta x)}], \tag{7}$$

where δ is the total coefficient of absorption minus the Compton-scattering coefficient in cm^{-1} for the given photon.

Variations of the absorbed energy per gram of tissue throughout the organ are, therefore, neglected, and the value of the effective energies given in Table 25 are strictly applicable only to tissues at the center of the organ. This leads to an overestimation, which is small in most cases, of the average energy, because the absorbed energy at points away from the center will be lower. The approximation becomes important when the effective energy of the radionuclide arises principally from γ radiation. In the case of children, for whom the effective radius (x) is smaller than for adults, the effective energy per distintegration of a radionuclide, and consequently the dose, will be correspondingly lower.

For the corpuscular radiations, which in tissues have ranges that are usually less than the linear dimensions of the organs concerned, practically all of the emitted energy will be absorbed in the organ where the material is deposited. Furthermore, provided that the radioactive material is uniformly distributed, the energy distribution within the organ will also be uniform, except at distances from the periphery of the organ that are less than the maximum range of the particles concerned, where it will be lower. However, if the distribution of the radioactive material is not uniform, the absorbed dose will also not be uniform, being highest at the sites of greatest concentration. Although this is known to be true in many cases, it is neglected here, and an average energy is calculated assuming uniform distribution of the material. The effect of nonuniform distribution of a radionuclide emitting α and β rays is also neglected by ICRP-2, with notable exceptions, namely, those radionuclides that are deposited in bone. In these cases the maximum permissible body burdens recommended by ICRP-2 are calculated by a comparison with radium. In the method of calculation adopted, the effective energy due to α and β radiations or from recoil atoms is increased by a factor 5, if the radionuclide is not an isotope of radium. This factor 5, known as the "Relative-Damage Factor," reflects the possibility, among others, that the radionuclide concerned may be more nonuniformly distributed than radium in bone. The values for E for bone in Table 25 are appropriate for the calculation of the average dose to the skeleton and do not include this

* It is to be noted that ICRP-2 always considers the intake of the parent nuclide alone, except in the case of inhalation of the gases radon and thoron, when some of the daughter products are assumed to be present.

factor 5. In the case of pure β emitters, they were obtained by dividing by a factor 5 the values of E that are given for bone in the ICRP-2 tables. For radionuclides that also emit γ rays, the values of E used were either those given in ICRP-2 for any organ having the same effective radius as bone, 5 cm, or they were calculated from first principles.

DOSE TO THE GASTROINTESTINAL TRACT

When radioactive materials are administered by mouth, parts of the gastrointestinal (G.I.) tract are irradiated. The average doses to various parts of the G.I. tract following a 1-μCi drink of different radionuclides in both soluble and insoluble form are shown in Table 25. A simple model of the G.I. tract, taken from the Recommendations of the International Commission on Radiological Protection,[6] was used in the calculation of these doses. In this it is assumed that ingested material passes through the tract in the following manner. First it remains in the stomach (S) for a period of one hour. During the following four hours it passes through the small intestine (SI), although, if it is in a soluble form that is taken up by the body, part of it may cross the intestinal wall and enter the blood stream. Any material remaining after traversing the small intestine passes through the upper large intestine (ULI) during the next eight hours, and finally traverses the lower large intestine (LLI) in the final eighteen hours before being excreted.

Table 25 shows the average dose to only one portion of the tract following the ingestion of 1 μCi of either the soluble or insoluble forms of any radionuclide. This is that portion of the tract which receives the highest average dose following the ingestion. With three exceptions, it is also that part of the tract which, according to ICRP-2, receives the highest local dose during conditions of chronic intake of the radionuclide. The three exceptions are for the soluble forms of ^{18}F, ^{24}Na and ^{82}Br, and they call for special comment.

Dose to the Small Intestine

In the earlier recommendations of ICRP (1955), where average doses throughout any portion of the tract were used as criteria, it was stated that the dose to the small intestine never limited the intake of radioactive materials, since another portion of the tract always received a higher average dose. In the present recommendations of ICRP-2, the maximum local dose in any portion of the tract limits the intake of radionuclides, and in some cases, including soluble forms of ^{18}F, ^{24}Na and ^{82}Br, this is said to occur in the small intestine. However, in the case of these three radionuclides a major reason for this is undoubtedly the high value recommended by ICRP-2 for that part of the effective energy (E) which is due to the γ rays emitted by them. In the case of the small intestine, this has been calculated on the assumption that the critical portion is at the center of a sphere with a radius of 30 cm. This undoubtedly high estimate for the effective radius of the small intestine, as well as the assumption that the maximum dose occurs near the center of the hypothetical sphere, overestimates E, and consequently the dose. There is, therefore, some doubt about the validity of the assumptions made by ICRP-2 in calculating the dose to the small intestine in the case of γ emitters. For these reasons, the stomach probably receives a higher average dose than the small intestine in the cases of ^{18}F, ^{24}Na and ^{82}Br, and values for this portion of the gastrointestinal tract are shown in Table 25.

Dose to the Stomach

If 1 μCi is ingested and remains in the stomach for a period of one hour, the integrated activity in the stomach, I_t, is given by

$$I_t \text{ (in μCi-days)} = \int_0^{1/24} e^{(-0.693t/T_r)} \, dt, \tag{8}$$

where T_r is the radioactive half-life in days.

$$\therefore I_t \text{ (in μCi-days)} = T_r[1 - e^{(-0.693/24T_r)}]/0.693. \tag{9}$$

If the average mass of the stomach contents is m grams, the dose within them is given by substituting the above value for I_t in equation (2). The average dose to the stomach wall, however, will be half the value so obtained, since this is irradiated over a solid angle of 2π only, and the dose to the stomach wall, D_s, is given by

$$D_S \text{ (in rems)} = 37 T_r E[1 - e^{(-0.693/24T_r)}]/m. \tag{10}$$

Dose to the Large Intestine

If a fraction (f) of the radioactive material reaches the large intestine, the integrated activity, I_t, in either the LLI or the ULI following a drink of 1 µCi is given by

$$I_i \text{ (in µCi-days)} = f \int_{h_0}^{h_1} e^{(-0.693t/T_r)} \, dt, \tag{11}$$

where h_0 and h_1 respectively are the times in days after the drink was given when the material enters and leaves the portion considered.

$$\therefore I_t \text{ (in µCi-days)} = f T_r [e^{(0.693h_0/T_r)} - e^{(-0.693h_1/T_r)}]/0.693. \tag{12}$$

For convenience let

$$A = 0.693 H / [e^{(-0.693h_1/T_r)} - e^{(0.693h_1/T_r)}] T_r,$$

where H is the time spent by the material in the portion of the large intestine considered (i.e., $H = h_1 - h_0$), which is 18/24 days for the LLI and 8/24 days for the ULI. (It is worth noting that A reduces to 1 for large values of T_r).

$$\therefore I_t \text{ (in µCi-days)} = fH/A. \tag{13}$$

The dose within the contents of a portion of the large intestine of mass m grams is then given by substituting for I_t in equation (2). The dose to the intestinal walls is half of this, and is, therefore, given by

$$D_L \text{ (in rems)} = 25.6 f H E / Am. \tag{14}$$

Values of m, the mass of the contents, were taken from ICRP-2. For the stomach, m was 250 g; for the lower large intestine, 150g; and for the upper large intestine, 135 g. The values of E that were used, taken from ICRP-2 when available, are shown in Table 25. For those radionuclides in Table 25 that are not listed in ICRP-2, E was calculated from first principles. For α particles (^{224}Ra is the only radionuclide concerned in Table 25), only 1 percent of the energy is included, in order to make some allowance for the shielding afforded by mucus.

For those radionuclides in Table 25 that have radioactive daughters, excepting ^{149}Pm, it was assumed that their radioactive daughters were in equilibrium with their parents at the time of ingestion and throughout their passage through the tract. The values of E, taken from ICRP-2, for parent and duaghter were, therefore, added, to find the value of E shown in Table 25. In the case of ^{147}Pm, the daughter, ^{147}Sm, is so long-lived that the contribution it makes to the effective energy per disintegration of its parent will be zero in any practical circumstances.

The value of f, the fraction of the ingested material reaching the large intestine, is shown under the column headed f_w in Table 25. It is equal to $(1 - f_l)$, where f_l is taken from ICRP-2, and is the fraction of ingested material that enters the blood stream. In accordance with the practice adopted by ICRP-2 for those radionuclides where f_l is unity, the value of f_w shown in Table 25 is arbitrarily fixed at 0.05; perhaps this is to make some allowance for material that, having entered the blood stream, is reexcreted into the gut. The need to make an arbitrary choice about the value of f_w in these cases only emphasizes the uncertainty of our knowledge about the fate of ingested material.

CONCLUSION

It will be clear from the foregoing that there are many gaps in our knowledge concerning the behavior of radioactive materials in the body. There is, in addition, considerable uncertainty about their effects on body tissues, and it is, therefore, doubly important to exercise the greatest caution in any experiments involving the use of radioactive materials on humans. Nevertheless, in the interests of progress it is important that the medical profession should not be denied this most useful new tool for diagnosis and investigation. This paper has been written, therefore, in the hope that it will assist those concerned to make a preliminary assessment of the radiation doses involved, so that they may balance the possible harm against the ultimate benefit of their investigation.

SUMMARY

When planning the administration of radionuclides for medical purposes, it is important to estimate the radiation doses that will be received by several organs of the body, as well as that received by the target organ. In order to make this estimate for any radionuclide, information is required about its uptake and retention in different organs of the body following its administration to the patient, and about the amount of energy absorbed in those organs following each of its disintegrations. Average values for these parameters for adult males are available in the Report of Committee 2 of the International Commission on Radiological Protection. This information has been used to compute values of the average radiation doses received by several organs of adult males following the ingestion or injection of 1 µCi of most of the radionuclides at present used in hospitals. Details of the method of calculation and of the approximations employed are discussed, so that the values of the radiation doses derived for adult males may be modified for other individuals—for example females and children—whenever the relevant metabolic data are available. Values of the parameters required in the calculation and of the radiation doses for adult males are tabulated.

AUTHORS' ACKNOWLEDGMENT

It is a great pleasure to acknowledge our indebtedness to the Director of the Radiological Protection Service, Mr. W. Binks, who suggested that we should write this paper, and who has, in many helpful discussions, materially assisted in its preparation.

ADDENDUM

Since writing this paper, the authors' attention has been drawn to a publication of the International Atomic Energy Agency, namely, *Regulations for the Safe Transport of Radioactive Materials*.[7] In Appendix II of this publication, values are given for the radiation dose received by the "critical organ" following the entry into the body, either through a wound or by inhalation, of 1 µCi of various radionuclides. In this context the term "critical organ" actually means the organ receiving the highest dose. The values of the radiation dose for entry through a wound that are given in the IAEA document should, therefore, be identical with those shown in Table 25 for the organ receiving the highest dose following an injection of 1 µCi of the same radionuclide. The fact that this is not always so is due principally to two factors. First, although the values of radiation dose given in IAEA are based on information from ICRP-2, as are values shown in Table 25, the two calculations have different starting points and, therefore, include different approximations. In the present paper the radiation dose is calculated from the basic information on metabolic behavior, effective energy and mass given in ICRP-2. In the IAEA document the dose is obtained from the value recommended in ICRP-2 for the maximum permissible body burden of the appropriate radionuclide. This is defined as the amount of radionuclide in the whole body that delivers certain specified maximum permissible dose rates to the critical organ. In ICRP-2 the values of these maximum permissible body burdens are given to one significant figure only.

The second important point of difference concerns the calculation of dose to bone. The values given in Table 25 are for the average dose, but in the IAEA document they are average doses only in the cases of radium and its isotopes and of pure γ emitters. For all other radionuclides the doses given for bone are multiples, varying between one and five, of the average doses. As mentioned in the preceding paper, this is due to the fact that in ICRP-2 the maximum permissible body burden for bone-seeking radionuclides is based on a comparison with radium. For all except the radioisotopes of radium, the effective energies of α and β radiations are increased by a factor 5, but there is no such increase for the energies of γ rays.

The values given in the IAEA document also have a limited usefulness when applied to the injection of radionuclides for medical purposes, since no information is given about the doses received by organs of the body other than that receiving the highest radiation dose. No information is given in the IAEA document concerning the radiation dose following ingestion of radionuclides.

REFERENCES

1. HMSO, *Radiological Hazards to Patients*. Second Report of the Committee, 1960.
2. International Commission on Radiological Protection, ICRP-2, Recommendations of the International Commission on Radiological Protection. *Report of Committee 2 on Permissible Dose for Internal Radiation*. Pergamon Press, Macmillan Co., New York, 1959.
3. Krogman, W., Growth of Man, *Tabulae Biologicae*, Vol. 20. W. Junk, Holland, 1941.
4. Ford, M., *Private Communication*. Health Physics Division, Oak Ridge National Laboratory, Oak Ridge, Tennessee, 1960.
5. Loevinger, R., Holt, J. G., and Hine, G. J., in *Radiation Dosimetry*, Hine, G. J. and Brownell, G. L., eds. Academic Press, New York, 1956.
6. International Commission on Radiological Protection, Recommendations of the International Commission on Radiological Protection. *Brit. J. Radiol.*, Suppl. 6, 1955.
7. International Atomic Energy Agency, *Regulations for the Safe Transport of Radioactive Materials*. Safety Series No. 7, International Atomic Energy Agency, Vienna, Austria, 1961.

DOSE FROM INGESTION OR INHALATION OF SOLUBLE RADIONUCLIDES

D. F. Bunch

At the National Reactor Testing Station (NRTS), a great number of calculations are performed by the various contractors each year to estimate the radiological consequences from operational releases, minor and major incidents, and potential accidental releases. Much of this is repetitious in that each person making the calculation must perform the necessary mathematics to solve the various equations used in calculating dose. In almost all cases the mathematical and biological parameters are those recommended by the International Commission on Radiological Protection (ICRP)[1,2] for the "standard man". To eliminate the need for this repetition, and to establish more uniform practices in calculations, a computer program was written, and estimates of dose were prepared for the isotopes and major organs listed in the ICRP reports. These estimates take the form of dose conversion factors, so that a rapid and reasonable estimate of dose may be made. This chapter gives information for dose from the ingestion or inhalation of soluble radionuclides, although the tables in the source publication have been severely abridged; only the dose conversion factors applying to the thyroid are given, to illustrate the principle of the calculation.

CALCULATION OF DOSE

Since, for the most part, close accuracy is not desired, or even warranted, the parameters used are those recommended by the ICRP for continuous exposure. The general expression is

$$\text{Dose} = \frac{AfET_E}{m}\left(1 - \exp\frac{-1.26 \times 10^4}{T_E}\right) \times \frac{1.6 \times 10^{-6} \times 3.2 \times 10^{15}}{0.693 \times 10^2}, \tag{1}$$

where

f = fractional uptake by ingestion or inhalation to the organ of interest;
E = effective energy = $\Sigma\, EF(\text{RBE})n$;
T_E = effective half-time of material in organ of interest;
m = mass of organ;
1.6×10^{-6} = erg/Mev;
3.2×10^{15} = dis/day/curie;
0.693×10^2 = erg/g/rad.

The exponential term assumes a 50-year post-exposure period, to correct for certain isotopes that do not reach equilibrium in this time. The term "A" is defined as:

$A = 1$ to calculate rem per curie inhaled or ingested; (2)

$A = 1B$ to calculate rem per curie-sec/m^3, where B = breathing rate in m^3/sec (3)
(curie-sec/m^3 is the time-integrated concentration of airborne radioactivity);

$A = 1/f$ to calculate rem per curie in the organ. (4)

The deviation, assumptions, and limitations of equation (1) have been discussed in detail in the preceding paper and in References 1 and 3, and these should be referred to for more detailed information. It should be emphasized that these calculations should not be applied to the general population and, further, that derived doses are only approximations.

Application

Sample Calculation 1.

If it is known that one microcurie of ^{131}I has been inhaled, the estimation of dose may be made as follows.

1. If the thyroid is the organ of interest, the conversion factor from curies inhaled to dose in rem is 1.48×10^6 (see Table 27), or $1.48E + 06$ (see Table 28) rem/curie inhaled.
2. Dose $= 1.48 \times 10^6 \dfrac{\text{rem}}{\text{Ci inhaled}} \times 10^{-6}$ curie $= 1.48$ rem ≈ 1.5 rem.

Sample Calculation 2.

If the air concentration is $10^{-6} \mu$Ci/cc of ^{135}I, and the individual will be exposed for eight hours, the estimation of dose may be made as follows:

1. If the exposure is occupational, assume the high breathing-rate factor (Ci-sec/m^3).
2. If the thyroid is the organ of interest, the conversion factor is $4.28E + 01$ or 4.28×10^1 rem/Ci-sec/m^3.
3. $10^{-6} \mu$Ci/cc $= 10^{-6}$ Ci/m^3.
4. Dose $= 10^{-6}$ Ci/m$^3 \times 8$ hours $\times 3600 \dfrac{\text{sec}}{\text{hour}} \times \dfrac{4.28 \times 10^1 \text{ rem}}{\text{Ci-sec/m}^3} \approx 1.2$ rem.

For the short-lived isotopes, the use of a single exponential model may grossly overestimate the actual dose. The conversion factors for these isotopes should be corrected by $T_r(T_r + T_u)$, where T_r is the radiological half-life, and T_u is the half-time for uptake (0.25 day for iodine). The corrected iodine factors are shown in Table 27. It can be seen that many of these isotopes would not constitute a significant hazard, if mixed fission products were released. In addition, for most of the organs a few isotopes constitute 80 to 90+ percent of the exposure. Therefore, for estimating dose from mixed fission products, it is not necessary to make a calculation for the contribution of every dose. Future reports will deal further with this subject, as well as present conversion factors for dose to the gastrointestinal tract.

Notes on Format for Table 28.

Column 1—the identification format is Z.A., as 01.003 for tritium. (Z = atomic number; A = atomic weight; M = metastable.)

Column 2—this is ΣEF(RBE)n, expressed in Mev.

Column 3—this is T_E expressed in days.

Column 4—rem/curie inhaled.

Column 5—rem/curie-sec/m^3 for breathing rate typical of active portion of day, 10 m^3/8 hours.

Column 6—rem/curie-sec/m^3 for average breathing rate, 20 m^3/24 hours.

Column 7—rem/curie in organ.

Column 8—rem/curie ingested.

Weight of the thyroid (20 g) assumes the "standard man".

Alternate Data

Since it may be desirable to use parameters other than those used in these conversion factors, a nomogram has been prepared and included as Figure 2, so that any or all of the parameters may be varied to obtain different conversion factors. This nomogram also is based on equation (1), with the assumption that $e^{-1.26} \times 10^{-4} T_E$ is near zero. The nomogram may be used as follows.

Table 27. CALCULATION OF IODINE CONVERSION FACTORS APPLYING TO THE THYROID

Isotope	ICRP-2			Corrected for Uptake Decay			
	rem/Ci Inhaled	rem/Ci-sec/m^3 (2.32×10^{-4} m^3/sec)	rem/Ci-sec/m^3 (3.47×10^{-4} m^3/sec)	rem/Ci Organ	rem/Ci Inhaled	rem/Ci-sec/m^3 (2.32×10^{-4} m^3/sec)	rem/Ci-sec/m^3 (3.47×10^{-4} m^3/sec)
^{131}I	1.48×10^6	343.0	514.0	6.30	1.44×10^6	330.00	500.0
^{132}I	5.35×10^4	12.4	18.5	0.23	1.50×10^4	3.50	5.2
^{133}I	4.00×10^5	92.8	139.0	1.80	3.10×10^5	72.00	110.0
^{134}I	2.50×10^4	5.8	8.7	0.11	3.30×10^3	0.75	1.1
^{135}I	1.24×10^5	28.8	43.0	0.54	6.60×10^4	15.00	23.0

$T_r(T_r + T_u)$, where $T_u = 0.25$ day

Table 28. DOSE CONVERSION FACTORS FOR THE THYROID

Identification, Z.A.	Energy, Mev	$T_{1/2}$, days	Inh. Dose, rem/Ci	Dose/Conc., rem/Ci-sec/m^3, Hi-Rate	Dose/Conc., rem/Ci-sec/m^3, Lo-Rate	Organ Dose, rem/Ci	Ing. Dose, rem/Ci
24.051	8.40E − 03	2.66E + 01	1.90E + 02	6.57E − 02	4.29E − 02	8.26E + 05	3.72E − 00
49.114M	9.20E − 01	7.20E − 00	2.45E + 03	8.47E − 01	5.53E − 01	2.45E + 07	1.96E + 01
49.115M	1.60E − 01	1.90E − 01	1.12E + 01	3.89E − 03	2.53E − 03	1.12E + 05	8.99E − 02
49.115	1.70E − 01	8.40E − 00	5.28E + 02	1.82E − 01	1.19E − 01	5.28E + 06	4.22E − 00
50.113	1.60E − 01	4.30E + 01	7.12E + 02	2.46E − 01	1.60E − 01	2.54E + 07	1.27E + 02
50.125	9.30E − 01	8.40E − 00	8.09E + 02	2.79E − 01	1.82E − 01	2.89E + 07	1.44E + 02
51.122	5.90E − 01	1.60E − 00	2.79E + 01	9.66E − 03	6.30E − 03	3.49E + 06	3.14E − 00
51.124	5.70E − 01	3.80E − 00	6.41E + 01	2.21E − 02	1.44E − 02	8.01E + 06	7.21E − 00
51.125		4.00E − 00	1.66E + 01	5.76E − 03	3.75E − 03	2.08E + 06	1.86E − 00
52.125	1.10E − 01	7.80E − 00	1.20E + 03	4.17E − 01	2.72E − 01	3.17E + 06	7.93E + 02
52.127M	3.00E − 01	8.30E − 00	3.50E + 03	1.21E − 00	7.90E − 01	9.21E + 06	2.30E + 03
52.127	2.40E − 01	3.70E − 01	1.24E + 02	4.31E − 02	2.81E − 02	3.28E + 05	8.21E + 01
52.129M	6.80E − 01	7.10E − 00	6.78E + 03	2.34E − 00	1.53E − 00	1.78E + 07	4.46E + 03
52.129	6.00E − 01	5.10E − 02	4.30E + 01	1.48E − 02	9.70E − 03	1.13E + 05	2.83E + 01
52.131M	6.90E − 01	1.10E − 00	1.06E + 03	3.69E − 01	2.40E − 01	2.80E + 06	7.02E + 02
52.132	7.40E − 01	2.40E − 00	2.49E + 03	8.63E − 01	5.63E − 01	6.57E + 06	1.64E + 03
53.126	1.60E − 01	1.21E + 01	1.64E + 06	5.69E + 02	3.71E + 02	7.16E + 06	2.14E + 06
53.129	6.80E − 02	1.38E + 02	7.98E + 06	2.76E + 03	1.80E + 03	3.47E + 07	1.04E + 07
53.131	2.30E − 01	7.60E − 00	1.48E + 06	5.14E + 02	3.35E + 02	6.46E + 06	1.94E + 06
53.132	6.50E − 01	9.70E − 02	5.36E + 04	1.85E + 01	1.21E + 01	2.33E + 05	6.99E + 04
53.133	5.40E − 01	8.70E − 01	3.99E + 05	1.38E + 02	9.02E + 01	1.73E + 06	5.21E + 05
53.134	8.20E − 01	3.60E − 02	2.51E + 04	8.69E − 00	5.66E − 00	1.09E + 05	3.27E + 04
53.135	5.20E − 01	2.80E − 01	1.23E + 05	4.28E + 01	2.79E + 01	5.38E + 05	1.61E + 05
75.183	3.40E − 02	2.90E − 00	1.27E + 03	4.41E − 01	2.88E − 01	3.64E + 05	1.27E + 03
75.186	3.60E − 01	1.70E − 00	7.92E + 03	2.74E − 00	1.78E − 00	2.26E + 06	7.92E + 03
75.187	1.20E − 02	3.00E − 00	4.66E + 02	1.61E − 01	1.05E − 01	1.33E + 05	4.66E + 02
75.188	8.00E − 01	5.70E − 01	5.90E + 03	2.04E − 00	1.33E − 00	1.68E + 06	5.90E + 03
85.211	6.10E + 01	3.00E − 01	1.80E + 06	6.22E + 02	4.05E + 02	7.82E + 07	2.34E + 06

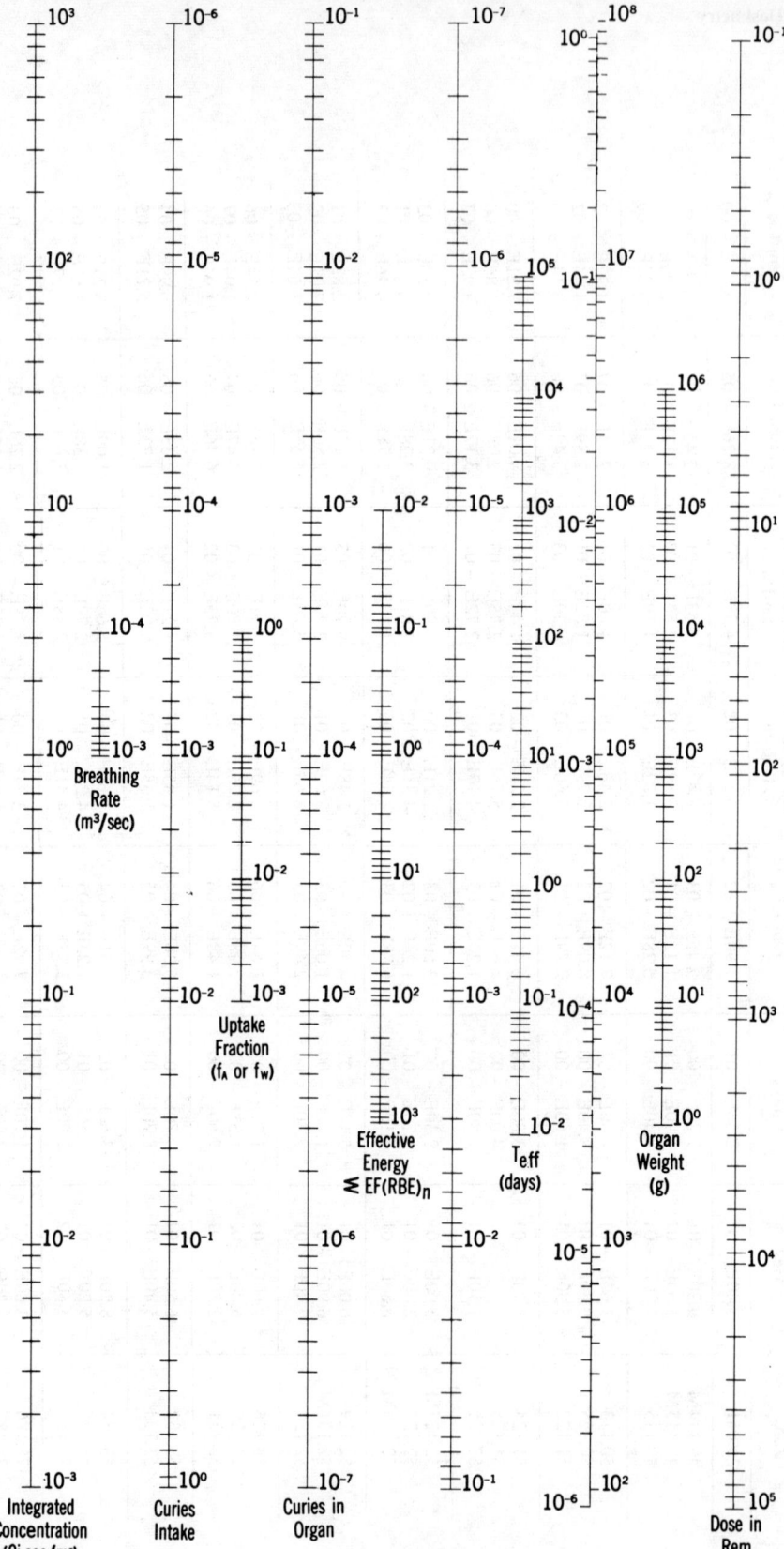

FIGURE 2. Nomogram for Infinity Dose Calculations.

1. rem/Ci-sec/m^3.
   a) Draw a line from the time-integrated concentration through the assumed breathing rate, to derive curies inhaled.
   b) Draw a line from curies inhaled through f_A, to derive curies in organ.
   c) Draw a line from curies in organ through the effective energy, mark the point, and draw a line through the effective half-life to the next column.
   d) Draw a line from this last column through the organ weight, to derive dose in rem.
2. rem/curies inhaled.
   Done in the same manner as 1., except that step a) is eliminated; begin at Column 3; to determine rem/curie ingested, substitute f_W for f_A.
3. rem/curie in organ.
   Same as 1., except eliminate steps a) and b), and begin at Column 5.

LEGAL NOTICE

This report was prepared as an account of Government-sponsored work. Neither the United States, nor the Commission, nor any person acting on behalf of the Commission
   a) makes any warranty or representation, express or implied, with respect to the accuracy, completeness, or usefulness of the information contained in this report, or that the use of any information, apparatus, method, or process disclosed in this report may not infringe on privately owned rights; or
   b) assumes any liabilities with respect to the use of, or for damages resulting from the use of, any information, apparatus, method, or process disclosed in this report.

As used in the above, "person acting on behalf of the Commission" includes any employee or contractor of the Commission, or employee of such contractor, to the extent that such employee provides access to any information pursuant to his employment or contract with the Commission, or his employment with such contractor.

ACKNOWLEDGMENT

This chapter is reprinted, with modifications, from *Dose to Various Body Organs from Inhalation or Ingestion of Soluble Radionuclides*, prepared in the Health and Safety Division of the U.S. Atomic Energy Commission, Idaho Operations Office, AEC Research and Development Report, Health and Safety, TID-4500, issued August 1966. It is reprinted in its entirety in the *Handbook of Laboratory Safety*, pp. 270–282. N.V. Steere, ed., published by The Chemical Rubber Co., Cleveland, Ohio, in 1967.

REFERENCES

1. International Commission on Radiological Protection, ICRP-2, Recommendations of the International Commission on Radiological Protection. *Report of Committee 2 on Permissible Dose for Internal Radiation.* Pergamon Press, Macmillan Co., New York, 1962.
2. International Commission on Radiological Protection, ICRP-6, 1959, revised 1962. *Recommendations o, the International Commission on Radiological Protection.* Pergamon Press, Macmillan Co., New York, 1962.
3. Loevinger, R., Holt, J. G., and Hine, G. J., Internally Administered Isotopes. *Radiation Dosimetry*, Chapter 17, Hine, G. J. and Brownell, G. L., eds. Academic Press, New York, 1956.
4. Markee, E. H., Jr., A Simplified Method of Estimating Environmental Hazards from Accidental Airborne Release of Radioactive Materials. *NRTS Meteorological Information Bulletin No. 2*, 1966.

PART V
BIOCHEMISTRY

STANDARDS OF ACTIVITY

Reprinted by permission of
The Radiochemical Centre, Amersham, Bucks, England.

GENERAL CONSIDERATIONS

Most users of radioactive substances need to know the approximate activity of their material, if only to abide by safety regulations or to be able to purchase the minimum activity that will make it possible to obtain useful results. Relatively few users need an accurate knowledge of activity; for example, nearly all tracer investigations rely on comparative measurements. However, for such applications as the determination of the yield of nuclear reactions, the measurements of cross sections, and for all applications where radiation dose is important, an accurate knowledge of the activity is essential. For the calibration of laboratory instruments to be used for measuring radioactive materials it is always desirable to use standards of known accuracy.

Radioisotope suppliers do not usually state or guarantee the accuracy of their activity measurement on routine supplies; when known accuracy is required, it is necessary to purchase specially measured and certified material.

DEFINITION OF STANDARDS OF ACCURACY

The activity of a quantity of radioactive material is the number of nuclear disintegrations that occur in this quantity in unit time.* The special unit of activity is the *curie*, which is defined by the ICRU as exactly 3.7×10^{10} disintegrations per second. Activity can, therefore, be quoted in curies or its subunits (millicuries—10^{-3}, microcuries—10^{-6}, nanocuries—10^{-9}, picocuries—10^{-12}), or in disintegrations per second or other suitable units of time.

Activity is not constant with time, but decays at a rate that is determined by the radionuclide present. A statement of activity is, therefore, only true at a specified time. The activity at any other time can be calculated, provided that the amount of each radionuclide present and the individual rates of decay are known to a sufficient accuracy.

Standards of activity must, therefore, be based on an accurate determination of the number of nuclear disintegrations occurring in a known quantity of material within a known period of time at a specified reference time. Because of uncertainties in purity and half-life, the accuracy to which the activity can be calculated for other times becomes less as the period from the reference time increases. In consequence, a standard of activity has only a limited period of usefulness.

The ephemeral nature of standards of activity has caused some misgiving about the use of the word *standard* in this context. An alternative is to describe such standards as *absolutely standardized radioactive material* or as *radioactive reference material*, according to the method of measurement.

STATEMENT OF ACCURACY

One of the most important qualities of a standard of activity is the accuracy to which the disintegration rate has been determined. It is important to distinguish between the accuracy (which refers to the possible deviation of the quoted value from the true value) and the precision

---

* The International Commission on Radiological Units and Measurements (ICRU), in its 1962 report (NBS Handbook 86), uses the term *activity* in preference to *radioactivity*, and gives a precise definition for it. In chemical practice the term *activity* is used as a measure of the effective concentration of ions.[1] In this article the term *activity* is used as defined by the ICRU.

(which, in English, refers only to the reproducibility of the measurement). Inconsistencies in terminology and in practice occur between laboratories; it is, therefore, essential to understand exactly what is meant by the statement of accuracy that is supplied with a standard of activity.

The assessment of accuracy is complicated, because both random and systematic errors must be considered. The random errors can be determined by repeated measurements, and are usually represented by the standard error, σ, (i.e., the standard deviation of the quoted mean of the experimental results); $\pm 3\sigma$ closely represents the maximum uncertainty in the result due to random errors. Systematic errors are more difficult to determine objectively. The figures quoted in the stated result must represent the maximum remaining uncertainties due to systematic errors after the best possible attempt has been made to eliminate all such errors. The extent of these errors depends on the magnitude of the correction factors applied to the experimental result, on the decay scheme and radiation characteristics of the nuclide, and on the techniques used for determining the disintegration rate. The results of intercomparisons with other laboratories must be taken into account. In the case of standards based on comparative measurements, the systematic errors will usually be predominantly the maximum overall errors of the standard source used in the comparison.

The combination of random and systematic errors has been the subject of much discussion. If they are combined to provide values for the maximum overall uncertainties in the quoted value of disintegration rate, the separate components should always be available for the user to combine in alternative ways or to use separately.

TYPES OF STANDARDS AVAILABLE
Solutions

The most satisfactory physical form for a standard of activity is a solution in a flame-sealed ampule. Carrier content and chemical form must be such that all the activity remains in solution over an extended period of time. For use, the ampule tip must be capable of being readily broken, and must allow aliquots to be taken easily and accurately for the preparation of counting sources. The whole solution may sometimes be used for the calibration of instruments such as ionization chambers. In this case, in order to maintain consistent geometry, the user may prefer to receive the standardized solution in exactly the same physical form as he normally receives his routine supplies of radioactive material—e.g. in a capped glass vial.

Absolutely Standardized Solutions are those for which the disintegration rate has been determined very accurately without reference to any other standard of activity. These were originally called *primary standards*, but this term is very unsuitable for a material whose important measured property is not constant with time.[2]

Absolute standardizations require the use of complex and reliable counting equipment, followed by expert interpretation of the results. They are, therefore, very expensive, unless many solutions can be distributed from each standardization. Such solutions are, therefore, not usually available at short notice (except for long-lived nuclides, which can be held in stock), but are prepared according to a published program. The common practice is to dispense a large number of ampules from a stock solution by accurate gravimetric dispensing, and to dispatch these to the users some days prior to the intended reference date. User and standardizing laboratory can then make measurements on or near the reference date. This reduces the error caused by uncertainty in half-life.

Reference Solutions have their activity measured by comparison, directly or indirectly, with one or more absolutely standardized solutions. These were originally called *secondary standards*, but the terms *reference*, *laboratory*, or *working standards* are preferable.[2]

Reference solutions can be prepared relatively cheaply, provided that suitable equipment is available for relating total activity of a solution to that of previous absolutely standardized solutions. For gamma-emitting nuclides, this can be done with a suitable reentrant ionization chamber, using a long-lived radiation source to link the activities. For pure beta-emitting nuclides it is necessary to use a "liquid" Geiger counter, or liquid scintillation counter, or to prepare a solid counting source. For these latter measurements, the ampule of solution must be opened. Alternatively, a stock solution can be dispensed into the ampules, and an aliquot taken for measurement. The problems of accurate direct comparison of pure beta-emitting solutions intact in their ampules, using, for example, bremsstrahlung or Cerenkov counters, have not yet been solved.

Standardized solutions are available from some national standardizing laboratories (e.g., National Bureau of Standards, Washington, D.C., in the United States, and National Physical Laboratory, Teddington, in England), but are increasingly becoming available from the normal suppliers of radioactive materials. The national laboratories and certain international bodies—such as the International Commission on Radiological Units, the International Bureau of Weights and Measures in Paris, France, and the International Atomic Energy Agency in Vienna, Austria—have initiated or taken part in many international comparisons of standards of activity over the last fifteen years. There is, therefore, some measure of international agreement on standards of activity for a limited number of nuclides. The firms or institutions specializing in the supply of radioactive material do not normally take part in these international comparisons, and they, therefore, have the responsibility for ensuring that their standards are consistent with those that are supplied by their own national laboratory, and hence with international standards.

Solid Sources

A solid source that is to be used as a standard of activity must be very thin, so that the effects of absorption and scattering within the source are very small. The source may thus be very fragile, unstable with time, hygroscopic, liable to cause contamination, and unsuitable for normal mechanical handling or for conveyance by normal transport. Absolute standards of activity are, therefore, not usually supplied as solid sources. Such sources are best prepared from an absolutely standardized solution in the user's laboratory immediately before use, and disposed of after use.

Commercially available solid sources have to be designed and constructed in rugged form in order to withstand usage. This usually means that the radiation suffers appreciable self-absorption and scattering within the source. They are, therefore, of only limited value as standards of activity, but may be of great value as standards of radiation emission. Such reference sources are widely available. Alpha-emitting sources are often not sealed, and have to be treated very carefully to prevent loss of activity and spread of contamination. Other types of sources are usually sealed, and have passed a test for absence of contamination and leakage of activity. Calibration of reference sources is usually in terms of the radiation emission, but this may be expressed as an "effective activity" or as an observed count rate in a particular counter. The effect of absorption and scattering of gamma radiation within a source may be small, and calibration of gamma reference sources can often be usefully expressed in terms of activity.

Mock Standards

It is sometimes considered desirable to have a long-lived source or solution with radiation characteristics very similar to those of a short-lived nuclide. If the long-lived material is made up of a mixture of nuclides, it is referred to as a *mock standard*. A mixture of ^{133}Ba and ^{137}Cs, in a ratio of approximately 12 to 1 by activity, simulates the radiation emission of ^{131}I, provided that the effect of the excess low-energy photons is eliminated by a suitable filter or detection system. Such a mixture is available as mock iodine-131. Mock standards are essentially radiation standards, and not activity standards, although they may be calibrated in terms of "effective activity" of the short-lived nuclide that they are meant to reproduce. Such a calibration is not absolute, and depends on the measuring system used. Mock standards should, therefore, not be used for accurate calibrations, but they may be useful for maintaining a calibration that was originally performed with a short-lived absolute standard, or for purely comparative purposes.

It should be noted that a source of a single long-lived nuclide having only a limited similarity of radiation emission to the short-lived nuclide is often as useful as an accurately prepared mixture of nuclides. For example, it is unnecessary to use a mixture of 12 parts ^{133}Ba and 1 part ^{137}Cs for maintaining the calibration of a gamma-ray spectrometer with a "window" set at about 360 kev (the gamma energy of ^{131}I), because the ^{137}Cs will have little or no effect on the count rate; a source of barium-133 alone is sufficient.

Bulk Materials

There is an increasing requirement for supplies of natural materials (such as milk, meat, bone, water, etc.) or derived materials (such as dried milk, bone ash, etc.) containing accurately known amounts of activity of various nuclides, particularly those occurring in fallout.

This requirement can be met in two ways: first, by adding a known amount of activity to the natural material (for this a standardized solution can be used, after suitable dilution); second, by incorporating an unknown amount of activity into the material, possibly by a metabolic route, and then measuring the resulting activity in the natural material. The former method is rarely reliable, because of the difficulty of incorporating the added material into the natural material with an accurately known efficiency; the latter method is, thus, preferable.

The supplier of this type of standard therefore requires special low-background laboratories, where the activity can be introduced into the natural material in a satisfactory manner and where measurements on the resulting material can be made to a sufficient accuracy.

Customer's Material

Some standardizing laboratories are prepared to make accurate measurements on material supplied to them. There may be several restrictions. For example, nominal activity and volume may have to be within specified limits, and no assessment may be made of the purity of the material. If purity is not known, an expensive measurement of activity may be useless.

CERTIFICATION OF STANDARDS OF ACTIVITY

The use of an activity standard is not always simple, and in order to assist the user, the supplier issues a certificate that not only records the results of his measurements, but also provides relevant associated data.

Ideally, a certificate for a standardized solution should provide the following information.
1. Measurement method.
2. Measurement result, quoted as a radioactive concentration in microcuries (or disintegrations per second) of the stated principal radionuclide per gram of solution.
3. Reference time and date for which the quoted measurement result was determined.
4. Recommended value for half-life of the principal nuclide, including, if possible, an assessment of this value.
5. Radioactive concentration of any daughter or grand-daughter activity present in the solution at the reference time.
6. Total mass of solution and/or the total activity of the principal nuclide in the solution. (Specific gravity may also be useful.)
7. Chemical composition of the solution.
8. Recommended diluent.
9. Essential data on the decay schemes of the nuclides present, including, if possible, an assessment of their uncertainties.
10. Assessment of the purity of the solution; radioisotopic purity is usually of greater relevance than radiochemical or chemical purity. (If purity data cannot be provided, the production method should be stated.)

11. Assessment of the accuracy of the quoted measurement result. Both random and systematic errors should be considered, together with information on the consistency of the present and past standardizations. The terms used should be clearly defined.
12. Assessment of the effect on the final result of making small changes in any assumption, including the decay scheme and half-life, that was necessarily made during the standardization.

In practice, the more information the certificates contain, the more expensive is the standard, and the more delay there may be in sending out the certificates. Nevertheless, users of absolutely standardized solutions should expect to be given most of the above information; however, this is not yet general practice.[3] Certificates for reference solutions, for which the emphasis is on ready availability and low cost, may reasonably be less detailed.

Certificates for reference sources must make it quite clear whether activity (i.e., content) or radiation emission rate is being quoted. Calibrated reference sources should be accompanied by a certificate with at least the following information.

1. Measurement method.
2. Measurement result.
3. Reference time and date.
4. Recommended half-life.
5. Description of the radioactive material used in the source.
6. Assessment of the accuracy of the quoted measurement result.
7. A statement whether the source is open or sealed; if sealed, the results of a contamination and leakage test should be quoted.

USE OF STANDARDS OF ACTIVITY

Calibration of Laboratory Instruments

The principal use of standards of activity is for the calibration of measuring instruments which, in turn, will be used to determine activity or radiation emission of laboratory samples. The calibration of such instruments must usually be maintained over a long period of time, either by adjustment of sensitivity or by applying a varying correction factor. This can be done in two ways.

Method 1. Using an accurately measured standard each time the calibration is to be checked.

If the half-life of the nuclide is short, it will be necessary to obtain a new standard at frequent intervals.

If the half-life is long, the same standard may be used repeatedly, provided that it can be stored without loss of activity. A new counting source can, for example, be prepared from an absolutely standardized solution for each calibration, provided that the solution is kept sealed (to prevent evaporation losses), that the chemical composition of the solution is such that it is stable with time and does not lose activity to the walls of the container, that the radioisotopic purity is high, and that the half-life is accurately known.

If the calibration is not in terms of activity, but in terms of radiation detection efficiency, a sealed reference source with a calibration in terms of radiation emission can be used.

Method 2. Using an accurately measured standard only once to establish the calibration and at the same time determine the response of an otherwise uncalibrated source of long half-life, then repeating measurements with the latter source only whenever the calibration is to be checked.

In this method only one expensive standard is required, even if the half-life is short. The standard can be prepared from an accurately standardized solution or can be a calibrated reference source. The uncalibrated source of long half-life should be sealed and have approximately the same radiation emission as the accurate standard. The extent to which the uncalibrated source and the accurate standard may be allowed to differ in the quality of their emissions depends on the characteristics of the detection system used and on the accuracy required.

Precalibrated Instruments. An alternative approach to using standards of activity is purchase of a measuring instrument that has already been calibrated. For example, the National Physical Laboratory has developed an ionization chamber for which it has published calibration data.[4,5] If such an instrument is used, it is important to ensure that the mechanical design is very rugged and that the electronic measuring equipment is very stable with time. If this is in doubt, the calibration must be checked periodically with suitable standards.

Precautions in the Use of Standardized Solutions

If full use is to be made of the high accuracy associated with absolutely standardized solutions, the following points must be considered.

1. In preparing these solutions, the issuing laboratory takes every care that the chemical form is such that the activity will remain uniformly in solution and will not deposit on the walls of the ampule or bottle. If further dilution of the solution is necessary, the recommended diluent should be used, to ensure that unwanted deposition does not occur.
2. Bacterial growth can sometimes preferentially absorb the activity from a solution. Solutions should, therefore, be heat-sterilized or contain a bacteriostat to prevent such growth. If a standardized solution is diluted and stored in the laboratory, similar precautions should be taken.
3. When an ampule or bottle is opened, evaporation begins. If the original radioactive concentration is to be maintained, it is essential to keep the solution closed from the atmosphere, except for the very short periods necessary to take aliquots.
4. Solutions may be in a chemical form that is not suitable for immediate deposition on aluminum supports—e.g., strong hydrochloric acid, which will dissolve aluminum and may cause a change in counting efficiency.
5. When sources are prepared from some solutions by desposition and evaporation, there may be a loss arising from volatilization of the resulting solid. If this is liable to occur, source preparation by precipitation (e.g., as sulfide), followed by evaporation, may be necessary.
6. The activity per gram of solution is commonly certified. If the emission rate of any type of radiation is required, allowance must be made for the decay scheme (branching ratios, internal conversion, L/K capture ratio, fluorescence yield, etc.). If a counting source is prepared, allowance must be made for absorption and scattering, as well as for the mass of solution used.
7. The time of use is normally not the same as the reference time for which the activity of the solution is certified. The activity at the time of use will, therefore, have to be calculated from the certified activity. The accuracy with which the half-life is known may then affect the accuracy of the calculated activity. In addition, the effects of shorter-lived impurities have become less. Allowance must also be made for growth or decay of any radioactive daughter products. Only when a daughter nuclide has a very much shorter half-life than the parent is an equilibrium reached in which the parent and daughter activities are equal. When the half-life of the daughter is comparable to that of the parent, the activity of the daughter may exceed that of the parent.[6] For example, in the transient equilibrium of ^{140}Ba (half-life 12.8 days) and ^{140}La (half-life 40.2 hours) the activity of the shorter-lived ^{140}La is about 15 percent greater than that of the parent ^{140}Ba.

Precautions in the Use of Reference Solutions

1. The rate of emission of particulate (or photon) radiation from the surface of a solid source is not necessarily the same as the activity (i.e., disintegra-

tion rate). In the first place, there will be absorption and scattering of the radiation within the source. Secondly, the decay scheme of the nuclide may be complex, so that a single particle or photon of specified type may not be produced at each disintegration. To relate the activity of a solid source to the radiation emission, it is necessary to know the decay scheme of the nuclide and the magnitude of internal absorption and scattering.

2. Variations in size, shape, or composition between the reference source and the source to be measured may caused differences in the absorption and scattering of the radiation either internally or externally to the source. The detection sensitivity may then be different for the two sources. Such effects should be minimized and taken into account in assessing the accuracy of measurement of the unknown source.

3. If a source is calibrated in terms of an "effective activity", it is essential to know and understand the exact definition of this term in the particular context. If the calibration is in terms of a specified counter, the geometric arrangement and the counting efficiency may need to be known.

4. The emission at the time of use will have to be calculated from the certified emission. The accuracy with which the half-life and purity are known may affect the accuracy of the calculated value.

REFERENCES

1. Glasstone, S., *Textbook of Physical Chemistry*, 2nd ed. Van Nostrand, New York, 1946.
2. McNish, A. G., Nomenclature for Standards of Radioactivity. *Int. J. Appl. Radiat. Isotop.*, 8(2/3): 145–146, 1960.
3. Kolde, H. E. and Karches, G. J., Some Limitations of Available Radioactivity Standards. *Health Phys.*, 10(9): 635–641, 1964.
4. Dale, J. W. G., Perry, W. E., and Pulfer, R. F., A Beta-Gamma Ionization Chamber for Substandards of Radioactivity. I. Uses and Calibration. *Int. J. Appl. Radiat. Isotop.*, 10(2/3): 65–71, 1961.
5. Dale, J. W. G., A Beta-Gamma Ionization Chamber for Substandards of Radioactivity. II. Instrument Response to Gamma Radiation. *Int. J. Appl. Radiat. Isotop.*, 10(2/3): 72–78, 1961.
6. Evans, R. D., *The Atomic Nucleus*. McGraw-Hill, New York, 1955.

SELECTED ADDITIONAL REFERENCES

Aglintsev, K. K., Bochkarev, V. V., Grabevskii, V., and Karavaev, F. M., Methods of Radioactivity Metrology in USSR. *Soviet J. At. Energy*, 8(4): 304–309, 1961.

Attix, F. H. and Ritz, V. H., A Determination of the Gamma-Ray Emission of Radium. *J. Res. Nat. Bur. Stand.*, 59: 293–305, 1957.

Bryant, J., Anticoincidence Counting Method for Standardizing Radioactive Materials. *Int. J. Appl. Radiat. Isotop.*, 13: 273–276, 1962.

Campton, P. J., Taylor, J. G. V., and Merritt, J. S., The Efficiency Tracing Technique for Eliminating Self-Absorption Errors in $4\pi\beta$ Counting. *Int. J. Appl. Radiat. Isotop.*, 8(1): 8–19, 1960.

Gandy, A., *Préparation et Étalonnage des Sources Radioactives de Référence*. (IAEA Review Series No. 14.) International Atomic Energy Agency, Vienna, Austria, 1961.

Gorsuch, T. T., Jenkins, D. J., and Ludbrook, C. A., Use of Complex Ions in the Efficiency Tracing Method for the Standardization of Pure Beta-Emitting Nuclides. *Nature*, 199: 368, 1963.

Harper, P. V., Siemens, W. O., Lathrop, K. A., and Endlich, H., Production and Use of Iodine-125. *J. Nucl. Med.*, 4(4): 277–289, 1963.

International Atomic Energy Agency, *Metrology of Radionuclides. Proceedings of a Symposium Organized by the IAEA, Vienna, Austria, October 1959*. International Atomic Energy Agency, Vienna, Austria, 1960.

Lyon, W. S., Availability and Use of Radioactivity Standards. *Anal. Chem.*, 36(2): 31A–39A, 1964.

Mann, W. B. and Seliger, H. H., *Preparation, Maintenance and Application of Standards of Radioactivity*. (NBS Circular No. 594.) United States Department of Commerce, National Bureau of Standards, Washington, D.C., 1958.

O'Kelley, G. D., Detection and Measurement of Nuclear Radiation. *National Research Council Report NAS-NS 3105* National Academy of Sciences, National Research Council, Washington; D.C., 1962.

Rapkin, E., Liquid Scintillation Counting 1957–63: A Review. *Int. J. Appl. Radiat. Isotop.*, 15(2): 69–87, 1964.

Sharpe, J. *Nuclear Radiation Detectors*, 2nd ed. Methuen, London, England, 1964.

Siegbahn, K., (ed.), *Alpha-, Beta- and Gamma-Ray Spectroscopy*, Vols. 1 and 2. North-Holland Publishing Company, Amsterdam, The Netherlands, 1965.

Snell, A. H., (ed.), *Nuclear Instruments and Their Uses*, Vol. 1. John Wiley and Sons, New York, 1962.

Steyn, J. and Hahne, F. J. W., Absolute Disintegration Rate Measurement of Beta-Emitters by Application of Efficiency Tracing to 4π Liquid Scintillation Counting. Paper presented at the *National Nuclear Energy Conference, Pretoria, Transvaal, Republic of South Africa, 1963*.

United States Department of Commerce, National Bureau of Standards, *Report of the International Commission on Radiological Units and Measurements* (ICRU) *1959*. (Handbook 78.) United States Department of Commerce, National Bureau of Standards, Washington, D.C., 1961.

United States Department of Commerce, National Bureau of Standards, *A Manual of Radioactive Procedures*. Recommendations of the National Committee on Radiation Protection and Measurement. (Handbook 80; NCRP Report No. 28.) United States Department of Commerce, National Bureau of Standards, Washington, D.C., 1963.

Watt, D. E. and Ramsden, D., *High Sensitivity Counting Techniques*. Pergamon Press, Oxford, England, 1964.

Young, M. E. J. and Batho, H. F., Dose Tables for Linear Sources Calculated by an Electronic Computer. *Brit. J. Radiol.*, 37(433): 38–44, 1964.

RADIOACTIVE ISOTOPE DILUTION ANALYSIS

Reprinted by permission of
The Radiochemical Centre, Amersham, Bucks, England.

Classical methods of chemical analysis generally rely either on the quantitative isolation of the component to be determined or on the measurement of some specific chemical or physical property of that component which is not obscured by other substances present. These methods often fail with more complex mixtures, particularly when the quantity to be determined is small or when the mixture contains chemically similar components. The difficulties stem from the basic conflict between the need for a quantitative recovery and the need to obtain a pure product; as the complexity increases or the scale of working decreases, it becomes progressively more difficult to reconcile these two requirements. It is for the analysis of mixtures of this kind that the group of elegant and versatile methods known as isotope dilution analysis is particularly well suited. These methods are simple in concept, straightforward in practice, and do not have excessive requirements for apparatus and equipment. They resolve the problem of the conflicting requirements by removing the need for a quantitative recovery. As a result they are capable of a precision and sensitivity often unobtainable by any other method, and they offer unique advantages in handling materials that are likely to decompose during separation and purification.

LABORATORY AND MATERIAL REQUIREMENTS

There is a widespread belief that any work with radioactive material involves the use of expensive equipment and restrictive safety precautions. This is not so.

For work on the analytical scale, any good-quality chemical laboratory can be made suitable with only very minor modifications, and these are more in the methods of working than in the equipment of the laboratory. Careful working is certainly required when handling radioactive material, but this is, in any event, an essential part of analytical work. There are certain points of technique and laboratory discipline that are peculiar to handling radioactive chemicals, but they are largely a matter of common sense, and at the low levels of radiation involved in isotope dilution procedures, none of them is particularly onerous. Full details of methods of working can be found in textbooks on the use of radioisotopes in analysis.[1,2]

The counting and monitoring equipment needed will cost far less than many other items that are today considered normal, or even essential, in a well-equipped analytical laboratory.

In addition to an orderly laboratory and suitable measuring equipment, the third requirement is a supply of radioactive labeled compounds of good purity. The effect of impure material is described later, but the need for radiochemically pure compounds must be stressed.

PRINCIPLE

The basic principle of radioactive isotope dilution analysis is very simple. It depends on the fact that, if a radioactive tracer is mixed with the corresponding unlabeled compound, the amount of activity per gram of the substance will be reduced; in other words, the radioactive material will be diluted with inactive material. If the reduction in activity per gram can be measured, the amount of diluting material added can be calculated.

If one considers the tracer compound C* with an activity A_0 and a mass W_0, one can calculate the specific activity, S_0, of the compound from these two first-mentioned values:

$$S_0 = \frac{A_0}{W_0}. \tag{1}$$

The inactive form of the same compound (C) has no activity, but it has a mass, W_u. If C* and C are mixed, the total activity is still A_0, but the mass is $W_0 + W_u$. The new specific activity is therefore

$$S_1 = \frac{A_0}{W_0 + W_u}. \tag{2}$$

Provided that A_0 and W_0 are known and that S_1 can be measured, the value of W_u can then be calculated:

$$W_u = W_0 \left(\frac{S_0}{S_1} - 1 \right). \tag{3}$$

The great advantage of this method is that it is not necessary to separate the whole of $W_0 + W_u$ to measure S_1. If the labeled compound and the inactive component are in the same chemical form and have been thoroughly mixed, then the specific activity (S_1) is independent of the amount of material used to measure it. As long as some of the component can be separated in a pure state, the amount of inactive material in the sample can be determined.

Example 1.

Problem. To determine the amount of naphthalene in 10 gram of coal tar.

Procedure.

1. Take some radioactive-labeled naphthalene of known weight (W_0) and activity (A_0); let $W_0 = 10$ mg, and $A_0 = 1{,}000{,}000$ cpm.
2. Add the labeled naphthalene to the coal tar.
3. Ensure that the naphthalene in the coal tar and the labeled naphthalene are thoroughly mixed; in the cited case warming and stirring will be sufficient.
4. Separate some of the naphthalene in a pure state.
5. Weigh the separated naphthalene (W_1); let $W_1 = 120$ mg.
6. Measure the activity of the separated naphthalene (A_1); let $A_1 = 240{,}000$ cpm.

We now have the following measurements:

$A_0 = 1{,}000{,}000$ cpm, $\qquad A_1 = 240{,}000$ cpm,
$W_0 = 10$ mg, $\qquad W_1 = 120$ mg,
$S_0 = 100{,}000$ cpm/mg, $\qquad S_1 = 2{,}000$ cpm/mg,

and by substituting these values in equation (2),

$$\text{naphthalene in the sample} = 10 \left(\frac{100{,}000}{2{,}000} - 1 \right) = 490 \text{ mg or } 4.9\%.$$

In order to ensure that the separated naphthalene was pure, it was possible to sacrifice more than three fourths of it during purification without affecting the result.

In carrying out an analysis of this kind, there are several aspects to be considered.

1. Purity of the Tracer. It is essential that the chemical and radio-chemical purity of the compound used should be known and be as high as possible. If it contains an unknown inactive impurity, the weight (W_0) will be higher than it should be; if it contains an unknown active impurity, the activity A_0 will be higher than it should be; in either case the calculated value of S_0 will be wrong, and so will the result.

One convenient method of checking the purity of the labeled compound is by reverse isotope dilution analysis. In this procedure a large amount of inactive pure compound is added to the radioactive material and thoroughly mixed. The specific activity of the mixture is measured, and part of the diluted material is then exhaustively purified. The specific activity of the purified product is also measured. The radiochemical purity of the labeled compound is then given by

$$\% \text{ radiochemical purity} = \frac{\text{specific activity of purified compound}}{\text{specific activity of original mixture}} \times 100. \quad (4)$$

2. Complete Equilibration. Unless the labeled compound and the material in the sample are fully mixed in step 3, the pure substance separated in step 4 may not be truly representative, and may therefore give the wrong value for S_1.

3. Purity of the Separated Material. Unless the material separated in step 4 is pure, the value obtained for S_1 will again be wrong. The separation and purification can be carried out by any suitable procedure (crystallization, distillation, chromatography, etc.), and, as the yield is relatively unimportant, rapid but wasteful methods of purification can be used. A derivative may be isolated and measured, if this assists in purification; the principle of the method remains unchanged.

4. Mass Determination. Although W_0 and W_u have been expressed in milligrams, it does not follow that it is always necessary to determine them by weighing. It is perfectly possible to work with microgram amounts of material and to carry out the mass determinations by spectrophotometry or other techniques. Vitamin B_{12} is routinely measured in dilution analysis by its light absorption at 361 and 548 mµ.

5. Unstable Substances. Since the analysis is essentially measurement of a dilution ratio at the time of mixing, it does not matter if the substance decomposes to some extent during purification. So long as a small pure sample is finally obtained, the result is not affected, because labeled and unlabeled molecules are decomposed equally.

DEVELOPMENTS IN DILUTION ANALYSIS

There are a number of situations in which direct dilution analysis proves unsuitable or inconvenient for one or more reasons.

Direct dilution analysis requires the determination of the specific activity of the separated material, and when the amount recovered falls to a fraction of a microgram, the accuracy of the determination falls off sharply.

When determinations are required on a number of related complex substances, it may not be possible to obtain pure radioactive tracers for each of them. Similarly, if tracers are available, it may be inconvenient to store stock solutions of many different tracer compounds for, perhaps, only occasional use.

These drawbacks in the basic method have led to a number of new developments in the field of dilution analysis.

Derivative Dilution

In this procedure the compound being determined is made to react quantitatively with a active reagent of known specific activity, and the radioactive product is separated and purified. The amount of activity in the pure product indicates the amount of radioactive reagent it contains, and as the stoichiometry of the reaction between reagent and compound is known, the amount of compound present can be calculated.

Example 2.

Problem. It is necessary to determine one component in the presence of several others.

Procedure.
1. Add an excess of a radioactive reagent of known specific activity (S_0, expressed in microcuries per micromole) to the mixture; the reagent reacts with the compound being determined and with some other components of the mixture; the specific activity of the radioactive products, when expressed in microcuries per micromole, will be the same as that of the radioactive reagent itself.
2. Separate the required component product from the other constituents of the mixture; it is particularly important to remove all traces of the excess reagent and of other radioactive products formed.
3. Measure the radioactivity of the separated product (A_1, expressed in microcuries); the amount of the original product can then be determined directly from

$$W_u = \frac{A_1}{S_0}, \tag{5}$$

where W_u is expressed in micromoles.

It is clear that there are two critical requirements in this procedure. The first is to ensure that the reagent and the required component react together quantitatively, or at least with a high and reproducible yield; this is achieved by careful selection of the reagent, and by close attention to the reaction conditions. The second is to achieve a successful separation of the required product from the other components of the mixture, most particularly from other radioactive products and from the excess of the reagent; this can be done in three ways: 1) by direct quantitative separation of the reaction product itself, 2) by adding macro amounts of carrier and correcting for the losses during processing, or 3) by adding small amounts of carrier labeled with a second radioactive nuclide and using the decrease in this to correct for purification losses. This can be demonstrated by a further example.

Example 3.

The mixture of amino acids obtained by hydrolysis of small amounts of proteins can be treated with p-iodobenzene sulfonyl chloride (pipsyl chloride) labeled with ^{131}I to give the radioactive pipsyl derivatives of the individual amino acids. This has been used as a basis for the determination of amino acids in such mixtures.[3,6] In this work three different separation procedures were used, each making use of a different principle. These can be demonstrated by assuming that it is necessary to determine glycine in a small amount of protein hydrolysate (Figure 1).

A. An overwhelming excess of unlabeled pipsyl glycine is added to the mixture, and equilibrated. The mixture of active and inactive pipsyl glycine is then separated and purified by repeated recrystallization and filtration through charcoal. When all the radioactive contaminants have been removed, the overall recovery is obtained by measuring the yield of carrier. Measurement of the ^{131}I activity in the recovered material then shows how much of the original glycine is present. See equation (5).

B. Purification of the derivative by recrystallization is a tedious procedure, whereas paper chromatography is simple and rapid. However, separation on paper is not suitable for large amounts of material, so the addition of carriers for checking recoveries is not feasible. If quantitative paper chromatograms can be carried out, the results can be calculated without the need for any correction for yield.

C. Although it is not feasible to add large amounts of inactive carrier for checking the recovery during paper chromatography, it is possible to use a radioactive tracer for this purpose. A small weight of ^{35}S-labeled pipsyl glycine can be added to the mixture before application to the paper, and the percentage recovered can be measured by counting. The ^{35}S and ^{131}I can be readily measured separately, as the radiations emitted by them differ considerably.

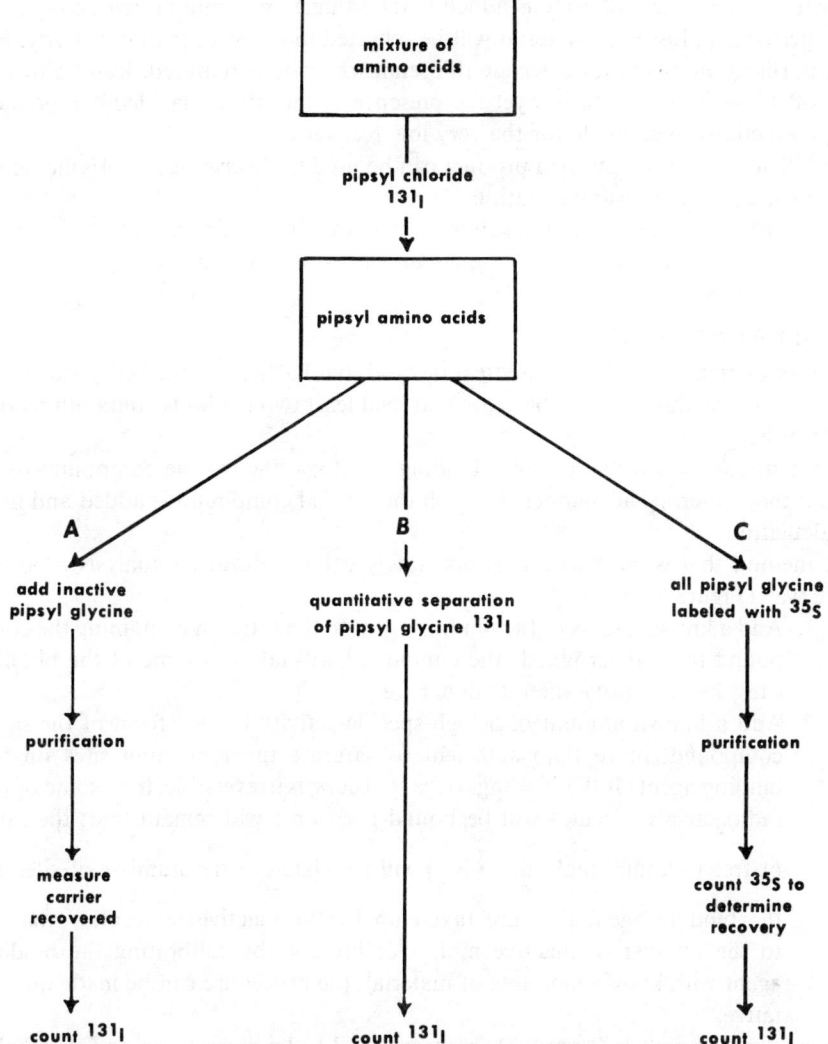

FIGURE 1. Separation Procedures Based on Three Different Principles.

Measurements of this kind were among the first applications of derivative dilution analysis, and although they clearly demonstrate the principle, they do not form an extreme test of the method. Recent developments in the determinatian of ketosteroids show the extraordinary sensitivity of which this technique is capable.[7]

The method described for the determination of testosterone in peripheral plasma depends upon a double-label technique using tritium-labeled testosterone and ^{35}S-labeled thiosemicarbazide. The stages in a determination are as follows.

1. Add a small known amount of tritium-labeled testosterone of high specific activity to a sample of peripheral plasma; equilibrate the active and inactive species thoroughly.
2. Separate the steroids from the plasma chemically.
3. React the steroids with ^{35}S-labeled thiosemicarbazide, which will convert the ketosteroids to the semicarbazones.
4. Purify the testosterone thiosemicarbazone by repeated chromatography, conversion to a derivative, and further chromatography.
5. Measure the tritium content.
6. Measure the ^{35}S content.

As the tritium-labeled testosterone is added to the sample and equilibrated before any separations are performed, losses of material will be reflected in losses of tritium activity. Extremely rigorous purification of the testosterone thiosemicarbazone is required, leading to the loss of the bulk of the separated material; yet the presence of the tritium-labeled compound enables accurate corrections to be made for the very low recoveries.

The ^{35}S activity in the purified product can be used to determine the original testosterone content in the usual way, using equation (5).

The sensitivity of this method depends on the availability of pure radioactive materials of very high specific activity, and levels of one part of testosterone in ten billion parts of sample have been measured.

Saturation Analysis

Another extremely sensitive analytical method that has been described is called *saturation analysis*. In practice, this term has been applied to at least two methods, and both modifications will be described here.

Both methods involve the use of a binding agent specific for the compound being determined, but they differ in the manner in which the labeled compound is added and in the form of the calculation.

The method that is, perhaps, most accurately called saturation analysis proceeds in the following way (Figure 2).

1. Add a known excess of the binding agent to the solution containing the compound to be determined; the compound will take up some of the binding sites, but will leave the remainder free.
2. Add a known amount of a high-specific-activity labeled form of the same compound, more than sufficient to saturate the remaining sites on the binding agent; if the binding, once it occurs, is irreversible, then some of the radioactive molecules will be bound and some will remain free; the ratio of free to bound molecules $\left(R\frac{f}{b}\right)$ will be related to the number of sites on the binding agent that are taken up by the inactive molecules—that is, to the number of inactive molecules present; by calibrating the binding agent with known amounts of material, the procedure can be made quantitative.

In the other procedure (Figure 3) the compound to be determined and its labeled tracer are mixed together before the binding agent is added. This means that, instead of all the inactive molecules being bound and the active molecules distributing between the remaining sites and the free state, the active and inactive molecules compete for the binding sites on an equal basis. This does not require irreversible reaction between the molecule and the binding agent, and is in many ways similar to the substoichiometric dilution analysis described later in reference to inorganic applications.

Most applications of these techniques have used specific binding agents of high molecular weight and of biological origin. With these agents very sensitive determinations of physiological importance have been carried out, such as insulin[8] and vitamin B_{12}[9] in plasma.

INORGANIC APPLICATIONS

Most of the applications of radioactive isotope dilution analysis that have been described were in organic chemistry, but that does not mean that the technique is limited to this field. Direct isotope dilution analysis can be applied equally well to the many situations where quantitative separation of an inorganic compound is difficult.

With many inorganic systems the tracer nuclide is available at very high specific activity, and the weight of tracer added to give reasonable count rates is very small compared with the weight of the element being determined. Under these circumstances it is useful to rearrange

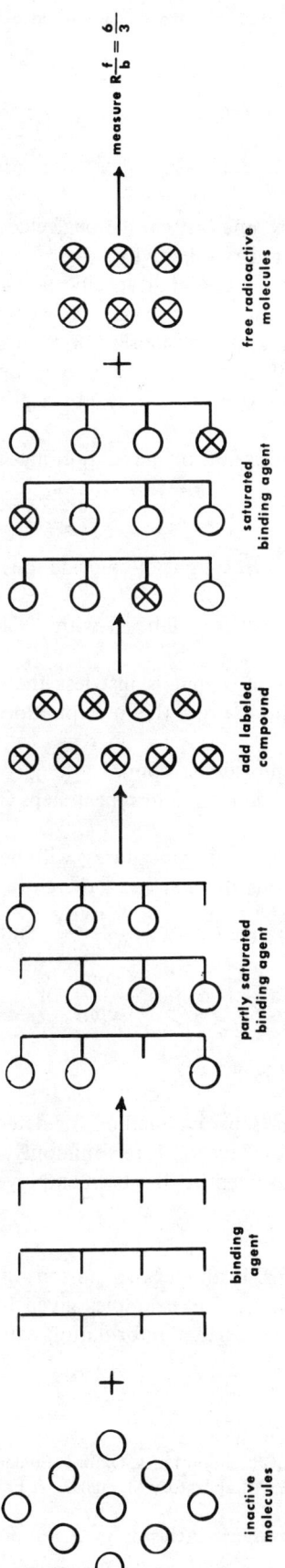

FIGURE 2. Saturation Analysis, Method 1.

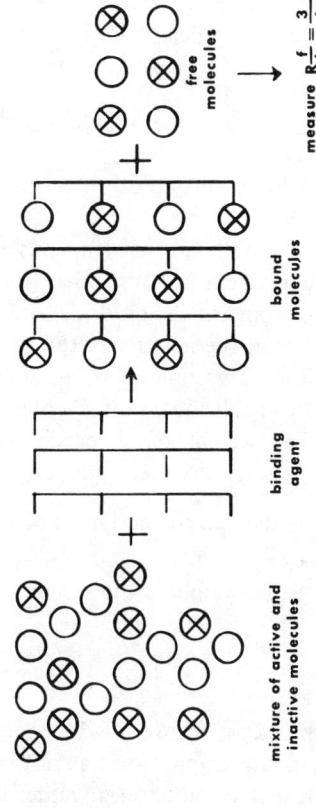

FIGURE 3. Saturation Analysis, Method 2.

equation (3): $W_u = W_0\left(\dfrac{S_0}{S_1} - 1\right)$, but $S = \dfrac{A}{W}$, therefore

$$W_u = W_0\left(\frac{A_0 W_1}{A_1 W_0} - 1\right), \quad \text{or} \quad W_u = \frac{A_0 W_1}{A_1} - W_0. \tag{6}$$

When the weight of active material added (W_0) is very small, it can be neglected, and a simple correction can be applied on the basis of the recovered radioactivity.

In inorganic analysis advantage has also been taken of the high specific activities to increase the sensitivity of the determinations. As the amounts being measured become less, the quantity W_0 in equation (6) can again not be neglected, and the calculation reverts to the form given in equation (3). The difficulty, then, is to measure the specific activity on small fractions of a microgram of material. This problem has been overcome neatly by a procedure known as *substoichiometric dilution analysis*.

A very simple example of this technique is the determination of microgram amounts of chloride, using ^{36}Cl-labeled chloride as the tracer.[10]

Example 4.

1. Take a volume of solution containing an unknown weight of chloride ions (W_u).
2. Add to this solution a known weight of chloride ions labeled with ^{36}Cl (W_0).
3. Mix well, and add a known amount of silver ions that is just less than equivalent to W_0 microgram of chloride; this causes the precipitation of some of the chloride ions.
4. Count the activity in the precipitate of silver chloride (X cpm).
5. Take a volume of solution containing no chloride ions, and repeat steps 2 to 4; call the count Y cpm.
6. The weight of silver chloride obtained in the two procedures will be identical, as there was an excess of chloride ions in each case; therefore, as the specific activities of the precipitates are given by $S = \dfrac{\text{Activity}}{\text{Weight}}$, and as the weight is the same in each case, the ratio $\dfrac{S_0}{S_1}$ becomes $\dfrac{Y \text{ cpm}}{X \text{ cpm}}$, or

$$W_u = W_0\left(\frac{Y}{X} - 1\right). \tag{7}$$

This type of procedure has been developed into a very sensitive method for determining metals, although precipitation has not been used at very low levels. More commonly, a complexing agent such as dithizon, together with an extraction procedure, has been employed.[11,12]

CONCLUSION

Radioactive isotope dilution, in its many modifications, is an elegant and sensitive technique for the determination of many substances in mixtures. The examples given here can represent only a small fraction of the possible applications; further information can be obtained from the references listed below.

REFERENCES

1. Cook, G. B. and Duncan, J. F., *Modern Radiochemical Practice*. Clarendon Press, Oxford, England, 1952.
2. Lambie, D. A., *Techniques for the Use of Radioisotopes in Analysis*. A Laboratory Manual. E. & F. N. Spon, 1952.
3. Keston, A. S. and Cannon, R. K., Microanalysis of Mixtures (Amino Acids) in the Form of Isotopic Derivatives. *J. Am. Chem. Soc.*, 68: 1390, 1946.

4. Keston, A. S., Udenfriend, S., and Cannon, K., Method for the Determination of Organic Compounds in the Form of Isotopic Derivatives. I. Estimation of Amino Acids by the Carrier Techniques. *J. Am. Chem. Soc.*, 71: 249–257, 1949.
5. Keston, A. S., et al. *J. Am. Chem. Soc.*, 69: 3151–3152, 1947.
6. Keston, A. S., Udenfriend, S., and Levy, M., Determination of Organic Compounds as Isotopic Derivatives. II. Amino Acids by Paper Chromatography and Indicator Technique. *J. Am. Chem. Soc.*, 72: 748–753, 1950.
7. Riondel, A., Tait, J. F., Gut, M., Tait, S. A. S., and Joachim, E., Estimation of Testosterone in Human Peripheral Blood, Using ^{35}S Thiosemicarbazide. *J. Clin. Endocrinol. Metab.*, 23: 620–628, 1963.
8. Hales, C. N. and Randle, P. J., Immunoassay of Insulin with Insulin-Antibody Precipitate. *Biochem. J.*, 88: 137–146, 1963.
9. Barakat, R. M. and Ekins, R. P., Assay of Vitamin B_{12} in Blood. *Lancet*, 2: 25–26, 1961.
10. Johannesson, J. K., Radiochemical Determination of Microgram Amounts of Chloride. *Analyst*, 86: 60–61, 1961.
11. Růžička, J. and Starý, J., Isotopic Dilution Analysis by Solvent Extraction. I. Principle and Theory of the Method. *Talanta*, 8: 228–234, 1961.
12. Růžička, J. and Starý, J., Isotopic Dilution Analysis by Solvent Extraction. II. Highly Selective Determination of Trace Amounts of Mercury. *Talanta*, 8: 535–538, 1961.

ADDITIONAL REFERENCES

Ashton, G. C. and Foster, M. C., Isotope Dilution Technique for Determining Benzylpenicillin in Fermentation Liquors. *Analyst*, 80: 123–132, 1955.

Bacher, F. A., Boley, A. E., and Shonk, C. E., Radioactive Tracer Assay for Vitamin B_{12} and Other Cobalamins in Complex Mixtures. *Anal. Chem.*, 26: 1146–1149, 1954.

Bayly, R. J., *Proceedings of the Conference on the Use of Radioisotopes in the Physical Sciences and Industry, Copenhagen, September 6–17, 1960*, 2: 305–308. International Atomic Energy Agency, Vienna, Austria, 1962.

Craig, J. T., Measuring the Active Ingredients in an Insecticide. *Nucleonics*, 14(5): 60–61, 1956.

Fremlin, J. H., Hardwick, J. L., and Suthers, J., Measurements of Small Quantities of Fluoride with the Help of Fluorine–18. *Nature*, 180: 1179–1181, 1957.

Hill, R., Jones, A. G., and Palin, D. E., The Determination of Gamma Isomer in Crude Benzene Hexachloride by a Carbon–14 Isotope Dilution Method. *Chemy. Ind.*, Feb. 6: 162–163, 1954.

Jones, L. and Trenner, M. B., Ampoule-Combustion Isotope Dilution Technique for Organic Nitrogen. *Anal. Chem.*, 28: 387–390, 1956.

Ralph, W. D., Sweet, R. T., and Mencis, I., Determination of Small Amounts of Cobalt by Isotope Dilution Analysis. *Anal. Chem.*, 34: 92–94, 1962.

Rosenblum, Ch., Principles of Isotope Dilution Assays. *Anal. Chem.*, 29: 1740–1744, 1957.

Schayer, R. W., Kobayashi, Y., and Smiley, R. L., Determination of Histamine as an Isotopic Derivative. *J. Biol. Chem.*, 212: 593–598, 1955.

Sorensen, P., Determination of Carboxylic Acids, Acid Chlorides and Anhydrides by Chlorine–36 Isotope Dilution Method. *Anal. Chem.*, 28: 1318–1320, 1956.

Weiler, H., The Accuracy of Isotope Dilution Methods. *Int. J. Appl. Radiat. Isotop.*, 12: 49–52, 1961.

Whitehead, J. K., Determination of Amino Acids by Double Isotope Dilution Techniques. *Biochem. J.*, 68: 662–668, 1958.

IMPROVED SOLUBILIZATION PROCEDURES FOR LIQUID SCINTILLATION COUNTING OF BIOLOGICAL MATERIALS

D. L. Hansen and E. T. Bush
Nuclear-Chicago Corporation, Des Plaines, Illinois.

Reprinted by permission from
Anal. Biochem., 18(2): 320–332, 1967.

Simplicity and economy of equipment and time are two important advantages associated with the direct-solution method of preparing labeled materials for liquid scintillation counting. In this method the sample material may be placed directly in the counting vial with a solubilizing agent; after dissolution a scintillator solution that dissolves the digest is added to the vial, and the sample is ready for counting. The search for a solubilizer capable of handling a diversity of biological materials has been in progress for at least ten years.[17-] Alcoholic Hyamine* hydroxide and alcoholic KOH have been the most generally successful solubilizers, but have provided less than adequate sensitivity for samples of low specific activity. The amount of sample material that can be incorporated into the standard 20-ml counting vial is limited by the solubility of the digest in a toluene or dioxane scintillator solution. It is customary to employ additional secondary solvents (alcohol or ether) with Hyamine hydroxide or KOH digests to increase their solubilities, but this reduces counting efficiency. The quenching introduced by the alcoholic Hyamine hydroxide and KOH solutions themselves is not negligible.

Two kinds of sensitivity are important to users of liquid scintillation counting. The minimum detectable disintegrations per minute per gram, called the *concentration sensitivity*, is the quantity of importance for materials of low specific activity when the size of the sample to be counted is not limited by the amount of the material available. The minimum detectable disintegrations per minute, or *absolute sensitivity*, is important for small samples of any specific activity. When the sample digest is less quenching, both kinds of sensitivity are increased; if, in addition, the digest is more soluble in the scintillator, concentration sensitivity is further improved.

The present study is primarily concerned with improving the concentration sensitivity or the figure of merit, defined as the product of the sample weight and the counting efficiency. The majority of the troublesome problems of sample preparation are associated with low-specific-activity samples collected from *in-vivo* tracer experiments. The results of this study, however, are applicable to all sample types that can be solubilized by a strong organic base, including whole tissue, purified tissue extracts, and biological fluids. Extensive experimentation was used to determine optimal digestion conditions, sample size, and scintillation solvent composition with alcoholic Hyamine hydroxide and KOH, as well as with the new toluene-soluble quaternary ammonium base, NCS; the latter is a mixture of bases in the molecular-weight range of 250 to 600, produced from a commercial mixture of chlorides of the type formula $R_2R'_2NCl$, where R is methyl and R' is a straight chain varying from C_6 to C_{20}, with an average of about 12 carbon atoms.

* Trademark of Rohm & Haas Co.

MATERIALS AND METHODS

Materials and Instrumentation

2,5-diphenyloxazole (PPO) was obtained from Pilot Chemicals, Inc., 2-butoxyethanol was obtained from Aldrich Chemical Co. Other solvents were reagent grade or the best available comparable grade.

A bleaching solution was prepared by adding 6 to 7 g Eastman Grade benzoyl peroxide to 30 ml toluene at 60°C, cooling rapidly to room temperature, and allowing to stand 1 hour before filtering. This saturated solution was prepared immediately prior to use.

The methanolic KOH was $2N$, as determined by titration. Hyamine hydroxide (0.9 1.0N in methanol) was used as obtained from Nuclear Enterprises, Ltd.

NCS is produced by Nuclear-Chicago Corporation, and was used as obtained, at $0.6N$ base strength in a toluene solution.

Dry sample materials were purchased in purified powdered form, except for a laboratory preparation of dried ground chicken muscle. Radioactive and control tissues were prepared through the cooperation of the staff of the Christ Hospital Institute of Medical Research, Cincinnati, Ohio, and included the following: liver, skeletal muscle, plasma, and blood of rabbits injected intraperitoneally daily from the ninth through the second day before sacrifice with either 150 μc of L-leucine-4, 5-^{3}H or 30 μc of D-glucose-^{14}C (uniformly labeled). Tissues were ground in a glass grinder, using one part thawed ground liver or muscle to four parts 0.85% NaCl solution. Labeled and control rat tissues (brain, spleen, liver, kidney, small intestine, and abdominal muscle) and human plasma were used in other experiments.

Digestion and counting were done in 20-ml glass scintillation vials whose foil-lined caps were provided with inert liners cut from 0.005"-thick Teflon sheet. (The metal foil is attacked by basic reagents, and serious quenching results from dissolution of the cap liner in the sample.) Digestion was carried out in a constant-temperature water bath or at room temperature, as well as in a Crest Corporation Model 511 ultrasonic tank driven by a 100-watt generator. Nuclear-Chicago Corporation Models 6725 and 6860 liquid scintillation spectrometers were used, with the counting chambers at about 10°C.

Method

Background samples for labeled tissue samples were prepared by performing identical procedures on unlabeled tissue from control animals. Most samples and backgrounds were prepared in duplicate. In the experiments that involved determining the counting efficiencies of samples prepared from unlabeled material, samples were internally standardized ("spiked") with 100 μl of tritiated toluene immediately prior to counting.

Digests were examined for completeness of dissolution, and diluted samples for evidence of precipitation, by holding vials against a dark background and over a shaded light source in such a way that only light scattered in the solution was observed. In this way it was possible, with a gentle swirling of the vial, to see very fine particles and transparent gelatinous pieces that could not be observed under ordinary room lighting or by transmitted light. Digestion was usually continued until no change in appearance was observed by this method.

The scintillator solutions added to the digests contained amounts of PPO that would yield a final concentration of 6 g per liter at each dilution. NCS samples were diluted with a 3:1 mixture of toluene:2-butoxyethanol + PPO, except that the first dilution of each digest to 5 ml was made with a 48% solution of 2-butoxyethanol in toluene. Most samples were counted successively at a series of volumes between 3 and 20 ml, diluting the samples with scintillator solution in increments of 5 ml or less. A figure of merit for each concentration could then be calculated. Counting efficiencies were determined from the known activities of spiked samples or from channels-ratio[8] counting of unspiked radioactive tissue samples.

In addition to preliminary experimentation to establish, qualitatively, the best method of sample preparation with each basic digestant, the following quantitative experiments were performed.

1. Determination of Optimal Sample Composition for Certain Dry Sample Materials. The weight of dry sample material and the volume of added water that resulted in the maximal figure of merit were determined for the following purified powdered substances: fibrinogen, collagen, RNA, glycogen, sucrose, tyrosine, histidine, methionine, and glycine. A series of samples of increasing weights, usually in increments of 20 or 30 mg, was weighed out in counting vials in triplicate. To each of the triplicate samples was added a different volume of water, in the range of 100 to 350 µl, depending on the weight of solid. It was necessary to test different amounts of water, because too little resulted in prolonged or incomplete digestion, while too much resulted in phase separation when the digest was diluted with scintillator. When the sample was completely wetted, 2 ml of digestant were added, and digestion was carried out at room temperature or 50°C (70°C in the case of some KOH samples). Sample size was increased until a weight was reached that would not dissolve in 2 ml of base. Collagen, RNA, and glycogen were digested in this manner in all three bases; the remaining samples were digested with NCS only. The cooled digests were diluted with scintillator solution and spiked with 100 µl toluene-^{3}H.

2. Comparison of Figure of Merit and Calculated Specific Activity for Tissues and Fluids with the Three Basic Digestants. The radioactive and control rabbit liver and muscle were thawed, and three different weights of each, in the range from 130 to 600 mg, were digested with 2 ml of each of the three bases; after standing overnight at room temperature, samples were heated for 20 hours at 50°C.

Homogenates of one part tissue in four parts 0.85% NaCl solution were dispensed by pipet; 0.15-ml and 0.3-ml samples were digested with 2 ml of each base overnight at room temperature, and then at 50°C until digestion appeared to cease; heating times varied from 2 to 10.5 hours for NCS, from 4 to 24 hours for Hyamine hydroxide, and from 24 to 32 hours for KOH. Tritium-labeled plasma and heparinized ^{14}C-labeled blood were pipetted into vials for digestion with 2 ml of each base; the samples were shaken vigorously; 0.35 ml of benzoyl peroxide solution was added for each 0.1 ml of blood. After digestion at 50°C for 0.5 hour, the samples were cooled for 10 minutes, and diluted for counting.

3. Study of Effects of Digestion Temperature and Duration on Figure of Merit and Calculated Specific Activity. 0.3-ml samples of homogenates of ^{3}H-labeled rabbit liver and muscle were pipetted into vials and digested with 2 ml NCS, Hyamine hydroxide or KOH at each of four temperatures: 25, 35, 50, and 70°C. Quadruplicate samples were run for each tissue with each base at each temperature. When digestion appeared to cease, i.e., as soon as the condition of the samples appeared static, two of each quartet of samples were cooled to 25°C and diluted with scintillator solution. The other pair of each set was digested twice as long; this period varied with each solvent and tissue. Figures of merit and specific activities were calculated for those with counting efficiencies greater than 4 percent. The same procedure was followed with 50-mg samples of RNA and dried chicken muscle wetted with 200 µl of water, except that figures of merit only were obtained.

4. Study of Effect of Ultrasonic Agitation on Digestion Time. Samples of human plasma were digested in NCS at 25°C with and without ultrasonic agitation. Unlabeled rat liver, muscle, and kidney homogenates were digested in NCS at 50°C, both with and without ultrasonics. These samples were periodically examined for 17 hours, at which time all except two samples were in solution. All samples were diluted with scintillator solution and spiked with toluene-^{3}H.

RESULTS AND DISCUSSION

For a fixed weight of sample, the highest efficiency and figure of merit were obtained, as expected, by maximal dilution of the sample. All results reported here are for samples that remained in solution during dilutions to 20 ml. In general, the highest efficiencies were obtained by using the minimal amount of basic digestant required to dissolve the sample. Close to the limit of solubility, however, this sometimes resulted in an excessive heating time and in greater quenching than was produced by the use of more base.

When sufficient sample material is available, particularly if it does not have a high specific activity, the optimal sample size should be determined. The counting rate increases with increasing weight of sample material dissolved until this is offset by decreased counting efficiency due to quenching produced by the sample material. The highest figure of merit is obtained at the highest counting rate. In order to provide a means for comparison, all figures of merit reported here are based on the weight of sample material that could be contained in a 20-ml sample of the same composition; for instance, if the volume of the counting sample was 5 ml, the actual sample weight (in mg) was multiplied by 20/5, and then by the counting efficiency in percent, to calculate the figure of merit.

It was found from preliminary experiments that, when dry sample material was to be dissolved, there were several important advantages to be gained from adding a small amount of water before adding base. Water increased the amount of dry sample that could be dissolved, reduced the digestion time, and increased the counting efficiency. The optimal amount of water was found to be the maximum that would stay in solution when the sample was diluted with scintillator. Whole tissue was found to contain a sufficient quantity of water for good digestion. The presence of water in the sample usually eliminated the phosphorescence associated with alkaline digests.[9] Normal background rates were reached within 30 minutes after samples were placed in the counter at 10°C, except with Hyamine samples of blood. Neutralization of the digests was unnecessary.

The relatively high solubility of water solutions in NCS permitted the use of a pure toluene scintillator to dilute NCS digests, thus minimizing quenching. If any phase separation did occur, better counting efficiency was achieved by adding more NCS than by adding alcoholic secondary solvents to restore homogeneity. Figure 4 shows the solubility of water in toluene solutions of NCS at 0°C. This curve was also found to predict the solubility of the 1:4 tissue homogenates. Water solubility was decreased in the presence of large amounts of solid sample material or of benzoyl peroxide.

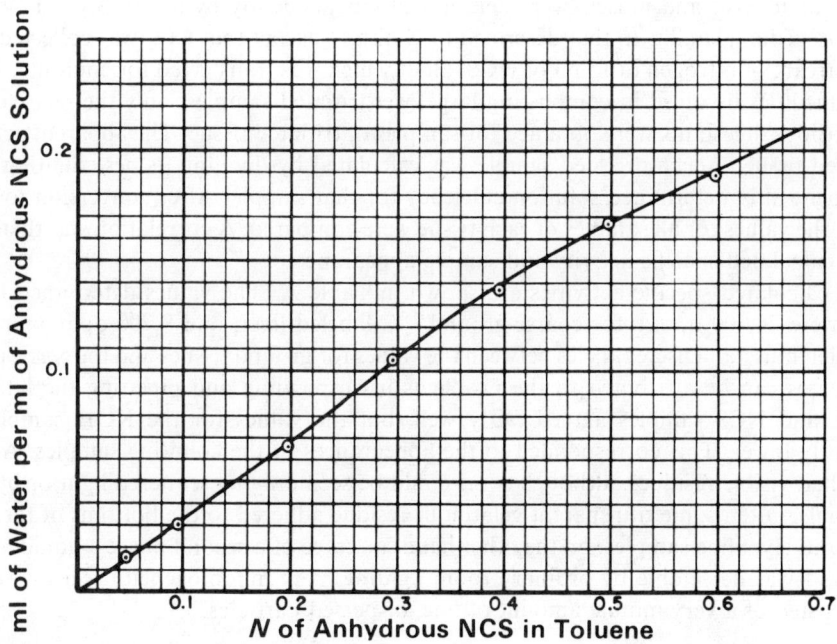

FIGURE 4. Solubility of Water in Toluene Solutions of NCS at 0°C.

The $2N$ methanolic KOH-butoxyethanol system was suggested by Petroff et al.[7] Butoxyethanol was also used as a secondary solvent in diluting Hyamine hydroxide digests with toluene. Comparisons with the "diotol" system of Herberg,[4] with a dioxane-naphthalene system, and with mixtures of toluene and Cellosolve[10] or methyl Cellosolve indicated that the butoxyethanol-toluene system could be expected to give comparable or better figures of merit with both KOH and Hyamine digests.

1. Dry Materials. Table 1 shows results for some of the dry materials tested. Hyamine hydroxide samples almost always required longer heating times than NCS samples. Alcoholic KOH appeared to be a very ineffective solvent, even when heated to 70°C, as suggested by Petroff et al. Some residue usually remained in KOH samples, even after heating times of up to 50 hours, and this residue did not dissolve in the scintillator solution. With the two solvents that completely dissolved the sample materials, the best figure of merit was obtained with NCS. The samples shown in the table are those that gave the highest figure of merit. Smaller weights of material would require proportionally smaller volumes of base and water, and would count with higher efficiencies, if diluted to the same volume of 20 ml.

In another experiment, with 50-mg samples of dry solids wet with 0.1 ml water, it was found that elastin, myoglobin, and dextrose dissolved in 2 ml of Hyamine hydroxide or NCS; casein hydrolyzate, keratin, and dextrin were almost completely dissolved; and starch and chitin did not dissolve. These samples were not diluted with scintillator solution.

2. Tissues and Fluids. An important objective of this study was the establishment of a visual test for completeness of digestion. Digestion may be termed complete when the digest of a labeled material can be diluted with a scintillator solution and counted with the efficiency determined independently to be the *solution efficiency* for that counting sample. Such independent efficiency determinations are commonly made by channels-ratio counting or external standardization. If some labeled sample material remains undissolved, the efficiency for this portion is likely to be lower than that for the portion in solution, especially with a very weak beta emitter such as tritium. To determine the efficiency with which a labeled sample material is actually being counted, it is necessary to know its activity. The specific activities of the ^{14}C-labeled ground liver and muscle were determined independently by a Van Slyke-Folch combustion procedure, similar to that described by Smith,[11] except that CO_2 was collected from a flowing stream of nitrogen in 15 ml of NCS, all of which was transferred for counting. Because of the difficulties involved in combusting large, wet tritiated samples, only the relative results for the various digestants were obtained for samples of tritiated tissue. The apparent activity of the labeled tissue in each digested sample was calculated by dividing its net counting rate by the independently determined solution efficiency for that sample. After conversion to specific activites, the values for all samples of each tissue were compared. A correlation was then sought between calculated specific activity and sample appearance.

The calculated specific activities are shown in Table 2. (The values determined by combustion were 24,400 dpm/g for coarse-ground ^{14}C-labeled liver, and 9,700 dpm/g for ground ^{14}C-labeled muscle.) Discrepancies between the values for ground tissues and the corresponding homogenates can be attributed to the problems in suspending and pipetting the latter. The Hyamine and NCS samples agreed fairly well, but the values for the KOH samples were significantly lower. This corresponded to the appearances of the counting samples. All KOH samples had visible residues, although some could be seen only by scattered light or by inverting the vials. Often some transparent gelatinous residue adhered to the bottoms of these vials. In NCS and Hyamine samples, on the other hand, no trace of a second phase—liquid, solid, or gelatinous—was detectable by ordinary room lighting. Any inhomogeneity seen by scattered light consisted of a very minute amount of fine, dispersed particles.

Table 1. OPTIMAL SAMPLE COMPOSITION FOR COUNTING ^{3}H IN DRY MATERIALS

Sample	Solvent	Volume, ml		Optimal Sample Weight, mg	Concentration, mg sample/mg solvent		% ^{3}H Efficiency	Figure of Merit
		Solvent	Water		Optimum	Maximum		
RNA	NCS	2.7	0.3	280	105	105	7.2	2,020
	HyOH*	2.0	0.3	270	135	150	3.7	1,000
Glycogen	NCS	8.0	1.4	600	75	120	16.8	10,100
	HyOH	2.7	0.5	320	120	120	8.4	2,700
Collagen	NCS	4.0	0.7	420	105	135	13.4	5,600
	HyOH	2.7	0.5	360	135	135	5.1	1,800
Fibrinogen	NCS	4.0	0.7	160	40	40	16.5	2,600
Sucrose	NCS	6.7	0.5	670	100	100	30.1	20,200
Tyrosine	NCS	2.7	0.2	120	45	45	16.4	2,000
Histidine	NCS	2.7	0.3	160	60	60	4.0	640
Methionine	NCS	8.0	1.4	720	90	90	23.4	16,800
Glycine	NCS	6.7	0.8	300	45	45	32.6	9,800

* Hyamine hydroxide.

Table 2. CALCULATED SPECIFIC ACTIVITIES IN VARIOUS DIGESTS
(dpm/g tissue or dpm/ml plasma)

Tissue	NCS	Hyamine Hydroxide	KOH
^{14}C-Labeled Liver			
Coarse-Ground	24,800 ± 870 (4)	22,800 ± 800 (2)	12,900 ± 1,000 (2)
Homogenate	26,500 ± 150 (4)	26,500 ± 140 (2)	19,200 ± 0 (2)
^{14}C-Labeled Muscle			
Coarse-Ground	9,500 ± 270 (5)	9,000 ± 0 (3)	5,700 ± 500 (3)
Homogenate	8,900 ± 80 (4)	8,200 ± 65 (2)	7,700 ± 240 (2)
^{3}H-Labeled Muscle			
Coarse-Ground	317,000 ± 9,000 (5)	330,000 ± 3,850 (3)	244,000 ± 16,000 (3)
Homogenate	293,000 ± 1,700 (16)	306,000 ± 4,500 (16)	265,000 ± 5,400 (16)
^{3}H-Labeled Plasma	463,000 ± 2,500 (2)	473,000 ± 1,000 (2)	344,000 ± 1,000 (2)

Values for the 1 : 4 homogenate were calculated back to the wet weight of tissue. Standard deviations of the mean were calculated as

$$\sqrt{\frac{\sum (x - \bar{x})^2}{n(n - 1)}}.$$

Values for n (the number of replicate samples) are given in parentheses.

It was possible to achieve this clarity in NCS digests with all tissues tested, except spleen, which apparently left a residue of ferric hydroxide. Digests in Hyamine were not always clear, but they dissolved when diluted, except those of blood and spleen. In KOH it was never possible to obtain clear digests or counting samples. The "diotol" solvent described by Herberg[4] was tested, but it did not dissolve the KOH digest.

The extent to which incomplete digestion affects the counting of a sample must depend on the extent to which its radioactive species are directly soluble in the scintillator solution. The homogenates of tritiated rabbit liver and muscle used in this experiment were tested for tritium lability by successively extracting portions of them with water, alcohol, and toluene. At least 25 percent of the tritium remained in the insoluble fraction of the muscle, and at least 30 percent in the liver.

The optimal sample compositions for homogenates and ground tissue are shown in Table 3. (Considerable variation in these values can be expected with different animals, even for the same tissue; among the three rabbits used in these experiments a substantial difference in color and ease of digestion was noted for both liver and muscle.) The amount of water in the homogenates limits the sample size. With tissues of low inherent quenching, better figures of merit can be obtained by counting coarse-ground material. The liver samples were not bleached, although this would have improved efficiency somewhat. Blood samples were bleached. The background samples of blood in Hyamine had counting rates several times normal after cooling to 10°C, but other backgrounds were normal. It was necessary to digest blood (but not plasma) samples for about 30 minutes at 50°C to prevent formation of a precipitate after dilution.

Benzoyl peroxide was the most satisfactory bleaching agent tested (others were hydrogen peroxide, sodium borohydride, and methyltrioctylammonium borohydride); it was the least quenching, and did not cause evolution of gas. Experimentation with blood samples showed that the highest counting efficiencies and lowest background rates were obtained by adding benzoyl peroxide before digestion. Maximal bleaching (and counting efficiency) without phase separation at 10°C was obtained by using 0.35 ml benzoyl peroxide per 0.1 ml blood.

3. Effects of Digestion Temperature and Duration. Complete dissolution was obtained in Hyamine hydroxide and NCS at 25, 35, 50, and 70°C, but digestion times varied from several days at 25°C to 1 hour at 70°C. The range from 35 to 50°C was the best for the sample materials

Table 3.

OPTIMAL SAMPLE COMPOSITION FOR COUNTING ^{3}H AND ^{14}C IN WET TISSUE

Tissue	Solvent	Solvent Volume, ml	Sample Wet Weight or Volume	% Efficiency	Figure of Merit
^{3}H-Labeled Muscle Coarse-Ground	NCS HyOH*	5.0 2.1	950 mg 450 mg	18 14	17,000 6,300
1:4 Homogenate	NCS HyOH	7.6 3.7	280 mg 110 mg	20 10.6	5,700 1,170
^{14}C-Labeled Muscle Coarse-Ground	NCS HyOH	11.5 5.4	2,300 mg 1,150 mg	47 42	100,000 48,000
1:4 Homogenate	NCS HyOH	15.3 7.0	560 mg 210 mg	58 47	32,200 9,800
^{3}H-Labeled Liver Coarse-Ground	NCS HyOH	0.7 0.4	75 mg 40 mg	18 19.2	1,350 770
1:4 Homogenate	NCS HyOH	4.2 2.0	125 mg 60 mg	19.2 12.2	2,400 730
^{14}C-Labeled Liver Coarse-Ground	NCS HyOH	1.9 1.3	220 mg 130 mg	42 41	9,200 5,300
1:4 Homogenate	NCS HyOH	9.3 4.5	280 mg 135 mg	51 42	14,300 5,700
^{3}H-Labeled Blood Whole	NCS HyOH	1.8 0.7	0.18 ml 0.11 ml	18 14.1	3.23 1.55
^{14}C-Labeled Blood Whole	NCS HyOH	4.5 1.5	0.45 ml 0.25 ml	45 48	20.20 12.10
^{3}H-Labeled Plasma	NCS HyOH	8.0 2.7	1.20 ml 0.50 ml	22 16.4	26.40 8.20

* Hyamine hydroxide.

studied in this experiment. The results of this and other experiments indicated that excessive quenching is produced in Hyamine and NCS digests at temperatures above 60°C, and that loss of sample from the screw-cap vial becomes more probable. In the extended periods of time required for good digestion at 25°C in this experiment more quenching was produced than in a few hours at 35 or 50°C. On the other hand, materials that are easily dissolved at 25°C, such as plasma, show less quenching when allowed to sit for a few hours or overnight at room temperature. KOH samples were best digested at 70°C, but never appeared completely dissolved.

Extension of digestion time beyond that required to give complete dissolution by visual test did not change either the behavior of the sample during dilution or the calculated specific activity.

4. Ultrasonic Agitation. For the four different sample materials examined, the use of ultrasonic agitation during digestion did not appear to offer any significant advantage. It did not materially affect the rate of sample digestion, final sample condition, background count rate, or tritium counting efficiency.

Sensitivity Comparison

It is of interest to compare the figures of merit obtained by basic digestion with those obtained by combustion and liquid scintillation counting of $^{14}CO_2$ and tritiated water.

Combustion has been used to improve detection sensitivity for low-specific-activity and highly quenching sample materials. For oxygen flask combustion a direct comparison can be made only in the case of dry samples. Wet combustion can be used for ^{14}C samples, but for wet tritium samples it is necessary to use high-temperature (Pregl tube) combustion. Sample size is limited in the case of ^{14}C by the amount of CO_2 that can be counted in 20 ml; this is about 8 millimoles by the best previous solution techniques, or 10 millimoles using NCS or $Ba(OH)_2$. Following Woeller[12] and Kalberer and Rutschmann,[13] it may be estimated that 8 millimoles CO_2, representing the combustion product of about 100 mg of an average dry tissue plus some 80 mg of cellulose wrapper, can be counted with an efficiency of 55 percent for a figure of merit of 5,500. For wet-tissue combustion the ^{14}C figure of merit, following Smith,[11] is approximately 12,000 on the basis of wet weight, in this case limited by the necessity of discarding two thirds of the trapping solution. This could be increased to about 40,000 by trapping the CO_2 in $0.6N$ NCS. 15 ml of the NCS bicarbonate can be counted at about 60 percent efficiency for ^{14}C.

For wet tritiated tissue the method of Knoche and Bell[14] gives a figure of merit of 17,000 for combustion of 1 g wet weight. This might be increased by using a higher counter temperature to increase water solubility. On the same weight basis, the figure of merit for plasma + NCS in Table 3 would be 26,400. For oxygen flask combustion the ^3H figure of merit may be estimated as about 5,600 on the basis of a dry-tissue weight of 400 mg combusted in a 2-liter flask.

It is not surprising that the combustion method shows a decided advantage in sensitivity when the sample material is highly colored (blood, liver) or when deep color is developed during basic digestion (RNA, tyrosine, histidine). Use of a bleaching agent with the latter samples might have reduced this advantage considerably. Combustion is less advantageous for ^{14}C than for ^3H, because of the slower decrease of ^{14}C counting efficiency with increasing quencher concentration, and because of the limitation of CO_2 solubility. The use of improved digestion procedures is seen to give figures of merit approaching or surpassing the figures for combusted samples in a significant number of cases.

SUMMARY

Hyamine hydroxide and NCS appear to be capable of solubilizing soft tissues and many purified biological materials, so that low-energy beta emitters (^3H and ^{14}C) can be counted with true solution efficiencies as determined by internal standardization, channels ratio, etc. Alcoholic KOH does not. Highest figures of merit are obtained with NCS, and in a number of cases these compare favorably with those attainable by combustion techniques. Improved digestion, bleaching, and dilution procedures are described.

ACKNOWLEDGMENT

The authors gratefully acknowledge the assistance of Lawrence J. Calicchio and Michael J. Gezing in preparing and counting many of the samples.

REFERENCES

1. Vaughan, M., Steinberg, D., and Logan, J., Liquid Scintillation Counting of ^{14}C- and ^3H-Labeled Amino Acids and Protein. *Science*, 126: 446, 1957.
2. Radin, N. S., in *Liquid Scintillation Counting*, p. 108, Bell, C. G., and Hayes, F. N., eds. Pergamon Press, Macmillan Co., New York, 1958.

3. Kinnory, D. S., Kanabrocki, E. L., Greco, J., Veatch, R. L., Kaplan, E., and Oester, Y. T., in *Liquid Scintillation Counting*, p. 223, Bell, C. G. and Hayes, F. N., eds. Pergamon Press, Macmillan Co., New York, 1958.
4. Herberg, R. J., Determination of Carbon-14 and Tritium in Blood and Other Whole Tissues. Liquid Scintillation Counting of Tissues. *Anal. Chem.*, 32: 42, 1960.
5. Gjone, E., Vance, H., and Turner, D. A., Direct Liquid Scintillation Counting of Plasma and Tissues. *Int. J. Appl. Radiat. Isotop.*, 8: 95, 1960.
6. O'Brien, R. D., Nitric Acid Digestion of Tissues for Liquid Scintillation Counting. *Anal. Biochem.*, 7: 251, 1965.
7. Petroff, C. P., Patt, H. H., and Nair, P. P., A Rapid Method for Dissolving Tissues for Liquid Scintillation Counting. *Int. J. Appl. Radiat. Isotop.*, 16: 599, 1965.
8. Bush, E. T., General Applicability of the Channels-Ratio Method of Measuring Liquid Scintillation Counting Efficiencies. *Anal. Chem.*, 35: 1024, 1963.
9. Herberg, R. J., Phosphorescence in Liquid Scintillation Counting of Protein. *Science*, 128: 199, 1958.
10. Bruno, G. A. and Christian, J. A., Determination of Carbon-14 in Aqueous Bicarbonate Solution by Liquid Scintillation Counting Technique. Application of Biological Fluids. *Anal. Chem.*, 33: 1216, 1961.
11. Smith, G. N., Ludwig, P. D., Wright, K. C., and Bauriedel, J. R., Simple Apparatus for Combustion of Samples Containing ^{14}C- Labeled Pesticides for Residue Analysis. *J. Agr. Food Chem.*, 12: 172, 1964.
12. Woeller, F. H., Liquid Scintillation Counting of $^{14}CO_2$ with Phenthylamine. *Anal. Biochem.*, 2: 508, 1961.
13. Kalberer, F. and Rutschmann, J., Eine Schnellmethode zur Bestimmung von Tritium, Radiokohlenstoff und Radioschwefel in beliebigen organischen Probematerial mittels des Flüssigkeit Scintillations Zählers. *Helv. Chim. Acta (Switzerland)*, 44: 1956, 1961.
14. Knoche, H. W. and Bell, R. M., Tritium Assay by Combustion with a Novel Oxygen Train and Liquid Scintillation Techniques. *Anal. Biochem.*, 12: 49, 1965.

SAMPLE PREPARATION FOR LIQUID SCINTILLATION COUNTING

J. C. Turner

Reprinted by permission of
The Radiochemical Centre, Amersham, Bucks, England.

INTRODUCTION

Liquid scintillation counting has become the generally preferred method of counting weak beta emitters and, to a lesser extent, alpha and gamma emitters. Why this should be so is made clear by considering the basis of the liquid scintillation counting process. The counting sample consists of three components: the radioactive material, an organic solvent or solvent mixture, and one or more organic phosphors. A particle or radiation emitted by the sample material is absorbed in, and its energy transferred to, the solvent, and thence to the phosphor, which emits a burst or scintillation of light photons. These photons are then absorbed by the photocathode of a photomultiplier tube, which converts them into an electronic pulse. The pulse, after suitable amplification, is registered as a count corresponding to the emission of the particle or radiation. The phosphor may thus be considered to be the first link in the detector assembly. Since the radioactive sample material in most methods of sample preparation is in intimate contact or in actual solution together with the phosphor, detection of the emitted particles or radiations is highly efficient, and may approach 100 percent. Problems of self-absorption of the emissions are thus absent, or considerably less than those associated with planchette counting of solid samples. This is of particular importance for the measurement of beta emitters of low energy, such as tritium, ^{14}C, and ^{35}S. Indeed, in the case of tritium the advent of the liquid scintillation counting technique has been one of the main factors in the rapid expansion of its use as a tracer. The technique has the particular advantage over other sensitive methods (such as gas counting) that sample changing can conveniently be mechanized to deal with large numbers of samples.

The apparent simplicity of the method can be misleading. Errors originating in the sample may arise from inhomogeneity, chemical quenching or color quenching, and from a variety of other causes. These sources of error, and the way in which they can affect the results, are discussed later in this chapter. Other problems are involved in the design of the counter itself, and although these are primarily the concern of the instrument manufacturer, they should be appreciated by the user. A variety of liquid scintillation counters are in use today, ranging from the simplest single-photomultiplier-tube type to the most sophisticated twin-photomultiplier-tube types with coincidence circuitry, multichannel counting, automatic determination of counting efficiency, and electronic calculators. The essentials of any liquid scintillation counter are the photomultiplier tube, a pulse amplifier of some sort, and a scaler to record the pulses. With this basic assembly, counting can be carried out, but there are at once several disadvantages. Because there is no means of discriminating between electronic pulses of different energies, only samples containing one isotope may be counted. The assembly suffers from a high background count rate, which has three main contributions: high-energy pulses due to cosmic radiation, low-energy pulses due to possible chemiluminescence of the sample or fluorescence of the sample bottle glass, and low-energy pulses generated by the random thermal emission of electrons from the photocathode of the photomultiplier tube. This last factor may be considerably reduced by cooling the tube to low temperatures, but this, in turn, leads to reduced sample solubility in the liquid scintillant, and the range of liquid scintillants available for use is restricted to those that do not freeze at low temperatures.

The more sophisticated counters possess additional electronic equipment that overcomes the disadvantages mentioned above. Pulse-height analyzers with discriminators that may be set at various energy levels allow setting of lower and upper limits to the energy of pulses accepted and passed on to the scaler; in this way most of the undesired pulses are eliminated. The pulses due to thermal emission of the photocathodes may largely be eliminated by using two photomultiplier tubes operated in a coincidence circuit; this circuit accepts only those pulses that arise simultaneously from both photomultiplier tubes, initiated by radioactive decay in the sample. Pulses due to random emission of the thermal electrons in the individual tubes are not simultaneous, except—rarely—by chance, and are rejected. The use of electronic circuits with very short resolving times ensures that two pulses arriving within a short time are not accepted as one pulse. The use of such coincidence circuits allows the photomultiplier tubes to be operated at temperatures above freezing point or at room temperature.

Further expansion of the counter assembly may be made with additional electronic channels with facilities for pulse-height analysis. More than one isotope may then be counted simultaneously in the same sample, and automatic methods of determining counting efficiency, such as the channels-ratio and external-standard methods, are possible. With the addition of an electronic calculator the operator is provided with a complete set of data for the sample, from which the true disintegration rate may be easily calculated. In some instruments this is also carried out by the calculator.

Generally the more advanced instruments have increased counting efficiencies, and under under the most favorable conditions this may reach 95 percent for ^{14}C and nearly 60 percent for tritium. The efficiencies attainable in routine use with the majority of quenched samples are lower. Nowadays nearly all available counters give satisfactory counting efficiencies, and unless this factor is particularly important to the potential user, the other capabilities of the system should be assessed first. An interesting paper by Ayrey and Mazza[1] describes the quite advanced counting work that may be carried out with relatively simple equipment, although at the cost of taking more time and trouble.

One of the disadvantages of liquid scintillation counting in its application to the measurement of sample materials is that the organic solvents employed in the liquid scintillant mixture are mostly of the aromatic hydrocarbon type, such as xylene and toluene. These solvents are used because they have favorable properties for the transfer of energy from the emitted particles or radiations to the phosphor. Unfortunately the range of materials that can be dissolved in such solvents is severaly limited. Considerable effort has been expended in the search for other suitable liquid scintillation solvents of wider application and for methods of introducing the sample material into the liquid scintillant as a colloidal mixture or suspension, by direct solution, by complexing or salt formation to form soluble compounds, or by conversion or degradation of the sample material to form soluble compounds. The literature reporting the results of this work is widely dispersed; despite the publication of several books and review articles[2-9] on various aspects of the subject, it is not always easy to get the information required, or even to select the most helpful publications. It is hoped that this review will be an aid to workers who wish to obtain rapid information on various practical aspects of liquid scintillation counting.

SELECTIVE LITERATURE REVIEW OF METHODS OF SAMPLE PREPARATION FOR LIQUID SCINTILLATION COUNTING

This review is not exhaustive; the methods of sample preparation that have been selected are thought to be of more general interest and application. Some of the advantages and disadvantages of various methods are mentioned, but no "best methods" are recommended, as these depend on the particular needs of the individual worker. The material of the review is arranged in sections, each of which considers a particular field of liquid scintillation counting, such as counting in solution and counting suspensions. Within these sections, sample preparation methods for classes of compounds or materials are considered.

Liquid Scintillation Solvents and Solutes

Solvents. The most efficient liquid scintillation solvents are the alkyl benzenes, such as xylene and toluene,[3] the latter being most widely used. Ethers, such as anisole (methoxybenzene) and 1,4-dioxane, have also been used.[3] Their efficiency is 20 to 30 percent less than that of the alkyl benzenes, but some have useful solvent properties, particularly 1,4-dioxane, because it is miscible with water. Many other solvents have been used; however, their function is not as primary solvents, but as secondary solvents to improve the solubility of the sample to be measured.

Since the solvent forms the largest part of the liquid scintillant, purity is important in order to maintain high efficiency. Tanielian et al.[10] have shown that this is not so troublesome with commercially available toluene, but may be so with 1,4-dioxane, which tends to form strongly quenching peroxides on standing.

Solutes. Scintillation solutes are classified as primary or secondary solutes. The primary solute accepts transferred energy from the solvent and emits light scintillations. The secondary solute absorbs these scintillations and reemits them at a longer wavelength, closer to the wavelength of maximum response of the photomultiplier tube. A higher detection efficiency results, and this gives significant improvement for tritium measurements. For isotopes with higher energy emissions this improvement is smaller, and a secondary solute can often be omitted from the liquid scintillant with no great loss in counting efficiency. Table 4 lists some of the more commonly used scintillation solutes. The abbreviations shown in the table will be used for the scintillation solutes throughout this chapter.

Table 4. SCINTILLATION SOLUTES

Solute	Abbreviation	Type	Fluorescence Maximum, mμ
p-Terphenyl	—	Primary	344
2, 5-Diphenyloxazole	PPO	Primary	363
2-(4-Biphenyl)-5-phenyl-1,3,4-oxadiazole	PBD	Primary	361
1,4-Bis-(5-phenyloxazol-2-yl) benzene	POPOP	Secondary	430
1,4-Bis-(4-methyl-5-phenyloxazol-2-yl) benzene	DM–POPOP	Secondary	430
2-(4'-t-Butylphenyl)-5-(4'-biphenyl)-1,3,4-oxadiazole	Butyl–PBD	Secondary	366
Naphthalene	—	—	—

PBD and p-terphenyl are now little used, because of limited solubility at low temperatures and in the presence of water. PPO is widely used, and has satisfactory solubility. Butyl-PBD is a new solute that has good solubility, is less affected by quenching than other solutes, and does not need a secondary solute even for tritium measurement. POPOP and the more recently introduced DM-POPOP are similar in their characteristics, except that DM-POPOP has improved solubility. Naphthalene has also been included in Table 4; its function is to improve the transfer of energy from the solvent to the phosphor in 1,4-dioxane-based liquid scintillants, which are used for counting aqueous samples; a marked increase in counting efficiency results.

Counting in Solution

1. Water and Aqueous Solutions.

A variety of liquid scintillant mixtures has been evolved for the counting of tritiated water and of aqueous solutions containing different radioisotopes. Toluene-based scintillants have been diluted with more polar solvents, such as ethanol, methanol, 1,4-dioxane and 2-ethoxyethanol, to improve the miscibility with water or aqueous solutions. 1,4-dioxane-based scintillants, although miscible with water, have been diluted with various glycols and alcohols to lower the freezing point of the solution. The recent introduction of liquid scintillation counters that operate at room temperature makes the use of pure 1,4-dioxane scintillants more attractive.[12] Baxter et al.[13] evaluated various liquid scintillant mixtures used for counting aqueous samples containing ^{14}C and tritium at room temperature, at 0 to 5°C, and at $-15°C$; a figure of merit was used that was the product of the percent total water content and the counting efficiency.

Probably the most widely used and most satisfactory liquid scintillant for counting large amounts of water and aqueous sample is that due to Bray,[14] which consists of naphthalene (60 g), PPO (4 g), POPOP (0.2 g), methanol (100 ml), ethylene glycol (20 ml), and 1,4-dioxane, giving a final volume of 1 liter. A recent paper[15] describes a liquid scintillant for use at room temperature based on 2-phenylethylamine, which is reported to incorporate up to 60 percent by volume of water or aqueous solutions of proteins and salts, although with rather low counting efficiencies. The liquid scintillant Triton* X-100, introduced by Patterson and Greene[16] for the emulsion counting of aqueous solutions, has recently, at the Radiochemical Centre, been found to be satisfactory for the incorporation of large amounts of aqueous solutions of highly polar ^{14}C compounds into homogeneous solution, with high counting efficiencies.

Calculations have shown that aqueous solutions of beta emitters with energy greater than 0.26 Mev will emit light in the spectral region from 300 mμ to 700 mμ.[17] This light, called *Cherenkov radiation*, can be detected by the photomultiplier tube of a liquid scintillation counter. Aqueous solutions of such beta emitters may, therefore, be counted without the use of a liquid scintillant. Counting efficiencies are very low at about 0.3 Mev, but become worthwhile above 1 Mev for beta emitters such as ^{24}Na and ^{32}P,[18] and ^{56}Mn.[19] The counting efficiency of lower-energy beta emitters can be much improved by adding a water-soluble fluorescent compound to the aqueous sample.[20]

2. Inorganic Salts.

Inorganic salts have been solubilized in liquid scintillant mixtures by two main methods. The first employs dilution with alcohols or ethers, as mentioned above for aqueous samples. The second achieves solubilization by the formation of toluene-soluble complexes or salts of the inorganic anions or cations. Table 5 lists some of the radioisotopes that have been counted by these methods.

Many metallic radioisotopes have been solubilized as complexes or salts of alkyl phosphates (^{147}Pm;[30] uranium and thorium;[31] ^{95}Zr and ^{95}Nb;[32] plutonium isotopes[33]) and of 2-ethylhexanoic acid (cadmium, lead, bismuth, and uranium;[34] ^{90}Sr and ^{90}Y;[34] ^{45}Ca;[35] ^{87}Rb [37]). The literature of radiochemical separation and processing contains many examples of complexing agents—either generally applicable or specific for a particular element—that are suitable for extracting metallic radioisotopes into toluene or other organic solvents suitable for liquid scintillation counting.

3. Biological Materials.

Solubilization Methods. Biological material is in many ways the most difficult type of sample material to prepare for counting in solution. Table 6 gives some methods for the solubilization of a variety of materials. Usually, strongly basic conditions with warming are employed. These conditions break down the polymeric materials to yield soluble products of lower molecular weight. The worker should be aware of the possibility of loss of volatile active products during such degradation procedures.

Hansen and Bush[38] have recently shown that the solubilizing agents NCS and Hyamine 10X are efficient in the solubilization of soft tissues and many purified biological materials to give homogeneous-solution counting samples. A figure of merit that was the product of the

* Trademark of Rohm & Haas Co.

Table 5. COUNTING OF RADIOISOTOPES IN INORGANIC FORM

Isotope	Chemical Form	Liquid Scintillant	Application	Reference
^{45}Ca	Chloride	Toluene/ethanol	Large amounts (∼100 mg/sample) of ^{45}Ca.	21
			Suitable for biological material after preliminary ashing and treatment.	22
	Nitrate	Toluene/ethanol/ethylene glycol/nitric acid	Large amounts (∼100 mg/sample) of ^{45}Ca.	21
			Suitable for biological material after preliminary ashing and treatment.	22
	Di-n-butyl phosphate complex	Toluene	Large amounts of calcium in plant material, soil, bone, and milk after ashing and conversion to the chloride.	27
	Tri-n-butyl phosphate complex	Toluene	Calcium in tissue and other biological material after ashing and conversion to perchlorate.	28
^{110m}Ag	Nitrate	Toluene/ethanol	Small amounts (0.25 ml of aqueous solution) of salts of metallic radioisotopes.	23
^{22}Na	Chloride or bromide	Toluene/ethanol	Small amounts (0.25 ml of aqueous solution) of salts of metallic radioisotopes.	23
^{147}Pm	Chloride	Toluene/ethanol	Small amounts (0.25 ml of aqueous solution) of salts of metallic radioisotopes.	23
^{204}Tl	Nitrate	Toluene/ethanol	Small amounts (0.25 ml of aqueous solution) of salts of metallic radioisotopes.	23
^{125}I	Sodium iodide	Bray's scintillant	Large amounts of aqueous iodide solution.	24
^{129}I	Sodium iodide	Bray's scintillant	Large amounts of aqueous iodide solution.	24
^{131}I	Sodium iodide	Bray's scintillant	Large amounts of aqueous iodide solution.	24
^{32}P	Phosphmolybdate	Toluene/ethanol/Hyamine 10X/methanol	Food samples after ashing; inorganic phosphate.	25
^{35}S	Primene* 81–R sulfate	Toluene/methanol	Sulfur as benzidine sulfate after preliminary treatment; inorganic sulfate.	26
Uranium, plutonium, transplutonium elements, and lanthanides	Tri-n-octylphosphine oxide complexes	Dioxane	Counting of alpha emitters with high efficiency and accuracy.	29

* Trademark of Rohm & Haas Co.

sample weight and the counting efficiency was used. In nearly all cases the highest figures of merit were obtained with NCS. The use of alcoholic potassium hydroxide was also investigated, and was found to be unsatisfactory, resulting in incomplete solubilization in many cases. Improved digestion, bleaching, and dilution procedures are described.

Table 6.
SOLUBILIZATION OF BIOLOGICAL MATERIALS FOR COUNTING IN SOLUTION

Biological Material	Solubilizing Agent	Remarks	Reference
Whole tissue, tissue extracts, and biological fluids	NCS (mixture of organic quaternary ammonium base)	Very efficient, widely applicable; used with toluene scintillant.	38
Bacterial cells	Formamide	Toluene/ethanol scintillant.	39
Dry tissue, protein, fresh tissue, serum	N potassium hydroxide/ Hyamine 10X chloride	Xylene scintillant.	40
Tissue	$2N$ methanolic potassium hydroxide	Toluene/ethylene glycol monobutyl ether scintillant.	41
Protein and amino acids	Hyamine 10X hydroxide	Toluene scintillant; up to 20 mg various amino acids or 10 mg protein per sample.	42
Nucleic acids	Hydrochloric acid, followed by Hyamine 10X hydroxide	Dioxane scintillant; 10 mg of RNA or 3 mg of DNA per sample.	43
Protein-nucleic acid	NCS	Toluene scintillant.	44
Protein	88% formic acid	Toluene/2-ethoxyethanol scintillant; specific activity of the solubilized protein can be determined by the biuret procedure.	45
Blood, tissue fluids, tissue, bone	60% perchloric acid 30% hydrogen peroxide	Toluene/ethylene glycol monoethyl ether scintillant; easily oxidized, or volatile ^{14}C or ^{3}H may be lost.	46
Plasma	Hyamine 10X hydroxide	Toluene/Hyamine/ethylene glycol monoethyl ether scintillant; large quantities of plasma are solubilized.	47
Serum	—	Direct solubilization in dioxane/methanol/ ethylene glycol; used for estimation of ^{45}Ca, but causes serum proteins to precipitate.	48

Combustion and Oxidation Procedures for Biological Material. Many biological samples are deeply colored or give colored solutions after solubilization by methods described above. This sometimes leads to severe quenching, and low counting efficiencies result. Some workers have preferred to adopt combustion or oxidation procedures, which, although generally more time-consuming, yield samples that are all in the same chemical form and have higher counting efficiencies. In this way more reproducible and reliable samples are obtained. Many methods of combustion or oxidation have been used, such as wet oxidation, sealed-tube combustion, oxygen train combustion, oxygen bomb combustion, and oxygen flask combustion. A review of these methods as applied to biological samples containing tritium, ^{14}C, ^{35}S, or ^{32}P has been published.[49] The oxidation products may be absorbed in liquid scintillation mixtures to yield homogeneous samples for counting in solution. Water may be incorporated into liquid scintillant, as described under the heading "Water and Aqueous Solutions". Again the worker should be aware of the possibility of loss of volatile active oxidation or combustion products, and guard against such loss.

Carbon-14 Dioxide and Sulfur-35 Dioxide. These gases may be absorbed in a variety of bases. The absorption solution may then be incorporated into a liquid scintillant. Table 7 lists some of the bases that have been used and their particular application to a variety of $^{14}CO_2$ and $^{35}SO_2$ estimations.

Table 7.
COUNTING CARBON-14 DIOXIDE AND SULFUR-35 DIOXIDE IN SOLUTION

Base Used for Absorption	Liquid Scintillant	Application of Method	Reference
Phenylethylamine	Toluene/methanol	Measurement of enzymic decarboxylation reactions.	50
		Quantitative absorption of large amounts (up to 5 mM) of $^{14}CO_2$.	54
		$^{14}CO_2$ and $^{35}SO_2$ from oxygen flask combustion.	60
	Toluene/ethanol/dioxane	$^{14}CO_2$ and $^{35}SO_2$ from combustion in the counting vial.	59
Primene 81-R	Toluene	Bicarbonate/^{14}C carbonate in plasma or aqueous solution.	51
	Toluene/methanol	Absorption of $^{14}CO_2$ from gas stream, and determination of specific activity.	52
Hyamine 10X hydroxide	Toluene/methanol	$^{14}CO_2$ collection in Warburg flask.	53
		Absorption of $^{14}CO_2$ from Van Slyke wet oxidations.	55
Ethanolamine	Toluene/methanol	Absorption of $^{14}CO_2$ from a gas stream.	56
		$^{14}CO_2$ and $^{35}SO_2$ from oxygen flask combustion.	61
	Toluene/2-methoxy-ethanol/ethanol	Determination of specific activity of ^{14}C barium carbonate ($^{14}CO_2$ regenerated and absorbed).	58
20% Potassium hydroxide	Toluene/methanol	$^{14}CO_2$ collection in Warburg flask.	57

Decolorization of Biological Samples for Counting in Solution. An alternative approach to the total combustion or oxidation of colored biological material to provide colorless samples for counting is the addition of a bleaching agent to remove the color. This may often be done in the counting vial, thereby reducing losses due to transfer, although here also the loss of volatile active material from the vial is a possibility. Herberg[62] used hydrogen peroxide to bleach solubilized samples of tissue or whole blood. Shneour et al.[63] decolorized carotenoids by the addition of chlorine water, while Walter and Purcell[64] found that benzoyl peroxide was to be preferred for this purpose. Fales[65] recommended the use of sodium borohydride for general bleaching purposes. Hansen and Bush[38] tested hydrogen peroxide, sodium borohydride, methyl trioctyl ammonium borohydride, and benzoyl peroxide for the bleaching of blood samples. Benzoyl peroxide was found to be the most satisfactory, and was least quenching among those examined.

4. Solution Counting of a Wide Range of Labeled Compounds and Materials with the Toluene/Triton X-100 Scintillant.[16]

Nonpolar compounds, such as lipids and fatty acids, may be readily dissolved in almost any of the available liquid scintillant mixtures. More polar or insoluble compounds present greater difficulty, and some approaches to this problem have been described on preceding pages for various classes of compounds or materials. Most water-soluble compounds may be satisfactorily counted as aqueous solutions incorporated in Bray's scintillant.[14] The toluene/

Triton X-100 scintillant of Patterson and Greene[16] has been found more useful at the Radiochemical Centre, and this is described in more detail below.

The toluene/Triton X-100 scintillant was originally introduced for the purpose of emulsion counting of aqueous samples. Several scintillant mixtures were described for various applications; the liquid scintillant mixture of composition toluene/Triton X-100, 2:1 v/v was used in the work described in the following paragraphs. The basic counting characteristics of the system are described under the heading "Emulsion Counting".

It was found that a wide variety of labeled compounds—ranging from polar compounds such as amino acids to nonpolar compounds such as hexadecane and glyceryl tristearate—can satisfactorily be incorporated and counted in the scintillant. In general, polar compounds should be counted as 0.5 to 1.0 ml of aqueous solution in 10 ml of scintillant. Hydrochloric acid ($2N$), sulfuric acid ($2N$), and nitric acid ($2N$) may also be used, although nitric acid causes more quenching, as might be expected. Buffer solutions, salts, and 1-ml quantities of serum and urine are also satisfactorily incorporated. Nonpolar compounds may be incorporated as the "neat" compound or dissolved in an organic solvent. The incorporation of the various types of aqueous sample does not lower the counting efficiency very much. Under similar conditions, counting efficiency in Bray's scintillant[14] is much more affected. Providing that the above conditions of sample preparation are adhered to, clear counting samples will result, which are homogeneous solutions. The system has excellent solubilizing properties; 25-mg quantities of various amino acids and other polar compounds were counted in 10 ml of liquid scintillant with high efficiency. The counting samples were stable for extended periods of time, with no evidence of crystallization or precipitation of the sample material. The results of counting samples in this system may be directly compared with results obtained from counting in other homogeneous systems using the channels-ratio method[66,67] of determining counting efficiency. The system is suitable for use at low temperatures (4°C) or at room temperature, and has the further advantage that the components are cheap and do not deteriorate on storage, in contrast to dioxane-based liquid scintillants.

Counting Suspensions

Solid samples of materials that are insoluble in liquid scintillation mixtures may be counted by suspending the material as a powder in the liquid scintillant. Originally the counting sample was simply shaken and then counted, allowance being made for the settling that occurred.[68] However, more satisfactory results are obtained when a gelling agent is incorporated with the liquid scintillant to prevent settling. Cab-O-Sil*, a finely divided silica with a very large surface area,[69] is frequently used. It is a thixotropic gelling agent, the liquid scintillant being fluid when shaken, but a firm gel when at rest. Table 8 gives some examples of materials that have been counted by the suspension technique, which has several advantages. Large quantities (up to 2 or 3 g) of heavy powders may be counted with good efficiency in one sample. Quenching is minimized, because the material is not in solution in the liquid scintillant. However, the material should be white or colorless, to avoid light absorption by the suspension. The material should not dissolve in the scintillant, since dissolved material will be counted at an efficiency different from suspended material.

If the results obtained by suspension counting are to be compared with those obtained by counting in solution, the losses due to self-absorption in the suspended solid particles should be determined in some way. To minimize such losses, the suspended particles should be as fine as possible. The material that is most frequently counted by this method is ^{14}C-labeled barium carbonate, and Cluly[70] gives full experimental details. Some characteristics of the suspension counting of ^{14}C-labeled barium carbonate and ^{14}C-labeled crystalline material are discussed in the section dealing with theoretical aspects of liquid scintillation counting, under the heading "Suspension Counting".

* Trademark of Cabot Corporation.

Table 8. SOME EXAMPLES OF THE USE OF SUSPENSION COUNTING

Isotope	Chemical Form of Suspended Material	Application	Reference
^{14}C	Barium carbonate	Specific activity of barium carbonate.	70
	Sodium carbonate	Determination of $^{14}CO_2$ in aqueous solutions of ^{14}C carbonate.	74
^{55}Fe + ^{59}Fe	Insoluble white ferri-phosphate complex	Determination of iron in blood.	71
	Insoluble white ferric benzene phosphinate	Determination of iron in blood.	72
^{45}Ca + ^{90}Sr	Whole algal cells	Determination of calcium and strontium in unicellular algae.	73
^{37}Cs	Cesium perchlorate	Direct counting of inorganic salts.	75
^{90}Sr	Strontium sulfate	Direct counting of inorganic salts.	75
^{131}I	Silver iodide	Direct counting of inorganic salts.	75

Counting Paper and Thin-Layer Chromatograms, and Samples Dried onto Paper or Glass Fiber

Paper chromatograms or electrophoretograms of active samples, after development, may be dried and cut up into pieces, placed in vials containing liquid scintillant, and counted. In this way the amount of activity that migrates as the pure compound may be compared with the total activity present on the chromatogram or electrophoretogram, and thus the radiochemical purity of the sample may be derived. In the same way the relative proportions of various components of a mixture may be readily estimated. The method is usually as rapid as scanning—that is, passing the paper strip—through a windowless gas flow Geiger counter, and in most cases is considerably more sensitive. The reproducibility of the liquid scintillation counting method is usually good for ^{14}C and the more energetic beta emitters, but there are some complications in the measurement of tritium, which are discussed below.

Some controversy exists in the literature as to whether the orientation in the counting vial of pieces of paper or glass fiber containing adsorbed samples affects the results obtained. Generally the variations in counting efficiency observed are small or negligible for ^{14}C and the more energetic beta emitters, but may be important for tritium. These effects may be minimized by arranging the piece of paper or glass fiber flat on the bottom of the counting vial. Counting efficiency is decreased slightly, but reproducibility is increased.[76]

There are two main approaches to this type of counting. The first aims to ensure that all sample material is eluted from the support and therefore counted in homogeneous solution. Takahashi et al.[77] used Hyamine 10X hydroxide in methanol to elute amino acids, protein hydrolysates and sugars from paper in the counting vial, the eluate then being mixed with toluene scintillant. This approach has the advantage that no problems of orientation or of self-absorption in the support are encountered, but it is often difficult to elute quantitatively small amounts of polar materials adsorbed onto paper. For this reason some workers have preferred to combust the paper plus sample, and then count the combustion products in homogeneous solution.[78]

If it can be ensured that all active sample material remains firmly adsorbed to the support, the method can be considered as an extension of the suspension counting technique. Disadvantages of this method, as discussed previously, are the problems of orientation and self-absorption. Advantages include the ability to count materials that are difficult to dissolve in liquid scintillant mixtures, diminished quenching by the sample material, as it is not in solution in the liquid scintillant, and increased ease of sample preparation.

Originally this technique was applied to the counting of paper chromatograms, but thin-layer chromatograms can also be counted. The spots or zones are scraped into a counting vial, thixotropic gelling agent—usually Cab-O-Sil[69]—and liquid scintillant are added, and the sample is shaken to disperse the particles of adsorbent. Here too, the sample, for best results, should either be eluted completely from the support or remain firmly bound to it.[79]

Because of the advantages of counting samples on paper, some workers have preferred to dry aqueous solutions of material onto paper for counting. Blair and Segal[80] counted substantial amounts of ^{14}C-labeled potassium-D-gluconate in this way. More recently, the use of glass-fiber paper has been preferred for this purpose, because a higher counting efficiency is obtained than with cellulose paper. Davies and Cocking[81] have described this technique for counting amino acids, protein hydrolysates, proteins, and nucleotides. The sample in aqueous solution (acid or basic solutions may also be used) is soaked into a glass-fiber disc, dried, and then counted in a small volume of toluene scintillant with high efficiency. A linear increase of determined activity was found with increasing sample volume, and the applied sample material remained firmly bound to the glass-fiber discs. The amount of solution that may be counted in this way is limited to about 0.1 ml, but several discs may be stacked together in a counting vial for larger samples. Hutchinson[82] has shown that self-absorption becomes apparent for ^{45}Ca for the larger sample quantities on glass-fiber paper.

With the qualifications discussed above (with regard to problems of orientation, self-absorption, and distribution of sample material between support and liquid scintillant), the technique of counting samples adsorbed onto an inert support is a valuable one. The glass-fiber-disc technique of Davies and Cocking[81] appears to be particularly useful.

Counting Emulsions

The emulsion counting technique was used by Patterson and Greene[16] in 1965 for the counting of large quantities of aqueous samples of ^{14}C- and tritium-labeled materials. (The aqueous phase formed up to 40 percent of the counting sample.) Three liquid scintillant mixtures, containing different quantities of the surface-active agent Triton X-100 for various applications, were described. Counting samples consisting of fluid emulsions or rigid gels were prepared, and gave good counting efficiencies for ^{14}C and fair counting efficiencies for tritium. The emulsions were said to be unaffected by strong acid, alkali, and concentrated phosphate buffer in the aqueous phase, and were of the water-in-oil type. This type of system has advantages similar to the suspension counting system in that quenching agents in the sample are in a different phase to the liquid scintillant and exert less effect. Since the aqueous phase is separate, it should be possible to count aqueous solutions of highly polar material; self-absorption would also be small, if the droplet size of the emulsion were small in comparison to the range of the emitted beta particles. However, the system is not a simple water-in-oil emulsion. Recently Benson[83] and van der Laarse[84] have demonstrated the complexity of the system toluene/Triton X-100 (2:1 v/v) when used for counting tritiated water, and are in agreement that close attention to the experimental detail is necessary to obtain reliable results. Turner at the Radiochemical Centre investigated the characteristics of the same system, and showed that the system moved from one of homogeneous solution through a transition region of unstable emulsions to stable emulsions with increasing aqueous-phase content.

It is evident that, although the system is capable of being used for emulsion counting of large quantities of aqueous samples, careful selection of experimental conditions is essential, if reliable results are to be obtained. The characteristics of such systems will be further discussed in the section covering theoretical aspects of liquid scintillation counting, under the heading "Emulsion Counting".

Counting Gases and Vapors

Gases may be counted by absorption in various solutions (see, for example, Table 7), and also by simple physical solution in the liquid scintillant. Vapors from gas–liquid chromatography columns can also be counted, but in this case the main difficulty has been to achieve quantitative trapping of the vapors. A brief summary of the methods of trapping and counting gases and vapors is given in Table 9.

Table 9. COUNTING GASES AND VAPORS

Gas or Vapor	Method of Trapping	Method of Counting	Reference
^{222}Rn	Gas condensed into evacuated ampule containing liquid scintillant cooled in liquid nitrogen.	Ampule sealed, thawed, and counted.	85
^{131m}Xe	Gas condensed into evacuated ampule containing liquid scintillant cooled in liquid nitrogen.	Ampule sealed, thawed, and counted.	85
^{85}Kr	Gas condensed into evacuated ampule containing liquid scintillant cooled in liquid nitrogen.	Ampule sealed, thawed, and counted.	85
^{35}S-labeled hydrogen sulfide	Gas injected into sealed evacuated ampule.	Toluene scintillant injected into ampule.	87
^{3}H	Gas injected into sealed evacuated ampule.	Toluene scintillant injected into ampule.	86
^{85}Kr	Gas injected into sealed evacuated ampule.	Toluene scintillant injected into ampule.	86
^{14}C-labeled fatty-acid methyl esters	Cartridge of anthracene crystals coated with silicone oil.	Cartridge counted directly, to give cumulative record of effluent activity.	88
^{14}C-labeled fatty-acid methyl esters	Millipore filters.	Sample washed from filter into toluene scintillant.	89
^{14}C-labeled fatty-acid methyl esters	Cartridges of anthracene crystals coated with silicone oil.	Cartridges counted directly, to give activity of effluent fractions.	90

Counting Flowing Aqueous Solutions

The continuous counting of flowing aqueous solutions—in the form of effluent from, for example, chromatography columns—presents some difficulty. Mixing of the effluent with liquid scintillant is not practicable as a method for counting. Instead, the effluent may be passed through a flow cell that contains solid or plastic scintillating material and is connected to the photomultipliers of a liquid scintillation counter. Strictly speaking, this should not be classified as a liquid scintillation process, as the sample material is not incorporated into liquid scintillant for counting. However, the method is a very useful one, and is included here for completeness. The solid scintillant used is usually pure crystalline anthracene, which gives reasonable counting efficiencies for aqueous solutions of ^{14}C, ^{35}S, and tritium in contact with it. (Higher counting efficiencies are, of course, obtained with beta emitters of higher energy.) Schram and Lombaert[91,92] and Elwyn[95] have described such cells and their application to the analysis of labeled amino acids in effluents from chromatography columns. Plastic scintillant that contains organic phosphors in a plastic medium, such as polyvinyltoluene, has also been used. The scintillant may be molded or machined to any desired shape, and is not affected by aqueous solutions, dilute acids, alkalis and salts, or the lower alcohols. Schram and Lombaert[93] have described a plastic scintillant cell in the form of a flattened tube between twin photomultiplier tubes. The system has good properties for the continuous counting of ^{14}C and ^{35}S in aqueous column effluents, although the efficiency is fairly low. Tkachuk[94] has described a packed column of finely divided plastic scintillant for the same purpose with much improved counting efficiencies.

SOME THEORETICAL ASPECTS OF LIQUID SCINTILLATION COUNTING

Determination of Counting Efficiency

The efficiency with which a radioactive sample is counted in a liquid scintillant usually varies from sample to sample. The quenching—that is, the decrease in counting efficiency—that occurs is produced by processes that interfere with the production of light in the liquid scintillant and its transmission to the photomultiplier tube of the liquid scintillation counter. Quenching may take two forms: chemical quenching, and color quenching. In chemical quenching, compounds in solution in the liquid scintillant interfere with the transfer of energy from the emitted particle or radiation to the organic phosphor, and the energy is degraded by processes that do not produce emission of light. In color quenching, colored materials in the liquid scintillant absorb light emitted by the organic phosphor and prevent it from being detected by the photomultiplier tube. Many materials, including dissolved oxygen, act as quenchers in the liquid scintillant, and it is rarely possible to predict the counting efficiency of a given sample. Some means of determining the counting efficiencies of samples is, therefore, essential to the liquid scintillation counting technique. Many methods have been evolved for this purpose, but nowadays only three main methods are used: internal standard, channels ratio, and external standard. A refinement of the external-standard method whereby the channels ratio of the gamma-generated counts rather than the gross count rate is determined is said to have several advantages. Better reproducibility due to increased influence of sample–gamma-source geometry and of sample volume is claimed. Peng[96] and Rapkin[97] have reviewed methods of determining counting efficiency. Bush[98] has described the general applicability and statistics of the channels-ratio method, and Rogers and Moran[99] have compared the internal-standard, channels-ratio and external-standard methods. The various methods will not be discussed at length here, as the subject is adequately covered in the literature. Instead, Table 10 summarizes the advantages and disadvantages of each method as far as they are reported in the literature.

Table 10. METHODS OF DETERMINING COUNTING EFFICIENCY

Method	Advantages	Disadvantages
Internal standard	Simple and reliable for singly labeled samples; with very careful handling, it is the most accurate method, especially for severely quenched samples; chemical and color quenching are corrected for.	Possible errors due to sample handling and measurement of small amounts of internal standards; sample cannot be recounted and is contaminated with activity; time-consuming, as two counts are needed.
Channels ratio	Only one count is needed; rapid, requires no manipulation of sample; composition of sample unaltered; accurate for moderate degrees of quenching and for moderate or high count rates; corrects for chemical and color quenching for tritium; independent of sample volume over a wide range; nonuniformly dispersed samples are acceptable.	Accuracy falls off for highly quenched samples; samples of low activity may need long counting times for adequate statistics to be gained; strong color quenching for ^{14}C not adequately corrected for; two counting channels are needed.
External standard	Faster than internal-standard method; composition of sample unaltered; avoids long counts on samples of low activity, which are necessary with the channels-ratio method.	Least accurate of the methods, particularly for highly quenched samples; variations in geometry between external source and sample and changes in sample volume introduce errors; two counts are needed; sample must be uniformly dispersed in counting vial.

Counting in Solution—Some Sources of Error

Once a sample is incorporated into homogeneous solution for counting, there are several possible sources of error that should be borne in mind. When, for example, an aqueous solution of a polar compound is mixed with liquid scintillant, the compound will sometimes precipitate or crystallize from solution. Because of the small chemical scale usually involved, it is difficult to predict whether a particular compound will come out of solution, or when, as this may be affected by slight temperature changes, by the presence of dust particles that may act as nuclei for precipitates or crystals, and by other factors. This problem has been noted at the Radiochemical Centre even with compounds present in sub-milligram quantities in 10 ml of liquid scintillant. Counting errors of 20 to 30 percent may arise, because the solid material is counted with a different efficiency from that remaining in solution, due to the self-absorption losses that occur in the solid particles. The process of precipitation or crystallization is often difficult to detect with small quantities of sample material, but a fall in counting rate and changes in the channels ratio (if this method of determining the counting efficiency is used) of the sample usually occur.

A further source of error in solution occurs with certain sample materials that become adsorbed onto the surface of the glass counting vial. This has been noted with fatty acids, benzoic acid, some polymeric materials, sodium phosphate, and calcium chloride. The problem is particularly acute with high-specific-activity material in a few parts per million chemical concentration, when the whole sample may become adsorbed onto the glass. In this case the counting geometry changes from 4π (homogeneous solution) to essentially 2π (flat surface), and counting losses of up to 50 percent may occur, which are not accounted for in the counting efficiency determination. (The worker, of course, also has the problem of adsorption onto pipets, graduated flasks, etc., which further complicates measurement.) The problem has been solved in several ways. Carrier sample material may first be added to the counting vial, thereby filling the adsorption sites. The counting vial may be siliconized by treatment with dimethyl dichlorosilane, or polyethylene counting vials may be used instead of glass. These last two methods have been found to be effective in preventing adsorption of ^{32}P-labeled sodium phosphate and ^{45}Ca-labeled calcium chloride.[100] Sub-micron silica (Cab-O-Sil[69]) has also been used, and in this case the silica is mixed with the liquid scintillant before the active sample is added. If adsorption occurs, 99 percent of it takes place onto the suspended silica particles; 4π geometry is retained, since the silica particles are so small that negligible self-absorption of ^{14}C and ^{35}S beta particles occurs.[101] The silica does not affect the counting efficiency of materials that are not adsorbed onto it or the glass of the counting vial.

A source of error when counting homogeneous samples obtained by the oxygen flask combustion method has been mentioned by Conway and Grace[102] and by Baggett et al.[103] An internal standard was used to determine counting efficiency. Since the samples were saturated with oxygen after combustion and trapping process, some oxygen escaped when they were opened for the introduction of internal standard. Dissolved oxygen quenches strongly, and since the samples had different oxygen contents when the internal standard was counted, this yielded a counting efficiency that was incorrect for the original samples. Use of the channels-ratio or external-standard methods of determining counting efficiency is satisfactory in this case, since the sample is not opened. Variations in oxygen quenching are not generally a large source of error in day-to-day measurements using the internal-standard method of samples in equilibrium with air. However, it should be appreciated by the worker that any manipulation of the sample increases the possibility of error.

Some other sources of error in counting samples in solution have been summarized in Table 10, and are mentioned by several of the authors listed in the references.[96-99]

Suspension Counting

As mentioned earlier, the suspension method has been widely used for the liquid scintillation counting of insoluble materials, particularly ^{14}C-labeled barium carbonate. The technique is useful and accurate if certain limitations are known and taken into account. The most important of these is that the direct determination of counting efficiency is impossible, because some of the emitted radiation is lost without interacting with the liquid scintillant, due to self-absorption in the suspended solid particles. The result of this self-absorption loss is that results obtained by using the suspension counting technique may only be compared directly with those obtained by counting homogeneous solutions when the fraction of counts lost through self-absorption is determined in some way. The counting efficiency of the radiation that does interact with the liquid scintillant may be determined by any of the methods used for homogeneous-solution counting. For tritium the self-absorption losses are so large that the technique loses much of its advantage. For ^{14}C and ^{35}S the losses, while not large, are appreciable and have to be taken into account. For the more energetic beta emitters such losses are probably very small for fine suspensions and may be neglected. (It may be noted that, although no reports of suspension counting of alpha emitters have appeared in the literature, the very short range of alpha particles in solid material would indicate that the technique is of little value in this case.)

Practical methods of determining self-absorption losses usually involve the suspension counting of a material of known specific activity and the estimation of the losses by difference.[70] Unfortunately such losses are not reproducible from sample to sample (except in certain cases, such as ^{14}C-labeled barium carbonate, discussed below), unless precautions are taken to ensure that particle size or size distribution remain more or less constant. This may involve grinding and sieving procedures, which are laborious and time-consuming and not suited to the manipulation of large numbers of samples. For crystalline materials it is found, as might be expected, that the larger the particle size, the greater the self-absorption losses. Data from work at the Radiochemical Centre illustrating this point for crystalline ^{14}C-labeled D-glucose (U) are given in Table 11.

Table 11.
SELF-ABSORPTION LOSSES IN THE SUSPENSION COUNTING OF CRYSTALLINE ^{14}C-LABELLED D-GLUCOSE (U)*

Sample Preparation Method	Specific Activity Found, μCi/mM (± S.D.)	Ratio (f) of Suspension Counting / Solution Counting	% Self-Absorption, Loss = $(1-f) \times 100$
Crystals as suspensions	0.96 ± 0.04	0.55 ± 0.02	45 ± 2
Coarse powder as suspension	1.27 ± 0.03	0.72 ± 0.02	28 ± 2
Fine powder as suspension	1.49 ± 0.09	0.84 ± 0.05	16 ± 5
In homogeneous solution†	1.77 ± 0.03	(1.00)	(0)

* Liquid scintillant solvent used for the suspensions was pure toluene, to ensure that negligible amounts of the samples dissolved in the scintillant.
† Aqueous solution of ^{14}C-labeled D-glucose (U) mixed with ethanol/dioxane/toluene scintillant.

The suspension counting method is, therefore, not accurate for crystalline materials, unless the worker is prepared to take the precautions outlined above. In the case of ^{14}C-labeled barium carbonate (and possibly for other amorphous materials, although this has not been tested in the laboratory) the gross particle size of the suspension does not affect the results obtained. Of course, self-absorption losses are still present, and have to be taken into account, but these losses are constant from sample to sample. Indeed, it has been observed that flakes of material 2 mm in diameter and the finest powder that could be obtained by grinding had the same self-absorption losses. It is thought that the aggregates of ^{14}C-labeled barium carbonate are thoroughly penetrated by liquid scintillant and are essentially transparent to the emitted scintillations. They may be composed of small units of regular size, independent of the method of sample preparation. Grinding of the samples to ensure fine or regular particle size is, therefore, unnecessary. The channels-ratio method of determining counting efficiency was found to be quite applicable to the gel suspensions prepared in this work.

Emulsion Counting

Some characteristics of the emulsion counting system toluene/Triton X-100 (2:1 v/v) of Patterson and Greene[16] have been described elsewhere in this chapter, and the point was made that this system is not a simple water-in-oil emulsion at all concentrations of aqueous sample. The counting characteristics of this system have been studied by Turner at the Radiochemical Centre for ^{14}C and tritium aqueous samples, and the results of this work are described below.

A series of samples of ^{14}C-labeled D-glucose (U) in aqueous solution were mixed with the scintillant to give counting samples containing between 0.2 and 3 g of aqueous phase per 10 ml of scintillant. The samples were counted, and the radioactive concentration in μ Ci/g of the ^{14}C-labeled D-glucose (U) was calculated. These results were compared as a percentage value with those obtained by counting in homogeneous solution, and a graph of percentage value versus weight of aqueous phase was plotted (Figure 5, ^{14}C). A series of samples of tritiated water were treated in the same way as the ^{14}C samples described above, and a similar graph of percentage value of result from emulsions, as compared with counting in homogeneous solution, versus weight of aqueous phase was plotted (Figure 5, tritium).

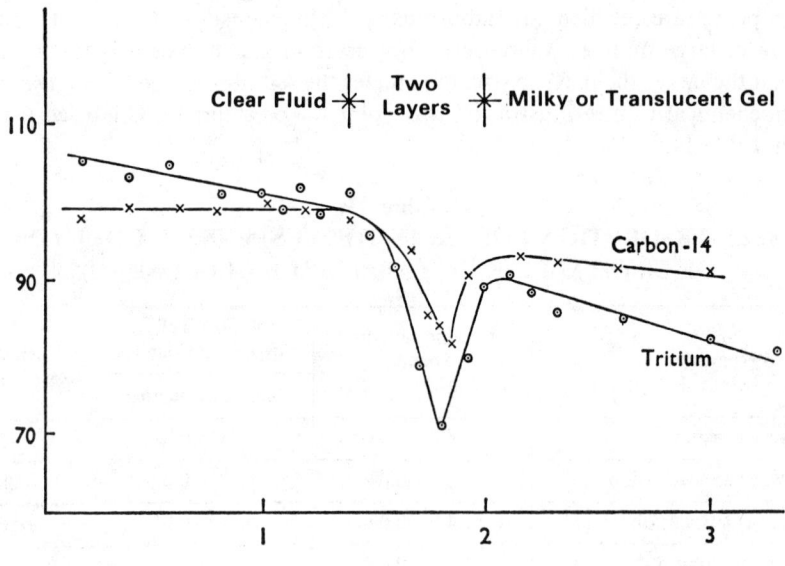

FIGURE 5.
Emulsion Counting ^{14}C-Labeled D-Glucose (U) and Tritiated Water,
Using Toluene/Triton X-100 (2 : 1 v/v).

The results obtained and their interpretation, together with recommended conditions of counting, are conveniently summarized in Table 12. It will be seen that, with certain qualifications, the emulsion method of counting is excellent for aqueous ^{14}C samples. In addition to this work, the application of this system to the counting of a wide variety of ^{14}C-labeled compounds has been investigated at the Radiochemical Centre, and the results have been summarized on preceding pages in this chapter. Accurate reproducible results are obtained, and their comparison with the results obtained from counting in homogeneous solution is direct and reliable. The system is also useful for counting large quantities (2 to 3 g) of aqueous ^{14}C sample, although less accurate. Unfortunately, for aqueous tritium samples the system seems capable of giving only approximate results, and is not recommended for accurate work.

Table 12.
CHARACTERISTICS OF THE TOLUENE/TRITON X–100 (2:1) SCINTILLANT FOR EMULSION COUNTING*

Characteristics	Weight of Aqueous Sample per 10 ml of Scintillant†					
	0–1.4 g		1.4–2.0 g		2.0–3.0 g	
	^{14}C	^{3}H	^{14}C	^{3}H	^{14}C	^{3}H
Percent value of result, compared with counting in homogeneous solution	100	105→100 (variable values)	100→82→93	100→71→90	93→91	90→82
Explanation of results	Counting sample is in homogeneous solution.		Fall of results due to self-absorption in large droplets of unstable collapsing emulsion; increase of results due to increasing viscosity of sample mixture, which hinders coalescence of emulsion droplets.		Decline of values caused by increasing percentage of aqueous sample that is in emulsion form (and has self-absorption), as compared with constant amount in solution.	
Appearance and physical state of sample mixture	Clear fluid, stable for extended periods.		Initially cloudy fluid, separating into clear fluid upper layer and opaque viscous lower layer within 1 hour.		Milky or translucent viscous gels, stable for at least 24 hours.	
Microscopic appearance	One phase (droplets of 0.1 μ diameter would have been clearly visible).		Not examined.	Not examined.	Many extremely fine droplets dispersed in a continuous phase.	
Distribution of activity in sample mixture	Homogeneous.	Homogeneous.	Clear upper layer 35%; opaque lower layer 65%.‡		Homogeneous in the continuous phase + droplets of aqueous phase more or less uniformly dispersed.	
Sample reproducibility (i.e., constancy of result with varying weight of aqueous sample)	Excellent.	Poor.	Poor.	Poor.	Moderate.	Poor.
Type of counting work possible with this system	Accurate.	Approximate (with precautions to minimize sample weight variations).	None—system too variable.		Approximate (with precautions to minimize weight variations).	
Comparison of results with those obtained in homogeneous solution	Direct, accurate.	Not advisable; absolute reproducibility poor.	None—system too variable.		Indirectly; self-absorption loss must first be calculated.	Not advisable; absolute reproducibility poor.

* Method of sample preparation: aqueous sample mixed with scintillant, vigorously hand-shaken for a few seconds, allowed to cool to counter temperature, reshaken, and counted.

† These values increased by 0.15 g of aqueous sample on increasing the temperature from 4°C to 20°C.

‡ These percentages refer to the radioactive concentration and are therefore independent of volumes. The opaque lower layer is, thus, richer in water. The same results were obtained with both ^{14}C and ^{3}H samples.

REFERENCES

1. Ayrey, G. and Mazza, R. J. *J. Labelled Compounds*, 3(1): 24–37, 1967.
2. Hayes, F. N. *Int. J. Appl. Radiat. Isotop.* 1: 46–56, 1956.
3. Davidson, J. D. and Feigelson, P. *Int. J. Appl. Radiat. Isotop.*, 2: 1–18, 1957.
4. Rapkin, E. *Int. J. Appl. Radiat. Isotop.*, 15: 69–87, 1964.
5. Bell, C. G. and Hayes, F. N., eds., *Liquid Scintillation Counting. Proceedings of a Conference Held at Northwestern University, August 20–22, 1957*. Pergamon Press, Oxford, England, 1958.
6. International Atomic Energy Agency, *Tritium in the Physical and Biological Sciences*, Vols. 1 and 2. International Atomic Energy Agency, Vienna, Austria, 1962.
7. Schram, E., *Organic Scintillation Detectors. Counting of Low-Energy Beta Emitters*. Elsevier, Amsterdam, Holland, 1963.
8a. Rothchild, S., ed., *Advances in Tracer Methodology*, Vol. 1. Plenum Press, New York, 1963.
8b. Rothchild, S., ed., *Advances in Tracer Methodology*, Vol. 2. Plenum Press, New York, 1965.
8c. Rothchild, S., ed., *Advances in Tracer Methodology*, Vol. 3. Plenum Press, New York, 1966.
9. Birks, J. B., *The Theory and Practice of Scintillation Counting*. Pergamon Press, Oxford, England, 1964.
10. Tanielian, C., Coche, A., Deluzarche, A., Laustriat, G., and Maillard, A. *Int. J. Appl. Radiat. Isotop.*, 15: 11–15 and 17–23, 1964.
11. Scales, B. *Int. J. Appl. Radiat. Isotop.*, 18: 1–6, 1967.
12. Hall, T. C. and Weiser, C. J. *Anal. Biochem.*, 17: 294–299, 1966.
13. Baxter, J. A., Fanning, L. E., and Swartz, H. A. *Int. J. Appl. Radiat. Isotop.*, 15: 415–418, 1964.
14. Gray, G. A. *Anal. Biochem.*, 1: 279–285, 1960.
15. Francis, G. E. and Hawkins, J. D. *Int. J. Appl. Radiat. Isotop.*, 18: 223–230, 1967.
16. Patterson, M. S. and Greene, R. C. *Anal. Chem.*, 37: 854–857, 1965.
17. Belcher, E. H. *Proc. Roy. Soc., Ser. A (England)*, 216: 90–102, 1953.
18. Braunsberg, H. and Guyver, A. *Anal. Biochem.*, 10: 86–95, 1965.
19. De Volpi, A. and Porges, K. G. A. *Int. J. Appl. Radiat. Isotop.*, 16: 496 and 498, 1965.
20. Haberer, K. and Kölle, W. *Atompraxis*, 11: 664–665, 1965.
21. Carr, T. E. F. and Parsons, B. J. *Int. J. Appl. Radiat. Isotop.*, 13: 57–62, 1962.
22. Sarnat, N. and Jeffay, M. *Anal. Chem.*, 34: 643–646, 1962.
23. Dyer, A., Fawcett, J. M., and Potts, D. U. *Int. J. Appl. Radiat. Isotop.*, 15: 377–380, 1964.
24. Rhodes, B. A. *Anal. Chem.*, 37: 995–997, 1965.
25. Ellis, M. K., Wampler, S. N., and Yager, R. H. *Anal. Chim. Acta*, 34: 169–174, 1966.
26. Radin, N. S. and Fried, R. *Anal. Chem.*, 30: 1926–1928, 1958.
27. Hardcastle, J. E., Hannapel, R. J., and Fuller, W. H. *Int. J. Appl. Radiat. Isotop.*, 18: 193–199, 1967.
28. Humphreys, E. R. *Int. J. Appl. Radiat. Isotop.*, 16: 345–348, 1965.
29. Ihle, H. R., Karayannis, M., and Murrenhoff, A. P., *Radioisotope Sample Measurement Techniques in Medicine and Biology*, pp. 485–503. International Atomic Energy Agency, Vienna, Austria, 1965.
30. Ludwick, J. D. *Anal. Chem.*, 36: 1104–1106, 1964.
31. Axtmann, R. C. and Cathey, L. *Int. J. Appl. Radiat. Isotop.*, 4: 261, 1959.
32. Ludwick, J. D. *Anal. Chem.*, 32: 607–610, 1960.
33. Horrocks, D. L. and Studier, M. H. *Anal. Chem.*, 30: 1747–1750, 1958.
34. Ronzio, A. R. *Int. J. Appl. Radiat. Isotop.*, 4: 196–200, 1959.
35. Uyesugi, G. S. and Greenberg, A. E. *Int. J. Appl. Radiat. Isotop.*, 16: 581–587, 1965.
36. Lutwak, L. *Anal. Chem.*, 31: 340–343, 1959.
37. Flynn, K. F. and Glendenin, L. E. *Phys. Rev.*, 116: 744–748, 1959.
38. Hansen, D. L. and Bush, E. T. *Anal. Biochem.*, 18: 320–332, 1967.
39. Neujahr, H. Y. and Ewaldsson, B. *Anal. Biochem.*, 8: 487–494, 1964.
40. Brown, W. O. and Badman, H. G. *Biochem. J.*, 78: 571–578, 1961.
41. Petroff, C. P., Patt, H. H., and Nair, P. P. *Int. J. Appl. Radiat. Isotop.*, 16: 599–601, 1965.
42. Vaughan, M., Steinberg, D., and Logan, J. *Science*, 126: 446–447, 1957.
43. Hattori, T., Aoki, H., Matsuzaki, I., Maruo, B., and Takahashi, H. *Anal. Chem.*, 37: 159–161, 1965.
44. Moorhead, J. F. and McFarland, W. *Nature*, 211: 1157–1159, 1966.
45. Bartley, J. C. and Abraham, S. *Atomlight*, No. 49: 1–8, 1965.
46. Mahin, D. T. and Lofberg, R. T. *Anal. Biochem.*, 16: 500–509, 1966.
47. Bloom, P. M. and Nelp, W. B. *J. Lab. Clin. Med.*, 65: 1030–1033, 1965.
48. Kumar, M. A. *Int. J. Appl. Radiat. Isotop.*, 17: 556–557, 1966.
49. Ragland, J. B. *Nucleus (Chicago)*, No. 20: 2–11, 1966.
50. Aures, D. and Clark, W. G. *Anal. Biochem.*, 9: 35–47, 1964.
51. Moss, G. *Int. J. Appl. Radiat. Isotop.*, 11: 47–48, 1961.
52. Oppermann, R. A., Nystrom, R. F., Nelson, W. O., and Brown, R. E. *Int. J. Appl. Radiat. Isotop.*, 7: 38–42, 1959.

53. Snyder, F. and Godfrey, P. *J. Lip. Res.*, 2: 195, 1961.
54. Woeller, F. H. *Anal. Biochem.*, 2: 508–511, 1961.
55. Edwards, B. and Kitchener, J. A. *Int. J. Appl. Radiat. Isotop.*, 16: 445–446, 1965.
56. Baggiolini, M. *Experientia (Switzerland)*, 21: 731–733, 1965.
57. Yardley, H. J. *Nature (England)*, 204: 281, 1964.
58. Kornblatt, J. A., Bernath, P., and Katz, J. *Int. J. Appl. Radiat. Isotop.*, 15: 191–194, 1964.
59. Gupta, G. N. *Anal. Chem.*, 38: 1356–1359, 1966.
60. Dobbs, H. E. *Anal. Chem.*, 35: 783–786, 1963.
61. Kalberer, F. and Rutschmann, J. *Helv. Chim. Acta*, 44: 1956–1966, 1961.
62. Herberg, R. J. *Anal. Chem.*, 32: 42–46, 1960.
63. Shneour, E. A., Aronoff, S., and Kish, M. R. *Int. J. Appl. Radiat. Isotop.*, 13: 623–627, 1962.
64. Walter, W. M. and Purcell, A. E. *Anal. Biochem.*, 16: 466–473, 1966.
65. Fales, H. M. *Atomlight*, No. 25: 8, 1963.
66. Baillie, L. A. *Int. J. Appl. Radiat. Isotop.*, 8: 1–7, 1960.
67. Bruno, G. A. and Christian, J. E. *Anal. Chem.*, 33: 650–651, 1961.
68. Hayes, F. N., Rogers, B. S., and Langham, W. H. *Nucleonics*, 14(3): 48–51, 1956.
69. Ott, D. G., Richmond, C. R., Trujillo, T. T., and Foreman, H. *Nucleonics*, 17(9): 106–108, 1959.
70. Cluley, H. J. *Analyst*, 87: 170–177, 1962.
71. Eakins, J. D. and Brown, D. A. *Int. J. Appl. Radiat. Isotop.*, 17: 391–397, 1966.
72. Graber, S., McKee, L. C., and Heyssel, R. M. *J. Lab. Clin. Med.*, 69: 170–176, 1967.
73. Yarbrough, J. D., Findeis, A. F., and O'Kelley, J. C. *Int. J. Appl. Radiat. Isotop.*, 17: 453–458, 1966.
74. Harlan, J. W., in *Advances in Tracer Methodology*, 1: 115–118, Rothchild, S., ed. Plenum Press, New York, 1963.
75. Germai, G. *Bull. Soc. Roy. Sci. Liège (Belgium)*, 33: 672–677, 1964.
76. Davidson, J. D., *United States Atomic Energy Commission Report TID-7612*, pp. 232–238, 1961.
77. Takahashi, H., Hattori, T., and Maruo, B. *Anal. Biochem.*, 2: 447–462, 1961.
78. Baxter, C. F. and Senoner, I. *Anal. Biochem.*, 7: 55–61, 1964.
79. Snyder, F. *Anal. Biochem.*, 9: 183–196, 1964.
80. Blair, A. and Segal, S. *Anal. Biochem.*, 3: 221–229, 1962.
81. Davies, J. W. and Cocking, E. C. *Biochim. Biophys. Acta*, 115: 511–513, 1966.
82. Hutchinson, F. *Int. J. Appl. Radiat. Isotop.*, 18: 136–137, 1967.
83. Benson, R. H. *Anal. Chem.*, 38: 1353–1356, 1966.
84. van der Laarse, J. D. *Int. J. Appl. Radiat. Isotop.*, 18: 485–491, 1967.
85. Horrocks, D. L. and Studier, M. H. *Anal. Chem.*, 36: 2077–2079, 1964.
86. Curtis, M. L., Ness, S. L., and Bentz, L. L. *Anal. Chem.*, 38: 636–637, 1966.
87. Gordon, B. E., Lukens, H. R., and Ten Hove, W. *Int. J. Appl. Radiat. Isotop.*, 12: 145–146, 1961.
88. Karmen, A. and Tritch, H. R. *Nature (England)*, 186: 150–151, 1960.
89. Hajra, A. K. and Radin, N. S. *J. Lip. Res.*, 3: 131–134, 1962.
90. Karmen, A., Giuffrida, L., and Bowman, R. L. *J. Lip. Res.*, 3: 44–52, 1962.
91. Schram, E. and Lombaert, R. *Arch. Int. Physiol. Biochim.* 68: 845, 1960.
92. Schram, E. and Lombaert, R. *Anal. Biochem.*, 3: 68–74, 1962.
93. Schram, E. and Lombaert, R. *Biochem. J.*, 66: 20P, 1957.
94. Tkachuk, R. *Can. J. Chem.*, 40: 2348–2356, 1962.
95. Elwyn, D. H., in *Advances in Tracer Methodology*, 2: 115–122, Rothchild, S., ed. Plenum Press, New York, 1965.
96. Peng, C. T., in *Advances in Tracer Methodology*, 3: 81–94, Rothchild, S., ed. Plenum Press, New York, 1966.
97. Rapkin, E. *Picker Lab. Scintillator*, 11(49): 1–12, 1966.
98. Bush, E. T. *Anal. Chem.*, 35: 1024–1029, 1963.
99. Rogers, A. W. and Moran, J. F. *Anal. Biochem.*, 16: 206–219, 1966.
100. Petroff, C. P., Nair, P. P., and Turner, D. A. *Int. J. Appl. Radiat. Isotop.*, 15: 491–494, 1964.
101. Blanchard, E. A. and Takahashi, I. T. *Anal. Chem.*, 33: 975–976, 1961.
102. Conway, W. D. and Grace, A. J. *Anal. Biochem.*, 9: 487–489, 1964.
103. Baggett, B., Presson, T. L., Presson, J. B., and Coffey, J. C. *Anal. Biochem.*, 10: 367–370, 1965.

THE STABILITY OF LABELED ORGANIC COMPOUNDS

Reprinted by permission of
The Radiochemical Centre, Amersham, Bucks, England

INTRODUCTION

This chapter is written not for the student of radiation chemistry, but for the user of radioactive tracer compounds whose interest in radioactivity is probably limited to using it in tracer experiments as a means of detecting and measuring particular components in a system. Other effects of radiation will, for him, be useless and unwanted; one such unwanted effect is the degradation of the radioactive tracer compounds he wishes to use.

The radiation chemistry of organic compounds is now a very popular subject of study, prompted by the use of organic structural materials, solvents and coolants to be used in intense radiation fields, by the possible technological uses of radiation, by the basic importance of the subject to radiobiology, and (to a much lesser degree) by the need of the tracer worker to know more about the stability of his reagents.

The user of radioactive tracer compounds will not, however, find all his questions answered in the voluminous literature of radiation chemistry. Theoretical studies concentrate on a limited number of relatively simple compounds under ideal or model conditions (for example, at extremely high chemical purity); empirical studies have mostly been with substances of special interest in reactor or technological development, such as solvents, coolants, and polymers. Most work is done with X or gamma rays, some with alpha or energetic beta particles, little with the common soft-beta emitters used so widely in tracer work. This is important because of the difference in effect between different types of radiation, independent of the energy absorbed in the sample.

Tracer workers, by contrast, are interested in an extraordinary variety of compounds, most of them too complex to encourage theoretical study at present. These compounds are prepared under conditions that make it impossible to ensure freedom from small amounts of chemical impurities that can greatly affect the stability; they are used in minute quantities that introduce the complicating factor of purely chemical decomposition, which may be independent of radioactivity.

The approach adopted here is, therefore, frankly empirical. Wherever possible, it uses experience of the actual storage of radioactive compounds, mostly at the Radiochemical Centre, and it attempts to answer the kind of question that is known to arise in using tracer compounds. Finally, the whole subject is approached from the point of view of the representative user (or, indeed, preparer) of labeled compounds, regarding self-decomposition as a troublesome phenomenon, to be studied only in order to reduce its magnitude.

DEFINITIONS

For newcomers to the subject, definitions of some of the terms in this chapter may be helpful.

Chemical purity. The proportion of a material in a specified chemical form, regardless of any isotopic substitution. It is commonly expressed as a percentage and by weight.

Radiochemical purity. The proportion of the radioisotope that is present in the specified chemical form. It may be necessary to specify the molecular form in detail of configuration (for example, optical isomerism) and position of labeling.

Note that this differs from the older use of the term, to define the proportion of the radioactivity that is in the form of the specified radionuclide. For this concept, which is not involved in this discussion, the term *radioisotopic purity* is preferred.

Radiochemical purity is most often determined by paper or other chromatographic methods, and by reverse isotope dilution analysis. A rather more extended discussion may be found in the "Radiochemical Manual" of the Radiochemical Centre.[1]

Radiation yield. This is conventionally expressed as a G- value: for example,

$G(-H_2)$ = the number of molecules of hydrogen formed per 100 ev of energy absorbed.

$G(-M)$ = the number of molecules permanently transformed per 100 ev of energy absorbed.

Specific activity. For pure compounds, this is conveniently expressed as curies/mole, or decimal values of those units.

Radioactive concentration. This term is preferred for solutions of various kinds, and is usually expressed in curies/milliliter, curies/gram, or decimal values.

RADIATION DECOMPOSITION

When tracer compounds first came into use in the 1940's, specific activities were rather low, and decomposition was correspondingly less rapid. It was not until 1953 that the first general observations were published by Tolbert et al.[2] This publication aroused much interest at the time.

Table 13 presents, in concise form, decomposition observed in various radioactive compounds, reported in Tolbert's paper and elsewhere. The reader should compare the decomposition observed with the specific activities and the times and conditions of storage.

Table 13. DECOMPOSITION OF RADIOACTIVE COMPOUNDS

Compound	Specfic Activity, mCi/mM	Age, days	Storage	Decomposition	Reference
Cholesterol-4-^{14}C	2.5	540	Solid in air	40%	2
Choline-chloride methyl-^{14}C	1.8	270	Solid in air	63%	2
Dextran-^{14}C sulfate (20 glucose units per molecule)	3	21	Solid in air	100%	3
L-Methionine-^{35}S	100	60	Solid in air (dry)	20%	4
L-Phenylalanine-^{14}C(U)	304	105	In 0.01N HCl	14%	4
Succinic acid-2,3-T	58,000	30	Solid	100%	4
9,10-Dimethylbenz-anthracene-9-^{14}C	9	30	Benzene solution	20%	4
9,10-Dimethylbenz-anthracene-T(G)	3,250	390	Benzene solution	37%	4

THE UNDERLYING FACTORS

A summary of the protective measures that can be taken would be misleading at this point, unless the underlying factors are recognized.[3]

Primary (Internal) Effect

When a radioactive atom disintegrates, it leaves the wreckage of the molecule behind as a "recoil fragment". If the individual molecule contains two or more radioactive atoms, this process will create radioactive recoil fragments that will, in fact, be radioactive impurities.

The magnitude of this effect is nearly (but not quite) always negligible; the reader may be referred to some examples of much theoretical interest.[6,7] It is fortunate that the effect is usually small, as it cannot possibly be controlled. With macromolecules and relatively short-lived isotopes the impurities produced may be significant,[3] and it is well to be aware of the effect.

Primary (External) Effect

This is defined as the transformation of a molecule, brought about by interaction with a nuclear particle. If the molecule affected is labeled, then a radioactive impurity will be produced, and the effect is therefore observed with singly labeled as well as with multiple-labeled molecules.

It is rather more serious in magnitude than the primary (internal) effect; one beta particle from ^{14}C may destroy up to 5,000 molecules. Fortunately, some degree of control is possible. If the nuclear particle can be intercepted and its energy harmlessly dissipated, or if it can be allowed to escape without interaction with a labeled molecule, the labeled-molecule "target" will be protected from damage. The stability of the compound depends, in fact, on its environment and physical form.

The simplest protective measure, in theory, is to spread the compound in a layer so thin that virtually all the particles escape. This has been done with very good results with vitamin B_{12}, for which the regular production ampules contain 1 µg or less in an area of about 0.1 cm^2; it has also been successful with ^{14}C-labeled chlorophyll at almost 300 mCi/mM, stored for some years. But with soft-beta emitters the method can be very inconvenient because of the large areas required. For tritium it is of no use. Supports such as glass fiber, powdered silica or charcoal, which provide large surface areas, may easily do more harm (because of their chemical reactivity or powerful adsorption) than good. Even for milligram quantities of organic compounds the containers should be perfectly clean, neutral, and chemically inert.

It is usually more practical to use a diluent rather than a thin layer. One may add "carrier" (pure unlabeled compound), thus lowering the specific activity (see Figure 6), or a different substance, chosen to make it easier to separate the labeled compound again when required. In the common case, when isotopic abundances are less than 100 percent, this last situation is illustrated more accurately by Figure 7.

For many uses a reduction of specific activity is not acceptable, but it should not be dismissed without thought. Quite low specific activities are adequate for some purposes, and they have advantages in easy preparation and purification. Primary (external) radiation damage is proportional to the molar specific activity.

The use of a different substance as solvent or diluent is more generally applicable, and has the advantage that it may be selected for convenience in use and stability to radiation. This subject is developed more fully below. It must be noted here that the radiation damages labeled and unlabeled molecules indiscriminately. Damage to labeled molecules is more important, because it produces labeled (radiochemical) impurities. Unlabeled (chemical) impurities are usually of no importance in themselves, but only insofar as they bring about further chemical reactions or the secondary radiation effects discussed below.

Secondary Effects

These may be defined as transformations of a molecule of labeled compound, brought about by chemically reactive species produced by the primary radiation effect; they account for most of the damage in storage of radioactive compounds.

Much may be done to reduce this damage. It will be proportional to the fraction of the energy absorbed in the sample, which (for soft-beta emitters) is approximately proportional to the radioactive concentration, but depends also on the shape (geometry) of the sample. It depends even more on the character of the molecules that surround the labeled molecule. As with the primary (external) effect, a diluent may be used to shield the labeled molecules. The possible effect of small amounts of chemical impurities must not be forgotten. In fact, Figure 8

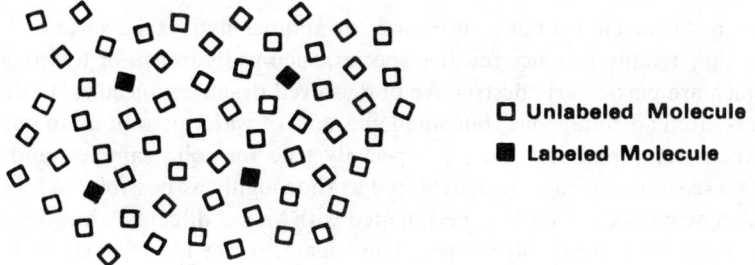

FIGURE 6. Reduction of Specific Activity by Addition of Pure Unlabeled Compound.

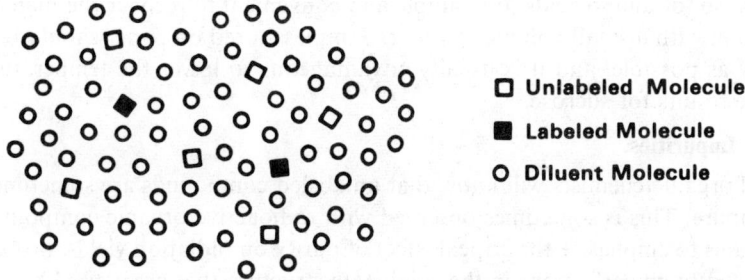

FIGURE 7. Reduction of Specific Activity by Addition of Diluent.

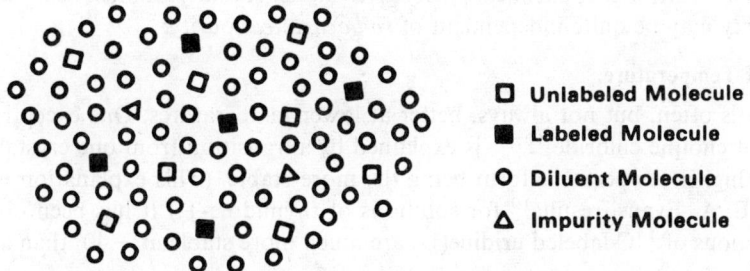

FIGURE 8. Reduction of Specific Activity by Dilution and Chemical Impurity.

is a more complete diagram of a radioactive molecule in a solution intended to protect it from self-irradiation.

Secondary radiation effects, being purely chemical, are temperature-dependent, and simple theory would predict that they will be reduced as the temperature is lowered, disappearing at absolute zero. This is true in a general sense, but there are exceptions that arise from other effects of temperature in the environment.

Another possible type of protective measure is use of a "scavenger" compound to react preferentially with the harmful radiation products. The most obvious are free radical scavengers.

Let us now consider some of these points in more detail.

Choice of Diluent.

Aromatic compounds such as benzene can dissipate absorbed radiation energy without transferring it to solute molecules.[8] Benzene is, therefore, the first solvent to be considered. It is chemically inert, readily obtained in a highly pure state, and easy to separate from a non-volatile dissolved compound, when required, by evaporation. Unluckily, many compounds are insoluble in benzene. A proportion of a secondary solvent, such as alcohol or dioxane, may then be used without too much disadvantage.

Often one has little choice but to use water or another hydroxylic solvent. These are not ideal, because they readily produce reactive species, such as hydrogen or hydroxyl radicals or peroxides, which are particularly destructive of dissolved organic molecules. In dilute solution the net result is often advantageous; but small amounts of water present as an impurity nearly always accelerate decomposition. This is especially true for solid samples, and it is recommended that these should always be dehydrated as thoroughly as possible.

The Radiochemical Centre has experimented with "solid diluents", but the only one that has established practical use is filter paper. This, again, is not ideal; Bayly and Weigel have shown[3] that the radiation yield (G value) for sugars is actually higher in this form than as anhydrous solids. The gain from dilution is, however, enough to compensate for this effect, and the net result is an improvement in stability. Paper storage has proved especially successful for sugars, less so for amino acids. It is simple and convenient to recover the material from the paper by elution with a small volume of water. Samples stored on paper should be dehydrated as completely as possible, and it is usually advantageous to lower the temperature. Table 14 presents some results for sucrose.

Effect of Impurities.

Practical organic chemists will know that unlabeled compounds are sometimes less stable if they are impure. This is sometimes observed with radioactive organic compounds. Tolbert[9] has been at pains to emphasize the critical effect of purity on radiation yields, probably because they are affected by imperfections in the solid-state structure that are caused by impurities.

If an impurity acts as a "scavenger" of reactive species, it may be beneficial; but this is exceptional. As a rule, higher chemical purity gives higher stability. It must be remembered that *chemical* purity may be quite independent of *radiochemical* purity.

Effect of Temperature.

Stability is often, but not always, better at lower temperatures. One exception, the well-known one of choline chloride,[10,11] is explained by a transition from one crystalline form to another, the higher-temperature form being the more stable. (This explanation is unlikely to account for E. A. Evans' results[12] for solutions of thymidine-T.) It has been observed[4] that aqueous solutions of ^{14}C-labeled uridine(U) are much more stable at $-40°$ than at $0°$.

Methods of Handling and Sampling.

One reason why freezing may be less effective in checking secondary reactions than one would expect has been advanced; long-lived radicals or reactive chemical species may be frozen temporarily, but are liberated when the solution is thawed, and have then a destructive effect.

It is not good practice to freeze and thaw a solution repeatedly. For experiments requiring small successive quantities it is better to store small amounts, so that all of the material is continuously frozen for its entire storage life. Another practical difficulty observed is sublimation of the solvent. If there is a small temperature difference between different parts of the ampule, solvent will slowly condense in the coldest part. This can leave the solute as a solid residue. In freezing aqueous solutions (in particular), care must be taken to avoid fractures of ampules from expansion.

Use of Free Radical Scavengers.

The more commonly used free radical scavengers have not been examined very much at the Radiochemical Centre; the few experiments made have not been very successful. There is a fundamental difficulty in that users of tracer compounds dislike the introduction of foreign chemical compounds into their experiments.

Some evidence is accumulating that simple primary alcohols, such as ethyl alcohol, are quite effective in low concentrations.[4] These are less objectionable than, for example, diphenyl-picrylhydrazyl or iodine and seem likely to be used increasingly to improve the stability of aqueous solutions.

Table 14. COMPARISON OF EFFECTS OF DIFFERENT STORAGE CONDITIONS

Compound	Tube No.	Storage Conditions			Impurity Observed, %	Initial Decomposition per Year per mCi/mM, %	Average Mass per Unit Area (freeze dried), mg/cm^2	Area of Paper, cm^2	Energy Absorbed by Sugar or Sugar Syrup (F × 100), %	G(−M) Value
		Form	Temp.	Pressure						
Sucrose	1	Freeze-dried	Room	Vac.	16.4	0.071	0.35	—	16	4.6
	2	Freeze-dried	−80°C	Vac.	15.1	0.065	0.35	—	16	4.2
	3	On paper	Room	Atm.	15.7	0.068	—	21	0.3	234
	4	On paper	−80°C	Atm.	4.9	0.020	—	21	0.3	68
	5	On paper	Room	Vac.	2.4	0.010	—	21	0.3	33
	6	On paper	−80°C	Vac.	1.8	0.007	—	21	0.3	25

Chain Reactions.

G(–M) values for organic compounds vary with their chemical character and environment, and with the type and energy of radiation; some examples are given in Table 15, and readers may be referred to a recent review by Wyant.[25] For self-irradiation of ^{14}C and tritium compounds, G(–M) values of 1 to 20 are fairly typical, but sometimes a chain reaction is initiated, and G(–M) values of hundreds—or even thousands—are found. Choline chloride is a well-known example of such high radiation sensitivity,[10,11] and vinyl monomers show similar values. For this reason it has not been possible to prepare and distribute ^{14}C-labeled acrylonitrile at useful specific activities; at specific activities of 30 to 50 mCi/mM, ^{14}C-labeled acetylene polymerizes at a steady rate to give a cuprene-like solid. An extreme example[5] has already been quoted in Table 13.

Table 15.

RELATION OF G(–M) VALUES TO TYPE AND ENERGY OF RADIATION

Compound	G(–M)	Type of Radiation	Reference
Methane	8.6	6 Mev α	15
Saturated hydrocarbons	4–9	—	16
Ethylene	20	—	16
Methanol	5.8	27 Mev α	17, 18, 19
Alcohols	3–12	—	16
Cholesterol	3.5	2–4 Mv e$^-$	20
Choline chloride	178	1.3 Mev γ	21, 22
	19	2–4 Mv e$^-$	16
Choline iodide	7	1.3 Mev γ	16
Aromatic hydrocarbons	1	—	23
Amino acids	2–35	^{137}Cs γ	24

Effect of Radiation Characteristics.

It will be seen from Table 15 that comparison of radiation yields on a basis of absorbed energy alone requires caution. G(–M) values found, for example, with gamma radiation from ^{60}Co cannot be used with any confidence to estimate stability to self-irradiation with ^{14}C.

Table 16.

OBSERVED DECOMPOSITION RATE OF ^{14}C-LABELED AMINO ACIDS IN STORAGE

Amino Acid	Storage Time, weeks	Decomposition Observed, %	No. of Carbon Atoms	Initial Weekly Rate of Decomposition, %	Initial Weekly Decomposition Rate Divided by No. of Carbon Atoms
Arginine	5	1	6	0.2	3.4×10^{-2}
Leucine	11	4	6	0.38	6.4×10^{-2}
iso-Leucine	6	1.5	6	0.25	4.4×10^{-2}
Lysine	11	1	6	0.1	1.6×10^{-2}
Phenylalanine	14	14	9	1.0	11.6×10^{-2}
Proline	10	nd*	5	< 0.05	$< 1 \times 10^{-2}$
Serine	11	nd	3	< 0.05	$< 2 \times 10^{-2}$
Threonine	10	1	4	0.1	2.4×10^{-2}
Tyrosine	14	4	9	0.3	3.4×10^{-2}
Valine	16	3	5	0.2	4.0×10^{-2}

* No detectable decomposition ($< 0.5\%$)

Table 16 illustrates this. It shows the observed decomposition on storage of ^{14}C-labeled amino acids (0.5-mCi quantities) in $0.01N$ aqueous HCl. The "initial weekly rate" of decomposition ranges from < 0.06 to 1.0 percent for a dose of approximately 0.01 megarad in a week. These samples—in exactly the same conditions of solvent, sample size, concentration, etc.—had been examined under ^{60}Co irradiation for a dose of 2.0 megarads, and in no case was the decomposition greater than about 3 percent.[4] The work of R. M. Lemmon et al.[10,11] on crystalline choline chloride provides another example; there were pronounced differences in G(–M) values between self-irradiation with ^{14}C (1500—1700), gamma irradiations with ^{60}Co (160—200), and 2 to 4 Mev (14—23).

Decomposition Independent of Radiation

Purely chemical decomposition (that is to say, decomposition that would occur even if there were no radioactivity there) is often overlooked. It is, however, very often troublesome, because labeled compounds are used at extremely low concentrations or in extremely small quantities, and under these conditions chemical reactions become apparent that would not be observed without such a sensitive method of detecting them. For example, a research worker had found cholesterol-4-^{14}C to be very unstable; he was dealing with an aqueous solution at a very low concentration, which had been left in daylight on the bench.[26]

Similar difficulties have been reported from time to time with radioactive thymidine. Research workers have sometimes stored this in very dilute aqueous solution for periods of weeks, or even months, withdrawing samples occasionally. The purity often falls, usually unpredictably. The probable reason for most of these troubles is microbiological infection, even at a very low level, which destroys the small quantities of thymidine present. The stability of thymidine has been reviewed by Evans and Stanford[12] and by Apelgot.[13]

A bacteriostatic agent may help; benzyl alcohol is sometimes used as one of the least likely to interfere with experimental results, but it may be necessary to verify by experiment that it has no unwanted physiological effects.

A third problem reported to the Radiochemical Centre some years ago concerned ^{14}C-labeled glucose. The experimenter using this was concerned because, he said, when he acidified it and passed air through the solution, he obtained quite a large proportion of radioactive carbon dioxide; he suspected that the glucose contained ^{14}C-labeled carbonate. This, however, was quite impossible. It appeared on investigation that the solution was extremely dilute and was heated rather strongly (to about 70° or 80°C); under these conditions, minute quantities of glucose are oxidized. This does not become apparent with conventional chemical quantities, for which the proportion oxidized would be undetectable by ordinary chemical means. Anyone investigating a suspected case of radiation decomposition should automatically consider whether purely chemical effects may be partly or wholly responsible for the observed decomposition. As with the comparable phenomenon of artefacts in paper chromatography,[27] experiments with added carrier will give evidence on this point.

The unexpected behavior of small quantities is sometimes regarded as "abnormal". That is inaccurate; it is abnormal only in the classical laboratory sense; in the quantities and concentrations in question it is the *normal* behavior, not previously observed because no method of detection was available.

Summary

At this point the factors in studying or observing decomposition from self-irradiation may be recapitulated.

The first is the chemical character of the molecule, which partly determines the liability to primary (external) and secondary effects.

The second is the "environment" of the labeled molecule, using the term in a very wide sense, including, for example, purity, solvents or diluents, concentration, temperature, specific activity, physical shape, and method of sampling.

Purely chemical effects may be a third factor.

The fourth is the type and energy of the radiation from the incorporated nuclide. Examples have been quoted above.

SPECIAL CASES

These are a few of the cases observed at the Radiochemical Centre, for which the findings have been unexpected or have some particular interest.

9,10-Dimethylbenzanthracene labeled with ^{14}C in position 9 underwent 20 percent decomposition in 30 days at 9 mCi/mM in the crystalline state. This corresponds to a quite high G(–M) value, about 280. Normally one would expect solution in benzene to protect such a material, but in fact benzene solutions were not much more stable than the solid form. 9,10-Dimethylbenzanthracene labeled generally with tritium showed similar behavior.

Diisopropylfluorophosphonate (DFP) labeled with ^{32}P at 40 mCi/mM undergoes about 10 percent decomposition in a week in the undiluted state. Dilution in propylene glycol or arachis oil has little effect on the stability.[4] The dose rate in such solutions is about 4×10^{-2} Mr/week. A much larger dose of gamma irradiation from ^{60}Co (2.5 Mr, enough to sterilize the solution) may, however, be given without much effect on the amount of decomposition, which remains the same as in a control sample not gamma irradiated.

L-Leucine-4,5-T has been prepared at 7 curies/mM. After 10 months in neutral aqueous solution at 0°C it was only 70 percent pure, and was repurified. Analysis by paper chromatography was satisfactory, but dilution analysis with L-carrier[4] revealed the presence of 18 percent of the D-form, which had been shown to be absent from the original labeled preparation. This racemization may prove to be a common phenomenon.

DL-*Methionine-*^{35}*S* was prepared regularly for many years at specific activities up to 100 mCi/mM and was found to be unstable in the presence of small amounts of water, as one might expect.[14] The anhydrous crystalline solid underwent little or no decomposition when stored in a good vacuum in a completely dry state over phosphorus pentoxide.

Naturally the same method of storage was used for L-methionine-^{35}S at similar specific activities, when this was prepared later. This behaved quite differently. There was decomposition of up to 20 percent in two months when it was stored as the anhydrous solid amino acid. This was somewhat reduced at –30°C. By storage in dilute aqueous solution, decomposition reached 8 percent in twelve weeks, but was not significant at six months at –30°C.

Solid L-methionine-^{35}S seemed, therefore, very much more sensitive to self-irradiation than the DL-methionine-^{35}S. In solution, there was probably no significant difference.

There was no obvious difference in purity between these two preparations, but the syntheses did, in fact, use different routes. It seemed, therefore, possible that a small and undetected trace of impurity in one or the other was affecting its stability. At first it seemed unlikely that DL- and L-amino acids would differ so much in their behavior merely because of their optical configuration. The synthesis of DL-methionine-^{35}S was therefore repeated, but using the same synthetic route as for L-methionine. This confirmed the observations already made that solid DL-methionine-^{35}S is more stable than the L-form. It is thought to be improbable that an impurity is the primary cause of these phenomena; it seems much more likely to be related to the differences in crystal structure between DL- and L-forms, and here the differences may well be magnified by traces of impurity.

PRECAUTIONS

One may now summarize briefly the precautions that may be advisable in handling or storing radioactive organic compounds.

1. Review the origin, preparation, age, and history of the compound, its specific activity, and its chemical character.
2. If in doubt, reanalyze it for radio-chemical purity, paying special attention to impurities that will interfere in the experimental use.

3. Remember the possibility of unexpected behavior at low concentrations or in small quantities, which may also obscure analytical results.
4. Be alert to the risks of chemical decomposition (for example, by heat, light or microbiological infection) independent of radiation.
5. Ensure that storage vessels or ampules are perfectly clean, neutral, and chemically inert.
6. Dissolve the compound in benzene, or in a solvent consisting largely of benzene, if that is possible; the radioactive concentration should be low, and it is probably wise to free it from oxygen.
7. If benzene is unsuitable, consider the use of alcohol, water, or other such solvents, or of filter paper, which should then be dehydrated as thoroughly as possible; again, freedom from oxygen is advisable.
8. If the sample is stored in the dry solid state (and this is often the best form, especially at moderate specific activities), the quantities should be as small as convenient, to reduce self-absorption; for ^{14}C compounds, quantities of a few milligrams or less are suggested; the expedient is, of course, valueless for tritium compounds; the material should be dehydrated and freed from oxygen as completely as possible.
9. Store samples at the lowest convenient temperature, unless there are known reasons to the contrary.

DECOMPOSITION DURING EXPERIMENTS

The possible significance of radiation decomposition affecting a tracer material in the course of an experiment is sometimes asked.

It may confidently be affirmed that only under exceptional conditions will there be any effect other than the uncontrollable primary (internal) decomposition, and this will usually be insignificant in itself. Radiochemical impurities do not arise from singly labeled molecules, and with small multiple-labeled molecules the magnitude of the effect will usually be negligibly small, even by the standards of tracer work. With really large macromolecules, such as proteins and nucleic acids, primary internal decomposition could become quite appreciable, even at moderate specific activities, and in some circumstances the nonradioactive recoil fragment may be deleterious. Genetic damage is the most obvious example.

In most other circumstances the labeled compound, during an experiment, will, in effect, be in extremely dilute solution for a time that is short compared with the ordinary storage of the compound on the shelf. Under these conditions external and secondary effects will not operate appreciably. The chemical behavior at such concentrations, *in vivo* as well as *in vitro*, may be unexpected, but will not be abnormal, and the radioactivity has nothing to do with it, except to make it detectable.

REFERENCES

1. Radiochemical Centre, *The Radiochemical Manual. Part 2*, 2nd ed., p. 163. The Radiochemical Centre, Amersham, Bucks, England, 1965.
2. Tolbert, B. M., et al. *J. Amer. Chem. Soc.*, 75: 1867–1868, 1953.
3. Bayly, R. J. and Weigel, H. *Nature (England)*, 188: 384–387, 1960.
4. Radiochemical Centre, *Unpublished Observations*. The Radiochemical Centre, Amersham, Bucks, England.
5. Murray, A. and Williams, D. L., *Organic Syntheses with Isotopes, Part 2*, pp. 1661–1663. Interscience Publications, John Wiley and Sons, New York, 1958.
6. Wolfgang, R. L., et al. *J. Chem. Phys.*, 24: 16–23, 1956.
7. Manning, P. G. and Monk, C. B. *J. Chem. Soc.*, pp. 2573–2576, 1962.
8. Burton, M. and Lipsky, S. *J. Phys. Chem.*, 61: 1461–1465, 1957.
9. Tolbert, B. M., et al., in *Proceedings of the Conference on Methods of Preparing and Storing Marked Molecules, Brussels, November 13–16, 1963*, pp. 671–698. European Atomic Energy Community, Brussels, Belgium, 1964.

10. Lemmon, R. M., et al. *J. Amer. Chem. Soc.*, 80: 2730–2733, 1958.
11. Lemmon, R. M., et al. *J. Amer. Chem. Soc.*, 77: 4139–4142, 1955.
12. Evans, E. A. and Stanford, F. G. *Nature (England)*, 199: 762–765, 1963.
13. Apelgot, S., et al., in *Proceedings of the Conference on Methods of Preparing and Storing Marked Molecules, Brussels, November 13–16, 1963*, pp. 221–234. European Atomic Energy Community, Brussels, Belgium, 1964.
14. Kopoldová, J., et al. *Nature (England)*, 182: 1074–1076, 1958.
15. Meisels, G. G., et al. *J. Phys. Chem.*, 61: 1456–1461, 1957.
16. Tolbert, B. M. and Lemmon, R. M. *Radiat. Res.*, 3: 52–67, 1955.
17. Adams, G. E. and Baxendale, J. H. *J. Amer. Chem. Soc.*, 80: 4215–4219, 1958.
18. Blase, E. F. *Diss. Abstr.*, 19: 965–966, 1958.
19. Meshitsuka, G. and Burton, M. *Radiat. Res.*, 8: 285–297, 1958.
20. Dauben, W. G. and Payot, P. H. *J. Amer. Chem. Soc.*, 79: 6577, 1957.
21. Collin, R. L., *J. Amer. Chem. Soc.*, 79: 6086, 1957.
22. Lemmon, R. L., et al. *J. Amer. Chem. Soc.*, 80: 2730–2733, 1958.
23. Tolbert, B. M. *Nucleonics*, 18(8): 74, 1960.
24. Tolbert, B. M., et al., in *Proceedings of the Conference on Methods of Preparing and Storing Marked Molecules, Brussels, November 13–16, 1963*, pp. 575–581. European Atomic Energy Community, Brussels, Belgium, 1964.
25. Wyant, R. E., in *Effects of Radiation on Materials and Components*, pp. 204–276, Kircher, J. F. and Bowman, R. E., eds. Reinhold, New York, 1964.
26. Hais, I. M. and Myant, N. B. *Biochem. J.*, 94: 85–90, 1965.
27. Tsuk, R. G., et al., in *Proceedings of the Conference on Methods of Preparing and Storing Marked Molecules, Brussels, November 13–16, 1963*, pp. 497–509. European Atomic Energy Community, Brussels, Belgium, 1964.

STORAGE AND STABILITY OF COMPOUNDS LABELED WITH RADIOISOTOPES. I.

R. J. Bayly and *E. A. Evans*

Reprinted by permission of
The Journal of Labelled Compounds

INTRODUCTION

Most users of compounds labeled with radioisotopes recognize that such compounds decompose on storage and that the decomposition is accelerated by self-irradiation. The degree of the decomposition in relation to the storage conditions of the compound and the measures that can be taken to control and minimize the rate of self-radiolysis are perhaps not always so well known. This review summarizes present knowledge of the decomposition of labeled compounds and methods of reducing it.

Information on this subject is largely empirical. The increasing sensitivity of methods for the analysis and measurement of radioactive chemicals is making more users aware of the problem, and its importance is being increasingly recognized. Not only are a large number of labeled compounds—particularly organic compounds—extensively used as tracers, but many applications demand a very high purity. Fractions of a percent radiochemical impurity can sometimes lead to wrong deductions being made from a tracer investigation, and under these conditions the problem of decomposition by self-irradiation becomes a very serious one. An example is the need for tyrosine labeled with tritium or ^{14}C at a very high radiochemical purity for the estimation of tyrosine hydroxylase. The method measures the formation of 3,4-dihydroxyphenylalanine, which is itself a product of the self-irradiation of the tyrosine in aqueous solution.[1] Another example is the need for ^{14}C-labeled glucose of high radiochemical purity when it is used for an insulin bioassay. Sonksen[2] has shown that fractions of a percent of impurity have a marked effect on the "no-tissue blank", and that this can be avoided if a purified labeled sugar is used.

RADIONUCLIDES

Compounds labeled with the pure beta-emitting radioisotopes—^{14}C, tritium, ^{35}S, ^{32}P, and ^{36}Cl—are most commonly used in tracer investigations. Compounds labeled with the gamma-emitting radioisotopes—such as ^{125}I, ^{131}I, ^{57}Co, ^{58}Co, and ^{75}Se—have special application in medicine. Some properties of these radionuclides are shown in Table 17.

Decomposition depends in part on the amount of energy absorbed by the compound during its useful life, so that, for a given amount of activity, the radiation energy emitted should be a guide to the seriousness of the problem. The problem of decomposition by self-irradiation might be expected to increase in magnitude as the series of pure beta emitters in Table 17 is descended, but in fact almost the reverse is true. This is largely for three reasons.

1. The fraction of energy absorbed is much less than unity for the more energetic beta emitters such as ^{32}P; on the other hand, almost complete total absorption of the beta energy occurs with tritium compounds. Gamma energy is, in general, little absorbed by the compound itself or by its immediate environs.

Table 17. PHYSICAL PROPERTIES OF SOME RADIONUCLIDES

Radionuclide	Half-Life	Beta Energy, Mev		Specific Activity, mCi/mA		Daughter Nuclide (stable)
		Max.	Mean	Maximum	Common Values for Compounds	
^{3}H	12.26 years	0.018	0.0057	2.9×10^4	10^2–10^4	^{3}He
^{14}C	5700 years	0.159	0.050	64	1–10^2	^{14}N
^{35}S	87.2 days	0.167	0.049	1.5×10^6	1–10^2	^{35}Cl
^{36}Cl	3.03×10^5 years	0.714	0.3	1.2	10^{-3}–10^{-1}	^{36}Ar
^{32}P	14.3 days	1.71	0.69	9.3×10^6	10–10^2	^{32}S
^{131}I	8.04 days	0.81	0.19	1.7×10^7	10^2–10^4	^{131}Xe
^{125}I	60 days	Electron capture		2.2×10^6	10^2–10^4	^{125}Te
^{57}Co	270 days	Electron capture		4.9×10^5	10^3–10^5	^{57}Fe
^{58}Co	71 days	Electron capture + β^+		1.9×10^6	10^3–10^5	^{58}Fe
^{75}Se	121 days	Electron capture		1.1×10^6	10–10^3	^{75}As

2. The decomposition also depends on the specific activity of the compound; as can be seen from Table 17, the specific activities of tritiated compounds in current use are usually much higher than those for compounds labeled with other pure beta-emitting radionuclides.
3. The absorbed energy decreases exponentially with time; this is an important factor for compounds labeled with radionuclides having a short half-life, such as ^{131}I or ^{32}P.

HOW AND WHY DO RADIOISOTOPICALLY LABELED COMPOUNDS DECOMPOSE?

The reason why labeled compounds decompose is not difficult to understand: the radiation energy will be commonly absorbed by the compound itself or by its environs. If the former occurs, then the excited molecules may break up in some manner; if the latter occurs, the radiation energy can produce free radicals and other reactive species, which may then cause destruction of the molecules of the labeled compound.

The observations of Calvin and his colleagues[3] in 1953 showed that extensive self-radiolysis of compounds labeled with the long-lived ^{14}C can occur, from which beta particles of quite modest mean energy are emitted. This shattered any illusions that radioisotopically labeled compounds were as stable as their unlabeled counterparts. However, surprisingly few publications that deal specifically with this important subject have subsequently appeared. On the other hand, much has been published concerning the effects of external radiation—such as gamma or X irradiation, for example—on compounds of all types.[4-8] Unfortunately the information from such experiments provides only a rough approximation as to the expected decomposition of labeled compounds by self-irradiation.

The modes by which the decomposition of labeled compounds can arise were classified by Bayly and Weigel[9] in 1960; this classification is summarized in Table 18.

Primary (internal) decomposition is the production of an impurity due to the disintegration of the unstable nucleus. A decomposition fragment so produced will be radioactive only if the molecule that is decomposing contains two or more radioactive atoms. For many investigations compounds that contain quite a small proportion of doubly labeled molecules are used, and for such compounds the radioactive impurities from primary (internal) decomposition can usually be neglected.

Table 18. MODES OF DECOMPOSITION OF LABELED COMPOUNDS

Mode of Decomposition	Cause	Method for Control
Primary (internal)	Natural isotopic decay.	None, for a given specific activity.*
Primary (external)	Direct interaction of the radioactive emission (alpha, beta, or gamma) with molecules of the compound.	Dispersal of the labeled molecules.
Secondary	Interactions of excited products with molecules of the compound.	Dispersal of active molecules; cooling to low temperatures; scavenging of free radicals.
Chemical	Thermodynamic instability of compounds; poor choice of environment.	Cooling to low temperatures; removal of harmful agents.

* Note that dilution with the inactive form of the compound subsequent to preparation is not beneficial in this case.[9]

All the effects of radiation decomposition are dependent upon specific activity, and in order to make comparisons between compounds, it is the molar specific activity that should be considered—that is, mCi/mM rather than mCi/mg. This is particularly important for macromolecules labeled, for example, with ^{14}C, where a modest isotopic abundance may be associated with quite a high molar specific activity.

Secondary decomposition is commonly the most damaging and the most difficult to control. It is also the mode most susceptible to minor variations of the environmental conditions.

In considering all the possible radiation effects causing decomposition, ordinary chemical decomposition of the compound is often overlooked. Smith[10] gives a brief, but very illuminating, account of factors (such as oxidation, hydrolysis, biological reactions, etc.) that influence the chemical decomposition of medical preparations. Chemical decomposition arising from such factors is even more likely to occur with radioactive chemicals, because these are often used in solution at very low chemical concentration or are prepared in such small chemical amounts that it is difficult to ensure complete freedom from inactive impurities that might be harmful. Even hard-glass surfaces have been found to adversely affect the stability of radioactive carbohydrates, although the nonneutrality of the glass could only be demonstrated by extraction with boiling water.[9] It is also essential to guard against photochemical or microbiological decomposition of the compound.

PERCENTAGE DECOMPOSITION IN RELATION TO THE G(–M) VALUE

In radiation chemistry it is usual to express decomposition in terms of G values—the yield in number of molecules (atoms, ions, etc.) per 100 ev absorbed by the system. In investigations of self-radiolysis it is sometimes useful, for comparisons between compounds, to calculate G(–M) values, the –M representing molecules of the starting compound irreversibly altered by the radiation process. The system may be the labeled compound itself, when it is stored in the pure state, or a solution of it in water or other solvents. G(–M) values are very dependent on the storage conditions for the compound.

Given a G(–M) value for a particular compound under specified conditions, it is possible to calculate the magnitude of the self-decomposition from the following equation (1):

$$P_d = (f \bar{E} S_a)(5.3 \times 10^{-9})G(-M), \tag{1}$$

where P_d is the initial percentage decomposition per day, f the fraction of the radiation energy absorbed by the system, Ē the mean energy of the emission in electron volts, and S_a the initial specific activity of the compound in millicuries per millimole. Inserting the appropriate E values, equation (1) reduces to

$$P_d = (3.0 \times 10^{-5})G(-M)S_a \quad \text{for tritium,} \tag{2}$$

$$P_d = (2.65 \times 10^{-4})G(-M)S_a f \quad \text{for } {}^{14}C, \tag{3}$$

$$P_d = (2.6 \times 10^{-4})G(-M)S_a f \quad \text{for } {}^{35}S. \tag{4}$$

To calculate the degree of decomposition for a particular time, equation (1) can be used in its simple linear form, providing the magnitude of decomposition is modest (e.g., $<10\%$) and the storage time short compared to the half-life of the isotope. If these restrictions are not met, the exponential forms need to be used.[9] For example, for ^{35}S the "equivalent storage time" (i.e., the time at which it can be regarded as being stored at its initial specific activity) is given by $126(1 - e^{t/126})$ days, where t is the actual time of storage in days. A similar adjustment must be made for the extent of decomposition when it exceeds about 10 percent.

For tritium compounds the value of f can be taken as unity because of the low penetrating power of the weak beta radiation; but for compounds labeled with other radioisotopes and stored under favorable conditions the value of f may be considerably less than unity. Unfortunately it is difficult to calculate or obtain a reasonable estimate for the value of f, except in the simplest of cases.[9] Consequently it has been the usual practice to assume a value of unity for f for calculation purposes; however, for practical purposes this is far from the truth, and the possible implications of this should not be forgotten.

It cannot be stressed too strongly that minor variations in the storage conditions of a compound can exert a major effect on the rate of decomposition and on the G(-M) value, and the tabulated results in this review should be treated only as a rough guide. It should also be remembered that determination of the percentage decomposition—and hence the G(-M) value—is related to the sensitivity of the methods that are employed for the detection and measurement of the radioactive impurities. Such methods have undoubtedly been improved in their sensitivity over the past few years, and this should be borne in mind when considering the significance of a few percent decomposition of very old samples that have been reanalyzed using perhaps a more sensitive method for detecting radioactive impurities than was originally used.

Rochlin[11] summarizes the methods that have been used for the calculation of absorbed dose, percentage decomposition, and G(-M) value.

ANALYSIS OF LABELED COMPOUNDS

Before describing some of the results of decomposition studies, it is useful to consider briefly the methods used for measuring the amounts of radiochemical impurities.

Some of the pitfalls in the analysis of ^{14}C-labeled compounds are discussed in detail by Catch,[12] and those for tritium compounds by Evans,[13] but the following points are of general interest.

1. In general, physical methods of analysis—such as melting point, boiling point, refractive index, and ultraviolet or infrared spectrophotometry—are inadequate and not sufficiently sensitive, to measure radiation decomposition.
2. Reverse isotope dilution analysis is a highly desirable method for the determination of the purity of a labeled compound. For detecting decomposition, however, it is less useful, unless one is concerned only with a few specific impurities.
3. Chromatographic or electrophoretic methods for detecting decomposition have generally been preferred. However, neither of these methods are without

pitfalls. For example, paper chromatography must not result in decomposition of the product on the paper, such as that observed with ^{14}C-labeled ribulose and ribulose diphosphate.[14] Gas–liquid chromatographic methods must also be geared for detecting radioactivity,[15] and must be quantitative for accurate interpretation.

4. The visual appearance of a labeled compound can sometimes give a misleading indication of radiation decomposition. While the presence of color, for example, in a normally colorless compound may indicate some impurity, it can often represent quite a negligible amount, and this may not be radioactive. Some examples are the red-violet (iodine) coloration of tritiated or ^{14}C-labeled methyl iodide[16] (and other alkyl or aryl iodides[17]), and the straw color of labeled benzene or acetic anhydride.[17] It is possible for a solid labeled compound to be deeply colored, due to deformations in the crystal lattice, without signifying the presence of any impurity. The dark-green color produced on storage of solid ^{35}S-labeled thiosemicarbazide is an example of this phenomenon.

One cannot rely upon one method of analysis only; for example, a paper chromatographic method used for determining the radiochemical purity of a L-amino acid will not indicate whether racemization has occurred—reverse dilution analysis is also required. Volatile impurities cannot be detected quantitatively by paper or thin-layer chromatographic methods. With tritium compounds it is also necessary to check for "labile" tritium, particularly when the compounds are stored in solution. Such "labile" tritium may be formed by slow exchange of the tritium atoms of the compound itself, or of its decomposition products, with the hydrogen atoms of the solvent.

EXPERIMENTAL OBSERVATIONS

For easy reference, studies of decomposition by self-irradiation on certain classes of organic compounds labeled with various radioisotopes are collected in Table 19.

Chlorine-36 Compounds

There does not appear to be any published information concerning the decomposition by self-irradiation of compounds labeled with ^{36}Cl. This is perhaps not surprising in view of the very long half-life of this radionuclide and the very low specific activity of the compounds labeled with it (see Table 17).

Sulfur-35 Compounds

Little information has been published concerning the self-radiolysis of compounds labeled with ^{35}S; in fact, only ^{35}S-labeled methionine has been previously studied in any detail.[32] Some results obtained by Dr. J. R. Ogle[17] are given in the Table 20, and it is seen that only a few of the wide variety of compounds studied show any marked decomposition, in spite of the fact that many of the compounds were stored in bulk at high specific activity.

Sodium salts of long-chain aliphatic sulfates show a marked sensitivity to self-radiolysis, which is similar to that observed for the decomposition of some long-chain aliphatic carboxylic acids labeled with ^{14}C (see Table 22).

The stability of ^{35}S-labeled thiosemicarbazide, a compound extensively used in the analysis of ketosteroids,[53] has been examined under several storage conditions. However, there is little to choose between storage in aqueous solution at $-30°$C, as a solid at $-30°$C, or at room temperature *in vacuo* over phosphorus pentoxide; but storage in methanol, even at $-30°$C, results in more rapid decomposition.

^{35}S-labeled DL-methionine at 100 mCi/mM is unstable at room temperature in the presence of moisture, but the anhydrous material is quite stable as a crystalline solid stored *in vacuo*. However, ^{35}S-labeled L-methionine undergoes up to 10 percent decomposition per month when stored as the anhydrous solid amino acid at a comparable specific activity; the decomposition rate is somewhat reduced at $-30°C$ (see Table 20). ^{35}S-labeled L-methionine, therefore, appears to be more sensitive to self-radiolysis than ^{35}S-labeled DL-methionine in the solid state. Their different behavior is probably due to the difference in the crystalline form of the two compounds,[54] perhaps similar to the radiation sensitivity of the two forms of choline

Table 19.
SELF-RADIOLYSIS OF ORGANIC COMPOUNDS
LABELED WITH RADIOISOTOPES

Compound Class	Radioisotopes	Table No.	References
Aliphatic acids, esters and salts	^{14}C	22	3, 18
	^{3}H	26	13, 19–22
Aliphatic alcohols	^{14}C	22	23
	^{3}H	26	13, 19, 20, 23
Aliphatic hydrocarbons	^{14}C	22	—
	^{3}H	—	13, 20, 24
Alkyl halides	^{14}C	22	16
	^{3}H	—	13, 20
Aromatic compounds	^{14}C	22	3
	^{3}H	26	13, 20, 25–27
Amino acids	^{14}C	22, 23	3, 28
	^{3}H	26	13, 19, 20, 26, 29–31
	^{35}S	20	32
	^{125}I, ^{131}I	27	71
Carbohydrates	^{14}C	24	9, 33–35
	^{3}H	—	13, 20
Heterocyclic compounds	^{14}C	22	3
	^{3}H	26	13, 20, 36
	^{35}S	20	—
	^{32}P	21	—
	^{57}Co, ^{58}Co	28, 29	37, 38
Macromolecules	^{14}C	—	9
	^{3}H	—	39
Nucleosides	^{14}C	25	40
	^{3}H	26	13, 20, 26, 40–46
	^{125}I, ^{131}I	27	47
Nucleotides	^{32}P	21	—
	^{3}H	26	—
Proteins and peptides	^{3}H	—	39, 48
Steroids	^{14}C	22	3, 49, 73
	^{3}H	26	13, 20, 50–52, 72, 74

Table 20. SELF-DECOMPOSITION OF COMPOUNDS LABELED WITH SULFUR-35

Compound	Specific Activity, mCi/mM	Storage Condition	Temperature, °C	Storage Time, days	Decomposition, %	G(−M)*
AET (2-aminoethyl-isothiouronium bromide hydrobromide)	4	Solid *in vacuo* over P_2O_5.	20	354	nd†	—
Chlorpromazine	7	Solid *in vacuo* over P_2O_5.	20	265	nd	—
	10	Solid *in vacuo* over P_2O_5.	20	105	nd	—
Cysteamine	2	Solid *in vacuo* over P_2O_5.	20	321	nd	—
Cysteamine dihydrochloride	25	Solid *in vacuo* over P_2O_5.	20	245	nd	—
	29	Solid *in vacuo* over P_2O_5.	20	164	nd	—
	30	Solid *in vacuo* over P_2O_5.	20	341	nd	—
	50	Solid *in vacuo* over P_2O_5.	20	203	nd	—
L-Cystine	184	Solid *in vacuo* over P_2O_5.	20	157	nd	—
s-Ethyl-L-cystine	1.5	Solid *in vacuo* over P_2O_5.	20	149	nd	—
L-Homocystine	7	Solid *in vacuo* over P_2O_5.	20	148	nd	—
6-Mercaptopurine	7	Solid *in vacuo* over P_2O_5.	20	126	nd	—
L-Methionine	96	Solid *in vacuo* over P_2O_5.	−30	33	5	7.0
	96	Solid *in vacuo* over P_2O_5.	−30	77	10	7.3
	96	Water (2.03 mCi/ml).	−30	77	nd	—
	135	Water (4.6 mCi/ml).	−30	152	2	0.7
	207	Water (7.6 mCi/ml).	−30	75	2	0.7
	46	Water†† (1.2 mCi/ml).	20	84	6	8.3
	58	Water†† (3.0 mCi/ml).	20	126	7	6.0
Potassium ethylxanthate	18	Solid *in vacuo* over P_2O_5.	20	190	nd	—
Sodium dodecylbenzenesulfonate	8	Water (3.3 mCi/ml).	20	175	12	65
Sodium ethane sulfonate	2–25	Solid *in vacuo* over P_2O_5.	20	128	nd	—
Sodium hexadecylsulfonate	56	Solid *in vacuo* over P_2O_5.	20	120	11	10.3
Sodium lauryl sulfate	6	Solid *in vacuo* over P_2O_5.	20	126	nd	—
	26	Solid *in vacuo* over P_2O_5.	20	183	10	16.1
	45	Solid *in vacuo* over P_2O_5.	20	78	7	10.5
Sodium octyl sulfate	10	Solid *in vacuo* over P_2O_5.	20	179	3 (sulfate)	—
Sodium sulfite	11	Freeze-dried solid under N_2.	20	29	26	406
	11	Water (5.5 mCi/ml) under N_2.	20	29	86	2650
Sodium thiosulfate (labeled at "inner" or "outer" S atom)	27	Solid *in vacuo* over P_2O_5.	20	63	nd	—
	37.5	Solid *in vacuo* over P_2O_5.	20	91	nd	—
	37.5	Solid *in vacuo* over P_2O_5.	20	129	nd	—
	37	Solid *in vacuo* over P_2O_5.	20	202	nd	—

Table 20. (*Continued*)
SELF-DECOMPOSITION OF COMPOUNDS LABELED WITH SULFUR-35

Compound	Specific Activity, mCi/mM	Storage Condition	Temperature, °C	Storage Time, days	Decomposition %	G(−M)*
Sulfanilic acid	160	Solid *in vacuo* over P_2O_5.	20	50	<2‡	<1.2
	140	Solid *in vacuo* over P_2O_5.	20	268	<2‡	<0.5
Taurine	30	Solid *in vacuo* over P_2O_5.	20	153	nd	—
Tetramethylthiuram disulfide	17	Solid *in vacuo* over P_2O_5.	20	209	nd	—
Thiamine (vitamin B_1)	25	Solid *in vacuo* over P_2O_5.	20	112	nd	—
Thioacetamide	2	Solid *in vacuo* over P_2O_5.	20	153	nd	—
Thiopentone ("Pentothal" Sodium**)	12	Solid *in vacuo* over P_2O_5.	20	65	nd	—
Thiosemicarbazide	160	Solid *in vacuo* over P_2O_5.	20	54	5‡	2.8
	155	Solid *in vacuo* over P_2O_5.	20	20	2‡	2.7
	155	Solid *in vacuo* over P_2O_5.	20	74	15‡	7.2
	220	Solid *in vacuo* over P_2O_5.	20	35	4‡	2.3
	191	Solid *in vacuo* over P_2O_5.	20	211	7‡	1.4
	196	Solid *in vacuo* over P_2O_5.	−30	18	4	4.7
	196	Solid *in vacuo* over P_2O_5.	−30	63	6	2.4
	196	Solid *in vacuo* over P_2O_5.	−30	112	8	2.2
	196	Water (2.0 mCi/ml).	−30	18	5	6.0
	196	Water (2.0 mCi/ml).	−30	63	7	2.8
	196	Water (2.0 mCi/ml).	−30	112	8	2.2
	196	MeOH (2.2 mCi/ml).	−30	18	15	19.0
	196	MeOH (2.2 mCi/ml).	−30	63	20	8.5
	196	MeOH (2.2 mCi/ml).	−30	112	30	9.4
Thiourea	20	Solid *in vacuo* over P_2O_5.	20	134	nd‡‡	—
	35	Solid *in vacuo* over P_2O_5.	20	101	nd‡‡	—
	46	Solid *in vacuo* over P_2O_5.	20	134	nd‡‡	—

* Calculated from equation (4), assuming f is unity.
** Trademark of Abbott Laboratories.
† No detectable deterioration.
†† Sterilized by autoclaving at 120°C (15 psi) for 20 minutes.
‡ Accompanied by considerable discoloration.
‡‡ Slight discoloration of samples.

chloride.[55] There is no significant difference between the rate of decomposition of ^{35}S-labeled DL- and L-methionine when stored in aqueous solution; storage at low temperature (−30°C) is best for both. At −30°C, in aqueous solution, less than 0.02 percent racemization of the L-isomer is observed during 152 days.[17]

Although the yields of radiolysis products are different, there is a similarity between the compounds produced by self-radiolysis of labeled methionine and those produced by gamma or X irradiation.[32,56−58] Products arise mainly through demethylation, deamination, and oxidation to the sulfone.

^{35}S-labeled sodium sulfite is very sensitive to decomposition, particularly when dissolved in water; unfortunately, the decomposition rates reported in Table 20 were not checked for chemical decomposition against unlabeled controls.

Radiation damage to ^{35}S-labeled barium sulfate, producing a highly "active" surface, is offered as an explanation for the observed increased solubility of ^{35}S-labeled barium sulfate, compared with the unlabeled sulfate.[59] Similar effects were not observed with ^{35}S-labeled lead sulfate.

Phosphorus-32 Compounds

^{32}P labeled compounds are normally used within four weeks of their preparation because of the short half-life (14.3 days) of the radionuclide. The high energy of the beta particles results in only a small fraction of the radiation energy being self-absorbed by the compound. Therefore, in general, compounds labeled with ^{32}P are best stored in their natural form as thin films. The use of solvents for these compounds has not, in general, proved beneficial.

Some results obtained by R. Monks[17] are recorded in Table 21. The presence of oxygen and chemical impurities are observed to increase the rate of decomposition.

^{32}P-labeled diisopropylphosphorofluoridate (DFP) is one of the more sensitive compounds, and dilution to the material in propylene glycol or arachis oil has little effect on the stability of the compound. The rate of decomposition is about 10 percent per week for solutions with an initial radioactive concentration of 300 µCi/ml; this corresponds to an internal dose of approximately 6×10^4 rads in the first week. Solutions of DFP in propylene gylcol can be sterilized by gamma irradiation (2.5 megarads) with less than 5 percent additional decomposition.

^{32}P-labeled sodium pyrophosphate in aqueous solution at pH 10 can be heated for 1 hour at 100°C in a sealed tube without detectable decomposition. At pH 10, solutions remain stable over a period of about 4 weeks,[60] but at pH values below 8.5, decomposition rates increase progressively as solutions are made more acidic. It is not certain whether this decomposition in acidic solutions is accelerated by self-radiolysis.

Carbon-14 Compounds

Decomposition of compounds labeled with ^{14}C is not, in general, an insuperable problem, provided suitable conditions are used for their storage.

Rochlin[11] has adequately reviewed the published information concerning the decomposition of ^{14}C compounds up to April 1965, which includes a detailed summary of the early work of Tolbert et al.[3] It would be superfluous to resummarize this information for the present review; instead, some results obtained at the Radiochemical Centre for a wide variety of compounds are listed in Table 22.

The storage of selected ^{14}C compounds in aqueous solution may not be as unsatisfactory as was once thought.[12] Deep-frozen ($-40°C$) solutions of thymidine-2-^{14}C, for example,[40] keep quite well (less than 1 percent decomposition per year at 18.3 mCi/mM, 0.1 mCi/ml), and even solutions stored at room temperature are satisfactory for compounds that are not readily oxidized and whose specific activity is moderate. For example, DL-3-phenylalanine-1-^{14}C (10.8 mCi/mM), glycine-^{14}C(U) (6.62 mCi/mM), orotic acid-6-^{14}C (11.6 mCi/mM), adenine-8-^{14}C sulfate (28.3 mCi/mM), and thiamine-(thiazole-2-^{14}C) hydrochloride (26.7 mCi/mM) can all be stored at room temperature in aqueous solution at a radioactive concentration of 25 µCi/ml with less than 2 percent decomposition per year.[17] However, until proven satisfactory for the compound under investigation, one must be cautious. Sodium acetate-1-^{14}C (37.3 mCi/mM) undergoes 14 percent decomposition on storage in aqueous solution (25 µCi/ml) during one year at room temperature, orotic acid-6-^{14}C (32.6 mCi/mM) 25 percent decomposition per annum, and sodium pyruvate-1-^{14}C rapidly decomposes in aqueous solution at room temperature.[18] In general, storage of solutions at low temperatures ($-40°C$ or below) or the addition of "protecting agents" (see below) are wise precautions.

Silverstein and Boyer[18] showed that the decomposition of pyruvate-^{14}C solution at 25°C during 35 days of storage is considerably reduced (from 84 percent to 4 percent) by the addition of benzene. The benzene had no protective effect in frozen solutions, probably because it freezes out of solution. Korff[61] also found ^{14}C-pyruvate to be unstable in solution, but ^{14}C-pyruvic acid stored at $-30°C$ is stable for at least 3 months.

Self-radiolysis of ^{14}C-labeled acetic acid in aqueous solution yielding nonionic products (approximately 2.4 percent) is suggested[62] to explain the deviation from the second-order law during a study of the kinetics of the esterification of ethanol with ^{14}C-labeled acetic acid.

Table 21.
SELF-DECOMPOSITION OF COMPOUNDS LABELED WITH PHOSPHORUS-32

Compound	Specific Activity, mCi/mM	Storage Condition	Temperature, °C	Storage Time, weeks	Observed Decomposition, %
Adenosine triphosphate-^{32}P (tetraammonium salt)	800	Water (0.3% solution).	−30	4	<2
Aze-TEPA-^{32}P (P,P-bis(1-aziridinyl)-N-ethyl,N-1,3,4-thia-(diazol-2-yl) phosphinic amide)	4.5	Sealed tube under air.	20	7	11
o-Butylethylphosphonothiolothionate (ammonium salt)	2	Sealed tube under air.	20	6	nd*
2-Cyanoethylphosphate (barium salt)	600	Sealed tube under air.	20	4	<1
Diazoxon	5	Sealed tube under air.	20	4	3†
Diisopropylphosphorofluoridate (DFP)	50–100 50–100	Sealed tube under air. Propylene glycol *in vacuo*.	20 20	1 1	10–20 10
Dipterex‡ (O,O-dimethyl 2,2,2-trichloro-1-hydroxyethylphosphonate)	2.5 2.5	Sealed tube under air.	20 20	4 10	2 6
Diethylstilbestrol diphosphate disodium salt	80	Sealed tube under air.	20	2	<5
Di-Syston‡ (O,O-diethyl-2-ethyl-thioethyl phosphorodithioate)	11	Sealed tube under air.	20	8	<1
Endoxan (N,N-bis(2-chloro-ethyl-N,O-trimethylene phosphorodiamidate)	1.5	Sealed tube under air.	20	7	4
Thymidine-5′-monophosphate (ammonium salt)	992 992 992 992	Freeze-dried solid. Water (3.4 mCi/ml). Water (3.4 mCi/ml). Water (3.4 mCi/ml).	0 20 20 0	7.5 1.5 7.5 7.5	8 2 25 20
Thio-TEPA (triethylene thiophosphoramide)	4	Sealed tube under air.	20	7	3
Trimethyl phosphite	5 10	*In vacuo.* In air.	20 20	8 2	nd 30
Sodium hypophosphite	100	Sealed tube under air.	20	2	<2

* No detectable decomposition.
† Decomposition rate sensitive to chemical impurities, particularly diethylphosphorochloridate.
‡ Trademark of Chemagro Corporation.

Table 22.
SELF-DECOMPOSITION OF COMPOUNDS LABELED WITH CARBON-14

Compound	Specific Activity, mCi/mM	Storage Condition	Temperature, °C	Storage Time, months	Decomposition, %	G(–M)*

Amino Acids (specifically labeled)

Compound	Specific Activity, mCi/mM	Storage Condition	Temperature, °C	Storage Time, months	Decomposition, %	G(–M)*
DL-Alanine-^{14}C	4.4	Solid.	20	36	nd†	—
	21.7	Freeze-dried solid.	20	12	1	0.5
2-Aminoisobutyric acid-1-^{14}C	6.4	Freeze-dried solid under N_2 or *in vacuo*.	20	36	nd	—
	14.6	Freeze-dried solid under N_2 or *in vacuo*.	20	12	1	0.7
L-Citrulline-(carboxyl-^{14}C)	22	Solid *in vacuo*.	20	17	nd	—
DL-Cystine-3-^{14}C hydrochloride	18.7	Freeze-dried solid *in vacuo*.	20	23	1	0.3
DL-3(3,4-Dihydroxyphenyl)-alanine-2-^{14}C	5.7	Solid.	20	23	2	1.8
Folic acid-2-^{14}C	31.4	Freeze-dried solid under N_2.	−40	6	nd	—
Glycine-1-^{14}C	7.9	Freeze-dried solid under N_2.	20	24	nd	—
DL-Histidine-2-^{14}C	21.6	Freeze-dried solid *in vacuo*.	−40	21	nd	—
DL-5-Hydroxytryptophan-(methylene-^{14}C)	7.45	Freeze-dried solid *in vacuo*.	−40	32	nd	—
	2.34	Freeze-dried solid *in vacuo*.	−40	32	1	1.4
DL-Leucine-1-^{14}C	36.4	Freeze-dried solid under N_2.	20	2	1	1.7
DL-Lysine-1-^{14}C	6.58	Freeze-dried solid *in vacuo*.	20	22	4	3.4
DL-3-Phenyl(alanine-1-^{14}C)	21	Freeze-dried solid under N_2.	20	12	2	1.0
DL-3-Phenyl(alanine-2-^{14}C)	4.6	Solid.	20	65	3	1.2
	4.5	Freeze-dried solid *in vacuo*.	20	47	nd	—
DL-3-(2-Thienyl)-alanine-1-^{14}C	0.78	Solid.	20	50	1	3.4
DL-Tryptophan-(benzene ring-^{14}C(U))	3.81	Freeze-dried solid under N_2.	20	18	nd	—
DL-Tryptophan-(methylene-^{14}C)	32.5	Freeze-dried solid under N_2.	−40	6	nd	—
	32.5	Freeze-dried solid under N_2.	−40	13	1	0.3
DL-Serine-3-^{14}C	5.23	Freeze-dried solid *in vacuo*.	20	36	nd	—

Table 22. (Continued)
SELF-DECOMPOSITION OF COMPOUNDS LABELED WITH CARBON-14

Compound	Specific Activity, mCi/mM	Storage Condition	Temperature, °C	Storage Time, months	Decomposition, %	G(–M)*
DL-Tyrosine-2-^{14}C	4.2 15.8	Freeze-dried solid in air. Freeze-dried solid under N_2.	20 20	63 12	nd nd	— —
DL-Valine-1-^{14}C	4.8	Freeze-dried solid *in vacuo*.	20	33	nd	—
DL-Valine-4-^{14}C	1.53	Solid under air.	20	96	2	1.4

Aliphatic Compounds

Compound	Specific Activity, mCi/mM	Storage Condition	Temperature, °C	Storage Time, months	Decomposition, %	G(–M)*
Acetic anhydride-^{14}C	32.2	Liquid‡ *in vacuo*.	20	2	2	3.7
Acetone-1,3-^{14}C	19.5	Liquid‡ *in vacuo*.	20	14	nd	—
Acetone-2-^{14}C	5.4	Liquid‡ *in vacuo*.	20	29	3	2.3
Acetonitrile-2-^{14}C	3.8	Liquid‡ *in vacuo*.	20	36	nd	—
Acetylene-^{14}C(U)	7	Gas *in vacuo*.	20	13	1	1.4
Acetyl bromide-1-^{14}C	2.77	Liquid *in vacuo*.	−40	12	10	39
Adipic acid-1,6-^{14}C	5.8	Solid under air.	20	31	3	2.0
Bromoacetic acid-1-^{14}C	2.36	Sealed tube under air.	20	36	nd	—
Bromoacetic acid-2-^{14}C	4.87	Sealed tube under air.	20	23	nd	—
Carbon tetrachloride-^{14}C	7.2	Liquid‡ *in vacuo*.	20	3	nd	—
Cetane-1-^{14}C (n-hexadecane-1-^{14}C)	4.45 10.7	Benzene *in vacuo*. Benzene *in vacuo*.	20 20	37 3	15 2	12 7.6
Cetyl alcohol-1-^{14}C	6.19	Sealed tube under air.	20	43	4	1.9
Chloroacetic acid-1-^{14}C	0.47 1.64	Solid in air. Solid in air.	20 20	36 36	nd 3	— 6.3
Choline chloride-(methyl-^{14}C)	5.7 37.6	On paper. On paper.	−80 −80	11 18	nd nd	— —
Creatinine-1-^{14}C hydrochloride	2.55	Freeze-dried *in vacuo*.	20	22	1	2.1
Cyclohexane-^{14}C(U)	5.8	Liquid *in vacuo*.	20	75	nd	—
Cyclohexane-1-carboxylic acid-(carboxyl-^{14}C) sodium salt	1.2	Freeze-dried *in vacuo*.	20	75	nd	—
trans-Cyclohexane-1,2-diamine tetracetic-2-^{14}C acid	7.69	Solid under air.	20	60	5	1.4

Table 22. (Continued)
SELF-DECOMPOSITION OF COMPOUNDS LABELED WITH CARBON-14

Compound	Specific Activity, mCi/mM	Storage Condition	Temperature, °C	Storage Time, months	Decomposition, %	G–(M)*
Decamethonium bromide-(methyl-^{14}C)	2.11 5.03	Solid in air. Solid in air.	20 20	53 12	10 1	11 2.1
n-Decane-1-^{14}C	3.3	Liquid under air.	20	35	nd	—
n-Decanoic acid-1-^{14}C	3.5	Solid under N_2.	20	24	nd	—
Dieldrin-^{14}C	20.7	Benzene *in vacuo*.	20	33	nd	—
Diethyl malonate-1-^{14}C	1.5 4.7	Liquid under air. Liquid *in vacuo*.	20 −40	25 11	7 nd	24 —
Diethyl malonate-2-^{14}C	6.06 4.36	Liquid under air. Liquid *in vacuo*.	20 −40	42 11	14 nd	7.3 —
N,N-Dimethyl(cetyl-1-^{14}C) amine	4.5	Benzene *in vacuo*.	20	34	3	2.3
Ethyl acetate-1-^{14}C	10.2	Liquid‡ in air.	20	36	5	1.7
Ethylene diamine-^{14}C (U) hydrochloride	4.8	Solid *in vacuo*.	20	24	nd	—
Ethylene diamine tetra-(acetic acid-2-^{14}C) sodium salt	4.2 25	Solid in air. Solid in air.	20 20	48 49	nd 25	— 2.8
Ethyl iodide-1-^{14}C	5.6	Liquid *in vacuo*.	−40	18	4	4.8
Ethyl iodide-2-^{14}C	4.28	Liquid *in vacuo*.	−40	33	3	4.4
Fumaric acid-1,4-^{14}C	4.9 23	Solid in air. Solid in air.	20 20	11 11	2 2	4.5 1
Fumaric acid-2,3-^{14}C	6.15	Solid in air.	20	22	nd	—
Glycerol-1-^{14}C	1.6	Liquid *in vacuo*.	−40	24	3	9.8
Glyceryl tri(stearate-1-^{14}C)	14.5	Benzene *in vacuo*.	20	55	nd	—
Glyceryl tri(stearate-2-^{14}C)	1.36	Benzene *in vacuo*.	−40	24	2	7.6
Glyceryl tri(oleate-1-^{14}C)	9.8	Benzene *in vacuo*.	20	16	nd	—
n-Hendecane-1-^{14}C	2.5	Solid in air.	20	42	nd	—
n-Hendecanoic acid-1-^{14}C	2.35	Solid in air.	20	90	4	2.1
Lauric acid-1-^{14}C	2.6	Benzene *in vacuo*.	20	22	1	2.1
Lauryl alcohol-1-^{14}C	2.5 10.7	Benzene under N_2. Benzene under N_2.	20 20	24 12	nd 3	— 2.9
Linoleic acid-1-^{14}C	24.7	Benzene under N_2.	20	12	nd	—

Table 22. (*Continued*)
SELF-DECOMPOSITION OF COMPOUNDS LABELED WITH CARBON-14

Compound	Specific Activity, mCi/mM	Storage Condition	Temperature, °C	Storage Time, months	Decomposition, %	G(−M)*
Linoleic acid-^{14}C(U)	39	Benzene *in vacuo*.	20	24	2	0.3
Linolenic acid-^{14}C(U)	139	Benzene under N_2.	20	29	nd	—
	139	Benzene *in vacuo*.	20	16	8	0.4
Malathion-^{14}C	2.2	Liquid in air.	20	24	13	33
Methanol-^{14}C	10.3	Liquid *in vacuo*.	20	11	nd	—
Methyl cyanoacetate-2-^{14}C	1.03	Liquid in air.	20	42	8	2.3
Mevalonic lactone-1-^{14}C	1.19	Benzene *in vacuo*.	20	33	nd	—
Mevalonic lactone-2-^{14}C	3.35	Benzene *in vacuo*.	20	12	1	3
Maleic anhydride-2,3-^{14}C	0.39	Solid *in vacuo*.	20	21	nd	—
	1.78	Solid *in vacuo*.	20	21	5	16.3
Methylamine-^{14}C hydrochloride	2.65	Solid *in vacuo*.	20	33	nd	—
Methyl bromoacetate-2-^{14}C	2.9	Liquid in air.	20	24	7	12.5
	4.7	Liquid in air.	20	12	nd	—
Methyl iodide-^{14}C	15.2	Liquid *in vacuo*.	−40	8	4	4.1
	25.2	Liquid *in vacuo*.	−40	23	nd	—
n-Octadecane-1-^{14}C	25.5	Solid under N_2.	20	11	nd	—
Oleic acid-1-^{14}C	24.6	Benzene *in vacuo*.	20	3	1	0.2
Oleic acid-^{14}C(U)	88	Benzene *in vacuo*.	20	22	nd	—
Oxalic acid-^{14}C(U)	14.7	Solid in air.	20	12	1	0.7
Palmitic acid-^{14}C(U)	76	Benzene *in vacuo*.	20	31	nd	—
	93	Benzene under N_2.	20	25	3	0.2
Potassium cyanide-^{14}C	0.43	Solid *in vacuo*.	20	33	nd	—
Potassium thiocyanate-^{14}C	4.4	Freeze-dried solid under air.	20	24	3	3.5
iso-Propyl iodide-1,3-^{14}C	1.9	Liquid *in vacuo*.	−40	36	nd	—
Sodium acetate-2-^{14}C	10	Freeze-dried solid in air.	20	6	nd	—
Sodium acetate-^{14}C (U)	16.7	Freeze-dried solid under N_2.	20	16	nd	—
	7.1	Freeze-dried solid *in vacuo*.	20	45	2	0.8
Sodium n-butyrate-1-^{14}C	11.9	Solid *in vacuo*.	20	30	10	3.5

Table 22. (*Continued*)
SELF-DECOMPOSITION OF COMPOUNDS LABELED WITH CARBON-14

Compound	Specific Activity, mCi/mM	Storage Condition	Temperature, °C	Storage Time, months	Decomposition, %	G–(M)*
Sodium isobutyrate-1-^{14}C	2.4	Solid in air.	20	84	20	13.5
Sodium isocaproate-1-^{14}C	2.06	Solid in air.	20	94	20	14
Sodium cyanide (alkaline)-^{14}C	16.3	Solid *in vacuo*.	20	24	2	0.6
Sodium cyanoacetate-2-^{14}C	4.7	Solid in air.	20	31	nd	—
Sodium formate-^{14}C	6.1	Freeze-dried solid *in vacuo*.	20	51	2	0.8
	17.4	Freeze-dried solid *in vacuo*.	20	12	nd	—
Sodium glyoxalate-1-^{14}C	4.8	Solid under N_2.	20	10	1	2.4
Sodium n-heptanoate-1-^{14}C	2.6	Solid in air.	20	88	20	11.8
Sodium n-hexanoate-1-^{14}C	2.87	Solid in air.	20	86	10	5.2
Sodium 2-ketoglutarate-5-^{14}C	3.05	Solid in air.	−40	24	10	17.7
	6.93	Solid in air.	−40	12	1	1.5
Sodium DL-lactate-2-^{14}C	4.88	Solid *in vacuo*.	−40	26	2	1.9
Sodium n-nonanoate-1-^{14}C	1.94	Solid in air.	20	36	nd	—
Sodium n-octanoate-1-^{14}C	3.3	Solid in air.	20	35	nd	—
Sodium propionate-1-^{14}C	5.6	Freeze-dried solid in air.	20	60	3	1.1
Sodium propionate-2-^{14}C	3.7	Freeze-dried solid under N_2.	20	24	3	4.2
	4.3	Freeze-dried solid *in vacuo*.	20	35	7	5.7
Sodium pyruvate-1-^{14}C	16	Freeze-dried solid under N_2.	−40	6	1	1.3
	5	Freeze-dried solid under N_2.	−40	9	1	2.7
Sodium pyruvate-2-^{14}C	2.7	Freeze-dried solid *in vacuo*.	−40	22	7	15
Sodium pyruvate-3-^{14}C	2.1	Freeze-dried solid *in vacuo*.	−40	24	5	13
Sodium pyruvate-^{14}C (U)	2.1	Freeze-dried solid *in vacuo*.	−40	22	nd	—
	2.14	Freeze-dried solid under N_2.	−40	8	2	13.6

Table 22. (*Continued*)
SELF-DECOMPOSITION OF COMPOUNDS LABELED WITH CARBON-14

Compound	Specific Activity, mCi/mM	Storage Condition	Temperature, °C	Storage Time, months	Decomposition, %	G(–M)*
Stearic acid-^{14}C(U)	92	Benzene *in vacuo*.	20	31	nd	—
Stearyl alcohol-1-^{14}C	1.89	Benzene *in vacuo*.	20	47	nd	—
Succinic acid-2,3-^{14}C	13.9	Solid *in vacuo*.	20	72	3	0.4
	4.25	Solid in air.	20	24	2	2.4
Succinyl bis(choline-^{14}C iodide	8.6	Solid in air.	20	19	2	1.4
DL-Tartaric acid-1,4-^{14}C	3.4	Solid in air.	20	24	nd	—

Aromatic Compounds

Compound	Specific Activity, mCi/mM	Storage Condition	Temperature, °C	Storage Time, months	Decomposition, %	G(–M)*
Acetyl salicylic acid-(carboxyl-^{14}C)	1	Solid in air.	20	80	nd	—
DL-nor-Adrenaline-(carbinol-^{14}C) DL-bitartrate	10.9	Freeze-dried solid *in vacuo*.	20	19	nd	—
Aniline-^{14}C(U) sulfate	5.1	Solid under N$_2$.	20	44	2	1.1
	5.1	Solid *in vacuo*.	20	32	nd	—
Benzaldehyde-^{14}C (carbonyl-^{14}C)	1.3	Sealed tube under air.	20	28	4	13.6
1,2-Benzanthracene-9-^{14}C	5.52	Benzene *in vacuo*.	20	66	nd	—
Benzene-^{14}C(U)	10.4	Liquid‡ *in vacuo*.	–40	36	nd	—
γ-Benzenehexachloride-^{14}C(U))	9.4	Benzene *in vacuo*.	20	36	nd	—
Benzoic (acid-ring-^{14}C(U))	5.1	Solid in air.	20	33	1	0.7
Benzylpenicillin-^{14}C (potassium salt)	24.7	Freeze-dried solid *in vacuo*.	–40	6	2	1.7
	24.7	Freeze-dried solid *in vacuo*.	–40	9	12	7
Bromobenzene-^{14}C(U)	2.7	Liquid in air.	20	21	3	6.5
Chlorobenzene-^{14}C(U)	6.9	Liquid in air.	20	24	3	2.2
4-Chloro-2-methylphenoxy-(acetic-1-^{14}C) acid	2.55	Solid under air.	20	75	nd	—
DDT-(phenyl-^{14}C)	4.4	Benzene *in vacuo*.	20	23	nd	—
1,2,3,4-Dibenzanthracene-9-^{14}C	2.8	Benzene *in vacuo*.	20	58	30	27

Table 22. (*Continued*)
SELF-DECOMPOSITION OF COMPOUNDS LABELED WITH CARBON-14

Compound	Specific Activity, mCi/mM	Storage Condition	Temperature, °C	Storage Time, months	Decomposition, %	G(–M)*
2,4-Dichlorophenoxy-(acetic-1-^{14}C) acid	2.94 14.1	Benzene *in vacuo*. Benzene *in vacuo*.	20 20	27 36	nd 8	— 2
γ-,3-Dichlorophenoxy-(butyric-1-^{14}C) acid	12.6	Benzene *in vacuo*.	20	61	nd	—
Diethyl stilbestrol-1-^{14}C	10.9	Benzene *in vacuo*.	20	32	nd	—
9,10-Dimethyl-1,2-benzanthracene-9-^{14}C	9.3 9.3	Solid in air. Solid in air.	0 0	4 21	2 6	6.8 3.7
2,4-Dinitrochlorobenzene-^{14}C(U)	6.84	Solid in air.	20	24	4	3
1-Fluoro-2,4-dinitrobenzene-^{14}C(U)	4.37	Liquid under N_2.	20	15	nd	—
Naphthalene-1-^{14}C	2	Solid under N_2.	20	73	nd	—
1-Naphthoic acid-(carboxyl-^{14}C)	5.8	Solid in air.	20	36	nd	—
2-Naphthoic acid-(carboxyl-^{14}C)	4.6	Solid in air.	20	112	2	0.5
Nitrobenzene-^{14}C(U)	2.4	Liquid under air.	20	44	nd	—
2-Methyl-^{14}C-naphthaquinone	1.63 1.63	Solid under air. Solid under air.	20 −40	43 50	3 nd	5.4 —
Phenanthrene-9-^{14}C	1.93	Solid in air.	20	36	5	9
Phenol-^{14}C(U)	1.5	Solid under air.	−40	24	10	36
Terephthalic acid-(carboxyl-^{14}C)	4	Solid in air.	20	64	20	10.7
o-Toluic acid-(carboxyl-^{14}C)	6.75	Solid in air.	20	33	nd	—
p-Toluic acid-(carboxyl-^{14}C)	2.27	Solid in air.	20	26	1	2.1
m-Toluic acid-(carboxyl-^{14}C)	1.84	Solid in air.	20	66	2	2.1

Heterocyclic Compounds

Compound	Specific Activity, mCi/mM	Storage Condition	Temperature, °C	Storage Time, months	Decomposition, %	G(–M)*
Adenine-8-^{14}C	31.3	Solid under N_2.	20	12	1	0.3
D-Biotin-(carbonyl-^{14}C)	2.65 32.4	Solid *in vacuo*. Solid *in vacuo*.	20 −40	24 20	2 1	3.9 0.2
Guanine-8-^{14}C sulfate	15.4	Solid in air.	20	24	nd	—
Hypoxanthine-8-^{14}C	12.2	Solid under N_2.	20	9	1	1.1

Table 22. (Continued)
SELF-DECOMPOSITION OF COMPOUNDS LABELED WITH CARBON-14

Compound	Specific Activity, mCi/mM	Storage Condition	Temperature, °C	Storage Time, months	Decomposition, %	G(–M)*
5-Hydroxytryptamine-3′-^{14}C creatinine sulfate	11.4 32	Solid *in vacuo*. Solid under N_2.	–40 20	12 12	nd 1	— 0.4
Indole (acetic-2-^{14}C) acid	5.3	Solid in air.	20	24	3	2.9
3-Indolyl (acetonitrile-1-^{14}C)	11.6	Benzene *in vacuo*.	20	37	1	0.3
Nicotinamide-(carbonyl-^{14}C)	10.7	Solid in air.	20	24	nd	—
Nicotinic acid-(carboxyl-^{14}C)	10.7	Solid *in vacuo*.	20	72	nd	—
iso-Nicotinic hydrazide-(carbonyl-^{14}C)	9.8	Solid *in vacuo*.	–40	31	nd	—
Orotic acid-6-^{14}C	32.6	Solid in air.	20	12	1	0.3
Thiamine-2-^{14}C hydrochloride	26.7	Solid under N_2.	20	25	1	0.2
Thymidine-2-^{14}C	18.3	Solid *in vacuo*.	–40	60	1	0.1
Thymine-2-^{14}C	9.5	Solid under air.	20	47	2	0.5
Uracil-2-^{14}C	37.5	Solid under N_2.	20	12	2	0.5
Uric acid-2-^{14}C	8.23	Solid under N_2.	20	24	3	1.4

Steroids

Compound	Specific Activity, mCi/mM	Storage Condition	Temperature, °C	Storage Time, months	Decomposition, %	G(–M)*
Cholesterol-4-^{14}C	13.7	Benzene *in vacuo*.	20	20	1	0.5
Cholesterol-26-^{14}C	24	Benzene *in vacuo*.	20	37	1	0.2
Cholestenone-4-^{14}C	19.3	Benzene *in vacuo*.	20	64	nd	—
Cholesteryl linoleate-1-^{14}C	10.3 10.9	Benzene *in vacuo*. Benzene under N_2.	20 20	12 12	40 7	51 6.6
Cholesteryl palmitate-1-^{14}C	11	Benzene *in vacuo*.	20	18	nd	—
Cholesteryl-4-^{14}C palmitate	21.2	Benzene *in vacuo*.	20	24	nd	—
Cortisol-4-^{14}C	22.3	Benzene under N_2.	20	9	3	1.8
	22.3	Benzene/2% EtOH under N_2.	20	12	4	1.8
	22.3	Benzene/5% MeOH under N_2.	20	12	7	3.3
	28.6	Benzene/10% MeOH under N_2.	20††	12	34	15
	29.2	Benzene/5% MeOH under N_2.	–40	12	8	2.9
	29.2	EtOH under N_2.	0	12	13	5
	29.2	Benzene/5% MeOH under N_2.	20	12	22	8.7
	29.2	Benzene/10% MeOH under N_2.	0	12	27	11.1
	29.2	Benzene/10% MeOH under N_2.	20	12	32	13.6

Table 22. (*Continued*)
SELF-DECOMPOSITION OF COMPOUNDS LABELED WITH CARBON-14

Compound	Specific Activity, mCi/mM	Storage Condition	Temperature, °C	Storage Time, months	Decomposition, %	G(−M)*
Cortisone-4-^{14}C	22.6	Benzene *in vacuo*.	20	12	7	3.2
	25	Benzene *in vacuo*.	20	48	nd	—
Cortisone-4-^{14}C acetate	17.1	Benzene *in vacuo*.	20	46	nd	—
Progesterone-4-^{14}C	26.1	Benzene *in vacuo*.	20	18	25	7.5
Pregnenolone-4-^{14}C	19.8	Benzene *in vacuo*.	20	12	2	1
Testosterone-4-^{14}C propionate	10.2	Benzene *in vacuo*.	20	69	nd	—

* Calculated from equation (3), assuming f is unity.
† No detectable composition.
‡ Or as a vapor.
†† Sample left in the laboratory in occasional sunlight.

As the radioactive concentration is only 15.45 µCi/ml, and the duration of the experiment only 170 hours, it would seem unlikely that radiation decomposition during the experiment is the complete explanation; impurities in the ^{14}C-labeled acetic acid sample used is a more likely possibility, particularly if solutions had been kept for several days or weeks before use.

Mizon, Boulanger and Osteux[63] draw attention to the possible decomposition of amino acids during the actual purification procedures. The ^{14}C-labeled acids, arginine, ornithine, lysine, aspartic and glutamic acids, proline, and pipecolic acid may undergo from 1 to 5 percent decomposition if left during 8 days at room temperature in the presence of air. It is necessary to guard against such chemical decomposition effects.

Examination of Table 22 shows that the rate of decomposition of cortisol-4-^{14}C stored in benzene solution is accelerated by the addition of 5 percent or more of methanol.[17] However, a few percent alcohol can often have a protective action when added to aqueous solutions of labeled compounds.[26, 28, 30] The protective effect of small amounts of ethanol[28] on solutions of ^{14}C-labeled amino acids(U) at high specific activity is seen from Table 23.

If ethanol is unacceptable to the user of the labeled compound, the stability of the amino acids in aqueous solution is increased by storage at −40°C.

In general, ^{14}C-labeled phenylalanine(U) is the most sensitive to self-radiolysis of the ^{14}C-labeled amino acids; it also has the highest molar specific activity for the same isotopic abundance.

Reducing sugars are among the labeled compounds more susceptible to the influence of their own radiation. Two reasons for this are the effect of oxidizing agents produced in solution and the difficulty of obtaining anhydrous sugars for storage in the dry state. Some results are recorded in Table 24.

The best method of storage for the sugars will vary with the user's requirement, but the following might be generally recommended for ^{14}C-labeled glucose:
1. Specific activity less than 0.1 mCi/mM: crystalline, anhydrous, with precautions to ensure that samples remain in a dry atmosphere.
2. Specific activity 0.1 to 3 mCi/mM: freeze-dried, sealed under vacuum, and spread out as much as possible over the surface of the tube (i.e., freeze-dried, from as large a volume as possible).
3. Specific activity above 3 mCi/mM: storage on paper (sealed under vacuum) is probably the best method, but this may sometimes be found inconvenient; aqueous ethanolic solutions are a possible alternative, but this is still under investigation.[17]

Table 23.
SELF-RADIOLYSIS OF AMINO ACIDS UNIFORMLY LABELED WITH CARBON-14

^{14}C-Labeled Amino Acid(U)	Specific Activity, mCi/mM	Storage Condition*	Temperature, °C	Storage Time, weeks	Decomposition, %	f, %	G(–M)†
L-Alanine	15	Freeze-dried in air.	20	104	1	27	1.5
	85	Water under N_2.	25	31	3	100	0.6
L-Arginine	35	Freeze-dried in air.	20	108	22	45	7.8
	193	Water under N_2.	25	33	21.7	100	2.1
L-Aspartic acid	106	Water under N_2.	25	30	2.8	100	0.5
L-Glutamic acid	27	Freeze-dried in air.	20	100	16	42	8.4
	148	Water‡ under N_2.	25	35	13.4	100	1.5
Glycine	67	Water under N_2.	25	29	1.4	100	0.4
L-Leucine	30	Freeze-dried in air.	20	117	3	5	9.7
	170	Water under N_2.	25	31	17.8	100	2
L-iso-Leucine	174	Water under N_2.	25	37	11.7	100	1
L-Phenylalanine	282	Water under N_2.	25	39	39.9	100	2.5
	282	Water under N_2.	20	8	8.4	100	2.1
	282	Water under N_2.	2	8	6.1	100	1.5
	282	Water under N_2.	–20	8	1	100	0.2
	282	Water/3% EtOH under N_2.	20	8	<0.3	100	<0.1
	282	Water/3% EtOH under N_2.	–20	8	<0.3	100	<0.1
L-Serine	17	Freeze-dried in air.	20	95	20	63	11.4
	87	Water under N_2.	25	33	5.2	100	1
L-Threonine	26	Freeze-dried in air.	20	87	8	32	6.2
	133	Water under N_2.	25	38	8.5	100	0.9
L-Tyrosine	240	Water under N_2.	25	33	11.8	100	0.9
L-Valine	26	Freeze-dried in air.	20	95	3	26	2.6

* Aqueous solutions autoclaved for 30 minutes at 120°C.
† Calculated from equation (3).
‡ Cannot be autoclaved without decomposition; sterile filtered.

The ^{14}C-labeled ribonucleosides(U) are best stored in aqueous solution at –40°C (or below); some results are shown in Table 25.

If storage at room temperature is essential, the rate of decomposition of the ^{14}C-labeled ribonucleosides(U) may be reduced by the addition of a few percent ethanol, but, as with other ^{14}C-labeled compounds, the ethanol does not give increased stability at –40°C.[28] Current work at the Radiochemical Centre indicates that the deoxyribonucleosides have stability similar to that of the ribonucleosides, provided they are carefully purified from chemical impurities. No information is available concerning the stability of the ^{14}C-labeled nucleotides.

Decomposition by free-radical reactions can sometimes proceed at great speed. Thus ^{14}C-labeled isobutene, even at 2.5 mCi/mM, polymerizes within a few minutes on freezing in liquid nitrogen. Similarly, acrylonitrile-1-^{14}C at 1.55 mCi/mM is largely converted into a solid polymer on standing at room temperature in the dark overnight.[17] N-(methyl-^{14}C)-N-nitroso-p-toluenesulfonic acid (11 mCi/mM) and N-ethyl (maleimide-2,3-^{14}C) (2.6 mCi/mM) are two other compounds that undergo considerable decomposition in the course of a few weeks, even at –40°C.[17]

Table 24.
SELF-RADIOLYSIS OF CARBOHYDRATES UNIFORMLY LABELED WITH CARBON-14

^{14}C-Labeled Amino Acid(U)	Specific Activity, mCi/mM	Storage Condition*	Temperature, °C	Storage Time, weeks	Decomposition, %	f, %	G(–M)†
D-Fructose	86	On paper *in vacuo*.	–40	69	1.5	0.3‡	45
	26	Freeze-dried *in vacuo*.	20	30	8	40	14
	26	Freeze-dried *in vacuo*.	–80	30	5.9	40	10
D-Glucose	42	Freeze-dried *in vacuo*.	20	34	13.1	10	53
	42	Freeze-dried *in vacuo*.	–80	34	3.6	10	14
	42	On paper *in vacuo*.	20	34	1.2	0.05‡	900
	42	On paper *in vacuo*.	–80	34	0.7	0.05‡	526
	42	Crystalline solid *in vacuo*.	20	34	9.7	100	3.8
	42	Crystalline solid *in vacuo*.	–80	34	5.7	100	2.2
	4.8	Water (0.5 mCi/ml).	2	40	2	100	5.6
	4.8	Water (0.5 mCi/ml).	–40	40	1.1	100	3.1
	80	Water (0.5 mCi/ml).	2	40	15.4	100	2.8
	80	Water (0.5 mCi/ml).	–40	40	7	100	1.2
	80	Water (0.05 mCi/ml).	2	40	12.4	100	2.2
	80	Water (0.05 mCi/ml).	–40	40	5.1	100	0.9
Sucrose	149	Freeze-dried *in vacuo*.	20	88	16.4	16	4.6
	149	Freeze-dried *in vacuo*.	–80	88	15.1	16	4.2
	149	On paper in air.	20	88	15.7	0.3‡	234
	149	On paper in air.	–80	88	4.9	0.3‡	68
	149	On paper *in vacuo*.	20	88	2.4	0.3‡	33
	149	On paper *in vacuo*.	–80	88	1.8	0.3‡	25
	171	Water (0.05 mCi/ml).	2	40	16.8	100	1.4
	171	Water (0.5 mCi/ml).	2	40	27.7	100	2.5

* Solutions sterilized by autoclaving for 30 minutes at 120°C.
† Calculated from equation (3).
‡ Absorbed energy by the sugar only; i.e., excluding the paper from the system.

Table 25.
SELF-RADIOLYSIS OF NUCLEOSIDES UNIFORMLY LABELED WITH CARBON-14

Compound	Specific Activity, mCi/mM	Storage Condition*	Temperature, °C	Storage Time, weeks	Decomposition, %	G(–M)†
Adenosine-^{14}C(U)	307	Water *in vacuo*.	37	17	3.5	0.4
	307	Water *in vacuo*.	–40	17	1.7	0.2
Cytidine-^{14}C(U)	256	Water *in vacuo*.	37	17	5.5	0.7
	256	Water *in vacuo*.	–40	17	1.3	0.2
Guanosine-^{14}C(U)	281	Water *in vacuo*.	37	19	7.6	0.8
	281	Water *in vacuo*.	–40	19	0.8	0.1
Uridine-^{14}C(U)	251	Water *in vacuo*.	37	22	9	0.9
	251	Water *in vacuo*.	–40	22	2.3	0.2

* Solutions sterilized by autoclaving for 30 minutes at 120°C; activity concentration 0.5 mCi/ml.
† Calculated from equation (3), assuming f as unity.

Tritium Compounds

The almost complete absorption of the beta-radiation energy and the very high specific activities that can be attained have made the control of self-radiolysis for tritium compounds much more difficult than for compounds labeled with other radioisotopes. The decomposition of tritium compounds has consequently been studied in rather more detail than other labeled compounds (see Table 19).

Fairly comprehensive reviews concerning the decomposition of organic compounds labeled with tritium have already been published,[13,19–26] and it is perhaps only necessary for this review to highlight some of the observations of general importance. These are:

1. The rate of decomposition may vary markedly between different preparations of the same compound, stored apparently under identical conditions at the same molar specific activity.
2. There is normally little variation between the observed decomposition of samples of the compound (dispensed into ampules at one time) prepared from a single batch and stored under identical conditions.
3. Racemization of optically active amino acids may occur on storing aqueous solutions of tritiated amino acids at $+2°C$.[26] This has not yet been observed for ^{14}C- or ^{35}S-labeled amino acids stored in aqueous solution.
4. Self-radiolysis of unsaturated compounds dissolved in benzene solution have not yet been observed to result in any significant cis-trans isomerism. Less than 5 percent trans-isomer is found during the storage of oleic acid-9,10-T or glyceryl-2-T trioleate (see Table 26), although about 10^5 rads of radiation energy were absorbed by the solutions. It is interesting to note that benzene solutions of some unsaturated compounds have been observed to undergo cis-trans isomerism when irradiated externally at comparable radiation doses.[65,66]
5. A slight variation in the conditions of storage can sometimes markedly affect the $G(-M)$ value. Thus, slowly frozen solutions of tritiated thymidine, for example, undergo a more rapid decomposition than solutions stored at $+2°C$.[40–45]
6. Some compounds at high specific activity (for example, orotic acid-5-T) can be stored in aqueous solution apparently without significant radiolysis. However, a more detailed analysis indicates that the molecular structure is destroyed, giving rise to "labile" tritium (which exchanges to form tritiated water) and unlabeled chemical impurities. This possibility should always be borne in mind when analyzing tritium compounds.

Osinski[52] has recently studied the comparative decomposition rates for nine tritiated steroids stored in several solvents. The sensitivity of self-radiolysis increases as the number of oxygen atoms in the steroid molecule is increased.

Some recent results obtained at the Radiochemical Centre for a miscellany of tritiated compounds are shown in Table 26.

Selenium-75 Compounds

Compounds labeled with the gamma emitter ^{75}Se would not be expected to undergo serious self-radiolysis. ^{75}Se-labeled L-seleno-methionine, the selenium analogue of the naturally occurring amino acid methionine, is the only compound studied at present. Storage of this compound at 400 mCi/mM in aqueous solution at $-30°C$ and a radioactive concentration of 8 mCi/ml, or at room temperature at 1 mCi/ml, results in no detectable decomposition over a period of six months.[17]

Table 26. SELF-RADIOLYSIS OF SOME TRITIUM-LABELED COMPOUNDS

Compound	Specific Activity, mCi/mM	Storage Condition*	Temperature, °C	Storage Time, months	Decomposition, %	G(–M)*
Adenine	2,900	Water (1 mCi/ml).	2	9	22	1.0
Adenosine-T(G)-5'-monophosphate (lithium salt)	644	Water (0.5 mCi/ml).	2	3	nd†	—
S-Adenosylmethionine-(methyl-T)	200	Aqueous sulfuric acid at pH 4 (1 mCi/ml).	20	0.17	10	350
	1,580	Aqueous sulfuric acid at pH 4 (1 mCi/ml).	20	0.23	10	32
	200	Aqueous sulfuric acid at pH 4 (1 mCi/ml).	−80	1.5	8	30
	1,580	Aqueous sulfuric acid at pH 4 (1 mCi/ml).	−80	1.5	8	3.7
	1,580	Aqueous sulfuric acid at pH 4 (1 mCi/ml).	−196	2	7	2.4
d(+)-Aldosterone-1,2-T	1,940	Benzene + 10% EtOH (0.2 mCi/ml).	2	6	<5	<0.5
	10,400	Benzene + 10% EtOH (0.2 mCi/ml).	2	6.5	<5	<0.1
Benzyl-T(G) penicillin	189	Freeze-dried in vacuo.	−40	5	5	5.7
	189	Water (1 mCi/ml).	2	5	90	260
3,4-Benzpyrene-T(G)	1,760	Benzene (6.7 mCi/ml).	−40	13	4	0.2
Cholesterol-1α-T	506	Benzene (1.4 mCi/ml).	20	6	2	0.7
	7,150	Benzene (1 mCi/ml).	20	6	8	0.2
Cholesterol-7α-T	3,450	Benzene (10 mCi/ml).	20	4.5	nd	—
Cytidine-T(G)	2,300	Water (1 mCi/ml).	2	13	20	0.8
Dehydroepiandrosterone-7α-T	18,300	Benzene (2.2 mCi/ml).	20	13	2	0.01
Dehydroepiandrosterone-7α-T acetate	1,460	Benzene (1.5 mCi/ml).	20	15	6	0.3
Deoxycytidine-5-T	8,300	Water (1 mCi/ml).	2	8	25	0.5
Deoxyuridine-5,6-T	5,160	Water (1 mCi/ml).	2	9	35	1
Dimethylaniline-T(G)	95	Liquid under air.	2	32	nd	—
9,10-Dimethyl-1,2-benzanthracene-T(G)	189	Freeze-dried solid in vacuo.	−80	5	nd	—
	15,100	Freeze-dried solid in vacuo.	−80	4.5	40	0.8
1-Fluoro-2,4-dinitrobenzene-3,5,6-T	10,400	Benzene (34 mCi/ml) in the dark.	20	8	nd	—
5-Fluorouracil-6-T	624	Water (1 mCi/ml).	2	12	<5	<0.8
	624	Water (1 mCi/ml).	−196	12	<2	<0.3
Glycerol-2-T	71	Liquid in air.	2	33	<2	<1.1
Glyceryl-2-T trioleate	104	Benzene (4.4 mCi/ml).	20	31	<2	<0.7

Table 26. (Continued)
SELF-RADIOLYSIS OF SOME TRITIUM-LABELED COMPOUNDS

Compound	Specific Activity, mCi/mM	Storage Condition*	Temperature, °C	Storage Time, months	Decomposition, %	G(–M)*
Glyceryl tri(stearate-9,10-T)	4,300	Benzene (30 mCi/ml).	20	9	8	0.2
DL-Leucine-4,5-T	785 23,000	Water (1 mCi/ml). Water (1 mCi/ml).	2 2	11 3	13 10	1.8 0.2
D-Leucine-4,5-T	500 16,100	Water (1 mCi/ml). Water (1 mCi/ml).	2 2	5 5	5 10	2.2 0.1
L-Leucine-4,5-T	23,000	Water (1 mCi/ml).	2	5	10	0.1
L-Methionine-(methyl-T)‡	1,610 1,610 1,610 1,610 1,610	Water (1 mCi/ml). Water (2.9 mCi/ml). Water + 1% sodium formate (2.9 mCi/ml). Water + 0.01% sodium formate (2.9 mCi/ml). Water + 0.1% ethanol (2.9 mCi/ml).	2 2 2 2 2	3.5 3.5 3.5 3.5 3.5	13 15 15 15 15	2.7 3.1 3.1 3.1 3.1
Nicotine-T (G)	761	Liquid in vacuo.	–40	14	45	6.3
Oleic acid-9, 10-T	730 2,480	Benzene (2.5 mCi/ml). Benzene (2.5 mCi/ml).	20 20	9 5	8 4	1.4 0.3
Orotic acid-5-T	4,600	Water (1 mCi/ml).	2	8	71††	—
Palmitic acid-9, 19-T	850	Benzene (1.9 mCi/ml).	20	8.5	5	0.8
L-Phenylalanine-(ring-4-T)	1,000 2,000 9,600	Water (1 mCi/ml). Water (1 mCi/ml). Water (1 mCi/ml).	2 2 2	5 12 3.5	5 25 10	1.1 1.4 0.3
β-Propiolactone-T	202	Liquid + 5% ether.	2	5	60	97
Succinic acid-2,3-T	330 649	Solid in air. Solid in air.	–40 2	15 14	10 16	2.4 2.2
Thymine-T(G)	12,200	Water (1 mCi/ml).	2	4.5	18	0.4
DL-Tryptophan-T(G)	1,460	Water (1.2 mCi/ml).	–40	15	15	0.9
L-Tyrosine-3,5-T	1,300	Water (1 mCi/ml).	2	6	15	2.2
Uracil-5,6-T	1,750	Water (1 mCi/ml).	2	5	12	1.6
Uridine-T(G)	1,760	Water (5.4 mCi/ml) + 0.01% sodium formate.	2	3.25	15	3
Uridine-5-T-5'monophosphate (ammonium salt)	572 7,320	Water (1 mCi/ml). Water (1 mCi/ml).	2 2	3 3	nd nd	— —

Benzene solutions sealed under vacuum.

* Calculated from equation (2).
† No detectable decomposition.
‡ No racemization detected in stored solutions of this compound.
†† Detailed analysis showed 66% labile tritium and 5% nonvolatile radiochemical impurity; only 30 to 33% of orotic acid remained, as determined by UV light absorption measurements.

Iodine-125 and Iodine-131 Compounds

The number of systematic investigations concerning the stability of compounds labeled with radioactive iodine is even smaller than for those labeled with most other isotopes. Many of these compounds are used *in vivo*, and under these circumstances the important parameter to measure has usually been taken to be the liberation of free iodide, rather than the overall drop in radiochemical purity.[67,71] Clearly, there is good reason behind this viewpoint, but it can be misleading; a faulty diagnosis due to the use of an impure radiochemical can be at least as damaging to a patient as an unwanted radiological dose to his thyroid.

Table 27 gives some results obtained on the decomposition of these compounds at room temperature.[17]

Table 27.
SELF-RADIOLYSIS OF SOME COMPOUNDS LABELED WITH IODINE-125 AND IODINE-131

Compound	Specific Activity, mCi/mM	Storage Condition	Radioactive Concentration, mCi/ml	Storage Time, days	Impurity, %	Total Iodide, %
4-Iodoantipyrine-^{125}I	126	Neutral aqueous solution.	0.075	104	nd*	nd
4-Iodoantipyrine-^{131}I	950	Neutral aqueous solution.	1.5	9	—	15
5-Iodo-2′-deoxyuridine-^{125}I	920	Aqueous solution.	0.126	30	—	1
5-Iodo-2′-deoxyuridine-^{131}I	280 5,800	Aqueous solution. Aqueous solution.	0.79 0.46	14 14	9 10	— —
o-Iodohippuric acid-^{125}I (sodium salt)	3.4	Aqueous solution.	0.28	92	nd	nd
o-Iodohippuric acid-^{131}I (sodium salt)	25	Aqueous solution.	3.75	8	—	1
Triiodothyronine-^{125}I	3,250	50% v/v aqueous propylene glycol.	0.17	40	—	1.2
Triiodothyronine-^{131}I	25,200	50% v/v aqueous propylene glycol.	0.62	28	15	4
Thyroxine-^{125}I	3,900	50% v/v aqueous propylene glycol.	0.19	40	—	1.5
Thyroxine-^{131}I	41,500	50% v/v aqueous propylene glycol.	0.48	21	21	7

* No detectable decomposition.

As with compounds labeled with other radioisotopes, the decomposition rates depend on the exact storage conditions; in addition, the problem of reliable analysis is met in an acute form for the iodothyronines.

Although the decomposition rates quoted in Table 27 are in some cases quite high, they do not represent examples of exceptionally labile compounds; the case of ^{131}I-labeled triiodothyronine reported in Table 27 has a G(–M) value of less than 0.06.

Compounds labeled with ^{125}I are more stable than those correspondingly labeled with ^{131}I, as would be expected from the lack of beta emission from ^{125}I.

Fluorinated xenon compounds have been isolated during the decay of ^{131}I in ^{131}I-labeled iodine pentafluoride.[68]

Radionuclides of iodine are also used to "tag" molecules such as proteins, which cannot be conveniently labeled by the incorporation of an appropriate isotope. In these cases it is not possible to describe a radiochemical impurity in the same way as one can do for "normally" labeled compounds. One can only speak of a certain percentage of the iodinated compound behaving in an identical manner to the compound for which it is serving as a tracer *under given test conditions*. For example, one could measure the stability of iodinated insulin; one value would be obtained for its behavior *in vivo*, and quite another might be obtained by measurement of the amount of activity bound to antibody—a parameter of some importance to those wishing to use the labeled protein for radioimmunoassay. However, many of the general remedial measures can be applied equally well for minimizing the decomposition of such iodinated materials—storage at very low temperatures, for example, and the addition of an inactive protein as a protective agent.

Cobalt-57 and Cobalt-58 Compounds

The only compound labeled with these radioisotopes that has been studied to any extent is cyanocobalamin (vitamin B_{12}), whose instability was first reported by Smith.[37] A further series of observations on the decomposition of labeled cyanocobalamins in aqueous solution was later reported by Rosenblum.[38]

For cyanocobalamin labeled with ^{58}Co at moderate specific activity the best method of storage appears to be as a thin film of freeze-dried solid, as illustrated by the results[17] shown in Table 28.

Table 28.
SELF-RADIOLYSIS OF CYANOCOBALAMIN LABELED WITH COBALT-58

Compound	Specific Activity, curies/mM	Storage Condition	Temperature, °C	Storage Time, weeks	Decomposition, %
Cyanocobalamin-^{58}Co	10	Aqueous solution.	2	10	45
	10	Aqueous solution.	−40	10	15
	10	Freeze-dried solid.	2	10	nd*

* No detectable decomposition.

Cyanocobalamin labeled with ^{57}Co is not as stable as the compound labeled with ^{58}Co when freeze-dried, and neither is it stable when dissolved in aqueous solution. However, the addition of 0.9 percent benzyl alcohol to the solution considerably reduces the rate of decomposition, even when the cyanocobalamin is at high specific activity. Benzyl alcohol can be used for stabilizing solutions of the compound labeled with ^{58}Co, and these results are summarized in Table 29.

Table 29.
PROTECTIVE EFFECT OF BENZYL ALCOHOL ON SOLUTIONS OF LABELED CYANOCOBALAMINS

Compound	Specific Activity, curies/mM	Storage Condition	Temperature, °C	Storage Time, weeks	Decomposition, %
Cyanocobalamin-^{57}Co	390	Freeze-dried* solid.	2	12	39
	390	Aqueous solution + 0.9% benzyl alcohol.	2	12	nd†
Cyanocobalamin-^{58}Co	400	Aqueous solution + 0.9% benzyl alcohol.	2	11	nd

* Included for comparison.   † No detectable decomposition.

The main product from the self-irradiation decomposition of ^{57}Co- or ^{58}Co-labeled cyanocobalamin has been identified as hydroxocobalamin.[17]

Care must be taken with cyanocobalamin to ensure ordinary chemical stability by storing the compound in a cool (or cold), dark place, free from noxious contaminants.

CONTROL OF SELF-IRRADIATION DECOMPOSITION

Fortunately, the decomposition of radioactive compounds by self-irradiation can be controlled and minimized to a tolerable level for most uses of these compounds. As already discussed, the problem is usually quite small for compounds labeled with ^{36}Cl, ^{32}P, ^{35}S, or with pure gamma emitters.

There are three main methods for the control of self-radiolysis.
 1. Dispersion of the active molecules.
 2. Reduction of the temperature of storage.
 3. Scavenging of the reactive species, such as free radicals.

In practice, a combination of all three methods is often used.

1. Dispersion of the Active Molecules. Homogeneous dilution of the labeled material with the same unlabeled compound (inactive carrier) serves to disperse the active molecules, but also reduces the specific activity of the compound, which is often unacceptable on other grounds. It is a wise precaution to dilute the radioactive compound to the lowest convenient molar specific activity for use.

From Tables 22, 23, 24 and 25 it is observed that the dispersal of molecules labeled with ^{14}C on paper, in aqueous solution, as freeze-dried solids, or by dissolution in organic solvents are all satisfactory methods for reducing their rate of decomposition.

For tritium compounds, dispersal of the compound as a thin film on paper and other supports is much less effective in reducing their decomposition rate.[19,23,29,69] Using such supporting materials as charcoal, benzanthracene, silica gel, iron oxides, or cellulose powder,[23] or sand,[29] offers to the research worker only more complications for recovery of the labeled compound ready for use in return for only a modest reduction in the G(–M) value. The use of clathrates has been tried by Guarino and his colleagues,[19,69] who recognized first the difficulty in preparing a suitable clathrate, and second the problem of recovering a chemically and radio-chemically pure compound from the clathrate. Labeled compounds, although perhaps representing only a small fraction of the expenditure on a research project, are valuable, and their preparation may represent many months' work; recoveries of such compounds from any method of storage must be high.

In general, a solution of the radioactive compound in a suitable solvent is preferred to dispersal on foreign solid supporting materials.

2. Reduction of the Temperature of Storage. The rate of self-radiolysis is often reduced by cooling the sample to as low a temperature as possible. For solutions, the dispersal of active molecules must remain homogeneous; otherwise a more rapid decomposition may occur than at higher temperatures, due to the formation of local "pockets" of radiation. This is particularly true for tritium compounds and has been observed, for example, with frozen aqueous solutions of thymidine-T[40–45] and with solutions of tetrasodium 2-methyl-1,4 naphthaquinol-T diphosphate.[27]

Lemmon et al.[70] showed that the rate of self-radiolysis of the radiation-sensitive compound choline chloride-^{14}C-methyl was reduced by several orders of magnitude by storing the compound at $-196°C$ (liquid nitrogen). Storage of tritium compounds at this very low temperature was first investigated by Apelgot et al.[32] for thymidine-T, and it has now been shown to reduce greatly the rate of decomposition for many tritium compounds at high specific activity.[26]

3. Scavenging of the Reactive Species. The protective action of benzyl alcohol,[26,40] cysteamine,[42] ethyl alcohol,[26,29] and formate[26] against self-radiolysis of tritium compounds stored in aqueous solution has been demonstrated. Ethanol in concentrations of only a few percent effectively protects amino acids uniformly labeled at high specific activity with ^{14}C[28] (see Table 23), and preliminary results at the Radiochemical Centre indicate that similar amounts of alcohol are effective in reducing the rate of decomposition of solutions of carbohydrates and nucleosides labeled with ^{14}C.

Provided the scavenger does not interfere in the tracer experiment, the use of such protective agents promises to be a most useful, practical, and convenient method for minimizing the decomposition of labeled compounds stored in solution. However, it should not be automatically assumed that the addition of, for example, 1 to 2 percent alcohol to an aqueous solution of a radioactive compound will result in a reduced rate of self-radiolysis of the compound. It can be seen from Table 26 that, for some compounds, scavenging does not always have any effect; an example is L-methionine-(methyl-T).

Other Precautions. Sometimes the rate of self-radiolysis can be reduced by changing the actual chemical form of the compound. Thus, while DL-noradrenaline-(^{14}C-carbinal) DL-bitartrate is quite stable, ^{14}C-noradrenaline hydrochloride is not.[17] The difference in radiation sensitivity between choline chloride and choline iodide is, of course, another example.[64]

Precautions should always be taken against ordinary chemical decomposition. This will include the use of scrupulously clean containers, storage at reduced temperature and in the absence of light, protection against microbiological attack, and—except where protective agents are used—storage as free from contaminants as possible.

ACKNOWLEDGEMENTS

The authors thank Dr. John R. Catch for his interest, and many colleagues at the Radiochemical Centre for their helpful contributions. We also thank the United Kingdom Atomic Energy Authority for permission to publish this review.

REFERENCES

1. Nagatsu, T., Levitt, M., and Udenfriend, S. *J. Biol. Chem.* 239: 2910, 1964.
2. Sonksen, P., *Private Communication.* Middlesex Hospital Medical School, London, England.
3. Tolbert, B. M., Adams, P. T., Bennett, E. L., Hughes, A. M., Kirk, M. R., Lemmon, R. M., Noller, R. M., Ostwald, R., and Calvin, M. *J. Amer. Chem. Soc.*, 75: 1867, 1953.
4. Swallow, A. J., *Radiation Chemistry of Organic Compounds.* Pergamon Press, Macmillan Co., New York, 1960.
5. Bolt, R. and Carroll, J. R., eds., *Radiation Effects on Organic Material.* Academic Press, New York, 1963.
6. Nosworthy, J. and Swallow, A. J., in *Atomic Energy Review*, 2: 35. International Atomic Energy Agency, Vienna, Austria, 1964.
7. Tolbert, B. M., Stansfield, R., and Krinks, M. H., in *Proceedings of the Conference on Methods of Preparing and Storing Marked Molecules, Brussels, November 13–16, 1963*, p. 575. European Atomic Energy Community, Brussels, Belgium, 1964.
8. Tolbert, B. M., Krinks, M. H., and Stevens, Cl. O., in *Proceedings of the Conference on Methods of Preparing and Storing Marked Molecules, Brussels, November 13–16, 1963*, p. 671. European Atomic Energy Community, Brussels, Belgium, 1964.
9. Bayly, R. J. and Weigel, H. *Nature (England)*, 188: 384, 1960.
10. Smith, G. *Pharm. J.*, 194: 219, 1965.
11. Rochlin, P. *Chem. Rev.*, 65: 685, 1965.
12. Catch, J. R., *Carbon-14 Compounds*, p. 68. Butterworths, London, England, 1961.
13. Evans, E. A., *Tritium and Its Compounds.* Butterworths, London, England, 1966.
14. Kauss, H. and Kandler, O. *Z. Naturforsch.*, 19(b): 439, 1964.
15. Muhs, M. A., Bastin, E. L., and Gordon, B. E. *Int. J. Appl. Radiat. Isotop.*, 16: 537, 1965.
16. Wagner, C. D., and Guinn, V. P. *J. Amer. Chem. Soc.*, 75: 4861, 1953.
17. Radiochemical Centre, *Unpublished Observations.* The Radiochemical Centre, Amersham, Bucks, England.
18. Silverstein, E. and Boyer, P. D. *Anal. Biochem.*, 8: 470, 1964.
19. Ciranni, G., Guarino, A., Pizzella, R., Possagno, E., Rabe, B., and Rabe, G., *Euratom Report EUR 2452 e.* European Atomic Energy Community, Brussels, Belgium, 1965.
20. Evans, E. A., and Sandford, F. G. *Nature (England)*, 197: 551, 1963.
21. Guarino, A., Pizzella, R., and Possagno, E. *J. Labelled Compounds*, 1: 10, 1965.

22. Haigh, W. G. and Hanahan, D. J. *Biochim. Biophys. Acta*, 98: 640, 1965.
23. Rabe, J. G., Guarino, A., and Rabe, B. *Int. J. Appl. Radiat. Isotop.*, 14: 571, 1963.
24. Aliprandi, B., Cacace, F., and Guarino, A., Preparation and Bio-Medical Applications of Labelled Molecules. *Euratom Report EUR 2200 e*, p. 35. European Atomic Energy Community, Brussels, Belgium, 1964.
25. Andrews, K. J. M., Bultitude, F., Evans. E. A., Gronow, M., Lambert, R. W., and Marrian, D. H. *J. Chem. Soc.*, p. 3440, 1962.
26. Evans, E. A. *Nature (England)*, 209: 169, 1966.
27. Evans, E. A. *Nature (England)*, 209: 196, 1966.
28. Bayly, R. J. and Shrimpton, S. The Radiochemical Centre, Amersham, Bucks, England, to be published.
29. Hempel, K., in *Proceedings of the Conference on Methods of Preparing and Storing Marked Molecules, Brussels, November 13-16, 1963*, p. 1009. European Atomic Energy Community, Brussels, Belgium, 1964.
30. Nouvertine, W. and Hempel, K., in *Euratom Report EUR 1828 d*. European Atomic Energy Community, Brussels, Belgium, 1965.
31. Winand, M., Bricteux-Gregoire, S., and Verly, W. G., Preparation and Bio-Medical Applications of Labelled Molecules. *Euratom Report EUR 2200 e*, p. 17. European Atomic Energy Community, Brussels, Belgium, 1964.
32. Kolousek, J., Liebster, J., and Babicky, A. *Nature (England)*, 179: 521, 1957; *Coll. Czech. Chem. Communs.*, 22: 874, 1957.
33. Phillips, G. O., Criddle, W. J., and Moody, G. J. *J. Chem. Soc.*, p. 4216, 1962.
34. Phillips, G. O., and Davies, K. W. *J. Chem. Soc.*, p. 2654, 1965.
35. Baker, N., Gibbons, A. P., and Shipley, R. A. *Biochim. Biophys. Acta*, 28: 579, 1958.
36. Bertino, J. R., Johns, D. G., Almquist, P., Hollingsworth, J. W., and Evans, E. A. *Nature*, 206: 1052, 1965.
37. Smith, E. L. *Lancet*, 7069: 387, 1959.
38. Rosenblum, C., *Vitamin B_{12} Intrinsic Factor*, 2: 294. European Symposia, Hamburg, Germany, 1961.
39. Baeyens, W., Charles, P., Davila, C., Huart, R., Ledous, L., and Zamorani, G., in *Euratom Report EUR 2419 f*. European Atomic Energy Community, Brussels, Belgium, 1965.
40. Evans, E. A. and Stanford, F. G. *Nature*, 199: 762, 1963.
41. Apelgot, S., Ekert, B., and Bouyat, A. *J. Chem. Phys.*, 60: 505, 1963.
42. Apelgot, S., Ekert, B., and Tisne, M. R., in *Proceedings of the Conference on Methods of Preparing and Storing Marked Molecules, Brussels, November 13-16, 1963*, p. 939. European Atomic Energy Community, Brussels, Belgium, 1964.
43. Apelgot, S., Ekert, B., and Frilley, M. *Biochim. Biophys. Acta*, 103: 503, 1965.
44. Apelgot, S. and Ekert, B. *J. Chem. Phys.*, 62: 845, 1965.
45. Apelgot, S. and Frilley, M. *J. Chem. Phys.*, 62: 838, 1965.
46. Petersen, D. F., Murray, A., Hayes, F. N., and Magee, M., in *Los Alamos Scientific Laboratory Report LAMS-2627*, p. 57. January–June, 1961.
47. Hughes, W. L., Commerford, S. L., Gitlin, D., Krueger, R. C., Schultze, B., Shah, V., and Reilly, P. *Fed. Proc.*, 23(3): 640, 1964.
48. Agishi, Y. and Dingman, J. F. *Biochem. Biophys. Res. Communs.*, 18: 92, 1965.
49. Dauben, W. G. and Payot, P. H. *J. Amer. Chem. Soc.*, 78: 5657, 1956.
50. Pearlman, W. H. *J. Biol. Chem.*, 236: 700, 1961.
51. Osinski, P. A., and Deconinck, J. M., in *Proceedings of the Conference on Methods of Preparing and Storing Marked Molecules, Brussels, November 13-16, 1963*, p. 931. European Atomic Energy Community, Brussels, Belgium, 1964.
52. Osinski, P., in *Euratom Report EUR 2435 f*. European Atomic Energy Community, Brussels, Belgium, 1965.
53. Horton, R. and Tait, J. F., in *Proceedings of the 2nd International Congress of Endocrinology, London, August 1964*, p. 262, Taylor, S., ed. Int. Congress Series No. 83, Excerpta Medica Foundation, London, England, 1965.
54. Radiochemical Centre, *Review Series No. 3*. The Radiochemical Centre, Amersham, Bucks, England, April, 1965.
55. Shanley, P. and Collin, R. L. *Radiat. Res.*, 16: 674, 1962.
56. Kopoldova, J., Kolousek, J., Babicky, A., and Liebster, J. *Nature (England)*, 182: 1074, 1958.
57. Kumta, U. S., Gurnani, S. U., and Sahasrabudhe, M. B. *J. Sci. Ind. Res. (India)*, 15(C): 45, 1956.
58. Kumta, U. S., Gurnani, S. U., and Sahasrabudhe, M. B. *J. Sci. Ind. Res. (India)*, 16(C): 25, 1957.
59. Bovington, C. H. *J. Inorg. Nucl. Chem.*, 27: 1975, 1965.
60. Rega, A. F., Caro, R. A., and Radicella, R. *Radiochim. Acta*, 2: 218, 1964.
61. Korff, R. W. *Anal. Biochem.*, 8: 171, 1964.
62. Dorabialska, A., Swiatkowski, W., and Tilk, S. *Nukleonika*, 8: 673, 1963.
63. Mizon, J., Boulanger, P., and Osteux, R. *Bull. Soc. Chim. Biol.*, 46: 759, 1964.
64. Tolbert, B. M. and Lemmon, R. M. *Radiat. Res.*, 3: 52, 1955.
65. Cundall, R. B. and Griffiths, P. A. *J. Amer. Chem. Soc.*, 85: 1211, 1963.
66. Nosworthy, J. *Trans. Faraday Soc.*, 61: 1138, 1965.

67. Wang, Y., Society of Nuclear Medicine 11th Annual Meeting, Berkley, California, U.A. CONF-640609-4. *Nucl. Sci. Abstr.*, 19: 19731, 1965.
68. Murin, A. N., Nefedvo, V. D., Kirin, I. S., Leonov, V. V., Zaitsev, V. M., and Akulov, G. P. *Radiokhimiya*, 7: 629, 1965.
69. Ciranni, G., Guarino, A., Pizzella, R., and Possagno, E. *J. Labelled Compounds*, 1: 1, 1965.
70. Lemmon, R. L., Parsons, M. A., and Chin, D. M. *J. Amer. Chem. Soc.*, 77: 4139, 1955.
71. Kato, S., Kurata, K., and Sugisawa, Y. *J. Pharm. Soc. Japan*, 85: 935, 1965.
72. von Rabitzsch, G. and Herzmann, H. *Justus Liebigs Ann. Chem.*, 685: 261, 1965.
73. Taylor, W. and Scratched, T. *Biochem. J.*, 97: 89, 1965.
74. Thompson, G. R., Lewis, B., and Booth, C. C. *J. Clin. Invest.*, 45: 94, 1966.

STORAGE AND STABILITY OF COMPOUNDS LABELED WITH RADIOISOTOPES. II.

R. J. Bayly and *E. A. Evans*
The Radiochemical Centre, Amersham, Bucks, England

INTRODUCTION

Observations and improvements concerning the storage and the stability of compounds labeled with radioisotopes are continually being assessed. There are currently about 1,500 radiochemicals labeled with the isotopes ^{14}C and tritium used in research. It is imperative that information derived from the use of these compounds should be valid, and not misleading due to the use of impure radiochemicals. Both users and suppliers of radiochemicals need to be aware of the problems involved in order to minimize not only the rate of self-radiolysis, but the chance of drawing the wrong conclusions from experimental data.

Existing knowledge concerning the decomposition of labeled compounds by self-irradiation was reviewed earlier[1] and summarized in the preceding chapter. Because of the great interest and importance of this subject, the need to keep the research worker informed of recent advances, pitfalls and developments—particularly those concerning methods for controlling decomposition by self-radiolysis—is evident. This chapter presents further data on the self-decomposition of particular compounds, which, though they are often empirical observations, may give some guidance to research workers as to the rates of decomposition to expect.

Readers are referred to the preceding chapter for a general introduction to the problem, for methods of calculating G(–M) values, and for other basic information. Two points that were made there might bear repetition. The first is that many of the data reported there, and many of those given here, are taken from analyses of single batches. This means that, in addition to the usual risk of errors and inaccuracies in analysis, batch-to-batch variation must be considered; this can be considerable, especially when the bulk of the self-decomposition is secondary in nature, as such decomposition is notoriously dependent[2] on even small amounts of impurities, which, if they are inactive, are very difficult to detect. It will be noted that in many of the tables a G(–M) value has been calculated. It might seem that this is almost a misuse of this parameter, as—apart from the potential inaccuracies already referred to—there are other, more fundamental, errors. For simplicity, complete absorption of the radiation energy by the system has been assumed, and the same assumption was made in many cases in the previous chapter.[1] However, for freeze-dried solids this assumption is often very erroneous; indeed, use can be, and is, made of this discrepancy in practice. Also, one implies in calculating a G(–M) value that decomposition is entirely due to primary or secondary radiation effects. But this is certainly not the case, and this source of error is most clearly seen in comparing two batches of a compound with very different specific activities. Particularly, if the compound is relatively unstable in its inactive form, it will often be found that the G(–M) value for the higher-specific-activity batch will be much lower than that for the lower-specific-activity batch. It should also be remembered that, in storing a high-specific-activity compound at a similar radioactive concentration as a low-specific-activity one, the two solutions will have very different chemical concentrations, and this will also affect the G(–M) values. In spite of these several severe limitations, these G(–M) have continued to be calculated to give some means of comparing compounds and different storage condition. Readers should bear these limitations in mind, when using the tables.

The second point to stress is the importance of the participation of the labeled-compound user in ensuring that the purity of the compounds he uses is adequate for his purpose. Whenever possible, the user should consider in advance what his purity requirements are. What particular impurities will be likely to be troublesome, and what are the maximum permissible levels of these compounds? What is the minimum acceptable purity for the labeled compound? In cases where he has special requirements he will then need to ensure that these are met, usually by conducting his own analyses. Indeed, some form of simple control analysis has a lot to commend it, if only because it provides a safeguard against unexpectedly large amounts of decomposition.

There is still a very large gap in our knowledge concerning the indentity of the products produced by self-radiolysis of labeled compounds in the various physical states of storage. The study of each labeled compound is a major research undertaking, and it is regrettable, but understandable, why this information is accumulated rather slowly.

SELF-RADIOLYSIS OF COMPOUNDS IN AQUEOUS SOLUTION

Many radiochemicals used in biochemical research are required in a suitable form for immediate use—e.g., in aqueous solution—and consequently many of the recent studies are concerned with the stability of radiochemicals stored in this condition.

The action of ionizing radiation on water is well known to produce hydrogen, hydrogen radicals, hydroxyl radicals, peroxides, and hydrated electrons as "reactive species".[3,4] These reactive species cause self-radiolysis of the radioactive compound through secondary radiation effects.[1]

Because of the necessity to supply compounds in aqueous solution, current investigations are primarily concerned with methods for reducing the damage caused by these reactive species. Basically such methods reduce to lowering the temperature of the solution, or to the use of scavengers that preferentially react with the reactive species.

Temperature Effects. Many research workers are uncertain as to whether aqueous solutions of tritiated or ^{14}C-labeled compounds should be kept frozen. In general, solutions of ^{14}C-labeled compounds are best kept frozen at temperatures of $-20°C$ or below; all the evidence available suggests that the lower the temperature of storage, the lower the decomposition will be, although the gain from going much below $-20°C$ is often small. In contrast, solutions of tritiated compounds should be kept just above the freezing point of the solution, unless facilities are available for storage at $-196°C$. Storage of tritiated compounds in frozen solutions at temperatures above $-100°C$ (or thereabouts) can often actually cause an acceleration in the rate of self-radiolysis. This unexpected effect was first observed with solutions of tritiated thymidine;[5] at least a partial explanation of it is the heterogeneity of the frozen solution, demonstrated by Apelgot and Frilley.[6] These authors also showed that, if 10 percent glycerol is present in the solution, the heterogeneity of the solute molecules on freezing is reduced considerably.

A difficulty in understanding what is happening when labeled compounds are stored in frozen solution is that there are at least two mechanisms whereby the labeled molecules can be decomposed. This either can occur by solute–reactive-species interaction at the low temperatures,[7] or it can be due to some build-up of certain of these reactive species with time and subsequent reaction of these with the labeled solute when the solution is thawed prior to its use or analysis.

Any complete explanation of the behavior of labeled compounds in frozen solution must account for:
    a) the increase in decomposition frequently found on freezing solutions of tritium-labeled compounds;
    b) the increased stability of such frozen solution at $-196°C$;
    c) the contrast between the behavior of ^{14}C- and tritium-labeled compounds.

Perhaps a clue to the second point is to be found in the behavior of the radicals produced in water on irradiation at about −160°C. Henriksen[8] has shown that around this temperature the diffusion of the electron-spin resonance centers, which are present at lower temperatures, becomes quite significant, and secondary reactions take place quite readily.

If, as is often the case, molecular clustering of the solute occurs on freezing, then it is quite probable that for tritiated compounds such clusters will soon be in juxtaposition to relatively high concentrations of reactive species because of the low range of the beta particle emitted by this radionuclide. Such a concentration will result in increased decomposition, whether that decomposition occurs in the frozen state or on thawing. It seems feasible that the reactive species will be more dispersed in the case of ^{14}C-labeled compounds, though this is probably a gross oversimplification of the mechanism.

Table 30 shows some results obtained by irradiating thymidine-2-^{14}C in tritiated water at +2°C and in the frozen state at −40°C. At the lower temperature solute–radical interaction is substantially reduced, only 10 percent of the thymidine-^{14}C being decomposed, compared with 100 percent at +2°C. With tritiated thymidine at the same chemical concentration one is probably observing the effect of molecular clustering coupled with the localization of reactive species on the rate of self-radiolysis.

Table 30.
DECOMPOSITION OF RADIOACTIVE THYMIDINE WITH BETA RADIATION FROM TRITIATED WATER

Thymidine-2^{14}C at 35.9 mCi/mM.

Weight of Labeled Thymidine, μg	Volume of Solution, ml	Tritium Activity, mCi	Storage Temperature, °C	Storage Time, months	Dose, megarads	Decomposition, %
Thymidine-2-^{14}C						
67.5	1	0 (control)	+2	16	0.012	nd*
67.5	1	10	+2	16	1.34	100
67.5	1	1	+2	16	0.14	23
67.5	1	10	−40	4	0.35	3
67.5	1	10	−50	16	1.34	10
Thymidine-T						
67.5	1	1	+2	16	0.13	18
67.5	1	1	−40	10	0.08	20

* No detectable impurities.

From a practical standpoint, it is clear that more attention needs to be paid to the rate of freezing of solutions of labeled compounds—especially tritiated ones—because of the effect on the heterogeneity of such solutions. Another subject meriting more study is the storage of labeled compounds at −196°C.

Attempts to rapidly freeze tritium compounds by immersion of their aqueous solutions in liquid nitrogen, followed by storage at −40°C, were not really successful; in many cases decomposition was still faster than at +2°C.[9,10]

Actual storage of labeled compounds at −196°C has been shown to reduce self-radiolysis considerably in a number of cases.[9-13] Some further results obtained by storing compounds at −196°C are given in Table 31; comparison of these results with those given elsewhere (see Table 39 and the preceding chapter) will emphasize that this method of storage is not a universal panacea.

Table 31. SELF-RADIOLYSIS OF LABELED COMPOUNDS AT −196°C

Compound	Specific Activity, mCi/mM	Storage Condition	Radioactive Concentration, mCi/ml	Storage Time, months	Decomposition, %	G(−M)
S-Adenosylmethionine-(methyl-T)	1,580	Aqueous solution, pH 4.	1.0	12	10	0.6
DL-Adrenaline-(ring-T(G))	200	Freeze-dried solid under vacuum.	—	10	10	5.9
DL-nor-Adrenaline-7-T hydrochloride	1,240	Aqueous solution.	0.9	4	10	2.2
Cortisol-4-^{14}C	30.4	Ethanolic solution.	0.006	10	3	1.2
	30.4	Methanolic solution.	0.006	10	2	0.8
Folic acid-3′,5′-T	25,100	Freeze-dried solid.	—	9	15	0.08
L-Leucine-4,5-T	7,600	Aqueous solution.	4.8	16	25	0.3
L-Methionine (methyl-T)		see Table 33				
Methotrexate-3′,5′-T	6,150	Freeze-dried solid.	—	4	7	0.3
L-Phenylalanine-^{14}C (U)	282	Aqueous solution.	0.18	22	2.3	0.05
	282	Aqueous solution containing 30% ethanol.	0.18	22	1.4	0.03
	282	Ethanolic solution containing 5% water.	0.18	22	4.4	0.09
Thymidine-6-T(n)	14,800	Aqueous solution.	5.6	23	22	0.08
DL-α-Tocopherol-(5-methyl-T)	872	Ethanolic solution.	1.0	7	6	1.1
DL-Tryptophan-T(G)	3,020	Aqueous solution.	1.0	9	2	0.08
Vitamin A(carbinol-^{14}C)	2.8	Benzene solution containing antioxidant.*	0.2	19	10	24

* Antioxidant was 0.05% butylated hydroxyanisole +0.05% butylated hydroxytoluene.

Effect of Scavengers. It is often inconvenient to store aqueous solutions of labeled compounds at liquid-nitrogen temperature (−196°C), particularly when relatively large volumes are involved; there is always the danger of the ampule cracking, with the possible loss of an expensive radiochemical. The difficulty of minimizing self-radiolysis of compounds in solution has been largely overcome by using reactive-species scavengers. These have included sodium formate, benzyl alcohol, and ethanol,[1,11,13] and also cysteamine;[9,10] in fact, many other compounds have been tried, and almost any substance that effectively reacts with radicals—hydroxyl radicals in particular—would be a possibility.

Ethanol is perhaps one of the best substances to use as a scavenger, as it is easily removed. It has been found especially useful for minimizing the decomposition of amino acids uniformly labeled with ^{14}C at high specific activities.[1] Some recent results obtained under these storage conditions are included in Table 39, and the effect of this stabilizer on the storage of tritium compounds is further illustrated in Table 38.

It should not be assumed that the self-radiolysis of a labeled compound will always be reduced by the addition of ethanol to its solution, although it is recommended frequently as being good practice, and to date no authenticated cases are known where it has aggravated the problem. One case that did not respond favorably to this addition is that of L-methionine-(methyl-^{14}C). As can be seen from Table 32, ethanol affords little benefit in minimizing the decomposition of ^{14}C-labeled methionine stored in aqueous solution, in agreement with results previously reported for the storage of the corresponding compound labeled with tritium.[1]

Table 32. SELF-RADIOLYSIS OF METHIONINE(METHYL-^{14}C)

All samples had a specific activity of 25 mCi/mM
and a radioactive concentration of 50 µCi/ml.

Storage Condition	Temperature, °C	G(–M)*
In deoxygenated water	20	14.9
	–20	10.3
In deoxygenated water containing 2% ethanol	20	14.0
	–20	8.8

* Calculated from the decomposition rates observed in samples stored for 6 months and for 12 months.

In practice, L-methionine (methyl-^{14}C) is best stored in the freeze-dried state (see Table 39), but just why ethanol and other scavengers fail to afford much protection to ^{14}C- or tritium-labeled methionine is not understood. It was thought that hydrated electrons may be the principal cause of self-radiolysis of the methionine.[14] To test this hypothesis, solutions of tritiated methionine(methyl-T) were saturated with nitrous oxide and stored for several weeks. Nitrous oxide is known to convert hydrated electrons into hydroxyl ions and hydroxyl radicals. The results are summarized in Table 33. Nitrous oxide accelerates the decomposition of the labeled

Table 33.
EFFECT OF TEMPERATURE, NITROUS OXIDE, AND SODIUM FORMATE ON THE SELF-RADIOLYSIS OF L-METHIONINE(METHYL-T)

Specific Activity   8.6 curies/mM
Radioactive Concentration   5.5 mCi/ml
Storage Time   7 weeks*

Addition to Aqueous Solution	Decomposition Observed, %			
	at 20°C	at 2°C	at –40°C	at –196°C
None	31	26	42	6
1% Sodium formate	23	20	29	4
Saturated with N$_2$O	43	37	27	7
Saturated with N$_2$O + 1% sodium formate	23	20	24	6

* Values given at 20°C and 2°C are extrapolations of measurements made after 6 weeks of storage.

methionine in the nonfrozen solutions, but the addition of sodium formate as scavenger reduces the rate of decomposition to the same as that of the controls containing formate alone. This suggests that hydroxyl radicals are causing most of the damage. It is interesting to note the considerable reduction in the rate of self-radiolysis that is affected by storing any of the solutions at –196°C, and it is possible that there may be some significance in the relatively low value for the solution saturated with nitrous oxide and stored at –40°C.

Although the presence of 1 percent ascorbic acid minimizes the decomposition of tritiated noradrenaline (29 curies/mM) in aqueous solution,[15] it is observed to accelerate the decomposition of methionine (methyl-T) in solution, as seen from Table 34.

Table 34. EFFECT OF ASCORBIC ACID

Compound	Specific Activity, mCi/mM	Storage Condition*	Activity Concentration, mCi/ml	Storage Time, months	Decomposition, %
L-Methionine-(methyl-T)	4330	Water.	1	4	35
	4330	Water + 1% ascorbic acid.	1	4	90
DL-nor-Adrenaline-7-T hydrochloride	3680	Water.	1	3	15
	3680	Water + 1% ascorbic acid.	2	3	6

* All solutions stored at 2°C.

There is no knowledge of any reports suggesting that the addition of compounds such as ethanol to solutions of labeled compounds interferes in their tracer uses. Often, of course, the solution supplied is well diluted before use, and the scavenger concentration is then very low indeed. However, it should be borne in mind that some effects are possible, and any necessary control experiments should be conducted to find this out. It should also be noted that, in order to minimize decomposition of the labeled compound in the diluted solution on storage, it may be necessary to raise the scavenger concentration to that of the original solution.

Cases of tritiated L-amino acids undergoing racemization on storage have been observed and reported.[11] The mechanism of this is far from clear, and even the facts themselves are uncertain, as considerable variation has been observed between batches that should be similar. However, it can be stated that no racemization has been observed in solutions of tritiated amino acids stored in the presence of small amounts of ethanol. No racemization has, as far as is known, ever been observed to occur as a result of self-radiolysis of amino acids labeled with ^{14}C. This includes a case in which some L-phenylalanine-^{14}C(U) was examined after it had been stored in a purely aqueous solution and had produced 92 percent of radiochemical impurity; even so, no racemization was detected by reverse isotope dilution analysis.[16]

SELF-RADIOLYSIS OF LABELED COMPOUNDS IN NONAQUEOUS SOLVENTS

The mechanism of the decomposition of labeled organic compounds by self-radiolysis on storage in nonaqueous solvents is not fully understood. The transfer and absorption of the radiation energy is undoubtedly quite different and produces forms of reactive species different from those produced in aqueous solution.

Commonly used solvents include benzene, ethanol, and methanol, and it is important that they should be pure. Frankel and Nalbandov[17] have investigated the deleterious effects on the stability of steroids by using solvents of varying purity, in particular when evaporating very dilute solutions of the labeled steroids; this problem has also been commented on by others.[18]

Classes of compounds that are often stored in nonaqueous solvents include steroids, long-chain aliphatic fatty acids, and hydrocarbons. Some recent observations on tritiated compounds are given in Tables 35 and 36; results for ^{14}C-labeled compounds stored in this way are included in Table 39.

In addition to the results given in Table 35, tritiated anthracene (54 curies/mM), 9,10-dimethyl-1,2-benzanthracene (19 curies/mM), 3,4-benzpyrene (24 curies/mM), and 20-methylcholanthrene (44 curies/mM), stored for 6 months in benzene solution in the dark at 0 to 3°C, are reported to undergo self-radiolysis at the rate of 12 to 20 percent per annum.[19]

The storage of labeled organic acids at low chemical concentrations in predominantly alcoholic solutions can often result in the formation of esters, especially if additional acid is present. Hempel[20] observed this occurrence on storage of tritiated phenylalanine, lysine,

Table 35.
SELF-RADIOLYSIS OF POLYCYCLIC AROMATIC HYDROCARBONS LABELED WITH TRITIUM

Compound	Specific Activity, mCi/mM	Storage Condition	Activity Concentration, mCi/ml	Storage Temperature, °C	Storage Time, months	Decomposition, %	G(–M)
3,4-Benzpyrene-T(G)	451	Benzene.	2	20	35	1	0.1
	560	Benzene.	10	20	8	4	1.0
	1,330	Benzene.	5.5	20	9	5	0.5
	1,760	Benzene.	6.7	–40	13	4	0.2
	3,730	Benzene.	1	20	8	5	0.2
	451	Solid.	—	–40	36	3	0.2
1,2,3,4-Dibenzanthracene-T(G)	151	Benzene.	2.3	20	18	2	0.9
1,2,5,6-Dibenzanthracene T(G)	834	Benzene.	5	20	8	2	0.3
9,10-Dimethyl-1,2-benzanthracene-T(G)	3,250	Benzene.	15	20	14	37	1.2
	361	Solid.	—	–80	7	10	4.7
20-Methylcholanthrene-T(G)	346	Benzene.	2	20	12	5	1.4
	400	Benzene.	1	20	7	7	2.9
	471	Benzene.*	1.1	20	7	18	6.8
	3,860	Benzene.	1	20	7	7	0.3
	3,980	Benzene.*	1.1	20	7	45	2.4

* Solution unprotected from light; all other solutions kept in dark.

Table 36. SELF-RADIOLYSIS OF SOME STEROIDS LABELED WITH TRITIUM

Compound	Specific Activity, mCi/mM	Storage Condition	Activity Concentration, mCi/ml	Storage Temperature, °C	Storage Time months	Decomposition, %	G(–M)
D(+)-Aldosterone-1,2-T	3,680	Benzene : ethanol (9 : 1).	0.1	0	5	nd*†	—
	7,800	Benzene : ethanol (4 : 1).	0.1	0	4	nd†	—
Cholesterol-7α-T	3,450	Benzene.	4	20	14	2	0.05
Cortisol-1,2-T	2,000	Benzene : methanol (1 : 1).	1	20	13	15	0.7
	30,900	Benzene : methanol (1 : 1).	2	20	13	23	0.07
Estradiol-6,7-T	500	Ethanol.	1	–20	5	3	1.3
	29,200	Ethanol.	2	0	2	9	0.2
	8,200	Ethanol.	1	2	1	nd	—
Estradiol-6,7-T-17β-acetate	36,300	Benzene : ethanol (9 : 1).	2	20	8	35	0.2
Pregnenolone-7α-T	454	Benzene.	1	20	17	1	0.1
	2,500	Benzene.	1	20	17	1	0.03

* No detectable decomposition.
† No isoaldosterone was observed to form on storage under these conditions.

α-aminoadipic acid, and leucine in 80 percent ethanol containing $0.1N$ hydrochloric acid. It is inadvisable to have both free hydrochloric acid and high concentrations of ethanol present together, unless the solutions are kept frozen.

ANALYTICAL PROBLEMS IN THE DETECTION OF IMPURITIES IN LABELED COMPOUNDS

Consignments of radiochemicals from commercial suppliers are normally accompanied by a technical data sheet describing the purity of a particular compound and the methods that have been employed for the analyses. It should be remembered that no method of analysis is infallible, and the more independent checks are used for evaluating the purity, the nearer one gets to the "absolute" value. Reliance on a single chromatographic system is normally unwise, and even much more evidence than this can often be misleading, as is illustrated by the following example concerning a batch of tritiated cholesterol.

Analysis of Cholesterol-T

1. *By Paper Chromatography.**
    a) Descending elution by (100–120°C) petroleum ether on paper preequilibrated by 80 percent methanol (this system separates double-bond isomers).
    b) Reverse phase on paper saturated with 10 percent paraffin in ethoxyethanol : n-propanol : methanol : water (70 : 20 : 60 : 50) saturated with paraffin.
    c) Reverse phase on paper saturated with 10 percent paraffin in 84 percent acetic acid (saturated with paraffin).
    d) Reverse phase on paper saturated with 10 percent phenoxyethanol in n-hexane saturated with phenoxyethanol.

2. *By Thin-Layer Chromatography on Silica Gel G.**
    a) In cyclohexane : ethyl acetate (60 : 40) (R_f 0.53).
    b) In methylene chloride : acetone (80 : 20) R_f 0.67.
    c) Reverse phase TLC on silica gel G layers predipped in 10 percent paraffin in ether, in 90 percent acetic acid saturated with liquid paraffin (R_f 0.2).

* None of these systems separates cholesterol and cholestanol.

In four paper chromatographic systems and three thin-layer systems one radioactive peak was obtained for the cholesterol-1α-T. Impregnation of the silica gel with silver nitrate (Figure 9),

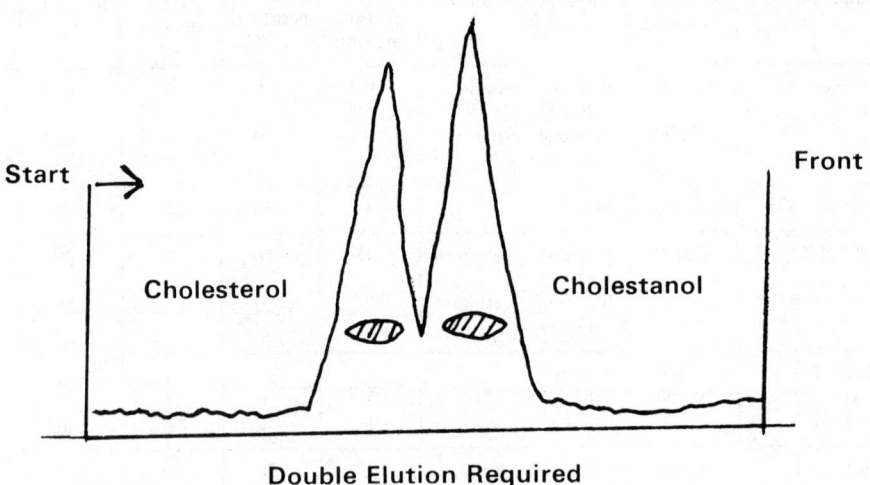

Double Elution Required
R_f^2 Cholesterol 0.53
Cholestanol 0.63

FIGURE 9. Analysis of Cholesterol Containing Cholestanol.
Thin-layer chromatography on silver nitrate–silica gel in chloroform : acetone (98 : 2). (30 g Merck silica gel G slurried with 7.5 g silver nitrate dissolved in 60 ml water.)

followed by elution with chloroform : acetone, demonstrated that the cholesterol-1α-T was impure and was contaminated with cholestanol-T.

Reverse isotope dilution analysis with cholesterol, even with four or more recrystallizations, failed to separate or show the presence of an impurity. It was necessary to prepare the 5,6-dibromocholesterol and recrystallize that several times to obtain the true radiochemical purity.

Reverse isotope dilution analysis of some deoxyadenosine that contained adenine as an impurity was another case in which recrystallization was found to be insufficient to separate the impurity.

Another case of chromatography failing to detect an impurity has recently been reported[21] in the case of some aldosterone-4-^{14}C.

EFFECT OF IMPURITIES ON THE TRACER USE OF RADIOCHEMICALS

It is seldom possible to predict the effect of impurities (both chemical and radiochemical) on the results obtained from a tracer experiment using an impure radiochemical. It is made even more difficult by the fact that the products of self-radiolysis of compounds on storage are known in only a few examples. At present, the only practical method is to isolate the impurities and to actually test these under the experimental conditions, as a type of control experiment. An example of this approach is described by Want, Zeuthen and Evans,[22] who isolated the self-radiolysis products of tritiated thymidine and studied their behavior with synchronized cells of *Tetrahymena pyriformis*. Results showed that these decomposition products did not label DNA, but were rapidly and efficiently incorporated into macromolecular structures in the cytoplasm. Treatment of the cells with DNase or RNase did not remove the labeling by these impurities. For comparison, a sample of thymidine-T irradiated with gamma rays was also tested. The results of this test demonstrate quite clearly how misleading results can be obtained and suggest that nonspecific labeling of the cytoplasm in the tracer use of thymidine labeled with tritium (or ^{14}C) for cytological studies must be interpreted with caution and related to the purity of the radioactive thymidine used for the experiments.

In the use of L-histidine(ring-2-^{14}C) to measure histidine decarboxylase activity, Thunberg[23] had trouble due to the formation of small amounts of histamine(ring-2-^{14}C) by self-decomposition. Although the levels were extremely low, they affected the assay, which involved measuring the histamine produced. Fortunately it was not difficult to remove this contaminant from the L-histidine, but he did find it necessary to do this every two to three weeks if the full sensitivity of the assay was to be maintained.

Recently Tait and co-workers[21] have demonstrated the potential source of inaccuracy in the measurement of aldosterone secretion rates, which is due to the presence of labeled 17-isoaldosterone in labeled aldosterone used for this purpose. Not only does this impurity arise fairly readily, both in the preparation of labeled aldosterone and in the preparation of a solution of that steroid suitable for injection, but the different handling of these two compounds by the body magnifies the effect of the impurity in the usual secretion rate determinations.[24, 25] At least it is now possible to state that very little, if any, of the "iso" compound is formed on storage of aldosterone-1,2-T in benzene solution.

It would be wrong to conclude from these foregoing examples that impurities will always be more trouble than their quantitative presence suggests. It has been pointed out[26] that, although the overall decomposition rate of ^{35}S-labeled thiosemicarbazide is greater in methanol than when it is stored dry or in aqueous solution,[1] the compound stored in methanolic solution behaves quite satisfactorily in steroid double-isotope-derivative analysis throughout its useful life. It seems likely that in this storage condition formaldehyde is produced by the action of the radiation on the methanol, and that this then reacts with the thiosemicarbazide; provided that this reaction is not too extensive, it appears not to interfere in the subsequent analytical procedure. In this and similar applications users may well be able to ignore the "recommended" storage conditions in order to gain convenience, provided they check the labeled compound under the conditions of their use of it. Indeed, it could well be that on occasions a "worse" method of storage (as determined by the supplier, using total impurity produced as his criteria) may be actually better for a particular application, if the impurities formed are different from those in the "recommended" method of storage and interfere less in the procedure.

The presence of radioactive molecules not labeled in the position specified for a given tracer compound can present a different kind of impurity problem. This problem is extremely unlikely to be met with in the use of ^{14}C-labeled compounds, where the position of the ^{14}C atom(s) is known with certainty from the method used for preparing the labeled compound. However, because of the ready exchange of hydrogen atoms with tritium atoms under conditions used for the preparation of tritiated compounds,[13] many "specifically" labeled tritiated compounds have as an impurity some radioactive molecules that are labeled in other positions. An example is the use of uridine-5-T as a specific precursor of RNA. A number of investigators have commented on the nonspecificity of uridine-5-T as a tracer for RNA.[27,28] When uridine-5-T is transformed into thymidine, the tritium label at the 5-position is lost. If the uridine-5-T contains any uridine-6-T, then methylation does not displace all the tritium. Analysis of various preparations of uridine-5-T prepared by catalyzed halogen–tritium replacement[13] has shown that the uridine-5-T may have up to 3 percent of uridine-6-T present. This is just one of many possible examples, and it pinpoints the importance of knowing the degree of specificity of labeling in specifically labeled compounds.

ADDITIONAL TABLES OF DECOMPOSITION RATES

Tritium Compounds

In addition to the data already presented in Tables 31, 33, 34, 35 and 36, two further tables are included, which deal with tritiated nucleotides (Table 37) and with the storage of a number of compounds in aqueous solution containing a few percent of ethanol (Table 38).

Table 37. SELF-RADIOLYSIS OF NUCLEOTIDES LABELED WITH TRITIUM

Compound*	Specific Activity, mCi/mM	Storage Time, months	Decomposition, %	G(–M)
Adenosine-T(G)-5′-monophosphate, lithium salt	644	12	<5	<0.7
Cytidine-5-T-5′-monophosphate, ammonium salt	500	7	3	0.9
	6,750	7	3	0.07
Deoxycytidine-5-T-5′-monophosphate, ammonium salt	9,200	5	35	1.0
Deoxyuridine-5-T-5′-monophosphate, ammonium salt	2,650	4	20	2.4
Thymidine-(methyl-T)-5′-monophosphate, ammonium salt	2,200	7	<5	<0.4
Uridine-5-T-5′-monophosphate, sodium salt	572	12	<3	<0.5
Uridine-5-T-5′-monophosphate, ammonium salt	6,630	6	<3	<0.1

* All these compounds were stored in sterilized aqueous solutions at +2°C, at a radioactive concentration of 1 mCi/ml.

Table 38.
SELF-RADIOLYSIS OF SOME TRITIATED COMPOUNDS IN AQUEOUS SOLUTION CONTAINING A FEW PERCENT ETHANOL

Compound	Specific Activity, mCi/mM	Storage Condition	Radioactive Concentration, mCi/ml	Storage Temperature, °C	Storage Time, months	Decomposition, %	G(–M)
DL-Adrenaline-7-T	1,400	Aqueous solution.	1	2	3.5	100	high
	1,400	Aqueous solution containing 1% ethanol.	1	2	3.5	20	5.0
	1,400	Aqueous solution containing 0.1% sodium formate.	1	2	3.5	100	high
Deoxycytidine-5-T	500	Aqueous solution containing 2% ethanol.	1	2	7	2	0.6
	14,700	Aqueous solution containing 2% ethanol.	1	2	5	nd*	—
L-3(3,4-Dihydroxyphenyl)alanine-(ring-2,5,6-T)	28,700	Aqueous solution containing 1% ethanol.	1	2	3.5	5	0.06
DL-Phenylalanine-T(G)	2,000	Aqueous solution.	1	−40	11	20	1.1
	2,000	Aqueous solution containing 1% ethanol.	1	−40	11	12	0.7
Serotonin-T(G)†	5,650	Aqueous solution.	1.1	2	2.5	25	2.2
	5,650	Aqueous solution containing 2% ethanol.	1.1	2	2.5	5	0.4
L-Tyrosine-3,5-T	1,000	Aqueous solution containing 1% ethanol.	1	2	3	nd	—
	42,300	Aqueous solution containing 1% ethanol.	1	2	3	nd	—
Uracil-5,6-T	5,600	Aqueous solution containing 1% ethanol.	0.5	2	3.5	2	0.1

* No detectable decomposition. † As creatinine sulfate complex.

Carbon-14 Compounds

Table 39 lists decomposition data for a number of compounds. For convenience, they are grouped by classes of compounds as follows.

- Amino acids
- Aliphatic compounds
- Aromatic compounds
- Heterocyclic compounds (including nucleosides and nucleotides)
- Carbohydrates
- Steroids

Table 39.
SELF-DECOMPOSITION OF SOME COMPOUNDS LABELED WITH CARBON-14

Compound	Specific Activity, mCi/mM	Storage Condition	Temperature, °C	Storage Time, months	Decomposition, %	G(–M)
Amino Acids						
S-Adenosyl-methionine (methyl-^{14}C)	23	Aqueous solution, pH 3, containing 2% ethanol.	−40	8	2	1.3
DL-Alanine-1-^{14}C	21.7	Freeze-dried solid.	20	47	3	0.3
L-Alanine-^{14}C(U)	20.2	Freeze-dried solid.	20	37	1	0.2
	85	Aqueous solution containing 2% ethanol.	−40	34	nd*	—
L-Arginine-^{14}C(U)	324	Aqueous solution containing 2% ethanol.	−40	3	nd	—
L-Asparagine-^{14}C(U)	102	Aqueous solution containing 2% ethanol.	−40	11	5	0.6
	102	Aqueous solution containing 2% ethanol.	−40	6	nd	—
L-Aspartic acid-^{14}C(U)	6.1	Freeze-dried solid.	20	23	2	1.8
	106	Aqueous solution containing 2% ethanol.	−40	20	nd	—
L-Citrulline (carbamyl-^{14}C)	35.9	Freeze-dried solid.	−40	13	5	1.3
DL-Cystine-3-^{14}C hydrochloride	39.5	Freeze-dried solid.	20	25	1	0.1
L-Cystine-^{14}C(U)	324	Aqueous solution containing 2% ethanol.	−20	4	nd	—
Folic acid-2-^{14}C, potassium salt	31.4	Freeze-dried solid.	−40	13	nd	—
L-Glutamic acid-^{14}C (U)	130	Aqueous solution containing 2% ethanol.	−40	24	nd	—
L-Glutamine-^{14}C(U)	32.2	Freeze-dried solid.	−40	23	4	0.7
	38.8	Freeze-dried solid.	−20	3	nd	—
Glycine-1-^{14}C	2	Aqueous solution, 10 μCi/ml.	20	15	nd	—
Glycine-2-^{14}C	31.7	Freeze-dried solid.	20	12	nd	—
Glycine-^{14}C(U)	8.1	Freeze-dried solid.	20	20	nd	—
	38	Aqueous solution containing 2% ethanol.	−40	12	nd	—
L-Histidine (ring-2-^{14}C)	41.5	Freeze-dried solid.	−40	11	nd	—
DL-5-Hydroxytryptophan(methylene-^{14}C)	21.8	Freeze-dried solid.	−40	47	3	0.3
L-Leucine-1-^{14}C	8.3	Freeze-dried solid.	20	15	nd	—
L-iso-Leucine-^{14}C(U)	8.7	Freeze-dried solid.	20	24	nd	—

Table 39. (*Continued*)
SELF-DECOMPOSITION OF SOME COMPOUNDS LABELED WITH CARBON-14

Compound	Specific Activity, mCi/mM	Storage Condition	Temperature, °C	Storage Time, months	Decomposition, %	G(–M)
DL-Lysine-1-^{14}C hydrochloride	15.5	Freeze-dried solid.	20	25	4	1.3
L-Lysine-^{14}C(U) hydrochloride	7.5 324	Freeze-dried solid. Aqueous solution containing 2% ethanol.	20 –40	14 2	1 nd	1.2 —
L-Methionine(methyl-^{14}C)†	25 56.8	Freeze-dried solid under nitrogen. Freeze-dried solid under nitrogen.	–20 –20	18 4	nd nd	— —
DL-3-Phenyl(alanine-1-^{14}C)	44.2	Freeze-dried solid.	20	12	nd	—
DL-3-Phenyl(alanine-2-^{14}C)	14.7	Freeze-dried solid.	20	23	2	0.7
L-3-Phenylalanine-^{14}C(U)	7	Freeze-dried solid.	20	12	nd	—
L-Proline-^{14}C(U)	8.2 120	Freeze-dried solid. Aqueous solution containing 2% ethanol.	20 –40	23 11	1 nd	0.7 —
Sarcosine-1-^{14}C	8.1	Freeze-dried solid.	20	83	nd	—
L-Serine-^{14}C(U)	87.4	Aqueous solution containing 2% ethanol.	–40	28	nd	—
L-Threonine-^{14}C(U)	5.5	Freeze-dried solid.	20	27	2	1.7
L-Thyroxine-^{14}C [L-4(3,5-diiodo-4-hydroxyphenoxy)-3,5-diiodo(phenylalanine-^{14}C(U))]	23.2	Freeze-dried solid.	–20	3	2	3.5
DL-Tryptophan(benzene ring-^{14}C(U))	19.8	Freeze-dried solid.	20	11	5	2.8
DL-Tryptophan(methylene-^{14}C)	32.9	Freeze-dried solid.	20	11	nd	—
L-Tyrosine-^{14}C(U)	5.5 225	Freeze-dried solid. Aqueous solution containing 2% ethanol.	20 –40	11 16	nd nd	— —
DL-Valine-^{14}C	4.8	Freeze-dried solid.	20	22	nd	—
L-Valine-^{14}C(U)	6.9	Freeze-dried solid.	20	23	1	0.8

Aliphatic Compounds

Compound	Specific Activity, mCi/mM	Storage Condition	Temperature, °C	Storage Time, months	Decomposition, %	G(–M)
(Acetyl-1-^{14}C) choline chloride	6.4	Freeze-dried solid.	–40	26	nd	—
Acetyl choline(methyl-^{14}C) chloride	10.4	Freeze-dried solid.	–40	23	nd	—

Table 39. (*Continued*)
SELF-DECOMPOSITION OF SOME COMPOUNDS LABELED WITH CARBON-14

Compound	Specific Activity, mCi/mM	Storage Condition	Temperature, °C	Storage Time, months	Decomposition, %	G(–M)
Adipic acid-1,6-^{14}C	11.1	Freeze-dried solid.	20	11	1	1.0
Bromoacetic acid-2-^{14}C	37.5	Liquid.	20	11	3	0.9
Calcium DL-glycerate-1-^{14}C	70.2	Freeze-dried solid.	–40	17	nd	—
Carbon tetrachloride-^{14}C	7.2	Liquid.	20	35	nd	—
Chloracetic acid-2-^{14}C	20.6	Solid.	0	12	15	8.2
Choline(methyl-^{14}C) chloride	32	Dried on paper.	–80	11	nd	—
DL-Citric acid-1,5-^{14}C	15.4	Freeze-dried solid.	–40	10	nd	—
Creatine-1-^{14}C	7.8	Freeze-dried solid.	20	23	nd	—
n-Decane-1-^{14}C	3.3	Liquid.	20	65	nd	—
Ethan-1-ol-2-amine-2-^{14}C hydrochloride	4.9	Freeze-dried solid.	20	12	nd	—
Ethyl-1-^{14}C iodide	7.1	Liquid.	20	17	8	8.5
Ethylene diamine-1,2-^{14}C dihydrochloride	4.8	Freeze-dried solid.	20	59	nd	—
Ethylene diaminetetra (acetic-2-^{14}C) acid, sodium salt	21.6	Freeze-dried solid.	20	12	nd	—
Formaldehyde-^{14}C	13.5	Aqueous solution.	20	12	19	16.1
Formic acid-^{14}C	21.8	Liquid.	20	23	nd	—
Fumaric acid-1,4-^{14}C	12	Solid.	20	12	nd	—
Fumaric acid-2,3-^{14}C	10.9	Solid.	20	19	nd	—
Glycerol-2-^{14}C	8	Liquid.	–40	11	nd	—
Glycerol-^{14}C(U)	14.3	Liquid.	–40	11	nd	—
Glyceryl-1-^{14}C tripalmitate	12.5	Benzene solution.	20	25	nd	—
n-Hexadecane-1-^{14}C	19.4	Benzene solution.	20	12	nd	—
Iodoacetic acid-2-^{14}C	6.5	Solid.	–40	17	nd	—
Lauric acid-1-^{14}C	21	Benzene solution.	20	36	nd	—
Linoleic acid-1-^{14}C	37.8	Benzene solution.	20	11	5	1.5
Linoleic acid-^{14}C(U)	442	Benzene solution.	20	3	100	high
Malathion-^{14}C [0,0-dimethyl S-(1,2-di(ethoxycarbonyl)ethyl-1,2-^{14}C) phosphorodithioate]	2.2	Benzene solution.	20	16	nd	—

Table 39. (*Continued*)
SELF-DECOMPOSITION OF SOME COMPOUNDS LABELED WITH CARBON-14

Compound	Specific Activity, mCi/mM	Storage Condition	Temperature, °C	Storage Time, months	Decomposition, %	G(–M)
Methyl bromoacetate-1-^{14}C	5.5	Liquid under nitrogen.	20	11	9	19.2
Methyl bromoacetate-2-^{14}C	5.4 40.5	Liquid under nitrogen. Liquid under nitrogen.	20 20	5 5	nd 14	— 9.2
Methylamine-^{14}C hydrochloride	34.5	Freeze-dried solid.	20	12	nd	—
Myristic acid-1-^{14}C	15.4	Benzene solution.	20	24	nd	—
n-Octadecane-1-^{14}C	25.5	Benzene solution.	20	24	nd	—
Palmitic acid-^{14}C(U)	495	Benzene solution.	20	3	nd	—
Potassium cyanate-^{14}C	15.8	Freeze-dried solid.	20	12	nd	—
iso-Propyl iodide-1, 3-^{14}C	7.5	Liquid.	20	23	1	0.7
Putrescine-^{14}C [tetramethylene diamine-1,4-^{14}C dihydrochloride]	7.8	Freeze-dried solid.	20	26	nd	—
Sodium n-butyrate-1-^{14}C	16	Freeze-dried solid.	20	22	1	0.4
Sodium cyanoacetate-2-^{14}C	14.7	Freeze-dried solid.	20	23	4.5	1.7
Sodium formate-^{14}C	10.8 42.8	Freeze-dried solid. Freeze-dried solid.	0 0	14 8	nd 2	— 0.7
Sodium glyoxalate-2-^{14}C	4.6	Freeze-dried solid.	20	23	3	3.5
Sodium DL-3-hydroxybutyrate-3-^{14}C	5.7	Freeze-dried solid.	−40	25	nd	—
Sodium DL-3-hydroxybutyrate-4-^{14}C	8.6	Freeze-dried solid.	−40	51	nd	—
Sodium DL-lactate-1-^{14}C	25.6	Freeze-dried solid.	−40	11	2.5	1.1
Sodium DL-lactate-2-^{14}C	35.6	Freeze-dried solid.	−40	13	nd	—
Sodium L-lactate-^{14}C (U)	19.4	Freeze-dried solid.	−20	16	nd	—
Sodium malonate-1-^{14}C	14.5	Freeze-dried solid.	20	3	nd	—
Sodium propionate-1-^{14}C	10.2	Freeze-dried solid.	20	23	2	1.1
Sodium propionate-2-^{14}C	9.9	Freeze-dried solid.	20	11	1	1.2

Table 39. (Continued)
SELF-DECOMPOSITION OF SOME COMPOUNDS LABELED WITH CARBON-14

Compound	Specific Activity, mCi/mM	Storage Condition	Temperature, °C	Storage Time, months	Decomposition, %	G(–M)
Stearyl alcohol-1-^{14}C	16	Benzene solution.	20	23	nd	—
Succinic acid-1,4-^{14}C	5.7	Solid.	20	11	nd	—
Succinic acid-2,3-^{14}C	4	Solid.	20	11	nd	—
Succinyl bis(choline (methyl-^{14}C) iodide)	8.6	Freeze-dried solid.	20	31	nd	—
DL-Tartaric acid-1,4-^{14}C	5.1	Freeze-dried solid.	20	23	nd	—
Thiourea-^{14}C	21.6	Solid.	20	23	3	0.7
Trichloroacetic acid-1-^{14}C	6.8	Solid.	–40	12	2	3.0
Trichloroacetic acid-2-^{14}C	10.5	Solid.	–40	11	nd	—
Urea-^{14}C	15.4	Freeze-dried solid.	20	25	nd	—
	39	Freeze-dried solid.	–40	22	nd	—

Aromatic Compounds

Compound	Specific Activity, mCi/mM	Storage Condition	Temperature, °C	Storage Time, months	Decomposition, %	G(–M)
DL-Adrenaline(carbinol-^{14}C)DL-bitartrate	7.3	Freeze-dried solid.	–20	9	nd	—
	10.9	Freeze-dried solid.	20	27	nd	—
DL-nor-Adrenaline(carbinol-^{14}C(DL-bitartrate	17.2	Freeze-dried solid.	20	9	nd	—
Aldrin-^{14}C	81	Benzene solution.	20	5	1	0.3
Aniline hydrogen sulfate-^{14}C(U)	37	Freeze-dried solid.	20	23	nd	—
Benzaldehyde-(carbonyl-^{14}C)	4.1	Liquid under nitrogen.	20	30	14	15.1
Benzene-^{14}C(U)	26.5	Liquid.	20	12	nd	—
γ-Benzene hexachloride-^{14}C(U)	34	Benzene solution.	20	20	nd	—
Benzoic acid(carboxyl-^{14}C)	23.3	Solid under nitrogen.	20	23	nd	—
Benzoic acid(ring-^{14}C(U))	48.2	Solid.	20	11	nd	—
Benzyl alcohol(carbinol-^{14}C)	3	Liquid.	20	23	2	3.6
Dieldrin-^{14}C	70.4	Benzene solution.	20	6	1	0.3
2,4-Dichlorophenoxy (acetic acid-1-^{14}C)	12.1	Solid.	20	11	nd	—

Table 39. (*Continued*)
SELF-DECOMPOSITION OF SOME COMPOUNDS LABELED WITH CARBON-14

Compound	Specific Activity, mCi/mM	Storage Condition	Temperature, °C	Storage Time, months	Decomposition, %	G(–M)
9,10-Dimethyl-1,2-benzanthracene-9-^{14}C	9.3 9.3	Solid. Solid.	20 20	8 32	1 6	1.7 2.5
2,4-Dinitrophenylhydrazine-^{14}C(U)	4.1	Solid.	20	59	5	2.6
2-(Methyl-^{14}C)naphthalene	5.1	Solid.	20	24	nd	—
1-Naphthol-1-^{14}C	2.3	Solid.	20	23	nd	—
Neostigmine iodide [tri(methyl-^{14}C)(N,N-dimethyl-m-carbamatophenyl) ammonium iodide]	5.3	Freeze-dried solid.	–40	40	nd	—
Salicylic acid(carboxyl-^{14}C)	15.5	Solid.	20	15	nd	—
Vitamin A (carbinol-^{14}C)	2.8	Benzene solution containing antioxidant.††	–20	19	8	19.0
Vitamin A (carbinol-^{14}C) acetate	2.9	Benzene solution containing antioxidant.††	–20	30	12	17.0

Heterocyclic Compounds

Compound	Specific Activity, mCi/mM	Storage Condition	Temperature, °C	Storage Time, months	Decomposition, %	G(–M)
Adenine-8-^{14}C	31.5	Freeze-dried solid.	20	12	nd	—
S-Adenosyl-methionine (methyl-^{14}C)	23	Aqueous solution, pH 3, containing 2% ethanol.	–40	8	2	1.3
Adenosine-8-^{14}C	28.4	Aqueous solution.	–40	20	nd	—
Adenosine-8-^{14}C-5′-monophosphate, ammonium salt	21.4	Freeze-dried solid.	–40	13	nd	—
Adenosine-^{14}C(U)-5′-monophosphate, ammonium salt	255	Sterile aqueous solution under vacuum.	–40	13	3	0.1
Adenosine-8-^{14}C-5′-triphosphate, sodium salt	4.8	Freeze-dried solid.	–20	7	2	7.0
Benzyl penicillin-^{14}C potassium [potassium 6-phenyl(acet-1-^{14}C)-amidopenicillanate]	15.7	Freeze-dried solid.	–40	17	2	0.0
D-Biotin(carbonyl-^{14}C)	31.5	Solid.	20	11	nd	—
Cytidine-^{14}C(U)-5′-monophosphate, ammonium salt	193	Aqueous solution.	–40	14	2	0.1

Table 39. (*Continued*)
SELF-DECOMPOSITION OF SOME COMPOUNDS LABELED WITH CARBON-14

Compound	Specific Activity, mCi/mM	Storage Condition	Temperature, °C	Storage Time, months	Decomposition, %	G(−M)
Cytosine-2-^{14}C sulfate	21.5	Freeze-dried solid.	20	39	nd	—
Deoxyadenosine-^{14}C (U)	275	Aqueous solution.	−40	6	nd	—
Deoxycytidine-^{14}C(U)	245	Aqueous solution.	−40	7	nd	—
Deoxyguanosine-^{14}C (U)	272	Aqueous solution.	−40	19	1	0.02
Diquat(ethylene-^{14}C) dibromide [N,N'-ethylene-^{14}C (U)-2,2'-bipyridylium monohydrate]	10.6	Freeze-dried solid.	20	14	nd	—
6-Furfurylaminopurine-8-^{14}C	16.5	Solid.	0	12	1	0.6
Guanine sulfate-8-^{14}C	36.1	Solid.	20	37	nd	—
5-Hydroxytryptamine-3'-^{14}C creatinine sulfate	39.6	Freeze-dried solid.	−40	12	nd	—
Hypoxanthine-8-^{14}C	9.6	Solid.	20	23	nd	—
Nicotinamide (carbonyl-^{14}C)	13.2	Freeze-dried solid.	20	12	nd	—
Nicotinic acid (carboxyl-^{14}C)	27.9	Freeze-dried solid.	20	23	nd	—
Nicotinic acid-6-^{14}C	26.2	Freeze-dried solid under nitrogen.	20	4	nd	—
Orotic acid-6-^{14}C	44.5	Freeze-dried solid.	20	11	nd	—
Paraquat(methyl-^{14}C) chloride [bis(N-methyl-^{14}C)-4,4'-bipyridylium chloride]	10.1	Freeze-dried solid.	20	21	nd	—
Quinolinic acid-6-^{14}C	29.3	Freeze-dried solid under nitrogen.	20	12	nd	—
2-Thiouracil-2-^{14}C	29.9	Freeze-dried solid.	20	23	1	0.2
Thymidine-^{14}C(U)	250	Aqueous solution.	−40	7	nd	—
Thymine-2-^{14}C	58.3	Freeze-dried solid.	20	4	nd	—
Uracil-2-^{14}C	61	Freeze-dried solid.	20	4	nd	—
Uric acid-2-^{14}C	20.8	Solid.	20	24	3	0.7
Uridine diphospho(glucose-^{14}C(U)), ammonium salt	76	Aqueous sodium phosphate buffer solution (0.05M, pH 6.4) containing 2% ethanol.	−20	10	nd	—

Table 39. (*Continued*)
SELF-DECOMPOSITION OF SOME COMPOUNDS LABELED WITH CARBON-14

Compound	Specific Activity, mCi/mM	Storage Condition	Temperature, °C	Storage Time, months	Decomposition, %	G(–M)
Uridine-4-^{14}C-5'-monophosphate, ammonium salt	16.6	Freeze-dried solid under nitrogen.	–20	9	1	0.8
Uridine-^{14}C(U)-5'-monophosphate, ammonium salt	223	Aqueous solution.	–40	12	1	0.05
Uridine-4-^{14}C-5'-triphosphate, ammonium salt	30.4	Aqueous solution containing 2% ethanol.	–20	7	2	1.2

Carbohydrates††

Compound	Specific Activity, mCi/mM	Storage Condition	Temperature, °C	Storage Time, months	Decomposition, %	G(–M)
D-Arabinose-^{14}C(U)	3.9	Freeze-dried solid.	–40	25	2	2.5
2-Deoxy-D-ribose-^{14}C(U)	1.7	Freeze-dried solid.	–40	15	nd	—
Dulcitol-1-^{14}C	8.2	Freeze-dried solid.	–20	5	nd	—
Erythritol-^{14}C(U)	2.3	Freeze-dried solid.	–20	25	nd	—
D-Fructose-^{14}C(U)	86.2	Dried on paper.	–40	14	nd	—
D-Galactosamine-1-^{14}C hydrochloride	3.8	Freeze-dried solid.	–40	16	nd	—
D-Galactose-1-^{14}C	3.8	Freeze-dried solid.	–40	30	nd	—
	35.4	Dried on paper.	–40	11	1	0.3
D-Glucosamine-1-^{14}C hydrochloride	4.1	Freeze-dried solid.	–40	15	3	6.0
D-Glucose-1-^{14}C	39.7	Dried on paper.	–40	8	nd	—
	39.7	Aqueous solution containing 3% ethanol.	–40	6	nd	—
	57.5	Aqueous solution containing 3% ethanol.	–40	3	nd	—
D-Glucose-2-^{14}C	21.1	Dried on paper.	–40	17	1	0.3
	33.4	Aqueous solution containing 3% ethanol.	–40	5	nd	—
D-Glucose-6-^{14}C	27.5	Aqueous solution containing 3% ethanol.	–40	6	nd	—
	35.2	Aqueous solution containing 3% ethanol.	–40	8	nd	—
D-Glucose-^{14}C(U)	194	Dried on paper.	–40	12	nd	—
	3.9	Aqueous solution.	–40	10	5	15.8
	65.7	Aqueous solution containing 3% ethanol.	–40	15	2	0.3
	196	Aqueous solution containing 3% ethanol.	–40	7	nd	—
L-Glucose-1-^{14}C	2.9	Freeze-dried solid.	–40	27	nd	—

Table 39. (*Continued*)
SELF-DECOMPOSITION OF SOME COMPOUNDS LABELED WITH CARBON-14

Compound	Specific Activity, mCi/mM	Storage Condition	Temperature, °C	Storage Time, months	Decomposition, %	G(−M)
D-Glucose-1-^{14}C-6-phosphate, disodium salt	3	Freeze-dried solid.	−40	30	nd	—
D-Glucose-^{14}C(U)-6-phosphate, disodium salt	127	Dried on paper.	−40	23	nd	—
Lactose-1-^{14}C	7.5	Freeze-dried solid.	−40	11	nd	—
Maltose-^{14}C(U)	4.5	Freeze-dried solid.	−40	10	nd	—
D-Mannose-1-^{14}C	3.5	Freeze-dried solid.	−40	23	2	3.1
	31.2	Dried on paper.	−40	36	nd	—
D-Mannose-^{14}C(U)	4.9	Freeze-dried solid.	−40	35	2	1.4
Methyl (α-D-gluco-^{14}C(U)) pyranoside	131	Dried on paper.	−40	35	nd	—
D-Ribose-1-^{14}C	8.2	Freeze-dried solid.	−40	11	nd	—
	26.8	Dried on paper.	−40	22	1	0.2
D-Ribose-^{14}C(U)	2	Freeze-dried solid.	−40	16	nd	—
Sorbitol-1-^{14}C(D-glucitol-1-^{14}C)	3.2	Freeze-dried solid.	−20	12	nd	—
L-Sorbose-^{14}C(U)	2.5	Freeze-dried solid.	−40	12	nd	—

Steroids

Compound	Specific Activity, mCi/mM	Storage Condition	Temperature, °C	Storage Time, months	Decomposition, %	G(−M)
Cholesterol-4-^{14}C	55.8	Benzene solution.	20	3	nd	—
Cholesteryl(linoleate-1-^{14}C)	16	Benzene solution.	20	11	2	1.4
Cholesteryl-4-^{14}C linoleate	20.9	Benzene solution.	20	23	nd	—
Cholesteryl(oleate-1-^{14}C)	10.4	Benzene solution.	20	23	2	1.0
Cholesteryl-4-^{14}C palmitate	21.2	Benzene solution.	20	22	2	0.5
Corticosterone-4-^{14}C	32.4	Benzene solution containing 2% ethanol.	20	14	nd	—
Cortisol-4-^{14}C‡	30.4	Benzene solution containing 10% ethanol.	20	13	1	0.3
Cortisone-4-^{14}C	25.2	Benzene solution containing 10% methanol.	20	8	4	2.5
	29.7	Benzene solution containing 10% methanol.	20	8	9	4.9
Dehydroepiandrosterone-4-^{14}C	27.5	Benzene solution.	20	23	nd	—

Table 39. (Continued)
SELF-DECOMPOSITION OF SOME COMPOUNDS LABELED WITH CARBON-14

Compound	Specific Activity, mCi/mM	Storage Condition	Temperature, °C	Storage Time, months	Decomposition, %	G(–M)
17-α-Hydroxyprogesterone-4-^{14}C	35.9	Benzene solution.	20	36	nd	—
17-α-Hydroxyprogesterone-4-^{14}C caproate	38.3	Benzene solution.	20	26	nd	—
Estradiol-4-^{14}C	54.4	Benzene solution containing 5% methanol.	20	3	nd	—
Prednisolone-4-^{14}C	25.4	Benzene solution containing 5% ethanol.	20	3	3	4.9
Δ^5-Pregnenolone-4-^{14}C	24	Benzene solution.	20	24	nd	—
Progesterone-4-^{14}C	36.1	Benzene solution.	20	24	nd	—
Testosterone-4-^{14}C	29.2	Benzene solution.	20	24	nd	—
19-nor-Testosterone-4-^{14}C	29.6	Benzene solution.	20	11	1	0.4
	50.2	Benzene solution.	20	4	nd	—

* No detectable decomposition.

† Also see Table 32.

‡ Also see Table 31.

‡‡ Nucleosides and nucleotides are listed under heterocyclic compounds

†† Antioxidant was 0.05% butylated hydroxyanisole + 0.05% butylated hydroxytoluene.

Readers are reminded that, because of the difficulty in many cases of estimating the fraction of its own energy absorbed, this has been assumed to be unity for the purposes of calculating G(–M) values. However, with freeze-dried ^{14}C-labeled compounds it is possible to keep this fraction relatively low by storing it as spread-out as possible.

Sulfur-35 Compounds

Some additional results[29] on the storage of compounds labeled with ^{35}S are given in Table 40. The discrepancy between the two results for sodium cholesteryl sulfate-^{35}S is of interest, as the two batches were prepared by quite different methods. This is evidently another case of susceptibility to impurity, though it is worth noting that both batches were carefully recrystallized, and in fact the more stable batch had a significantly lower initial radiochemical purity.

Results of a more detailed study[29] on the self-decomposition of DL-cysteine-^{35}S are given in Table 41.

Selenium-75 Compounds

L-Selenomethionine-^{75}Se, having an initial specific activity of 1 curie/mM, exhibited no detectable decomposition or racemization when stored for four months at 20°C and a radioactive concentration of 1 mCi/ml, or when stored for a similar period at −30°C and 20 mCi/ml.[29]

Table 40.
SELF-DECOMPOSITION OF SOME COMPOUNDS LABELED WITH SULFUR-35

Compound	Specific Activity, mCi/mM	Storage Condition	Temperature, °C	Storage Time, weeks	Decomposition, %	G(−M)
Captan-^{35}S [N-(trichloromethyl-thio)-tetrahydrophthalamide]	7 9	Solid *in vacuo* over P_2O_5. Solid *in vacuo* over P_2O_5.	20 20	44 27	<4 nd*	<19 —
Dibenzyl sulfoxide-^{35}S	5	Solid *in vacuo* over P_2O_5.	20	29	nd	—
Dimethyl sulfoxide-^{35}S	23	Liquid.	20	14	<4	<10
p-Iodobenzenesulfonyl chloride-^{35}S	160 175 175	Solid. Solid. Solid.	−30 −30 −30	6 15 56	5 nd 11	3.4 — 2.1
L-Methionine-^{35}S	206 281 206 206	Aqueous solution, 5 mCi/ml initially. Aqueous solution, 10.1 mCi/ml initially. Sterile aqueous solution, 5 mCi/ml initially. Sterile aqueous solution, 5 mCi/ml initially.	−30 −30 20 20	5 25 4 10	nd 6† 1 5	— 0.9 0.7 1.8
Mustard gas-^{35}S [bis-2-chloroethyl sulfide-^{35}S]	50 280 280	Liquid. Solution in diethyl ether, 0.8 mg/ml. Solution in diethyl ether, 0.8 mg/ml.	−30 −30 −30	12 30 28	nd 8‡ 24	— 1.1 3.8
Parathion-^{35}S [O,O-diethyl O-p-nitrophenyl phosphorothionate-^{35}S]	2	Liquid.	20	5	4	250**
Phenyl isothiocyanate-^{35}S	235 235	Acetonic solution, 23 mCi/ml initially. Acetonic solution, 7 mCi/ml initially.	−30 −30	5 5	1 nd	0.5 —
Potassium thiocyanate-^{35}S	12	Solid *in vacuo* over P_2O_5.	20	23	nd	—
Sodium cholesteryl sulfate-^{35}S††	5 3.3	Solid *in vacuo* over P_2O_5. Solid *in vacuo* over P_2O_5.	20 20	17 26	nd 34	— 500
Sodium dehydroepiandrosterone sulfate-^{35}S	5	Solid *in vacuo* over P_2O_5.	20	22	<3	<26
Sodium ethanesulfonate-^{35}S	11	Solid *in vacuo* over P_2O_5.	20	51	<4	<12
Sodium hexadecyl sulfate-^{35}S	5.5	Solid *in vacuo* over P_2O_5.	20	25	<3	<22
Sodium octadecyl sulfate-^{35}S	3	Solid *in vacuo* over P_2O_5.	20	45	<3	<33
Sodium pregnenolone sulfate-^{35}S	5	Solid *in vacuo* over P_2O_5.	20	16	nd	—

Table 40. (Continued)
SELF-DECOMPOSITION OF SOME COMPOUNDS LABELED WITH SULFUR-35

Compound	Specific Activity, mCi/mM	Storage Condition	Temperature, °C	Storage Time, weeks	Decomposition, %	G(−M)
Sodium testosterone sulfate-^{35}S	5	Solid in vacuo over P_2O_5.	20	23	nd	—
Sodium tetradecyl sulfate-^{35}S	2.2	Solid in vacuo over P_2O_5.	20	45	<3	<46
2-Sulfanilamido-3-methoxypyrazine-^{35}S	3.5	Solid in vacuo over P_2O_5.	20	39	nd	—
Tetramethylthiuram disulfide-^{35}S	11.5	Solid in vacuo over P_2O_5.	20	26	2	7
Thiopentone-^{35}S [5-ethyl-5-(1-methylbutyl)-2-thiobarbituric acid-^{35}S, sodium salt]	6	Solid in vacuo over P_2O_5.	20	27	nd	—
Thiosemicarbazide-^{35}S	297	Solid in vacuo over P_2O_5.	20	4	4	2.1
	367	Solid in vacuo over P_2O_5.	20	11	2	0.4
	150	Solid in vacuo over P_2O_5.	0	9	1	0.5
	150	Aqueous solution, 1 mg/ml.	0	9	6	3.2
	150	Methanolic solution, 1 mg/ml.	0	9	12	6.6
Thiouracil-^{35}S	2	Solid in vacuo over P_2O_5.	20	16	nd	—
	8	Solid in vacuo over P_2O_5.	20	44	<4	<17
Triethylamine-N-sulfonate-^{35}S	35	Solid in vacuo over P_2O_5.	20	8	20	54
p-Toluene sulfonic acid-^{35}S	170	Solid.	20	10	<4‡‡	<2
	210	Solid.	−30	20	3	0.7

* No detectable decomposition.
† D-Methionine >0.03%.
‡ About one half of the impurity was thiodiglycol.
†† See text.
** This value is liable to more than usual error, in view of the low time, decomposition, and specific activity from which it is calculated.
‡‡ After storage, 36% of the activity became insoluble in ether, although it had the same radiochromatographic purity and infrared spectrum as the soluble fraction.

Table 41.
PRODUCTION OF CYSTINE FROM DL-CYSTEINE-^{35}S HYDROCHLORIDE ON STORAGE IN AQUEOUS SOLUTION

Initial Specific Activity   39.5 mCi/mM
Initial Radioactive Concentration   3.9 mCi/ml
Solutions Stored under Nitrogen at 20°C

Time of Storage, days	Cystine* Present, %	Cystine Formed After 2nd Day, %	G(Cystine)
0	1.9†	—	—
2	7.8†	—	—
15	9.4	1.6	13
33	13.5	5.7	21
70	14.6	6.8	13
126	18.5	10.7	14

* Determined polarographically.
† After the initial sample had been taken, the solutions were sterilized by autoclaving; the bulk of the cystine produced in the first two days is due to this treatment.

Phosphorus-32 Compounds

In the previous chapter DFP-^{32}P (diisopropylphosphorofluoridate-^{32}P was described as having a high rate of self-decomposition.[1] This was an error due to a faulty analytical method. Use of two alternative methods of analysis have made it possible to revise this figure, and from three series of experiments[30] the initial rate of decomposition for this compound in propylene glycol can now be quoted as about 1 percent per week. This is for an initial specific activity of about 80 m/mM and a radioactive concentration of about 300 µCi/ml. The impurity produced by sterilization (2.5 megarads) is now found to be less than 0.2 percent.

ACKNOWLEDGEMENTS

The authors thank many colleagues at the Radiochemical Centre for providing data and for useful discussions. We also thank Dr. John R. Catch for his interest and advice, and the United Kingdom Atomic Energy Authority for permission to publish this review.

REFERENCES

1. Bayly, R. J. and Evans, E. A. *J. Labelled Compounds*, 2: 1–34, 1966.
2. Geller, L. and Silberman, N. *Steroids*, 9: 157–161, 1967.
3. Scholes, G., Shaw, P., Wilson, R. L., and Ebert, M., in *Pulse Radiolysis*, pp. 151–164, Ebert M., Keene, J. P., Swallow, A. J., and Baxendale, J. H., eds. Academic Press, London, England, 1965.
4. Bacq, Z. M. and Alexander, P., *Fundamentals of Radiobiology*, 2nd ed. Pergamon Press, London, England, 1961.
5. Evans, E. A. and Stanford, F. G. *Nature (England)*, 199: 762–765, 1963.
6. Apelgot, S. and Frilley, M. *J. Chem. Phys.*, 62: 838–844, 1965.
7. Sanner, T. *Radiat. Res.*, 25: 586–600, 1965.
8. Henriksen, T. *Radiat. Res.*, 17: 158–172, 1962.
9. Apelgot, S. and Ekert, B. *J. Chem. Phys.*, 62: 845–852, 1965.
10. Apelgot, S., Ekert, B., and Tisne, M. R., in *Proceedings of the Conference on Methods of Preparing and Storing Marked Molecules, Brussels, November 13–16, 1963*, pp. 939–952. European Atomic Energy Community, Brussels, Belgium, 1964.
11. Evans, E. A. *Nature (England)*, 209: 169–171, 1966.
12. Evans, E. A. *Nature (England)*, 209: 196–197, 1966.
13. Evans, E. A., *Tritium and Its Compounds*. Butterworth, London, England, 1966.
14. Willix, R. L. S. and Garrison, W. M., University of California Nuclear Chemistry Division Annual Progress Report *UCRL-16580*, 1965.
15. Hesselbo, T. and Long, R. F., in *2nd International Conference on Methods of Preparing and Storing Labelled Molecules, Brussels, November 28–December 3, Abstracts*, p. 83, 1966.
16. Bayly, R. J., in *Proceedings of the Conference on the Use of Radioisotopes in the Physical Sciences and Industry, Copenhagen, September 6–12, 1960*, 2: 305–308. International Atomic Energy Agency, Vienna, Austria, 1962.
17. Frankel, A. I. and Nalbandov, A. V. *Steroids*, 8: 749–764, 1966.
18. Idler, D. R., Kimball, N. R., and Truscott, B. *Steroids*, 8: 865–876, 1966.
19. Emmerich, H. and Schmialek, P. *Z. Naturforsch.*, 21(b): 855–858, 1966.
20. Hempel, K., in *Proceedings of the Conference on Methods of Preparing and Storing Marked Molecules, Brussels, November 13–16, 1963*, pp. 1009–1019. European Atomic Energy Community, Brussels, Belgium, 1964.
21. Flood, C., Pincus, G., Tait, J. F., Tait, S. A. S., and Willoughby, S. *J. Clin. Invest.*, 46: 717–727, 1967.
22. Wand, M., Zeuthen, E., and Evans, E. A. *Science*, 157: 436–438, 1967.
23. Thunberg, R., *Private Communication*.
24. Peterson, R. E. *Recent Progr. Hormone Res.*, 15: 231–274, 1959.
25. Siegenthaler, W. E., Dowdy, A., and Luetscher, J. A. *J. Clin. Endocrinol. Metab.*, 22: 1092, 1962.
26. Lobotsky, J., *Private Communication*.
27. Comings, P. E. *Exp. Cell Res.*, 41: 677, 1966.
28. Winter, G. C. B. and Yoffey, J. M. *Exp. Cell Res.*, 43: 84, 1966.
29. Ogle, J. R. and Phare, L., *Unpublished Results*. The Radiochemical Centre, Amersham, Bucks, England.
30. Monks, R., *Unpublished Results*. The Radiochemical Centre, Amerhsam, Bucks, England.

SYNTHESIS OF LABELED COMPOUNDS

Reprinted, with modifications, from
The Radiochemical Manual, 2nd ed., pp. 31–55, 268–284, 1966.

By permission of
The Radiochemical Centre, Amersham, Bucks, England.

INTRODUCTION

A labeled compound has one of its atoms or larger structural units substituted in a way that distinguishes this atom or unit from others. This chapter is concerned in particular with radioisotopically substituted compounds, but a molecule may be labeled with stable isotopes or with characteristic chemical groupings; Knoop[1] used phenyl radicals to label long-chain aliphatic acids even before isotopes were known, and similar labeling methods that may seem crude by present-day practice have also been effectively used in the past. Many of the features discussed in this chapter apply to substances labeled with stable isotopes or with characteristic chemical groups. The term "labeled" compounds seems to be coming into general use in the English language, superseding the alternatives, "tagged" and "marked" compounds. These two latter terms are used in the same sense. The practice of describing an isotopically substituted compound by the prefix "heavy", "radio-", or "hot" is in many ways unsatisfactory and is declining.

Interest in labeled compounds is almost exclusively due to their use as tracers. It is dominated by the immense value of tracers in biological research, and particularly in biochemistry. Interest is therefore concentrated on those elements that are either ubiquitous or specially important in biological systems: carbon, hydrogen, nitrogen, oxygen, sulfur, phosphorus, and iodine. It is unfortunate that there are no useful radioactive isotopes of oxygen and nitrogen that are sufficiently long-lived for most tracer experiments, and the relatively low sensitivity of measurement of the stable isotopes of carbon, hydrogen, nitrogen and oxygen has discouraged tracer work with these elements. Ingenious attempts have been made[2] to improve the sensitivity of tracer work with ^{18}O by activation of chromatograms to give the radioactive ^{18}F;[3] a good example of this technique, used to study intracellular phosphate turnover, may be of interest to readers. These methods are unavoidably more cumbersome than conventional tracer work and are at present not widely used.

Labeled compounds are not necessarily covalent compounds, nor are they necessarily organic, but the uses made of them have caused attention to be concentrated on these classes. From a tracer point of view, compounds that ionize lose their identity in solution and should then be thought of as labeled ions.

The wide and growing use of labeled compounds in medical diagnosis is at present mostly concerned with location of organs or structures by selective uptake and external scanning, a rather specialized kind of tracer work. Labeled compounds also find applications in analysis, in organic chemical research and process investigation, and in physicochemical studies; examples are in determinations of vapor pressure and solubility. Although such uses are still overshadowed in volume by uses in academic biochemistry, they remain of great interest, and many of them promise to develop considerably.

Whatever use is to be made of a labeled compound, there are some general considerations that must be continually borne in mind if results are to be correctly interpreted.

"Isotopic" and "Nonisotopic" Labeling

It is, first, important to remember that isotopic labeling identifies an atom; it does not identify a molecule or a physical phase in the absence of further chemical and physical knowledge. The measurement of radioactivity at the end of a tracer experiment shows in itself nothing but the presence and proportion of the tracer atom introduced at the beginning of the experiment in the sample measured. No other inference can be drawn without associated chemical and physical evidence, which may be quite simple or extremely complicated. The labeling of carbon dioxide is a simple, but instructive, example. This might be labeled either with a carbon isotope or an oxygen isotope. The carbon tracer will give a precise reflection of the fate of the carbon atom; in an animal experiment (given by inhalation or by injection as bicarbonate) it will, for example, reveal carboxylation reactions, such as that of Wood and Werkman,[4] to give labeled acids of the citric acid cycle. Since these reactions are relatively inefficient, the carbon tracer will also give, over a short period of time, fairly accurate evidence of carbon dioxide–bicarbonate equilibria, circulation and turnover. Labeling the molecule with an oxygen isotope is quite useless for either of these purposes, since the oxygen of carbon dioxide exchanges rapidly with bicarbonate, and hence with body water. To the biochemist this proposition that isotopic labeling is essentially atomic is so obvious that he would hardly formulate it consciously; a large part of biochemistry is essentially study of the transformation of molecules. Yet even biochemists, having this implicit understanding of the concept, have been known to fall into error—for example, in some attempts to use tritium as a tracer for carbon without due caution.

Labeling of a compound may be either "isotopic" or "nonisotopic". The terminology of tracer work, in this as in other aspects, is unhappily not yet standardized. As used here, "isotopic" labeling refers to a compound in which an atom is replaced by an isotope of the same element, in the same position, and without any other change in the molecule. It refers also to compounds in which more than one atom is replaced under the same conditions. Except for isotope effects, which are not often large enough to prejudice practical use of the compound (tritium is the most notable exception), the evidence obtained from this form of labeling is the most precise and reliable obtainable. To take another simple example, phenol is used as a solvent in refining hydrocarbon oils, and is recovered and recycled; losses of phenol resulting in incomplete recovery could be studied by a tracer method, using phenol labeled with ^{14}C. It could perhaps be assumed—or, if necessary, demonstrated by investigation—that the phenol ring does not break down under the operating conditions, and that being so, the ^{14}C-labeled phenol would be an infallible tracer for phenol itself. Phenol labeled with tritium in one or another of the ring hydrogens might possibly be a valid tracer, but this would depend on the degree to which the carbon–tritium bond is preserved throughout the processes. It will be observed that in both examples the radioactive measurement is only part of the evidence, the other part being a knowledge of the chemistry. Isotopic labeling is, unfortunately, not always attainable in practice. In the example given above it would be much too expensive to use ^{14}C-labeled phenol in experiments on the refinery scale. The labeling of proteins, and indeed of many other natural products not amenable to synthesis, is often not practicable with isotopes of carbon, or even with isotopes of hydrogen.

"Nonisotopic" labeling is not very easy to define, as it would include a number of practices that may or may not be acceptable. It often substitutes a convenient radioactive isotope of an element foreign to the compound—as, for example, when 2,4-dichloro-5(^{131}I)phenoxyacetic acid is used as a tracer for 2,4-dichlorophenoxyacetic acid. This is one of the less justifiable uses, since there is no real need to adopt this expedient; labeling with ^{14}C or tritium, or with ^{36}Cl, is quite practicable. There are much better reasons for the widespread practice of labeling proteins with iodine isotopes, and the conditions under which this can give valid results have been carefully defined.[5-7] To return to a comparable example, orthobromophenol-^{82}Br has been used in an attempt to determine phenol losses in oil refinery practice, but the results were inevitably qualitative in many respects, since phenol and orthobromophenol are quite distinct substances and will, for example, have different partition coefficients. With proper precautions and in suitable circumstances, nonisotopic labeling can be of great value, but it always requires special caution, since chemical, physical or biological processes may distinguish the "foreign" element or compound and produce a startlingly erroneous result. The errors that can arise from working with proteins too heavily substituted with iodine isotopes are perhaps the best known.[8]

Sheppard[9] has proposed the terms "perfect" and "imperfect" labeling, but observes that even isotopic substitution cannot be considered entirely perfect, because there are differences of mass and reaction rate between isotopic atoms and isotopically labeled molecules. Isotopic labeling is, however, as nearly perfect as most tools of scientific investigation can be and stands clearly apart from nonisotopic labeling, which can be very imperfect indeed. An even less refined form of labeling is to use a tracer substance that differs from the material being traced not only in one of the atoms in the molecule, but in its whole chemical or physical nature. Examples are the use of ^{24}Na to trace leaks in water mains, of ^{82}Br to examine water flow and circulation in reservoirs, and of radon or ^{85}Kr to trace air and gas flow. Radioactive tracers of this kind may be classed with dyestuffs such as fluorescin, used to trace water flow, and odorous compounds such as crude thiophene, added to town gas to make it detectable. If properly selected and used, they are entirely satisfactory and useful tracers, but they always require special care in evaluating the chemical and physical evidence associated with the tracer measurements.

Isomerism

Isotopic labeling creates a new element of isomerism in chemical compounds. It is most obvious in compounds of carbon and hydrogen isotopes, but other isotopes show it also; examples are iodine labeling in thyroxine and sulfur labeling in sodium thiosulfate. The term "specific labeling" generally denotes labeling in one specified position. Since carrier-free isotopes are rarely used in compound labeling, this means that a specifically labeled compound is essentially a mixture of one radioactive species substituted in one specified position with unlabeled molecules. To take a simple example, propionic acid can exist in three forms specifically (singly) labeled with ^{14}C, and in three forms specifically labeled with tritium (the tritium, of course, being readily exchangeable in the hydroxyl position):

$$^{14}CH_3.CH_2.COOH \qquad CH_3.^{14}CH_2.COOH \qquad CH_3.CH_2.^{14}COOH$$
$$CH_2T.CH_2.COOH \qquad CH_3.CHT.COOH \qquad CH_3.CH_2.COOT$$

Multiple labeling may arise from introduction of isotopes of two different elements, different isotopes of the same element, or the same isotope of the same element in two or more different positions. The term "double labeling" is used for all of these when only two positions or two isotopes are involved. It is most commonly used for labeling with two different isotopes, not necessarily in only two specific positions. A distinction immediately arises in that multiple labeling may be truly molecular, so that one molecule is substituted in more than one position, or it may be only statistical (i.e., the overall labeling pattern of a large number of molecules is that described, although any individual molecule may show only one element of the pattern). This may be illustrated diagrammatically:

$$n \underset{O \diagup \diagdown OH}{\overset{^{35}S \diagdown \diagup OH}{S}} \; + \; n \underset{O \diagup \diagdown OH}{\overset{S \diagdown \diagup OH}{^{35}S}} \quad \text{is equivalent for most purposes to} \quad 2n \underset{O \diagup \diagdown OH}{\overset{^{35}S \diagdown \diagup OH}{^{35}S}}$$

Only very rarely does the application of the tracer compound call for strictly molecular multiple labeling; examples are the study of primary (internal) radiation effects and the formation of labeled recoil products,[10,11] kinetics of diffusion, and possibly some aspects of isotope effects. But for nearly all purposes statistically uniform labeling is all that is needed. This is an important practical point, because multiple labeling is very often most easily attained by mixing the separately prepared singly labeled molecules—a particularly convenient method when the two labels are nuclides of differing half-life, as the relative specific activities may then be adjusted as required. A synthesis that on paper leads to a compound labeled with two nuclides (whether of the same or different elements) does not mean that every molecule is substituted with both isotopes. The probability of the molecule being doubly labeled depends on the isotopic abundance of each isotope used at the stage in the synthesis at which labeling arose, and as these abundances are commonly quite low, the product usually consists of a large proportion of unlabeled molecules, a small proportion of molecules labeled with one or the other isotope, but not both, and a very small proportion of molecules labeled with both isotopes.

Isotopic labeling can make apparent a form of biological asymmetry (sometimes loosely, but incorrectly, called "isotopic asymmetry") that is otherwise undetectable and, indeed, was not recognized until fairly recently. It is really a manifestation of the asymmetry of an enzyme, which may be able to distinguish completely between two functional groups in the molecule that appear chemically identical. The best known example is of D- and L-citric acid-1-^{14}C.[12,13]

The reader will observe that, if the molecule is bound to an enzyme by three points, such as (1)(2)(3) or (2)(3)(4), the two terminal carboxyls (2) and (4) are no longer equivalent and are open to specific, selective reaction.

The terms "uniform", "general", and "random" labeling call for comment. They are most relevant to compounds of carbon and hydrogen isotopes. They are not precise terms, and there is no consensus of opinion of their exact meanings, but they are in use, and two of them, at least, can serve as convenient abbreviations. "Uniform" or "U" may be taken as meaning that the labeling is distributed with statistical uniformity throughout all the atoms of the element concerned, without further definition of the molecular species present. "General" or "G" may imply, though not always, labeling in all atoms concerned, which is known or suspected to be nonuniform. The term "random" appears to be used in both senses and might well be dispensed with.

Isotopic substitution raises problems of formal nomenclature, which have been discussed,[14] but which have not been resolved by general agreement in practice between the various national and international bodies interested in the subject. The Chemical Society of London and the Biochemical Society have collaborated in formulating recommendations for use in their publications.[15] The problem is most acute with carbon and tritium labeling. Until there is general agreement on terminology, everyone who names labeled compounds should strive to adopt a consistent and explicit convention and avoid at least the worst ambiguities; irritating examples of these, such as tryptophan-α-^{14}C or pentobarbitone-2-^{14}C, are so easily avoided or clarified, even without an international convention, that there is no excuse for using them.

Tryptophan
(2-indolylalanine)

Pentobarbitone
(5-methyl-5-(1-methylbutyl)barbituric acid)

Synthesis by Chemical and Biological Methods

When chemical synthesis is possible, it is usually the method of choice for preparation of a labeled compound; but, unfortunately, an efficient synthesis is not always available. Synthetic methods generally give the greatest control over yield, position of labeling, and purification of the product. The use of purified enzymes to carry out specific reactions may logically be classified with chemical synthetic methods. If care is taken to look out for possible rearrangement, chemical synthesis usually gives completely specific labeling with ^{14}C, and (with some reservations) even with tritium and other isotopes. The complexity of biochemical reactions is such that labeling patterns from biological preparations should never be readily assumed.

Chemical syntheses that give pure stereoisomeric forms are the exception rather than the rule, and resolution of racemic mixtures into pure optical isomers on a small scale is not easy by classical methods. Enzyme reactions are not, of course, subject to these disadvantages. Uniform labeling, with the high specific activity that can be a consequence (particularly with ^{14}C), is not, as a rule, readily attained by synthesis; only occasionally, with comparatively simple compounds, do symmetrical intermediates make uniform labeling possible. Should this not be clear, the reader may refer to the example given of the labeling of pyruvic acid. As explained there, uniform labeling is conveniently attained for virtually all practical purposes by mixing the single-labeled forms. To obtain the highest possible specific activity, however, it would be necessary to carry out the following series of reactions, using isotopes of the highest available specific activity at each stage.

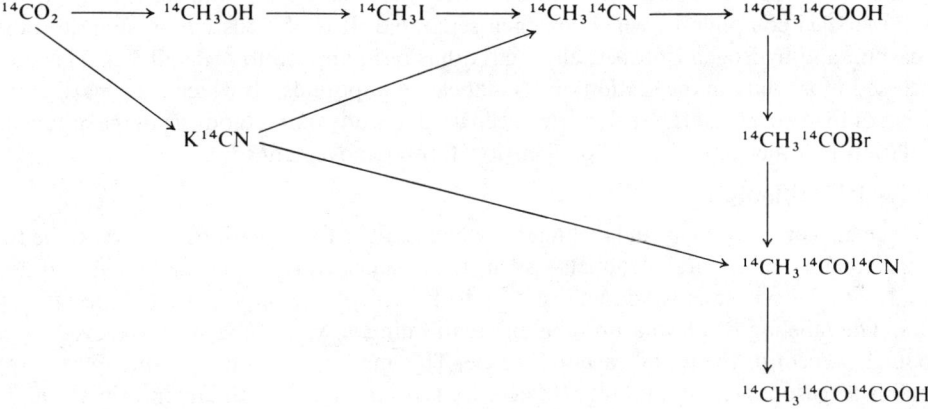

The risks of poor yields or accidental loss in such a reaction sequence are much greater than in the separate syntheses.

Biosynthetic methods, which at present apply to few isotopes other than ^{14}C, are often available when chemical synthesis is not possible. They very often have the advantages of giving uniform labeling, if this is desired, of high specific activity, and of naturally occurring stereoisomeric forms. Biosynthetic methods at present are limited in scope, lacking in control of product and positional labeling, and can be very inefficient. They also, as a rule, require isolation of a product from a very complex mixture, which makes purification particularly difficult. These disadvantages often do not apply to reactions carried out by specific enzymes.

Radiation Synthesis

Radiation synthesis (activation, recoil, or "hot-atom" methods), which is sometimes superficially promising, is rarely useful in practice for preparing tracer compounds, since the yields and specific activities are usually very low and purification is often extremely difficult.[16] Only for tritium labeling is this approach of much use, with some notable success to its credit; but even with tritium the results are not, at present, predictable with any confidence.

An interesting example of a useful hot-atom synthesis with ^{14}C has been given by A. P. Wolf,[17] but this labeling was a rather favorable case. The compound concerned, isonicotinic hydrazide,

$$\underset{\underset{N}{\bigcirc}}{\overset{\overset{O}{\parallel}}{C-NH.NH_2}}$$

is more radiation-resistant than most organic compounds, and the specific activity achieved (0.575 mµCi/mg carbon—a factor of 10^5 lower than might be achieved by synthesis) was sufficient for the intended use of the compound. Most biological experiments require much higher specific activities and are concerned with more sensitive compounds. Wolf is, however, probably justified in believing that the method could be used with ^{14}C to meet some other needs.

"Shot Gun" Methods

Many radiation syntheses with ^{14}C or tritium could be classed under this heading, as they produce a relatively large quantity of labeled by-products. The most widely known method, however, is "isotope farming".

The idea of isotope farming is simple and attractive. The isotope, in a cheap and readily attainable form, is given to a living organism that can assimilate it and turn it into a variety of useful labeled compounds, which are then separated. It is obviously best suited to labeling with carbon and hydrogen isotopes, although it has been applied to ^{32}P and ^{35}S. This method is discussed more fully in the section on ^{14}C-labeled compounds, to which it primarily applies. Labeling of living organisms has also often been used to study the metabolism of the organisms,[18] assimilation by other organisms,[19] or their location or movement.[20]

Unsolved Problems

There are some problems in labeling, for which satisfactory solutions have yet to be found. Isotopic labeling of natural products—such as proteins, adrenocortical steroids, and plant alkaloids—by biosynthesis is so inefficient as to be acceptable only in rather special circumstances. The labeling of plasma proteins of small animals with ^{14}C can be achieved in yields of about 1 percent of the initial carbon dioxide. This makes the product rather expensive, but it remains a reasonable undertaking. It becomes less reasonable with larger animals, and presents special difficulties, for example, for human plasma proteins. Production of any of the less abundant animal proteins, such as insulin, by the same method (feeding the animal with labeled amino acids) would make the product unacceptably expensive, merely because the conversion of amino acids into insulin would be so poor. Continual advances in synthetic methods—for example, the recent improvements in methods of preparing peptides—will gradually solve some of these problems, but it will be a long time before all of them are disposed of. It is for this reason that nonisotopic labeling—with iodine, for example—has been extensively used with proteins, and with much success.

Natural products such as coal tar or petroleum, or the complex mixtures of partially refined products obtained from them, cannot be labeled in a strictly representative way. This is simply because the natural preparation cannot be reproduced artificially. It seems peculiarly frustrating that a tracer method cannot be applied to such obvious problems as the determination of residual petroleum solvents in paint films or in synthetic resins. Even tritium labeling by exchange is unlikely to give a reliable result, since it is very unlikely that the hydrogen labeling would be anything like uniform among all the components. The best that can be done is to label one or more of the pure known components of the petroleum fraction by synthesis and study these individually.

Even when labeling is theoretically possible, the efficiency of the preparation may be too low and the labor too great to make it acceptable. ^{14}C labeling of aromatic compounds in specific positions is an obvious example. It would be possible to work out on paper syntheses of, for example, dihydroxyphenylalanine, starting from carbon dioxide, leading to specific labeling in the various carbon atoms of the ring. Some ingenious and patient work has, indeed, been directed to syntheses of this kind,[21, 22] but the many stages and poor yields only serve to underline the difficulty of providing such compounds as tracers at useful specific activities and at a reasonable cost.

The Role of the Isotope Supplier

Most people prefer to buy labeled compounds, rather than make them, despite what appear to be high prices. Provision of high specific activities and ready availability is often possible only by centralizing production in some way.

Even when preparations that do not seem technically feasible are set aside, the demand for labeled compounds outstrips in variety their production by commercial suppliers, and it seems inevitable that it will always do so. This may also be said of fine chemicals for research in general, and for the same obvious reason—that research work is always more advanced and more diversified than production for general supply can ever be. Only a small fraction of known organic compounds, for example, is called for in labeled forms, but this is still a very large number, which is multiplied by the isomerism arising from positional labeling.

Many of the inquiries received at the Radiochemical Centre are for labeled compounds of quite specialized interest, which are perhaps needed for only one investigation. As the preparation of the tracer compound can be more tedious and laborious than the tracer experiment in which it is to be used, any serious attempt to supply all these compounds would, in effect, be equivalent to taking over a significant fraction of the world's research work in many fields of science. The commercial supply of labeled compounds is growing vigorously, but it will always be necessary for individual research workers to supplement it with their own efforts. Progress in many departments of biochemistry, and even in medicine, may come to demand that more effort be given to the preparation of labeled compounds.

Tradition in Labeling

Before the advent of isotopic tracers, attempts to study the selective uptake of chemical compounds by living tissues had to use other methods of detection, such as color or color reactions. The long-standing use of selective dyes in histology and the sensitivity of spectrophotometric methods caused attention to be concentrated on materials that were "colored" in the visible or ultraviolet regions. When tracer isotopes arrived, those that were most readily available were naturally most readily used. ^{131}I and ^{32}P became especially prominent, because of their successful applications in medicine. The use of tritium and, to a lesser extent, of ^{14}C was inhibited by the earlier difficulties of measurement.

These traditions still betray their influence on the approach to research problems. The study of tissue-localizing agents still tends to concentrate on dyestuffs, X-ray-contrast agents, and similar materials, admittedly with some justification, since much is known about them already; but insufficient attention is perhaps given to compounds of other classes, which can now be studied equally well by tracer methods. Isotopes of iodine are naturally of particular value, due to the chemical character of the element and the ready availability and cheapness of the isotopes; nevertheless, one feels that their use as tracers in preference to other isotopes sometimes arises from tradition rather than for any other reason. An instructive recent review by Silver[23] shows at first glance a remarkable range of work in progress on medical uses of isotopes, but on closer study evidence can be found of limitations of the kind referred to.

The technical applications of dilution analysis (as apart from the biological applications) also show a disappointing tendency to "run in grooves". Of the inquiries received at the Radiochemical Centre, for example, a large proportion is limited to interest in a particular published application, down to the last detail of the compound in question and the isotope used to label it. It appears that the versatility of this method of analysis, drawing on the wide range of tracer isotopes now available, will not be recognized until many more examples have been published. It is not a convincing explanation that the labeled compounds required are not available. For this application the requirements are particularly easy to meet. There are no reasons to specify a particular isotope, a particular position of labeling (it need not even be known completely), or a high specific activity. Even the purity is less exacting than in many biochemical tracer applications.

CARBON-14 COMPOUNDS

Requirements and Basic Factors

The intensive study of biochemical processes during the last two decades has been greatly assisted by the use of carbon isotopes as tracers, and has come to rely heavily upon them. ^{14}C has virtually replaced the stable isotope ^{13}C, whose measurement is by comparison rather clumsy and insensitive, and the radioactive isotope ^{11}C, whose half-life of twenty minutes is too short to allow extended experiments.

^{14}C has a particularly favorable combination of properties as a tracer isotope. It emits only beta particles, which are sufficiently energetic to make measurements fairly simple, but weak enough to make shielding unnecessary and to permit fairly good definition in autoradiography. Its half-life of 5,760 years makes it unnecessary to correct for decay. It is a reactor-produced isotope, which may be obtained in adequate quantities and at high specific activity. Its toxicity is relatively low and is not a practical handicap in most current applications, because organic compounds generally are metabolized and excreted with a short biological half-life.

^{14}C is nowadays produced by neutron irradiation of beryllium nitride or aluminum nitride, and although a variety of one-carbon compounds can be extracted directly from the target material,[24] the only form readily available in practice is carbon dioxide. This is the most distinctive feature that sets carbon isotope synthesis apart from ordinary synthetic organic chemistry, but two other factors are perhaps of even greater importance. They are the cost of the isotope and the common need for high specific activities. Because of the low cross section of the reaction $^{14}N(n,p)^{14}C$ and the long half-life of ^{14}C, large amounts of target material must be exposed to high fluxes over long periods to produce useful quantities. This makes ^{14}C expensive, compared to many other reactor-produced isotopes.

The ordinary worker with tracers, setting out to prepare a ^{14}C-labeled compound, is fortunately no longer limited to carbon dioxide as a starting material. During the past fifteen years, a number of laboratories have specialized in the synthesis of ^{14}C-labeled compounds, and many hundreds of these are now readily available, including many useful synthetic intermediates. For further information, the catalogs of individual suppliers, or reference books such as the *Isotope Index*[25] or the directory of the International Atomic Energy Agency,[26] should be consulted.

Specific Activities

The need for high specific activities in ^{14}C-labeled compounds arises from the high biological potency or low normal physiological concentration of the compounds to be studied. ^{14}C, if it were of 100 percent isotopic abundance, would have a specific activity of 4.6 mCi/mg element, equivalent to 64 mCi/milliatom; the isotopic abundances currently available are upwards of 60 percent, or about 40 mCi/milliatom. The lower limit at which organic synthesis ceases to be practicable is commonly 2 to 5 mM. This means that preparative batches at the highest available specific activity are much larger in radioactive quantity than any individual worker normally requires. Work with such small masses and large amounts of expensive isotope is therefore rather specialized, and this is one reason why it has tended to concentrate in a small number of laboratories.

The preparation of ^{14}C-labeled compounds is fairly well documented. Calvin's *Isotopic Carbon* remains an excellent basic textbook.[27] Murray and Williams' *Organic Syntheses with Radioisotopes* is a compendium of many hundreds of preparative methods for syntheses with ^{14}C and other isotopes;[28] it does not, however, give any attention to biological preparations. The bibliographies of Nevenzel,[29] although unfortunately not continued beyond 1954, are useful and thorough. *Carbon-14 Compounds*[14] is a rather more recent review and guide to the literature. The continuous and vigorous development of synthesis and use of ^{14}C-labeled compounds inevitably means that none of these texts now gives a complete and current conspectus of the subject.

Chemical Scale

Although small-scale work sometimes has advantages (such as the use of paper chromatography on a preparative scale), it is on the whole more troublesome and tedious than conventional bench-scale work of 0.05 to 0.1 mole or more, and anyone making a ^{14}C-labeled compound for his own use should not aim at a specific activity higher than he really needs. Many research workers are still using comparatively insensitive methods of measurement. Use of more sensitive equipment would often economize substantially in expenditure on isotope and effort, and extend the scope of work generally; in particular, it sometimes permits reduction of the necessary specific activity for the starting material. The pronounced extension in the use of liquid scintillation counting is undoubtedly leading to greater economy in the use of tracers and is extending the range and precision of tracer methods.

Laboratories engaged in an organized supply of ^{14}C-labeled compounds must, in any event, try to attain the highest specific activities for compounds that are in wide demand, simply because many users demand it. For such compounds no reduction in cost will be attained by carrying out separate preparations at lower levels, because the dilution of the existing material at high specific activity is even easier. Experience shows that users sometimes call for specific activities higher than they really need, but for many others the highest levels are genuinely necessary. The synthesis of a compound for a specific purpose—one that does not meet a general need—will be made more difficult, slower, and more costly, if the specific activity called for is higher than necessary.

Chemical Synthesis

Synthesis of ^{14}C, in which one might logically include the use of specific enzymes, has the advantage of being relatively efficient and controllable; it gives a high degree of choice over the product and position of labeling. Synthetic chemicals are also, as a rule, more easily purified than biological products. In the nature of things, starting from a one-carbon compound such as carbon dioxide, synthesis generally results in labeling in a specific carbon atom, although multiple or uniform labeling may effectively result by passing through a symmetrical intermediate such as benzene or succinic acid.

Synthesis from carbon dioxide is necessarily systematic. There are only twelve compounds or types of compound that are attainable in one stage. Of these, only the first six are of much importance, and the last five find practically no use.

1. Carboxyl-labeled acids
2. Methanol
3. Cyanide
4. Carbides (acetylene)
5. Formic acid
6. Cyanamides
7. Carbon monoxide
8. Urea (from ammonium carbonate)
9. Methane
10. Formaldehyde
11. Elementary carbon
12. Alkyl carbonates

The carboxyl-labeled acids are important not only in themselves, but as synthetic intermediates. Some examples of their use are shown in Table 42 and Table 43.

Table 42.
CARBON-14 COMPOUNDS—SYNTHESES VIA CARBOXYL-LABELED ACIDS

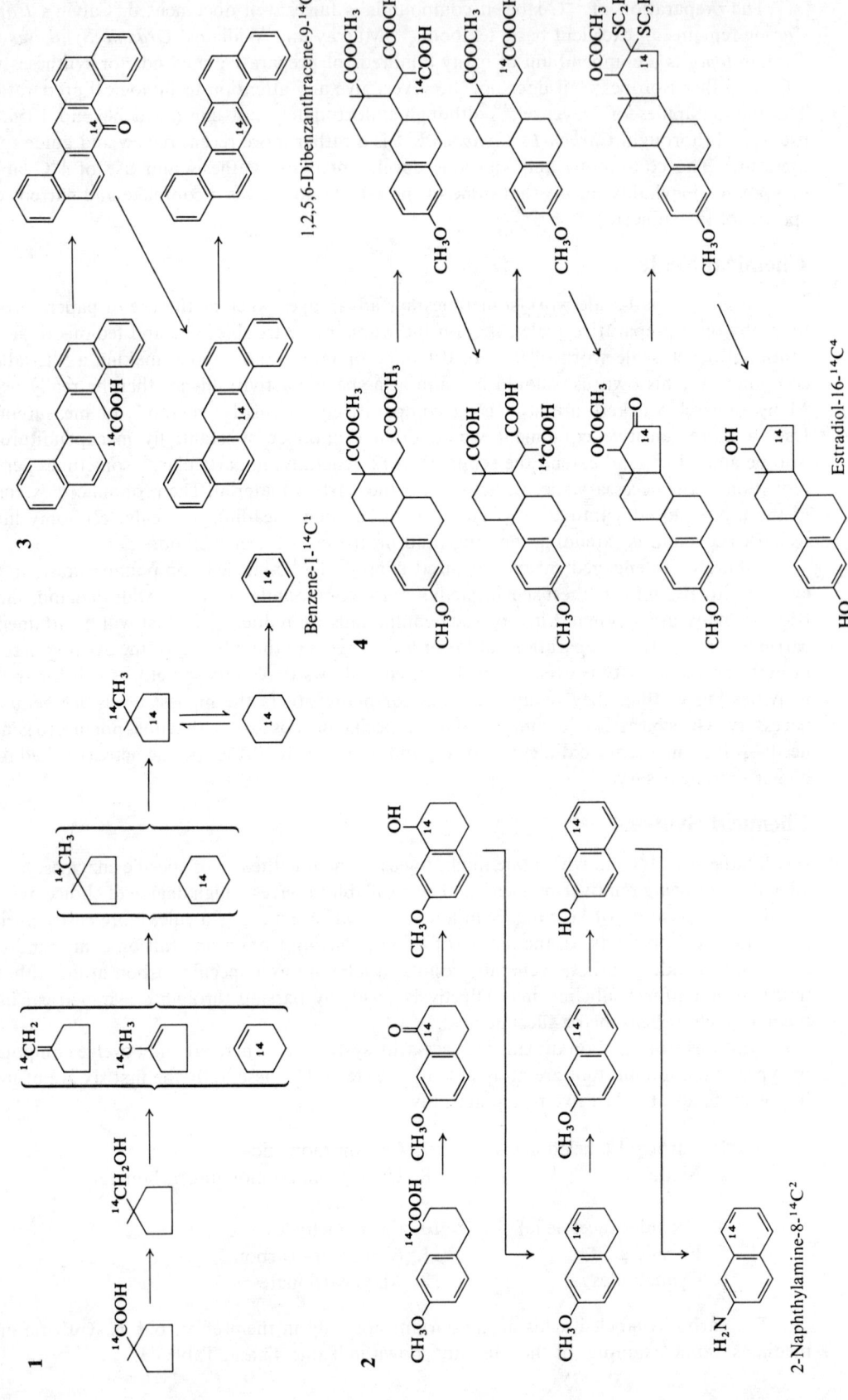

Table 42. (*Continued*)
CARBON-14 COMPOUNDS—SYNTHESES VIA CARBOXYL-LABELED ACIDS

5 $^{14}COOH$-naphthalene → $^{14}CH_3$-naphthalene → 2(Methyl-^{14}C)-1,4-naphthoquinone[5,6]

6 Phenyl-CH(CH_3)-CH$_2$-$^{14}COOH$ → tetralone (C=O labeled ^{14}C, with CH_3) → 2-methylnaphthalene (^{14}C labeled) → 2-Methyl-1,4-(naphthoquinone-4-^{14}C)[7]

7 $HOO^{14}C$-C$_6$H$_4$-CH$_3$ with propyl chain → methyl-tetralone (^{14}C at C=O) → 2-methylnaphthalene → 2-Methyl-1,4-(naphthoquinone-8-^{14}C)[8]

8 $^{14}COOH$-benzene → $^{14}CH_2OH$-benzene → $^{14}CH_2Cl$-benzene ; C_6H_5-$^{14}CH_2$-CH(NH$_2$)-COOH 3-Phenyl (Alanine-3-^{14}C)[9]

9 C_6H_5-CH_2-$^{14}COOH$ → C_6H_5-CH_2-$^{14}CH_2OH$ → C_6H_5-CH=$^{14}CH_2$ Styrene-β-^{14}C[10,11]

10 $(CH_3)_2CH^{14}COOH$ → $(CH_3)_2CH^{14}CH_2OH$ → $[(CH_3)_2CH^{14}CH_2-]_2Cd$ + cholesteryl acetate COCl → acetate-ketone intermediate → Cholesterol-24-^{14}C[12]

Table 43. CARBON-14 COMPOUNDS—SYNTHESES VIA ACETIC ACID

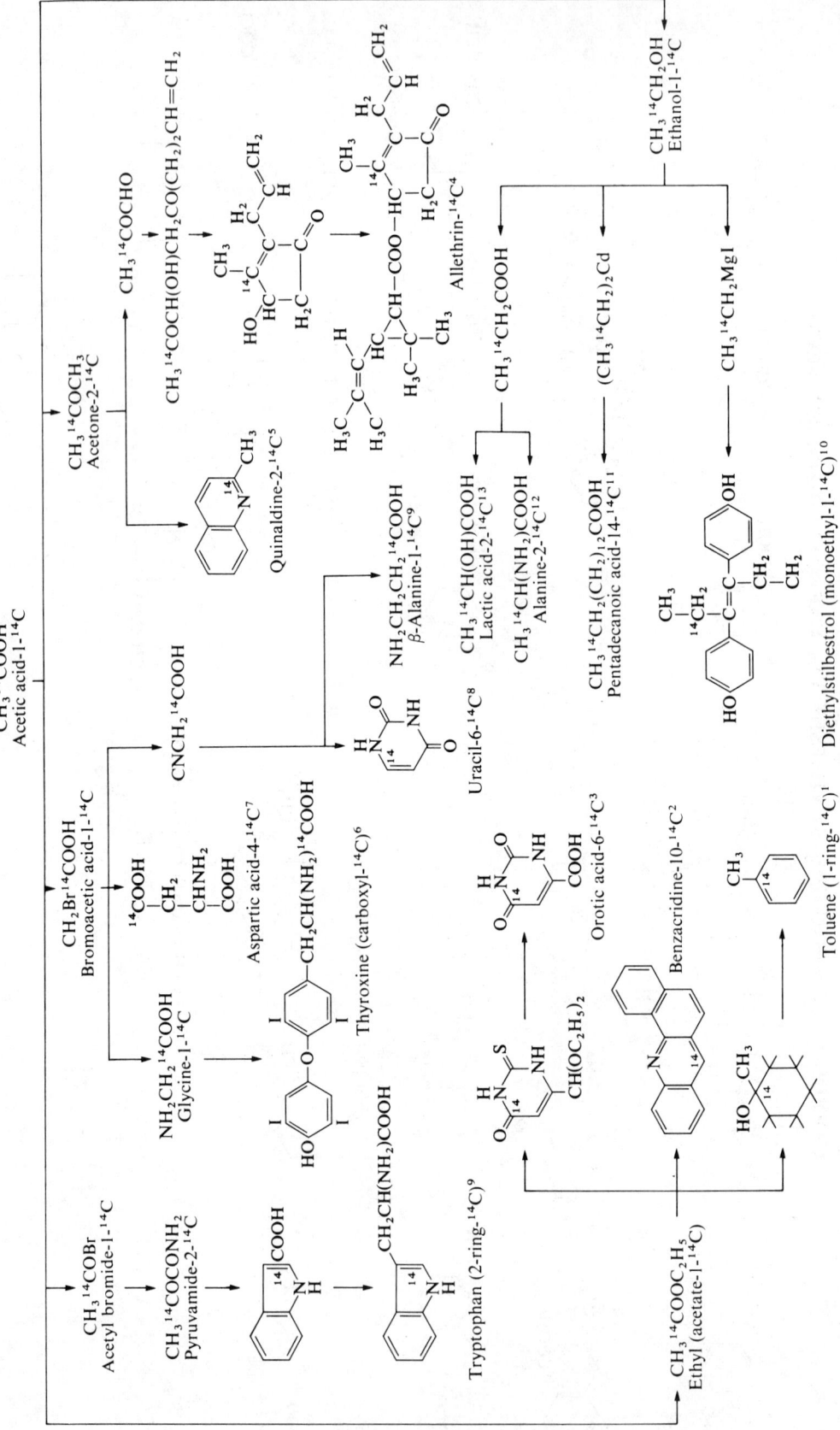

Methanol, cyanide, acetylene, formic acid, and cyanamide are chiefly of importance as starting materials for syntheses. The reader will perhaps best gain an overall view of this kind of synthesis from Tables 44 to 48, which show the remarkable variety of carbon-labeled compounds fairly rapidly derived from these five starting materials. These routes are necessarily only a small selection of the recorded preparations; for a more thorough and documented account of the primary reactions and the labeled compounds to which they lead, the reader may refer to *Isotopic Carbon*[27] and *Carbon-14 Compounds*.[14]

Labeling by synthesis is highly specific. Unsuspected rearrangements can result in the appearance of activity in other than the predicted carbon atom, but in practice this is uncommon—so much so that, when degradation appears to show a rearrangment, it is prudent to check that the degradation reaction itself is not at fault.[30] With the labeling limited to one carbon atom in each molecule, the specific activity of the compounds is correspondingly limited. It is a disadvantage of the synthetic approach to ^{14}C-labeled compounds that, with rare exceptions, it gives racemic forms and mixtures of epimers, so that separation of the desired isomer is often both troublesome and inefficient.

The synthesis of a compound from carbon dioxide often passes through many successive stages. In checking such a synthesis it is essential to check it all the way through, from carbon dioxide to product, exactly as the final preparation will be done. It is not sufficient to check each successive stage, using pure reagents. In practice, in small-scale preparation, it cannot be assumed that intermediates will always be as pure as reagents off the shelf, and a small amount of impurity can have a catastrophic effect on a stage yield. This trouble may, however, be reduced, if—as is normally the case—some use can be made of inactive carriers to dilute the specific activity. By judicious use of carrier at appropriate stages good yields can often be assured. It must, however, not be forgotten that any labeled impurities will remain present at higher specific activity, unless removed by appropriate purification.

Biosynthesis

Biosynthesis methods with ^{14}C fall into several classes. The most interesting, and the most potentially valuable, is the use of specific enzyme reactions with substrates labeled either by synthesis or biosynthesis. Such enzyme methods are really akin to chemical synthesis in the degree of control over yield and position of labeling they afford. A number are now in use, examples being in the preparation of thymidine[31] and other nucleosides and nucleotides, but they are only a fraction of those known to be available. The practical difficulty is in developing a demonstrated possibility into an efficient preparative method.

Other biosynthetic methods do not lend themselves to systematic exposition. The presently most useful depend on photosynthesis. Detached plant leaves can produce excellent yields of the more common carbohydrates—sucrose, glucose, fructose, and starch.[32–35] Direct photosynthetic methods for less familiar carbohydrates have hardly been developed. The other major biosynthetic source is the green alga *Chlorella*, which can be used to produce a high yield of protein[36] or of lipid.[37] Other microorganisms can, of course, be used; examples are sulfur bacteria grown on carbon dioxide and yeasts, or bacteria grown on acetate or carbohydrates, but these are on the whole of minor importance.

Much attention was at one time paid to a more glamorous aspect of isotope farming—the growth of intact plants, from seedling to maturity, in ^{14}C dioxide. This approach has, indeed, been used to obtain interesting natural products, such as cardiac glycosides and alkaloids uniformly labeled with ^{14}C. For some problems in labeling it is undoubtedly the only possible solution, but it has serious practical disadvantages. The technique of closed culture of plants is difficult and expensive, the yields of desired products are often exceedingly low, and the long period of growth seriously limits the specific activities attainable. Plants begin to show pronounced abnormalities at 1 mCi/g of carbon (12 µCi/milliatom), whereas short-term photosynthesis in detached leaves and cultures of ^{14}C-labeled *Chlorella* may be conducted at upwards of 30 to 40 mCi/milliatom under favorable conditions, and synthetic preparations are often carried out at this level also. It is, therefore, impossible to attain really high specific activities in this type of isotope farming, and it has consequently proved to be of only limited value.

Table 44. CARBON-14 COMPOUNDS—SYNTHESES VIA METHANOL

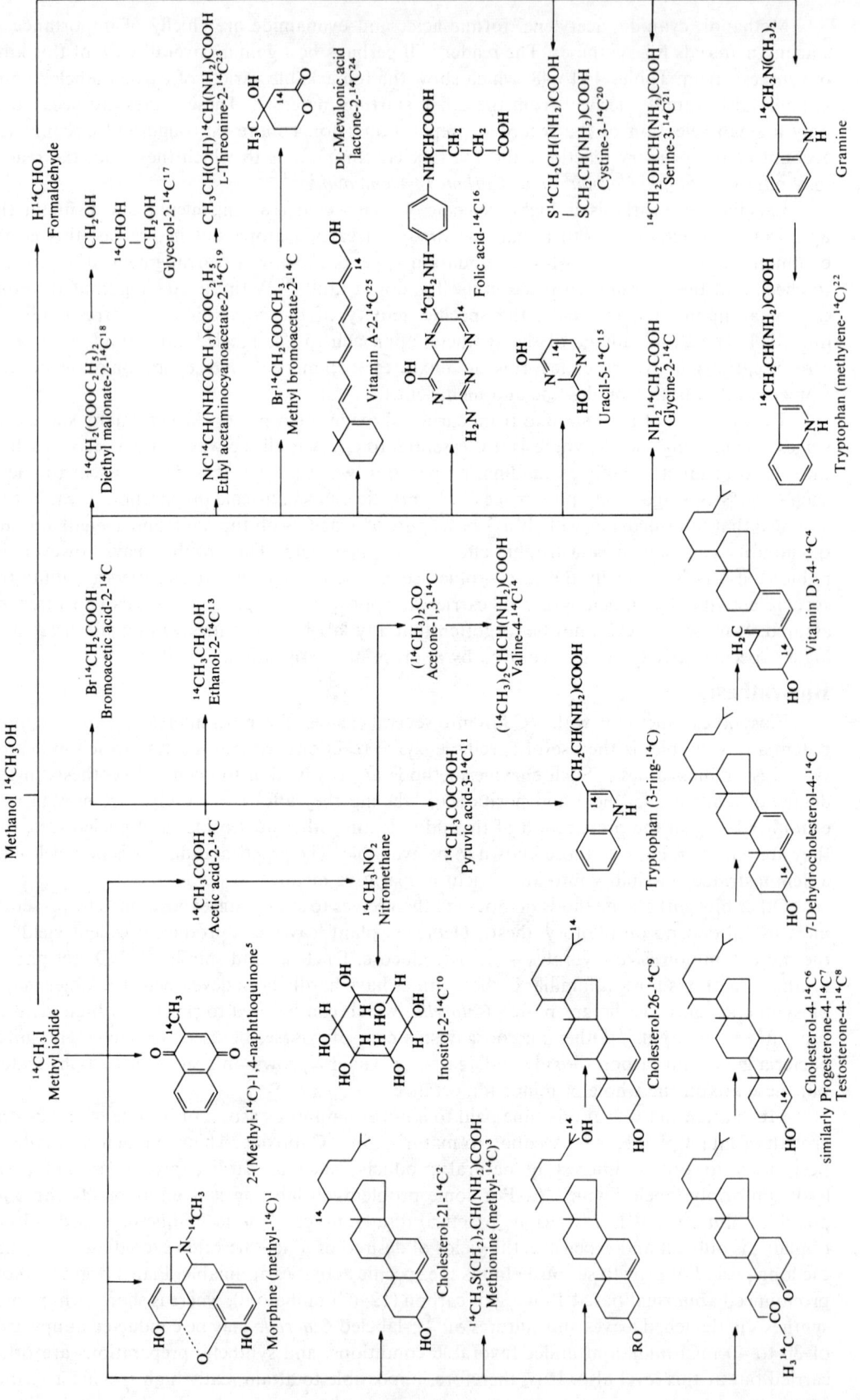

Table 45. CARBON-14 COMPOUNDS—SYNTHESES VIA CYANIDE

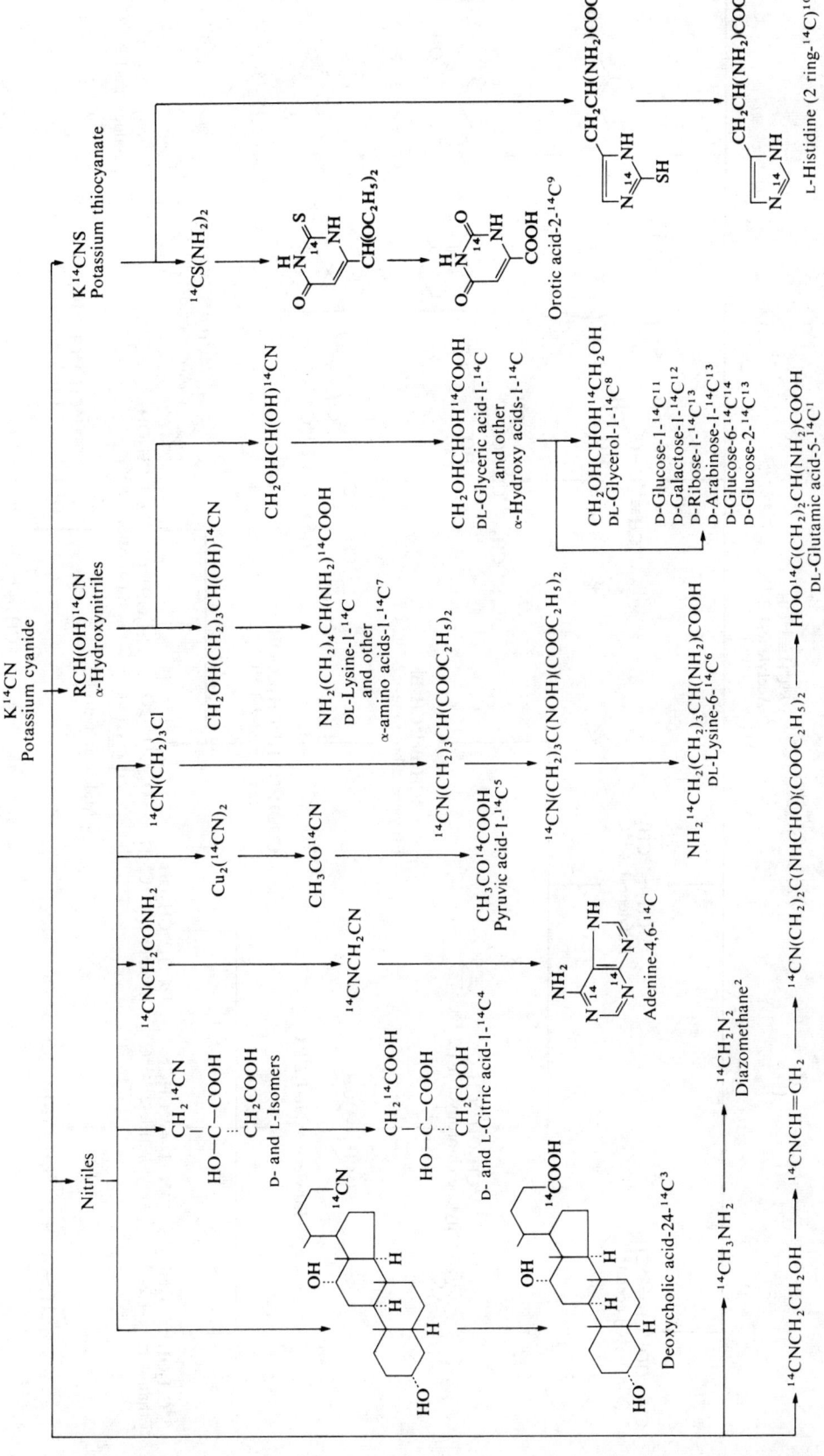

Table 46. CARBON-14 COMPOUNDS—SYNTHESES VIA ACETYLENE

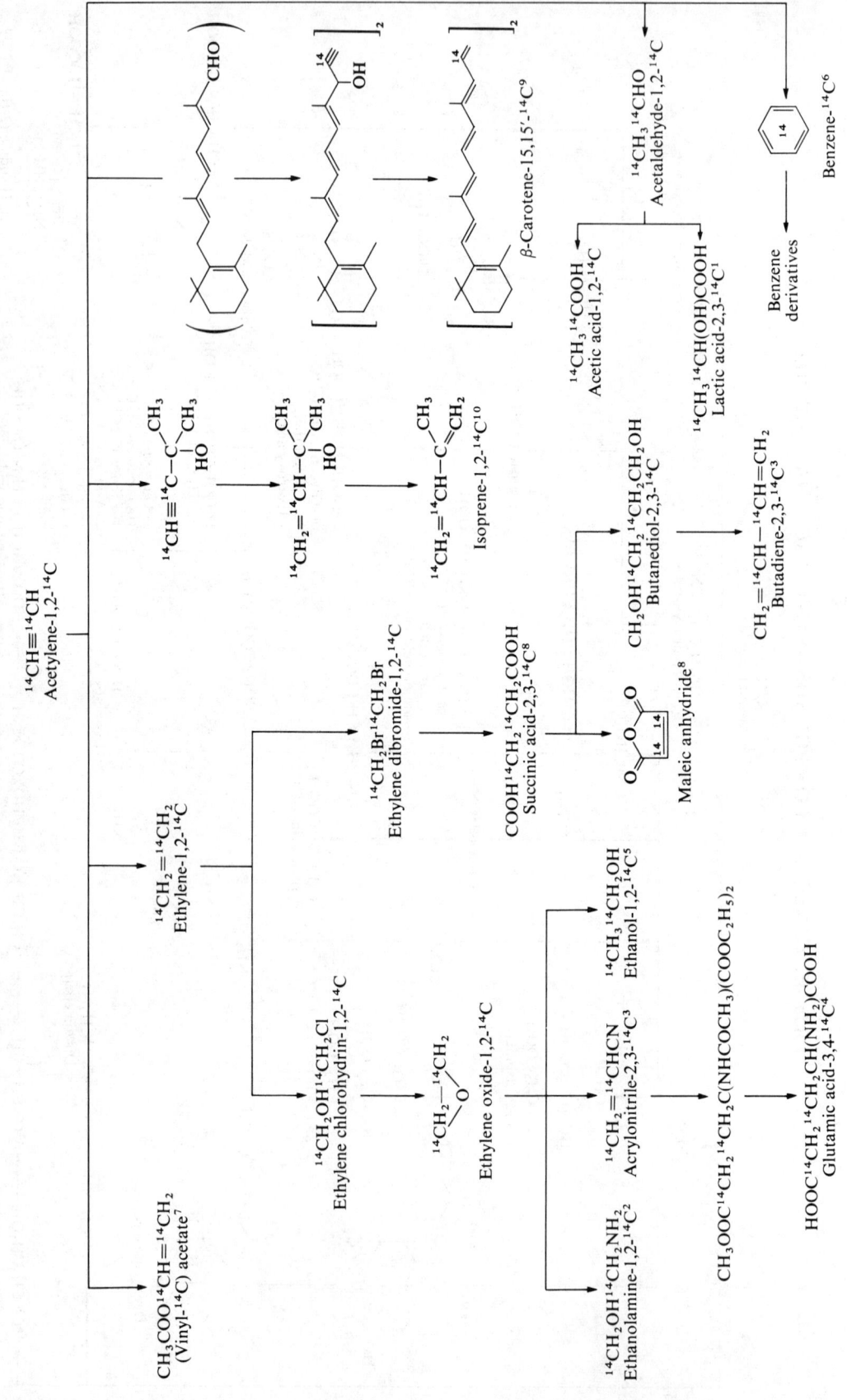

Table 47. CARBON-14 COMPOUNDS—SYNTHESES VIA FORMIC ACID

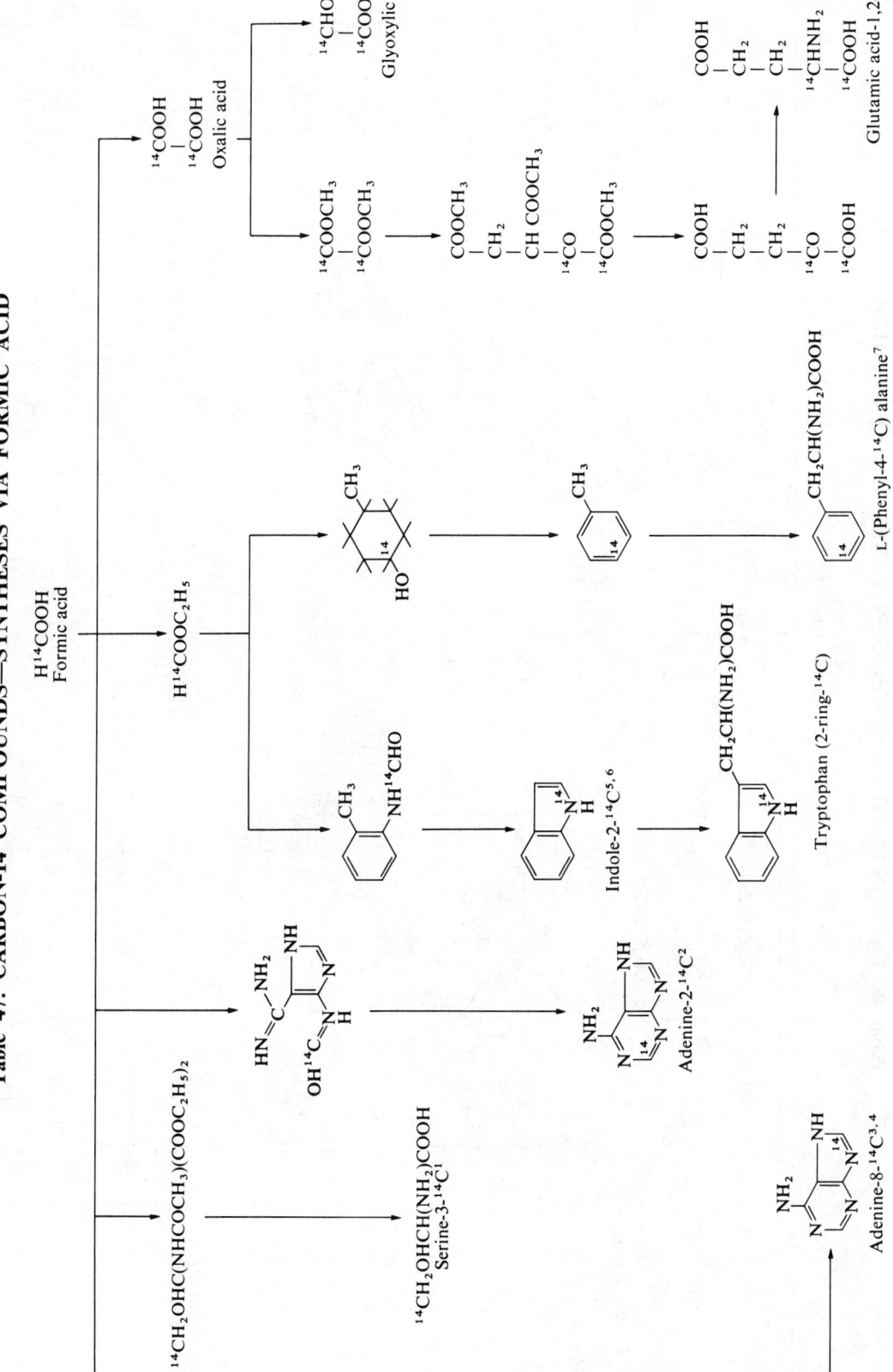

Table 48. CARBON-14 COMPOUNDS—SYNTHESES VIA CYANAMIDE

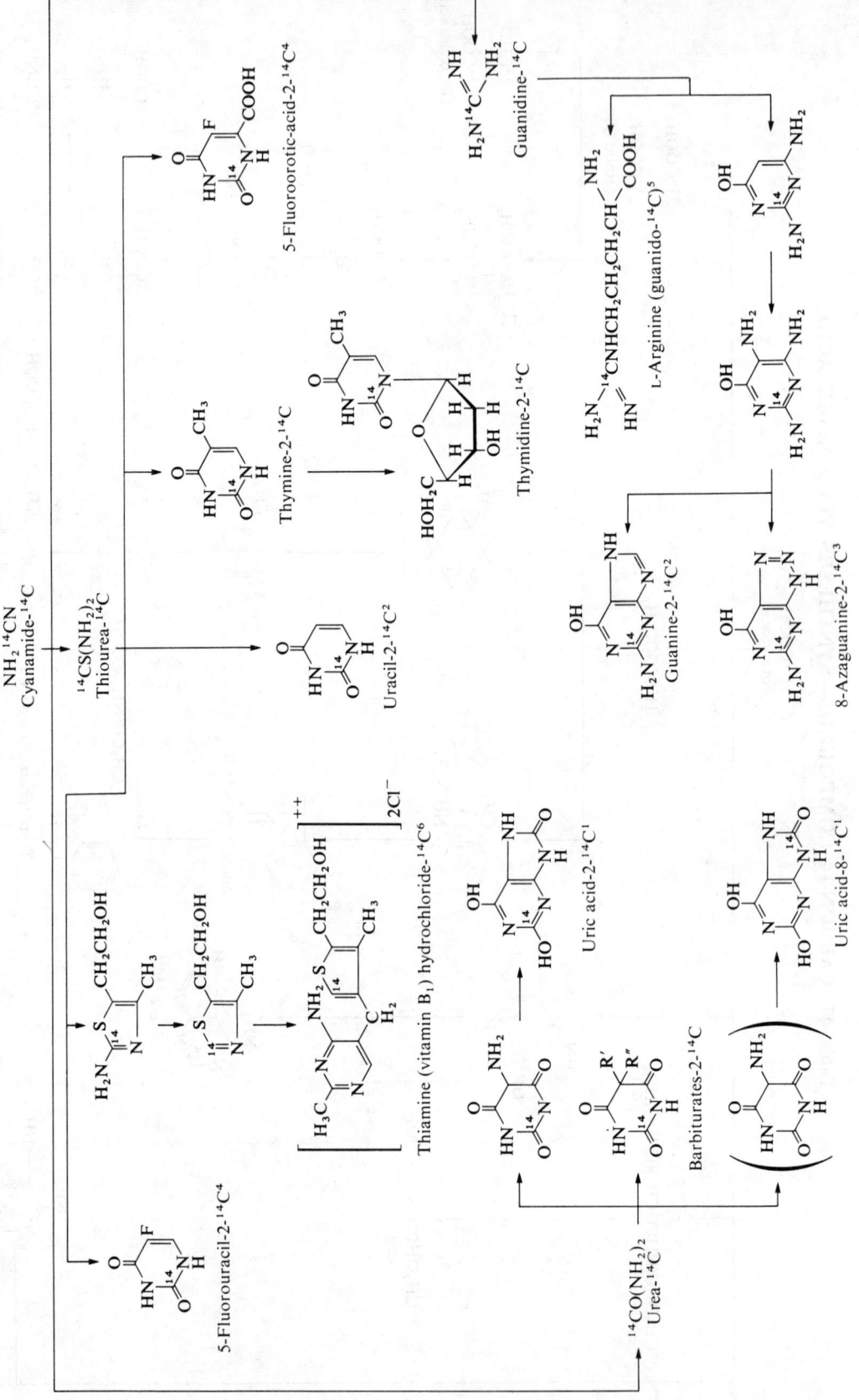

The Argonne National Laboratory in the U.S.A. has specialized in this work, which has been well described in the *Proceedings of the International Conference on the Peaceful Uses of Atomic Energy*, Geneva, 1955,[38] and some more recent advances have been described by Stepka.[39] A substantial bibliography[40] has been compiled of research work, much of it based on products of isotope farming.

Many compounds of animal origin that would be of great interest (for example, ^{14}C-labeled insulin and other protein hormones) have hitherto appeared to be beyond the reach of isotopic labeling at useful specific activities, for much the same reasons. Theoretically, insulin might be labeled by feeding animals with labeled amino acids or protein; this has, indeed, been done successfully—although rather expensively—for the more plentiful proteins, such as serum proteins.[41] The yield of a specific protein such as insulin would, however, be so small that this method of labeling could hardly be acceptable. Recent developments in methods of efficient synthesis for peptides may be expected to overcome many of the difficulties experienced with compounds of this class.

Labeling by biosynthesis can have its particular advantages. It gives the naturally occurring isomers of compounds, and not racemic mixtures or unwanted epimers. The photosynthetic systems, since they achieve multiple labeling, can give extremely high molecular specific activities and also uniform labeling. *Chlorella* has the special advantage that residues can be readily recovered by oxidation to carbon dioxide without appreciable waste by dilution with ^{12}C. Against these advantages one must set the undeniably greater difficulty of purifying biologically produced compounds and the incompatibility of extraction methods for compounds of different kinds. Superficially it would appear that labeled *Chlorella*, for example, would be a veritable mine of labeled compounds. In practice the methods for extraction and purification are often tedious and difficult, out of all proportion to the quantity of product to be obtained; and the most efficient methods, even for groups of major importance (such as nucleic acids and amino acids), do not always go well together.

Radiation Synthesis

Attempts have been made to use "hot-atom" reactions to prepare ^{14}C-labeled compounds. At present they remain of academic interest; the yields and specific activities are much too low to meet the needs of tracer workers. An additional problem is the difficulty of purifications of the products. The interested reader may be referred to published work,[16] and particularly to the specific example of a useful radiation synthesis with ^{14}C mentioned earlier.

Peculiar Features

Organic compounds substituted with isotopes have a number of characteristic features that affect their behavior and use, and they need to be clearly understood if they are to be used without the risk of error. These features are particularly apparent with carbon isotope labeling, because of the element of positional isomerism referred to above.

When dealing with multiple-labeled compounds, it may be necessary to bear in mind their actual molecular composition; this has already been referred to briefly. To take a simple example: what is meant by pyruvic acid labeled in all three carbon atoms? Since the isotope is not yet available at an isotopic abundance of 100 percent, it is not possible to prepare a compound in which every labeled molecule contains only ^{14}C atoms. A suitable method of preparation will give pyruvic acid in which the labeling is statistically uniform, but will in fact be made up of molecules labeled variously in one, two, or—in a minor degree—three atoms. Fortunately nearly all uses of ^{14}C-labeled compounds depend only on the statistical probability of labeling, and not on the actual molecular make-up. A synthesis that gives formal double labeling is, therefore, very rarely justifiable on any grounds whatsoever, and it is very often much more satisfactory to mix the single-labeled molecules. As an actual example, one may prepare pyruvic acid in the three single-labeled forms:

$$CH_3COBr + Cu_2(^{14}CN)_2 \rightarrow CH_3CO^{14}CN \rightarrow CH_3CO^{14}COOH,$$
$$CH_3{}^{14}COBr + Cu_2(CN)_2 \rightarrow CH_3{}^{14}COCN \rightarrow CH_3{}^{14}COCOOH,$$
$$^{14}CH_3COBr + Cu_2(CN)_2 \rightarrow {}^{14}CH_3COCN \rightarrow {}^{14}CH_3COCOOH;$$

by mixing these, all the multiple-labeled forms are available at choice. Synthesis with multiple labeling would certainly be more troublesome, and less efficient.

The novel element of positional isomerism introduced by isotopic labeling complicates the production of ^{14}C-labeled compounds. Even in a simple molecule such as stearic acid the difficulties of labeling many of the specific carbon atoms would be considerable, and specific labeling of anything as complex as a steroid is really only practicable in a few positions that happen to be open to reasonably efficient partial synthesis. This extension of specific labeling to more difficult positions is a challenge to chemical ingenuity.

The specification of labeling position and of molecular composition raises particular problems in the nomenclature of ^{14}C-labeled compounds. Ambiguities in nomenclature arise because numbering systems for organic molecules do not always allow for numbering where substitution (in the classical sense) is not possible; there is still no uniformity of convention on details such as the form and relative positions of isotope symbols and numbering or lettering, and there is a deep-rooted preference for familiar or trivial names. Most users of carbon isotope compounds are not primarily organic chemists and are unlikely to give up the use of trivial names such as acetone, tryptophan and cholesterol. The whole problem of carbon-isotope nomenclature is too complex to deal with here. Everyone who uses labeled compounds should, however, strive to adopt a consistent and explicit convention and avoid at least the obvious ambiguities, even if this means supporting a trivial name by a systematic name or formula. Before leaving this subject, one may comment that any attempt to develop a systematic notation to describe the degrees as well as the positions of isotopic substitution seems likely to be so cumbrous that it cannot be expected to come into common use.

Isotope effects—differences in reaction rates between, for example, ^{14}C and ^{12}C—are well known and documented by extensive literature.[42-45] They have been observed with biological systems[46-49] as well as with isolated synthetic reactions; rate constants may differ by as much as 10 percent. They do not, however, seem to prejudice tracer work, which is still too preoccupied with major conclusions to be concerned with moderate quantitative errors. In synthetic and degradative work with ^{14}C they are rarely serious enough to be troublesome. Reactions are usually selected to be as nearly quantitative as possible, which reduces the likelihood of an isotope effect becoming noticeable. When an unexplained change in specific activity is found, it is wise to refrain from postulating an isotope effect until all other evidence—on purities, for example—has been thoroughly sifted.

Carbon isotopes have been very useful in the study of organic reaction mechanisms, a subject covered by extensive literature.[50-53] Many rearrangements have been demonstrated or elucidated with the use of carbon tracers, and conversely, it is important to anyone preparing carbon-labeled compounds to be alert to the possibility of a rearrangement producing an unexpected labeling pattern. Although there are many examples of these rearrangements, they remain exceptional rather than common in ordinary preparative work and have, in practice, not very often given rise to errors in tracer work.

The use of carbon tracers has revealed that enzymes can distinguish completely between two apparently identical groupings in a conventionally symmetrical molecule—for example, the terminal carboxyl groups of citric acid,

$$\begin{array}{c} CH_2COOH \\ | \\ HO-C-COOH, \\ | \\ CH_2COOH \end{array}$$

which is the best known example of this phenomenon. This concept is now generally well known.¹³ It is not produced by the isotope, and is not an isotope effect in any way; the isotopic label only makes the asymmetry apparent. It must be a perfectly general phenomenon and may be found in examples other than the few that have hitherto been investigated.

The need to be cautious in interpreting degradation reactions of labeled compounds has already been referred to. A simple—but quite typical—example observed at the Radiochemical Centre may be quoted. Acetamide, prepared by heating acetic acid-2-¹⁴C with urea,

$$^{14}CH_3COOH + NH_2CONH_2 \xrightarrow{heat} {}^{14}CH_3CONH_2 + NH_3 + CO_2,$$

was degraded by the Hofmann method,

$$^{14}CH_3CONH_2 + NaOBr \xrightarrow[88\%]{NaOH} {}^{14}CH_3NH_2 + NaBr + CO_2,$$

and this revealed about 3 percent of the activity in the liberated carbon dioxide; but hydrolysis to acetic acid, followed by a Schmidt azide degradation,

$$^{14}CH_3COOH \xrightarrow[96\%]{HN_3, H^+} {}^{14}CH_3NH_2 + N_2 + CO_2,$$

gave carbon dioxide that was completely inactive. Clearly, alkaline hypobromite effects some oxidation of the methyl group. The need to be sure of the chemical reactions is obvious, and it is really remarkable that there are not more numerous examples of such misleading evidence. It may be suspected that they often arise, but are not observed because they do not affect or are not relevant to the conclusions from the experiment. The cautious experimenter will always bear in mind that organic chemical reactions are rarely as simple in fact as the equations used to notate them on paper. Studies of the Willgerodt reaction offer a well-known published example:³⁰

$$C_6H_5COCH_3 \xrightarrow[C_5H_5N]{NH_4OH, S} \begin{array}{l} (I) \ C_6H_5CH_2CONH_2 \ (\text{major product}) \\ (II) \ C_6H_5CH_2COOH \ (\text{minor product}). \end{array}$$

Shantz and Rittenberg,⁵⁴ labeling the carbonyl group in acetophenone with ¹³C, found no ¹³C in the carboxyl group of the phenylacetic acid, (II), and concluded that no rearrangement occurred.

A similar study with ¹⁴C showed a small amount of isotope in the carbamido carbon of (I), and a large proportion in the carboxyl carbon of (II). This prompted a suggestion that (II) arose by an alternative reaction sequence involving a rearrangement, but it was later shown³⁰ that in the isotopic analysis of (II) by thermal decarboxylation the carbon dioxide produced arises in some measure from the phenylmethyl residue. Decarboxylation of (II) by the cleaner Hofmann reaction gave carbon dioxide devoid of activity.

TRITIUM COMPOUNDS

Essential Features

Tritium-labeled compounds are proving second in importance only to ¹⁴C-labeled compounds in tracer work, and may rival them in versatility. The value of tritium as a specific hydrogen tracer in studying hydrogen transfer reactions has been increasingly recognized during recent years, but it is still true to say that tritium in labeled compounds is more frequently used as ancillary tracer for a carbon atom or for a molecule or molecular fragment than as a hydrogen tracer. This is a most important point; it means that much tritium tracer work depends on some knowledge of the integrity of a carbon–hydrogen bond.

If this consideration is given due weight, tritium is a most useful tracer. It is readily available in quantities, and extremely high specific activities (up to 29 curies/milliatom) can be attained. Although the very soft beta emission has in the past made for difficulties in measurement, it has great advantages in high-resolution autoradiography. As with ¹⁴C, tritium has favorable properties as a bioligocal tracer. Its half-life (12.26 years) is conveniently long, there are no shielding problems, the biological half-life is short, and the toxicity is among the lowest in radioactive materials. An extensive review of present knowledge of tritium and tritium compounds has recently been published.

Until quite recently, the difficulty of measuring tritium was a serious handicap to its wide use. Although solid sources can be measured in a windowless proportional counter, this method has not proved reproducible enough or sensitive enough for most laboratories; it has its value as an approximate method of measurement for monitoring. Proportional counting and ion-current measurement of gaseous samples were virtually the only satisfactory methods, and not everyone has the skill or the patience to operate them successfully. Experimental use of liquid scintillation counting goes back many years, but only comparatively recently has good equipment appeared on the market. However, the method is now well established and is for most people the only tolerable one for tritium measurement.[57-62]

The introduction or subsititution of hydrogen isotopes into an organic molecule is much easier than carbon isotope substitution, and tritium-labeled compounds are correspondingly more readily prepared. Conversely, the prudent user will remember that the label is also more readily lost. The primary chemical forms are elementary tritium (isotopic abundances up to 100 percent), equivalent to 2.6 curies/ml at standard temperature and pressure, and tritium oxide. Tritiated water is supplied from the Radiochemical Centre at up to 5 curies/ml without special arrangements; above this level it is necessary to take into account the formation of "electrolytic gas" by radiolysis, and arrangements for packaging and transport may need to be made according to the specific activity and time.

All methods of labeling with tritium involve ultimately either exchange, substitution, or addition reactions, which may be supplemented by further synthetic processes. The primary labeling processes may be classified as

a) exchange with heterogeneous catalysis,
b) exchange with homogeneous catalysis,
c) exchange catalyzed by radiation,
d) substitution by chemical reduction,
e) addition (hydrogenation).

Exchange with Heterogeneous Catalysis

This is generally the easiest and quickest method of tritium labeling. The unlabeled organic substance is heated with, for example, tritium-labeled water or acetic acid and a catalyst such as palladium or platinum. After disposal of the excess solvent, labile tritium is removed by repeated equilibration with water or other appropriate solvents, and the product is purified by suitable methods. Chemically catalyzed exchange can give quite high specific activities (1 to 20 curies/mM). It results in labeling that is general, but rarely uniform, and the precise determination of labeling is usually so laborious that it is not often attempted. The method is, of course, not applicable to compounds that are unstable under the conditions used, and it is not often possible to obtain a theoretical equilibrium concentration without excessive breakdown of the starting material. Purification of the products is not usually too difficult, and the method, although rather unpredictable, is generally useful. The erratic results of these exchange reactions are suspected to be due to variation in the activity of the catalyst used, and greater reliability could, no doubt, be attained by more thorough investigation.

Exchange with Homogeneous Catalysis

Hydrogen atoms in "labile" positions undergo rapid exchange and equilibration with tritium oxide. Examples are 2-naphthoic acid,

$$\text{Naphthalene-COOH} + T_2O \rightleftharpoons \text{Naphthalene-COOT} + \text{THO,}$$

and malonic acid,

$$CH_2(COOH)_2 + 3T_2O \rightleftharpoons CHT(COOT)_2 + 3THO.$$

The products themselves are not of much interest, because the label will exchange off again equally readily in aqueous solution; but by reacting the substituted naphthoic acid with diazomethane, followed by hydrolysis,

$$\text{Naphthyl-COOT} + CH_2N_2 \longrightarrow \text{Naphthyl-COOCH}_2T + N_2$$

$$\downarrow NaOH$$

$$CH_2TOH$$

tritium-labeled methanol is readily obtained. Similarly, by decarboxylation of the labeled malonic acid,

$$CHT(COOT)_2 \xrightarrow{heat} CHT_2COOT + CO_2$$

$$\downarrow NaOH$$

$$CHT_2COOH \longleftarrow CHT_2COONa + THO,$$

followed by removal of the labile tritium, one obtains acetic acid labeled with tritium in the methyl group. Exchanges of this kind are necessarily possible only in particular cases, but—as in the examples quoted—can be useful practical methods.

Hydrogen exchanges catalyzed by acid reagents, although extensively studied for their theoretical interest, are not much used in practical tritium labeling. This, no doubt, is because the conditions required for a high degree of substitution are rather drastic, so that relatively few compounds will survive them; for those that do, such as the simpler aromatic compounds, more efficient alternative methods of labeling are already available.

Exchange Catalyzed Radiation

Radiation-induced labeling, as described by Wilzbach and others,[63,64] often appeals to the beginner in tracer work because of its superficial simplicity. In its simplest form (there are elaborations of the method), the finely divided organic compound is exposed to elementary tritium. Hydrogen exchange occurs in some measure with most compounds, but is complicated by side reactions, such as additions to unsaturated centers,[65-67] and above all by extensive radiation decomposition. This decomposition seriously limits the specific activities attainable and, unfortunately, is most marked with large and sensitive molecules, for which the method would otherwise be particularly valuable. As a consequence, a very complex mixture of labeled compounds results, and special care is necessary in purification.[68,69] On really rigorous purification and analysis it will often be found that the desired product is present only in very small concentration, compared with labeled impurities. Observations published at the Brussels conference in 1963,[70] two of them dealing with lysozyme, show interesting differences of opinion on the success of the Wilzbach labeling of protein materials and suggest that much care is necessary in using it. The success of labeling by the Wilzbach method or its variants is very unpredictable, although it can occasionally score a pronounced success, as in the example of atropine.[67] The method is, however, at least relatively simple to try, although it is not often that of choice.

Labeling by recoil tritons—for example, by neutron irradiation of a mixture of an organic substance with a lithium salt—may also be mentioned under this heading. As a means of preparing labeled compounds it is less useful than the Wilzbach method and is, at present, of academic interest only. This academic interest is considerable, but it should not mislead the reader into overestimating the value of this approach to tritium labeling for practical purposes.

Substitution by Chemical Reduction

Reduction of a halogen compound by a metal, such as zinc, in the presence of tritium oxide or tritium-labeled acetic acid, of the general type

$$RX + M^{++} + T_2O \rightarrow RT + MOTX,$$

will produce a compound with a tritium atom in place of the original halogen.

Reductions of this kind, although sometimes used for labeling, require a large excess of tritium in the water, acetic acid, or other comparable solvent used. This large excess is avoided by converting the halide RX into an organometallic compound, such as a Grignard reagent, and reacting this with an equivalent quantity of tritium oxide:

$$RMgX + T_2O \rightarrow RT + MgOTX.$$

The method obviously lends itself to high specific activities and to work on a small scale. A comparable method of chemical reduction uses sodium borotritide for reduction of carbonyl compounds,

$$4R_2CO + NaBT_4 + 2H_2O \rightarrow 4R_2CTOH + NaBO_2,$$

while reductions with tritiated lithium borohydride are often used for labeling carbohydrates.[71] These chemical reductions can usually be relied upon to give completely specific labeling, although isotope effects and exchange reactions often reduce the specific activity of the product, as it would be predicted from the simple equation.

Another variant is reduction by using elementary tritium and a catalyst such as platinum or palladium:

$$RX + T_2 \xrightarrow{Pd \text{ or } Pt} RT + TX.$$

This is often a very efficient method, but it should not be assumed that the labeling by this means is completely specific, since hydrogen migration can occur in the presence of catalysts. As in labeling by addition of tritium, care is needed in the selection of solvents and other conditions, such as pH.[72] Elementary tritium in the presence of platinum or palladium exchanges rapidly with water, ethanol, acetic acid, and many other solvents that are used for catalytic hydrogenations or reductions. Ethers, particularly dioxan, and esters have proved particularly useful as solvents, since they exchange less readily.

Addition (Hydrogenation)

Addition to unsaturated centers (particularly carbon–carbon double and triple bonds), catalyzed by platinum or palladium, is a generally useful and efficient method of tritium labeling. There is always some risk that exchange or hydrogen migration will occur at the same time, and it cannot be assumed that tritium will be located only at places indicated by simple addition to the double or triple bond. Another source of trouble is tritium exchange with solvent, which has been discussed in the preceding paragraph.

Specific Activities

One of the advantages of tritium as a tracer is that the theoretical specific activity of 29 curies/milliatom can often be approached in practice. It is only to be expected that compounds substituted at such levels will be very unstable. Acetylene, for example, polymerizes rapidly and spontaneously,[73] and 2-methyl naphthoquinol diphosphate, in aqueous solution at room temperature and at a specific activity of 70 to 80 curies/mM, decomposes to the extent of 5 to 10 percent per day. Present methods of reducing radiation decomposition are not really adequate at this level, and it is therefore often difficult to make full use of the high specific activities that tritium can provide.

A feature of tritium-labeling procedures that is in particular contrast to ^{14}C labeling is the relatively low yields based on the isotope. This may be partly because the isotope is too cheap to encourage economy, but it is to a greater degree the result of the methods used. For catalyzed exchange reactions it may be necessary to use some hundreds of curies of tritium in order to obtain some hundreds of millicuries of product. The isotope is, of course, normally recovered, repurified if necessary, and used again. The formation of labeled impurities and the ready incorporation of tritium into labile positions, from which it must then be removed, occur, however, quite generally and make it necessary to work with very much larger quantities of tritium than are customary with ^{14}C. Large quantities of tritium-labeled water and similar compounds make for problems in control of contamination and personal uptake by the operator, and much care is necessary, even in the 10-curie range. The determination of the body burden by urine analysis is fortunately fairly easy and gives meaningful results.

Biosynthesis

Enzymatic syntheses, which—as in ^{14}C work—are really akin to chemical synthetic methods in specificity and control, have proved to be valuable, for example, for producing the important compound thymidine at high specific activity.[74] In any kind of isotope farming, however, tritium is at a serious disadvantage, compared with ^{14}C, simply because water makes up a predominant part of the composition and environment of all living cells. If, for example, *Chlorella* is grown in tritium oxide, the cell receives the radiation from the layer of water surrounding it; with carbon labeling, the ^{14}C dioxide in the aqueous phase at any time is relatively less, while a large proportion of the beta energy of disintegrations occurring within the cell (diameter of about 5 μ) is dissipated in the surrounding aqueous medium. *Chlorella* shows a gross reduction in viability in tritium-labeled water at 20 mCi/ml,[75] which is equivalent to about 0.18 mCi/milliatom of hydrogen, a disappointingly low level for useful labeling. The conversion of the isotope will obviously also be very inefficient. These disadvantages will apply to all biological preparations of this kind, and although they have been tried at relatively low levels, they are not of much importance. Another disadvantage of such preparations will be the difficulty of predicting the distribution of labeling. The large isotope effects observed with tritium and the varying stability of hydrogen atoms in different chemical forms make it unlikely that the purified products will ever be uniformly labeled.

Biosynthesis from specific precursors is more promising: one example is the conversion of progesterone-16-T to aldosterone-16-T and other steroids,[76] and another is the biosynthesis of actinomycin from tritium-labeled amino acids.[77] Tritium-labeled amino acids should also prove useful for protein labeling by biosynthesis, in the same way as ^{14}C,[7,41] but some caution may need to be exercised; McFarlane[78] has shown that the tritium label in glycine is lost almost entirely, if its use for this purpose is attempted. The reader may also wish to remember the known instability of the alpha-hydrogen of amino acids.[79]

Stability of Labeling

The increasing use of tritium as a tracer is revealing gaps in our knowledge of the stability of carbon–hydrogen bonds in all circumstances, and most people who have worked with tritium compounds are aware of tritium losses and migrations that are not readily explained.[80,81] This is true even for professed organic chemists; but most users of tritium compounds are not primarily organic chemists and can easily be misled by unjustified assumptions about the stability of a hydrogen atom. The Radiochemical Centre is often asked if assurance of the stability of the tritium label in an organic compound can be given. This is a very difficult question to answer. The hydrogen in some functional groups, such as hydroxyl and amino groups, is well known to be rapidly exchangeable in water or hydroxylic solvents, and this is, indeed, a standard procedure in purifying tritium-labeled compounds. At the other extreme, tritium in a benzenoid structure, for example, may show no detectable loss whatsoever. For many compounds and many positions of labeling the loss of tritium will depend very much upon the pH. One starts, therefore, with rather imperfect knowledge, even under the simpler conditions. More uncertainties are introduced when the possible effects of chemical reaction in synthesis or degradations are considered. The loss or exchange of a hydrogen atom is often predictable with certainty, but the retention of a tritium atom much less so. If the conditions that may obtain in a biological tracer experiment are taken into account, the stability becomes even more problematical. It is usually possible to get some evidence about stability under the actual conditions by direct experiment, and this is the only course that can be recommended at present.[82]

SULFUR-35 COMPOUNDS

^{35}S is the only practical radioisotopic label for sulfur atoms. In many tracer applications, however, it is required only to follow whole molecules, or large groups, containing sulfur. For these purposes, if tritium labeling is unacceptable, there is often a choice between ^{14}C and ^{35}S. ^{14}C and ^{35}S emit soft beta radiations of practically the same maximum energy, so that the counting techniques employed are essentially the same, and the same degree of resolution is obtained in autoradiography. The essential differences are those of half-life (87.2 days for ^{35}S, compared with 5,760 years for ^{14}C), price, and specific activity. Most ^{35}S-labeled compounds are prepared to fill individual orders, and—once the initial costs have been met—they are much cheaper than ^{14}C-labeled compounds.

The Pattern of Demand

The main groups of chemicals in which ^{35}S-labeled compounds are of interest are the following.

1. The simpler groups of chemical intermediates that are usually required for further synthetic work. This group includes such compounds as chlorosulfonic acid, sulfur trioxide and its complexes, hydrogen sulfide, sulfur dioxide, sodium sulfide and sulfite, carbon disulfide, and thiourea.
2. Compounds of biochemical importance, especially the naturally occurring amino acids L-cysteine, L-cystine and L-methionine.
3. Labeled drugs to study the distribution in the body, the metabolism, and the mode of action of the corresponding unlabeled compounds. This is an important group. A large number of widely used drugs contain sulfur, typical and important examples being the sulfonamides, the phenothiazines (chlorpromazine is of particular interest here), some antimetabolites (for instance, 2-thiouracil and 6-mercaptopurine), the thiobarbiturates (for instance, thiopentone), and diaminodiphenyl sulfone (or dapsone, as it is usually known). The ease with which information on the behavior of pharmaceuticals can be obtained by the use of the corresponding labeled compounds is reflected in the steadily increasing demand for the latter.
4. Surface-active agents, such as the xanthates (flotation agents), alkyl sulfosuccinates (wetting agents), long-chain alkylbenzene sulfonates, long-chain alkyl sulfates, and sodium lauroyl methyltauride (detergents). These ^{35}S-labeled compounds, used in conjunction with the autoradiographic technique, can give much information on the physical behavior of this class of compound, as well as provide a sensitive analytical method for studying the distribution of the materials in complex effluent systems and their pattern of degradation during sewage disposal.
5. Rubber additives, especially accelerators, such as thiocarbanilide, mercaptobenzothiazole, dialkyl dithiocarbamates, and tetraalkylthiuram disulfides. These labeled compounds are used in studying the mixing of rubber stocks and the mechanism of the vulcanization process.
6. Labeled reagents, such as phenyl isothiocyanate and pipsyl chloride for

sequence and end group determinations in proteins, and thiosemicarbazide, pipsyl chloride and toluene-p-sulfonic anhydride for the clinical determination of steroids in body fluids. Using a double-isotope-derivative technique, nanogram quantities of steroids can be estimated by using these reagents.

7. The use of radioactive tracer techniques in studying the mode of action, distribution, and persistence of insecticides or fungicides is now common, and ^{35}S has proved valuable in this field. Typical examples include the use of ^{35}S-labeled malathion in large-scale field studies and the use of labeled Captan* and TMTD (tetramethylthiuram disulfide) in distribution studies within plants.

The Approach to Synthesis

^{35}S can be obtained by the nuclear reaction $^{34}S(n,\gamma)^{35}S$. The production of ^{35}S-labeled compounds by direct neutron irradiation is therefore possible and has been achieved. However, the specific activity is limited, there is extensive decomposition under irradiation, a large amount of ^{32}P is concurrently formed, and, because of the Szilard-Chalmers effect, the products cannot be relied upon to be radiochemically pure. As a method on which to base the supply of ^{35}S-labeled compounds, direct irradiation is impracticable.

^{35}S is usually extracted as sulfate at high specific activity from neutron-irradiated potassium chloride resulting from the nuclear reaction $^{35}Cl(n,p)^{35}S$. The half-life allows extended syntheses, and also—to a limited extent—stock holding of labeled compounds. But few ^{35}S-labeled compounds, except the simple intermediates and the amino acids, are in sufficient demand to justify the holding of stocks.

Successful synthesis of a range of ^{35}S-labeled compounds depends on the reliable preparation of simple intermediates, such as concentrated sulfuric acid, chlorsulfonic acid, sulfur, sulfur dioxide, hydrogen sulfide, and thiourea. The routes to some of these key intermediates are shown in Table 49.

The major item in the cost of a ^{35}S-labeled compound is almost always that of labor; it is necessary to devise rapid and reliable methods of synthesis for a large number of compounds and to take into account—among other things—the need to balance the cost of effort against the yield that is required. This consideration does not eliminate the need to aim for high yields, but it often rules out time-consuming operations such as the recovery of "second crops". The synthetic routes from a typical intermediate, sulfur dioxide-^{35}S, are shown in Table 50. The routes are generally straightforward, but the techniques are complicated by the fact that many of the early stages involve handling the intermediate on a very small chemical scale; thus, 1 mCi of sulfur dioxide-^{35}S at a specific activity of 20 mCi/mM occupies only 1 ml at standard temperature and pressure. As with ^{14}C syntheses, the vacuum manifold technique is extremely valuable. The syntheses of sulfur-containing amino acids and their related compounds have required the development of special synthetic routes. For these benzyl mercaptan is a key intermediate, and it is prepared by the action of elementary sulfur on benzyl magnesium chloride. Some examples of the routes to DL-homocysteine and the S-alkyl-DL-homocysteines, including DL-methionine, are illustrated in Table 51. The ready availability of key intermediates is an advantage to a research worker who wishes to carry out his own synthetic work and for whom it is uneconomical to acquire the equipment and experience necessary for the preparation of simple intermediates. But with these available, he can put his own special skills to good use in a particular field of synthetic work.

Biosynthesis has been used for ^{35}S-labeled compounds, notably methionine, cystine, and glutathione. At low specific activity, and especially for the individual preparative worker, it has advantages, notably provision of the L-forms of optically active compounds. For an organized supply, it seems unlikely to compete with chemical synthesis, and its scope is, of course, limited.

---

* Trademark of Natural Gas Odorizing, Division of Helmerich & Payne, Inc.

Table 49. SULFUR-35 COMPOUNDS—ROUTES TO INTERMEDIATES

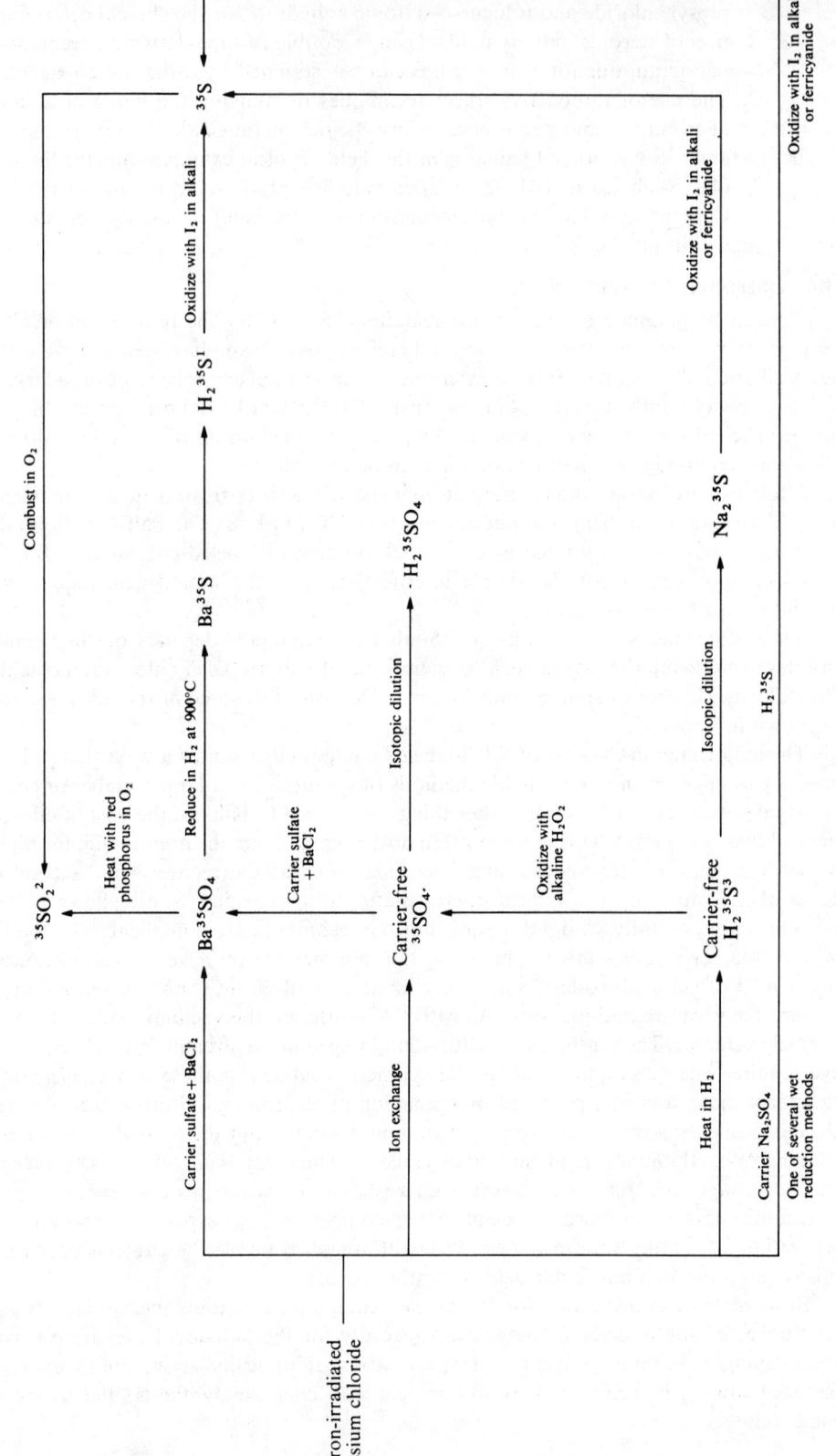

Table 50. SULFUR-35 COMPOUNDS—SYNTHESES VIA SULFUR DIOXIDE

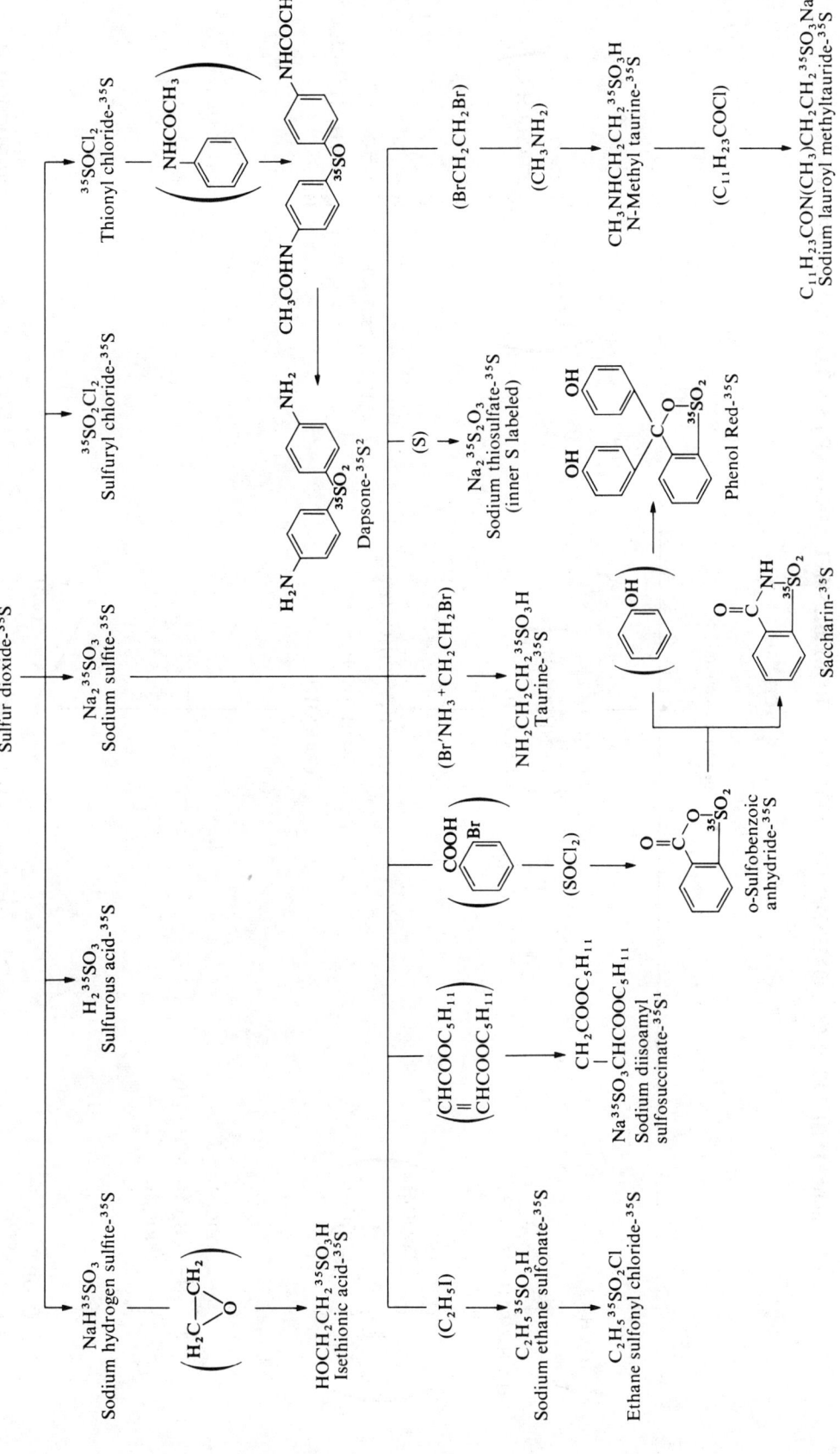

Table 51. SULFUR-35 COMPOUNDS—ROUTES FROM BENZYL MERCAPTAN TO DL-METHIONINE

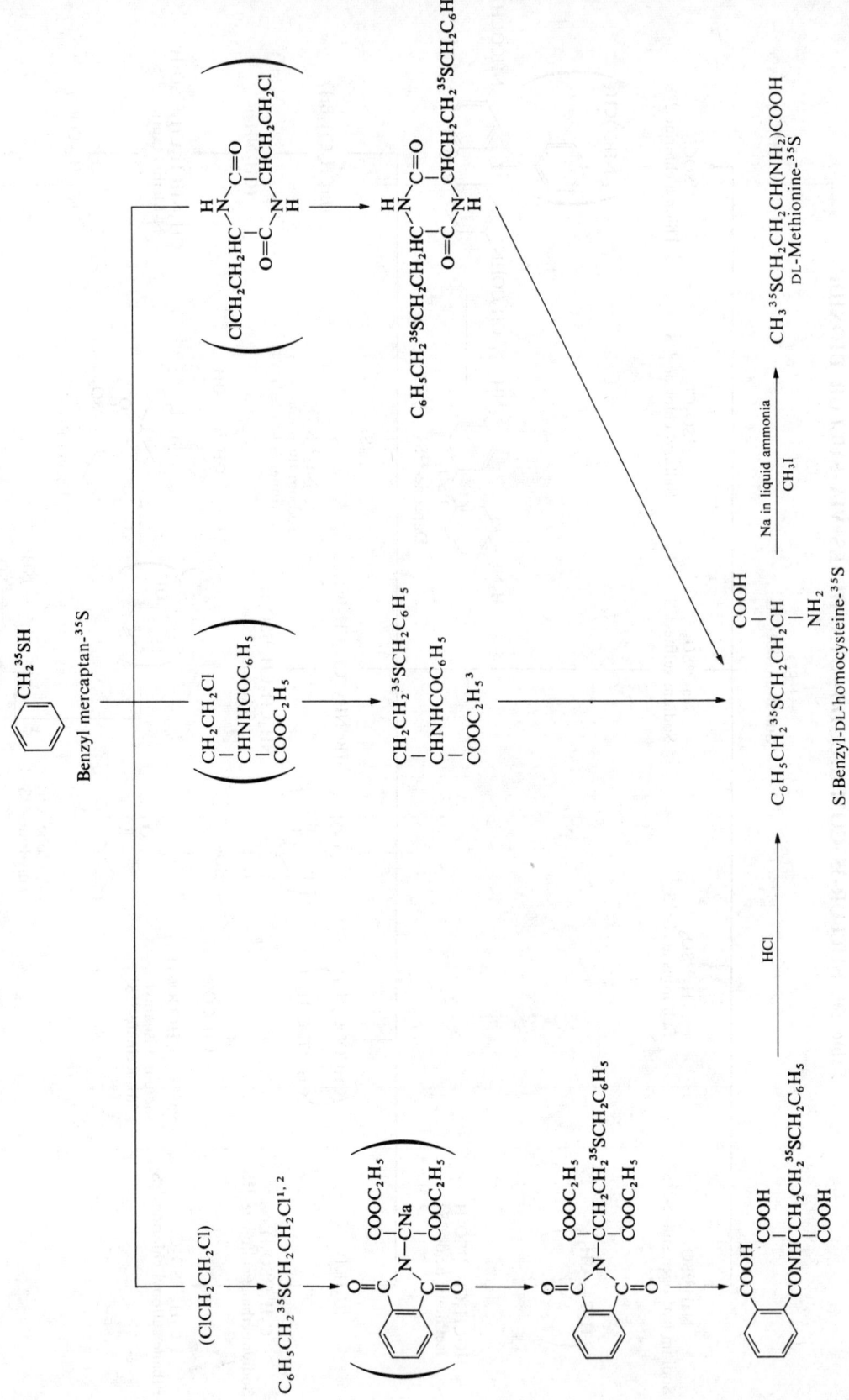

The Specific Activity and Stability of Compounds Labeled with Sulfur-35

One of the great advantages of ^{35}S labeling is the high specific activity that can be attained. The maximum theoretical specific activity is 1,500 curies/milliatom, compared with 64 mCi/milliatom for ^{14}C. The specific activities that may be achieved in practice are usually limited by radiation decomposition and by the small chemical scale. Radiation decomposition must be considered from two points of view: as it affects the storage of the final product, and as it may affect intermediates during synthesis. An interesting example of the latter is the preparation of 2-mercaptoethanol, which may be prepared by reacting ethylene oxide hydrate with hydrogen sulfide. The reaction takes many days to achieve completion, but, with inactive materials, it gives an excellent yield. When hydrogen sulfide labeled with ^{35}S at 10 mCi/mM is used under the same conditions, no product is obtained. Another striking illustration is afforded by the preparation of high-specific-activity toluene-p-sulfonic anhydride for steroid determination. The nonradioactive reagent can be obtained relatively easily by several methods, but at specific activities of 200 to 300 mCi/mM, radiation decomposition during the synthesis has necessitated substantial modifications to the method employed.

However, as experience has grown, it has proved possible to make some ^{35}S-labeled compounds at extremely high specific activities. For example, at the Radioactive Centre hydrogen sulfide, sulfur dioxide, pipsyl chloride, thiosemicarbazide, toluene-p-sulfonic acid, L-cystine, and L-methionine are now produced routinely at 150 to 200 mCi/mM, and individual preparations of mustard gas at 300 mCi/mM and of sulfonic acid at 1,000 mCi/mM have been performed. Here the limits are set by the chemical scale of working, since—unless a very high activity is required—the chemical scale must be reduced in order to maintain the level of the specific activity. This difficulty may, in part, be overcome when a compound is in sufficient demand to be made periodically in large batches, rather than for individual orders.

PHOSPHORUS-32 COMPOUNDS

^{32}P is readily produced either by the direct neutron irradiation of elementary red phosphorus or by separation as carrier-free phosphate from irradiated sulfur. The hard beta emission, having a maximum energy of 1.7 Mev, makes for ease of detection and measurement, but—on the other hand—poses problems of local shielding, With a half-life of 14.3 days, the need for speed in synthesis is not as pressing as with ^{131}I; nevertheless, stock holding is not possible, and for those compounds in frequent demand it is necessary to carry out preparations at suitable intervals. Many ^{32}P-labeled compounds are difficult to make and purify, and those that are produced to individual order necessarily incur high initial charges; the unit price per millicurie of product is low thereafter.

The following groups are distinguishable in the pattern of demand for ^{32}P-labeled compounds.

1. Inorganic fertilizers for agricultural research; superphosphate-^{32}P is a typical example.
2. Organophosphorus insecticides for studies of their own distribution and metabolism.
3. Compounds used in medical therapy and diagnosis: ^{32}P colloids for therapy, and diisopropyl phosphorofluoridate-^{32}P(DFP-P32) for the labeling of blood cells.
4. Labeled pharmaceutical substances for distribution and metabolic studies; cytotoxic drugs such as thiotepa-^{32}P and cyclophosphamide-^{32}P are important examples.
5. Biochemicals, especially nucleotides, such as adenosine-5′-triphosphate-^{32}P.
6. Industrial chemicals, such as plasticizers and solvent extraction reagents.
7. Intermediates for further synthetic work: phosphorus halides and sulfides, and phosphorylating agents, such as 2-cyanoethyl phosphate-^{32}P.

The Approach to Synthesis

^{32}P may be produced by the nuclear reaction $^{31}P(n,\gamma)^{32}P$, and it is therefore possible to make ^{32}P-labeled compounds by direct irradiation in the reactor. However, because of the radiation decomposition of the target and the operation of the Szilard-Chalmers effect, this method is not trustworthy. The reaction is, however, a useful source of elementary red ^{32}P. If the target material is carefully prepared, then, after irradiation, the ^{32}P activity is present only as phosphorus element, so that conversion of the irradiated material to phosphorus halides and sulfides gives radiochemically pure products. These, in turn, may be used for the synthesis of many organophosphorus compounds.

The direct irradiation of phosphorus, however, provides only moderate specific activities, which, while adequate for a large proportion of ^{32}P-labeled compounds, are inadequate for many others. For these it is possible to utilize the nuclear reaction $^{32}S(n,p)^{32}P$, which is used to make carrier-free phosphate-^{32}P. This product, after the addition of orthophosphate carrier, is a starting point for many labeled-compound syntheses, especially those where high specific activity is required.

While a required specific activity may determine the synthetic route to a product, more often the factors of speed and reliability influence the choice of a particular route, since—as with ^{35}S—the major item in the cost of a ^{32}P-labeled compound is that of labor. For a similar reason, purity of the end product is of more importance than yield, as retreatment of impure material wastes valuable time. To this end, the provision of pure intermediates by reliable methods is essential (see Table 52).

In many organophosphorus compounds it is difficult to remove impurities, and recourse is often made to the technique of preparative thin-layer chromatography. Table 53 shows the routes to a number of organophosphorus insecticides, and serves to illustrate some syntheses deriving from a typical intermediate, phosphorus trichloride-^{32}P.

Enzymatic syntheses are used for the production of certain ^{32}P-labeled biochemicals, notably adenosine-5'-triphosphate-^{32}P, and these methods are likely to acquire wider application in the field of ^{32}P-labeled nucleotides.

RADIOIODINE COMPOUNDS

Two major radioisotopes of iodine are now available for the synthesis of labeled compounds. ^{131}I is in plentiful supply from neutron-irradiated tellurium and from fission products, and its extraction in high purity and at high specific activity is simple and relatively cheap. The 8-day half-life, however, limits the time available for synthesis and quality control, and precludes stock holding of ^{131}I-labeled compounds. This and the problems of radiation protection and remote handling determine the type of chemistry feasible with this isotope. ^{125}I has become available relatively recently, and is obtained by the neutron irradiation of xenon. It can be produced at high specific activity, and the amount of shorter-lived ^{126}I (produced by successive nuclear reactions) can be brought to a low level by allowing a suitable decay period. The higher cost of producing ^{125}I, as compared to ^{131}I, is offset in many applications by the longer half-life. The low-energy radiation and 60-day half-life render radiation protection less difficult, make longer syntheses possible, and permit stocks of labeled compounds to be held,

The Pattern of Demand

The major demand for radioiodine-labeled compounds falls largely into four groups.

1. Those biologically active materials, such as L-thyroxine, 3',3,5-triiodo-L-thyronine and 3,5-diiodo-L-tyrosine, in which iodine is used as a true isotopic label in compounds at the higher specific activity.
2. Iodinated organic compounds used in medical diagnosis; many in this group derive from earlier applications of inactive iodine-containing compounds as X-ray contrast media and from the use of iodine-containing dyes in

Table 52. PHOSPHORUS-32 COMPOUNDS—ROUTES TO INTERMEDIATES

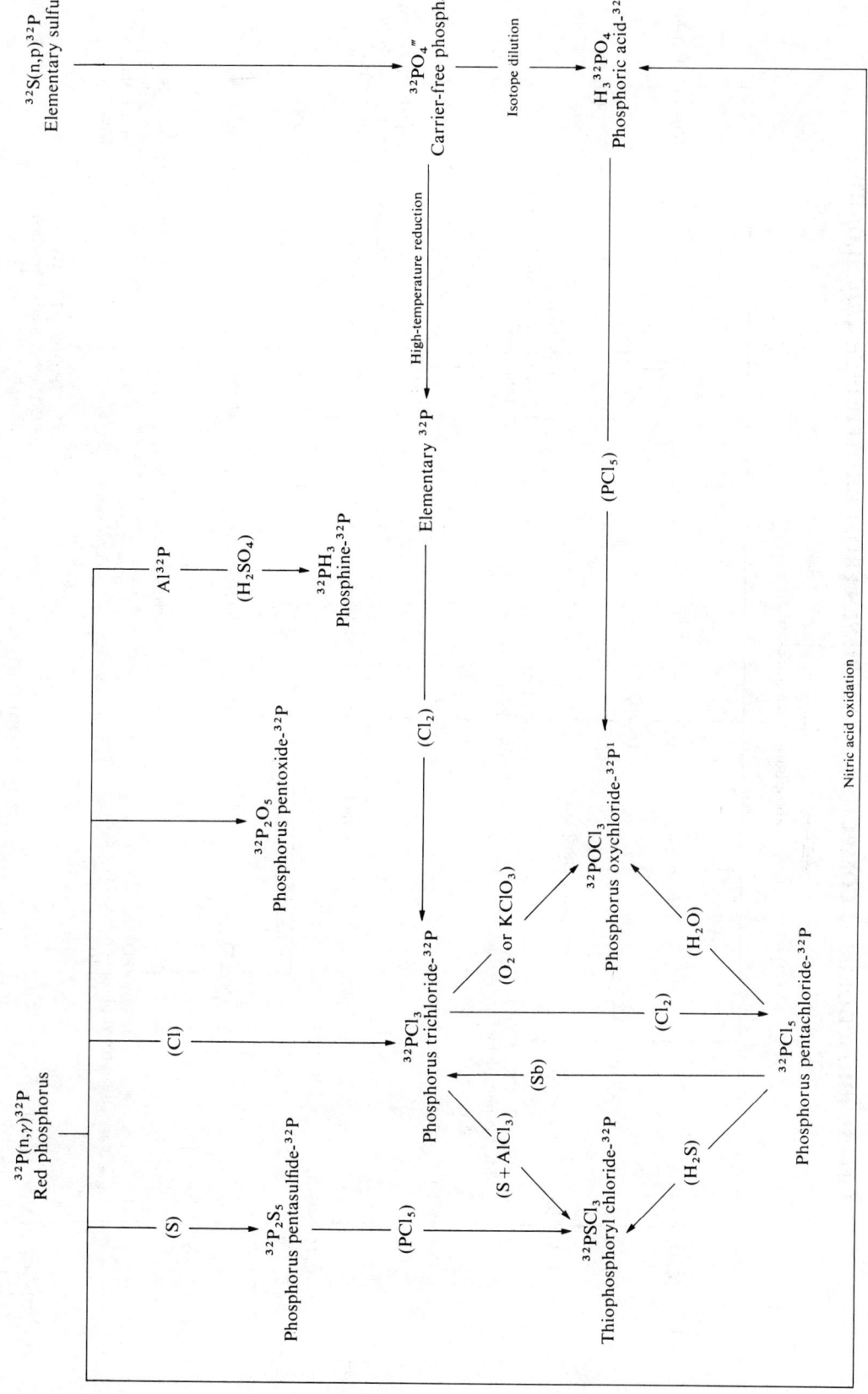

Table 53. PHOSPHORUS-32 COMPOUNDS—SYNTHESES FROM PHOSPHORUS TRICHLORIDE

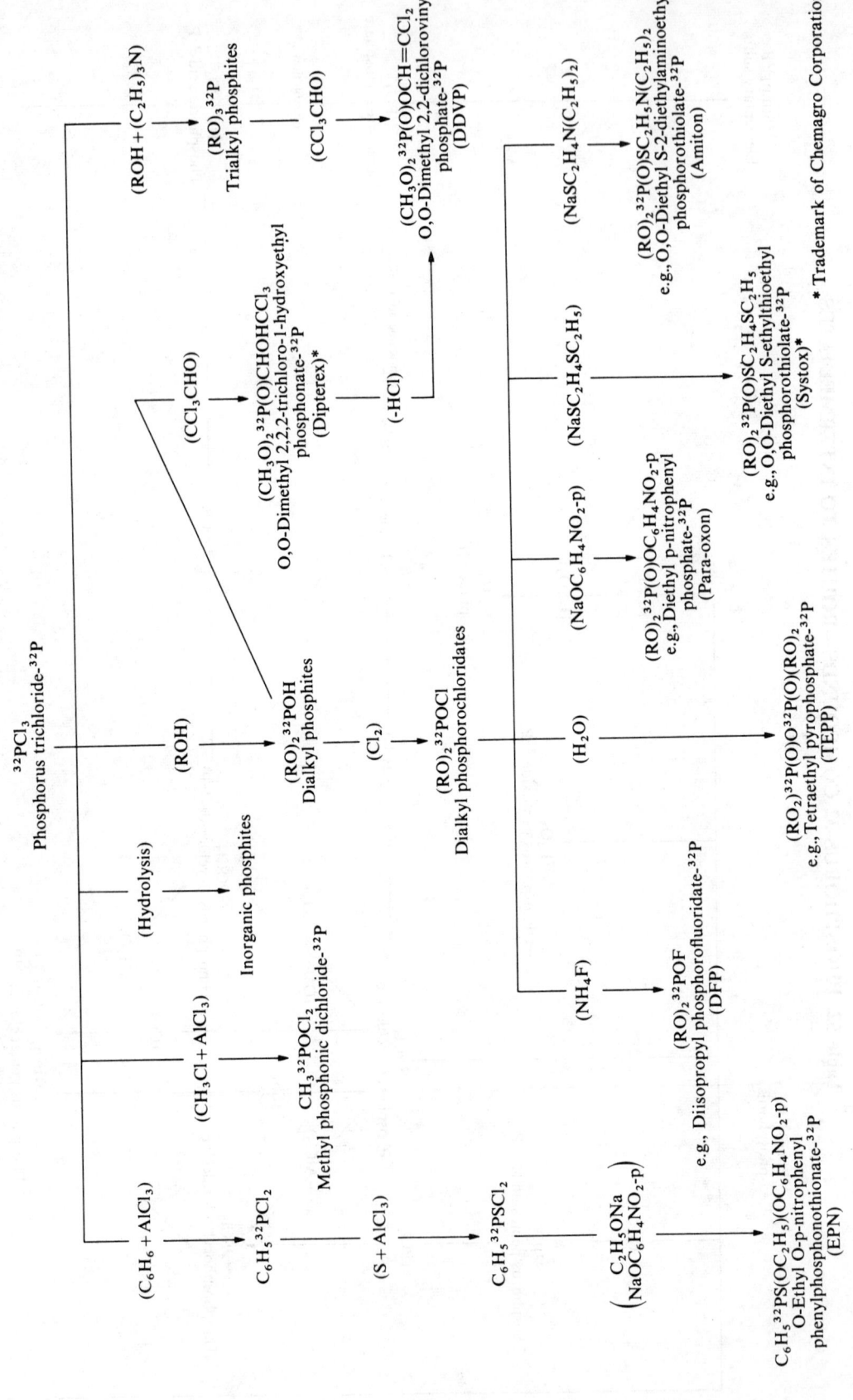

diagnosis. Compounds such as Rose Bengal (tetrachloro(p)tetraiodo(R)-fluorescin), sodium o-iodohippurate, and sodium acetrizoate (sodium, 2,4,6-triiodo-3,5-diacetylaminobenzoate) are typical examples.

3. Compounds formed by the iodination of large molecules, in which the iodine is introduced as a foreign label; the molecule so labeled may then be used as a tracer. This group includes iodinated human serum albumin, iodinated polyvinylpyrrolidone, and iodinated triolein.
4. Of growing interest is a fourth group, in which larger quantities of ^{131}I are combined in a convenient chemical form for therapeutic use. An example is iodinated triolein-^{131}I in Lipiodol,* in which the radioactivity is in the form of a stable oil; by suitable diffusion into the lymphatic system, irradiation of specific parts of the body is accomplished.

The majority of ^{131}I-labeled compounds are in regular demand, and it is the practice of suppliers to synthesize them at fixed and frequent intervals, usually weekly. Many ^{125}I-labeled compounds are sufficiently stable to be delivered from stock.

The Approach to Synthesis

The short half-life of ^{131}I makes it necessary to seek rapid and simple synthetic routes that can be reliably reproduced. The synthesis of most radioiodine-labeled compounds follows conventional methods; esterification, iodination, addition of iodine monochloride to double bonds, and replacement of the diazo group are important.

The biologically active compounds that are required at extremely high specific activity, up to 20 curies/mM, are often best synthesized by exchange reactions:

$$RI + I^{\blacktriangle} \rightleftharpoons RI^{\blacktriangle} + I.$$

Synthesis by exchange, however, must be undertaken with caution. It is frequently necessary, in order to obtain a favorable radiochemical yield, to employ amounts of inorganic radio-iodide that are small by comparison with organic iodide. As conditions for exchange are also often conditions for iodination, two interfering effects may arise. Impurities in the organic iodide may be iodinated, and it is therefore essential to ensure the purity of the reagents; also, the organic iodide itself may undergo further iodination: for example, L-thyroxine is formed as an impurity in the exchange between triiodo-L-thyronine and ^{131}I. Rigorous analytical control is necessary when compounds are labeled by exchange, and the failure to recognize the pitfalls of the method is apparent in some published work.

In some tracer experiments it may be necessary to follow the behavior of material in bulk, or of large molecules. Provided that the bulk properties or the molecular properties are not altered significantly and that the limitations of the method are clearly recognized, it is possible to label the compound with a "foreign" label. ^{131}I is commonly used in this way in such compounds as iodinated human serum albumin, iodinated insulin, and iodinated oils and fats. If care is taken to alter the properties of the original molecule as little as possible during the process of labeling (and this is often the whole art of foreign labeling), the information that is obtained from the radioiodinated product can be of considerable value; this is particularly true in comparative tests. A common pitfall in foreign labeling with radioiodine is the use of too small a quantity of the iodinating agent in comparison with the amount of material to be labeled. Attempts to achieve high specific activity in this way may result in the iodination of minor impurities present in the original material or in the solvent; the product then contains mainly a labeled impurity, rather than the desired labeled compound.

The Stability of Radioiodine Compounds

The use of ^{131}I as a radioactive label poses some problems of stability. The starting material for all ^{131}I syntheses is carrier-free iodide in aqueous solution. This solution deteriorates rapidly on storage because of the readiness with which iodine is radiolytically oxidized to

* Trademark of Laboratoires André Guerbet.

iodate and periodate. Oxidation can be prevented by the addition of a reducing agent such as sodium thiosulfate, but this is not acceptable for all applications—for example, in the iodination of proteins.

Thyrosine and related substances labeled with ^{131}I, since they are prepared at high specific activity, undergo considerable radiation decomposition in aqueous solution; a complex distribution of radioactive impurities is formed, including radioiodide.

The use of ^{125}I as an alternative label has considerable advantages in this respect. This isotope decays by electron capture, with associated low-energy X radiation, and the radiolytic effect is much smaller than with ^{131}I. Solutions of carrier-free iodide-^{125}I can be stored for some weeks, in the absence of reducing agents, without significant oxidation. High-specific-activity L-thyroxine-^{125}I and related compounds form only minor quantities of impurities over some months.

The most objectionable impurity in radioiodine-labeled compounds is often radioiodide; this may be formed in aqueous solution by radiation decomposition, or by chemical decomposition during storage or autoclaving. In some cases it may be removed as it is formed by including in the container a suitable ion-exchange resin.

A further example of radiolytic decomposition is found in aqueous solutions of sodium o-iodohippurate-^{131}I; a brown coloration is formed, which is not associated with a significant amount of radioactive impurity and does not affect its physiological properties.

The Choice of Isotope

As a tracer for investigating chemical and biological systems, ^{125}I has clear advantages. The long half-life enables the system to be kept under observation over long periods. Considerable confidence may be placed in the purity of the labeled compound used, as more time is available for thorough purification and analysis, and radiation damage is generally negligible. Several successive experiments may be carried out with the same stock of labeled compound, thus eliminating possible errors due to varying batches. Handling is easier, as no radiation protection is required for tracer quantities. ^{125}I is measured with high efficiency by crystal scintillation counting.

In the field of medicine and biology the characteristics of ^{131}I are often more valuable. The gamma emission enables its labeled compounds to be traced by external counting, and this has rendered them valuable as medical diagnostic agents. The uptake of ^{131}I-labeled compounds in various organs of the body can be detected by placing a suitable counter over the particular organ, and the distribution of such compounds can be determined by gamma-scanning techniques. Rapid clearance of the isotope from the system under observation is ensured by the short half-life, and this permits successive tests to be performed without interference. The beta emission of ^{131}I facilitates its use also in larger quantities, as a therapeutic agent.

More sophisticated techniques combine the use of ^{125}I- and ^{131}I-labeled compounds. The two isotopes can be counted at the same time, using a crystal scintillation counter in conjunction with a pulse-height analyzer. This permits two measurements to be performed simultaneously on a single system, and extends the versatility of the tracer technique.

Other Iodine Isotopes

In addition to ^{125}I and ^{131}I, other radioisotopes of iodine are available, including ^{124}I and ^{132}I. The latter has the merit of a very short half-life (2.3 hours)—of particular importance when carrying out serial measurements on a patient and in minimizing the total radiation dose. ^{124}I is a positron emitter and lends itself to the special technique of positron scanning.

CHLORINE-36 COMPOUNDS

^{36}Cl has a long half-life (3×10^5 years) and is a pure beta emitter of medium energy (0.714 Mev). Because of the long half-life, its production is expensive and its specific activity is very low; the maximum theoretical specific activity is approximately 1.1 mCi/milliatom. Until recently, the highest specific activity readily available was about 25 µCi/milliatom, and the use of ^{36}Cl in tracer work was consequently severely limited. It is, on the other hand, the only long-lived isotopic label for the element chlorine, an important industrial chemical; it is fairly readily detected and measured, and it is a convenient isotope for carrying out synthetic work. The increasing availability at higher specific activities, up to 350 µCi/milliatom, has roused greater interest in this isotope than previously existed. In many ways ^{36}Cl is an ideal radioisotope for use in the teaching of practical radiochemistry.

^{36}Cl is made by the neutron irradiation of potassium chloride, when several nuclear reactions take place, of which the most important are $^{35}Cl(n,\gamma)^{36}Cl$ and $^{35}Cl(n,p)^{35}S$.

Commercially, ^{36}Cl and ^{35}S are produced concurrently; since long irradiation is necessary in order to obtain ^{36}Cl of reasonable specific activity, it is convenient to interrupt the irradiation periodically in order to extract the ^{35}S and then recycle the potassium chloride. For the same reason it is advantageous to recover the ^{36}Cl from the residues obtained in synthetic work and, after purification, reprocess these also.

Some of the basic routes to ^{36}Cl-labeled compounds are set out in Table 54, and the example below explains more fully some of the points involved.

In one of the published syntheses of carbon tetrachloride-^{36}Cl, the first step is the preparation of aluminum chloride-^{36}Cl—either by the action of elementary chlorine on aluminum or by heating a mixture of silver chloride and aluminum metal. All residues at this stage are at the original specific activity and may be reprocessed for their ^{36}Cl. The aluminum chloride-^{36}Cl is then allowed to undergo isotopic exchange with inactive tetrachloride. As the system is a nonhomogeneous one, it is difficult to get the exchange to proceed to complete equilibrium. After separation of the product, the remaining aluminum chloride is of a considerably lower specific activity than it was initially, and this material may be recovered, converted to potassium chloride, and irradiated once more.

At present the demand for ^{36}Cl-labeled compounds is small, and individual users tend to carry out their own special syntheses. For this reason the main interest in ^{36}Cl-labeled compounds is for useful intermediates—such as elementary chlorine, hydrogen chloride, thionyl chloride, sulfuryl chloride, and anhydrous aluminum chloride—and for simple chlorinated solvents. In this respect the pattern of demand resembles the early demand for ^{14}C, and—as with ^{14}C—the growth of an increasingly diverse interest will result in more varied and complex ^{36}Cl-labeled compounds being offered for sale.

Table 54. CHLORINE-36 COMPOUNDS

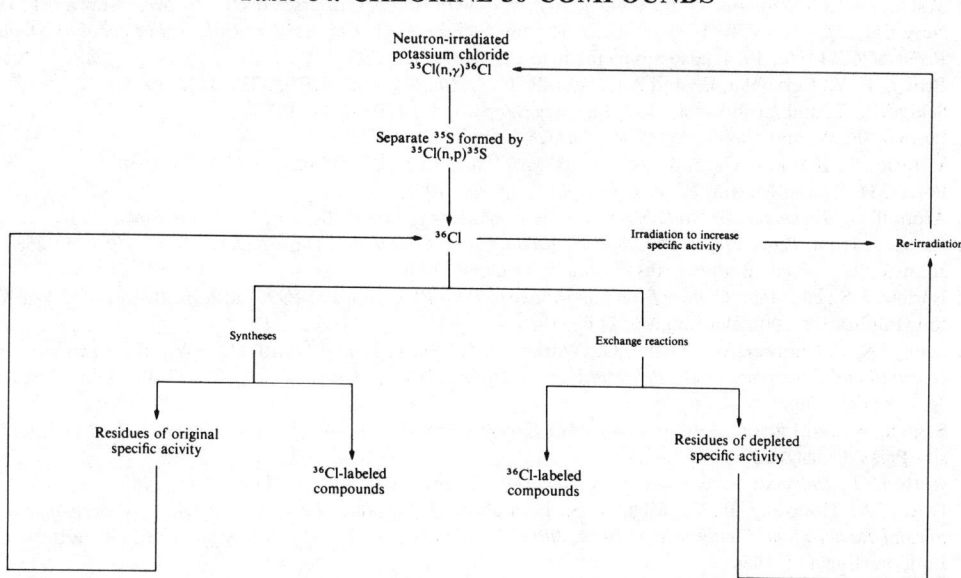

REFERENCES

General

1. Knoop, F. *Beitr. Chem. Physiol. Pathol.*, 6: 150–162, 1904.
2. Fogelstrom-Fineman, I., Holm-Hansen, O., Tolbert, B. M., and Calvin, M. *Int. J. Appl. Radiat. Isotop.*, 2: 280–286, 1957.
3. Fleckenstein, A., Janke, J., and Marmier, P., in *Isotopes in Experimental Pharmacology*, pp. 33–45, Roth, L. J., ed. University Press, Chicago, 1965.
4. Wood, H. G., Werkman, C. H., Hemingway, A., and Nier, A. O. *J. Biol. Chem.*, 139: 483–484, 1941.
5. Campbell, R. M., Cuthbertson, D. P., Matthews, C., and McFarlane, A. S. *Int. J. Appl. Radiat. Isotop.*, 1: 66–84, 1956.
6. Cohen, S., Holloway, C. C., Matthews, C., and McFarlane, A. S. *Biochem. J.*, 62: 143–154, 1956.
7. McFarlane, A. S. *Nature (England)*, 182: 53, 1958.
8. McFarlane, A. S. *Biochem. J.*, 62: 135–154, 1956.
9. Sheppard, C. W., *Basic Principles of the Tracer Method. Introduction to Mathematical Tracer Kinetics*. John Wiley and Sons, New York, 1962.
10. Bayly, R. J. and Weigel, H. *Nature (England)*, 188: 384–387, 1960.
11. Wolfgang, R. L., Anderson, R. C., and Dodson, R. W. *J. Chem. Phys.*, 24: 16–23, 1956.
12. Wilcox, P. E., Heidelberger, C., and Potter, V. R. *J. Am. Chem. Soc.*, 72: 5019–5024, 1950.
13. Krebs, H. A., in *Chemical Pathways of Metabolism*, 1: 115–125, Greenberg, D. M., ed. Academic Press, New York, 1954.
14. Catch, J. R., *Carbon-14 Compounds*, pp. 60–80. Butterworths, London, England, 1961.
15. Chemical Society. *J. Chem. Soc.*, 4201–4205, 1953.
16. Nesmeyanov, A. N., Chemical Effects of Nuclear Transformation. *Proceedings of the Symposium on Chemical Effects of Nuclear Transformations, Prague, October 24–27, 1960*, 2: 269–290. International Atomic Energy Agency, Vienna, Austria, 1961.
17. Wolf, A. P., Preparation and Biomedical Application of Labeled Molecules. Proceedings of a Symposium, Venice, August 23–29, 1964. *Euratom Report EUR 2200 e*, pp. 423–425. European Atomic Energy Community, Brussels, Belgium, 1964.
18. Kuzin, A. M. and Tokarskaya, V. I. *Biochemistry*, 24: 71–77, 1959.
19. Sorokin, Yu. I. and Meshkov, A. N. *Dokl. Akad. Nauk. SSSR*, 118(1): 205–207, 1958.
20. Suss, A. and Weigand, G. *Atompraxis*, 7: 326–327, 1961.
21. Swan, G. A. and Wright, D. *J. Chem. Soc.*, 1549–1557, 1956.
22. Dische, R. and Rittenberg, D. *J. Biol. Chem.*, 211: 199–212, 1954.
23. Silver, S. *Nucleonics*, 23(8): 106–111, 1965.
24. Yankwich, P. E. and Cornman, W. R. *J. Am. Chem. Soc.*, 78: 1560–1562, 1956.
25. Sommerville, J. L., ed., *The Isotope Index 1963–64*. Scientific Equipment Co., Indianapolis, 1963.
26. International Atomic Energy Agency, *International Directory of Isotopes*, 3rd ed. International Atomic Energy Agency, Vienna, Austria, 1964.
27. Calvin, M., Heidelberger, C., Reid, J. C., Tolbert, B. M., and Yankwich, P. E., *Isotopic Carbon. Techniques in Its Measurement and Chemical Manipulation*. John Wiley and Sons, New York, 1949.
28. Murray, A. and Williams, D. L., *Organic Syntheses with Isotopes*. Interscience Publishers, New York, 1958.
29. Nevenzel, J. C., Riley, R. F., Hinton, D. R., and Steinberg, G. *University of California School of Medicine Report UCLA–316*, 1954; also supplement to above, *UCLA–395*, 1957.
30. Brown, E. V., Cervonka, E., and Anderson, R. C. *J. Am. Chem. Soc.*, 73: 3735–3738, 1951.
31. Sekiguchi, T. and Yoshikawa, H. *J. Biochem. (Japan)*, 46: 1505–1511, 1959.
32. Putman, E. W. and Hassid, W. Z. *J. Biol. Chem.*, 196: 749–752, 1952.
33. Vittorio, P., Krotkov, G., and Reed, G. B. *Proc. Soc. Exp. Biol. Med.*, 74: 775–776, 1950.
34. Porter, H. K. and Martin, R. V. *J. Exp. Bot.*, 3: 326, 1952.
35. Aronoff, S., *Techniques of Radiobiochemistry*, p. 66. Iowa State College Press, Ames, Iowa, 1956.
36. Catch, J. R., in *Proceedings of the Second Radioisotope Conference, Oxford, July 19–23, 1954.*, 1: 258–261, Johnston, J. E., ed. Butterworths, London, England, 1954.
37. Burlew, J. S., ed., *Algal Culture from Laboratory to Pilot Plant*, pp. 290–299. Carnegie Institute of Washington (Publication 600), Washington, D.C., 1953.
38. Scully, N. J., Chorney, W., Kostal, G., Watanabe, R., Skok, J., and Glattfeld, J. W., in *Proceedings of the International Conference on the Peaceful Uses of Atomic Energy, Geneva, 1955.*, 12: 377–385. United Nations, New York, 1956.
39. Stepka, W. and Larson, P. S., in *Isotopes in Experimental Pharmacology*, pp. 7–13, Roth, L. J., ed. University Press, Chicago, 1965.
40. Roth, L. J., *University of Chicago Department of Pharmacology Report AECU–4618*, 1960.
41. Dovey, A., Holloway, R. C., Piha, R. S., Humphrey, J. H., and McFarlane, A. S., in *Proceedings of the Second Radioisotope Conference, Oxford, July 19–23, 1954.*, 1: 337–345, Johnston, J. E., ed. Butterworths, London, England, 1954.
42. Ropp, G. A. *Nucleonics*, 10(10): 22–27, 1952.

43. Yankwich, P. E. *Annu. Rev. Nucl. Sci.*, 3: 235–248, 1953.
44. Comar, C. L., *Radioisotopes in Biology and Agriculture. Principles and Practice*, pp. 76–79. McGraw-Hill, New York, 1955.
45. Ropp, G. A. and Hodnett, E. M. J. *J. Chem. Phys.*, 25: 587–588, 1956.
46. Nier, A. O. and Gulbransen, E. A. *J. Am. Chem. Soc.*, 61: 697–698, 1939.
47. Weigl, J. W., Warrington, P. M., and Calvin, M. *J. Am. Chem. Soc.*, 73: 5058–5063, 1951.
48. van Norman, R. W. and Brown, A. H. *Plant Physiol. (England)*, 27: 691–709, 1952.
49. Buchanan, D. L., Nakao, A., and Edwards, G. *Science*, 117: 541–545, 1953.
50. Ropp, G. A. and Neville, O. K. *Nucleonics*, 9(2): 22–37, 70, 1951.
51. Semenow, D. A. and Roberts, J. D. *J. Chem. Educ.*, 33: 2–14, 1956.
52. Harris, G. M. *Rev. Pure Appl. Chem.*, 2(1): 57–64, 1952.
53. Burr, J. G., *Tracer Applications for the Study of Organic Reactions*. Interscience Publishers, New York, 1957.
54. Shantz, E. M. and Rittenberg, D. *J. Am. Chem. Soc.*, 68: 2109–2110, 1946.
55. Dauben, W. G., Reid, J. C., Yankwich, P. E., and Calvin, M. *J. Am. Chem. Soc.*, 68: 2117, 1946.
56. Evans, E. A., *Tritium and Its Compounds*. Butterworths, London, England. 1966.
57. Kelly, R. G., Peets, E. A., Gordon, S., and Buyske, D. A. *Anal. Biochem.*, 2: 267–273, 1961.
58. Jones, J. R. and Monk, C. B. *Lab. Pract.*, 11: 675–677, 1962.
59. Foskett, A. C., in *Atomic Energy Research Establishment, Harwell Report AERE-Bib-132*. Atomic Energy Research Establishment, Harwell, Berks, England, 1961.
60. Martin, L. E. and Harrison, C. *Biochem. J.*, 83: 18P, 1962.
61. Kaufman, W. J., Nir, A., Parks, G., and Hours. R. M., Tritium in the Physical and Biological Sciences. *Proceedings of the Symposium on the Detection and Use of Tritium in the Physical and Biological Sciences, Vienna, May 3–10, 1961.*, 1: 249–261. International Atomic Energy Agency, Vienna, Austria, 1962.
62. Mantescu, C. and Genunche, A., in *Institul de Fizica Atomica (Roumania) Report IFA/CO/29*. Institul de Fizica Atomica, Bucharest, Roumania, 1965.
63. Dorfman, L. M. and Wilzbach, K. E. *J. Phys. Chem.*, 63: 799–801, 1959.
64. Wilzbach, K. E., Tritium in the Physical and Biological Sciences. *Proceedings of the Symposium on the Detection and Use of Tritium in the Physical and Biological Sciences, Vienna, May 3–10, 1961.*, 2: 3–10. International Atomic Energy Agency, Vienna, Austria, 1962.
65. Dutton, H. J., Jones, E. P., Mason, L. H., and Nystrom, R. F. *Chemy Ind.*, 36: 1176–1177, 1958.
66. Bradlow, H. L., Fukushima, D. K., and Gallagher, T. F. *Atomlight*, 9: 2–3, 1959.
67. Evans, E. A. *Chemy Ind.*, 51: 2097, 1961.
68. Jellinck, P. H. and Smyth, D. G. *Nature (England)*, 182: 46, 1958.
69. Chada, S., Woeller, F. H., and Lemmon, R. M., in *University of California Radiation Laboratory Report UCRL-10032*, pp. 94–101, 1962.
70. European Atomic Energy Community., *Proceedings of the Conference on Methods of Preparing and Storing Marked Molecules, Brussels, November 13–16, 1963*. European Atomic Energy Community, Brussels, Belgium, 1964.
71. Isbell, H. S., Frush, H. L., and Moyer, J. D. *J. Res. Nat. Bur. Stand.*, 64A: 359–362, 1960.
72. Nystrom, R. F. *Atomlight*, 23: 5–8, 1962.
73. Ronzio, A. R., in *Los Alamos Scientific Laboratory Report LADC-1475*, 1953.
74. Crowter, D. G., Evans, E. A., and Lambert, R. W. *Chemy Ind.*, 29: 899, 1960.
75. Porter, J. W. and Watson, M. S. *Am. J. Bot.*, 41: 550–555, 1954.
76. Ayres, P. J., Pearlman, W. H., Tait, J. F., and Tait, S. A. S. *Biochem. J.*, 70: 230–236, 1958.
77. Ciferri, O., Fraccaro, M., Albertini, A., Cassini, G., Mannini, A., and Tiepolo, L., Preparation and Biomedical Application of Labeled Molecules. Proceedings of a Symposium, Venice, August 23–29, 1964. *Euratom Report EUR 2200 e*, pp. 147–163, Sirchis, J., ed. European Atomic Energy Community, Brussels, Belgium, 1964.
78. McFarlane, A. S., *Private Communication*.
79. Evans, E. A., Green, R. H., Spanner, J. A., and Waterfield, W. R. *Nature (England)*, 198: 1301–1302, 1963.
80. Crawhall, J. C. and Smyth, D. G. *Biochem. J.*, 69: 280–286, 1958.
81. Glascock, R. F. and Reinius, L. R. *Biochem. J.* 62: 529–534, 1956.
82. Okita, G. T. and Spratt, J. L., Tritium in the Physical and Biological Sciences. *Proceedings of a Symposium on the Detection and Uses of Tritium in the Physical and Biological Sciences, Vienna, May 3–10, 1961.*, 2: 85–92. International Atomic Energy Agency, Vienna, Austria, 1962.

Table 42.

1. Turner, H. S. and Warne, R. H. *J. Chem. Soc.*, 789–795, 1953.
2. Catch, J. R. Huggill, H. P. W., and Somerville, A. R., *J. Chem. Soc.*, 3028–3030, 1953.
3. Catch, J. R. and Evans, E. A. *J. Chem. Soc.*, 2796–2800, 1957.
4. Levitz, M. *J. Am, Chem. Soc.*, 75: 5352–5355, 1953.
5. Murray, A. and Ronzio, A. R., in *Los Alamos Scientific Laboratory Report LADC-1051 (AECU-1667)*, 1951.
6. Solvonuk, P. F., et al. *Proc. Soc. Exp. Biol. Med.*, 79: 597–604, 1952.

7. Liang Li and Elliott, W. H. *J. Am. Chem. Soc.*, 74: 4089–4090, 1952.
8. Collins, C. J. *J. Am. Chem. Soc.*, 73: 1038–1039, 1951.
9. Bergel, F., Burnop, V. C. E., and Stock, J. A. *J. Chem. Soc.*, 1223–1230, 1955.
10. Bobalek, E. G., et al. *Anal. Chem.*, 28: 906–908, 1956.
11. Pajaro, G. *Energia Nucl.*, 6: 273–276, 1959.
12. Kurath, P., Ganis, F. M., and Radakowich, N. *Helv. Chim. Acta*, 40: 933–936, 1957.

Table 43.

1. Fields, M., et al. *J. Am. Chem. Soc.*, 74: 5498–5499, 1952.
2. Daudel, P., et al. *Bull. Soc. Chim. France*, 19: 86–89, 1952.
3. Heidelberger, C. and Hurlbert, R. B. *J. Am. Chem. Soc.*, 72: 4704–4706, 1950.
4. Winteringham, F. P. W., Harrison, A., and Bridges, P. M. *Biochem. J.*, 61: 359–367, 1955.
5. Dauben, W. G. and Vaughan, C. W. *J. Am. Chem. Soc.*, 75: 4651–4655, 1953.
6. Wang, S. C., Hummel, J. P., and Winnick, T. *J. Am. Chem. Soc.*, 74: 2445, 1952.
7. Wang, S. C., Winnick, T., and Hummel, J. P. *J. Am. Chem. Soc.*, 73: 2390, 1951.
8. Fritzson, P. *Acta Chem. Scand.*, 9: 1239–1240, 1955.
9. Leete, E. Marion, L., and Spenser, I. D. *Can. J. Chem.*, 33: 405–410, 1955.
10. Twombly, G. H. and Schoenewaldt, E. F. *Cancer (USA)*, 4: 296–302, 1951.
11. Chapman, D. D., Chaikoff, I. L., and Dauben, W. G. *J. Biol. Chem.*, 222: 363–372, 1956.
12. Ostwald, M., Adams, P. T., and Tolbert, B. M. *J. Am. Chem. Soc.*, 74: 2425–2427, 1952.
13. Hughes, D. M., Ostwald, R., and Tolbert, B. M. *J. Am. Chem. Soc.*, 74: 2434, 1952.

Table 44.

1. Rapoport, H., Lovell, C. H., and Tolbert, B. M. *J. Am. Chem. Soc.*, 73: 5900, 1951.
2. Kurath, P. and Capezzuto, M. *J. Am. Chem. Soc.*, 78: 3527–3529, 1956.
3. Melville, D. B., Rachele, J. R., and Keller, E. B. *J. Biol. Chem.*, 169: 419–426, 1947.
4. Hunziker, F. *Helv. Chim. Acta*, 38: 917–925, 1955.
5. Murray, A. Ronzio, A. R. *J. Am. Chem. Soc.*, 74: 2408–2409, 1952.
6. Dauben, W. G. and Eastham, J. F. *J. Am. Chem. Soc.*, 73: 4463–4464, 1951.
    Belleau, R. and Gallagher, T. F. *J. Am. Chem. Soc.*, 73: 4458–4460, 1951.
    Dauben, W. G., Eastham, J. F., and Micheli, R. A. *J. Am. Chem. Soc.*, 73: 4496, 1951.
7. Fujimoto, G. I. and Prager, J. *J. Am. Chem. Soc.*, 75: 3259–3261, 1953.
8. Fujimoto, G. I. *J. Am. Chem. Soc.*, 73: 1856, 1951.
9. Dauben, W. G. and Bradlow, H. L. *J. Am. Chem. Soc.*, 72: 4248–4250, 1950.
    Ryer, A. I., Gebert, W. H., and Murrill, N. M. *J. Am. Chem Soc.*, 72: 4247–4248, 1950.
10. Posternak, Th., Schopfer, S. H., and Hugeunin, R. *Helv. Chim. Acta.*, 40: 1875–1880, 1950.
11. Thomas, R. C., Wang, C. H., and Christensen, B. E. *J. Am. Chem. Soc.*, 73: 5914, 1951.
    Anker, H. S. *J. Biol. Chem.*, 176: 1333–1335, 1948.
12. Leete, E., Marion, L., and Spenser, I. D. *Can. J. Chem.*, 33: 405–410, 1955.
13. Cox, J. D. and Turner, H. S. *J. Chem. Soc.*, 3176–3180, 1950.
14. Burnop, V. C. E., *Unpublished Report*.
15. Fritzson, P. *Acta Chem. Scand.*, 9: 1239–1240, 1955.
16. Weygand, F. and Swoboda, O. P. *Chem. Ber.*, 89: 18–21, 1956.
17. Karnovsky, M. L. and Gidez, L. I. *Fed. Proc.*, 10: 205, 1951.
18. Catch, J. R., *Unpublished Report*.
19. Fields, M., Walz, D. E., and Rothchild, S. *J. Am. Chem. Soc.*, 73: 1000–1002, 1951.
20. Arnstein, H. R. V. and Crawhall, J. C. *Biochem. J.*, 55: 280–285, 1953.
21. Cramer, R. D. and Kistiakowsky, G. B. *J. Biol. Chem.*, 137: 549–555, 1941.
22. Weygand, F. and Linden, H. *Z. Naturforsch*, 9b: 682–683, 1954.
23. Krasna, A. I., Peyser, P., and Sprinson, D. B. *J. Biol. Chem.*, 198: 421–426, 1952.
24. Cornforth, J. W., et al. *Biochem. J.*, 66: 10P, 1957.
    Tavormina, P. A., Gibbs, M. H., and Huff, J. W. *J. Am. Chem. Soc.*, 78: 4498–4499, 1956.
25. Garbers, C. F. *J. Chem. Soc.*, 3234–3236, 1956.
    Wolf, G., Johnson, B. C., and Kahns, S. G., in *Radioisotope Conference 1954*, 1: 283–286. Butterworths, London, England, 1954.

Table 45.

1. Tiedemann, H. *Biochem. Z.*, 326: 511–514, 1955.
2. Jones, A. R. and Skraba, W. J. *Nucleonics*, 7(3): 53–54, 1950.
3. Mahowald, T. A., et al. *J. Biol. Chem.*, 225: 781–793, 1957.
4. Wilcox, P. E., Heidelberger, C., and Potter, V. R. *J. Am. Chem. Soc.*, 72: 5019–5024, 1950.
5. Gould, R. G., et al. *J. Biol. Chem.*, 177: 727–731, 1949.
    Wood, H. G., Christensen, B. E., and Anker, H. S. *Nucleonics*, 7(3): 60, 1950.

6. Olynyk, P., et al. *J. Org. Chem.*, 13: 465–470, 1948.
   Rothstein, M. and Claus, C. J. *J. Am. Chem. Soc.*, 75: 2981–2982, 1953.
7. Borsook, H. et al. *J. Biol. Chem.*, 184: 529–543, 1950; 187: 839–848, 1950.
   Gaudry, R. *Can. J. Res.*, 26B: 387–392, 1948.
   Rothstein, M. *J. Am. Chem. Soc.*, 79: 2009–2011, 1957.
8. Doershuk, A. P. *J. Am. Chem. Soc.*, 73: 821–822, 1951.
9. Weed, L. L. and Wilson, D. W. *J. Biol. Chem.*, 189: 435–442, 1951.
10. Schayer, R. W. *J. Am. Chem. Soc.*, 74: 2440–2441, 1952.
11. Isbell, H. S., et al. *J. Res. Nat. Bur. Stand.*, 48: 163–171, 1952.
12. Isbell, H. S., Frush, H. L., and Holt, N. B. *J. Res. Nat. Bur. Stand.*, 53: 217–220, 1954.
13. Isbell, H. S., Frush, H. L., and Schaffer, R. *J. Res. Nat. Bur. Stand.*, 54: 201–203, 1955.
14. Schaffer, R. and Isbell, H. S. *J. Res. Nat. Bur. Stand.*, 56: 191–195, 1956.

Table 46.

1. Sakami, W., Evans, E. W., and Gurin, S. *J. Am. Chem. Soc.*, 69: 1110–1112, 1947.
2. Pilgeram, L. O., et al. *J. Biol. Chem.*, 204: 367–377, 1953.
3. Nystrom, R. F., et al. *Nucleonics*, 7(3): 61, 1950.
4. Pichat, L., et al., in *Radioisotope Conference 1954.*, 1: 245–257. Butterworths, London, England, 1954.
5. Cox, J. D. and Warne, R. J. *J. Chem. Soc.*, 1893–1896, 1951.
6. Pichat, L. and Baret, C. *Tetrahedron*, 1: 269, 1957.
7. Roberts, J. D., Lee, C. C., and Saunders, W. H. *J. Am. Chem. Soc.*, 76: 4501–4510, 1954.
8. Brown, S. A. and Neish, A. C. *Can. J. Biochem. Physiol.*, 32: 170–177, 1954.
9. Inhoffen, H. H., et al. *J. Am. Chem. Soc.*, 77: 1053–1054, 1955.
10. McNall, L. R. and Eby, L. T. *Anal. Chem.*, 29: 951–954, 1957.

Table 47.

1. Elwyn, D., et al. *J. Biol. Chem.*, 213: 281–295, 1955.
2. Paterson, A. R. P. and Zbarsky, S. H. *J. Am. Chem. Soc.*, 75: 5753, 1953.
3. Clark, V. M. and Kalckar, H. M. *J. Chem. Soc.*, 1029–1030, 1950.
4. Abrams, R. and Clark, L. *J. Am. Chem. Soc.*, 73: 4609–4610, 1951.
5. Leete, E. and Marion, L. *Can. J. Chem.*, 31: 1195–1202, 1953.
6. Pichat, L., Audinot, M., and Monnet, J. *Bull. Soc. Chim. France*, 21: 85–88, 1954.
7. Dische, T. and Rittenberg, D. *J. Biol. Chem.*, 211: 199–212, 1954.
8. Krupka, R. M. and Towers, G. H. N. *Nature (England)*, 181: 335–336, 1958.
9. Anker, R. M. and Boehne, J. W. *J. Am. Chem. Soc.*, 74: 2431–2432, 1952.

Table 48.

1. Canellakis, E. S. and Cohen, P. P. *J. Biol. Chem.*, 213: 379–384, 385–395, 1955.
2. Bennett, L. L. *J. Am. Chem. Soc.*, 74: 2432–2433, 1952.
3. Skipper, H. E. and Bennett, L. L. *Nucleonics*, 7(4): 49, 1950.
4. Murray, A., in *Los Alamos Scientific Laboratory Report LA-2145*, 1957.
5. Turba, F. and Leismann, A. *Angew. Chem.*, 65: 535, 1953.
6. Williams, D. L. and Ronzio, A. R. *J. Am. Chem. Soc.*, 74: 2409–2410, 1952.

Table 49.

1. Tarver, H. *Advance. Biol. Med. Phys.*, 2: 281–311, 1951.
2. Johnson, R. E. and Huston, J. L. *J. Am. Chem. Soc.*, 72: 1841–1842, 1950.
3. Koski, W. S. *Nature (England)*, 165: 565–566, 1950.

Table 50.

1. Salley, D. J., et al. *Proc. Roy. Soc., Ser. A*, 203: 42–55, 1950.
2. Saraiya, P. R., Khanolkar, V. R., and Gopal-Ayengar, A. R., in *Proceedings of the International Conference on the Peaceful Uses of Atomic Energy, Geneva, 1955.*, 10: 487–489. United Nations, New York, 1956.

Table 51.

1. Tarver, H. and Schmidt, C. L. A. *J. Biol. Chem.*, 130: 67–80, 1939.
2. Seligman, A. M., Rutenburg, A. M., and Banks, H. *J. Clin. Invest.*, 22: 275–279, 1943.
3. Tarver, H. and Schmidt, C. L. A. *J. Biol. Chem.*, 146: 69–84, 1942.
4. Wood, J. L. and Gutmann, H. R. *J. Biol. Chem.*, 179: 535–542, 1949.

Table 52.

1. Axelrod, B. *J. Biol. Chem.*, 176: 295–298, 1948.

PART VI
RADIONUCLIDES FOR MEDICAL APPLICATION*

* Research carried out under the auspices of the U.S. Atomic Energy Commission.

PART VI

RADIONUCLIDES FOR MEDICAL APPLICATION

CEREBRONERVOUS APPLICATIONS
I. MEASUREMENT OF CEREBRAL BLOOD FLOW

Y. Wang, M.D., D.Sc.(Med.)

Measurement of cerebral blood flow can be obtained by rapid serial carotid angiography. Serial angiography is a traumatic procedure, particularly in the case of repeated studies. Furthermore, the Ketz-Schmidt nitrous oxide method[1] has been widely applied for brain blood flow measurement and has derived most of the present knowledge of human brain blood flow. However, the complexity of these techniques has precluded their routine use clinically.[2]

The study of cerebral blood flow can be simplified by using a radioactive gas. Therefore, numerous methods using radioactive tracers have been developed. The techniques are classified as intracarotid, inhalation, and intravenous. Each technique is subdivided further, according to the use of nondiffusible or diffusible gases.

General Comments on Intracarotid Injection Techniques.

The nuclide preparations injected are pharmacologically completely inert to the brain circulation. The technique is safer than angiography, because a smaller needle is used. It can be performed at the same time as carotid angiography, and the brain circulation time can be determined simultaneously, using the tracer technique and a serial carotid angiogram. The exact quantitative recording of timing and blood flow can be easily reproduced.

General Comments on Intravenous Techniques.

These procedures are less traumatic than the intracarotid injection techniques. An intravenous diffused preparation must become equilibrated with lung gas and tissue water before arriving in the brain. The bolus is distorted and considerably streamed out. ^{133}Xe and ^{85}Kr equilibrate rapidly with lung gas and are completely eliminated on one passage through the lungs; therefore they cannot be used for intravenous techniques. An asymmetry of blood flow in either cerebral hemisphere can easily be established by intravenous injection of ^{131}I antipyrine.

At present, all of these procedures for cerebral blood flow measurements are still in the evaluation stage. The technique for adequate quantitation is the carotid injection of liquid ^{85}Kr, ^{133}Xe or ^{131}I antipyrine, or inhalation of ^{133}Xe or ^{85}Kr, with determination of subsequent clearance rate from the brain. All these techniques using radioactive preparations give quantitative evaluation of cerebral blood flow, and this assesses the cerebral hemodynamics. The cerebral hemodynamics is not routinely available from other techniques. In the future, an intravenous or intracarotid injection of a diffusible tracer should allow the display and measurement of regional blood flow pattern whenever a rapid, high-resolution detecting system can be used on the head.

Table 1A.
CEREBRAL BLOOD FLOW MEASUREMENTS, INTRACAROTID INJECTION TECHNIQUE
Nondiffusable Tracer

Nuclides	Techniques	Results	Comments
[131]I-hippurate[3] ^{131}I-albumin	Following carotid injection, continuous external monitoring is carried out, with scintillation probes at the lateral aspect of the cranium. A relative index of blood flow can be established by recording the clearance rate of the tracers from the blood pool.	An asymmetry of the clearance rate is reflecting hemisphere-perfusing rate.	A relatively simple procedure, with nonspecific information.
[32]P-erythrocytes[4]	Following a rapid carotid injection, serial jugular blood samples are collected, count rate recorded, and dilution curves plotted. The flow rate is calculated from the curves.	Brain blood pool volume is about 132 ml. Circulation time is about 8.5 seconds.	The technique requires multiple great-vessel punctures and a serial blood sample collection.
[125]I-hippurate[5] [203]Hg-neohydrin [197]Hg-neohydrin	After a craniotomy, scintillation detectors in position over selected cortical sites on the exposed surface of the brain monitor the regional blood flow following an intracarotid injection of the low-energy gamma emitters. The count rates are recorded, and transformed into circulatory curves. This has permitted analysis of the contribution of the flow pattern derived from arteries, cortex and veins.	Anterior Sylvania Artery / Post-Central Cortex / Vein / Tumor	A selective study of blood flow patterns in arteries, cortex and veins shows abnormalities in regional flow in cerebral vascular occlusal disease, in A–V shunt, and in glioma.
[131]I-Sodium[6]	Following intracarotid injection, the nuclide circulation is continuously monitored externally at the parieto-parasagittal region. The time between the injection and the peak count rate is the nuclide circulation time.	6 to 8 seconds.	

Table 1B.
CEREBRAL BLOOD FLOW MEASUREMENTS, INTRACAROTID INJECTION TECHNIQUE
Diffusable Tracer

Nuclides	Techniques	Results	Comments
^{85}Kr[7] ^{133}Xe[8] ^{131}I-antipyrine[9]	Following carotid injection, the count rates are monitored externally at the various sites of the cranium for about 10 minutes. The decreased count rates are recorded, and a clearance curve is established. The clearance curves consist of two components, which represent a fast and slow type of flow, probably corresponding to the blood flow through the gray and white matter.	Gross different perfusion between the white and gray matter is about 1:4. Average perfusion rate is 60 ml/100g.	A local circulatory disturbance can be quantitatively measured with multiple probes used simultaneously. Abnormalities of blood flow in the regions not primarily involved by the lesion can also be studied. The method is also used to detect shunts in a lesion.

Table 2.
CEREBRAL BLOOD FLOW MEASUREMENTS, INTRAVENOUS TECHNIQUES

Type of Tracer	Nuclides	Techniques	Results
Nondiffusible	^{131}I-hippurate[10] ^{131}I-albumin[11] ^{51}Cr-erythrocytes[12]	A bolus of radioactive preparation is injected rapidly into a tightly constricted antecubital vein with a pressure cuff, and the cuff is suddenly released. For about 7 seconds the bolus head enters the brain circulation. Multiple detectors are monitoring the count rate externally at the various sites of the head. A clearance curve is plotted according to the count rate changes. The interval between the positive and negative peaks is the most common circulation time. The interval of the peaks is prolonged in cerebrovascular disease.	The range of normal circulation time is 6 to 11 seconds. This may be prolonged—14 to 26 seconds—in cases of occlusive disease. The adequacy of collateral supply is best assessed by the shape of the curve.
Diffusible	^{131}I-antipyrine[9]	The actual technique for performing the procedure is similar to that used with nondiffusible preparations. Two or more detectors may be used for monitoring the clearance rate after an antecubital venous injection. A clearance curve can be plotted according to the count rate changes.	A large A–V shunt can be detected. The test can be used for screening purposes.

Table 3.
CEREBRAL BLOOD FLOW MEASUREMENTS
INHALATION TECHNIQUE
Diffusible Tracer

Nuclides	Techniques	Results	Comments
$^{85}Kr^{13}$ $^{133}Xe^{14}$	After 10 to 15 minutes inhalation of the radioactive preparations, the brain count rate is externally monitored to determine the nuclide washing-out rate. The technique does not require any vascular punctures.	The average cerebral perfusion rate is 35 ml/100 g/min.	The procedure is simple and atraumatic. However, the result of this inhalation technique is essentially difficult to evaluate in situations where both ventilation and the peripheral blood flow change, because the nuclide clearance in a number of noncerebral compartments influences the cerebral washing out.

REFERENCES

1. Ketz, S. S., and Schmidt, C. F., The Nitrous Oxide Method for Determination of Cerebral Blood Flow in Man: Theory, Procedures and Normal Values. *J. Clin. Invest.*, 27: 476, 1948.
2. Oldendorf, W. H., Measurement of Cerebral Blood Flow. *Nuclear Medicine*, pp. 422–429, Blahd, W. H., ed. McGraw-Hill, New York, 1965.
3. Fazio, C., Fieschi, C., and Agnoli, A., Direct Common Carotid Injection of Radioisotopes for the Evaluation of Cerebral Circulation Disturbances. *Neurology*, 13: 561, 1963.
4. Nylin, G., Silfverskiold, B. P., Lofstedt, S., Regnstrom, O., and Hedlund, S., Studies on Cerebral Blood Flow in Man, Using Radioactive Labelled Erythrocytes. *Brain*, 83: 293, 1960.
5. Feindel, W., Garretson, H., Yamamoto, Y. L., Perot, P., and Rumin, N., Blood Flow Patterns in the Cerebral Vessels and Cortex in Man, Studied by Intracarotid Injection of Radioisotopes and Coomassie Blue Dye. *J. Neurol. Surg.*, 23: 12–22, 1965.
6. Greitz, T., A Radiologic Study of the Brain Circulation by Rapid Serial Angiography of the Carotid Artery. *Acta Radiol.* (*Sweden*), Suppl. 140: 123, 1956.
7. Ingvar, D. H. and Lassen, N. N., Methods for Cerebral Blood Flow Measurements in Man. *Brit. J. Anesth.*, 37: 216, 1965.
8. Glass, H. I. and Harper, A. M., Measurement of Regional Blood Flow in Cerebral Cortex of Man Through Intact Skull. *Brit. Med. J.*, 1: 593, 1963.
9. Oldendorf, W. H. and Kitano, M., The Free Passage of ^{131}I Antipyrine Through Brain as Indication of A–V Shunting. *Neurology*, 14: 1078, 1964.
10. Oldendorf, W. H. and Kitano, M., Isotope Study of Brain Blood Turnover in Vascular Disease. *Arch. Neurol.*, 12: 30, 1965.
11. Lang, E. K. and Hann, E. C., Angiographic and Isotope Pool Circulation Study of the Cerebral Hemispheres After Internal Carotid Occlusion. *Diseases Chest*, 48: 278, 1965.
12. Ljungren, K., Nylin, G., Berggren, B., Hedlund, S., and Regnstrom, O., Observations on the Determination of Blood Passage Times in Brain by Means of Radioactive Erythrocytes and Externally Placed Detectors. *Int. J. Appl. Radiat.*, 12: 53, 1961.
13. Lassen, N. A. and Klee, A., Cerebral Blood Flow Determined by Saturation and Desaturation with ^{85}Kr. *Circul. Res.*, 16: 26, 1965.
14. Mallett, B. L. and Veall, N., Investigation of Cerebral Blood Flow in Hypertension, Using Radioactive Xenon Inhalation and Extracranial Recording. *Lancet*, 1: 1081, 1963.

CEREBRONERVOUS APPLICATIONS
II. BRAIN SCANNING

Y. *Wang*, M.D., D.Sc.(Med.)

INTRODUCTION

Brain scanning in the past decade has proved that radionuclide techniques are of value in the diagnosis of intracranial lesions.[1-5] The brain scan is relatively atraumatic, and it gives direct indication as to the site of the lesion. The results of brain scanning have shown an accuracy comparable to the other methods.[3,6]

Historically, Moore and colleagues[7] first introduced the use of radioisotopic technique for the detection of intracranial lesions in 1948. By 1951, their work had resulted in correct localization of seventeen out of twenty-six verified brain tumors.[8] Since then, the development of the field has resulted in several desirable radioactive preparations and new scanning devices.

Brain scanning depends on the breakdown of the so-called blood–brain barrier, where certain elements are normally not taken up by the central nervous system.

MECHANISM

The following are the three most probable causes for the formation of an abnormal brain scan.

1. An interruption or breakdown of the so-called blood–brain barrier between brain stroma and cerebral capillaries by various pathologic conditions (such as trauma, infection, necrosis, or cellular alterations).
2. Increased vascularities, either in the form of abnormally large or numerous blood vessels (as in the case of hemangioma) or of arteriovenous malformation.
3. A changing rate of cellular growth or degree of cellular differentiation.

In addition, molecular size, electric charge, and biochemical factors are implicated in the regulation of the blood–brain barrier. A positive-scan study reveals a higher differential ratio of concentration of radioactivity in the diseased area to concentration in normal brain tissue, or a higher target to non-target ratio.

RADIOACTIVE NUCLIDES

Many radioactive nuclides are being tested for brain scanning. Some of the proposed radioactive preparations[1-29] are: ^{131}I-diiodofluorescein, ^{131}I-serum albumin, ^{131}I-octoiodofluorescein, ^{131}I-tetracycline, ^{131}I-antibody, ^{131}I-triiodothyronine, ^{131}I-polyvinylpyrrolidone, ^{131}I-globulin, ^{203}Hg-neohydrin, ^{197}Hg-neohydrin, ^{99m}Tc, ^{74}As, ^{64}Cu, ^{64}Cu-porphyrin, ^{206}Bi, ^{82}Br, ^{24}Na, ^{68}Ga, ^{42}K, and ^{18}F-tetrafluoroborate.

The most widely reported preparations are ^{99m}Tc, ^{203}Hg-neohydrin, ^{197}Hg-neohydrin, ^{131}I-serum albumin, and ^{131}I-diiodofluorescein as gamma emitters; ^{74}As, ^{64}Cu, ^{68}Ga, and ^{18}F are used as positron emitters.

METHODOLOGY

Commonly, a gamma-emitting nuclide preparation is used; a positron-emitting nuclide preparation is also used in a limited number of laboratories. This discussion will be limited

Table 4. USE OF GAMMA-EMITTING PREPARATIONS FOR BRAIN SCANNING

Radioactive Preparations	Physical Properties			Biological Properties	Techniques	
	Emission	Half-Life	Supply Sources		Preparation of the Patient	Administration
^{131}IHSA[1,2,7-13] ^{131}I-labeled preparations	610 kev β (87%) 360 kev γ (87%)	8.05 days	Most radiopharmaceutical companies.	^{131}I concentrates in the thyroid gland. ^{131}IHSA has a long biological half-life; about 80 to 90% of the radioactivity remain in the body after 24 to 48 hours; about 40 to 50% remain in the blood.	Administration of 15 drops of Lugol's solution 4 times on the day prior to the administration of RISA, to block the thyroid.	5 μCi/kg body weight of RISA, intravenously.
^{99m}Tc[15-20]	140 kev γ (100%)	6.0 hours	A pertechnetate product from a ^{99m}Tc generator; the eluting procedure from the generator is given with each shipment.	The gastrointestinal absorption and tissue distribution are similar to those of ^{131}I iodide; it becomes concentrated in the thyroid, in the salivary glands, and in the gastric mucosa; the thyroid uptake of ^{99m}Tc pertechnetate is maximal 1 to 2 hours after administration; 3 to 4% of the administered dose are present within the gland at this time, as measured by external counting; the effective half-life, by external assessment, is about 24 hours.	Administration of 1 g thiocyanate orally about 2 to 3 hours before the administration of ^{99m}Tc, to block the thyroid.	100 μCi/kg body weight of ^{99m}Tc, either orally or intravenously.
^{197}Hg-neohydrin[3-6,21-24] ^{203}Hg-neohydrin[3-6,21-24]	70 kev γ (99%) 210 kev β (100%)	2.7 days 47 days	Most radiopharmaceutical companies.	Approximately 10 to 15% of the administered dose is bound firmly by the kidneys, with an effective half-life approaching the physical half-life; on the basis of 10 μCi/kg bodyweight, whole-body radiation from ^{203}Hg-neohydrin is about 290 mrads, and 35 to 40 rads to the kidneys; ^{196}Hg-neohydrin reportedly will reduce the kidneys' dose to 1 rad.	Administration of 1 ml mercuhydrin 12 to 24 hours prior to the administration, to block the kidneys from taking up radioactive neohydrin.	30 μCi/kg body weight, intravenously. 10 μCi/kg body weight, intravenously.

to the technique using the gamma-emitting isotopes; however, a recently developed detecting system may prove application of positron-emitting nuclide preparations to be advantageous.

The currently employed gamma-emitting nuclide preparations for brain scanning can be classified into three groups: ^{131}IHSA, ^{99m}Tc, and ^{203}Hg- and ^{197}Hg-neohydrin. The methodology for each preparation is described in Table 4.

INSTRUMENTS

Most scanners have a highly sensitive scintillation detecting system with adequate shielding and collimation, a stable gamma spectrometer, and both a dot and photographic system for information display. Until recently, most of the scanners were equipped with sodium iodide (Tl-activated) crystals 3 inches in diameter and 2 inches thick. Within the past few years, however, larger sodium iodide crystals have been developed, with diameters of 5, 8, and 11 inches, along with a large assembly of multiple small crystals and other kinds of detecting systems. These kinds of large-dimension crystals have enhanced the sensitivity, efficiency and speed of the scanning system. An appropriate collimator, pulse-height analyzer setting, and scanning factors are selected according to the type of scanner and radioactive preparations. For the ordinary 3-inch scanner, a 19-inch focusing collimator is used for ^{131}I and ^{203}Hg, and a low-energy collimator may be used for ^{99m}Tc and ^{197}Hg. Anger's camera, Picker's Dynapix, and other stationary detection devices can perform one view of the brain scan in a few minutes, while a mechanical scanner with a 3- to 5-inch crystal will take about 10 to 20 minutes.

Table 5 shows the recommended time interval between administration and start of scanning for each of the gamma-emitting preparations.

Table 5.
ADMINISTRATION–SCANNING INTERVALS
FOR GAMMA-EMITTING NUCLIDE PREPARATIONS

^{131}IHSA ^{131}I-Labeled Preparation	^{99m}Tc	^{197}Hg ^{203}Hg-Neohydrin
Scanning can start at 8-, 12-, 24-, or 36-hour intervals after injection of radioactive tracer, but usually starts at 24 hours.	Scanning usually starts after about 1 hour, but it can start at any time within 3 hours after administration of the radioactive substance.	Scanning starts about 2 to 3 hours after injection of the radioactive substance.

For detecting vascular abnormality, scanning should start immediately and be followed by a delayed scan.

The patient is placed on a comfortable table; his head is immobilized to prevent head movements during the scanning. At least two views are usually required, but three or four views are desirable. This can be easily performed when the camera scanning system is employed. By comparing the immediate scan with the follow-up or delayed scan, a more accurate diagnosis in detecting the presence of a lesion and in making characteristic differentiation of the lesion can often be obtained.

When the scan is completed, it should clearly delineate the contour of the skull. To aid in recognizing anatomical features, certain landmarks should be drawn on the dot scan. For the lateral scan these landmarks are the external auditory meatus, the external canthus, and the nasion; for the anterior scan the orbital canthus is marked, and for the posterior scan the landmarks are the mastoid processes and occipital tuberosity.

Any questionable or unsatisfactory scan should be repeated, and additional views taken, if indicated. The proper selection of scanning factors is essential for a successful scan. Many times a lesion will be missed or be highly questionable on one scan, but be well defined on a repeat scan using slightly different scanning factors. On the other hand, an abnormal finding on a technically unsatisfactory scan may incorrectly be interpreted as a significant finding.

INTERPRETATION

The Normal Brain Scan

Specific normal structures in the endocranium, which are presented as areas of high and low activity, should be delineated. It is imperative for diagnosis that these structures be recognized. Several workers[1,19,30] have described the appearance of the normal structure of a normal brain scan. The overall shape of the skull is ordinarily well delineated on the scan and shows good agreement with skull roentgenograms.

Anatomic considerations of normal brain scans are illustrated in Figures 1 to 3.

Figure 1A is a typical anterior-up view of a normal brain scan. The formation of the characteristic pattern of a brain scan is illustrated as to the significance of various anatomical structures in Figure 1B.

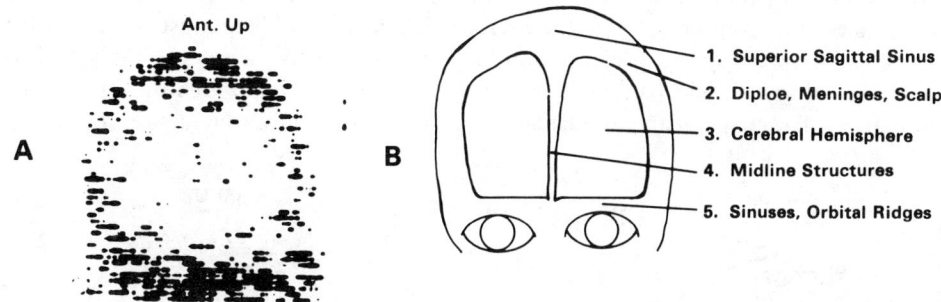

FIGURE 1. Anterior View of Normal Brain Scan.

Figure 2A is a typical lateral-up view of a normal brain scan. Figure 2B illustrates the formation of the characteristic pattern of a brain scan, showing the significance of various anatomical features.

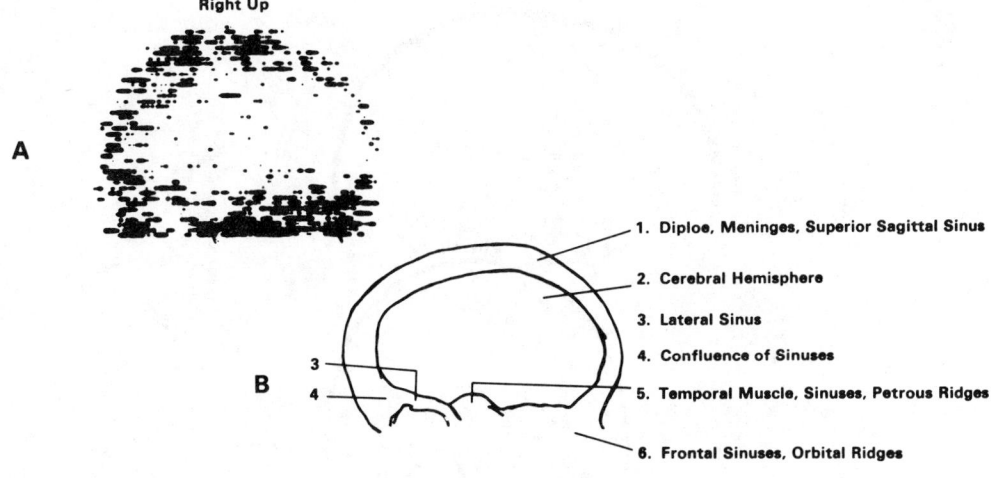

FIGURE 2. Lateral View of Normal Brain Scan.

Figure 3A shows a normal posterior-up view of a brain scan. Figure 3B illustrates the formation of the characteristic pattern of a brain scan, showing the corresponding anatomic features.

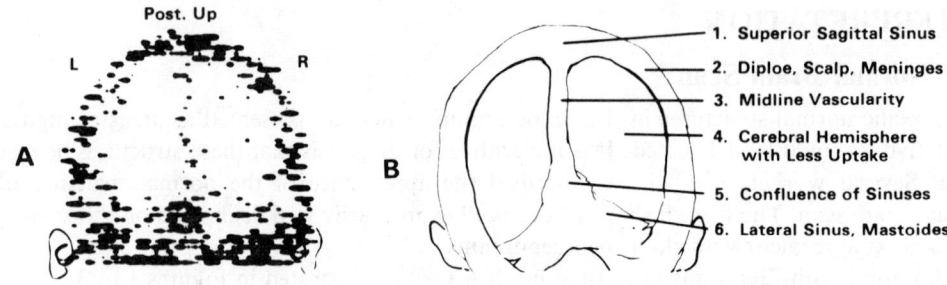

FIGURE 3. Posterior View of Normal Brain Scan.

The Abnormal Brain Scan

An abnormal or positive scan can be defined as one that shows increased activity on at least three lines in a given area, using the present technique. This area must be demonstrated on at least two views. If the suspected lesion fails to meet the above criteria, the scan is considered equivocal and should be repeated. A negative scan shows no area of abnormally increased activity. Examples of the abnormal (positive) scan are given in Figures 4 and 5.

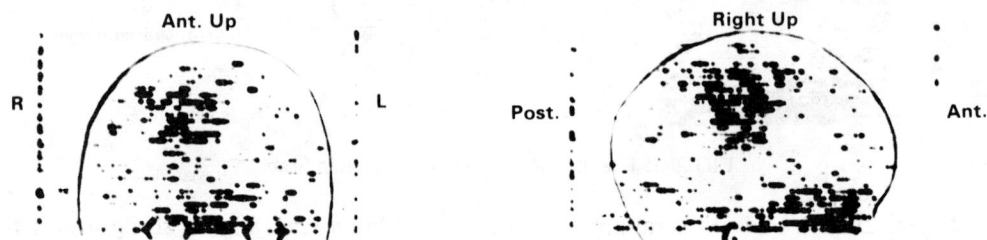

FIGURE 4. Abnormally Increased Activity in Neoplastic Disease.

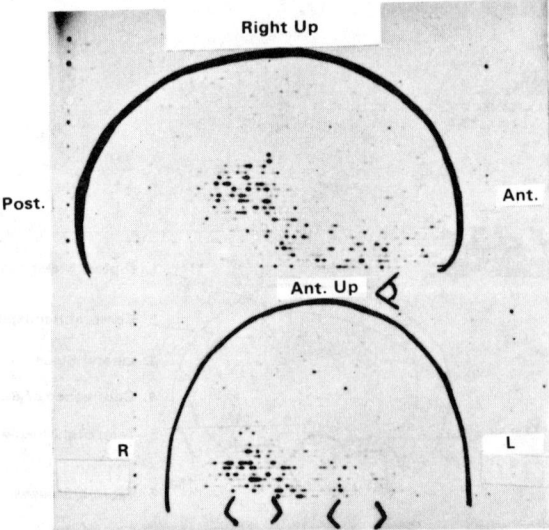

FIGURE 5. Abnormally Increased Activity in Nonneoplastic Disease.

USEFULNESS

The brain scan is of value in detecting a variety of neoplastic and nonneoplastic intracranial disease processes. The neoplastic processes include glioblastoma, glioma, meningioma, astrocytoma, metastatic tumor, hemangioma, medulloblastoma, melanoma, oligodindroglioma, ependymoma, sarcoma, pinealoma, pituitary tumor, acoustic neurinoma, craniopharyngioma, gangliocytoma, neurofibroma, spongioblastoma, teratoma, and cylindroma. Figure 4 shows a positive scan of a case with neoplastic disease. The nonneoplastic processes include abscess, cyst, aneurysm, arteriovenous malformation, encephalitis, encephalomalaria, infarction, thrombosis, hematoma, trauma, and granuloma. Figure 5 illustrates a positive scan of a case with nonneoplastic disease.

It is generally consented that the overall accuracy of brain scanning for detecting intracranial lesions is between 70 and 90 percent.

Comparison of Various Neurological Diagnostic Tests

Several groups of workers[1,3,6,21] have reported that brain scanning fulfills the need for an atraumatic screening procedure for diagnosis of intracranial lesions. It has the advantage, in contrast to carotid angiography and pneumoencephalography, of being free from complications or morbidity. Furthermore, the lesion or pathologic tissue is demonstrated by the scan, and the size and location of the lesion are directly established. Detection of intracranial lesions, therefore, does not depend upon the distortion of cerebral vasculatures or on an air-filled ventricular system. Brain scanning has been proved to be significantly more accurate than routine skull roentgenography, electroencephalography, or echoencephalography.

The brain scan should be used as a routine screening procedure in all patients with suspected intracerebral lesion before special neuroloradiological procedures are contemplated.

The brain scan is more comfortable and atraumatic to the patients, and less time-consuming to the physician, compared to other procedures. It can be done on an inpatient or outpatient basis. Therefore, follow-up scanning is also useful for evaluating the course of patients with intracranial lesion after surgical intervention or therapeutic measurements.

REFERENCES

1. McAfee, J. G. and Taxdal, D. R., Comparison of Radioisotopic Scanning with Cerebral Angiography and Air Studies in Brain Tumor Localization. *Radiology*, 77: 207, 1961.
2. Budabin, M., RISA Brain Scanning. *J. Mt. Sinai Hosp.*, 32: 527, 1965.
3. Wang, Y., *Clinical Radioactive Scanning*. Charles C Thomas, Springfield, Ill., 1967.
4. Feindel, W., Yamamoto, Y. L., and Rumin, N., Comparison of RISA and ^{203}Hg for Brain Scanning. *J. Neurosurg.*, 21: 1, 1964.
5. Spencer, R., Scintiscanning in Space-Occupying Lesions of the Skull. *Brit. J. Radiol.*, 38: 1, 1965.
6. Wang, Y., Shea, F. J., and Rosen, J. A., Comparison of the Accuracy of Brain Scanning and Other Procedures Used for Brain Tumor Detection. *Neurology*, 15: 1117, 1965.
7. Moore, G. E., The Use of Radioactive Diiodoflurescein in the Diagnosis and Localization of Brain Tumors. *Science*, 107: 569, 1948.
8. Chou, S. N., Aust, J. B., Peyton, W. T., and Moore G. E., Radioactive Isotopes in Localization of Intracranial Lesions. A Survey of Various Types of Isotopes and "Tagged Compounds" Useful in the Diagnosis and Localization of Intracranial Lesions, with Special Reference to the Use of Radioactive Iodine-Tagged Human Serum Albumin. *Arch. Surg.*, 63: 554, 1951.
9. Raimondi, A. J., Localization of Radioiodinated Serum Albumin in Human Glioma. *Arch. Neurol.*, 11: 74, 1964.
10. Pitlyk, P. J., Tauxe, W. N., Kerr, F. W. L., Sedlack, R. E., and Svien, H. T., Location of Brain Tumors With Polyvinylpyrrolidone-^{131}I. *Arch. Neurol.*, 9: 437, 1963.
11. Bale, W. F., Spar, I. L., and Goodland, R. L., Experimental Radiation Therapy of Tumor, Using ^{131}I-Carrying Antibodies to Fibrin UR567. *University of Rochester Atomic Energy Project*, 1960.
12. Tator, C. H., Morley, T. P., and Olszewski, J., A Study of the Factors Responsible for the Accumulation of Radioactive Iodinated Human Serum Albumin (RISA) by Intracranial Tumors and Other Lesions. *J. Neurosurg.*, 22: 60, 1965.
13. Rosenthall, L., Human Brain Scanning with Radioiodinated Macroaggregates of Human Serum Albumin. *Radiology*, 85: 110, 1965.

14. McAfee, J. G., Fueger, C. F., Stern, H. S., Wagner, H. N., Jr., and Migita, T., ^{99m}Tc Pertechnetate for Brain Scanning. *J. Nucl. Med.*, 5: 811, 1964.
15. Harper, P. V., Beck, R., Charleston, D., and Lathrop, K. A., Optimization of a Scanning Method Using ^{99m}Tc. *Nucleonics*, 22: 50, 1964.
16. Harper, P. V., Lathrop, K. A., McCardle, R. J., and Andros, G., The Use of ^{99m}Tc as Pertechnetate for Thyroid, Liver, and Brain Scanning in Medical Scanning. *Proceedings of the Athens Symposium*. International Atomic Energy Agency, Vienna, Austria, 1964.
17. Smith, E. M., Properties, Uses, Radiochemical Purity, and Calibration of ^{99m}Tc. *J. Nucl. Med.*, 5: 871, 1964.
18. Quinn, J. L., Ciric, I., and Hauser, W. H., Analysis of 96 Abnormal Brain Scans Using ^{99m}Technetium. *J. Amer. Med. Ass.*, 194: 157, 1965.
19. Webber, M. M., ^{99m}Technetium Normal Brain Scans and Their Anatomic Features. *Amer. J. Roentgenol.*, 94: 315, 1965,
20. Witcofski, R., Maynard, D., and Meschan, I., The Utilization of ^{99m}Technetium in Brain Scanning. *J. Nucl. Med.*, 6: 121, 1965.
21. Wang, Y. and Rosen, J. A., Positive Brain Scan in Non-Space-Occupying Lesion. *Amer. J. Roentgenol.*, 94: 816, 1965.
22. Mealey, J., Jr., Dehner, J. R., and Reese, I. C., Clinical Comparison of Two Agents Used in Brain Scanning: Radioiodinated Serum Albumin vs. Chlormerodrin-^{203}Hg. *J. Amer. Med. Ass.*, 189: 260, 1964.
23. Bucy, P. C. and Ciric, I. S., Brain Scans in Diagnosis of Brain Tumors (Scanning with Chlormerodrin-^{203}Hg and Chlormerodrin-^{197}Hg). *J. Amer. Med. Ass.*, 191: 437, 1965.
24. Overton, M. O., III, Otte, W. K., Beentjes, L. B., and Haynic, T. P., A Comparison of ^{197}Mercury and ^{203}Mercury Chlormerodrin in Clinical Brain Scanning. *J. Nucl. Med.*, 6: 28, 1965.
25. Morrison, R. T., Afifi, A. K., VanAllen, M. W., and Evans, T. C., Scintiencephalography for the Detection and Localization of Non-Neoplastic Intracranial Lesions. *J. Nucl. Med.*, 6: 7, 1965.
26. Blau, M. and Bender, M. A., Clinical Evaluation of ^{203}Hg-Neohydrin and ^{131}I-Albumin in Brain Tumor Localization. *J. Nucl. Med.*, 1: 106, 1960.
27. Sklaroff, D., Cerebral Scanning with Radioactive ^{203}Chlormerodrin. *Neurology*, 13: 79, 1963.
28. Rhoton, A. L., Carlson, A. M., and Ter-Pogassion, M. M., Brain Scanning with Chlormerodrin-^{197}Hg and Chlormerodrin-^{203}Hg. *Arch. Neurol.*, 10: 369, 1964.
29. Brownell, G. L., and Sweet, G. L. Scanning of Positron-Emitting Isotopes in Diagnosis of Intracranial and Other Lesions. *Acta Radiol.*, 56: 425, 1956.
30. DiChiro, G., New Radiographic and Isotopic Procedures in Neurological Diagnosis. *J. Amer. Med. Ass.*, 188: 524, 1964.
31. Dugger, G. S. and Pepper, F. D., The Reliablity of Radioisotope Encephalography. *Neurology*, 13: 1042, 1963.

RADIONUCLIDES AND THE ENDOCRINE SYSTEM

H. L. Atkins, M.D.
Medical Research Center, Brookhaven National Laboratory, Upton, New York.

INTRODUCTION

Thyroid function studies were among the earliest applications of radioactivity, even before the atomic age began. The striking affinity of the thyroid for iodine and the role of iodine in the production of thyroid hormone made it only natural that this would be so. Following World War II and the introduction of reactor-produced ^{131}I, this radioisotope rapidly became the most commonly used nuclide in medicine. Its low cost and practical shelf-life have continued to maintain its position as the radioisotope of choice for thyroid function studies. For several reasons—e.g., long physical half-life relative to the time in which observations must be made, and the presence of high-energy gamma rays and beta particles—recent attention has been focused on other radioisotopes of iodine, which can provide more information with less radiation to the patient.

The use of radionuclides for the study of other endocrine glands in addition to the thyroid is somewhat limited at present. Recent advances have included the use of high-specific-activity ^{131}I-labeled protein and peptide hormones in an immunoassay of circulating levels of hormone in humans. The method is described by Berson et al.[1] Other aspects of radionuclides in the endocrinopathies are the use of ^{75}Se-selenomethionine for parathyroid scanning and the use of ^{90}Y seeds for pituitary irradiation. Newer labeled compounds and the use of other nuclides, such as ^{11}C, may open up vast new areas of clinical usefulness of radioactive isotopes in the study of endocrine disease in the near future.

RADIOISOTOPES OF IODINE AND TECHNETIUM

Myers[2] has given an excellent summary of the radioiodines that could be useful. In addition to radioiodine, ^{99m}Tc has aroused some interest in the study of thyroid physiology.

Of the many isotopes of iodine that exist, four are presently of greatest interest. ^{131}I has been the most widely used and is most readily available. Its 8.05-day half-life is much longer than is required for most thyroid function studies and, together with its high-energy beta emission, contributes unnecessarily to the patient radiation dose. In addition, the presence of gamma rays with energy higher than the major one of 364 kev is a hindrance in scanning procedures.

For short-term studies ^{132}I has been utilized, particularly in England,[3,4,5,6] The gamma-ray spectrum, however, is not suitable for scanning. Sixty-day ^{125}I has been useful for double labeling. The low-energy gamma emission is suitable for scanning,[7] but can cause difficulties when the thyroid is substernal or when looking for metastatic thyroid carcinoma. The long half-life can be a disadvantage. It is useful for radioautography.

^{123}I, although not yet widely available, has aroused considerable interest. The 13.3-hour half-life, virtual absence of beta emission, and low gamma energies allow studies to be done conveniently with low patient dose. It is particularly suitable for scanning procedures.

In addition, ^{129}I has a high thermal-neutron cross section and could be useful in activation analysis.

The following Table 6 offers a list of radioactive isotopes of iodine.

Table 6. RADIOISOTOPES OF IODINE

Mass Number	Chief Decay Modes, Mev	$T_{1/2}$	Chief γ rays, Mev
117		6.5 min	0.34, 0.51
118		13.9 min	0.51, 1.15
119	β⁺	19.5 min	0.26, 0.73
120	4.0 β⁺	1.3 hr	0.62, 1.52
121	EC, 1.1 β⁺	2.1 hr	0.028, 0.21
121m	I.T.	80.0 μsec	0.06, 0.19
122	EC, 3.1 β⁺	3.5 min	0.56, 3.1
123	100% EC	13.3 hr	0.028, 0.16
124	EC, 2.2 β⁺	4.2 d	0.028, 1.69
125	100% EC	60.0 d	0.028, 0.035
126	EC, 0.87 β⁺	13.3 d	0.39, 0.67
?126m		2.6 hr	
128	2.12 β⁻	25.0 min	0.45, 0.54
129	0.15 β⁻	16×10^6 yr	0.029, 0.038
130	1.02 β⁻	12.5 hr	0.74, 1.15
131	0.61 β⁻	8.0 d	0.36, 0.64
132	1.60 β⁻	2.3 hr	0.76, 1.41
133	1.3 β⁻	21.0 hr	0.53, 0.85
134	2.5 β⁻	53.0 min	0.85, 1.8
135	1.4 β⁻	6.7 hr	1.28, 1.69
136	5.6	86.0 sec	1.32, 3.2
137	β⁻(n)	24.0 sec	
138	β⁻(n)	6.3 sec	
139	β⁻(n)	2.0 sec	

The following data, given in Tables 7 to 11, were obtained from Myers.[2]

Table 7. EMISSIONS OF 8.05-DAY ^{131}I

Beta Particles		Gamma Rays		
Energy, kev	Abundance, %	Energy, kev	Abundance, %	Calculated Narrow-Beam Half-Thickness in Water, cm
815	0.7	722	3.0	8.5
608	87.2	637	9.0	8.0
335	9.3	364	80.0	6.3
250	2.8	284	5.3	5.7
		163	0.7	4.7
		80	2.2	3.8

Table 8. EMISSIONS of 2.3-HOUR ^{132}I

Beta Particles		Gamma Rays	
Energy, Mev	Abundance, %	Energy, Mev	Abundance, %
0.73	15	0.520	30
0.97	20	0.620	6
1.16	23	0.670	100
1.60	24	0.760	93
2.13	18	0.970	23
		1.140	9
		1.410	13
		1.900	5
		2.200	~2

Table 9. CHIEF EMISSIONS FROM 60-DAY ^{125}I

Energy, kev	Type	Number per 100 Decays
Photons		
~28	K X rays	137
~35	Gamma rays	7
~4	L X rays	~7
Monoenergetic Electrons		
~34	Conversion	2
~31	Conversion	11
~23	Auger	23
~4	Conversion	80
~3	Auger	48

Table 10.
^{123}I GAMMA RAYS AND X RAYS USABLE IN DIAGNOSIS

Photons	Energy, kev	Number Emitted per 100 Disintegrations
Gamma rays	159	84
X rays	28 (av.)	92

X Ray Type	Energy, kev	% of X Rays	Number Emitted per 100 Disintegrations
Kα_1	27.5	54.4	50.0
Kα_2	27.2	27.9	25.6
Kβ_1	31.0	14.7	13.5
Kβ_2	31.7	3.0	2.8

Table 11. DECAY SCHEME OF ^{99m}Tc

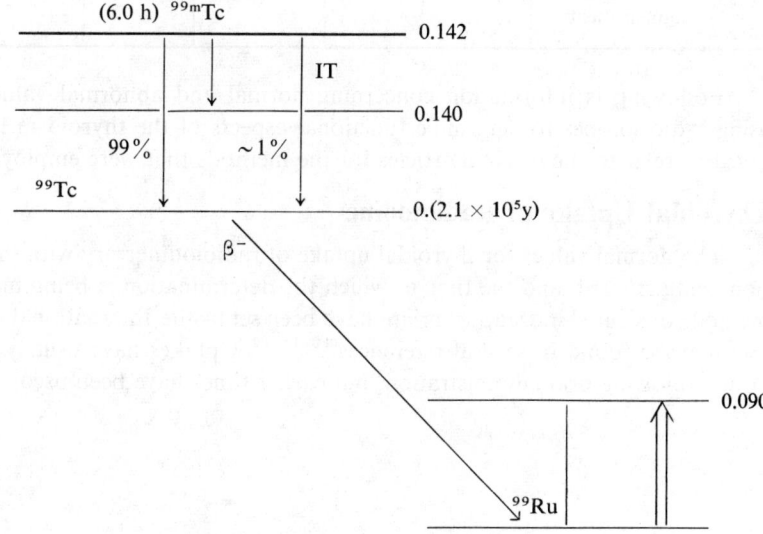

THYROID METABOLISM AND FUNCTION TESTS

Quantitative aspects of thyroid metabolism have been discussed by Riggs.[9] Below, in Table 12, is a partial listing of some values useful in interpreting isotopic tests of thyroid function.

Table 12. SOME THYROID METABOLISM VALUES

Quantity	Units	Condition				
		Normal	Acute Iodine Deficiency	Chronic Iodine Deficiency	Thyrotoxicosis	Thyrotoxicosis; Acute Block of Hormonal Synthesis
Rate of secretion of hormonal iodine	μg/day	70	70	70	597	597
Daily intake of iodide	μg/day	150	15	15	250	250
Rate of renal excretion of iodide	μg/day	144	53.1	9	199	796
Rate of excretion of organic iodine (chiefly fecal)	μg/day	6	6	6	51.1	51.1
Total iodide in compartment of inorganic iodide	μg	75	27.8	4.69	103	664
Organic iodine in thyroid	μg	8000	8000	3000	5500	4310
Extrathyroidal organic iodine	μg	1200	1200	1200	3360	3360
Thyroidal plasma clearance of iodide	ml/min	16.2	16.2	259	100	0
Renal plasma clearance of iodide	ml/min	33.3	33.3	33.3	33.3	33.3
Uptake of iodide in the thyroid	proportion	0.327	0.327	0.886	0.750	0.000
Iodide excreted	proportion	0.673	0.673	0.114	0.250	1.000
Biological half-life of ^{131}I in						
inorganic iodide compartment	hours	5.8	5.8	1.0	2.2	13.9
thyroid (initial)	days	79	79	30	6.4	5.0
thyroid (final)	days	113	(non-equilibrium)	156	20.3	(non-equilibrium)
extrathyroidal organic iodine compartment	days	11.9	11.9	11.9	3.9	3.9

Following is information concerning normal and abnormal values for various studies using radioisotopes to determine functional aspects of the thyroid in humans. The reader is urged to refer to the original articles for the methods that were employed.

Thyroidal Uptake of Radioiodine

The normal values for thyroidal uptake of radioiodine vary with the method, the population being studied, and the time at which the determination is being made. Standards for the method considered most appropriate have been set by the International Atomic Energy Agency and may be found in several references.[10,11,12] Uptakes have usually been performed at 24 hours following oral administration, but earlier times have been used.

Table 13.
NORMAL VALUES OF THYROIDAL RADIOIODINE UPTAKE AT 24 HOURS

Source	Normal Value, % dose
Sisson[13]	10.5–38.9
Gaffney et al.[14]	20–50
Werner et al.[15]	10–40 (90% of all euthyroids)

Table 14.
EFFECT OF THYROID POOL ON 24-HOUR THYROIDAL UPTAKE OF RADIOIODINE[13]

Condition	Iodide Pool, µg	Thyroid Uptake at 24 Hours	
		µg	% ^{131}I
Normal	280	70	25
Iodide deficiency	100	70	70
Iodide excess	2800	70	2.5
Iodide excess	28000	70	0.25
TSH in iodide excess	28000	105	0.38

Table 15. UPTAKES AT OTHER THAN 24 HOURS, % DOSE

Source	Condition	2 Hours	6 Hours	8 Hours	24 Hours	48 Hours
Adams and Purves[16]	Thyrotoxic			39–96	26–93	13–90
	Euthyroid			8–62	12–67	12–63
McConahy et al.[17]	Exophthalmic goiter		33.2–89.6		39.8–88.6	
	Euthyroid		7.5–25		11.9–46.5	
	Myxedema		1.7–6.7		0.6–18.1	
Vanderlaan[18]	Hyperthyroid	34.8 ± 2.2				
	Euthyroid	11.2 ± 0.9				

Table 16. ONE-HOUR UPTAKE STUDIES

Source	Condition	% Dose in Thyroid at 1 Hour
Vanderlaan[18]	Hyperthyroid	26.9 ± 1.9
	Euthyroid	9.8 ± 0.5
	Hypothyroid	3.8 ± 0.2
Crispell et al.[19]	Hyperthyroid	20–57
	Euthyroid	7–19 (2 patients had 20, 25% uptake)
	Hypothyroid	4–14
Kohler and Wynn[20]	Hyperthyroid	9–60
	Euthyroid	1–9
	Nontoxic nodular goiter	3–12
	Hypothyroid	1–8
Kriss[21] (intravenous)	Hyperthyroid	18–72
	Probable hyperthyroid	13–17
	Euthyroid	4–11
	Nontoxic nodular goiter	4–11
	Myxedema	2

Table 17.
SUMMARY OF
THYROIDAL RADIOACTIVE IODINE UPTAKE CHANGES[13]

Type of Radioiodine Uptake Disturbance	Concomitant PBI
Elevated Values	
Hyperthyroidism	high
Hyperthyroidism Adequately Treated	normal
Low Iodide Pool	normal
Thyroid Diseases	
Hashimoto's (early)	normal or high
Subacute thyroiditis (recovery)	low or normal
Intrathyroidal biochemical disturbances	low or normal
Rebound from Inhibition of Hormonogenesis	
After iodide therapy	normal
After antithyroid drug therapy	low or normal
Extrathyroidal Diseases	
Cirrhosis	low or normal
Nephrosis	low
Choriocarcinoma	high
Depressed Values	
Hypothyroidism	low
Large Iodide Pool	
Iodide administration	normal or high
Organic iodine administration	high
Thyroid Diseases	
Hashimoto's (late)	low
Subacute thyroiditis (early)	normal or high
Intrathyroidal biochemical disturbances	low or normal
Thyroid hormones	depends on type
Antithyroid drugs	low or normal
Adrenal corticosteroids	low
Extrathyroidal Diseases	
Congestive heart failure	normal

Use of ^{132}I in Diagnosis.

The reader is referred to reports by Halnan,[5] Halnan and Pochin,[4] Goolden and Mallard,[3] and Hobbs et al.[6] for descriptions of the clinical use of ^{132}I. Halnan and Pochin used a 2-hour neck/thigh ratio. Their results are as follows:

$$\begin{array}{ll} \text{Euthyroid} & 3.0 \pm 1.0 \\ \text{Thyrotoxicosis} & 43 \pm 22 \\ \text{Thyroid ablation} & 1.5 \pm 0.2 \end{array}$$

Table 18. STIMULATION TESTS OF THYROID UPTAKE[13]

References	Subjects (Number)	TSH* Dose, μ	TSH* Frequency of Injection	Baseline, % dose	Radioiodine Uptake Time Tracer, (hours after TSH)	Radioiodine Uptake Time Count, (hours after tracer)	After TSH Values Difference (after−before), mean ± SD, or range	After TSH Values % Difference ($\frac{after-before}{before} \times 100$)
Multiple Injections of TSH								
Schneeberg et al.[22]	Normals (6)	10	qd × 3	<15	?	24	+29, +15 to +37	
	Normals (18)	10	qd × 3	>15		24	+18, + 6 to +35	
	Pituitary Insufficiency (12)	10	qd × 3			24	+34, +20 to +49	
	Primary Hypothyroidism							
	Thyroidectomy (5)	10	qd × 3			24	−6 to +11	
	Spontaneous or Cretinism (17)	10	qd × 3			24	−8 to + 5	
Taunton et al.[23]	Normals (10)	5	qd × 3	17	18	24	±18	
	Pituitary Insufficiency							
	Sheehan's (3)	5	qd × 3	1	18	24	± 1	
	Other (5)	5	qd × 3	5	18	24	±11	
	Primary Hypothyroidism (15)	10	qd × 3	4	18	24	± 3	
Single Injection of TSH								
Fletcher and Besford[24]	Normals (14)	10		<30†	8	24		>49
	Pituitary Insufficiency (12)	10			8	24		>49
	Primary Hypothyroidism (5)	10			8	24		0
Jefferies et al.[25]	Normals (12)	5		28	21	3	+20.4 ± 7.8	
Taunton et al.[23]	Normals (11)	5			18	24	+25 ± 11	
	Pituitary insufficiency							
	Sheehan's (3)	5		1	18	24	+2 ± 2	
	Other (5)	5		4	18	24	+10 ± 6	

* Thytropar, Armour Pharmaceutical Co.

† 30% baseline resulted in smaller responses to TSH.

Table 19. TYPES OF THYROID SUPPRESSION TESTS USING THE THYROID UPTAKE[13]

Reference	Subjects (Number)	Method			Results (24-Hour Uptake ^{131}I)		
					% of Dose		RAIU After ÷ RAIU Before
		Hormone	Daily Dose, mg	Days Given	Before Hormone	After Hormone	
Werner and Spooner[26]	Euthyroid (41)	T_3	0.075	8	37 ± 2.2*	12 ± 1.4†	
	Hyperthyroidism (20)	T_3	0.075	8	56 ± 3.4	62 ± 3.4	
	Hyperthyroidism (28)	T_3	0.150	8	63 ± 9.0	63 ± 11.1†	
Dresner and Schneeberg[27]	Euthyroid (35)	T_3	0.300	2	48.8	26.2	0.533 ± 0.088‡
	Hyperthyroidism (18)	T_3	0.300	2	81.1	75.3	0.930 ± 0.086‡
Cassidy and Jagiello[28]	Euthyroid (56)	Des. thyroid	180	21		<20 in all except 1	<0.65 in all except 1
	Hyperthyroidism (26)	Des. thyroid	180	21		<10 in all except 4 <20 in all except 1	>0.66 in all††

* Mean ±SE.
† All but one euthyroid <20%, and all euthyroid <20% at 150 µg/d; all hyperthyroid >35%.
‡ 0.073 will separate hyper- and euthyroid groups, so there is no overlap of means ±2 SE.
‡‡ 0.65 gives good separation of hyper- and euthyroid groups.

Table 20. URINARY EXCRETION OF RADIOIODINE[29]

Condition	24 Hours, %	48 Hours, %
Normal	60.6 ± 0.7	65.9 ± 0.7
Thyrotoxicosis	15.5 ± 0.8	17.4 ± 0.9
Thyrotoxicosis (diagnostic problems)	26.6 ± 1.2	29.0 ± 1.3
Nontoxic nodular goiter	59.9 ± 1.7	66.6 ± 2.2
Nontoxic nodular goiter (diagnostic problems)	59.6 ± 1.6	64.6 ± 1.6
Myxedema	61.4 ± 2.3	83.6 ± 2.1

Table 21. THYROID AND RENAL CLEARANCE OF RADIOIODINE

Condition	Thyroid Clearance, ml/min/1.73 m²		Renal Clearance, ml/min/1.73 m²	
	Mean	Range	Mean	Range
Hyperthyroid	210.5	74.5–512.0	30.3	5.2–53.3
Euthyroid	17.7	3.7– 41.0	35.6	11.1–58.0
Hypothyroid	2.0	0.0– 4.1	26.8	9.1–40.0

The method assumes constancy of the iodide space in the first half-hour following injection.

Table 22.
RENAL AND EXTRARENAL CLEARANCE OF RADIOIODIDE[31]

$$\text{Renal Clearance} = \frac{U'_f}{P'_0} r; \quad \text{Extrarenal Clearance} = \frac{r(1 - U'_f)}{P'_0}$$

U'_f = estimated asymptotic value of urinary excretion, fraction of administered dose.
r = exponential rate constant.
P'_0 = concentration of inorganic ^{131}I in blood at time zero by extrapolation.

Condition	Clearance of Serum, ml/min			
	Renal		Extrarenal	
	Normal Function	Impaired Function	Range	Mean
Hyperthyroid	32.7 ± 1.0		49–479	154.6 ± 19.5
Euthyroid	33.3 ± 3.1	(8.7 ± 1.6)	10–64 (3–11)	26.3 ± 5.2 (6.5 ± 0.2)
Myxedema	17.9 ± 1.6		3–8	4.7 ± 0.7

Table 23.
IODIDE CONCENTRATING ABILITY OF THE THYROID[32]

Thyroidal iodide space and thyroid/plasma ratios, binding blocked.

Condition	Thyroidal Iodide Space, liters	Thyroid/Plasma Ratio	Thyroidal Iodide Clearance, ml/min
Euthyroid: mean	1.0	44	15.8
SE	0.05	1.4	1.0
range	0.6–1.6	28–63	7.0–25.0
Hyperthyroid: mean	6.1	125	420.9
SE	0.3	22	94.9
range	1.7–18.2	49–455	43.6–1822.0

Table 24.
TECHNETIUM CONCENTRATING ABILITY OF THE THYROID[33]

Condition	Uptake, %	Thyroidal Tc Space, ml	Thyroid/Plasma Ratio
Euthyroid	0.5–4.0	92–608	3.7–31.1
Hyperthyroid	4.0–28.4	405–4850	16.3–58.4

Table 25.
FACTORS INFLUENCING RADIOIODINE UPTAKE IN THE THYROID[34]

A. Factors Causing Low 24-Hour Uptake

    Excess Iodine or Iodide
    Diet
        Foods
        Iodized salt
    Radiographic Media
    Goitrogenic Foods
        Brassica family
            Turnips
            Cabbage
            Kale
            Rape
        Milk from cattle eating chou-moullier and certain other fodders
    Thiourylene Derivatives
        Thiouracil
        Propylthiouracil
        Methimazole
    Aromatic Antithyroid Drugs
        Resorcinol
        Aminobenzenes
            Para-aminosalicylic acid
            Para-aminobenzoic acid
        Sulfonylureas
    Cardiac Decompensation
    Sedatives
        Thiopental
        Bromides, if contaminated with iodine
    Miscellaneous Drugs
        Monovalent anions
        Perchlorate
        Thiocyanate
        Others, vitamin A, phenylbutazone, salicylates
    Hormones
        Adrenocortical steroids
            Endogenous steroids activated by adrenocorticotrophic hormone
            Cortisone and related compounds
            Desoxycorticosterone
        Thyroid
        Progesterone
        Testosterone

B. Potential Causes of Low Uptake (unconfirmed in man)

    Heavy Metal Administration
        Mercury
        Cobalt
        Arsenic
        Bismuth
    Goitrogenic Foods
        Peaches
        Pears
        Strawberries
        Spinach
        Carrots
        Peanuts
    Antibiotics
        Penicillin
        Chlortetracycline
        Sulfonamides
    Morphine
    Epinephrine

Table 25. (*Continued*)
FACTORS INFLUENCING RADIOIODINE UPTAKE IN THE THYROID[34]

C. **Clinical and Experimental Factors that May Increase Uptake**

    Iodine Deficiency
    Chronic Liver Disease with Dietary Iodine Deficiency
    Rebound Phenomenon Following Withdrawal of Antithyroid Drugs
    Nephrosis
    Estrogens
    Low Serum Chlorides
    Soybean Meal, Cellulose and Bulk Producer

D. **Factors Apparently Not Influencing Uptake**

    Mercurial Diuresis
    Malabsorption Syndrome
    Increased Glomerular Filtration
    Decreased Glomerular Filtration
    Sedatives
        Reserpine
        Pentobarbital
        Tribromethanol

Table 26.
COMPARISON OF 6-HOUR AND 24-HOUR THYROIDAL UPTAKE OF ^{131}I BEFORE AND AFTER THE ADDITION OF POTASSIUM IODIDE[35]

Carrier Iodide, μg	Time, hours after dose	Euthyroids, % difference*	Graves' Disease, % difference*
0	6	6	0
	24	4	4
300	6	36	21
	24	15	5
500	6	18	43
	24	11	29
750	6	19	57
	24	16	53
1000	6	27	58
	24	27	47
1500	6	33	—
	24	36	72
2000	6	48	81
	24	37	77
3000	6	53	67
	24	46	49

* Percentile difference = $\dfrac{\text{mean difference of percent uptake}}{\text{mean uptake before added carrier iodide}} \times 100$

Table 27.
THYROID UPTAKES IN STUDIES USING VARIOUS ^{131}I-LABELED PREPARATIONS[35]

Compound	Thyroid Uptake at 24 Hours, %
^{131}I-Triolein	7.2–18.5
^{131}I-Hippuran	0.0–2.6
^{131}I-Rose Bengal	0.0–7.9
^{131}I-RHSA	5.8–9.0

Table 28. THYROID BINDING OF RADIOIODINE

Condition	48-Hours PBI-131[36]	72-Hours PBI-131[37]	Conversion Ratio*[38] at 24 Hours
Euthyroid	<0.2%/1	<0.3%/1	3–41%
Hyperthyroid	0.7–2.8%/1	>0.3%/1	>45%

* C.R. = $\frac{\text{protein bound }^{131}\text{I}}{\text{total plasma }^{131}\text{I}} \times 100$

Table 29.
PROTEIN BINDING OF TRIIODOTHYRONINE
(IN-VITRO TESTS)

Hamolsky, RBC Binding[39]

Condition	Female	Male
Euthyroid	11.0–17.0%	11.8–19.0%
Hyperthyroid	17.0–35.0%	19.5–37.9%
Hypothyroid	6.1–11.0%	5.5–11.6%

Resin Uptake of T_3, Expressed as % of Normal [40]

Condition	Mean	Range
Euthyroid	99.1 ± 4.3	91–106
Hyperthyroid	122.0 ± 9.4	108–135
Hypothyroid	82.0 ± 4.9	70–90
Pregnancy	71.0 ± 8.8	55–83

Table 30.
EXAMPLES OF CHANGES IN THYROXIN-BINDING GLOBULIN CAPACITIES
AND T-3-TEST RESULTS[13]

A. Increased TBG Capacities and Decreased T-3-Test Results

    Pregnancy
    Estrogen Administration (including the common ovulatory suppressants—oral contraceptives—as norethyndrol*)
    Acute Liver Disease and Cirrhosis
    Familial or Hereditary
    Idiopathic

B. Decreased TBG Capacities and Increased T-3-Test Results

    Androgen Administration
    Anabolic-Hormone Administration
    Nephrosis
    Advanced Cirrhosis
    Familial or Hereditary
    Idiopathic

* Enovid, G. D. Searle and Company, Chicago.

THERAPY WITH IODINE-131

The status of ^{131}I in the therapy of hyperthyroidism is still undergoing change. More recently the concern has been associated with the late development of hypothyroidism at an incidence hitherto unsuspected. No conclusive proof of carcinogenic action has been demonstrated, but further follow-up is required. A number of factors must be taken into account when deciding among surgery, radioiodine, and chemotherapy for any individual patient.

A scheme of suggested form of treatment is given by Silver.[42]

Original Treatment

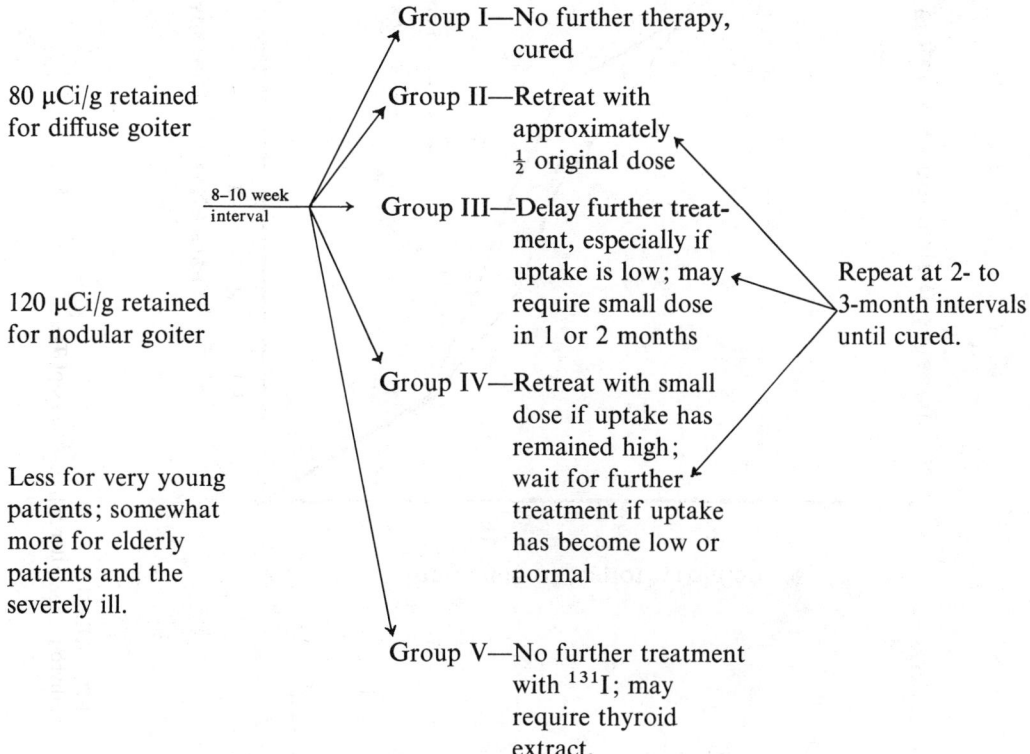

With this scheme, the following treatment was required in a series of 1600 patients[43] (see Table 31).

Table 31.
SUMMARY OF IODINE-^{131} TREATMENT IN A SERIES OF 1600 PATIENTS

	Average Number of Treatments per Patient	Average Total Dose of Radioactive Iodine Administered per Patient, mCi
Diffuse toxic thyroids	1.9	7.2
Solitary toxic nodules	1.8	7.4
Multinodular toxic goiters	2.3	10.8
Nodular of all types	2.2	10.3
Total for entire series	2.0	8.13

The late complications of the radioiodine treatment of hyperthyroidism, particularly the continuously rising incidence of hypothyroidism, was pointed out by Beling and Einhorn,[44] and confirmed by others.[45,46]

Figure 6 shows the incidence of hypothyroidism after ^{131}I treatment of 791 patients controlled for 2 to 8 years. Each incidence is based on the number of patients followed up.[44]

Tables 32 to 40 offer extensive data on the effects of ^{131}I therapy in a variety of disease states.

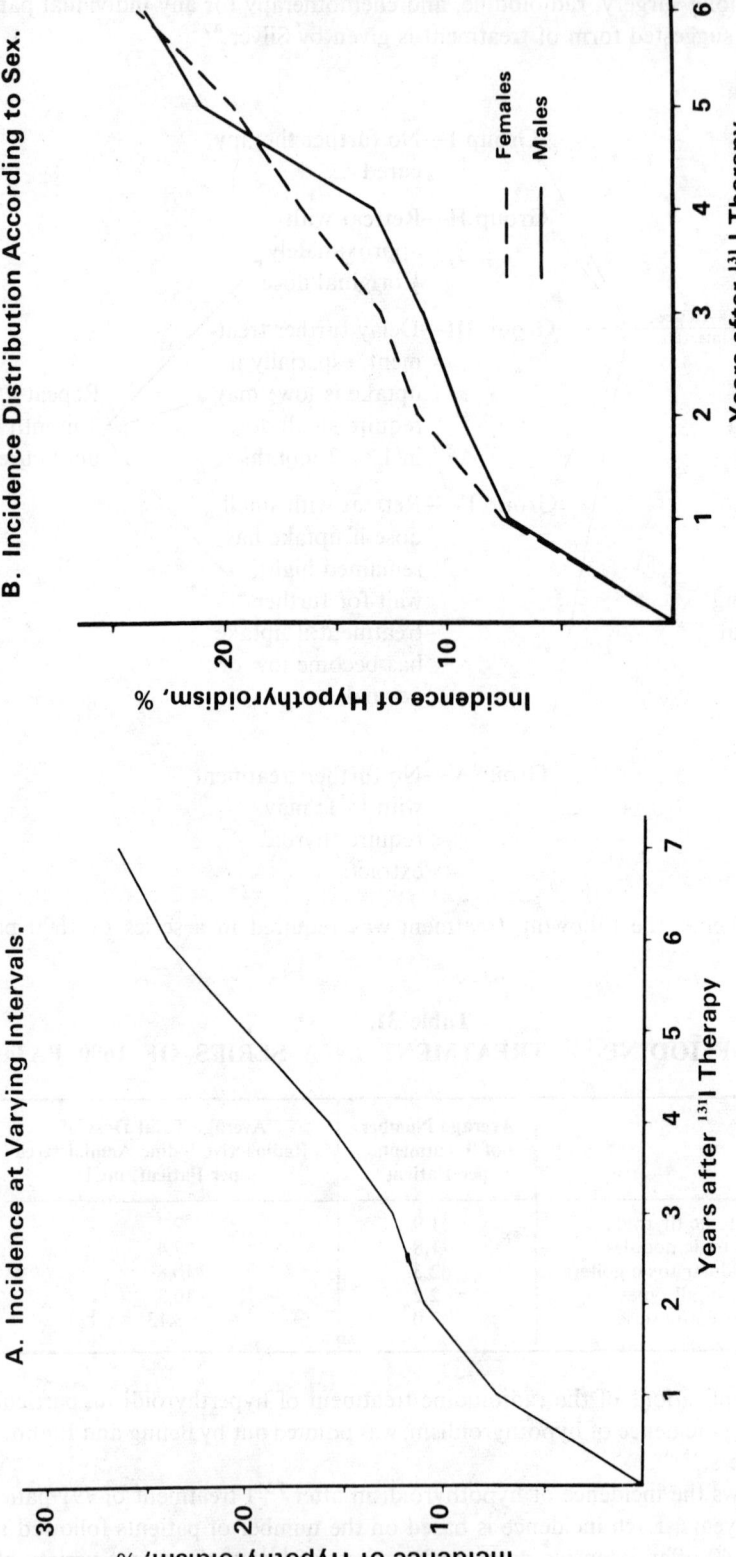

FIGURE 6. Incidence of Hypothyroidism After ^{131}I Therapy of Thyroid Hyperfunction.

Table 32.
INCIDENCE OF RECURRENT HYPERTHYROIDISM FOLLOWING ^{131}I THERAPY[44]

Source of Data	Number of Patients Treated	Recurrence of Hyperthyroidism, number of cases
Chapman and Maloof, 1955	520	1
Clark and Rule, 1955	628	3
McCullagh, 1956	1235	2 (8)
Werner et coll., 1957	525	6
Cassidy and Astwood, 1959	200	0
Eller et coll., 1960	1308	12
Beling and Einhorn, 1961	791	3

Table 33. CUMULATIVE INCIDENCE OF HYPOTHYROIDISM

Treatment and Interval, years	Number Observed at Start of Interval and Euthyroid	Number Lost to Observation During Interval	Patients Becoming Hypothyroid During Interval		Cumulative Percent Hypothyroid
			Number	Percent*	
Sodium Iodide-^{131}I					
0–1/12	848	49	2	0.24	0.24
1/12–1/4	797	12	55	6.95	7.18
1/4–1/2	730	56	197	28.06	33.23
1/2–3/4	477	14	41	8.72	39.05
3/4–1	422	17	12	2.90	40.82
1–2	393	28	24	6.33	44.57
2–3	341	41	14	4.37	46.99
3–4	286	33	15	5.57	49.94
4–5	238	21	9	3.96	51.92
5–6	208	24	13	6.63	55.11
6–7	171	21	8	4.98	57.34
7–8	142	21	12	9.12	61.24
8–9	109	22	9	9.18	64.80
9–10	78	20	5	7.35	67.38
10–	53	16	3	6.67	69.56†
Surgery					
0–1/12	121	6	5	4.24	4.24
1/12–1/4	110	8	10	10.20	14.01
1/4–1/2	92	6	8	9.64	22.30
1/2–3/4	78	1	1	1.29	23.30
3/4–1	76	2	2	2.67	25.34
1–2	72	4	1	1.43	26.41
2–3	67	1	0	0.00	26.41
3–4	66	5	1	1.57	27.57
4–5	60	9	2	4.30	30.68
5–6	49	7	0	0.00	30.68
6–7	42	9	1	3.51	33.12
7–8	32	6	2	8.69	38.93
8–9	24	3	0	0.00	38.93
9–10	21	2	0	0.00	38.93
10–	19	4	1	5.88	42.52‡

* Number becoming hypothyroid/(number euthyroid beginning of interval)—1/2 (number lost to observation during interval).
† Standard error, 5.7%; effective sample size 635.
‡ Standard error, 5.0%; effective sample size 97.

Table 34.
LEUKEMIA FOLLOWING RADIOIODINE TREATMENT OF THYROTOXICOSIS[47]

Country	Replies from Main Clinics	Patients Treated	Patient Years	Leukemia Rate, d/m/yr*	Cases Expected by Chance	Cases Observed in Treated Patients
United Kingdom	All (47)	13,300	38,000	73	2.8	3
U.S.A.	Most (67)	31,400	142,000	106	15.1	10
Canada	All (34)	13,900	40,000	80	3.2	4
Austria	All (3)	600	1,700	(73)	0.1†	1
Total		59,200	221,900		21.2 (S.D. 4.6)	18

* dead/million/year.
† based on U.K. rates.

Table 35.
OBSERVED AND EXPECTED NUMBERS OF LEUKEMIA CASES AMONG PATIENTS TREATED WITH RADIOIODINE[48]

Types of Leukemia Cases	Male			Female			Total		
	Obs.*	Exp.†	Pr.‡	Obs.*	Exp.†	Pr.‡	Obs.*	Exp.†	Pr.‡
All leukemias	8	3.66	0.014††	2	10.10	0.003††	10	13.76	0.195
Acute myelogenous	6	0.58	<0.001††	1	2.48	0.287	—	—	—
Acute, all other	—	0.40	0.670	—	1.56	0.210	—	—	—
Chronic lymphatic	—	1.40	0.247	—	2.47	0.085	—	—	—
Chronic, all other	2	1.28	0.143	1	3.59	0.126	—	—	—
Acute	6	0.98	0.001††	1	4.04	0.092	7	5.02	0.133
Chronic	2	2.68	0.494	1	6.06	0.017††	3	8.74	0.025††

* Observed number of cases.
† Expected number of cases.
‡ Probability number of cases is as great (or as small) as that observed for given expected number.
†† Statistically significant with stated value of Pr.

Table 36.
OBSERVED AND EXPECTED NUMBERS OF LEUKEMIA CASES BY INTERVAL BETWEEN RADIOIODINE TREATMENT AND DIAGNOSIS OF LEUKEMIA[48]

Years ^{131}I Rx to Dx Leukemia	Expected Cases, Total	Observed Cases		
		Total	Acute	Chronic
Totals	13.7	10	7	3
0–	3.5	1	—	1
1–	3.1	3	3	—
2–	2.6	2	1	1
3–	2.0	2	2	—
4–	1.3	1	1	—
5–	0.8	—	—	—
6–	0.3	1	—	1
7+	0.1	—	—	—

Table 37.
SUMMARY OF RESULTS IN ^{131}I TREATMENT OF ANGINA PECTORIS IN EUTHYROID PATIENTS[49]

Result	Segal et al.		Jaffe		Blumgart		Wolferth		Chapman		Goldman		Blumgart (Combined from 50 Clinics)	
	No.	%	No.	%	No.	%	No.	%	No.	%	No.	%	No.	%
Excellent	15	23	53	56	24	39	12	67	4	44	5	38	284	39
Good	23	35	35	37	20	32	4	22	5	56	3	24	257	36
Fair and poor	19	30	6	7	18	28	2	11					179	25
Died	8	12*	16†								5	38		
Total	65		110		62		18		9		13		720	

* Three other deaths are included in the excellent, good, and fair-and-poor categories.
† Jaffe included seven of these deaths in his excellent categories, seven in the good, and two in fair and poor.

Table 38.
RESULTS OF ^{131}I TREATMENT OF THYROID CARCINOMA IN 200 PATIENTS AT 14 YEARS[50]

Group 1. ^{131}I in lymph node metastases.
Group 2. ^{131}I in distant metastases.
Group 3. ^{131}I in thyroid bed only—normal tissue presumed.
Group 4. Residual thyroid carcinoma by palpation or X ray; no concentration of ^{131}I in metastases.

Category of Patients	No.	Mean Age, years	Well Diffused Carcinoma, %	Mean Total Dose, mCi	Mean Follow-Up Interval, years
Group 1.					
Clinically free of disease	35	28	94	148	5.1
Clinical evidence of residual disease	14	30	78	304	3.5
Dead	3	53	0	122	1.7
Group 2.					
Clinically free of disease	16	27	94	366	7.0
Clinical evidence of residual disease	7	44	86	533	5.4
Dead	7	63	100	316	3.4
Group 3.					
Clinically free of disease	72	35	97	136	4.0
Clinical evidence of residual disease	2	42	50	376	1.5
Dead	5	47	60	174	3.4
Group 4.					
Clinically free of disease	3	43	100	259	4.5
Clinical evidence of residual disease	6	46	67	169	2.8
Dead	30	60	30	93	1.3

Table 39. UPTAKE OF RADIOIODINE AND SURVIVAL IN PATIENTS WITH THYROID CARCINOMA HAVING DISTANT METASTASES AT TIME OF TREATMENT WITH RADIOIODINE[51]

Survival After Radioiodine	Number of Patients	Uptake of Radioiodine in Metastases		
		Good	Some	None
Died within 6 months	14	—	4	10
Died at 6 months to 1 year	3	—	2	1
Died at 1 to 2 years	3	2	—	1
Died at 2 to 5 years	4	2	1	1
Alive at 2 to 5 years	3	2	1	—
Alive over 5 years	3	3 9, 10, 15½	—	—
Total	30	9	8	13

Table 40. SURVIVAL OF 67 PATIENTS WITH THYROID CANCER[52]

Age of Patients	Number of Patients	Living		Dead	
		No Cancer	Cancer	No Cancer	Cancer
Patients less than 40 years old	35	26	8	1	0
Patients more than 40 years old	32	9	4	3	16
Total	67	35	12	4	16

Positive localization: 19/48 patients in metastases
26 patients in thyroid remnant
3 no localization
Neck freed of metastases, 6/13 patients; lung or bone, 0/6 patients

PITUITARY IRRADIATION

Ablation of the pituitary for treatment of advanced hormonally sensitive malignancies as well as for primary pituitary conditions has been attempted with various radionuclides. Radon(^{222}Rn), ^{32}P, ^{198}Au, and ^{90}Y have been among the radioactive sources used in various implantation procedures. The most popular at the present time is ^{90}Y, because of its pure beta emission (2.26 Mev) and its susceptibility to roentgenologic control during implantation. The decay scheme[53] is as follows:

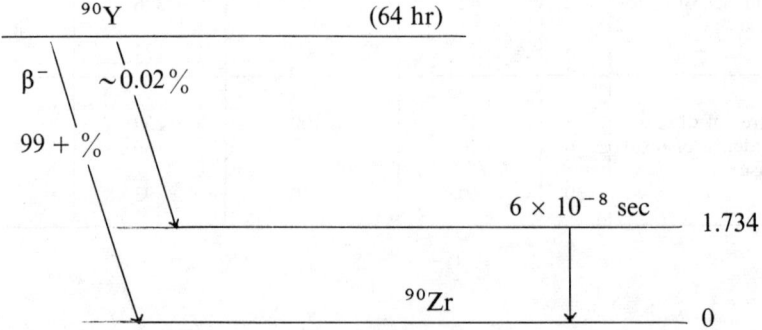

Figure 7 illustrates the pattern of radiation intensity from a given source in relation to distance.

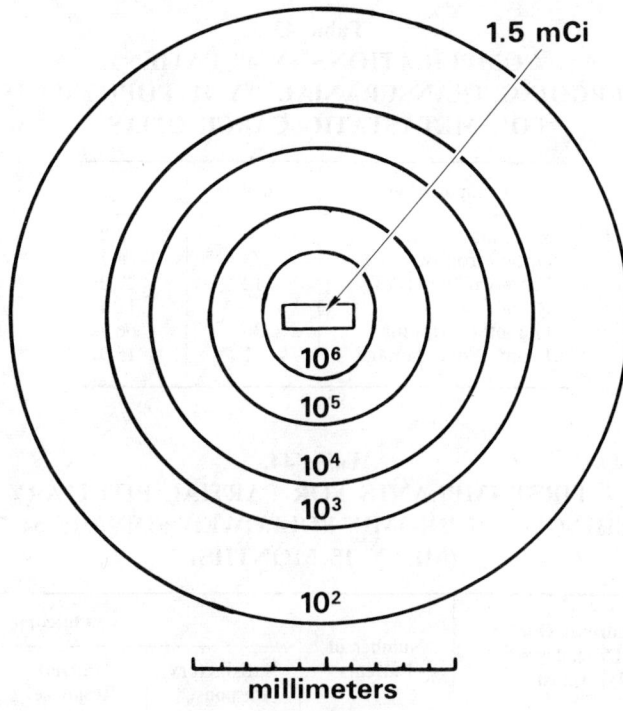

FIGURE 7.
Isodose Curves Surrounding Y_2O_3 Pellet in Lucite Phantom.
(Numbers indicate rads at complete decay.)

The necrotic effects of radiation dosages on various tissues were established in experiments conducted by Rasmussen et al.,[55] and the data are given in Table 41.

Table 41.
RADIATION DOSAGE REQUIRED TO PRODUCE NECROSIS

Tissue	Rads
Anterior lobe of hypophysis	110,000 to 190,000
Hypothalamus	60,000 to 120,000
Optic chiasm and tract	60,000 to 140,000
Oculomotor nerve	

Data are from experiments on monkeys. The figures are probably somewhat high.

A comparison of visual-field complications from implants of yttrium and radon was made by Harrington;[56] Table 41 indicates why yttrium is preferred.

Table 42.
FIELD DEFECTS IN PATIENTS TREATED WITH RADON AND YTTRIUM

Treatment	Normal Fields	Field Defects	Total
Radon	28	8	36
Yttrium	23	4	27
Total	51	12	63

Additional data on the effects of pituitary irradiation are shown in Tables 43 to 46.

Table 43.
COMPLICATIONS IN 48 PATIENTS UNDERGOING TRANSCRANIAL ^{90}Y HYPOPHYSECTOMY FOR METASTATIC CARCINOMAS[57]

Complication	Number	%
Anosmia	4	8.9
Optic atrophy	3	6.7
Extraocular palsies	12	27.2
Rhinorrhea	5	11.4
Diabetes insipidus	30	66.7
Frontal lobe damage	7	15.9

Table 44.
RESULTS OF FIRST IMPLANTS FOR PARTIAL PITUITARY ABLATION FOR CUSHING'S SYNDROME, FOLLOWED FOR 3 to 54 MONTHS (MEAN 15 MONTHS)[58]

Cushing's Syndrome Due to Pituitary-Dependent Adrenal Hyperplasia	Number of Patients	Outcome		
		Satisfactory Response*	Partial Response†	Unchanged
With obvious pituitary tumor	6	0	4	2
Others				
^{198}Au only	7	5	0	2
^{198}Au + ^{90}Y	7	3	4	0
Total	20	8	8	4

* Satisfactory response = symptom-free, facies normal, glucose tolerance test normal or improved, urinary 17-oxygenic steroid normal or improved.
† Partial response = partial loss of clinical signs and symptoms, or correction of urinary 17-oxygenic steroid or response only temporary.

Table 45.
RESULTS OF PARTIAL PITUITARY ABLATION BY ^{90}Y AND/OR ^{198}Au FOR ACROMEGALY, FOLLOWED FOR 4 TO 48 MONTHS (MEAN 19 MONTHS)[59]

Isotope Implanted	Number of Patients	Change in Acromegaly		
		Satisfactory Response*	Partial Response	Unchanged
After first implant (^{198}Au	15	5	5	5
(^{198}Au + ^{90}Y	6	1	2	3
(^{90}Y	1	1	—	—
After second implant, various	9	6	3	0
Final result	22	13	6	3

* Satisfactory response = regression of acromegalic facies, symptom-free, insulin tolerance test normal, glucose tolerance test normal or improved.

Table 46.
PITUARY ^{90}Y IMPLANTATION VS. ADRENALECTOMY
PLUS OOPHORECTOMY[60]

Method	Number of Patients	Objective Improvement		Not Assessed
		Number	Duration of Regression, weeks	
Pituitary ^{90}Y implantation	22	7	36+ (13–63)	3
Adrenalectomy + oophorectomy	22	8	42+ (13–70)	2

RADIOIMMUNOASSAY OF HORMONES

Radioimmunoassay has been developed for a number of protein and peptide hormones, including insulin, human growth hormone (HGH), parathyroid hormone, adrenocorticotropic hormone (ACTH), luteinizing hormone (LH), thyrotropic hormone (TSH), and glucagon. The method has extremely high sensitivity for minute quantities present in plasma, and has enabled investigators to study new aspects of human metabolism and disease. The basic principles have been outlined by Berson, et al.,[1] and the results of a few studies are presented here in Figures 8 to 12 and in Tables 47 to 52. Undoubtedly, in the future there will be a much greater utilization of the method in clinical diagnosis as well as in more basic studies of mechanism of disease.

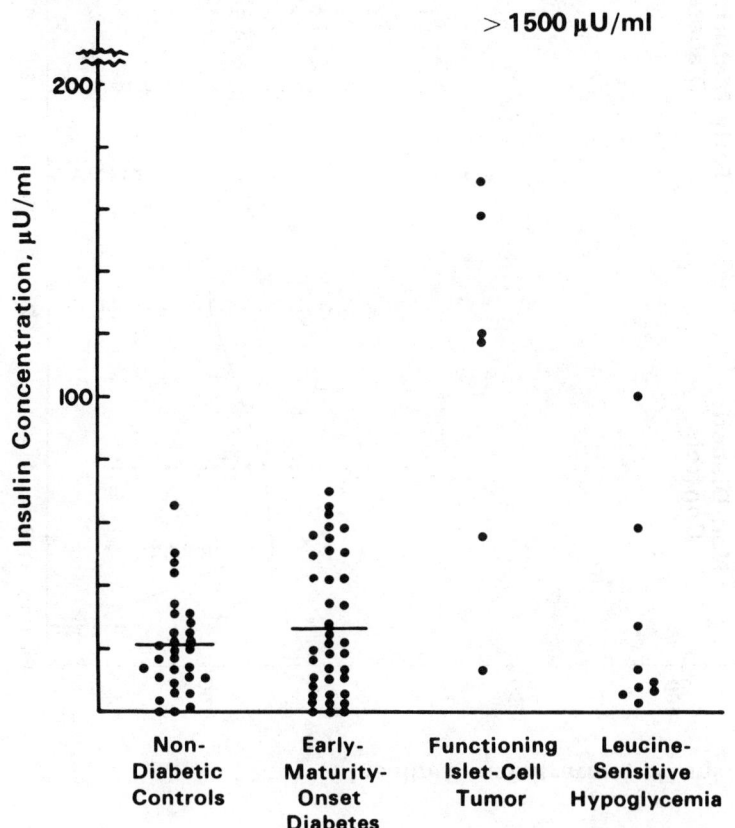

FIGURE 8. Fasting Plasma Insulin Concentrations in Various Groups of Subjects.[61]

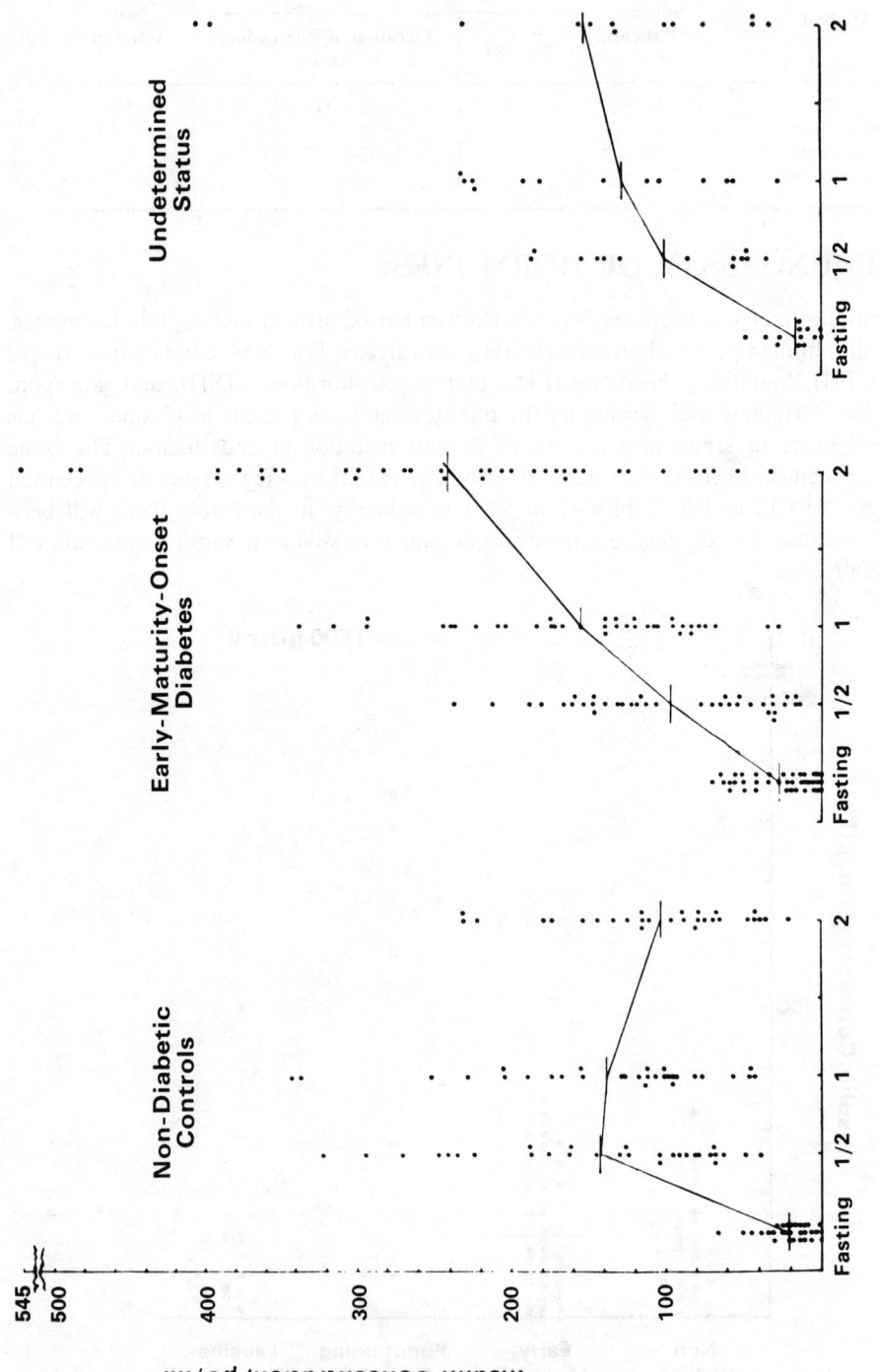

FIGURE 9. Plasma Insulin Concentration During Standard 100-g (p.o.) Glucose-Tolerance Test in Various Groups of Subjects.[61]

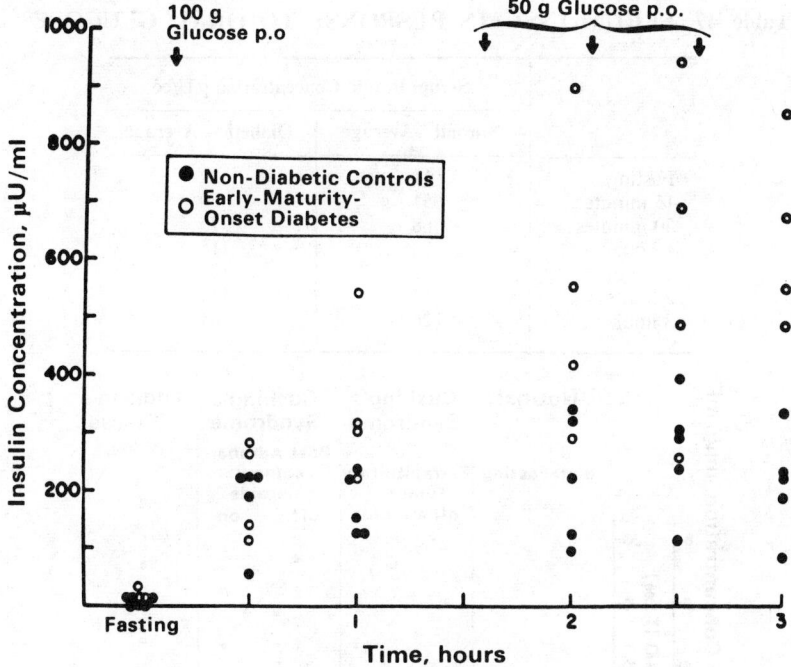

FIGURE 10.
Plasma Insulin Concentrations During Heavy-Glucose-Loading Experiments in Diabetic and Nondiabetic Subjects.[61]

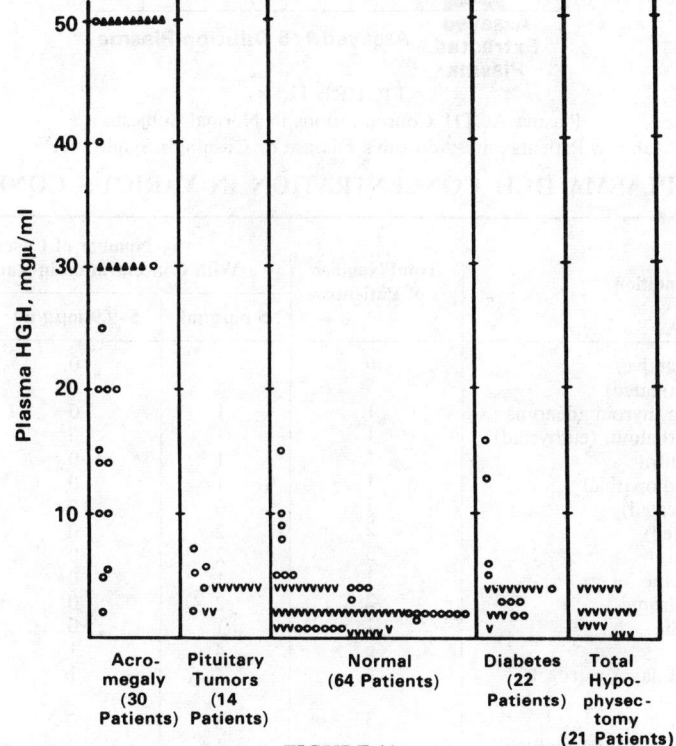

FIGURE 11.
HGH Concentration in Plasma Samples Randomly Obtained from 150 Adult Subjects.
Since the sensitivity of the assay increased during the course of these studies, HGH concentrations are plotted according to the lower limit of sensitivity of the particular assay:
V = less than; ▲ = more than; O = equal to.

Table 47. SERUM INSULIN RESPONSE TO ORAL GLUCOSE[62]

	Serum Insulin Concentration μU/cc	
	Normal—Average	Diabetic—Average
Fasting	11	22
15 minutes	51	50
30 minutes	68	75
1 hour	78	115
1½ hours	72	121
2 hours	53	115
3 hours	26	81

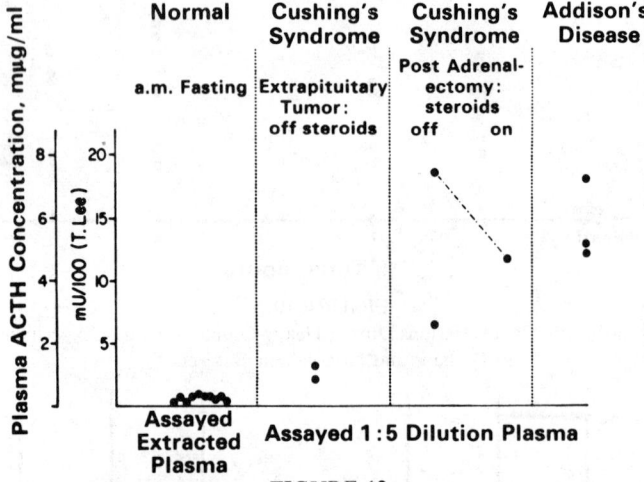

FIGURE 12.
Plasma ACTH Concentrations in Normal Subjects and in Patients with Addison's Disease or Cushing's Syndrome.[65]

Table 48. PLASMA HGH CONCENTRATION IN VARIOUS CONDITIONS

Condition	Total Number of Patients	Number of Cases With Concentration in Range Indicated		
		5 mμg/ml	5–9.9 mμg/ml	10 mμg/ml
Grave's disease (active)	6	6	0	0
Grave's disease (treated)	5	4	0	1
Hyperfunctioning thyroid adenoma	1	1	0	0
Unilateral exophthalmus (euthyroid)	1	1	0	0
Hypoparathyroidism	1	1	0	0
Hypercalciuria (idiopathic)	1	1	0	0
Myxedema (untreated)	1	0	0	1
Myxedema (treated)	2	2	0	0
Anorexia nervosa	3	2	0	1
Sheehan's syndrome	1	1	0	0
Klinefelter's syndrome	2	2	0	0
Turner's syndrome	1	1	0	0
Pregnancy	6	4	1	1
Post-partum (0–3 days), estrogen treated	5	4	1	0
Normal lactation	3	3	0	0
Post-partum (1 month), normal lactation	1	1	0	0
Cervical carcinoma	3	1	2	0
Endometrial hyperplasia	5	3	1	1
Endometrial carcinoma	8	1	0	2
Other	11	9	2	0
Totals	62	48	7	7

Table 49.
VALUES FOR THE CONCENTRATION OF GROWTH HORMONE IN PLASMA OBTAINED BY IMMUNOASSAY[64]

The concentrations of growth hormone, mµg/ml plasma, are given as means ±SD; the numbers of subjects studied are given in parentheses.

Source of Plasma	Concentration of Growth Hormone, mµg/ml	Comment
Cord blood	49.7 ± 57.8 (10)	—
Normal children	18.6 ± 12.5 (10)	Age <1 year old
Normal children	8.5 ± 7.8 (26)	Age 1–8 years
Normal children	2.8 ± 2.8 (14)	Age 9–10 years
Normal children	14.0 ± 13.4 (34)	Age 11–17 years
Adults in hospital	0.55 ± 0.68 (18)	Age 24–68 years
Women with breast growth	1.5 ± 2.5 (35)	Benign and malignant
Men with lung cancer	1.3 ± 1.4 (7)	Age 42–66 years
Women in lactation	4.5 ± 2.1 (11)	1–6 weeks post partum
Women with galactorrhea	7.0 ± 3.4 (7)	—
Acromegalic subjects	72.3 ± 69.3 (23)	Active and inactive

Table 50. LUTEINIZING HORMONE—NORMAL VALUES IN PLASMA

	Bagshawe et al.[66]	Odell et al.[67]
Children, 5–10 years of age	<25 I.U./day	0.5–2.0 mµg/ml
Postmenopausal females	>50 I.U./day	4.6–12.2 mµg/ml
Adult males	25–100 I.U./day	0.8–2.8 mµg/ml

Table 51. THYROID-STIMULATING-HORMONE PLASMA LEVELS[68]

Euthyroid (50)*	
Hypophysectomized (10)	<3 mµg TSH/ml plasma
Pregnant (12)	
Euthyroid (11)	0.6–2.8 mµg TSH/ml plasma
Hyperthyroid (10)	<3 mµg TSH/ml plasma
Primary myxedema (13)	7–156 mµg TSH/ml plasma

* Numbers in parentheses indicate number of patients in each category.

Table 52. INFLUENCE OF TAPAZOLE ON SERUM TSH IN MAN[69]

Patient	Beginning of Tapazole Treatment		End of Tapazole Treatment		End of Tapazole Treatment Schedule
	Serum TSH, mµg/ml	PBI, µg/100 ml	Serum TSH, mµg/ml	PBI, µg/100 ml	
C.M.	<5	—	16	—	30 mg/day for 14 days
P.L.	<3	4.8	14	3.4	45 mg/day for 14 days
J.W.	<3	—	5	—	30 mg/day for 10 days
M.A.	<3	5.9	4	4.9	30 mg/day for 7 days
S.H.	<3	—	<3	—	45 mg/day for 14 days
B.E.	<3	4.1	<3	3.8	45 mg/day for 28 days
G.D.	<3	—	<3	—	30 mg/day for 14 days
J.G.	<3	4.6	<3	4.2	45 mg/day for 14 days

Immunoassay of bovine and human parathyroid hormone has been achieved,[70] but clinical data are not available at this time.

PARATHYROID LOCALIZATION

Localization of hyperfunctioning parathyroid tissue has been performed with two isotopic preparations: ^{75}Se-selenomethionine[71,72] and ^{57}Co-labeled vitamin B_{12}.[73]

The results of Potchen and Awwad[72] with ^{75}Se-selenomethionine are shown in Table 53 as follows:

Table 53. PARATHYROID SCAN–SURGICAL FINDINGS

Result	Total	Adenoma	Hyperplasia	Carcinoma	Negative Exploration
Complete correlation	18	10	3	2	3
Partial correlation	4	2	2	0	0
False positive	3	0	0	0	3
False negative	3	2	1	0	0
Unsatisfactory	2	2	0	0	0
Totals	30	16	6	2	6

Operative localization of parathyroid adenomas with a surgical probe following administration of ^{57}Co- and ^{58}Co-labeled vitamin B_{12} has yielded the following results:[73]

Table 54. SOME RESULTS OF PARATHYROID LOCALIZATION STUDIES

Isotopic Preparation	Total Patients	Scan (+)	Scan (±)	Scan (−)
^{57}Co-B_{12}	7	6	1	0
^{58}Co-B_{12}	3	2	0	1
Totals	10	8	1	1

REFERENCES

1. Berson, S. A., Yalow, R. S., Glick, S. M. and Roth, J., Immunoassay of Protein and Peptide Hormones. *Metabolism*, 13: 1135–1153, 1964.
2. Myers, W. G., Radioisotopes of Iodine in Radioactive Pharmaceuticals. *Proceedings of a Symposium Held at the Oak Ridge Institute of Nuclear Studies, November 1–4, 1965*, pp. 217–243. U.S. Atomic Energy Commission, Division of Technical Information, 1966.
3. Goolden, A. W. G. and Mallard, J. R., The Use of Iodine-132 in Studies of Thyroid Function. *Brit. J. Radiol.*, 31: 589–595, 1958.
4. Halman, K. E. and Pochin, E. E., The Use of Iodine-132 for Thyroid Function Tests. *Brit. J. Radiol.*, 31: 581–588, 1958.
5. Halman, K. E., The Radioiodine Uptake of the Human Thyroid in Pregnancy. *Clin. Sci.*, 17: 281–290, 1958.
6. Hobbs, J. R., Bayliss, R. I. S., and Maclagan, N. F., The Routine Use of ^{132}I in the Diagnosis of Thyroid Disease. *Lancet*, 1: 8–13, 1963.
7. Endlich, H., Harper, P., Siemans, W., and Lathrop, K., The Use of ^{125}I to Increase Isotope Scanning Resolution. *Amer. J. Roentgenol. Radium Ther. Nucl. Med.*, 81: 148–155, 1962.
8. *Radiological Health Handbook*, p. 291. United States Department of Commerce, Office of Technical Services, Washington, D.C., 1960.
9. Riggs, D. S., Quantitative Aspects of Iodine Metabolism in Man. *Pharmacol. Rev.*, 4: 284–370, 1952.
10. Consultants Meeting 1960, Convened by the International Atomic Energy Agency, Calibration and Standardization of Thyroid Radioiodine Uptake Measurements. *Acta Radiol.*, 58: 233–240, 1962.
11. Consultants Meeting 1960, Convened by the International Atomic Energy Agency, Calibration and Standardization of Thyroid Radioiodine Uptake Measurements. *Phys. Med. Biol.*, 6: 533–540, 1962.
12. Consultants Meeting 1960, Convened by the International Atomic Energy Agency, Calibration and Standardization of Thyroid Radioiodine Uptake Measurements. *Brit. J. Radiol.*, 35: 205–210, 1962.
13. Sisson, J. C., Principles of, and Pitfalls in, Thyroid Function Tests. *J. Nucl. Med.*, 6: 853–901, 1965.

14. Gaffney, G. W., Gregerman, R. I., and Shock, N. W., Relationship of Age to the Thyroidal Accumulation, Renal Excretion, and Distribution of Radioiodide in Euthyroid Man. *J. Clin. Endocrinol. Metab.*, 22: 784–794, 1962.
15. Werner, S. C., Hamilton, H. B., Leifer, E., and Goodwin, L. D., An Appraisal of the Radioiodine Tracer Technic as a Clinical Procedure in the Diagnosis of Thyroid Disorders. *J. Clin. Endocrinol. Metab.*, 10: 1054–1076, 1950.
16. Adams, D. D. and Purves, H. D., The Change in Thyroidal ^{131}I Content Between 8 and 48 Hours as an Index of Thyroid Activity. *J. Clin. Endocrinol. Metab.*, 17: 126–132, 1957.
17. McConahy, W. M., Owen, C. A. J., and Keating, F. R., A Clinical Appraisal of Radioiodine Tests of Thyroid Function. *J. Clin. Endocrinol. Metab.*, 16: 724–734, 1956.
18. Vanderlaan, W. P., Accumulation of Radioactive Iodine. Observations on Its Early Phase in Hyperthyroid, Euthyroid and Hypothyroid Subjects. *New Engl. J. Med.*, 257: 752–756, 1957.
19. Crispell, K. R., Parson, W., and Sprinkle, P., A Simplified Technique for the Diagnosis of Hyperthyroidism, Utilizing the One-Hour Uptake of Orally Administered ^{131}I. *J. Clin. Endocrinol. Metab.*, 13: 221–224, 1953.
20. Kohler, P. O. and Wynn, J., One-Hour Thyroid Uptake of Radioactive Iodine. A Screening Test for Hyperthyroidism. *Arch. Intern. Med.*, 116: 177–182, 1965.
21. Kriss, J. P., Uptake of Radioactive Iodine After Intravenous Administration of Tracer Doses. *J. Clin. Endocrinol. Metab.*, 11: 289–296, 1951.
22. Schneeberg, N. G., Perloff, W. H., and Levy, L. M., Diagnosis of Equivocal Hyperthyroidism, Using Thyrotropin Hormone (TSH). *J. Clin. Endocrinol. Metab.*, 14: 223–231, 1954.
23. Taunton, O. D., McDaniel, H. G., and Pittman, J. A., Jr., Standardization of TSH Testing. *J. Clin. Endocrinol. Metab.*, 25: 266–277, 1965.
24. Fletcher, R. F. and Besford, H., A Test of Thyroid and Pituitary Function, Using Thyrotropin. *Clin. Sci.*, 17: 113–120, 1958.
25. Jefferies, W. McK., Kelley, L. W., Jr., Levy, R. P., Cooper, G. W., and Prouty, R. L., The Significance of Low Thyroid Reserve. *J. Clin. Endocrinol. Metab.*, 16: 1438–1455, 1956.
26. Werner, S. C. and Spooner, M., A New and Simple Test for Hyperthyroidism, Employing l-Triiodothyronine and the Twenty-four Hour ^{131}I Uptake Method. *Bull. N.Y. Acad. Med.*, 31: 137–145, 1955.
27. Dresner, S. and Schneeberg, N. G., Rapid Radioiodine Suppression Test Using Triiodothyronine. *J. Clin. Endocrinol. Metab.*, 18: 797–799, 1958.
28. Cassidy, C. E. and Jagiello, G., Test of Thyroid Function. *Clinical Endocrinology*, pp. 659–660, Astwood, I. E. B., ed. Grune & Stratton, New York, 1960.
29. Skanse, B., Radioactive Iodine in the Diagnosis of Thyroid Disease. *Acta Med. Scand.*, Suppl. 235: 1–186, 1949.
30. Berson, S. A., Yalow, R. S., Sorrentino J., and Roswit, B., The Determination of Thyroidal and Renal Plasma ^{131}I Clearance Rates as a Routine Diagnostic Test of Thyroid Dysfunction. *J. Clin. Invest.*, 31: 141–158, 1952.
31. McConahy, W. M., Keating, F. R., Jr., and Power, M. H., An Estimation of the Renal and Extrarenal Clearance of Radioiodide in Man. *J. Clin. Invest.*, 30: 778–780, 1951.
32. Ingbar, S. H., Simultaneous Measurement of the Iodide-Concentrating and Protein-Binding Capacities of the Human Thyroid Gland. *Trans. Amer. Goiter Ass.*, pp. 387–401, 1963.
33. Atkins, H. L. and Richards, P., Assessment of Thyroid Function and Anatomy with Technetium-99m as Pertechnetate. *J. Nucl. Med.*, 9: 7–15, 1968.
34. Grayson, R. R., Factors Which Influence the Radioactive-Iodine Thyroidal-Uptake Test. *Amer. J. Med.* 28: 397–415, 1960.
35. Feinberg, W. D., Hoffman, D. L., and Owen, C. A., Jr., The Effects of Varying Amounts of Stable Iodide on the Function of the Human Thyroid. *J. Clin. Endocrinol. Metab.*, 19: 567–582, 1959.
36. Wang, Y., Thyroid Uptake in Studies Using Various ^{131}I-Labeled Preparations. *J. Nucl. Med.*, 5: 349, 1964 (abstract).
37. Macgregor, A. G., Miller, H., Blaney, P. J., and Whimster, W. S., Diagnosis of Thyrotoxicosis by a Simple Outpatient Radioactive-Iodine Technique. *Brit. Med. J.*, 2: 21–22, 1953.
38. Silver, S., Yohalem, S., and Newberger, R. A., Pitfalls in Diagnostic Use of Radioactive Iodine. *J. Amer. Med. Ass.*, 159: 1–5, 1955.
39. Sheline, G. E. and Clark, D. E., Index of Thyroid Function: Estimation by Rate of Organic Binding of ^{131}I. *J. Lab. Clin. Med.*, 36: 450–455, 1950.
40. Hamolsky, M. W., Goloditz, A., and Freedberg, H. S., The Plasma Protein–Thyroid Hormone Complex in Man. III. Further Studies on the Use of the *in-vitro* Red Blood Cell Uptake of ^{131}I-1-Triiodothyronine as a Diagnostic Test of Thyroid Function. *J. Clin. Endocrinol. Metab.*, 19: 103–116, 1959.
41. Nava, M. and DeGroot, L. J., Resin Uptake of ^{131}I-Labeled Triiodothyronine as a Test of Thyroid Function. *New Engl. J. Med.*, 266: 1307–1310, 1962.
42. Silver, S., *Radioactive Isotopes in Medicine and Biology: Medicine*, 2nd ed., p. 124. Lea & Febiger, Philadelphia, 1962.
43. Silver, S., *Radioactive Isotopes in Medicine and Biology: Medicine*, 2nd ed., pp. 125–126. Lea & Febiger, Philadelphia, 1962.

44. Beling, U. and Einhorn, J., Incidence of Hypothyroidism and Recurrences Following ^{131}I Treatment of Hyperthyroidism. *Acta Radiol.*, 56: 275–288, 1961.
45. Dunn, J. T. and Chapman, E. M., Rising Incidence of Hypothyroidism After Radioactive-Iodine Therapy in Thyrotoxicosis. *New Engl. J. Med.*, 271: 1037–1042, 1964.
46. Nofal, M. N., Beierwaltes, W. H., and Patno, M. E., Treatment of Hyperthyroidism with Sodium Iodide-^{131}I. *J. Amer. Med. Ass.*, 197: 605–610, 1966.
47. Pochin, E. E., Leukaemia Following Radioiodine Treatment of Thyrotoxicosis. *Brit. Med. J.*, 2: 1545–1550, 1960.
48. Werner, S. C., Gittelsohn, A. M., and Brill, A. B., Leukemia Following Radioiodine Therapy of Hyperthroidism. *J. Amer. Med. Ass.*, 177: 646–648, 1961.
49. Segal, R. L., Silver, S., Yohalem, S. B., and Newburger, R. A., Use of Radioactive Iodine in the Treatment of Angina Pectoris. *Amer. J. Cardiol.*, 1: 671–681, 1958.
50. Haynie, T. P., Nofal, M. N., and Beierwaltes, W. H., Treatment of Thyroid Carcinoma with ^{131}I: Results at 14 Years. *J. Amer. Med. Ass.*, 183: 303–306, 1963.
51. Smithers, D. W., Howard, N., and Trott, N. G., Treatment of Carcinoma of the Thyroid with Radioiodine. *Brit. Med. J.*, 2: 969–974, 1965.
52. Saenger, E. L., Barrett, C. M., Passino, J. W., Seltzer, R. A., and Dooley, W. D., Experiences with ^{131}I in the Management of Carcinoma of the Thyroid. *Radiology*, 83: 892–901, 1964.
53. *Radiological Health Handbook*, p. 280. United States Department of Commerce, Office of Technical Services, Washington, D.C., 1960.
54. Harper, P. V., Moseley, R. D., Jr., Ironside, W. S., Kelly, W. A., Fenge, W., and DeVos, W., Experiences with Yttrium-90 Hypophysectomy. *Argonne Cancer Research Hospital Semiannual Report to the USAEC*, 9: 1–11, 1958.
55. Rasmussen, T., Harper, P. V., and Kennedy, T., The Use of a Beta-Ray Point Source for Destruction of the Hypophysis. American College of Surgeons, *Surgical Forum 1953*, 4: 681–686. W. B. Saunders Co., London, England, 1954.
56. Harrington, R. W., Pituitary Ablation and the Visual Fields. *Endocrine Aspects of Breast Cancer*, pp. 54–56. E. & S. Livingstone Ltd., London, England, 1958.
57. Evans, J. P., Fenge, W., Kelly, W. A., and Harper, P. V., Jr., Transcranial Yttrium-90 Hypophysectomy. *Surg. Gynecol. Obstet.*, 108: 393–405, 1959.
58. Hartog, M., Doyle, F., Fotherby, K., Fraser, R., and Joplin, G. F., Partial Pituitary Ablation with Implants of Gold-198 and Yttrium-90 for Cushing's Syndrome with Associated Adrenal Hyperplasia. *Brit. Med. J.*, 2: 392–395, 1965.
59. Hartog, M., Doyle, F., Fraser, R., and Joplin, G. F., Partial Pituitary Ablation with Implants of Gold-198 and Yttrium-90 for Acromegaly. *Brit. Med. J.*, 2: 396–398, 1965.
60. Forrest, A. P. M., Blair, D. W., Morris, S. R., Peebles-Brown, D. A., Sandison, A. T., Valentine, J. S., and Illingworth, C. F. W., Treatment of Advanced Breast Cancer by Yttrium-90 Implantation of the Pituitary. *Endocrine Aspects of Breast Cancer*, pp. 46–54. E. &. S. Livingstone Ltd., London, England, 1958.
61. Yalow, R. S. and Berson, S. A., Immunoassay of Endogenous Plasma Insulin in Man. *J. Clin. Invest.*, 39: 1157–1175, 1960.
62. Meade, R. C. and Klitgaard, H. M., A Simplified Method for Immunoassay of Human Serum Insulin. *J. Nucl. Med.*, 3: 407–416, 1962.
63. Glick, S. M., Roth, J., Yalow, R. S., and Berson, S. A., Immunoassay of Human Growth Hormone in Plasma. *Nature*, 199: 784-787, 1963.
64. Hunter, W. M. and Greenwood, F. C., A Radioimmunoelectrophoretic Assay for Human Growth Hormone. *Biochem. J.*, 91: 43–56, 1964.
65. Yalow, R. S., Glick, S. M., Roth, J., and Berson, S. A. Radioimmunoassay of Human Plasma ACTH. *J. Clin. Endocrinol. Metab.*, 24: 1219–1225, 1964.
66. Bagshawe, K. D., Wilde, C. E., and Orr, A. H., Radioimmunoassay for Human Chorionic Gonadotropin and Luteinizing Hormone. *Lancet*, 1: 1118–1121, 1966.
67. Odell, W. D., Ross, G. T., and Rayford, P. L., Radioimmunoassay for Human Luteinizing Hormone: Preliminary Report. *Metabolism*, 15, 287–289, 1966.
68. Odell, W. D., Wilber, J. F., and Paul, W. E., Radioimmunoassay of Thyrotropin in Human Serum. *J. Clin. Endocrinol. Metab.*, 25: 1179–1188, 1965.
69. Wilber, J. F. and Odell, W. D., Influence of Tapazole upon Serum TSH. *J. Clin. Endocrinol. Metab.*, 25: 1407–1408, 1965.
70. Berson, S. A., Yalow, R. S., Aurbach, G. D., and Potts, J. T., Jr., Immunoassay of Bovine and Human Parathyroid Hormone. *Proc. Nat. Acad. Sci.*, 49: 613–617, 1963.
71. Potchen, E. J., Adelstein, S. J., and Dealy, J. B., Jr., Radioisotope Localization of the Overactive Human Parathyroid. *Amer. J. Roentgenol. Radium Ther. Nucl. Med.*, 93: 955–961, 1965.
72. Potchen, E. J. and Awwad, H. K., Parathyroid Scanning: The Biochemical Approach. *Recent Advances in Nuclear Medicine*, pp. 207–211, Croll, M. N. and Brady, L. W., eds. Appleton-Century-Crofts, New York, 1966.
73. Workman, J. B., Parathyroid Localization Studies: Clinical Aspects. *Recent Advances in Nuclear Medicine*, pp. 212–217, Croll, M. N. and Brady, L. W., eds. Appleton-Century-Crofts, New York, 1966.

RADIONUCLIDE DIAGNOSIS OF CARDIOVASCULAR DISEASE

L. Rosenthall, M.D., Director, Division of Nuclear Medicine
Montreal General Hospital, Montreal, P.Q., Canada.

CARDIAC OUTPUT

Methods for measuring cardiac output involve either the Fick principle or the dilution principle.

The Fick principle necessitates the measurement of oxygen consumption and of arteriovenous differences. It is a time-consuming method, requiring cardiac catheterization.

The dilution technique[1] employs a sudden single intravenous injection of a dye (Evans' blue; indocyanine green) or a radiopharmaceutical (radioiodinated human serum albumin) and determination of the changing concentration of the test agent in the arterial blood as the initial bolus passes through.

Table 55. CARDIAC OUTPUT MEASUREMENT WITH RADIOPHARMACEUTICALS

Radiopharmaceutical	Technique	Results	Comments
Erythrocytes labeled ^{32}P[2] ^{131}I-human serum albumin[3]	Rapid injection of test agent into the venous system or through cardiac catheter; serial sampling or continuous monitoring of arterial blood; dilution curve is plotted; flow rate is calculated from curves (see Figure 13).		Requires an arterial tap.
^{131}I-human serum albumin[4-8]	Rapid intravenous injection of radioiodinated human serum albumin and precordial monitoring with a scintillation crystal detector; activity is recorded on a strip chart; blood volume is determined 10 minutes after injection of ^{131}I-HSA; cardiac output is calculated from curves.	8.02 ± 2.13 l/min at rest. 12.74 ± 3.43 l/min following exercise.	Simple procedure, requiring an intravenous injection and precordial monitoring.
Myocardial Blood Flow			
^{131}I-human serum albumin ^{131}I-diodrast[10]	Rapid intravenous injection with precordial monitoring, using scintillation detectors, rate meters, and strip chart recorders; three peaks are described, which are attributed to the right heart, left heart, and myocardial blood flow emanating from the descending limb of the left ventricular curve; calculations of myocardial flow are based on the area encompassed by this third peak.		There are theoretical objections to this technique;[11] the third peak could not be obtained by other investigators.[12]

Table 55. (*Continued*)
CARDIAC OUTPUT MEASUREMENT WITH RADIOPHARMACEUTICALS

Radiopharmaceutical	Technique	Results	Comments
^{131}I-hippurate [13]	Intravenous injection of radioiodinated hippurate while the precordium and head are monitored with scintillation detectors; the output is recorded simultaneously by a dual-channel digital count computer; a semilogarithmic plot is made of head and precordial counts, and the half-disappearance times of each determined; a coronary flow index (CFI) is defined as $\dfrac{T_{1/2} \text{ left heart}}{T_{1/2} \text{ brain}}$.	Normal: 1.62 ± 0.23 CFI Coronary disease: 1.06 ± 0.11 CFI	Simple procedure, requiring about 10 minutes to perform; necessitates expensive digital computing and recording systems.
^{131}Xenon[14] ^{85}Krypton[15]	A bolus of radioxenon or radiokrypton gas dissolved in saline is injected into the coronary arterial circulation; the activity over the heart is monitored by a precordial scintillation detector, rate meter, and strip chart recorder; the rate of disappearance of activity is a measure of myocardial perfusion; half-disappearance time is determined from the semilogarithmic plot of the curve (see Figure 14).	Values obtained are slightly lower than those obtained by the nitrous oxide method. Right coronary A: 50/ml/100 g/min. Left coronary A: 60/ml/100 g/min.	The procedure and calculations are simple; procedure requires coronary artery catheterization.
^{84}Rubidium[16] ^{86}Rubidium[17]	Still largely experimental; different techniques using either a single bolus injection or a constant infusion of rubidium have been tried; the methods can be grouped into intravenous injections and coronary-artery catheterization combined with precordial external monitoring.[17]		Major difficulty has been separation of heart muscle from surrounding structures; also, changes in activity of the blood in the cardiac chambers or coronary arteries may interfere with counts recorded from heart muscles alone; solutions to these problems have been suggested.[16,18]

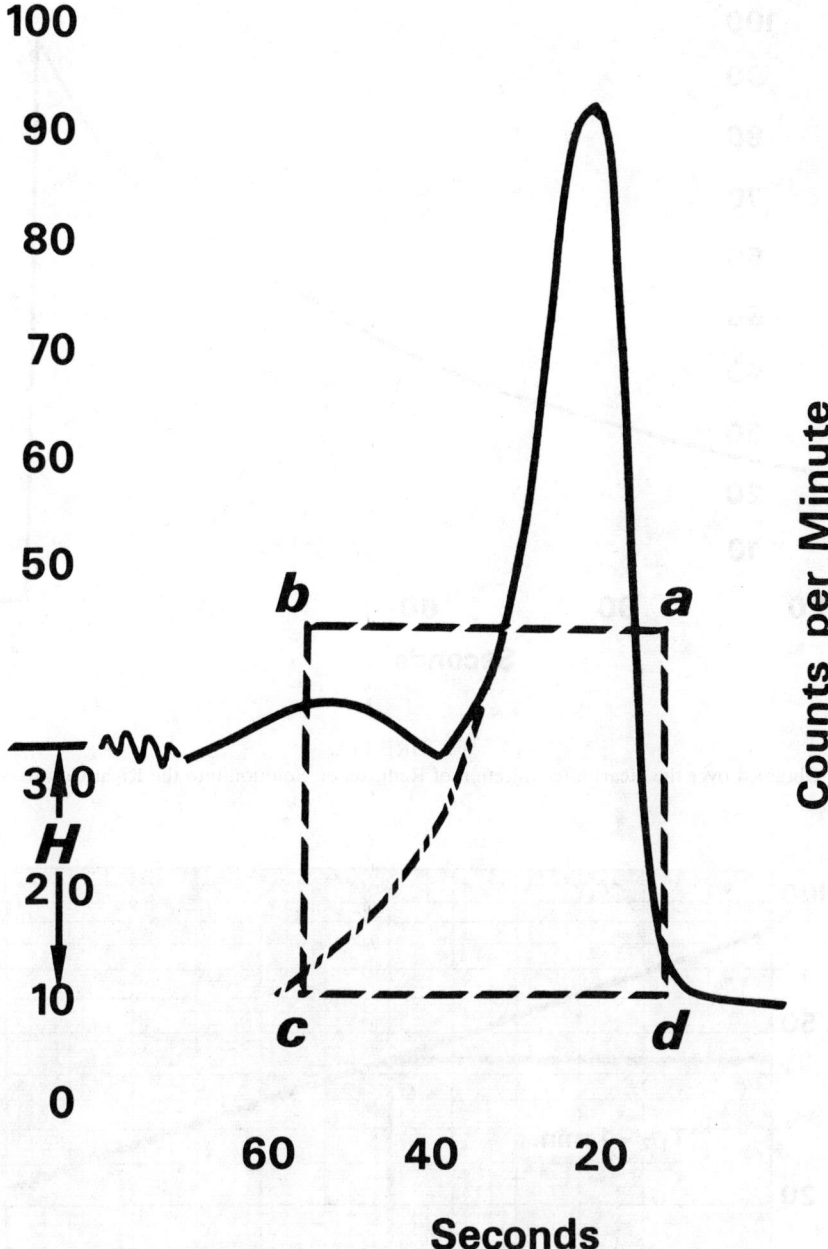

FIGURE 13. Typical Curve Obtained with ^{131}I-Human Serum Albumin.

The down slope of the first bolus passing through is extrapolated to the base, and the area under the resultant curve is measured with a planimeter; *abcd* is the equivalent area, and its height (*ad* or *bc*) represents the average blood concentration of ^{131}I-HSA.

$$\text{Cardiac output} = \frac{\text{amount of test agent injected}}{\text{average concentration} \times \text{elapsed time}} = \frac{C_f}{C_{av} \times \text{elapsed time}} = \frac{H \times BV \times S}{A},$$

where C_f = final concentration,
C_{av} = average concentration of initial bolus,
H = height of curve above base line 10 minutes after injection,
A = area under curve,
S = chart speed.

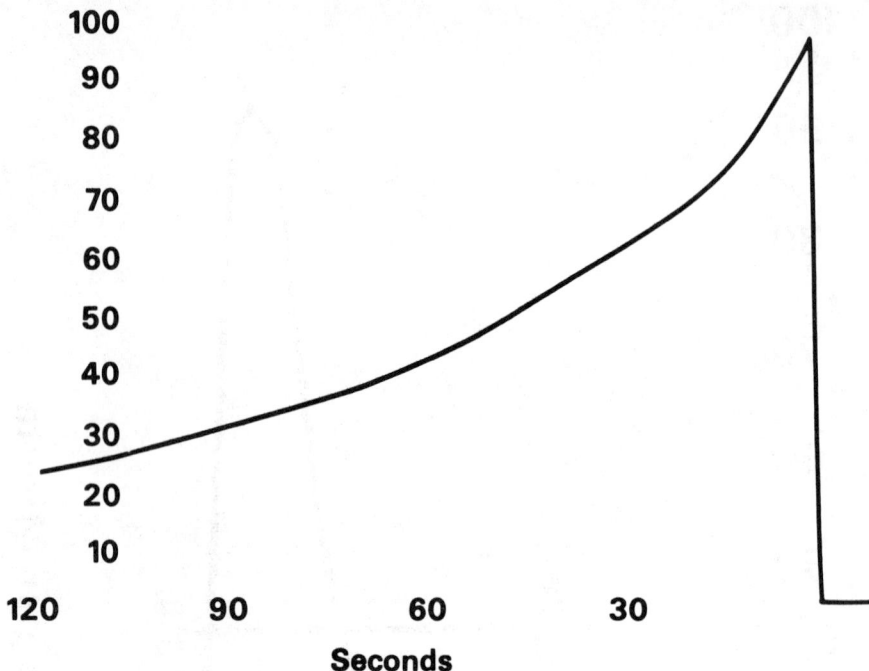

FIGURE 14A.
Curve Obtained over the Heart after Injection of Radioxenon Solution into the Right Coronary Artery.

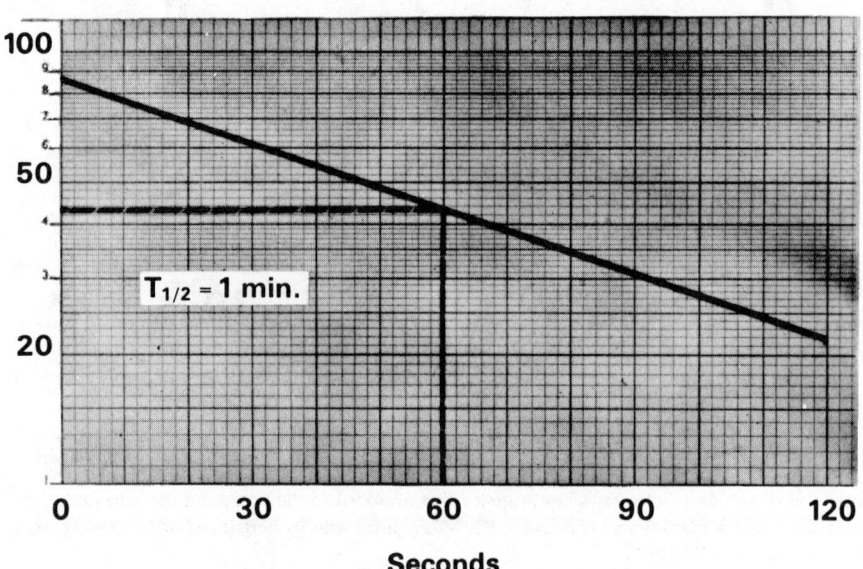

FIGURE 14B. Semilogarithmic Plot of Above Curve.

$$F = \text{flow}/100 \text{ g myocardium/min.} = \frac{69.3\, p}{T_{1/2} \times s} = \frac{(69.3)(0.72)}{(1)(1.05)} = 47.4,$$

where p = partition coefficient,
s = specific gravity of myocardium,
$T_{1/2}$ = half-disappearance of radioxenon.

RADIONUCLIDE ANGIOGRAPHY

The gamma-ray scintillation camera[19] is a rapid imaging device that monitors every point within the field of view continuously, in contrast to rectilinear scanners, which pass over each point momentarily. This feature permits a bolus of radioactivity to be followed through the large arteries (Figures 15 and 16), veins, heart, lung, brain (Figure 17), and kidneys (Figure 18).[20-25]

FIGURE 15.

Aortogram Showing Complete Obstruction of the Abdominal Aorta Below the Origin of the Renal Arteries.

Corresponding ^{99m}Tc Pertechnetate Blood Flow Study Depicting the Aorta Down to Its Site of Obstruction and the Kidneys.

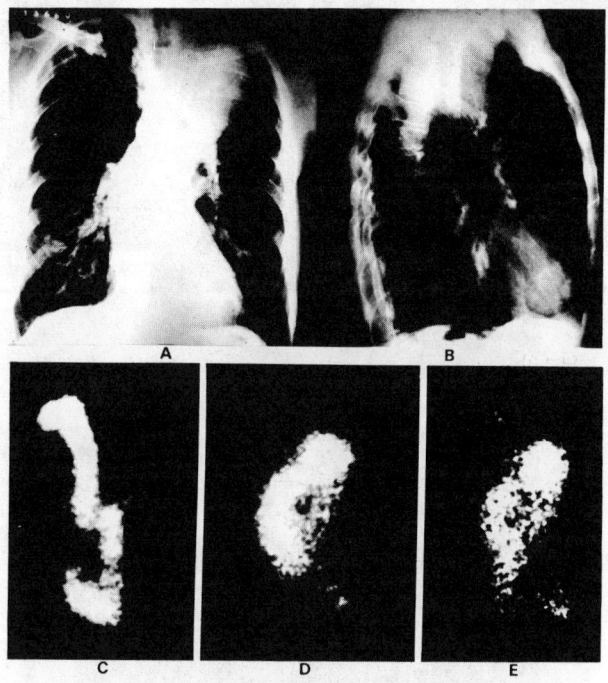

FIGURE 16.
(From: Rosenthall, L. J. Can. Ass. Radiol., 18: 270, 1967.)

A. Frontal Chest Roentgenogram Showing a Mass in the Superior Mediastinum Extending to the Left.
B. Lateral Chest Roentgenogram Showing Aneurysmal Dilatation of the Ascending Limb of the Aorta.
C. 0- to 5-Second Scintiphoto Following a Right Antecubital-Vein Injection, Exhibiting a Narrowed—But Not Obstructed—Superior Vena Cava.
D. 10- to 14-Second Scintiphoto Demonstrating an Unfolded Aorta with Dilatation of the Ascending Limb and an Aneurysm in the Arch of the Aorta.
E. 15- to 20-Second Scintiphoto Showing the Aneurysm in the Arch of the Aorta After a Good Deal of the ^{99m}Tc Has Left the Field.

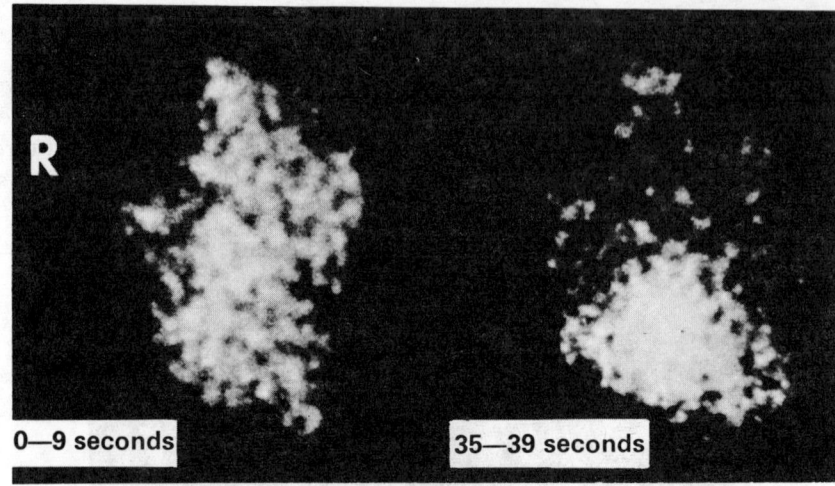

FIGURE 17.

^{99m}Tc-Pertechnetate Cerebral Blood Flow Study in a Patient
Who Suffered a Recent Cerebral Vascular Accident on the Right.

Following an antecubital-vein injection, the 0- to 9-second scintiphoto exhibits reduced arterial blood flow in the right hemisphere; at 35 seconds this disparity is no longer seen, and the conventional scan, obtained a half hour later, was normal.

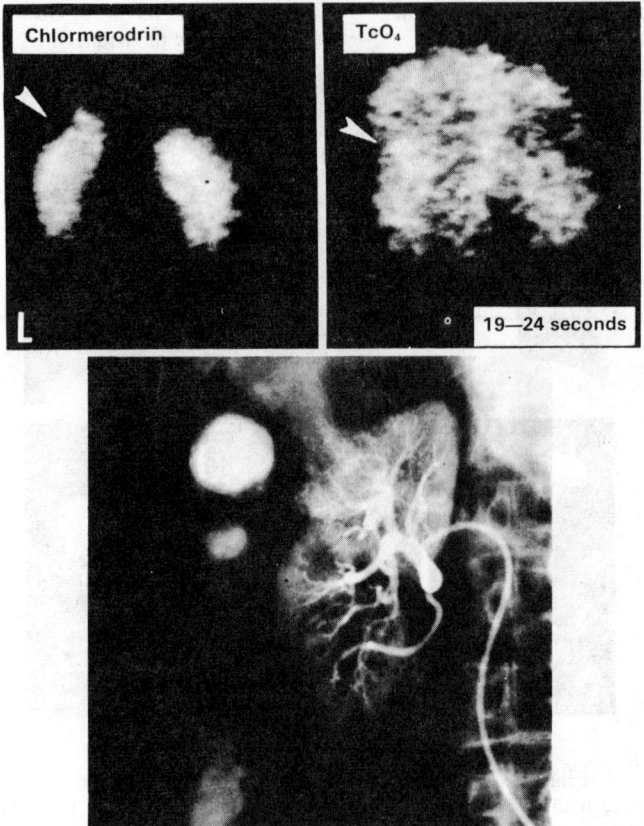

FIGURE 18.

Verification of Renal Carcinoma by Radionuclide Angiography.

The chlormerodrin-^{203}Hg renal study showed a lesion on the lateral margin in the upper half of the left kidney (arrow); a ^{99m}TcO$_4$ renal blood flow series demonstrated this defect as having a vascular supply; the selective renal ateriogram verified the presence of a renal carcinoma.

Obstructions, compressions, and aneurysms of the major vessels can be demonstrated. Pericardial effusions, lung perfusion defects, and differences in renal blood flow may be discerned. Renal cysts and neoplasms can also be distinguished.

Total-body absorbed dose is about 120 millirads per 10 millicuries $^{99m}TcO_4$.

Technique

A rapid injection of 10 to 15 millicuries of ^{99m}Tc– pertechnetate into an appropriate vein is made, and serial 1- to 5-second scintiphotos are obtained over the area of interest.

CIRCULATION TIME

Any radionuclide or radiopharmaceutical can be used to determine circulation time between two or more points. It is preferable to use a test agent with a short effective half-life in order to reduce total-body radiation exposure. The earliest reported circulation time studies were performed in 1927, using radium C′, a Geiger-Mueller type of tube, and a pen galvanometer connected to a vacuum tube amplifier;[26] an arm-to-arm time of 18 seconds was obtained.

The technique differs somewhat from author to author, but basically involves the placement of a scintillation detector, which is linked to a rate meter and strip chart, over an area of interest. A rapid intravenous injection of test agent is made, and the time between the injection and the arrival over the area is measured.

The arm–femoral circulation time and arm–heart circulation time can be obtained by placing one scintillation detector over the femoral artery in the groin and another over the precordium. Using $Na^{131}I$,[2] the values shown in Table 56 are obtained.

Table 56.
ARM-TO-FEMORAL AND ARM-TO-HEART CIRCULATION TIMES

	Condition	Mean, sec.	Range, sec.
Arm to femoral	Normal	11.2	5.5–18.5
	Cardiac failure	35.1	19.0–67.0
Arm to heart	Normal	1.8	0.8– 5.0
	Cardiac failure	4.3	0.5–11.0
Decholin	Normal	14.4	8.5–24.0
	Cardiac failure	43.1	19.5–78.0

Techniques using radiopharmaceuticals to evaluate peripheral circulation are described in Table 57.

DIAGNOSIS OF CONGENITAL HEART DISEASE

Radioisotopes can be used to detect left-to-right and right-to-left shunts.

Left-to-Right Shunts

Inhalation Technique. Radioactive ^{85}Kr gas is inhaled; during the first two minutes a cardiac catheter samples blood from the pulmonary artery, right ventricle, right atrium, and venae cavae if an anomalous pulmonary return is suspected.[34,35]

The presence of a shunt is established when an abnormally high concentration of radioactive gas is obtained; its location is the most proximal chamber exhibiting this high concentration.

In a large series of patients with proven left-to-right shunts,[34] the activity in the pulmonary-artery blood sample was equal to or greater than 13 percent of the systemic blood activity.

Table 57.
PERIPHERAL-CIRCULATION STUDIES WITH RADIOPHARMACEUTICALS

Radiopharmaceutical	Technique	Results	Comments
Intravenous Injection			
^{32}Phosphorus[28] ^{24}Sodium[29]	About 200 µCi of ^{32}P or 100 µCi of ^{24}Na are injected rapidly into an antecubital vein, and build-up curves are obtained over the foot, using the appropriate detector.	Steep build-up of ^{32}P in tissues of patients with arteriosclerosis and arteriosclerosis obliterans, flat curves in impending gangrene, and no build-up in gangrenous extremities. Studies with ^{24}Na show similar appearance times in normal patients and in those with peripheral vascular disease, but the latter has a reduced total accumulation after one hour.	The two electrolytes are freely diffusible into the extracellular space; the steep build-up of ^{32}P in patients with arteriosclerosis is attributed to capillary dilatation and early diffusion into the tissues; both the ^{32}P and ^{24}Na techniques have been used to evaluate drug therapy and sympathectomy.
^{131}I-human serum albumin[30]	About 100 µCi of ^{131}I-HSA are injected rapidly into an arm vein; a scintillation detector is applied against the sole of each foot, and the build-up of activity is monitored.	The rate of rise and the height of the plateau are reduced in gangrenous extremities, arterial and venous obstruction, and with cooling of the extremity; heating the extremity increases the slope and height of the curve.	
Local Injection			
^{85}Krypton ^{131}Iodine ^{24}Sodium[31] ^{133}Xenon[32] ^{86}Rubidium[33] ^{22}Sodium ^{42}Potassium	A volume of about 0.1 ml of the test agent is injected into the muscle or skin with a very thin (26 or 27 gauge) needle; the rate of disappearance from the site is monitored with a scintillation detector that is linked to a rate meter and strip chart recorder, or with a scaler.	The rate of clearance is exponential, and can be expressed by a half-disappearance time.	The noble gases ^{85}Kr and ^{133}Xe are freely diffusible across cell membranes, and their clearance is supposedly a measure of regional blood flow; I, K, Na, and Rb are not freely diffusible, and their clearance is determined by other factors in addition to blood flow.

Injection Technique. ^{85}Kr is dissolved in saline and injected into the left side of the heart. The expired gas is monitored with a thin-window Geiger-Mueller tube inserted into the expiratory line. Activity is recorded continuously on a strip chart. In the absence of a shunt, the radioactive gases pass through the systemic circulation and reach the lung in ten seconds or more. If the injection of ^{85}Kr is made proximal to the left-to-right shunt, the expired air contains activity between two and five seconds after administration.

External Pulmonary Monitoring. A scintillation detector probe is placed over the lung, and a rapid antecubital-vein injection of ^{131}I-diodrast[37] or ^{131}I-hippurate is made. The curve obtained on the strip chart exhibits an ascending and a descending limb. A prolongation of the descending limb indicates delayed clearance of the radiopharmaceutical due to a left-to-right shunt.

Right-to-Left Shunts

^{85}Kr dissolved in saline is injected via cardiac catheter into the veins, right atrium, right ventricle, and pulmonary artery successively. Systemic arterial blood is sampled after each selective injection.[36] If a given volume of arterial blood is obtained over a fixed interval of time, and if the activity collected is expressed as a percentage of the injected dose, the magnitude of the shunt can be estimated. The method also locates the site of the shunt.

LOCALIZATION OF THE PLACENTA

The placenta is a large blood pool, and any radiopharmaceutical that remains long enough in the vascular system to be counted or scanned and renders an acceptably low fetal radiation exposure is satisfactory.

Table 58 lists techniques using various radiopharmaceuticals to locate the placenta.

Table 58.
LOCALIZATION OF THE PLACENTA WITH RADIOPHARMACEUTICALS

Radiopharmaceutical	Technique	Results	Comments
^{131}I-human serum albumin[38-44] ^{132}I-human serum albumin[45]	About 10 drops of Lugol's solution is administered prior to the intravenous injection of 5 to 10 μCi ^{131}I; 10 minutes are allowed for adequate mixing; the abdomen is divided into approximately 12 equal sections; counts are obtained over each section for one minute; the precordial count is used as a reference—100%.	Those sections with the highest count rate reflect the location of the placenta; if the maximum count rate is less than 70% of the reference, it is presumed that the placenta is located on the posterior wall of the uterus.	An accuracy of 95% is obtained in the third trimester, compared to 78% with roentgenographic placentography; 5 μCi of ^{131}I-HSA yield a total-body fetal dose of 0.005 to 0.07 rads, compared to 1.5 to 3 rads for the radiographic technique.
^{51}Cr-erythrocytes[46]	The dose is 10 to 20 μCi of ^{51}Cr; the technique is the same as with ^{131}I-HSA or ^{125}I-HSA.		Fetal exposure between 0.002 and 0.008 rads.
^{99m}Tc-human serum albumin[47] ^{113m}Indium—protein complex	250 mg potassium perchlorate are given one hour before the test to block the thyroid trapping of free or released ^{99m}Tc; 1 mCi ^{99m}Tc-albumin is intravenously injected, and the scan of the abdomen is started immediately afterwards from the pubis cephalad; 1 to 2 mCi of ^{113m}In can be used without pretreatment.	The organ is visualized and localized relative to pubic fundal, umbilical and perhaps intravaginal markers.	Fetal absorbed dose is about 14 mrads per mCi of ^{99m}Tc-albumin. The organ is more clearly delineated and more accurately localized than with the surface count technique.
^{99m}Tc-pertechnetate[49] ^{87m}Strontium[50]	Same technique as with ^{99m}Tc-albumin, but a rapid imaging device, such as a gamma-ray scintillation camera, is required, to obtain scintiphotos before the activity reaches the urinary vesicle and the extravascular build-up reduces the target to background ratio (see Figure 19). Dosage: 300 μCi to 1 mCi of ^{99m}TcO$_4$; 1 mCi of ^{87m}Sr.		For ^{99m}Tc, the fetal absorbed dose is about 10 to 12 mrads per mCi of ^{99m}Tc; the total fetal dose of ^{87m}Sr is 12 to 25 mCi. The organ is more clearly delineated and more accurately localized than with the surface count technique.

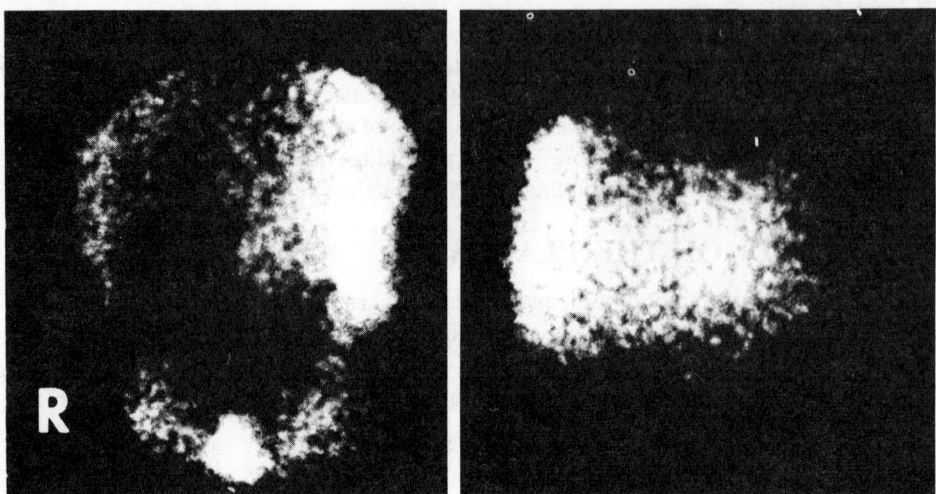

FIGURE 19.

Frontal and Left Lateral Views of Placenta,

Obtained with ^{99m}Tc-Pertechnetate and a Gamma-Ray Scintillation Camera. Patient was 29 weeks pregnant, and admitted to the hospital for investigation of vaginal bleeding; the placenta is localized on the left lateral wall of the uterus, remote from the radioactive marker placed over the symphysis pubis.

REFERENCES

1. Hamilton, W. F., Moore, J. W., Kinsman, J. M., and Spurling, R. C., Studies on Circulation. IV. Further Analysis of the Injection Method and of Changes in Hemodynamics under Physiological and Pathological Conditions. *Amer. J. Physiol.*, 99: 534, 1931.
2. Nylin, G. and Celander, H., Determination of Blood Volume in Heart and Lung and the Cardiac Output through the Injection of Radiophosphorus. *Circulation*, 1:76, 1950.
3. MacIntyre, W. J., Pritchard, W. H., Eckstein, R. W., and Friedell, H. L., The Determination of Cardiac Output by a Continuous Recording System Utilizing Iodinated (^{131}I) Human Serum Albumin. *Circulation*, 4: 552, 1951.
4. Prinzmetal, M., Corday, E., Spritzler, R. J., and Flieg, W., Radiocardiography and Its Clinical Applications. *J. Amer. Med. Ass.*, 139: 617, 1949.
5. Huff, R. L., Feller, D. D., Judd, O. J., and Bogardus, G. M., Cardiac Output of Men and Dogs Measured by *in-vivo* Analysis of Iodinated (^{131}I) Human Serum Albumin. *Circul. Res.*, 3: 564, 1955.
6. Shackman, R., Radioactive Isotope Measurements of Cardiac Output. *Clin. Sci.*, 17: 317, 1958.
7. Duffy, B. J., Howley, J. R., Casina, J., and Hufnagel, C. A., Cardiac Output in Man Determined by External Counting of RISA. *Clin. Res. Proc.*, 5: 94, 1957.
8. Glick, G., Schreiner, B. F., Luria, M. N., and Yu, P. N., Determination of Cardiac Output by Means of Radioisotope Dilution Technique. *Progr. Cardiov. Dis.*, 4: 50, 1962.
9. Blahd, W. H., ed., *Nuclear Medicine*, p. 403. McGraw-Hill, New York, 1965.
10. Seveluis, G. and Johnson, P. C., Myocardial Blood Flow Determined by Surface Counting and Ratio Formula. *J. Lab. Clin. Med.*, 54: 669, 1959.
11. Conn, H. L., Use of External Counting Technics in Studies of the Circulation. *Circul. Res.*, 10: 505, 1962.
12. Mena, I., Kattus, A. A., Greenfield, M. A., and Bennett, L. R., Effect of Coronary Blood Flow on Radioisotope Dilution Curves Measured by Precordial Scintillation Detection. *Circul. Res.*, 9: 911, 1961.
13. Bishop, H. A., Kattus, A. A., Bennett, L. R., and Mena, I., Evaluation of an Isotope Coronary-Potency Test by Angiography. *Radiology*, 81: 428, 1963.
14. Ross, R. S., Ueda, K., Lichtlen, P. R., and Rees, J. R., Measurement of Myocardial Blood Flow in Animals and Man by Selective Injection of Inert Gas into the Coronary Arteries. *Circul. Res.*, 15: 28, 1964.
15. Herd, J. A., Hollenberg, M., Thorburn, G. D., Kopald, H. H., and Barger, A. C., Myocardial Blood Flow Determined with ^{85}Krypton in Unanesthetized Dogs. *Amer. J. Physiol.*, 203: 122, 1962.
16. Cohen, A., Zaleski, E. J., Luebs, E. D., and Bing, R. J., The Use of Positron Emitter in the Determination of Coronary Blood Flow in Man. *J. Nucl. Med.*, 6: 651, 1965.
17. Nolting, D., Mack, R., Lutly, E., Kirsch, M., and Hogancamp, C., Measurement of Coronary Flow and Myocardial Rubidium Uptake with ^{86}Rubidium. *J. Clin. Invest.*, 37: 921, 1958.

18. Donato, L., Bartolmei, G., and Giordani, R., Evaluation of Myocardial Blood Perfusion in Man with Radioactive Potassium and Rubidium and Precordial Counting. *Circulation*, 29: 195, 1964.
19. Anger, H. O., Gamma and Positron Scintillation Camera. *Nucleonics*, 21: 56, 1963.
20. Rosenthall, L., Applications of the Gamma-Ray Scintillation Camera to Dynamic Studies in Man. *Radiology*, 86: 634, 1966.
21. Rosenthall, L., Radionuclide Diagnosis of Renal Cysts and Neoplasms, Using the Gamma-Ray Scintillation Camera: Preliminary Work. *J. Can. Ass. Radiol.*, 17: 85, 1966.
22. Rosenthall, L., Radionuclide Venography, Using Technetium-99m Pertechnetate and the Gamma-Ray Scintillation Camera. *Amer. J. Roentgenol., Radium Therapy, Nucl. Med.*, 97: 874, 1966.
23. Rosenthall, L., Thoracic Radionuclide Angiograph with the Gamma-Ray Scintillation Camera. *J. Can. Ass. Radiol.*, 18: 270, 1966.
24. Rosenthall, L., Detection of Altered Cerebral Blood Flow, Using Technetium-99m Pertechnetate and the Gamma-Ray Scintillation Camera. *Radiology*, 88: 713, 1967.
25. Powell, M. R., and Anger, H. O., Blood Flow Visualization with the Scintillation Camera. *J. Nucl. Med.*, 7: 729, 1966.
26. Blumgart, H. L. and Weiss, S., Studies on the Velocity of Blood Flow. II. The Velocity of Blood Flow in Normal Resting Individuals and a Critique of the Method Used. *J. Clin. Invest.*, 4: 15, 1927.
27. Shipley, R. A. and Clark, R. E., Measurement of Circulation Times with Na-^{131}I. *Circul. Res.*, 4: 456, 1956.
28. Friedell, M. T., Schaffner, F., Pickett, W. J., and Hammon, I. W., Radioactive Isotopes in the Study of Peripheral Vascular Disease. I. Derivation of a Circulatory Index. *Arch. Intern. Med.*, 83: 608, 1949.
29. Smith, B. C. and Quimby, E. H., The Use of Radioactive Sodium as a Tracer in the Study of Peripheral Vascular Disease. *Radiology*, 45: 335, 1945.
30. MacIntyre, W. J., Storaasli, J. P., Krieger, H., Pritchard, W., and Friedell, H. L., ^{131}I-Labeled Serum Albumin: Its Use in the Study of Cardiac Output and Peripheral Vascular Flow. *Radiology*, 59: 849, 1952.
31. Kety, S. S., Measurement of Regional Circulation by Local Clearance of Radioactive Sodium. *Amer. Heart J.*, 38: 321, 1949.
32. Lassen, N. A., Lindbjerg, I. F., and Munck, O., Measurement of Blood Flow Through Skeletal Muscle by Intramuscular Injection of Xenon-133. *Lancet*, 1: 686, 1964.
33. Bauer, F. K., Cassen, B., Youtcheff, E., and Shoop, L., Jet Injection of Radioisotopes. *Amer. J. Med. Sci.*, 225: 374, 1953.
34. Braunwald, E., Goldbatt, A., Long, R. T. L., and Morrow, A. G., The Krypton-85 Inhalation Test for the Detection of Left-to-Right Cardiac Shunts. *Brit. Heart J.*, 24: 47, 1962.
35. Braunwald, E., Morrow, A. G., and Folse, R., The Use of Radioisotopes in Clinical Studies of the Central Circulation. *Progr. Cardiovas. Dis.*, 4: 543, 1962.
36. Long, R. T. L., Braunwald, E., and Morrow, A. G., Intracardiac Injection of Radioactive Krypton. Clinical Applications of New Methods for Characterization of Circulatory Shunts. *Circulation*, 21: 1126, 1960.
37. Folse, R. and Braunwald, E., Pulmonary Vascular-Dilution Curves Recorded by External Detection in Diagnosis of Left-to-Right Shunts. *Brit. Heart J.*, 166, 1962.
38. Hutchinson, D. L., Bennett, L. R., and Gean, D. A., Isotopic Localization of the Placenta in Placenta Previa. *Surg. Gynecol. Obstet.*, 107: 370, 1958.
39. Heavy, F. C. and Swartz, D. P., Localizing the Placenta with Radioactive Iodinated Human Serum Albumin. *Radiology*, 76: 936, 1961.
40. Shapiro, B. J. and Shaul, D., Radioisotope Localization of the Placenta. *Can. Med. Ass. J.*, 97: 218, 1967.
41. Rosenthall, L., Placental Localization, Using Radioiodinated Human Serum Albumin and the Gamma-Ray Scintillation Camera. *J. Can. Ass. Radiol.*, 17: 221, 1966.
42. Visscher, R. D. and Baker, W. S., Isotope Localization of the Placenta in Suspected Cases of Placenta Previa. *Amer. J. Obst. Gynecol.*, 80: 1154, 1960.
43. Wheeler, P. C. and Dolan, K. D., A Modified Method of Radioisotopic Placental Localization. *Radiology*, 81: 989, 1963.
44. Durfee, R. B. and Howieson, J. L., Localization of the Placenta with RISA. *Amer. J. Obstet. Gynecol.*, 84: 577, 1962.
45. Hibbard, B. M., Placental Localization, Using Radioiodinated Serum Albumin (RISA). *J. Obstet. Gynaecol. Brit. Comm.*, 68: 481, 1961.
46. Paul, J. D., Gahres, E. E., Albert, S. N., Terrell, W. D., and Dodek, S. M., Placental Localization, Using ^{51}Cr-Tagged Erythrocytes. *Obstet. Gynecol.*, 21: 33, 1963.
47. McAfee, J. G., Stern, H. S., Fueger, G. S., Baggish, M. S., Holzman, G. B., and Zolle, I., ^{99m}Tc-Labeled Serum Albumin for Scintillation Scanning of the Placenta. *J. Nucl. Med.*, 5: 936, 1964.
48. Stern, H. S., Goodwin, D. A., and Wagner, H. N., Cardiac and Placental Scanning with Indium-113m. *J. Nucl. Med.*, 8: 351, 1967 (Abstract).
49. Rosenthall, L., Radionuclide Visualization of the Placenta with the Gamma-Ray Scintillation Camera. *Can. Med. Ass. J.*, 97: 212, 1967.
50. Fish, M. B., Lavine, D., Pollycove, M., and Khentigan, A., Rapid Placental Imaging and Localization, Utilizing Ionic Technetium-99m. *J. Nucl. Med.*, 8: 350, 1967 (Abstract).

RADIONUCLIDES IN RESPIRATORY-SYSTEM STUDIES

H. N. Wagner, Jr., R. A. Holmes, V. Lopez-Majano, and *D. E. Tow*

INTRODUCTION

A major contribution of radioactive tracers is their ability to make possible the measurement of *regional* pulmonary function, supplementing the tests of *total* function. Although bronchospirometry is still used in measuring the ventilation and oxygen uptake of each lung, that method is often too complex for routine clinical or research use, particularly with severely ill patients. Knipping and his colleagues[1] first suggested the use of radioactive gases, an important technical advance that has been greatly extended since that time. In addition to radioactive gases, radioactive particles have also been used to study the regional circulation in a variety of diseases and physiological conditions. At times, external radiation detectors permit the diagnosis of disease of certain regions before impairment of total function can be detected, or areas of greatest involvement in generalized disease of the lungs can be located. This information is not only of diagnostic value; it can also help the surgeon decide which parts of the lung may need to be removed.

Also, the effectiveness of some of the nonrespiratory functions of the lungs—such as coughing, mucociliary activity, and alveolar phagocytosis—can be measured with radioactive tracers. In general, radioactive tracers are an aid in the study of the lungs in four ways: 1) in the measurement of physiological abnormalities in precise quantitative terms, 2) in locating areas of malfunction, 3) in the planning of the optimum management of a patient's illness, and 4) in the objective evaluation of the results of treatment.

RADIOACTIVE GASES

The function of various regions of the lungs can be determined without the use of radioactive tracers. Sampling of the gases from individual lobes of the lung can be carried out by means of radiopaque polyvinyl catheters placed into selected lobar bronchi,[2] or a triple-lumen catheter can be used to study gaseous exchange in the upper and lower lobes.[3] But both these methods are difficult for the patient and have not been used widely.

The first use of radioactive gases to study pulmonary disease was by Knipping et al.;[4] they were concerned primarily with the early diagnosis of bronchial carcinoma, and developed a procedure called "radioxenon thorakography". By placing radiation detectors over various regions of the lung and observing the change in radioactivity as the patient breathed radioactive xenon through a closed system, they could detect impairment in regional ventilation caused by local bronchial stenosis or pleural thickening. In intrapulmonary diseases not resulting in bronchial narrowing there was only slight impairment of regional ventilation. These earlier methods, although novel and ingenious, were qualitative rather than quantitative; they have been greatly improved by subsequent modifications.

The three principal respiratory gases—oxygen, carbon dioxide, and nitrogen—can be made radioactive by deuteron bombardment of nitrogen, boron, or carbon in a cyclotron. The use of ^{15}O, ^{11}C, and ^{13}N for lung function studies was introduced by West et al.,[5] Dollery,[6] and Dyson et al.[7] Unfortunately these nuclides cannot be produced in appreciable quantities in a nuclear reactor, and they decay within minutes; for these reasons, their clinical usefulness is limited.

Radioactive carbon dioxide was prepared by passing ^{15}O through copper tubing as the radioactive gas evolved from a cyclotron. Despite the 2.5-minute physical half-life of ^{15}O, it was possible to measure the initial concentration and rate of disappearance of inhaled radioactive carbon dioxide by means of two pairs of crystal scintillation detectors, that were arranged in front of and behind the chest. By moving the detectors from one position to another, it was possible to record the rate of clearance of the radioactive gas from different parts of the lungs, and to determine the relative distribution of ventilation and perfusion in different regions.

Since ^{15}O and its derivatives $C^{15}O_2$ and $C^{15}O$ were available only in medical units located near cyclotrons, the application of the ^{133}Xe method was modified and extended by Ball et al.[8] This radioactive gas is readily available and inexpensive, and it can be injected in aqueous solution intravenously. The patient holds his breath during the injection; when the xenon reaches the pulmonary capillary bed, it diffuses into the air within the alveoli. If the patient holds his breath for 5 to 10 seconds after the injection, intrapulmonary distribution of radioactivity is determined primarily by regional pulmonary blood flow, which can be measured with external radiation detectors.

After the initial distribution has been measured, a second determination is made after a period of rebreathing, during which time the gas is distributed evenly throughout the alveoli and thus provides an indication of the lung volume being viewed by the collimated detector.

An example of physiological information obtained with radioactive gases is the finding that perfusion of the lungs in normal persons is usually greater in the lower regions of the lung—compared to the upper portions—when the person is in an upright position, but that this difference disappears when he is supine.

Instead of injection of the xenon solution (as in the perfusion studies) to measure the ventilation of different regions of the lungs, the patient can be instructed to take a single breath of the radioactive gas and hold his breath for 10 to 15 seconds. Scintillation detectors located in front of or behind the chest record the time course of radioactivity in various regions of the lungs in a manner similar to the regional perfusion studies. During inspiration the radioactive gas enters various regions of the lung at a rate directly related to ventilation. A modification of the single-breath technique consists of measuring the time required for the radioactivity to reach equilibrium when the patient breathes from a closed system.[9,10] In contrast to the single-breath method, which measures primarily dead-space ventilation, the equilibration-time method gives a better indication of alveolar ventilation.

In the measurement of both ventilation and perfusion, an important advantage of the xenon method is that it lets the air within the lungs become uniformly labeled with the radioactive gas. This provides an indication of the volume of lung within the field of the detector and makes it possible to determine the ventilation and perfusion per unit volume of lung.

RADIOACTIVE PARTICLES

Instead of using radioactive gases, regional pulmonary arterial blood flow can be measured by the *particle distribution method*. The technique is easy, relatively inexpensive, and free of significant danger or discomfort to the patient. The method is based on a principle somewhat similar to that used in measurement of regional blood flow by determination of the fractional distribution of injected ^{42}K or ^{131}I-antipyrine.[11] However, the use of particulate rather than soluble indicators obviates the problem of recirculation, and the detection instruments are simpler. With the particle distribution method, the particles are injected under the particular physiological condition to be studied, and measurements can be made at a later time. The fractional distribution within the lungs of particles that were injected intravenously is determined by the regional pulmonary arterial blood flow at the time of injection, and this can be quantified by radioisotope scanning. Macroaggregates of human serum albumin (MAA) labeled with ^{131}I have been the most widely used radioactive particles.[12]

An important advantage of the use of metabolized macroaggregates of albumin—rather than the ceramic microspheres that were used formerly[13]—is that the albumin particles can be safely administered to man. An extensive testing for potential toxicity was carried out before the substance was used in man, and subsequent experience with thousands of patients has confirmed the safety of the method.

The absence of hemodynamic effects of macroaggregates in this size range (10 to 50 microns) is due to the small amount injected and to the structural characteristics of the lung. Weibel[14] established that the human lung contains about 280 billion capillary segments that arise from arterioles at about the twenty-eighth consecutive branching of the pulmonary artery. In carrying out a lung scan, fewer than one million particles are injected.

Gold and McCormack[15] measured pulmonary function of eleven patients before and after lung scans; they found that scans produced no effect on lung volumes, pulmonary diffusing capacity, wasted ventilation, arterial oxygen and carbon dioxide pressures and pH, or the cardiopulmonary response to exercise. Lung scanning was a safe procedure even in the presence of preexisting pulmonary vascular disease.

Validity of the Particle Distribution Method

The particle distribution method is based on the assumptions that 1) the particles are uniformly mixed in the blood in the course of their passage from the point of injection to the pulmonary artery, 2) hemodynamic and gravitational forces affect the distribution of the particles in a manner similar to that of red blood cells, 3) the particles are almost completely extracted from the pulmonary circulation in a single passage through the lung, 4) the particles—in the small quantities administered—do not themselves alter the distribution of blood flow, 5) the particles are not metabolized so rapidly that the initial distribution is significantly altered before their detection by the external radiation detectors, and 6) proper calibration can be made to correct for the effects of variations in chest wall thickness, lung volume, and other geometric factors affecting the quantification of radioactivity by the scanning method.

Use of Indium-113m-Labeled Particles

Recently, particles of iron hydroxide labeled with radioactive indium have been used.[16] Among the advantages of ^{113m}In are that it decays with a half-life of 1.7 hours and emits 390-kev gamma rays without emission of beta particles, thus resulting in a lower radiation dose than ^{131}I. Carrier-free quantities of ^{113m}In can be obtained from ^{113}Sn, which has a half-life of 118 days.

Iron hydroxide particles labeled with ^{113m}In are made as follows. Ferric chloride in acidified solution is added to an aliquot from the ^{113}Sn column so that the final ferric ion concentration is ~ 10 μg per ml. The solution is titrated with sodium hydroxide to a pH ranging from 10.5 to 11.5. Gelatin (20-percent solution) is added so that the final concentration of gelatin is 20 mg per ml. The pH is adjusted to 7.5 with dilute hydrochloric acid. The final product is sterilized by autoclaving for twenty minutes. The majority of the particles range from 20 to 40 microns in diameters, with no particles in excess of 60 microns.

In addition to providing excellent lung scans, ^{113m}In-labeled iron hydroxide has the following advantages.

1. The absorbed radiation dose to the lungs from 1 mCi of ^{113m}In is about 0.75 rads, instead of 2 to 4 rads from a 300-μCi dose of ^{131}I-labeled MAA.
2. The radionuclide is available from a long-lived parent.
3. The particles are easy to prepare.
4. The final product can be sterilized by autoclaving.
5. The larger dose that can be administered provides increased numbers of emitted photons and decreases image distortion resulting from statistical fluctuation.

EXTERNAL DETECTION

The regional distribution of radioactivity in the lungs of humans after the injection of labeled particles can be measured by multiple stationary detectors, by a camera device, or by a single detector that scans the thorax in a rectilinear pattern. Usually the patient lies supine during the scanning procedure, with the detector mounted beneath the subject and capable of moving at a relatively high speed. Patients can be scanned with an over-the-table detector, but occasionally they have difficulty in lying prone for the posterior view.

Quantification of the regional distribution of radioactivity can be accomplished in several ways. One method is to count the number of dots printed automatically on paper at a rate proportional to the counting rate. The method is tedious, and thus is not often used. A second method is to use the density of the film on which the scanning image is obtained for quantification; this can be done if the system is calibrated to correct for the nonlinear response of the film. Another method is recording of the counting rates by two scalers, each recording the activity from one lung. The radioactivity in particular segments of lung is usually expressed as a percentage of the total radioactivity in the lungs.

A rectilinear scanner can be used to measure the distribution of radioactive particles within the lung, but it cannot be used to measure the rapid changes in radioactivity that occur with the use of radioactive gases. With the gases, three types of instrument systems have been used. The first is a battery of two to six detectors arranged over the front or back of the chest; the output of the detectors can be stored in an intermediate system, such as magnetic or paper punched tape, and then played back one channel at a time through a single or dual recording system. A second method is to use a single pair of counters that rapidly scan the chest in two vertical lines from the bottom to the top of the lungs in a period of about seven seconds; the time response of the detection and recording system must be fast (about a third of a second), because changes in activity occur in a matter of seconds. The third method of measuring the distribution of rapidly changing concentrations of radioactivity in the lungs is to use one of the newer stationary imaging devices or cameras; a suitable instrument should cover the entire lung field and have a spatial resolution of about 1 cm, as well as fast time response.

Up to the present time, radioactive xenon has been used primarily to study the ventilatory physiology of the lungs in normal persons and in patients with obstructive lung disease. It is predictable that, with simplification of the techniques, the radioactive gases will be used more and more to solve clinical problems. Justification for simpler methods arise from basic investigations, which have revealed that the degree of spatial resolution obtained with six or more detectors is usually not necessary for the solution of most clinical problems, particularly since identification of the detailed size and shape of the blood flow defects can be determined by scanning after the injection of radioactive particles. To solve a diagnostic problem, it is necessary to determine whether ventilation to a particular area is normal or reduced. This can be accomplished with equipment that is readily available in most departments of nuclear medicine. The most readily adaptable equipment is the dual-detector system widely used for the detection of unilateral renal disease. If the temporal response is fast enough (measurement must be made at 1-second intervals), regional ventilation can readily be determined.

The technique is as follows. The patient lies quietly in a supine position, with the two detectors positioned first over the upper halves of the lungs. A 10-liter spirometer is filled with oxygen or room air and several millicuries of ^{133}Xe; the ^{133}Xe can be introduced into the spirometer from a shielded tank containing oxygen (or compressed air) and xenon, or by injecting aqueous solutions of the dissolved radioactive gas. The patient's nostrils are occluded with a padded clamp, and he breathes through a two-way valve connected to the spirometer; after several practice breaths, the patient takes a single deep breath of the xenon-and-oxygen mixture and holds his breath for 10 seconds. The time course of radioactivity in the two regions of the lungs is recorded at 1-second intervals, and the counting rates are graphed on linear graph paper. In the second phase of the study, the patient breathes at a normal tidal volume; after a period of time, the valve connecting the mouthpiece to the spirometer is turned so that the patient breathes from the closed system containing the xenon. The system contains a carbon dioxide absorber, and rebreathing continues until equilibrium has been reached, or for at least ten minutes. The counting rates are recorded from each of the detector sites at 1-second intervals and are graphed as a function of time on linear graph paper.

The lung volume in the field of view of the detector can be estimated from the counting rate at the end of the equilibration period. The plateau values after the single breath are related to both ventilation and the lung volume viewed by the detector. If the lung volumes are the same, differences in the single-breath plateau values can be used to determine the relative ventilation of the two regions. The rate with which equilibration is reached is also used as an index of regional ventilation. In addition, the time required to reach 5, 50, and 90 percent of the final equilibration activity parameter is measured.

In practice, the regional perfusion of the lung is first assessed by scanning, and then ventilation is measured. The finding of decreased perfusion with relatively normal ventilation may help in distinguishing pulmonary embolism from chronic obstructive pulmonary disease.

PULMONARY ARTERIAL BLOOD FLOW IN NORMAL PERSONS

A typical lung scan made with the detector in front of the chest (anterior view) usually shows the outline of the heart as an area without radioactivity. Respiratory motion may give a serrated appearance to the lower borders, but the remainder of the lung fields usually have smooth convex borders. Concavities at the borders, often seen in pulmonary embolism, should be carefully noted. The lower lateral borders of the scan often have a rounded contour, even in a normal person.

In the posterior view the cardiac image is less prominent, and the two lungs appear more symmetrical. The apices may have slightly less radioactivity than the bases, although this is usually not pronounced if the injection was made with the patient in a supine position. In the posterior view the medial borders of the lungs appear straight, except in such conditions as cardiomegaly, distention of the pulmonary arteries, and hilar adenopathy.

The lateral views are very helpful and should be performed routinely; at the very least they should be performed whenever an abnormality is noted in the anterior or posterior view. In the left lateral view the heart may produce a concave indentation of the anterior borders of the lung, particularly if the heart is enlarged. On both lateral views the inferior margin of the lung is usually concave, because of the diaphragm; respiratory movements may produce serrations of this concave border. To avoid an image that is difficult to interpret, the lateral views require the utmost care in ensuring that the patient's position is truly lateral, rather than oblique.

ARTIFACTS AND FREQUENT NORMAL VARIANTS

Multiple views of the lungs are needed because of the limited depth of focus of most focusing collimators. In one instance, despite the small infiltrate in the left mid-lung field on the radiograph, the posterior scan suggested that the abnormality included the entire mid-lung field. The left lateral view, however, showed the lesion to involve the posterior segment of the upper lobe.

The particle distribution technique has confirmed the results of the radioactive gas studies, i.e., the regional shifts in pulmonary blood flow occur as the result of minor changes, such as shifting the body position.

Awareness of the sensitivity of the method in detecting changes in regional pulmonary blood flow is helpful in interpreting lung scans. Unilateral pleural effusion may also result in marked reduction in blood flow, particularly if the fluid is free and impinges upon the pulmonary vessels when the patient is lying down.

Another variant is the decrease in regional blood flow resulting from an enlarged heart. The area of decreased activity does not usually extend to the periphery of the lung, except in cases of extreme enlargement. The effects of cardiomegaly should not be interpreted as pulmonary lesions. The left lateral view is particularly helpful in determining whether there is a pulmonary lesion in the left lower field in patients with cardiomegaly. Lesions of the left lower lung will often be missed if only an anterior view is obtained.

Developmental (or acquired) abnormalities of the chest wall, thoracic spine or sternum—such as hemithorax, kyphoscoliosis, pectus excavatum, and pectus carinatum—may alter the appearance of the lung scan. These conditions emphasize one of the first principles of lung scanning: one should never interpret a scan without examining the patient and the chest radiograph.

Some congenital disorders alter the pulmonary circulation in ways that provide diagnostic information. Proper interpretation of lung scans in patients with congenital cardiac defects requires an understanding of the relationship between the pulmonary and bronchial circulation. Except when there is an abnormal communication between the right and left side of the heart, such as a ventricular septal defect, the bronchial circulation does not contain radioactive particles after intravenous injection; the distribution of the particles within the lungs is a function of the pulmonary arterial blood flow only. However, if right-to-left shunting is present, and especially if the bronchial circulation is increased, particles can reach the lungs by way of both the pulmonary and the bronchial circulations. For example, in patients with total pulmonary atresia all the particles are shunted into the systemic circulation and reach the lungs only via the bronchial circulation.

Friedman and Braunwald (unpublished) have proposed that serial lung scans after intravenous injection of particles can be a simple means of evaluating the natural history of a palliative subclavian anastomosis. If the distribution of particles becomes more symmetrical after intravenous injection of particles, this is evidence that the anastomosis has become smaller or has closed completely.

Serial scans also permit estimation of the patency of superior-vena-cava–pulmonary anastomoses. If the shunt is working properly, particles injected into a vein of the upper arm will produce a scan of the lung on the same side as the anastomosis; the opposite lung will be visualized after injection into a lower extremity.

DIFFERENTIAL LUNG SCAN CHARACTERISTICS
Obstructive Pulmonary Disease

Although imbalance between ventilation and perfusion may occur, resulting in inefficient perfusion and deficient oxygenation of blood, the body is usually quite efficient in balancing changes in regional ventilation by adjustments in the regional pulmonary blood flow. Thus, regional decreases in blood flow will usually be found in bronchial and alveolar diseases as well as in primary disorders of the pulmonary circulation. Bronchial obstruction often produces demonstrable perfusion abnormalities that are detectable by the particle distribution method. In such cases the defect can be further characterized by the study of regional ventilation with radioactive xenon.

Bentivoglio and his associates[9] used the xenon method to study patients with acute asthma and found regional defects in ventilation even when the patients were free of symptoms between acute asthmatic attacks.

Ball and his associates,[8] using ^{133}Xe, found focal abnormalities of both ventilation and perfusion in patients with emphysema. Delineation of these areas may be of practical value in indicating those patients whose total lung function might be improved by surgery. Lopez-Majano et al.[17] used lung scans to study 62 patients with severe pulmonary emphysema, and they confirmed the high incidence of perfusion abnormalities.

Bronchiectasis is frequently focal, and in such areas it results in decreased perfusion, as seen in the lung scan.

In the early diagnosis of pulmonary infections, such as pneumonia or tuberculosis, the sensitivity of lung scanning—as compared to chest radiography—has not yet been determined.

The diminished radioactivity corresponds to the infiltrate in the radiograph; this does not necessarily indicate a decrease in total perfusion of this region, since it is likely that the bronchial circulation is adequate, or even increased.

In lung scans of 67 patients with chronic pulmonary tuberculosis[18] the most significant finding was a decrease in the pulmonary arterial blood flow to the diseased areas, which were usually larger than had been expected from the size of the lesion in the chest radiograph. Thus, the scans were of value in delineating the extent or severity of disease.

The use of lung scanning in the early diagnosis of bronchogenic carcinoma was first reported by Wagner et al.,[19] who found that avascular areas in the lung scan were far larger than the radiograph abnormalities; the scans were markedly abnormal in a few patients with unexplained hemoptysis and positive cytology at a time when the radiograph was entirely within normal limits. Characteristically, the vascular defects in carcinoma of the lung did not change significantly for periods up to a month or more. This helps to distinguish these abnormalities from the changing defects characteristic of pulmonary embolism. The persistence of the defect in the scanning image of a patient with a bronchogenic carcinoma suggests the possibility that the defect might be a tumor rather than an embolus, as suspected on clinical evidence.

Pulmonary Embolism

Lung scanning is a simple and effective procedure for the diagnosis of pulmonary embolism. Although it is not sufficiently specific by itself, it provides the right answer when the proper question is asked. Lung scanning indicates the regional distribution of pulmonary arterial blood flow. It clearly delineates areas where pulmonary arterial blood flow is impaired—a universal finding in massive pulmonary embolism. However, if pulmonary arterial blood flow is reduced generally, as in multiple small embolism or in acute pulmonary edema, pulmonary blood flow will retain an essentially normal distribution. An advantage of the scanning procedure is that, because of its safety and simplicity, it can be performed readily and repeatedly. It can also provide information that cannot be obtained by other means, such as the determination of the rate of return of blood flow to affected areas—either with or without therapy.

It is true that many other diseases besides pulmonary embolism result in regional decreases in pulmonary arterial blood flow. A lung scan does not have the specificity of arteriography, which often shows encroachment of an embolus upon a pulmonary artery, but the scan does provide information concerning the perfusion of the pulmonary capillary bed. Thus, lung scanning and pulmonary arteriography should always be considered as complementary rather than competitive procedures.

At times, if the findings meet certain criteria, the lung scan can be relatively specific for pulmonary embolism. First, it is important to look for large areas of greatly diminished pulmonary arterial blood flow in regions that do not appear affected in the chest radiograph; this is particularly helpful, because pneumonia and other infiltrative lesions that that been found to be associated with a decrease in pulmonary arterial blood flow can usually be seen in the radiograph. If the scan shows the characteristic pattern of avascularity, one can usually be certain of the diagnosis. However, if the avascular areas correspond to opacities on the chest X-ray, one cannot distinguish between primary vascular disease and secondary involvement of the pulmonary vasculature. Particularly troublesome lesions are lung cysts and bullae, which may appear as areas of increased radiolucency in a manner similar to pulmonary emboli. Second, the lateral borders of the scan must be examined for concave defects. These characteristic defects correspond to the hemispherical lesions found at the lung periphery on post-mortem examination;[20] presumably they represent areas in which the collateral circulation is inadequate. Since large areas of the lung must be compromised before pulmonary hypertension and right ventricular failure result, search for large defects on the scan must be made before severe systemic hypotension can be attributed to massive embolism. The scan is often helpful in ruling out massive embolism in a patient who develops hypotension following surgery or delivery. If the scan is normal, the cause of the hypotension must be found elsewhere.

Pulmonary Hypertension

In pathological conditions, such as severe mitral stenosis and left ventricular failure, when the pulmonary venous pressure is raised, there is a reversal of the normal distribution of flow when the patient is erect. In normal persons in erect position the blood flow per gram of lung is less in the apex than in the base.[21] In severe pulmonary hypertension this is reversed, and a distinct pattern of distribution of flow is found associated with advanced mitral-valve disease. This was confirmed by Dollery and West,[6] using radioactive oxygen, carbon dioxide, and carbon monoxide, and later by Dawson et al.[22] with the ^{133}Xe method.

Friedman and Braunwald[23] studied the alterations in regional pulmonary blood flow in mitral-valve disease with the particle distribution technique. When left-to-right shunting was present in patients with pulmonary hypertension, the ratio of radioactivity in the upper zones of the lung to that in the lower zones was significantly higher than when pulmonary arterial pressure was normal. This is also true in patients with primary pulmonary hypertension.

Diseases of the Pleura

The effect of pleural effusion on the distribution of pulmonary arterial blood flow, as seen in the lung scan, has recently been studied by Tow and others,[24] who observed a diminished blood flow to the involved lung in cases of large pleural effusions. They concluded from their studies that the diminished perfusion was probably the result of hilar pulmonary artery compression by the accumulated free fluid in the paravertebral gutter.

Other Diseases

Included in this group are pneumothorax; disease entities with "alveolar capillary block"; diffusion abnormalities, including sarcoidosis and interstitial pulmonary fibrosis (Hamman-Rich syndrome); pneumoconioses, such as silicosis, asbestosis, and berylliosis; connective-tissue diseases (dyscollagenoses); and diseases of the mediastinum and diaphragm, including injuries of the heart and great vessels. In all these conditions lung scanning has been used infrequently, and its exact role has not yet been defined.

CONCLUSIONS

Present scientific knowledge of the pulmonary circulation and the mucociliary activity of the tracheobronchial tree is based to a large extent on observations made on anesthetized animals, on isolated lung preparations, and on patients with cardiopulmonary diseases. Although such studies have produced fundamental contributions to our understanding, the circumstances of study have usually involved altered, impaired, or absent function. Even studies in normal and conscious humans frequently necessitate abnormal conditions—such as bronchospirometry, cardiac catheterization, and arterial cannulation. A significant advantage of the particle distribution method, employing radioisotope scanning for delineation, is the fact that there is minimal disturbance of the patient's normal physiology by the technique itself. Also, quantification of findings can be made readily, although some spatial geometric problems are encountered. The techniques are comparatively easy and relatively inexpensive, and they can be repeated in the same person.

The studies of weightlessness and of patients with acute asthma illustrate a unique advantage of the particle distribution method—its applicability under difficult experimental conditions. The primary requirement is that the injection of the particles be made under the conditions of the experiments. If it can be established that the internal mixing is completed during the experiment, one can derive the distribution of regional pulmonary blood flow that was present at the time of the injection, despite the fact that the fractional distribution is determined after a period of time (up to 1.5 hours or more).

Furthermore, use of the scanning technique with injected particles has confirmed the data obtained from studies with radioactive gases—the changes that occur in the distribution of pulmonary blood flow when a person stands up or lies down. It has been found in thousands of patients that regional diseases of the lungs result in a shift of pulmonary arterial blood away from the diseased area. The mechanism is not fully known, but mechanical obstruction, obliteration of the vascular bed, and functional changes such as alveolar hypoxia are probable factors.

Vasoactive substances, such as serotonin and acetylcholine, have been demonstrated to have a significant vasomotor effect when perfused through various parts of the pulmonary vascular bed. The particle distribution technique has made it quite easy to demonstrate vasomotor changes, because the problem of monitoring the pressure gradient across the pulmonary vascular bed can be avoided. The effect on regional flow is measured, rather than changes in pressure. The technique is a simple and effective way to study the circulation and other physiological activities in various regions of the lungs of healthy persons and of patients with disease.

REFERENCES

1. Knipping, H. W., Bolt, W., Venrath, H., Valentin, H., Ludes, H., and Endler, P., Eine neue Methode zur Prüfung der Herz- und Lungenfunktion: die regionale Funktionsanalyse in der Lungen- und Herzklinik mit Hilfe des radioaktiven Edelgases Xenon-133. *Deut. Med. Wochensch.*, 80: 1146, 1955.
2. Martin, C. J., Cline, F., Jr., and Marshall, H., Lobar Alveolar Gas Concentrations: Effects of Body Position. *J. Clin. Invest.*, 32: 617, 1953.
3. Mattson, S. B. and Carlens, E., Lobar Ventilation and Oxygen Uptake in Man: Influence of Body Position. *J. Thorac. Surg.*, 30: 676, 1955.
4. Knipping, H. W., Bolt, W., Valentin, H., Venrath, H., and Endler, P., Regionale Funktionsanalyse in der Kreislauf- und Lungenklinik mit Hilfe der Isotopenthorakographie und der selektiven Angiographie der Lungengefässe. *Münchener Med. Wochensch.*, 99: 1, 1957.
5. West, J. B., Hollard, R. A. B., Dollery, C. T., and Matthews, C. M. E., Interpretation of Radioactive-Gas Clearance Rates in the Lung. *J. Appl. Physiol.*, 17: 14, 1962.
6. Dollery, C. T. and West, J. B., Regional Uptake of Radioactive Oxygen, Carbon Monoxide and Carbon Dioxide in the Lungs of Patients with Mitral Stenosis. *Circul. Res.*, 8: 765, 1960.
7. Dyson, N. A., Hugh-Jones, P., Newbery, G. R., Sinclair, J. D., and West, J. B., Studies of Regional Lung Function, Using Radioactive Oxygen. *Brit. Med. J.*, 1: 231, 1960.
8. Ball, W. C., Stewart, P. B., Newsham, L. G. S., and Bates, D. V., Regional Pulmonary Function Studied with Xenon-133. *J. Clin. Invest.*, 41: 519, 1962.
9. Bentivoglio, L. G., Beerel, F., Bryan, A. C., Stewart, P. B., Rose B., and Bates, D. V., Regional Pulmonary Function Studied with Xenon-133 in Patients with Bronchial Asthma. *J. Clin. Invest.*, 42: 1193, 1963.
10. Bentivoglio, L. G., Beerel, F., Stewart, P. B., Bryan, A. C., Ball W. C., and Bates, D. V., Studies of Regional Ventilation and Perfusion in Pulmonary Emphysema, Using Xenon-133. *Amer. Rev. Resp. Dis.*, 88: 315, 1963.
11. Sapirstein, L. A., Regional Blood Flow by Fractional Distribution of Indicators. *Amer. J. Physiol.*, 193: 161, 1958.
12. Taplin, G. V., Johnson, D. E., Dore, E. K., and Kaplan, H. S., Lung Photoscans with Macroaggregates of Human Serum Radioalbumin. Experimental Basis and Initial Clinical Trials. *Health Phys.*, 10: 1219, 1964.
13. Ariel, I. M., Quoted in Highlights of the Society of Nuclear Medicine Meeting. *J. Amer. Med. Ass.*, 183: 32, 1963.
14. Weibel, E. R., *Morphometry of the Human Lung.* Academic Press, New York, 1963.
15. Gold, W. M. and McCormack, K. R., Pulmonary-Function Response to Radioisotope Scanning of the Lungs. *J. Amer. Med. Ass.*, 197: 146, 1966.
16. Stern, H. S., Goodwin, D. A., Wagner, H. N., Jr., and Kramer, H. H., In-113m—a Short-Lived Isotope for Lung Scanning. *Nucleonics*, 24: 57, 1966.
17. Lopez-Majano, V., Tow, D. E., and Wagner, H. N., Jr., Regional Distribution of Pulmonary Arterial Blood Flow in Emphysema. *J. Amer. Med. Ass.*, 197: 81, 1966.
18. Lopez-Majano, V., Wagner, H. N., Jr., Tow, D. E., and Chernick, V., Radioisotope Scanning of the Lungs in Pulmonary Tuberculosis. *J. Amer. Med. Ass.*, 194: 1053, 1965.
19. Wagner, H. N., Jr., Lopez-Majano, V., Tow, D. E., and Langan, J. K., Radioisotope Scanning of Lungs in Early Diagnosis of Bronchogenic Carcinoma. *Lancet*, 1: 344, 1965b.
20. Hampton, A. O. and Castleman, B., Correlation of Post-Mortem Chest Teleroentgenograms with Autopsy Findings. *Am. J. Roentgenol. Radium Ther. Nucl. Med.* 43: 305, 1940.
21. West, J. B., *Ventilation/Blood Flow and Gas Exchange.* F. A. Davis, Philadelphia, 1965.
22. Dawson, A., Kaneko, K., and McGregor, M., Regional Lung Function in Patients with Mitral Stenosis, Studied with Xenon-133 During Air and Oxygen Breathing. *J. Clin. Invest.*, 44: 999, 1965.
23. Friedman, W. F. and Braunwald, E., Alterations in Regional Pulmonary Blood Flow in Mitral-Valve Disease Studied by Radioisotope Scanning. *Circulation*, 34: 363, 1966.
24. Tow, D. E., Mishkin, F. S., Wagner, H. N., Jr., Baker, R. B., and Jensen, A. D., Effect of Free Pleural Fluid on the Pulmonary Circulation. (*Unpublished Data.*)

NUCLEAR-MEDICINE TECHNIQUES IN GASTROENTEROLOGIC DIAGNOSIS

M. R. Powell, M.D.
University of California Medical Center, San Francisco, California.

Radioisotope counting and distribution detection techniques have assumed a relatively minor role in diagnosis of diseases of the gut and related organs. This is a paradox in that the gut, liver, pancreas, and other related organs have primary roles in a vast variety of metabolic processes that should be readily susceptible to introduction of radioactive tracer labels. The problems encountered are similar to those in other areas: internal labeling of metabolites is required to preserve their normal metabolism and is for the most part restricted to ^{14}C and tritium labeling. This limits counting techniques to those suitable for weak beta radiation: either liquid scintillation counting *in vitro* or, in the case of ^{14}C, breath analysis. Both counting techniques have been slow to be commonly used in clinical nuclear medicine, but will no doubt see increasing use in gastroenterologic diagnosis.

The clinical procedures in common use at this time employ gamma-radiation detection. Table 58 lists the present tests usually available for diagnosis of gastroenterologic diseases. The organ visualization tests will be discussed first, in the order listed, with a discussion of the other tests following.

Table 58. RADIOISOTOPIC TESTS IN GASTROENTEROLOGIC DIAGNOSIS

Organ Visualization

Organ	Radiopharmaceutical	Labeling Mechanism
Liver	^{198}Au-colloid	Phagocytosis by reticuloendothelial cells (Kupffer Cells)
	^{99m}Tc-colloid	
	^{113m}In-colloid	
	^{131}I-microaggregated albumin	
	^{131}I-Rose Bengal	Polygonal-cell concentration
Gall bladder	^{131}I-Rose Bengal	Excretion in bile
Pancreas	^{75}Se-methionine	Participation in protein anabolism
Salivary glands	^{99m}Tc-pertechnetate	Secretion in saliva

Counting Studies

Diagnostic Test	Radiopharmaceutical	Specimen Counted
Fat hydrolysis	^{131}I-triolein	Feces or serum
Fat absorption	^{131}I-oleic acid	Feces or serum
Vitamin B_{12} absorption	^{57}Co- or ^{60}Co-vitamin B_{12}	24-hour urine or serum
G. I. blood loss	^{51}Cr-tagged erythrocytes	Feces
Protein-losing enteropathy	^{51}Cr-labeled albumin	Feces
Liver function	^{131}I-Rose Bengal	Liver (probe) or serum
	^{131}I-iodipamide*	Liver (probe)
Biliary obstruction	^{131}I-Rose Bengal	Feces

* Cholegrafin.

LIVER IMAGING
Liver Labeling

Liver scanning is the most frequently performed procedure discussed in this chapter. The liver was one of the first organs imaged in rectilinear radioisotope scanning. Cassen developed the first scanning apparatus in 1951, and participated with Stirrett and Yuhl in application of the instrument and technique to imaging liver defects in 1954.[1,2] Their scans were done using ^{198}Au-colloid to label the reticuloendothelial cells of the liver. ^{131}I-Rose Bengal was soon suggested as a suitable label for the polygonal cells.[3] It was not until 1964 that a major advance in liver-labeling radiopharmaceuticals was made with Atkin's report of liver, spleen and marrow scanning with ^{99m}Tc-colloid.[4] This colloid was the first to provide for high-data-density liver imaging, with considerable improvement in image quality, but no increase of radiation exposure. More recently ^{113m}In-colloid has been introduced as another short-lived liver label, and ^{131}I-labeled microaggregated human serum albumin has been reintroduced by being made commercially available.[5,6] Each of these radiopharmaceuticals has some advantages unique to it and will continue to enjoy clinical utilization.

Of the colloid preparations, colloidal gold is inexpensive and has a physical half-life long enough to allow stocking in laboratories where liver scans are rather infrequent. The radiation dose to the patient is highest among the usual doses of various liver phagocyte labels summarized in Table 59. The commercially available microaggregated human serum albumin is relatively more expensive, but has a physical half-life even more favorable for storage. It also has the advantage of being the only reticuloendothelial-cell label that is metabolized after phagocytosis; this considerably reduces the radiation dose to the reticuloendothelial system. The thyroid should be blocked by administration of Lugol's solution during the time in which free iodide is liberated from the microaggregated albumin—approximately two days.

Table 59. LIVER PHAGOCYTE LABELS

Agent	Physical Half-Life	Photon Energy, kev	Dose Administered, µCi	Liver Dose, rads	Whole-Body Dose, rads	Reference
^{198}Au-colloid	2.7 d	411	200	8.0	0.4	7
^{131}I-aggregated albumin	8.1 d	364	500	0.3	0.6	8*
^{99m}Tc-sulfide colloid	6.0 h	140	3,000	1.0	0.05	9
^{113m}In-colloid	1.7 h	390	3,000	1.7		5

* Estimated from data by Taplin and associates.

The ^{99m}Tc- and ^{113m}In-colloids, which have short half-lives, provide high count rates with reduced radiation doses. ^{99m}Tc-sulfide colloid is now the most generally used agent in larger laboratories. Harper and his colleagues have been very active in developing radiopharmaceuticals labeled with ^{99m}Tc, including sulfur colloid.[10] A rapid preparation method for ^{99m}Tc-sulfur colloid has been published by Patton, and further assessed by Larson.[11,12] Preparation must be done on the day of use, because of the 6-hour half-life of ^{99m}Tc. This has led most laboratories to prepare their own colloid, although preparation kits and prepared colloid are being introduced commercially. The original preparation methods required use of H_2S gas, a highly toxic and noxious substance. Furthermore, the original methods required approximately 60 to 90 minutes preparation time, used principally in the H_2S–$^{99m}TcO_4^-$ reaction and subsequent purging of all H_2S from the solution with nitrogen gas. The methods of Patton and Larson utilize reaction between thiosulfite ($S_2O_3^=$) and pertechnetate ($^{99m}TcO_4^-$), with stable perrhenate (ReO_4^-) as a carrier. Rhenium, like technetium, is a member of the manganese subgroup VII B of the Periodic Table of the Elements. Rhenium is apparently nontoxic in man in doses hundreds of times greater than those used in this preparation.[13]

The only problems of drug reaction encountered with these radiopharmaceuticals appear related to colloid-stabilizing agents used in most of the preparation methods, according to a national survey by Smith.[14] The survey showed three minimal reactions in 3,200 studies using a gelatin stabilizer, and eleven reactions—many rather severe—in the 13 studies employing a dextran stabilizer. The typical severe reaction resembles a histamine release, showing first a cutaneous flushing, then unstable blood pressure with periods of sudden severe hypotension and diaphoresis, and late colicky abdominal pains. There have been some complaints of mild pruritis, but no urticaria, and only minimal bronchospasm. Response to pressor, antihistaminic, and steroid therapy has been good, with no recorded death or severe morbidity to date. The author, however, has been sufficiently impressed with the two reactions in his experience with high-molecular-weight dextran stabilization of colloid to discontinue using dextran in the preparation. By simply trying to do liver colloid injections within an hour of preparation, significant lung labeling due to colloid instability can be avoided. This has worked out quite well in practice, with no further reactions occurring in the more than 500 studies conducted since the dextran was omitted. Liver labeling has been sufficiently greater than any lung labeling, and thus is entirely satisfactory. The small amount of lung labeling that may occur does interfere with visualization of bone marrow in the ribs, if the ^{99m}Tc-colloid is to be used for this purpose.

The other type of radiopharmaceutical for liver labeling shows uptake by the polygonal cells and eventual biliary excretion. The dye in common clinical use for this purpose, ^{131}I-Rose Bengal, has a long history of use in clinical liver function testing prior to its being labeled with radioiodide for use in nuclear-medicine diagnostic tests.[15] Since its transit time through the liver is dependent both upon normal polygonal-cell function and upon normal biliary drainage, the radiation dose to the subject may show wide variations. Assuming the worst circumstances, i.e., no excretion after uptake of the dye by the liver (as might occur in neonatal jaundice), liver radiation exposure is estimated at 4 rads for 3-µCi/kg-body-weight dose of the dye, gonadal dose at 0.4 rads, and whole-body dose at 70 mrad.[16]

Usual Rose Bengal liver radiation doses when the dye is normally excreted would be less than the radiation doses seen with ^{99m}Tc-colloid studies, since the usual radioactive doses of these two radiopharmaceuticals are quite different. The Rose Bengal liver scan is a lower-data-density scan, more analogous to the gold-colloid scans in data density. Since relatively small amounts of the dye are injected, drug reactions have not been a problem.

Of the liver labels discussed, the ^{99m}Tc-colloid is the one most commonly used in those laboratories doing rather large numbers of liver visualization studies, but ^{131}I-Rose Bengal will certainly deserve considerable further attention as the best available label for the polygonal cells and for its ability to label the biliary system even when radiographic dyes may not. It should be pointed out that combined studies may be performed, wherein liver labeling by colloid is compared with labeling by dye. If this is done, it is best to image first the isotope having lower gamma energy and then the one with the higher energy; e.g., ^{99m}Tc-colloid, with a 140-kev photopeak, should be imaged first, and ^{131}I-Rose Bengal, having a 364-kev photopeak, should be imaged second. This avoids accepting Compton scatter radiation in a window set for a lower-energy photopeak after prior study with an isotope of higher gamma-ray energy. The scattered radiation would relate to the wrong radiopharmaceutical, and its detected position of origin would be different from the origin of its primary gamma ray, causing apparent loss of resolution in the image.

Liver Anatomy by Gamma Imaging Techniques

The largest organ in the body is suspended from the diaphragm superiorly and posteriorly, but has considerable mobility with change of body position, respiratory excursions of the diaphragm, and with external pressure caused by change in the shape or size of adjacent viscera. The liver is not only mobile, but might in addition be described as plastic: it normally has good ability to respond to external pressure by deformity of its own shape. Such deformity may be temporary with body position change, or may constitute a constant defect in the liver anatomy. Liver fibrosis may either prevent local liver response to pressure, causing the whole organ to respond with position change, or it may serve to make a defect in normal liver contours permanent. Surprising ability of the liver lobes to return to their normal contours after partial surgical excision has often been noted, such regeneration apparently depending upon the regenerating lobe being otherwise normal.[17] The regeneration that occurs in a cirrhotic liver is typically nodular and results in further distortion of already abnormal liver contours.

Gamma imaging defines the general external contours of the liver, any prominent interlobar fissures, and, if these are distinguishable, the major liver lobes.[18] The right and left lobes are usually considered distinguishable in the anterior view. The anterior view may also define what appears to be an intermediate lobe, either the caudate or quadrate; most likely it is the latter, since it tends to be larger and is anteriorly located, where its outlines are more susceptible to definition by either focusing or nonfocusing collimation systems. Other views of the liver seldom define such lobar anatomy, except that in the lateral views the left lobe tends to protrude anteriorly, and in particular its inferior margin may be seen extending beyond the right lobe. Definition of detail at such a distance from the detector usually occurs only with non-focused collimation systems or may be seen with "coarse-focusing" long-focal-length (5 inches) collimators. While the foregoing description of liver lobe anatomy is that generally taught with regard to scan techniques, it appears that it is not entirely correct in view of newer anatomic knowledge from vinyl-cast studies of hepatic vascular and bile duct anatomy.[19] The correct morphologic division between the right and left lobes is the lobar fissure, visible only on the inferior (visceral) surface of the liver and corresponding to a line extending from the gall bladder inferiorly to the inferior vena cava fossa superiorly. What was previously thought to be the fissure between the right and left lobes in the area of the falciform ligament superiorly and the ligament venosum posteriorly is now considered to be a fissure between lateral and medial segments of the left lobe, and the right lobe is found to have anatomic anterior and posterior segments. The latter are not ordinarily recognizable in gamma images. The caudate lobe remains an anatomic third lobe, while the quadrate lobe is the inferior area of the medial segment of the left lobe, which is sometimes seen prolonged inferiorly as an anatomic variant. These considerations of newer liver anatomy assume major importance with reference to liver scanning as evaluations for partial hepatectomy are more frequently required. Figure 20 illustrates normal variations of liver lobe outlines. A prolongation of the right lobe tip inferiorly is referred to as a "Reidel's lobe".

Scintillation camera views of several patients are illustrated. The livers were labeled with ^{99m}Tc-sulfur colloid, given intravenously in doses of 3.5 millicuries shortly before the study. In the illustration the left-hand column shows anterior views, the center column shows right lateral views, and the right-hand column shows right posterior views.

A, B, C.

Typical liver lobe outlines in a mildly obese, short-statured individual with a rather horizontal liver. The margin of picture A goes across the spleen image. In view B the spleen is dimly seen through the body of the patient, posterior to the liver image and adjoining it. View C shows the attenuation of gamma radiation from the left lobe of the liver as it passes through the vertebral column and associated soft tissues. The marker in view A indicates the gall bladder fossa, the inferior extent of the main interlobar fissure (see text).

D, E, F.

Liver lobe outlines in an individual of average stature. The marker in view E shows in this slightly anterior right lateral view how the left lobe of the liver protrudes anterior to the right lobe in most individuals.

G, H, I.

Liver appearance in an asthenic individual with a pointed tip of the right lobe, which was described clinically as a "Reidel's lobe". Here there is a prominent raphe between the lateral segment of the left lobe and the remainder of the liver, associated with the falciform

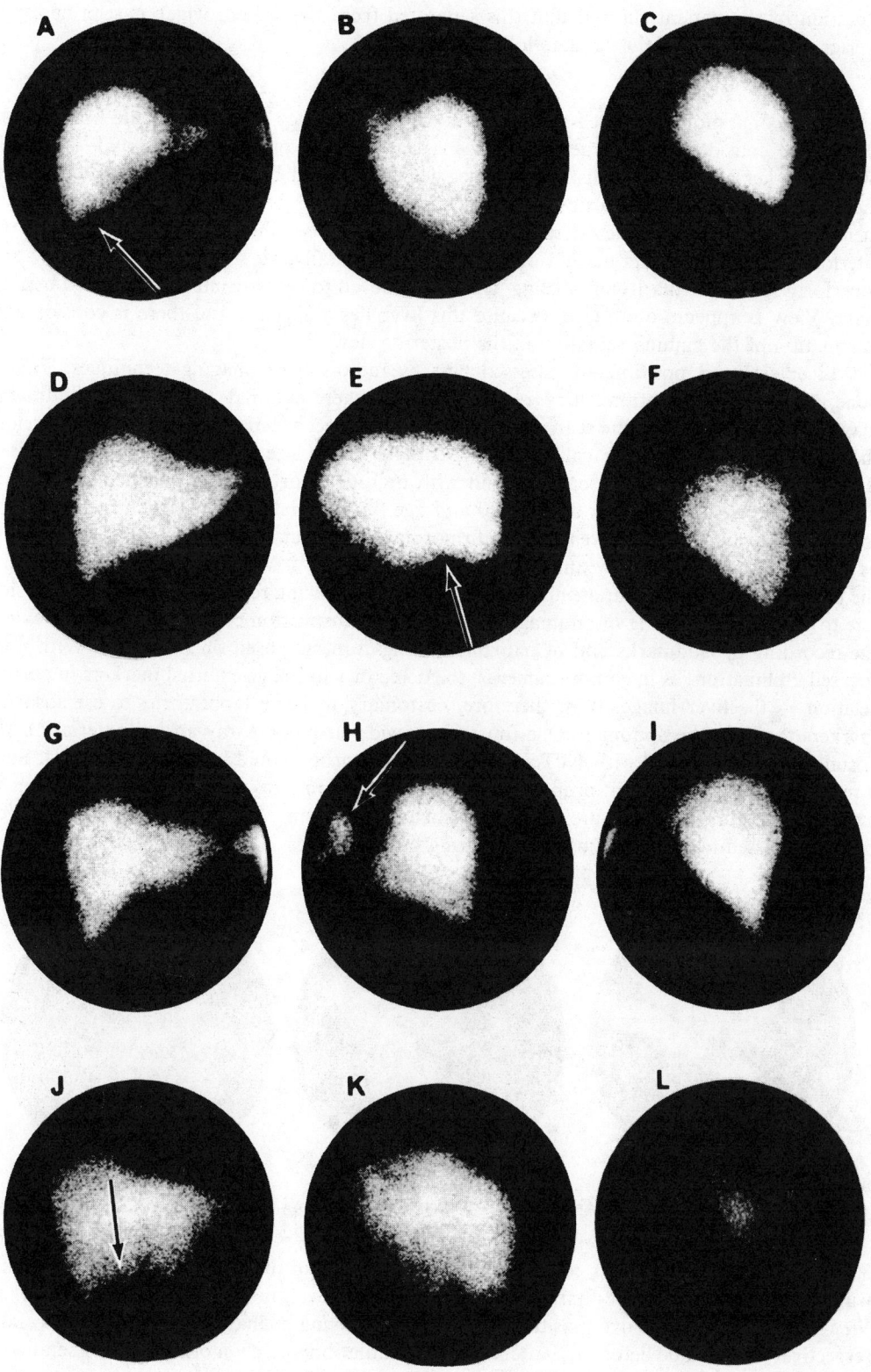

FIGURE 20. Normal Variations of Liver Lobe Outlines.

ligament. View H shows what is believed to be a caudate lobe, as indicated by the marker. Rotation of the patient showed that this separated from the spleen, which is seen here just nferior and posterior to the caudate lobe.

J, K, L.

Liver congestion associated with severe aortic insufficiency and rather acute heart failure has caused considerable enlargement of the lateral segment of the left lobe and—indicated by the marker in view J—a distinct lobe segment that could be termed a quadrate lobe or, more correctly, would be described as the medial segment of the left lobe. The falciform ligament lies between it and the lateral segment of the left lobe; the gall bladder fossa and the interlobar fissure are at its margin on the right. Incidentally, view K shows posteriorly and superiorly some irregularity of labeling, which is believed to be associated with dilated hepatic veins. View L appears quite dim, because this liver lies anteriorly and there is considerable attenuation of the gamma radiation in the posterior view.

Liver size and position are assessed best by radioisotope imaging techniques. This is done easily by a conventional 1 : 1 rectilinear scan, where external and palpable landmarks may be drawn directly on the scan image. If the image is recorded by electronic transmission, then the instrument must be calibrated to correlate image size with the actual size. This is the case with various gamma cameras and with multidetector scanners producing miniature images. No correlation of liver dimensions and age, body weight, height, or surface area was found in individuals over 30 years old with normal livers, but maximum diameters are given as 18.3 ± 1.7 cm horizontally and 16.7 ± 2.1 cm vertically.[18] More meaningful, certainly, is the use of scan image measurements on serial scans in following response of liver and of lesion size to therapy. Recording of an image by electronic transmission complicates to some extent the recording of landmarks and of palpable findings superimposed on the image. With nonfocused collimation, as in gamma cameras, the detector will image external markers in correct relation to the liver image. It is, therefore, customary in some laboratories to use absorber markers taped to the abdomen. One-fourth-inch wide strips of X-ray apron material on the costal margins serve well with ^{99m}Tc, but a heavier absorber would be required for ^{131}I. Such a marker absorbs a sufficient proportion of the ^{99m}Tc gamma rays to appear in the images as a dark line (Figure 21). These are used only in an anterior supine view, so that the liver position may be assessed in the same position as during palpation.

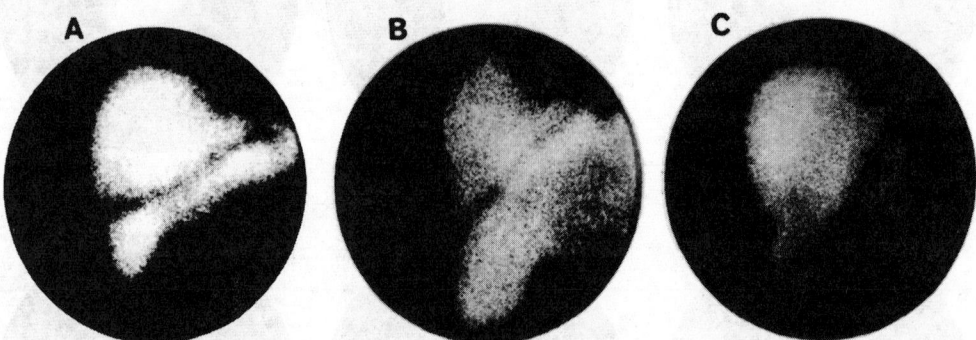

FIGURE 21. Absorber Marker on the Liver.

Figure 21 represents anterior-view scintiphotos in two patients after liver labeling with 3.5 millicuries of ^{99m}Tc-sulfur colloid. The costal margin and the xiphoid process are marked with X-ray apron strips one-fourth-inch wide, taped to the skin over the area to be marked. View A demonstrates the liver position to be rather low-lying in an elderly man. Views B and C respectively were made before and after combined radiation and chemotherapy for seminoma metastatic to the liver. The costal-margin marker in view B demonstrates the position of the bulk of the liver below the level of the costal margin, and in view C, after a dramatic response to the combined therapy, above it.

Small surface depressions may be noted in the liver image where the costal margin flares. These are generally best seen in an anterior view. A Reidel's lobe frequently will show an area of decreased labeling above the tip of the lobe; this is just thinning of the lobe above the tip. Sometimes the gall bladder impression is observed in anterior liver views. If such a defect appears unusual, it may be studied directly by a Rose Bengal liver imaging procedure conducted so as to show the gall bladder area after labeling of the gall bladder has occurred, normally within 30 minutes post injection.[20] These are generally the only surface defects observed in a normal liver image. Figure 22 illustrates some of these surface depressions and indicates the surrounding viscera that are directly related to liver surfaces. Internal normal anatomy of the liver that may be seen is principally related to the porta hepatis, where anteriorly there is a considerable unlabeled volume related to the portal vein, hepatic artery, and major bile ducts; farther posteriorly there is the similar large unlabeled volume of the hepatic veins and the inferior vena cava impression upon the posterior surface of the liver. All these contribute to an ill-defined labeling defect often seen in anterior gamma-image views in the central areas of the image (Figure 22).

The views illustrated in Figure 22 are all of liver labeling with 3.5-millicurie doses of ^{99m}Tc-colloid injected intravenously, study D, E, F being a series of rapid-sequence views during injection of the colloid.

A, B, C.

View A is anterior, View B right lateral, and C posterior, showing the effect of external pressure by a costal margin that was sharply flared in an 83-year-old male. This is the same study illustrated in Figure 21A, where a costal-margin marker is shown to correspond with the defect.

D, E, F.

Selected gamma photos from a series of rapid-sequence, 10-second exposures after peripheral intravenous injection of the usual dose of technetium colloid for liver labeling. The patient is in upright position, and the camera is viewing the area of the superior aspect of the left lobe and the cardiac impression upon the liver. In this view, any collection of pericardial or pleural fluid and any subdiaphragmatic abscess would be evident as areas in which there is no labeling during the entire sequence. Accurate measurements of the difference between the margins of the cardiac-chamber images, the lungs, and the liver have shown that there is always less than the equivalent of a one-inch difference in each of these measurements in normals. View D shows a predominately right heart and pulmonary arterial arborization pattern, with the left ventricular region showing very little labeling; view E shows lung and left-heart labeling, with the aorta indicated by the marker; view F shows a late phase, after considerable labeling of the left lobe of the liver had occurred. This is a normal perfusion study.

G, H, I.

Views G and H respectively are anterior and right lateral views in a patient with liver congestion due to right heart failure and a loop of large bowel around the right lobe of the liver. This bowel is distended and apparently caused the unusual appearance of the external contour of the liver image, since view I, taken two weeks later, showed a considerable decrease of liver size in the anterior view after correction of the heart failure and relief of the bowel distention. The cause of the small labeling defect is unexplained, although it may relate to the dilation of the porta hepatis. In particular, the change in appearance in the left-lobe region should be noted, which either shows a cut-off or a disproportionate decrease in size and labeling. No explanation for this change was obtained clinically. The size calibration of the pictures was unchanged in the interval between the two studies.

J, K, L.

These show in an anterior, right lateral, and posterior view the effect of an abscess that is

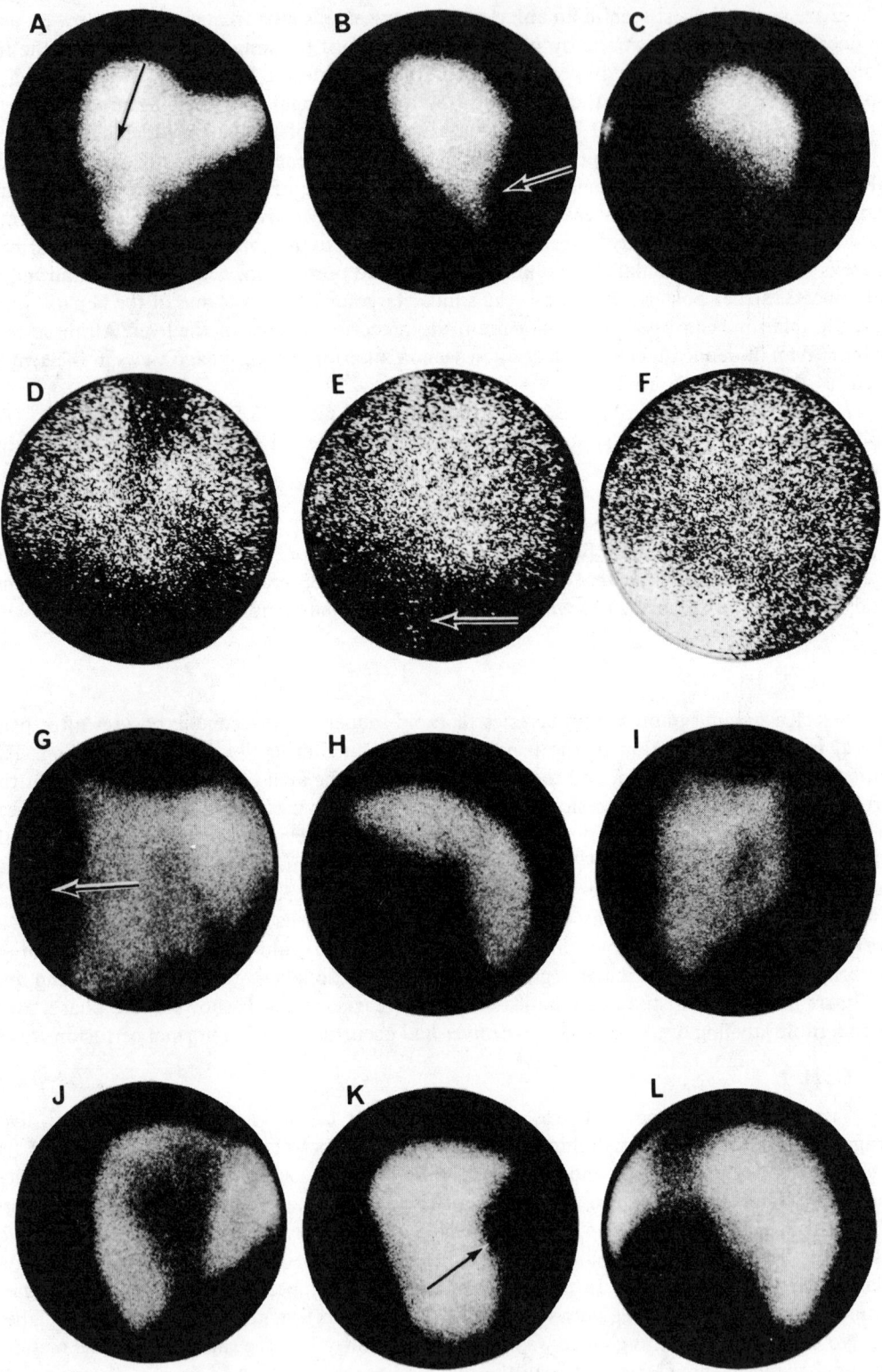

FIGURE 22.
Surface Impressions and Relation to Surrounding Viscera.

extrahepatic, but sufficient in size to cause attenuation of gamma rays and an apparent intrahepatic defect in the anterior view. The posterior view shows considerable liver labeling in this area, and in the lateral view it can be seen that the abnormal area is quite far anterior.

Liver Imaging Techniques

Colloid labeling of the liver occurs quite rapidly after intravenous injection, so that gamma imaging may be started soon after administration. If scanning techniques are used with ^{99m}Tc or ^{113m}In, the scan should be programmed so that image data density approximates 2,000 counts per square centimeter in at least half the image area. This restricts the scan to a scanning speed approximate to available count rate with the collimator used. ^{198}Au-colloid, ^{131}I-Rose Bengal, and ^{131}I-micoaggregated albumin provide lower available count rates, and usually a compromise must be made between desirable data density in readout and available count rate.[21] Data density of approximately 600 counts per square centimeter is usually considered acceptable in these scans with lower available count rate.

The patient should be positioned for a minimum of target motion during the procedure.[22] This may involve use of an abdominal support belt for abdominal breathers as well as patient instruction. The instrument : patient orientation must also be carefully fixed prior to beginning a view. If an instrument with a detector that will view upwards is used, there is theoretical advantage in using the detector in this position. The considerable weight of the liver is sufficient to shift more of the organ toward the detector. This is important with the relatively short effective focal lengths of focused collimation systems; it is also important with gamma cameras, where resolution falls off with distance from the target. Such positioning would assume particular importance in studying individuals with rigid chests, increased anteroposterior diameter, and flared costal margins, which would all serve to introduce a considerable distance between the liver and detector in the anterior view of a supine patient.

Numerous authors have emphasized the importance of multiple views in thorough examination of the liver by scanning and by scintiphotography.[23,24] It is usually possible to obtain anterior, right lateral, and posterior views in each study. With scintiphotography, the spleen should be imaged as well by a left posterior view, and some effort should be made to assess marrow uptake of the colloid in other than the vertebral marrow, which may be seen in the posterior liver view.

If a Rose Bengal liver imaging procedure is used, it is best to begin scanning 10 minutes after intravenous injection of the dye. Scanning is begun below the tip of the right lobe of the liver, so that the gall bladder area will be passed prior to heavy labeling. Where differential diagnosis of jaundice is a primary concern, scanning may be programmed from the diaphragm level caudad and repeated at appropriate intervals to several days.[25] Scintiphotography is usually performed as a dynamic study, with a series of pictures showing regional liver and biliary function.[20] Picture intervals are usually 5 minutes, but simultaneous longer exposures or photographic multiple-exposure copying of the 5-minute scintiphotos should be used to provide better data density. Rose Bengal liver scanning, in addition to showing polygonal-cell function and biliary drainage, serves to differentiate the left lobe of the liver and the spleen, since the spleen is not labeled by this dye, as it is by colloids.[18]

Interpretation of Liver Gamma Images

The general aspects of a normal liver image have been discussed in the section on normal liver anatomy, as imaged by gamma radiation. Abnormalities of labeling with colloid are usually seen as defects of labeling. If large and distinct, focal defects are readily related to autopsy, laparotomy. or needle biopsy findings; such defects imply local destruction, displacement, or devascularization of liver reticuloendothelial cells (Kupffer cells). If a defect is spherical, it implies cyst, noninfiltrating neoplasm (often sarcoma), or abscess without significant surrounding cellulitis (e.g., amebic abscess). Irregular margins of defects suggest neoplastic disease, but as the margin of a defect of labeling and the decrease of labeling become less distinct, the probability becomes greater that hepatocellular disease is responsible for a given defect. Hepatocellular disease typically causes irregularity of labeling throughout the liver image. This must be distinguished from "mottling" due to poor count rate statistics in a scan.[21]

A certain amount of deformity of liver outlines is observed in cirrhosis as scarring progresses; changes in liver size occur with enlargment early, particularly of the left lobe, and decrease late;[26] there is decreased liver uptake of colloid labels, despite known Kupffer cell increase in cirrhosis.[27] This implies shunting of portal venous return or other cause of reduced opportunity for colloid phagocytosis by these cells. Castell documented the decrease of liver labeling with portal–systemic shunting.[28] The shunting results in increased splenic labeling, and eventually in increased marrow labeling. In Castell's study the degree of spleen labeling, and after that the degree of marrow labeling, had a good correlation to ammonia intolerance, which was used as a measure of portal–systemic shunting.

Table 60 outlines findings that may be observed by gamma imaging of the liver.

Table 60. FINDINGS IN LIVER GAMMA IMAGES

Liver Outline
    Size, relative lobe size
    Position
    Extrinsic pressure defects
    Developmental anomalies

Liver Structure
    Nonuniform labeling
    Focal defects of labeling

Liver Hemodynamics
    Liver labeling relative to spleen, marrow labeling
    Lobular defects due to infarction

Polygonal-Cell Function (Rose Bengal studies only)

Biliary Drainage (Rose Bengal studies only)

Attention is also directed to the data of Root, where otherwise normal livers in moribund patients given gold colloid several hours prior to death were found to contain 60 to 94 percent of the colloid administered, the spleens 1.2 to 16 percent, and the marrow approximately the same as the spleen range (10 percent).[29] Rate of disappearance of the colloid particles is a function of particle size, smaller particles disappearing more slowly, and of hepatic blood flow.[30,31] The normal rate of radiogold disappearance from the blood is reported at 14 to 27 percent per minute, suggesting that most scans may be started only a few minutes post injection, but that better opportunity for localization should be allowed in severe cirrhosis.[32,33]

A, B, C.

Views A and B show different exposure intensities of an anterior view of an individual, with active viral hepatitis, and view C the same individual two weeks later, after some improvement had occurred. A patchy decrease of labeling is seen in the earlier study, particularly in the low-intensity exposure, and this patchy decrease of labeling seems to have cleared to some extent in the later study.

D, E, F.

These are Rose Bengal studies done on the same patient as the illustrations above. View D shows a half-hour exposure after injection of 200 µCi of ^{131}I-labeled Rose Bengal. Excretion into the small intestine is seen at the marker, but there is no labeling of the gall bladder, which should be seen as a very bright focus just lateral to the area of intestinal labeling. View E, taken 5 hours later, shows loss of labeling from the liver and bright labeling within the bowel. View F, at 24 hours post injection, shows labeling of the hepatic flexure of the colon and the transverse colon, with a faint residual labeling of the liver. These studies indicate that there is nonfunction of the gall bladder, but no extrahepatic biliary obstruction, to account for the episode of jaundice.

G, H, I.

Views G and H respectively are an anterior and right lateral view of a liver involved by carcinoid metastases, which appear as discrete spheroidal areas of decreased liver labeling. There is apparently very little involvement of the liver beyond the margin of each focal area of decreased labeling. Also, there is surprisingly little liver enlargement for the amount of volume displaced. The arrows indicate the margins of a faintly visualized large metastasis, as seen anteriorly. View H shows the large metastasis in a lateral view as a very distinct posteriorly located defect. View I is an anterior view of a normal to slightly enlarged liver with a focal defect in the left-lobe area. This is the same liver that was shown earlier in the blood flow study (Figure 20, D, E, F) and shows that the reason for this patient's abdominal pain was not a suspected subdiaphragmatic abscess, but rather an intrahepatic cellulitis and abscess in the left lobe.

J, K, L.

View J shows focal lesions in the grossly enlarged liver of a patient with polycystic disease of the liver with considerable hepatic pain. These lesions have discrete margins with rounded contours and little evidence of involvement of surrounding liver, except by contiguous defects. Views K and L are anterior views of a patient with massive involvement of the liver by a giant hemangioma. View K is of the right upper quadrant, with the arrow showing the approximate position of the xiphoid process; view L is of the left upper quadrant, showing the spleen and better labeling of the left lobe of the liver than of the right lobe, which is more involved in the hemangioma.

M, N, O.

View M shows the anterior view of a liver with involvement by metastatic melanoma. Focal lesions are difficult to identify, but there is an apparently infiltrative process involving the lower half of the liver image. Views N and O are identical, showing different exposure intensities in a left lateral view. These bring out a definite round lesion lying in the anterior surface of the right-lobe image, as viewed from the left lateral. The marker indicates the position of the lesion.

P, Q, R.

View P is an anterior view of the right lung base, the diaphragmatic area, and the superior aspect of the right lobe of the liver of a patient with a giant septic abscess involving the right lobe, as indicated by the marker. Slight lung labeling often occurs with technetium colloid. It should be noted that there is no interruption between the slight amount of labeling in the lung base and the subdiaphragmatic part of the right liver lobe, indicating that there is no subdiaphragmatic collection or abscess and no pleural effusion seen in this upright position. View Q shows a right lateral view of the large abscess occupying the superior aspect of the right lobe. View R was obtained several weeks later, after surgical drainage and antibiotic therapy had greatly reduced the size of the abscess; the lesion, seen as an ill-defined area of decreased labeling, is indicated by the marker.

S, T, U.

These are anterior views in a patient with scleroderma, a stasis malabsorption syndrome, and marked elevation of alkaline phosphatase and BSP retention without more than mild change in SGOT and other liver function tests. Views S and T, obtained with a parallel-hole multichannel collimator, present exposures of different intensity and show many irregularities of labeling; view U was obtained with divergent-hole collimation and shows the whole liver and

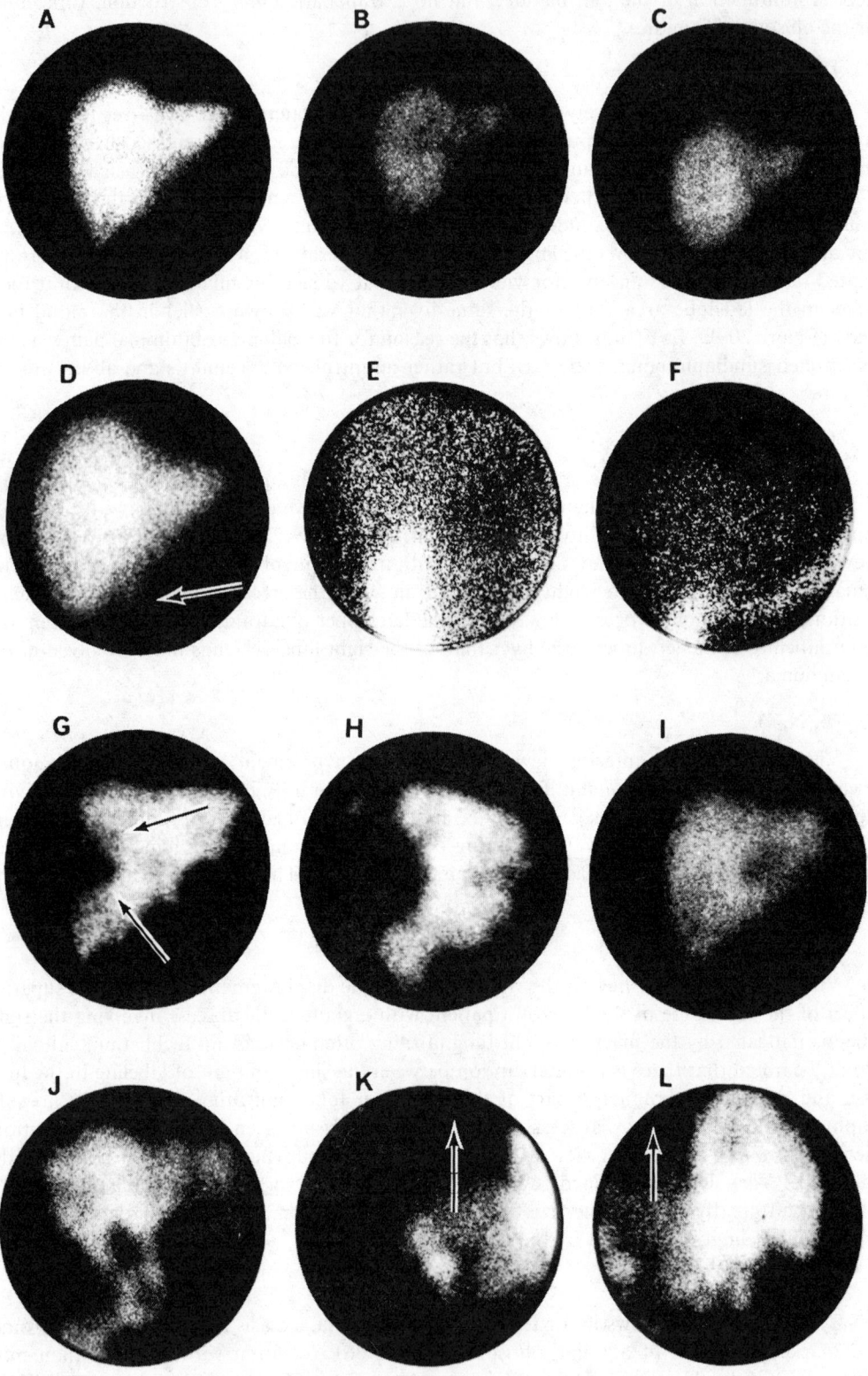

FIGURE 23.
Illustration of a Variety of Liver Labeling Abnormalities.

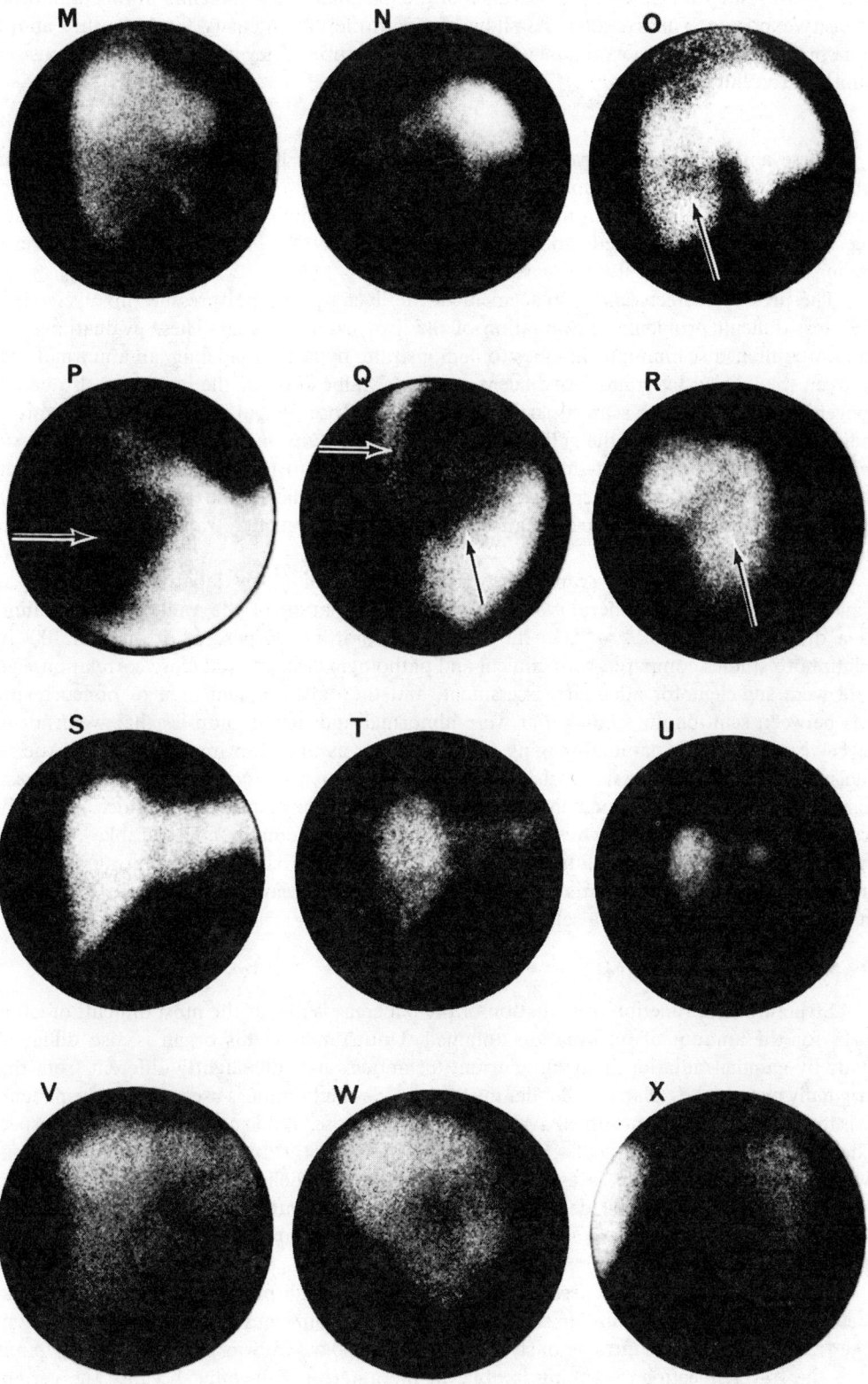

FIGURE 23. (*Continued*)
Illustration of a Variety of Liver Labeling Abnormalities.

spleen with somewhat different appearance of the irregularities of labeling in this diminished-intensity exposure. The irregularities show best with the low-intensity film exposure and are more prominent than they would be with ordinary exposures. They are thought to be consistent with hepatocellular disease.

V, W, X.

These anterior, right lateral, and posterior views show hepatomegaly, slightly irregular labeling, a prominent porta hepatis, and no focal defects in a liver with chronic passive congestion due to rheumatic heart disease and heart failure. The liver has overall decreased labeling, and the spleen increased labeling, as is seen in view X. This is a frequently observed finding in the presence of liver disease.

The pressure defects and displacement of the liver by external pressure involve some of the most difficult problems in evaluation of the liver gamma image. These evaluations often call for combined scanning techniques to demonstrate, by lack of labeling, an abnormal space between the labeled liver and an adjacent organ.[34] Malposition of the liver may also be on a congenital basis; the liver scan is particularly useful if Rose Bengal is used to demonstrate the major lobe of the liver and the gall bladder on the left in *situs inversus*, or on the right in *situs solitus* associated with a right-sided cardiac apex. The first of these conditions, *situs inversus*, is associated with about 5 percent incidence of cardiac anomalies, while the second, *situs solitus*, is usually associated with cardiac anomalies, often a corrected transposition, and other problems.[35]

Correlations of colloid scan abnormalities with findings by laparotomy, biopsy, and clinical evaluation have generally shown strong confirmation of the value of liver scanning as a diagnostic method.[18,26,36] In the author's laboratory, 90 percent of the first 100 liver scintiphoto studies compared with clinical and pathologic data showed close correlation where data were sufficient for adequate assessment, and the only frequent area of noncorrelation was between scintiphoto studies that were abnormal and needle biopsies that were normal. Liver visualization by gamma-imaging techniques has assumed an important diagnostic role and is devoid of patient risk or discomfort. It is efficient in its contribution to diagnosis of right-upper-quadrant mass or pain, liver metastasis, and mechanism of jaundice. Serial liver scans are useful in following response to therapy. There remains considerable latitude for improvement in minimal resolution available with current procedures. Promising approaches are offered by Kuhl and Edwards[37] in cylindrical and section scanning, and by Cavalieri et al.[38] in the use of ^{75}Se-selenite as a selective label of neoplasms.

PANCREAS IMAGING

Structural and functional evaluation of the pancreas is one of the most difficult of clinical tasks for the amount of information obtained. Unfortunately this organ is also difficult to study by gamma-radiation imaging. Current techniques are only slightly different from those originally proposed by Blau and Bender in 1962.[39] ^{75}Se-methionine is used to label the pancreas. It is the only radiopharmaceutical available for the purpose, and is not ideal in several respects. The 128-day physical half-life of ^{75}Se is not proportional to the duration of the study, although whole-body radiation dose is estimated at only 0.6 rad, delivered over several months.[39] Another disadvantage is the degree to which the selenomenthionine enters all pathways of amino acid metabolism. There is labeling of many structures around the pancreas, particularly the liver.

The technique originally used was to first provide a high-protein, fat-free breakfast, and then—two hours later—stimulate secretion of pancreatic enzymes by injecting pancreozymin. The ^{75}Se label was given intravenously an hour later, and a scan was performed thirty minutes after the tracer injection. Scanning is done in the anterior view only, because the vertebral column causes sufficient gamma-radiation absorption to interfere with posterior views. Pancreas visualization was obtained in approximately 75 percent of the cases originally reported, and the technique appeared to have some clinical utility.[39-42] Some improvements in technique have evolved, so that most pancreas studies are now conducted without use of adjuncts, either pancreozymin or anticholinergics; the label is simply injected one half-hour after breakfast, and scanning is started immediately.[43-46] Difficulty with overlap of the liver edge and pancreas image has been minimized by a variety of techniques. Sodee proposes elevating the left side of the patient 6 inches, so as to cause change of the liver edge position.[46] Inclination of the detector of the gamma camera so that the parallel-hole collimator direction of view is cephalad and 5° to the patient's right will usually cause effective separation of liver and pancreas images.[44] Image subtraction techniques have also been used to eliminate liver interference.[47]

Gamma imaging is continued until a satisfactory normal study is obtained or until adequate definition of abnormality is obtained. There is some change in appearance of the images obtained at different intervals after label injection, because both accretion and loss of label from the pancreas occur as enzymes are synthesized and secreted. The average study requires ninety minutes, enough time for two scans or nine scintiphotos. Detector : patient orientation should be kept as constant as possible through the study, so that images may be compared and photographically superimposed for greater data density.

The normal pancreas image varies considerably in outline, but in general a head and a tail may be distinguished. The tail appears smaller than the head in scans.[45] Gamma photos show prominent tail labeling, particularly late in the study.[44] The difference may be related to the fact that the tail is often too distant from the focal point of scanning collimators for efficient imaging. The brighter labeling of the tail probably is related to slower drainage of enzymes from the tail, possibly due to a degree of obstruction in the thin neck area where the abdominal aorta crosses the pancreas; this neck area is demonstrated as an area of narrowing and diminished pancreas image width, often with vertical lateral borders, that can be shown to correspond with the aorta position by rapid-sequence aortic-flow-pattern studies.

Examples of pancreas imaging are shown in Figure 24. These scintiphotos were obtained by injection of 100 to 150 µCi of ^{75}Se-labeled methionine one half-hour after a high-protein, low-fat breakfast. The pancreas is photographed in an anterior view; multiple exposures are obtained in sequence without change of the camera : patient orientation, so that several images may be photographically superimposed for production of a higher-data-density image.

A, B, C.

Views A and B are early and late 10-minute exposures during a 90-minute observation of accumulation and loss of ^{75}Se-methionine label from a normal pancreas. View C is a photographic superimposition of nine views obtained during the entire study, providing a data density of somewhat over 500,000 counts in the one scintiphoto. The identifiable anatomy of the pancreas consists of the head and the tail, separated by a somewhat attenuated neck. The neck area is normally less well visualized than the head and the tail. The marker indicates the neck and is aligned in the approximate position of the abdominal aorta. In late scintiphotos the tail often appears to be somewhat more brightly labeled than the head of the pancreas, probably due to better drainage of secretions from the head than from the tail. The labeled muscles in the patient's flank, the bladder, and labeling in the liver are also imaged.

D, E, F.

View D shows the pancreas in a patient with clinically documented chronic pancreatitis. The anatomy is grossly distorted with several dark areas, which may relate to pseudocysts. Since the patient has done relatively well for two years following this study, the diagnosis of malignancy is thought to be excluded, although laparotomy was not performed. Views E and F demonstrate anatomy related to the pancreas in view D, the bright spot marker being on the xiphoid process; view E shows the position of the left heart and abdominal aorta during injection of technetium colloid, view F the position of the left lobe of the liver after localization of this label. Since camera : patient orientation was unchanged throughout, views D, E, and F may be compared.

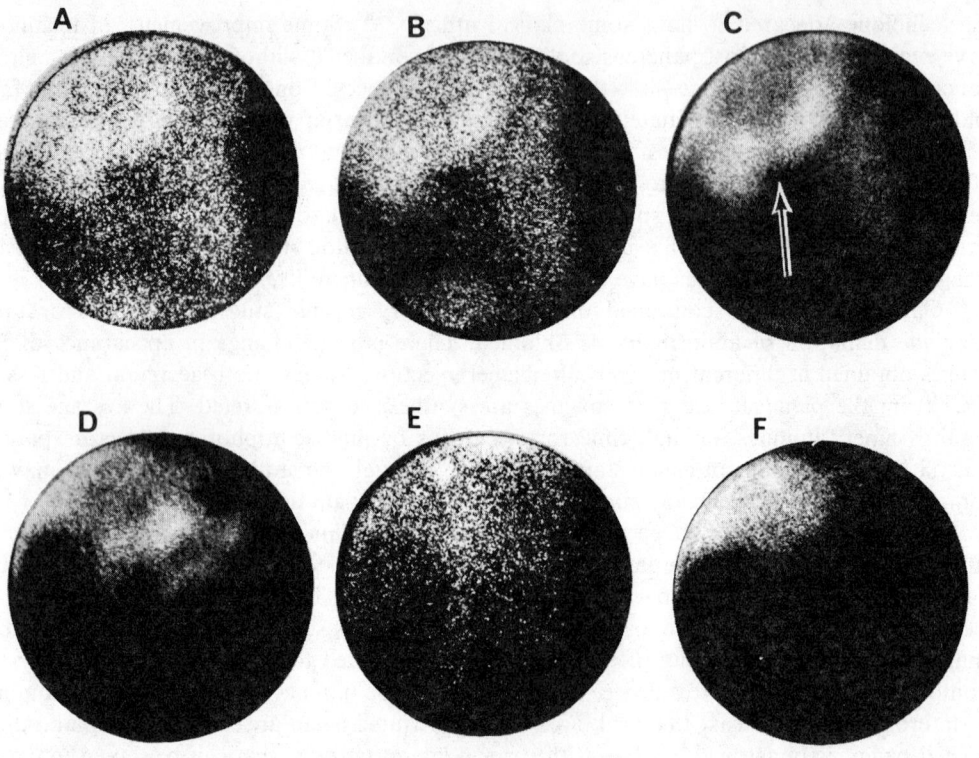

FIGURE 24. Pancreas Scintiphotography.

While the technique does provide useful information in experienced hands, diagnostic efficiency is limited.

SALIVARY-GLAND IMAGING

Recent evaluation of ^{99m}Tc-pertechnetate labeling of the salivary glands during brain scanning suggests that scanning of these glands may occasionally prove to be a very useful study.[48] Pertechnetate is concentrated by the salivary glands, just as is iodide. The particular case in point would be that of the painful, enlarged salivary gland. A pertechnetate scan might provide useful diagnostic data by positive labeling of function in an obstructed gland, or by absence of labeling in an area of neoplasm. It should be remembered, however, that palpable salivary-gland induration may be localized inflammation and, though "cold" in an initial scan, may prove to return later to normal function. Salivary-gland labeling is readily identified in scintiphotos, but conventional scanning or single-pinhole-collimation gamma-camera techniques would be suggested for best correlation of palpable anatomy with the gamma image.

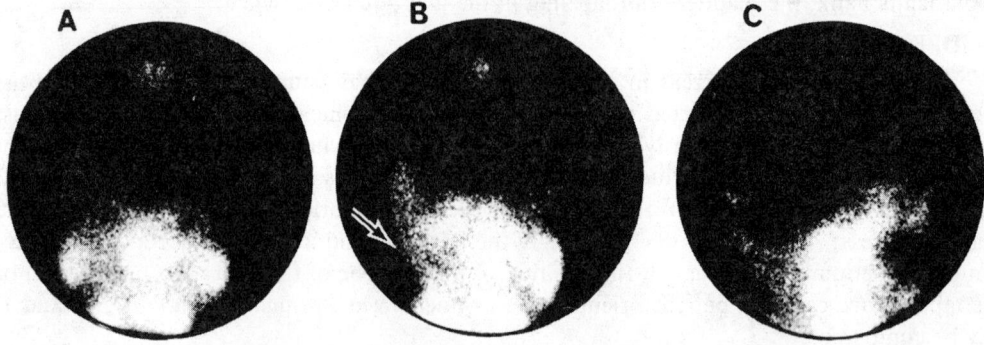

FIGURE 25. Salivary Glands Visualized with ^{99m}Tc-Pertechnetate.

Figure 25 demonstrates the appearance of the various salivary glands after pertechnetate labeling. Although salivary glands are visualized most often during brain imaging studies using 10 to 15 millicuries of ^{99m}Tc-pertechnetate, adequate studies may be performed with considerably less label immediately post injection. View A is an anterior view showing the parotid, submaxillary, and sublingual salivary glands, as well as nasal and other midline labeling. Views B and C show an abnormal study, with involvement of the right parotid gland by neoplasm and complete absence of labeling.

IN-VITRO TESTS IN GASTROENTEROLOGIC DIAGNOSIS

This section is devoted to tests that have found rather general application in diagnostic nuclear medicine. Such tests are restricted in most laboratories to evaluation of the following: intestinal absorption of three substances, i.e., fatty acids, fat, and vitamin B_{12}; gastrointestinal loss of two materials, namely labeled red blood cells and albumin; and liver uptake of dyes, principally Rose Bengal. As discussed in the introduction to this chapter, a vast diversity of clinical tests of intermediary metabolism involving the G.I. tract and accessory organs may be expected as newer technology is utilized for medical diagnosis. The organ-visualization era of nuclear medicine may be expected to diversify to an era in which tests will become more oriented to tracing specific metabolites during absorption and utilization, and gastroenterology may be expected to become particularly involved. This will occur as whole-body counting, breath analysis for $^{14}CO_2$, and even liquid scintillation counting become better adapted and evaluated for general diagnostic use. Beta counting vastly broadens the scope of available labeled metabolites, and in some instances leads to utilization of better radiopharmaceuticals than those in general use for gamma-counting techniques, as will be commented upon in discussion of fat absorption studies.

Fat Absorption

Decreased fat absorption leads in many patients to striking abnormalities of the stool. Lesser degrees of steatorrhea present more of a diagnostic problem. Normally, less than 5 g of the 50- to 100-g dietary fat content will be found in the stool. Many diverse problems can contribute to steatorrhea: failure of neutral-fat hydrolysis, small-intestinal mucosal abnormality, lack of bile salts necessary for fatty acid absorption, decreased transit time, and anatomic "short-circuiting" of the small bowel as a result of surgery or fistulae. Steatorrhea is also observed in conjunction with exudative enteropathy in which the more prominent finding is loss of serum albumin through the bowel mucosa, and not the malabsorption occurring with the disease. Even suppression of intestinal flora with antibiotics is associated with mild steatorrhea.

Fat absorption tests using radioactive labeled fats have long been performed in efforts to both quantitate and characterize steatorrhea. Stanley and Thannhauser used radioiodinated olive oil to study fat absorption in patients with essential hyperlipemia.[49] Absorption of ^{131}I-iodinated oleic acid was first evaluated by Hoffman[59] in 1953, and absorption of ^{131}I-iodinated glyceryl trioleate by Ruffin et al.[51] in 1956. Actually, iodination changes oleic acid to stearic acid, but since the chemical amount iodinated is small, and since the literature refers to the radiopharmaceuticals as "oleic acid" or "triolein", these designations are used here. Commercial preparations of ^{131}I-labeled triolein and oleic acid were soon available and have been the subject of many evaluations.[52,53] Absorption of triolein could be measured by study of retention through whole-body counting, by determining the amount excreted unabsorbed in the stool, or by following blood levels of radioactivity during absorption and utilization of the fat. Convenience of the determination led to frequent use of the blood-counting methods in earlier studies.[51] Normal triolein absorption is followed by a peak of blood radioactivity 3 to 6 hours post administration of the fat, the peak being approximately 8 to 17 percent of the administered dose. Estimation or determination of the blood volume are required to calculate this percentage of the dose. Several hourly determinations of the blood counts must be obtained for best results, since peak time varies considerably with gastrointestinal motility. Administration of the fat in a commercially prepared test capsule with milk as a test meal appears to give satisfactory results.[52] The thyroid is blocked with Lugol's solution during the test. Despite the appeal of counting blood specimens, and despite careful attention to procedural details, the blood levels of radioactivity have never been found entirely acceptable as reflections of fat absorption and are very infrequently used.[52]

Stool radioactivity measurements offer much better correlation with chemical determinations of fat absorption. Stool measurements of unabsorbed radioactivity usually require a 72-hour collection, and the specimen must be uncontaminated by urine. Careful attention must be given to sample preparation for counting. The stool is a two-phase system of oil and water; uniform emulsification must be accomplished prior to aliquoting for well counting, so whole-stool counting is the more usual technique. Normal fecal levels of radioactivity are considered to be less than 5 percent of the administered dose, and usually are 1 or 2 percent.[52] Although a 72-hour stool collection usually suffices, a charcoal marker may be used to signal complete collection.

Once steatorrhea is detected by triolein absorption abnormality, the defect may be further evaluated by determining whether absorption is corrected; this is done by presenting the small intestine with hydrolyzed fat. Two general methods have been used. Most frequently, the triolein test (or other neutral-fat test) is followed by a similar absorption test, using radioiodinated oleic acid;[50,54,55] if normal absorption ensues, this is evidence for defective hydrolysis of triglyceride. The other method sometimes used is more direct; the triolein test is repeated with addition of pancreatic extract; if this corrects triolein malabsorption, then the defect is attributed to pancreatic-lipase insufficiency;[56] pancreatic lipase has a pH optimum of 8.0 and may appear to be lacking in the presence of uncompensated gastric hypersecretion. Again, it should be emphasized that decreased fat absorption occurs with biliary obstruction not involving the pancreatic secretions, since bile salts are necessary for absorption. The various tests assume a normal intake of 50 g of fat per day.

There have been reports concerning lack of correlation between blood radioactivity levels and fecal fat content in milder cases of steatorrhea.[57,58] Commercial ^{131}I-labeled triolein and oleic acid have been shown to contain methyl esters, monoglycerides and diglycerides as impurities in significant amounts—as high as 60 percent of the label in some preparations. Since monoglycerides participate with fatty acids in solution in bile salt micelles during the mucosal phase of fat absorption, and di- and triglycerides do so only to a slight extent (2 to 9% vs 9 to 24%), the monoglyceride absorption could account for considerable effect on blood radioactivity counts.[60] Measuring the unabsorbed residual in the stool might show less evidence of steatorrhea in the presence of monoglyceride impurity. Use of ^{14}C-labeled fat, which is available in a much purer preparation, has been shown to demonstrate minimal steatorrhea much more readily than ^{131}I-labeled triolein.[61] ^{14}C-labeled carbon dioxide may be measured in expired air by various techniques of breath analysis and reflects data similar to the familiar blood radioactivity levels.[62]

The usual test employing commercially available ^{131}I-labeled fat and stool counting is useful and sensitive if performed with attention to detail, but probably does not offer as much sensitivity in distinguishing minimal steatorrhea as chemical analysis of stool fat. When accurate chemical analysis is not available, the radioactive tracer tests have well-documented utility in measuring steatorrhea and some usefulness in distinguishing lipase deficiencies from other causes of decreased fat absorption.

Vitamin B_{12} Absorption

Measurement of vitamin B_{12} absorption is discussed in detail in the chapter on hematologic aspects of nuclear medicine. It would serve the purpose of this discussion of gastroenterologic aspects to comment briefly that, in addition to classical pernicious anemia due to deficient gastric secretion of intrinsic factor, there are a number of conditions with defective vitamin B_{12} absorption and varying manifestations of vitamin B_{12} deficiency. While fish tapeworm (*Diphyllobothrium latum*) infestation and sometimes sprue can have clinical pictures very difficult to distinguish from pernicious anemia, many conditions involving decreased vitamin B_{12} absorption are predominately oriented in the area of gastroenterology or areas other than hematology.[63,64] These other malabsorptive conditions generally do not show any correction of vitamin B_{12} absorption by intrinsic factor. Vitamin B_{12} malabsorption may occur in regional ileitis, small-bowel diverticulosis, blind-loop syndrome, multiple small-bowel strictures, total gastrectomy, and even in myxedema.[65] In addition, there is evidence that in pancreatic insufficiency there may be vitamin B_{12} malabsorption that improves with either alkalinization of duodenal contents or administration of pancreatic extract.[66] Thus, vitamin B_{12} malabsorption is observed most commonly in a broad range of gastroenterologic conditions where bowel organisms or parasites utilize the vitamin in amounts sufficient to cause deficiency or interfere with the vitamin B_{12}: in intrinsic-factor complex, in diseases of the ileum with block of mucosal absorption, and even in relation to pancreatic insufficiency.

Gastrointestinal Clearance of Plasma Albumin and of Circulating Red Blood Cells

Having considered the three tests commonly used to evaluate malabsorption by use of radioisotope tracer techniques, the similar measurements of loss of substances via the gastrointestinal tract are even more restricted in most laboratories, since they are limited to measuring the stool content of labeled plasma proteins and red blood cells. Gastrointestinal loss of plasma proteins may be increased above normal protein clearance rates in a variety of diseases, many of which have no other prominent gastrointestinal features. The protein-losing enteropathies do have a common denominator of enteric mucosal abnormality; the list of prominent conditions showing plasma protein loss includes giant gastric mucosal hypertrophy, gastric carcinoma, sprue, Whipple's disease, regional ileitis, dysgammaglobulinemic (infectious) enteritis, lymphoma of the stomach or intestine, congestive heart failure (especially constrictive pericarditis), nephrosis, intestinal lymphangiectasia, allergic gastroenteritis, and ulcerative colitis. Evaluation of plasma protein loss in these conditions requires stable labeling of a plasma protein, no alteration of the biologic characteristics of the protein, no enteric loss of the protein by any normal process, and absence of reabsorption of the labeled protein or free label if introduced into the gastrointestinal tract. Chromated albumin does not have the same biologic half-life in the plasma as iodinated albumin, which is thought to have a more normal half-time, but is the most satisfactory radiopharmaceutical for measurement of enteric protein loss in other respects, according to extensive studies by Waldemann and his colleagues.[67,68] Simultaneous use of ^{125}I-labeled albumin and ^{51}Cr-labeled albumin is proposed to study plasma albumin degradation rate and fecal loss of albumin label respectively.[69] Normal loss of ^{51}Cr-labeled albumin by the gut is given as 1.73 ± 0.63 g per day (equivalent to 34 ± 12 ml serum).[70] These values for albumin loss are slightly higher than some that have been reported, since collections of feces were delayed until all albumin compartments were labeled, starting four to seven days after intravenous injection of the label; they, therefore, measure loss from the total albumin pool, which is in excess of 300 g—about twice the intravascular albumin.[70]

Similarly, as with plasma albumin loss determination, ^{51}Cr tagging provides for most laboratories the best means to quantitate red-blood-cell loss via the gut. Lacking access to a whole-body counter for measurement of rate of loss of ^{59}Fe after cohort labeling of red cells *in vivo*, red-cell tagging with ^{51}Cr and measurement of loss in the stools provide adequate data to quantitate blood loss by the gastrointestinal tract.[71] As with measuring the loss of ^{51}Cr-labeled albumin, the stool collection should be accomplished when the rate of attrition of labeled red blood cells has become constant, at least four days post administration of the labeled cells. Loss of ^{51}Cr-labeled red blood cells in normal subjects is given as 0.44 ± 0.30 ml erythrocytes per day (equivalent to 1.0 ± 0.7 cc of whole blood).[70]

After intravenous injection of ^{51}Cr-labeled albumin or red blood cells, separation of the label will not greatly affect the results of stool counting. More than half of ^{51}CrCl$_3$ given intravenously appears in the urine, and as much as 14 percent of ^{51}Cr-labeled albumin given intravenously to normal individuals will appear in a 3-day urine collection, but neither exceeded

1 percent in the 4-day stools.[71] Since the ^{51}Cr-labeled-albumin and red-blood-cell measurements of gastroenteric loss of these substances are considerably more satisfactory than those obtained with other generally available labels, combined studies of albumin and red-blood-cell loss are probably best done sequentially, using the ^{51}Cr-label for each measurement.

Liver Function Evaluation with Radiopharmaceuticals

Liver function evaluation can be approached from several aspects with radiopharmaceuticals. Any liver uptake of a radioactive label reflects liver hemodynamics as well as liver cellular function. The same liver labels used in scanning may be utilized in these studies to accomplish labeling of either the reticuloendothelial cells or the polygonal cells. The classical technique of Dobson and Jones[31] employed ^{32}P-colloid to measure liver blood flow, and this was improved by Vetter and his associates[32] by use of gamma-emitting ^{198}Au-labeled colloid for external detection. Since extraction ratios are quite high for colloids (well over 90% on the first pass), the colloids provide considerable information as to liver blood flow, despite difficulties introduced by some trapping of labeled particles in the lungs, particle size variation, and any cause for changed phagocytic function in disease.

Dye studies are used more frequently in evaluation of liver function. Two radioiodinated dyes are available commercially: ^{131}I-labeled Rose Bengal and ^{131}I-labeled iodipamide. The iodipamide has a longer biologic half-life, with various measurements of its clearance by the liver showing rather poor discrimination between normal subjects and individuals with hepatocellular jaundice.[73] This radiopharmaceutical has found more use in cardiovascular scanning and cardiac-output determination.

Rose Bengal dye has found some use in liver function evaluation. Various tests have employed probe counting of the liver, head (blood pool), mid abdomen (bowel), as well as blood levels, Taplin and co-workers having been the first to describe the method.[74] Two approaches have been used sufficiently to require brief description. Lowenstein[75] used single-probe counting over the right liver lobe, but away from the gall bladder, to obtain an uptake : excretion curve, which was plotted on semi-log paper; liver uptake and excretion half-times were derived by graphical analysis, as well as a fractional liver blood volume; quantitative uptake remained normal in obstructive jaundice, but was reduced in hepatocellular diseases, while both rate of uptake and excretion were reduced in obstructive jaundice as well as in hepatocellular disease. Nordyke[76] used a somewhat simpler approach, comparing 20-minute post-injection ^{131}I-Rose Bengal blood levels with 5-minute levels. Normal individuals had ratios of 39 to 51 percent, with a mean of 46 percent, and elevated levels correlated well with BSP-retention testing. Unfortunately, a generally accepted standard method of description of Rose Bengal clearance has not evolved, and there is no ready comparison of Rose Bengal data and BSP data. The use of Rose Bengal for gamma-radiation distribution imaging as well as functional analysis offers an approach with less uncertainty as to the significance of the delay in clearance.[20]

Any discussion of Rose Bengal clearance evaluation must include its use in differentiation of various forms of neonatal jaundice. Ghadimi and Sass-Kortsak[16] report that failure to clear more than 5 percent in a 72-hour stool collection is diagnostic of biliary atresia or other cause of complete extrahepatic obstruction, such as choledochal cyst. The test is quite demanding in that there can be absolutely no admixture of urine in the stool specimens if accurate evaluation is to be obtained. Again, the procedure should be combined with interval scanning or scintiphotography to define any intrahepatic tumor, collection of labeled bile at site of obstruction, or bowel drainage, if this occurs.

REFERENCES

1. Cassen, B., Curtis, L., Reed, C., and Libby, R., Instrumentation for ^{131}I Medical Studies. *Nucleonics*, 9: 46, 1951.
2. Stirrett, L. A., Yuhl, E. T., and Cassen, B., Clinical Applications of Hepatic Radioactive Surveys. *Amer. J. Gastroenterol.*, 21: 310, 1954.

3. Taplin, G. V., Meredith, O. M., and Kade, H., The Radioactive (^{131}I-Tagged) Rose Bengal Uptake Excretion Test for Liver Function, Using External Gamma-Ray Scintillation Counting Techniques. *U.S. Atomic Energy Commission Report UCLA-319.* University of California at Los Angeles, 1954.
4. Atkins, H. L., Richards, P., and Schiffer, L., Scanning of Liver, Spleen, and Bone Marrow with Colloidal ^{99m}Tc. *U.S. Atomic Energy Commission Report BNL-9210.* Brookhaven National Laboratory, Upton, New York, 1964.
5. Goodwin, D. A., Stern, H. S., Wagner, H. N., Jr., and Kramer, H. H., ^{113m}In: A New Radiopharmaceutical for Liver Scanning. *Nucleonics*, 24: 65, 1966.
6. Benacerraf, B., Biozzi, G., Halpern, B. N., Stiffel, C., and Morton, D., Phagocytosis of Heat-Denatured Human Serum Albumin Labelled with ^{131}I, and Its Use as a Means of Investigating Liver Blood Flow. *Brit. J. Exp. Pathol.*, 38: 35, 1957.
7. Quimby, E. H., Some Problems Regarding Permissable Doses with Radioisotopes. *J. Nucl. Med.*, 1: 14, 1960.
8. Taplin, G. V., Johnson, D. E., Dore, E. K., and Kaplan, H. S., Suspensions of Radioalbumin Aggregates for Photoscanning the Liver, Spleen, Lung, and Other Organs. *J. Nucl. Med.*, 5: 259, 1964.
9. Smith, E. M., Internal-Dose Calculation for ^{99m}Tc. *J. Nucl. Med.*, 6: 231, 1965.
10. Harper, P. V., Lathrop, K. A., and Gottschalk, A., Pharmacodynamics of Some ^{99m}Tc Preparations. *U.S. Atomic Energy Commission Symposium Series #6, Radioactive Pharmaceuticals,* p. 335. U.S. Atomic Energy Committee, Division of Technical Information, Oak Ridge, Tennessee, 1966.
11. Patton, D. P., Garcia, E. N., and Webber, M. M., Simplified Preparation of ^{99m}Tc-Sulfur Colloid for Liver Scanning. *Amer. J. Roentgenol. Radium Ther. Nucl. Med.*, 97: 880, 1966.
12. Larson, S. M. and Nelp, W. B., Radiopharmacology of a Simplified ^{99m}Tc-Colloid Preparation for Photoscanning. *J. Nucl. Med.*, 7: 817, 1966.
13. Marsh, F., Lustok, M. J., and Cohen, P. P., Physiologic Studies of Rhenium Compounds. *Proc. Soc. Exp. Biol. Med.*, 45: 576, 1940.
14. Smith, E. M., Smoak, W. M., and Gilson, A. J., Results of Survey of Reactions to ^{99m}Tc-Sulfur Colloid (Letter to Editor). *J. Nucl. Med.*, 8: 896, 1967.
15. Delprat, G. D., Epstein, N. N., and Kerr, W. J., A New Liver Function Test: The Elimination of Rose Bengal, When Injected into the Circulation of Human Subjects. *Arch. Intern. Med.*, 34: 533, 1924.
16. Ghadimi, H. and Sass-Kortsak, A., Evaluation of the Radioactive Rose Bengal Test for the Differential Diagnosis of Obstructive Jaundice in Infants. *New Engl. J. Med.*, 265: 351, 1961.
17. Parker, J. J. and Siemsen, T. K., Liver Regeneration Following Hepatectomy; Evaluation by Scintillation Scanning. *Radiology*, 88: 342, 1967.
18. McAfee, J. G., Ause, R. J., and Wagner, H. G., Jr., Diagnostic Value of Scintillation Scanning of the Liver. *Arch. Intern. Med.*, 116: 95, 1965.
19. Michels, N. A., Newer Anatomy of the Liver and Its Variant Blood Supply and Collateral Circulation. *Amer. J. Surg.*, 112: 337, 1966.
20. Burke, G. and Halko, A., Dynamic Clinical Studies with Radioisotopes and the Scintillation Camera. II. Rose Bengal-^{131}I Liver Function Studies. *J. Amer. Med. Ass.*, 198: 608, 1966.
21. Christie, J. H., MacIntyre, W. J., Gomez-Crespo, G., and Koch-Weser, D., Radioisotope Scanning in Hepatic Cirrhosis. *Radiology*, 81: 455, 1963.
22. Gottschalk, A., Harper, P. V., Jiminez, F. F., and Petasnick, J. P., Quantification of the Respiratory Motion Artifact in Radioisotope Scanning with the Rectilinear Focused-Collimator Scanner and the Gamma Scintillation Camera. *J. Nucl. Med.*, 7: 243, 1966.
23. Czerniak, P., Lubin, E., Dzaldetti, M., and de Vries, A., Scintillographic Follow-Up of Amebic Abscesses and Hydatid Cysts of the Liver. *J. Nucl. Med.*, 4: 35, 1963.
24. Gottschalk, A., Liver Scanning. *J. Nucl. Med.*, 8: 372, 1967.
25. Eyler, W. R., Schuman, B. M., Du Sault, L. A., and Hinson, R. E., Radioiodinated Rose Bengal Liver Scan as Aid in Differential Diagnosis of Jaundice. *Amer. J. Roentgenol.*, 94: 469, 1965.
26. Wang, K. S., Fish, M. B., and Pollycove, M., Evaluation of Hepatic Photoscanning with Radioactive Colloidal Gold. *J. Nucl. Med.*, 6: 494, 1965.
27. Leevy, C. M., Ten Hove, W., and Howard, M., Mesenchymal-Cell Proliferation in Liver Disease of the Alcoholic. *J. Amer. Med. Ass.*, 187: 598, 1964.
28. Castell, D. O. and Johnson, R. B., The ^{198}Au Scan: An Index of Portal–Systemic Collateral Circulation in Chronic Liver Diseases. *New Engl. J. Med.*, 275: 188, 1966.
29. Root, S. W., Andrews, G. A., Kinseley, R. M., and Tyer, M. P., The Distribution and Radiation Effects of Intravenously Administered Colloidal ^{198}Au in Man. *Cancer*, 7: 856, 1956.
30. Zilversmit, D. B., Boyd, G. A., and Brucer, M., The Effect of Particle Size on Blood Clearance and Tissue Distribution of Radioactive Gold Colloids. *J. Lab. Clin, Med.*, 40: 255, 1952.
31. Dobson, E. L. and Jones, H. B., The Behavior of Intravenously Injected Particulate Material: Its Rate of Disappearance from the Blood Stream as a Measure of Liver Blood Flow. *Acta Med. Scand.*, 144, Suppl. 273: 1, 1952.

32. Vetter, H., Falkner, R., and Newmayr, R., The Disappearance Rate of Colloidal Radiogold from the Circulation, and Its Application to the Estimation of Liver Blood Flow in Normal and Cirrhotic Subjects. *J. Clin. Invest.*, 33: 1594, 1954.
33. Wakisaka, G. and Fujita, T., ^{198}Au Clearance and Liver Cell Clearance in Hepatic Injuries. *Jap. Circ. J.* (English ed.), 28: 7, 1964.
34. Brown, D. W., Lung–Liver Radioisotope Scans in the Diagnosis of Subdiaphragmatic Abscess. *J. Amer. Med. Ass.*, 197: 728, 1966.
35. Elliott, L. P., Sue, K. L., and Amplatz, K., A Roentgen Classification of Cardiac Malpositions. *Invest. Radiol.*, 1: 17, 1966.
36. Nagler, W., Bender, M. A., and Blau, M., Radioisotope Photoscanning of the Liver. *Gastroenterology*, 44: 36, 1963.
37. Kuhl, D. E. and Edwards, R. Q., Cylindrical and Section Radioisotope Scanning of the Liver and Brain. *Radiology*, 83: 926, 1964.
38. Cavalieri, R. R., Scott, K. G., and Sairenji, E., Selenite-^{75}Se as a Tumor-Localizing Agent in Man. *J. Nucl. Med.*, 7: 197, 1966.
39. Blau, M. and Bender, M. A., ^{75}Se-Methionine for Visualization of the Pancreas by Isotope Scanning. *Radiology*, 78: 974, 1962.
40. Haynie, T. P., Svoboda, A. C., and Zuidema, G. D., Diagnosis of Pancreatic Disease by Photoscanning. *J. Nucl. Med.*, 5: 90, 1964.
41. Sodee, D. B., Radioisotope Scanning of the Pancreas with Selenomethionine (^{75}Se). *Radiology*, 83: 910, 1964.
42. Burke, G. and Goldstein, M. S., Radioisotope Photoscanning in the Diagnosis of Pancreatic Disease. *Amer. J. Roentgenol.*, 92: 1156, 1964.
43. Burdine, J. A. and Haynie, T. P., Diagnosis of Pancreatic Carcinoma by Photoscanning. *J. Amer. Med. Ass.*, 194: 979, 1965.
44. Powell, M. R., Miale, A., Jr., and Anger, H. O., Pancreas Visualization with the Scintillation Camera. *J. Nucl. Med.*, 7: 372, 1966.
45. Rodriguez-Antunez, A., Filson, E. J., Sullivan, B. H., Jr., and Brown, C. H., Photoscanning in Diagnosis of Carcinoma of the Pancreas. *Ann. Intern. Med.*, 65: 730, 1966.
46. Sodee, D. B., Pancreatic Scanning. *Radiology*, 87: 641, 1966.
47. Kaplan, E., Ben-Porath, M., Fink, S., Clayton, G. D., and Jacobson, B., Evaluation of Pancreatic Disease by Dual-Channel Scanning. *J. Nucl. Med.*, 8: 349, 1967.
48. Grove, A. S., Jr. and Di Chiro, G., Salivary-Gland Scanning with ^{99m}Tc-Pertechnetate. *Amer. J. Roentgenol.* 102: 109, 1968.
49. Stanley, M. M. and Thannhauser, S. B., The Absorption and Disposition of Orally Administered ^{131}I-Labeled Natural Fat in Man. *J. Lab. Clin. Med.*, 34: 1634, 1949.
50. Hoffman, M. C., Radioactive-Iodine-Labeled Fat. *J. Lab. Clin. Med.*, 41: 521, 1953.
51. Ruffin, J. M., Stringleton, W. W., Baylin, G. J., Hymans, J. C., Isley, J. K., Sanders, A. P., and Soluner, M. F., Jr., ^{131}I-Labeled Fat in the Study of Intestinal Absorption. *New Engl. J. Med.*, 225: 594, 1956.
52. Pimparker, C. D., Tulsky, E. G., Kalser, M. H., and Bockus, H. L., Correlation of Radioactive and Chemical Fecal-Fat Determinations in the Malabsorption Syndrome. I. Studies in Normal Man and in Functional Disorders of the Gastrointestinal Tract. *Amer. J. Med.*, 30: 910, 1961.
53. Rufin, F., Blahd, W. H., Nordyke, R. A., and Grossman, M. I., Reliability of ^{131}I-Triolein Test in the Detection of Steatorrhea. *Gastroenterology*, 41: 220, 1961.
54. Malin, J. R., Reemtsma, K., and Barker, H. G., Comparative Fat and Fatty-Acid Absorption Test Utilizing Radioiodine-Labeling Results in Normal Subjects. *Proc. Soc. Exp. Biol.* 92: 471, 1956.
55. Reemtsma, K., Malm, J. R., and Barker, H. G., The Comparative Absorption of Labeled Fat and Fatty Acid in the Study of Pancreatic Disease. *Surgery*, 42: 22, 1957.
56. Polachek, A. A. and Williard, R. F., ^{131}I-Labeled Fat and Pancreatin as a Differential Absorption Test in Patients with Steatorrhea. *Ann. Intern. Med.*, 52: 1195, 1960.
57. Shingleton, W. W., Isley, J. K., Floyd, R. D., Saunders, A. P., and Baylin, G. J., Studies on Post-Gastrectomy Steatorrhea, Using Radioactive Triolein and Oleic Acid. *Surgery*, 42: 12, 1957.
58. Grossman, M. and Jordon, P. H., The Radioiodinated-Triolein Test for Steatorrhea. *Gastroenterology*, 34: 892, 1958.
59. Tuna, M., Mangold, H. K., and Mosser, D. G., Reevaluation of the ^{131}I-Triolein Absorption Test. Analysis and Purification of Commercial Radioiodinated Triolein, and Clinical Studies with Pure Preparation. *J. Lab. Clin. Med.*, 61: 620, 1963.
60. Hofmann, A. F. and Borgstrom, B., Intraluminal Phase of Fat Digestion in Man: Lipid Content of Micellar and Oil Phases of Intestinal Content Obtained During Fat Digestion and Absorption. *J. Clin. Invest.*, 43: 247, 1964.
61. Rothfeld, B. and Rabinowitz, J. L., Comparison of Measurement of Fat Absorption, Using ^{131}I- and ^{14}C-Labeled Fats. *Amer. J. Digest. Dis.*, 9: 263, 1964.

62. Von Schuching, A. L. and Abt, A. F., ^{14}C-Fat Oxidation Test: A New Method for Measuring Fat Utilization in the Human. *Advances in Tracer Methodology*, Vol. 2, Rothchild, S., ed. Plenum Press, New York, 1965.
63. Nyberg, W., Absorption and Excretion of Vitamin B_{12} in Subjects Infected with *Diphyllobothrium Latum* and in Noninfected Subjects Following Oral Administration of Radioactive Vitamin B_{12}. *Acta Haematol.*, 19: 90, 1958.
64. Reisner, E. H., Jr., Gilbert, J. P., Rosenblum, C., and Morgan, M. C., Application of the Urinary Tracer Test of Schilling as an Index of Vitamin B_{12} Absorption. *Amer. J. Clin. Nutr.*, 4: 134, 1956.
65. Tudhope, G. R. and Wilson, G. M., Anemia in Hypothyroidism. *Quart. J. Med.*, 29: 513, 1960.
66. Veeger, W., Abels, J., Hellemans, N., and Nieweg, H. O., Effect of Sodium Bicarbonate and Pancreatin on the Absorption of Vitamin B_{12} and Fat in Pancreatic Insufficiency. *New Engl. J. Med.*, 267: 1341, 1962.
67. Waldemann, T. A., Gastrointestinal Protein Loss Demonstrated by ^{51}Cr-Labeled Albumin. *Lancet*, 2: 121, 1961.
68. Waldemann, T. A., Protein-Losing Enteropathy. *Gastroenterology*, 50: 422, 1966.
69. Kerr, R. M., DuBois, J. J., and Hoh, P. R., Use of ^{125}I- and ^{51}Cr-Labeled Albumin for the Measurement of Gastrointestinal and Total Albumin Catabolism. *J. Clin. Invest.*, 46: 2064, 1967.
70. Beekin, W. L., Clearance of Circulating Radiochromated Albumin and Erythrocytes by Gastrointestinal Tract of Normal Subjects. *Gastroenterology*, 52: 35, 1967.
71. Owen, C. A., Jr., Bollman, J. L., and Grindlay, J. H., Radiochromium-Labeled Erythrocytes for Detection of Gastrointestinal Hemorrhage. *J. Lab. Clin. Med.*, 44: 238, 1954.
72. Mabry, C. C., Greenlaw, R. H., and De Vore, W. D., Measurement of Gastrointestinal Loss of Plasma Albumin: A Clinical and Laboratory Evaluation of ^{51}Cr-Labeled Albumin. *J. Nucl. Med.*, 6: 93, 1965.
73. Freiman, H. D., Cohn, W. E., and Sklaroff, D. M., The Inadequacy of Radioiodinated (^{131}I) Iodipamide as a Liver Function Test. *J. Nucl. Med.*, 3: 63, 1962.
74. Taplin, G. V., Meredith, O. M., Jr., and Kade, H., The Radioactive (^{131}I-Tagged) Rose Bengal Uptake–Excretion Test for Liver Function, Using External Gamma-Ray Scintillation Counting Techniques. *J. Lab. Clin. Med.*, 45: 665, 1955.
75. Lowenstein, J. M., Radioactive-Rose Bengal Test as a Quantitative Measure of Liver Function. *Proc. Soc. Exp. Biol. Med.*, 93: 377, 1956.
76. Nordyke, R. A., Radioiodinated Rose Bengal in Liver and Biliary-Tract Function Testing: A Reappraisal. *Gastroenterology*, 39: 258, 1960.

THE USE OF RADIOISOTOPES IN THE OSSEOUS AND CARTILAGINOUS SYSTEM

Robert D. Ray, M.D., Ph.D., and *Riad Barmada*, M.D.
University of Illinois College of Medicine, Chicago, Illinois.
Research and Educational Hospitals, Chicago, Illinois.
Presbyterian-St. Luke's Hospital, Chicago, Illinois.

INTRODUCTION

Among the problems confronting the human biologist or clinician concerned with the skeletal system and with quantitating skeletal metabolism are the identification of normal processes (does calcium enter the cell?) and the patterns of metabolism and cell activity (life history of cells). Until 1946, radioisotopes were available to only a few investigators, but in recent years, as a result of the large-scale expansion of work in nuclear energy, they have become generally available; this has resulted in a burst of research activity in the fields of medicine and biology. It is now widely accepted that radioisotope techniques provide a unique and powerful research tool for the medical scientist and an increasing range of potentially valuable diagnostic procedures for clinicians in general.

In using the radioactive emitters to study the osseous or cartilaginous systems, one of the first steps is to select an appropriate isotope. Among the factors that must be considered are:

1. The physiological process to be studied; mineral metabolism may be studied with "bone-seeking" isotopes, such as those of calcium, strontium or phosphorus; organic-matrix metabolism may be studied with sulfur and carbon or with tritium-labeled compounds; the cells may be labeled with tritium-tagged thymidine; and the circulation of bone may be investigated with tagged red cells or with iodinated albumin.
2. The radiation characteristics of the isotope; radioactivity depends on the emission of energy in the form of radiation due to instability of the nucleus, and this emission may be in one of the following forms:
   a) *alpha particles:* the alpha-ray emitters are useful in autoradiographic techniques, but they are not advisable for clinical studies.
   b) *beta particles:* the beta-ray emitters are useful for autoradiographic techniques, and also for kinetic studies; identification of beta radioactivity can be carried out with liquid scintillation, Geiger, proportional, or gas-flow counters.
   c) *gamma rays:* gamma-ray emitters have been used for kinetic studies and lend themselves to surface-monitoring techniques with scintillation detectors.

Some of the ways in which radioactive isotopes have been used to study problems related to the osseous and cartilaginous system will be reviewed below. These may be summarized under the following general headings:

1. Tracer Distribution Studies
2. Quantitative Distribution Studies
3. Dilution Studies
4. Kinetic Studies
5. Circulatory Studies

TRACER DISTRIBUTION STUDIES

One of the earliest applications of the tracer distribution studies was in 1938, when Bulliard, Grundland and Moussa, using autoradiograms, demonstrated uptake of ^{32}P by the adrenal gland. In 1940, Hamilton, Soley and Eichorn demonstrated uptake of ^{131}I by the thyroid. Subsequently, studies of the skeletal uptake and distribution of various ions—such as ^{45}Ca, ^{32}P, and ^{35}S—were done; still later, compounds such as glycine and proline labeled with ^{3}H or ^{14}C were used. Recent studies have involved labeling cells and their progeny with tritiated thymidine, which is incorporated into the desoxyribonucleic acid (DNA) fraction of the cell nucleus. Once the nucleus is labeled, the cell can be identified by autoradiographic techniques, and the percentage of cells undergoing desoxyribonucleic acid synthesis at the time of isotope injection can be determined. In addition, as the labeled cell divides, the concentration of tritium in the nucleus is reduced, thus making it possible to follow cell division. This method has been useful in studying the life cycle of the precursors of the osteoblast, osteoclast, and osteocyte, and also in the study of the origin of new bone cells (whether from graft or host) following bone-grafting procedures. In the autoradiographic studies, alpha- and beta-ray emitters are used. Gamma-ray emitters are not used, because of the long distance these rays may travel before undergoing absorption, and because of the resultant poor quality of the autoradiograms produced. For this reason, one should take into account the range of the particles in the tissues, as well as the specific activity and the physical half-life of the radioisotope in the sample, before undertaking any autoradiographic study.

Tracer distribution studies are performed by the following method. After administration of the radioisotope to an organism, the tissue sample is fixed, dried, embedded, and sectioned. The sections should be thin, on the order of 6 to 8 microns. They are placed in close contact with a photographic film or are coated with an emulsion for a desired time, which should be long enough to allow exposure of the emulsion. After development, all the areas corresponding to the location of the radioisotope are darkened; in this fashion an image is produced that provides visualization of the location of the radioactivity in the sample, as is shown in Figure 26.

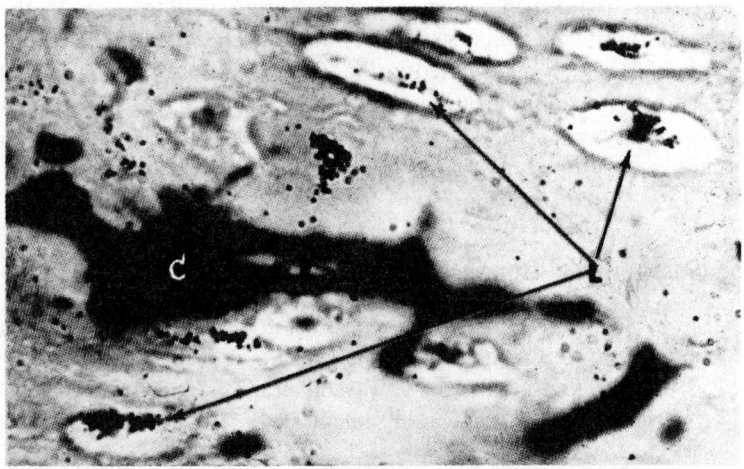

FIGURE 26.
Autoradiographic Distribution and Localization of ^{45}Ca
in Undecalcified Fresh and Devitalized Rat Bone.
(Autografts by Carl D. Salomon, M.D. and Robert D. Ray, M.D., Ph.D.
J. Bone and Joint Surg., 48A: 1575–1584, 1966.)

There are a few problems that must be kept in mind when preparing and interpreting autoradiograms, including loss of histological details and dislodgment or displacement of the tracer with respect to the tissue or cell, autoradiographic artefacts caused by chemical reactions between tissue and emulsion, background radiation and static electricity (especially when long exposures are required), distortion of the emulsion, and localization of the isotope with reference to its exposure image. In conclusion, it should be pointed out that the information obtained by autoradiograms is mainly of a qualitative nature. However, the densities of the autoradiograms may be measured with a densitometer or with high-resolution techniques on a microscopic scale. Quantitative information may be obtained by counting alpha or beta tracks, or by counting the numbers of darkened grains in the emulsion or the numbers of labeled cells.

QUANTITATIVE DISTRIBUTION STUDIES

Investigations involving radioisotope techniques lend themselves to quantitative studies. If one is not interested in histological detail, direct radioassay of representative tissues may be carried out with radiosensitive detectors of various types. Selection of the method and type of detector to be employed depends entirely on the nature of the sample and on the type of activity of the radioisotope used. Fluid samples—such as blood, plasma, and urine—can be counted directly; solid or semisolid samples may have to be homogenized or ashed by either the dry or wet method. Various types of counting containers are available; in choosing one suitable for the material to be counted, it must be kept in mind that the geometry of the system must be reproducible.

Quantitative distribution studies can determine the length of time required for an isotope to reach its destination or the amount of isotope accumulated in bone. Such a study was done by Ray, Thomson, Wolf and LaViolette (Toxicity and Metabolism of Radioactive Strontium (^{90}Sr) in Rats. *J. Bone and Joint Surg.*, 38A: 160-174, 1956), where the plasma concentration of ^{90}Sr and its accumulation in bone as well as the loss of the isotope from the body in the urine and feces were determined by sacrificing groups of animals at different intervals and counting the radioactivity in tissues and excreta. These investigators concluded from their studies that the metabolism of ^{90}Sr following intraperitoneal injection of the dose is very similar to that of ^{89}Sr and ^{45}Ca. The ^{90}Sr was quickly absorbed from the peritoneal cavity into the blood stream and was at first distributed in part to the soft tissue. However, by the end of three hours 75 to 80 percent of the injected dose had been deposited in the skeleton. Up to 15 percent of the injected dose of isotope was excreted in the urine, and up to 8 percent in the feces, during the first twenty-four hours; mobilization of the isotope from the skeleton was not appreciable during the five-day experimental period. Another example is illustrated in Figure 27.

Surface-counting techniques may also be classified under the quantitative distribution studies.

Radioisotope Scanning of Bone

Scanning of body surfaces for detection of radioisotopes can be useful under a variety of circumstances. It has been used for measurements over the thyroid after administration of ^{131}I as well as in cases of suspected metastases of thyroid tumors, and to study a wide variety of other isotopes in various problems, including hematology, cardiology, and liver and kidney metabolism.

Scanning techniques for the study of problems in bone metabolism were used by Bauer and Ray;[1] they were able to follow the distribution of an isotope in the body by surface counting. These investigators used a collimated scintillation crystal for this study and combined the surface counting with assay of the radioactivity of blood, urine, and feces.

The surface-scanning technique requires emission of penetrating radiation, and therefore radioisotopes that are primary emitters of gamma radiation are usually selected. The isotopes ^{47}Ca and ^{85}Sr have been used in studies of localized skeletal lesions;[2] for this purpose a scintillation detector, mounted on a collimation head in such a way that various lead collimators can be added, can be used. The detector is placed with the collimator close to the skin of the area to be counted. When the measurements are made over the limbs, the corresponding sites on the right and left sides are usually both counted to permit comparison of a diseased region with a normal region. In this fashion the counting rate can be obtained and corrected for background. Activity is usually higher over bone lesions such as metastatic tumors, Paget's disease, primary bone tumors, infection, and fractures.

Whole-Body Counting

This also is a type of surface counting where the amount of the retained radioactivity is measured. It eliminates the necessity for counting the activity lost in the urine and feces and subtracting this from the initial dose to arrive at a figure for retained activity. The whole-body counter has the advantage of greater sensitivity and accuracy; much smaller doses of radioisotopes can be administered when this method is used. Unfortunately, whole-body counters are not easily available.

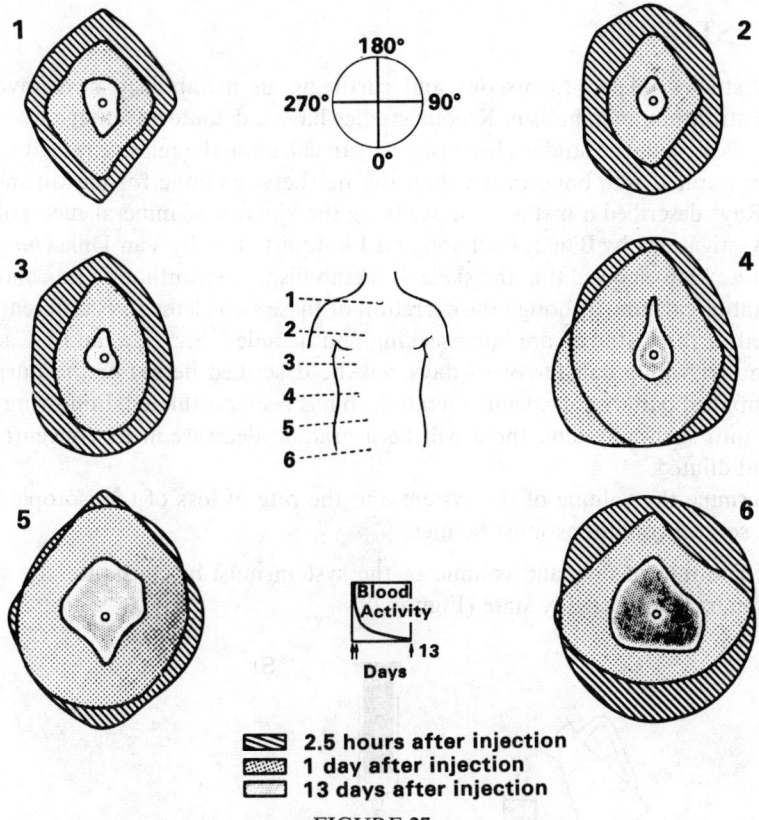

FIGURE 27.
Kinetics of Strontium Metabolism in Man.[1]
(Bauer, G. C. H. and Ray, R. D. *J. Bone and Joint Surg.*, 40A: 175, 1958.)

DILUTION STUDIES

Dilution techniques used to be performed with various dyes. The introduction of radioisotopes into these techniques made them much easier. ^{131}I-tagged albumin, ^{51}Cr-tagged red cells, and ^{32}P-labeled diisopropylphosphate have been used to study plasma and red-cell volumes. Total-body water, sodium, iron, and uric acid have been studied using the isotope dilution methods.

Dilution studies are performed by adding to the system containing the substance to be analyzed a small quantity of the same substance tagged with a radioactive isotope; after allowing some time for mixing, a sample is isolated and analyzed for the tracer. The unknown volume could be calculated from the formula

$$V = \frac{A}{C},$$

where V is the volume of the system or amount of the substance to be measured, A is the amount of the tracer administered, and C its final concentration.

When using the dilution studies, certain assumptions should be considered.
1. It is assumed that the system is in a steady state.
2. The tracer is uniformly distributed after mixing is complete.
3. There is no loss of isotope during the interval between its introduction and its assay.

KINETIC STUDIES

Kinetic studies related to osseous and cartilaginous metabolism have involved mainly calcium and strontium metabolism. Kinetic studies have a definite advantage over the classical calcium metabolic-balance studies, because one can calculate the relative rate at which calcium is being incorporated into bone rather than the net between bone formation and resorption. Bauer and Ray[1] described a method for studying the kinetics of mineral metabolism in man. Previous investigations by Bauer, Carlsson, and Lindquist[4] and by Van Dilla (*Internat. J. Appl. Radiat. Isotop.*, 1956) showed that the skeletal metabolism of strontium in rats and man closely simulates that of calcium, although the excretion of these two elements is different.[5] The radio-isotopes used in these studies are bone-seeking and include ^{85}Sr, ^{45}Ca, and ^{47}Ca; use of ^{85}Sr, a gamma emitter with a half-life of 64 days, will be described here. For the purpose of these studies a simplified pattern for calcium metabolism has been postulated. Following introduction of a tracer into such a system, there will be a gradual decrease in its concentration as it is both lost and diluted.

To determine the volume of the system and the rate of loss of the isotope and its stable hemologue, several conditions must be met:
1. The rate of flow and volume of the system must be constant—the system must be in a steady state (Figure 28).

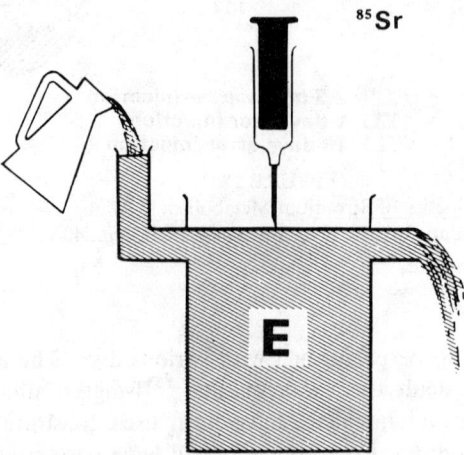

FIGURE 28.
Premise of Isotope Dilution Studies:
Constant Volume–Constant Turnover (Steady State).

2. The mixing or equilibration time must be relatively short in comparison to the time interval in which the rate of loss from the system is calculated.
3. There must be no return or feedback of tracer during the interval under study.

If these assumptions are valid, the concentration of isotope in the system at any time may be described by the following relationship: $C_t = C_0 e^{-Ka}$, where C_t is the concentration at time t, C_0 the concentration at zero time (obtained by extrapolating the curve for concentration, plotted semilogarithmically against time), and K the fractional removal rate. The rate constant K may be determined from the half-life of the tracer in the system as follows:

$$K = \frac{\ln 2}{t^{1/2}} = \frac{0.693}{t^{1/2}},$$

where $t^{1/2}$ is the amount of time required for removal of half of the injected tracer, and presumably the time required for removal of half of the unlabeled hemologue. In the case of bone salt metabolism it would appear that one is dealing with a situation similar to the model described above, except that there are two routes of loss of isotope (excretion and skeletal accretion) rather than one. Following intravenous administration, there is a rapid drop in plasma concentration of a radioactive bone-seeking tracer. Three processes appear to account for this.

1. Diffusion of the isotope from the blood stream into the body fluids and into bone, with establishment of a steady state between the body fluids and those portions of the bone salt in rapid exchange with the body fluids.
2. Loss from the blood stream and body fluids by way of urinary and fecal excretion.
3. Loss of the isotope by incorporation (accretion) into the nonexchangeable fraction of the bone salt.

Once mixing and equilibration have occurred, the subsequent fall in plasma concentration is due to the last two processes alone. At a later interval the bone in which isotope was incorporated by accretion is resorbed or remodeled, and isotope previously deposited is mobilized and recirculated. At this point a break in the plasma concentration curve occurs, due to reentry of isotope into the circulation. However, during the interval between the initial establishment of a steady state and recirculation of the isotope the conditions outlined for the theoretic model are approximated, and the plasma concentration curve approximates a straight line when plotted semilogarithmically against time. In clinical studies the amount of isotope lost from the body can be determined indirectly by measurement of the urine and feces, or directly by whole-body counting.

For purposes of these studies, the isotope remaining in the body may be considered as partitioned between two fractions—one in the exchangeable pool and the other lost from the pool by skeletal accretion. Two unknowns remain: the volume of the exchangeable space, and the rate of loss of isotope by skeletal accretion. The details of the various methods proposed for solving these two unknowns have been presented previously by Bauer, Carlsson and Lindquist,[6] Bauer and Ray,[1] and others.

In performing kinetic studies, the procedure described below is applied. Carrier-free ^{85}Sr, 10 to 20 μCi in 2 or 3 cc of saline, is injected into the antecubital vein of the patient. A plasma activity curve is developed by taking blood samples every 12 hours for 7 days. The heparinized plasma is placed in 2-cc ampules and counted with a sodium iodine scintillation crystal counter or any other suitable counter. By direct comparison with a standard, the concentration of isotope in the plasma in terms of percent of injected dose may be determined. The standard is prepared by diluting a measured amount of ^{85}Sr in a known volume of 1 M NaCl. A 2-cc aliquot of this solution is counted under the same conditions as the plasma samples. The plasma content of ^{85}Sr as a percentage of the injected dose is determined from the plasma concentration by multiplying the value for the latter with that for the plasma volume (determined independently by the Evans blue dye dilution method). Twenty-four-hour urine samples are collected in 1,000-cc plastic bottles. The stools for 2- or 4-day periods are homogenized, diluted to volume, and counted in 2,000-cc glass bottles. The data obtained are expressed in the graphs represented in Figures 29, 30, and 31.

Figure 29 shows the interval of time during which the rate of loss of isotope from the plasma is constant. Extrapolation of this part of the plasma curve, expressed as percentage of injected dose per plasma volume, to the ordinate at zero time should give the theoretic concentration of isotope at the instant of injection (P_{t_0}), thereby correcting for mixing and equilibration. This concentration of isotope is assumed to represent the theoretical concentration at zero time throughout the exchangeable compartment. Although this is not strictly accurate, it simplifies the calculations greatly and should not alter the results significantly. When the injected dose of isotope (100%) is divided by the zero-time concentration of isotopes, the quotient is the theoretic size of the exchangeable pool (E), expressed in plasma volumes. The product of this number and the isotope concentration of the plasma at any time following injection is the total amount of isotope present in the exchangeable compartment at that time ($E \times P_{t_0} = E_t$).

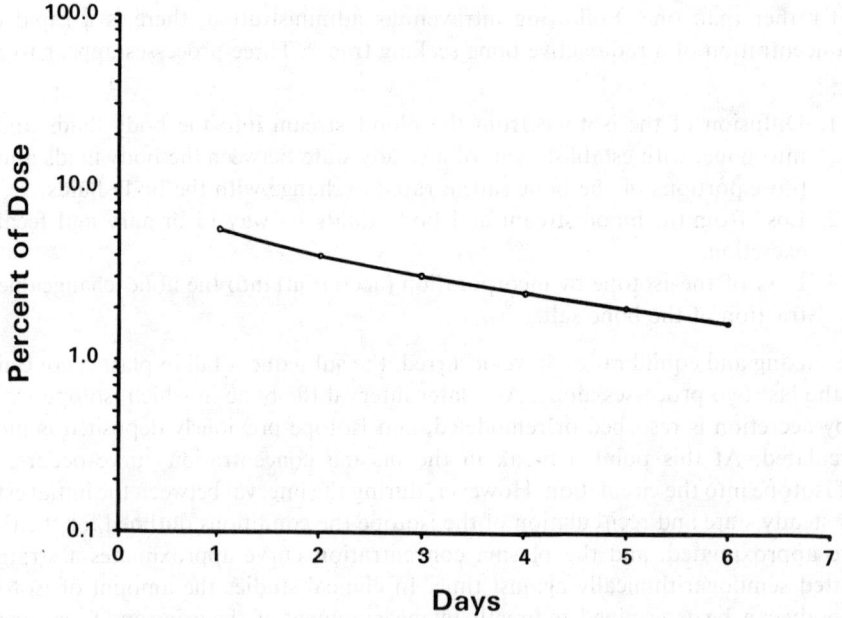

FIGURE 29.
Plasma Concentration of ^{85}Sr Following Intravenous Injection.[3]
(Ray, R. D., Meltzer, W., Lyon, I., and Mensen, E. D.,
Radioisotope Studies of Generalized Skeletal Disorders.
Clin. Orthop., 17: 269–287, 1960.)

The cumulative excretion curves for urine and feces (U) are presented in Figure 30. Subtracting the cumulative excreted isotope, expressed as percentage of dose from 100 percent (the injected dose), will give the amount of isotope retained (R), or the sum of the amount of isotope in the exchangeable compartment (E) and the amount of isotope in the nonexchangeable (or "accreted") compartment (A). The equation may be written as follows:

$$R = \text{injected dose} - U = E + A.$$

If one plots the rate of excretion (percentage of dose excreted per unit of time) against the plasma concentration for the corresponding interval (Figure 31), the excretory clearance rate of isotope (u) may be determined from the slope of the line. By taking two different times on the plasma curve and the figure for the excretory clearance rate, one can calculate the accretion

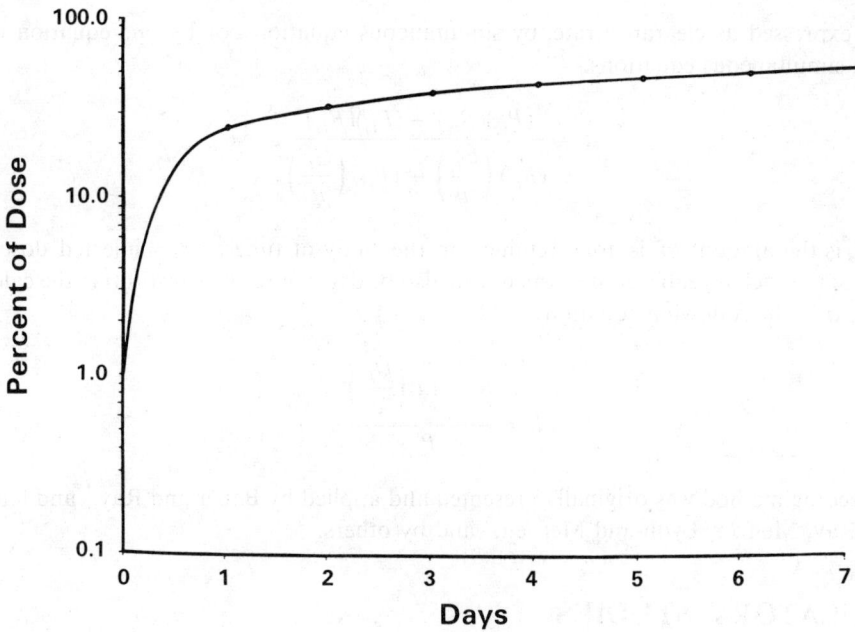

FIGURE 30.
Cumulative Urinary and Fecal Excretion of ^{85}Sr
Following Intravenous Injection.
(Ray, R. D., Meltzer, W., Lyon, I., and Mensen, E. D.,
Radioisotope Studies of Generalized Skeletal Disorders.
Clin. Orthop., 17: 269–287, 1960.)

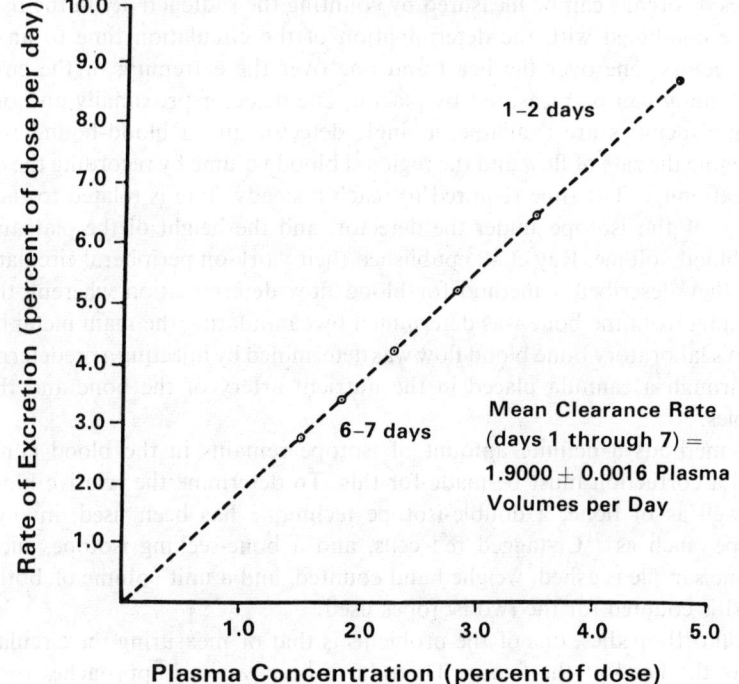

FIGURE 31.
Plasma Clearance of Strontium by Urinary and Fecal Excretion.[3]
(Ray, R. D., Meltzer, W., Lyon, I., and Mensen, E. D.,
Radioisotope Studies of Generalized Skeletal Disorders.
Clin. Orthop., 17: 269–287, 1960.)

rate (a), expressed as clearance rate, by simultaneous equations, or by one equation derived from the simultaneous equations.

$$a = \frac{(P_{t_1})(R_{t_2}) - (P_{t_2})(R_{t_1})}{(P_{t_1})\left(\frac{U_{t_2}}{u}\right) - (P_{t_2})\left(\frac{U_{t_1}}{u}\right)},$$

where R_t is the amount of isotope retained in the body at time t (R_t = injected dose $- U_t$). The size of the exchangeable compartment can also be determined by substituting the calculated value for a in the following equation:

$$E = \frac{R_t - (a)\left(\frac{U_{t_1}}{u}\right)}{P_{t_1}}.$$

The preceding method was originally presented and applied by Bauer and Ray,[1] and later used also by Ray, Meltzer, Lyon and Mensen,[3] and by others.

CIRCULATORY STUDIES

Some of the principles discussed earlier, in particular those related to distribution and dilution of isotopes, have also been applied to measurement of circulation time, local bone blood volume, and flow. If a blood-bound radioactive isotope, such as ^{131}I-tagged albumin or ^{51}Cr-tagged red cells, is administered, and if sufficient time is allowed to permit uniform mixing, the concentration of the tagging agent in the plasma and the volumes of plasma in various tissues or organs can be measured by counting the radioactivity in the tissues. Volume studies can be combined with the determination of the circulation time to an extremity by using two detectors, one over the heart and one over the extremity, or the circulation time within an extremity can be measured by placing one detector proximally and one distally. If rapid-recording facilities are available, a single detector and a blood-bound isotope can be used to determine the rate of flow and the regional blood volume by recording the early build-up curve in an extremity. The time required to reach a steady state is related to the volume rate of blood flow of the isotope under the detector, and the height of the plateau is related to the regional blood volume. Ray et al.[7] published their work on peripheral circulation and bone growth, and they described a method for blood flow determination whereby the amount of isotope returning from the bone was determined by cannulating the main metaphyseal effluent vein. In Copp's laboratory bone blood flow was determined by injecting a predetermined amount of isotope through a cannula placed in the nutrient artery of the bone and then collecting venous samples.[8]

In both methods a definite amount of isotope remains in the blood contained in the marrow, and a correction must be made for this. To determine the relative blood volume of marrow as well as of bone, a double-isotope technique has been used, involving a blood-bound isotope, such as ^{51}Cr-tagged red cells, and a bone-seeking isotope, such as ^{45}Ca or ^{85}Sr. The bone sample is ashed, weighed and counted, and a unit volume of both arterial and venous blood is counted for the two isotopes used.

In clinical orthopedics, one of the problems is that of measuring the circulation of bone, particularly of the head of the femur. There have been various approaches to this problem. One of the first was that of Boyd and Calandruccio,[9] who injected ^{32}P into the circulatory system. Since ^{32}P is a bone-seeking isotope and was not bound to red cells, it diffused out of the bloodstream fairly rapidly. The isotope was injected some time before counts were made of the head of the femur at surgery. The counters used for the purpose were end-window Geiger counters about the size of a Steinmann pin; they were introduced into the head of the femur, and the counts were registered. This was repeated in the trochanteric area in order to obtain a ratio. One disadvantage of this particular technique is that dead bone will pick up a bone-seeking isotope—either strontium, calcium, or ^{32}P. If one tibia of a rat is killed by repeated freezing and thawing and reintroduced into the rectus sheath, the count in such a tibia—compared with the normal tibia, with its blood supply intact—reveals that in the adult rat 70 percent as much isotope is accumulated by the dead as by the living bone. It is, therefore, difficult to draw conclusions when using bone-seeking isotopes that diffuse rapidly out of the circulatory system. The mechanism of isotope uptake by dead tissues is still an open question, whether by accretion or adsorption on the surfaces, by an exchange process between crystals and the surrounding body fluids, or by diffusion, but it is still an observed fact.

A second method for measuring the circulation of the head of the femur was that of Stein, Beller and Nedwich,[10] who placed a counter in the head of the femur and then injected an isotope to determine the circulation time to the head of the femur as well as the rate of accumulation. This gives a little more information than the previous technique, but it still is subject to interpretation errors.

Another method was that proposed by Ferguson and Laing, who injected ^{23}Na into the head of the femur and measured the rate of disappearance. This is a method of assessing venous return and is similar to older venography techniques.

Woodall tried to measure the oxygen tension in the head of the femur, using a platinum electrode; Massie used tagged red cells, injected at the time of surgery, to obtain a ratio between the head of the femur and the trochanteric area, a figure related to the relative blood volumes. One interesting result of this study was that a reduction in blood volume of the head of the femur at the time of surgery was not necessarily correlated with the clinical outcome; however, studies carried out a week later showed a high correlation between delayed and nonunion and avascular necrosis and reduced blood volume, but this, of course, is not much help clinically.

In summary, one must admit that, at this stage of our knowledge, there is no suitable isotope, nor a technique sufficiently accurate, to determine at the time of surgery whether the head of the femur is viable or, not after fracture.

REFERENCES

1. Bauer, G. C. H. and Ray, R. D., Kinetics of Strontium Metabolism in Man. *J. Bone and Joint Surg.*, 40A: 171–186, 1958.
2. Bauer, G. C. H. and Wendeberg, B., External Counting of ^{47}Ca and ^{85}Sr in Studies of Localized Skeletal Lesions in Man. *J. Bone and Joint Surg.*, 41B: 558–580, 1959
3. Ray, R. D., Meltzer, W., Lyon, I., and Mensen, E. D., Radioisotope Studies of Generalized Skeletal Disorders. *Clin. Orthop.*, 17: 269–287, 1960.
4. Bauer, G. C. H., Carlsson, A., and Lindquist, B., A Comparative Study on the Metabolism of ^{90}Sr and ^{45}Ca. *Acta Physiol. Scand.*, 35: 56–66, 1955.
5. Comar, C. L., Wasserman, R. H., Ulberg, S., and Andrews, G. A., Strontium Metabolism and Strontium–Calcium Discrimination in Man. *Proc. Soc. Exp. Biol. Med.*, 95: 386–391, 1957.
6. Bauer, G. C. H., Carlsson, A., and Lindquist, B., Evaluation of Accretion, Resorption and Exchange Reactions in the Skeleton. *Kungl. Fysiogr. Sallsk. Lund Forhandl.*, 25: 1–16, 1955.
7. Ray, R. D., Kawabata, M., and Galante, J., Peripheral Circulation and Bone Growth. *Clin. Orthop. Relat. Res.*, 54: 175–185, 1967.
8. Copp, D. H. and Shim, S. S., Quantitative Studies of Bone Blood Flow in Dogs and Rabbits. *J. Bone and Joint Surg.*, 46B: 781–782, 1964.
9. Boyd, H. B. and Calandruccio, R. G., Further Observation on the Use of Radioactive Phosphorus (^{32}P) to Determine the Viability of the Head of the Femur: Correlation of Clinical and Experimental Data in 130 Patients with Fractures of the Femoral Neck. *J. Bone and Joint Surg.*, 45A: 445–449, 1963.
10. Stein, I., Beller, M. L., and Nedwich, A., Isotopic Measurement of Blood Supply to Upper Femur. *Proceedings of the Conference on Aseptic Necrosis of the Femoral Head, St. Louis, Missouri, January 1964*, pp. 265–296. National Institutes of Health, United States Public Health Service, 1964.

ADDITIONAL REFERENCES

Bertram, V. A. and Low-Beer, M. D., *The Clinical Uses of Radioactive Isotopes*. Charles C. Thomas, Springfield, Illinois, 1950.

Brounner, F., Harris, R. S., Maletskos, C. J., and Benda, L. E., Studies in Calcium Metabolism: The Fate of Intravenously Injected Radiocalcium in Human Beings. *J. Clin. Invest.*, 35: 78–88, 1956.

Budy, A. M., The Use of Radioisotopes in Orthopaedics. II. Application of Radioactive-Tracer Techniques to Bone. *J. Bone and Joint Surg.*, 45A: 1073–1083.

Cohen, J., Maletskos, C. J., Marshall, J. H., and Williams, S. M., Radioactive-Calcium Tracer Studies in Bone Grafts. *J. Bone and Joint Surg.*, 39A: 561–577, 1957.

Copp, D. H., Seminar on Bone Disease, Calcium and Phosphorus Metabolism. *Amer. J. Med.*, 22: 275–285, 1957.

Dow, E. C. and Stanbury, J. B., Strontium and Calcium Metabolism in Metabolic Bone Disease. *J. Clin. Invest.*, 39: 885–902, 1960.

Dymling, J. F., Accretion and Excretory Clearance Rates and Exchangeable Spaces Measured in Man with ^{47}Ca and ^{85}Sr Under Normal and Pathological Conditions. *Medical Uses of ^{47}Ca*. International Atomic Energy Agency, Vienna, Austria, 1962.

Edelman, I. S., James, A. H., Baden, H., and Moore, F. D., Electrolyte Composition of Bone and the Penetration of Radiosodium and Deuterium Oxide into Dog and Human Bone. *J. Clin. Invest.*, 33: 122–131, 1954.

McLean, F. C., The Use of Isotopes in Orthopaedics. I. Instructional Course Lecture, The American Academy of Orthopaedic Surgeons. *J. Bone and Joint Surg.*, 45A: 1067–1073, 1963.

MacLean, F. C. and Budy, A. M., Radioisotopes in the Study of Bone. *Clin. Orthop.*, 24: 178–197, 1962.

Post, M. and Shoemaker, W. C., Method for Measuring Bone Metabolism *in vivo*. *J. Bone and Joint Surg.*, 46A: 111–120, 1964.

Ray, R. D., LaViolette, O., Buckley, H. D., and Mosiman, R. S., Studies of Bone Metabolism: A Comparison of the Metabolism of ^{90}Sr in Living and Dead Bone. *J. Bone and Joint Surg.*, 37A: 143–155, 1955.

Ray, R. D. and Mueller, K. H., The Use of Radioisotopes in Orthopaedics. III. Instructional Course Lecture, The American Academy of Orthopaedic Surgeons. *J. Bone and Joint Surg.*, 47A: 417–425, 1965.

Ray, R. D., Mueller, K. H., Sankaran, B., Mensen, E. D., and Schwartz, T. B., Metabolic Diseases of Bone: Kinetic Studies. *Med. Clin. N. Amer.*, 49: 241–258, 1965.

Ray, R. D., Stevens, J., Lyon, I., and Mensen, E. D., Calcium Metabolism. *Fed. Proc.*, 20: 119–122, 1961.

Salomon, C. D. and Ray, R. D., Concomitant Microradiographic, Autoradiographic and Histologic Observations on Undecalcified Bone. *Stain Technol.*, 39: 373–380, 1964.

Siri, W. E., *Isotopic Tracers and Nuclear Radiations with Applications to Biology and Medicine*, pp. 388–402. McGraw-Hill, New York, 1949.

Tonna, E. A., The Cellular Complement of the Skeletal System Studied Autoradiographically with Tritiated Thymidine (^{3}HTDR) During Growth and Aging. *J. Biophys. Biochem. Cytol.*, 9: 813–824, 1961.

Wendeberg, B., Mineral Metabolism of Fractures of the Tibia in Man Studied with External Counting of ^{85}Sr. *Acta Orthop. Scand.*, Suppl. 52, 1961.

RADIOISOTOPES AND THE HEMATOPOIETIC SYSTEM

August Miale, Jr., M.D., Director, Division of Nuclear Medicine
Georgetown University Hospital, Washington, D.C.

INTRODUCTION

Classical morphological and biochemical studies provided the cornerstone for progress in understanding the physiology, chemistry and pathology of blood. The introduction of radiotracer techniques to modern hematology advanced the understanding of normal erythropoiesis and of disease states even further. The work of Hahn and co-workers[1] stands as a significant example of the utilization of a radiotracer such as radioiron in the understanding of iron metabolism in man. Subsequently other tracers were applied to hematological research, with a remarkable outpouring of knowledge. The applications of some specific radioisotope techniques described in the research literature were designed for specialized purposes, which are impractical in most hospital laboratories. In this chapter emphasis has been directed to those methods that best answer the need for practicality, specificity, and clinical usefulness.

VOLUME OF PLASMA, RED-CELL MASS, AND TOTAL BLOOD VOLUME

Principles

Three approaches are possible: 1) direct measurement to each compartment of blood separately by a tracer designed to measure each, 2) measurement of either plasma or red cells and extrapolation of the other to get total volume, and 3) measurement of both simultaneously. In all cases the dilution principle is utilized. Simply stated, the dilution of a known quantity of tracer is proportional to the volume of diluent.

Methodology

I. Red-Cell Mass and Blood Volume with ^{51}Cr-Labeled RBC (Ascorbic-Acid Technique).

Preparation of Dose Tubes.

1. To sterile Vacutainer tubes (yellow top)* add 2 ml of sterile low-glucose ACD† solution and label.
2. Dilute ^{51}Cr (as sodium radiochromate) with sterile normal saline to give about 35 µCi per 0.5 ml of solution.
3. Draw 0.5 ml of the ^{51}Cr solution into a disposable syringe; cap needle and label the syringe.
4. Draw 10 ml of the patient's blood into the dose tube containing the 2 ml ACD solution (draw another tube for a blank or background calculation if the patient had previous radioisotope study).
5. Add the 0.5 ml of ^{51}Cr to the dose tube and gently invert several times; incubate at room temperature for 15 minutes, with occasional inversion.
6. Add 50 mg of ascorbic-acid solution to the dose tube, and invert gently

---
* Becton-Dickinson, Rutherford, New Jersey.
† Abbot Laboratories, North Chicago, Illinois.

several times; five minutes after the addition of the ascorbic acid withdraw exactly 4 ml of the labeled blood into a sterile, heparinized disposable glass syringe; make sure that the dose has been inverted gently to insure uniform distribution of the red cells.

Procedure.
1. Inject 4 ml of the tagged blood intravenously; extreme care should be taken, to insure that there is no possibility of infiltration.
2. 10, 20, and 30 minutes post injection withdraw sufficient blood with a heparinized syringe or heparinized Vacutainer tube to yield enough whole blood and plasma samples for counting (about 5 ml each).
3. Do duplicate microhematocrits on the tagged whole blood (dose tube) and on the patient's whole-blood sample.
4. Make a 1:100 dilution on the tagged whole blood (dose tube); centrifuge the remaining tagged whole blood (dose tube) and make a 1:100 dilution of the plasma; invert the flask 30 times before withdrawing duplicate 3-ml samples.
5. Withdraw duplicate 3-ml samples of the patient's whole blood; centrifuge the rest of the patient's whole-blood sample and withdraw duplicate 3-ml samples of plasma.
6. Count each 3-ml sample for 4 minutes and record the results on a suitable worksheet.
7. Plot the counts on linear graph paper, with time along the abscissa; draw the best straight line through the three points and read true counts on the ordinate at zero time; use the true-counts value to derive *average Patient's WB cpm/ml value*.

Calculation.

$$TBV = \frac{\text{Vol Inj} \times (\text{av. Std WB cpm/ml}) \times (\text{Dil Fact}) - (\text{av. Std Plas cpm/ml} \times \text{Dil Fact} \times \text{Plascrit})}{(\text{av. Patient's WB cpm/ml}) - (\text{av. Patient's Plasma cpm/ml} \times \text{Plascrit})}$$

RCV = TBV × Body Hematocrit   Body Hematocrit = Venous Hematocrit × 0.91
TPV = TBV × Plasmacrit   Plasmacrit = 1 − Body Hematocrit

II. Blood Volume with ^{51}Cr-Labeled Red Cells (Washed-Cell Technique).

Preparation of Multiple Stock Dose Tubes.
1. Inject about 350 µCi of sodium radiochromate aseptically into a sterile Vacutainer tube (yellow top).
2. With sterile technique, inject 1 ml of heparin (1:100) into the Vacutainer tube containing the sodium radiochromate.
3. Add enough normal saline to this Vacutainer tube to make the total volume of the solution 5 ml. (Note: this will give a concentration of about 35 µCi in 0.5 ml of solution.)

Procedure.
1. Withdraw 0.5 ml of the solution and inject this into nine sterile Vacutainer tubes (yellow top); pipet 0.5 ml of solution and label an expiration date at 14 days (always maintain sterile technique).
2. Draw 10 ml of blood from the patient, avoiding stasis, and carefully expel the blood into the sterile dose tube.
3. Incubate for 45 minutes at 37°C, inverting the tube every few minutes, or at least three times, during incubation.
4. Centrifuge the tube at 3,000 rpm for 10 minutes; withdraw the plasma by aseptic technique, using a sterile 18-gauge spinal needle and 20-ml syringe;

avoid sucking up any red cells; a small film of plasma over the level of red cells is permissible.
5. Reconstitute the blood to the original volume with cold isotonic saline, retaining aseptic technique; centrifuge as before, after slowly inverting the tube a few times.
6. Withdraw the supernatant fluid and repeat the above steps (1—5).
7. Inject exactly 5 ml of this reconstituted blood intravenously. (Note: the syringe should be heparinized.)
8. After 10, 20, and 30 minutes of mixing time withdraw 5 ml of blood from the opposite arm of the patient, avoiding stasis.
9. Perform duplicate hematocrits of the last sample; body hematocrit is approximated by multiplying the venous hematocrit by 0.91.
10. Count a 3-ml sample of the whole blood from each drawing in a well-type scintillation counter.
11. Plot the counts on linear graph paper, with time along the abscissa; draw the best straight line through the three points and read true counts on the ordinate at zero time; use the true-counts value to derive *corrected cpm of WB sample*.
12. Dilute 1 ml of the original suspension of washed RBC's (step 6) to 100 ml with water in a 100-ml volumetric flask (this is the *standard*); the cells hemolyze to ensure complete mixing of the labeled hemoglobin in the solution.
13. Count a 3-ml sample of this standard in a well-type scintillation counter; use blood-type pipets, and be as precise as possible.

Calculation.

$$TBV = \frac{(\text{corrected cpm of Standard}) \times (\text{Dil Fact}) \times (\text{Vol Inj})}{\text{corrected cpm of WB Sample}}$$

$RCV = TBV \times \text{Body Hematocrit} \qquad TPV = TBV - RCV$
$\text{Body Hematocrit} = \text{Venous Hematocrit} \times 0.91$

III. Plasma and Blood Volume Determination Using RISA (^{131}I) (Well Counter Method).

Preparation of Dose Bottles.

1. Using aseptic technique, withdraw 1 ml from a sterile 20-ml bottle of normal saline.
2. Add 1 ml of sterile normal human serum albumin to the bottle and mix well.
3. After mixing, add 40 µCi of RISA (^{131}I), giving a concentration in the dose bottle of 2 µCi per 3 ml of solution; label and refrigerate.
4. This dose bottle should be prepared twice weekly, as the RISA tends to dissociate after a few days.

Preparation of Standard.

1. Using aseptic technique, withdraw exactly 3 ml of solution from the dose bottle and inject into a volumetric flask of 1 liter capacity; add sufficient tap water to bring to volume; mix thoroughly by inverting the flask several times; label and date the flask; this can be used as a standard until a new dose bottle is made up.
2. Count a 3-ml aliquot from this standard with each blood volume determination; always mix the standard solution thoroughly before withdrawal of the aliquot for counting.

Procedure.

1. Using a sterile syringe and needle, withdraw 3 ml of the RISA solution from the dose bottle and place the plastic cap over the needle.

2. Before injecting the dose into the patient, withdraw a 10-ml sample of blood from the patient into a heparinized or oxalated tube; invert the test tube several times to prevent clotting; label as blank.
3. Inject the 3-ml dose of RISA into the patient intravenously.
4. At 10, 20, and 30 minutes after injection of the dose withdraw and label 10 ml of venous blood into a heparinized test tube; invert several times to prevent clotting.

Counting of Samples.
1. Count a 3-ml aliquot of the standard solution for 3 minutes, or to 3% counting error; correct for background and record a net counts per minute of standard.
2. Pipet 3 ml of the whole blood from the blank into a test tube and count for 3 minutes; correct for background and record as net counts per minute of blank.
3. Pipet 3 ml of whole blood from each sample into a test tube and count for 3 minutes; correct for background and record as net counts per minute per sample.
4. Plot the counts on linear graph paper, with time along the abscissa; draw the best straight line through the three points and read true counts on the ordinate at zero time; use the true-counts value to derive *corrected Sample cpm*.

Calculation.

$$TBV = \frac{(\text{corrected Standard cpm}) \times (\text{Dil Fact}) \times (\text{Vol Inj})}{(\text{corrected Sample cpm}) - (\text{corrected Blank cpm})}$$

RCV = TBV × Body Hematocrit   Body Hematocrit = Venous Hematocrit × 0.91
TPV = TBV × Plasmacrit   Plasmacrit = 1 − Body Hemactocrit

IV. Plasma and Blood Volume Determination Using RISA (^{131}I) (Volemetron* and Hemo-litre† Method).

These instruments are suitable for determination of plasma or red-cell volumes individually or of both combined by following directions published by the respective manufacturers. Certain modifications are required to allow simultaneous red-cell measurement with ^{51}Cr and plasma volume with ^{125}I-RISA. It is important to remember that the machine measures only the dilution factor and does not correct the values for body hematocrit.

V. Simultaneous Red-Cell Mass and Plasma Volume Determination.

Procedure.
Add 5 μCi of ^{125}I-RISA to 40 μCi of ^{51}Cr-labeled washed RBC's; inject 20 ml of mixture; measure hematocrit (H) of remaining mixture; dilute the mixture (A) and spun supernatant (E) by 100 volumes; count each at the ^{51}Cr and ^{125}I photopeaks respectively; after mixing in the body, withdraw a heparinized sample (S) at 20 minutes; count whole blood at the ^{51}Cr peak (C), and plasma at the ^{125}I peak (F).

     H = hematocrit of injected mixture
     A = cpm/ml of mixture
     E = cpm/ml of supernatant of A
     S = hematocrit of whole venous blood (after mixing) × 0.91
     C = cpm/ml whole blood (after mixing) at ^{51}Cr peak
     F = cpm/ml whole blood (after mixing) at ^{125}I peak

* Ames Co., Elkhart, Indiana.
† Picker Nuclear Corp., White Plains, New York.

Calculation.

$$RCV = \frac{A \times 2000 \times S}{C}$$

$$TPV = \frac{E \times (1 - H) \times 2000}{F}$$

Clinical Interpretation

The values for normal total blood volume can best be determined in each laboratory, but this is impractical. The Tulane table[3] is widely used and is sufficiently accurate for clinical purposes (see Tables 61 and 62). Normal values for red-cell volume are 28 to 32 ml/kg, and for plasma volume 35 to 45 ml/kg.

The value of measuring red-cell mass, plasma volume, or both is related to the underlying condition and the clinical problem to be solved. The diagnosis of polycythemia vera is best accomplished by measuring both in order to show a relative increase of red-cell mass with normal plasma volume. Blood loss postoperatively or from acute trauma is reflected in the lowered hematocrit, but the adequacy of replacement blood can be ascertained by a total-blood-volume measurement to avoid hypervolemia. Clinical shock may result from lowered total volume due to plasma deficit alone, as in severe dehydration or other causes of plasma loss. A normal hematocrit alone does not ensure adequate blood volume, since red-cell-to-plasma ratio may remain fairly constant early in cases of hemorrhage. Conversely, a decreased hematocrit does not prove a red-cell deficit, but may be due to dilution from any cause.

IRON METABOLISM AND RED-CELL FORMATION

Principle

Radioiron follows the same metabolic pathway as nonlabeled iron and permits quantification of its progress from plasma to bone marrow and from bone marrow to mature red cell. A tracer dose given intravenously combines with transferrin and begins to clear from the plasma immediately. The rate of clearance is an index of the turnover of iron in the plasma, the avidity of storage sites in the liver and elsewhere, and of active metabolic sites in marrow. Ultimately, the radioiron reaching cells in the process of hemoglobin formation actually "labels" newly formed hemoglobin biosynthetically and is incorporated into new red cells or reticulocytes in a manner indistinguishable from nonlabeled hemoglobin. These cells are subsequently released into the circulating blood, or destroyed before release in certain diseases. The released cells dilute with nonlabeled cells, and the ratio or percent of labeled versus nonlabeled cells rises steadily for several days and then reaches a plateau. The latter indicates the point where normal destruction and formation are balanced. When observed as a function of time, the relative concentration of radioiron in liver, spleen, bone marrow and blood of normal individuals shows an overwhelming predilection for the rapid uptake of iron only in marrow, with incorporation of label in red cells over a period of 8 to 10 days.

Methodology

Combined Ferrokinetic Study (Plasma Iron Disappearance, Plasma Iron Turnover, Red-Cell Utilization, and Organ Localization).

Preparation.

Radioiron (^{59}Fe) is obtained commercially as ferrous citrate; other salts, such as ferrous sulfate and ferrous gluconate, are satisfactory; the specific activity should be greater than 500 µCi/mg Fe.

Table 61. BLOOD VOLUME (ml) IN NORMAL MEN (Predicted from Height and Weight)

Weight, lbs	\multicolumn{19}{c}{Height, inches}	Weight, kg
	59	60	61	62	63	64	65	66	67	68	69	70	71	72	73	74	75	76	77	
90	3155	3219	3285	3353	3424	3497	3572	3649	3729	—	—	—	—	—	—	—	—	—	—	40.8
95	3288	3292	3358	3427	3497	3570	3645	3722	3802	3884	3969	4056	4146	4238	4333	4430	4530	4633	4738	43.1
100	3302	3365	3431	3500	3570	3643	3718	3795	3875	3957	4042	4129	4219	4311	4406	4503	4603	4706	4812	45.4
105	3375	3439	3505	3573	3643	3716	3791	3868	3948	4030	4115	4202	4292	4384	4479	4576	4676	4779	4885	47.6
110	3448	3512	3578	3646	3716	3789	3864	3942	4021	4104	4188	4275	4365	4457	4552	4649	4750	4852	4958	49.9
115	3521	3585	3651	3719	3790	3862	3937	4015	4095	4177	4261	4349	4438	4530	4625	4723	4823	4926	5031	52.2
120	3594	3658	3724	3792	3863	3936	4011	4088	4168	4250	4335	4422	4511	4604	4698	4796	4896	4999	5104	54.4
125	3667	3731	3797	3865	3936	4009	4084	4161	4241	4323	4408	4495	4584	4677	4771	4869	4969	5072	5177	56.7
130	3741	3804	3870	3939	4009	4082	4157	4234	4314	4396	4481	4568	4658	4750	4845	4942	5042	5145	5251	59.0
135	3814	3878	3944	4012	4082	4155	4230	4307	4387	4469	4554	4641	4731	4823	4918	5015	5115	5218	5324	61.2
140	3887	3951	4017	4085	4155	4228	4303	4381	4460	4543	4627	4714	4804	4896	4991	5088	5189	5291	5397	63.5
145	3960	4024	4090	4158	4229	4301	4376	4454	4534	4616	4700	4787	4877	4969	5064	5162	5262	5365	5470	65.8
150	4033	4097	4163	4231	4302	4374	4450	4527	4607	4689	4773	4861	4950	5042	5137	5235	5335	5438	5543	68.0
155	4106	4170	4236	4304	4375	4448	4523	4600	4680	4762	4847	4934	5023	5116	5210	5308	5408	5511	5616	70.3
160	4180	4243	4309	4378	4448	4521	4596	4673	4753	4835	4920	5007	5097	5189	5284	5381	5481	5584	5690	72.6
165	4253	4317	4383	4451	4521	4594	4669	4746	4826	4908	4993	5080	5170	5262	5357	5454	5554	5657	5763	74.8
170	4326	4390	4456	4524	4594	4667	4742	4820	4899	4981	5066	5153	5243	5335	5430	5527	5627	5730	5836	77.1
175	4399	4463	4529	4597	4668	4740	4815	4893	4972	5055	5139	5226	5316	5408	5503	5601	5701	5803	5909	79.4
180	4472	4536	4602	4670	4741	4813	4888	4966	5046	5128	5212	5300	5389	5481	5576	5674	5774	5877	5982	81.6
185	4545	4609	4675	4743	4814	4887	4962	5039	5119	5201	5286	5373	5462	5555	5649	5747	5847	5950	6055	83.9
190	4618	4682	4748	4817	4887	4960	5035	5112	5192	5274	5359	5446	5356	5628	5723	5820	5920	6023	6128	86.2
195	4692	4755	4822	4890	4960	5033	5108	5185	5265	5347	5432	5519	5609	5701	5796	5893	5993	6096	6202	88.5
200	4765	4829	4895	4963	5033	5106	5181	5259	5338	5420	5505	5592	5682	5774	5869	5966	6066	6169	6275	90.7
205	4838	4902	4968	5036	5107	5179	5254	5332	5411	5494	5578	5665	5755	5847	5942	6039	6140	6242	6348	93.0
210	4911	4975	5041	5109	5180	5252	5327	5405	5485	5567	5651	5739	5828	5920	6015	6113	6213	6316	6421	95.3
Height, cm	149.9	152.4	154.9	157.5	160.0	162.6	165.1	167.6	170.2	172.7	175.3	177.8	180.3	182.9	185.4	188.0	190.5	193.0	195.6	

Table 61. (Continued)
BLOOD VOLUME (ml) IN NORMAL MEN (Predicted from Height and Weight)

Weight, lbs	Height, inches																			Weight, kg
	59	60	61	62	63	64	65	66	67	68	69	70	71	72	73	74	75	76	77	
215	4984	5048	5114	5182	5253	5326	5401	5478	5558	5640	5725	5812	5901	5994	6088	6186	6286	6389	6494	97.5
220	5057	5121	5187	5256	5326	5399	5474	5551	5631	5713	5798	5885	5975	6067	6162	6259	6359	6462	6567	99.8
225	5131	5194	5260	5329	5399	5472	5547	5624	5704	5786	5871	5958	6048	6140	6235	6332	6432	6535	6641	102.1
230	5204	5268	5334	5402	5472	5545	5620	5697	5777	5859	5944	6031	6121	6213	6308	6405	6505	6608	6714	104.3
235	5277	5341	5407	5475	5545	5618	5693	5771	5850	5933	6017	6104	6194	6286	6381	6478	6579	6681	6787	106.6
240	5350	5414	5480	5548	5619	5691	5766	5844	5924	6006	6090	6177	6267	6359	6454	6552	6652	6755	6860	108.9
245	5423	5487	5553	5621	5692	5765	5840	5917	5997	6079	6164	6251	6340	6433	6527	6625	6725	6828	6933	111.1
250	5496	5560	5626	5694	5765	5838	5913	5990	6070	6152	6237	6324	6413	6506	6600	6698	6798	6901	7006	113.4
255	5570	5633	5699	5768	5838	5911	5986	6063	6143	6225	6310	6397	6487	6579	6674	6771	6871	6974	7080	115.7
260	5643	5707	5773	5841	5911	5984	6059	6136	6216	6298	6383	6470	6560	6652	6747	6844	6944	7047	7153	117.9
265	5716	5780	5846	5914	5984	6057	6132	6210	6289	6372	6456	6543	6633	6725	6820	6917	7018	7120	7226	120.2
270	5789	5853	5919	5987	6058	6130	6205	6283	6362	6445	6529	6616	6706	6798	6893	6991	7091	7193	7299	122.5
275	5862	5926	5992	6060	6131	6203	6279	6356	6436	6518	6602	6690	6779	6871	6966	7064	7164	7267	7372	124.7
280	5935	5999	6065	6133	6204	6277	6352	6429	6509	6591	6676	6763	6852	6945	7039	7137	7237	7340	7445	127.0
285	6008	6072	6138	6207	6277	6350	6425	6502	6582	6664	6749	6836	6926	7018	7113	7210	7310	7413	7519	129.3
290	6082	6146	6212	6280	6350	6423	6498	6575	6655	6737	6822	6909	6999	7091	7186	7283	7383	7486	7592	131.5
295	6155	6219	6285	6353	6423	6496	6571	6649	6728	6810	6895	6982	7072	7164	7259	7356	7456	7559	7665	133.8
300										6884	6968	7055	7145	7237	7332	7430	7530	7632	7738	136.1
	149.9	152.4	154.9	157.5	160.0	162.6	165.1	167.6	170.2	172.7	175.3	177.8	180.3	182.9	185.4	188.0	190.5	193.0	195.6	
	Height, cm																			

Table 62. BLOOD VOLUME (ml) IN NORMAL WOMEN (Predicted from Height and Weight)

Weight, lbs	Height, inches																			Weight, kg
	58	59	60	61	62	63	64	65	66	67	68	69	70	71	72	73	74	75	76	
80	2524	2584	2646	2710	2776	2845	2915	2988	3063	3141	—	—	—	—	—	—	—	—	—	36.3
85	2599	2659	2721	2785	2852	2920	2991	3063	3139	3216	3296	3378	3462	3549	3639	3731	3826	3923	4022	38.6
90	2675	2735	2797	2861	2927	2995	3066	3139	3214	3291	3371	3453	3538	3625	3714	3806	3901	3998	4098	40.8
95	2750	2810	2872	2936	3002	3070	3141	3214	3289	3366	3446	3528	3613	3700	3789	3881	3976	4073	4173	43.1
100	2825	2885	2947	3011	3077	3146	3216	3289	3364	3442	3521	3603	3688	3775	3864	3957	4051	4148	4248	45.4
105	2900	2960	3022	3086	3152	3221	3291	3364	3439	3517	3596	3679	3763	3850	3940	4032	4126	4223	4323	47.6
110	2975	3035	3097	3161	3228	3296	3367	3439	3514	3592	3672	3754	3838	3925	4015	4107	4201	4299	4398	49.9
115	3051	3110	3172	3237	3303	3371	3442	3515	3590	3667	3747	3829	3914	4001	4090	4182	4277	4374	4474	52.2
120	3126	3186	3248	3312	3378	3446	3517	3590	3665	3742	3822	3904	3989	4076	4165	4257	4352	4449	4549	54.4
125	3201	3261	3323	3387	3453	3521	3592	3665	3740	3817	3897	3979	4064	4151	4240	4332	4427	4524	4624	56.7
130	3276	3336	3398	3462	3528	3597	3667	3740	3815	3893	3972	4055	4139	4226	4316	4408	4502	4599	4699	59.0
135	3351	3411	3473	3537	3603	3672	3742	3815	3890	3968	4048	4130	4214	4301	4391	4483	4577	4675	4774	61.2
140	3426	3486	3548	3612	3679	3747	3818	3890	3966	4043	4123	4205	4289	4376	4466	4558	4653	4750	4849	63.5
145	3502	3562	3624	3688	3754	3822	3893	3966	4041	4118	4198	4280	4365	4452	4541	4633	4728	4825	4925	65.8
150	3577	3637	3699	3763	3829	3897	3968	4041	4116	4193	4273	4355	4440	4527	4616	4708	4803	4900	5000	68.0
155	3652	3712	3774	3838	3904	3973	4043	4116	4191	4269	4348	4430	4515	4602	4691	4784	4878	4975	5075	70.3
160	3727	3787	3849	3913	3979	4048	4118	4191	4266	4344	4423	4506	4590	4677	4767	4859	4953	5050	5150	72.6
165	3802	3862	3924	3988	4055	4123	4194	4266	4341	4419	4499	4581	4665	4752	4842	4934	5028	5126	5225	74.8
170	3878	3937	3999	4064	4130	4198	4269	4342	4417	4494	4574	4656	4741	4828	4917	5009	5104	5201	5301	77.1
175	3953	4013	4075	4139	4205	4273	4344	4417	4492	4569	4649	4731	4816	4903	4992	5084	5179	5276	5376	79.4
180	4028	4088	4150	4214	4280	4348	4419	4492	4567	4644	4724	4806	4891	4978	5067	5159	5254	5351	5451	81.6
185	4103	4163	4225	4289	4355	4424	4494	4567	4642	4720	4799	4882	4966	5053	5143	5236	5329	5426	5526	83.9
190	4178	4238	4300	4364	4430	4499	4569	4642	4717	4795	4875	4957	5041	5128	5218	5310	5404	5502	5601	86.2
195	4253	4313	4375	4439	4506	4574	4645	4717	4793	4870	4950	5032	5116	5203	5293	5385	5480	5577	5676	88.5
	147.3	149.9	152.4	154.9	157.5	160.0	162.6	165.1	167.6	170.2	172.7	175.3	177.8	180.3	182.9	185.4	188.0	190.5	193.0	
	Height, cm																			

Table 62. (Continued)
BLOOD VOLUME (ml) IN NORMAL WOMEN (Predicted from Height and Weight)

Weight, lbs	\multicolumn{19}{c}{Height, inches}	Weight, kg
	58	59	60	61	62	63	64	65	66	67	68	69	70	71	72	73	74	75	76	
200	4329	4389	4451	4515	4581	4649	4720	4793	4868	4945	5025	5107	5192	5279	5368	5460	5555	5652	5752	90.7
205	4404	4464	4526	4590	4656	4724	4795	4868	4943	5020	5100	5182	5267	5354	5443	5535	5630	5727	5827	93.0
210	4479	4539	4601	4665	4731	4800	4870	4943	5018	5096	5175	5257	5342	5429	5518	5611	5705	5802	5902	95.3
215	4554	4614	4676	4740	4806	4875	4945	5018	5093	5171	5250	5333	5417	5504	5594	5686	5780	5877	5977	97.5
220	4629	4689	4751	4815	4882	4950	5021	5093	5168	5246	5326	5408	5492	5579	5669	5761	5855	5953	6052	99.8
225	4705	4764	4826	4891	4957	5025	5096	5169	5244	5321	5401	5483	5568	5655	5744	5836	5931	6028	6128	102.1
230	4780	4840	4902	4966	5032	5100	5171	5244	5319	5396	5476	5558	5643	5730	5819	5911	6006	6103	6203	104.3
235	4855	4915	4977	5041	5107	5175	5246	5319	5394	5471	5551	5633	5718	5805	5894	5986	6081	6178	6278	106.6
240	4930	4990	5052	5116	5182	5251	5321	5394	5469	5547	5626	5709	5793	5880	5970	6062	6156	6253	6353	108.9
245	5005	5065	5127	5191	5257	5326	5396	5469	5544	5622	5702	5784	5868	5955	6045	6137	6231	6329	6428	111.1
250	5080	5140	5202	5266	5333	5401	5472	5544	5620	5697	5777	5859	5943	6030	6120	6212	6307	6404	6503	113.4
255	5156	5216	5278	5342	5408	5476	5547	5620	5695	5772	5852	5934	6019	6106	6195	6287	6382	6479	6579	115.7
260	5231	5291	5353	5417	5483	5551	5622	5695	5770	5847	5927	6009	6094	6181	6270	6362	6457	6554	6654	117.9
265	5306	5366	5428	5492	5558	5627	5697	5770	5845	5923	6002	6084	6169	6256	6345	6438	6532	6629	6729	120.2
270	5381	5441	5503	5567	5633	5702	5772	5845	5920	5998	6077	6160	6244	6331	6421	6513	6607	6704	6804	122.5
275	5456	5516	5578	5642	5709	5777	5848	5920	5995	6073	6153	6235	6319	6406	6496	6588	6682	6780	6879	124.7
280	5532	5591	5653	5718	5784	5852	5923	5996	6071	6148	6228	6310	6395	6482	6571	6663	6758	6855	6955	127.0
285	5607	5667	5729	5793	5859	5927	5998	6071	6146	6223	6303	6385	6470	6557	6646	6738	6833	6930	7030	129.3
290	—	—	—	—	—	—	—	—	—	—	6378	6460	6545	6632	6721	6813	6908	7005	7105	131.5
	147.3	149.9	152.4	154.9	157.5	160.0	162.6	165.1	167.6	170.2	172.7	175.3	177.8	180.3	182.9	185.4	188.0	190.5	193.0	
	\multicolumn{19}{c}{Height, cm}	

Procedure.
1. Draw from the patient the following samples:
   a) two sterile (yellow top) Vacutainer tubes of heparinized blood, to be centrifuged for plasma.
   b) two tubes unclotted blood, for serum iron and unsaturated iron-binding capacity;
   c) one oxalate tube, for a whole-blood and plasma blank; perform a hematocrit on this sample.
2. Centrifuge the sterile tubes and draw off plasma, using sterile technique; add 10 ml of this plasma to another sterile tube.
3. To this plasma add 40 µCi of ^{59}Fe; incubate with agitation for 20 minutes at room temperature.
4. Calibrate the counting system 0 to 2 Mev; count a long-lived standard in a geometry that will approximate the organ counting sites (an isotope that approximates the gamma energies of ^{59}Fe should be used—^{60}Co is satisfactory for this purpose).
5. After the counting sites (spleen, liver, sacrum) have been localized and marked, and after a mark has been placed on the collimator to allow reproduction of the geometry, take background counts over each organ site.
6. Place an intracath or Rochester needle in an antecubital vein; attach a 3-way stopcock and a 2-ml syringe filled with heparin on the catheter; the system must be heparinized after insertion and after each sample is drawn.
7. Draw into a syringe sufficient labeled plasma to give the patient a dose of 20 µCi.
8. Inject this dose into the opposite arm from the intracath; start the interval timer.
9. Start organ counts immediately, counting one minute at each site; record elapsed time for each count; during the first hour, obtain as many organ counts as possible; after the first hour, count the sites every 30 minutes until 4 hours after injection.
10. Draw blood samples at 15, 30, 60, 120, and 180 minutes after the dose is injected, and place them in oxalated tubes; the intracath may be removed after the 180-minute sample is obtained.
11. The patient should return daily for the first two days for organ counting; obtain a 10-ml oxalated blood sample at this time; the patient must return every other day after this time, for a total of 14 days, for blood samples and organ counting.
12. A ^{51}Cr blood volume determination should be performed on the patient at the end of the procedure.

Sample Preparation—Plasma Iron Disappearance Half-Time (PID-$T_{1/2}$).
1. Using the samples obtained in step 10, measure a hematocrit.
2. Centrifuge the samples and withdraw 3-ml aliquots of the plasma.
3. Place in counting tubes and count with a scintillation well-type counting system calibrated 0 to 2 Mev.
4. Plot cpm/ml versus time on semi-logarithmic graph paper and graphically determine the PID-$T_{1/2}$

Sample Preparation—Percent ^{59}Fe Utilization by RBC.
1. Prepare a 1:1000 dilution from the dose tube for a standard.
2. Prepare duplicate 3-ml samples from the standard.
3. Using the samples obtained in step 11,
   a) prepare 3-ml aliquots of whole blood in counting tubes; measure microhematocrit of each sample; hemolyze, using 15 drops of 5% saponin;

b) count in a well counter, using the same settings as for the PID-$T_{1/2}$; correct each count for variation in hematocrit;
c) calculate the percent ^{59}Fe utilization by RBC for each sample, using the formula

$$\% \ ^{59}\text{Fe Utilization by RBC} = \frac{\text{cpm/ml} \times \text{RBC Vol} \times 100}{\text{Vol Inj} \times \text{cpm/ml Standard}};$$

d) plot the percent utilization as a function of time in days.

Plotting the *in-vivo* Organ Counts.

1. Divide the decay-corrected cpm of each organ at given times by the highest count obtained at any time for that organ.
2. Plot this ratio versus time in minutes for the initial day of the test and versus time in days for the remainder of the test on the same sheet of linear graph paper; use 1.0 as the base; values less than 1.0 may be obtained; this represents a normalization of the data, so that one organ can be compared to another in terms of relative concentration of activity.

Calculations.

1. Plasma Iron Turnover (PIT), in mg/kg/day:

$$\text{PIT} = \frac{0.693 \times \text{Plas Vol} \times \text{Serum Iron (mg/ml)} \times 24 \text{ hrs}}{\text{PID-}T_{1/2} \text{ (hrs)} \times \text{body weight (kg)}}.$$

2. Red Blood Cell ^{59}Fe Turnover:
   RBC ^{59}Fe Turnover = PIT × % ^{59}Fe Utilization.

Clinical Interpretation

The plasma iron turnover value (normal 27 to 40 mg/day or 0.46 to 0.78 mg/kg/day) is helpful when elevated, because it is indicative of adequate or increased erythropoiesis. Low values indicate poor red-cell production. The utilization of red-cell iron is very helpful in hemolytic anemias, since the initial incorporation is normal, with a premature plateau or truncation of the curve indicating an equilibrium between production and destruction of new cells. Figure 32 shows a case of congenital spherocytosis with shortened red-cell survival and reduced iron utilization.

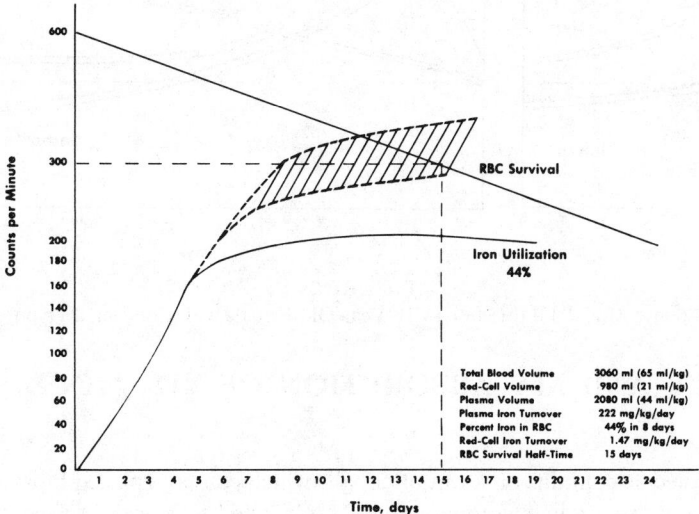

FIGURE 32. Congenital Spherocytosis.

The organ concentration of labeled red cells reflects initially, or during the first three hours, the most active site of erythropoiesis. Normally this is the sacrum, but extramedullary sites can be detected by showing a decrease in bone marrow (sacrum) and increase elsewhere (spleen). Figure 33 shows a typical plot of organ data in a normal individual. The rapid rise in relative radioiron concentration over the sacrum with a rapid decrease during the first four days is typical of normal iron incorporation and release of labeled cells in marrow. Figure 34 is a plot of data from a case of congenital hypoplastic anemia, which shows reduced marrow activity relative to other organs, with uptake in liver suggesting some erythropoiesis there and gradual uptake of labeled cells in the spleen indicating splenic sequestration.

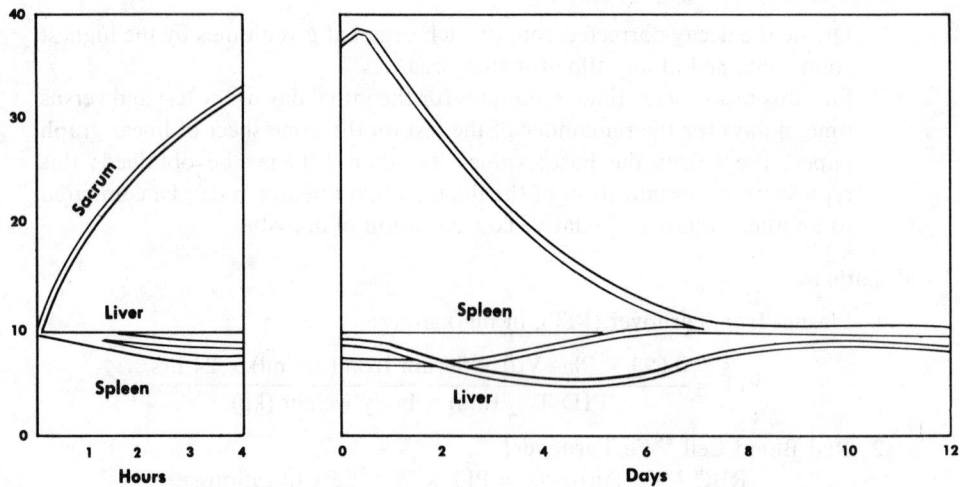

FIGURE 33. *In-vivo* Organ Counts After ^{59}Fe Injection in Normal Subject.

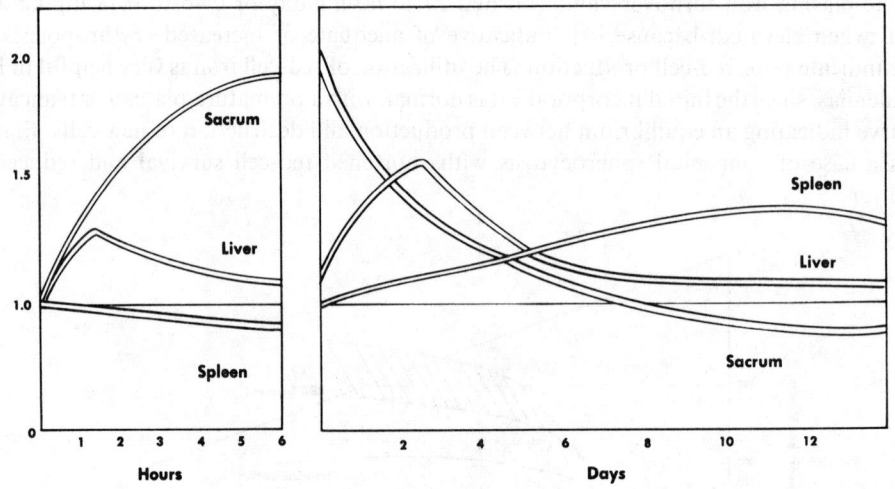

FIGURE 34.
In-vivo Organ Counts After ^{59}Fe Injection in 16-Year Old Female with Congenital Hypoplastic Anemia.

DEFICIENCY AND MALABSORPTION OF VITAMIN B_{12}

Principle

The B_{12} molecule contains on atom of cobalt, which can be replaced biosynthetically with radiocobalt; any of three cobalt isotopes—^{57}Co, ^{58}Co, and ^{60}Co—can be used. ^{57}Co is favored because of its shorter physical half-life and relative safety. The labeled B_{12}, given orally in absorbed carrier molecules of unlabeled B_{12}, is absorbed to a degree commensurate with the ability of the mucosa of stomach to produce intrinsic factor (I.F.) and form a complex of B_{12} + I.F. About 70 percent of the B_{12} as a complex is absorbed in the terminal ileum and transported to body stores, primarily in the liver. If a flushing dose of unlabeled B_{12} is injected intramuscularly, then a large percentage of the labeled B_{12} will tend to be displaced from storage or prevented from remaining bound to protein by dilution. This "free" labeled B_{12} is rapidly filtered by the kidney and excreted. The amount excreted in a 24-hour period is directly proportional to the amount originally absorbed. Thus, if absorption is impaired by disease, then very little is "flushed" into the urine.

Methodology

1. B_{12}-^{55}Co, containing 0.25 µCi of ^{55}Co as radiocyanocobalamin and about 0.75 µg of B_{12}, is administered to a fasting patient with an empty urinary bladder.
2. Two hours later a "flushing dose" of B_{12} in the form of 1000 mg of B_{12} solution is administered intramuscularly, after which the patient may resume normal diet.
3. The urine is collected for the next 24 hours and returned to the laboratory; an aliquot of the urine is counted, and the percent of the total dose is determined:

$$\frac{\text{cpm of 3 ml Urine} \times \text{Vol of Urine} \times 100}{\text{cpm of 3 ml Standard} \times \text{Dil Fact}} = \% \text{ excreted.}$$

4. Normal range is 10 to 15 percent of administered dose excreted in urine in the first 24 hours; if less than 10 percent is excreted, the examination is repeated in 4 days, with the following change of procedure.
5. Intrinsic factor (30 mg) is given orally with another dose of B_{12}-^{57}Co.

Clinical Interpretation

To prove the presence of pernicious anemia (PA) after an abnormal Schilling test is obtained, the test is repeated, with the patient ingesting intrinsic factor as well as B_{12}-^{57}Co. If the patient has PA, the repeat test will have a normal result; if he has a different malabsorption syndrome, the result will again be abnormal. Renal disease may also give false positive results. In such cases excretion of B_{12}-^{57}Co is characteristically low and not affected by intrinsic-factor administration.

HEMOLYTIC DISEASE AND HYPERSPLENISM

Principle

Red-cell life span is shortened in acquired hemolytic disorders, blood loss, and in diseases where defective red cells are produced. Measurement of the red-cell survival half-time is readily made by labeling the red cells with ^{51}Cr. This binds to hemoglobin, and remains with the living circulating cell until it is destroyed. A small elution or leakage of ^{51}Cr occurs, but in most clinical situations this does not seriously affect the data. The labeled cells for evaluating the survival of transfused or known normal cells by testing the patient's ability to destroy them may be prepared from patient's or donor's blood. The spleen normally sequesters aged, damaged, or defective red blood cells from the circulation. Increased splenic sequestration may result from disease, and produce a shortening of the life span of normal cells.

Methodology

I. Red-Cell Survival Half-Time with ^{51}Cr.

Procedure.

1. Labeled red blood cells are prepared as described earlier.
2. 1 ml of labeled cells is retained as a standard, and the rest is injected; whole-blood samples are obtained three times weekly for four weeks.

Calculation.

The heparinized blood samples are hemolyzed with saponin and counted in a well-type scintillation counter; the initial blood sample (100%) is retained and counted with the other blood samples, so that no physical-decay correction is required; the radioactivity of each blood sample is related to this initial sample and is expressed as percent of the initial activity:

$$\% \text{ Survival of } {}^{51}\text{Cr-labeled RBC's at any day (t)} = \frac{\text{cpm/ml of Blood Sample on day t}}{\text{cpm/ml of Initial 24-hour Sample}} \times 100.$$

For most clinical purposes the test is ended when the half-time period is reached; normal values for apparent half-time survival of red cells in human subjects is 32 ± 2 days in most laboratories utilizing the 100-percent sample at 24 hours.

II. Red-Cell Survival with DF-^{32}P.

Procedure.

1. In usual clinical practice, DF-^{32}P is administered intravenously in a single dose of 0.5 to 1.0 mg; venous blood samples are withdrawn three times a week for 1 to 2 weeks, and then at approximately weekly intervals for the rest of the study.
2. The erythrocytes are separated initially by differential centrifugation.
3. After centrifugation the red cells are washed three times in cold physiologic saline to remove remaining plasma, leukocytes and platelets; the erythrocytes are heavily labeled with DF-^{32}P, and further purification is usually unnecessary; if leukocyte and platelet DF-^{32}P studies are to be done simultaneously, their further purification must be carried out.
4. 1-ml aliquots of washed red cells are pipetted into disposable boats of aluminum foil and dried at 70°C; the boats are sealed in cellophane, and counted with a Geiger-Mueller tube; the counting error with this method is less than 3 percent for approximately 90 days, and then it is in the range of 5 to 10 percent.

III. Splenic Sequestration.

Procedure.

1. Body surface counting is begun 30 minutes after injection of ^{51}Cr-labeled erythrocytes; the precordium is used as a measure of blood activity; surface projections of the liver and spleen are determined by physical examination, and their approximate centers are outlined with an indelible material.
2. A wide-angle scintillation probe is placed just at the skin surface in a standardized position, with the patient supine; this position approximates a parallelism with the body surface overlying the organ.
3. A background count is determined by placing the scintillation counter over the back of the thigh.
4. The counts over each organ are plotted after subtracting the background.
5. Body counts are performed at 6 hours and at 24 hours after injection of the radioactivity on patients in whom severe hemolytic states are suspected; subsequently, body surface counting is performed three times weekly until the ^{51}Cr half-time is reached; in most clinical situations significant information regarding potential sites of red-cell sequestration can be gathered within one week; the normal ratio of liver:spleen counting is 1.2:1; this relationship does not change during the entire period of the survival study;

the precordial radioactivity is always greater than that in the liver or spleen of a normal subject and slowly declines with time, roughly approximating the determined red-cell life span.

Clinical Interpretation

McCurdy and Rath[2] have devised the following calculation:

$$\text{Splenic Localization Index (SLI)} = \frac{\frac{S/P}{S/P_0} \times 10}{d\ max},$$

where S = spleen counts,

P = precordial counts,

S/P = the maximum change in ratio,

S/P_0 = the initial ratio,

d max = the day on which the maximum ratio occurs.

It was found that significant splenic sequestration could be anticipated if the SLI was 1.0 or more and if the maximum S/P ratio exceeded 1.5.

Splenic-sequestration measurements by either of the above techniques are reliable. A good correlation exists between positive results and improvement after splenectomy, but negative results do not preclude possible clinical improvement from splenectomy.

Mild anemias of uncertain etiology are occasionally encountered, in which knowledge of red-cell survival may be useful. If normal, then a hemolytic disorder or significant bleeding can be ruled out; if shortened, then the question of red-cell defect or splenic hyperactivity or both must be faced.

REFERENCES

1. Hahn, P. T., Bale, W. F., Lawrence, E. O., and Whipple, G. H., Radioactive Iron and Its Metabolism in Anemia. *J. Exp. Med.*, 69: 739, 1939.
2. McCurdy, P. R. and Rath, C. E., Splenectomy in Hemolytic Anemia: Results Predicted by Body Scanning After Injection of ^{51}Cr-Tagged Red Cells. *New Engl. J. Med.*, 259: 459, 1958.
3. Nadler, S., Hidalgo, J., and Block, T., Prediction of Blood Volume in Normal Human Adults. *Surgery*, 51: 22, 1962.

RADIONUCLIDE TECHNIQUES APPLIED TO THE GENITOURINARY SYSTEM

L. Rosenthall, M.D.,
Montreal General Hospital and McGill University, Montreal, P.Q., Canada.

KIDNEY-FUNCTION STUDIES

Radioiodinated Orthoiodohippurate Renogram

When ^{131}I-orthoiodohippurate was introduced in 1960,[1,2] the renogram became widely accepted as a modality to assess the function of each kidney individually by external monitoring with paired scintillation detectors. The resultant curve from each kidney is composed of three components: tracer appearance, renal transit time, and drainage. Renal transit time is the interval between appearance of the activity subtended by the scintillation detector and the peak of the curve (Figure 35)[3]. The ascending slope of the curve during the transit time is reported to reflect the blood flow rather than tubular function.

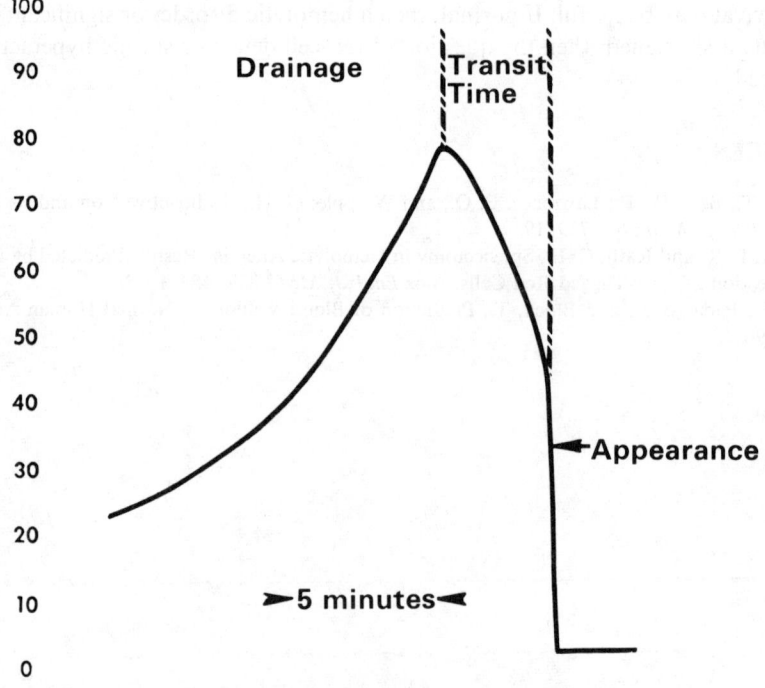

FIGURE 35.
Normal Radiohippurate Renogram Curve Exhibiting Three Phases:
Appearance of the Bolus, Transit Time or Passage of the Radiopharmaceutical Through the Kidney as Viewed by the Detector, and Drainage.

Technique.

The patient should be comfortably immobilized. Most investigators favor seating the patients, but some have them lying supine, others prone or semiprone.

A dose of about 50 microcuries is rapidly injected into an antecubital vein, and the kidneys are monitored with a matched pair of scintillation detectors; each unit is linked to a rate meter and strip chart recorder. It is essential that the field of view of the detectors encompass the kidneys. Accurate localization can be obtained by first injecting a small dose (5 microcuries) of ^{131}I-hippurate or radiochlormerodrin and then placing the detectors over the area of maximum count rate.

Some authors favor doing the renogram first in a dehydrated state, then repeating the test with water loading;[4] others omit special preparation.[5]

Comments.

The renogram is eminently suitable for depicting a difference in function between the kidneys. Contrary to the initial enthusiasm, the curves are nonspecific, and a clinical interpretation is only possible in light of the clinical findings and pathoanatomical information derived from such examinations as the intravenous urogram, aortogram, and perhaps the renal scan (Figure 36).

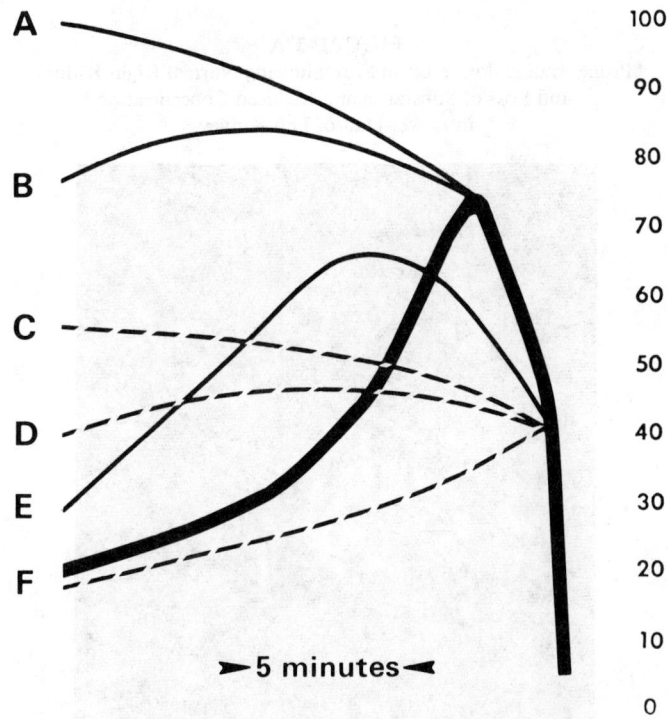

FIGURE 36.
Renogram Curves of Normal and Abnormal Kidney Function.

The heavy black line is a typical normal radiohippurate renogram
- A. Acute complete obstructive uropathy
- B. 1. Partial acute obstructive uropathy
    - a) Lithiasis, blood clot, tissue
    - b) Extrinsic compression by tumor or abdominal contents in the prone position; kinking of the ureter in sitting position
  2. Renal-artery stenosis
  3. Large extrarenal pelvis
- C. 1. Complete obstructive uropathy of about 3 to 10 days duration
  2. Moderate to severe renal-artery stenosis
  3. Primary renal disease
  4. Acute obstructive uropathy superimposed on primary renal disease
- D. 1. Severe primary renal disease
  2. Severe renal-artery stenosis
  3. Dehydration or hypovolemia
  4. Prolonged partial obstruction
- E. 1. Renal-artery stenosis
  2. Primary renal disease
- F. Absent renal function due to disease or surgical removal

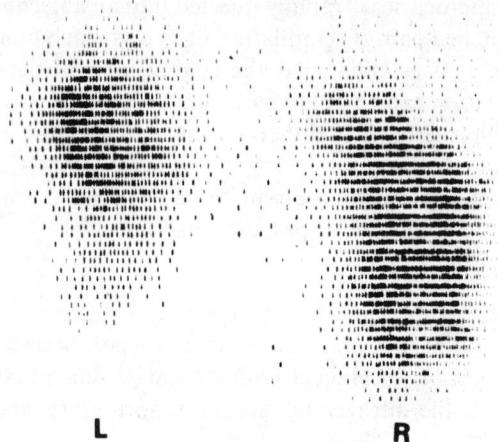

FIGURE 37A.
Prone Radiochlormerodrin Scan Showing Normal Right Kidney
and Loss of Substance and Reduced Concentration
in Lower Half of Left Kidney.

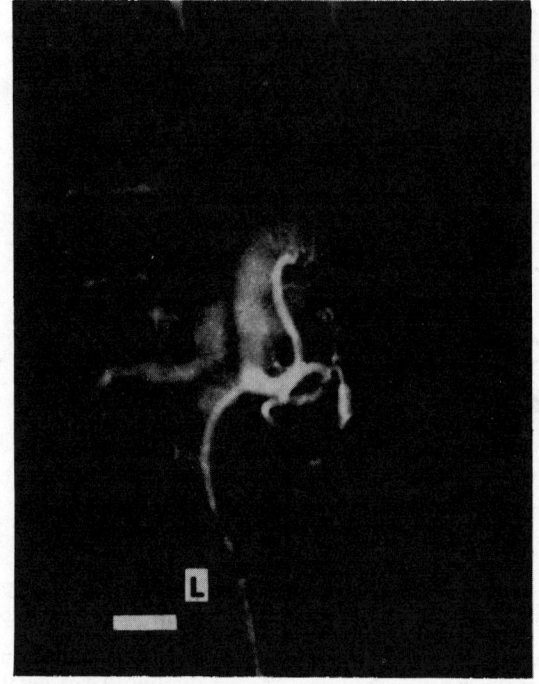

FIGURE 37B.
Selective Left Renal Arteriogram Demonstrating an Infarction
and Small Intrarenal Aneurysm.
This patient suffered an acute left-flank pain, and
two months later developed severe bouts of headache; his blood pressure was elevated at the time;
the excretory urogram was interpreted as normal. In
this particular case, the combination of a normal excretory urogram and a grossly abnormal scan was specific
for focal ischemia. (After Rosenthall, L., The Role of
Radioisotope Renal Scanning in the Assessment of
Renal Disease. *Can. Med. Ass. J.*, 90: 999, 1964.)

Radiochlormerodrin Renal Uptake

The rate of accumulation of radiochlormerodrin by the kidneys is purported to reflect the integrity of the renal blood flow and renal tubular efficiency, and to be influenced little by urine flow.[6,7] There is experimental work, however, that has shown that the urine flow rate does influence the radiochlormerodrin uptake.[8]

When the accumulation test is followed by a renal scan, an accuracy of 95 percent is claimed for uncovering hypertensive patients with unilateral renal disease.

Technique.

The patient is comfortably immobilized, preferably in the supine position,[8] and a pair of scintillation detectors are carefully placed to encompass each kidney within the field of view. No special preparation is necessary. A scanning dose of 100 to 150 microcuries ^{197}Hg- or ^{203}Hg-chlormerodrin is administered intravenously, and the accumulation curve for each kidney is recorded on a strip chart for one hour.

Results.

If R_t represents the ratio of the count rate at t minutes to the count rate at 5 minutes on the right renal accumulation curve, and L_t represents the corresponding ratio on the left renal accumulation curve, then the normal 2-standard-deviation range of $\frac{R_t}{L_t}$ is 0.85 to 1.05. This is valid between 20 and 60 minutes after injection. The value of $\frac{R_t}{L_t}$ is less than 0.85 with right renal disease, and greater than 1.05 with left renal disease.

Comments.

The test reveals only the side of predominant renal disease and is nonspecific. In spite of the declared accuracy,[7,8] it is not widely used, probably because it takes about one hour to perform, whereas the radiohippurate renogram is complete in 15 to 20 minutes.

RENAL SCANNING

Radiochlormerodrin Renal Imaging

^{197}Hg- or ^{203}Hg-chlormerodrin is concentrated in the proximal tubules of the kidney and remains there sufficiently long to be scanned.[10] Any lesion that affects enough cortex to be resolved will be seen as an area of reduced or absent concentration of activity.[11-16]

Technique.

There is no special preparation of the patient. An intravenous dose of 100 to 200 microcuries of ^{197}Hg- or ^{203}Hg-chlormerodrin is administered. Scanning is performed between one and four hours.

Comments.

The finding of an area of reduced or absent activity is generally nonspecific. Cysts, malignant neoplasms, infarctions (Figure 37), and intrarenal aneurysms cannot, for the most part, be distinguished. It is essential that the renal outline derived from the radiochlormerodrin scan be compared to the silhouette on the roentgenogram, because a lack of congruence would indicate an abnormality.

The renal scan is a useful adjunct to the excretory urogram. It can confirm, negate, or augment the X-ray findings.[16] The scan is particularly well suited to determine whether the renal " bulge " is a normal fetal lobulation or a lesion. Lesions that do not comprise the cortex will not be detected; these include such entities as medullary sponge kidneys, diverticulae, peripelvic cysts, and pelvic tumors. There is no interference from overlying gas and fecal material, and the scan will often visualize the kidney when the degree of azotemia is too high to permit sufficient concentration of the usual doses of radiographic contrast materials for proper renal assessment. Qualitatively, the radiochlormerodrin scan is more sensitive than radiography as an indicator of renal function.

Radiohippurate Renal Imaging

Radiohippurate is excreted too rapidly by the normal kidney to permit visualization with the conventional rectilinear scanning apparatus, although it can be imaged with the gamma-ray scintillation camera.[17] In the presence of renal failure the transit time of radiohippurate is sufficiently prolonged to enable the kidneys to be seen as much as 72 hours after administration.[18]

Technique.

No patient preparation is required. 200 microcuries of ^{131}I-iodohippurate are intravenously injected. The first scan of the kidneys is obtained at 15 minutes post dose, and is repeated at 45 minutes if adequate visualization is not obtained. A 3-hour and 24-hour scan should be secured in the absence of sufficient concentration at 45 minutes.

The urinary vesical is imaged after each renal study to rule out obstructive uropathy.

Although the scanning procedure can be performed with conventional rectilinear devices, it is greatly facilitated by more sensitive equipment, such as the gamma-ray scintillation camera (Nuclear-Chicago) and Dynapix (Picker-Nuclear).

Comments.

The radiohippurate method may visualize the kidney when the degree of failure precludes conventional radiographic and radiochlormerodrin identification (Figures 38 and 39). Calyceal, pelvic, and ureteral configuration can be demonstrated with an infusion intravenous urogram in the presence of renal failure, but the radiohippurate technique is simpler, safer, and will demonstrate the renal outline, size and disease more clearly and, as a rule, earlier (Figure 40). The contrast infusion method, however, is superior for studying the ureters and bladder. When the kidneys fail to show hippurate concentration, the patient is usually a candidate for dialysis.

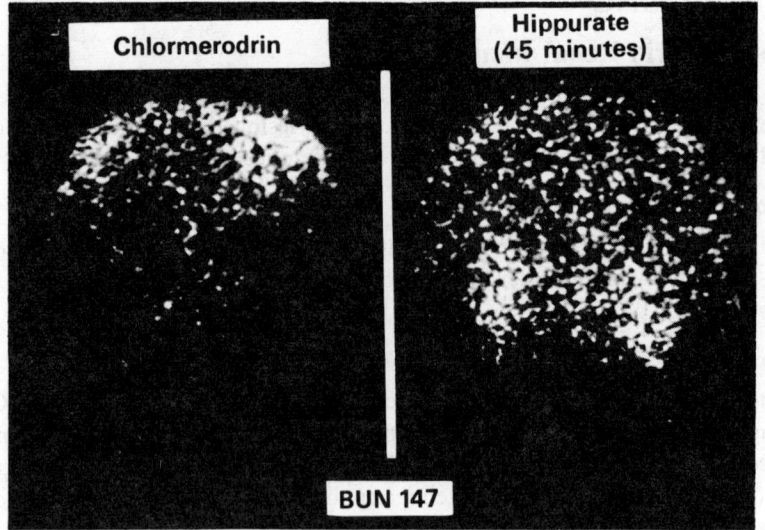

FIGURE 38.
Comparison of Radiochlormerodrin and Radiohippurate Renal Scans.
This 13-year-old girl had a nephrotic syndrome and a BUN of 147
milligram percent. Left: the radiochlormerodrin scan failed to
visualize the kidneys. Right: the radiohippurate scan portrayed a
pair of large kidneys with uniform distribution of activity.

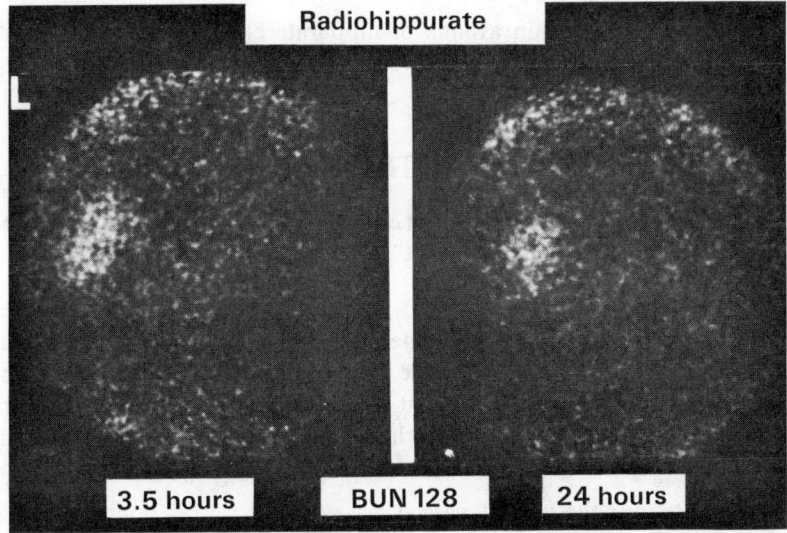

FIGURE 39.
Visualization of Kidney by Radiohippurate Method in Patient with High Degree of Renal Failure.
This 30-year-old white male was admitted to the medical ward in renal failure. The BUN was 128 milligram percent. A radiochlormerodrin scan did not visualize the kidneys, and most of the activity was concentrated in the liver. The radiohippurate renal scan exhibited a solitary small left kidney at 3½ hours and at 24 hours post injection. At post mortem, pyelonephritis was observed on the left, and an absent kidney on the right.

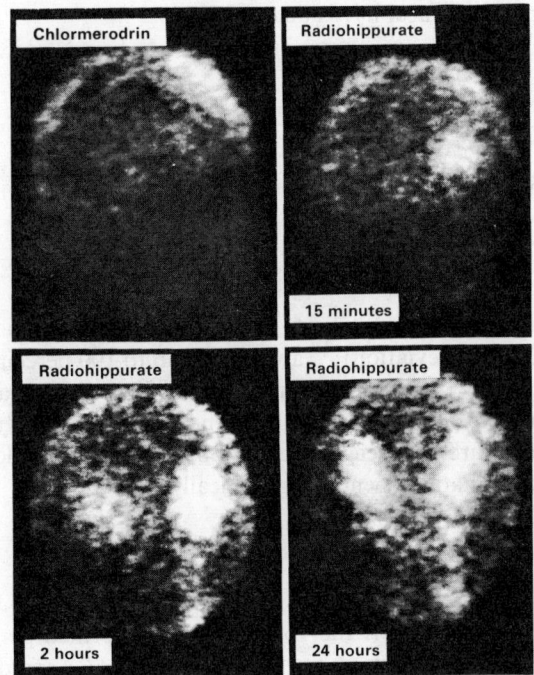

FIGURE 40.
Activity Concentration in Kidneys Observed in a Series of Radiohippurate Scans
after Radiochlormerodrin Scan Failed to Identify the Organs.
Renal failure developed in this male octogenarian with carcinoma of the urinary vesical. An infusion intravenous urogram demonstrated a dilated right upper collecting system at 2½ hours. At no time was the left kidney seen. The radiochlormerodrin scan failed to visualize either kidney. A radiohippurate series followed, and it showed concentration in the right kidney at 15 minutes and in the left kidney at 2 hours. At 2 hours and 24 hours, a right ureteral dilatation due to obstruction in the urinary bladder was seen. The radiohippurate series demonstrated bilateral renal viability, whereas the infusion pyelogram exhibited right renal function only.

Although the liver is the main avenue of hippurate elimination when the kidneys are defective, the degree of concentration in the liver is not enough to obscure the right kidney, as may be the case with radiochlormerodrin.

RESIDUAL URINE DETERMINATION

Residual urine may be quantitated without catheterization by excretory urography[19,20] and radionuclide methods.[21,22]

Technique.

Any radioactive substance that is eliminated primarily via the renal system may be used. Radiohippurate is most suitable. It is preferable to give the patient 250 ml of water immediately before the test. While in supine position, about 25 microcuries of radiohippurate are administered intravenously. Counts are obtained with a 20°-divergent collimator at a crystal–skin distance of 20 cm one minute after injection.

1. Count rate over the urinary vesical, u.
2. Count rate over the face, centered on the nose, n_1.

Ratio of the backgrounds, $\dfrac{u}{n_1}$.

The patient is instructed not to void for at least one hour, preferably two hours.

3. Pre-voiding count rate over urinary vesical, P_1.
4. Patient voids immediately afterward, and the volume, v, is measured.
5. Post-voiding count rate over the urinary vesical, P_2, is obtained.
6. Count rate over the face, n_2, is determined.

Calculation.

$$\text{Residual urine volume} = \left(\frac{v}{P_1 - P_2}\right)\left(P_2 - \frac{un_2}{n_1}\right)$$

If the one-minute background ratio is not obtained, a good approximation of $\dfrac{u}{n_1}$ is 1.5, and if the heart is used instead of the face, 0.5.

Comments.

The radionuclide technique can quantitate the residual urine volume with an accuracy of about 12 percent (one standard deviation);[22] a variation up to 100 percent is obtained with the roentgenographic technique.[19,22] There is no hazard of iodinated contrast materials, and the technique can be combined with the radiohippurate renogram.[4] Residual urine can be obtained in the presence of renal failure, whereas there may be insufficient concentration of opaque medium to make the determination roentgenographically.

RENAL-CLEARANCE METHODS

A continuous infusion of a test agent is made until equilibrium between arterial and peripheral venous blood is obtained. If the volume of urine formed per minute (V) and the concentrations of the test agent in the urine (U) and plasma (P) are determined, the renal clearance of the test agent is given by $\dfrac{UV}{P}$.

Para-aminohippuric acid is handled very efficiently by glomerular filtration and tubular excretion. When used as the test agent in the continuous-infusion technique, a measurement of the effective renal plasma flow is obtained. On the other hand, inulin is completely filtered by the glomeruli, and therefore it measures glomerular filtration.

Radionuclide-labeled test agents can be substituted to overcome the tedious biochemical determinations of para-aminohippuric acid and inulin concentrations in the urine and plasma (see Tables 63 and 64).

Table 63. RADIOPHARMACEUTICALS USED FOR RENAL CLEARANCE

Glomerular Filtration	Effective Renal Plasma Flow
Meglumine diatrizoate-^{131}I (Renografin)	Orthoiodohippurate-^{131}I
Cyanocobalamin-^{57}Co	Orthoiodohippurate-^{125}I
Sodium iothalamate-^{131}I (Conray)	Iodopyracet-^{131}I
^{125}I-allyl inulin	

Table 64. COMPARATIVE STUDIES

Renografin / Inulin	1.04[23] Range: 0.76 — 1.86
Cyanocobalamin-^{57}Co / Inulin	0.99 ± 0.09[24,25] Range: 0.76 — 1.86
Iothalamate-^{131}I / Inulin	1.06[26]
^{125}I-allyl inulin / Inulin	0.97[27]
Orthoiodohippurate / PAH	0.88 (r = 0.99)[28] 0.96 ± 0.10[24]
Iodopyracet-^{131}I / PAH	1.02[29]

A single-injection technique of radiohippurate, that obviates urine collection has been developed for measuring effective renal blood flow by serial plasma sampling.[30]

The effective renal plasma flow of each kidney can be obtained from the radiohippurate renogram by analyzing the slopes during the transit phase and measuring the percentage of radiopharmaceutical excreted in the urine in 15 minutes.[31]

The renografin-^{131}I renogram can be quantitated, so that the glomerular filtration of each kidney is measured individually without the necessity of urinary-excretion determinations.[32,33]

Radioxenon[34] or radiokrypton[35] dissolved in saline and injected into the renal artery can be monitored by an external scintillation detector over the kidney. The washout curve consists of at least two exponential components. Cortical blood flow in ml/g renal tissue/min is calculated from the half-disappearance time of the fast component.

REFERENCES

1. Tubis, M., Posnick, E., and Nordyke, R. A., Preparation and use of ^{131}I-Labeled Sodium Iodohippurate in Kidney-Function Tests. *Proc. Soc. Exp. Biol. Med.*, 103: 497, 1960.
2. Nordyke, R. A., Tubis, M., and Blahd, W. H., Use of Radioiodinated Hippuran for Individual-Kidney-Function Test. *J. Lab. Clin. Med.*, 56: 438, 1960.
3. Dore, E. K., Taplin, G. V., and Johnson, D. E., Current Interpretation of the Sodium Iodohippurate-^{131}I Renocystogram. *J. Amer. Med. Ass.*, 185: 925, 1963.
4. Burbank, M. K., Hunt, J. C., Tauxe, W. N., and Maher, F. T., Radioisotopic Renography: Diagnosis of Renal Arterial Disease in Hypertensive Patients. *Circulation*, 27: 328, 1963.

5. Winter, C. C., A Kidney-Function Test Performed with Radioisotope-Labeled Agents. *Radioisotope Renography*. Williams and Wilkins, Baltimore, 1963.
6. Reba, R. C., Wagner, H. N., and McAfee, J. G., Measurement of ^{203}Hg-Chlormerodrin Accumulation by the Kidneys for Detection of Unilateral Disease. *Radiology*, 79: 134, 1962.
7. Reba, R. C., McAfee, J. G., and Wagner, H. N., Radiomercury-Labeled Chlormerodrin Accumulation by the Kidneys for Detection of Unilateral Disease. *Radiology*, 79: 134, 1962.
8. Johnston, G. S., Wettlaufer, J. N., and Murphy, G. P., Clinical and Experimental Studies Using ^{197}Hg-Neohydrin as a Measure of Renal Blood Flow. *J. Urol.*, 92: 378, 1964.
9. Reba, R. C., Wagner, H. N., and McAfee, J. G., Radioactive-Chlormerodrin Kidney Uptake Test. *Nuclear Medicine*, Blahd, W. H., ed. McGraw-Hill, New York, 1965.
10. McAfee, J. G. and Wagner, H. N., Visualization of Renal Parenchyma by Scintiscanning with ^{203}Hg-Neohydrin. *Radiology*, 75: 820, 1960.
11. Cohen, M. B., Pearman, R. O., Mims, M. M., and Blahd, W., Radioisotope Photoscanning of the Kidneys in Urologic Disease. *J. Urol.*, 89: 360, 1963.
12. Haynie, T. P., Stewart, B. H., Nofal, M. M., Carr, E. A., and Beierwaltes, W. H., Diagnosis of Renal Vascular Disease and Renal Tumor by Photoscanning. *J. Amer. Med. Ass.*, 179: 137, 1962.
13. Izenstark, J. L., Burden, J. J., Mardis, H. K., and Varella, R., Clinical Indications for Kidney Scanning. *J. Amer. Med. Ass.*, 188: 136, 1964.
14. Rosenthall, L., The Usefulness of Renal Scanning as an Adjunct to Excretory Urography. *J. Can. Ass. Radiol.*, 14: 76, 1963.
15. Rosenthall, L., The Role of Radioisotope Renal Scanning in the Assessment of Renal Disease. *Can. Med. Ass. J.*, 90: 999, 1964.
16. MacEwen, D. W. and Rosenthall, L., Assessment of Excretory Urography and Radioisotope Renal Scanning in Diseases of the Kidneys. *Radiology*, 86: 1010, 1966.
17. Burke, G. and Halko, A., Scintillation Camera Renography in the Study of Prolonged Renal Transit Time. *Radiology*, 88: 704, 1967.
18. Rosenthall, L., Orthoiodohippurate-^{131}I Kidney Scanning in Renal Failure. *Radiology*, 87: 298, 1966.
19. Bretland, P. M., Relation of Bladder Shadow to Bladder Volume on Excretion Urography. *J. Fac. Radiol.*, 9: 152, 1958.
20. Hershman, H. A., New Method of Determining Bladder Residual-Urine Volume. *J. Urol.*, 83: 283, 1960.
21. Mulrow, P. J., Huvos, A., and Buchanan, D. L., Measurement of Residual Urine with ^{131}I-Labeled Diodrast. *J. Lab. Clin. Med.*, 57: 109, 1961.
22. Rosenthall, L., Residual-Urine Determination by Roentgenographic and Isotope Means. *Radiology*, 80: 454, 1963.
23. Morris, A. M., Elwood, C., Sigman, E. M., and Catanzaro, A., Renal Clearance of ^{131}I-Labeled Meglumine Diatrizoate (Renografin) in Man. *J. Nucl. Med.*, 6: 183, 1965.
24. Cutler, R. E. and Glatte, H., Simultaneous Measurement of Glomerular Filtration Rate and Effective Renal Plasma Flow with Cyanocobalamin-^{57}Co and Hippuran-^{125}I. *J. Lab. Clin. Med.*, 65: 1041, 1965.
25. Nelp, W. B. and Wagner, H. N., Use of Radioactive Vitamin B_{12} to Measure Glomerular Filtration Rate (G.F.R.). *Proc. Amer. Fed. Clin. Res.*, 11: 93, 1963.
26. Sigman, E. M., Elwood, C., Reagan, M. E., Morris, A. M., and Catanzaro, A., Renal Clearance of ^{131}I-Labeled Sodium Iothalamate in Man. *Invest. Urol.*, 2: 432, 1965.
27. Concannon, J. P., Summers, R. E., Brewer, R., Cole, C., Weil, C., and Foster, W. D., ^{125}I-Allyl Inulin for the Determination of Glomerular Filtration Rate. *Amer. J. Roentgenol. Radium Ther. Nucl. Med.*, 92: 302, 1964.
28. Tauxe, W. N., Burbank, M. K., Maher, F. T., and Hunt, J. C., Renal Clearances of Radioactive Orthoiodohippurate and Diatrizoate. *Proc. Staff Meetings Mayo Clin.*, 39: 761, 1964.
29. Elwood, C. M., Armenia, J., Orman, D., Morris, A. M., and Sigman, E. M., Measurement of Renal Plasma Flow by Iodopyracet-^{131}I. *J. Amer. Med. Ass.*, 193: 771, 1965.
30. Wagoner, R. D., Tauxe, W. N., Maher, F. T., and Hunt, J. C., Measurement of Effective Renal Plasma Flow with Sodium Orthoiodohippurate-^{131}I. *J. Amer. Med. Ass.*, 187: 811, 1964.
31. Taplin, G. V., Dore, E. K., and Johnson, D. E., The Qualitative Radiorenogram for Total and Differential Renal Blood Flow Measurements. *J. Nucl. Med.*, 4: 404, 1963.
32. Meschan, I., Watts, F. C., Maynard, C. D., Schultz, J. L., Bolliger, T. T., and Morris, M. L., The Quantitation of the Renografin-^{131}I Renogram for Renal-Clearance Determination., *J. Nucl. Med.*, 7: 442, 1966.
33. Meschan, I., Watts, F. C., Lathem, E., Boyce, W. H., Schmid, H. E., Maynard, C. D., Roper, T., and Hosick, T. A., Simultaneous PAH, Inulin and Renografin-^{131}I Renal-Clearance Determinations and a Method for Calculating Renografin Clearance from Renograms in Patients. *Amer. J. Roentgenol. Radium Ther. Nucl. Med.*, 97: 909, 1966.
34. Lewis, H. D. and Bergentz, S. E., Renal Blood Flow Measurement with Xenon-133 at the Time of Operation for Renal-Artery Stenosis. *Surgery*, 59: 1043, 1966.
35. Kemp, E., Hoedt-Rasmussen, K., Bjerrum, J. K., Fahrenkrug, A., and Ladefoged, J., A New Method for Determination of Divided Renal Blood Flow in Man. *Lancet*, 1: 1402, 1963.

PART VII
RADIONUCLIDES FOR INDUSTRIAL APPLICATIONS

P. S. Baker
Isotopes Information Center
Oak Ridge National Laboratory

Prepared at the request of Division of Isotopes Development of USAEC for inclusion in the CRC Handbook of Radioactive Nuclides.

PART VII

RADIONUCLIDES FOR INDUSTRIAL APPLICATIONS

CHARACTERISTIC EFFECTS OF RADIATION

In previous sections of this handbook it has been shown that the basis for their utilization —whether it be in chemistry and biochemistry, biology, or medicine—is the fact that radioisotopes continuously give off radiations. The industrial uses follow the same pattern. Figure 1 illustrates the basic principles of the industrial applications.

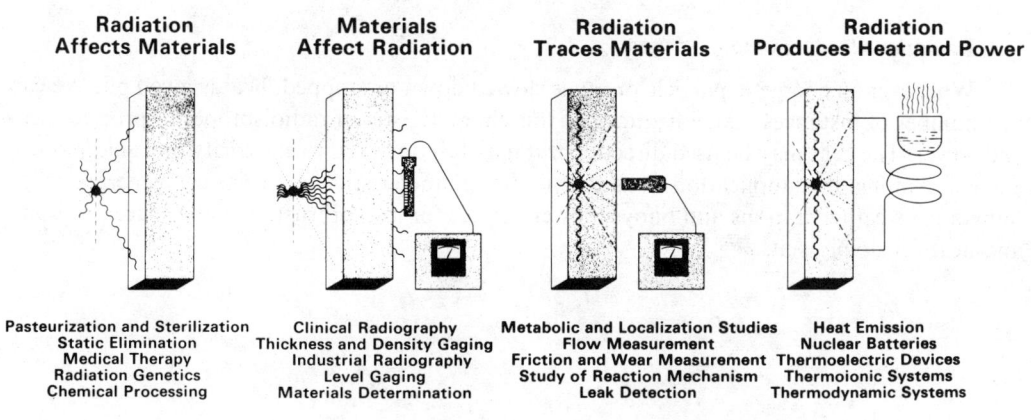

FIGURE 1.
Basic Principles of Radioisotope Utilization.

Radiation Affects Materials.

This particular characteristic can be compared to a suntan in that the radiation affects to various degrees any materials exposed to it. This permits such applications as pasteurization and sterilization of foods, polymerization of organic compounds, sterilization of medical supplies, and elimination of static electricity. In general, we can think of these effects as resulting from the properties of "ionizing radiation."

Materials Affect Radiation.

Carrying through the suntan analogy mentioned above, one can conveniently think of this effect as similar to the use of sunglasses, where the intensity of the sun's rays is reduced by the use of thicker (or darker) glasses: the intensity of nuclear radiation is reduced by thicker (or denser) materials introduced into the path of the radiation. This characteristic is responsible for such applications as radiography (picture taking *through* objects), locating or controlling hidden levels of solids and liquids (especially helpful when the liquid is viscous, hot, corrosive, or under pressure), and determining thicknesses of materials. This property also permits the user to protect himself from radiation by placing suitable "shielding" around the radioisotopes or around himself, as discussed in the next chapter.

Radiation Traces Materials.

Radioisotopes differ from their stable brothers and sisters in that they are able to "announce" their presence through the radiations that are given off. Except for this fact, however, their chemical behavior is identical to that of the stable species. Thus, not only do the radionuclides take part in any reaction or process, but they continuously identify their exact location by the signals they give off; all that is necessary is some sort of device that can "hear" or detect the signals; fortunately many such devices are available, as discussed in Part III. This particular characteristic is the basis for the largest variety of isotope applications. It permits us to test wear, to locate leaks, to trace fluid flow, to evaluate detergent efficiency, etc. It should be remembered, however, that the tracing principle is not a prerogative of any particular field; it is equally at home in any industrial environment.

Neutron activation *in situ* provides a variation of the tracer technique and is rapidly growing in popularity. It is useful in all industries for determining trace elements and has the advantage of not requiring the protection and precautions ordinarily required for radioisotopes added to systems. Neutron generators are used to activate short-lived isotopes in process streams or analytical samples, with the interpretations made immediately. Furthermore, the results can often be fed back into the system to provide automation.

Radiation Produces Heat.

Whenever an energetic particle or ray is slowed down or stopped, heat is given off. We can in a number of instances take advantage of this characteristic of radioisotopes to produce heat and power. The heat may be used directly, or it may be converted to electricity or to mechanical energy. Among the applications based on this characteristic are electrical generators for unmanned weather stations and buoys, power devices for microthrusters in the space program, and heat for diving suits.

AVAILABILITY OF ISOTOPES

Although isotope production is covered in Part I, something should be said about the availability of suitable isotopes for industrial applications. A few radionuclides are naturally available, radium and radon probably being the best-known. ^{40}K occurs in small amounts in the human body; ^{14}C occurs in nature, as do ^{87}Rb, ^{113}Cd, ^{187}Re, ^{147}Sm, ^{124}Sn, and ^{50}V. However, their half-lives are generally so long, or their abundances so low, that they are unsatisfactory for most isotope uses. Hence it becomes necessary to manufacture most of the isotopes that are to be used. Thus, most of the radioisotopes used industrially are man-made and are referred to as *artificial isotopes*. These may be produced in reactors, in accelerators such as cyclotrons, or may occur as a result of the fissioning of uranium or plutonium in reactors. At the present time, approximately 200 radionuclides can be considered commercially available; on the other hand, perhaps 400 more have suitable properties and may well be useful if satisfactory methods of production can be developed.

ISOTOPES FROM REACTORS

Reactor production of isotopes depends upon the interaction of neutrons (which are available in excess in any reactor) with nuclei of target atoms. Table 1 lists the types of neutron reactions that are commercially involved, along with examples of specific reactions in each category. The most important of these by far is the (n,γ) reaction, involving so-called "thermal" neutrons, which are captured with coincident formation of gamma rays. Also shown are some common reactions involving "fast" neutrons—that is, sufficiently energetic to simultaneously remove other particles as they penetrate the nuclei. In general, the more energetic the impinging neutron, the more extensive the damage to the nucleus (i.e., the more mass is knocked out). Table 2 lists some of the common reactor-produced isotopes and shows the method of production.

ISOTOPES FROM ACCELERATORS

In a manner similar to the way in which fast neutrons can affect nuclei, high-energy charged particles from an accelerator can also affect nuclei by "knocking out" particles. Accelerators generally produce what are known as "neutron-deficient" isotopes, whereas the reactors, for the most part, produce "neutron-excess" isotopes. Table 3 lists some charged-particle reactions involved in the production of radioisotopes, and Table 4 some common accelerator-produced isotopes.

ISOTOPES FROM FISSION

Table 1 includes the reaction (n, fission), which is responsible for a number of isotopes, such as ^{131}I, ^{137}Cs, ^{144}Ce, ^{90}Sr, ^{99}Tc, ^{133}Xe, and ^{91}Y. There are two significant characteristics of the fission process: the large quantities of resultant radioisotopes, and the fact that they are produced whether we want them or not. Megacurie quantities are available as "nuclear wastes" from the numerous reactors around the world; a single batch may easily involve hundreds of thousands of curies. Except for ^{60}Co, and perhaps ^{192}Ir, the production of reactor and accelerator isotopes is limited to quantities ranging from millicuries to perhaps a few hundred curies. Furthermore, the fission-product isotopes tend to be rather cheap, with the price per curie often lower than the price per millicurie for reactor and accelerator isotopes.

Table 1. NEUTRON-INDUCED REACTIONS

Type	Example
(n,γ)	$^{46}Ca(n,\gamma)^{47}Ca$ $^{197}Au(n,\gamma)^{198}Au$
(n,fission)	$^{235}U(n, fission)^{131}I$ $^{235}U(n, fission)^{90}Sr$
(n,p)	$^{32}S(n,p)^{32}P$ $^{14}N(n,p)^{14}C$
(n,α)	$^{40}Ca(n,\alpha)^{37}Ar$ $^{6}Li(n,\alpha)^{3}H$
(n,γ) $\xrightarrow{decay}$	$^{124}Xe(n,\gamma)^{125}Xe \xrightarrow{ec} {}^{125}I$ $^{198}Pt(n,\gamma)^{199}Pt \xrightarrow{\beta} {}^{199}Au$

Table 2. ROUTINELY AVAILABLE RADIOISOTOPES

Isotope	Production Method	Isotope	Production Method
^{37}Ar	$^{40}Ca(n,\alpha)^{37}Ar$	^{125}I	$^{124}Xe(n,\gamma)^{125}Xe \xrightarrow{ec} {}^{125}I$
^{45}Ca	$^{44}Ca(n,\gamma)^{45}Ca$	^{131}I	$^{238}U(fission)^{131}I$
^{47}Ca	$^{46}Ca(n,\gamma)^{47}Ca$	^{55}Fe	$^{54}Fe(n,\gamma)^{55}Fe$
^{14}C	$^{14}N(n,p)^{14}C$	^{59}Fe	$^{58}Fe(n,\gamma)^{59}Fe$
^{144}Ce	$^{238}U(fission)^{144}Ce$	^{32}P	$^{32}S(n,p)^{32}P$
^{137}Cs	$^{238}U(fission)^{137}Cs$	^{42}K	$^{41}K(n,\gamma)^{42}K$
^{198}Au	$^{197}Au(n,\gamma)^{198}Au$	^{35}S	$^{35}Cl(n,p)^{35}S$

Table 3. IMPORTANT REACTIONS FOR CYCLOTRON-PRODUCED RADIOISOTOPES

Reaction	Example	Reaction	Example
p,n	$^{7}Li(p,n)^{7}Be$	p,pn	$^{48}Ca(p,pn)^{47}Ca$
d,n	$^{72}Ge(d,n)^{73}As$	d,p	
d,2n	$^{52}Cr(d,2n)^{52}Mn$	p,d	could be used if needed
d,α	$^{24}Mg(d,\alpha)^{22}Na$	p,γ	
p,α	$^{25}Mg(p,\alpha)^{22}Na$	α,n	

Table 4. CYCLOTRON-PRODUCED ISOTOPES

Radionuclide	Half-Life	Radionuclide	Half-Life	Radionuclide	Half-Life
^{7}Be	53 days	^{67}Ga	77.9 hours	^{123}I	13 hours
^{18}F	1.87 hours	^{68}Ge	280 days	^{139}Ce	140 days
^{22}Na	2.58 years	^{74}As	18 days	^{143}Pm	265 days
^{44}Ti	~1,000 years	^{84}Rb	33 days	^{144}Pm	380 days
^{48}V	16.1 days	^{85}Sr	64 days	^{146}Pm	1,600 days
^{51}Cr	27.8 days	^{87}Sr	2.8 hours	^{181}W	145 days
^{52}Mn	5.7 days	^{86}Y	15 hours	^{195}Au	180 days
^{54}Mn	314 days	^{87}Y	80 hours	^{196}Au	6.2 days
^{55}Fe	2.6 years	^{88}Y	105 days	^{197}Hg	65 hours
^{56}Co	77.3 days	^{95m}Tc	60 days	^{200}Tl	26 hours
^{57}Co	267 days	^{103}Pd	17 days	^{201}Tl	73 hours
^{65}Zn	245 days	^{109}Cd	1.3 years	^{202}Tl	12 days
^{68}Ga	^{68}Ge cow, 280 days			^{207}Bi	28 years

RADIOISOTOPE UTILIZATION

Rather than discussing the industrial uses of isotopes under the four principles outlined at the beginning of this section, it may be more meaningful to relate these basic areas to specific industrial usage. For the purposes of the discussion we are using, in part, a breakdown suggested by A. D. Little, Inc.,* and are including eight general categories of industries: metals, electrical, transportation, chemical process, consumer products, crude petroleum and natural gas, mining, and utilities. In addition, agriculture (other than food processing) has been treated as an industry, thus allowing consideration of applications of radioisotopes to plants and animals and to the products from both, and so have aerospace and environmental uses in order to include some of the newer applications. In each case the order followed is: 1) tracing; 2) gaging; 3) radiography; 4) ionizing radiation; 5) heat and power. It should be noted that gaging and radiography both represent the principle "materials affect radiation"; however, both uses are important enough to warrant separate discussion.

On the other hand, no separate acknowledgement is given to firms providing research, engineering, and testing services, although they are a major segment of the industrial users of radioisotopes; the isotope applications in this industry are very similar to those found in the manufacturing companies and are covered sufficiently there. The largest group within this service industry is the considerable number of radiography laboratories, which offer nondestructive testing services to a variety of manufacturing firms. Other services include tracer studies, gaging, radiation research, and the rapidly growing field of activation analysis.

No attempt will be made to identify all of the hundreds of specific uses; however, enough typical examples are included to permit the reader to have a relatively complete insight into the subject. It should be pointed out that the basic principles are applicable to many different problems, so that a single representative photograph or drawing may fit many situations. However, there are instances where a specific principle of utilization that is widely used in one industry may be of little consequence in another.

Before considering the specific applications, a few comments should be made concerning the need for and the advantages and disadvantages of isotope methods. In general, the principal advantages of isotopes lie in their ability to get things done better, or faster, or more cheaply than by other methods. In some cases isotopes provide ways to do things that could not be accomplished otherwise (e.g., tracing complicated metabolic processes, or studying photosynthesis). However, it does not follow that the use of isotopes, *per se*, will always be better than conventional methods. In each case the relative merits must be weighed against each other.

The disadvantages of radioisotopes revolve primarily around the radiations themselves, since they might be harmful to the ultimate user of a product. However, now that more and more short-lived nuclides are becoming available, this disadvantage is becoming much less serious. A second possible drawback to the use of radioisotopes is the "red tape" involved in licensing procedures. However, this precautionary regulatory measure should be considered, in the light of safety and protection, as a public benefit—not a hindrance. A third possible disadvantage of radionuclides is the cost of equipment. However, radioisotopes are not a universal panacea, and one must judge each application on its own economic merits: if it can't be justified in terms of dollars saved, or a better product, or better customer relations, then it shouldn't be used.

---

* "Isotopes in Industry," USAEC Report NYO-3337-16 (September 1965).

TRACING

The use of radioisotopes to trace the flow of materials is probably the most widely used industrial application of isotopes—largely because there is usually no other satisfactory way to do the job. Since the isotopes do not lose their identity during chemical or physical processing operations, the path of a material can be followed despite any changes in its chemical structure.

Since very sensitive radiation detectors are available, even such slow processes as friction and wear can be continuously and effectively measured. There is no other known way to measure these effects "on the job".

GAGING

The principal advantages of radioisotopes for thickness gaging are related to the fact that the measurements are nondestructive and can be made continuously; furthermore, the signal can be used to provide the input for automation of the system. For density and level gaging, the isotopes allow measurement of hidden materials as well as of corrosive, abrasive, high-temperature, or high-pressure materials. There are no real disadvantages to these uses, since the sources are sealed and no radiocontamination of the product results.

RADIOGRAPHY

The radioisotope "X-ray" devices have the advantages of portability, freedom from need for an external power source, and applicability to odd and complex shapes. The large number of available radioisotopes provides a broad spectrum of energies. On the other hand, a possible disadvantage is the fact that the devices can't be "turned off".

IONIZING RADIATION

The advantages of radioisotopes as sources of ionizing radiation are just starting to be exploited on a large scale. Although not necessarily uniquely applicable, isotopes eliminate the need for catalysts in initiating chemical reactions, sterilize materials (by killing microorganisms) without the inherent disadvantages of chemical treatment or autoclaving, and change the molecular structure of substances to provide, for example, cross-linking of polymers or to induce mutations in plants. Radiation sterilization of foods is cheap, rapid, and effective, and does not require heat. The principal disadvantage, at present, is the high cost of a large-scale facility. However, this capital investment can be readily amortized if used effectively.

HEAT AND POWER

Perhaps the major advantage of radioisotopes as sources of heat and power lies in the fact that the energy is provided for long periods of time without need for "recharging" the energy source. This permits operation of remote unmanned weather stations, buoys, etc., as well as the supplying of heat and power to satellites, and makes self-contained life-support systems possible. The disadvantage, at present, is the high cost of the units.

APPLICATIONS IN THE METALS INDUSTRIES

The metals industries comprise *primary metals* and *fabricated metal products*, including machinery. The industry as a whole is dominated by steel, since in the U.S. nearly twenty times as much steel is produced as all other primary metals combined, with aluminum being its nearest rival. It is interesting to note that, with the exception of these two, the metal industries do not make significant use of isotopes—even in proportion to their sizes.

TRACING

In blast furnace operations, ^{59}Fe has been used to study residence time and distribution of constituents in the various metallurgical processes. Control of dust losses from blast furnaces and improvement of mixing during alloy formation are aided by the use of radionuclides. Other tracer studies have compared methods of chemically cleaning copper and stainless-steel parts, have evaluated tin-plating techniques, and have added to the knowledge concerning the structure of electroplated coatings. Isotopes have been used to evaluate the effects of the diffusion of gases into metals on embrittlement. Tracer studies yield valuable information regarding the rate of tool wear in punching and machining operations (Figure 2) and in the rate of wear of gear trains and other metal moving parts, thereby allowing manufacturers to evaluate cutting-oil stability and the characteristics of alloys, to specify optimal operating conditions, and to improve design methods. Engine pistons and cylinders, piston rings, bearings, engine gaskets, and lubricating oil can all be evaluated with the tracer technique by irradiating, for example, piston rings and cylinder liners to produce the isotope ^{59}Fe, or by irradiating bearings to produce ^{64}Cu and ^{65}Zn, and then observing the growth of radioactivity in the lubricant (Figure 3). Perhaps most important is the fact that systems can be tested while running (Figure 4), i.e., without dismantling, so that effects of starting, stopping, idling, dust feed (i.e., size of dust particles), speeding, torque, coolant temperature, and sudden temperature changes can be related to friction and wear; these tests take only a fraction of the time required with conventional methods. It has been found in the case of one particular single-cylinder diesel engine that the piston rings were fully broken in within six hours, and that the wear rate then dropped sharply and remained almost constant. The advantage of chrome plate over unplated cast iron for diesel compression rings has also been proved by the tracer method.

Nucleonic equipment has found a very profitable application in a system called the "coke ray" for controlling the alignment of the "pusher" and "receiving car" in discharge operations from batteries of coke ovens. It is critically important to make sure that the receiving car is exactly in position before the coke is pushed out of the oven. A link-up system using a cobalt source on one end and a radiation detection device at the other end has been found to be the most effective system under the extremely rugged conditions surrounding the oven operation.

Of interest to the electroplating sector of the metals industry is a project, now in progress, to determine the efficiency of chromium removal from chromium-plating solutions by ion exchange. A minute quantity of ^{51}Cr is added to the plating baths; when the capacity of the ion-exchange columns to remove chromium from the spent bath is exceeded, radioactivity is detected in the effluent from the columns, and the solution is then automatically routed to fresh ion-exchange columns. Thus the amount of chromium inadvertently dumped into sewers can be minimized, thereby reducing the contaminants fed to streams in industrial waste water.

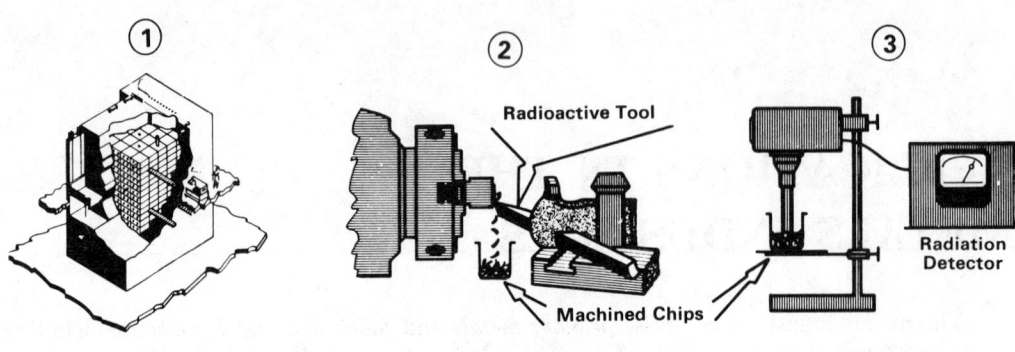

Advantages: more reproducible and sensitive than other tests
faster and more efficient
yield knowledge of wear process

FIGURE 2.
Measuring Cutting-Tool Wear and Life by Radioactivity Tests.

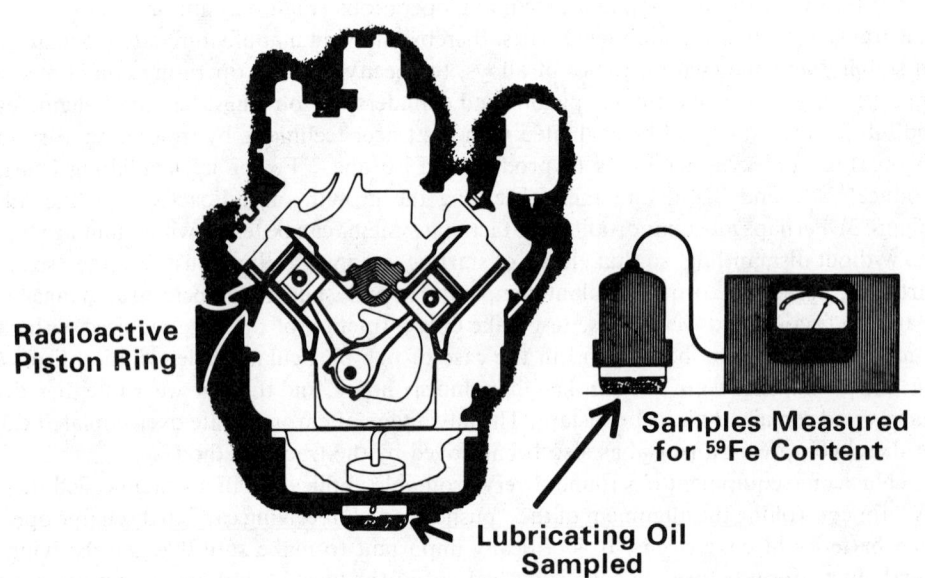

Advantages: transfer of metal measured to 1/100,000 ounce
oil sampled during operation of motor
developed film shows location of wear

FIGURE 3.
Radioactive Iron (^{59}Fe) for Friction and Lubrication Studies.

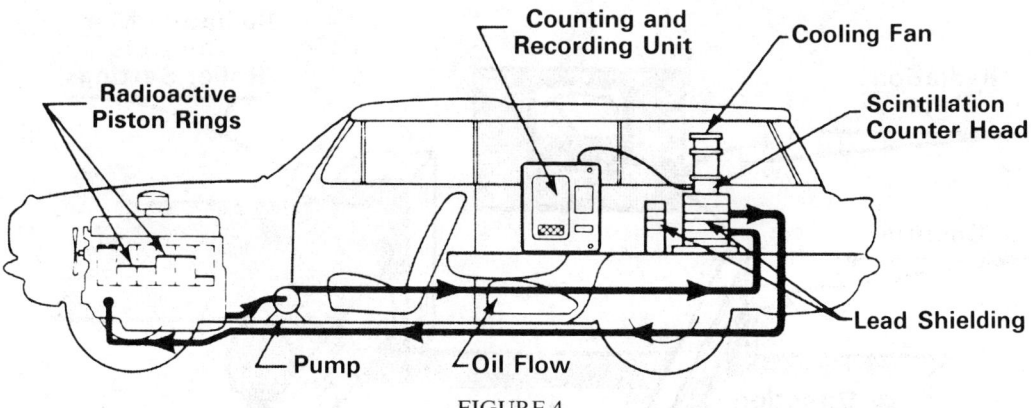

FIGURE 4.
Radioisotopes Permit Measurements During Operation.

GAGING

Gaging techniques are based, in general, upon the measurement of the mass of material between a radioisotope source and a detector; hence the "sunglasses" effect is one involving a thickness–density relationship, making possible the measurement and control of thickness, density, and levels of liquids and solids.

Thickness Gaging

The merits of radioisotope gaging are well recognized by industry, since the gages permit much closer tolerances in film thickness. In the past, continuous control of the uniformity of thickness of various kinds of sheets was difficult. In some cases it was necessary to shut down a machine long enough to make a measurement, but this resulted in lost production. In other cases the sheet moved continuously between the rollers of a gage, but this often streaked the material, especially hot metals. It can be seen that, for a given density, a wide range of thicknesses can be readily measured with great accuracy by choice of appropriate energies in beta or gamma sources, since the thickness is the only variable in the thickness–density relationship. An important advantage of the isotope gage is that it makes continuous monitoring over large areas of the material possible without contact. Furthermore, the system can be completely automated, so that the response to thickness changes is used to actuate the rollers, thereby providing closer control than would otherwise be possible and upgrading the quality of the product. One major steel manufacturer alone has 75 nuclear thickness gages, and the aluminum industry uses them for quality control in foil production. Figure 5 shows the principle of a thickness gage.

A rather interesting application involves the measurement of wall thickness of hollow turbine blades through injection of a radioactive liquid into the blades. Thickness of firebrick lining can also be gaged with radioisotopes, thereby providing a measure of wear (Figure 6).

Related to this, and almost as important as transmission gaging, is "backscatter gaging" (Figure 7). When radiation enters matter, some of it is reflected or backscattered, and the amount backscattered can be used as a measure of the thickness, weight, or density of the material. Thickness of boiler shells, pipe walls, or ships' hulls up to 3/4 inch in thickness can therefore be measured from one side in order to assess the effects of corrosion. By choice of an isotope source of suitable energy, the thickness of protective platings of many metals may be measured continuously, thus obviating the need for destructive assaying of samples. For example, a major application that has been widely accepted is in the manufacture of galvanized sheeting; beta-ray reflectance gages monitor zinc-coating thicknesses on both sides of the steel; improved uniformity and reduced zinc usage are benefits that have accrued. A similar application is the use of beta backscatter gages to measure the weight of vinyl coating applied to tinplate in tin cans.

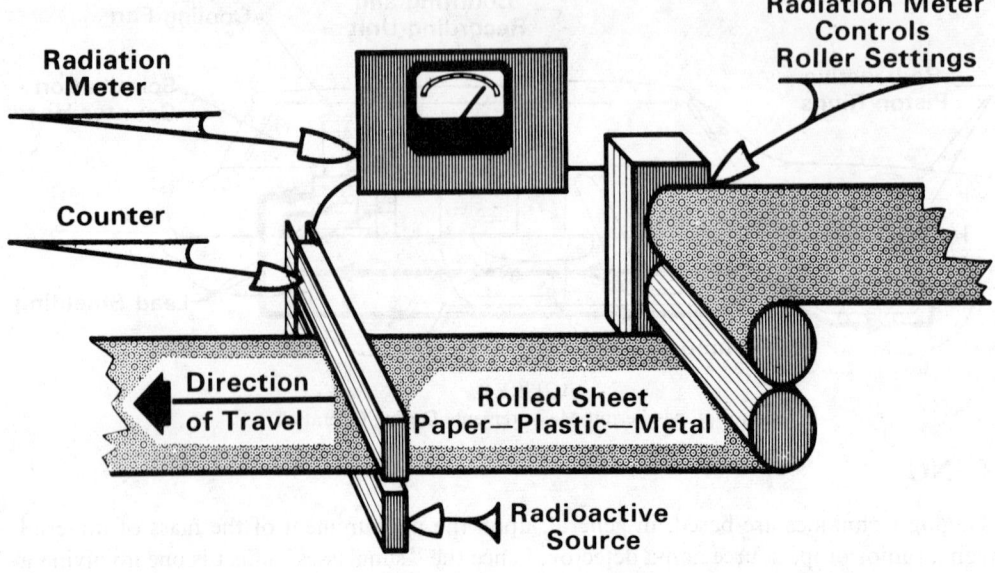

**Advantages: radiation source selected to suit material
no contact, no tearing, no marking of material
rapid and reliable**

FIGURE 5.
Radioactive Source for Gaging Thickness.

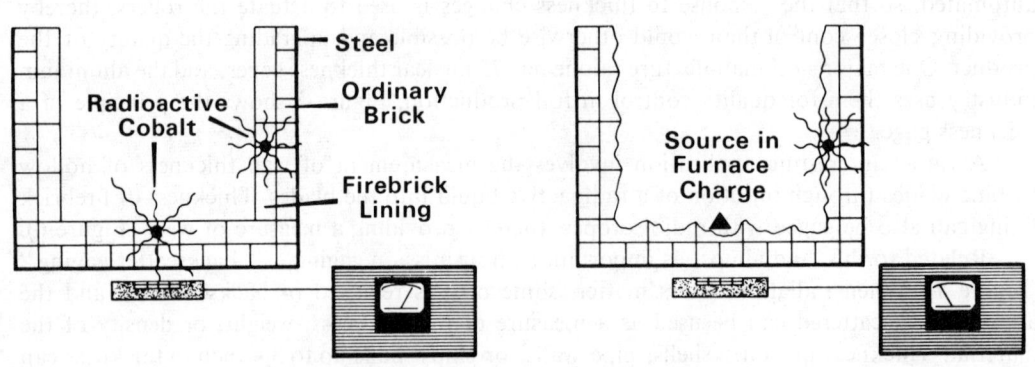

**Advantages: permits normal operation
warns of incipient failure**

FIGURE 6.
Measuring Wear of Firebrick Lining,
Using Radioactive Cobalt (^{60}Co).

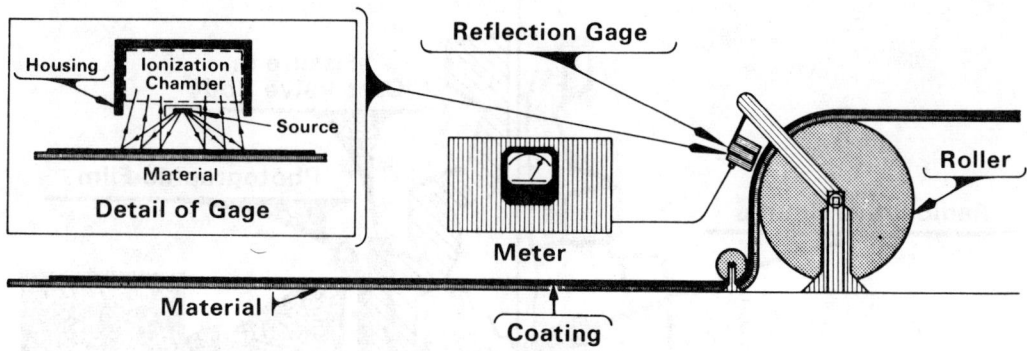

**Advantages: can measure thickness of coating and/or material
measurement made from one accessible side
can measure a variety of materials with one calibration**

FIGURE 7. Radioactive Source for Reflection (Backscattering) Thickness Gage.

Density Gaging

Like thickness gages, density gages are another important type of process control instrument (where the dimensions of a plate or cylinder of liquid or solid are kept constant) to measure densities of a variety of liquids, slurries, powders, and granular solids—usually by virtue of having the radioisotope source and detector mounted on opposite sides of the material being measured. An important application is the determination of the bulk density of coal in connection with the elimination of mechanical damage that results from the excess expansion of an overloaded coke oven. In the aluminum industry the devices are used for determining densities of slurries of bauxite in water.

A new gaging development involves the use of beta-ray backscattering devices to determine the hardness of steels. The effect, discovered empirically, seems to depend on the relationship between backscattering and grain size, which is in turn related to hardness.

Level Gaging

Level gages are used to determine liquid-metal levels in continuous-casting equipment and the level of burden in a blast furnace during production of pig iron.

RADIOGRAPHY

The major advantages of radioisotope radiography over conventional X-ray inspection are lower investment cost, portability, absence of wires, and ability to make exposures with the source of radiation placed inside a complex shape (e.g., a valve body).

The use of ^{60}Co for flaw detection in masses of metal was one of the earliest applications of isotopes. Since the density of a casting, for example, is relatively uniform, any variations in thickness will decrease or increase the amount of radiation passing through the casting, depending upon whether a particular area is thicker or thinner. The principle is illustrated in Figure 8. Most large foundry operations maintain a selection of isotope sources, which usually include ^{60}Co, ^{192}Ir, and ^{137}Cs, but ^{155}Eu, ^{144}Ce, ^{170}Tm, ^{182}Ta, ^{169}Yb, and others have also been used successfully. X-ray machines are still used extensively, but isotopes are required where the shape and accessibility of the casting make the X-ray technique ineffective.

The portability of radioisotope sources is particularly advantageous for checking weldments. For example, in a large boiler fabrication plant (manufacturing boilers for high-pressure chemical reactors and both nuclear and steam generators) every inch of weld is radiographed because of the high stresses expected in service; for a unit 15 feet in diameter and 75 feet long, this involves as much as 500 feet of welded seams up to 8 inches thick. Both ^{60}Co and

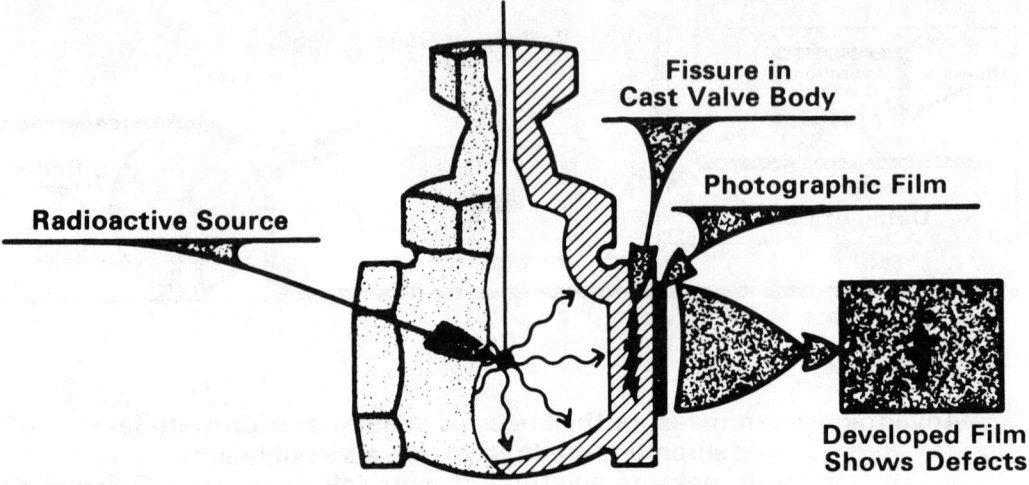

Advantages: versatile and reliable inspection
inspection made without dismantling
sources of desired shape and size
very-high-activity sources available at low cost

FIGURE 8. Radioactive Cobalt (^{60}Co) for Radiography Testing.

^{192}Ir are used in the plant, with a 2500-curie ^{60}Co source being the workhorse, since it is much better than even a 1-Mev X-ray machine. In another instance a firm used ^{60}Co on the premises to inspect 90,000-pound steel castings for defects, but used a scintillation counter and a simple strip chart instead of a photographic film to record the information; with X-rays, inspection would have required 600 exposures on $14'' \times 17''$ film after moving the castings to a site 200 miles away.

Machinery manufacturers—particularly producers of earth-moving equipment, crushers, and other devices subjected to high levels of impact and stress—use radioisotopes to detect flaws before it is too late to repair them. For example, subsurface defects that would make a final machined part useless can be detected and repaired by welding prior to machining. In the manufacture of turbines, ^{60}Co sources are indispensable for radiographing parts and detecting subsurface defects where shapes of the critical parts do not lend themselves to X-ray inspection.

Table 5 shows a few typical uses of radioisotopes in the metals industry.

Table 5. TYPICAL ISOTOPES IN THE METALS INDUSTRY

Applications	Isotopes
Tracing blast furnace operations	^{59}Fe
Study of hydrogen embrittlement	^{3}H
Study of piston-ring and bearing wear	^{59}Fe, ^{64}Cu, ^{65}Zn, ^{51}Cr, ^{63}Ni
Control of discharge in coke ovens	^{60}Co
Thickness gaging	^{90}Sr, ^{85}Kr, ^{170}Tm, ^{155}Eu, ^{144}Ce, ^{137}Cs
Measuring wear of firebrick linings	^{60}Co
Detecting thickness variation and defects in castings; weld inspection	^{60}Co, ^{192}Ir, ^{137}Cs, ^{145}Sm, ^{153}Gd, ^{155}Eu, ^{144}Ce, ^{170}Tm, ^{182}Ta, ^{169}Yb

APPLICATIONS IN THE ELECTRICAL INDUSTRY

In the electrical-equipment industry, radioisotopes are involved to a limited extent in the manufacturing of instruments and controls, appliances, and heavy electrical apparatus; they are used to a greater extent in various areas of research. Both radioisotopes and ionizing radiation are important to the manufacturers of electronic components in production control as well in research applications.

TRACING

The routine use by the electronics industry of ^{85}Kr leak testing has, between 1958 and 1966, grown from virtually nothing to the point where it was estimated that twelve companies employed this technique in about twenty facilities. The test procedure involves exposing a batch of the components to ^{85}Kr at 100 to 120 psi for 2 to 16 hours, depending on the specific activity of the gas; during this pressurization period any leaky components are at least partially filled with the radioactive gas. After evacuation to clean the surfaces of adsorbed radioactivity, the leaky components can be detected by the residual radioacitvity. In another instance the components tested are computer-grade transistors, microwave diodes, and integrated circuits; three hundred components are chosen at random, pressurized with ^{85}Kr for 2 hours, and then inspected in groups of four, to provide an acceptance procedure for the daily output of a transistor welder.

The use of tracers in production control has been employed by a manufacturer of printed circuit boards. Incorporation of tracers into masking agents and etch solutions permits, after cleaning of the boards, detection of minute residual quantities of these compounds, which are eventually responsible for defects in solder joints. The tracer technique is advantageous because it is nondestructive and yields results in a shorter time than the aging tests.

Some other uses of isotope tracers to solve specific problems are: 1) the investigation of stress-induced corrosion of steel in calcium chloride solutions by relating stress gradients to the diffusion rate of chloride ion in iron oxide films (this permitted an improved understanding of the corrosion mechanism, and ultimately the development of a method for inhibiting the corrosion); 2) a study of adsorption and desorption of mercury by glass surfaces in mercury switches; 3) studies of corrosion of silver contacts by a fused salt; 4) the development of a high-integrity compression seal; 5) evaluation of methods for cleaning metal surfaces prior to electroplating or enameling; 6) wear testing of shaft bearings; 7) determination of lubrication-seal characteristics; 8) investigation of the mechanisms of the diffusion of doping agents in semiconducting materials.

GAGING

The application of isotopic gaging to the electrical industries is rather limited. However, a successful use is that of a special isotope density gage for automatic control of the weight of lead oxide pasted on the grids of lead-acid storage batteries. Improved battery performance results from the increased uniformity of the plates, and in addition there is a reduction in oxide used.

RADIOGRAPHY

As was the case in the metal industries, radioisotopes are used to check the integrity of welds on structural components of heavy industrial electrical equipment.

IONIZING RADIATION

Isotope sources are used by manufacturers of precision balances for static elimination. Tritium sources are placed in these instruments to eliminate the small forces due to static and to the collection of dust on knife-edge pivots, thereby increasing balance accuracy.

Certain kinds of fire-detection equipment use alpha-emitting isotopes in an ionization chamber, taking advantage of the fact that trace quantities of smoke decrease the ion chamber current considerably, thereby indicating a potential fire.

A manufacturer of office copying equipment is experimenting with sealed sources of ^{85}Kr to eliminate static in copying machines. A significant market could develop in the office copy and document reproduction field if the static eliminators work satisfactorily.

In the instrument field, radioisotope-activated phosphors—using beta emitters such as tritium, ^{147}Pm and ^{85}Kr—are being used on dials and hands of timepieces and meters, since these luminous sources are considerably brighter than radium-activated phosphors and emit less external radiation.

The ionization of gases by radioisotopes is important to producers of gas-filled electronic tubes, where operation is stabilized by the presence of small amounts of isotopes in the tubes, which keep the gas in a state of continual ionization; as a result the performance is unaffected by variations in the intensity of light or cosmic rays. In a typical application of this principle, a company manufacturing microwave tubes puts a few microcuries of ^{60}Co inside the assembly, which give a rapid discharge within the gas upon excitation of the device. The use of ^{63}Ni is also being considered for increasing the efficiency of spark plugs.

HEAT AND POWER

There is considerable interest on the part of the electrical industry in the use of isotopes to replace batteries and related power sources. For the most part the applications have been of major interest to the space program and are discussed more completely under that heading. However, a demonstration of a generator for commercial use has been announced as a joint venture between private industry and the Federal Government; a 60-watt ^{90}Sr generator is providing power for flashing navigational lights and operating a foghorn on an unmanned offshore oil platform in the Gulf of Mexico; the expected five-year lifetime of the generator is a distinct advantage over diesel generators and battery systems now used, both of which require frequent maintenance at high cost.

Beta radiation from ^{147}Pm can be used to produce electrical power by interaction with n/p silicon solar cells; although the problem of radiation degradation of the diode is not completely solved, attractive conversion efficiencies have resulted.

A few examples of the uses of radioisotopes in the electrical industries are shown in Table 6.

Table 6. TYPICAL ISOTOPES IN THE ELECTRICAL INDUSTRY

Applications	Isotopes
Leak testing	^{85}Kr
Mercury-switch studies	^{197}Hg
Luminous dials	^{3}H, ^{85}Kr, ^{147}Pm
Pre-ionization of gases in electronic tubes	^{60}Co, ^{63}Ni
Power for navigational lights	^{90}Sr

APPLICATIONS IN TRANSPORTATION EQUIPMENT INDUSTRIES

The transportation industry in the U.S. grosses over 25 billion dollars a year, with approximately 80 percent split almost evenly between railroads and trucks; airlines and buses account for most of the balance. In addition to this, however, there is a 35-billion-dollars-a-year industry involving transportation equipment. There are a number of applications of radioisotopes in these areas.

TRACING

The measurement of the wear of bearings and other moving parts is important in the manufacture of all kinds of vehicles. Data can be obtained more rapidly by use of isotope methods, and the results are cheaper and more reliable. The evaluation of piston-ring wear has already been mentioned. One aircraft manufacturer has eliminated the problem of "bucking bars" (heavy metal pads that are placed behind rivets as they are driven into place) being left in the aircraft by placing microcurie quantities of ^{137}Cs in the bars. Prior to this the bars were frequently left in the aircraft and became a major hazard in flight. The same principle is applicable to metal tools, not only to eliminate hazards, but also to combat pilferage.

Radioactive gold has been employed in Austria in tests aimed at prolonging the life of railway wheels and tracks used by fast trains. It permits selection of an oil film only a millionth of a millimeter thick, which is spread automatically from a moving engine onto wheels and rails.

Although not directly concerned with the manufacture of transporation equipment, the problems of measuring and controlling both the flow of river sediments and the underwater movement of silt and shingle under the influence of tides and estuary currents are of importance to the overall transportation picture. Suitably tagged portions of the sediment can be traced; the results can be used in predicting the rate of filling up of natural and artificial waterways and can lead to large savings in the cost of dredging dock and harbor approaches.

GAGING

Although gaging is of limited importance in the actual manufacture of transportation equipment, suppliers make considerable use of this technique—for example, the gaging of sheet steel used in the manufacture of automobiles and of the plastic fabrics used for their interiors. Another application is in the quality control of lead-acid-battery separators. In the railroad industry there is interest in the possibilities of nuclear weighing, a technique whereby the mass in moving loaded railroad cars can be determined by the attenuation of gamma radiation from a source in the roadbed. The railroad industry has also done significant studies on the structural strength of timbers for bridges and road ties and on the composition of the soil underneath the road by use of nuclear gaging techniques.

RADIOGRAPHY

A major application for isotopes in manufacturing transportation equipment is in the radiography of welds and structural members. Radiographic equipment using ^{60}Co or ^{192}Ir is much more readily portable and can be used in areas that are inaccessible or inappropriate for X-ray equipment. This technique is well suited to the shipbuilding industry, and naval construction—particularly of submarines—takes advantage of isotope radiography.

A major airline is now using a ^{60}Co inspection gage for the routine maintenance inspection of the inner cores of jet engines; it considers that it has already saved a half-million dollars by introduction of this technique.

IONIZING RADIATION

The aircraft industry appears to have taken a strong lead in the use of radioisotopes for illumination. Illuminated emergency signs use a tritium- or ^{85}Kr-activated phosphor, are independent of external power supplies, and have a lower maintenance cost than the conventional battery–filament-bulb systems. The railroad industry is considering ^{85}Kr-activated phosphors for switch lamp signals in isolated locations, and the automobile industry has been licensed to use tritium, ^{85}Kr, and ^{147}Pm to illuminate door locks and ignition lights.

Typical uses in the transportation industry are listed in Table 7.

Table 7. TYPICAL ISOTOPES IN THE TRANSPORTATION INDUSTRY

Applications	Isotopes
Measurement of wear in pistons and bearings	^{55}Fe, ^{65}Zn
Studies of sediment and sand movement	^{133}Xe
Evaluation of rail life	^{198}Au
Gaging of automobile sheet steel	^{60}Co, ^{192}Ir, ^{137}Cs
Luminous locks and dials	^{3}H, ^{85}Kr, ^{147}Pm

APPLICATIONS IN CHEMICAL PROCESSING

The chemical industry realizes more than 30 billion dollars yearly; when one adds to this approximately 40 billion dollars for petroleum refining, 5 billion dollars for pharmaceuticals, 15 billion dollars for paper, and 10 billion dollars each for rubber and for stone, clay and glass, the overall chemical-process industries account for a large portion of the gross national product. As would be expected, the use of radioisotopes is widespread and includes all conceivable categories. Of special importance are the applications to process control and the use of tracer techniques in the optimization of production methods, in the design of new processes, and in research.

Petroleum refineries were among the first industrial operations to use radioisotopes. Refineries are characterized by having fluids not only as raw materials, but also as in-plant inventory and products; this provides a wide opportunity for instrumentation and makes refinery operations suitable for automatic computerized control; hence, radioisotopes find significant application in this industry. In the pharmaceutical industry, on the other hand, although sealed sources are used for density and level measurements in a few production operations, by far the most extensive use of iosotpes is as tracers in the research and development of new drugs.

TRACING

One of the original uses of isotope tracers is the location of obstructions in underground pipe lines, and this application remains a useful technique. Isotope sources, mounted in a rubber ball or other suitable carrier, have been used frequently to locate blockage in drains and pipe lines and mark their location, and to find "go-devils" lost during pipe line cleaning. In a recent case, a pipe line extending 25 miles from a compressor station to a polymer processing plant had been installed 30 years earlier, but had not been used for about 25 years; when a new use for the line was proposed, a cleaning tool with a sealed source containing 300 millicuries of radioactive gold attached to it was propelled down the pipe by air pressure to remove debris; when it struck an obstruction somewhere in the middle of the line, a crew with a Geiger counter located the obstruction within 25 feet and was able to remove it; without this technique the line would have been abandoned.

Leak testing is another one of the older "tried-and-true" applications of radioactive tracers. The leak is located by virtue of an accumulation of activity at the site of the leak (Figure 9). For leaks in underground piping, ^{131}I, ^{82}Br, ^{24}Na, etc., are often used. Leaks in heat exchangers can be located by the use of tritiated water or steam.

Radioisotope tracers are used by the various industries to study catalyst efficiency and the efficiency of separation procedures, and to measure flow patterns and mixing and residence times of fluids, gases, and solids in plant-kinetics studies designed to optimize the operation of industrial plants (Figures 10 and 11). For example, the flow of acids and other heavy chemicals in a series of reactors was studied with ^{86}Rb in order to optimize product yield. In many of these techniques activation analysis is of some advantage, since no radiation protection is needed and analyses can be performed separately.

In the petroleum industry, a major use of radioisotopes relates to the determination of mass balances on the various sections of the refinery. Injection of a labeled compound at an appropriate point in the fluid stream, along with the applications of sampling and measuring techniques

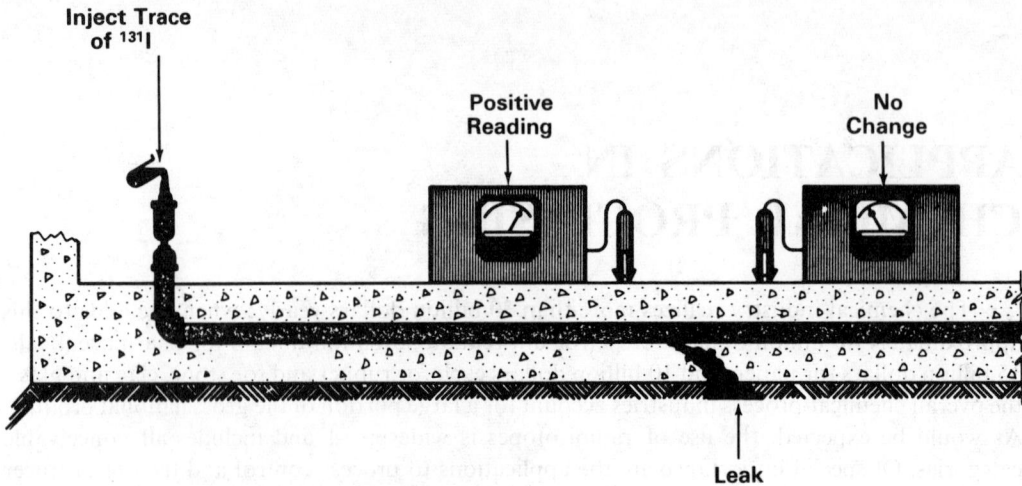

FIGURE 9. Leak Detection.

at other appropriate points, permits determination of residence times in various parts of the flow system and also indicates whether or not part of the product is being lost or mismetered. In one case, a mass-balance experiment using triphenylstibine labeled with ^{124}Sb pointed out a malfunction of one of the conventional flowmeters, that had been indicating a loss of $600 per day. Many refineries use ^{14}C to study reaction mechanisms, and other applications are concerned with determining the rates of loss of additives from polymers. Some refineries use isotope techniques for measuring gas flow velocities, primarily for use in the associated processing of petrochemicals. Very good results have been obtained by injecting radioactive krypton or xenon into the gas stream and measuring the time required for the activity to move a fixed distance down the pipe line. Neutron-activation analysis can be used to determine trace amounts of vanadium and other catalyst poisons in petroleum feed stocks to catalytic cracking towers.

Since the greatest portion of petroleum products is used by the transportation industry—in automobiles, trucks, locomotives, ships, and aircraft—as fuels and as lubricants, the refining industry has considerable interest in the technology of engines. To this end, radioisotopes are

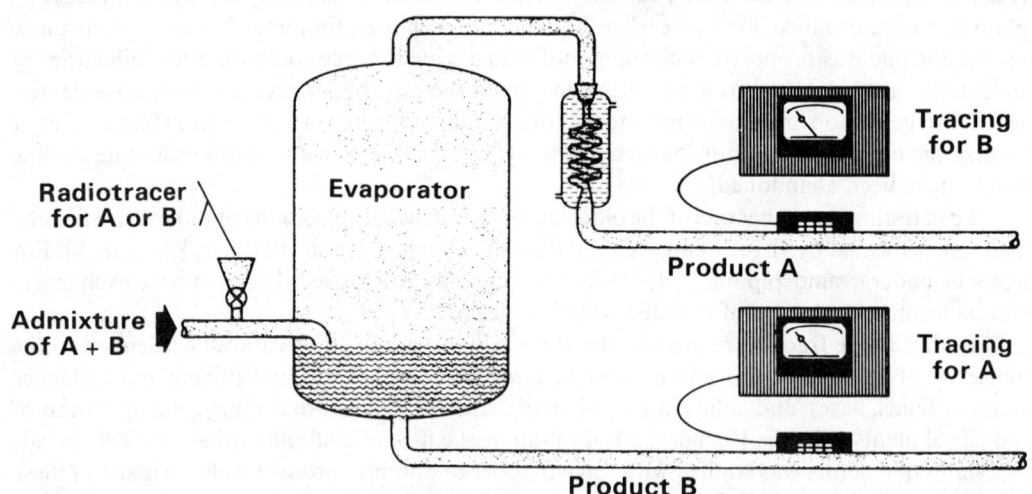

FIGURE 10. Measuring Efficiency of Separation, Using Radioactive Tracer.

used in work relating to this area—usually by irradiating pistons or cylinder liners in a reactor and then reassembling them into an engine for study of engine wear in the laboratory. This can be followed by monitoring the appearance of radioactivity in the lubricant. The points of greatest wear can be located by plating chromium over various parts of the piston and separately monitoring for radioisotopes of chromium and iron.

Several years ago, those components of an inexpensive diesel fuel that were responsible for the increased wear in diesel locomotive engines were determined by a technique similar to that used for automobile pistons, and this yielded results of direct economic benefit; it permitted the cheaper fuel to be treated, and its subsequent use saved the railroad some four hundred thousand dollars a year.

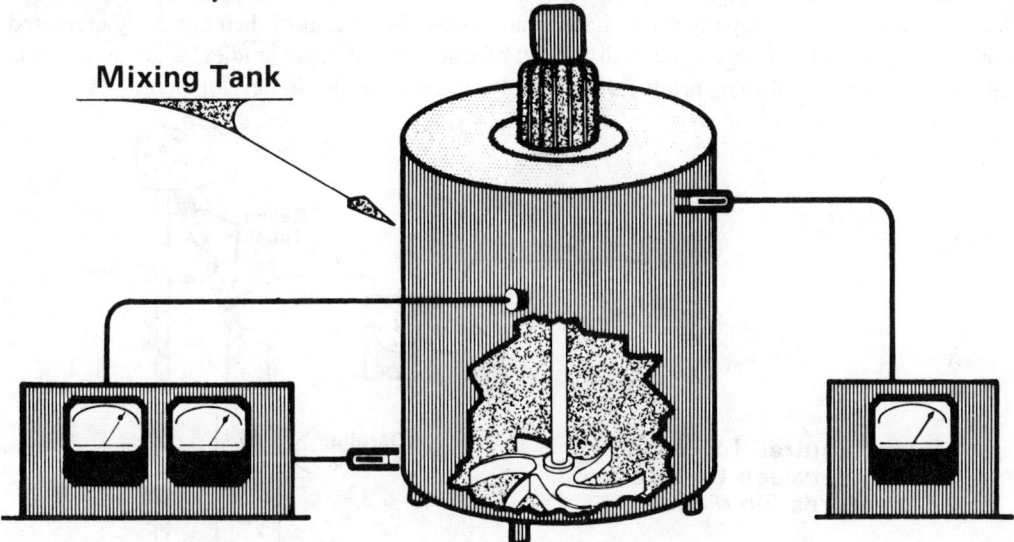

Advantages: uniformity of mixing easily assured excessive mixing time eliminated

FIGURE 11. Radioactive Isotopes for Determining Thoroughness of Mixing.

Also of interest to the petroleum industry is the study of catalytic cracking operations. Catalyst beads are tagged with some suitable gamma emitter, such as ^{60}Co or ^{95}ZrNb, and inserted into the fluid bed (Figure 12). It is possible to determine flow rate, particle distribution within the chamber, residence time, and effects of chamber geometry on the results of the cracking operation. Corresponding information for the gas stream can be obtained by using ^{133}Xe or ^{85}Kr.

In fuels research, radioisotopes are used to determine the component responsible for carbon deposits; this facilitates the formulation of more efficient fuel combinations on a nonempirical basis. Gaseous radioisotopes such as ^{85}Kr have also been used to study air-flow patterns in carburetors.

In the pharmaceutical industry, isotopes are never used in production equipment; production-oriented tracers are, therefore, limited. A tracer use indirectly related to production was the measurement of the extraction of zinc stearate from plastic bags containing saline solution, using radioactive zinc as the label, in order to confirm that the grade of plastic used was suitable for containing the product for use in transfusions.

On the other hand, in pharmaceutical research the tracer technique finds perhaps its greatest support. The significance of radioisotopes in drug research is related to two properties of labeled compounds: the extreme sensitivity of analytical methods to radiation, and the fact that the determination is independent of the actual chemical structure of the labeled compound.

These properties permit the drug researcher to detect, isolate, and measure not only the original labeled compound, but also its derivatives. Thus considerable time is saved in the investigation of the metabolic fates of drugs, where the original compounds 1) may travel several biochemical routes through the living system, 2) often degrade to smaller chemical fragments, and 3) are oxidized, reduced, or in other ways changed from the original chemical structures. Much of the pharmaceutical tracer work involves ^{14}C or tritium, since these can be readily used to tag complex organic molecules; ^{32}P and ^{35}S are also used, and in special cases, such as routine assay of vitamin B_{12}, ^{60}Co or ^{58}Co is the tracer atom.

Particularly significant is the importance of radioisotopes in helping to provide the detailed knowledge of the physiological disposition of a drug in connection with application to the FDA for approval. At one company it was stated that at least 50 percent of their currently marketed ethical drugs had been developed with the aid of tracer metabolism studies, whereas 10 years ago only 10 percent of their products had involved radioisotopes in development.

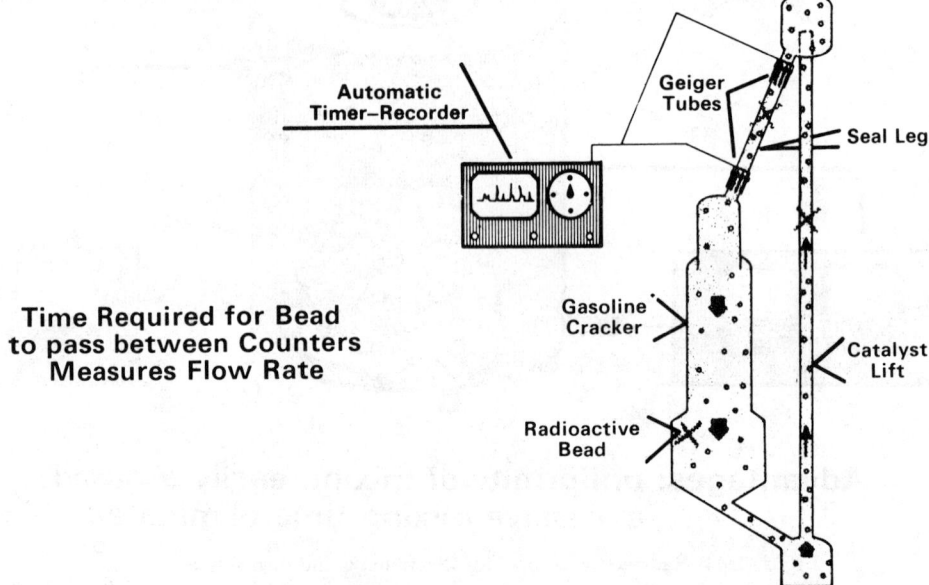

FIGURE 12.
Control of Catalyst Flow Rate, Using Radioactive Beads.

Radioisotope-labeled drugs are very important in studies involving veterinary medicines for administration to animals that are eventually to be eaten; it is absolutely essential to establish the time required for complete elimination of the drug from edible tissues, so that there will be no chemical residue for human consumption.

In many cases analytical techniques involving radioisotopes are far superior to the more conventional methods, and much time and expense can be saved by using them. Tracers are also used during process design to study mixing efficiency, residence time in chemical reactors, and flow rates and patterns in chromatographic columns.

In the rubber industry, some companies use radioisotopes to study the mixing of carbon black, zinc oxide, and other pigments and fillers that are used in the rubber formulation. This makes it possible to determine the degree of dispersion of these components and to correlate the physical properties of the rubber with the effectiveness of the mixing. Studies of the vulcanizing reaction have been carried out using ^{35}S, and internal frictional forces in rubber under dynamic stresses have been studied by using ^{14}C-labeled molecules. Migration of the components in a rubber formulation and its permeability to gases have also been studied by using isotopes.

Paper mills and pulp plants occasionally use tracers to optimize processes. Radioactive wood chips, for example, can be traced through continuous digesters to study the dwell time and rate of passage of the chips as well as the extent of spiraling and channeling through the vessel; this also provides valuable information in connection with the digestion process.

Paper companies are using radioisotopes to investigate the flow patterns in a waste outlet prior to the expansion of a paper mill. In process streams, the rate of dilution of mill wastes in a river is determined by the introduction of an isotope in the mill effluent and counting numerous water samples downstream. In another instance, lanthanum is introduced into wood pulp, and the amount of pulp discharged into various streams is determined by taking water samples, activating the lanthanum in a reactor, and then measuring it.

In one case a paper mill used a neutron-activated impeller in a high-speed blender to study the abrasiveness of filler materials for paper; the amount of radioactivity picked up by a slurry of the filler material was correlated with the results of standard gravimetric wear tests. This method was a good measure of abrasiveness and was faster, more sensitive, and more convenient than the usual gravimetric techniques. Furthermore, the technique could be used to measure the abrasiveness of fibrous materials, such as asbestos, which gave inconsistent results with the conventional method.

In the stone, clay, and glass industry, tracers are used to study the flow of molten glass from the furnace through the machinery for making glass bottles. In research, radioisotope tracers have been very helpful in studies on the diffusion of chloride, calcium, and sodium ions into glass over a wide range of temperatures, up to the softening point of the glass.

With the increasing importance of pollution control, industries are experimenting with isotopes to determine the quantitative magnitude of their specific problems in waste disposal for both liquids and gases. Often there are holdup lagoons or ponds into which waste is discharged; radioisotopes provide a means of determining the residence time of pollutants in the settling ponds. In other instances, where plant effluents are diluted in rivers and streams, radioisotopes have been helpful in evaluating dilution rates.

Atmospheric pollution presents another problem of increasing significance. Smog, which is one result of our growing concentration of population, automobiles, and industry, has as some of its most irritating components substances that exist in such small quantities that ordinary analysis and detection methods do not give sufficient information as to the source of the smog or as to conditions that make the smog toxic. The extreme sensitivity of radioisotope tracer methods can provide answers to problems of smog formation and other types of air pollution. More specifically, isotopes have also been used in experiments at a refinery to determine the extent of air pollution attributable to open oil tanks or oil–water separators, or that of the pollution (especially SO_2) coming from a coal- or oil-burning steam plant. The concentration of hydrocarbons in the vapor is determined by mass spectrometry, and the residence time of the air is obtained from the rate of loss of an isotope such as ^{85}Kr or ^{133}Xe. Such information is frequently very important from a public-relations standpoint.

Other pollution work involving isotopes concerns studies of rate of degradation of detergents subjected to microbial action.

GAGING

There are many instances in the chemical and allied industries where radioisotope gaging devices are widely used.

Thickness Gaging

In the rubber industry, in order to achieve the close tolerances required in the fabrication of tires capable of long service under high-speed driving conditions, very precise quality control is essential in the manufacturing process. A most important step is the regulation of the thickness of the rubber coating on the nylon or rayon ply used in the tires. All five of the largest tire manufacturers in the U.S. have installed radioisotope gages on their rubber calenders for this purpose, with 100 or so of these gages presently in use by the industry.

A second major application of gaging is in regulating the thickness of plastic films. Most rubber companies produce films of plastic or rubber either as a primary product or as a component of some manufactured article. Both here and in the plastics industry, beta gages are standard equipment for quality control of film thickness.

In the paper industry, isotope gaging is widely used to measure thickness or basic weight of paper—usually with ^{85}Kr or ^{90}Sr beta gages. Often the gage is an integral part of a system to provide automatic control of the pulp feed; close control reduces "giveaway" without sacrificing quality. Furthermore, since the gage operates continuously, there is no downtime for sampling. Lastly, the gage reduces the time necessary for weight changes. These gages are also used by paper companies in coating and impregnating operations. Two gages are usually used, one before and one after the coating operation, and the outputs are subtracted and/or divided to provide control signals.

Another application of beta gages for paper-coating control is in the extrusion coating of both paper and board with polyethylene resin. In a typical installation the coating thickness varies from 0.3 mil to 8 mils on substrates ranging from 25 pounds per 3000-square-foot ream to 30-mil board. Since many of the products are made on a custom basis with relatively short runs, the major advantages of the two beta gages in this mill are in reducing startup time and rapidly attaining a flat coating profile, and in continuously controlling coating at high speed.

One of the most successful and widespread uses of isotope thickness gages has been in the manufacture of coated abrasive materials, such as sandpaper and emery cloth, where it is difficult to control uniformly the weight of adhesive and abrasive by others means. Gages are installed before and after each coating operation, and the signal representing difference in weight is used to automatically adjust the calenders (Figure 13). Users of this technique emphasize that increased uniformity of the product is the greatest advantage of this gaging system, which allows their product to command a premium price. Figure 14 shows a comparison of coating uniformity without and with isotope gaging control.

Thickness gages are not at all widely used in the stone, clay, and glass industry. One exception is the measurement of the thickness of gypsum wallboard at the wet end of the process.

Density Gaging

In many of the chemical industries the severe environmental conditions in manufacturing operations (high temperatures, abrasion, dust, etc.) favor instrumentation that can operate without physically contacting the material being processed.

Density gages are used in the paper industry to measure and control several aspects of the pulping operation. The measurement of moisture in wood chips is important to the economic operation of a digester; a combination of gamma density gaging and neutron moisture gaging has been successfully employed to continuously measure the moisture of wood-chip feed, allowing adjustment of liquor addition to maintain the digestion process at a rate sufficient to meet specifications in a fixed digestion cycle. The continuous isotope gaging method yields a reported 2 percent increase in pulp production with the same input of cordwood and digestion chemicals, representing savings of nearly $150,000 per year for a pulp plant producing 400 tons per day. The cost of the gage is approximately $24,000, including installation. Another important application of nuclear density gages to the pulping process is measurement and feedback control of the solids present in the feed slurry in order to give a satisfactory lime-mud thickness. Variations in the solids content of the filter cake can produce caking conditions in the limekiln and wasteful consumption of fuel. These variations can be avoided if the filter feed slurry is maintained at a constant ratio of solids to weak liquor. This is accomplished by continuously measuring the density of the feed stream to the filters with a gamma gage and using the gage output in a feedback loop to control the rate of underflow from the mud thickener. Other

Radionuclides for Industrial Applications 525

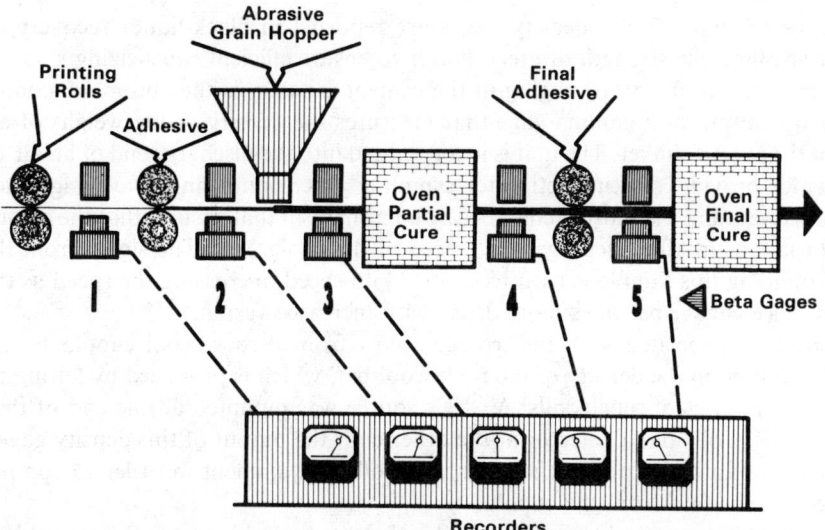

FIGURE 13. Use of Multiple Beta Gages.

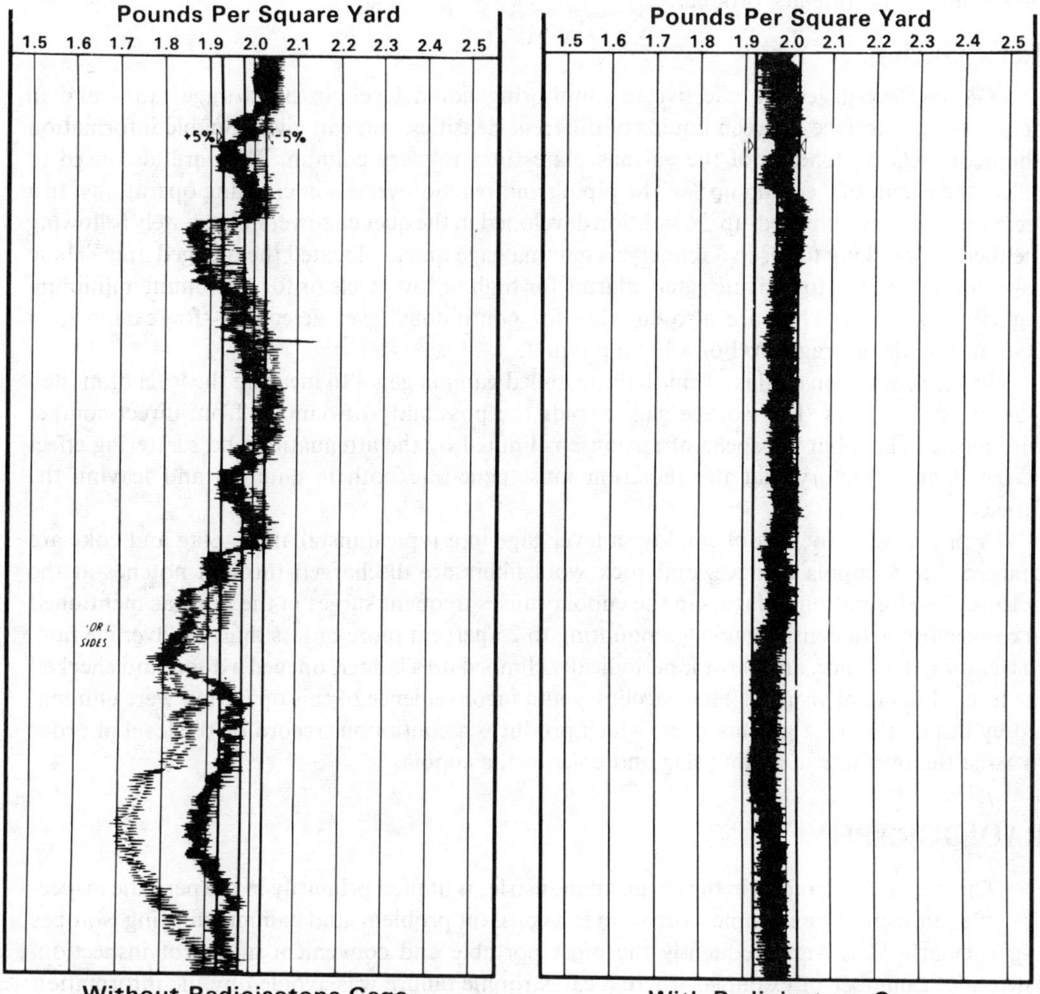

FIGURE 14.
Effect of Isotope Gaging on Coating Uniformity
(Courtesy of Industrial Nucleonics)

applications of radioisotope density gages are reported in black-liquor recovery operations and in controlling the strength of green liquor to ensure efficient causticizing.

An application of isotope gaging in the cement industry is the automatic control of kiln speed by the output of a gamma gage that measures the density or literweight of a classified sample of the cooled clinker. The unit is incorporated into the discharge end of an air-quenching cooler, which provides shaking action for sampling, rescreening, and conveying. The screened sample is conveyed to the sensing chamber by a mechanism that ensures that the chamber is full despite variations in kiln discharge rate, speed of the cooler, etc. The signal from the density gage surrounding this sample is used to control kiln speed, increasing the speed as the density increases, since clinker becomes more dense with increased fusion.

A problem associated with the irregular flow from a rock wool cupola furnace is the resulting variation in the density of the rock wool bat, which is produced by felting the cooled material with a series of pinch rolls. A ^{137}Cs source was mounted at one end of the gap in a densifying pinch roll, and a detector at the other end; the output of this density gage has been used to control automatically the speed of the rolling equipment in order to compensate for fluctuations in the feed from the cupola.

Density gages have rarely been used in the pharmaceutical industry, one possible exception being the installation of two density gages in conjunction with a volume flowmeter to control the flow rate of a proprietary product.

Level Gaging

Gamma-ray gages are effective in monitoring liquid levels in oil storage tanks and in detecting the interface between liquids of different densities, and can give valuable information about the relative loading of the various plates in a refinery column. They are also used to reflect the extent of "coking up" of the piping and reaction vessels in cracking operations. In a recent example, a "plugged-up" condition developed in the quench tower immediately following the thermal cracking tower in a refinery; a gamma gage quickly located the plugged area. These gages are often used to activate safety alarms for high or low levels or for maximum–minimum regulation systems. They are also suitable for continuous level detection—for example, in determining the average level of a boiling liquid.

In the production of glass, firms have installed gamma gages to measure the level of molten glass inside furnaces. The isotope gage avoids the possibility of damage from direct contact with molten glass, but the speed of response is limited by the attenuation and scattering effect of the thick refractory that the radiation must penetrate both in entering and leaving the furnace.

A producer of rock wool employs a level gage in a typical installation. Slag and coke are charged into a cupola furnace, and rock wool fibers are discharged through notches at the bottom. The formation of layers in the cupola causes frequent surges in the flow, as mentioned in connection with density gaging, amounting to 25 percent more or less than the average rate. In traditional practice, an operator periodically climbed up a ladder, opened a door, and checked the level of material visually. The inaccuracy and inconvenience of this operation were eliminated by installation of a gamma gage, which produces a continuous record of the level in order to assist the operator in adding slag and coke to the cupola.

RADIOGRAPHY

The use of radiography in the chemical industries is limited primarily to the periodic inspection of equipment. For example, corrosion is a constant problem, and gamma-emitting isotopes —particularly ^{192}Ir—are frequently the most portable and convenient means of inspection. There are a number of examples where a catastrophic failure was avoided by the information provided by isotopic radiographs, although this type of economic asset cannot easily be estimated quantitatively.

IONIZING RADIATION

The use of ionizing radiation in the various chemical-process and related industries includes static elimination, initiation of chemical reactions (including polymerization), and radiation sterilization.

In industries where powders, plastics, etc., are used or manufactured, there are serious hazards from dust explosions resulting from static sparks. The radiations from many isotopes have the property of ionizing the media through which they pass, including air. Ionized air will neutralize static electricity, thereby eliminating the dangers of static discharges (Figure 15).

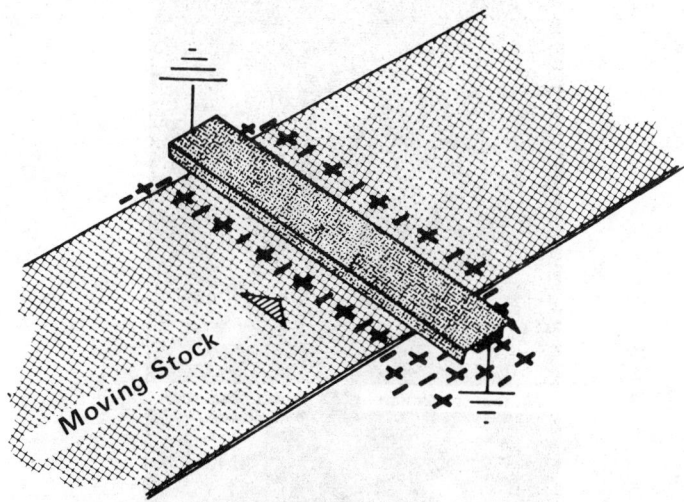

Static Eliminator

Radiation Preionizes Air, Providing Exit Path for Static

FIGURE 15.
Ionized Air Neutralizes Static Electricity,
Eliminating the Dangers of Static Discharges.

One of the first commercial-scale applications of radiation to chemical processing was initiated by the Dow Chemical Company in 1962 to produce ethyl bromide. A 3100-curie ^{60}Co source is used to bring about the reaction between ethylene and hydrogen bromide; an annual capacity of approximately one million pounds is reported, and the process is economically attractive because of higher yields and cheaper raw materials.

With the importance of biodegradable detergents being emphasized, it is significant that the radiation-induced sulfoxidation of hydrocarbons was developed by Esso Research at Linden, N. J., in a pilot-scale operation, and later carried on commercially by Farbwerke Hoechst in Germany. Oxidation of paraffins on a large scale is also tied to the use of gamma radiation.

Perhaps the most active area of radiation processing has been in the initiation of polymerization. Polymerization of ethylene is being studied intensively, using ^{60}Co, and the process appears to be suitable for a continuous-flow operation. Advantages of using ionizing radiation include elimination of catalyst impurities in the product, better control of rates of production and of molecular weight distribution during operation, and production of both high- and low-density polyethylene in a single plant by simply altering process conditions.

The successful manufacture of radiation-cross-linked polyethylene products by several firms is one of the major milestones in commercial radiation use. At present, however, electron beams from accelerators rather than radiations from isotopes are used for the most part. On the other hand, much of the development work involved ^{60}Co (Figure 16). The principal products are electrical insulation and packaging film. In the case of insulation, the chief advantage of the cross-linked product over the ordinary polymer is improved mechanical strength at elevated temperature; this material competes with fluorocarbon polymers. However, the shrink "memory" of the irradiated insulation is also important; stretched tubing can be heat-shrunk over electrical connectors.

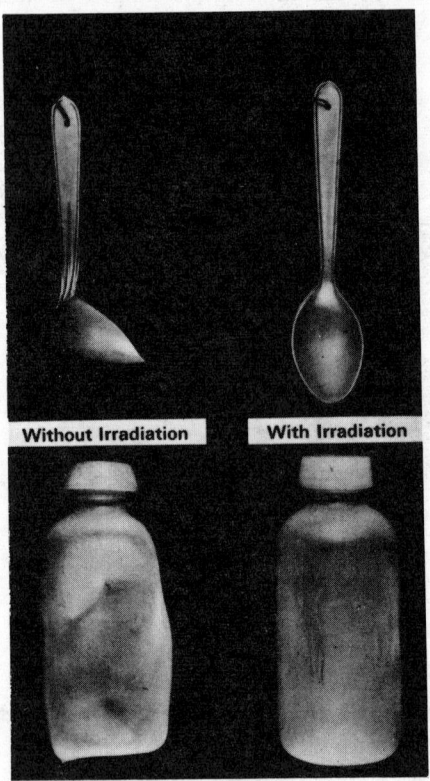

FIGURE 16.
Effect of ^{60}Co-Radiation-Induced Cross-Linking on Melting Point of Polyethylene.

The polymerization of various monomers within the structure of wood has recently been commercialized. Wood is placed under vacuum, then the pores are filled with a monomer (for example, monomethylmethacrylate), and the wood–monomer system is exposed to the radiation from ^{60}Co. Not only is the monomer polymerized, but a small amount of it appears to be "grafted" to the cellulose of the wood, thereby enhancing some of the wood properties (e.g., hardness, compression strength and abrasion resistance, dimensional stability, shear and static-bending strength). These "wood plastics" appear to be particularly suitable for use as veneers, furniture, and flooring as well as for other special products, including musical instruments, rifle stocks, golf-club heads, tool handles, and salad bowls. The product retains the natural wood grain and color, or can be artificially colored throughout; it can be sawed, drilled, turned, and sanded with conventional equipment to a hard, satin-smooth finish. However, the economic advantages of the irradiation process over heat-catalyst methods still require further evaluation.

A research use of radioisotopes by glass companies is also related to plastic products. Radiation-induced vinyl polymerization in thin films is being investigated by using a glass plate containing a beta-emitting isotope as a constituent of the glass; this arrangement allows containment of the isotope without excessive attenuation of the beta radiation. Films of monomer are deposited on the glass, and polymerization is initiated by the beta rays.

Radiation sterilization is included here as part of the pharmaceuticals industry, although it might equally well be covered in relation to consumer products. It has found a major use in the medical and hospital supply business, but not in the production of drugs. One medical-supply company subsidiary operates six plants that use gamma radiation from a total of nearly 800 kilo curies of ^{60}Co to sterilize bandages, surgical sutures, plastic blood-donor kits, catheters, disposable hypodermic syringes, rubber gloves, and surgical blades, needles and instruments. Initially the process depended upon high-energy (about 6 Mev) electrons from linear accelerators; now ^{60}Co is used, because of the advantages of radiation over conventional heat treatment, which include elimination of the need for aseptic handling prior to packaging of the product. Heat sterilization requires that the absorbable type of suture known as catgut be

Table 8. TYPICAL ISOTOPES IN THE CHEMICAL-PROCESSING INDUSTRY

Applications	Isotopes
Efficiency of separation	^{35}S, ^{3}H
Thoroughness of mixing	^{198}Au, ^{131}I, ^{24}Na, ^{56}Mn, ^{82}Br, ^{51}Cr
Leak location	^{131}I, ^{82}Br, ^{24}Na, ^{3}H
Gaging of liquid or solid levels	^{60}Co, ^{137}Cs
Study of process-stream flow patterns	^{86}Rb
Location of pipe obstructions	^{198}Au
Study of mass balances in refinery stream	^{124}Sb
Study of reaction mechanisms	^{14}C, ^{3}H
Measurement of gas-flow velocities	^{133}Xe, ^{85}Kr
Catalyst flow studies	^{95}ZrNb, ^{60}Co
Study of carbon deposits in fuel research	^{14}C
Drug-metabolism studies	^{32}P, ^{58}Co, ^{60}Co, ^{14}C
Study of vulcanizing process and tire wear	^{35}S
Study of frictional forces in rubber	^{14}C
Evaluation of plastic blood bags	^{65}Zn
Study of diffusion in glass	^{45}Ca, ^{24}Na, ^{38}Cl
Determination of air pollution from refinery	^{85}Kr, ^{133}Xe
Control of rubber thickness on tire ply	^{90}Sr, ^{85}Kr
Control of rock wool production	^{137}Cs
Initiation of chemical reactions; effecting of polymerization	^{60}Co, ^{137}Cs
Elimination of static	^{90}Sr, ^{85}Kr
Sterilization of medical supplies	^{60}Co, ^{137}Cs
Thickness gaging of paper and plastics	^{90}Sr, ^{85}Kr

carefully dehydrated and then heated in an open tube containing a high-boiling mixture of hydrocarbons; after sterilization the container must be emptied of this liquid, filled with a sterile tubing fluid, and sealed. All of these operations following the heating had to be carried out under aseptic conditions in order to prevent recontamination of the sutures, and this requires workers to observe the same precautions as in an operating room. Furthermore, the method is only practical as a batch process. Radiation sterilization, on the other hand, is accomplished after the suture is sealed in an aluminum foil package, which is irradiated to a dose of about 2.5 megarads; the package is conveniently opened by tearing, whereas the glass vial previously used had to be broken in the operating room.

Other reported advantages of the isotope radiations are economy of operation because of fewer personnel, less downtime for repair and maintenance, and the ability to sterilize products and package geometries that could not be penetrated by the electron beam. Formerly the sutures were belt-conveyed in shallow trays under the beam; the thickness was limited to an inch or two. Now the sutures are irradiated in racks, each containing four boxes of sutures, with a typical thickness of one foot.

Typical chemical-processing industry uses of isotopes are shown in Table 8.

APPLICATIONS IN CONSUMER-PRODUCTS INDUSTRIES

This particular group of applications has arbitrarily been broken down to include food, tobacco, textiles, and miscellaneous. Actually, medical supplies could also be logically included here, but this subject has already been covered.

TRACING

Radioisotopes are not used in tracer studies of production operations in the food industry because of regulations against the practice, although small quantities of short-lived isotopes would be well suited for determining the mixing of ingredients in sausage meat, for example. To avoid this problem, stable isotope tracers can be used, followed by neutron-activation analysis. But two categories of tracer applications, in particular, are worthy of mention in connection with the food-processing industry: one has to do with the study of pesticide removal from fruits and vegetables, and the other is concerned with food additives and food purity. We have arbitrarily chosen to cover these here rather than under agriculture.

Pesticide Removal. For a number of years the National Canners Association has been investigating pesticide residue on marketed raw food and has developed methods for studying the effectiveness of pesticide removal from fruits and vegetables by washing with various wetting agents and detergents. Tritium and tritium-labeled DDT were used in most of the studies.

Food Additives and Food Purity. Many chemical products get into foods in a variety of ways: insecticides and herbicides often remain on fruits and vegetables; chemicals frequently are added to foods to retard deterioration or to enhance appearance; still other additives are used to produce better-tasting products that stay fresh longer; or food wrappings may contain materials that can be transferred to the foods. Unfortunately, some of the additives can be cumulatively carcinogenic or toxic when ingested regularly in more than trace amounts. Under the new federal food and drug laws, chemical manufacturers or food processors must conclusively demonstrate the harmlessness of any toxicant or additive that is introduced into food products. Since analyses of these chemicals often involve identification of trace amounts of material (too small for ordinary analytical techniques), radioisotope tracers make it possible not only to detect the additives, but also to follow their metabolic fate.

Thus, one of the most useful and economically advantageous applications of isotopes in the food industry is in metabolic studies. Here—in connection with FDA certification studies, for example—experiments to determine the metabolic pathways of a given chemical intended as a food additive, or perhaps present as a plasticizer, antioxidant, or antistatic agent in a wrapping material, can be carried out by labeling with ^{14}C or, less frequently, with tritium. The chemical is fed to a laboratory animal; an accurate mass balance is kept of the labeled products exhaled and issued in its waste products; after a suitable period of time the animal is sacrificed, and the distribution of isotopes in its body is measured. This permits rapid and unambiguous determination of the distribution of a given chemical compound in the body. Prior to the availability of isotopes, these experiments usually required at least a year; then it was necessary to make morphological examinations of the sacrificed animal in order to determine whether or not any organic anomalies had developed. The isotope technique takes only about six weeks and is only about one-tenth as costly; for example, a major meat packer accustomed to carrying out two or three of these studies a year estimates that use of radioisotopes saves one hundred thousand dollars annually.

A rather indirect—but widespread—use of isotopes in the industry is in the tritium detection of vapor chromatographs; these instruments are widely used, particularly in companies manufacturing food products that are highly dependent on flavor and odor quality. In the USSR, the brewing industry has introduced an improved production process: by finding the lifetime of labeled lactic-acid bacteria, it has been possible to reduce the fermentation process from the previous five or six days to one or one-and-a-half days; the quality of the beer is also improved. A related field is the use of dating techniques to determine the true age of brandies for tax purposes.

In the tobacco industry, tracing is limited almost entirely to research, where ^{14}C and ^{3}H are used 1) in determination of yields at various stages of analytical procedures, 2) in studies of the biosynthesis of tobacco, and 3) in studies of the mechanism of formation of various pyrolytic constituents of tobacco—often with the help of gas chromatography. Studies of tobacco biosynthesis usually involve the growing of plants in an atmosphere containing ^{14}C-labeled carbon dioxide; in this way the distribution of the label within the structure of the plant and among the chemical compounds isolated from plant tissue can be determined. Labeled nutrients added to the soil are used for the same purpose. In many of the pyrolysis studies, the labeled compounds used in isotope dilution analyses of the constituents of smoke permit evaluations that could not be performed without isotope tracers. The development of analytical methods for trace compounds in smoke is an arduous procedure, and isotope tracers have been extremely valuable in determining the extent of the severe losses encountered in isolation techniques. Once a compound is identified, it is often more easily determined quantitatively by isotope dilution analysis than by conventional techniques.

Tracer techniques have been frequently reported in research papers on textile processes and materials. For example, the drawing of web has been studied with ^{32}P-tagged wool fibers by using two Geiger counters to measure the time required for passage of the material between two sets of rollers. In another case, the crimp of wool has been observed during various stages of processing by taking autoradiographs of fibers tagged with ^{32}P. The uniformity of uptake of various substances, such as oil in the oiling of yarn and resist salt in a padding operation, has also involved radiotracers. Others have investigated the effect of detergents in the washing process and of sequestering agents in the dyeing process. Finally, many tracer investigations deal with the nature of fibers, especially of natural fibers, and with the mechanism of absorption processes.

Another tracer application in the textile industry is the control of dye carry-over from one color box to a second of different color in the printing of fabric. To avoid off-colors, the contamination in the second dye box must be kept below a certain level; radioactive tracers readily monitor this level. There is some objection to this application, however, because of the possible radiation hazards to personnel and to the public.

Two interesting applications to miscellaneous consumer products have to do with the evaluation of detergent efficiency and wear characteristics of floor wax. How can a person *really* tell how good a washing machine is, or how effective a particular soap or detergent is? One way is to simply use bacteria labeled with a radioisotope to soil the test clothes and then measure the residual activity after laundering (Figure 17). This procedure makes it very easy to compare the cleansing action of different kinds of soaps and detergents. Similarly, if a small amount of radioisotope is incorporated into a test wax, then the amount of wear caused by scruffing can be determined quite easily—either by measuring the amount of activity removed by the scuffing, or by measuring the activity in the residual wax (Figure 18).

GAGING

As in all processing industries involving fluids (including flowing solids), radioisotope density and level gages also find application in food processing—e.g., for the control of the density of such products as tomato paste and fruit-juice concentrates or of water content in

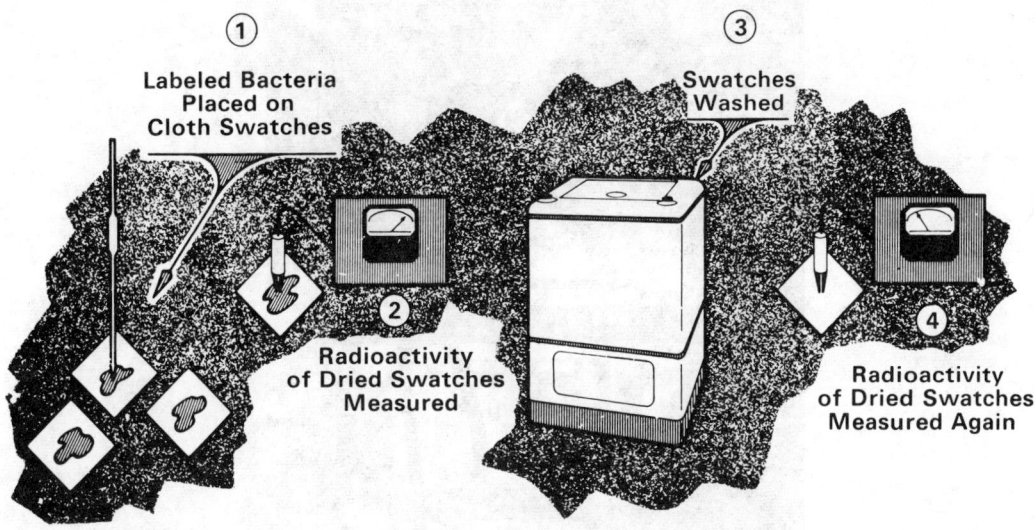

Demonstrates: efficiency of washing procedures for removing bacteria
comparative efficiencies of various detergents
amount of cross contamination

FIGURE 17.
Test for Washing Efficiency, Using Radioactive Phosphorus (^{32}P).

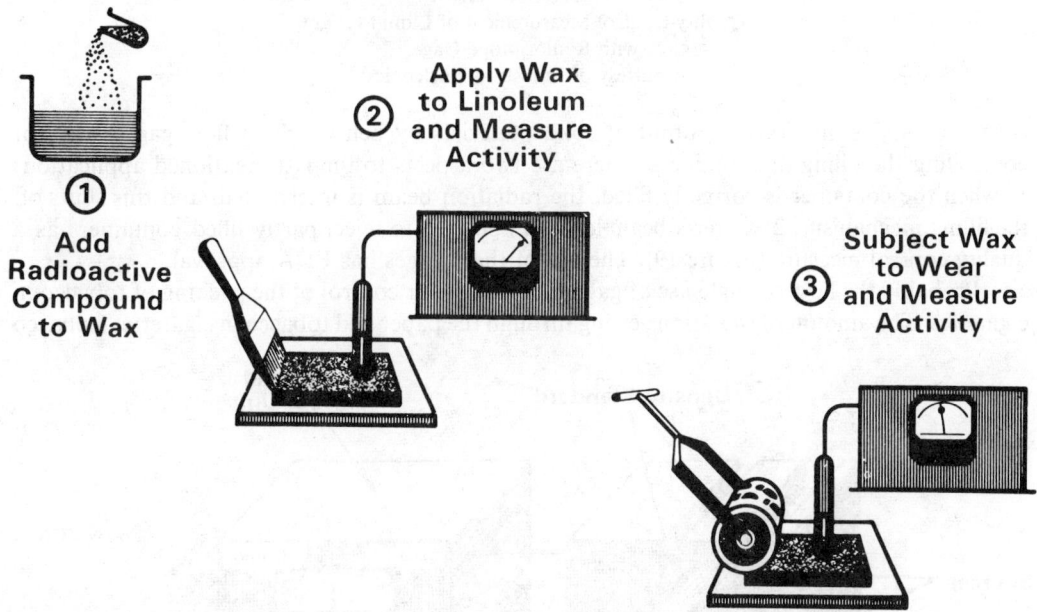

Advantages: quantitative measurement of wear
measures early phases of wear
measures extremely minute amounts

FIGURE 18.
Radioactive Isotopes for Testing Wear Resistance of Floor Wax.

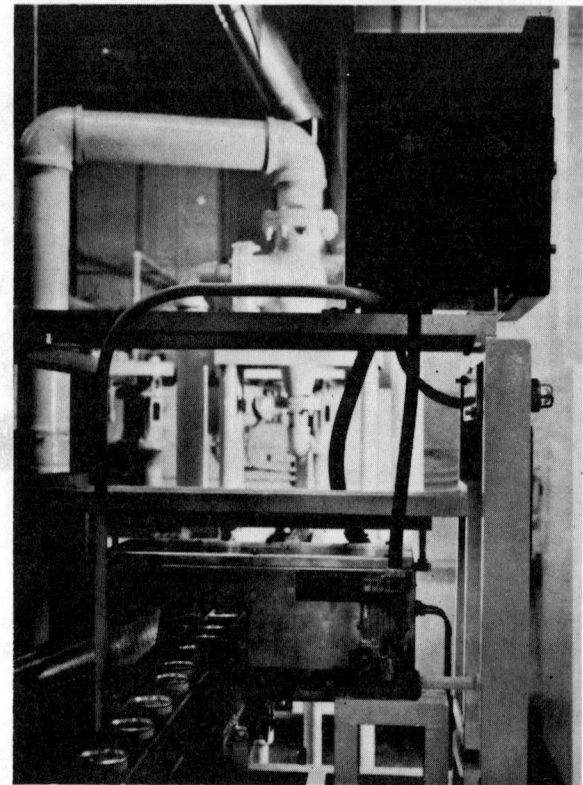

FIGURE 19.
Quality-Control Measurement of Liquid Height
with Radioisotope Gage.
(Courtesy of Industrial Nucleonics)

corn syrups, for monitoring output of solid materials by belt-weighing flow gages, and for controlling the filling of containers. There are two aspects to the last-mentioned application: 1) when the container is correctly filled, the radiation beam is interrupted, and this shuts off the filling mechanism; 2) a preset beam level can be used to reject partly filled containers as a quality control measure (Figure 19). The use of these gages has FDA approval.

Probably the largest single use of gaging devices is for control of the packing of tobacco in cigarettes. The amount of radiation getting through the paper and tobacco in cigarettes (tobacco

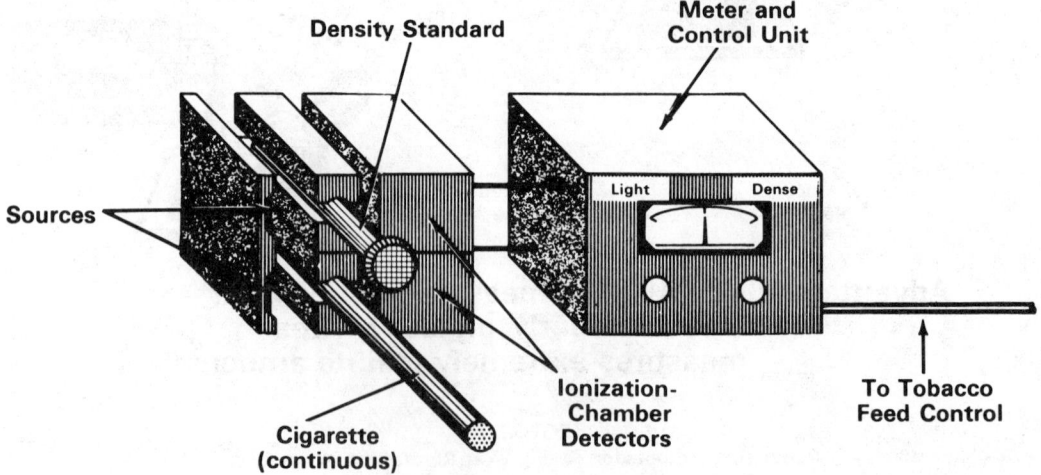

FIGURE 20. Cigarette-Firmness Gage.

rod) is a measure of the amount of tobacco in the cigarette (the paper, being very thin, contributes very little "resistance" to the radiation). Hence, if it has been decided what a desirable firmness is, the gaging device can be set to that degree of packing the tobacco in the cigarette. If too much radiation starts to come through, the machine is "told" to add more tobacco; on the other hand, if too little radiation penetrates the rod, the tobacco stuffing machine is automatically told to ease up a little. More than 2,500 density gages, each containing 15 to 20 mCi of ^{90}Sr, are employed to gage nearly 90 percent of the cigarettes made in the U.S.A.; it is estimated that each gage can save about $5,000 per year by elimination of off-specification rod and increased productivity of the machines (Figure 20).

The textile industry does not yet use nuclear gages for control in typical textile manufacturing operations—primarily for economic reasons. The amount of material processed per textile line is small (when compared, for example, to the paper industry), and many more machines are needed; hence the cost of equipping each machine with a gage cannot be justified on the basis of the possible savings that can be envisaged.

IONIZING RADIATION

Perhaps the most exciting area of potential application of radioisotopes in the 100-billion-dollar-per-year food industry is in the use of their ionizing radiations for processing foods —either by sterilization or by pasteurization. The fact that the U.S. Army Quartermaster Corps (sterilization) and the U.S. Atomic Energy Commission (pasteurization) together spent approximately 35 million dollars in the years from 1950 to 1968 to develop this area is evidence of its significance.

The Army emphasis has been on complete sterilization of foods, making possible their preservation under the variety of environmental conditions that may be experienced in military operations. Relatively high doses of radiation have been applied to a variety of food products, particularly meats, including bacon, ham, pork, beef, and chicken. The program has been successful in preserving foods without thermal canning or refrigeration, and the products themselves are adequately palatable, although sometimes an off-taste develops.

Some of the Army's procurement plans that have recently been completed or announced are:

1. thirty thousand pounds of radiation-sterilized bacon procured in June and July 1966;
2. four hundred thousand pounds of irradiated white potatoes procured during the winter of 1966;
3. two hundred thousand pounds of radiation-sterilized ham procured in the fiscal year 1967, or as soon as FDA and USDA clearances are obtained;
4. one hundred thousand pounds of pasteurized marine products procured following FDA clearance;
5. two hundred thousand pounds of irradiated wheat flour procured during the fiscal year 1967.

The USAEC program has concentrated more on the pasteurization of foods, the objective being to retard spoilage, particularly during passage through marketing channels and immediately after purchase, thereby extending marketing time. The procedures differ from sterilization in that they involve much lower doses of radiation and therefore have a much lesser effect on texture and palatability. The most successful work has been done with seafood and fruit; in particular, the effects of radiation on *Cl. botulinum* and on *Salmonellae* have been subjects of much concern. The shelf-life of refrigerated seafoods can be extended from one to several weeks by radiation pasteurization, and surveys in the fishing industry show that extra costs of the order of 1 to 3 cents per pound would be acceptable for such products. In 1964, an irradiation facility was opened in Gloucester, Mass., for semicommercial-scale production of irradiation-pasteurized seafoods. In 1967, twelve irradiator units were being used in the USAEC's development program, including an on-board-ship irradiator (Figure 21) and a mobile irradiator (Figure 22).

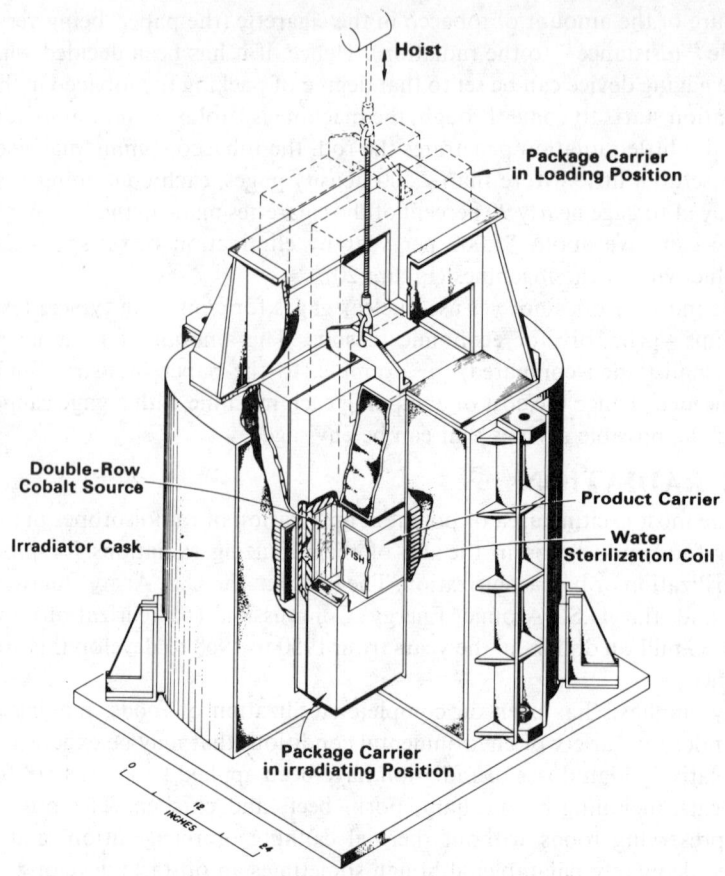

FIGURE 21. On-Board-Ship Irradiator.

FIGURE 22. Mobile Irradiator.

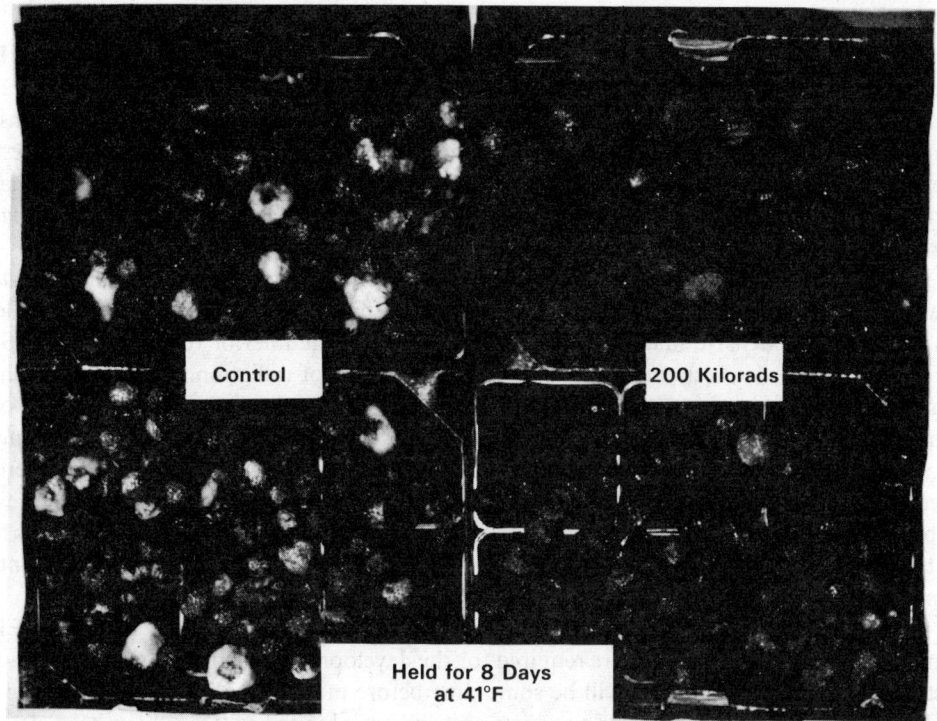

FIGURE 23.
Retardation of Spoilage of Strawberries by Irradiation.

FIGURE 24.
Inhibition of Sprouting of Potatoes by Irradiation.
(The potato at top left is unirradiated; the remaining potatoes have received increasing doses; potatoes at lower right show that too much radiation causes an undesirable effect.)

Interestingly, a number of individual food companies are actively cooperating with the USAEC in carrying out important program aspects that must precede commercial use. For example, commercial transportation channels were used to send irradiated and nonirradiated haddock fillets to principal seafood buyers of interested chain stores and other industry members located hundreds of miles away from the Gloucester irradiator; the fillets were tested by experienced people under commercial quality-control conditions; eight major food chains participated in these tests, and all reported that the irradiated fillets continue to be acceptable for 6 to 14 days after the nonirradiated fillets are no longer marketable.

Consumer tests have been carried out for a variety of fish, clams, shrimp, and crabs, and all were acceptable. These developments are extremely promising, since they might well lead to a much more extensive market for seafood products in the interior regions of the U.S.A.

Certain fruits also respond well to pasteurization doses of radiation. Strawberries, which are notoriously subject to spoilage during marketing, can be preserved by irradiation with only slight, but apparently acceptable, changes in texture (Figure 23). Retardation of the ripening of bananas by pasteurizing doses of radiation also holds out hope of developing into a commercial process. The radiation disinfestation of grains and the sprout inhibition of potatoes (Figure 24) are both quite successful, and the future appears promising for sprout inhibition of onions and for treatment of tropical fruits (e.g., mangoes and papayas) to destroy insects for quarantine purposes.

One of the major problems of the food-irradiation program is that of obtaining FDA clearance. Since two-year studies are required for the development of sufficient data to show the safety of the irradiated foods, it will be some time before most of the items are commercially available. Sterilized grain and grain products and sprout-inhibited potatoes have been cleared. Table 9 shows the 1969 status of some of the petitions before the FDA. All of the food petitions have been withdrawn pending accumulation of additional data required by the FDA.

Table 9. SUMMARY OF PETITIONS PRESENTLY BEFORE THE FDA

Product	Source	Dose, krads	Date Filed
Nylon	Gamma	1,000	9-8-65
Polyester (Saran-coated)	Gamma	1,000	7-30-65
Polypropylene (Saran-coated)	Gamma	1,000	7-30-65

In the textile industry, the effect of nuclear radiation on fibers is generally detrimental, reducing the strength and inducing degradation. In rare cases, such as Dacron polyester filament, where cross-linking of the polymer is induced, the improvements in the fiber properties are small, and commercial exploitation does not appear worthwhile. On the other hand, the grafting of monomers to various fibers through a "delayed radiation curing" offers considerable promise in the manufacture of permanent-press and soil-release fabrics, although the actual applications are very limited at the present time. However, electron accelerators may be more efficient in this area than radioisotopes. Materials that are relatively unstable toward radiation—such as cotton, rayon, cellulose esters, polyamides, polyesters, and polypropylene—show the best grafting results. Present research efforts are directed toward achieving a larger number of short-chain grafts instead of fewer grafts of long chains.

Also of interest to the textile industry is static elimination. The radiation from radioisotopes has been used commercially to ionize air, thus rendering it conductive to electricity and allowing the dissipation of static charges that accumulate in weaving of textiles (particularly in those made of synthetic fibers) or in the polishing or handling of plastic materials. An application of proven value involves a small source to remove the static charges in a loom while it is stationary, thus preventing the attraction of dust to the fibers; this trouble occurs especially when the atmosphere is dirty or foggy and gives rise to dirty marks on the fabric ("fog-marking").

Table 10 lists a few isotope uses in the consumer-products industries.

Table 10. TYPICAL ISOTOPES IN CONSUMER-PRODUCTS INDUSTRIES

Applications	Isotopes
Floor-wax wear	^{106}Ru-^{106}Rh
Detergent studies; laundering efficiency	^{14}C, ^{32}P
Cigarette firmness	^{90}Sr
Pesticide-removal studies	^{14}C, ^{3}H
Metabolic studies of food additives	^{14}C
Tobacco biosynthesis	^{14}C, ^{3}H
Measurement of movement of textile layers	^{32}P
Control of solid and liquid levels of foods and beverages	^{60}Co, ^{137}Cs
Food sterilization and pasteurization	^{60}Co, ^{137}Cs
Dye migration in printing	^{32}P

APPLICATIONS IN CRUDE-PETROLEUM AND NATURAL-GAS INDUSTRIES

Most of the isotope applications in the crude-petroleum and natural-gas industry involve tracing and gaging, although, of course, radiography is used for weld inspection, as in other industries.

TRACING

The tracing techniques that are used are both direct and indirect and are related to the location, recovery, and transmission of petroleum and natural gas. As a matter of fact, radioisotopes are so well suited to these applications that natural radioisotopes were being used prior to World War II. The use of isotopes permits reliable estimates of the extent of the recoverable sources and also makes valuable contributions to the techniques of getting the oil out.

Nuclear well logging is probably the most frequently used technique. It involves the generation of a signal inside the borehole of a well and monitoring the return signal, which has been attenuated by the surrounding strata; the operation is carried out as a continuous function of depth. Information about the composition and physical properties of the strata can be deduced from the log and used to estimate the probability of oil or gas being present. Of approximately 50,000 wells logged each year, nearly three fourths use nuclear methods. Most applications involve the lowering of a neutron source (or occasionally a gamma source) into the well and subsequent detection of either neutron-induced or gamma-induced emission (Figure 25).

Increasing oil yield by water-flooding of wells is a common technique that is made more efficient by the use of radioisotopes. Much-needed information about flow patterns in the subterranean rock formations and about hydrodynamic relations between oil-bearing geological formations is obtained by use of ^{3}H or ^{131}I tracers, and the economic benefit runs into hundreds of millions of dollars.

Information about the quantity of oil in a stratum can be obtained by using tritiated water in the bentonite mud that is pumped around the drilling bit. As a result of isotope dilution, the residual radioactivity in core samples taken from the drilling can be used as a measure of the water content of the stratum, and ultimately of the oil:water ratio and hence of the potential resources of the oil field.

The measurement of trace quantities of elements—often by activation analysis—is very helpful in estimating the age of rock samples, thereby giving clues as to the origin of the rock, and also in providing a more or less empirical correlation with the probability of its bearing oil.

Tritiated methane and ^{85}Kr are used in natural-gas fields to study permeability, porosity, and degree of saturation. The degree of saturation is of particular significance in relation to the need for repressurization of oil-bearing strata by the natural gas in order to avoid loss of the large quantities of gas that would be flared away during the development of a field. Similarly, the usefulness of natural underground "warehouses" as storage and distribution points for natural gas depends upon a knowledge of their boundaries and pressure-bearing capacities.

One of the oldest uses of isotopes is in the definition of the interfaces between various types of fluids being pumped through an oil pipe line. A given pipe line must be used for a variety of products (as many as 30), and since the economic value of the product is dependent on its purity, the mixed products at the interface clearly represent an economic loss. The extent of

FIGURE 25.
Complete Mobile Units Designed to Bring Radioisotope Facilities and Services to Oil Fields Are in Routine Use in the U.S.A. Unit Shown is for Logging Oil-Well Production.

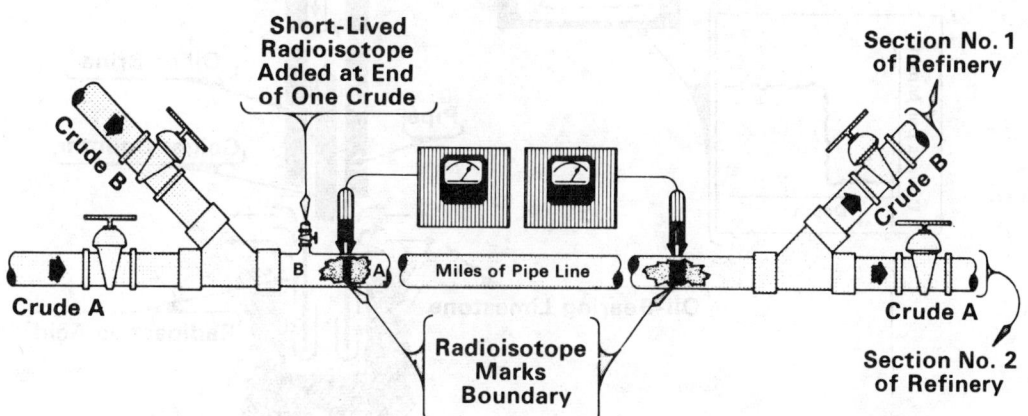

FIGURE 26.
Radioactive Isotopes for Tracing Oil Flow in Pipe Lines.

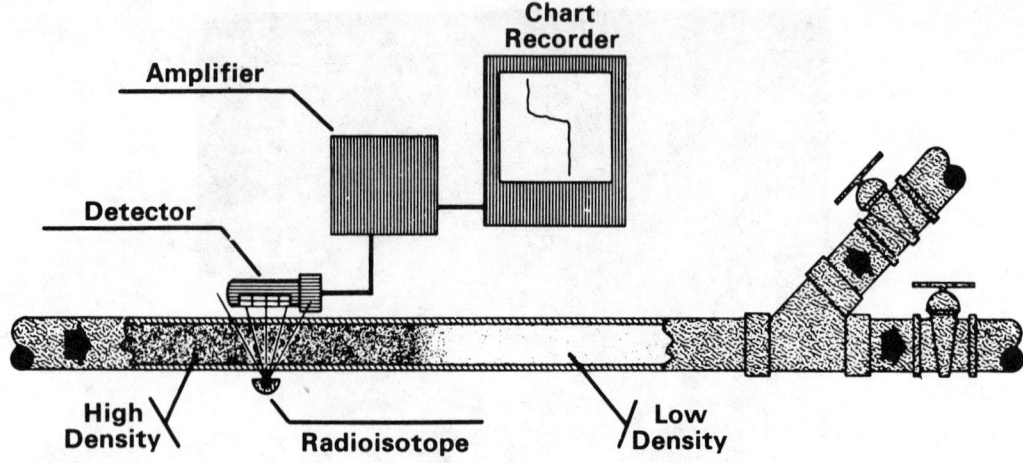

**Advantages: permits separation of liquids with minimal loss
method is quick and requires no sampling**

FIGURE 27.
Detection of Interfaces in Pipe-Line Flow, Using a Density Gage.

mixing and uncertainty in the position of the interface can be minimized if a radioactive tracer is used in the leading part of the incoming fluid (Figure 26). A major oil company has used this technique successfully to distinguish between grades of lube oil and considers that it has saved three hundred thousand dollars over ten years in this way. It should be mentioned, however, that the interface can also be detected by a density gage that responds to the change in density of the two adjoining liquids as the interface passes a source and detector. (Figure 27). Either of these radioisotope techniques lends itself to automation of the pipe-line operation.

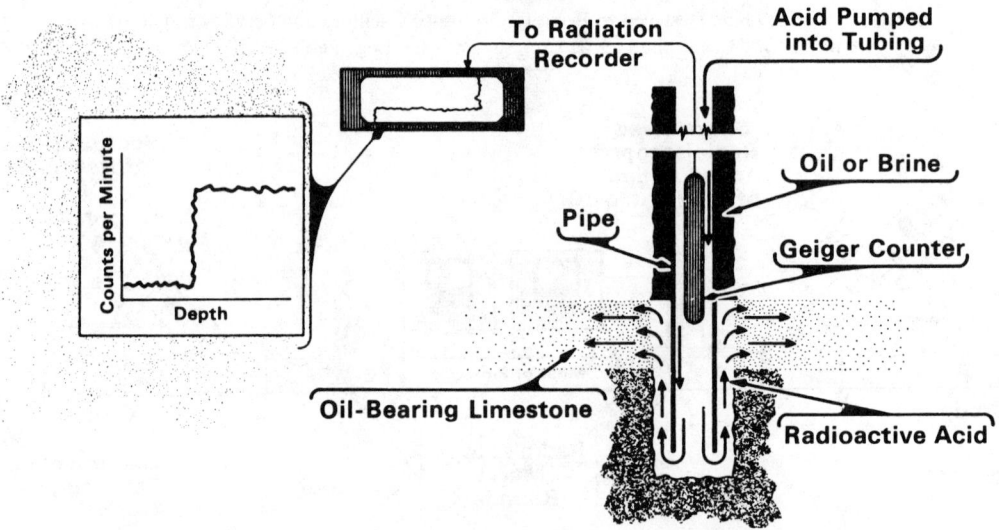

**Advantages: permits controlling site of acid action
increases efficiency of oil production
saves time and money
less hazardous than removing pipe**

FIGURE 28.
Radioactive Isotopes for Control of Oil-Well Acidizing.

Another use of the tracer technique is in the control of oil-well acidizing (Figure 28). Where oil-bearing limestone strata are involved in oil production, hydrochloric acid is often used to dissolve the limestone and release the oil. The use of a small amount of tracer in the acid permits a detector to locate and—indirectly—control the site of the acid action.

GAGING

The usual gaging equipment for determining densities, levels, etc., can also be found in the crude-petroleum and natural-gas industry.

HEAT AND POWER

A potential use of isotopes—particularly the relatively inexpensive fission products—is in heat sources for fluidizing oil wells and oil shales. These isotopes provide a relatively cheap source of heat, but problems of transportation and shielding have been deterrents to this application.

Four uses in the crude-petroleum and natural-gas industries are listed in Table 11.

Table 11.
TYPICAL ISOTOPES IN THE CRUDE-PETROLEUM AND NATURAL-GAS INDUSTRIES

Applications	Isotopes
Pipe-line flow	^{124}Sb, ^{60}Co, ^{137}Cs
Obtaining information about subterranean flow patterns in oil-bearing strata	^{3}H, ^{131}I
Permeability studies in gas fields	^{3}H, ^{85}Kr
Gaging (level, density, etc.)	^{60}Co, ^{137}Cs

APPLICATIONS IN MINING

The mining industry includes the producers of coal, metal ores, and nonmetallic minerals. The operations involve not only extracting the material from the ground, but also beneficiating it to upgrade its value. In coal mining, the operations include cleaning, removal of foreign matter, grading according to size, and classification according to ash content. In the case of metal ores, processing involves such operations as crushing, grinding, flotation, magnetic separation, leaching, and sintering, in order to concentrate the metal values so that metal recovery in the smelter or blast furnace is economically feasible. The use of radioisotopes is related almost entirely to these processing operations, in which gamma density gages are employed to measure operating variables critical to the performance of process equipment. Tracer applications are largely dependent upon neutron activation.

TRACING

Tracer studies of ore-processing variables have made use of radiosotopes, as have the investigations of the mechanisms of coal-cleaning processes. Gamma sources attached to pipe-cleaning devices are used to determine their location and velocity. The identification and assay of beryllium ores is accomplished by the use of ^{124}Sb, whose gamma ray interacts with ^{9}Be to produce neutrons.

The continuous assay of coal or ores for moisture and ash content, and of slurries for individual constituents, has been worked out through the use of neutron-activation analysis. In this technique, short-lived radioisotopes are produced *in situ* rather than being added (Figure 29). A neutron generator bombards the moving coal, ore, or slurry process stream with neutrons; then detectors and analyzers scan the activated material for radiations characteristic of the product stream; the results are then interpreted with respect to such things as moisture and ash content in the coal or ore or the individual components of a slurry. This on-line application is not only faster, since it is continuous, but also results in greater uniformity of product, because the solid stream can be appropriately diverted according to ash and/or moisture content, or the slurry according to any desired component.

Pollution of lakes, streams, and the atmosphere is often serious in the mining industry. It is being handled in much the same way as in the other industries.

GAGING

An important application of gamma density gages in ore processing is for the measurement and control of "pulp" density in the underflow from thickeners. This is particularly valuable in connection with the need for economizing on water usage. In one case, in a copper mine, successful water conservation is based on the use of two ^{137}Cs gamma density gages to measure the density of the underflow from the thickener; feedback from the gages controls diaphragm valves that maintain the water content to the tailing pond. Radioisotope density gages were chosen to control this pulp density because they do not require physical contact with the abrasive slurry. The measurement and control system based on these gamma gages aids in saving nearly 5 million gallons of water per day.

Another use of nuclear density gages in copper mining is exemplified by the design of a large new copper-concentrating plant. The central control room for this installation includes over 100 control loops depending on 82 primary measurements with isotope density gages.

One of the major innovations in this concentrator is the use of autogenous grinding, in which the material being ground also acts as a grinding medium—i.e., copper-ore pebbles grind the finer ore. The performance of these mills is very sensitive to loading, and gamma gages are used to maintain a constant recirculating load. The measurement of load by these gages on the twelve pebble mills is used either to control the pebble feed or to regulate output of the preceding rod mill; either way, the recirculating load is kept steady for optimum performance of the autogenous grinders.

In the mining and processing of iron ore, gamma gages are used to assure the quality of fired ore pellets by maintaining close control over the filtration of finely ground magnetite. The gages measure the density of magnetite slurry being pumped to the filters, and the amplified output of the gage controls a water valve connected to the underflow line at the preceding thickener; this automatic control of the filter-feed moisture results in higher quality green and fired pellets than were obtained with manual moisture control.

FIGURE 29.
Neutron Generator for On-Line Analysis of Coal.
(Courtesy of Texas Nuclear Corporation)

In the coal-mining industry, in order to hold ash content to an acceptable percentage, anthracite coal is separated from slate, stone, and other foreign materials by maintaining the density of a liquid medium so that the coal floats and the foreign matter sinks. The liquid medium used for this process, a slurry of finely ground magnetite and water, has an optimum specific gravity in the vicinity of 1.70, which must be maintained within very narrow limits in order to prevent losses of coal or excess ash content. Through the use of radioisotope density gages, the closed-loop automatic control system is able to maintain the specific gravity to ± 0.003 of the optimum preset value, as compared to only ± 0.02 by the conventional manual technique. As a consequence the ash content averages 10 to 11 percent, while previously, with the cleaning methods based on manual process control, it ranged from 12 to 16 percent.

IONIZING RADIATION

An interesting use of ionizing radiation in the mining industry is in the operation of a methane detector. The early detection of methane in mines is very important as a safety precaution, providing warning against possible explosions. Two foils, each consisting of a compound of tritium and titanium (TiT_2), are mounted parallel to each other; they are connected to the proper instrumentation to measure the ionization of the gas between the plates, which is caused by the radiations from the tritium. The detection system is set so that it will show any changes in gas composition. Of the usual components of mine gas, only a change in methane concentration will affect the detector. This change can be used to actuate a safety signal, either a warning light or an audible signal, if the concentration reaches a dangerous level.

Three typical mining-industry applications are shown in Table 12.

Table 12. TYPICAL ISOTOPES IN THE MINING INDUSTRY

Applications	Isotopes
Methane detection	^3H
Assay of beryllium ores	^{124}Sb
Control of water flow in slurry operations	^{137}Cs

APPLICATIONS IN THE UTILITIES

The electric, gas, and water-management utilities use radioisotopes primarily in process-control instrumentation, although tracer applications, radiography, and ionizing radiation are occasionally involved.

TRACING

Radioisotopes can be used to measure gas-flow rates through distribution pipe lines; the information is useful in designing pipe line extensions and new facilities as well as in indicating deterioration of existing pipe lines. Radioisotope methods are also superior to others for calibrating permanently installed flow meters. Radioisotopes have been used to determine the efficiency of turbines in hydroelectric plants in order to establish as precisely as possible the flow of water through the turbines. In England, isotopes have been used in civil engineering for acceptance tests on large-flow systems, such as water-cooling facilities at power stations. Leak-testing with radiotracers is very valuable in locating leaks in water and gas mains.

Related to water supplies are the movement of water in soils, the measurement and control of snowmelt, leaks in reservoirs, and stream pollution. Measurement of water movement in soils involves the introduction of a radioisotope tracer (e.g., ^{3}H, ^{82}Br) into ground water at one location and following it as it moves through the ground; sometimes this involves the use of deep holes dug at various spots around the point of injection of the radioisotopes. ^{32}P and ^{35}S can be used in conjunction with an autoradiographic technique to locate microfissures in rock, thereby helping avoid such accidents as dam breakage. Also, in certain areas, it is possible to use natural radon gas to determine if potentially dangerous faults or fissures exist in places where dams are proposed.

Measuring the density of mountain snowpacks requires both gaging and tracing. The Forest Service of the USDA and the USAEC have cosponsored work in this area. It has been found that knowledge of the thickness and density of snow layers high in the mountains allows prediction of the quantity and rate of water formation in warmer weather, thereby permitting institution of proper flood-control measures as well as better management and use of the resulting melt water. The snow density is determined by measuring the amount of radiation transmitted from a source (^{60}Co) to a detector. Associated with this work are studies of ground water flow, using tritium as a tracer, and of the control of evaporation of water from the snow surface itself and through vegetation, using hexadecanol labeled with the ^{14}C isotope.

The source of reservoir leakage and of stream pollution can often be identified by the use of radioisotope tracers. If the reservoir leakage is appreciable, there is usually an underground stream or spring that can be associated with it by noting the appearance of a radioisotope tracer previously put into the reservoir. In the case of stream pollution, radioisotopes permit pinpointing the sources of trouble, since the various suspects can be tagged and identified.

As for atmospheric pollution, significant amounts of SO_2 are introduced into our atmosphere primarily through the burning of coal. The sulfur content of bituminous coal ranges from 0.5 to 6 percent, with an average value of 2.5 percent. Thus a 150-megawatt utility plant burning 1,800 tons of coal per day will produce about 90 tons of SO_2 per day. The overall problem involves continuous monitoring of the SO_2 content of effluent stack gases, measuring the distribution of the stack-gas plume over the neighboring areas, and determining the sulfur content of coal on rapidly moving belts. Isotope tracers are involved in the first two operations, and tritium-excited X-ray-fluorescence analysis of the sulfur content of coal appears entirely feasible. Relating pollution to a particular industrial smokestack can be carried out by injecting a fine spray of cobalt sulfate solution into the base of a stack (e.g., at a power station), taking air samples over a large area, and neutron-activating the samples to determine the stack's "share" of the pollution.

GAGING

The use of radioisotope gages in process control is illustrated by the example of a power generation plant recently constructed and equipped with on-line computer analysis of the operating variables. Coal is brought from a storage pile by a single variable-speed conveyor belt to the crusher, which discharges onto two parallel belts into the generating station. Since the residence time of coal in the plant is only about four hours, nuclear level gages provide the most reliable method of indicating coal weight in the silos to the computer, and gamma gages serve the silos and chutes feeding coal to the furnaces. Belt speed is controlled by analysis of the coal level in the silo as indicated by the isotope gages, the first-stage steam pressure, and the output of strain gages on the belts from the crusher.

Another application of gamma density gages to conventional power generation is the measurement of the ratio of pulverized coal to air at the burner feed inlet. Automatic adjustment of the fuel–air mixture increases furnace efficiency and contributes to safe operation of the burner.

A relatively early use of gamma gages in electric-generation stations was the installation of a pair of units on each duct feeding coal to the cyclone burners. Under certain conditions the pulverized coal bridges across the duct, causing a constriction; the flow stops completely, cold air is drawn into the furnace, and a hot gas mixture may feed back up the coal passage, creating an explosion hazard. The source and detector of the gamma gage are mounted outside the standpipe, thus suffering no damage from the flow of coal particles, as occurred with the internally mounted capacitance probes formerly used. An audio-visual warning in the event of blockage signals an operator to start a hopper vibrator for clearing the plugged condition

Physical inventory of coal piles is becoming increasingly important as generating stations and annual coal consumption rates grow larger. The procedure involves measuring total volume, density, and moisture content of a pile; the use of a combination density–moisture gage based on isotopic sealed sources has improved the accuracy of these measurements. Moisture content is derived from the total hydrogen concentration of the coal, which in turn is determined by the output of a slow-neutron detector in conjunction with an isotopic source of fast neutrons. Density is determined by backscattered gamma radiation. Variations in both moisture and density with position are rapidly and accurately observed *in situ*, and the results lead to a more precise measurement of coal inventory than was previously possible.

RADIOGRAPHY

Radiographic inspection of boiler tubes has been accomplished with gamma sources where X-ray equipment would be impossible to locate properly. In one case nearly 1,600 14" × 17" films were exposed around the circumference of a large cyclone furnace whose cylindrical geometry was appropriate for radiographic inspection. In spite of the insulation and furnace shell between the tubes and the film, corroded areas were clearly visible and were selectively replaced.

IONIZING RADIATION

The use of ionizing radiation in the preparation of plastics for the closely related electrical industry has already been mentioned. Another application involves the use of radioisotope radiations for sterilization of water and sewage.

A few of the applications in the field of utilities are listed in Table 13.

Table 13. TYPICAL ISOTOPES IN THE UTILITIES

Applications	Isotopes
Studies of ground water flow	^{3}H, ^{82}Br
Studies of turbine flow	^{198}Au, ^{82}Br
Stream-current studies	^{131}I
Study of water evaporation from snowpacks	^{14}C
Density measurement of snowpacks	^{60}Co
Tracing of reservoir leakage	^{3}H, ^{131}I
Pollution studies	^{35}S, ^{85}Kr, ^{133}Xe
Location of microfissures in rock	^{32}P, ^{35}S
Density gaging	^{60}Co, ^{137}Cs

APPLICATIONS IN AGRICULTURE

Whether agriculture can truly be considered an industry is perhaps academic. However, since the applications are not being considered elsewhere, they are being taken up here.

TRACING

In agriculture, tracers are increasing the production and variety of food and other farm products. They are providing more information about the ways in which fertilizers, growth hormones, weed killers, insecticides, and other agricultural chemicals do their important work. The use of tracers in the agricultural industry represents one of the earliest of all radioisotope applications.

Fertilizers.

^{32}P was being used for the study of fertilizer uptake in plants (plant nutrition) in the late 1930's. It has become possible to answer questions about the efficiency of use of added fertilizers —i.e., when, where, and how much of what form (e.g., orthophosphates as compared to other phosphates) to add—as well as about the part that soil exchange plays in the fertilization process, the relative advantages of root and leaf application (Figure 30), the way the fertilizers move in the plants, and moisture and temperature effects.

Metabolism and Photosynthesis.

A number of studies of the biochemical changes (apart from the combination of chromosomes) taking place in flowers during fertilization and in growing fruit have used isotopes. The effects of growth stimulators such as gibberellins and auxins, the effect of plant-growth regulators (Figure 31), and the roles of certain proteins and nucleic acids have been evaluated with the help of the isotope tracer technique.

A tremendous amount of effort has been spent investigating plant metabolism of proteins and amino acids, of nitrogen, sulfur, and phosphates, and of 2,4-D. The effect of chelates on plant nutrients has been reported, and the uptake and distribution of inorganic ions and micronutrients have been the subjects of a number of papers. Related to this latter problem is the study of the ion-exchange properties of soils and clays.

One of the early applications of radioisotopes was in the study of the important process of photosynthesis (Figure 32). Calvin, Bassham, et al. at the University of California, among others, have for the most part used ^{14}C in their studies, which began as early as 1948; still others have used ^{18}O. An interesting result of these studies is the discovery that all of the oxygen released in the process of photosynthesis comes from the water and that the chloroplasts evidently act as photocatalysts.

Plant Diseases.

Throughout history, the epidemic spread of plant diseases caused by microbes has been a major agricultural problem. The Irish famines in the 1840's, for example, resulted from potato blight. In the U.S. alone the crop losses from plant diseases run into billions of dollars annually. The best way to fight this problem (aside from developing resistant strains of plants) is by chemical control of these fungus diseases. With radioisotope tracers (e.g., ^{35}S) it now becomes possible to measure chemical uptake in single spores and to follow the chemicals through the plant. Now we can study the life cycle of the microorganism, learn about the importance of

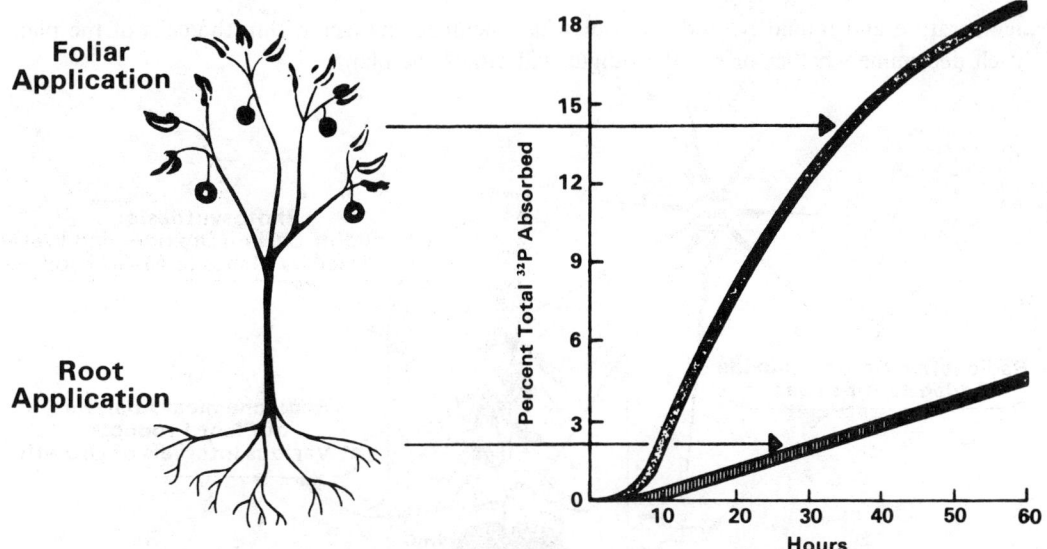

FIGURE 30.
Fertilizer Uptake by Roots and Foliage
Using Radioactive Phosphorus (^{32}P).

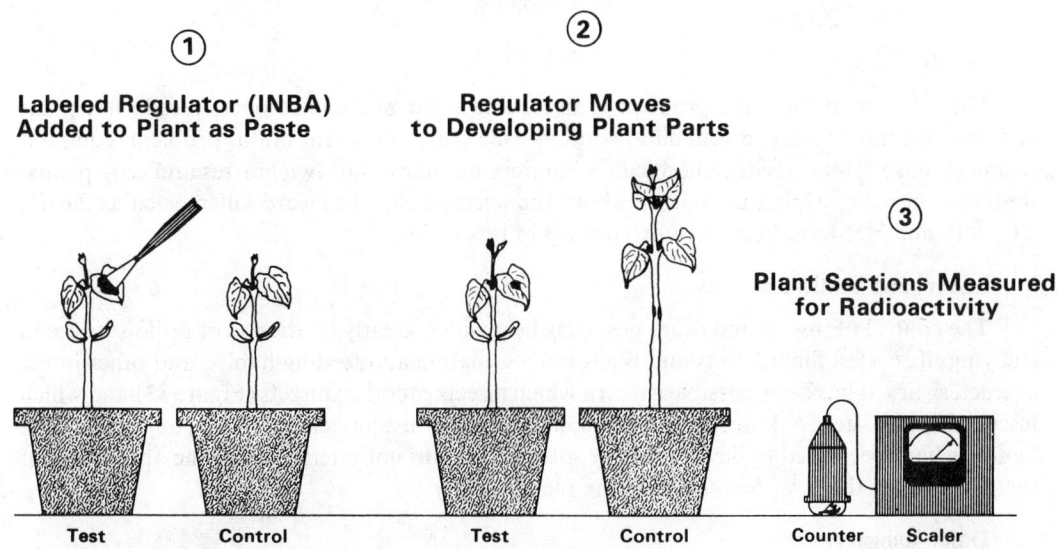

FIGURE 31.
Radioactive Iodine (^{131}I) for Studying Movement of Plant-Growth Regulators.

temperature and humidity, and find out what chemical changes within the cells of the plant itself determine whether or not the fungus will attack the plant.

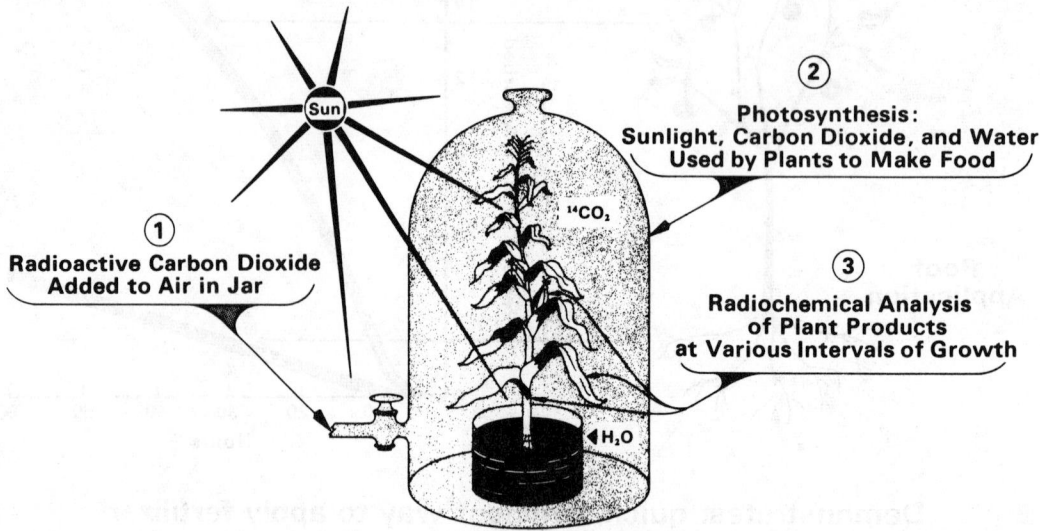

FIGURE 32.
Radioactive Carbon (^{14}C)
for Studying Food Production by Plants
(Photosynthesis).

Weeds.

The solution of the weed problem is also being aided by radioisotopes. Weeds, like plant diseases, cost this country several billion dollars each year. Treatment of this problem requires a chemical that kills the weeds (a herbicide), but does not harm the lawn grasses and crop plants. Radioisotopes are helping us to learn about the selective chemical weed killers such as 2,4-D; ^{14}C, ^{32}P, and ^{131}I have been used extensively in this regard.

Insects and Pesticides.

The control of insects and other pests has been aided greatly by the use of radioisotopes in studying life cycles, flight habits and flight ranges, mating and feeding habits, and other insect characteristics. It has been possible to learn which insects eat other insects (Figure 33) and which insects are parasites. A knowledge of this enables us to use insecticides to better advantage. Isotopes have been used to develop better solvents that do not interfere with the absorption of the pesticide by the pest, but still coat the plants evenly.

Dairy Industry.

Many radioisotopes have been used in studying body processes and habits of animals. The effectiveness of various kinds of feeds in causing weight gain, the importance of vitamins and trace elements in diets, and both the effect of hormones in fattening meat animals and the use of tranquilizers to reduce weight loss during shipping have been studied with radioactive tracers; ^{14}C, ^{35}S, ^{32}P, and ^{45}Ca have been used.

A rather interesting use of radioisotopes has been the study of the activity of the thyroid gland as it relates to milk production in cows and to egg-laying in chickens. The relation suggests that a farmer may be able to screen his stock and flock for increased productivity simply by measuring thyroid activity. Other tracer uses include studies of cheese fermentation (^{3}H, ^{14}C, and ^{35}S) to improve quality and flavor, of oxidation flavors of milk (^{64}Cu), and of cleaning surfaces in contact with dairy products (^{32}P) in order to evaluate the efficiency of cleaning compounds.

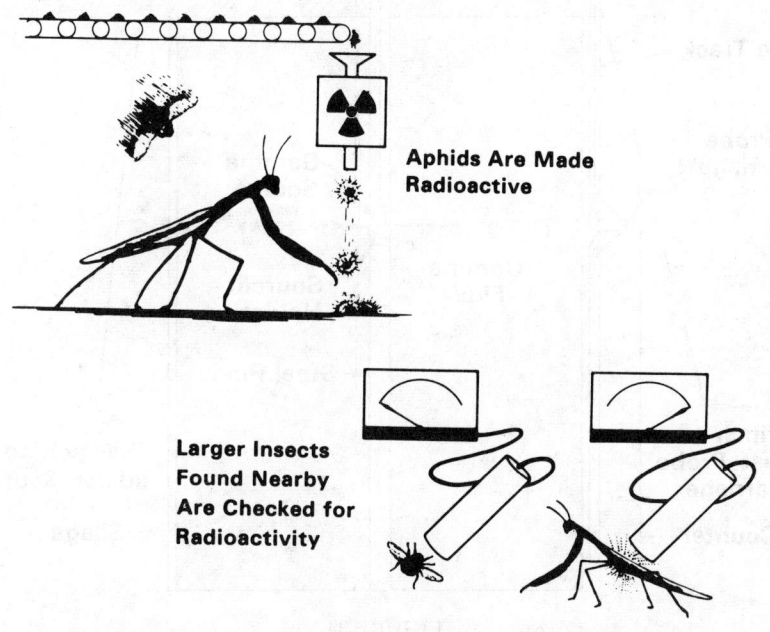

FIGURE 33.
Identifying Predators That Destroy Unwanted Insects.

GAGING

The previously described technique of automatically controlling levels by the use of radioisotopes is being used for filling milk cans and paper cartons. Density gages are used to control the density of condensed milk and for on-line regulation of the air content of ice cream. The measurement of eggshell thickness is being used for selection of thick-shelled eggs in order to decrease breakage during shipping. A rather unusual use of the gaging technique is in the measurement of silage density (Figure 34).

Radioisotopic neutron sources (e.g., Ra-Be) are used to determine soil moisture, which must be known for efficient irrigation; gamma-ray sources are used to determine soil density, which is related to space in the soil available for water and oxygen.

RADIOGRAPHY

Sanitary pipe lines in dairy plants are checked to determine the integrity of welds. Autoradiography is very commonly used in conjunction with the tracer technique in fertilizer-uptake and plant-metabolism studies.

IONIZING RADIATION

The agricultural applications of ionizing radiation comprise primarily three areas: plant mutations, sterilization of goat hair, and sex sterilization of insects. In addition, soils for use in hothouses are being sterilized on a small scale to provide potting material that is free of weeds.

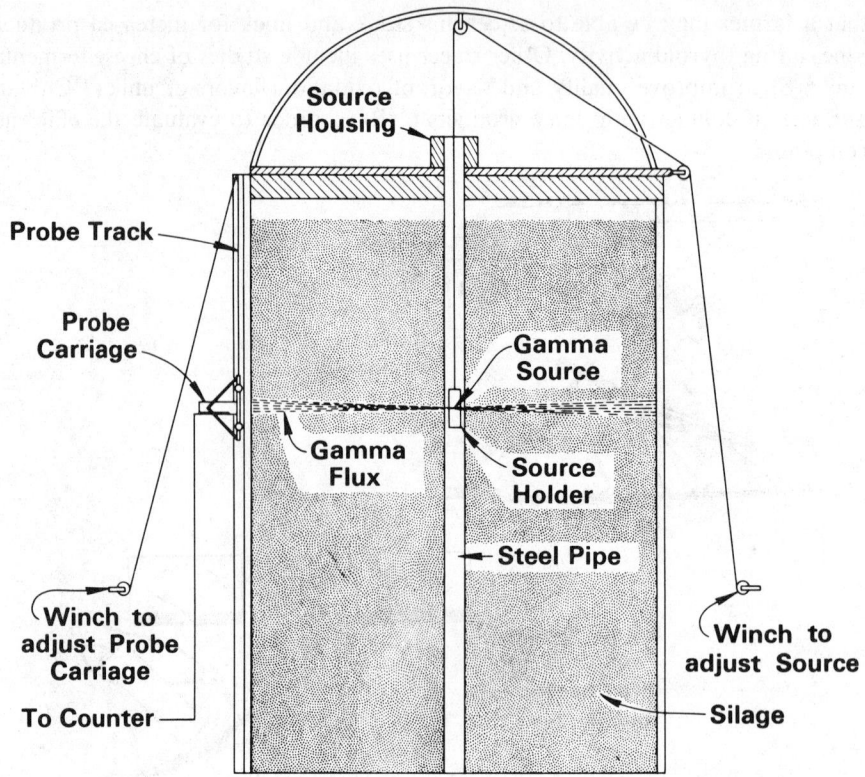

FIGURE 34.
Measurement of Silage Density with Isotope Gage.

Plant Mutations.

One of the most exciting aspects of atomic energy is the prospect of creating new plant varieties. Although it is not possible to control or predict what will happen when this invisible energy is poured into seeds or buds, there have been numerous instances where radiation has caused hereditary changes. At Brookhaven National Laboratory's "gamma garden", thousands of plants are being exposed to ^{60}Co gamma radiation. It should be remembered, however, that, although any feature of a plant subject to hereditary control can be altered by radiation, the efficiency of mutation is poor, and that, furthermore, most changes are undesirable—they interfere with the normal functions of the plant; on the other hand, a very small percentage of mutations results in improvement, although often not for several generations. Table 14 identifies some of these improvements.

Sterilization of Goat Hair.

One of the world's largest radiation sources (2 million curies ^{60}Co), located at Dandenong in Australia, is used for the disinfestation of goat hair. Just as radiation is used to disinfest grains, it is also effectively employed to kill the anthrax spores associated with most goat hair.

Sex Sterilization of Insects.

An ingenious application of the radiations from radioisotopes concerns eradication of the screwworm fly, which inhabits extensive areas of the southern U.S., Mexico, and the Caribbean. The fly lays its eggs in open wounds of livestock, and even in navels of newborn animals; eventually the maggots burrow into the animals and kill them. In the southeastern U.S., this damage has amounted to as much as 25 million dollars annually. A fortunate combination of circumstances made this technique (often referred to as the "sterile-male technique") successful: the male flies can be readily sterilized in the pupal stages by low doses of radiation (approximately 2500 roentgens); the insect can be reared easily in large numbers; sterile males compete equally with normal males for mates, but, of course, the sterile eggs do not hatch; and incidentally, but helpfully, females mate only once. By saturating an area with matured sterilized males (50 million at a time), it was possible to rapidly reduce to zero the number of eggs that hatched from the normal native flies. The program was continued for 18 months, during which more than two billion flies were released; at the end of this period the insect had been completely eliminated.

It is of interest to note, in passing, that oriental scientists are using gamma rays instead of conventional heating to kill silkworms inside cocoons.

Table 15 summarizes some of the application of isotopes in agriculture.

Table 14. RADIATION-INDUCED PLANT MUTATIONS

Crop	Variety Name	Place of Release	Date of Release	Improved Features
Tobacco	Chlorina Mutant	Indonesia	1930's	Leaf color and quality.
White mustard	Primex	Sweden	1950	Greater seed yield and oil content.
Summer oil rape	Regina II	Sweden	1953	Seed yield and oil content.
Pea	Weibull Stralart	Sweden	1953	5–10% higher yield.
Winter barley	Jutta	Germany	1953	Winter hardiness; yield of grain; lodging resistance.
Pea bean	Sanilac	Michigan	1957	Bush type; disease resistance.
Spring barley	Pallas	Sweden	1958	Tolerance to heavy nitrogen fertilization.
Peanut	N.C. 4X	North Carolina	1959	Tougher hull (better for shipping).
Oats	Florad	Florida	1960	Resistance to Victoria blight.
Pea bean	Seaway	Michigan	1960	Bush type; virus resistance.
Spring barley	Mari	Sweden	1960	Early maturity (8 days); stiffer straw.
Oats	Alamo-X	Texas	1961	Disease resistance.
White mustard	Seco	Sweden	1961	Seed yield; crude fat; early maturity; stiff stalk.
Bread wheat	NP 836	India	1961	Awned to resist bird damage; higher yield without irrigation.
Spring barley	VR_z	Czechoslovakia	1962	Higher yield of grain; stiff straw.
Pea bean	Gratiot	Michigan	1963	Better seed type.
Barley	Pennrad	Pennsylvania	1963	Improved winter hardiness.
Carnation	Yukon-1	Connecticut	1963	Fewer petals; greater durability.
Bread Wheat	Sinvalocho Gama	Argentina	1965(?)	Resistance to stem and leaf rusts.

(By courtesy of T. S. Osborne, University of Tennessee, Agricultural Experiment Station, March 1967.)

Table 15. TYPICAL ISOTOPES IN AGRICULTURE

Applications	Isotopes
Study of the movement of growth regulators	^{131}I
Investigation of sulfur damage to citrus fruits	^{35}S
Study of potassium uptake in fertilizer	^{39}K, ^{86}Rb
Study of cheese fermentation	^{3}H, ^{14}C, ^{35}S
Study of the translocation of labeled sucrose in sugar beet	^{14}C
Determination of nutritional factors influencing uptake of iron and other elements in plant tissues	^{55}Fe, ^{65}Zn
Study of the effect of pH, soil cations, and the soil nutrients on uptake and translocation of iron	^{55}Fe
Comparison of phosphorus uptake from various forms of phosphate fertilizer	^{32}P
Comparison of fertilizer uptake through roots with uptake as result of topical application	^{32}P
Evaluation of the efficiency of cleaning compounds	^{32}P
Study of oxidation flavors in milk	^{64}Cu
Study of decomposition rates of organic materials in soil	^{14}C
Measurement of eggshell thickness	^{106}Ru-Rh
Determination of the leach rate of calcium and sulfur in soils	^{35}S, ^{45}Ca
Study of flight patterns of aphids	^{32}P
Study of the metabolism of insecticides in roaches	^{32}P
Study of the movement of insect larvae through soil	^{60}Co
Location of hibernation sites of beetles	^{32}P
Study of metabolic changes induced by virus infection in plants	^{14}C
Determination of the effect of various elements, plant characteristics, and environmental conditions on iron metabolism under conditions conducive to chlorosis	^{55}Fe, ^{59}Fe
Study of mineral metabolism in animals	^{32}P, ^{35}S, ^{45}Ca, ^{55}Fe, ^{60}Co, ^{64}Cu, ^{65}Zn, ^{89}Sr, ^{99}Mo, ^{131}I, ^{134}Cs, ^{182}Ta
Determination of blood volume in farm animals in relation to climatic factors and milk yield	^{51}Cr
Study of the physiological availability of diet additives	^{131}I, ^{35}S
Study of the metabolism of the insecticide parathion in goats	^{32}P
Sterilization of goat hair; sex sterilization of insects; plant mutations	^{60}Co, ^{137}Cs
Soil density measurements	^{137}Cs

APPLICATIONS IN AEROSPACE AND OTHER ENVIRONMENTAL USES

Although the uses of isotopes to aerospace and other environmental applications seem to revolve principally around heat and power, there are some very important uses in other areas. Table 16 lists some typical application.

Table 16. RADIOISOTOPES IN AEROSPACE

Applications	Isotopes
Measurement of atmospheric density	^{144}Ce; ^{75}Se; ^{241}Am; ^{85}Kr
Helicopter close-order formation-keeping system	^{85}Kr; ^{90}Sr
Prompt miss distance measurement (missiles vs. targets)	^{60}Co
Fuel gaging	^{60}Co; ^{137}Cs; ^{85}Kr
Oil gaging	Any gamma source
Lighting	^{85}Kr; ^{3}H; ^{147}Pm
Extraterrestrial life detection	^{14}C
Radiographic inspection	^{60}Co; ^{192}Ir
Measurement of strain in metal components	^{85}Kr
Measurement of turbine-block temperature	^{85}Kr

TRACING

Radioisotope tracers have proved very helpful in at least one instance in evaluating detergent effectiveness in cleaning out LOX (liquid oxygen) lines; "tagged" dirt in the lines was used to compare the cleaning action of various detergents.

Isotopes have been used to study fuel mixing, particularly for solid-fueled missiles.

The proposed use of radioactive ^{14}C to detect life on other planets involves introduction of soil samples from the planet into culture media containing ^{14}C; if there are any growing organisms, the culture will release radioactive ^{14}CO$_2$, which can be detected; the information can be telemetered back to earth.

^{85}Kr in the form of "kryptonates" (solids into which ^{85}Kr has been incorporated physically—i.e., without undergoing chemical reactions) is a very useful tracer for measuring the temperature to which turbine blades are exposed in jet engines. The measurement is based upon the fact that a kryptonate has a sort of "memory" in that, when heated to any given temperature below the melting point, it loses a fixed fraction of its initial activity; furthermore, no additional activity is lost until that temperature is exceeded. Thus, if a turbine blade is "kryptonated", run for a time, removed, and then heated in an oven, no loss of ^{85}Kr occurs until the highest operating temperature of the turbine blade has been reached.

Other tracing applications parallel those already mentioned in previous chapters.

GAGING

One problem in satellites and rockets is the fact that they operate essentially in a "zero-gravity" situation; thus the usual types of fuel and oil gages are useless. Isotope transmission gages, one of which is called a "Nucleonic Fuel Gage", can be used to give the average mass of fuel in a tank. Another technique uses radioactive ^{85}Kr gas in the void space of a fuel tank; as the fuel is forced out by the introduction of pressurized nonradioactive gas, the ^{85}Kr is diluted; the measurement of the concentration by isotope detectors can be related directly to the amount of residual fuel.

The tracking of satellites or aircraft during launch or takeoff can be aided by the use of an isotope such as ^{60}Co in the space vehicle. This can then be followed with ground-based detectors.

An interesting future application still in the research stage is the radionuclide altitude gage, which uses radiation attenuation from ground sources to guide aircraft in instrument-approach landings.

In research applications, the aerospace manufacturers have made use of devices resembling nuclear thickness gages to study the rate of ablation of nose cones under simulated reentry conditions.

RADIOGRAPHY

Most radiographic applications are related to quality control of manufactured parts, which has already been covered.

IONIZING RADIATION

In the presence of suitable phosphors, beta-emitting isotopes can cause luminescence, making possible illuminated warning signs, landing markers (Figure 35), and watch dials—all of which are important in the aerospace industry. The use of ionizing radiation in plastics manufacture has already been covered.

FIGURE 35.
^{85}Kr-Activated Light Sources.
(Courtesy of American Atomics)

HEAT AND POWER

As was mentioned earlier, when the radiations from an isotope are slowed down or stopped, much of the energy is converted to heat. This heat can be used directly, or it can be converted to electricity by suitable thermoelectric devices or to mechanical energy by appropriate conversion systems; we then have a kind of nuclear power generator, which will provide energy for long periods of time without need for recharging or replenishing the fuel. Figure 36 shows a ^{242}Cm pellet glowing from its own heat of self-absorption.

FIGURE 36.
^{242}Cm Pellet Glowing from Its Own Heat of Self-Absorption.

Among the important applications of isotopic power are the SNAP (Systems for Nuclear Auxiliary Power) devices listed in Table 17. This table, based on data supplied by the AEC, is a revision of information published in *Nuclear Industry*, p. 30, June 1965.

Isotopic power provides a reliable primary energy source that has the advantages of compactness, long life, no need for separate power storage, excess heat for other systems, and applicability to any environment. Some of the problems associated with large-scale use of isotopes as power sources are comparative economics with competing power supplies such as solar cells, current limited availability of some of the isotopic power "fuels", objectionable radiation (in some cases), and safety requirements.

Recently some of the other transuranic elements (the man-made elements beyond uranium in the Periodic Chart of the Elements) have become very attractive for SNAP devices because they have relatively high power-to-weight ratios, and because they are primarily alpha emitters and do not require much shielding. On the other hand, they are relatively expensive.

Table 18 lists characteristics of some of the fuels now being used or considered for isotopic power applications.

Thermal Applications

The direct use of heat from radioisotope decay, although potentially suitable in a number of ways, has so far been quite limited. However, it is not difficult to imagine many applications for isotopic heat. A device for heating diving suits is soon to be tested; it supplies 400 to 500

Table 17. "SNAP" ISOTOPE-FUELED GENERATORS

Designation	Application	User	Contractor	Output, watts	Design Life, years
SNAP-3A	Transit 4A (navigation satellite)	Navy	Martin	2.7	5
SNAP-3A	Transit 4B	Navy	Martin	2.7	5
"SENTRY"	Weather station	Weather Bureau	Martin	5	2
SNAP-7A	Light buoy	Coast Guard	Martin	10	2
SNAP-7B	Land light	Coast Guard	Martin	60	2
SNAP-7C	Weather station	Navy	Martin	10	2
SNAP-7D	Floating weather station	Navy	Martin	60	2
SNAP-7E	Undersea acoustic beacon	Navy	Martin	8	2
SNAP-7F	Offshore oil rig	Phillips Petroleum	Martin	60	5
SNAP-9A	DOD satellite	Navy	Martin	25	5
SNAP-9A	DOD satellite	Navy	Martin	25	5
SNAP-9A	DOD satellite	Navy	Martin	25	5
SNAP-11	Surveyor* (lunar surface use)	NASA	Martin	25	0.3
SNAP-15A	Nuclear weapons	AEC	General Atomic	0.001	5
SNAP-15C	Communications	AEC	Minnesota Mining	0.1	Classified
SNAP-17A	Developmental†	Undetermined	Martin	30	3–5
SNAP-17B	Developmental	Undetermined	General Electric	30	3–5
SNAP-19	Nimbus B weather satellite	NASA	Martin	30	5
SNAP-21	Advanced deepsea	Undetermined	Minnesota Mining	10–60	5
SNAP-23	Advanced terrestrial	Undetermined	Minnesota Mining	25–100	5
SNAP-27	Apollo Lunar Surface Experiment Package (ALSEP)	NASA	General Electric	50	1
SNAP-29	Space application	Defense Department	Martin	400	0.25

* According to the AEC, the success of the solar cells powering the recently launched Surveyor experiment makes it unlikely that SNAP-11 will be used for a Surveyor mission, unless problems are encountered with solar cells during later missions.

IN OPERATION AND UNDER DEVELOPMENT

Weight, pounds	Fuel		Status as of 6/6/66
	Isotope	Quantity	
5	^{238}Pu	95 g	Launched June 1961; operating at an undetermined lower power level.
5	^{238}Pu	95 g	Launched November 1961; shutdown by power system failure in June 1962.
1,680	^{90}Sr	19 kCi	Installed August 1961 on Axel Heiberg Island, Canada; removed October 1965 because of problems with nonnuclear equipment; now on display at Oak Ridge Nat'l Lab; future use undetermined.
1,870	^{90}Sr	41 kCi	Installed December 1961 in Curtis Bay, Md; withdrawn from operation in June 1962; to be removed for examination, probably next month.
4,600	^{90}Sr	225 kCi	Installed May 1964 in Baltimore harbor lighthouse; still operating; scheduled to replace the defective SNAP-7F generator in an offshore oil rig in the Gulf of Mexico, probably next month.
1,870	^{90}Sr	41 kCi	Installed February 1962 at McMurdo Sound; still operating.
4,600	^{90}Sr	225 kCi	Installed January 1964 in the Gulf of Mexico; still operating.
6,000	^{90}Sr	31 kCi	Installed July 1964 off Bermuda; still operating.
4,600	^{90}Sr	225 kCi	Removed last October from an oil rig in the Gulf of Mexico, after swelling in the biological shield shorted part of its thermoelectric system; Martin and Oak Ridge National Laboratory are investigating.
27	^{238}Pu	~1 kg	Launched September 1963; operating at an undisclosed lower power level.
27	^{238}Pu	~1 kg	Launched December 1963; operating at an undisclosed lower power level.
27	^{238}Pu	~1 kg	Launched April 1964; burned up on reentry.
30	^{242}Cm	Classified	Fueled generator scheduled to be tested this summer at Oak Ridge National Laboratory.
1	^{239}Pu	2–3 g	Advanced units under test at General Atomic.
Classified	^{238}Pu	Classified	Fueled units under test at Minnesota Mining.
30	^{90}Sr	Classified	Contractor completed engineering design studies in November 1964; life evaluation tests of thermoelectric modules by subcontractor, RCA, are continuing; work now limited to module testing, with the results to be applied to future SNAP generator projects.
30	^{90}Sr	Classified	Contractor completed engineering design studies in November 1964; life evaluation tests of thermoelectric modules by subcontractor, Minnesota Mining, are continuing; work now limited to module testing, with the results to be applied to future SNAP generator projects.
30	^{238}Pu	Classified	Two electrically heated units are to be delivered to GE's Missile and Space Division this month; two fueled units will be sent to GE next February, with two additional fueled backup units to follow in April 1967.
500 (10-W unit)	^{90}Sr	Classified	Minnesota Mining has begun development, which is to lead to the delivery of fueled units in about three years.
1,000 (60-W unit)	^{90}Sr	Classified	AEC is evaluating Minnesota Mining design study and may soon award a development contract.
30	^{238}Pu	Classified	Under development; first fueled units to be delivered next April, and flight hardware to follow in July 1967.
400	^{210}Po	Classified	Under development; schedule calls for delivery of developmental prototype units in approximately two to three years.

† The Air Force is no longer interested in using SNAP-17 for communication satellites applications; it is interested in a 250-watt ^{90}Sr or ^{238}Pu unit, on which Martin, General Electric and Lockheed are working.

Table 18. ISOTOPIC POWER FUELS

Isotope	Half-Life, years	Predominant Radiation	Power Density, watts/cm³	
			Theoretical	Practical
^{238}Pm	87.5	α,n	4.8	3.6
^{244}Cm	18.4	α,n	26.4	22.5
^{242}Cm	0.45	α,n	947	150
^{210}Po	0.38	α,n	1,150	150
^{147}Pm	2.62	β	2.3	2
^{144}Ce	0.78	β	24.4	21.9
^{90}Sr	27.7	β	1.1–1.7	0.85–1.5
^{60}Co	5.24	γ	120	27–54

thermal watts for up to 24 months from a ^{170}Tm-^{171}Tm source; the source heats water, which is then circulated through a specially designed diver's undergarment. Figure 37 shows the design of a warm-up station for an underseas laboratory; the heating unit consists of a 500-watt ^{90}Sr source.

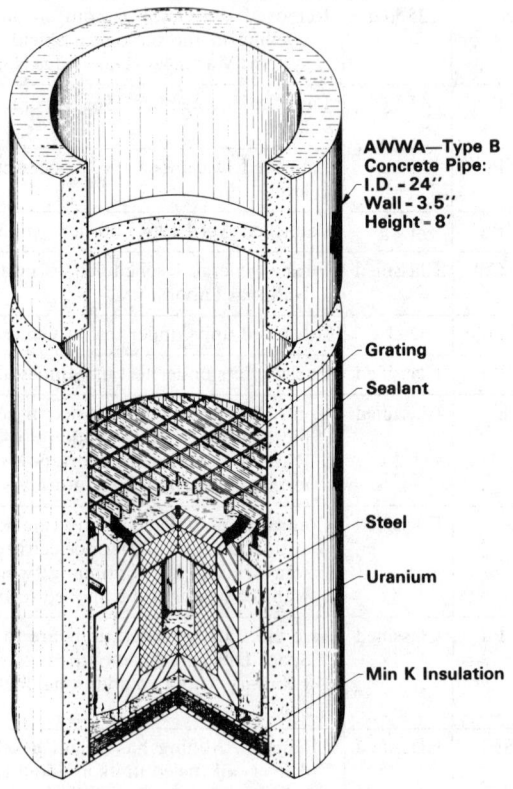

FIGURE 37.
500-Watt Underseas Warm-Up Station.

Other possible applications might be the heating of small huts on trails between major stations in the Arctic and Antarctic or in other unusually cold regions and the warming of diesel fuel oil pans in extremely cold weather. In the space program, heat sources could be used in satellites to ensure thermal control and to process human waste. In one study, for example, a ^{238}Pu-powered water-recovery system successfully recovered 15 pounds of water per day for an extended period of time.

Thermoelectric Devices

Various thermoelectric conversion systems are potentially useful for a number of applications. Figure 38 illustrates several of these.

Thermoelectric Junctions.

The earliest isotope-powered thermoelectric devices used thermojunctions (thermocouples) to convert the isotopic heat to electricity, since the principle was well known and had been applied for many years in heat-measuring instruments. These devices are based on the fact that, when wires of two dissimilar metals are connected at both ends and then subjected to different temperatures, an electrical potential is set up between them. Although each junction usually develops only a few milliwatts, a large number of such junctures can develop sizeable voltages and useful currents. The first such device, called "SNAP-3", was built for the U.S. Atomic Energy Commission and used on board a navigational satellite that was launched in 1961; it was still operating in 1967; the source consisted of 95 g of ^{238}Pu. In August 1961, an 18-kilocurie ^{90}Sr thermoelectric device was placed at Axel Heiberg Island in northern Canada to power an unmanned weather station, which reported wind velocity, temperature, and barometric pressure daily for two years (Figure 39); the "fuel" was in the form of the compound strontium titanate ($SrTiO_3$). Figure 40 is a photograph of trays of $SrTiO_3$ pellets suitable for this type of application.

Figures 41 and 42 show a diagram and photograph of a flashing-light navigational buoy, also fueled with ^{90}Sr. Figure 43 shows a cutaway of a 250-kilocurie generator being used to operate a shorelight at Curtis Bay, Maryland.

A very important, but low-level (approximately 100 microwatts), thermoelectric application is the cardiac pacemaker, which provides a regular electrical stimulus to maintain the heartbeat. Hopefully, an isotopic source can be implanted in the chest cavity for years at a time; ^{238}Pu is presently being considered as a power source for this purpose.

Thermionic Devices.

In its simplest form, a thermionic energy converter consists of two electrodes separated by a vacuum; one electrode—the emitter—is kept hot enough to emit electrons thermionically, and the other electrode—the collector—is kept cool. The load through which the electric current passes is connected externally between the two electrodes. Since heat is supplied to one electrode and part of it is removed from the other, the converter may be considered a heat engine.

Dynamic Systems.

Several types of cycles have been considered for converting heat to electricity by generating steam and then using the steam in a turbogenerator to produce the electricity. The Sterling, Rankine, and Brayton cycles are all feasible, but the systems development is largely an engineering problem, independent of the particular isotope, and so will not be considered here. The applications will, in general, be in the area of high-energy requirements.

Thermophotovoltaic Systems and Beta Batteries.

The principles of these devices are shown in Figure 38. Actual applications will probably be limited to very low powers (i.e., microwatts) and can be expected to use low-energy sources, such as tritium. These devices can also be miniaturized, since the low levels of activity require a minimum of shielding.

Mechanical-Energy Devices

Heat from radioisotopes can also be converted to mechanical energy for different uses; for example, the indirect use of the heat from radioisotopes in propulsion and thrusting devices for spacecraft has drawn a good deal of attention. In such applications the thermal energy supplied

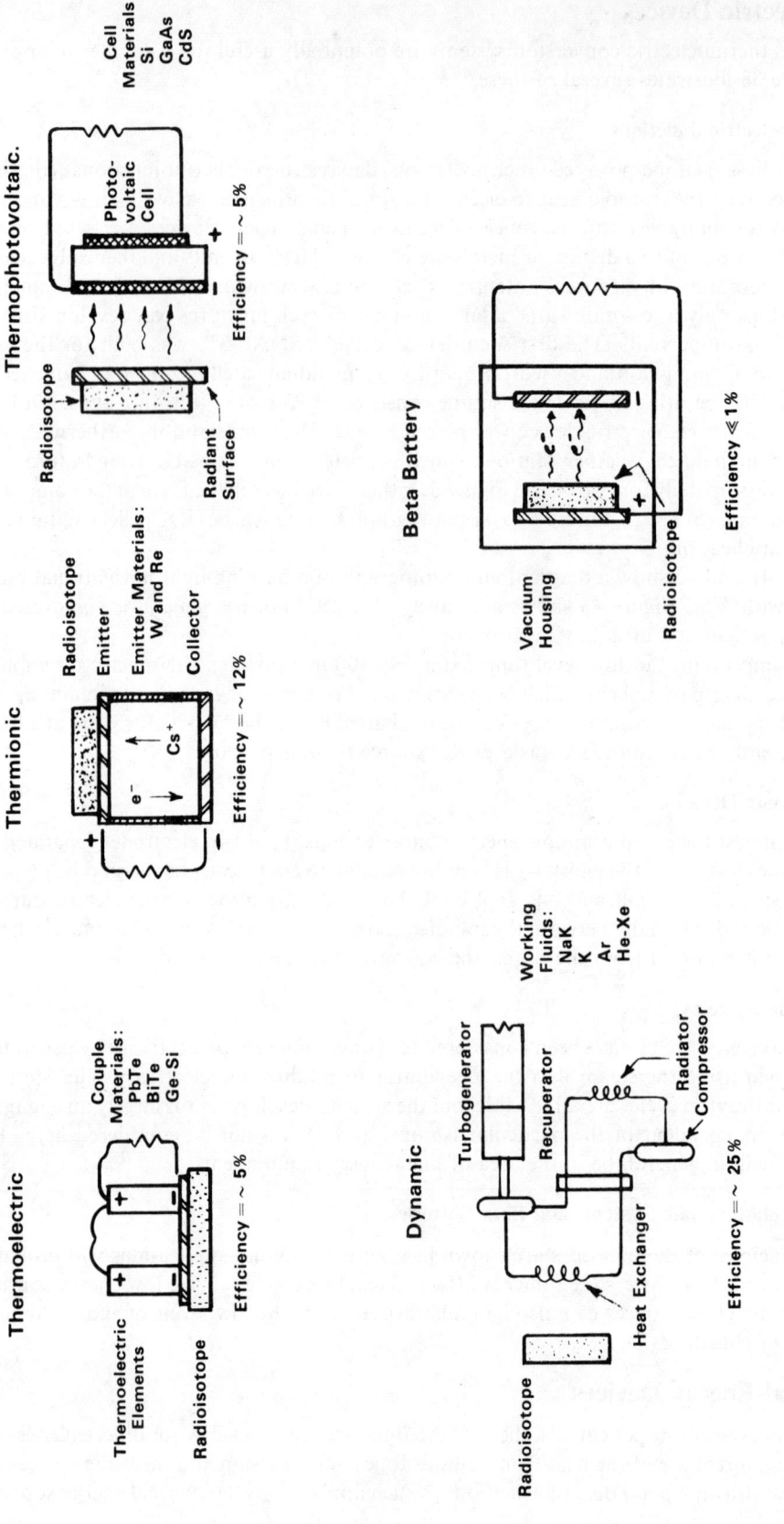

FIGURE 38. Electrical Conversion Systems.

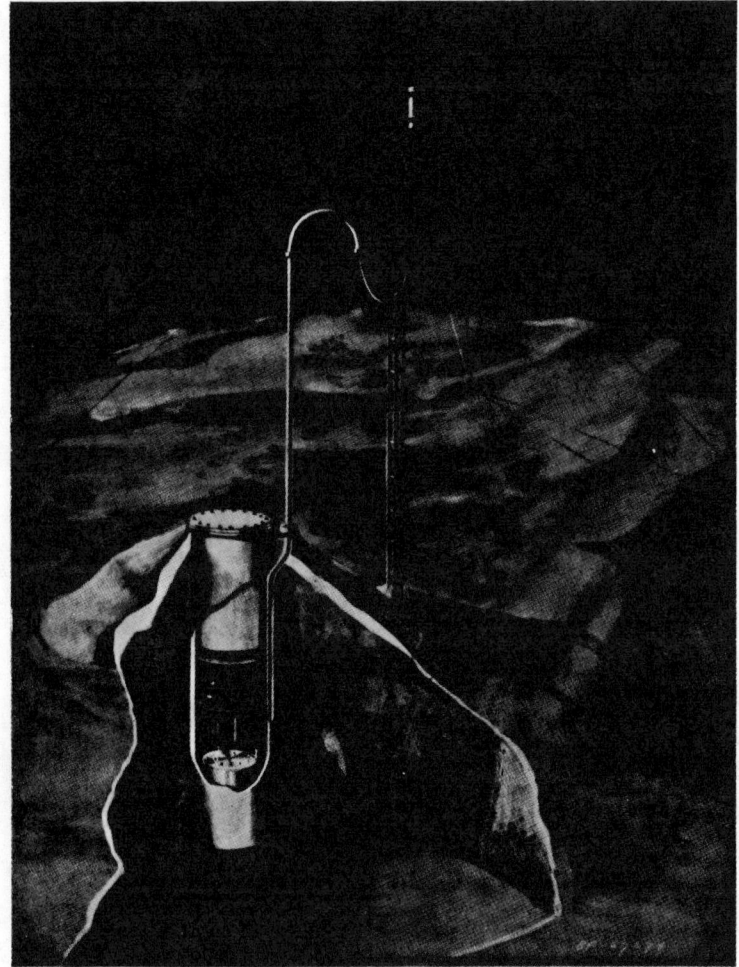

FIGURE 39.
Schematic Drawing of the World's
First Radioisotope-Powered Weather Station.
Located on remote Axel Heiberg Island in the Canadian Arctic, the unattended station began transmitting wind velocity, temperature and barometric pressure in August 1961; power is supplied by a ^{90}Sr thermoelectric generator capable of 10 years continuous service.

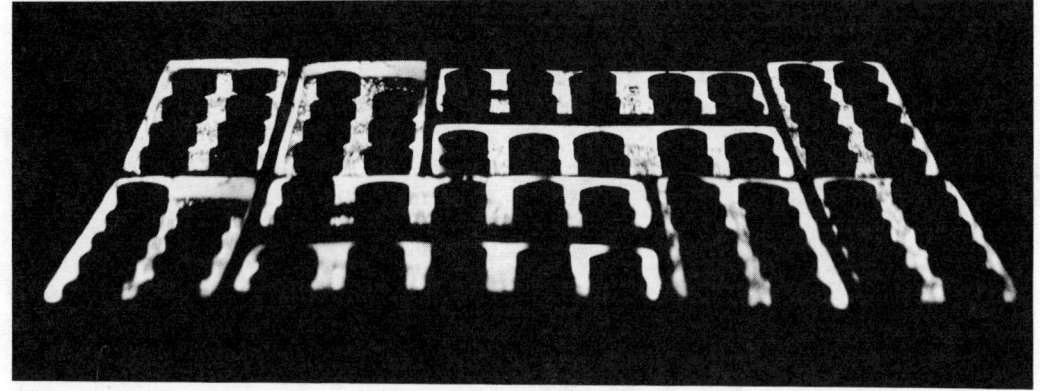

FIGURE 40.
Strontium Titanate Pellets
Used for Thermoelectric Generators.

FIGURE 41.
Conceptual View of Flashing-Light Buoy.

FIGURE 42.
Flashing-Light Buoy Near Baltimore,
Powered by a SNAP-7A Radioisotope Generator.

FIGURE 43.
Cutaway of Isotope-Powered Generator for Shorelight.

by the radioisotope serves to heat a working fluid, which is then expanded through a nozzle. The specific uses envisioned include the low-acceleration transfer of objects from a low orbit to a higher orbit (e.g., large television satellites) and low-energy thrusters for control of the attitude and position of space vehicles (Figure 44). For much of this type of application ^{210}Po, ^{238}Pu, and ^{147}Pm are being considered.

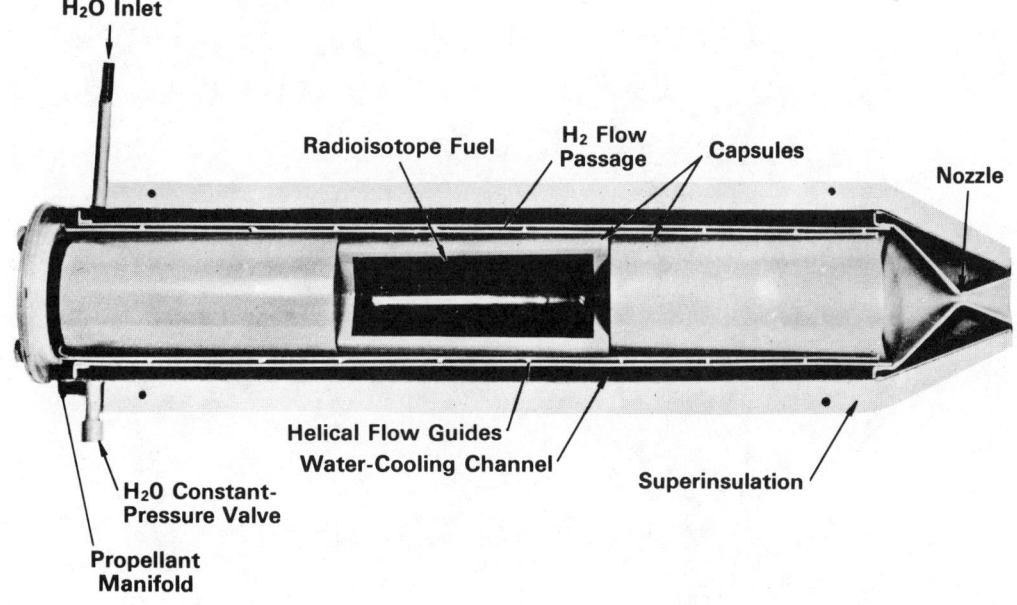

FIGURE 44.
Isotope-Powered Low-Energy Thruster for Space Vehicles.

Another application under this category is the development of an artificial heart. Studies have shown that the number of individuals who could benefit from heart assist and replacement devices was sufficiently large to justify a major effort on a national basis. Data indicate that approximately one hundred thousand of the seven hundred thousand Americans who will die from heart disease this year might be helped if circulatory systems were available. National Heart Institute studies indicate that radioisotopes, along with the biological fuel cell, offer the most promising means for supplying power for a fully implantable circulatory support system. Isotopes under consideration are ^{238}Pu and ^{147}Pm.

GLOSSARY

Accelerator—a device capable of accelerating charged particles to high velocities (e.g., cyclotron, Van de Graaff).
Activation—the formation of a radioisotope from a stable isotope by changing the neutron:proton ratio.
Activation analysis—identifying elements by virtue of the radiations given off *subsequent to activation* of a material.
Alpha emitter—a radioisotope that gives off alpha particles.
Alpha particle or ray (α)—a helium nucleus consisting of two protons and two neutrons.
Atom—the smallest unit of an element; consists of protons and neutrons in the nucleus and electrons in outer "shells".
Atomic number—a number equal to the number of protons in the nucleus of an atom; it is a characteristic of an element.
Beta particle or ray (β^+, β^-)—a positive or negative electron; if positive, it is usually called a "positron".
Compound—a chemical combination of two or more elements.
Cross section—a measure of the ability of a nucleus to accept or "capture" bombarding particles (e.g., neutrons or protons).
Curie—a measure of the activity of a radioisotope; defined as that quantity of an isotope that is equal to 3.7×10^{10} disintegrations per second (approximately the same as the activity of 1 g radium).
Cyclotron—a device for accelerating charged particles.
Decay—the rearrangement of a radioactive nucleus to a more stable configuration, during which radiations are given off.
Electron—a negatively charged particle with a mass about 1/1840 of that of a proton.
Element—a substance in which all of the atoms have the same number of protons.
Fission—the division of a heavy nucleus into two approximately equal parts, with simultaneous liberation of neutrons and energy.
Flux—a measure of neutron or charged-particle concentration; used for purposes of calculating radioisotope production rates.
Gamma emitter—a radioisotope that emits gamma rays.
Gamma ray (γ)—an electromagnetic radiation emitted by a nucleus.
Half-life—the time ($T_{1/2}$) it takes for one-half of any starting amount of radioisotope to decay.
Irradiation—the impinging of radiation upon a material.
Isotopes—atoms of an element differing from each other only in the number of neutrons in the nucleus.
Labeling—the attaching of radioactive isotopes to a compound.
Neutron—an uncharged nuclear particle, of approximately the same mass as a proton, that adds mass to an atom without much effect on the chemical properties of the atom.
Neutron capture—the acceptance of a neutron by a nucleus; often results in formation of a radioactive isotope.
Neutron-deficient isotope—a radioisotope in which the neutron : proton ratio is *below* the stability ratio.
Neutron-excess isotope—a radioisotope in which the neutron : proton ratio is *above* the stability ratio.
Neutron source—a radioisotope or accelerator that produces neutrons; the isotope produces neutrons through decay, the accelerator as a result of a nuclear reaction.

Nuclide—see Isotopes.
Photon—a quantity of electromagnetic radiation.
Positron—a positive electron.
Proton—a positively charged nuclear particle of mass number 1 and charge equal, but opposite, to that of a negative electron.
Radiation—for the purpose of this review, the rays and particles resulting from the decay of radioisotopes or produced in accelerators.
Radiation pasteurization—the use of radiation to pasteurize foods.
Radiation sterilization—the use of a sufficient quantity of radiation to sterilize a substance.
Radioactivity—spontaneous nuclear rearrangement with emission of radiation.
Radiography—the taking of "pictures" *through* objects by means of radiation, usually electromagnetic, but may include even neutrons.
Radioisotope—an isotope whose nucleus has a neutron : proton ratio that is either too high or too low, and which is therefore radioactive; may be natural or artificial (man-made).
Radionuclide—a radioisotope.
Reactor fuel—a material, such as uranium or plutonium, that will undergo fission, thereby producing neutrons and energy.
Short-lived isotope—a radioisotope with a relatively short half-life (e.g., a few hours, for most practical purposes).
Source—radioactive material that can be used for its radiation or heat.
Sterile-male technique—the use of radiation to control insect population by sterilizing the males and eventually eliminating the species through inability to reproduce.
Stringer—a device for introducing targets into a reactor core, often for the purpose of making radioisotopes.
Target—a sample to be bombarded with neutrons or charged particles, often to produce radioisotopes.
Teletherapy—treatment with radiation, usually with gamma rays or X rays.
Thermoelectric device—a device capable of converting heat (as from a radioisotope source) to electricity.
Tracer—a radioisotope that is mixed with or attached to a given substance and used to determine the location or path of that substance; often used to follow industrial process streams, metabolic fates, etc.
Transmutation—the changing of one element to another through nuclear rearrangements.
Transuranic element—any element with an atomic number greater than that of uranium (92).
X ray—electromagnetic radiation resulting from electron shifts in the "electron shells" of atoms; when an electron moves from an "outer" shell to a shell nearer the nucleus, an X ray is given off, with energy equivalent to the difference between the energy levels of the two shells involved.

PART VIII
RADIATION PROTECTION AND REGULATIONS

Allen Brodsky, Sc.D.
University of Pittsburgh
Graduate School of Public Health

Francis J. Bradley, Ph.D.
New York State Department of Labor

BASIC UNITS
OF RADIATION MEASUREMENT

Allen Brodsky, Sc.D., C.H.P.

RELATIONSHIP BETWEEN PHYSICAL QUANTITIES AND BIOLOGICAL EFFECTS

As discussed in the sections on biological effects of radiation, a number of acute and long-term effects on animal and human species have been related to the physical energy absorbed from various types of ionizing radiation. However, the relative effectiveness of each type of radiation per unit energy absorbed in biological tissue has been found to vary not only with the type of radiation and its quantum energy, but also with the rate at which the energy is delivered, the kind of tissue, age and species of animal, the biological effect under consideration, and other experimental and epidemiologic variables. For mammals, beta radiation and the recoil electrons ejected from atoms by X or gamma radiation generally produce approximately the same order of magnitude of biological effects. On the other hand, heavier particles, such as the alpha particles emitted by certain radionuclides, lose their energy at higher rates of linear energy transfer (LET) along their paths and seem to produce somewhat higher damage per unit energy absorbed. In the case of alpha-emitting radionuclides, of course, since the characteristic alpha particles emitted do not have sufficient range to penetrate the dead layer of skin, biological damage is produced only when the radioactive material itself is distributed within an organ by inhalation or ingestion.

Thus, for purposes of radiation hazard evaluation, several units of radiation exposure and dose must be introduced to account for the several methods of measuring and assessing the effects of different types of radiation. Since most radiations of a given type, as emitted by most radionuclides, have average LET's within a narrow range, and consequently seem to have relative biological effectivenesses (RBE's) within a narrow range, characteristic simplifications can be introduced to limit the number of new units and the definitions required for most applications. In this chapter, only the most useful definitions and data are presented, and some basic references are given for further details.

BASIC UNITS OF RADIOACTIVITY, RADIATION MEASUREMENT, AND RADIATION DOSE

The following definitions of quantities and units will suffice in dealing with most problems in radiation protection (health physics) and dosimetry:

roentgen (r)—a unit for expressing exposure from X or gamma radiation in terms of the ionization produced in air, which can be measured by appropriate air ionization chambers and electrical instruments. It is defined as an exposure "such that the associated corpuscular emission per 0.001293 g of air produces, in air, ions carrying one electrostatic unit of quantity of electricity of either sign."[1] This simply means that an exposure (formerly called "exposure dose"[2]) of one roentgen generates recoil electrons per 0.001293 g of dry air within a small volume surrounding the point of measurement to the extent that these electrons produce 1 esu (electrostatic unit) of positive charge and 1 esu of negative charge (as ion pairs) in air. Since many electrons will leave the element of volume before losing all of their energy, a "standard air chamber" is necessary in order to most accurately measure the roentgen (see Figure 1). This chamber is specially designed to achieve an "electronic equilibrium" condition in which the ionization lost by electrons leaving that volume is compensated for by an equilibrium number of electrons entering the volume from a preceding volume of air.[1] Thus, the roentgen is a unit that expresses a point quantity, essentially the ionization density produced near the point of measurement in air. When applied to human exposure in a large beam or field of radiation, either the field must be uniform or the exposure in r must be measured at each point in the field. In summary, the roentgen may be remembered for practical purposes as the X or gamma exposure producing 1 esu (+ or −) per cc of air (at STP 0°C, 760 mm Hg).

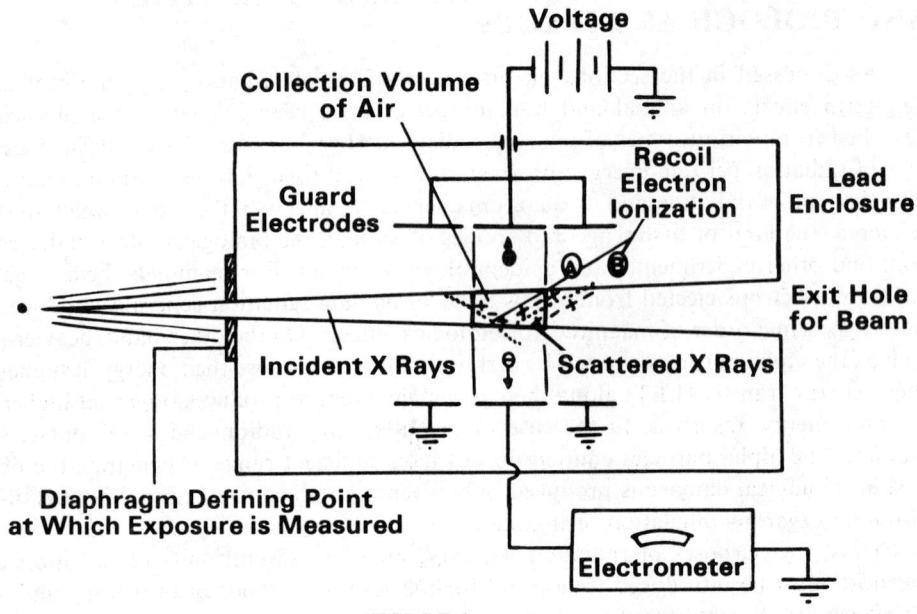

FIGURE 1.
Schematic Diagram of a "Standard Air Chamber".

rad—a unit of "absorbed dose" (D) in any medium for any kind of ionizing radiation.[1,2] It is simply defined as:

$$1 \text{ rad} = 100 \text{ ergs/gram}.$$

Integral absorbed dose—the integral $\int D dm$ over an organ or the whole body, where D is the variable absorbed dose within mass element dm. It is useful in obtaining an idea of the total energy absorbed by a region of tissue, which is more closely related to certain macroscopic biological changes than the dose only near one point. Common units are gram-rads, and 1 gram-rad = 100 ergs. The integral absorbed dose may be divided by the total mass of tissue in the region of interest to obtain the average tissue dose $\bar{D}$ in rads.

rep—a unit formerly used for similar purposes as the rad, but defined for any type of radiation in terms of the energy absorbed in tissue equivalent to that which would be absorbed from 1 r of X or gamma radiation. 1 rep has been defined variously as the absorption of energy ranging from 84 ergs/cm³ to 93 ergs/gram of tissue.[3,4]

rem—a unit of "RBE-Dose" or "Dose Equivalent" (DE), used to express the estimated equivalent of any type of radiation that would produce the same biological end point as 1 rad delivered by X or gamma radiation. Thus,

$$\text{DE (in rem)} = \text{Dose (in rad)} + \text{QF} \times \text{DF},$$

where QF (formerly called RBE) accounts for the relative biological effectiveness of the radiation compared to X radiation for radiation protection purposes, and the DF (formerly designated as n) is the "relative damage factor"[5] or "distribution factor"[1,2] used to account for differences in the distribution of the rad dose to the organ of concern as a result of uneven uptake of the radionuclide, etc., as opposed to the QF or RBE factors reserved for more intrinsic biological characteristics of the emitted radiations.

gram-rem—the unit for integral absorbed dose when several types of radiation are involved and the equivalent doses of each are to be added after multiplication by appropriate relative biological effectiveness factors.

man-rem—a term used in estimating expected frequencies of disease in a population by determining the equivalent integral absorbed dose over the population. For somatic effects, the number of persons in the exposed population might be multiplied by the average dose equivalent to each person; for genetic effects, the average gonadal dose would be multiplied by the number of people exposed.

kerma (K)—the quotient E_K/m, where E_K is the sum of all kinetic energies of charged particles liberated by indirectly ionizing particles (e.g., neutrons) in a volume element, and m is the mass of matter in that volume element.[6]

Energy fluence (F)—as defined by the ICRU,[6] the quotient $\Delta E_F/\Delta a$, where ΔE_F is the sum of all the energies, exclusive of rest energies, entering a sphere of cross-sectional area Δa (e.g., in units of Mev/cm^2).

Energy flux density (intensity I)—the quotient $\Delta F/\Delta t$, where ΔF is the energy fluence in a small time interval Δt.

Particle fluence (Θ)—the quotient $\Delta N/\Delta a$, the number of particles flowing into a small sphere per unit cross-sectional area. This is the same quantity as nvt, used in the case of neutron-diffusion theory, where n = neutron density in neutrons/cm^3, v = average neutron velocity in cm/sec, and t = time in seconds over which the fluence is integrated. Another way of envisioning the meaning of nvt would be to imagine the total distance in cm traveled by all the neutrons present at a given moment per cm^3 of volume.

Particle flux density (ϕ)—the quotient $\Delta\Phi/\Delta t$, where $\Delta\Phi$ is the particle fluence in time Δt; units may be particles/cm^2-sec.

Mass attentuation coefficient (μ/ρ)—the fractional number of incident particles (or photons) interacting with a given material per unit mass thickness that they pass through; i.e., $\mu/\rho = dN/N\rho dl$, where dN/N is the probability of interaction per unit thickness dl of density ρ. Common units are cm^2/g (i.e., fractional number/g-cm^{-2}).

Mass energy-absorption coefficient (μ_{en}/ρ)—the fractional energy removed from incident indirectly ionizing particles (or photons) per unit mass thickness; i.e., $\mu_{en}/\rho = dE/E\rho dl$, where dE is the energy removed from the incident particles (not including rest energy or energy reirradiation as bremsstrahlung), E is the sum of the energies (excluding rest energies) of the incident particles, and ρdl is the mass thickness in g/cm^2, as above. Common units are again cm^2/g (i.e., fractional energy/g-cm^{-2}).

Curie (Ci)—the most common unit used to express the radioactivity (A) of a material; it is the amount of any radionuclide (or combination of radionuclides) in which there are 3.7×10^{10} nuclear transformations or disintegrations per second, or 2.22×10^{12} disintegrations per minute (dpm). Combined with appropriate prefixes, the symbol is commonly used for the following other scientific units:

1 megacurie (MCi) = 10^6 Ci
1 kilocurie (kCi) = 10^3 Ci
1 millicurie (mCi) = 10^{-3} Ci
1 microcurie (μCi) = 10^{-6} Ci
1 nanocurie (nCi) = 10^{-9} Ci
1 picocurie (pCi) = 10^{-12} Ci = μμCi.

The above designations are widely used, since we deal with a wide range of quantities of radioactivity in evaluating radiation exposure potentials.

In addition to the basic quantities and units defined above, other derived units will be introduced in subsequent sections of this chapter. Table 1 presents the other fundamental quantities and units listed by the ICRU.[2,6]

Table 1. QUANTITIES AND UNITS

No.	Name	Symbol	Dimensions	Units mksa	Units csg	Special
4	Energy imparted (integral absorbed dose)	—	E	J	erg	g-rad
5	Absorbed dose	D	EM^{-1}	J kg^{-1}	erg g^{-1}	rad
6	Absorbed-dose rate	—	$EM^{-1} T^{-1}$	J kg^{-1} s^{-1}	erg g^{-1} s^{-1}	rad s^{-1}, etc
7	Particle fluence or fluence	Φ	L^{-2}	m^{-2}	cm^{-2}	
8	Particle flux density	φ	$L^{-2} T^{-1}$	m^{-2} s^{-1}	cm^{-2} s^{-1}	
9	Energy fluence	F	EL^{-2}	J m^{-2}	erg cm^{-2}	
10	Energy flux density or intensity	I	$EL^{-2} T^{-1}$	J m^{-2} s^{-1}	erg cm^{-2} s^{-1}	
11	Kerma	K	EM^{-1}	J kg^{-1}	erg g^{-1}	
12	Kerma rate	—	$EM^{-1} T^{-1}$	J kg^{-1} s^{-1}	erg g^{-1} s^{-1}	
13	Exposure	X	—	—	—	r (roentgen)
14	Exposure rate	—	$QM^{-1} T^{-1}$	C kg^{-1} s^{-1}	esu g^{-1} s^{-1}	r s^{-1}, etc.
15	Mass attenuation coefficient	$\frac{\mu}{\rho}$	$L^2 M^{-1}$	m^2 kg^{-1}	cm^2 g^{-1}	
16	Mass energy-transfer coefficient	$\frac{\mu K}{\rho}$	$L^2 M^{-1}$	m^2 kg^{-1}	cm^2 g^{-1}	
17	Mass energy-absorption coefficient	$\frac{\mu_{en}}{\rho}$	$L^2 M^{-1}$	m^2 kg^{-1}	cm^2 g^{-1}	
18	Mass stopping power	$\frac{S}{\rho}$	$EL^2 M^{-1}$	J m^2 kg^{-1}	erg cm^2 g^{-1}	
19	Linear energy transfer	LET	EL^{-1}	J m^{-1}	erg m^{-1}	kev (um)$^{-1}$
20	Average energy per ion pair	W	E	J	erg	ev.
22	Activity	A	T^{-1}	s^{-1}	s^{-1}	Ci (curie)
23	Specific gamma-ray constant	Γ	$QL^2 M^{-1}$	C m^2 kg^{-1}	esu cm^2 g^{-1}	r m^2 h^{-1}, etc.
	Dose equivalent	DE	—	—	—	rem

From *NBS HANDBOOK 87*, p. 43.

PHYSICAL MEASUREMENTS OF DOSE AND THE BRAGG-GRAY PRINCIPLE

Since the standard air chamber described earlier is a relatively large, expensive, and sensitive instrument, and since measurements of the roentgen apply directly only to X or gamma rays, many other field and laboratory instruments have been devised for measuring the absorbed dose or dose equivalent from various types of radiation more directly, although in some cases less precisely. Instruments have also been designed to measure over wide ranges of intensity and energy. Some of the methodology has been reviewed in References 7 to 12.

Small-cavity chambers with air-equivalent or tissue-equivalent walls have often been used as secondary-standard instruments for measuring radiation exposure or absorbed dose. By use of the modified Bragg-Gray principle,[11] the ionization collected by an electrode within a small cavity in a suitably designed chamber can be related to the energy deposited per gram of wall material; also, the relationship may hold constant for a wide range of energies as long as the Bragg-Gray conditions are fulfilled. In its simplest form, the Bragg-Gray principle states that the ratio of the energy absorbed per gram of a medium to the energy absorbed per gram of gas in a small cavity in the medium is constant (almost independent of the initial energy of the recoil electrons produced in the medium). Since the energy to produce an ion pair in a gas (W_g) is also apparently independent of energy, we have the Bragg-Gray principle[13]

$$E_m = S_g^m \times W_g \times J_g,$$

where E_m is the energy absorbed per gram of wall material, S_g^m represents the relative mass stopping power ratio $(dE/\rho dl)_m/(dE/\rho dl)_a$ for electrons in the material and in air (i.e., the ratio of the rates of energy loss per unit path measured in g/cm^2), W_g expresses the average energy required to produce an ion pair in the gas (now usually taken as 34 ev/ion pair[11]), and J_g is the number of ion pairs produced per gram of gas in the cavity. Average mass stopping power ratios are given in Table 2 for electrons of various initial kinetic energies.[13,14]

The conditions to be met for the Bragg-Gray relation to hold[13] are briefly:

a) cavity dimensions must be small compared to the ranges of most secondary ionizing particles;
b) most ionizing particles should originate in the chamber walls, and very few primary interactions should occur in the gas;
c) the fluence of primary and secondary particles should be nearly uniform across the cavity;
d) the wall of the cavity should be thick enough so that all recoil charged particles traversing the cavity originate within the wall material, but the wall must not be so thick that it attenuates the primary nonionizing radiation appreciably.

Cavity chambers may be made with walls having an average atomic number $\overline{Z}$ simulating that of soft tissue, so that they can measure the rad dose more directly. More often, however, r chambers having air-equivalent walls are used, with appopriate wall thicknesses for the X- or gamma-ray energies to be measured (see Table 3 and Figure 2). Then the exposure or "air dose" measured in roentgens (r) can be converted to the appropriate absorbed dose in tissue, expressed in rads, by the equation[13]

$$D_{tissue}(rads) = 0.877 \times \frac{(\mu_{en}/\rho)_{tissue}}{(\mu_{en}/\rho)_{air}} \times R = f \times R.$$

Some f factors and values of $m\mu_{en} = \mu_{en}/\rho$ for various materials and gamma-ray energies are given in Table 4; for X-ray spectra, see Figure 3.

When exposure to X or gamma radiation is measured at or near the surface of the body with an air wall chamber of "equilibrium" thickness, the dose in rads at various depths in tissue must be corrected for attenuation by the use of depth dose curves or tables. However, the fractional dose at each depth depends not only on the quantum energy of the radiation, but on the area of the beam at the body surface, on the distance from source to skin, and on other factors.[2,13] A few representative depth dose data taken from Johns[13] are presented in Tables 5 to 13; the equivalent kilovoltage (effective kev) corresponding to each of the HVL specifications may be obtained from Figure 4. Table 14 provides data for converting photon fluences

Table 2.
MEAN MASS STOPPING POWER RATIOS RELATIVE TO AIR (S_m)

$$S_m = \frac{1}{T} \int_0^{T_0} S_m \, dT$$

Initial Electron Kinetic Energy (T_0), Mev	Including Density Effect		
	C	Water	Tissue
0.002	1.070	1.238	1.216
0.003	1.064	1.226	1.216
0.004	1.060	1.220	1.199
0.005	1.058	1.215	1.195
0.006	1.055	1.212	1.191
0.007	1.054	1.208	1.188
0.008	1.052	1.206	1.186
0.009	1.051	1.203	1.183
0.01	1.050	1.202	1.182
0.02	1.044	1.191	1.172
0.03	1.041	1.185	1.166
0.04	1.039	1.181	1.163
0.05	1.038	1.179	1.160
0.06	1.037	1.177	1.159
0.07	1.036	1.175	1.157
0.08	1.035	1.174	1.156
0.09	1.034	1.173	1.155
0.1	1.034	1.172	1.154
0.2	1.030	1.166	1.148
0.3	1.027	1.163	1.145
0.4	1.024	1.161	1.143
0.5	1.022	1.159	1.141
0.6	1.020	1.158	1.140
0.7	1.017	1.156	1.138
0.8	1.016	1.154	1.136
0.9	1.014	1.152	1.134
1	1.012	1.150	1.132
2	1.001	1.139	1.121
3	0.985	1.121	1.103
4	0.976	1.110	1.093
5	0.968	1.108	1.084
6	0.961	1.093	1.076
8	0.950	1.080	1.063
10	0.940	1.069	1.052

From: Johns, H. E., *The Physics of Radiology*, 2nd ed., p. 703. Charles C Thomas, Springfield, Illinois, 1964.

Table 3.
EQUILIBRIUM WALL THICKNESSES AND f FACTORS FOR SEVERAL PHOTON SPECTRA UP TO 3 Mev

Radiation	Peak Energy, Mev	Mean Energy, Mev	Equilibrium Wall, cm	f—rads/roentgens		
				Water	Bone	Muscle
^{137}Cs	0.66	0.66	0.2–0.3	0.975	0.933	0.966
^{60}Co	1.25	1.25	0.4–0.6	0.974	0.928	9.965
X rays	2.00	0.67	0.3–0.4	0.975	0.933	0.966
X rays	3.00	1.00	0.4–0.6	0.974	0.927	0.965

From: Johns, H. E., *The Physics of Radiology*, 2nd ed., p. 293. Charles C Thomas, Springfield, Illinois, 1964.

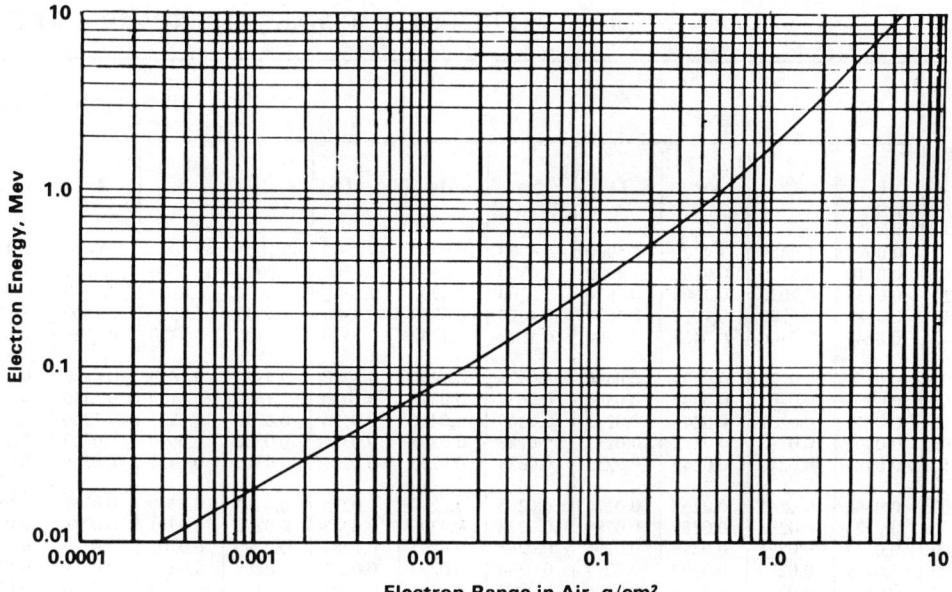

FIGURE 2.
Electron Range in Air (in g/cm²)
as a Function of the Electron Energy.
(From: Johns, H. E., *The Physics of Radiology*,
2nd ed., p. 291. Charles C Thomas,
Springfield, Illinois, 1964.)

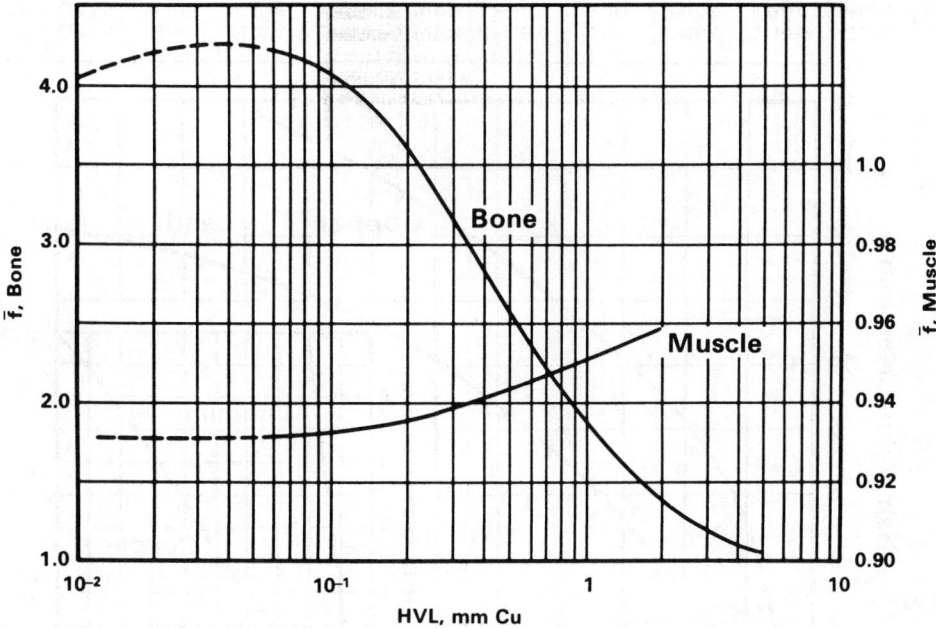

FIGURE 3.
Conversion Factors $\bar{f}$ for X-Ray Spectra
of Various Half-Value Layers of Copper,
to Convert Roentgens to Rads for Bone and Muscle.
(From: Johns, H. E., *The Physics of Radiology*,
2nd ed., p. 282. Charles C Thomas,
Springfield, Illinois, 1964.)

Table 4. VALUES OF THE MASS ENERGY-ABSORPTION

Mass Energy-Absorption

Photon Energy, Mev	H	C	N	O	Na	Mg	Al	P	S	Ar	K
0.010	0.0099	1.9400	3.4200	5.5000	15.4000	20.9000	26.5000	40.1000	49.7000	62.0000	77.0000
0.015	0.0110	0.5170	0.9160	1.4900	4.4300	6.0900	7.6500	11.9000	15.2000	19.4000	24.6000
0.020	0.0133	0.2030	0.3600	0.5870	1.7700	2.4700	3.1600	5.0000	6.4100	8.3100	10.5000
0.030	0.0186	0.0592	0.1020	0.1630	0.4820	0.6840	0.8800	1.4500	1.8500	2.4600	3.1200
0.040	0.0280	0.0306	0.0465	0.0700	0.1940	0.2740	0.3510	0.5700	0.7310	0.9740	1.2500
0.050	0.0270	0.0226	0.0299	0.0410	0.0996	0.1400	0.1760	0.2820	0.3610	0.4840	0.6260
0.060	0.0305	0.0203	0.0244	0.0304	0.0637	0.0845	0.1040	0.1660	0.2140	0.2840	0.3670
0.080	0.0362	0.0201	0.0248	0.0239	0.0369	0.0456	0.0536	0.0780	0.0971	0.1240	0.1580
0.100	0.0406	0.0213	0.0222	0.0232	0.0288	0.0334	0.0372	0.0500	0.0599	0.0725	0.0909
0.150	0.0485	0.0246	0.0249	0.0252	0.0258	0.0275	0.0282	0.0315	0.0351	0.0368	0.0433
0.200	0.0530	0.0267	0.0267	0.0271	0.0265	0.0277	0.0275	0.0292	0.0310	0.0302	0.0339
0.300	0.0573	0.0288	0.0289	0.0289	0.0278	0.0290	0.0283	0.0290	0.0301	0.0278	0.0304
0.400	0.0587	0.0295	0.0296	0.0296	0.0283	0.0295	0.0287	0.0290	0.0301	0.0274	0.0299
0.500	0.0589	0.0297	0.0297	0.0297	0.0284	0.0293	0.0287	0.0288	0.0300	0.0271	0.0294
0.600	0.0588	0.0296	0.0296	0.0296	0.0283	0.0292	0.0286	0.0287	0.0297	0.0270	0.0291
0.800	0.0573	0.0288	0.0289	0.0289	0.0276	0.0285	0.0278	0.0280	0.0287	0.0261	0.0282
1.000	0.0555	0.0279	0.0280	0.0280	0.0267	0.0275	0.0269	0.0270	0.0280	0.0252	0.0272
1.500	0.0507	0.0255	0.0255	0.0255	0.0243	0.0250	0.0246	0.0245	0.0254	0.0228	0.0247
2.000	0.0464	0.0234	0.0234	0.0234	0.0225	0.0232	0.0227	0.0228	0.0235	0.0212	0.0228
3.000	0.0398	0.0204	0.0205	0.0206	0.0199	0.0206	0.0201	0.0204	0.0210	0.0193	0.0208
4.000	0.0351	0.0184	0.0186	0.0187	0.0184	0.0191	0.0188	0.0192	0.0199	0.0182	0.0199
5.000	0.0316	0.0170	0.0172	0.0174	0.0173	0.0181	0.0180	0.0184	0.0192	0.0176	0.0193
6.000	0.0288	0.0160	0.0162	0.0166	0.0166	0.0175	0.0174	0.0179	0.0187	0.0175	0.0190
8.000	0.0249	0.0145	0.0148	0.0154	0.0158	0.0167	0.0169	0.0175	0.0184	0.0172	0.0190
10.000	0.0222	0.0137	0.0142	0.0147	0.0154	0.0163	0.0167	0.0174	0.0183	0.0173	0.0191

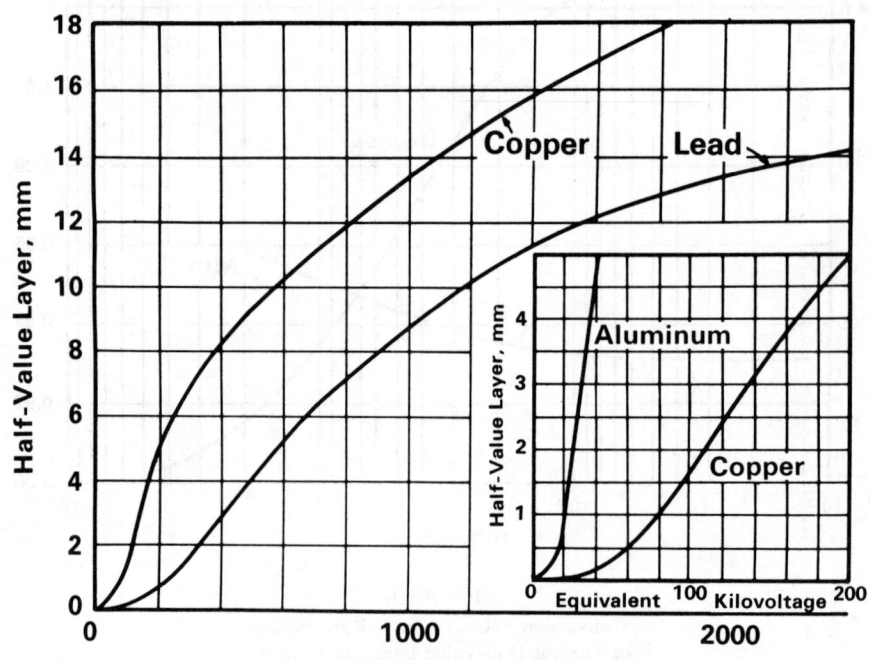

FIGURE 4.
Relationships Between HVL and Effective Kilovoltage.
(From: Johns, H. E., *The Physics of Radiology*,
2nd ed., p. 256. Charles C Thomas,
Springfield, Illinois, 1964.)

COEFFICIENTS AND THE FACTOR f

Coefficient $(m\mu_{en})$ cm²/g									$f = 0.87_7 \left[\dfrac{(m\mu_{en}) \text{ medium}}{(m\mu_{en}) \text{ air}} \right]$		
Ca	Poly-styrene	Lucite	Poly-ethylene	Bakelite	Water	Air	Compact Bone	Muscle	$\dfrac{\text{Water}}{\text{Air}}$	$\dfrac{\text{Compact Bone}}{\text{Air}}$	$\dfrac{\text{Muscle}}{\text{Air}}$
89.8000	1.7900	2.9200	1.6600	2.4300	4.8900	4.6600	19.0000	4.9600	0.92_0	3.5_8	0.93_3
28.9000	0.4780	0.7880	0.4440	0.6510	1.3200	1.2900	5.8900	1.3600	0.89_7	4.0_0	0.92_5
12.5000	0.1880	0.3110	0.1760	0.2570	0.5230	0.5160	2.5100	0.5440	0.88_7	4.2_7	0.92_5
3.7500	0.0561	0.8920	0.0534	0.0743	0.1470	0.1470	0.0743	0.1540	0.87_7	4.4_3	0.91_9
1.5200	0.0300	0.0426	0.0295	0.0368	0.0647	0.0640	0.3050	0.0677	0.88_7	4.1_8	0.92_8
0.7640	0.0229	0.0288	0.0232	0.0259	0.0394	0.0384	0.1580	0.0409	0.90_6	3.6_1	0.93_4
0.4480	0.0211	0.0243	0.0218	0.0226	0.0304	0.0292	0.0979	0.0312	0.91_3	2.9_4	0.93_7
0.1910	0.0213	0.0226	0.0224	0.0217	0.0253	0.0236	0.0520	0.0255	0.94_0	1.9_3	0.94_5
0.1110	0.0228	0.0235	0.0241	0.0227	0.0252	0.0231	0.0386	0.0252	0.95_7	1.4_7	0.95_7
0.0488	0.0264	0.0267	0.0280	0.0261	0.0278	0.0251	0.0304	0.0276	0.97_1	1.0_3	0.96_4
0.0367	0.0287	0.0289	0.0305	0.0283	0.0300	0.0268	0.0302	0.0297	0.98_2	0.98_2	0.97_7
0.0319	0.0310	0.0311	0.0329	0.0305	0.0320	0.0288	0.0311	0.0317	0.97_7	0.94_7	0.96_3
0.0308	0.0317	0.0319	0.0337	0.0312	0.0329	0.0296	0.0316	0.0325	0.97_3	0.93_8	0.96_3
0.0304	0.0319	0.0320	0.0339	0.0314	0.0330	0.0297	0.0316	0.0327	0.97_4	0.93_3	0.96_8
0.0301	0.0318	0.0319	0.0338	0.0313	0.0329	0.0296	0.0315	0.0326	0.97_3	0.93_6	0.96_3
0.0290	0.0310	0.0311	0.0329	0.0305	0.0321	0.0289	0.0306	0.0318	0.97_4	0.92_9	0.96_5
0.0279	0.0300	0.0301	0.0319	0.0295	0.0311	0.0280	0.0297	0.0308	0.97_4	0.92_7	0.96_5
0.0253	0.0274	0.0275	0.0291	0.0270	0.0283	0.0255	0.0270	0.0281	0.97_3	0.92_9	0.96_6
0.0234	0.0252	0.0252	0.0267	0.0247	0.0260	0.0234	0.0248	0.0257	0.97_4	0.92_9	0.96_6
0.0213	0.0219	0.0220	0.0232	0.0216	0.0227	0.0205	0.0219	0.0225	0.97_1	0.93_7	0.96_3
0.0204	0.0197	0.0198	0.0208	0.0194	0.0205	0.0186	0.0199	0.0203			
0.0200	0.0181	0.0183	0.0191	0.0179	0.0190	0.0173	0.0186	0.0188			
0.0198	0.0170	0.0172	0.0178	0.0168	0.0180	0.0163	0.0178	0.0178			
0.0197	0.0153	0.0156	0.0160	0.0153	0.0165	0.0150	0.0165	0.0163			
0.0201	0.0144	0.0147	0.0149	0.0144	0.0155	0.0144	0.0159	0.0154			

to exposures expressed in roentgen units. Depth dose data and conversion factors for many other conditions may be found in Johns[13] and other references cited.

QUANTITATION OF DOSE-EFFECT RELATIONSHIPS

The analysis of possible mechanisms relating radiation dose delivered to tissue and the observed biological effects has been reviewed by Burch,[15] who emphasizes the need for quantitative methods of predicting long-term as well as early biological effects resulting from radiation exposure. Upton and Kimball[16] review data indicating the relationships between exposure and effects in animals, and Wald[17] reviews the various biomedical effects observed in humans exposed to radiation. Many other comprehensive reviews of radiobiological effects are available; they are cited in References 15 to 17 and, notably, in the United Nations Scientific Committee Reports.[18] The information now available shows certain definite biomedical effects of radiation on humans at high doses and dose rates, but continuing investigation is underway to provide improved understanding and predictability of the shape of the probabilistic dose–response relationships at low doses and dose rates. Therefore, recommended limits of long-term occupational and population exposure must be based on estimated upper limits of the probabilities of disease determined by conservative extrapolation of dose–response curves to low dose or dose rate regions. Since a comprehensive treatment of dose–response relations is beyond the scope of this handbook, only some examples of possible types of response relationships are presented below.

All survival after irradiation is often plotted as the logarithm of the surviving fraction versus dose (see Figure 5). These curves are often fitted to more or less smoothly varying data, showing that for large numbers of experimental units, such as bacterial or mammalian cells, an underlying probabilistic dose–response relationship can be explicitly and accurately defined.

Table 5.
DEPTH DOSE PERCENTAGES FOR CIRCULAR FIELDS FOR 70- AND 120-KVP X RAYS, VARIOUS FILTRATIONS AND FOCAL-TO-SKIN DISTANCES (FSD)

A. HVL 1.0 mm Aluminum (Approximately 70 kvp with Inherent Filtration)

FSD = 15 cm

Area, cm^2		0	3.1	7.0	12.5	28.3	50.0	100.0
Diameter, cm		0	2	3	4	6	8	11.3
Depth, cm	0.0	100	100	100	100	100	100	100
	0.5	61	74	79	81	84	86	87
	1.0	42	56	61	63	66	67	69
	2.0	23	32	36	39	41	42	44
	3.0	13	19	22	24	26	27	29
	4.0	8	12	13	15	17	19	20
	8.0	2	2	3	3	4	4	5

FSD = 20 cm

Area, cm^2		0	3.1	7.0	12.5	28.3	50.0	100.0
Diameter, cm		0	2	3	4	6	8	11.3
Depth, cm	0.0	100	100	100	100	100	100	100
	0.5	62	75	80	82	84	86	88
	1.0	44	58	63	65	67	68	70
	2.0	24	34	38	41	43	44	45
	3.0	14	20	23	25	28	29	31
	4.0	9	13	15	16	18	20	21
	8.0	2	3	3	4	4	5	6

FSD = 30 cm

Area, cm^2		0	3.1	7.0	12.5	28.3	50.0	100.0
Diameter, cm		0	2	3	4	6	8	11.3
Depth, cm	0.0	100	100	100	100	100	100	100
	0.5	63	76	81	83	85	88	89
	1.0	45	60	64	66	68	70	71
	2.0	25	36	40	42	44	46	48
	3.0	16	22	25	27	30	31	33
	4.0	10	14	16	18	20	22	23
	8.0	2	3	4	4	5	6	7

B. HVL 2.0 mm Aluminum (Approximately 120 kvp with Inherent Filtration)

FSD = 15 cm

Area, cm^2		0	3.1	7.0	12.5	28.3	50.0	100.0
Diameter, cm		0	2	3	4	6	8	11.3
Depth, cm	0.0	100	100	100	100	100	100	100
	0.5	71	82	85	87	88	89	90
	1.0	52	65	69	72	74	76	77
	2.0	31	42	47	49	53	55	56
	3.0	20	28	32	34	38	40	42
	4.0	14	19	22	24	27	30	32
	8.0	3	5	6	7	9	10	11

Table 5. *(Continued)*
DEPTH DOSE PERCENTAGES FOR CIRCULAR FIELDS FOR 70- AND 120-KVP X RAYS, VARIOUS FILTRATIONS AND FOCAL-TO-SKIN DISTANCES (FSD)

FSD = 20 cm

Area, cm²		0	3.1	7.0	12.5	28.3	50.0	100.0
Diameter, cm		0	2	3	4	6	8	11.3
Depth, cm	0.0	100	100	100	100	100	100	100
	0.5	72	83	86	88	89	90	91
	1.0	54	66	71	73	76	77	78
	2.0	33	44	49	51	55	57	58
	3.0	22	30	34	36	40	42	44
	4.0	15	21	24	26	30	32	34
	8.0	4	6	7	8	10	11	13

FSD = 30 cm

Area, cm²		0	3.1	7.0	12.5	28.3	50.0	100.0
Diameter, cm		0	2	3	4	6	8	11.3
Depth, cm	0.0	100	100	100	100	100	100	100
	0.5	73	84	87	88	89	91	92
	1.0	55	68	73	74	77	79	80
	2.0	35	47	51	54	57	60	61
	3.0	24	33	37	39	43	45	47
	4.0	17	23	27	29	32	35	37
	8.0	5	7	8	9	11	13	15

C. HVL 3.0 mm Aluminum (Approximately 120 kvp, 1 mm Aluminum Filter)

FSD = 15 cm

Area, cm²		0	3.1	7.0	12.5	28.3	50.0	100.0
Diameter, cm		0	2	3	4	6	8	11.3
Depth, cm	0.0	100	100	100	100	100	100	100
	0.5	75	85	87	88	89	90	90
	1.0	58	70	74	76	77	78	80
	2.0	37	48	53	56	59	60	62
	3.0	24	33	37	41	45	46	48
	4.0	17	23	27	30	34	35	37
	8.0	4	6	8	9	11	13	14

FSD = 20 cm

Area, cm²		0	3.1	7.0	12.5	28.3	50.0	100.0
Diameter, cm		0	2	3	4	6	8	11.3
Depth, cm	0.0	100	100	100	100	100	100	100
	0.5	76	86	88	89	90	91	91
	1.0	60	72	75	77	79	80	81
	2.0	39	51	55	58	62	63	65
	3.0	27	35	40	43	47	49	51
	4.0	19	25	29	32	36	38	40
	8.0	5	7	9	10	12	14	16

Table 5. (Continued)
DEPTH DOSE PERCENTAGES FOR CIRCULAR FIELDS FOR 70- AND 120-KVP X RAYS, VARIOUS FILTRATIONS AND FOCAL-TO-SKIN DISTANCES (FSD)

FSD = 30 cm

Area, cm²		0	3.1	7.0	12.5	28.3	50.0	100.0
Diameter, cm		0	2	3	4	6	8	11.3
Depth, cm	0.0	100	100	100	100	100	100	100
	0.5	77	86	88	90	91	92	92
	1.0	62	74	77	79	81	82	83
	2.0	41	54	58	61	65	66	67
	3.0	29	39	43	46	51	53	55
	4.0	21	28	32	35	40	42	44
	8.0	6	9	10	12	14	17	19

D. HVL 4.0mm (Approximately 140 kvp, 2.0 mm Aluminum Filter)

FSD = 15 cm

Area, cm²		0	3.1	7.0	12.5	28.3	50.0	100.0
Diameter, cm		0	2	3	4	6	8	11.3
Depth, cm	0.0	100	100	100	100	100	100	100
	0.5	78	87	89	90	91	92	93
	1.0	62	74	77	79	80	81	84
	2.0	40	52	56	59	62	63	67
	3.0	27	37	41	44	47	49	53
	4.0	19	26	30	32	36	38	42
	8.0	5	8	9	10	12	14	17

FSD = 20 cm

Area, cm²		0	3.1	7.0	12.5	28.3	50.0	100.0
Diameter, cm		0	2	3	4	6	8	11.3
Depth, cm	0.0	100	100	100	100	100	100	100
	0.5	79	88	89	90	92	93	94
	1.0	63	76	78	80	82	83	86
	2.0	43	55	59	62	64	66	70
	3.0	30	40	44	46	49	52	56
	4.0	21	29	32	35	38	41	45
	8.0	6	9	10	12	14	16	19

FSD = 30 cm

Area, cm²		0	3.1	7.0	12.5	28.3	50.0	100.0
Diameter, cm		0	2	3	4	6	8	11.3
Depth, cm	0.0	100	100	100	100	100	100	100
	0.5	80	90	91	92	93	94	95
	1.0	65	78	81	82	83	84	87
	2.0	45	58	62	65	68	69	73
	3.0	32	43	47	50	54	56	60
	4.0	24	32	36	38	42	45	49
	8.0	7	11	12	14	17	19	22

From: Johns, H. E., *The Physics of Radiology*, 2nd ed., pp. 705–706. Charles C Thomas, Springfield, Illinois, 1964.

Table 6.
DEPTH DOSE PERCENTAGES FOR CIRCULAR FIELDS, X RAYS OF HVL = 0.5 AND 1 mm COPPER

A. HVL 0.5 mm Copper, FSD 40 cm

Depth, cm	Area of Field, cm^2								
	0	20	35	50	80	100	150	200	400
0	100.0	100.0	100.0	100.0	100.0	100.0	100.0	100.0	100.0
1	74.6	91.7	93.6	94.7	96.4	97.0	98.0	98.6	99.3
2	56.5	78.1	81.5	83.4	86.0	86.9	88.8	89.9	91.9
3	43.2	64.8	68.9	71.6	74.6	76.0	78.4	80.0	83.4
4	33.3	52.9	57.7	60.5	64.2	65.6	68.1	69.7	73.9
5	25.8	43.3	47.8	50.9	54.6	56.2	59.0	61.0	65.1
6	20.0	35.4	39.3	42.4	46.0	47.5	50.5	52.8	57.0
7	15.5	28.9	32.6	35.6	38.8	40.1	43.2	45.4	49.8
8	12.1	23.7	27.1	29.5	32.5	34.0	36.8	39.0	43.5
9	9.4	19.4	22.3	24.7	27.3	28.7	31.4	33.4	37.5
10	7.4	16.1	18.4	20.5	23.0	24.3	26.6	28.5	32.7
11	5.8	13.2	15.3	17.0	19.3	20.5	22.5	24.3	28.2
12	4.6	10.8	12.8	14.3	16.3	17.4	19.2	20.8	24.5
13	3.7	8.8	10.7	12.0	13.7	14.7	16.3	17.6	21.1
14	2.9	7.3	8.9	10.0	11.5	12.3	13.9	15.3	18.3
15	2.4	6.0	7.4	8.3	9.7	10.4	11.8	13.0	15.7
16	1.9	4.9	6.1	6.9	8.2	8.8	10.1	11.1	13.6
17	1.5	4.1	5.1	5.8	6.9	7.4	8.6	9.6	11.7
18	1.2	3.4	4.2	4.8	5.8	6.3	7.3	8.2	10.1
19	1.0	2.8	3.5	4.0	4.9	5.3	6.2	7.0	8.7
20	0.8	2.3	2.9	3.4	4.1	4.5	5.3	5.9	7.5

B. HVL 0.5 mm Copper, FSD 50 cm

Depth, cm	Area of Field, cm^2								
	0	20	35	50	80	100	150	200	400
0	100.0	100.0	100.0	100.0	100.0	100.0	100.0	100.0	100.0
1	75.3	92.3	94.3	95.4	97.1	97.7	98.7	99.3	100.0
2	55.7	79.0	82.5	84.4	87.0	88.0	89.9	91.0	93.0
3	44.5	66.0	70.2	72.9	76.0	77.4	79.8	81.5	84.9
4	34.5	54.3	59.2	62.1	65.9	67.3	69.9	71.6	75.9
5	27.0	44.7	49.3	52.5	56.3	58.0	60.9	62.9	67.2
6	21.1	36.7	40.8	44.0	47.7	49.3	52.4	54.8	59.1
7	16.5	30.1	34.0	37.1	40.4	41.8	45.0	47.3	51.9
8	13.0	24.8	28.3	30.8	34.0	35.5	38.5	40.8	45.2
9	10.1	20.4	23.4	25.9	28.6	30.1	32.9	35.0	39.4
10	8.0	16.9	19.4	21.6	24.2	25.6	28.0	30.0	34.4
11	6.3	13.9	16.2	18.0	20.4	21.6	23.8	25.7	29.8
12	5.1	11.4	13.5	15.1	17.2	18.4	20.3	22.0	25.9
13	4.1	9.4	11.3	12.7	14.5	15.6	17.3	18.7	22.4
14	3.3	7.7	9.4	10.6	12.2	13.1	14.8	16.2	19.4
15	2.6	6.4	7.8	8.8	10.3	11.1	12.6	13.8	16.7
16	2.1	5.3	6.5	7.4	8.7	9.4	10.8	11.8	14.5
17	1.7	4.3	5.4	6.2	7.3	7.9	9.2	10.2	12.5
18	1.4	3.6	4.5	5.2	6.2	6.7	7.8	8.7	10.8
19	1.1	3.0	3.8	4.3	5.2	5.7	6.6	7.5	9.3
20	0.9	2.4	3.1	3.6	4.4	4.8	5.6	6.4	8.1

Table 6. *(Continued)*
DEPTH DOSE PERCENTAGES FOR CIRCULAR FIELDS, X RAYS OF HVL = 0.5 AND 1 mm COPPER

C. HVL 1.0 mm Copper, FSD 40 cm

Depth, cm	Area of Field, cm²								
	0	20	35	50	80	100	150	200	400
0	100.0	100.0	100.0	100.0	100.0	100.0	100.0	100.0	100.0
1	78.3	93.5	96.2	97.5	99.2	100.1	101.3	101.9	102.3
2	61.7	82.1	87.2	89.0	92.0	93.0	94.7	95.6	97.1
3	49.0	71.1	75.9	79.0	83.1	84.7	87.1	88.9	91.4
4	39.0	60.5	65.5	68.8	73.2	75.2	78.2	80.3	84.2
5	31.1	50.9	55.8	59.3	63.9	65.6	69.1	71.3	75.5
6	25.0	42.8	47.4	50.7	55.1	57.1	60.3	62.6	67.4
7	20.0	35.8	40.1	43.2	47.4	49.3	52.7	55.1	59.9
8	16.1	29.8	33.7	36.5	40.5	42.6	45.7	48.1	53.1
9	13.0	24.9	28.5	31.0	34.7	36.7	39.9	41.9	46.9
10	10.4	20.8	24.9	26.4	29.6	31.4	34.4	36.4	41.5
11	8.4	17.4	20.3	22.4	25.3	27.0	29.6	31.6	36.4
12	6.7	14.6	17.1	19.0	21.5	23.1	25.6	27.5	31.8
13	5.4	12.2	14.4	16.0	18.4	19.7	22.0	23.9	27.8
14	4.4	10.2	12.2	13.6	15.7	16.9	19.0	20.7	24.3
15	3.5	8.5	10.2	11.5	13.5	14.5	16.3	17.8	21.3
16	2.8	7.1	8.6	9.7	11.5	12.4	14.0	15.4	18.6
17	2.3	6.0	7.2	8.3	9.8	10.6	12.1	13.3	16.3
18	1.9	5.0	6.1	7.0	8.3	9.0	10.4	11.5	14.3
19	1.5	4.2	5.2	5.9	7.1	7.8	8.9	9.9	12.5
20	1.2	3.5	4.4	5.0	6.1	6.7	7.7	8.5	10.9

D. HVL 1.0 mm Copper, FSD 50 cm

Depth, cm	Area of Field, cm²								
	0	20	35	50	80	100	150	200	400
0	100.0	100.0	100.0	100.0	100.0	100.0	100.0	100.0	100.0
1	79.0	94.2	96.9	98.2	99.9	100.8	102.0	102.6	103.0
2	63.0	83.2	88.3	90.2	93.2	94.2	95.9	96.9	98.4
3	50.5	72.5	77.4	80.5	84.7	86.3	88.8	90.6	93.5
4	40.5	62.0	67.2	70.6	75.1	77.1	80.2	82.4	86.4
5	32.5	52.5	57.5	61.1	65.9	67.6	71.2	73.5	77.8
6	26.3	44.4	49.1	52.5	57.1	59.2	62.5	64.9	69.8
7	21.3	37.3	41.8	45.0	49.4	51.4	54.8	57.3	62.3
8	17.3	31.2	35.2	38.2	42.4	44.6	47.8	50.3	55.5
9	14.0	26.1	29.9	32.5	36.4	38.5	41.8	43.9	49.3
10	11.3	21.9	25.2	27.8	31.2	33.1	36.2	38.3	43.6
11	9.1	18.3	21.4	23.7	26.7	28.5	31.3	33.4	38.5
12	7.4	15.4	18.2	20.1	22.8	24.4	27.1	29.1	33.8
13	5.9	12.9	15.3	17.0	19.5	20.9	23.4	25.3	29.5
14	4.8	10.8	13.0	14.4	16.7	17.9	20.2	21.9	25.8
15	3.9	9.1	10.8	12.2	14.3	15.4	17.4	18.9	22.7
16	3.2	7.6	9.1	10.3	12.2	13.2	14.9	16.4	19.8
17	2.6	6.4	7.7	8.8	10.4	11.3	12.9	14.2	17.3
18	2.1	5.3	6.5	7.4	8.9	9.6	11.1	12.3	15.2
19	1.7	4.5	5.5	6.3	7.6	8.3	9.5	10.6	13.3
20	1.4	3.7	4.7	5.4	6.5	7.1	8.2	9.1	11.6

From: Johns, H. E., *The Physics of Radiology*, 2nd ed., pp. 707–708. Charles C Thomas, Springfield, Illinois, 1964.

Table 7.
DEPTH DOSE PERCENTAGES FOR CIRCULAR FIELDS, X RAYS OF HVL = 3 mm COPPER

A. HVL 3.0 mm Copper, FSD 50 cm

Depth, cm	Area of Field, cm²								
	0	20	35	50	80	100	150	200	400
0	100.0	100.0	100.0	100.0	100.0	100.0	100.0	100.0	100.0
1	82.3	94.7	96.5	97.4	98.6	99.0	100.0	100.5	101.4
2	68.0	85.8	88.2	89.8	91.7	92.7	94.3	95.4	97.6
3	56.2	75.0	78.8	81.0	84.1	85.4	87.5	89.2	92.4
4	46.4	64.8	69.1	71.8	75.4	77.0	79.8	81.8	85.9
5	38.6	56.0	60.0	63.0	66.8	68.6	71.6	73.9	78.4
6	32.0	47.7	52.0	54.9	58.8	60.9	64.0	66.4	71.0
7	26.5	40.8	44.8	47.8	51.8	54.0	56.9	59.4	61.4
8	22.0	34.9	38.7	41.5	45.5	47.6	50.4	53.0	58.2
9	18.4	29.7	33.3	36.0	39.8	41.7	44.6	47.2	52.2
10	15.4	25.3	28.6	31.1	34.7	36.6	39.5	41.8	46.8
11	12.8	21.7	24.6	26.9	30.3	32.0	34.8	37.2	41.9
12	10.7	18.5	21.1	23.2	26.4	27.9	30.6	32.7	37.3
13	9.0	15.7	18.2	20.0	22.9	24.4	26.9	28.8	33.3
14	7.5	13.4	15.7	17.3	19.9	21.2	23.6	25.4	29.5
15	6.3	11.5	13.4	15.0	17.3	18.5	20.7	22.4	26.3
16	5.3	9.8	11.5	12.9	15.0	16.4	18.2	19.7	23.4
17	4.5	8.4	9.9	11.2	13.4	14.0	15.9	17.4	20.8
18	3.7	7.2	8.5	9.6	11.1	12.2	14.0	15.4	18.5
19	3.1	6.1	7.3	8.3	9.9	10.7	12.3	13.6	16.5
20	2.6	5.2	6.3	7.2	8.6	9.3	10.8	11.9	14.6

B. HVL 3.0 mm Copper, FSD 60 cm

Depth, cm	Area of Field, cm²								
	0	20	35	50	80	100	150	200	300
0	100.0	100.0	100.0	100.0	100.0	100.0	100.0	100.0	100.0
1	82.9	95.3	97.1	98.0	99.2	99.5	100.6	101.1	102.0
2	68.8	86.7	89.2	90.8	92.7	93.7	95.3	96.4	98.7
3	57.3	76.2	80.1	82.3	85.4	86.8	88.9	90.6	93.9
4	47.5	66.1	70.5	73.2	76.8	78.5	81.3	83.3	87.4
5	39.8	57.5	61.4	64.5	68.3	70.2	73.2	75.5	80.1
6	33.2	49.1	53.4	56.4	60.3	62.5	65.6	68.1	72.8
7	27.6	42.2	46.2	49.3	53.3	55.6	58.6	61.1	66.2
8	23.1	36.2	40.0	42.9	47.0	49.1	52.0	54.7	60.0
9	19.4	30.9	34.6	37.3	41.2	43.2	46.2	48.9	54.0
10	16.3	26.4	29.8	32.3	36.1	38.0	41.0	43.3	48.5
11	13.6	22.7	25.7	28.0	31.5	33.3	36.2	38.6	43.4
12	11.4	19.4	22.1	24.2	27.5	29.1	31.9	34.1	38.8
13	9.6	16.5	19.1	20.9	24.0	25.5	28.1	30.1	34.8
14	8.1	14.2	16.5	18.1	20.9	22.2	24.7	26.6	30.9
15	6.8	12.2	14.2	15.7	18.2	19.5	21.8	23.6	27.6
16	5.8	10.4	12.2	13.6	15.8	17.0	19.2	20.8	24.6
17	4.9	8.9	10.5	11.8	13.9	14.8	16.8	18.4	21.9
18	4.1	7.6	9.1	10.2	12.1	12.9	14.8	16.3	19.6
19	3.5	6.5	7.9	8.9	10.5	11.4	13.1	14.4	17.5
20	2.9	5.6	6.8	7.7	9.2	9.9	11.5	12.7	15.6

Table 7. (Continued)
DEPTH DOSE PERCENTAGES FOR CIRCULAR FIELDS, X RAYS OF HVL = 3 mm COPPER

C. HVL 3.0 mm Copper, FSD 80 cm

Depth, cm	Area of Field, cm²								
	0	20	35	50	80	100	150	200	400
0	100.0	100.0	100.0	100.0	100.0	100.0	100.0	100.0	100.0
1	83.8	95.9	97.8	98.6	99.7	100.1	101.1	101.6	102.5
2	70.0	88.0	90.3	92.0	93.9	94.8	96.5	97.6	99.7
3	58.6	77.9	81.7	83.8	86.9	88.2	90.4	92.1	95.4
4	49.0	68.1	72.3	75.2	78.7	80.4	83.3	85.3	89.5
5	41.3	59.5	63.4	66.5	70.4	72.3	75.3	77.7	82.3
6	34.7	51.1	55.5	58.6	62.6	64.8	68.0	70.5	75.1
7	29.2	44.1	48.3	51.4	55.5	57.8	60.9	63.6	68.9
8	24.5	38.1	42.0	44.9	49.1	51.4	54.4	57.1	62.6
9	20.7	32.7	36.5	39.2	43.3	45.3	48.4	51.2	56.4
10	17.5	28.1	31.5	34.1	38.0	40.0	43.1	45.7	50.9
11	14.7	24.2	27.3	29.7	33.4	35.2	38.2	40.7	45.8
12	12.5	20.8	23.6	25.8	29.2	30.9	33.8	36.1	41.0
13	10.5	17.8	20.4	22.2	25.5	27.1	29.9	31.9	36.8
14	8.9	15.3	17.7	19.4	22.3	23.7	26.3	28.2	32.8
15	7.6	13.2	15.2	16.9	19.4	20.8	23.2	25.0	29.4
16	6.4	11.3	13.2	14.7	17.0	18.2	20.5	22.1	26.3
17	5.4	9.7	11.4	12.8	14.9	15.9	18.1	19.7	23.5
18	4.6	8.3	9.9	11.1	13.0	13.9	15.9	17.5	21.0
19	3.9	7.2	8.5	9.6	11.4	12.3	14.1	15.5	18.8
20	3.3	6.2	7.4	8.4	9.9	10.7	12.4	13.7	16.7

D. HVL 3.0 mm Copper, FSD 100 cm

Depth, cm	Area of Field, cm²								
	0	20	35	50	80	100	150	200	400
0	100.0	100.0	100.0	100.0	100.0	100.0	100.0	100.0	100.0
1	84.0	96.4	98.1	99.0	100.1	100.5	101.5	101.9	102.8
2	70.7	88.8	91.1	92.7	94.5	95.5	97.0	98.2	100.3
3	59.6	79.0	82.7	84.8	87.9	88.2	91.4	93.0	96.3
4	50.1	69.3	73.5	76.3	79.8	81.5	84.4	86.4	90.6
5	42.4	60.8	64.7	67.8	71.7	74.6	76.7	79.1	83.7
6	35.7	52.4	56.8	59.9	63.9	66.1	69.3	71.8	76.6
7	30.1	45.5	49.6	52.8	56.9	59.3	62.3	65.0	70.3
8	25.4	39.3	43.4	46.3	50.6	52.8	55.8	58.6	64.1
9	21.6	33.9	37.7	40.6	44.7	46.7	49.8	52.6	58.0
10	18.3	29.2	32.7	35.4	39.3	41.4	44.5	47.0	52.4
11	15.4	25.2	28.4	30.9	34.6	36.5	39.5	42.1	47.3
12	13.1	21.7	24.6	26.9	30.4	32.1	35.0	37.4	42.4
13	11.1	18.6	21.3	23.4	26.6	28.2	31.0	33.1	38.0
14	9.5	16.1	18.5	20.4	23.3	24.7	27.4	29.4	34.0
15	8.1	13.8	16.0	17.8	20.4	21.7	24.2	26.1	30.5
16	6.9	11.9	13.9	15.4	17.8	19.0	21.4	23.4	27.3
17	5.8	10.3	12.1	13.5	15.6	16.7	18.9	20.5	24.4
18	5.0	8.9	10.5	11.7	13.7	14.6	16.7	18.2	19.6
19	4.2	7.6	9.2	10.2	12.0	12.9	14.7	16.2	19.6
20	3.6	6.6	7.9	8.9	10.5	11.3	13.0	14.3	17.5

From: Johns, H. E., *The Physics of Radiology*, 2nd ed., pp. 714–715. Charles C Thomas, Springfield, Illinois, 1964.

Table 8.
DEPTH DOSE PERCENTAGES FOR CIRCULAR FIELDS, X RAYS OF HVL = 4 mm COPPER

A. HVL 4.0 mm Copper, FSD 50 cm

Depth, cm	Area of Field, cm²								
	0	20	35	50	80	100	150	200	400
0	100.0	100.0	100.0	100.0	100.0	100.0	100.0	100.0	100.0
1	83.1	94.4	96.0	96.8	97.7	98.0	98.8	99.3	100.0
2	69.3	85.9	87.8	89.1	90.8	91.6	93.0	93.9	96.0
3	57.8	75.6	78.8	80.7	83.3	84.3	86.2	87.6	90.1
4	48.2	65.5	69.5	71.8	75.0	76.4	78.9	80.5	84.2
5	40.7	56.6	60.4	63.2	66.6	68.2	71.2	73.4	77.1
6	34.3	48.5	52.7	55.5	58.9	60.8	63.8	66.1	70.2
7	28.9	41.6	45.6	48.4	51.8	53.7	56.8	59.4	63.8
8	24.4	35.7	39.5	42.0	45.5	47.3	50.5	53.1	57.8
9	20.5	30.6	34.0	36.5	39.8	41.6	44.8	47.3	51.8
10	17.3	26.3	29.4	31.6	35.0	36.7	39.7	42.0	46.6
11	14.6	22.6	25.4	27.4	30.6	32.3	35.1	37.4	41.8
12	12.4	19.4	21.9	23.7	26.8	28.4	30.9	33.1	37.5
13	10.5	16.7	19.0	20.6	23.4	24.9	27.3	29.2	33.6
14	8.9	14.3	16.4	17.9	20.4	21.8	24.1	25.8	30.0
15	7.5	12.3	14.1	15.5	17.8	19.0	21.2	22.8	26.7
16	6.4	10.6	12.2	13.5	15.6	16.7	18.7	20.1	23.8
17	5.4	9.1	10.6	11.7	13.6	14.6	16.4	17.7	21.2
18	4.6	7.8	9.1	10.2	11.8	12.8	14.4	15.7	18.9
19	4.0	6.7	7.9	8.8	10.3	11.2	12.6	13.8	16.9
20	3.4	5.8	6.8	7.7	9.0	9.7	11.1	12.2	15.1

B. HVL 4.0 mm Copper, FSD 80 cm

Depth, cm	Area of Field, cm²								
	0	20	35	50	80	100	150	200	400
0	100.0	100.0	100.0	100.0	100.0	100.0	100.0	100.0	100.0
1	84.6	95.6	97.2	97.9	98.8	99.1	99.8	100.3	101.0
2	71.4	88.0	89.7	91.1	92.8	93.6	95.0	95.9	97.9
3	60.2	78.5	81.5	83.4	86.0	87.0	88.9	90.4	93.2
4	50.9	68.8	72.7	75.0	78.2	79.7	82.2	83.9	87.5
5	43.5	60.2	63.9	66.7	70.2	71.9	74.9	77.1	80.9
6	37.2	52.2	56.3	59.2	62.7	64.8	67.8	70.1	71.2
7	31.7	45.3	49.2	52.1	55.7	57.8	60.9	63.6	67.9
8	27.1	39.3	43.1	45.7	49.4	51.3	54.6	57.3	62.1
9	23.1	34.0	37.5	40.1	43.6	45.4	48.8	51.5	56.0
10	19.7	29.5	32.7	35.0	38.6	40.4	43.6	46.1	50.8
11	16.8	25.6	28.5	30.6	34.0	35.8	38.8	41.3	45.9
12	14.4	22.2	24.8	26.7	29.9	31.6	34.4	36.8	41.4
13	12.3	19.3	21.6	23.4	26.4	27.9	30.6	32.7	37.4
14	10.5	16.6	18.8	20.4	23.1	24.7	27.2	29.1	33.6
15	9.0	14.4	16.3	17.8	20.3	21.6	24.0	25.9	30.1
16	7.7	12.5	14.2	15.6	17.8	19.1	21.3	23.0	27.0
17	6.6	10.8	12.4	13.7	15.7	16.8	18.9	20.3	24.2
18	5.7	9.4	10.8	12.0	13.8	14.8	16.7	18.1	21.7
19	4.9	8.1	9.4	10.5	12.1	13.1	14.8	16.1	19.5
20	4.2	7.0	8.2	9.1	10.6	11.5	13.1	14.3	17.6

From: Johns, H. E., *The Physics of Radiology*, 2nd ed., p. 716. Charles C Thomas, Springfield, Illinois, 1964.

Table 9.
DEPTH DOSE PERCENTAGES FOR ^{137}Cs FOR SEVERAL FIELD SIZES, SOURCE-TO-SKIN DISTANCES (SSD), AND DIAPHRAGM-TO-SKIN DISTANCES (DSD)

A. Source 1.5 cm, SSD 15 cm, DSD = 0 (Courtesy of *Brit. J. Radiol.*)

Depth, cm	Circular Fields, Diameter, cm					Rectangular Fields, cm × cm		
	0	1	6	8	10	4 × 6	6 × 8	6 × 10
0.15	100.0	100.0	100.0	100.0	100.0	100.0	100.0	100.0
0.5	94.0	94.6	95.0	95.2	95.6	95.2	97.0	98.0
1	84.6	86.5	87.6	88.5	89.2	88.6	90.5	91.3
2	66.3	73.4	75.2	76.5	77.0	75.5	77.7	78.5
3	54.6	62.0	64.2	66.0	66.7	64.4	67.0	67.7
4	44.5	52.4	54.8	56.8	57.6	54.5	57.5	58.3
5	36.6	44.4	47.2	49.0	50.0	46.3	49.5	50.2
6	30.3	37.5	40.4	42.4	43.3	40.2	42.5	43.3
7	25.4	32.0	34.6	36.5	37.5	34.3	36.5	37.2
8	21.2	27.5	29.8	31.5	32.4	29.5	31.5	32.0
9	17.7	23.6	25.7	27.5	28.6	25.3	27.3	28.0
10	15.0	20.4	22.5	24.2	25.3	21.9	23.8	24.4
11	12.8	17.4	19.5	21.3	22.4	19.0	20.9	21.5
12	11.0	15.2	17.0	18.8	19.7	16.7	18.4	18.9
13	9.4	13.2	14.9	16.6	17.6	14.6	16.3	16.7
14	8.0	11.4	13.1	14.7	15.6	12.8	14.4	14.7
15	6.8	9.9	11.5	13.0	13.8	11.3	12.6	12.9
16	5.9	8.6	10.1	11.5	12.3	9.8	11.1	11.3
17	5.1	7.6	8.9	10.2	10.9	8.6	9.8	10.0
18	4.4	6.6	7.8	9.0	9.7	7.5	8.6	8.8
19	3.8	5.8	6.9	7.9	8.6	6.5	7.6	7.7
20	3.3	5.1	6.0	7.0	7.6	5.7	6.7	6.8

B. Source 2.7 cm, SSD 35 cm, DSD 6 cm (Courtesy of *Brit. J. Radiol.*)

Depth, cm	Area of Field, cm^2					
	0	25	50	100	200	400
0	0.0	35.0	50.0	65.0	75.0	80.0
0.15	100.0	100.0	100.0	100.0	100.0	100.0
1	88.1	93.4	94.2	94.5	94.9	95.0
2	76.4	85.1	86.4	87.1	87.6	88.2
3	66.4	76.8	78.8	80.0	80.7	81.6
4	57.7	69.1	71.6	73.2	74.2	75.3
5	50.1	62.1	64.8	66.7	68.0	69.3
6	43.6	55.6	58.4	60.6	62.2	63.6
7	37.9	49.7	52.5	54.7	56.7	58.3
8	33.0	44.4	47.2	49.3	51.5	53.3
9	28.8	39.6	42.4	44.5	46.7	48.7
10	25.3	35.3	38.1	40.2	42.3	44.5
11	22.3	31.5	34.2	36.3	38.5	40.6
12	19.7	28.1	30.7	32.8	35.0	37.2
13	17.3	25.1	27.5	29.6	31.9	34.0
14	15.2	22.4	24.6	26.7	29.0	31.0
15	13.4	20.0	22.0	24.1	26.3	28.3
16	11.8	17.8	19.7	21.7	23.9	25.8
17	10.4	15.9	17.7	19.6	21.7	23.6
18	9.2	14.2	15.9	17.7	19.7	21.6
19	8.1	12.7	14.3	16.0	17.9	19.8
20	7.2	11.3	12.8	14.4	16.2	18.0

Table 9. *(Continued)*
DEPTH DOSE PERCENTAGES FOR ^{137}Cs FOR SEVERAL FIELD SIZES, SOURCE-TO-SKIN DISTANCES (SSD), AND DIAPHRAGM-TO-SKIN DISTANCES (DSD)

C. Source, 3.2 cm, SSD 50 cm, DSD 10 cm (Courtesy of C. S. Simons et al. and *Am. J. Roentgenol.*)

Depth, cm	Area of Circular Field, cm²					
	0	25	50	75	100	150
0.15	100.0	100.0	100.0	100.0	100.0	100.0
0.5	95.8	98.0	98.2	98.4	98.8	98.6
1	90.1	94.4	95.2	95.6	96.0	95.7
2	80.2	87.0	88.0	88.4	88.8	89.6
3	70.6	79.6	81.2	81.8	82.4	83.3
4	62.6	72.0	74.2	75.3	76.0	76.8
5	55.5	65.2	67.6	68.5	70.0	71.0
6	49.3	58.4	61.4	62.5	64.0	65.3
7	43.9	52.8	55.6	57.0	58.4	59.8
8	38.9	47.4	50.4	51.9	53.4	54.8
9	34.7	42.8	45.3	47.3	48.6	50.2
10	30.8	38.4	41.2	42.8	44.4	46.1
11	27.5	34.4	37.2	38.7	40.4	41.6
12	24.4	30.6	33.6	35.6	36.4	38.0
13	21.8	27.4	30.2	31.8	33.4	34.6
14	19.4	24.8	27.2	29.0	30.2	31.4
15	17.3	22.4	24.6	26.3	27.4	28.7
16	15.5	20.0	22.2	24.0	24.8	26.1
17	13.9	17.8	20.0	21.7	22.6	23.8
18	12.4	16.0	18.0	19.8	20.4	21.5
19	11.1	14.4	16.2	17.7	18.4	19.7
20	9.9	13.0	14.6	16.0	16.8	17.7

From: Johns, H. E., *The Physics of Radiology*, 2nd ed., pp. 719–720. Charles C Thomas, Springfield, Illinois, 1964.

Table 10.
DEPTH DOSE PERCENTAGES FOR ^{60}Co FOR SEVERAL FIELD SIZES AND SOURCE-TO-SKIN DISTANCES (SSD)

A. Average Photon Energy 1.25 Mev, HVL 11 mm Lead, SSD 50 cm

Depth, cm	Area of Field, cm^2					
	0	20	50	100	200	400
0.5	100.0	100.0	100.0	100.0	100.0	100.0
1	94.6	96.2	97.0	97.5	97.6	97.7
2	85.2	89.2	90.6	91.4	91.8	92.1
3	76.8	82.3	84.2	85.4	86.1	86.8
4	69.3	75.7	78.2	79.6	80.6	81.0
5	62.6	69.5	72.4	74.0	75.3	76.0
6	56.4	63.7	66.8	68.6	70.2	71.3
7	51.0	58.3	61.4	63.4	65.3	67.1
8	41.6	53.3	56.4	58.6	60.7	62.7
9	41.7	48.7	51.7	53.9	56.2	58.6
10	37.8	44.5	47.4	49.7	52.2	54.9
11	34.3	40.6	43.5	45.8	48.4	51.2
12	31.1	37.1	40.0	42.2	45.0	47.8
13	28.2	33.9	36.7	39.0	41.7	44.7
14	25.6	31.0	33.7	36.0	38.7	41.7
15	23.3	28.4	30.9	33.2	36.0	39.0
16	21.1	26.0	28.4	30.6	33.4	36.5
17	19.3	23.8	26.1	28.3	31.1	34.2
18	17.5	21.8	24.0	26.2	28.9	32.0
19	15.9	19.9	22.2	24.2	26.9	29.9
20	14.5	18.2	20.3	22.4	25.0	28.1

B. Average Photon Energy 1.25 Mev, HVL 11 mm Lead, SSD 60 cm

Depth, cm	Area of Field, cm^2					
	0	20	50	100	200	400
0.5	100.0	100.0	100.0	100.0	100.0	100.0
1	95.0	96.7	97.1	97.8	97.9	98.1
2	86.0	90.1	91.2	92.2	92.6	93.0
3	77.9	83.7	85.4	86.6	87.4	88.0
4	70.7	77.6	79.7	81.2	82.3	83.2
5	64.2	71.7	74.2	75.9	77.3	78.4
6	58.3	66.1	68.9	70.7	72.4	73.7
7	53.0	60.8	63.7	65.7	67.6	69.2
8	48.2	55.8	58.8	60.9	63.0	65.0
9	43.9	51.2	54.2	56.4	58.6	60.9
10	39.9	46.9	49.9	52.2	54.5	57.1
11	36.3	43.0	46.0	48.3	50.7	53.4
12	33.1	39.4	42.4	44.7	47.2	50.0
13	30.2	36.1	39.1	41.4	44.0	47.0
14	27.5	33.1	36.0	38.3	41.0	44.0
15	25.1	30.4	33.2	35.5	38.2	41.2
16	22.9	27.9	30.6	32.9	35.6	38.6
17	20.9	25.7	28.2	30.5	33.2	36.2
18	19.1	23.7	26.0	28.3	31.0	34.1
19	17.4	21.8	24.0	26.2	28.9	32.0
20	15.9	20.0	22.1	24.2	27.0	30.0

Table 10. *(Continued)*
DEPTH DOSE PERCENTAGES FOR ^{60}Co FOR SEVERAL FIELD SIZES AND SOURCE-TO-SKIN DISTANCES (SSD)

C. Average Photon Energy 1.25 Mev, HVL 11 mm Lead, SSD 80 cm

Depth, cm	Area of Field, cm^2					
	0	20	50	100	200	400
0.5	100.0	100.0	100.0	100.0	100.0	100.0
1	95.4	97.0	97.7	98.2	98.4	98.5
2	87.1	91.0	92.5	93.4	93.7	94.0
3	79.5	85.3	87.2	88.4	89.0	89.6
4	72.7	79.6	82.0	83.4	84.4	85.2
5	66.5	74.1	76.9	78.5	79.9	80.8
6	60.8	68.9	71.8	73.7	75.2	76.4
7	55.6	63.8	66.8	68.9	70.7	72.1
8	50.9	58.9	62.1	64.2	66.3	68.0
9	46.6	54.3	57.5	59.8	62.1	64.1
10	42.7	50.1	53.3	55.7	58.1	60.3
11	39.2	46.2	49.4	51.8	54.3	56.7
12	35.9	42.6	45.8	48.2	50.8	53.3
13	32.9	39.3	42.4	44.9	47.6	50.1
14	30.2	36.3	39.3	41.8	44.5	47.1
15	27.7	33.5	36.4	38.9	41.8	44.3
16	25.4	31.0	33.8	36.2	39.0	41.7
17	23.3	28.7	31.3	33.8	36.5	39.2
18	21.4	26.5	29.0	31.4	34.2	36.9
19	19.6	24.5	27.0	29.3	32.0	34.7
20	18.0	22.6	25.0	27.3	30.0	32.7

D. Average Photon Energy 1.25 Mev, HVL 11 mm Lead, SSD 100 cm

Depth, cm	Area of Field, cm^2					
	0	20	50	100	200	400
0.5	100.0	100.0	100.0	100.0	100.0	100.0
1	95.9	97.2	97.9	98.6	98.8	98.8
2	87.9	91.7	93.0	94.0	94.5	94.6
3	80.7	86.3	88.1	89.4	90.1	90.5
4	73.8	81.0	83.2	84.8	85.7	86.4
5	67.8	75.7	78.1	80.2	81.3	82.3
6	62.3	70.6	73.6	75.6	76.9	78.2
7	57.3	65.7	68.8	71.0	72.5	74.1
8	52.7	61.0	64.2	66.5	68.3	70.1
9	48.5	56.5	59.7	62.1	64.2	66.2
10	44.7	52.3	55.5	57.9	60.3	62.5
11	41.2	48.4	51.6	54.0	56.6	58.8
12	38.0	44.8	48.0	50.4	53.1	55.4
13	35.0	41.5	44.6	47.1	49.8	52.2
14	32.2	38.5	41.5	44.0	46.7	49.2
15	29.6	35.7	38.6	41.4	43.8	46.4
16	27.2	33.1	35.9	38.4	41.1	43.7
17	25.0	30.7	33.4	35.9	38.6	41.2
18	23.0	28.5	31.1	33.6	36.3	38.8
19	21.2	26.4	29.0	31.4	34.1	36.6
20	19.5	24.4	27.0	29.2	32.0	34.5

From: Johns, H. E., *The Physics of Radiology*, 2nd ed., pp. 721–722. Charles C Thomas, Springfield, Illinois, 1964.

Table
DEPTH DOSE PERCENTAGES FOR 3 mm COPPER FOR

Depth, cm	Rectangular									
	4 × 4	4 × 6	4 × 8	4 × 10	4 × 15	4 × 20	6 × 6	6 × 8	6 × 10	6 × 15
*	111.6	113.7	114.9	115.8	117.0	117.6	116.4	118.2	119.4	121.1
0	100.0	100.0	100.0	100.0	100.0	100.0	100.0	100.0	100.0	100.0
1	93.9	95.1	95.6	95.9	96.3	96.5	96.5	97.1	97.5	98.1
2	84.6	86.2	87.1	87.6	88.4	88.7	88.3	89.4	90.1	91.2
3	73.7	76.0	77.3	78.1	79.2	79.7	78.8	80.5	81.6	83.1
4	63.1	65.8	67.4	68.4	69.6	70.3	69.0	71.0	72.4	74.1
5	54.2	56.7	58.4	59.5	60.9	61.6	60.1	62.2	63.7	65.6
6	46.3	48.7	50.4	51.5	53.1	53.8	52.0	54.2	55.7	57.8
7	39.3	41.7	43.4	44.5	46.1	46.8	44.9	47.1	48.6	50.8
8	33.4	35.7	37.3	38.5	40.0	40.7	38.7	40.9	42.4	44.5
9	28.5	30.6	32.1	33.2	34.7	35.4	33.4	35.4	36.9	38.9
10	24.3	26.2	27.6	28.6	30.0	30.8	28.7	30.6	32.0	34.0
11	20.7	22.4	23.7	24.6	26.0	26.7	24.7	26.4	27.7	29.6
12	17.6	19.1	20.4	21.2	22.5	23.1	21.2	22.8	24.0	25.7
13	15.1	16.3	17.5	18.3	19.5	20.0	18.2	19.7	20.8	22.4
14	12.9	14.0	15.0	15.7	16.9	17.4	15.7	17.0	18.0	19.5
15	11.0	12.0	12.8	13.5	14.6	15.1	13.5	14.6	15.5	17.0
16	9.4	10.3	11.0	11.6	12.6	13.1	11.6	12.6	13.4	14.8
17	8.0	8.8	9.4	10.0	10.9	11.4	10.0	10.9	11.6	12.9
18	6.8	7.5	8.1	8.6	9.4	9.9	8.6	9.4	10.0	11.2
19	5.8	6.4	7.0	7.4	8.1	8.6	7.4	8.1	8.7	9.8
20	4.9	5.5	6.0	6.4	7.1	7.5	6.3	7.0	7.6	8.5

* The first line gives the dose at the maximum for 100 r of primary.

From: Johns, H. E., *The Physics of Radiology*, 2nd ed., p. 731. Charles C Thomas, Springfield, Illinois, 1964.

11.
RECTANGULAR FIELDS OF VARIOUS DIMENSIONS

Fields, cm × cm										
6 × 20	8 × 8	8 × 10	8 × 15	8 × 20	10 × 10	10 × 15	10 × 20	15 × 15	15 × 20	20 × 20
122.1	120.4	121.9	124.1	125.3	123.7	126.2	127.7	129.6	131.5	133.7
100.0	100.0	100.0	100.0	100.0	100.0	100.0	100.0	100.0	100.0	100.0
98.3	97.9	98.3	99.1	99.3	98.9	99.7	100.0	100.6	101.0	101.4
91.5	90.7	91.6	92.8	93.3	92.6	93.9	94.5	95.6	96.2	96.8
83.6	82.4	83.7	85.4	86.3	85.1	87.0	88.1	89.5	90.8	92.3
75.0	73.4	74.9	77.1	78.1	76.7	79.1	80.3	82.1	83.8	85.7
66.6	64.8	66.5	68.8	70.0	68.4	71.1	72.5	74.5	76.2	78.3
58.8	56.8	58.6	61.1	62.3	60.6	63.5	65.0	67.0	68.9	71.3
51.8	49.7	51.5	54.1	55.3	53.6	56.5	58.1	60.1	62.0	64.2
45.6	43.4	45.2	47.7	49.0	47.2	50.1	51.7	53.7	55.6	57.9
40.0	37.8	39.6	42.0	43.3	41.5	44.3	45.9	47.9	49.8	52.0
35.0	32.8	34.5	36.9	38.1	36.3	39.1	40.6	42.6	44.5	46.7
30.6	28.5	30.0	32.3	33.5	31.7	34.4	35.8	37.7	39.6	41.7
26.7	24.8	26.1	28.3	29.4	27.7	30.3	31.5	33.3	35.1	37.2
23.3	21.5	22.7	24.7	25.8	24.2	26.6	27.8	29.4	31.1	33.1
20.3	18.6	19.7	21.6	22.6	21.1	23.3	24.5	25.9	27.5	29.4
17.7	16.1	17.1	18.9	19.8	18.4	20.4	21.6	22.9	24.4	26.2
15.5	13.9	14.9	16.5	17.4	16.0	17.9	19.0	20.2	21.6	23.2
13.5	12.0	13.0	14.4	15.3	13.9	15.7	16.7	17.8	19.1	20.7
11.8	10.4	11.3	12.6	13.4	12.2	13.8	14.7	15.7	16.9	18.5
10.3	9.0	9.8	11.1	11.8	10.7	12.1	13.0	13.9	15.0	16.4
9.0	7.8	8.5	9.7	10.3	9.3	10.6	11.4	12.3	13.3	14.5

Table
DEPTH DOSE PERCENTAGES FOR
FOR 80 cm AND 100 cm

A. FSD 80 cm

Depth, cm									Rectangular	
	4 × 4	4 × 6	4 × 8	4 × 10	4 × 15	4 × 20	6 × 6	6 × 8	6 × 10	6 × 15
*	101.1	101.3	101.5	101.6	101.8	101.9	101.6	101.8	102.0	102.3
0								Surface dose 30 to 50%,		
0.5	100.0	100.0	100.0	100.0	100.0	100.0	100.0	100.0	100.0	100.0
1	96.8	97.0	97.2	97.3	97.4	97.4	97.4	97.6	97.7	97.8
2	90.6	91.2	91.5	91.6	91.8	91.8	91.9	92.2	92.5	92.7
3	84.7	85.5	85.9	86.1	86.4	86.4	86.5	86.9	87.3	87.6
4	79.0	79.9	80.4	80.6	81.0	81.1	81.1	81.7	82.1	82.5
5	73.5	74.5	75.1	75.3	75.7	75.9	75.9	76.6	77.0	77.5
6	68.1	69.2	69.9	70.1	70.5	70.7	70.7	71.5	71.9	72.5
7	62.9	64.1	64.8	65.1	65.5	65.7	65.7	66.5	67.0	67.6
8	58.0	59.2	59.9	60.3	60.8	61.0	60.8	61.7	62.2	62.9
9	53.5	54.7	55.3	55.8	56.3	56.6	56.2	57.1	57.7	58.5
10	49.3	50.5	51.1	51.6	52.2	52.5	52.0	52.9	53.5	54.4
11	45.5	46.6	47.3	47.8	48.4	48.6	48.4	49.0	49.6	50.5
12	41.9	43.0	43.7	44.2	44.8	45.4	44.5	45.4	46.0	46.9
13	38.6	39.7	40.4	40.9	41.4	41.8	41.1	42.0	42.7	43.6
14	35.6	36.6	37.3	37.8	38.4	38.7	38.0	38.9	39.6	40.5
15	32.9	33.8	34.5	35.0	35.6	35.9	35.2	36.1	36.7	37.6
16	30.4	31.3	32.0	32.4	33.1	33.4	32.6	33.5	34.1	35.0
17	28.1	29.0	29.6	30.0	30.7	31.0	30.2	31.1	31.6	32.6
18	26.0	26.9	27.4	27.9	28.5	28.8	28.0	28.8	29.4	30.0
19	24.0	24.9	25.4	25.9	26.5	26.8	26.0	26.7	27.4	28.2
20	22.1	22.9	23.5	23.9	24.5	24.8	24.0	24.8	25.4	26.2

B. FSD 100 cm

Depth, cm									Rectangular	
	4 × 4	4 × 6	4 × 8	4 × 10	4 × 15	4 × 20	6 × 6	6 × 8	6 × 10	6 × 15
*	101.1	101.3	101.5	101.6	101.8	101.9	101.6	101.8	102.0	102.3
0								Surface dose 30 to 50%,		
0.5	100.0	100.0	100.0	100.0	100.0	100.0	100.0	100.0	100.0	100.0
1	97.1	97.3	97.5	97.6	97.7	97.7	97.7	97.9	98.0	98.2
2	91.4	91.9	92.2	92.4	92.5	92.6	92.6	92.9	93.1	93.4
3	85.8	86.5	86.9	87.2	87.3	87.5	87.5	87.9	88.2	88.6
4	80.2	81.2	81.7	82.0	82.2	82.4	82.4	83.0	83.4	83.8
5	74.8	76.0	76.6	76.9	77.2	77.4	77.3	78.1	78.6	79.0
6	69.7	70.9	71.6	71.9	72.3	72.5	72.4	73.2	73.8	74.3
7	64.8	66.0	66.7	67.1	67.5	67.7	67.6	68.4	69.0	60.6
8	60.1	61.3	62.0	62.4	62.9	63.1	62.9	63.8	64.4	65.1
9	55.7	56.9	57.6	58.0	58.5	58.8	58.4	59.4	60.0	60.7
10	51.5	52.7	53.4	53.8	54.4	54.7	54.2	55.2	55.8	56.6
11	47.7	48.8	49.5	49.9	50.5	50.8	50.3	51.3	51.9	52.7
12	44.1	45.2	45.9	46.3	46.9	47.2	46.7	47.7	48.2	49.1
13	40.8	41.9	42.6	43.0	43.6	43.9	43.3	44.3	44.9	45.8
14	37.8	38.9	39.5	40.0	40.6	40.9	40.2	41.2	41.8	42.7
15	35.0	36.1	36.7	37.2	37.8	38.1	37.4	38.3	38.9	39.9
16	32.5	33.5	34.1	34.5	35.2	35.5	34.8	35.6	36.3	37.2
17	30.1	31.1	31.7	32.1	32.8	33.1	32.3	33.1	33.8	34.7
18	27.9	28.8	29.4	29.8	30.5	30.8	30.0	30.8	31.5	32.4
19	25.8	26.7	27.3	27.7	28.4	28.7	27.9	28.7	29.3	30.2
20	23.8	24.7	25.3	25.7	26.4	26.7	25.9	26.7	27.3	28.2

* The first line gives the dose at the maximum for 100 r of primary.

Note that percentages are relative to 100% at 0.5 cm, the approximate wall thickness of a ^{60}Co r chamber. Doses at the surface 0.5-cm depth at the surface of the phantom.

From: Johns, H. E., *The Physics of Radiology*, 2nd ed., pp. 734–735. Charles C Thomas, Springfield, Illinois, 1964.

12.
^{60}Co FOR RECTANGULAR FIELDS, FOCAL-TO-SKIN DISTANCES

Fields, cm × cm										
6 × 20	8 × 8	8 × 10	8 × 15	8 × 20	10 × 10	10 × 15	10 × 20	15 × 15	15 × 20	20 × 20
102.5	102.1	102.3	102.7	102.9	102.5	103.0	103.3	103.6	104.1	104.6
depending upon collimator										
100.0	100.0	100.0	100.0	100.0	100.0	100.0	100.0	100.0	100.0	100.0
97.8	97.8	98.0	98.1	98.1	98.2	98.3	98.3	98.4	98.4	98.4
92.8	92.7	93.0	93.2	93.3	93.3	93.6	93.6	93.9	93.9	94.0
87.7	87.6	87.9	88.3	88.5	88.3	88.8	88.9	89.3	89.4	89.6
82.7	82.5	82.9	83.4	83.6	83.4	84.0	84.2	84.7	84.9	85.2
77.7	77.4	77.9	78.5	78.8	78.5	79.2	79.5	80.1	80.4	80.8
72.7	72.4	73.0	73.7	74.0	73.6	74.4	74.7	75.4	75.8	76.4
67.9	67.5	68.1	68.9	69.2	68.8	69.8	70.1	70.8	71.4	72.1
63.3	62.7	63.4	64.3	64.7	64.1	65.2	65.7	66.5	67.2	68.0
58.9	58.2	58.9	59.9	60.4	59.7	60.9	61.4	62.3	63.1	64.0
54.8	54.0	54.8	55.8	56.3	55.6	56.9	57.4	58.4	59.2	60.2
51.0	50.1	50.9	52.0	52.5	51.7	53.1	53.7	54.7	55.6	56.6
47.4	46.5	47.3	48.4	49.0	48.1	49.5	50.2	51.2	52.1	53.2
44.1	43.2	44.0	45.1	45.7	44.8	46.2	46.9	47.9	48.8	50.0
41.0	40.1	40.9	42.0	42.6	41.8	43.1	43.9	44.9	45.8	47.0
38.1	37.2	38.0	39.2	39.8	38.9	40.3	41.0	42.0	43.0	44.2
35.5	34.5	35.3	36.5	37.1	36.2	37.6	38.8	39.3	40.3	41.5
33.1	32.1	32.8	34.0	34.6	33.7	35.1	35.8	36.8	37.8	39.0
30.8	29.8	30.5	31.7	32.3	31.4	32.8	33.5	34.5	35.5	36.7
28.7	27.7	28.4	29.6	30.2	29.2	30.7	31.4	32.3	33.4	34.6
26.8	25.7	26.4	27.6	28.2	27.2	28.6	29.4	30.3	31.4	32.6

Fields, cm × cm										
6 × 20	8 × 8	8 × 10	8 × 15	8 × 20	10 × 10	10 × 15	10 × 20	15 × 15	15 × 20	20 × 20
102.5	102.1	102.3	102.7	103.0	102.5	103.0	103.4	103.7	104.1	104.6
depending upon collimator										
100.0	100.0	100.0	100.0	100.0	100.0	100.0	100.0	100.0	100.0	100.0
98.2	98.1	98.3	98.5	98.5	98.6	98.8	98.8	99.0	98.9	98.9
93.4	93.3	93.6	93.9	93.9	93.9	94.3	94.3	94.6	94.6	94.7
88.6	88.4	88.9	89.3	89.3	89.3	89.8	89.8	90.2	90.3	90.5
83.9	83.7	84.2	84.7	84.8	84.7	85.3	85.4	85.9	86.1	86.3
79.2	78.9	79.6	80.1	80.3	80.1	80.8	81.0	81.6	81.9	82.2
74.5	74.2	74.9	75.6	75.8	75.5	76.3	76.6	77.3	77.7	78.1
69.9	69.5	70.2	71.0	71.3	70.9	71.8	72.2	73.0	73.5	74.0
65.4	64.9	65.6	66.5	66.9	66.4	67.4	67.9	68.7	69.3	70.0
61.1	60.5	61.2	62.1	62.6	62.0	63.1	63.7	64.5	65.2	66.1
57.0	56.3	57.0	58.0	58.6	57.8	59.0	59.7	60.6	61.3	62.3
53.2	52.4	53.1	54.2	54.8	53.9	55.2	55.9	56.9	57.7	58.7
49.6	48.7	49.5	50.7	51.2	50.3	51.7	52.4	53.4	54.3	55.3
46.3	45.4	46.1	47.3	47.9	47.0	48.4	49.1	50.2	51.1	52.1
43.2	42.3	43.0	44.2	44.8	43.9	45.3	46.0	47.1	48.1	49.1
40.3	39.4	40.1	41.3	41.9	41.0	42.4	43.1	44.2	45.2	46.2
37.7	36.7	37.4	38.6	39.2	38.3	39.7	40.4	41.5	42.5	43.5
35.2	34.2	34.9	36.1	36.7	35.8	37.2	37.9	39.0	40.0	41.0
32.9	31.9	32.6	33.8	34.4	33.5	34.9	35.6	36.7	37.6	38.6
30.7	29.7	30.5	31.6	32.3	31.3	32.7	33.4	34.5	35.4	36.4
28.7	27.7	28.5	29.6	30.2	29.3	30.6	31.3	32.4	33.3	34.4

are 30 to 50% of the peak dose. The actual peak dose obtained is given in the first line relative to an r chamber reading at the nominal

Table 13.
TUMOR–AIR RATIOS FOR RECTANGULAR FIELDS FOR X RAYS OF HVL 3 mm COPPER AND ^{60}Co, RELATIVE TO THE AIR DOSE MEASURED IN A PARALLEL BEAM (INFINITE FSD)

A. HVL 3 mm Copper, FSD ∞ (Courtesy of *Am. J. Roentgenol.*)

Depth, cm	Field Size at Axis of Rotation, cm × cm													
	4 × 4	4 × 6	4 × 8	4 × 10	4 × 15	6 × 6	6 × 8	6 × 10	6 × 15	8 × 8	8 × 10	8 × 15	10 × 10	10 × 15
0	1.120	1.140	1.150	1.160	1.180	1.160	1.180	1.200	1.220	1.210	1.230	1.240	1.240	1.260
1	1.100	1.120	1.140	1.160	1.170	1.150	1.180	1.200	1.220	1.210	1.230	1.250	1.250	1.280
2	1.000	1.050	1.070	1.090	1.110	1.100	1.130	1.150	1.180	1.170	1.200	1.230	1.220	1.260
3	0.895	0.946	0.964	0.988	1.010	1.000	1.040	1.070	1.100	1.107	1.130	1.160	1.150	1.200
4	0.798	0.844	0.868	0.888	0.917	0.898	0.936	0.970	1.000	0.988	1.030	1.070	1.060	1.120
5	0.705	0.752	0.776	0.798	0.830	0.809	0.850	0.877	0.915	0.902	0.943	0.990	0.978	1.040
6	0.624	0.668	0.697	0.721	0.750	0.728	0.771	0.798	0.836	0.820	0.860	0.905	0.902	0.959
7	0.549	0.588	0.615	0.641	0.670	0.645	0.686	0.712	0.750	0.731	0.776	0.822	0.815	0.876
8	0.476	0.546	0.543	0.565	0.595	0.566	0.607	0.634	0.674	0.655	0.700	0.745	0.736	0.796
9	0.419	0.450	0.479	0.500	0.529	0.500	0.539	0.565	0.600	0.580	0.622	0.669	0.660	0.722
10	0.363	0.394	0.417	0.438	0.468	0.439	0.473	0.497	0.535	0.513	0.549	0.596	0.587	0.645
11	0.317	0.345	0.366	0.384	0.412	0.382	0.417	0.436	0.473	0.450	0.489	0.531	0.520	0.574
12	0.273	0.299	0.319	0.337	0.364	0.335	0.365	0.381	0.417	0.398	0.434	0.474	0.461	0.512
13	0.239	0.261	0.279	0.296	0.320	0.294	0.320	0.337	0.370	0.350	0.381	0.419	0.407	0.455
14	0.206	0.228	0.244	0.259	0.281	0.257	0.281	0.298	0.326	0.308	0.335	0.371	0.360	0.405
15	0.182	0.199	0.214	0.228	0.248	0.226	0.247	0.260	0.288	0.270	0.295	0.329	0.318	0.361
16	0.158	0.174	0.187	0.198	0.218	0.198	0.217	0.230	0.255	0.238	0.259	0.290	0.281	0.320
17	0.138	0.152	0.165	0.175	0.192	0.174	0.190	0.202	0.225	0.210	0.228	0.258	0.249	0.287
18	0.120	0.133	0.143	0.153	0.169	0.151	0.166	0.177	0.200	0.185	0.200	0.228	0.220	0.256
19	0.105	0.116	0.126	0.135	0.149	0.133	0.146	0.156	0.176	0.163	0.176	0.202	0.194	0.226
20	0.092	0.102	0.110	0.118	0.132	0.117	0.129	0.137	0.156	0.143	0.157	0.180	0.172	0.201

Table 13. (Continued)
TUMOR–AIR RATIOS FOR RECTANGULAR FIELDS FOR X RAYS OF HVL 3 mm COPPER AND ^{60}Co, RELATIVE TO THE AIR DOSE MEASURED IN A PARALLEL BEAM (INFINITE FSD)

B. ^{60}Co, FSD ∞ (Courtesy of *Am. J. Roentgenol.*)

Depth, cm	4 × 4	4 × 6	4 × 8	4 × 10	4 × 15	6 × 6	6 × 8	6 × 10	6 × 15	8 × 8	8 × 10	8 × 15	10 × 10	10 × 15
0.5	1.011	1.013	1.014	1.016	1.019	1.016	1.019	1.020	1.023	1.021	1.023	1.026	1.026	1.029
1	0.993	0.998	1.000	1.010	1.020	0.990	0.996	1.000	1.005	1.005	1.010	1.115	1.020	1.025
2	0.950	0.958	0.961	0.965	0.968	0.966	0.973	0.978	0.984	0.981	0.986	0.992	0.992	0.998
3	0.910	0.916	0.925	0.930	0.936	0.928	0.941	0.946	0.952	0.948	0.954	0.962	0.959	0.968
4	0.866	0.878	0.884	0.889	0.893	0.893	0.903	0.908	0.916	0.913	0.920	0.926	0.927	0.936
5	0.823	0.835	0.845	0.850	0.854	0.852	0.869	0.871	0.878	0.872	0.882	0.890	0.888	0.900
6	0.780	0.794	0.802	0.808	0.815	0.811	0.822	0.829	0.839	0.836	0.844	0.854	0.853	0.867
7	0.739	0.752	0.761	0.768	0.774	0.769	0.783	0.788	0.800	0.794	0.804	0.815	0.811	0.829
8	0.693	0.708	0.716	0.723	0.732	0.726	0.740	0.748	0.759	0.754	0.764	0.778	0.775	0.792
9	0.653	0.668	0.676	0.683	0.694	0.685	0.700	0.708	0.723	0.716	0.726	0.740	0.736	0.756
10	0.613	0.628	0.637	0.644	0.653	0.646	0.661	0.670	0.683	0.676	0.687	0.703	0.699	0.717
11	0.577	0.591	0.600	0.608	0.616	0.608	0.622	0.632	0.649	0.639	0.650	0.669	0.662	0.682
12	0.541	0.555	0.563	0.570	0.580	0.573	0.587	0.597	0.611	0.602	0.614	0.632	0.628	0.648
13	0.509	0.521	0.531	0.538	0.549	0.538	0.552	0.565	0.580	0.568	0.580	0.599	0.595	0.615
14	0.476	0.489	0.498	0.505	0.516	0.507	0.522	0.532	0.547	0.537	0.549	0.568	0.563	0.584
15	0.448	0.461	0.469	0.476	0.488	0.476	0.491	0.503	0.518	0.505	0.518	0.536	0.533	0.553
16	0.420	0.432	0.440	0.448	0.459	0.449	0.463	0.473	0.487	0.478	0.489	0.508	0.503	0.525
17	0.398	0.409	0.416	0.421	0.433	0.421	0.436	0.448	0.460	0.450	0.460	0.480	0.475	0.499
18	0.371	0.383	0.391	0.397	0.408	0.398	0.411	0.421	0.434	0.425	0.436	0.454	0.449	0.472
19	0.351	0.361	0.370	0.375	0.386	0.376	0.387	0.398	0.410	0.399	0.411	0.430	0.421	0.448
20	0.330	0.340	0.348	0.354	0.364	0.354	0.366	0.374	0.386	0.378	0.387	0.404	0.399	0.425

Note the higher backscatter percentages for the lower-energy X rays.
From: Johns, H.E., *The Physics of Radiology*, 2nd ed. Charles C Thomas, Springfield, Illinois, 1964.

Table 14.
ENERGY FLUX PER ROENTGEN, PHOTON FLUX PER ROENTGEN, AND EXPOSURE DOSE RATE PER MILLICURIE AS A FUNCTION OF PHOTON ENERGY

hn Photon Energy, Mev	Energy-Absorption Coefficient $(\mu_{en})_{air}$, cm²/g	Energy Flux per Roentgen, ergs/cm²/r	Photon Flux per Roentgen, N/cm²/r	Roentgens per Hour per Millicurie at 1 cm
0.010	4.6600	18.8	11.7×10^8	9.020
0.015	1.2900	68.0	28.30	3.740
0.020	0.5160	170.0	53.10	2.000
0.030	0.1470	597.0	124.00	0.853
0.040	0.0640	1,370.0	214.00	0.495
0.050	0.0384	2,284.0	286.00	0.371
0.060	0.0292	3,003.0	313.00	0.339
0.080	0.0236	3,716.0	290.00	0.365
0.100	0.0231	3,796.0	237.00	0.447
0.150	0.0251	3,494.0	146.00	0.728
0.200	0.0268	3,272.0	102.00	1.090
0.300	0.0288	3,045.0	63.40	1.670
0.400	0.0296	2,963.0	46.30	2.290
0.500	0.0297	2,953.0	36.90	2.870
0.600	0.0296	2,963.0	30.80	3.440
0.800	0.0289	3,035.0	23.70	4.470
1.000	0.0280	3,132.0	19.60	5.420
1.500	0.0255	3,439.0	14.30	7.400
2.000	0.0234	3,748.0	11.70	9.050
3.000	0.0205	4,278.0	8.91	11.900
4.000	0.0186	4,715.0	7.36	14.400
5.000	0.0173	5,069.0	6.33	16.700
6.000	0.0163	5,380.0	5.60	18.900
8.000	0.0150	5,847.0	4.57	23.200
10.000	0.0144	6,090.0	3.80	27.900

From: Johns, H. E., *The Physics of Radiology*, 2nd ed., p. 694. Charles C Thomas, Springfield, Illinois, 1964.

These types of curves often are fitted by the n-target model

$$S_D = 1 - (1 - e^{-\lambda D})^n,$$

where n is termed the extrapolation (or target) number, and λ the probability that unit radiation dose D will "hit" any of the n targets.

Although useful for describing much of the accumulated data, this model is of limited use in describing the influence of many factors on radiosensitivity, as pointed out by Burch,[15] who has developed more elaborate multiple-hit models. Dose–response curves determined in terms of chromosome breaks in the plants *Tradescantia* and *Vicia* show that two-break aberrations increase nearly as the square of the dose for radiation of low LET at high intensity, but increase at a lower rate and almost linearly with the dose at low intensity.[16] For high LET radiation (see the neutron curve of Figure 6), the two-break aberrations increase almost linearly with dose and independently of intensity.

Figure 7 illustrates the type of dose–response relationships that may be observed in studying genetic mutations at the animal level. One may note that, although large confidence intervals must be attached to each data point as a result of statistical limitations in observing the relatively rare mutations with an animal population of limited size, a definite increase in mutation rate with dose is, nevertheless, observed. Also, high dose rates again appear more effective than low dose rates.

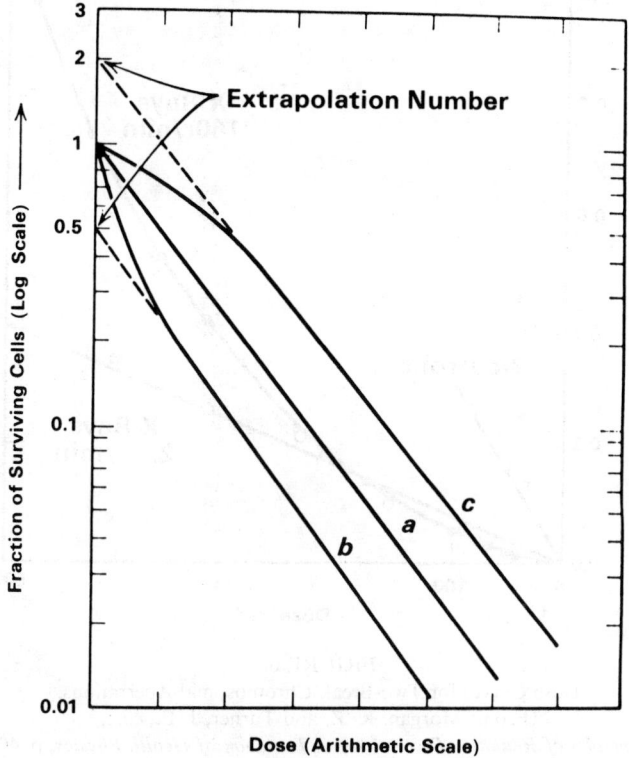

FIGURE 5.
Types of Mammalian Cell Survival Curve.
[Logarithm of fraction of cells S surviving the (acute) irradiation against dose D]
(From: Morgan, K. Z. and Turner, J. E., eds., *Principles of Radiation Protection—A Textbook of Health Physics*, p. 377. John Wiley and Sons, New York, New York, 1967.)
Curve a: $S = 2D$; observed when homogeneous cells are irradiated by particles (such as low-energy natural α particles) of high LET. Curve b gives an extrapolation number of less than unity (0.5 in example); observed when heterogeneous mixtures of cells (of low and high radiosensitivity) are irradiated. Curve c gives an extrapolation number of more than unity (2 in example); observed when most types of mammalian cells are irradiated by, for example, ^{60}Co γ rays—that is, by radiation of low average LET.

When a response such as mortality (1 − survival) of animals is plotted as a function of radiation dose in rad, curves such as those in Figure 8 are obtained. The S-shaped curve typical of many chemically toxic agents[19] is observed here for radiation.

When plotting this type of mortality data for mice exposed to high-energy protons (440 and 730 Mev) and to X rays, Bradley et al.[20] found that a probit mortality (or probability) versus log (dose) scale linearized the dose–response curves and made them approximately parallel for given strains exposed to different LET radiations (see Figure 9). The point where each line crosses the 50% mortality level is called the LD_{50} for that type of radiation, that age, strain and species of animal, and that time interval of observation. Thus, whatever the basic mechanisms at the cellular and subcellular levels, they may often combine to produce an effect represented at the animal or human level by the empirical log-normal dose–response function.[19,21] This is the functional shape that the probit-type plot is designed to linearize for convenience in bioassay of toxic agents or pharmaceuticals, and statistical methods of curve fitting and analysis are available for these functions.[19]

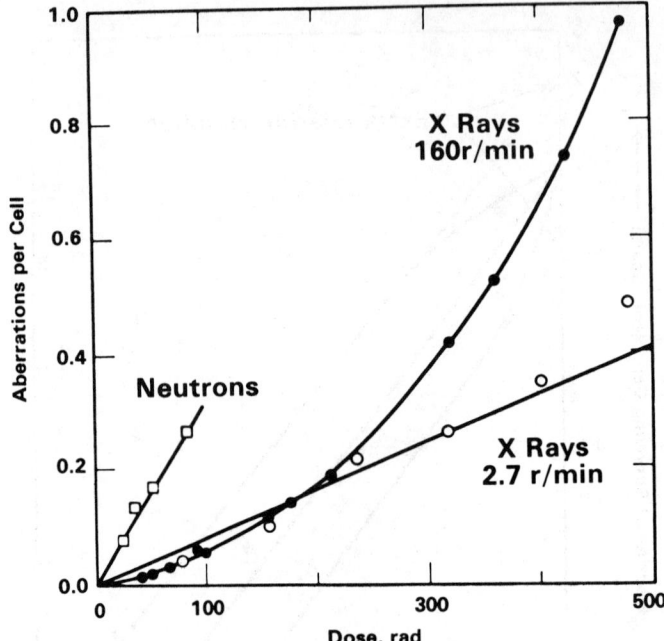

FIGURE 6.
Dose Curves for Two-Break Chromosomal Aberrations.
(From: Morgan, K. Z. and Turner, J. E., eds.,
Principles of Radiation Protection—A Textbook of Health Physics, p. 404.
John Wiley and Sons, New York, New York, 1967.)
The X-ray data are replotted from Sax (1941), the neutron data from Giles
(1943). The neutron doses were given in *n* units. and have been converted to rad by multiplication with the factor 2.5 (this conversion is very
approximate and should not be relied upon).

A summary of the log-normal function, its properties, and simplified procedures for its application are presented in Reference 21. Briefly, suppose that x is the dose of a population of animals and that $P(\ln X \leq \ln x)$ is the probability of animals having "sensitivities" or lethal effects at doses $X \leq x$; then P may be considered as the (expected) fraction of animals dying in a group administered a dose x, and a log-normally distributed P would be given by the function

$$P(\ln X \leq \ln x) = \frac{1}{\sqrt{2\pi}\sigma_g} \int_{\ln X = \ln 0 = -\infty}^{\ln X = \ln x} \exp\left[-(\ln X - \ln \mu_g)^2 / 2 \sigma^2_g\right] d(\ln X).$$

Here, σ_g is the standard deviation ($\sqrt{\mathrm{Var}(\ln X)}$) in the frequency distribution of $\ln X$, $\overline{\ln}$ is the mean value of $\ln X$, and μ_g is the "geometric mean of X". The above expression is the form for the cumulative integral over a normal distribution of $\ln X$, and has been shown to be a special case of a family of logarithmic distributions.[22] The LD_{50} value where a probit-type mortality plot (Figure 9) crosses the 50% mortality line would be an estimate of μ_g of the log-normal dose–response function (if the data indicate that a straight line on such a plot is applicable). The parameter σ_g is the standard deviation in $\ln X$ (the natural logarithm of the randomly distributed dose required to kill an animal), and this parameter may also be easily estimated from the probit plot. The estimated value of σ_g can be obtained from the equation

$$\sigma_g = \ln x_{0.8413} - \ln x_{0.50} = \ln(x_{0.8413}/x_{0.50}),$$

or

$$\sigma_g = \ln x_{0.50} - \ln x_{0.1587} = \ln(x_{0.50}/x_{0.1587}),$$

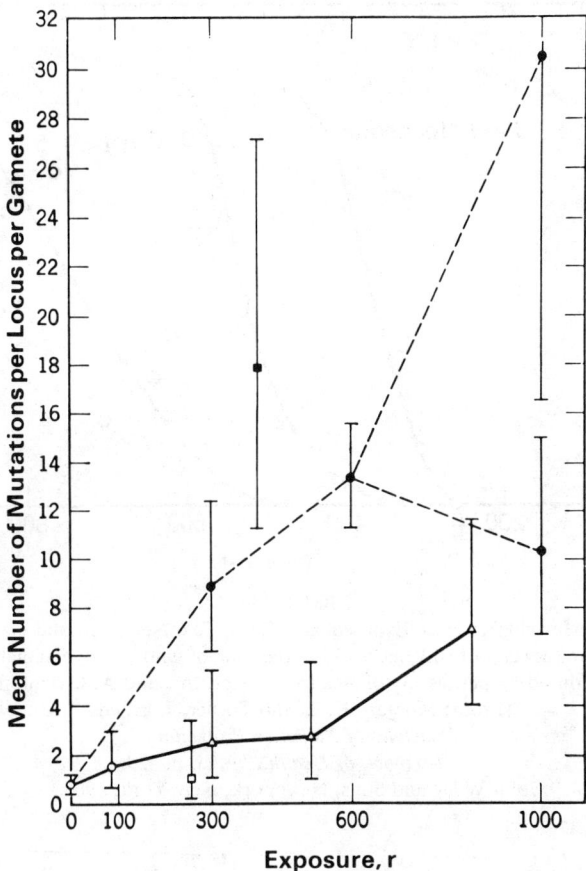

FIGURE 7.
Exposure Curves for Specific-Locus Mutation in the Mouse.
(From: Morgan, K. Z. and Turner, J. E., eds.,
*Principles of Radiation Protection—
A Textbook of Health Physics*, p. 416.
John Wiley and Sons, New York, New York, 1967.)
90% confidence intervals shown. Solid points represent
results with acute X rays (80 to 90 r/min). Open points
are chronic γ-ray results (triangles and square, 90/r week;
circles, 10 r/week). Squares are mutation rates
in females; all other points are mutation rates in males.
The point for zero dose is the sum of all male controls.
The top 1000-r point represents results of a single exposure;
the lower represents results of successive exposures
to 600 and 400 r respectively, separated by an
interval of 15 weeks. (Russel et al., 1960.)

where $x_{0.8413}$ is the value of x where the " % mortality versus log X line" crosses the 84.13% mortality value, and $x_{0.50} = \mu_g$ is the value of x where the line crosses the 50% mortality value. Now the "standard geometric deviation"[21] or geometric mean standard deviation[22] is defined as

$$s_g = x_{0.8413}/x_{0.50} = x_{0.50}/x_{0.1587},$$

and the average dose $\bar{x}$ required to kill an animal is then given by the relationship[21]

$$\log_{10} \bar{x} = \log \mu_g + 1.1513(\log_{10} s_g)^2.$$

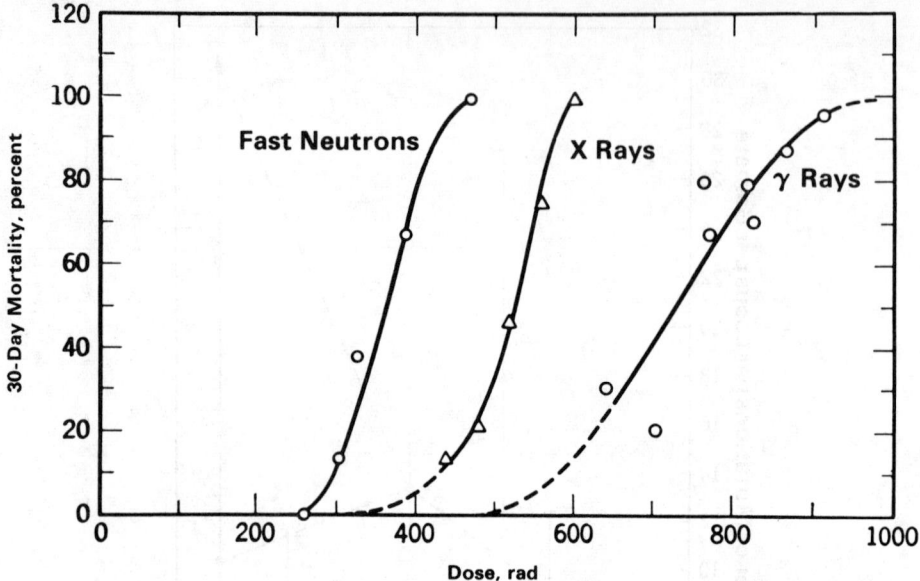

FIGURE 8.
Thirty-Day Mortality of Mice Exposed to X-Rays, Fast Neutrons, and Gamma Rays.
(The mice were 9 to 12 weeks old at the time of whole-body irradiation.)
Reproduced by permission of Academic Press, Inc. and A. C. Upton et al.
(From: Morgan, K. Z. and Turner, J. E., eds.,
*Principles of Radiation Protection—
A Textbook of Health Physics*, p. 429.
John Wiley and Sons, New York, New York, 1967.)

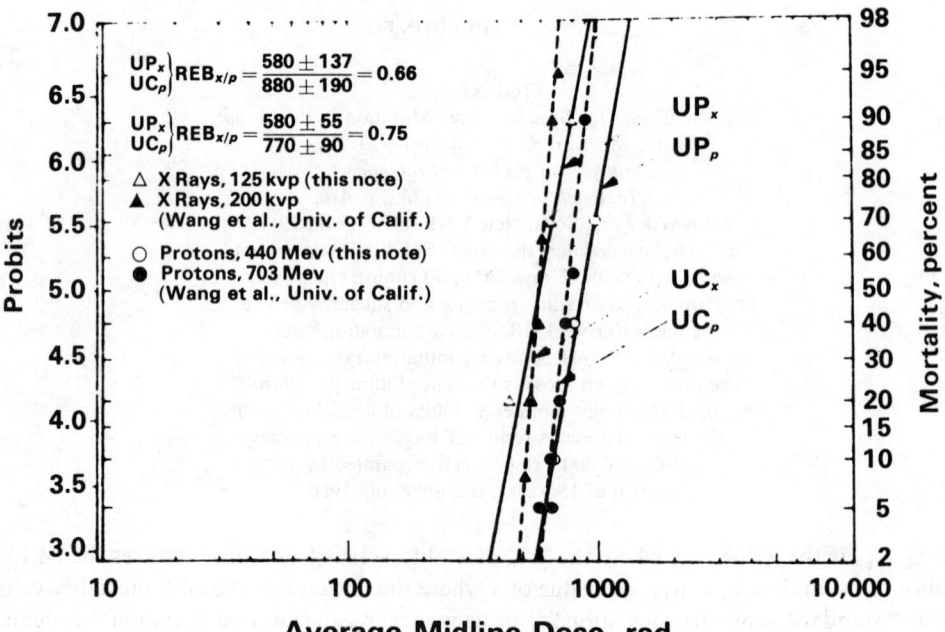

FIGURE 9.
Relationship Between Probit of Mortality and Average Midline Dose
of X and Proton Radiation to Mice.
(From: Bradley, F. J., Watson, J. A., Doolittle, D. P.,
Brodsky, A., and Sutton, R. B. *Health Phys.*, 10: 72, 1964.)

The particular significance of μ_g and s_g for a log-normal dose–response relationship is that $\mu \overset{x}{\div} s_g$, and not $\bar{x} \pm \sigma_x$, gives the interval in which 68 percent of the population "sensitivities" will lie, if X is interpreted as the dose to which a fraction of the population is sensitive enough to die (or manifest some other biological end point). The concept of s_g may also be useful, for example, in predicting that less than 2.5 percent of the population are likely to require doses greater than $s_g^2 \mu_g$ to die; in other words, if the log-normal function holds at higher doses, less than about 2.5 percent of the population would survive more than s_g^2 times the LD_{50}. On

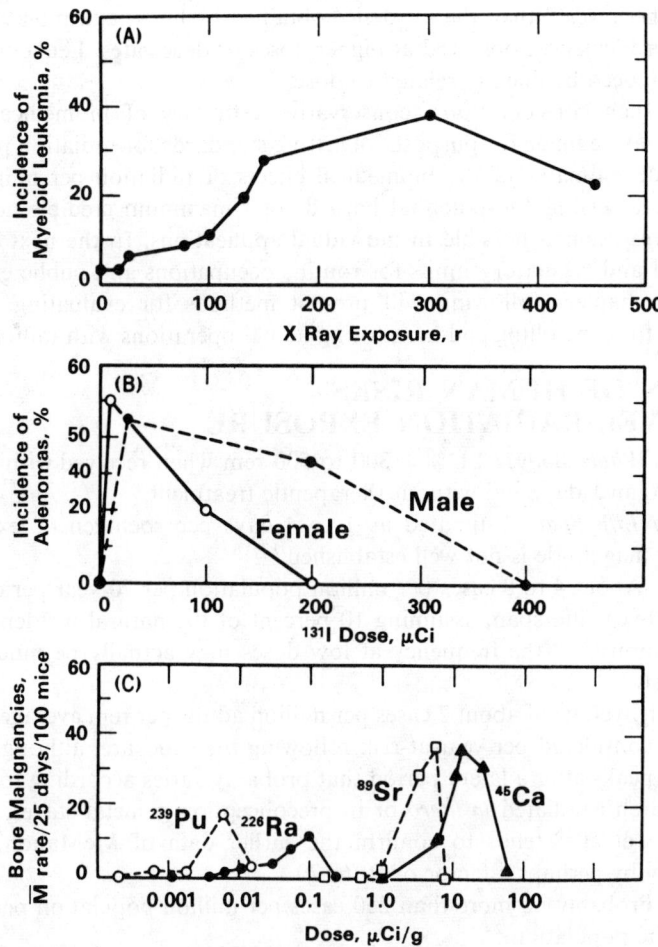

FIGURE 10.
Incidence of Various Neoplasms in Relation to Radiation Dose.
(From: Morgan, K. Z. and Turner, J. E., eds.,
*Principles of Radiation Protection—
A Textbook of Health Physics*, p. 434.
John Wiley and Sons, New York, New York, 1967.)

the other hand, 2.5 percent of the population would be affected by doses less than the LD_{50} divided by s_g^2. Thus, the S-shaped or log-normal dose–response relationship would predict that at low doses some finite, although small, effect is still possible.

The S-shaped curve has also been suggested by some of the animal data on carcinogenesis. However, when dose levels are so high that cell lethality is produced, the rising portion of the S-shaped curve may not be evident (see Figure 10). However, a stochastic two-stage model of carcinogenesis,[23] involving a second event conditioned on a specific prior event followed by the cell growth stage of Arley,[24] predicts a dose–response curve shape so similar to the log-normal that it could not be distinguished from it by any reasonably sized experiment. Thus, the variation in the dose X at which the biological end point (e.g., an observed tumor by time T) may occur could be interpreted as resulting largely from ordinary chance variations in the action of a radiation dose on an individual animal rather than from real individual variations in sensitivity between animals. Some purely stochastic variation of this kind, as well as real variations in individual susceptibility, may also be expected to contribute to the range of doses over which lethality may occur. In any case, however, the type of dose–response relationship represented by any of these S-shaped curves would yield a predicted incidence of cancer at low doses or dose rates that would be lower[23] than the incidence obtained by linear extrapolations through the origin from cancer incidences observed at higher doses or dose rates. Leukemogenesis, on the other hand, may indeed be linearly related to dose.[18]

Thus, from such considerations, conservative estimates of biomedical effects at the population level may be made for purposes of setting standards for radiation protection. Some order-of-magnitude estimates of the biomedical effects of radiation per unit dose are given below for use in assessing the potential hazards of "maximum credible accidents" or inadvertent exposures deemed possible in individual applications. In the next chapter some of the recommended and regulatory limits for routine occupations and public exposures will be summarized; the chapters following will present methods for evaluating radiation safety requirements and for controlling and monitoring actual operations with radioactive materials.

ESTIMATION OF HUMAN RISKS OF LOW-LEVEL RADIATION EXPOSURE

Lethal Dose (Whole Body). LD_{50}^{30} = 300 to 500 rem when received within a short period of time (say, less than 1 day) and without therapeutic treatment.[17]

Shortening of Life Span. Estimated at 1 to 4 days per roentgen of exposure early in life, but the exact magnitude is not well established.[17,18]

Bone Cancer. About 4 to 8 cases per million population per 70-year period per rem dose received over a 70-year life span, assuming 10 percent of the natural incidence is a result of background radiation;[15,18] the frequency at low doses may actually be much less than this upper-limit estimate.

Leukemia. An average of about 2 cases per million adults per rem average exposure to the entire population considered per year-at-risk following the exposure, although the incidence-versus-time curve peaks after a latent period that probably varies according to dose level;[23,25] however, for children irradiated *in utero* or in preconception gametal stages, recent evidence of Saxon Graham et al.[26] tends to confirm the earlier data of McMahon,[27,28] indicating a higher sensitivity by perhaps a factor of 50 to 70.

Cell Cancer. Probably no more than 250 cases per million population per 70-year period per rem dose to the population.[24]

Cataracts. More than 1,000 rads of X or gamma radiation to the lens of the eye are required, or more than 100 rads from neutrons, to cause an appreciable increase in cataract incidence.[17,18]

Genetic Effects. A single-generation "doubling dose" to double the natural mutation rate has been estimated to be 40 rems to the entire population (40 × 200,000,000 man-rems to the United States population, for example).[29] Considering the natural mutation rate, a doubling dose of 40 rems to every member of a population might cause about 1 out of every 200 births in the next generation to result in a death or failure to reproduce.[24] The total numbers of induced extinctions in all future generations might be a maximum of about 40 million for a constant population size of 200 million. If the 40-rem dose were delivered to the population each generation for much longer than the average life of a mutant gene, i.e., much longer than 40 generations or 1,200 years,[30] then at equilibrium there would be an additional 40 million abortive conceptions, stillbirths, neonatal deaths, or failures to reproduce per generation per 200,000,000 population. Of course, the proportion of each of these is uncertain. In addition, however, the surviving population would presumably carry a large pool of radiation-caused mutants who might have varying degrees of health impairment, as well as a possible small—but increasing—percentage of individuals evolving with superior characteristics as a result of selection in favor of the beneficial long-surviving mutant fraction.

Other Radiation Effects. Some additional nonspecific health impairment or loss of vitality might result from cell loss following somatic mutations, although, if the destruction of cells is at a low rate, regeneration may prevent organ failure or ill health.[31] These additional effects, particularly by degenerative cardiovascular and renal diseases,[15] may have a smaller relative increase per rem above natural incidence levels; but since they are more prevalent, they may somewhat exceed in absolute numbers the excess deaths from radiation-induced cancer. Nevertheless, the total excess mortality induced by radiation exposure would not be expected to be more than the order of magnitude produced by genetic and carcinogenic effects.[24]

REFERENCES

1. International Commission on Radiological Units and Measurements, Report of the ICRU. *NBS Handbook 78.* U.S. Government Printing Office, Washington, D.C., 1959.
2. International Commission on Radiological Units and Measurements, Clinical Dosimetry—Recommendations of the ICRU. *NBS Handbook 87*, p. 38. U.S. Government Printing Office, Washington, D.C., 1963.
3. Parker, H. M., Health Physics, Instrumentation, and Radiation Protection. *Advance. Biol. Med. Phys.*, 1: 243, 1948.
4. Roesch, W. C. and Attix, H. F., eds., *Radiation Dosimetry*, 2nd ed., Vol. 1, Ch. 1. Academic Press, New York, 1968.
5. International Commission on Radiological Protection, Report of Committee 2. *Health Phys.*, 3: 1–380, 1960.
6. International Commission on Radiological Units and Measurements, Report 10a, Radiation Quantities and Units. *NBS Handbook 84.* U.S. Government Printing Office, Washington, D.C., 1962.
7. Attix, F. H. and Roesch, W. C., eds., Instrumentation. *Radiation Dosimetry*, Vol. 2. Academic Press, New York, 1966.
8. Hine, G. J. and Brownell, G. L., eds., *Radiation Dosimetry*. Academic Press, New York, 1956.
9. Price, W. J., *Nuclear Radiation Detection*, 2nd ed. McGraw-Hill, New York, 1966.
10. Morgan, K. Z. and Turner, J. E., eds., *Principles of Radiation Protection—A Textbook of Health Physics.* John Wiley and Sons, New York, 1967.
11. National Committee on Radiation Protection and Measurements, NCRP Report No. 27, Stopping Powers for Use with Cavity Chambers. *NBS Handbook 79.* U.S. Government Printing Office, Washington, D.C., 1961.
12. National Committee on Radiation Protection and Measurements, Measurement of Neutron Flux and Spectra for Physical and Biological Applications, Recommendations of the NCRP. *NBS Handbook 72.* U.S. Government Printing Office, Washington, D.C., 1960.
13. Johns, H. E., *The Physics of Radiology*, 2nd ed. Charles C Thomas, Springfield, Illinois, 1964.
14. International Commission on Radiological Units and Measurements, Report of the ICRU. *NBS Handbook 62*, U.S. Government Printing Office, Washington, D.C., 1956.
15. Burch, P. R. J., Radiation Physics. *Principles of Radiation Protection—A Textbook of Health Physics*, pp. 366–397, Morgan, K. Z. and Turner, J. E., eds. John Wiley and Sons, New York, 1967.
16. Upton, A. C. and Kimball, R. F., Radiation Biology. *Principles of Radiation Protection—A Textbook of Health Physics*, pp. 398–447, Morgan, K. Z. and Turner, J. E., eds. John Wiley and Sons, New York, 1967.
17. Wald, N., Evaluation of Human Exposure Data. *Principles of Radiation Protection—A Textbook of Health Physics*, pp. 448–496, Morgan, K. Z., and Turner, J. E., eds. John Wiley and Sons, New York, 1967.
18. United National Scientific Committee on the Effects of Atomic Radiation, Various Reports. United Nations, New York, 1958, 1962, 1964, 1966. (Continuing reviews are underway.)
19. Finney, D. J., *Probit Analysis.* Cambridge University Press, New York, 1962.
20. Bradley, F. J., Watson, J. A., Doolittle, D. P., Brodsky, A., and Sutton, R. B. *Health Phys.*, 10: 71–74, 1964.
21. Schubert, J., Brodsky, A., and Tyler, S., The Log-Normal Function as a Stochastic Model of the Distribution of Strontium-90 and Other Fission Products in Humans. *Health Phys.*, 13: 1187–1204, 1967.
22. Espenscheid, W. F., Kerker, M., and Marijevic, E. *J. Phys. Chem.*, 68: 3093, 1964.
23. Brodsky, A., A Stochastic Model of Carcinogenesis and Its Implications in the Dose–Response Plane. Presented at the Health Physics Society Meeting in Los Angeles, 1966. Abstract in *Health Phys.*, 12: 1176, 1966. Detailed treatment of the model in Brodsky, A., *A Stochastic Model of Carcinogenesis as Applied to Skin Tumors in Mice* (Dissertation). University of Pittsburgh, 1966.

24. Brodsky, A. *Am. J. Public Health*, 55(12): 1971–1992, 1965.
25. Cobb, S., Miller, M., and Wald, N., On the Estimation of the Incubation Period in Malignant Disease, I. *J. Chron. Dis.* 9(4); 385–393, 1959.
26. Graham, L. S., Levine, M. L., Lilienfeld, A. M., Schuman, L., Gibson, R., David, J. E., and Hempleman, L. H., Preconception Intrauterine and Postnatal Irradiation as Related to Leukemia. Presented at the Meeting of the American Public Health Association, New York, October 8, 1964.
27. McMahon, B. *J. Nat. Cancer Inst.*, 28: 1173, 1962.
28. United Nations Scientific Committee on the Effects of Atomic Radiation, Report. United Nations, New York, 1962.
29. National Academy of Sciences, The Biological Effects of Atomic Radiation. *Summary Reports*. National Research Council, Washington, D.C., 1956.
30. Muller, H. J., Radiation and Human Mutations. *Sci. Amer.*, 193: 58–68, 1955.
31. Henshaw, P. S. *Health Phys.*, 1: 141–151, 1958.

RADIATION PROTECTION GUIDES AND REGULATORY LIMITS OF EXPOSURE

Allen Brodsky, Sc.D., C.H.P.

GUIDES FOR LIMITING ORGAN AND WHOLE-BODY DOSE

The history and rationale behind the limits of exposure recommended by the ICRP, NCRP and FRC have been reviewed recently by Morgan.[1] We shall list—for reference only—some of the more important guides for limiting exposure from sources external and internal to the body. These guides were generally arrived at by the consensus of individual scientists, considering the types of dose–response data illustrated in previous sections. As stated by Morgan,[2] "The goal is to avoid all unnecessary radiation exposure, to receive exposure even within the limits discussed above when—and only when—the expected benefits exceed the likely harmful consequences of radiation exposure." Of course, the "harmful consequences" may be estimated by the reader from the data in the preceding chapter; in most cases he will probably find the probabilities of harm small compared to those of many other occupations and the normal risks of daily life,[3] if radiation exposures are kept well below the recommended guide levels.

Table 15 presents the basic recommended limits of occupational exposure to various parts of the body for a period of 13 weeks and for one year, and cumulative limits according to age for certain organs. Although the quarterly limits may differ somewhat from those recommended by the various committees and agencies, the more important annual limits are practically identical. The most important rule to remember is that the whole-body dose should remain within the cumulative limit 5(N-18) rems, where N is the age after the 18th birthday. If this limit is met for workers exposed in a general field of external radiation, and if no appreciable additional extremity exposure or intake of radioactive material is received (which is often the case), then the other annual limits will automatically be met. If the exposure is fairly uniformly distributed throughout the year by administrative controls, then the quarterly limits are also easily met. In any case, regulatory limits must also be met as set forth in the Code of Federal Regulations, Title 10, Part 20, and in other applicable federal, state, and local regulations, which will be discussed further on.

A summary and comparison of recommended limits for workers and for the general public is presented in Table 16, as adopted from Morgan.[1] The groups B and C are no longer separately recommended by the ICRP.[13] Also, the safety factor of 1/3 for the average dose to the "critical" population group is not suggested, but left up to individual countries to decide under specific circumstances. (The factor 1/3 may be reasonable, in the absence of other data, for use in evaluating exposure distributions from radionuclides distributed to the environment; for weapons-test fallout, for example, less than 0.4 percent of the population in the 0-to-25 age group receive exposures from ^{90}Sr in excess of 3 times the average exposure.[16]) In general, no member of the public should receive more than 10 percent of the annual whole-body dose permitted for radiation workers.

Furthermore, when large populations may be exposed to radioactivity dispersed from the environment, consumer items, etc., the limits on genetic dose presented in Table 17 must be considered. As indicated in the previous chapter, the total number of mutants produced throughout all future generations is related to the man-rem product, regardless of how the

Table 15.
PRESENTLY (1967) RECOMMENDED DOSE EQUIVALENTS*
TO BODY ORGANS OF OCCUPATIONAL WORKERS
EXPOSED TO IONIZING RADIATION

Body Organ	Maximum Dose Equivalent In Any 13 Weeks, rem/13 weeks	Annual Permissible Dose Equivalent, rem/year	Accumulated Dose Equivalent To Age N, rem‡
Red bone marrow†	3^{4-6}	$5^{4,6,7}$	$5(N-18)^{4-7}$
Total body	$3^{4,6-9}$	$5^{5,6,8,9}$	$5(N-18)^{4,6-9}$
Head and trunk	$3^{4,6}$	$5^{4,5}$	$5(N-18)^{4,6,7}$
Gonads	$3^{4-6,8,9}$	5^{4-9}	$5(N-18)^{4-9}$
Lenses of eyes	$3^{4-6,8-11}$ 4^{12} 8^{13}	5^{4-11} 15^{13}	$5(N-18)^{4-11}$
Skin	$8^{5,9,11}$ $10^{4,9,15}$ 15^{13}	$30^{4-6,8,9,14,15}$	
Thyroid	$8^{5,9,11}$ 10^{4} 15^{13}	$30^{4-6,8,9}$	
Bone	15^{13}	$30^{8,9}$	
Feet, ankles, hands, forearms	20^{5} $25^{4,6,9}$ 38^{13}	$75^{4-6,9,14}$	
Other single organs	$4^{5,8}$ 5^{4} 8^{13}	$15^{4-6,8,9,14}$	

* The recommended permissible dose equivalent for the occupational worker is in addition to doses from medical and background exposure.

† Referred to in earlier reports as active "blood-forming organs."

‡ The 5(N − 18) accumulated dose equation may put a ceiling on the 13-week dose, but doses should be kept as far below even this cumulative limit as is feasible.

From: Morgan, K. Z., Maximum Permissible Exposure Levels—External and Internal. *Principles of Radiation Protection*, Ch. 14, Morgan, K. Z. and Turner, E. J., eds., 1967. By permission of John Wiley and Sons, New York.

Table 16.
DOSE-EQUIVALENT LEVELS RECOMMENDED IN EARLY REPORTS
OF THE ICRP, AND FACTORS TO BE APPLIED TO OCCUPATIONAL VALUES

Exposed Group	Dose Equivalent to Gonads or Total Body		Dose Equivalent to Most Other Organs	
	Dose Rate	Factor	Dose Rate	Factor
A. Occupational worker	3.00 rem/qtr 12.00 rem/yr or 5.00 rem/yr (av)	1	8.0 rem/qtr 15.0 rem/yr (av)	1
B. Work in vicinity of controlled area	1.50 rem/yr	3/10	1.5 rem/yr	1/10
C. Visit area occupationally	1.50 rem/yr	3/10	1.5 rem/yr	1/10
D. Whole population	0.50 rem/yr	1/10	1.5 rem/yr	1/10
	5.00 rem (av) to age 30 or 0.17 rem/yr (av)	1/30	15.0 rem (av) to age 30 or 0.5 rem/yr (av)	1/30
	0.05 rem/yr (av)	1/100		

From: Morgan, K. Z., Maximum Permissible Exposure Levels—External and Internal. *Principles of Radiation Protection*, Ch. 14, Morgan, K. Z. and Turner, E. J., eds., 1967. By permission of John Wiley and Sons, New York.

mutants may be distributed among immediately succeeding generations. Since the average occupational genetic dose is actually less than 0.3 rem per year distributed over more than 2×10^5 employees,[17] it represents fewer man-rem than the natural background genetic dose of about 0.1 rem per year received by 2×10^8 persons in the United States.[18] Thus, occupational exposure adds less than about 0.3 percent to the naturally received population genetic dose. However, as the nuclear-energy industry expands, and as radioisotopes are increasingly used in consumer products, it will become necessary to account for and limit all of the man-made sources of population exposure. A more detailed discussion of this problem and an illustrative method for such population exposure accounting have been presented elsewhere.[3] It has been suggested that all consumer items ever distributed together should not expose the public to an average genetic dose of more than about 0.01 rem per year (0.3 rem per generation of 30 years).[3]

Table 17.
PERMISSIBLE GENETIC DOSE TO THE POPULATION AT LARGE
SUGGESTED IN THE EARLY REPORTS OF THE ICRP TO SERVE AS A GUIDE
(Dose Equivalent in rem to Age 30)

Medical		4.5 rem/generation
Background		3.0 rem/generation
Other Sources		
Population at large		
Internal	1.5	
External	0.5	
Subtotal		2.0
Other groups		
Occupational	1.0	
Special groups	0.5	
Reserve	1.5	
Subtotal		3.0
Total from Other Sources	5.0	
Total from All Sources		12.5 rem/generation

From: Morgan, K. Z., Maximum Permissible Exposure Levels—External and Internal. *Principles of Radiation Protection*, Ch. 14, Morgan, K. Z. and Turner, E. J., eds., 1967. By permission of John Wiley and Sons, New York.

DETERMINATION OF PERMISSIBLE CONCENTRATIONS IN AIR AND WATER

By equating the absorbed dose from qf_2 microcuries in each body organ to the maximum permissible dose rate to that organ (q being the total activity in the body, and f_2 the fraction in the organ), the ICRP has calculated for each of the organs likely to absorb each radionuclide in "soluble" form a corresponding maximum permissible body burden q.[3,19] Then, from differential equations relating the rate of uptake of a radionuclide into an organ by inhalation or ingestion to the build-up of activity in the organ and to the biological and physical elimination from the organ, the ICRP has calculated maximum permissible concentrations (MPC's) for air and water for occupational exposures of 40 hours per week and 168 hours per week. These concentrations are calculated so that, during 50 years of occupational exposure, a "standard" 70-kilogram man would at no time exceed the permissible dose rate to the body organ receiving the highest fraction of its respective permissible dose rate. Calculations were also made, using an idealized lung model and G.I. tract model, to consider the doses to lung and G.I. tract from "insoluble" materials inhaled or ingested. Parameters were selected to yield MPC's$_{air}$ or MPC's$_{water}$ that would not, in any case, be dangerously high for the standard man. Since new physical and physiological data for many of the nuclides, as well as new nuclides, have become available since 1959, and since children or persons other than standard man may require special consideration, a summary of some of the basic equations for calculating maximum permissible body burdens (MPBB's) and MPC's will be given below. However, the simpler method of calculating the G.I. tract dose given in earlier ICRP reports will be presented for general radiation hazard evaluation purposes.

The following equations for permissible body burden have been condensed from References 19 and 20, where more detailed discussions and definitions of parameters may be found. Also, Reference 19 presents extensive tabulations of physical and physiological parameters, which are useful for (generally conservative) estimates of integrated exposures after single or continuous intake as well as in the equations. These data are continually being updated, and even more extensive tabulations of physical and physiological data for internal-dose evaluations are in preparation (1968) for publication by the Internal-Dose Evaluation Center.*
In addition, a committee† has been formed by the Society of Nuclear Medicine to recommend methods and data for more precise determination of organ exposure in medical diagnosis or treatment with radioisotopes. Recently their first publication[22] was issued, which includes methods of calculation and tabulations of energy deposited in various shapes and sizes of organs for various photon energies.

Maximum Permissible Concentration, Normal Occupational Exposure[19,20]

Body Burden of Radioisotopes for Which the Skeleton is the Critical Organ.

The maximum permissible skeletal content is generally derived by comparison with radium, for which the recommended maximum is 0.1 µCi in bone. On the basis of assumptions made by the National Commission on Radiation Protection (NCRP)-IRCP, such comparison leads to the following equation, calculated to give a dose rate to bone of 0.56 rem per week,

$$q = \frac{11}{f_2 \sum_i (E_i \times RBE_i \times N_i)} \mu Ci,$$

where q = body burden, in µCi;

f_2 = fraction of the amount in the body that is in the critical organ (bone);

E_i = effective energy released to the organ, in Mev per disintegration, weighted according to the relative biological effectiveness (RBE) of each particle or quantum emitted;

N_i = nonuniform-distribution factor, taken as 5 for elements other than radium, and as 1 for radium.‡

Body Burden to Give 0.3 rem per Week to Critical Organ.

$$q = \frac{8.4 \times 10^{-4} \, m}{f_2 \sum_i (E_i \times RBE_i \times N_i)} \mu Ci,$$

where m = mass of the critical organ, in grams;

f_2 = fraction in the critical organ as compared to the total body;

$E_i \times RBE_i$ = energy absorbed, in Mev per disintegration, weighted by RBE;

N_i = nonuniform-distribution factor, which is usually taken as $N = 1$ for organs other than bone.

* For additional information or assistance, contact Dr. W. S. Snyder, Health Physics Division, Oak Ridge National Laboratory, Oak Ridge, Tenn.
† For information, contact Dr. Edward M. Smith, Chairman, Medical Internal Radiation Dose Committee, Society for Nuclear Medicine.
‡ NOTE: A factor $N = 5$ has also been used for ^{90}Sr, based on animal experiments. An RBE of 1 is used for beta particles, and an RBE of 10 is used for alpha particles. The value of $(\sum E \times RBE \times N)$ for ^{226}Ra is taken as 110-Mev equivalent per disintegration, assuming $N = 1$ for radium and a 30% retention of ^{226}Ra daughter products in bone.

Radiation Protection and Regulations

Maximum Permissible Concentration, Based on Body Burden.

$$\text{MPC}_{\text{air}} = \frac{3.5 \times 10^{-8} \, qf_2}{Tf_a(1 - e^{-0.693t/T})},$$

where MPC_{air} = concentration, in μCi/cc, that will eventually build up to a dose rate of 0.3 rem per week if breathed continuously for a time t;

T = the effective half-life of the radionuclide in the body, in days;

t = period of inhalation, in days;

q = body burden, in μCi;

f_2 = fraction of material in the body that is in the critical organ;

f_a = fraction inhaled that arrives in the critical body organ.

$$\text{MPC}_{\text{water}} = \frac{3.1 \times 10^{-4} \, qf_2}{Tf_w(1 - e^{-0.693t/T})},$$

where $\text{MPC}_{\text{water}}$ = the concentration ingested for t days that will build up a burden of qf_2 μCi in the critical organ, giving a dose rate of 0.3 rem per week to the critical organ at time t;

f_w = the fraction ingested that reaches the critical organ.

Maximum Permissible Concentration, Based on Gastrointestinal Tract.

These equations have been superseded,[19] but will serve for making approximate estimates of G.I.-tract dose.

$$(\text{MPC})^{\text{G.I.}}_{\text{water}} = \frac{7.6 \times 10^{-7} \, m^1 G}{Hf_w \sum_i (E_i \times RBE_i)} \, \mu\text{Ci/cc},$$

and

$$(\text{MPC})^{\text{G.I.}}_{\text{air}} = \frac{8.4 \times 10^{-11} \, m^1 G}{Hf_a \sum_i (E_i \times RBE_i)} \, \mu\text{Ci/cc},$$

where $G = 0.693 \, H/(e^{-0.693 \, h_0/T_r} - e^{-0.693 \, h_1/T_r})T_r$;

m^1 = mass of contents of the G.I. tract in the section considered;

H = time (fraction of day) the radioactive material remains in the G.I. section considered (H = $h_1 - h_0$);

f_w and f_a = fractions of radioactive material arriving at the section considered ($f_w = (1 - f_1)$ and $f_a = 0.62(1 - f_1)$);

f_1 = fraction absorbed through the wall of the section;

h_0 = time of arrival in the section considered following inhalation or ingestion;

h_1 = time of elimination from section following inhalation or ingestion.

Necessary parameters are given in Table 18.

MPC of Noble Gases, Based on Submersion in Cloud.[20]

The 40-hour-per-week occupational MPC_{air}, based on the dose within a semi-infinite cloud, is

$$\text{MPC}_{\text{air}} = \frac{0.024 \, W}{\sum_i E_i} \rho_a \frac{P_a}{P_t} \, \mu\text{Ci/cm}^3,$$

where W = dose rate allowed, in rem/week;

$\rho_a = 0.0012$ g/cc;

P_a/P_t = stopping power of air relative to tissue, or 1/1.13 for betas or secondary electrons;

$\sum_i E_i$ = sum over energy, in Mev per disintegration of each isotope in question.

Combining parameters, the MPC for a dose rate of 0.1 rem in 40 hours becomes

$$\text{MPC}_{\text{air}} = \frac{2 \times 10^{-6}}{\sum_i E_i} \mu\text{Ci/cm}^3.$$

MPC for Insoluble Compounds in Lung.[19]

$$\text{MPC}_{\text{air}} = \frac{2 \times 10^{-9}}{\sum_i E_i(\text{RBE})_i N_i} \mu\text{Ci/cm}^3,$$

to give a dose rate of 0.3 rem per week to the lung, assuming continuous exposure with an equilibrium concentration in the lung and a biological half-life of 120 days; biological half-lives shorter and longer than this have been observed[24] for materials in various forms, but use of 120 days for prediction purposes is not likely to be wrong by more than a factor of 3 for most insoluble particulates. More detailed discussions of methods to estimate radiation doses from internal radionuclides and permissible concentrations are now available in several texts.[27-31]

Table 18. CONSTANTS FOR G.I.-TRACT CALCULATIONS*

Critical Tissue	m^1, grams	H, days	h_0, days	h_1, days	Value of T_r for G = 1.00 to 1.01
Lower large intestine, $T_r > 0.5$ day	150	18/24	13/24	31/24	$T_r > 64$ days
Upper large intestine, 0.08 day $< T_r < 0.05$ day	135	8/24	5/24	13/24	$T_r > 26$ days
Stomach, $T_r < 0.08$ day	250	1/24	0	1/24	$T_r > 1.5$ days

* See also slightly revised data in Reference 23, p. 153.
From: Morrow, P. E., Task Group Report, *Health Phys.*, 3: 1—380, 1960. Copyright Pergamon Press, New York.

REGULATORY LIMITS OF EXPOSURE

In regulatory agencies there has been a need to establish some legal boundaries to allowable limits of exposure, which may be inspected conveniently and which may serve as a basis for enforcement of remedial measures or legal action, if necessary. Thus, although regulations established by different nations and states will generally be consistent with ICRP recommendations (Table 15) in the degree of protection afforded, there may be slight differences in detailed requirements, which should be noted by the user of radioactive material. For example, for ease in examining the majority of licenses, the regulations of the United States Atomic Energy Commission, 10 CRF 20,[30] reduce the allowable quarterly limit of radiation exposure below the values in Table 15 when retrospective verification of an employee's previously accumulated exposure is inconvenient or impossible. This may be in some cases more lenient rather than more restrictive, when codified, than the recommendations of the ICRP, since the maximum recommended annual exposure given in Table 15 may in some cases already have been exceeded. Yet it could be unreasonable to insist that an employer determine previous exposure histories that may not readily be available. Thus, according to 10 CFR, Section 20.101 (a):

"(a) Except as provided in paragraph (b) of this section, no licensee shall possess, use, or transfer licensed material in such a manner as to cause any individual in a restricted area to receive in any period of one calendar quarter from radioactive material and other sources of radiation in the licensee's possession a dose in excess of the limits specified in the following table:

Rems per Calendar Quarter
1. Whole body; head and trunk;
   active blood-forming organs;
   lenses of eyes; or gonads   1¼
2. Hands and forearms; feet and ankles   18¾
3. Skin of whole body   7½ "

On the other hand, Section 20.101 (b) spells out additional record-keeping requirements for those employers who occasionally must avail themselves of the 3-rem-per-calendar-quarter limit, in order to ensure that the $5(N - 18)$ cumulative lifetime limit (Table 15) is not grossly exceeded in the long run.

Additional regulatory considerations resulted in the publication in 10 CFR 20 of limiting permissible concentrations of radionuclides in air or water, which are generally taken from ICRP recommendations,[23] but which may differ slightly for some nuclides. Although all the conditions to which the 10 CFR 20 concentration limits are referred cannot be reproduced here, Appendix B of 10 CFR 20 is presented here, modified for easy reference, as Table 19. The concentrations of various radionuclides, soluble (S) or insoluble (I) in body fluids, as given in the columns under Section A of this table, are generally used as the limiting average concentrations in air or drinking water to which an employee may be exposed 40 hours per week during a lifetime of employment; these values may be used for disposal of small quantities (less than 1 curie per year of radioactive waste into sanitary-sewerage systems). The values in Section B are reduced, representing maximum average concentrations in air and water that may occur in unrestricted areas. Additional restrictions and special conditions, as well as other Title 10 regulations, are also covered in 10 CFR 20.

Other requirements of 10 CFR 20 deal with the following: exposure limits for minors; permissible levels of radiation in unrestricted areas (two maxims—neither more than 2 mrem in one hour, nor more than 100 mrem in any seven consecutive days); orders requiring furnishing of bioassay services; surveys; personnel monitoring; caution signs, labels, and signals; instruction of personnel and posting of notices of employee rights; storage of licensed materials; disposal of waste; records of surveys, radiation monitoring, and disposal; reports and notification required for accidents of varying degrees of seriousness. Before ordering radioactive materials, copies of this regulation should be examined together with other pertinent state or local regulations. Additional specific requirements may be added by the AEC or other regulatory body, where necessary, to a license to use radioactive materials. Information on obtaining a license to use by-product materials, source materials containing fissionable elements, or special nuclear materials may be obtained in the United States from:

Director of Regulation
U.S. Atomic Energy Commission
Washington, D.C. 20545

In about 19 states of the U.S.A., agreements have been made between the respective states and the USAEC, so that each "agreement state" now regulates the use of materials originating from the federal atomic-energy program as well as other radiation sources not under AEC jurisdiction, such as radium, X-ray machines, and high-energy particle accelerators. However, state regulations are generally fairly consistent with those of the AEC under the conditions of the agreements between these bodies. Moreover, licensing control of the larger production and utilization facilities (including nuclear reactors) is retained by the AEC.

In addition to AEC and state regulations, the user of radioactive materials should also consult the regulations of the U.S. Department of Transportation (Code of Federal Regulations, Title 49) and recent regulations from the Department of Labor, which are related to the Walsh-Healey Act. There are, however, certain small and sufficiently safe quantities of various radioactive materials that have been exempted from regulatory control by the USAEC for use in teaching, demonstration, and calibration of sensitive radiation detectors. Information on exempt and/or "generally licensed" radioactive materials may be obtained from many nuclear chemical and pharmaceutical companies. Additional sources of information and supply may

Table 19.
MAXIMUM PERMISSIBLE CONCENTRATIONS OF RADIONUCLIDES IN AIR AND WATER[30]
(See Notes at End of Table)

Element	Atomic Number	Isotope*	A - In Air, µCi/ml	A - In Water, µCi/ml	B - In Air, µCi/ml	B - In Water, µCi/ml
Actinium	89	^{227}Ac				
		S	2×10^{-12}	6×10^{-5}	8×10^{-14}	2×10^{-6}
		I	3×10^{-11}	9×10^{-3}	9×10^{-13}	3×10^{-4}
		^{228}Ac				
		S	8×10^{-8}	3×10^{-3}	3×10^{-9}	9×10^{-5}
		I	2×10^{-8}	3×10^{-3}	6×10^{-10}	9×10^{-5}
Americium	95	^{241}Am				
		S	6×10^{-12}	1×10^{-4}	2×10^{-13}	4×10^{-6}
		I	1×10^{-10}	8×10^{-4}	4×10^{-12}	2×10^{-5}
		^{242m}Am				
		S	6×10^{-12}	1×10^{-4}	2×10^{-13}	4×10^{-6}
		I	3×10^{-10}	3×10^{-3}	9×10^{-12}	9×10^{-5}
		^{242}Am				
		S	4×10^{-8}	4×10^{-3}	1×10^{-9}	1×10^{-4}
		I†	5×10^{-8}	4×10^{-3}	2×10^{-9}	1×10^{-4}
		^{243}Am				
		S	6×10^{-12}	1×10^{-4}	2×10^{-13}	4×10^{-6}
		I	1×10^{-10}	8×10^{-4}	4×10^{-12}	3×10^{-5}
		^{244}Am†				
		S	4×10^{-6}	1×10^{-1}	1×10^{-7}	5×10^{-3}
		I	2×10^{-5}	1×10^{-1}	8×10^{-7}	5×10^{-3}
Antimony	51	^{122}Sb				
		S	2×10^{-7}	8×10^{-4}	6×10^{-9}	3×10^{-5}
		I	1×10^{-7}	8×10^{-4}	5×10^{-9}	3×10^{-5}
		^{124}Sb				
		S	2×10^{-7}	7×10^{-4}	5×10^{-9}	2×10^{-5}
		I	2×10^{-8}	7×10^{-4}	7×10^{-10}	2×10^{-5}
		^{125}Sb				
		S	5×10^{-7}	3×10^{-3}	2×10^{-8}	1×10^{-4}
		I	3×10^{-8}	3×10^{-3}	9×10^{-10}	1×10^{-4}
Argon	18	^{37}Ar				
		Sub	6×10^{-3}	—	1×10^{-4}	—
		^{41}Ar				
		Sub	2×10^{-6}	—	4×10^{-8}	—
Arsenic	33	^{73}As				
		S	2×10^{-6}	1×10^{-2}	7×10^{-8}	5×10^{-4}
		I	4×10^{-7}	1×10^{-2}	1×10^{-8}	5×10^{-4}
		^{74}As				
		S	3×10^{-7}	2×10^{-3}	1×10^{-8}	5×10^{-5}
		I	1×10^{-7}	2×10^{-3}	4×10^{-9}	5×10^{-5}
		^{76}As				
		S	1×10^{-7}	6×10^{-4}	4×10^{-9}	2×10^{-5}
		I	1×10^{-7}	6×10^{-4}	3×10^{-9}	2×10^{-5}
		^{77}As				
		S	5×10^{-7}	2×10^{-3}	2×10^{-8}	8×10^{-5}
		I	4×10^{-7}	2×10^{-3}	1×10^{-8}	8×10^{-5}

Table 19. (Continued)
MAXIMUM PERMISSIBLE CONCENTRATIONS OF RADIONUCLIDES IN AIR AND WATER[30]

(See Notes at End of Table)

Element	Atomic Number	Isotope*	A		B	
			In Air, µCi/ml	In Water, µCi/ml	In Air, µCi/ml	In Water, µCi/ml
Astatine	85	^{211}At				
		S	7×10^{-9}	5×10^{-5}	2×10^{-10}	2×10^{-6}
		I	3×10^{-8}	2×10^{-3}	1×10^{-9}	7×10^{-5}
Barium	56	^{131}Ba				
		S	1×10^{-6}	5×10^{-3}	4×10^{-8}	2×10^{-4}
		I	4×10^{-7}	5×10^{-3}	1×10^{-8}	2×10^{-4}
		^{140}Ba				
		S	1×10^{-7}	8×10^{-4}	4×10^{-9}	3×10^{-5}
		I	4×10^{-8}	7×10^{-4}	1×10^{-9}	2×10^{-5}
Berkelium	97	^{249}Bk				
		S	9×10^{-10}	2×10^{-2}	3×10^{-11}	6×10^{-4}
		I	1×10^{-7}	2×10^{-2}	4×10^{-9}	6×10^{-4}
		^{250}Bk†				
		S	1×10^{-7}	6×10^{-3}	5×10^{-9}	2×10^{-4}
		I	1×10^{-6}	6×10^{-3}	4×10^{-8}	2×10^{-4}
Beryllium	4	^{7}Be				
		S	6×10^{-6}	5×10^{-2}	2×10^{-7}	2×10^{-3}
		I	1×10^{-6}	5×10^{-2}	4×10^{-8}	2×10^{-3}
Bismuth	83	^{206}Bi				
		S	2×10^{-7}	1×10^{-3}	6×10^{-9}	4×10^{-5}
		I	1×10^{-7}	1×10^{-3}	5×10^{-9}	4×10^{-5}
		^{207}Bi				
		S	2×10^{-7}	2×10^{-3}	6×10^{-9}	6×10^{-5}
		I	1×10^{-8}	2×10^{-3}	5×10^{-10}	6×10^{-5}
		^{210}Bi				
		S	6×10^{-9}	1×10^{-3}	2×10^{-10}	4×10^{-5}
		I	6×10^{-9}	1×10^{-3}	2×10^{-10}	4×10^{-5}
		^{212}Bi				
		S	1×10^{-7}	1×10^{-2}	3×10^{-9}	4×10^{-4}
		I	2×10^{-7}	1×10^{-2}	7×10^{-9}	4×10^{-4}
Bromine	35	^{82}Br				
		S	1×10^{-6}	8×10^{-3}	4×10^{-8}	3×10^{-4}
		I	2×10^{-7}	1×10^{-3}	6×10^{-9}	4×10^{-5}
Cadmium	48	^{109}Cd				
		S	5×10^{-8}	5×10^{-3}	2×10^{-9}	2×10^{-4}
		I	7×10^{-8}	5×10^{-3}	3×10^{-9}	2×10^{-4}
		^{115m}Cd				
		S	4×10^{-8}	7×10^{-4}	1×10^{-9}	3×10^{-5}
		I	4×10^{-8}	7×10^{-4}	1×10^{-9}	3×10^{-5}
		^{115}Cd				
		S	2×10^{-7}	1×10^{-3}	8×10^{-9}	3×10^{-5}
		I	2×10^{-7}	1×10^{-3}	6×10^{-9}	4×10^{-5}
Calcium	20	^{45}Ca				
		S	3×10^{-8}	3×10^{-4}	1×10^{-9}	9×10^{-6}
		I	1×10^{-7}	5×10^{-3}	4×10^{-9}	2×10^{-4}
		^{47}Ca				
		S	2×10^{-7}	1×10^{-3}	6×10^{-9}	5×10^{-5}
		I	2×10^{-7}	1×10^{-3}	6×10^{-9}	3×10^{-5}

Table 19. *(Continued)*
MAXIMUM PERMISSIBLE CONCENTRATIONS OF RADIONUCLIDES IN AIR AND WATER[30]
(See Notes at End of Table)

Element	Atomic Number	Isotope*	A		B	
			In Air, µCi/ml	In Water, µCi/ml	In Air, µCi/ml	In Water, µCi/ml
Californium	98	^{249}Cf				
		S	2×10^{-12}	1×10^{-4}	5×10^{-14}	4×10^{-6}
		I	1×10^{-10}	7×10^{-4}	3×10^{-12}	2×10^{-5}
		^{250}Cf				
		S	5×10^{-12}	4×10^{-4}	2×10^{-13}	1×10^{-5}
		I	1×10^{-10}	7×10^{-4}	3×10^{-12}	3×10^{-5}
		^{251}Cf†				
		S	2×10^{-12}	1×10^{-4}	6×10^{-14}	4×10^{-6}
		I	1×10^{-10}	8×10^{-4}	3×10^{-12}	3×10^{-5}
		^{252}Cf				
		S	2×10^{-11}	7×10^{-4}	7×10^{-13}	2×10^{-5}
		I	1×10^{-10}	7×10^{-4}	4×10^{-12}	2×10^{-5}
		^{253}Cf				
		S	8×10^{-10}	4×10^{-3}	3×10^{-11}	1×10^{-4}
		I	8×10^{-10}	4×10^{-3}	3×10^{-11}	1×10^{-4}
		^{254}Cf†				
		S	5×10^{-12}	4×10^{-6}	2×10^{-13}	1×10^{-7}
		I	5×10^{-12}	4×10^{-6}	2×10^{-13}	1×10^{-7}
Carbon	6	^{14}C				
		S	4×10^{-6}	2×10^{-2}	1×10^{-7}	8×10^{-4}
		(CO_2) Sub	5×10^{-5}	—	1×10^{-6}	—
Cerium	58	^{141}Ce				
		S	4×10^{-7}	3×10^{-3}	2×10^{-8}	9×10^{-5}
		I	2×10^{-7}	3×10^{-3}	5×10^{-9}	9×10^{-5}
		^{143}Ce				
		S	3×10^{-7}	1×10^{-3}	9×10^{-9}	4×10^{-5}
		I	2×10^{-7}	1×10^{-3}	7×10^{-9}	4×10^{-5}
		^{144}Ce				
		S	1×10^{-8}	3×10^{-4}	3×10^{-10}	1×10^{-5}
		I	6×10^{-9}	3×10^{-4}	2×10^{-10}	1×10^{-5}
Cesium	55	^{131}Cs				
		S	1×10^{-5}	7×10^{-2}	4×10^{-7}	2×10^{-3}
		I	3×10^{-6}	3×10^{-2}	1×10^{-7}	9×10^{-4}
		^{134m}Cs				
		S	4×10^{-5}	2×10^{-1}	1×10^{-6}	6×10^{-3}
		I	6×10^{-6}	3×10^{-2}	2×10^{-7}	1×10^{-3}
		^{134}Cs				
		S	4×10^{-8}	3×10^{-4}	1×10^{-9}	9×10^{-6}
		I	1×10^{-8}	1×10^{-3}	4×10^{-10}	4×10^{-5}
		^{135}Cs				
		S	5×10^{-7}	3×10^{-3}	2×10^{-8}	1×10^{-4}
		I	9×10^{-8}	7×10^{-3}	3×10^{-9}	2×10^{-4}
		^{136}Cs				
		S	4×10^{-7}	2×10^{-3}	1×10^{-8}	9×10^{-5}
		I	2×10^{-7}	2×10^{-3}	6×10^{-9}	6×10^{-5}
		^{137}Cs				
		S	6×10^{-8}	4×10^{-4}	2×10^{-9}	2×10^{-5}
		I	1×10^{-8}	1×10^{-3}	5×10^{-10}	4×10^{-5}

Table 19. *(Continued)*
MAXIMUM PERMISSIBLE CONCENTRATIONS OF RADIONUCLIDES IN AIR AND WATER[30]
(See Notes at End of Table)

Element	Atomic Number	Isotope*	A		B	
			In Air, µCi/ml	In Water, µCi/ml	In Air, µCi/ml	In Water, µCi/ml
Chlorine	17	^{36}Cl				
		S	4×10^{-7}	2×10^{-3}	1×10^{-8}	8×10^{-5}
		I	2×10^{-8}	2×10^{-3}	8×10^{-10}	6×10^{-5}
		^{38}Cl				
		S	3×10^{-6}	1×10^{-2}	9×10^{-8}	4×10^{-4}
		I	2×10^{-6}	1×10^{-2}	7×10^{-8}	4×10^{-4}
Chromium	24	^{51}Cr				
		S	1×10^{-5}	5×10^{-2}	4×10^{-7}	2×10^{-3}
		I	2×10^{-6}	5×10^{-2}	8×10^{-8}	2×10^{-3}
Cobalt	27	^{57}Co				
		S	3×10^{-6}	2×10^{-2}	1×10^{-7}	5×10^{-4}
		I	2×10^{-7}	1×10^{-2}	6×10^{-9}	4×10^{-4}
		^{58m}Co				
		S	2×10^{-5}	8×10^{-2}	6×10^{-7}	3×10^{-3}
		I	9×10^{-6}	6×10^{-2}	3×10^{-7}	2×10^{-3}
		^{58}Co				
		S	8×10^{-7}	4×10^{-3}	3×10^{-8}	1×10^{-4}
		I	5×10^{-8}	3×10^{-3}	2×10^{-9}	9×10^{-5}
		^{60}Co				
		S	3×10^{-7}	1×10^{-3}	1×10^{-8}	5×10^{-5}
		I	9×10^{-9}	1×10^{-3}	3×10^{-10}	3×10^{-5}
Copper	29	^{64}Cu				
		S	2×10^{-6}	1×10^{-2}	7×10^{-8}	3×10^{-4}
		I	1×10^{-6}	6×10^{-3}	4×10^{-8}	2×10^{-4}
Curium	96	^{242}Cm				
		S	1×10^{-10}	7×10^{-4}	4×10^{-12}	2×10^{-5}
		I	2×10^{-10}	7×10^{-4}	6×10^{-12}	3×10^{-5}
		^{243}Cm				
		S	6×10^{-12}	1×10^{-4}	2×10^{-13}	5×10^{-6}
		I	1×10^{-10}	7×10^{-4}	3×10^{-12}	2×10^{-5}
		$^{244\alpha}$Cm				
		S	9×10^{-12}	2×10^{-4}	3×10^{-13}	7×10^{-6}
		I	1×10^{-10}	8×10^{-4}	3×10^{-12}	3×10^{-5}
		^{245}Cm				
		S	5×10^{-12}	1×10^{-4}	2×10^{-13}	4×10^{-6}
		I	1×10^{-10}	8×10^{-4}	4×10^{-12}	3×10^{-5}
		^{246}Cm				
		S	5×10^{-12}	1×10^{-4}	2×10^{-13}	4×10^{-6}
		I	1×10^{-10}	8×10^{-4}	4×10^{-12}	3×10^{-5}
		^{247}Cm				
		S	5×10^{-12}	1×10^{-4}	2×10^{-13}	4×10^{-6}
		I	1×10^{-10}	6×10^{-4}	4×10^{-12}	2×10^{-5}
		^{248}Cm†				
		S	6×10^{-13}	1×10^{-5}	2×10^{-14}	4×10^{-7}
		I	1×10^{-11}	4×10^{-5}	4×10^{-13}	1×10^{-6}
		^{249}Cm				
		S	1×10^{-5}	6×10^{-2}	4×10^{-7}	2×10^{-3}
		I	1×10^{-5}	6×10^{-2}	4×10^{-7}	2×10^{-3}

Table 19. *(Continued)*
MAXIMUM PERMISSIBLE CONCENTRATIONS OF RADIONUCLIDES IN AIR AND WATER[30]
(See Notes at End of Table)

Element	Atomic Number	Isotope*	A		B	
			In Air, µCi/ml	In Water, µCi/ml	In Air, µCi/ml	In Water, µCi/ml
Dysprosium	66	^{165}Dy				
		S	3×10^{-6}	1×10^{-2}	9×10^{-8}	4×10^{-4}
		I	2×10^{-6}	1×10^{-2}	7×10^{-8}	4×10^{-4}
		^{166}Dy†				
		S	2×10^{-7}	1×10^{-3}	8×10^{-9}	4×10^{-5}
		I	2×10^{-7}	1×10^{-3}	7×10^{-9}	4×10^{-5}
Einsteinium	99	^{253}Es				
		S	8×10^{-10}	7×10^{-4}	3×10^{-11}	2×10^{-5}
		I	6×10^{-10}	7×10^{-4}	2×10^{-11}	2×10^{-5}
		^{254m}Es				
		S	5×10^{-9}	5×10^{-4}	2×10^{-10}	2×10^{-5}
		I	6×10^{-9}	5×10^{-4}	2×10^{-10}	2×10^{-5}
		^{254}Es				
		S	2×10^{-11}	4×10^{-4}	6×10^{-13}	1×10^{-5}
		I	1×10^{-10}	4×10^{-4}	4×10^{-12}	1×10^{-5}
		^{255}Es				
		S	5×10^{-10}	8×10^{-4}	2×10^{-11}	3×10^{-5}
		I	4×10^{-10}	8×10^{-4}	1×10^{-11}	3×10^{-5}
Erbium	68	^{169}Er				
		S	6×10^{-7}	3×10^{-3}	2×10^{-8}	9×10^{-5}
		I	4×10^{-7}	3×10^{-3}	1×10^{-8}	9×10^{-5}
		^{171}Er				
		S	7×10^{-7}	3×10^{-3}	2×10^{-8}	1×10^{-4}
		I	6×10^{-7}	3×10^{-3}	2×10^{-8}	1×10^{-4}
Europium	63	^{152}Eu (T/2 = 9.2 hrs)				
		S	4×10^{-7}	2×10^{-3}	1×10^{-8}	6×10^{-5}
		I	3×10^{-7}	2×10^{-3}	1×10^{-8}	6×10^{-5}
		^{152}Eu (T/2 = 13 yrs)				
		S	1×10^{-8}	2×10^{-3}	4×10^{-10}	8×10^{-5}
		I	2×10^{-8}	2×10^{-3}	6×10^{-10}	8×10^{-5}
		^{154}Eu				
		S	4×10^{-9}	6×10^{-4}	1×10^{-10}	2×10^{-5}
		I	7×10^{-9}	6×10^{-4}	2×10^{-10}	2×10^{-5}
		^{155}Eu				
		S	9×10^{-8}	6×10^{-3}	3×10^{-9}	2×10^{-4}
		I	7×10^{-8}	6×10^{-3}	3×10^{-9}	2×10^{-4}
Fermium†	100	^{254}Fm				
		S	6×10^{-8}	4×10^{-3}	2×10^{-9}	1×10^{-4}
		I	7×10^{-8}	4×10^{-3}	2×10^{-9}	1×10^{-4}
		^{255}Fm				
		S	2×10^{-8}	1×10^{-3}	6×10^{-10}	3×10^{-5}
		I	1×10^{-8}	1×10^{-3}	4×10^{-10}	3×10^{-5}
		^{256}Fm				
		S	3×10^{-9}	3×10^{-5}	1×10^{-10}	9×10^{-7}
		I	2×10^{-9}	3×10^{-5}	6×10^{-11}	9×10^{-7}

Table 19. *(Continued)*
MAXIMUM PERMISSIBLE CONCENTRATIONS OF RADIONUCLIDES IN AIR AND WATER[30]

(See Notes at End of Table)

Element	Atomic Number	Isotope*	A		B	
			In Air, µCi/ml	In Water, µCi/ml	In Air, µCi/ml	In Water, µCi/ml
Fluorine	9	^{18}F				
		S	5×10^{-6}	2×10^{-2}	2×10^{-7}	8×10^{-4}
		I	3×10^{-6}	1×10^{-2}	9×10^{-8}	5×10^{-4}
Gadolinium	64	^{153}Gd				
		S	2×10^{-7}	6×10^{-3}	8×10^{-9}	2×10^{-4}
		I	9×10^{-8}	6×10^{-3}	3×10^{-9}	2×10^{-4}
		^{159}Gd				
		S	5×10^{-7}	2×10^{-3}	2×10^{-8}	8×10^{-5}
		I	4×10^{-7}	2×10^{-3}	1×10^{-8}	8×10^{-5}
Gallium	31	^{72}Ga				
		S	2×10^{-7}	1×10^{-3}	8×10^{-9}	4×10^{-5}
		I	2×10^{-7}	1×10^{-3}	6×10^{-9}	4×10^{-5}
Germanium	32	^{71}Ge				
		S	1×10^{-5}	5×10^{-2}	4×10^{-7}	2×10^{-3}
		I	6×10^{-6}	5×10^{-2}	2×10^{-7}	2×10^{-3}
Gold	79	^{196}Au				
		S	1×10^{-6}	5×10^{-3}	4×10^{-8}	2×10^{-4}
		I	6×10^{-7}	4×10^{-3}	2×10^{-8}	1×10^{-4}
		^{198}Au				
		S	3×10^{-7}	2×10^{-3}	1×10^{-8}	5×10^{-5}
		I	2×10^{-7}	1×10^{-3}	8×10^{-9}	5×10^{-5}
		^{199}Au				
		S	1×10^{-6}	5×10^{-3}	4×10^{-8}	2×10^{-4}
		I	8×10^{-7}	4×10^{-3}	3×10^{-8}	2×10^{-4}
Hafnium	72	^{181}Hf				
		S	4×10^{-8}	2×10^{-3}	1×10^{-9}	7×10^{-5}
		I	7×10^{-8}	2×10^{-3}	3×10^{-9}	7×10^{-5}
Holmium	67	^{166}Ho				
		S	2×10^{-7}	9×10^{-4}	7×10^{-9}	3×10^{-5}
		I	2×10^{-7}	9×10^{-4}	6×10^{-9}	3×10^{-5}
Hydrogen	1	^{3}H				
		S	5×10^{-6}	1×10^{-1}	2×10^{-7}	3×10^{-3}
		I†	5×10^{-6}	1×10^{-1}	2×10^{-7}	3×10^{-3}
		Sub	2×10^{-3}	—	4×10^{-5}	—
Indium	49	^{113m}In				
		S	8×10^{-6}	4×10^{-2}	3×10^{-7}	1×10^{-3}
		I	7×10^{-6}	4×10^{-2}	2×10^{-7}	1×10^{-3}
		^{114m}In				
		S	1×10^{-7}	5×10^{-4}	4×10^{-9}	2×10^{-5}
		I	2×10^{-8}	5×10^{-4}	7×10^{-10}	2×10^{-5}
		^{115m}In				
		S	2×10^{-6}	1×10^{-2}	8×10^{-8}	4×10^{-4}
		I	2×10^{-6}	1×10^{-2}	6×10^{-8}	4×10^{-4}
		^{115}In				
		S	2×10^{-7}	3×10^{-3}	9×10^{-9}	9×10^{-5}
		I	3×10^{-8}	3×10^{-3}	1×10^{-9}	9×10^{-5}

Table 19. *(Continued)*
MAXIMUM PERMISSIBLE CONCENTRATIONS OF RADIONUCLIDES IN AIR AND WATER[30]
(See Notes at End of Table)

Element	Atomic Number	Isotope*	A		B	
			In Air, μCi/ml	In Water, μCi/ml	In Air, μCi/ml	In Water, μCi/ml
Iodine	53	^{125}I†				
		S	5×10^{-9}	4×10^{-5}	8×10^{-11}	2×10^{-7}
		I	2×10^{-7}	6×10^{-3}	6×10^{-9}	2×10^{-4}
		^{126}I††				
		S	8×10^{-9}	5×10^{-5}	9×10^{-11}	3×10^{-7}
		I	3×10^{-7}	3×10^{-3}	1×10^{-8}	9×10^{-5}
		^{129}I				
		S	2×10^{-9}	1×10^{-5}	2×10^{-11}	6×10^{-8}
		I	7×10^{-8}	6×10^{-3}	2×10^{-9}	2×10^{-4}
		^{131}I				
		S	9×10^{-9}	6×10^{-5}	1×10^{-10}	3×10^{-7}
		I	3×10^{-7}	2×10^{-3}	1×10^{-8}	6×10^{-5}
		^{132}I				
		S	2×10^{-7}	2×10^{-3}	3×10^{-9}	8×10^{-6}
		I	9×10^{-7}	5×10^{-3}	3×10^{-8}	2×10^{-4}
		^{133}I				
		S	3×10^{-8}	2×10^{-4}	4×10^{-10}	1×10^{-6}
		I	2×10^{-7}	1×10^{-3}	7×10^{-9}	4×10^{-5}
		^{134}I				
		S	5×10^{-7}	4×10^{-3}	6×10^{-9}	2×10^{-5}
		I	3×10^{-6}	2×10^{-2}	1×10^{-7}	6×10^{-4}
		^{135}I				
		S	1×10^{-7}	7×10^{-4}	1×10^{-9}	4×10^{-6}
		I	4×10^{-7}	2×10^{-3}	1×10^{-8}	7×10^{-5}
Iridium	77	^{190}Ir				
		S	1×10^{-6}	6×10^{-3}	4×10^{-8}	2×10^{-4}
		I	4×10^{-7}	5×10^{-3}	1×10^{-8}	2×10^{-4}
		^{192}Ir				
		S	1×10^{-7}	1×10^{-3}	4×10^{-9}	4×10^{-5}
		I	3×10^{-8}	1×10^{-3}	9×10^{-10}	4×10^{-5}
		^{194}Ir				
		S	2×10^{-7}	1×10^{-3}	8×10^{-9}	3×10^{-5}
		I	2×10^{-7}	9×10^{-4}	5×10^{-9}	3×10^{-5}
Iron	26	^{55}Fe				
		S	9×10^{-7}	2×10^{-2}	3×10^{-8}	8×10^{-4}
		I	1×10^{-6}	7×10^{-2}	3×10^{-8}	2×10^{-3}
		^{59}Fe				
		S	1×10^{-7}	2×10^{-3}	5×10^{-9}	6×10^{-5}
		I	5×10^{-8}	2×10^{-3}	2×10^{-9}	5×10^{-5}
Krypton	36	^{85m}Kr				
		Sub	6×10^{-6}	—	1×10^{-7}	—
		^{85}Kr				
		Sub	1×10^{-5}	—	3×10^{-7}	—
		^{87}Kr				
		Sub	1×10^{-6}	—	2×10^{-8}	—

Table 19. *(Continued)*
MAXIMUM PERMISSIBLE CONCENTRATIONS OF RADIONUCLIDES IN AIR AND WATER[30]
(See Notes at End of Table)

Element	Atomic Number	Isotope*	A		B	
			In Air, μCi/ml	In Water, μCi/ml	In Air, μCi/ml	In Water, μCi/ml
Krypton *(cont.)*	36	^{88}Kr$^+$ Sub	1×10^{-6}	—	2×10^{-8}	—
Lanthanum	57	^{140}La				
		S	2×10^{-7}	7×10^{-4}	5×10^{-9}	2×10^{-5}
		I	1×10^{-7}	7×10^{-4}	4×10^{-9}	2×10^{-5}
Lead	82	^{203}Pb				
		S	3×10^{-6}	1×10^{-2}	9×10^{-8}	4×10^{-4}
		I	2×10^{-6}	1×10^{-2}	6×10^{-8}	4×10^{-4}
		^{210}Pb				
		S	1×10^{-10}	4×10^{-6}	4×10^{-12}	1×10^{-7}
		I	2×10^{-10}	5×10^{-3}	8×10^{-12}	2×10^{-4}
		^{212}Pb				
		S	2×10^{-8}	6×10^{-4}	6×10^{-10}	2×10^{-5}
		I	2×10^{-8}	5×10^{-4}	7×10^{-10}	2×10^{-5}
Lutatium	71	^{177}Lu				
		S	6×10^{-7}	3×10^{-3}	2×10^{-8}	1×10^{-4}
		I	5×10^{-7}	3×10^{-3}	2×10^{-8}	1×10^{-4}
Manganese	25	^{52}Mn				
		S	2×10^{-7}	1×10^{-3}	7×10^{-9}	3×10^{-5}
		I	1×10^{-7}	9×10^{-4}	5×10^{-9}	3×10^{-5}
		^{54}Mn				
		S	4×10^{-7}	4×10^{-3}	1×10^{-9}	1×10^{-4}
		I	4×10^{-8}	3×10^{-3}	1×10^{-9}	1×10^{-4}
		^{56}Mn				
		S	8×10^{-7}	4×10^{-3}	3×10^{-8}	1×10^{-4}
		I	5×10^{-7}	3×10^{-3}	2×10^{-8}	1×10^{-4}
Mercury	80	^{197m}Hg				
		S	7×10^{-7}	6×10^{-3}	3×10^{-8}	2×10^{-4}
		I	8×10^{-7}	5×10^{-3}	3×10^{-8}	2×10^{-4}
		^{197}Hg				
		S	1×10^{-6}	9×10^{-3}	4×10^{-8}	3×10^{-4}
		I	3×10^{-6}	1×10^{-2}	9×10^{-8}	5×10^{-4}
		^{203}Hg				
		S	7×10^{-8}	5×10^{-4}	2×10^{-9}	2×10^{-5}
		I	1×10^{-7}	3×10^{-3}	4×10^{-9}	1×10^{-4}
Molybdenum	42	^{99}Mo				
		S	7×10^{-7}	5×10^{-3}	3×10^{-8}	2×10^{-4}
		I	2×10^{-7}	1×10^{-3}	7×10^{-9}	4×10^{-5}
Neodymium	60	^{144}Nd				
		S	8×10^{-11}	2×10^{-3}	3×10^{-12}	7×10^{-5}
		I	3×10^{-10}	2×10^{-3}	1×10^{-11}	8×10^{-5}
		^{147}Nd				
		S	4×10^{-7}	2×10^{-3}	1×10^{-8}	6×10^{-5}
		I	2×10^{-7}	2×10^{-3}	8×10^{-9}	6×10^{-5}

Table 19. *(Continued)*
MAXIMUM PERMISSIBLE CONCENTRATIONS OF RADIONUCLIDES IN AIR AND WATER[30]
(See Notes at End of Table)

Element	Atomic Number	Isotope*	A		B	
			In Air, µCi/ml	In Water, µCi/ml	In Air, µCi/ml	In Water, µCi/ml
Neodymium *(cont.)*	60	^{149}Nd S I	 2×10^{-6} 1×10^{-6}	 8×10^{-3} 8×10^{-3}	 6×10^{-8} 5×10^{-8}	 3×10^{-4} 3×10^{-4}
Neptunium	93	^{237}Np S I ^{239}Np S I	 4×10^{-12} 1×10^{-10} 8×10^{-7} 7×10^{-7}	 9×10^{-5} 9×10^{-4} 4×10^{-3} 4×10^{-3}	 1×10^{-13} 4×10^{-12} 3×10^{-8} 2×10^{-8}	 3×10^{-6} 3×10^{-5} 1×10^{-4} 1×10^{-4}
Nickel	28	^{59}Ni S I ^{63}Ni S I ^{65}Ni S I	 5×10^{-7} 8×10^{-7} 6×10^{-8} 3×10^{-7} 9×10^{-7} 5×10^{-7}	 6×10^{-3} 6×10^{-2} 8×10^{-4} 2×10^{-2} 4×10^{-3} 3×10^{-3}	 2×10^{-8} 3×10^{-8} 2×10^{-9} 1×10^{-8} 3×10^{-8} 2×10^{-8}	 2×10^{-4} 2×10^{-3} 3×10^{-5} 7×10^{-4} 1×10^{-4} 1×10^{-4}
Niobium (Columbium)	41	^{93m}Nb S I ^{95}Nb S I ^{97}Nb S I	 1×10^{-7} 2×10^{-7} 5×10^{-7} 1×10^{-7} 6×10^{-6} 5×10^{-6}	 1×10^{-2} 1×10^{-2} 3×10^{-3} 3×10^{-3} 3×10^{-2} 3×10^{-2}	 4×10^{-9} 5×10^{-9} 2×10^{-8} 3×10^{-9} 2×10^{-7} 2×10^{-7}	 4×10^{-4} 4×10^{-4} 1×10^{-4} 1×10^{-4} 9×10^{-4} 9×10^{-4}
Osmium	76	^{185}Os S I ^{191m}Os S I ^{191}Os S I ^{193}Os S I	 5×10^{-7} 5×10^{-8} 2×10^{-5} 9×10^{-6} 1×10^{-6} 4×10^{-7} 4×10^{-7} 3×10^{-7}	 2×10^{-3} 2×10^{-3} 7×10^{-2} 7×10^{-2} 5×10^{-3} 5×10^{-3} 2×10^{-3} 2×10^{-3}	 2×10^{-8} 2×10^{-9} 6×10^{-7} 3×10^{-7} 4×10^{-8} 1×10^{-8} 1×10^{-8} 9×10^{-9}	 7×10^{-5} 7×10^{-5} 3×10^{-3} 2×10^{-3} 2×10^{-4} 2×10^{-4} 6×10^{-5} 5×10^{-5}
Palladium	46	^{103}Pd S I ^{109}Pd S I	 1×10^{-6} 7×10^{-7} 6×10^{-7} 4×10^{-7}	 1×10^{-2} 8×10^{-3} 3×10^{-3} 2×10^{-3}	 5×10^{-8} 3×10^{-8} 2×10^{-8} 1×10^{-8}	 3×10^{-4} 3×10^{-4} 9×10^{-5} 7×10^{-5}

Table 19. *(Continued)*
MAXIMUM PERMISSIBLE CONCENTRATIONS OF RADIONUCLIDES IN AIR AND WATER[30]
(See Notes at End of Table)

Element	Atomic Number	Isotope*	A		B	
			In Air, µCi/ml	In Water, µCi/ml	In Air, µCi/ml	In Water, µCi/ml
Phosphorus	15	^{32}P				
		S	7×10^{-8}	5×10^{-4}	2×10^{-9}	2×10^{-5}
		I	8×10^{-8}	7×10^{-4}	3×10^{-9}	2×10^{-5}
Platinum	78	^{191}Pt				
		S	8×10^{-7}	4×10^{-3}	3×10^{-8}	1×10^{-4}
		I	6×10^{-7}	3×10^{-3}	2×10^{-8}	1×10^{-4}
		^{193m}Pt				
		S	7×10^{-6}	3×10^{-2}	2×10^{-7}	1×10^{-3}
		I	5×10^{-6}	3×10^{-2}	2×10^{-7}	1×10^{-3}
		^{197m}Pt				
		S	6×10^{-6}	3×10^{-2}	2×10^{-7}	1×10^{-3}
		I	5×10^{-6}	3×10^{-2}	2×10^{-7}	9×10^{-4}
		^{197}Pt				
		S	8×10^{-7}	4×10^{-3}	3×10^{-8}	1×10^{-4}
		I	6×10^{-7}	3×10^{-3}	2×10^{-8}	1×10^{-4}
Plutonium	94	^{238}Pu				
		S	2×10^{-12}	1×10^{-4}	7×10^{-14}	5×10^{-6}
		I	3×10^{-11}	8×10^{-4}	1×10^{-12}	3×10^{-5}
		^{239}Pu				
		S	2×10^{-12}	1×10^{-4}	6×10^{-14}	5×10^{-6}
		I	4×10^{-11}	8×10^{-4}	1×10^{-12}	3×10^{-5}
		^{240}Pu				
		S	2×10^{-12}	1×10^{-4}	6×10^{-14}	5×10^{-6}
		I	4×10^{-11}	8×10^{-4}	1×10^{-12}	3×10^{-5}
		^{241}Pu				
		S	9×10^{-11}	7×10^{-3}	3×10^{-12}	2×10^{-4}
		I	4×10^{-8}	4×10^{-2}	1×10^{-9}	1×10^{-3}
		^{242}Pu				
		S	2×10^{-12}	1×10^{-4}	6×10^{-14}	5×10^{-6}
		I	4×10^{-11}	9×10^{-4}	1×10^{-12}	3×10^{-5}
		^{243}Pu				
		S	2×10^{-6}	1×10^{-2}	6×10^{-8}	3×10^{-4}
		I	2×10^{-6}	1×10^{-2}	8×10^{-8}	3×10^{-4}
		^{244}Pu†				
		S	2×10^{-12}	1×10^{-4}	6×10^{-14}	4×10^{-6}
		I	3×10^{-11}	3×10^{-4}	1×10^{-12}	1×10^{-5}
Polonium	84	^{210}Po				
		S	5×10^{-10}	2×10^{-5}	2×10^{-11}	7×10^{-7}
		I	2×10^{-10}	8×10^{-4}	7×10^{-12}	3×10^{-5}
Potassium	19	^{42}K				
		S	2×10^{-6}	9×10^{-3}	7×10^{-8}	3×10^{-4}
		I	1×10^{-7}	6×10^{-4}	4×10^{-9}	2×10^{-5}
Praseodymium	59	^{142}Pr				
		S	2×10^{-7}	9×10^{-4}	7×10^{-9}	3×10^{-5}
		I	2×10^{-7}	9×10^{-4}	5×10^{-9}	3×10^{-5}
		^{143}Pr				
		S	3×10^{-7}	1×10^{-3}	1×10^{-8}	5×10^{-5}
		I	2×10^{-7}	1×10^{-3}	6×10^{-9}	5×10^{-5}

Table 19. *(Continued)*
MAXIMUM PERMISSIBLE CONCENTRATIONS OF RADIONUCLIDES IN AIR AND WATER[30]

(See Notes at End of Table)

Element	Atomic Number	Isotope*	A		B	
			In Air, µCi/ml	In Water, µCi/ml	In Air, µCi/ml	In Water, µCi/ml
Promethium	61	^{147}Pm				
		S	6×10^{-8}	6×10^{-3}	2×10^{-9}	2×10^{-4}
		I	1×10^{-7}	6×10^{-3}	3×10^{-9}	2×10^{-4}
		^{149}Pm				
		S	3×10^{-7}	1×10^{-3}	1×10^{-8}	4×10^{-5}
		I	2×10^{-7}	1×10^{-3}	8×10^{-9}	4×10^{-5}
Protoactinium	91	^{230}Pa				
		S	2×10^{-9}	7×10^{-3}	6×10^{-11}	2×10^{-4}
		I	8×10^{-10}	7×10^{-3}	3×10^{-11}	2×10^{-4}
		^{231}Pa				
		S	1×10^{-12}	3×10^{-5}	4×10^{-14}	9×10^{-7}
		I	1×10^{-10}	8×10^{-4}	4×10^{-12}	2×10^{-5}
		^{233}Pa				
		S	6×10^{-7}	4×10^{-3}	2×10^{-8}	1×10^{-4}
		I††	2×10^{-7}	3×10^{-3}	6×10^{-9}	1×10^{-4}
Radium	88	^{223}Ra				
		S	2×10^{-9}	2×10^{-5}	6×10^{-11}	7×10^{-7}
		I	2×10^{-10}	1×10^{-4}	8×10^{-12}	4×10^{-6}
		^{224}Ra				
		S	5×10^{-9}	7×10^{-5}	2×10^{-10}	2×10^{-6}
		I	7×10^{-10}	2×10^{-4}	2×10^{-11}	5×10^{-6}
		^{226}Ra				
		S	3×10^{-11}	4×10^{-7}	3×10^{-12}	3×10^{-8}
		I	5×10^{-11}	9×10^{-4}	2×10^{-12}	3×10^{-5}
		^{228}Ra				
		S	7×10^{-11}	8×10^{-7}	2×10^{-12}	3×10^{-8}
		I	4×10^{-11}	7×10^{-4}	1×10^{-12}	3×10^{-5}
Radon	86	^{220}Rn				
		S	3×10^{-7}	—	1×10^{-8}	—
		I	—	—	—	—
		^{222}Rn				
		S	1×10^{-7}	—	3×10^{-9}	—
		I	—	—	—	—
Rhenium	75	^{183}Re				
		S	3×10^{-6}	2×10^{-2}	9×10^{-8}	6×10^{-4}
		I	2×10^{-7}	8×10^{-3}	5×10^{-9}	3×10^{-4}
		^{186}Re				
		S	6×10^{-7}	3×10^{-3}	2×10^{-8}	9×10^{-5}
		I	2×10^{-7}	1×10^{-3}	8×10^{-9}	5×10^{-5}
		^{187}Re				
		S	9×10^{-6}	7×10^{-2}	3×10^{-7}	3×10^{-3}
		I	5×10^{-7}	4×10^{-2}	2×10^{-8}	2×10^{-3}
		^{188}Re				
		S	4×10^{-7}	2×10^{-3}	1×10^{-8}	6×10^{-5}
		I	2×10^{-7}	9×10^{-4}	6×10^{-9}	3×10^{-5}
Rhodium	45	^{103m}Rh				
		S	8×10^{-5}	4×10^{-1}	3×10^{-6}	1×10^{-2}
		I	6×10^{-5}	3×10^{-1}	2×10^{-6}	1×10^{-2}

Table 19. *(Continued)*
MAXIMUM PERMISSIBLE CONCENTRATIONS OF RADIONUCLIDES IN AIR AND WATER[30]
(See Notes at End of Table)

Element	Atomic Number	Isotope*	A		B	
			In Air, µCi/ml	In Water, µCi/ml	In Air, µCi/ml	In Water, µCi/ml
Rhodium *(cont.)*	45	^{105}Rh				
		S	8×10^{-7}	4×10^{-3}	3×10^{-8}	1×10^{-4}
		I	5×10^{-7}	3×10^{-3}	2×10^{-8}	1×10^{-4}
Rubidium	37	^{86}Rb				
		S	3×10^{-7}	2×10^{-3}	1×10^{-8}	7×10^{-5}
		I	7×10^{-8}	7×10^{-4}	2×10^{-9}	2×10^{-5}
		^{87}Rb				
		S	5×10^{-7}	3×10^{-3}	2×10^{-8}	1×10^{-4}
		I	7×10^{-8}	5×10^{-3}	2×10^{-9}	2×10^{-4}
Ruthenium	44	^{97}Ru				
		S	2×10^{-6}	1×10^{-2}	8×10^{-8}	4×10^{-4}
		I	2×10^{-6}	1×10^{-2}	6×10^{-8}	3×10^{-4}
		^{103}Ru				
		S	5×10^{-7}	2×10^{-3}	2×10^{-8}	8×10^{-5}
		I	8×10^{-8}	2×10^{-3}	3×10^{-9}	8×10^{-5}
		^{105}Ru				
		S	7×10^{-7}	3×10^{-3}	2×10^{-8}	1×10^{-4}
		I	5×10^{-7}	3×10^{-3}	2×10^{-8}	1×10^{-4}
		^{106}Ru				
		S	8×10^{-8}	4×10^{-4}	3×10^{-9}	1×10^{-5}
		I	6×10^{-9}	3×10^{-4}	2×10^{-10}	1×10^{-5}
Samarium	62	^{147}Sm				
		S	7×10^{-11}	2×10^{-3}	2×10^{-12}	6×10^{-5}
		I	3×10^{-10}	2×10^{-3}	9×10^{-12}	7×10^{-5}
		^{151}Sm				
		S	6×10^{-8}	1×10^{-2}	2×10^{-9}	4×10^{-4}
		I	1×10^{-7}	1×10^{-2}	5×10^{-9}	4×10^{-4}
		^{153}Sm				
		S	5×10^{-7}	2×10^{-3}	2×10^{-8}	8×10^{-5}
		I	4×10^{-7}	2×10^{-3}	1×10^{-8}	8×10^{-5}
Scandium	21	^{46}Sc				
		S	2×10^{-7}	1×10^{-3}	8×10^{-9}	4×10^{-5}
		I	2×10^{-8}	1×10^{-3}	8×10^{-10}	4×10^{-5}
		^{47}Sc				
		S	6×10^{-7}	3×10^{-3}	2×10^{-8}	9×10^{-5}
		I	5×10^{-7}	3×10^{-3}	2×10^{-8}	9×10^{-5}
		^{48}Sc				
		S	2×10^{-7}	8×10^{-4}	6×10^{-9}	3×10^{-5}
		I	1×10^{-7}	8×10^{-4}	5×10^{-9}	3×10^{-5}
Selenium	34	^{75}Se				
		S	1×10^{-6}	9×10^{-3}	4×10^{-8}	3×10^{-4}
		I	1×10^{-7}	8×10^{-3}	4×10^{-9}	3×10^{-4}
Silicon	14	^{31}Si				
		S	6×10^{-6}	3×10^{-2}	2×10^{-7}	9×10^{-4}
		I	1×10^{-6}	6×10^{-3}	3×10^{-8}	2×10^{-4}
Silver	47	^{105}Ag				
		S	6×10^{-7}	3×10^{-3}	2×10^{-8}	1×10^{-4}
		I	8×10^{-8}	3×10^{-3}	3×10^{-9}	1×10^{-4}

Table 19. (*Continued*)
MAXIMUM PERMISSIBLE CONCENTRATIONS OF RADIONUCLIDES IN AIR AND WATER[30]

(See Notes at End of Table)

Element	Atomic Number	Isotope*	A		B	
			In Air, µCi/ml	In Water, µCi/ml	In Air, µCi/ml	In Water, µCi/ml
Silver (*cont.*)	47	^{110m}Ag				
		S	2×10^{-7}	9×10^{-4}	7×10^{-9}	3×10^{-5}
		I	1×10^{-8}	9×10^{-4}	3×10^{-10}	3×10^{-5}
		^{111}Ag				
		S	3×10^{-7}	1×10^{-3}	1×10^{-8}	4×10^{-5}
		I	2×10^{-7}	1×10^{-3}	8×10^{-9}	4×10^{-5}
Sodium	11	^{22}Na				
		S	2×10^{-7}	1×10^{-3}	6×10^{-9}	4×10^{-5}
		I	9×10^{-9}	9×10^{-4}	3×10^{-10}	3×10^{-5}
		^{24}Na				
		S	1×10^{-6}	6×10^{-3}	4×10^{-8}	2×10^{-4}
		I††	1×10^{-7}	8×10^{-4}	5×10^{-9}	3×10^{-5}
Strontium	38	^{85m}Sr				
		S	4×10^{-5}	2×10^{-1}	1×10^{-6}	7×10^{-3}
		I	3×10^{-5}	2×10^{-1}	1×10^{-6}	7×10^{-3}
		^{85}Sr				
		S	2×10^{-7}	3×10^{-3}	8×10^{-9}	1×10^{-4}
		I	1×10^{-7}	5×10^{-3}	4×10^{-9}	2×10^{-4}
		^{89}Sr				
		S	3×10^{-8}	3×10^{-4}	3×10^{-10}	3×10^{-6}
		I	4×10^{-8}	8×10^{-4}	1×10^{-9}	3×10^{-5}
		^{90}Sr				
		S	1×10^{-9}‡	1×10^{-5}	3×10^{-11}	3×10^{-7}
		I	5×10^{-9}	1×10^{-3}	2×10^{-10}	4×10^{-5}
		^{91}Sr				
		S	4×10^{-7}	2×10^{-3}	2×10^{-8}	7×10^{-5}
		I	3×10^{-7}	1×10^{-3}	9×10^{-9}	5×10^{-5}
		^{92}Sr				
		S	4×10^{-7}	2×10^{-3}	2×10^{-8}	7×10^{-5}
		I	3×10^{-7}	2×10^{-3}	1×10^{-8}	6×10^{-5}
Sulfur	16	^{35}S				
		S	3×10^{-7}	2×10^{-3}	9×10^{-9}	6×10^{-5}
		I	3×10^{-7}	8×10^{-3}	9×10^{-9}	3×10^{-4}
Tantalum	73	^{182}Ta				
		S	4×10^{-8}	1×10^{-3}	1×10^{-9}	4×10^{-5}
		I	2×10^{-8}	1×10^{-3}	7×10^{-10}	4×10^{-5}
Technetium	43	^{96m}Tc				
		S	8×10^{-5}	4×10^{-1}	3×10^{-6}	1×10^{-2}
		I	3×10^{-5}	3×10^{-1}	1×10^{-6}	1×10^{-2}
		^{96}Tc				
		S	6×10^{-7}	3×10^{-3}	2×10^{-8}	1×10^{-4}
		I	2×10^{-7}	1×10^{-3}	8×10^{-9}	5×10^{-5}
		^{97m}Tc				
		S	2×10^{-6}	1×10^{-2}	8×10^{-8}	4×10^{-4}
		I	2×10^{-7}	5×10^{-3}	5×10^{-9}	2×10^{-4}
		^{97}Tc				
		S	1×10^{-5}	5×10^{-2}	4×10^{-7}	2×10^{-3}
		I	3×10^{-7}	2×10^{-2}	1×10^{-8}	8×10^{-4}
		^{99m}Tc				
		S	4×10^{-5}	2×10^{-1}	1×10^{-6}	6×10^{-3}

Table 19. (*Continued*)
MAXIMUM PERMISSIBLE CONCENTRATIONS OF RADIONUCLIDES IN AIR AND WATER[30]

(See Notes at End of Table)

Element	Atomic Number	Isotope*	A		B	
			In Air, µCi/Ml	In Water, µCi/ml	In Air, µCi/ml	In Water, µCi/ml
Technetium (*cont.*)	43	^{99m}Tc I	1×10^{-5}	8×10^{-2}	5×10^{-7}	3×10^{-3}
		^{99}Tc S	2×10^{-6}	1×10^{-2}	7×10^{-8}	3×10^{-4}
		I	6×10^{-8}	5×10^{-3}	2×10^{-9}	3×10^{-4}
Tellurium	52	^{125m}Te S	4×10^{-7}	5×10^{-3}	1×10^{-8}	2×10^{-4}
		I	1×10^{-7}	3×10^{-3}	4×10^{-9}	1×10^{-4}
		^{127m}Te S	1×10^{-7}	2×10^{-3}	5×10^{-9}	6×10^{-5}
		I	4×10^{-8}	2×10^{-3}	1×10^{-9}	5×10^{-5}
		^{127}Te S	2×10^{-6}	8×10^{-3}	6×10^{-8}	3×10^{-4}
		I	9×10^{-7}	5×10^{-3}	3×10^{-8}	2×10^{-4}
		^{129m}Te S	8×10^{-8}	1×10^{-3}	3×10^{-9}	3×10^{-5}
		I	3×10^{-8}	6×10^{-4}	1×10^{-9}	2×10^{-5}
		^{129}Te S	5×10^{-6}	2×10^{-2}	2×10^{-7}	8×10^{-4}
		I	4×10^{-6}	2×10^{-2}	1×10^{-7}	8×10^{-4}
		^{131m}Te S	4×10^{-7}	2×10^{-3}	1×10^{-8}	6×10^{-5}
		I	2×10^{-7}	1×10^{-3}	6×10^{-9}	4×10^{-5}
		^{132}Te S	2×10^{-7}	9×10^{-4}	7×10^{-9}	3×10^{-5}
		I	1×10^{-7}	6×10^{-4}	4×10^{-9}	2×10^{-5}
Terbium	65	^{160}Tb S	1×10^{-7}	1×10^{-3}	3×10^{-9}	4×10^{-5}
		I	3×10^{-8}	1×10^{-3}	1×10^{-9}	4×10^{-5}
Thallium	81	^{200}Tl S	3×10^{-6}	1×10^{-2}	9×10^{-8}	4×10^{-4}
		I	1×10^{-6}	7×10^{-3}	4×10^{-8}	2×10^{-4}
		^{201}Tl S	2×10^{-6}	9×10^{-3}	7×10^{-8}	3×10^{-4}
		I	9×10^{-7}	5×10^{-3}	3×10^{-8}	2×10^{-4}
		^{202}Tl S	8×10^{-7}	4×10^{-3}	3×10^{-8}	1×10^{-4}
		I	2×10^{-7}	2×10^{-3}	8×10^{-9}	7×10^{-5}
		^{204}Tl S	6×10^{-7}	3×10^{-3}	2×10^{-8}	1×10^{-4}
		I	3×10^{-8}	2×10^{-3}	9×10^{-10}	6×10^{-5}
Thorium	90	^{228}Th S	9×10^{-12}	2×10^{-4}	3×10^{-13}	7×10^{-6}
		I	6×10^{-12}	4×10^{-4}	2×10^{-13}	1×10^{-5}
		^{230}Th S	2×10^{-12}	5×10^{-5}	8×10^{-14}	2×10^{-6}
		I	1×10^{-11}	9×10^{-4}	3×10^{-13}	3×10^{-5}

Table 19. (*Continued*)
MAXIMUM PERMISSIBLE CONCENTRATIONS OF RADIONUCLIDES IN AIR AND WATER[30]

(See Notes at End of Table)

Element	Atomic Number	Isotope*	A In Air, µCi/ml	A In Water, µCi/ml	B In Air, µCi/ml	B In Water, µCi/ml
Thorium (*cont.*)	90	^{232}Th				
		S	3×10^{-11}	5×10^{-5}	1×10^{-12}	2×10^{-6}
		I	3×10^{-11}	1×10^{-3}	1×10^{-12}	4×10^{-5}
		^{natural}Th				
		S	3×10^{-11}	3×10^{-5}	1×10^{-12}	1×10^{-6}
		I	3×10^{-11}	3×10^{-4}	1×10^{-12}	1×10^{-5}
		^{234}Th				
		S	6×10^{-8}	5×10^{-4}	2×10^{-9}	2×10^{-5}
		I	3×10^{-8}	5×10^{-4}	1×10^{-9}	2×10^{-5}
Thulium	69	^{170}Tm				
		S	4×10^{-8}	1×10^{-3}	1×10^{-9}	5×10^{-5}
		I	3×10^{-8}	1×10^{-3}	1×10^{-9}	5×10^{-5}
		^{171}Tm				
		S	1×10^{-7}	1×10^{-2}	4×10^{-9}	5×10^{-4}
		I	2×10^{-7}	1×10^{-2}	8×10^{-9}	5×10^{-4}
Tin	50	^{113}Sn				
		S	4×10^{-7}	2×10^{-3}	1×10^{-8}	9×10^{-5}
		I	5×10^{-8}	2×10^{-3}	2×10^{-9}	8×10^{-5}
		^{125}Sn				
		S	1×10^{-7}	5×10^{-4}	4×10^{-9}	2×10^{-5}
		I	8×10^{-8}	5×10^{-4}	3×10^{-9}	2×10^{-5}
Tungsten (Wolfram)	74	^{181}W				
		S	2×10^{-6}	1×10^{-2}	8×10^{-8}	4×10^{-4}
		I	1×10^{-7}	1×10^{-2}	4×10^{-9}	3×10^{-4}
		^{185}W				
		S	8×10^{-7}	4×10^{-3}	3×10^{-8}	1×10^{-4}
		I	1×10^{-7}	3×10^{-3}	4×10^{-9}	1×10^{-4}
		^{187}W				
		S	4×10^{-7}	2×10^{-3}	2×10^{-8}	7×10^{-5}
		I	3×10^{-7}	2×10^{-3}	1×10^{-8}	6×10^{-5}
Uranium	92	^{230}U				
		S	3×10^{-10}	1×10^{-4}	1×10^{-11}	5×10^{-6}
		I	1×10^{-10}	1×10^{-4}	4×10^{-12}	5×10^{-6}
		^{232}U				
		S	1×10^{-10}	8×10^{-4}	3×10^{-12}	3×10^{-5}
		I	3×10^{-11}	8×10^{-4}	9×10^{-13}	3×10^{-5}
		^{233}U				
		S	5×10^{-10}	9×10^{-4}	2×10^{-11}	3×10^{-5}
		I	1×10^{-10}	9×10^{-4}	4×10^{-12}	3×10^{-5}
		^{234}U				
		S	6×10^{-10}	9×10^{-4}	2×10^{-11}	3×10^{-5}
		I	1×10^{-10}	9×10^{-4}	4×10^{-12}	3×10^{-5}
		^{235}U				
		S	5×10^{-10}	8×10^{-4}	2×10^{-11}	3×10^{-5}
		I	1×10^{-10}	8×10^{-4}	4×10^{-12}	3×10^{-5}
		^{236}U				
		S	6×10^{-10}	1×10^{-3}	2×10^{-11}	3×10^{-5}
		I	1×10^{-10}	1×10^{-3}	4×10^{-12}	3×10^{-5}

Table 19. (Continued)
MAXIMUM PERMISSIBLE CONCENTRATIONS OF RADIONUCLIDES IN AIR AND WATER[30]

(See Notes at End of Table)

Element	Atomic Number	Isotope*	A		B	
			In Air, µCi/ml	In Water, µCi/ml	In Air, µCi/ml	In Water, µCi/ml
Uranium (*cont.*)	92	^{238}U				
		S	7×10^{-11}	1×10^{-3}	3×10^{-12}	4×10^{-5}
		I	1×10^{-10}	1×10^{-3}	5×10^{-12}	4×10^{-5}
		^{240}U†				
		S	2×10^{-7}	1×10^{-3}	8×10^{-9}	3×10^{-5}
		I	2×10^{-7}	1×10^{-3}	6×10^{-9}	3×10^{-5}
		natural U				
		S	7×10^{-11}	5×10^{-4}	3×10^{-12}	2×10^{-5}
		I	6×10^{-11}	5×10^{-4}	2×10^{-12}	2×10^{-5}
Vanadium	23	^{48}V				
		S	2×10^{-7}	9×10^{-4}	6×10^{-9}	3×10^{-5}
		I	6×10^{-8}	8×10^{-4}	2×10^{-9}	3×10^{-5}
Xenon	54	^{131m}Xe				
		Sub	2×10^{-5}	—	4×10^{-7}	—
		^{133}Xe				
		Sub	1×10^{-5}	—	3×10^{-7}	—
		^{133m}Xe†				
		Sub	1×10^{-6}	—	3×10^{-7}	—
		^{135}Xe				
		Sub	4×10^{-6}	—	1×10^{-7}	—
Ytterbium	70	^{175}Yb				
		S	7×10^{-7}	3×10^{-3}	2×10^{-8}	1×10^{-4}
		I	6×10^{-7}	3×10^{-3}	2×10^{-8}	1×10^{-4}
Yttrium	39	^{90}Y				
		S	1×10^{-7}	6×10^{-4}	4×10^{-9}	2×10^{-5}
		I	1×10^{-7}	6×10^{-4}	3×10^{-9}	2×10^{-5}
		^{91m}Y				
		S	2×10^{-5}	1×10^{-1}	8×10^{-7}	3×10^{-3}
		I	2×10^{-5}	1×10^{-1}	6×10^{-7}	3×10^{-3}
		^{91}Y				
		S	4×10^{-8}	8×10^{-4}	1×10^{-9}	3×10^{-5}
		I	3×10^{-8}	8×10^{-4}	1×10^{-9}	3×10^{-5}
		^{92}Y				
		S	4×10^{-7}	2×10^{-3}	1×10^{-8}	6×10^{-5}
		I	3×10^{-7}	2×10^{-3}	1×10^{-8}	6×10^{-5}
		^{93}Y				
		S	2×10^{-7}	8×10^{-4}	6×10^{-9}	3×10^{-5}
		I	1×10^{-7}	8×10^{-4}	5×10^{-9}	3×10^{-5}
Zinc	30	^{65}Zn				
		S	1×10^{-7}	3×10^{-3}	4×10^{-9}	1×10^{-4}
		I	6×10^{-8}	5×10^{-3}	2×10^{-9}	2×10^{-4}
		^{69m}Zn				
		S	4×10^{-7}	2×10^{-3}	1×10^{-8}	7×10^{-5}
		I	3×10^{-7}	2×10^{-3}	1×10^{-8}	6×10^{-5}
		^{69}Zn				
		S	7×10^{-6}	5×10^{-2}	2×10^{-7}	2×10^{-3}
		I	9×10^{-6}	5×10^{-2}	3×10^{-7}	2×10^{-3}

Table 19. (*Continued*)

MAXIMUM PERMISSIBLE CONCENTRATIONS OF RADIONUCLIDES IN AIR AND WATER[30]

(See Notes at End of Table)

Element	Atomic Number	Isotope*	A		B	
			In Air, µCi/ml	In Water, µCi/ml	In Air, µCi/ml	In Water, µCi/ml
Zirconium	40	^{93}Zr				
		S	1×10^{-7}	2×10^{-2}	4×10^{-9}	8×10^{-4}
		I	3×10^{-7}	2×10^{-2}	1×10^{-8}	8×10^{-4}
		^{95}Zr				
		S	1×10^{-7}	2×10^{-3}	4×10^{-9}	6×10^{-5}
		I	3×10^{-8}	2×10^{-3}	1×10^{-9}	6×10^{-5}
		^{97}Zr				
		S	1×10^{-7}	5×10^{-4}	4×10^{-9}	2×10^{-5}
		I	9×10^{-8}	5×10^{-4}	3×10^{-9}	2×10^{-5}
Any single radionuclide not listed above with decay mode other than alpha emission or spontaneous fission and with radioactive half-life less than 2 hours†						
		Sub	1×10^{-6}	—	3×10^{-9}	—
Any single radionuclide not listed above with decay mode other than alpha emission or spontaneous fission and with radioactive half-life greater than 2 hours†			3×10^{-9}	9×10^{-5}	1×10^{-10}	3×10^{-6}
Any single radionuclide not listed above that decays by alpha emission or spontaneous fission†			6×10^{-13}	4×10^{-7}	2×10^{-14}	3×10^{-8}

* S = soluble; I = insoluble; "Sub" means that values given are for submersion in a semispherical infinite cloud of airborne material.‡

† Added 30 FR 15801.

‡ Revised 30 FR 15801.

†† Revised 29 FR 14434.

NOTE: In any case where there is a mixture in air or water of more than one radionuclide, the limiting values for purposes of this table should be derived as follows:

1. If the identity and concentration of each radionuclide in the mixture are known, the limiting values should be derived as follows: determine for each radionuclide in the mixture the ratio between the quantity present in the mixture and the limit otherwise established in this table for the specific radionuclide when not in a mixture. The sum of such ratios for all the radionuclides in the mixture may not exceed-1 (i.e., unity).

Example: If radionuclides A, B, and C are present in concentrations C_A, C_B, and C_C, and if the applicable MPC's are MPC_A, MPC_B, and MPC_C respectively, then the concentrations shall be limited so that the following relationship exists:

$$\frac{C_A}{MPC_A} + \frac{C_B}{MPC_B} + \frac{C_C}{MPC_C} \leq 1$$

2. ‡ If either the identity or the concentration of any radionuclide in the mixture is not known, the limiting values for purposes of this table shall be:

   a) for purposes of Section A, Column 1—6×10^{-13};
   b) for purposes of Section A, Column 2—4×10^{-7};
   c) for purposes of Section B, Column 1—2×10^{-14};
   d) for purposes of Section B, Column 2—3×10^{-8}.

3. (26 FR 11046) If any of the conditions specified below are met, the corresponding values specified below may be used in lieu of those specified in paragraph 2 above:

   a) if the identity of each radionuclide in the mixture is known, but the concentration of one or more of the radionuclides in the mixture is not known, the concentration limit for the mixture is the limit specified in this table for the radionuclide in the mixture having the lowest concentration limit; or

   b) if the identity of each radionuclide in the mixture is not known, but it is known that certain radionuclides specified in this table are not present in the mixture, the concentration limit for the mixture is the lowest concentration limit specified in this table for any radionuclide that is not known to be absent from the mixture; or

   c) (30 FR 15801).

Table 19. (*Continued*)
MAXIMUM PERMISSIBLE CONCENTRATIONS OF RADIONUCLIDES IN AIR AND WATER[30]

Element	Atomic Number	Isotope*	A		B	
			In Air, μCi/ml	In Water, μCi/ml	In Air, μCi/ml	In Water, μCi/ml
		If it is known that ^{90}Sr, ^{125}I, ^{126}I, ^{129}I, ^{131}I, (^{133}I, Section B only), ^{210}Pb, ^{210}Po, ^{211}At, ^{223}Ra, ^{224}Ra, ^{226}Ra, ^{227}Ac, ^{228}Ra, ^{230}Th, ^{231}Pa, ^{232}Th, ^{natural}Th, ^{248}Cm, ^{254}Cf, and ^{256}Fm are not present	—	9×10^{-5}	—	3×10^{-6}
		If it is known that ^{90}Sr, ^{125}I, ^{126}I, ^{129}I, (^{131}I, ^{133}I, Section B only), ^{210}Pb, ^{210}Po, ^{223}Ra, ^{226}Ra, ^{228}Ra, ^{231}Pa, ^{natural}Th, ^{248}Cm, ^{254}Cf, and ^{256}Fm are not present	—	6×10^{-5}	—	2×10^{-6}
		If it is known that ^{90}Sr, ^{129}I, (^{125}I, ^{126}I, ^{131}I, Section B only), ^{210}Pb, ^{226}Ra, ^{228}Ra, ^{248}Cm, and ^{254}Cf are not present	—	2×10^{-5}	—	6×10^{-7}
		If it is known that (^{129}I, Section B only), ^{226}Ra, and ^{228}Ra are not present	—	3×10^{-6}	—	1×10^{-7}
		If it is known that alpha emitters and ^{90}Sr, ^{129}I, ^{210}Pb, ^{227}Ac, ^{228}Ra, ^{230}Pa, ^{241}Pu, and ^{249}Bk are not present	3×10^{-9}	—	1×10^{-10}	—
		If it is known that alpha emitters and ^{210}Pb, ^{227}Ac, ^{228}Ra, and ^{241}Pu are not present	3×10^{-10}	—	1×10^{-11}	—
		If it is known that alpha emitters and ^{227}Ac are not present	3×10^{-11}	—	1×10^{-12}	—
		If it is known that ^{227}Ac, ^{230}Th, ^{231}Pa, ^{238}Pu, ^{239}Pu, ^{240}Pu, ^{242}Pu, ^{244}Pu, ^{248}Cm, ^{249}Cf, and ^{251}Cf are not present	3×10^{-12}	—	1×10^{-13}	—

4. (25 FR 13952) If the mixture of radionuclides consists of uranium and its daughter products in ore dust prior to chemical processing of the uranium ore, the values specified below may be used in lieu of those determined in accordance with paragraph 1 above or those specified in paragraphs 2 and 3 above:

    a) for purposes of Section A, Column 1—1×10^{-10} μCi/ml gross alpha activity; or 2.5×10^{-11} μCi/ml natural uranium; or 75 μg/m^3 air natural uranium;

    b) for purposes of Section B, Column 1—3×10^{-12} μCi/ml gross alpha activity; or 8×10^{-13} μCi/ml natural uranium; or 3 μg/m^3 air natural uranium.

5. (26 FR 11046) For purposes of this note, a radionuclide may be considered as not present in a mixture if a) the ratio of the concentration of that radionuclide in the mixture (C_A) to the concentration limit for that radionuclide specified in Section B of this table (MPC$_A$) does not exceed 1/10 (i.e., C_A/MPC$_A \leq 1/10$), and b) the sum of such ratios for all the radionuclides considered as not present in the mixture does not exceed 1/4 (i.e., C_A/MPC$_A$ + C_B/MPC$_B$ + $\cdots \leq 1/4$).

Reprinted by permission of the U.S. Atomic Energy Commission from Title 10, *Code of Federal Regulations*, Part 20, Standards for Protection Against Radiation. Revised August 1966.

be found by consulting *Health Physics*, *Nucleonics* (before 1967), and other technical and scientific journals. Professional assistance in planning a facility for handling radioactive materials and establishing a radiation-safety program that will meet regulatory requirements may be obtained by writing to the authorities named below for lists of certified health physicists, industrial hygienists, or radiological physicists:

    Mr. Russell F. Cowing
    Executive Secretary
    Health Physics Society
    194 Pilgrim Road
    Boston, Massachusetts 02215

Dr. Henry F. Smyth, Secretary
American Board of Industrial Hygiene
c/o Graduate School of Public Health
University of Pittsburgh
Pittsburgh, Pennsylvania 15213

Dr. James Kereiakes
Radioisotope Laboratory
General Hospital
Cincinnati, Ohio 45229

RECOMMENDATIONS FOR LIMITING EXPOSURES IN EMERGENCIES

Recommendations for limiting the radiation exposure of employees as well as members of the public when unplanned and unexpected situations arise that release radioactive materials to uncontrolled areas have been promulgated by various national and international organizations, such as the ICRP, NCRP, FRC, and the British Medical Research Council. Some of the recommendations have been summarized in various texts.[21,27,33-36] The calculations used to estimate the acceptable emergency exposures to inhalation or land contamination by gross fission-product mixtures have been recalculated with more recent data[37] and more conservative biological assumptions. Many considerations are required in an emergency in order to balance the risk of exposure of the individual against the need for some emergency action to save either life or property, and thus comprehensive discussion of this subject within the scope of this handbook is difficult. Generally, additional references containing summarized data needed for making emergency decisions should be on hand.[21,23,31,33,35-38]

Morgan[1] summarized some limits for "planned special exposures and emergency exposures", giving the basic dose limits of the various organs of the body in situations where there are good reasons for allowing persons to exceed the limits recommended in previous sections of this chapter. The summarized recommendations, taken from ICRP,[13] include:

a) a limit of 2 R_{50} committed in any single event to any body organ for a planned special exposure, where R_{50} represents the annual permissible dose equivalent for the respective organ under consideration;

b) a limit of 5 R_{50} committed in a lifetime to any body organ in planned radiation exposures;

c) a maximum permissible intake of any radionuclide for planned special exposures corresponding to the intake that would result from breathing at the MPC for 2 years;

d) a maximum permissible intake from all planned special exposures equivalent to breathing a nuclide at MPC for 5 years;

e) in addition, planned special exposures are not permitted if a single exposure exceeding $R_{50}/2$ has been received in the previous 12 months or if the worker previously—at any time—received an abnormal exposure exceeding 5 R_{50}.

Also, planned special exposures "are not permitted to organs of reproductive capacity". They are not permitted to gonads, total body, or red bone marrow if—as a consequence—the individual's cumulative lifetime limit of $5(N - 18)$ rem would be exceeded.

Thus, the external planned special exposure limits in a single event for the respective body organs, rounded off to R_{50}, would be as follows.

1. Gonads, total body, and red bone marrow: 2 R_{50}, if taken as 12 rem, would correspond to the routinely permitted exposure if the individual has not exceeded $5(N - 18)$, and 5 R_{50} (equivalent to 25 rem) for planned special exposures over a lifetime.

2. Thyroid, skin, and bone: $R_{50} = 30$ rem, $2\ R_{50} = 60$ rem, and $5\ R_{50} = 150$ rem.
3. Hands, forearm, feet, and ankles: $R_{50} = 75$ rem, $2\ R_{50} = 150$ rem, and $5\ R_{50} = 375$ rem.
4. For all other body organs (including lens of the eye): $R_{50} = 15$ rem, $2\ R_{50} = 30$ rem, and $5\ R_{50} = 75$ rem.

After inhalation of fission-product mixtures, the revised acceptable emergency doses (AED's) chosen by Brodsky[37] are closer to those given above than to the previous values used by Cowan and Kuper[39] in the calculations quoted in NCRP Report No. 29.[31] The later calculations[37] should result in estimates of acceptable emergency exposures in curie-seconds per cubic meter, or in curies inhaled, that would "not be likely to produce clinically observable injury either acutely or after many years following an incident."[37] Since a considerable degree of synergism is known to occur between radiation exposures to different body organs, the individual-organ AED's were further chosen so that the fractional AED's contributed to separate parts of the body, as developed by the various nuclides, could be assumed to be roughly additive in determining the total fraction or multiple number of AED's contributed by a particular fission-product mixture.

Table 20 shows the comparison between the resulting dose estimates in rem/(curie-second/m^3) for various exposures as calculated by Cowan and Kuper[39] and in the revised paper (1 curie-second/m^3 is equivalent to an inhalation of 280 µCi of activity for standard man). Since the revised calculations indicated considerably higher exposures, only these results will be presented here for use in estimating the maximum permissible intakes of fission products.

Table 20.
EXPECTED EFFECTS OF EXPOSURE TO 1.0 Ci-sec/m^3 IN A RADIOACTIVE CLOUD CONTAINING MIXED FISSION PRODUCTS FROM 180-DAY OPERATION AND 2-DAY DECAY[37]

Type of Dose	Dose in rem/Ci-sec-m^{-3}	
	NCRP Report No. 29 Cowan and Kuper 1957	Proposed Revision A. Brodsky 1964
External gamma dose	0.28 R	0.28 R (for infinite cloud)
Lung beta dose	0.24 R	36 rem over several months
Bone dose from ^{89}Sr + ^{90}Sr	0.69 R	7 rem within several months, 23 rem within 50 years
Bone dose from ^{144}Ce + ^{144}Pr	0.53 R	8.3 rem within several months
Thyroid dose from radioisotopes of iodine	0.25 R	10 rem within several weeks
Dose to intestinal tract	0.28 R	7.2 rem within 2 days

From: Brodsky, A., Criteria for Acute Exposure to Mixed Fission-Product Aerosols, *Health Phys.*, 11: 1028, 1965. Copyright Pergamon Press, New York.

Figure 11 presents the resulting estimates of the number of multiples of an accepted emergency dose for the inhalation of various fission-product mixtures in relation to various reactor fuel elements, irradiation times, and various decay times after removal of the gross fission products from the reactor. In most real situations, of course, fractionation among fission products will occur, and the contributions from particular nuclides will have to be reduced according to the fractionation of each nuclide released in an accident.

Table 21 presents the estimated gross effects of exposure to full fission-product release for a 180-day operation and short decay in comparison with exposure to the release of all volatiles, but only 1 percent of strontium and 50 percent of other fission products. In order to determine appropriate permissible intakes for other fractionation conditions, the respective multiples of AED of each nuclide must be calculated from the known fraction of the respective nuclide in the inhaled mixture as compared to its amount in the original nuclear fuel.

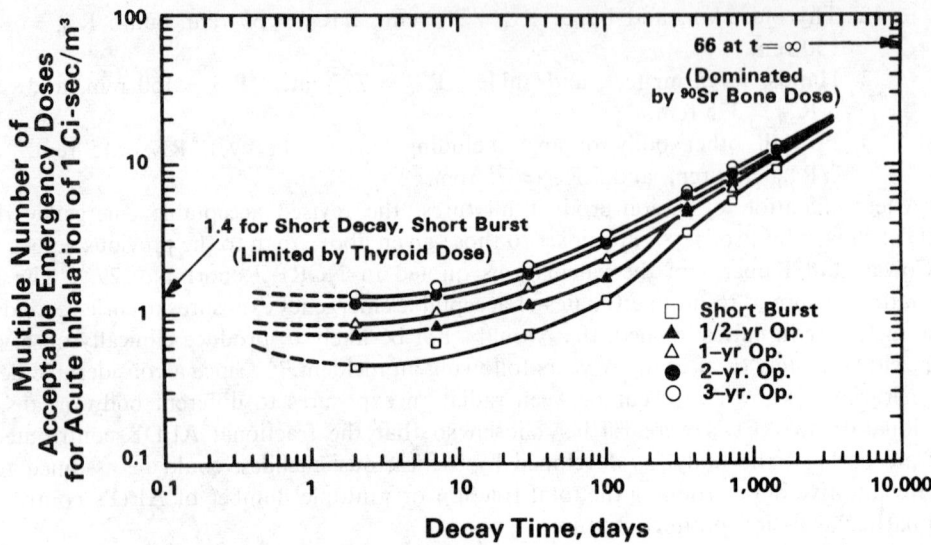

FIGURE 11.
Acceptable Emergency Doses of Gross Fission Products
for Acute Inhalation.[37]

(From: Brodsky, A., Criteria for Acute Exposure to Mixed Fission-Product Aerosols, *Health Phys.*, 11: 1027, 1965. Copyright Pergamon Press, New York.)

Figure 11 shows that an equivalent acceptable emergency dose does not differ from 1 Ci-sec/m³ by more than a factor of 2 for various operating times when the decay time is less than 30 days. Tables 22 and 23 present information on the relative exposure of separate parts of the body. Actually, the G.I.-tract dose per curie of fission product inhaled does not vary very strongly with the fission-product composition.[37] However, calculations for other fission-product mixtures may be made for the G.I. tract as well as for other organs, using the summarized fission-product inventory for reactor operations and for criticality bursts given in Tables 24 and 25. An interesting finding from these tables was that the sum of the activities of the biologically more important fission products for short criticality bursts or nuclear detonations follows the Way-Wigner total-beta-activity formula within a factor of 2 over decay times ranging from about 3 days to 2,000 days. This formula gives the total beta activity as $350\ t^{-1/2}$ curies per 100 megawatt-seconds.[37]

Table 21. EFFECTS OF FISSION-PRODUCT CLOUD EXPOSURES[37]

Category	Exposure to Full Fission-Product Release (Ci-sec/m³) for 180-Day Operation and Short Decay		Exposure to Release of All Volatiles, 1% Strontium, and 50% of Other Fission Products	
	Ref. 39	A. Brodsky	Ref. 39	A. Brodsky
A. Lethal exposure (possible)	>400	>10	>350–80	>10
B. Illness likely	400–90	10–5	350–80	10–5
C. Injury unlikely, but some expense may be incurred	90–10	5–1	80–10	5–2
D. Injury or expense unlikely	< 10	< 1	< 10	< 2

From: Brodsky, A., Criteria for Acute Exposure to Mixed Fission-Product Aerosols, *Health Phys.*, 11: 1028, 1965. Copyright Pergamon Press, New York.

Table 22.
MULTIPLES OF ACCEPTABLE EMERGENCY DOSES FOR INHALATION OF 1 Ci-sec/m^3

Mode of Exposure	Acceptable Emergency Dose Total	2-Day Decay	7-Day Decay	30-Day Decay	100-Day Decay	365-Day Decay	730-Day Decay	1460-Day Decay
1/2-Year Operation								
External gamma	25 R	0.0112	0.0112	0.0112	0.0112	0.0112	0.0112	0.0112
Lung beta	300 rem	0.1200	0.1340	0.1610	0.2290	0.5500	0.6450	0.5660
^{90}Sr bone	150 rem	0.1070	0.1330	0.2400	0.5670	2.5930	5.4470	10.6000
^{89}Sr bone	50 rem	0.1410	0.1670	0.2190	0.1990	0.0250	0.0004	—
^{144}Ce-^{144}Pr bone	50 rem	0.1660	0.2070	0.3440	0.6460	1.5840	1.4130	0.4900
Pu lung	150 rem	0.0080	0.0080	0.0080	0.0080	0.0080	0.0080	0.0080
I thyroid	300 rem	0.0330	0.0350	0.0100	—	—	—	—
G.I. tract	50 rem	0.1430	0.1450	0.1430	0.1210	0.0840	0.0830	0.1100
Total		0.7292	0.8402	1.1362	1.7812	4.8552	7.6076	11.7852
1-Year Operation								
External gamma	25 R	0.0112	0.0112	0.0112	0.0112	0.0112	0.0112	0.0112
Lung beta	300 rem	0.1060	0.1457	0.2107	0.2910	0.6284	0.6670	0.5740
^{90}Sr bone	150 rem	0.1870	0.2270	0.3670	0.7670	2.6800	5.2800	10.4800
^{89}Sr bone	50 rem	0.1400	0.1630	0.1940	0.1570	0.0150	0.0002	—
^{144}Ce-^{144}Pr bone	50 rem	0.2480	0.3030	0.5680	0.9490	1.7680	1.4780	0.5250
Pu lung	150 rem	0.0170	0.0170	0.0170	0.0170	0.0170	0.0170	0.0170
I thyroid	300 rem	0.0300	0.0320	0.0080	—	—	—	—
G.I. tract	50 rem	0.1410	0.1420	0.1370	0.1190	0.0795	0.0840	0.1110
Total		0.8802	1.0409	1.5129	2.3112	5.1991	7.5374	11.7182
2-Year Operation								
External gamma	25 R	0.0112	0.0112	0.0112	0.0112	0.0112	0.0112	0.0112
Lung beta	300 rem	0.1451	0.1690	0.2200	0.3080	0.6020	0.6126	0.5630
^{90}Sr bone	150 rem	0.3400	0.4130	0.6600	1.3270	3.9870	6.9400	11.5330
^{89}Sr bone	50 rem	0.1340	0.1540	0.1820	0.1410	0.0115	—	—
^{144}Ce-^{144}Pr bone	50 rem	0.3250	0.3940	0.5920	0.9450	1.5130	1.1160	0.3330
Pu lung	150 rem	0.0350	0.0350	0.0350	0.0350	0.0350	0.0350	0.0350
I thyroid	300 rem	0.0280	0.0300	0.0070	—	—	—	—
G.I. tract	50 rem	0.1380	0.1390	0.1340	0.1120	0.0840	0.0910	0.118
Total		1.1563	1.3452	1.8412	2.8792	6.2437	8.8058	12.5932
3-Year Operation								
External gamma	25 R	0.0112	0.0112	0.0112	0.0112	0.0112	0.0112	0.0112
Lung beta	300 rem	0.1550	0.1810	0.2340	0.3300	0.6000	0.6010	0.5700
^{90}Sr bone	150 rem	0.4930	0.6000	0.9400	1.8130	4.8470	7.8400	12.0130
^{89}Sr bone	50 rem	0.1300	0.1480	0.1720	0.1280	0.0090	—	—
^{144}Ce-^{144}Pr bone	50 rem	0.3540	0.4250	0.6270	0.9640	1.3760	0.9430	0.2590
Pu lung	150 rem	0.0520	0.0520	0.0520	0.0520	0.0520	0.0520	0.0520
I thyroid	300 rem	0.0280	0.0280	0.0070	—	—	—	—
G.I. tract	50 rem	0.1370	0.1370	0.1320	0.1100	0.0870	0.1220	0.1230
Total		1.3602	1.5822	2.1752	3.4082	6.9822	9.5792	13.0282

From: Brodsky, A., Criteria for Acute Exposure to Mixed Fission-Product Aerosols, *Health Phys*,. 11: 1022, 1965. Copyright Pergamon Press, New York.

Table 23.
MULTIPLES OF ACCEPTABLE EMERGENCY DOSES FOR INHALATION OF 1 Ci-sec/m³ SHORT-CRITICALITY BURSTS[37]

Mode of Exposure	Acceptable Emergency Dose Total	Zero Decay	2-Day Decay	7-Day Decay	30-Day Decay	100-Day Decay	365-Day Decay	730-Day Decay	1460-Day Decay
External gamma	25 R	0.0112	0.0112	0.0112	0.0112	0.0112	0.0112	0.0112	0.0112
Lung beta	300 rem	0.0100	0.0289	0.0838	0.1150	0.1760	0.5000	0.6750	0.5440
⁹⁰Sr bone	150 rem	0.0013	0.0067	0.0200	0.0533	0.1930	1.2530	3.2330	7.9000
⁸⁹Sr bone	50 rem	0.0071	0.0320	0.1030	0.1860	0.2680	0.0475	0.0009	—
¹⁴⁴Ce-¹⁴⁴Pr bone	50 rem	0.0039	0.0185	0.0633	0.1470	0.4660	1.6340	1.7790	0.7800
I thyroid	300 rem	1.1400*	0.1900	0.1770	0.0600	—	—	—	—
G.I. tract	50 rem	0.1450	0.1430	0.1790	0.1600	0.1310	0.0868	0.0756	0.0605
Total		1.3185	0.4303	0.6374	0.7325	1.2452	3.5317	5.7747	9.2957

* Allowing 10 minutes before uptake in thyroid.

From: Brodsky, A., Criteria for Acute Exposure to Mixed Fission-Product Aerosols, *Health Phys.*, 11: 1023, 1965. Copyright Pergamon Press, New York.

In lieu of utilizing the above data on inhalation of fission products, a simple rule of thumb may be utilized for making the most rapid decision in the event of an emergency release of fission products from a reactor accident or nuclear burst. The result of all the calculations in Reference 3 indicated that for decay times of less than about 30 days the simple rule of thumb that 1 Ci-sec/m³ (i.e., inhalation of at most 280 µCi of gross activity by a standard man) would be within a factor of 2 of the acceptable emergency dose that would " not be likely to produce clinically observable injury."[37] This can be seen by referring to Figure 11. The method for calculating single-intake doses for nuclides other than those included in considering gross fission products will be presented in the next two chapters. Exposure criteria similar to those outlined in References 1 or 37 may be used in determining the limiting intake in microcuries after the maximum dose per microcurie has been estimated. Alternatively, the limiting intake may be calculated from the ICRP rule presented above[1] that a planned special exposure should not exceed an intake equivalent to that obtained by breathing at the MPC for 2 years. Appropriate MPC's may be obtained from Table 19 or from Reference 23 for emergency purposes.

As is the case in deciding appropriate emergency levels of radiation exposure for people, many considerations regarding the uptake of particular chemical forms in which radionuclides are found in the environment and the special circumstances under which they are released and taken up in food supplies would determine the most appropriate limits of contamination in individual circumstances. Considerable literature is available on the subject,[40,41] which should be consulted when planning particular emergency procedures or environmental surveillance. However, there is sometimes an unexpected need for rapidly estimating the orders of magnitude or possible consequences of nuclear accidents. Thus, various guidelines have been developed for use in emergency situations.

In considering allowable emergency levels of land or surface contamination, one must estimate hazards resulting from either possible inhalation of activity reentrained in the air, possible routes of intake of radioactive material in foods, or—at higher contamination levels— external radiation exposure of individuals from materials on the ground. Only a few of the more useful data will be presented here for ready reference, since the methods of detailed analysis in cases of planned exposure may be obtained from the references already cited. Emergency actions for various ground contamination levels after reactor accidents releasing 180-day-irradiation fission products may be found in Table 26, taken from Reference 39. The estimates of Cowan and Kuper[39] in the case of ground contamination are considered adequate, although the inhalation values have been revised as described above.

Table 24. FISSION-PRODUCT INVENTORY (MCi/100 Mw)[37]

Nuclide	Half-Life, Days	2-Day Decay	7-Day Decay	30-Day Decay	100-Day Decay	365-Day Decay	730-Day Decay	1460-Day Decay
Reactor Operating Time = 1/2 Year								
^{89}Sr	50.5	3.31	3.10	2.31	0.884	2.35×10^{-2}	1.59×10^{-4}	7.27×10^{-9}
^{90}Sr	1.01×10^4	0.084	0.084	0.084	0.0836	0.0821	0.080	0.0762
^{90}Y	2.68	0.084	0.084	0.106	0.0836	0.0821	0.080	0.0762
^{91}Sr	0.403	0.15	0.000	0.000	—	—	—	—
^{91}Y	57.5	4.08	3.88	3.09	1.33	5.52×10^{-2}	6.85×10^{-4}	1.06×10^{-7}
^{95}Zr	63.3	4.50	4.25	3.34	1.59	9.39×10^{-2}	1.91×10^{-3}	7.96×10^{-7}
^{95}Nb	35.0	4.29	4.29	4.14	2.65	0.203	4.15×10^{-3}	1.728×10^{-6}
^{131}I	8.08	1.85	1.25	0.205	5.06×10^{-4}	—	—	—
^{133}I	0.87	1.11	0.02	—	—	—	—	—
^{133}Xe	5.27	4.69	2.50	0.123	—	—	—	—
^{137}Cs	1.1×10^4	0.0832	0.083	0.083	0.0826	0.081	0.079	0.0752
^{137}Ba	1.81×10^{-3}	0.0832	0.083	0.083	0.0826	0.081	0.079	0.0752
^{140}Ba	12.8	6.00	4.58	1.53	0.0345	1.95×10^{-8}	—	—
^{140}La	1.68	6.45	5.17	1.76	0.0397	2.24×10^{-8}	—	—
^{141}Ce	32	3.54	3.19	1.89	0.439	1.68×10^{-3}	8.00×10^{-7}	1.809×10^{-12}
^{143}Ce	1.33	1.81	0.15	—	—	—	—	—
^{143}Pr	13.7	4.73	3.83	1.20	0.0353	5.42×10^{-8}	—	—
^{144}Ce	290.0	1.54	1.52	1.44	1.139	0.597	0.247	0.042
^{144}Pr	1.18×10^{-2}	1.54	1.52	1.44	1.139	0.597	0.247	0.042
^{147}Nd	11.3	1.84	1.37	0.346	0.00431	—	—	—
^{147}Pm	920.0	0.26	0.26	0.28	0.25	0.21	0.16	0.093
Total		52.02	41.21	23.45	9.86	2.11	0.978	0.479
Reactor Operating Time = 1 Year								
^{89}Sr	50.5	3.59	3.37	2.51	0.964	2.56×10^{-2}	1.73×10^{-4}	7.92×10^{-9}
^{90}Sr	1.01×10^4	0.158	0.158	0.158	0.157	0.154	0.150	0.143
^{90}Y	2.68	0.158	0.158	0.196	0.16	0.154	0.150	0.143
^{91}Sr	0.403	0.15	—	—	—	—	—	—
^{91}Y	57.5	4.60	4.38	3.49	1.50	0.0621	7.72×10^{-4}	1.19×10^{-7}
^{95}Zr	63.3	5.20	4.98	3.85	1.83	0.108	2.20×10^{-3}	9.17×10^{-6}
^{95}Nb	35.0	5.15	5.12	4.88	3.08	0.226	4.76×10^{-3}	1.99×10^{-5}
^{131}I	8.08	1.85	1.25	0.205	5.06×10^{-4}	—	—	—
^{133}I	0.87	1.11	0.02	—	—	—	—	—
^{133}Xe	5.27	4.69	2.50	0.123	—	—	—	—
^{137}Cs	1.1×10^4	0.156	0.156	0.156	0.155	0.152	0.148	0.141
^{137}Ba	1.81×10^{-3}	0.156	0.156	0.156	0.155	0.152	0.148	0.141
^{140}Ba	12.8	6.00	4.58	1.53	0.0345	1.95×10^{-8}	—	—
^{140}La	1.68	6.45	5.17	1.76	0.0396	2.24×10^{-8}	—	—
^{141}Ce	32.0	3.55	3.19	1.90	0.440	1.69×10^{-3}	8.02×10^{-7}	1.81×10^{-12}
^{143}Ce	1.33	1.8	0.15	—	—	—	—	—
^{143}Pr	13.7	4.73	3.83	1.20	0.0352	5.4×10^{-8}	5.0×10^{-16}	5.0×10^{-32}
^{144}Ce	290.0	2.52	2.49	2.92	2.31	1.21	0.500	0.0853
^{144}Pr	1.18×10^{-2}	2.52	2.49	2.92	2.31	1.21	0.500	0.0853
^{147}Nd	11.3	1.84	1.37	0.35	0.00431	—	—	—
^{147}Pm	920.0	0.49	0.49	0.49	0.456	0.377	0.290	0.171
Total		56.87	46.00	28.79	13.63	3.83	1.89	0.909

Table 24. FISSION-PRODUCT INVENTORY (MCi/100 Mw)[37] (Continued)

Nuclide	Half-Life, Days	2-Day Decay	7-Day Decay	30-Day Decay	100-Day Decay	365-Day Decay	730-Day Decay	1460-Day Decay
Reactor Operating Time = 2 Years								
^{89}Sr	50.5	3.63	3.41	2.53	0.971	0.0258	1.75×10^{-4}	7.98×10^{-9}
^{90}Sr	1.01×10^4	0.306	0.306	0.306	0.305	0.299	0.292	0.278
^{90}Y	2.68	0.306	0.306	0.382	0.31	0.299	0.292	0.278
^{91}Sr	0.403	0.15	—	—	—	—	—	—
^{91}Y	57.5	4.68	4.45	3.55	1.52	0.0634	7.87×10^{-4}	1.21×10^{-7}
^{95}Zr	63.3	5.30	5.00	3.94	1.87	0.111	2.26×10^{-3}	9.4×10^{-7}
^{95}Nb	35.0	5.48	5.43	5.12	3.19	0.232	4.88×10^{-3}	2.04×10^{-6}
^{131}I	8.08	1.85	1.25	0.205	5.06×10^{-4}	—	—	—
^{133}I	0.87	1.11	0.02	—	—	—	—	—
^{133}Xe	5.27	4.69	2.50	0.123	—	—	—	—
^{137}Cs	1.1×10^4	0.31	0.31	0.31	0.308	0.302	0.294	0.280
^{137}Ba	1.81×10^{-3}	0.31	0.31	0.31	0.308	0.302	0.294	0.280
^{140}Ba	12.8	6.00	4.58	1.53	0.0345	2.0×10^{-8}	—	—
^{140}La	1.68	6.45	5.17	1.76	0.0397	2.3×10^{-8}	—	—
^{141}Ce	32.0	3.55	3.19	1.90	0.440	1.69×10^{-3}	8.02×10^{-7}	1.81×10^{-12}
^{143}Ce	1.33	1.81	0.15	—	—	—	—	—
^{143}Pr	13.7	4.73	3.83	1.20	0.0352	5.4×10^{-8}	5.0×10^{-16}	5.0×10^{-32}
^{144}Ce	290.0	3.51	3.47	3.27	2.58	1.35	0.559	0.0954
^{144}Pr	1.18×10^{-2}	3.51	3.47	3.27	2.58	1.35	0.559	0.0954
^{147}Nd	11.3	1.84	1.37	0.35	0.00431	—	—	—
^{147}Pm	920.0	0.86	0.86	0.84	0.800	0.662	0.508	0.300
Total		60.38	49.38	30.89	15.29	4.99	2.80	1.60
Reactor Operating Time = 3 Years								
^{89}Sr	50.5	3.63	3.41	2.53	0.971	0.0258	1.75×10^{-4}	7.98×10^{-9}
^{90}Sr	1.01×10^4	0.46	0.46	0.46	0.458	0.450	0.438	0.417
^{90}Y	2.68	0.46	0.46	0.58	0.46	0.450	0.438	0.417
^{91}Sr	0.403	0.15	—	—	—	—	—	—
^{91}Y	57.5	4.68	4.45	3.55	1.52	0.0634	7.87×10^{-4}	1.21×10^{-7}
^{95}Zr	63.3	5.3	5.00	3.94	1.87	0.111	2.26×10^{-3}	9.4×10^{-7}
^{95}Nb	35.0	5.48	5.43	5.12	3.19	0.232	4.88×10^{-3}	2.04×10^{-6}
^{131}I	8.08	1.85	1.25	0.205	5.06×10^{-4}	—	—	—
^{133}I	0.87	1.11	0.02	—	—	—	—	—
^{133}Xe	5.27	4.69	2.50	0.123	—	—	—	—
^{137}Cs	1.1×10^4	0.48	0.48	0.48	0.477	0.468	0.456	0.434
^{137}Ba	1.81×10^{-3}	0.48	0.48	0.48	0.477	0.468	0.456	0.434
^{140}Ba	12.8	6.00	4.58	1.53	0.0345	2.0×10^{-8}	—	—
^{140}La	1.68	6.45	5.17	1.76	0.0397	2.3×10^{-8}	—	—
^{141}Ce	32.0	3.55	3.19	1.90	0.440	1.69×10^{-3}	8.03×10^{-7}	1.81×10^{-12}
^{143}Ce	1.33	1.81	0.15	—	—	—	—	—
^{143}Pr	13.7	4.73	3.83	1.20	0.0352	5.4×10^{-8}	5.0×10^{-16}	5.0×10^{-32}
^{144}Ce	290.0	3.93	3.88	3.66	2.90	1.52	0.627	0.107
^{144}Pr	1.18×10^{-2}	3.93	3.88	3.66	2.90	1.52	0.627	0.107
^{147}Nd	11.3	1.84	1.37	0.346	0.00431	—	—	—
^{147}Pm	920.0	1.14	1.14	1.14	1.06	0.877	0.674	0.398
Total		62.15	51.13	32.66	16.83	6.18	3.72	2.31

From: Brodsky, A., Criteria for Acute Exposure to Mixed Fission-Product Aerosols, *Health Phys.*, 11: 1019—1020, 1965. Copyright Pergamon Press, New York.

Table 25. FISSION-PRODUCT INVENTORY, SHORT BURST (Ci/100 Mw-sec)[37]

Nuclide	0-Day Decay	2-Day Decay	7-Day Decay	30-Day Decay	100-Day Decay	365-Day Decay	730-Day Decay	1460-Day Decay
^{89}Sr	0.640	0.622	0.582	0.424	0.162	4.33×10^{-3}	2.93×10^{-5}	1.34×10^{-8}
^{90}Sr	0.0039	0.0039	0.0039	0.00389	0.00387	0.00380	0.00371	0.00353
^{90}Y	—	0.00236	0.00326	0.0038	0.0038	0.0038	0.0037	0.0035
^{91}Sr	97.8136	3.13	6.11×10^{-4}	—	—	—	—	—
^{91}Y	—	0.652	0.634	0.480	0.206	0.0086	1.06×10^{-4}	—
^{95}Zr	0.645	0.632	0.599	0.468	0.223	0.0132	2.68×10^{-4}	1.12×10^{-7}
^{95}Nb	—	0.0126	0.0796	0.244	0.289	0.0276	5.79×10^{-4}	2.43×10^{-7}
^{131}I	2.60	2.19	1.43	0.199	4.91×10^{-4}	—	—	—
^{133}I	54.0	10.9	0.200	—	—	—	—	—
^{133}Xe	9.5	6.19	8.84×10^{-4}	—	—	—	—	—
^{137}Cs	0.0043	0.0043	0.0043	0.00429	0.00427	0.00419	0.00409	0.00388
^{137}Ba	—	0.0043	0.0043	0.00429	0.00427	0.00419	0.00409	0.00388
^{140}Ba	3.35	2.93	2.29	0.656	0.0148	8.31×10^{-9}	—	—
^{140}La	—	1.02	1.32	0.695	0.0170	9.64×10^{-9}	—	—
^{141}Ce	1.23	1.18	1.06	0.656	0.152	5.86×10^{-4}	2.78×10^{-7}	6.28×10^{-14}
^{143}Ce	28.2	10.30	0.829	—	—	—	—	—
^{143}Pr	—	1.68	2.11	0.691	0.0203	—	—	—
^{144}Ce	0.143	0.142	0.141	0.133	0.112	0.0589	0.0243	0.00415
^{144}Pr	—	0.142	0.141	0.133	0.112	0.058	0.0243	0.00415
^{147}Nd	1.65	1.45	1.07	0.252	0.0031	—	—	—
^{147}Pm	—	0.00226	0.0067	0.0158	0.0178	0.0147	0.0113	0.00668
Total	198.77	43.18	12.50	5.06	1.34	0.2019	0.0765	0.0298

From: Brodsky, A., Criteria for Acute Exposure to Mixed Fission-Product Aerosols. *Health Phys.*, 11: 1021, 1965. Copyright Pergamon Press, New York.

The calculation of permissible land contamination levels for individual nuclides requires some knowledge of the relative uptake in plants and animals and of the discrimination factor in transferring this contamination through the food chain to man. These ecological factors are generally quite complex and subject to stochastic variations between different vicinities, populations, and time periods. A recent summary describes the log-normal variations in ^{90}Sr in milk and in human bone, and shows that the same type of variation may be expected with other radionuclide deposits from the atmosphere and that the uptake of one nuclide may not

Table 26. REACTOR ACCIDENTS, LIMITS OF EXPOSURE[39]

Category	Full Fission-Product Release, Ci/m^2	Volatile Fission-Product Release, Ci/m^2
A. Urgent evacuation necessary (within 12 hours)	> 0.2	> 0.1*
B. Evacuation necessary	>10.0–2.0	> 0.1
C. Severe restrictions on land use, possible temporary evacuation, restrictions on outdoor work	10^{-2}–10^{-3}	0.1–10^{-2}
D. Probable destruction of standing crops, restrictions on agriculture for first year	10^{-3}–10^{-4}	10^{-2}–10^{-3}
E. No expense likely	$<10^{-4}$	$<10^{-3}$

* Unless adequate shelter is available.
From: Cowan, F. P. and Kuper, J. B. *Health Phys.*, 1: 76, 1958. Copyright Pergamon Press, New York.

necessarily be correlated with that of any other.[16] However, upper-limit estimates of the hazard of a given level of area contamination for a given nuclide may still lead to conservative limits for use in carrying out emergency operations suitable for saving life and property. Generally, one can arrive at the order of magnitude at which the land contamination becomes serious by comparing nuclides with ^{90}Sr relative to the drinking water MPC's as given in Table 19. Experience with nuclear-weapons fallout would indicate that ^{90}Sr levels on the order of 100 millicuries per square mile would not result in annual bone doses of more than about 10 millirad per year, even after many years of living in the environment[16] (see Figure 12 and Table 27). Comparable contamination levels of other nuclides may be roughly estimated by dividing the drinking water MPC's by the MPC for ^{90}Sr in order to obtain a conservative estimate of emergency permissible levels of contamination, below which immediate danger to life would not be expected as a result of ingestion. This oversimplified assumption is useful as a result of

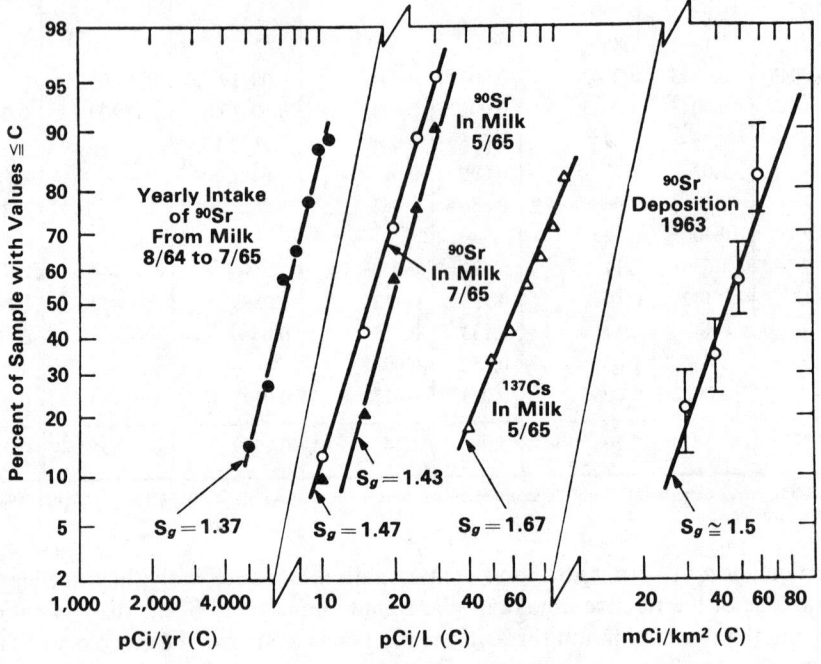

FIGURE 12.
Comparison of the Log-Normal Frequency Distribution of ^{90}Sr Deposition and ^{90}Sr and ^{137}Cs Concentrations in Milk in the U.S.[16]

(From: Schubert, J., Brodsky, A., and Tyler, S., The Log-Normal Function as a Stochastic Model of the Distribution of ^{90}Sr and Other Fission Products in Humans. *Health Phys.*, 13: 1198, 1967. Copyright Pergamon Press, New York.)

Table 27.
ESTIMATES OF PERCENTAGES OF UNITED STATES 0–25 AGE GROUP ABOVE VARIOUS ^{90}Sr LEVELS[16]

^{90}Sr Level	^{90}Sr Concentration		^{90}Sr Bone Marrow Dose, mrad/yr	Percent of U.S. in 0–25 Age Group with ^{90}Sr Levels Greater than Indicated
	pCi/g Ca	pCi/g bone		
Median	3.15	1.35	2.8	50.00
Average	3.47	1.49	3.1	40.00
2 × average	6.94	2.70	6.2	3.60
3 × average	10.41	4.05	9.3	0.36

From: Schubert, J., Brodsky, A., and Tyler, S., The Log-Normal Function as a Stochastic Model of the Distribution of ^{90}Sr and Other Fission Products in Humans. *Health Phys.*, 13: 1201, 1967. Copyright Pergamon Press, New York.

the fact that discrimination factors through the food chain are usually greater for other nuclides than they are for ^{90}Sr, except perhaps for the iodines, and because other nuclides generally have lower probabilities of absorption through the G.I. tract than those used in estimating the MPC_{water}.

In order to evaluate the probability of inhaling contamination resuspended from ground or laboratory surfaces, one may consider the fact that experience has shown—both in laboratory and field experiments and in routine operations with radionuclides—that the ratios of air concentrations in $\mu Ci/m^3$ divided by surface concentrations of loose material in units of $\mu Ci/m^2$ are generally below 10^{-4}, and usually average more like 10^{-5} to 10^{-6} or lower under ordinary conditions.[21,42] Thus, considerably higher contamination levels may be tolerated in the laboratory or the field on a temporary basis than would be considered suitable for long-term protection against the transfer of radioactive material to other areas, materials, or ecological systems. As a result of considerations of this kind, the limits of plutonium land contamination in the event of nuclear-weapons accidents or of criticality bursts involving plutonium have yielded recommended limits such as the following:

Air Force Limits for Plutonium[43]
0–100 $\mu g/m^2$—releasable to controlled area
for continuous residence;
1,000 $\mu g/m^2$—safe for continuous occupancy;
4,000 $\mu g/m^2$—safe for approximately 8 days;
5,000 $\mu g/m^2$—safe for 3 to 4 days.

Although these values have been used as a guide in weapons incidents, there are usually many other public-relations considerations that require the reduction of contamination levels to much lower values as soon as possible.[44] (The specific activity of plutonium is about 0.0617 Ci/g.) Under routine conditions in many atomic-energy laboratories such levels of plutonium would be unacceptable, of course, and in fact would be kept below values many orders of magnitude lower than those given above.

In the case of ^{131}I, experience after the Windscale, England, incident showed that iodine deposited on the grass was quickly carried into the milk. It was found that a grazing-area contamination level of 1 microcurie per square meter on grass yielded a level of about 0.1 microcurie per liter of milk, which was taken as the emergency limit based on a dose of 20 rads to an infant thyroid of 1.5 grams;[21] this amount of ^{131}I spread over a large area would produce a gamma dose rate of about 3 millirem per hour near ground level.[21] Of course, in the case of a nuclear criticality burst releasing only the most volatile fission products, this contamination limit would be more pertinent than those given in Table 26 for gross fission-product mixtures. For emergency estimation of gamma-ray dose rates from land contamination, one may use the above estimate of external radiation intensity from iodine contamination together with the rough rule of thumb that the roentgen intensity for a gamma emitter would be roughly proportional to the gamma photon energy for equivalent contamination levels in terms of gamma-emitting disintegrations per minute per square meter.

As an example of contamination limits for laboratory work, Table 28 presents contamination limits used by the French Atomic Energy Commission for alpha and beta emitters on equipment and working surfaces and on clothing and skin.[45] These levels will vary between and within countries, depending on the type of operation involved and on the danger of also contaminating different operations or materials. Often, under routine operations, levels may feasibly be kept much lower than those in Table 28, when suitable containment is provided.

Emergency guides for food and water may be obtained from the reports put out by various agencies.[1,16,21,41,46,47]

Table 29 gives some emergency food and water levels that could be ingested after the detonation of a nuclear weapon or a short-criticality burst of a nuclear reactor releasing fission products to the environment.[48-50] However, these levels are not recommended for

Table 28. MAXIMUM PERMISSIBLE LEVELS FOR SURFACE CONTAMINATION[45]

Radiotoxicity of Isotopes	Equipment and Working Places		Clothing	Skin
	in inactive areas	in active areas		
Very high	α emitters: 10^{-5} μCi/cm^2 β emitters: 10^{-4} μCi/cm^2	α emitters: 10^{-4} μCi/cm^2 β emitters: 10^{-3} μCi/cm^2	α emitters: 10^{-5} μCi/cm^2 β emitters: 10^{-4} μCi/cm^2	α emitters: 5.1^{-6} μCi/cm^2 β emitters: 5.1^{-5} μCi/cm^2
High Medium Low	10^{-4} μCi/cm^2	10^{-3} μCi/cm^2	10^{-4} μCi/cm^2	5.1^{-5} μCi/cm^2

From: *Note No. 172*, issued by the French Atomic Energy Commission, May 1956. By permission of the International Atomic Energy Agency, Vienna, Austria.

peacetime use, when uncontaminated food or water is readily available. For planned actions taken after an unexpected incident, it would be best to follow the ICRP- recommended exposure levels given in this section, estimating the internal dose from ingestion from the respective amounts of radioactivity equivalent to a quarterly or annual intake at maximum permissible concentrations. Whenever possible, the intake of radionuclides should be kept well below the recommended limits for routine exposure.[21,31,35,51]

Table 29. EMERGENCY FOOD AND WATER LEVELS[21]

Duration of Ingestion	μCi/cm^3	dis/min/m^3
Preferable 10 days	3.5×10^{-3}	7.7×10^3
Acceptable 10 days	9.0×10^{-2}	2.0×10^5
Preferable 30 days	1.1×10^{-3}	2.6×10^3
Acceptable 30 days	3.0×10^{-2}	7.0×10^4

From: Brodsky, A. and Beard, G. V., *A Compendium of Information for Use in Controlling Radiation Emergencies*, TID-8206, Rev. 1960. By permission of the U.S. Atomic Energy Commission Washington, D.C.

REFERENCES

1. Morgan, K. Z., Maximum Permissible Exposure Levels—External and Internal. *Principles of Radiation Protection—A Textbook of Health Physics*, Ch. 14, Morgan, K. Z. and Turner, J. E., eds. John Wiley and Sons, New York, 1967.
2. Morgan, K. Z., in *Principles of Radiation Protection—A Textbook of Health Physics*, Ch. 14, p. 532, Morgan, K. Z. and Turner, J. E., eds. John Wiley and Sons, New York, 1967.
3. Brodsky, A. *Amer. J. Public Health*, 55(12): 1971–1992, 1965.
4. Federal Radiation Council, Report No. 2, *Radiation Protection for Federal Agencies*. Federal Radiation Council, Washington, D.C., 1961.
5. International Commission on Radiological Protection, Recommendations of the International Commission on Radiological Protection. ICRP Publ. 2. Pergamon Press, London, England, 1959.
6. National Committee on Radiation Protection and Measurements, Maximum Permissible Radiation Exposures to Man. *Suppl. NBS Handbook 59*. U.S. Government Printing Office, Washington, D.C., 1958.
7. National Committee on Radiation Protection and Measurements, Maximum Permissible Radiation Exposures to Man. *NBS Handbook 59*. U.S. Government Printing Office, Washington, D.C., 1957.
8. International Commission on Radiological Protection, Report of Committee 2 on Permissible Dose for Internal Radiation. ICRP Publ. 2. Pergamon Press, London, England, 1959, and *Health Phys.*, 3: 1, 1960.,
9. National Committee on Radiation Protection and Measurements, Maximum Permissible Body Burden and Maximum Permissible Concentrations of Radionuclides in Air and Water for Occupational Exposure. *NBS Handbook 69*. U.S. Government Printing Office, Washington, D.C., 1959.
10. International Commission on Radiological Protection, Report on Decisions at the 1959 Meeting of the International Commission on Radiological Protection (ICRP), Addendum to ICRP Publ. 15. *Radiology*, 74: 116, 1960.

11. International Commission on Radiological Protection, Report of Committee 3 on Protection Against X Rays up to Energies of 3 Mev and Beta and Gamma Rays from Sealed Sources. *ICRP Publ. 3.* Pergamon Press, London, England, 1960.
12. International Commission on Radiological Protection, Recommendations of the International Commission on Radiological Protection. *ICRP Publ. 6.* Pergamon Press, London, England, 1966.
13. International Commission on Radiological Protection, Recommendations of the International Commission on Radiological Protection. *ICRP Publ. 9.* Pergamon Press, London, England, 1966.
14. National Committee on Radiation Protection and Measurements, Maximum Permissible Amounts of Radioisotopes in the Human Body and Maximum Permissible Concentrations in Air and Water. *NBS Handbook 52.* U.S. Government Printing Office, Washington, D.C., 1953.
15. National Committee on Radiation Protection and Measurements. *Radiology*, 75: 122, 1960.
16. Schubert, J., Brodsky, A., and Tyler, S., The Log-Normal Function as a Stochastic Model of the Distribution of Strontium-90 and Other Fission Products in Humans. *Health Phys.*, 13: 1187–1204, 1967.
17. Brodsky, A., *Unpublished Data.*
18. United Nations Scientific Committee on the Effects of Atomic Radiation, *Various Reports.* United Nations, New York, 1958, 1962, 1964, 1966. (Continuing reviews are underway.)
19. International Commission on Radiological Protection, Report of Committee 2. *Health Phys.*, 3: 1–380, 1960.
20. International Commission on Radiological Protection, *Recommendations of the International Commission on Radiological Protection.* Pergamon Press, Macmillan Co., New York, 1959.
21. Brodsky, A. and Beard, G. V., compilers and eds., *A Compendium of Information for Use in Controlling Radiation Emergencies, TID-8206* (Rev.). U.S. Atomic Energy Commission, Washington, D.C., 1960.
22. Society for Nuclear Medicine, MIRD Report. *J. Nucl. Med.*, Suppl. 1: 5–39, 1968.
23. International Commission on Radiological Protection, Report of Committee 2. *Health Phys.*, 3: 1–380, 1960.
24. International Commission on Radiological Protection, Report of the ICRP Task Group on Lung Dynamics. *Health Phys.*, 12: 173–207, 1966.
25. Fitzgerald, J. J., Brownell, G. L., and Mahoney, F. J., *Mathematical Theory of Radiation Dosimetry.* Gordon and Breach, New York, 1967.
26. Cember, H., *Introduction to Health Physics.* Pergamon Press, London, England, 1968.
27. Morgan, K. Z. and Turner, J. E., eds., *Principles of Radiation Protection—A Textbook of Health Physics.* John Wiley and Sons, New York, 1967.
28. Hine, G. J. and Brownell, G. L., eds., *Radiation Dosimetry*, 1st ed. Academic Press, New York, 1956.
29. Roesch, W. C. and Attix, F. H. *Radiation Dosimetry*, 2nd ed., Vol. 1, Ch. 1, Attix, H. F. and Roesch, W. C., eds. Academic Press, New York, 1968.
30. U.S. Atomic Energy Commission, *USAEC Rules and Regulations, Title 10, Code of Federal Regulations*, Part 20, Standards of Protection Against Radiation, Revised August 9, 1966. U.S. Government Printing Office, Washington, D.C.
31. National Committee on Radiation Protection and Measurements, Exposure to Radiation in an Emergency. *NCRP Report No. 29.* NCRP Publications Office, Washington, D.C., 1968.
32. National Committee on Radiation Protection and Measurements, Medical X-Ray and Gamma-Ray Protection for Energies up to 10 Mev. Equipment Design and Use, *NCRP Report No. 33.* NCRP Publications Office, Washington, D.C., 1968.
33. National Research Council, *Radiobiological Factors in Manned Space Flight*, Langham, W. H., ed. National Academy of Sciences, Washington, D.C., 1967.
34. Lanzl, L. H., Pingel, J. H., and Rush, J. H., *Radiation Accidents and Emergencies in Medicine, Research, and Industry.* Charles C Thomas, Springfield, Illinois, 1965.
35. U.S. Atomic Energy Commission, *The Effects of Nuclear Weapons.* U.S. Government Printing Office, Washington, D.C., 1962.
36. Martin, T. L. and Latham, D. C., *Strategy for Survival.* The University of Arizona Press, Tucson, Arizona, 1963.
37. Brodsky, A., Criteria for Acute Exposure to Mixed Fission-Product Aerosols. *Health Phys.*, 11: 1017–1032, 1965.
38. Bureau of Ships, Principles of Radiation and Contamination Control. *Procedures and Guidelines Relating to Nuclear-Weapons Effects*, Vol. 2. Department of the Navy, Washington, D.C.
39. Cowan, F. P. and Kuper, J. B. H. *Health Phys.*, 1: 76, 1958.
40. Adams, J. A. S. and Lowder, W. M., eds., *The Natural Radiation Environment.* University of Chicago Press, Chicago, Illinois, 1964.
41. Eisenbud, M., *Environmental Radioactivity.* McGraw-Hill, New York, 1963.
42. Brodsky, A., Determining Industrial Hygiene Requirements for Installations Using Radioactive Materials. *Amer. Ind. Hyg. Ass. J.*, 26: 249–310, 1965.
43. Meyer, A. F., Jr., *Handbook of Peacetime Nuclear Accidents.* Office of the Surgeon, HQ, Strategic Air Command, Offutt Air Base, Nebraska, 1958.

44. Langham, W. H., in a Lecture to the Western Pennsylvania Chapter of the Health Physics Society, 1967.
45. International Atomic Energy Agency, Safe Handling of Radioisotopes. *Safety Series No. 1*, p. 95. Intertional Atomic Energy Agency, Vienna, Austria, 1958.
46. National Research Council, *The Biological Effects of Atomic Radiation—Summary Reports*. National Academy of Sciences, Washington, D.C., 1956.
47. Federal Radiation Council, *Report No. 7, Protective Action Guides for Strontium-89, Strontium-90, and Cesium-137*, p. 13. Federal Radiation Council, Washington, D.C., 1965.
48. Hursh, J. B., Zizzo, S., and Dahl, A. H., Use of Commercially Available Portable Survey Meters for Emergency Fission-Product Monitoring of Water Supplies. *Report UR-180*. University of Rochester, 1951.
49. Federal Civil Defense Administration, Permissible Emergency Levels of Radiation in Water and Food. *Technical Bulletin TB 11-8*, 1952; reprinted by the Office of Civil and Defense Mobilization, 1959.
50. Federal Civil Defense Administration, Emergency Measurements of Radioactivity. *Technical Bulletin TB 11-9*, 1952.
51. Saenger, E. L., *Medical Aspects of Radiation Accidents*. U.S. Government Printing Office, Washington, D.C., 1963.

DATA AND METHODS FOR ESTIMATING RADIATION EXPOSURES FROM INTERNAL AND EXTERNAL RADIATION SOURCES

Allen Brodsky, Sc.D., C.H.P.

There is now abundant literature on methods for estimating radiation exposures from radioactive sources external or internal to the body.[1-13] For the most precise calculations of radiation dose to various body organs, a specialist should be employed who can use the detailed physical calculations and correction factors—such as organ-shape factors—given in the references and in other literature. However, it is often useful and adequate to make rapid estimates of exposures that are generally accurate to within perhaps a factor of 2, particularly for radiation control purposes, evaluation of safety requirements for facilities, emergency planning, and evaluation of hazards of radionuclides dispersed to the environment. For such purposes, the ordinary biological variations between individual radiosensitivities usually require the application of additional safety factors of 10 or more, which overcome the ordinary errors of estimating organ exposures based on standard physical and biological constants.[9,10,14,15] It is usually only after the individual has received a known exposure under known physical geometry and circumstances, or a known intake of a given body counting, that it is worthwhile, either for medical diagnosis or for treatment purposes, to estimate the dose to that particular individual as precisely as possible. Thus, only a summary of some of the general formulae and data useful for rapidly determining the general magnitudes of radiation exposures are given in this section.

The gamma and beta doses may be estimated from the following rules of thumb.[16]

Gamma Dose.

a) $1 \text{ r/hr} = \dfrac{5.6 \times 10^5}{E} = $ photons/cm²/sec, for E from 0.07 to 2.0 Mev, good within ±12 percent, assuming W_{air} = 34 ev/ion pair.

The above identity is equivalent to the equation:

$$I = \frac{5.64 \, C \sum_i E_i}{d^2},$$

where I = intensity in r/hr at d feet from a point source;
 C = curies of activity,
 $\sum_i E_i$ = sum of gamma energies in Mev per disintegration, if more than one gamma is emitted per disintegration,
 d = distance in feet.

Note: convert exposure in r to depth doses in rads, using tables in the first chapter of this section.

b) For any energy E in Mev and for its corresponding total minus Compton-scatter absorption coefficient $(\mu - \sigma_s)$,

$$1 \text{ r} = \frac{7.1 \times 10^4}{(\mu - \sigma_s)E} \text{ photons/cm}^2.$$

Beta Surface Dose in Air.

In air of density 0.001293 g/cc,

$$1 \text{ rad/hour} = \frac{6.1 \times 10^5}{S_a} \text{ betas/cm}^2/\text{sec},$$

where S_a is the average number of ion pairs produced per centimeter of path in air. This equation holds within ±6 percent for 0.01 to 2.0 Mev. Values of $(\mu - \sigma_s)$ are given in Table 30, and the theoretical values of $(\mu - \sigma_s)$ for calculating external exposure rates for various photon energies may be taken from Table 31.

Table 30. SPECIFIC IONIZATION OF ELECTRONS[21]

Energy, Mev	Sa, ion pairs/cm	Range in Air	
		mg/cm²	cm
0.05	250	3.9	3.02
0.10	175	14.0	10.80
0.20	96	42.0	32.50
0.30	76	77.0	59.60
0.50	60	158.0	122.00
1.00	53	400.00	310.00
1.50	47	680.0	526.00

From: Brodsky, A. and Beard, G. V., *A Compendium of Information for Use in Controlling Radiation Emergencies*, TID-8206, Rev. 1960. By permission of the U.S. Atomic Energy Commission, Washington, D.C.

Table 31. THEORETICAL VALUES OF $(\mu - \sigma_s)$ AND OF r/hr AT 1 m FROM A 1-CURIE SOURCE

Energy, Mev	$(\mu - \sigma_s)_{air}$, cm²/g	r/hr at 1 m from 1- Ci Source
0.02	0.500	0.160
0.04	0.078	0.063
0.06	0.035	0.042
0.08	0.025	0.040
0.10	0.023	0.047
0.20	0.026	0.110
0.40	0.029	0.230
0.60	0.029	0.350
0.80	0.028	0.450
1.00	0.027	0.550
2.00	0.023	0.930
4.00	0.019	1.500
6.00	0.017	2.100
8.00	0.016	2.600
10.00	0.015	3.000
20.00	0.013	5.300
40.00	0.013	11.000
60.00	0.014	17.000
100.00	0.015	30.000

From: Gray, *American Institute of Physics Handbook*, p. 8–257. Copyright 1957, by permission of McGraw-Hill Book Co., New York.

Internal doses for various nuclides taken into the body may be determined for various body organs by standard techniques given in the literature and by some of the equations presented in the two preceding chapters for single-intake situations. The total dose to the critical organ in 50 years per microcurie intake may be calculated, assuming instantaneous uptake in the critical organ from a single exposure of short duration (compared to the effective half-life of the radionuclide in the body), from the equation:

$$\text{Dose (rem/}\mu\text{Ci)} = \int_{t=0}^{t=50\times 365} I_0 e^{-0.693 t/T} dt = 1.44\, I_0 T[1 - \exp(1.265 \times 10^4/T)], \quad (1)$$

where I_0 = initial dose rate to the critical organ per microcurie intake, or $f_a R/qf_2$ (with the standard symbols from Reference 3);

q = body burden listed beside the corresponding critical organ in Table 1 of Reference 3;

T = effective half-life in the body in days, from Table 12 of Reference 3;

R = permissible dose rate for continuous exposure in rem per day fort he body organ concerned, obtained from Reference 3 as $0.1/7$ rem/day for irradiation of the whole body, 0.08 rem/day for bone, $0.1/7$ rem/day for the gonads, $0.6/7$ rem/day for the thyroid and skin, and $0.3/7$ rem/day for other parts of the body.

The expression in brackets in the above equation is essentially a factor of 1, except for the few bone-seeking radionuclides whose effective half-lives are not short compared to 50 years. For ^{90}Sr, with an effective half-life of 6.4×10^3 days, the factor in brackets becomes 0.861. For purposes of this chapter, the same relative dose from daughter products is assumed for single intake as for continuous exposure, where radionuclides that have radioactive daughters building up in the body are concerned. The contribution from daughters is thus taken into account by using the total body burdens, q, to give dose rates, R. These body burdens were calculated by the ICRP Committee, taking into account daughter products that build up in the body.

Since for materials insoluble in the lnug an average of about $12\frac{1}{2}$ percent of the material inhaled may remain in the lung, with a half-life of 120 days, the lung must also be taken into consideration as a possible critical organ. Deviations from this half-life in lung are described for various conditions,[10] and human data are in many cases still incomplete. However, the use of a 120-day half-life for lung exposure is reasonably conservative in the absence of other data. Rarely has lung clearance been more than an order of magnitude slower than 120 days, and it is often much less.[10] Single-intake doses based on the lung dose may be calculated from the equation

dose to the lung per microcurie inhaled (insoluble)

$$= \frac{\text{Permissible dose rate per week}}{(\text{MPC}_{air} \text{ based on continuous exposure to lung}) \times 1.4 \times 10^8 \text{ cc/week}}$$

$$= 2.14 \times 10^{-9}/\text{MPC}_{air} \text{ based on lung}, \quad (2)$$

since the equilibrium dose rate from continuous exposure is the same as the average dose rate from a series of single intakes of the same total quantity of radioactivity per week for effective half-lives that are short compared to 50 years.

Similar equations may be written for ingested radionuclides taken into the body by mouth. The f_a of equations (1) and (2) and the air intake in cc's per week should be replaced by the respective f_w, representing the fractional amount of the ingested material reaching the critical organ, or the respective MPC_{water} multiplied by an intake of approximately 1.75×10^4 cc's of water per week.

Some of the specific data in regard to lung clearance rates may be obtained from the compilation of the ICRP Task Group on Lung Dynamics, given in Table 32.[10] For approximate dose calculations, Tables 33, 34 and 35, giving parameters for standard man and for the simplified lung model previously outlined by the ICRP, are still useful.

In Table 36 some of the calculated maximum permissible body burdens of more frequently encountered radionuclides are given, based on the quantity of radioactivity (in microcuries) deposited in the organ of reference that delivers a radiation dose rate equal to the maximum permissible dose rate for continuous lifetime occupational exposure. Also represented are

Table 32.
BIOLOGICAL HALF-LIFE OF RADIOACTIVE SUBSTANCES IN THE LUNG[10]

Substance	Species	Biological Half-Life, days	Recommended Values for Man		Notes
			Single Exposure	Multiple Exposure	
Actinium (?)	Man	100–200	200		
Aluminum oxide (corundum)	Rats	100		100	Multiple
Antimony oxide	Rats	5	5		2-week study
Antimony chloride	Rats	94	94		
Arsenic trioxide	Man	16	16		
Barium sulfate	Rats	6 27½ and 10 10	10		
	Dogs	8			
Beryllium citrate	Rats	20–500 (?)	80		
Beryllium oxide	Rats	120 (?)	120		
Beryllium sulfate, carrier free	Rats	80	80		
Beryllium sulfate	Rats	80	80		
Calcium chloride, with carrier	Guinea pigs	18	18		
Cerium dioxide	Rats Dogs Rats	80 150 54	150		
Cerium (^{144}Pr) trichloride	Mice	58	60		
Cerium trichloride	Rats	63	60		Intratracheal injection
Cerous fluoride	Rats	200	200		
Cesium (?)	Man	89 109–149 140	90	150	Multiple Not only in lungs
Cesium chloride	Man	128			
Cesium sulfate	Man	78 (?)	80		Whole body (?),
Chromium sesquioxide	Dogs	>100	160		
Cobalt	Man	90			
Cobalt and cobalt oxide	Man	1.5–2.1 yrs	90	720	Multiple (?), mainly inhalation
Cobalt chloride	Mice	17	17		

Table 32. (*Continued*)
BIOLOGICAL HALF-LIFE OF RADIOACTIVE SUBSTANCES IN THE LUNG[10]

Substance	Species	Biological Half-Life, days	Recommended Values for Man		Notes
			Single Exposure	Multiple Exposure	
Europium nitrate	Mice	28	28		
Ferric oxide	Man Dogs	70 63	70		
Indium sesquioxide	Rats	70		70	90-day exposure
Iodine	Mice, rats	<5 hrs	<5 hrs		
Iridium metal	Rats	14	14		
Iron powder	Rabbits	14	14		
Manganese (?)	Man	185	185		Not only in lungs (?)
Manganic oxide	Dogs Man	40 60	60		
Mercuric oxide	Dogs	33	33		
Mercuric sulfide	Rats	30	30		
Mercury vapor	Rats	20	20		
Plutonium dioxide	Mice Dogs Rats Dogs Rats	460 >1,400 150 >400 300 180	500		Excretion analysis
Plutonium nitrate	Rats Dogs Rats	12 40 38	38		
Polonium hydroxide	Dogs Rats Rabbits	29 22 18–30 35 30	30	30	10-day multiple excretion analysis
Praseodymium tri-chloride	Mice	24	24		
Protoactinium (Pa$_2$O$_5$ or KPaO$_3$)	Man	1,200 ± 600	1,500		
Radium sulfate	Man	180	180		
Ruthenium dioxide	Mice	230	230		
Silica (glass spheres)	Rats	180	180		

Table 32. (*Continued*)
BIOLOGICAL HALF-LIFE OF RADIOACTIVE SUBSTANCES IN THE LUNG[10]

Substance	Species	Biological Half-Life, days	Recommended Values for Man		Notes
			Single Exposure	Multiple Exposure	
Silicon dioxide	Rats	30	30		
	Rats (Winkelman)	56			
	Rats (Sprague-Dawley)	31	40	200	
	Rats	>180			Multiple
Silver iodide	Mice	<5 hrs	<5 hrs		Based on iodine
Stannic phosphate	Dogs	59	59		
Strontium sulfate	Mice	56	56		
Strontium titanate	Man	500	500		
Strontium chloride	Mice	9	9		
Thallium chloride	Rats	>40–80 (1,400?)	$^{60}/_{1,400}$		
Thallous nitrate	Rats	<2 hrs	<2 hrs		
Thorium dioxide	Rats	>400	500		Intratracheal administration
Uranium dioxide	Rats	270			Multiple
		135			
		141	150		17 days multiple
		169			20 days multiple
		239			58 days multiple
		289			140 days multiple
	Man	120–150			
	Dogs	180			
Uranium dioxide, natural	Dogs	380			Multiple
Uranium dioxide, enriched	Man	300			Multiple
Uranium octoxide	Man	380			
		>120	120	380	
	Rats	300			
Uranium octoxide, fume	Dogs	120	120	380	
Uranyl nitrate	Rats	<1	<1		
Yttrium chloride, with carrier	Guinea pigs	23			Excretion analysis
Yttrium chloride, carrier free	Guinea pigs	12	23		Excretion analysis

Table 32. (Continued)
BIOLOGICAL HALF-LIFE OF RADIOACTIVE SUBSTANCES IN THE LUNG[10]

Substance	Species	Biological Half-Life, days	Recommended Values for Man		Notes
			Single Exposure	Multiple Exposure	
Yttrium chloride	Guinea pigs Mice	21 20			Lung analysis
Zinc phosphate	Dogs	15, 160	160		
Zirconium-niobium	Man	35			
Zirconium oxychloride, carrier free	Guinea pigs	23 30			Excretion analysis Lung analysis
Zirconium oxychloride, with carrier	Guinea pigs	23	35		Excretion analysis
Zirconium oxychloride	Guinea pigs	18			Lung analysis
Zirconium oxalate	Guinea pigs	12	12		Lung analysis

From: Report of the ICRP Task Group on Lung Dynamics. *Health Phys.*, 12: 173–207, 1966. Copyright Pergamon Press, New York.

Table 33. CHARACTERISTICS OF THE STANDARD MAN[16]

Water intake in food	1,000 cc/day
Water intake in fluids	1,200 cc/day
Oxidation	300 cc/day
Total water intake per day	2,500 cc/day
Water excretion from lungs	300 cc/day
Water excretion by feces	200 cc day
Water excretion by urine	1,400 cc/day
Water excretion by sweat	600 cc/day
Total water excretion per day	2,500 cc/day
Air inhaled during 8-hour work day	10^7 cc/day
Air inhaled during 16 hours not at work	10^7 cc/day
Total air inhaled per day	2×10^7 cc/day
Vital capacity of lungs	
Men	3 to 4 liters
Women	2 to 3 liters
Total surface area of respiratory tract	70 m^2
Total water in body	43 kg
Average life span of man	70 years
Occupational exposure of man	8 hours per day 40 hours per week 50 weeks per year
Minimum thickness of epidermis	0.07 mm (7 mg/cm^2)
Depth of blood-forming organs	5 cm

From: Brodsky, A. and Beard, G. V., *A Compendium of Information for Use in Controlling Radiation Emergencies, TID–8206* (Rev.). By permission of the U.S. Atomic Energy Commission, Washington, D.C.

Table 34. MASSES OF CRITICAL ORGANS[16]

Organ	Mass, g
Total body	70,000
Muscles	30,000
Fat	10,000
Bone	7,000
Blood	5,400
Skin	2,000
Liver	1,700
Lungs	1,000
Kidneys	300
Spleen	150
Thyroid	20

From: Brodsky, A. and Beard, G. V., *A Compendium of information for Use in Controlling Radiation Emergencies*, TID–8206 (Rev.). By permission of the U.S. Atomic Energy Commission, Washington, D.C.

maximum permissible concentrations of radionuclides of air and water for organ exposures of 40 hours per week, based on different body organs, and the single-intake doses per microcurie taken into the body by ingestion or by injections for particular radionuclides. In the case of the permissible body burden and concentrations, the more restrictive (smaller) values indicated for each respective nuclide should be used in order to limit the exposure to the critical organ (i.e., the part of the body exposed to the highest relative multiple of its permissible dose rate limit).

More detailed methods of estimating exposures, as applicable to various body organs, are presented in the references, particularly for radionuclides in bone;[11,17] a more detailed treatment of bone dosimetry would be beyond the scope of this handbook. Tables 37 and 38 present the specific dose rates from internal and external exposures to various nuclides, and the gamma-radiation levels from radioactive materials spread over wide land areas, for use in estimating exposures to populations from environmental radioactivity.[18]

Table 35.
PARTICULATE RETENTION IN THE RESPIRATORY TRACT[16]

Distribution	Readily Soluble Compounds, %	Other Compounds, %
Exhaled	25	25
Deposited in upper respiratory passages and swallowed	50	50
Deposited in the lungs (lower respiratory passages)	25 (this is taken up into the body)	25*

* Half of this is eliminated from the lungs and swallowed in the first 24 hours, making a total of 62½ percent swallowed. The remaining 12½ percent is retained in the lungs with a half-life of 120 days; it is assumed that this portion is taken up into body fluids.

From: Brodsky, A. and Beard, G. V., *A Compendium of Information for Use in Controlling Radiation Emergencies*, TID–8206 (Rev.). By permission of the U.S. Atomic Energy Commission, Washington, D.C.

Table 36. MPBB AND MPC VALUES FOR SOME RADIOISOTOPES[11]

Radionuclides and Type of Decay	Organ of Reference*	Maximum Permissible Burden in Total Body (q), µCi	Maximum Permissible Concentrations for a 40-Hour Week		Dose per µCi Administered†	
			MPC_{water}, µCi/cm³	MPC_{air}, µCi/cm³	By Ingestion, rem	By Injection, rem
³H (HTO or ³H₂O) β⁻, soluble	Body tissue Total body	1×10^3 2×10^3	0.1 0.2	5×10^{-6} 8×10^{-6}	2.1×10^{-4} 1.3×10^{-4}	2.1×10^{-4} 1.3×10^{-4}
³H₂, submersion	Skin			2×10^{-3}		
¹⁴C (CO₂), soluble β⁻	Fat Total body Bone	300 400 400	0.02 0.03 0.04	4×10^{-6} 5×10^{-6} 6×10^{-6}	2.4×10^{-3} 5.7×10^{-4} 5.7×10^{-4}	2.4×10^{-3} 5.7×10^{-4} 5.7×10^{-4}
¹⁴C, submersion	Total body			5×10^{-5}		
²²Na, soluble β⁺, γ	Total body G.I. (LLI)	10	1×10^{-3} 0.01	2×10^{-7} 2×10^{-6}	1.9×10^{-2} 3.4×10^{-3}	1.9×10^{-2}
²²Na, insoluble	Lung G.I. (LLI)		9×10^{-4}	9×10^{-9} 2×10^{-7}	6.8×10^{-2}	
²⁴Na, soluble β⁻, γ	G.I. (SI) Total body	7	6×10^{-3} 0.01	1×10^{-6} 2×10^{-6}	6.2×10^{-3} 1.7×10^{-3}	1.7×10^{-3}
²⁴Na, insoluble	G.I. (LLI) Lung		8×10^{-4}	1×10^{-7} 8×10^{-7}	4.8×10^{-2}	
³²P, soluble β	Bone Total body G.I. (LLI) Liver Brain	6 30 50 300	5×10^{-4} 3×10^{-3} 3×10^{-3} 5×10^{-3} 0.02	7×10^{-8} 4×10^{-7} 6×10^{-7} 6×10^{-7} 3×10^{-6}	3.8×10^{-2} 7.4×10^{-3} 2.1×10^{-2} 1.2×10^{-2} 2.4×10^{-3}	5.1×10^{-2} 9.8×10^{-3} 1.7×10^{-2} 3.2×10^{-3}
³²P, insoluble	Lung G.I. (LLI)		7×10^{-4}	8×10^{-8} 1×10^{-7}	8.4×10^{-2}	
³⁸Cl, soluble β⁻, γ	G.I. (S) Total body	9	0.01 0.3	3×10^{-4} 4×10^{-5}	4.9×10^{-3} 6.3×10^{-5}	6.3×10^{-5}

Table 36. (Continued) MPBB AND MPC VALUES FOR SOME RADIOISOTOPES[11]

Radionuclides and Type of Decay	Organ of Reference*	Maximum Permissible Burden in Total Body (q), μCi	Maximum Permissible Concentrations for a 40-Hour Week		Dose per μCi Administered†	
			MPC_{water}, μCi/cm³	MPC_{air}, μCi/cm³	By Ingestion, rem	By Injection, rem
^{38}Cl, insoluble	G.I. (S)		0.01	2×10^{-6}	4.9×10^{-3}	
	Lung			1×10^{-5}		
^{42}K, soluble β^-, γ	G.I. (S)		9×10^{-3}	2×10^{-6}	6.2×10^{-3}	8.8×10^{-4}
	Total body	10	0.02	3×10^{-6}	8.8×10^{-4}	1.5×10^{-3}
	Brain	20	0.04	6×10^{-6}	1.5×10^{-3}	1.5×10^{-3}
	Spleen	20	0.04	6×10^{-6}	1.5×10^{-3}	1.3×10^{-3}
	Muscle	20	0.04	6×10^{-6}	1.3×10^{-3}	6.8×10^{-4}
	Liver	50	0.08	1×10^{-5}	6.8×10^{-4}	
^{42}K, insoluble	G.I. (LLI)		6×10^{-4}	1×10^{-7}	5.9×10^{-2}	
	Lung			9×10^{-7}		
^{45}Ca, soluble β^-, γ	Bone	30	3×10^{-4}	3×10^{-8}	7.9×10^{-2}	1.3×10^{-1}
	Total body	200	2×10^{-3}	3×10^{-7}	8.8×10^{-3}	1.5×10^{-2}
	G.I. (LLI)		0.01	3×10^{-6}	4.4×10^{-3}	
^{45}Ca, insoluble	Lung			1×10^{-7}		
	G.I. (LLI)		5×10^{-3}	9×10^{-7}	1.1×10^{-2}	
^{47}Ca, soluble β^-, γ	Bone	5	1×10^{-3}	2×10^{-7}	1.9×10^{-2}	3.1×10^{-2}
	G.I. (LLI)		2×10^{-3}	5×10^{-7}	3.0×10^{-2}	7.2×10^{-3}
	Total body	10	4×10^{-3}	5×10^{-7}	4.3×10^{-3}	
^{47}Ca, insoluble	G.I. (LLI)		1×10^{-3}	2×10^{-7}	7.6×10^{-3}	
	Lung			2×10^{-7}		
^{59}Fe, soluble β^-, γ	G.I. (LLI)		2×10^{-3}	4×10^{-7}	3.3×10^{-2}	1.4×10^{-1}
	Spleen	20	4×10^{-3}	1×10^{-7}	1.4×10^{-2}	3.6×10^{-2}
	Total body	20	5×10^{-3}	2×10^{-7}	3.6×10^{-3}	9.9×10^{-2}
	Liver	30	6×10^{-3}	2×10^{-7}	9.9×10^{-3}	
	Lung	100	0.02	8×10^{-7}		
	Bone	100	0.03	1×10^{-6}	1.3×10^{-3}	1.3×10^{-2}

Table 36. (Continued) MPBB AND MPC VALUES FOR SOME RADIOISOTOPES[11]

Radionuclides and Type of Decay	Organ of Reference*	Maximum Permissible Burden in Total Body (q), µCi	Maximum Permissible Concentrations for a 40-Hour Week		Dose per µCi Administered†	
			MPC_{water}, µCi/cm³	MPC_{air}, µCi/cm³	By Ingestion, rem	By Injection, rem
^{59}Fe, insoluble	Lung			5×10^{-8}		
	G.I. (LLI)		2×10^{-3}	3×10^{-7}	3.7×10^{-2}	
^{85}Sr, soluble ε, γ	Total body	60	3×10^{-3}	2×10^{-7}	6.8×10^{-3}	2.3×10^{-2}
	Bone	70	4×10^{-3}	4×10^{-7}	1.3×10^{-2}	4.4×10^{-2}
	G.I. (LLI)		7×10^{-3}	2×10^{-6}	8.1×10^{-3}	
^{85}Sr, insoluble	Lung			1×10^{-7}		
	G.I. (LLI)		5×10^{-3}	9×10^{-7}	1.2×10^{-2}	
^{90}Sr, soluble β⁻	Bone	2	1×10^{-5}	1×10^{-9}		
	Total body	3	2×10^{-5}	2×10^{-9}		
	G.I. (LLI)		1×10^{-3}	3×10^{-7}		
^{90}Sr, insoluble	Lung			5×10^{-9}		
	G.I. (LLI)		1×10^{-3}	2×10^{-7}		
^{131}I, soluble β⁻, γ, e⁻	Thyroid	0.7	6×10^{-5}	9×10^{-9}	1.9	1.9
	Total body	50	5×10^{-3}	8×10^{-7}	3.5×10^{-3}	3.5×10^{-3}
	G.I. (LLI)		0.03	7×10^{-6}	1.5×10^{-3}	
^{131}I, insoluble	G.I. (LLI)		2×10^{-3}	3×10^{-7}	2.9×10^{-2}	
	Lung			3×10^{-7}		
^{132}I, soluble β⁻, γ, e⁻	Thyroid	0.3	2×10^{-3}	2×10^{-7}	2.0×10^{-2}	2.0×10^{-2}
	G.I. (SI)	10	0.01	3×10^{-6}	3.7×10^{-3}	
	Total body		0.1	2×10^{-5}	1.7×10^{-4}	1.7×10^{-4}
^{132}I, insoluble	G.I. (ULI)		5×10^{-3}	9×10^{-7}	4.1×10^{-3}	
	Lung			7×10^{-6}		
^{137}Cs, soluble β⁻, γ, e⁻	Total body	30	4×10^{-4}	6×10^{-8}		
	Liver	40	5×10^{-4}	8×10^{-8}		
	Spleen	50	6×10^{-4}	9×10^{-8}		
	Muscle	50	7×10^{-4}	1×10^{-7}		
	Bone	100	1×10^{-3}	2×10^{-7}		
	Kidney	100	1×10^{-3}	2×10^{-7}		
	Lung	300	5×10^{-3}	6×10^{-7}		
	G.I. (SI)		0.02	5×10^{-6}		

Table 36. (Continued) MPBB AND MPC VALUES FOR SOME RADIOISOTOPES[11]

Radionuclides and Types of Decay	Organ of Reference*	Maximum Permissible Burden in Total Body (q), μCi	Maximum Permissible Concentrations for a 40-Hour Week		Dose per μCi Administered†	
			MPC_{water}, μCi/cm³	MPC_{air}, μCi/cm³	By Ingestion, rem	By Injection, rem
^{137}Cs, insoluble	Lung			1×10^{-8}		
	G.I. (LLI)		1×10^{-3}	2×10^{-7}		
^{226}Ra, soluble α, β^-, γ	Bone	0.1	4×10^{-7}	3×10^{-11}		
	Total body		6×10^{-7}	5×10^{-11}		
	G.I. (LLI)	0.2	1×10^{-3}	3×10^{-7}		
^{226}Ra, insoluble	G.I. (LLI)		9×10^{-4}	2×10^{-7}		
^{235}U, soluble α, β^-, γ	G.I. (LLI)		8×10^{-4}	2×10^{-7}		
	Kidney	0.03	1×10^{-4}	5×10^{-10}		
	Bone	0.06	1×10^{-4}	6×10^{-10}		
	Total body	0.4	4×10^{-4}	2×10^{-9}		
^{235}U, insoluble	Lung			1×10^{-10}		
	G.I. (LLI)		8×10^{-4}	1×10^{-7}		
^{239}Pu, soluble α, γ	Bone	0.04	1×10^{-4}	2×10^{-12}		
	Liver	0.4	5×10^{-4}	7×10^{-12}		
	Kidney	0.5	7×10^{-4}	9×10^{-12}		
	G.I. (LLI)		8×10^{-4}	2×10^{-11}		
	Total body	0.4	1×10^{-3}	1×10^{-11}		
^{239}Pu, insoluble	Lung			4×10^{-11}		
	G.I. (LLI)		8×10^{-4}	2×10^{-7}		
^{241}Am, soluble α, γ	Kidney	0.1	1×10^{-4}	6×10^{-12}		
	Bone	0.05	1×10^{-4}	6×10^{-12}		
	Liver	0.4	2×10^{-4}	9×10^{-12}		
	Total body	0.3	4×10^{-4}	2×10^{-11}		
	G.I. (LLI)		8×10^{-4}	2×10^{-7}		
^{241}Am, insoluble	Lung			1×10^{-10}		
	G.I. (LLI)		8×10^{-4}	1×10^{-7}		

The abbreviations G.I., S, SI, ULI, and LLI refer to gastrointestinal tract, stomach, small intestine, upper large intestine, and lower large intestine respectively; critical organs are listed first.
* Data from the Report of ICRP Committee 2.
† Data from Vennart and Minski.

From: Spiers, F. W., *Radioisotopes in the Human Body: Physical and Biological Aspects*, pp. 320–323. Copyright 1968, Academic Press, New York.

Table 37.
SPECIFIC DOSE RATES FROM INTERNAL AND EXTERNAL EXPOSURE TO VARIOUS RADIONUCLIDES[18]

Radionuclide	Maximum Dose to Any Body Organ, rem per μCi inhaled	Maximum External Dose Rate from Point Source, r/hr at 1 meter per μCi
^{3}H (as HTO)	0.00038	negligible
^{14}C as (CO_2)	0.00052	$<1.00 \times 10^{-8}$
^{99}Tc	0.0030	$<1.00 \times 10^{-9}$
^{55}Fe	0.0069	$<2.00 \times 10^{-6}$
^{198}Au	0.021	2.50×10^{-7}
^{35}S	0.021	$<1.00 \times 10^{-8}$
^{60}Co	0.021	1.30×10^{-6}
^{140}La	0.035	9.50×10^{-7}
^{59}Fe	0.037	6.50×10^{-7}
Mixed gross fission products, 180-day irradiation time	0.040	$<7.00 \times 10^{-7}$
^{192}Ir	0.047	5.10×10^{-7}
^{137}Cs	0.058	3.60×10^{-7}
^{140}Ba	0.11	1.54×10^{-6}
^{147}Pm	0.17	$<1.00 \times 10^{-8}$
^{32}P	0.17	$<1.00 \times 10^{-8}$
^{170}Tm	0.23	4.00×10^{-9}
^{45}Ca	0.35	$<1.00 \times 10^{-8}$
^{144}Ce	1.09	2.00×10^{-7}
^{131}I	1.25	2.50×10^{-7}
^{210}Po	11	$<1.00 \times 10^{-8}$
^{90}Sr	38	$<1.00 \times 10^{-8}$
^{233}U (with 20 ppm ^{232}U)	56	2.00×10^{-10}
^{242}Cm	57	$<1.00 \times 10^{-8}$
^{238}U	80 (but very low specific activity)	$<2.00 \times 10^{-9}$
^{235}U + 1% ^{234}U	88	$<2.00 \times 10^{-9}$
U, natural	103 (but low specific activity)	$<2.00 \times 10^{-9}$
^{241}Am	2,280	3.90×10^{-8}
^{230}Th	5,320	9.00×10^{-9}
Th, natural	6,600 (but low specific activity)	$<2.00 \times 10^{-10}$
^{238}Pu	6,850	$<2.00 \times 10^{-8}$
^{239}Pu	7,370	$<1.00 \times 10^{-11}$
^{240}Pu	7,400	$<1.00 \times 10^{-9}$
^{85}Kr	see Reference 3	1.90×10^{-9}

From: Brodsky, A., Balancing Benefit versus Risk in Control of Consumer Items Containing Radioactive Material. *Amer. J Public Health*, 55(12):1971—1992, 1965.

Neutron dosimetry is particularly complicated, since it involves knowledge of the spectra of neutron radiation to which the individual is exposed as well as of the detailed absorption and scattering cross sections for all body elements as the functions of neutron energy.[7,19] Table 39 from the *NBS Handbook 75* presents maximum permissible neutron fluxes in neutron/cm^2 per second required to deliver 5 rems in a 2,000-hour work year. The assumed relative biological effectiveness of neutrons, as well as the flux, is given as a function of neutron energy. These values have been calculated for the multiple-collision tissue dose at the maximum of the depth dose curves,[7,19] taking into account the known neutron interactions, including secondary gamma radiation produced within the body. Snyder and Neufeld[20] have shown that the build-up ratio (maximum multiple-collision dose divided by first-collision dose) is 1.6 ± 0.2 for fast-neutron energies from 0.5 to 10 Mev.

Table 38.
GAMMA-RADIATION LEVELS FROM RADIOACTIVE MATERIALS SPREAD OVER WIDE LAND AREAS[18]

Radionuclides	Physical Half-Life	Activity in Curies per Square Mile to Produce 0.001 Roentgens per Hour at Three Feet Above Ground	
		Smooth Surface	Rough Soil
^{24}Na	15.1 hours	120	
^{59}Fe	45.0 days	340	
^{60}Co	5.2 years	168	
^{95}Zr-^{95}Nb	65.0 days / 37.0 days	—	250
^{103}Ru	40.0 days	—	360
^{106}Ru	1.0 year	—	770
^{131}I	8.1 days	884	
^{137}Cs	27.0 years	620	290
^{140}Ba-^{140}La	12.8 days	144	118
^{144}Ce-^{144}Pr	285.0 days	1,100	5,900
^{170}Tm	129.0 days	53,000	
^{192}Ir	74.0 days	434	
^{226}Ra + daughters	1,620.0 years	263	
^{239}Pu	24,000.0 years	1,310	
^{241}Am	470.0 years	5,660	

From: Brodsky, A., Balancing Benefit versus Risk in the Control of Consumer Items Containing Radioactive Material. *Amer. J. Public Health*, 55(12): 1971—1992, 1965.

Table 39.
AVERAGE YEARLY MAXIMUM PERMISSIBLE NEUTRON FLUX[19]

Neutron Energy, Mev	RBE and Flux	Flux, $ncm^{-2} sec^{-1}$
Thermal	3.0	670
0.0001	2.0	500
0.005	2.5	570
0.02	5.0	280
0.1	8.0	80
0.5	10.0	30
1.0	10.5	18
2.5	8.0	20
5.0	7.0	18
7.5	7.0	17
10	6.5	17
10 to 30		10*

*Suggested limit.

The calculation of dose from implanted radium needles or other implanted radionuclides is also discussed in detail in the literature on radiological physics.[8,13] Basically, the dose rates near multiple-needle arrays have been calculated from the fundamental equation for the dose rate at any point near a single linear source by adding the dose rate contributed by each needle at each point in tissue. In Johns[8] this dose rate is expressed by the equation

$$I \cong \frac{K\rho}{h}(\theta_2 - \theta_1)e^{-\mu d}$$

where I = the dose rate, in r/hour, at point P in Figure 13;
    K = the specific dose-rate constant for radium, 8.26 ± 0.05 R/hr/mg at 1 cm from a ^{226}Ra source filtered by 0.5 mm Pt;
    ρ = the linear density of activity, in mg Ra/cm of length;
    h = the distance from the source, in cm, in a perpendicular to the center line;
    θ_2 and θ_1 = the angles that the perpendicular makes with the lines drawn from P to the far and near ends of the active length of the source (as shown in Figure 13);
    μ = the average absorption coefficient of Pt, in cm^{-1};
    d = the average path length that a Ra gamma must traverse through the Pt filter before it escapes the needle in the direction of the point P.

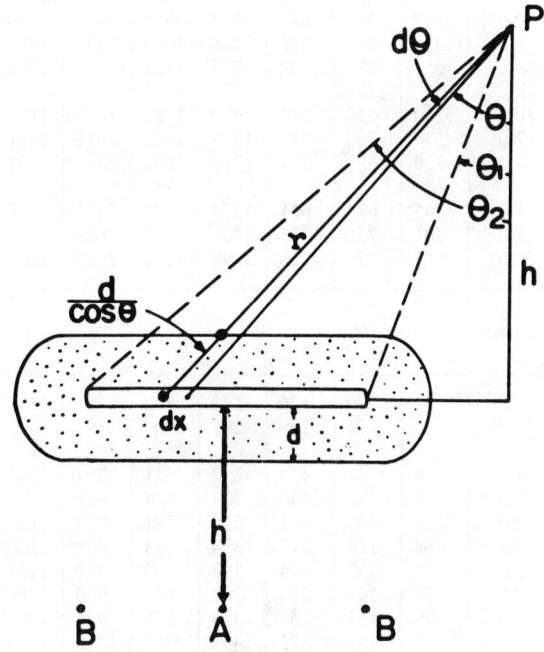

FIGURE 13.
Diagram Illustrating How the Dose May Be Calculated at Points Near a Linear Source.
(From: Johns, H. E., *The Physics of Radiology*, 2nd ed., p. 488.
Charles C Thomas, Springfield, Illinois, 1964.)

More precise calculations have been carried out integrating over the variable self-absorption path, and corrections are available for various thicknesses of Pt.[21] Table 40 gives the mg-hours to deliver 1000 rad at a point h centimeters from the center of a linear source of active length L for filtration of 1.0 and 0.5 mm of Pt. This table was taken from Johns,[8] who obtained these values in rads by converting the original data of Meredith[22] by multiplying the values by 1.055 to correct for the newer measurements on K for Ra and a conversion from roentgens to rads.

Table 40. LINEAR SOURCES[8]

Active Length, L	Treatment Distance, h cm						
	0.5	0.75	1.0	1.5	2.0	2.5	3.0

Filtration 1.0 mm Pt

0.0	35	78	139	314	557	870	1253
0.5	41	82	142	319	562	875	1257
1.0	50	90	151	327	571	882	1266
1.5	58	102	165	338	585	896	1282
2.0	69	118	181	359	604	915	1302
2.5	79	135	200	381	628	942	1329
3.0	92	151	223	407	656	975	1361
3.5	106	170	245	438	690	1011	1397
4.0	118	187	268	467	726	1050	1438
5.0	145	225	317	532	805	1136	1532
6.0	171	263	366	601	891	1231	1638
7.0	197	302	417	673	981	1336	1754
8.0	225	343	468	748	1075	1447	1874
9.0	251	384	522	823	1171	1560	1998
10.0	280	424	576	900	1266	1677	2131
12.0	334	506	683	1056	1467	1916	2411
14.0	390	589	791	1214	1669	2163	2701
16.0	445	672	901	1374	1878	2415	2996
18.0	501	755	1011	1536	2087	2669	3298
20.0	557	839	1118	1699	2298	2929	3603

Filtration 0.5 mm Pt

0.0	32	70	126	283	502	785	1130
0.5	35	74	128	287	504	788	1132
1.0	40	81	134	292	511	794	1142
1.5	50	91	146	303	524	806	1155
2.0	58	103	161	318	541	823	1175
2.5	68	118	177	338	564	844	1200
3.0	78	132	194	362	589	869	1227
3.5	89	148	214	388	615	900	1258
4.0	99	161	232	414	645	935	1289
5.0	122	193	273	467	712	1010	1365
6.0	146	226	313	525	783	1092	1454
7.0	168	257	356	583	857	1180	1551
8.0	190	289	401	646	936	1270	1654
9.0	211	324	444	708	1016	1363	1761
10.0	233	356	489	771	1099	1460	1872
12.0	277	423	576	900	1262	1660	2104
14.0	323	592	666	1030	1430	1866	2342
16.0	368	561	754	1161	1602	2079	2587
18.0	416	629	846	1291	1776	2290	2835
20.0	461	698	936	1426	1951	2502	3091

From: Johns, H. E., *The Physics of Radiology*, 2nd ed. Copyright 1964, Charles C. Thomas, Springfield, Illinois.

Other radionuclides have been used in implants, using the following conversion factors:

1 mCi ^{60}Co = 1.62 mg Ra;

1 mCi ^{137}Cs = 0.40 mg Ra;

1 mCi ^{182}Ta (half-life = 115 days) = 0.74 mg Ra.[21]

Radon gas needles and ^{198}Au needles are used for permanent implants, since they decay to negligible activity after a period of a month. Additional empirical graphs, tables, and rules of thumb are available for estimating the dose rate at any point near almost any given configuration of Ra needles.[8,13]

REFERENCES

1. International Commission on Radiological Units and Measurements, Clinical Dosimetry—Recommendations of the ICRU. *NBS Handbook 87*, p. 38. U.S. Government Printing Office, Washington, D.C., 1963.
2. Roesch, W. C. and Attix, F. H. *Radiation Dosimetry*, 2nd ed., Vol. 1, Ch. 1, Attix, F. H. and Roesch, W. C., eds. Academic Press, New York, 1968.
3. International Commission on Radiological Protection, Report of Committee 2. *Health Phys.*, 3: 1–380, 1960.
4. Attix, F. H. and Roesch, W. C., eds., *Radiation Dosimetry*, 2nd ed., Vol. 2. Academic Press, New York, 1968.
5. Hine, G. J. and Brownell, G. L., eds. *Radiation Dosimetry*, 1st ed. Academic Press, New York, 1956.
6. Morgan, K. Z. and Turner, J. E., eds., *Principles of Radiation Protection—A Textbook of Health Physics*. John Wiley and Sons, New York, 1967.
7. National Committee on Radiation Protection and Measurements, Measurement of Neutron Flux, and Spectra for Physical and Biological Applications, Recommendations of the NCRP. *NBS Handbook 72*. U.S. Government Printing Office, Washington, D. C., 1960.
8. Johns, H. E., *The Physics of Radiology*, 2nd ed. Charles C Thomas, Springfield, Illinois, 1964.
9. Society for Nuclear Medicine, MIRD Report. *J. Nucl. Med.*, Suppl. 1: 5–39, 1968.
10. International Commission on Radiological Protection, Report of the ICRP Task Group on Lung Dynamics. *Health Phys.*, 12: 173–207, 1966.
11. Spiers, F. W., *Radioisotopes in the Human Body: Physical and Biological Aspects*. Academic Press, New York, 1968.
12. International Atomic Energy Agency, Clinical Uses of Whole-Body Counting. *Proceedings of a Panel on the Clinical Uses of Whole-Body Counters*. International Atomic Energy Agency, Vienna, Austria, 1966.
13. Glasser, O., Quimby, E. H., Taylor, L. S., Weatherwax, J. L., and Morgan, R. H., *Physical Foundations of Radiology*. Hoeber Medical Division, Harper and Row, New York, 1963.
14. International Commission on Radiological Protection, Report of Committee 2. *Health Phys.*, 3: 1–380, 1960.
15. Fitzgerald, J. J., Brownell, G. L., and Mahoney, F. J., *Mathematical Theory of Radiation Dosimetry*. Gordon and Breach, New York, 1967.
16. Brodsky, A. and Beard, G. V., compilers and eds., *A Compendium of Information for Use in Controlling Radiation Emergencies, TID-8206* (Rev.). U.S. Atomic Energy Commission, Washington, D.C., 1960.
17. Saenger, E. L., *Medical Aspects of Radiation Accidents*. U.S. Government Printing Office, Washington, D.C., 1963.
18. Brodsky, A., Balancing Benefit versus Risk in the Control of Consumer Items Containing Radioactive Material. *Amer. J. Public Health*, 55(12): 1971–1992, 1965.
19. National Committee on Radiation Protection and Measurements, Measurement of Absorbed Dose of Neutrons, and of Mixtures of Neutrons and Gamma Rays. *NBS Handbook 75*. U.S. Government Printing Office, Washington, D.C., 1961.
20. Snyder, W. S. and Neufeld, J., Calculation of Depth Dose Curves in Tissue for Broad Beams of Fast Neutrons. *Brit. J. Radiol.*, 28: 342, 1955.
21. International Commission on Radiological Units and Measurements, Report of the ICRU. *NBS Handbook 62*. U.S. Government Printing Office, Washington, D.C., 1956.
22. Meredith, W. J., ed., *Radium Dosage, the Manchester Systems*. Williams and Wilkins, Baltimore, 1949.

DETERMINATION OF FACILITIES, EQUIPMENT, AND PROCEDURES REQUIRED FOR VARIOUS TYPES OF OPERATIONS

Allen Brodsky, Sc.D., C.H.P.

There are already many good references on general safety procedures and proper design considerations for laboratories or facilities using radioactive material or radiation sources.[1-5] This chapter will summarize only some of the general considerations in selecting appropriate facilities, equipment and procedures, including Table 41, which arranges the radionuclides according to intrinsic radiotoxicity along with the respective radioactivity levels above which various safeguards or combinations of several safeguards should be considered.[6]

A check list of considerations involved in determining the types of facilities, equipment, and procedures that might be required for handling radioactive materials in quantities large enough to be of some potential hazard to personnel is given below, presented in a format such as one might find in a hazard summary report prepared for a licensing review.

INFORMATION TO BE INCLUDED IN A HAZARD SUMMARY REPORT*

I. A Description of Operations or Applications Involving Radioactive Materials.

Include (as applicable to operations with radioactive material) a description of any chemical, physical, metallurgical, or nuclear processes to be carried out. The description should be detailed enough to permit evaluation of the radiation hazards involved. The forms and amounts of radioactivity to be handled in the proposed processes and any thermal energy likely to be generated should be given.

II. A Description of Facilities and Equipment.

Describe the design criteria for the facility as a whole and for those parts that are essential to the safe operation of the facility. The description should contain enough detail to allow an evaluation of the various methods proposed to minimize any chances of exposing persons on or off site to excessive amounts of radiation or radioactive materials. The description should also cover any activities (in addition to those involving radioactive material) that will be carried on in the building that will house the facility and on the balance of the site. The description should include, but not necessarily be limited to, such items as: shielding to be provided, including types of material, densities, dimensions, and attenuations expected; detailed descriptions of radiation monitoring systems and alarms and their sensitivities; features of the air ventilation and filtration systems that will prevent contamination of unrestricted areas or nearby farms and com-

---

* The items indicated are intended to serve as a relatively exhaustive check list; however, for individual installations many of these items may not be applicable or may require only brief answers. In some cases, items not indicated in this list may deserve mention.

Table 41.
RADIOTOXICITY VERSUS LEVELS ABOVE WHICH VARIOUS SAFEGUARDS MAY BE REQUIRED[6]
(See footnotes at end of table.)

Radionuclides	Physical Properties			Relative Radiotoxicity	
	Physical Half-Life, days	Specific Activity, Ci/g	External Gamma Dose Rate, r/hr at 1 meter per curie	Single Inhalation, in curies, to Give 15 rem to Critical Organ, Ci/15 rem	Single Inhalation, in curies, to Give 15 rem to Lung,[a] Ci/15 rem
Group I					
^{3}H	4.5×10^{3}	9.78×10^{3}	<0.0002	6.15×10^{-2}	—
^{14}C	2.0×10^{6}	4.61	<0.01	2.88×10^{-2}	—
Group II					
^{82}Br	1.5	1.06×10^{6}		7.47×10^{-3}	—
^{51}Cr	27.8	9.20×10^{4}		8.84×10^{-2}	5.30×10^{-3}
^{55}Fe	1.1×10^{3}	2.51×10^{3}		2.17×10^{-3}	2.30×10^{-3}
Group III					
^{35}S	87.1	4.28×10^{4}	<0.01	7.23×10^{-4}	6.90×10^{-4}
^{198}Au	2.7	2.44×10^{5}	0.25	7.25×10^{-4}	5.30×10^{-4}
^{47}Ca	4.9	5.90×10^{5}		2.59×10^{-4}	4.60×10^{-4}
^{132}I	0.097	1.05×10^{7}		4.50×10^{-4}	—
^{141}Ce	32	2.80×10^{4}		7.06×10^{-4}	4.20×10^{-4}
Mixed Fission* Products	*	$<4.00 \times 10^{11}$*		1.40×10^{-4}	*
^{85}Sr	65	2.37×10^{4}		2.00×10^{-3}	2.70×10^{-4}
^{140}La	1.68	5.61×10^{5}	0.95	4.20×10^{-4}	2.60×10^{-4}
^{95}Nb	35	3.93×10^{4}		3.60×10^{-3}	2.30×10^{-4}
^{65}Zn	245	8.20×10^{3}		2.60×10^{-4}	1.50×10^{-4}
^{58}Co	72	3.13×10^{4}		8.40×10^{-3}	1.30×10^{-4}
^{59}Fe	45.1	4.92×10^{4}	0.65	3.00×10^{-4}	1.30×10^{-4}
Group IV					
^{181}Hf	46	1.62×10^{4}		9.94×10^{-5}	1.92×10^{-4}
^{147}Pm	920	9.25×10^{2}	<0.01	8.90×10^{-5}	2.30×10^{-4}
^{32}P	14.3	2.85×10^{5}	<0.01	8.70×10^{-5}	2.10×10^{-4}
^{140}Ba	12.8	7.30×10^{4}	1.54	1.40×10^{-4}	8.60×10^{-5}
^{234}Th	24.1	2.32×10^{4}		8.50×10^{-5}	7.30×10^{-5}
^{85}Kr	3.9×10^{3}	39.6	0.0019	6.90×10^{-2}	5.80×10^{-5}
^{192}Ir	74.5	9.16×10^{3}	0.51	3.20×10^{-4}	6.90×10^{-5}
^{36}Cl	1.2×10^{8}	3.21×10^{-2}		2.70×10^{-3}	5.30×10^{-5}
^{91}Y	58	2.50×10^{4}		5.00×10^{-5}	7.30×10^{-5}
^{182}Ta	112	6.20×10^{3}		1.10×10^{-4}	5.00×10^{-5}
^{45}Ca	164	1.77×10^{4}	<0.01	4.30×10^{-5}	2.60×10^{-4}
^{89}Sr	50.5	2.77×10^{4}		4.00×10^{-5}	8.50×10^{-5}
^{137}Cs	1.1×10^{4}	98.5	0.36	2.60×10^{-4}	3.00×10^{-5}
^{60}Co	1.9×10^{3}	1.14×10^{3}	1.32	2.60×10^{-3}	2.20×10^{-5}
^{144}Ce	290	3.18×10^{3}	0.20	1.40×10^{-5}	1.50×10^{-5}
^{126}I	13.3	7.80×10^{4}		1.40×10^{-5}	7.30×10^{-4}

Table 41. (*Continued*)
RADIOTOXICITY VERSUS LEVELS
ABOVE WHICH VARIOUS SAFEGUARDS MAY BE REQUIRED[6]
(See footnotes at end of table.)

Radionuclides	Physical Properties			Relative Radiotoxicity	
	Physical Half-Life, days	Specific Activity, Ci/g	External Gamma Dose Rate, r/hr at 1 meter per curie	Single Inhalation, in curies, to Give 15 rem to Critical Organ, Ci/15 rem	Single Inhalation, in curies, to Give 15 rem to Lung,[a] Ci/15 rem
Group IV (*cont.*)					
^{154}Eu	5.8×10^3	1.45×10^2		1.30×10^{-5}	1.60×10^{-5}
^{131}I	8	1.24×10^5	0.25	1.20×10^{-5}	7.30×10^{-4}
^{170}Tm	127	6.08×10^3	0.004	3.80×10^{-5}	7.50×10^{-5}
Group V					
^{129}I	6.3×10^9	1.62×10^{-4}		2.30×10^{-6}	1.60×10^{-4}
^{99}Tc		1.71×10^{-2}		9.10×10^{-6}	
Group VI					
^{223}Ra	11.7	5.00×10^4		3.90×10^{-6}	5.30×10^{-7}
^{210}Po	138.4	4.50×10^3	<0.00005	1.30×10^{-6}	5.00×10^{-7}
^{227}Th	18.4	3.17×10^4		5.50×10^{-7}	4.60×10^{-7}
^{90}Sr	1.0×10^4	1.44×10^2	<0.01	3.90×10^{-7}	1.30×10^{-5}
^{210}Pb	7.1×10^3	88		3.20×10^{-7}	5.30×10^{-7}
^{242}Cm	162.5	3.34×10^3	<0.01	3.00×10^{-7}	4.60×10^{-7}
^{233}U	5.9×10^7	0.01 (with 80 ppm ^{232}U)	0.0002 (with 20 ppm ^{232}U)	7.00×10^{-7}	2.70×10^{-7}
^{235}U (+1% ^{234}U)	2.6×10^{11}	2.15×10^{-6}	<0.002	1.10×10^{-6}	2.60×10^{-7}
^{238}U+ ^{Natural}U	1.6×10^{12}	3.34×10^{-7}	<0.002	1.90×10^{-7}	3.00×10^{-7}
^{232}Th+ ^{Natural}Th	5.1×10^{12}	1.11×10^{-7}	<0.0002	2.25×10^{-9}	2.60×10^{-8}
Group VII					
^{147}Sm	4.8×10^{13}	1.95×10^{-8}		7.70×10^{-8}	6.90×10^{-7}
^{144}Nd	7.3×10^{17}	4.97×10^{-13}		7.70×10^{-8}	7.30×10^{-7}
^{226}Ra	5.9×10^5	1.00	0.826	4.90×10^{-8}	1.50×10^{-8}
^{244}Cm	6.7×10^3	82		1.10×10^{-8}	2.30×10^{-7}
Group VIII					
^{243}Am	2.9×10^6	1.85×10^{-1}		7.60×10^{-9}	2.70×10^{-7}
^{241}Am	1.7×10^5	3.21	0.039	6.60×10^{-9}	2.70×10^{-7}
^{237}Np	8.0×10^8	6.90×10^{-4}		5.20×10^{-9}	2.70×10^{-7}
^{227}Ac	8.0×10^3	72		3.00×10^{-9}	6.90×10^{-8}
^{230}Th	2.9×10^7	1.97×10^{-2}	0.009	2.80×10^{-9}	2.30×10^{-8}
^{242}Pu	1.4×10^8	3.90×10^{-3}		2.50×10^{-9}	8.50×10^{-8}
^{238}Pu	3.3×10^4	16.8	<0.02	2.20×10^{-9}	7.30×10^{-8}
^{240}Pu	2.4×10^6	0.227	<0.001	2.00×10^{-9}	8.50×10^{-8}
^{239}Pu	8.9×10^6	0.0617	<0.00001	2.00×10^{-9}	8.50×10^{-8}

Table 41. RADIOTOXICITY VERSUS LEVELS ABOVE WHICH VARIOUS SAFEGUARDS MAY BE REQUIRED[6] (*Continued*)

Radionuclides	Facilities and Equipment								Site
	Chemical Hood Required	Glove Box Required	Glove Box Inside Hot Cell or Cave[b]	1 Absolute Filter[c]	2 Absolute Filters in Series[d]	Continuous General Air Sampler with Alarm[e]	Continuous Exhaust Stack Monitor and Alarm[f]	Building Containment or Controlled Leak Rate[f]	Radius[g] of Low-Population Zone (X), meters
Group I									
^{3}H	1 Ci	10 Ci		10 Ci	10^4 Ci	10^4 Ci	10^5 Ci	10^6 Ci	0.47 $Q^{2/3}$
^{14}C	1 Ci	10 Ci		10 Ci	10^4 Ci	10^4 Ci	10^5 Ci	10^6 Ci	0.47 $Q^{2/3}$
Group II									
^{82}Br	0.1 Ci	1 Ci		1 Ci	10^3 Ci	10^3 Ci	10^4 Ci	10^5 Ci	2.2 $Q^{2/3}$
^{51}Cr	0.1 Ci	1 Ci		1 Ci	10^3 Ci	10^3 Ci	10^4 Ci	10^5 Ci	2.2 $Q^{2/3}$
^{55}Fe	0.1 Ci	1 Ci		1 Ci	10^3 Ci	10^3 Ci	10^4 Ci	10^5 Ci	2.2 $Q^{2/3}$
Group III									
^{35}S	10^{-2} Ci	0.1 Ci		0.1 Ci	10^2 Ci	10^2 Ci	10^3	10^4 Ci	10 $Q^{2/3}$
^{198}Au	↕	↕	4 Ci	↕	↕	↕	↕	↕	↕
^{47}Ca									
^{132}I									
^{141}Ce			1 Ci						
Mixed Fission* Products									
^{85}Sr			1 Ci						
^{140}La									
^{95}Nb									
^{65}Zn									
^{58}Co									
^{59}Fe	10^{-2} Ci	0.1 Ci	2 Ci	0.1 Ci	10^2 Ci	10^2 Ci	10^3	10^4 Ci	10 $Q^{2/3}$

Table 41. RADIOTOXICITY VERSUS LEVELS ABOVE WHICH VARIOUS SAFEGUARDS MAY BE REQUIRED[6] (*Continued*)

Radionuclides	Facilities and Equipment								Site
	Chemical Hood Required	Glove Box Required	Glove Box Inside Hot Cell or Cave[b]	1 Absolute Filter[c]	2 Absolute Filters in Series[d]	Continuous General Air Sampler with Alarm[e]	Continuous Exhaust Stack Monitor and Alarm[f]	Building Containment or Controlled Leak Rate[f]	Radius[g] of Low-Population Zone (X), meters
Group IV									
^{181}Hf	10^{-3} Ci	10^{-2} Ci		10^{-2} Ci	10 Ci	10 Ci	10^2 Ci	10^3 Ci	47 $Q^{2/3}$
^{147}Pm			100 Ci						
^{32}P			100 Ci						
^{140}Ba			0.5 Ci						
^{234}Th									
^{85}Kr			500 Ci						
^{192}Ir			2 Ci						
^{36}Cl									
^{91}Y									
^{182}Ta									
^{45}Ca			100 Ci						
^{89}Sr									
^{137}Cs									
^{60}Co			3 Ci						
^{144}Ce			1 Ci						
^{126}I			5 Ci						
^{154}Eu									
^{131}I			4 Ci						
^{170}Tm	10^{-3} Ci	10^{-2} Ci	250 Ci	10^{-2} Ci	10 Ci	10 Ci	10^2 Ci	10^3 Ci	47 $Q^{2/3}$
Group V									
^{129}I	10^{-4} Ci	10^{-3} Ci		10^{-3} Ci	1 Ci	1 Ci	10 Ci	10^2 Ci	220 $Q^{2/3}$
^{99}Tc	10^{-4} Ci	10^{-3} Ci		10^{-3} Ci	1 Ci	1 Ci	10 Ci	10^2 Ci	220 $Q^{2/3}$

Table 41. RADIOTOXICITY VERSUS LEVELS ABOVE WHICH VARIOUS SAFEGUARDS MAY BE REQUIRED[6] (*Continued*)

Radionuclides	Facilities and Equipment							Site
	Chemical Hood Required	Glove Box Required	Glove Box Inside Hot Cell or Cave[b]	1 Absolute Filter[c]	2 Absolute Filters in Series[d]	Continuous General Air Sampler with Alarm[e]	Continuous Exhaust Stack Monitor and Alarm[f]	Building Containment or Controlled Leak Rate[f]	Radius[g] of Low-Population Zone (X), meters
Group VI									
^{223}Ra	10^{-5} Ci	10^{-4} Ci		10^{-4} Ci	0.1 Ci	0.1 Ci	1 Ci	10 Ci	1,000 Q$^{2/3}$
^{210}Po	↔	↔	20,000 Ci	↔	↔	↔	↔	↔	↔
^{227}Th			100 Ci						
^{90}Sr									
^{210}Pb									
^{242}Cm			100 Ci						
^{233}U			5,000 Ci (500 kg)						
^{235}U (+1% ^{234}U)									
^{238}U+ Natural U									
^{232}Th+ Natural Th	10^{-6} Ci	10^{-4} Ci		10^{-5} Ci	0.1 Ci	0.1 Ci	1 Ci	10 Ci	1,000 Q$^{2/3}$
Group VII									
^{147}Sm	10^{-6} Ci	10^{-5} Ci		10^{-5} Ci	10^{-2} Ci	10^{-2} Ci	0.1 Ci	1 Ci	4,700 Q$^{2/3}$
^{144}Nd		↔			↔	↔	↔	↔	
^{226}Ra									
^{244}Cm	10^{-6} Ci	10^{-5} Ci		10^{-5} Ci	10^{-2} Ci	10^{-2} Ci	0.1 Ci	1 Ci	4,700 Q$^{2/3}$

Table 41. RADIOTOXICITY VERSUS LEVELS ABOVE WHICH VARIOUS SAFEGUARDS MAY BE REQUIRED[6] (Continued)

Radionuclides	Facilities and Equipment								Site
	Chemical Hood Required	Glove Box Required	Glove Box Inside Hot Cell or Cave[b]	1 Absolute Filter[c]	2 Absolute Filters in Series[d]	Continuous General Air Sampler with Alarm[e]	Continuous Exhaust Stack Monitor and Alarm[f]	Building Containment or Controlled Leak Rate[f]	Radius[g] of Low-Population Zone (X), meters

Group VIII

^{243}Am	10^{-7} Ci	10^{-6} Ci		10^{-6} Ci	10^{-3} Ci	10^{-3} Ci	10^{-2} Ci	0.1 Ci	22,000 $Q^{2/3}$
^{241}Am	↔	↔	25 Ci	↔	↔	↔	↔	↔	↔
^{237}Np									
^{227}Ac			100 Ci						
^{230}Th									
^{242}Pu			50 Ci						
^{238}Pu									
^{240}Pu									
^{239}Pu	10^{-7} Ci	10^{-6} Ci	1,000 Ci†	10^{-6} Ci	10^{-3} Ci	10^{-3} Ci	10^{-2} Ci	0.1 Ci	22,000 $Q^{2/3}$

Table 41. RADIOTOXICITY VERSUS LEVELS ABOVE WHICH VARIOUS SAFEGUARDS MAY BE REQUIRED[6] (Continued)

Radionuclides	Procedures						
	Personnel Monitoring and/or Appropriate Shielding vs. External Gamma Radiation	Occasional Excretion Radioassay Spot Checks of Operating Personnel	Routine Excretion Assay of All Operating Personnel	Emergency Dosimeters Worn to Measure High External Doses	Routine Environmental Monitoring of Site and Community	Preplanned Written Emergency Procedures and Drills	Written Routine Operating Procedures	Written Preoperational Analysis of Maximum Credible Accidents

Group I

^{3}H		10 Ci	10^{2} Ci		10^{3} Ci	10^{4} Ci	10^{6} Ci	10 Ci
^{14}C		10 Ci	10^{2} Ci		10^{3} Ci	10^{4} Ci	10^{6} Ci	10 Ci

Table 41. RADIOTOXICITY VERSUS LEVELS ABOVE WHICH VARIOUS SAFEGUARDS MAY BE REQUIRED[6] (*Continued*)

Radionuclides	Procedures							
	Personnel Monitoring and/or Appropriate Shielding vs. External Gamma Radiation	Occasional Excretion Radioassay Spot Checks of Operating Personnel	Routine Excretion Assay of All Operating Personnel	Emergency Dosimeters Worn to Measure High External Doses	Routine Environmental Monitoring of Site and Community	Preplanned Written Emergency Procedures and Drills	Written Routine Operating Procedures	Written Preoperational Analysis of Maximum Credible Accidents
Group II								
^{82}Br	0.50 Ci	1 Ci	10 Ci	50 Ci	10^2 Ci	10^3	10^5 Ci	1 Ci
^{51}Cr		1 Ci	10 Ci		10^2 Ci	10^3	10^5 Ci	1 Ci
^{55}Fe		1 Ci	10 Ci		10^2 Ci	10^3	10^5 Ci	1 Ci
Group III								
^{35}S	0.40 Ci	0.1 Ci	1 Ci	40 Ci	10 Ci	10^2 Ci	10^4 Ci	0.1 Ci
^{198}Au								
^{47}Ca								
^{132}I	0.10 Ci			10 Ci				
^{141}Ce								
Mixed Fission* Products								
^{85}Sr								
^{140}La								
^{95}Nb								
^{65}Zn								
^{58}Co								
^{59}Fe	0.20 Ci	0.1 Ci	1 Ci	20 Ci	10 Ci	10^2 Ci	10^4 Ci	0.1 Ci

Table 41. RADIOTOXICITY VERSUS LEVELS ABOVE WHICH VARIOUS SAFEGUARDS MAY BE REQUIRED[6] (Continued)

Radionuclides	Procedures							
	Personnel Monitoring and/or Appropriate Shielding vs. External Gamma Radiation	Occasional Excretion Radioassay Spot Checks of Operating Personnel	Routine Excretion Assay of All Operating Personnel	Emergency Dosimeters Worn to Measure High External Doses	Routine Environmental Monitoring of Site and Community	Preplanned Written Emergency Procedures and Drills	Written Routine Operating Procedures	Written Preoperational Analysis of Maximum Credible Accidents
Group IV								
^{181}Hf		10^{-2} Ci	0.1 Ci		1 Ci	10 Ci	10^3 Ci	10^{-2} Ci
^{147}Pm	10.00 Ci			1,000 Ci				
^{32}P	10.00 Ci			1,000 Ci				
^{140}Ba	0.05 Ci			5 Ci				
^{234}Th								
^{85}Kr	50.00 Ci			5,000 Ci				
^{192}Ir	0.02 Ci			20 Ci				
^{36}Cl								
^{91}Y								
^{182}Ta								
^{45}Ca	10.00 Ci			1,000 Ci				
^{89}Sr								
^{137}Cs								
^{60}Co	0.30 Ci			30 Ci				
^{144}Ce	0.10 Ci			10 Ci				
^{126}I	0.05 Ci			5 Ci				
^{154}Eu								
^{131}I	0.40 Ci			40 Ci				
^{170}Tm	25.00 Ci	10^{-2} Ci	0.1 Ci	2,500 Ci	1 Ci	10 Ci	10^3 Ci	10^{-2} Ci
Group V								
^{129}I	100.00 Ci	10^{-3} Ci	10^{-2} Ci	10,000 Ci	0.1 Ci	1 Ci	10^2 Ci	10^{-3} Ci
^{99}Tc		10^{-3} Ci	10^{-2} Ci		0.1 Ci	1 Ci	10^2 Ci	10^{-3} Ci

Table 41. RADIOTOXICITY VERSUS LEVELS ABOVE WHICH VARIOUS SAFEGUARDS MAY BE REQUIRED[6] *(Continued)*

Radionuclides	Procedures							
	Personnel Monitoring and/or Appropriate Shielding vs. External Gamma Radiation	Occasional Excretion Radioassay Spot Checks of Operating Personnel	Routine Excretion Assay of All Operating Personnel	Emergency Dosimeters Worn to Measure High External Doses	Routine Environmental Monitoring of Site and Community	Preplanned Written Emergency Procedures and Drills	Written Routine Operating Procedures	Written Preoperational Analysis of Maximum Credible Accidents
Group VI								
^{223}Ra		10^{-4} Ci ←────	10^{-3} Ci ←────		10^{-2} Ci ←────	0.1 Ci ←────	10 Ci ←────	10^{-4} Ci ←────
^{210}Po	10.00 Ci			1,000 Ci				
^{227}Th								
^{90}Sr	10.00 Ci			1,000 Ci				
^{210}Pb								
^{242}Cm	1.00 Ci			100 Ci				
^{233}U	50.00 Ci			5,000 Ci				
^{235}U (+1% ^{234}U)								
^{238}U + Natural U								
^{232}Th + Natural Th		10^{-4} Ci	10^{-3} Ci		10^{-2} Ci	0.1 Ci	10 Ci	10^{-4} Ci
Group VII								
^{147}Sm		10^{-5} Ci ←────	10^{-4} Ci ←────		10^{-3} Ci ←────	10^{-2} Ci ←────	1 Ci ←────	10^{-5} Ci ←────
^{144}Nd								
^{226}Ra		10^{-5} Ci	10^{-4} Ci		10^{-3} Ci	10^{-3} Ci	1 Ci	10^{-5} Ci
^{244}Cm								

Table 41. RADIOTOXICITY VERSUS LEVELS ABOVE WHICH VARIOUS SAFEGUARDS MAY BE REQUIRED[6] (Continued)

Radionuclides	Procedures							
	Personnel Monitoring and/or Appropriate Shielding vs. External Gamma Radiation	Occasional Excretion Radioassay Spot Checks of Operating Personnel	Routine Excretion Assay of All Operating Personnel	Emergency Dosimeters Worn to Measure High External Doses	Routine Environmental Monitoring of Site and Community	Preplanned Written Emergency Procedures and Drills	Written Routine Operating Procedures	Written Preoperational Analysis of Maximum Credible Accidents
Group VIII								
^{243}Am		10^{-6} Ci	10^{-5} Ci		10^{-4} Ci	10^{-3} Ci	0.1 Ci	10^{-6} Ci
^{241}Am	0.25 Ci	↔	↔	25 Ci	↔	↔	↔	↔
^{237}Np								
^{227}Ac								
^{230}Th	1.00 Ci			100 Ci				
^{242}Pu								
^{238}Pu	0.50 Ci			50 Ci				
^{240}Pu	10.00 Ci†			1,000 Ci†				
^{239}Pu		10^{-6} Ci	10^{-5} Ci		10^{-4} Ci	10^{-3} Ci	0.1 Ci	10^{-6} Ci

Note: The curie levels for protection against inhalation and contamination are intended for dry, dusty materials that may be easily dispersed in concentrated form. For liquids, or where active material will be diluted by other materials, the above safeguard levels may be raised by factors of 10 or more. For simple storage of stock solutions, or where operations are conducted only with materials of specific activity much less than 0.1 microcurie per milligram, multiply by 100 or more, depending on the nature of the material and the particular combination of safeguards selected (see text).

* For mixed gross fission products of 0-to-4-year operation, the overall relative hazard per curie changes relatively slowly with decay time for decay times shorter than 30 days, although for very short decay times (<1 day) the thyroid dose predominates (see Brodsky, A., Criteria for Acute Exposure to Mixed Fission-Product Aerosols, *Health Phys.*, 11: 1017–1032, 1965).

† Based on criticality.

a Insoluble materials.
b Based on gamma dose rates.
c In the exhaust from active atmosphere.
d In the active exhaust.
e In work areas.
f To protect public.
g Distance beyond which cloud dose is less than 15 rem for Q curies released, where Q = fc for C curies in process.

From: Brodsky, A., Determining Industrial Hygiene Requirements for Installations Using Radioactive Materials. *Amer. Ind. Hyg. Ass. J.*, 26: 294–310, 1965.

munities either during routine or accident conditions; hood or glove box construction details; doors and interlocks for preventing inadvertent entry into high-radiation areas; remote-handling equipment; storage facilities; and decontamination and waste disposal facilities in relation to expected levels of waste, including a description of instrumentation for monitoring all waste effluents.

III. A Description of the Site.

Include: a map of the area showing the location of the site and indicating the use to which the surrounding land is put (e.g., industrial, commercial, agricultural, residential); location of sources of potable or industrial water supply, watershed areas, and public utilities; and a scale plot plan of the site showing the proposed location of the facility in which the radioactive materials will be stored or used. Exclusion area boundaries should be well defined, and means of controlling access to the exclusion area should be specified. Meteorological, hydrological, geological, seismological, and population data needed to evaluate any possible radioactivity hazards to the public should also be included.

IV. A Description of the Organization and Qualifications of Personnel.

This should include: a chart showing the organizational relationships between managers, supervisors, operators, and radiation-safety personnel concerned with operations involving radioactive materials; the composition of cognizant committees and their responsibilities, authority, frequency of meetings, and extent of review of uses and users; operations for which detailed operating procedures will be written, and operating procedures themselves where appropriate; control of procurement, inventory, and records; and the qualifications of radiation-safety personnel and operating personnel.

V. A Description of Standard Procedures Affecting Radiation Safety.

This should include, but not necessarily be limited to: procedures for pre-operational checking of all facilities and equipment important to radiation safety; procedures for initiating dry runs on any new processes to be carried out; procedures for routine maintenance and calibration of radiation instruments, alarms, and emergency devices in relation to their required sensitivities or responses; programs for personnel training in radiation safety; procedures for preventing and controlling fires or explosions; procedures—such as locked controls or doors, check lists, and close supervision—for minimizing operational mishaps; and plans for investigating unusual or unexpected incidents. The planned radiation-safety program should be described, including any procedures for: environmental surveys of air, water, soil, and vegetation in the vicinity of the facility, both before and periodically after radioactive material is received; receiving and unloading material; storing radioactive materials; labeling and restricting radiation areas in accordance with 10 CFR, Part 20, or other applicable federal, state, or local regulations; monitoring radiation levels and concentrations of radioactive materials in air and water, both in controlled and uncontrolled areas; monitoring and recording cumulative amounts of radioactive materials in air and liquid effluents; external and internal dosimetry of personnel; decontamination of facilities and personnel; leak-testing of sources; packaging and shipping radioactive materials; traffic control of materials and personnel to and from contaminated areas; and disposing of radioactive wastes.

VI. A Description of Emergency Plans for Handling Radiation Incidents.

Plans to be carried out in the event of possible unexpected incidents should be presented in detail. These plans should be related to plausible incidents that could occur as a result of operational mistakes or equipment failures in the proposed operations, or as a result of fire, electric-power failure, flood, earthquake, storm, strike, riot, or air raid, as applicable to the proposed facility and site. The emergency plans should include, but not necessarily be limited to: procedures for detecting an incident and activating emergency plans; emergency organization and command responsibilities; coordination and communication between various emergency teams, such as fire, medical, health physics, and rescue; coordination with local civil authorities; procedures for evacuating personnel and processing them at a decontamination and first-aid center; procedures or facilities for preventing the dispersal of radioactivity to farms or communities in the vicinity; dosimetry and follow-up procedures on personnel exposed to high radiation doses or high concentrations of radioactivity; means for detecting and removing wound contamination; procedures for reentry and recovery of facilities; and instruction and drill of personnel in emergency procedures.

VII. An Analysis of Credible Accidents and Their Effects.

The possible effects of postulated extreme and unexpected accidents within the possibilities and limitations of the proposed operations should be analyzed in detail. This includes calculations of estimated doses received and numbers of persons affected, both in the facility and in the surrounding community. Alternate accidents should be described where they are possible and where the effects are not delimited by those of accidents already described. After each description of the effects of an incident, the procedures or facilities that will serve to prevent or minimize these effects should be indicated.

In most facilities, some of the considerations listed above would not be pertinent. Furthermore, for quantities of radionuclides below the respective amounts shown in Table 41 for various facilities, equipment, and procedures, the individual safeguards indicated in this table would probably not be required, even if the radionuclide were to be handled in its most hazardous, dispersible form. Appropriate multiplication factors are suggested in the footnote of Table 41 and in Reference 2 for multiplying the quantities in the table in cases where the radioactive material is in liquid form, or in more diluted forms where the possibility of inhaling hazardous quantities is reduced even in the event of an accident involving total release of the material.

It should be noted that the radionuclide groups covered in Table 41 range over about eight orders of magnitude of radiotoxicity,[6] rather than the four groups often recommended previously,[2] since it has been found that the order of the radionuclides in terms of maximum dose per microcurie inhaled, or in terms of MPC_{air}, ranges over about eight orders of magnitude in the same fashion.[6] For most of these radionuclides the specific activities of the pure radionuclide are so high that, in practice, the maximum dose per microcurie inhaled is indexed as the only fundamental way of ranking these nuclides without the addition of extrinsic considerations of the specific chemical forms and processes unique to each facility. Such an order automatically takes into account a certain amount of ingestion, as described earlier in the presentation of the ICRP lung models.[7,8]

The rationale behind the selection of the various levels of radioactivity above which the various safeguards might be required is outlined in Reference 6; it is based on a quantitative consideration of measurements of typical amounts of radioactivity resuspended from contaminated surfaces as well as on experience with the safety factors provided by various safeguards and meteorological dispersion. Thus, although some professional experience would be necessary in designing suitable facilities combining the various safeguards that may be required, the table may be used as a guide to help avoid expensive over-design as well as the possibility of overlooking needed safeguards for work with higher levels of potentially radiotoxic materials. Basically, the table serves as a starting point or baseline from which hazard evaluation and facility design and operations may be planned in a consistent and safe manner, even though the radionuclides vary widely in their fundamental radiotoxicity. Considerations of probability of intake, systematic absorption, and retention in critical organs may then be taken into account in order to reduce the requirements according to specific knowledge of operations to be conducted in the facility.[6]

The design of open hoods for handling radioactive materials is carried out according to the same principles of capturing contaminated air as that already developed for other industrial-hygiene purposes.[9,10] Generally, a face velocity of air at the opening of an ordinary laboratory hood would be about 150 feet per minute, to guard against backdrafts resulting from movement of materials within the hood. Face velocities should be checked regularly with a velometer. However, when materials are in a state of high radionuclide purity or specific activity, and when processes have a high potential for dispersing the material (e.g., dry, loose powder under mechanical agitation or pulverization, or experimental chemical procedures with a potential for volatilization or large energy release), completely enclosed operations within a "dry-box" or glove box have often been required for quantities in process above those given in the glove-box column of Table 41. Failure to provide complete containment when operating with loose radioactive materials above the levels indicated has on specific occasions resulted in significant personnel exposures. Limited experience with radiation-accident cases lends some validation to the semi-empirical methods of derivation of Table 41.[6]

A typical radioisotope hood is shown in Figure 14, and a sealed glove box with exhaust fan and an absolute filter in the intake is shown in Figure 15. For work with extremely hazardous quantities of radionuclides in loose form, an additional filter may also be placed in the dry-box exhaust preceding the blower. Also, provisions for filling the dry-box with a suitable inert gas are made when handling pyrophoric or potentially combustible materials. Special procedures are required when replacing the gloves of a glove box, or when "bagging" materials or wastes in or out of the box through the double airlock.[11,12]

The arrangement of hoods, dry-boxes, and other laboratory furnishings required in a radiochemical laboratory needs special planning, preferably with the assistance of an engineer experienced in operations with radioactive materials. Figure 16 shows the layout of a simple radiochemical laboratory, such as one might use, for example, for mixed-fission-product-separations chemistry below about 10 millicuries, or for plutonium chemistry with less than 1 microcurie in process;[6] it illustrates how the general air flow in the laboratory should be directed towards the hood, where higher amounts of radioactive contamination are likely to be found, and away from the areas that are to be kept free of radioactivity. Storage areas for radioactivity are shown, as well as waste containers for both radioactive and nonradioactive materials.

In selecting furniture for such a laboratory, consideration should be given to using surfaces that may be easily decontaminated when necessary.[13] Paints forming coatings that may be stripped from surfaces as necessary are commercially available. Floors should generally be covered with a paint or plastic coating free of cracks and may be covered with wax to simplify decontamination procedures.

Although federal and state regulations allow only very small amounts of radioactivity to be disposed of through sanitary sewers, plumbing should be planned to insure that there are no connections between radioactive waste lines and drinking water supplies. For production facilities where larger amounts of radioactive material may need to be washed down drains after an accident, or where storage of larger quantities of radioactive waste may be required, special waste-treatment facilities and waste-storage tanks must be designed.[14] Generally, sanitary engineers who have specialized in radioactive-waste treatment should be consulted in the design of chemical processes and equipment for handling larger amounts of wastes. Smaller amounts of waste may be stored in drums for collection by AEC-licensed commercial waste-disposal firms.

There are various items of special equipment for use in handling radioactive materials at a distance, even for the smaller laboratories, when the quantities of material in process emit sufficient beta or gamma radiation to require tongs, manipulators, or shields, to reduce external radiation exposures. Most of the items available for ordinary radiochemical laboratories are commercially available and may be found advertised in the journals servicing the nuclear professions.

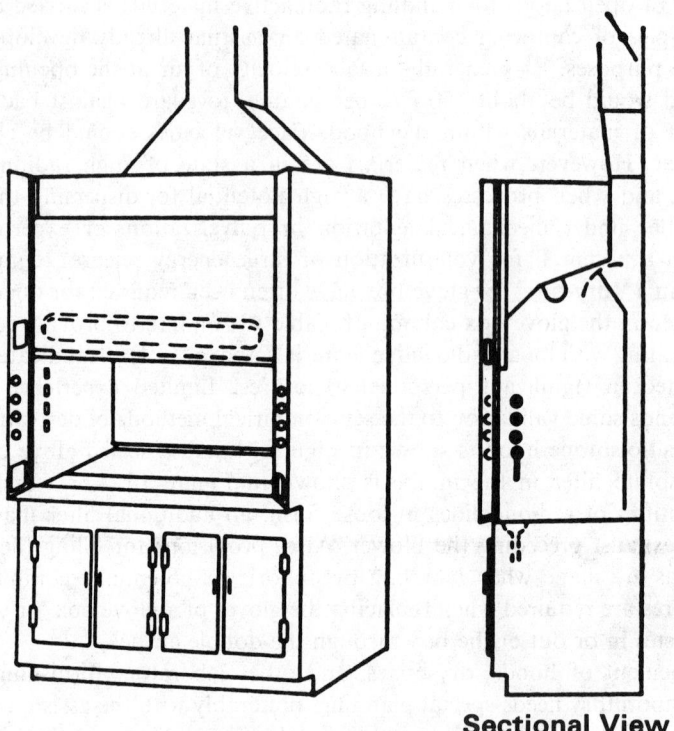

Sectional View

FIGURE 14. Radioisotope Hood.
(From: Blatz, H., ed., *Radiation Hygiene Handbook*,
p. 9-5. McGraw-Hill, New York, 1959.)

When the amounts of radioactive material emit bremsstrahlung or gamma radiation in such intensities that massive shielding would be required between the operator and the material in process, operations may be carried out inside a "hot cell" or "cave". In this case the operator would handle the materials and equipment through special mechanical manipulator arms that provide a sense of touch directly to the operator;[15] glove boxes may also be utilized within these hot cells, to avoid excessive contamination of the hot cell by personnel entering it from time to time in the event that processes must be changed. In some cases disposable dry-boxes may be utilized for economy in disposing of the contaminated wastes when the operations are completed.[16]

The provision of appropriate radiation shielding for facilities where large radiation levels are produced is a subject that has received intensive investigation by specialists. For shielding such complex radiation fields as those originating from a nuclear reactor, nuclear engineers must draw on complex theoretical considerations of radiation scattering and absorption, as

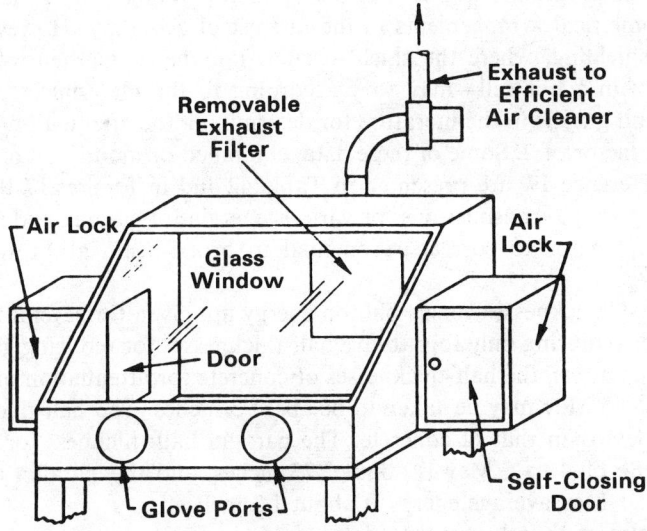

FIGURE 15.
Typical Glove-Box Design for Handling of Alpha and Beta Materials.
(From: Blatz, H., ed., *Radiation Hygiene Handbook*, p. 9-9.
McGraw-Hill, New York, 1959.)
Air flow = 50 cfm/sq. ft. of open-door area, entry loss = 0.25 VP plus
dirty-filter resistance, duct velocity = 3,500 fpm. Filters: 1) inlet
dust filters in doors, 2) prefilter at exhaust connection to hood, and
3) after cleaner for final air cleaning. All facilities totally enclosed
in hood; exterior controls may be advisable. Arm-length rubber gloves
are sealed to glove port rings; strippable plastic on interior and air
cleaner on exhaust may be used, to facilitate decontamination of the
system. Filter units may be installed in the doors, to allow the air
flow necessary for burners, etc. (American Conference of Governmental
Hygienists, *Industrial Ventilation Manual*.)

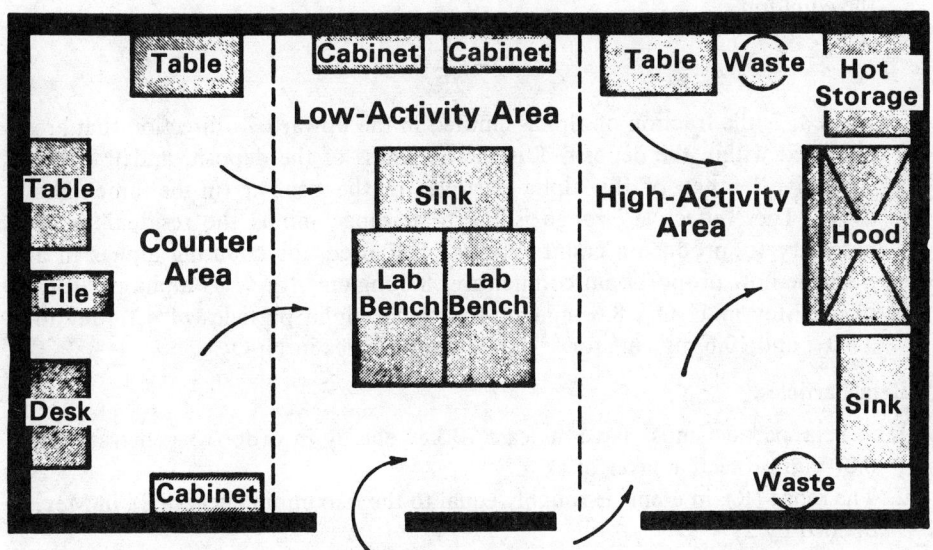

FIGURE 16.
Simple Radiochemical Laboratory.
(From: Blatz, H., ed., *Radiation Hygiene Handbook*,
p. 9-9. McGraw-Hill, New York, 1959.)

well as on computer programs, to arrive at the optimal combination of materials and thicknesses and of geometrical arrangements in the interest of economy. However, for designing most laboratory shielding, where the shield itself is not the major item of expense and its functions—such as in a hot cell—may vary according to the material in process, there are abundant tables and graphs in the literature for determining the attenuation of most shielding materials within a factor of 2. Some of these data, combined or modified from various sources in literature in Reference 17, are presented in Table 42 and in Figures 17 through 20. These tables and figures give the transmissions for various energies of gamma and neutron radiation where the build-up factor to correct exponential to broad-beam absorption is taken into account.

In Table 43 half-thicknesses versus photon energy are given for several materials, for use in designing shields requiring only four to five half-thicknesses for reducing gamma dose rates. For crude approximations, the half-thicknesses of concrete for attenuation of pile neutrons or fast neutrons up to 15 Mev may be taken to be about 3 inches. The half-thickness of water is about 10 percent less than that of concrete. The paraffin half-thickness for neutrons spectra from PuBe or RaBe of 4 to 5 Mev is about 2.73 inches, and the paraffin half-thickness for fission neutrons of 1 Mev average energy is about 1.3 inches.[17]

Additional rules of thumb useful in determining radiation exposure rates and required shielding protection are given below.

RULES OF THUMB

Alpha Particles.

1. An alpha particle of 7.5 Mev just penetrates the 0.07-mm minimum protective layer of skin.
2. Plutonium has an alpha activity of about 140,000 alphas per minute per microgram; natural-uranium activity is 1.5 alphas per minute per microgram; ^{238}U emits about 0.741 alpha per minute per microgram; ^{232}Th emits 0.247 alpha per minute per microgram; and ^{237}Np emits 1,519 alphas per minute per microgram.
3. Self-absorption of a thin layer of alpha-emitting deposit may be estimated by the equation

$$f_a = \frac{T}{2R},$$

where f_a is the fraction of alphas emitted in the upward 2π direction that are absorbed within the deposit, T is the thickness of the deposit, and R is the "effective" range of the alpha particles in the deposit (in the same units as T). The "effective" range is the total range minus the residual range necessary to produce a count. Values of R used for counting alphas in a methane-flow proportional counter are: 8.6 mg/cm^2 for ^{238}U alpha particles of 4.2 Mev in U_3O_8; 8.6 mg/cm^2 for ^{232}Th alpha particles of 4.1 Mev in ThO_2; and 10.6 mg/cm^2 for ^{237}Np alpha particles in NpO_2.

Beta Particles.

1. A beta particle must have at least 70 kev energy in order to penetrate the 0.07-mm protective layer of skin.
2. The range (R), in g/cm^2 is roughly equal to the maximum energy (E), in Mev, divided by 2:

$$= R\frac{E}{2}.$$

3. The range of beta particles in air is about 12 feet per Mev.

Table 42.
SHIELD THICKNESS VERSUS GAMMA-DOSE TRANSMISSION[17]

Broad-Beam Transmission	Shield Thickness, inches		
	Concrete* (147 lbs/cu. ft.)	Iron	Lead

Radium (11 principal gammas, 0.24 to 2.20 Mev)

0.1	10	3.1	1.6
0.01	19	6.2	3.5
0.001	28	9.1	5.5
0.0001	38	12.0	7.8
0.00001	47	15.3	10.2

Cobalt-60 (1.33 + 1.17 Mev per disintegration)

0.1	11	3.2	1.7
0.01	19	6.0	3.3
0.001	27	8.8	4.8
0.0001	35	11.4	6.5
0.00001	43	14.6	8.1

Cesium-137 (0.66 Mev)

0.1	8.5	2.6	0.85
0.01	15	4.7	1.7
0.001	22	6.8	2.5
0.001	28	8.9	3.4
0.00001	34	11.0	4.2

Iridium-192 (gammas from 0.13 to 0.87 Mev, averaging 0.3 Mev)

0.1	7		0.48
0.01	13		1.1
0.001	18.3		1.9
0.0001	24		2.6
0.00001	30		3.5

Gold-198 (0.41-, 0.68-, and 1.1-Mev gammas)

0.1	6.6		0.35
0.01	12.0		0.83
0.001	17.4		1.7
0.0001	22.6		2.8
0.00001	28.0		4.3

Iodine-131 (0.08 to 0.723 Mev, predominantly 0.36 Mev)

0.1	6		
0.01	12		
0.001	18		

* After several mean free paths, each ten inches of concrete reduce the radiation by another factor of 10.

Table 42. (Continued)
SHIELD THICKNESS VERSUS GAMMA-DOSE TRANSMISSION[17]

Barium-140 + Lanthanum-140 (0.030 to 2.5 Mev, averaging about 1.6 Mev)

Broad-Beam Transmission	Water	Aluminum	Iron	Lead	Uranium
0.1	25	9.8	3.4	1.6	0.87
0.01	44	18	6.4	3.4	2.0
0.001	64	27	9.2	5.2	3.0
0.0001	81	35	11.8	7.1	4.1
0.00001	104	44	14.8	9.0	5.2
0.000001		51			6.3

From: Brodsky, A. and Beard. G. V., *A Compendium of Information for Use in Controlling Radiation Emergencies*, TID-8206 (Rev.). By permission of the U.S. Atomic Energy Commission, Washington, D.C.

Table 43.
HALF-THICKNESS VERSUS PHOTON ENERGY FOR SEVERAL MATERIALS[17]

Photon Energy, Mev	Half-Thickness, inches			
	Water	Concrete	Iron	Lead
0.2	2.0	0.8	0.28	0.06
0.5	3.0	1.1	0.45	0.17
1.0	4.1	1.7	0.63	0.35
1.5	4.7	2.0	0.70	0.46
2.0	5.7	2.3	0.82	0.53
2.5	6.8	2.7	0.88	0.57
3.0	7.7	3.0	0.92	0.59
4.0	8.4	3.3	1.02	0.59
5.0	9.2	3.7	1.10	0.58

From: Brodsky, A. and Beard, G. V., *A Compendium of Information for Use in Controlling Radiation Emergencies*, TID-8206 (Rev.). By permission of the U.S. Atomic Energy Commission, Washington, D.C.

4. The air dose rate at 1 foot from a beta point source is about 200 C rads per hour, neglecting self-absorption and air absorption, where C is the number of curies. Variation with energy is small for most beta emitters.
5. Readings on Geiger-Müller tube survey meters calibrated with gamma radiation in mr/hr must be multiplied by 2 to measure beta radiation in mrad/hr with the window open.
6. Beta-ray surface dose rates for several materials are given in Table 44.
7. The fallout beta dose to the skin from contact with deposited fallout within 200 days after detonation is up to 150 times the gamma dose rate in rads per hour at 3 feet above ground. (Project 37.2, Operation Teapot, May 1955.)
8. The dose to a 30-g adult thyroid from 1 microcurie of ^{131}I is about 1 rad delivered within several weeks after intake.
9. The beta activity from fission products produced in a short-criticality burst is: beta activity = $294\, t^{-1.2}$ curies per watt-second, where t is the time in seconds since the burst, ranging from 10 seconds to 100 days. For a weapon in the kiloton-TNT range, this becomes: beta activity = $1.17 \times 10^{13}\, t^{-1.2}$ curies/kiloton TNT.

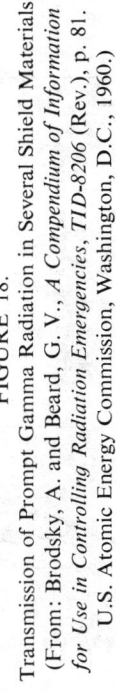

FIGURE 18.
Transmission of Prompt Gamma Radiation in Several Shield Materials.
(From: Brodsky, A. and Beard, G. V., *A Compendium of Information for Use in Controlling Radiation Emergencies*, *TID-8206* (Rev.), p. 81. U.S. Atomic Energy Commission, Washington, D.C., 1960.)

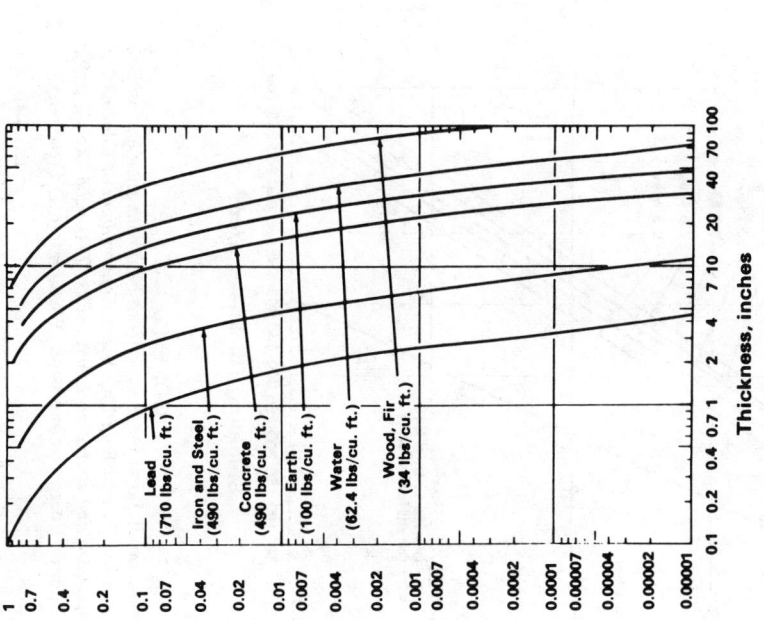

FIGURE 17.
Transmission of Fission Product Gamma Radiation in Several Shield Materials.
(From: Brodsky, A. and Beard, G. V., *A Compendium of Information for Use in Controlling Radiation Emergencies*, *TID-8206* (Rev.), p. 80. U.S. Atomic Energy Commission, Washington, D.C., 1960.)

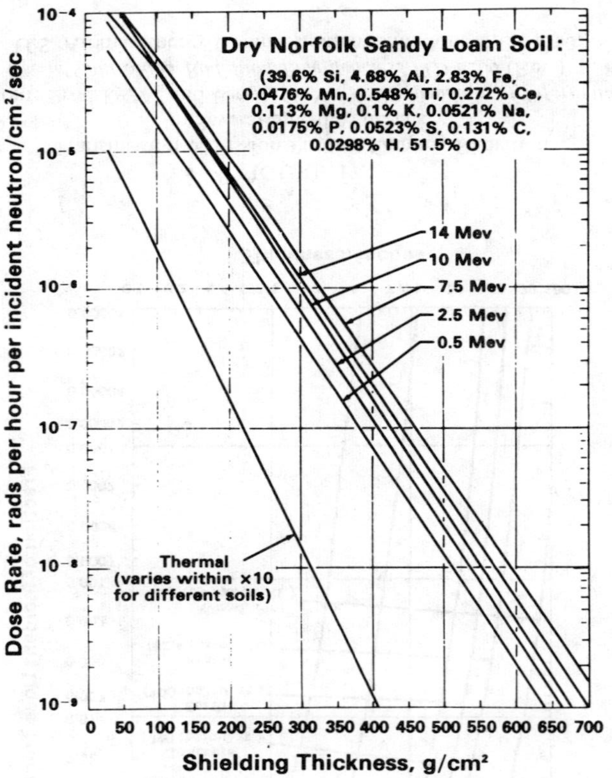

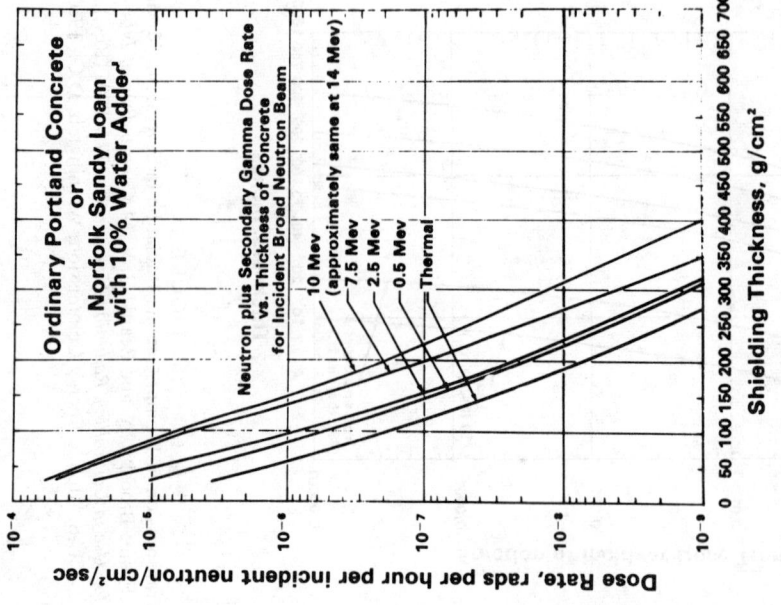

FIGURE 20.
Transmission of Total Dose Rate from Neutrons Incident on 10% Moist Soil and Concrete. (From: Brodsky, A. and Beard, G. V., *A Compendium of Information for Use in Controlling Radiation Emergencies, TID-8206* (Rev.), p. 83. U.S. Atomic Energy Commission, Washington, D.C., 1960.)

FIGURE 19.
Transmission of Total Dose Rate from Neutrons Incident on Soil. (From: Brodsky, A. and Beard, G. V., *A Compendium of Information for Use in Controlling Radiation Emergencies, TID-8206* (Rev.), p. 82. U.S. Atomic Energy Commission, Washington, D.C., 1960.)

Table 44. BETA-RAY SURFACE DOSE RATES[17]

Material	mrad/hr
Thorium, 4 to 5 years after separation	40
Tuballoy, D-38	200
Oralloy	
40%	180
93%	140
Plutonium-239	
nickel coated	360
uncoated	440
Uranium-233	
1-month ^{232}U build-up	7,000
1-year ^{232}U build-up	58,000
Uranium slug, natural	233
UO_2, brown oxide	207
UF_4, green salt	179
$UO_2(NO_3)_2 \cdot 6H_2O$	111
UO_3, orange oxide	204
U_3O_8, black oxide	203
UO_2F_2	176
$Na_2U_2O_7$	167

From: Brodsky, A. and Beard, G. V., *A Compendium of Information for Use in Controlling Radiation Emergencies*, TID-8206 (Rev.). By permission of the U.S. Atomic Energy Commission, Washington, D.C.

Bremsstrahlung.

1. The energy radiated as radiation per beta ray absorbed is:

$$B = 1.23 \times 10^{-4}(\bar{Z} + 3)E^2 \text{ Mev/beta,}$$

where E is the maximum beta energy, in Mev, and $\bar{Z}$ is the effective atomic number given by

$$\bar{Z} = \frac{\sum f_a Z_a^2}{\sum f_a Z_a},$$

where f_a is the fraction of the number of atoms of atomic number Z_a. The bremsstrahlung spectrum is given below, in Table 45.

Table 45. BREMSSTRAHLUNG SPECTRUM FROM BETA ABSORPTION[17]

Photon Energy Intervals in Fractions of the Maximum Beta Energy	Percent of Total Intensity Contributed by Photons in Energy Intervals
0.0 to 0.1	43.5
0.1 to 0.2	25.8
0.2 to 0.3	15.2
0.3 to 0.4	8.3
0.4 to 0.5	4.3
0.5 to 0.6	2.0
0.6 to 0.7	0.7
0.7 to 0.8	0.2
0.8 to 0.9	0.03
0.9 to 1.0	<0.01

From: Brodsky, A. and Beard, G. V., *A Compendium of Information for Use in Controlling Radiation Emergencies*. TID-8206 (Rev.). By permission of the U.S. Atomic Energy Commission, Washington, D.C.

2. When beta particles from a 1-curie source of ^{90}Sr-^{90}Y are absorbed in aluminum, the bremsstrahlung intensity is approximately equal to the gamma intensity from 12 mg of radium. The average bremsstrahlung energy is about 300 kev (Haybittle, *Phys. Med. Biol.*, 1(3): 270, 1956).
3. The bremsstrahlung from a 1-curie ^{32}P aqueous solution in a glass bottle is about 3 mr/hr at 1 meter.

Gamma Radiation.

1. The gamma-ray dose rates from various radiosotopes are given in Table 37 in the preceding chapter.
2. As a rule of thumb, accurate to ± 12 percent from 0.07 to 2.0 Mev, the dose rate at 1 foot from a point source of gamma radiation is: r/hr at 1 foot = 5.64 CE, where C is the number of curies of the parent nuclide, and E is the total gamma energy, in Mev, emitted per disintegration of the parent. Rounding off, the dose rate to within ± 20 percent from 0.07 to 4 Mev is: r/hr at 1 foot = 6 CE.
3. The dose rate versus distance from a 100-curie ^{60}Co source, taking into account inverse-square attenuation, air absorption, and build-up factor for air at normal temperature and pressure, is given in Table 46.

Table 46.
DOSE RATE VERSUS DISTANCE
FROM 100 CURIES ^{60}Co[17]

Distance, feet	Dose Rate, r/hr
1	1,500
10	15
50	0.6
100	0.15
400	0.0075
800	0.0012
1,000	0.0006

From: Brodsky, A. and Beard, G. V. *A Compendium of Information for Use in Controllling Radiation Emergencies, TID-8206* (Rev.). By permission of the U.S. Atomic Energy Commission, Washington, D. C.

4. The gamma activity of fission products produced in a nuclear burst may be expressed in equivalent curies according to the equation

gamma activity = $1.82 \, t^{-1.2}$ curies per watt-second,

where t is the time in seconds since the burst. This equation holds approximately between 10 seconds and 100 days following the burst. For kiloton-TNT-size bursts, this equation becomes

gamma activity = $7.3 \times 10^{12} \, t^{-1.2}$ curies/kiloton TNT.

5. A deposition of 1 microcurie per square meter of ^{131}I on grass gives a gamma radiation level of about 0.003 mr/hr near ground level and results in about 0.1 microcurie per liter of milk from cows grazing in the area. A deposition of 1 megacurie of fission products per square mile gives a gamma dose rate of about 4 r/hr at 3 feet above ground. A deposition of 1 microcurie of fission products per square meter gives 10.6 μr/hr at 3 feet above ground.
6. The dose rate in a foxhole due to air-scattered radiation from a fallout field is 2 percent of the open-field dose rate per steradian of the sky viewed from the foxhole.
7. The backscattered intensity of ^{60}Co gamma radiation from a thick wall per square meter of the wall visible at both source and detector may be approximated by the equation

$$\text{mr/hr/m}^2 \text{ for 1 kCi } ^{60}\text{Co} = \frac{4 \times 10^3}{D^2 d^2},$$

where D is the distance (in meters) of the source from the wall, and d is the distance (in meters) at which the backscattered dose rate is being measured. It is assumed that D and d are large compared to the dimensions of the irradiated area of the wall and that the gamma rays are scattered at angles greater than about 140°. For example, if a 10-kilocurie ^{60}Co source in a well-shielded collimator 10 meters from a thick concrete wall irradiated the wall with a beam area measuring 4 square meters at the wall, the dose rate at a point beside the collimator would be about 16 mr/hr from radiation scattered back from the wall.

Shelter Shielding.

The added shielding against fallout radiation to give an additional factor of 5 to a nominal basement shelter factor of 20, or a total factor of 100, is given in Table 47. The dose rate in a foxhole due to air-scattered radiation from a fallout field is 2 percent of the open-field dose rate per steradian of sky viewed from the foxhole.

Table 47.
SHELTER SHIELDING MATERIALS FOR ONE-FIFTH REDUCTION[17]

Material	Density, lbs/ft^3	Thickness, inches
Wood (birch, oak, maple, etc.)	40	17.5
Earth, loose	75	9.3
Sand, dry	100	7.0
Brick, common	110	6.4
Concrete block, solid	140	5.0

From: Brodsky, A. and Beard, G. V., *A Compendium of Information for Use in Controlling Radiation Emergencies*, TID-8206 (Rev.). By permission of the U.S. Atomic Energy Commission, Washington, D.C.

In operations requiring dry-boxes, usually one or more absolute filters will be required in the exhaust ventilation in order to insure a high degree of removal of radioactivity in exhaust air before its release to the environment. There are stringent federal and state regulations that require average concentrations in exhaust air to be lower than those specified in the table of regulatory maximum permissible concentrations for occupational exposure, presented earlier. These filters are generally designed to retain more than 99.97 percent of particulates of particle size greater than 0.3 microns and may be fairly fragile. Requirements for inspection, storage, handling, and installation have been proposed by the U.S. Atomic Energy Commission.[18] Figure 21 shows a filter damaged by moisture and by shipment.

The quantities of radionuclides requiring glove-box operations, absolute filters in the exhaust, and other special protective equipment and procedures are given in Table 41 for most of the radionuclides of concern. However, the values listed for mixed fission products pertain to a fuel irradiation time of 180 days and decay times of less than 30 days. Other quantities of fission products may be pertinent for longer operating times and longer decay times, where the radioactivity in curies represents longer half-lives and more radiotoxic fission, products. Estimates of requirements for the various fission product mixtures may be obtained with the assistance of Tables 22, 23, 24, and 25, and from Figure 11.[19]

In addition to absolute filters, larger chemical-processing installations or larger nuclear reactors may require continuous air samples, continuous exhaust-stack monitors, and even building containment or controlled leak rates, to protect the public and prevent a nuclear disaster in the event of an accident (see Table 41). Generally, a team of engineers, health physicists and industrial hygienists who have specialized in the design of such facilities must be consulted in order to deal appropriately with these problems. General recommendations may be found in more detail in the literature and in handbooks already available.[1] In some cases, a special site location may be required in the event of a release of radioactive material.

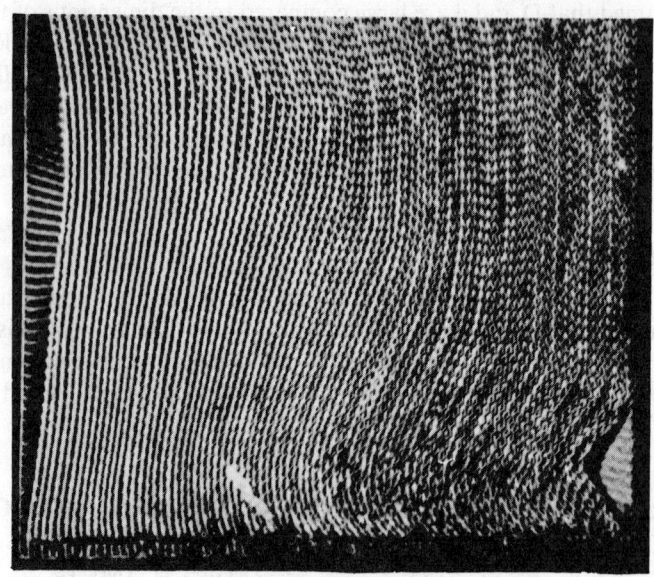

FIGURE 21.
Separators in Filter Units Absorbed Moisture;
Damage Aggravated by Shipment.
(From: Gilbert, H. and Palmer, G. H., *High-Efficiency Air Filter Units*.
U.S. Atomic Energy Commission, Washington, D.C., 1961.)

Table 41 gives formulae for calculating the site radius for the release of Q curies of mixed fission products in order to give no more than 15 rem to any body organ, even under adverse meteorological conditions. These formulae have been calculated for a release at ground level, using the diffusion equations from Reference 20. The use of proper containment, filtration, or other safeguards may diminish the site radius required, depending on considerations of population density in the vicinity and of the maximum quantity and type of radioactivity that could be released accidentally. The appropriate regulatory agency should be consulted in regard to its policies on these matters when planning high-level operations. For assistance in estimating upper limits to the hazards of releasing various radionuclides, curves are presented in Figures 22 and 23, from Reference 6, for determining, under adverse conditions and for a ground release, the maximum number of curies inhaled versus distance per curie released and the maximum ground contamination at distance x. These curves are based on conservative dispersion parameters and a wind speed of only 1 meter per second. The ground deposition in curies per square meter per curie released at various distances from the point of release and along the center trajectory in the wind direction assumes an average particle deposition velocity of about 2×10^{-3} meters per second. Figures 22 and 23, together with previously given data on external dose rates of fission products per curie per square meter, allow upper-limit estimates of the probable land contamination problems and exposure rates following a nuclear incident.

The general levels of activity at which personnel monitoring and/or shielding is necessary in order to meet regulatory requirements as well as recommendations of various committees are also presented in Table 41. These quantities were taken as those that could produce a dose rate of about 10 mr/hr at one meter. Suitable protection at lower or higher levels than this may be more appropriate in special circumstances, depending on the experience of operating

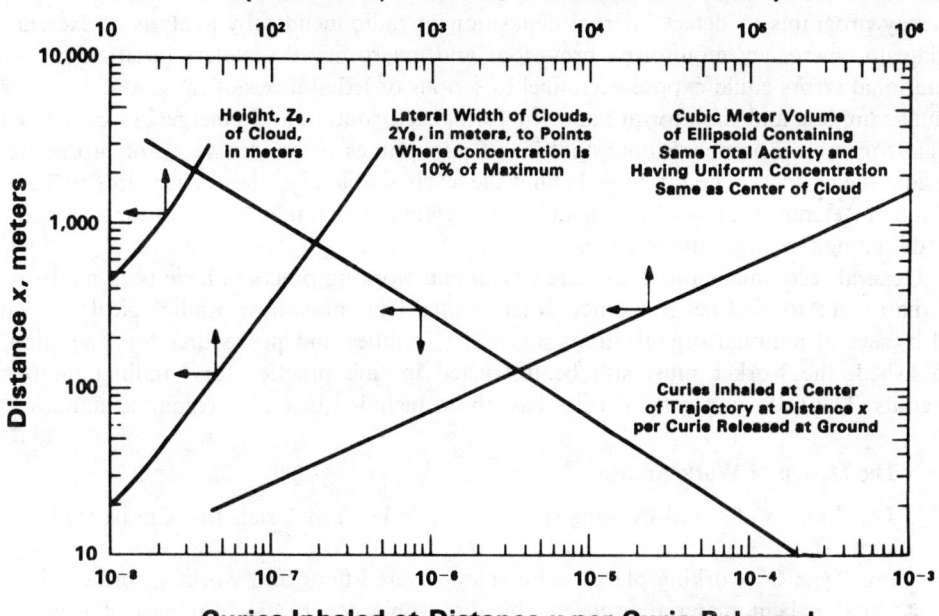

FIGURE 22.
Cloud Characteristics Under Inversion Conditions.
(From: Brodsky, A., Determining Industrial Hygiene Requirements
for Installations Using Radioactive Materials. *Amer. Ind. Hyg. Ass. J.*, 26: 300, 1965.)

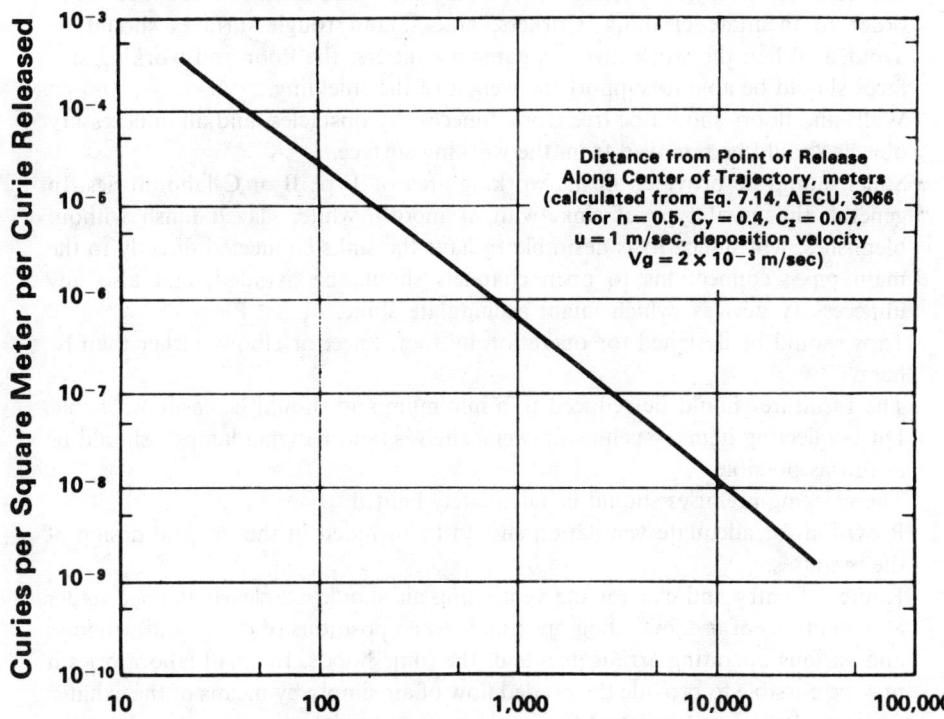

FIGURE 23.
Typical Deposition of Cloud Activity Plotted vs. Distance.
(From: Brodsky, A., Determining Industrial Hygiene Requirements
for Installations Using Radioactive Materials. *Amer. Ind. Hyg. Ass. J.*, 26: 300, 1965.)

personnel, the type of operation, possibility of rotating personnel, etc. Recommendations of levels above which various safeguards may be required are also given for carrying out routine bioassay programs to detect internal deposition of radionuclides by analysis of excreta; for setting up emergency monitoring provisions and emergency dosimeters in situations where operational errors could expose personnel to serious or lethal doses of radiation; for environmental sampling and monitoring; and for formalized routine and emergency procedures, as well as for written preoperational analysis of possibilities of incidents and of procedures to alleviate their effects. The rationale behind the levels selected for these safeguards in Table 41 is given in Reference 6, together with additional references that may be of assistance in planning and designing the larger installations.

General recommendations for safe design and working practices have been made by the International Atomic Energy Agency, International Commission on Radiological Protection, and by several national organizations. Once the facilities and procedures for operation are established, the worker must still be instructed in safe practice for handling radioactive materials. The following general rules have been included in IAEA recommendations.[21,22]

The Design of Work Areas.

The floors, walls, and working surfaces should be of materials that can be easily kept clean.

For Type C* working places, a linoleum-covered floor and working surface of nonabsorbent material protected by disposable covers are examples of what would be considered satisfactory. The working surfaces must be able to support the weight of the necessary shielding against gamma radiation.

For Type B* working places, the walls and the ceilings should be covered with a washable, hard, nonporous paint, and the floor with such materials as linoleum, rubber tiles, or vinyl. The junction of floors and walls should be rounded off in order to facilitate cleaning. Corners, cracks, and rough surfaces should be avoided. When the work involves gamma emitters, the floor and working surfaces should be able to support the weight of the shielding.

Walls and floors should be free from unnecessary obstacles, and all unnecessary objects should be removed from the working surface.

Sinks should be provided in the working area of Type B or C laboratories. In general, the usual type of sink, with a smooth, white, glazed finish without blemishes, will suffice. It is desirable to have the sinks connected directly to the main pipe; connections to open channels should be avoided, and also any unnecessary devices, which might accumulate slime.

Taps should be designed for operation by foot, knee, or elbow, rather than by hand.

The furniture should be reduced to a minimum and should be easily washable. Dust-collecting items—such as drawers, shelves, and hanging lamps—should be as few as possible.

The working premises should be adequately lighted.

Provision for adequate ventilation should be included in the original design of the premises.

Routes of entry and exit for the ventilating air should be clearly defined under all conditions of use, including open and closed positions of doors and windows and various operating arrangements of the fume hoods. In small laboratories it may be possible to provide the needed flow of air simply by means of the exhaust systems of the fume hoods, but in such a case special attention must be given to inflow of fresh air into the laboratory under all conditions, which can be

---

* Type C and Type B laboratories are defined in the IAEA publications.[21,22] However, this author would prefer defining them as laboratories that might only require a chemical hood or completely enclosed dry-box operation respectively, as indicated in Table 41.

accomplished by such means as providing adequate louvres in the doors of rooms. Consideration should be given to any need to treat or filter incoming air. In cool climates, the problem of heating the intake air for a large group of fume hoods should not be overlooked, as this may be a major problem.

The siting of inlet and exhaust vents should be such as to prevent any recirculation of exhausted air.

Fume hoods should produce a regular air flow without any eddies. The speed of the air flow should be such that there can be no escape of air into the working place from the fume hood under typical operating conditions, including opening of windows and doors and the suction of other fume hoods; this can be checked by smoke tests. It is recommended that the fan be placed on the exhaust side of any filter in the system. The gas, water, and electrical appliances should be operated from the outside of the fume hood. The inside of the hood and the exhaust ducts should be as easy to clean as possible.

Manipulations should be carried out over a suitable drip tray, or with some form of double container that will minimize the importance of breakages or spills. It is also useful to cover the working surfaces with absorbent material in order to soak up minor spills. The absorbent material should be changed when unsuitable for further work and be treated as radioactive waste.

Protective Clothing and Personal Protective Measures.

Protective clothing appropriate to the radioactive contamination risks should be worn by every person in the controlled area, even if only very small quantities of radioactive materials are manipulated.

In Type C working places, the personnel should wear simple protective clothing, such as ordinary laboratory coats or surgical coats. In Type A or B working places, protective clothing or devices should be provided according to the nature of the work. When working with experimental animals, clothing proof against teeth or claws may be desirable, and protection of the face against blood or body fluid splashings should be provided.

In Type A and B working places, the protective clothing should be clearly identified—for example, by a different color. It should not, in any case, be worn outside the controlled area.

The working clothes and street clothes should be kept in separate cubicles or changing rooms. When changing from one or the other, one should be careful to avoid cross-contamination risks.

Rubber gloves should be worn when working with unsealed radioactive substances.

Care should be taken to avoid needless contamination of objects by handling them with protective gloves—in particular, light switches, taps, door knobs, etc. The gloves should be either taken off, or a piece of noncontaminated material (paper), which should be disposed of later together with the contaminated residue, should be interposed.

Contaminated gloves should be washed before taking them off.

A method of putting on and removing rubber gloves without contaminating the inside of the gloves should be used. This procedure is such that the inside of the glove is not touched by the outside, nor is any part of the outside allowed to come in contact with the bare skin.

No unsealed radioactive sources should be manipulated with the unprotected hand.

No solution should be pipetted by mouth in any isotope laboratory.

It is recommended that special precautions be taken to avoid punctures or cuts, especially when manipulating the more dangerous isotopes.

Anyone who has an open skin wound below the wrist (whether protected by a bandage or not) should not work with radioactive isotopes without medical approval.

The use of containers, glassware, etc., with cutting edges should be avoided.

Care should be taken with contaminated animals to avoid bites or scratches.

An annual compendium of United States and international standards in the nuclear field, including radiation protection, is now available from the United States of America Standards Institute.[23]

PROTECTION IN MEDICAL PRACTICE

The basic principles of radiation protection for the patient and the physician or his assistant are the same as for those working in other facilities involving potential radiation exposure. Basically, diagnostic radiation exposure to the patient should be a minimum consistent with the best medical practice. In radiation therapy, the radiation dose to the tumor or to the pathogenic tissue should be optimized, and exposure to healthy tissues minimized. The principles of shielding the healthy tissues or minimizing the time they are exposed relative to exposure of the pathogenic tissues are utilized in the practice of beam collimation or by the use of lead aprons to shield the patient against scattered radiation, or, in rotational therapy, by reducing the exposure time of healthy tissues relative to the tumor that remains in the center of the beam.

In addition to the references previously mentioned,[24,25] there are a number of excellent reports and texts now available.[26-29] The National Council of Radiation Protection and Measurements (NCRP) is presently preparing a revised set of recommendations on the design of medical X-ray installations, but for the present the existing recommendations are sufficiently detailed to provide design criteria.[26,29]

For purposes of this handbook, some of the more frequently needed recommendations and data are summarized below from Reference 26.

General Guidelines in the Clinical Use of Radiation.

As a general principle, the exposure to the patient shall be kept to the practical minimum consistent with clinical objectives. To this end, the following recommendations are presented for the guidance of physicians and of others responsible for the exposure of patients.

The useful beam should be limited to the smallest area practicable and consistent with the objectives of the radiological examination or treatment.

The voltage, filtration, and source–skin distance (SSD) employed in medical radiological examinations should be as great as is practical and consistent with the diagnostic objectives for the study (for dental X-ray examinations, see NCRP Report No. 35).

Protection of the embryo or fetus during radiological examination or treatment of women known to be pregnant should be given special consideration.*

Suitable protective devices to shield the gonads of patients who are potentially procreative should be used when the examination or method of treatment may include the gonads in the useful beam, unless such devices interfere with the conditions or objectives of the examination or treatment.

Fluoroscopy should not be used as a substitute for radiography, but should be reserved for the study of dynamics or spatial relationships, or for guidance in spot-film recording of critical details.

X-ray films, intensifying screens, and other image-recording devices should be

---

* Ideally, abdominal radiological examination of a woman of childbearing age should be performed during the first few (approximately 10) days following the onset of menses, to minimize the possibility of irradiation of an embryo. In practice, medical needs should be the primary factors in deciding the timing of the examination.

as sensitive as is consistent with the requirements of the examination.
Film-processing materials and techniques should be those recommended by the X-ray film manufacturer, or those otherwise tested, to ensure maximum information content of the developed X-ray film; where practical, quality-control methods should be employed to ensure optimal results.

The section on design requirements for fluoroscopic equipment is quoted in the following paragraph, since it contains data that may be checked to insure the safety of equipment in use. NCRP Report No. 33 also gives performance criteria for the design of fluoroscopic and other diagnostic equipment.

FLUOROSCOPIC EQUIPMENT*

Design Recommendations.

A diagnostic-type protective tube housing shall be used (see definition in Appendix A of Reference 26).

The source–panel or source–tabletop distance shall be at least 12 inches (30 cm) and should not be less than 15 inches (38 cm). The source–skin distance of image-intensifier equipment should not be less than 15 inches (38 cm).†

The total filtration permanently in the useful beam shall be at least 2.5 millimeters aluminum equivalent. When the tabletop or panel surface is interposed between the source and the patient, its aluminum equivalent may be included as part of the total filtration (see comment under 3.2.2(a) in Reference 26).

The equipment shall be so constructed that, under conditions of normal use, the entire cross section of the useful beam is attenuated by a primary protective barrier permanently incorporated into the equipment. The exposure shall automatically terminate when the barrier is removed from the useful beam.

1. The lead equivalent of the barrier of conventional fluoroscopes shall be at least 1.5 millimeters for equipment capable of operating up to 100 kvp, at least 1.8 millimeters for equipment whose maximum operating potential is greater than 100 kvp and less than 125 kvp, and at least 2.0 millimeters for equipment whose maximum operating potential is 125 kvp or greater (see Reference 8 in Reference 26). Special attention shall be paid to the shielding of image intensifiers, so that neither the useful beam nor the scattered radiation from the intensifier itself or from the patient will produce significant radiation exposure to the operator or other personnel.

2. A collimator shall be provided to restrict the size of the useful beam to less than the area of the barrier. The X-ray tube and collimating system shall be linked with the fluorescent screen assembly, so that the useful beam at the fluorescent screen is confined within the barrier, irrespective of the panel–screen distance. For image intensifiers, the useful beam should be centered on the input phosphor, and during fluoroscopy or cine-recording it should not exceed the diameter of the input phosphor. (Ideally, for spot-film radiography with image-intensifier equipment, the shutters should automatically open to the required field size before each exposure.)

3. Collimators, adjustable diaphragms, and shutters shall provide the same degree of attenuation as is required of the tube housing.

---

* Including image-intensified fluoroscopic equipment.

† The greater the source–tabletop distance, the lower is the entrance dose (and, to a lesser extent, the integral dose) for a given screen luminance. Image unsharpness and image magnification are also reduced. However, other considerations place a practical upper limit on the source–tabletop distance. The heating load on the X-ray tube increases rapidly with distance, because greater tube current is required to maintain constant screen luminance. For the same reason it may be necessary to increase spot-film exposure time, resulting in greater motion unsharpness. From the standpoint of radiation safety, it appears that the source–tabletop distance is not critical within rather broad limits. For conventional fluoroscopes, a distance of 15 to 18 inches seems to be a reasonable compromise between the conflicting factors involved (see Reference 7 in Reference 26).

The fluoroscopic-exposure switch shall be of the dead-man type (see definition in Appendix A of Reference 26).

Provision shall be made to intercept the scattered X rays from the undersurface of the tabletop and other structures under the table. In most cases this may be accomplished either by a cone extending from the tube housing to the tabletop, or by a shield around the fluoroscope understructure, or both. The cone shall provide the same degree of attenuation as that required of the tube housing, with the incident angle of the useful beam taken into consideration.

A shielding device of at least 0.25 mm lead equivalent for covering the Bucky slot during fluoroscopy should be provided.

A shield of at least 0.25 mm lead equivalent—such as overlapping protective drapes, or hinged or sliding panels—should be provided to intercept scattered radiation that would otherwise reach the fluoroscopist and others near the machine.

A cumulative timing device, activated by the fluoroscope exposure switch, shall be provided. It shall indicate the passage of a predetermined period of irradiation either by an audible signal or by temporary interruption of the irradiation when the increment of exposure time exceeds a predetermined limit not exceeding five minutes.*

Devices that indicate the X-ray tube potential and current shall be provided. On image-intensified fluoroscopic equipment, such devices should be located in such a manner that the operator may monitor the tube potential and current during fluoroscopy (see footnote under *Guidelines for the Fluoroscopist*).

Image intensification shall always be provided on mobile fluoroscopic equipment. It shall be impossible to operate mobile fluoroscopic equipment unless the useful beam is intercepted by the image intensifier. Inherent provisions shall be made, so that the machine is not operated at a source–skin distance of less than 12 inches (30 cm).

Equipment to be operated in areas where explosive gases may be used should have the approval of Underwriters Laboratory for such use.†

Since the exposure rates in fluoroscopy and the exposure times may possibly be subject to considerable variation, the following guidelines are important in training fluoroscopists to minimize radiation exposure.[25]

Guidelines for the Fluoroscopist.

The exposure rate used in fluoroscopy should be as low as is consistent with the fluoroscopic requirements and shall not normally exceed 10 r/min (measured in air) at the position where the beam enters the patient. This recommendation applies to the use of image-intensifier equipment (with or without television cameras) as well as to conventional (direct-viewing) fluoroscopes (see comment under 3.1.2(a) in Reference 26).

The fluoroscopist should know the radiation characteristics of his equipment. Therefore, periodic measurements of tabletop or patient exposure rate shall be made. Patient-exposure measurements are especially necessary on apparatus employing image intensifiers in which the intensifier brightness is automatically controlled and the X-ray factors in use are not readily ascertained. ‡Such measurements necessitate the use of a phantom in the fluoroscopic beam.

---

* While the timer does not ensure safe operation, it is of value as a training device for physicians learning the techniques of fluoroscopy, and for the experienced fluoroscopist as a means for emphasizing the passage of time. The design should be such that the timer reset mechanism does not create a nuisance for the physician.

† Information may be obtained from Underwriters Laboratory, 207 East Ohio Street, Chicago, Illinois, 60611.

‡ Image intensifiers may significantly reduce both observation time and exposure rate when properly used, but do not inherently accomplish this reduction. In equipment with automatic brightness control, the tube potential and current may rise to high values without knowledge of the operator, particularly if the gain of the intensifier is diminished. It is important, therefore, for the operator to monitor tube current and potential on such equipment,

The smallest practical field sizes and shortest exposure times should be employed. The possibilities of reducing dose by techniques utilizing high tube potential and low current should be considered.

Fluoroscopy should not be used as a substitute for radiography, but should be reserved for the study of dynamics or spatial relationships or for guidance in spot-film recording critical details.

Medical fluoroscopy should be performed only by or under the immediate supervision of physicians properly trained in fluoroscopic procedures.

The fluoroscopist's eyes should be sufficiently dark-adapted for the visual task required, before commencing fluoroscopy.* Under no circumstances should he attempt to compensate for inadequate adaptation by increasing the exposure factors employed or by prolonging the fluoroscopic examination.

Extraneous light that interferes with the fluoroscopic examination shall be eliminated.

Special precautions, consistent with clinical needs, should be taken to minimize exposure of the gonads of potentially procreative patients and exposure of the embryo or fetus in patients known to be or suspected of being pregnant (see 2.4.3 and 2.4.4 in Reference 26).

In cineradiography, special care should be taken to limit patient exposure when —as is often the case—tube currents and potentials employed are higher than those normally used in fluoroscopy. The exposure rates to which patients are normally subjected shall be determined periodically.

Protective aprons of at least 0.25 mm lead equivalent should be worn in the fluoroscopy room by each person (except the patient) whose trunk is exposed to radiation fields of 5 mr/hr or more .†

The hand of the fluoroscopist should not be placed in the useful beam unless the beam is attenuated by the patient and a protective glove of at least 0.25 mm lead equivalent.

Only persons whose presence is needed should be in the fluoroscopy room during X-ray exposures.

Design recommendations for diagnostic X-ray machines, including appropriate filtration, diaphragms, cones, and collimators, are given in the following paragraphs.

FIXED RADIOGRAPHIC EQUIPMENT

Design Recommendations.

A diagnostic-type protective tube housing shall be used (see definition in Appendix A and Reference 6 in Reference 26).

Suitable devices (diaphragms, cones, adjustable collimators) capable of restricting the useful beam to the area of clinical interest shall be provided to define the beam and shall provide the same degree of attenuation as that required of the tube housing. Such devices shall be calibrated in terms of the size of the projected useful beam at specified source–film distances (see 3.2.2(b) in Reference 26). For chest photofluorographic equipment, the collimator shall restrict the beam to dimensions no greater than those of the fluorographic screen.

Radiographic equipment, particularly multipurpose machines, should be equipped with adjustable collimators containing light localizers that define the

---

* The perception of detail under conditions of scotopic vision requires retinal adaptation. The adaptation time necessary for the competent performance of a specific visual task depends upon the nature of the task itself, the preexposure luminance level and color, the conditions of adaptation, and on a number of other physiologic factors. While wearing red goggles for 10 minutes will usually satisfy adaptation requirements in fluoroscopy, no specific adaptation period can be recommended for all situations (see Reference 9 in Reference 26). Dark-adaptation normally is not necessary when using image intensifiers.

† A busy fluoroscopist is unlikely to operate a fluoroscope more than five hours per week. Therefore he would be unlikely to receive more than 1/4 the maximum permissible dose to the trunk of the body if the scattered radiation level is less than 5 mr per hour. However, the other sources of exposure also should be taken into account when deciding whether a protective apron is to be worn.

entire field. Rectangular collimators are usually preferable. Means to produce a visible indication of adequate collimation and alignment on the developed X-ray film should be provided. The field size indication on adjustable collimators shall be accurate to within one inch for a source–film distance of 72 inches. The light field shall be aligned with the X-ray field with the same degree of accuracy. The aluminum equivalent of the total filtration in the useful beam shall not be less than that shown in the following table (see also 3.2.2(a) in Reference 26). For dental radiography, see forthcoming NCRP Report No. 35.

Operating kvp	Minimum Total Filter (inherent plus added)
Below 50 kvp	0.5 mm aluminum
50–70 kvp	1.5 mm aluminum
Above 70 kvp	2.5 mm aluminum

A device that terminates the exposure at a preset time interval or exposure shall be provided. The operator should be able to terminate the exposure at any time. The exposure switch, except for those used in conjunction with spot-film devices in fluoroscopy, shall be so arranged that it cannot be conveniently operated outside a shielded area.

The control panel shall include a device (usually a milliammeter) to give positive indication of the production of X rays whenever the X-ray tube is energized. The control panel shall include devices (labeled control settings and/or meters) indicating the physical factors (such as kvp, mA, exposure time, or whether timing is automatic) used for the exposure.

Machines equipped with beryllium-window X-ray tubes* shall contain keyed filter-interlock switches in the tube housing and suitable indication of the control panel of the added filter in the useful beam if the total filtration permanently in the useful beam is less than 0.5 mm aluminum equivalent. The total filtration permanently in the useful beam shall be clearly indicated on the tube housing. Beryllium-window X-ray tubes should not be used on multipurpose radiographic equipment.

The aluminum equivalent of the tabletop when a cassette tray is used under the tabletop, or the aluminum equivalent of the front panel of the vertical cassette holder, shall not be more than 1 mm at 100 kvp.

Equipment to be operated in areas where explosive gases may be used should have the approval of Underwriters Laboratory for such use.†

The following guidelines for the use of diagnostic X-ray equipment are based on the common-sense application of the principles of minimizing radiation exposure through shielding, distance, or reducing time of exposure.[26]

Guidelines for the User. (See also 2.4 in Reference 26.)

Particular care should be taken to limit the useful beam to the smallest area consistent with the clinical requirements and to align the X-ray beam accurately with the patient and film (see also 3.2.2(b) in Reference 26).

Gonadal shielding should be used for the patient when appropriate (see 2.4.4 in Reference 26), but never as a substitute for adequate beam collimation and alignment.

When a patient must be held in position for radiography, mechanical supporting or restraining devices should be used. If the patient must be held by an individual,

---

* Beryllium-window X-ray tubes with no added filtration emit low-energy X rays at very high exposure rates. It is particularly important, therefore, that the operator be able to tell by a glance at the control panel how much added filter, if any, is present.
† Information may be obtained from Underwriters Laboratory, 207 East Ohio Street, Chicago, Illinois 60611.

that individual shall be protected with appropriate shielding devices such as protective gloves and apron, and he should be so positioned that no part of his body will be struck by the useful beam and that his body is as far as possible from the edge of the useful beam.

Only persons whose presence is necessary shall be in the radiographic room during exposure. All such persons shall be protected.

The radiographer shall stand behind the barrier provided for his protection during radiographic exposures.

Special care shall be taken to ensure adequate filtration in multipurpose machines.*

Particular care shall be taken to ensure adequate filtration in any machine equipped with a beryllium-window tube. Appropriate added filter is required, to provide the filtration values recommended (see 3.2.1(d), also 3.2.1(i) and (j) in Reference 26).

Other guidelines should be consulted for the design and use of mobile radiographic equipment and of X-ray therapy equipment.[26] Similar considerations of limiting the width of the beam to necessary dimensions, using appropriate filtration to reduce the soft component that would be absorbed in tissues before reaching the tissue of interest, and increasing the distance and shielding protection of medical personnel working in the vicinity of radiographic or therapy equipment are included. Particular items that may be checked most frequently by government agencies would include: the leakage radiation from the housing of therapy machines, which should not exceed 0.1 r/hr at 5 cm; the use of permanent diaphragms or cones for collimating the beam with the same degree of protection outside the beam as provided by the housing; the use of a suitable exposure-control device (e.g., an automatic timer or exposure meter) to terminate exposures after a preset time interval or preset exposure; emergency means of terminating the exposure; and various mechanical and electrical interlocks and warning lights to insure that operating personnel do not inadvertently expose themselves or the patients. NCRP Report No. 33 should be examined in detail before designing and establishing procedures for a therapy installation.[26]

Similar guidelines are given for gamma-beam therapy equipment that utilizes the radiation from radioactive materials. In addition, there are requirements for leak-testing these radioactive sources, to show that there is less than 0.005 μCi of transferable activity when surfaces of the devices are wiped clean with moistened cotton swabs or filter paper. Leak tests and the records of their results are usually made a legal requirement in the licensing of these radioactive sources and devices. Guidelines for the use of such equipment and for emergency procedures are usually incorporated in the inspection programs of regulatory agencies. Regular calibration of the quality and intensity of radiation from therapy equipment should be carried out. Annual calibrations may suffice, however, as long as spot checks are made at least once a month or after every 50 hours of operating time. A log shall be kept of all spot-check measurements.[26] Radiation protection surveys are required periodically, as well as appropriate warning signs;[26] these surveys and signs are often necessary in order to meet regulatory requirements.

In addition to safe equipment and procedures, it is necessary to have appropriate supervision of the overall radiation safety of the radiology department operations, as well as regular monitoring and recording of personnel exposures by the use of film badges or other devices worn by the medical staff and their assistants. NCRP recommendations should be consulted in detail, in addition to requirements of state and federal agencies.[26,30] Most of the state supervision of medical facilities is carried out by state health departments. The following general working conditions are recommended by NCRP and are generally considered good practice.[26]

---

* For soft-tissue radiography, such as mammography, operating potentials considerably below 50 kvp may be required. In performing such examinations on multipurpose machines, it is usually necessary to reduce the amount of filtration. It is important, however, that the appropriate filter be replaced before proceeding with exposures requiring normal filtration.

General.

The owner (see definition in Appendix A of Reference 26) is responsible for radiation safety. He is responsible for assuring that radiation sources under his jurisdiction are used only by persons competent to use them. He is responsible for providing the instruction of personnel in safe operating procedures and for promulgating rules for radiation safety.

Deliberate exposure of an individual to the useful beam for training or demonstration purposes shall not be permitted unless there is also a medical (or dental) indication for the exposure and the exposure is prescribed by a physician (or dentist).

Radiation-Protection Supervisor.

A radiation-protection supervisor (who may be the user himself) shall be designated for every installation to assume the responsibilities outlined below and to advise on the establishment of safe working conditions according to the recommendations of this report and in compliance with all pertinent federal, state, and local regulations. He should be familiar with the basic principles of radiation protection in order to properly discharge his responsibilities, although for details he may consult with appropriate qualified experts for advice.

Among the specific responsibilities of the radiation-protection supervisor or his deputy are the following.

1. To establish and supervise operating procedures, and to review them periodically to assure their conformity with the recommendations of this report.
2. To instruct personnel in proper radiation-protection practices.
3. To conduct or have conducted radiation surveys and source-leak tests where indicated (see Section 6 and 4.2.2(c) in Reference 26), and to keep records of such surveys and tests, including summaries of corrective measures recommended and/or instituted.
4. To assure that personnel-monitoring devices are used where indicated (see below) and that records are kept of the results of such monitoring.
5. To assure that interlock switches and warning signals are functioning and that signs are properly located.
6. To investigate each known or suspected case of excessive or abnormal exposure in order to determine the cause, and to take steps to prevent its recurrence.

Personnel-Monitoring.

Personnel-monitoring is valuable for checking the adequacy of the radiation-safety-program. It can be useful in disclosing inadequate or improper radiation-protection practices and potentially serious radiation-exposure situations. Personnel-monitoring may be of value also in documenting occupational exposure if proper consideration is given to the limitations of the monitoring system (see Reference 12 in Reference 26). Accordingly, the following recommendations are made.

1. Personnel-monitoring shall be performed in controlled areas for each occupationally exposed individual for whom there is reasonable possibility of receiving a dose exceeding 1/4 the applicable MPD (see Table 1, Appendix B in Reference 26).
2. A qualified expert should be consulted on establishing and evaluating the personnel-monitoring system. When feasible, the system should be tested periodically.

3. All reported cases of apparently high exposures shall be investigated by the radiation-protection supervisor, and his findings and conclusions should be made a part of the personnel-monitoring record.
4. Devices worn for the monitoring of occupational exposure shall not be worn by the individual when he is exposed as a patient for medical (or dental) reasons.
5. Monitoring devices used to estimate whole-body exposure normally should be worn on the chest or abdomen. When a protective apron is worn (e.g., during fluoroscopy), particular care should be taken in choosing the location of the monitoring device and in interpreting its reading. Devices worn on the inside of the apron will normally not provide a reliable indication of the radiation environment outside of the apron. Devices worn on the outside of the apron will usually provide only an upper limit for the estimation of the exposure of parts of the body covered by the apron. Accordingly, a qualified expert should be consulted in situations where the interpretation of the reading is highly dependent upon the conditions under which the monitoring device is used. For further information on the use of personnel-monitoring devices, see the forthcoming NCRP Report on Instrumentation and Monitoring Methods for Radiation Protection.
6. Blood counts shall not be used for personnel-monitoring (see Reference 13 in Reference 26).

Medical Examination.

A preplacement medical examination is recommended, to establish baseline values for the radiation worker and to reveal any physical condition that later might otherwise be attributed to radiation exposure. It should include medical history, radiation-exposure history, physical examination, and—at the discretion of the physician in charge—a complete blood count.

Whenever it is known or suspected that a person has received a dose substantially in excess of the MPD, the individual should be referred at once to a competent medical authority.

Vacations.

Vacations shall not be used as a substitute for adequate protection against exposure to radiation.

The emergency procedures to be used in case of failure of gamma-ray beam-control equipment depends on the individual installation, as pointed out by the NCRP.[26] However, since the procedure is brief and illustrates the general sequence of removing persons from direct intense exposure, notification of appropriate supervisory and radiation-protection personnel, planned careful entry to the area by radiation-protection personnel, utilization of appropriate survey materials to measure radiation intensity while exploring the causes, and possible rectification of the equipment of procedural failure, it is presented here.

Emergency Procedure in Case Beam Control Fails.

If the light signals indicate that the beam-control mechanism has failed to terminate the exposure at the end of the preset time (for example, if the red light stays on and/or the green signal does not light up), the source may still be in the "on" position. The following steps are to be carried out in a calm manner.

For the Radiation-Therapy Technician.

1. Open the door to the treatment room.

2. If the patient is ambulatory, direct him to get off the table and leave the room.
3. If the patient is not ambulatory, enter the treatment room, but avoid exposure to the useful beam; pull the treatment table as far away from the useful beam as possible; transfer the patient to a stretcher and remove him from the room.
4. Close the door.
5. Turn off the main switch at the control panel.
6. Notify the radiation therapist and radiation-protection supervisor at once.

For the Radiation-Protection Supervisor.

1. Secure a portable survey meter; check to see that the meter is functioning properly.
2. Turn the power on and open the door a few inches.
3. Stand behind the door and insert the survey meter into the door opening to test whether in fact the source is still in the "on" position.
4. If the source is still in the "on" position, enter the room and manually turn the source off as per manufacturer's instructions; avoid intercepting the useful beam with any part of your body.
5. Adjust the limiting diaphragm to the smallest field size.
6. Close the door to the treatment room; turn off the power; lock the control panel; post a sign warning people not to enter.
7. Notify the equipment manufacturer's representative.

Some useful data for estimating radiation exposure rates and design requirements for medical installations are given in Tables 48 to 56.

Table 48.
HALF-VALUE LAYERS AS A FUNCTION OF FILTRATION
AND OF THE TUBE POTENTIAL FOR DIAGNOSTIC UNITS[26]

Total Filtration, mm Al	Peak Potential, kvp									
	30	40	50	60	70	80	90	100	110	120
	Typical Half-Value Layers, mm of Al									
0.5	0.36*	0.47*	0.58	0.67	0.76	0.84	0.92	1.00	1.08	1.16
1.0	0.55	0.78	0.95	1.08	1.21	1.33	1.46	1.58	1.70	1.82
1.5	0.78	1.04	1.25*	1.42*	1.59*	1.75	1.90	2.08	2.25	2.42
2.0	0.92	1.22	1.49	1.70	1.90	2.10	2.28	2.48	2.70	2.90
2.5	1.02	1.38	1.69	1.95	2.16	2.37*†	2.58*†	2.82*†	3.06*†	3.30*†
3.0	—	1.49	1.87	2.16	2.40	2.62	2.86	3.12	3.38	3.65
3.5	—	1.58	2.00	2.34	2.60	2.86	3.12	3.40	3.68	3.95

Note: for full-wave rectified potential; derived from Reference 32 by interpolation and extrapolation.
* Recommended minimum HVL for radiographic units (see Section 3.2.2(a) in Reference 26.
† Recommended minimum HVL for fluoroscopes (see Section 3.1.2(b) in Reference 26.
From: Medical X-Ray and Gamma-Ray Protection for Energies up to 110 Mev—Equipment Design and Use. *NCRP Report No. 33.* By permission of the National Council on Radiation Protection and Measurements, Washington, D.C.

Table 49.
EFFECT OF TUBE POTENTIAL, DISTANCE, AND FILTRATION ON AIR EXPOSURE RATE AT PANEL OF FLUOROSCOPES[26]

Potential kvp	Source-to-Panel Distance		Equivalent Total Aluminum Filtration				
			1 mm	2 mm	2.5 mm	3 mm	4 mm
	cm	inches	roentgens per milliampere minute				
70	30	12	5.3	2.7	2.2*	1.8	1.3
	38	15	3.5	1.7	1.4†	1.2	0.8
	46	18	2.4	1.2	1.0	0.8	0.6
80	30	12	7.0	3.9	3.2*	2.6	2.0
	38	15	4.6	2.5	2.1†	1.7	1.3
	46	18	3.2	1.8	1.4	1.2	0.9
90	30	12	9.0	5.2	4.3*	3.6	2.8
	38	15	5.8	3.3	2.8†	2.3	1.8
	46	18	4.0	2.3	1.9	1.6	1.2
100	30	12	11.0	6.6	5.5*	4.7	3.7
	38	15	7.0	4.2	3.5†	3.0	2.3
	46	18	4.9	2.9	2.5	2.1	1.6
110	30	12	13.1	8.0	6.8*	5.9	4.6
	38	15	8.4	5.1	4.4†	3.8	3.0
	46	18	5.8	3.5	3.0	2.6	2.0
120	30	12	14.7	9.3	8.0*	7.0	5.5
	38	15	9.5	6.0	5.1†	4.5	3.6
	46	18	6.5	4.1	3.6	3.1	2.5
130	38	15	—	6.8	5.9†	5.2	4.2
	46	18	—	4.7	4.1	3.6	2.9
140	38	15	—	7.6	6.6†	5.9	4.8
	46	18	—	5.3	4.6	4.1	3.3
150	38	15	—	8.5	7.5†	6.7	5.4
	46	18	—	5.8	5.2	4.6	3.7

Note: typical exposure rates produced by equipment with medium length cables, derived from references 31 and 32 by interpolation and extraoplation. Filtration includes that of the tabletop and the X-ray tube with its inherent and added filter. As used above, panel means either panel or tabletop.
† See Section 3.1.2(a) in Reference 26.
From: Medical X-Ray and Gamma-Ray Protection for Energies up to 110 Mev—Equipment Design and Use. *NCRP Report No. 33.* By permission of the National Council on Radiation Protection and Measurements, Washington, D.C.

Table 50.
SCATTERED-RADIATION EXPOSURE RATE AT SIDE OF FLUOROSCOPY TABLE[26]

Distance from Source to Point of Measurement		Tube Potential						
		50 kvp	60 kvp	70 kvp	80 kvp	90 kvp	100 kvp	125 kvp
inches	cm	roentgens per 100 milliampere seconds						
12	30	1.8	2.8	4.2	5.8	8.0	9.8	15.2
18	46	0.8	1.3	1.8	2.5	3.4	4.2	6.7
24	61	0.4	0.7	1.1	1.4	1.9	2.3	3.8
39	100	0.2	0.3	0.4	0.5	0.7	0.9	1.4
54	137	0.1	0.1	0.2	0.3	0.4	0.5	0.7
72	183	0.1	0.1	0.1	0.2	0.2	0.3	0.4

Note: measured in air, with total filtration equivalent to 2.5 mm aluminum.
From: Medical X-Ray and Gamma-Ray Protection for Energies up to 110 Mev—Equipment Design and Use. *NCRP Report No. 33.* By permission of the National Council on Radiation Protection and Measurements, Washington, D.C.

Table 51.
EXPOSURE RATE THROUGH FLUOROSCOPIC SCREEN WITHOUT PATIENT[26]

X-Ray Tube Potential kvp	Source-to-Tabletop Distance		Lead Equivalent of Screen Protective Barrier*		
			1.5 mm	1.8 mm	2.0 mm
	inches	cm	Typical Exposure Rate, mr/h per r/min at tabletop		
80	12	30	10	4.5	2.5
	15	38	13	6	3.5
	18	46	15	7	4
90	12	30	12	6	3.5
	15	38	16	7.5	4.5
	18	46	19	9	5.5
100	12	30	15	7	4.5
	15	38	20	9	5.5
	18	46	23	11	7
110	12	30	19	9	5.5
	15	38	24	12	7
	18	46	29	14	8.5
120	12	30	23	11	7
	15	38	30	14	9
	18	46	35	17	10
130	15	38	35	17	10
	18	46	42	20	12
140	15	38	41	19	12
	18	46	49	23	14
150	15	38	46	20	12
	18	46	55	24	15

Total filtration: 3 mm aluminum equivalent
Tabletop-to-screen distance: 14 inches
Screen-to-chamber distance: 2 inches
Medium-length high-tension cables

Note: adapted from Reference 31 by interpolation and extrapolation; actual exposure rate values may differ from the typical values given above by ±15%, depending upon length of high-tension cables.
* See Section 3.1.1(d) and 3.1.2(c) of Reference 26.
From: Medical X-Ray and Gamma-Ray Protection for Energies up to 110 Mev—Equipment Design and Use. *NCRP Report No. 33.* By permission of the National Council on Radiation Protection and Measurements, Washington, D.C.

Table 52. PRIMARY PROTECTIVE-BARRIER REQUIREMENTS FOR 100 mr-week[21]

Tube Voltage, Constant Potential	WUT, mA-min per week	Primary Protective Barrier Requirements in mm of Lead at a Target Distance of						Primary Protective Barrier Requirements in cm of Concrete (2.35 g/cm^3) at a Target Distance of					
		TVL	1 m	2 m	3 m	5 m	10 m	TVL	1 m	2 m	3 m	5 m	10 m
50 kv	10,000	0.2	0.7	0.6	0.5	0.4	0.3	2	7.0	6.0	5.0	4.0	3.0
	3,000		0.6	0.5	0.4	0.3	0.2		6.0	5.0	4.0	3.0	2.0
	1,000		0.5	0.4	0.3	0.2	0.1		5.0	4.0	3.0	2.0	1.0
	300		0.4	0.3	0.2	0.2	0.1		4.0	3.0	2.0	2.0	1.0
	100		0.3	0.2	0.2	0.1	0.1		3.0	2.0	2.0	1.0	1.0
70 kv	10,000	0.5	1.6	1.3	1.1	0.9	0.7	4	14.0	12.0	10.0	8.0	6.0
	3,000		1.4	1.1	0.9	0.7	0.5		12.0	10.0	8.0	6.0	4.5
	1,000		1.1	0.9	0.7	0.5	0.4		10.0	8.0	6.0	4.5	3.5
	300		0.9	0.7	0.5	0.4	0.2		8.0	6.0	4.5	3.5	2.0
	100		0.7	0.5	0.4	0.3	0.1		6.0	4.5	3.5	2.5	1.0
85 kv	10,000	0.8	2.7	2.2	1.9	1.5	1.1	6.5	23.0	19.0	16.0	13.0	9.5
	3,000		2.3	1.8	1.5	1.2	0.8		19.5	15.5	13.0	11.0	7.0
	1,000		1.8	1.4	1.1	0.9	0.6		15.5	12.5	9.5	8.0	5.0
	300		1.4	1.1	0.8	0.6	0.4		12.5	9.5	7.0	5.0	3.5
	100		1.1	0.8	0.6	0.4	0.2		9.5	7.0	5.0	3.5	2.0
100 kv	10,000	0.85	3.3	2.8	2.5	2.1	1.6	7	26.5	22.0	20.0	17.0	13.0
	3,000		2.9	2.4	2.0	1.7	1.2		23.0	19.0	16.0	14.0	10.0
	1,000		2.5	2.0	1.6	1.3	0.8		20.0	16.0	13.0	10.5	6.5
	300		2.0	1.5	1.2	0.9	0.5		16.0	12.0	10.0	7.5	4.0
	100		1.6	1.1	0.8	0.6	0.3		13.0	9.0	6.5	5.0	2.5
125 kv	10,000	0.9	3.7	3.2	2.8	2.5	1.9	7	30.0	26.0	24.0	21.0	16.5
	3,000		3.3	2.7	2.4	2.0	1.5		27.0	23.0	20.0	17.0	13.0
	1,000		2.8	2.3	1.9	1.6	1.0		24.0	19.0	16.5	14.0	9.0
	300		2.4	1.8	1.5	1.1	0.7		20.0	16.0	13.0	10.0	6.0
	100		1.9	1.4	1.1	0.8	0.4		16.5	12.0	10.0	7.0	3.0

Table 52. PRIMARY PROTECTIVE-BARRIER REQUIREMENTS FOR 100 mr-week[21] (*Continued*)

Tube Voltage, Constant Potential	WUT, mA-min per week	Primary Protective Barrier Requirements in mm of Lead at a Target Distance of						Primary Protective Barrier Requirements in cm of Concrete (2.35 g/cm^3) at a Target Distance of					
		TVL	1 m	2 m	3 m	5 m	10 m	TVL	1 m	2 m	3 m	5 m	10 m
150 kv	10,000	0.9	3.9	3.4	3.1	2.7	2.1	8	33.0	29.0	26.0	23.0	18.0
	3,000		3.5	2.9	2.6	2.2	1.6		30.0	25.0	22.0	19.0	14.0
	1,000		3.0	2.5	2.2	1.7	1.2		25.5	21.0	19.0	14.5	10.5
	300		2.6	2.1	1.7	1.3	0.8		22.0	18.0	15.0	11.0	7.0
	100		2.2	1.6	1.3	0.9	0.5		19.0	14.0	12.0	8.0	4.0
200 kv	30,000	2	8.0	6.5	6.0	5.0	4.0	9	49.0	42.0	39.0	34.0	30.0
	10,000		7.0	5.5	5.0	4.2	3.3		44.0	37.0	34.0	30.0	25.0
	3,000		6.0	4.5	4.0	3.3	2.5		39.0	32.0	29.0	25.0	21.0
	1,000		5.0	3.8	3.3	2.7	1.8		34.0	27.0	25.0	22.0	16.0
	300		4.0	3.0	2.5	1.9	1.2		30.0	23.0	21.0	17.0	12.0
	100		3.3	2.4	1.9	1.3	0.9		25.0	20.0	17.0	13.0	8.0
250 kv	30,000	3	13.5	12.0	10.5	9.0	7.5	10	55.0	49.0	45.0	41.0	35.0
	10,000		12.0	10.5	9.0	7.5	6.0		50.0	45.0	40.0	35.0	30.0
	3,000		10.5	8.5	7.5	6.0	4.5		45.0	39.0	35.0	30.0	24.0
	1,000		9.0	7.0	6.0	5.0	3.5		40.0	34.0	30.0	26.0	20.0
	300		7.5	5.5	4.5	3.5	2.5		35.0	28.0	25.0	20.0	15.0
	100		6.0	4.5	3.5	2.5	1.5		30.0	25.0	20.0	15.0	10.0
300 kv	30,000	6	24.0	20.0	18.0	15.5	12.0	10	58.0	51.0	48.0	44.0	38.0
	10,000		21.0	17.0	15.0	12.5	9.5		53.0	46.0	43.0	39.0	33.0
	3,000		18.0	14.0	12.0	10.0	7.0		48.0	41.0	38.0	33.0	28.0
	1,000		15.0	11.5	10.0	7.5	5.0		43.0	36.0	33.0	29.0	23.0
	300		12.0	9.0	7.5	5.5	3.5		38.0	32.0	29.0	24.0	18.0
	100		9.5	7.0	5.5	4.0	2.5		33.0	28.0	24.0	19.0	15.0

Note: the tabulated values give the shielding required to reduce the exposure dose to 100 mr in a week, the value assumed for design purposes in controlled areas; to compute the shielding required outside controlled areas, it is necessary to add half a tenth-value layer to reduce the weekly exposure dose to about 30 mr, and one tenth-value layer to reduce it to 10 mr.

From: Recommendations of the ICRP. *Publication No. 3, Report of Committee 3.* Copyright 1960, Pergamon Press, New York.

Table 53.
SECONDARY PROTECTIVE BARRIER REQUIREMENTS FOR 100 mr-week[21]

Tube Voltage, Constant Potential	WUT mA-min per week	Secondary Protective Barrier Requirements in mm of Lead at Target Distances of					Secondary Protective Barrier Requirements in mm of Concrete (2.35 g/cm^3) at Target Distances of				
		1 m	2 m	3 m	5 m	10 m	1 m	2 m	3 m	5 m	10 m
50 kv[1]	10,000	0.35	0.25	0.2	0.1	0	3.5	2.5	2	1	0
	3,000	0.25	0.15	0.1	0.1	0	2.5	1.5	1	1	0
	1,000	0.2	0.1	0.1	0	0	2	1	1	0	0
	300	0.1	0	0	0	0	1	0	0	0	0
	100	0	0	0	0	0	0	0	0	0	0
70 kv[1]	10,000	0.9	0.7	0.5	0.3	0.1	7	5.5	4	2.5	1
	3,000	0.7	0.5	0.3	0.1	0	5.5	4	2.5	1	0
	1,000	0.5	0.3	0.1	0	0	4	2.5	1	0	0
	300	0.3	0.1	0	0	0	2.5	1	0	0	0
	100	0.1	0	0	0	0	1	0	0	0	0
85 kv[1]	10,000	1.4	1.0	0.8	0.4	0.2	12	8	6.5	4	2
	3,000	1.1	0.7	0.4	0.2	0	9	6	4	2	0
	1,000	0.8	0.4	0.2	0	0	6.5	4	2	0	0
	300	0.4	0.2	0	0	0	4	2	0	0	0
	100	0.2	0	0	0	0	2	0	0	0	0
100 kv[1]	10,000	1.6	1.1	0.9	0.5	0.2	13	10	7	4	2
	3,000	1.2	0.8	0.5	0.2	0	10	7	4	2	0
	1,000	0.9	0.4	0.2	0	0	7	4	2	0	0
	300	0.5	0.2	0	0	0	4	2	0	0	0
	100	0.2	0	0	0	0	2	0	0	0	0
125 kv[1]	10,000	1.8	1.4	1.0	0.5	0.2	14.5	11	8	4	2
	3,000	1.4	0.9	0.5	0.2	0	11	7.5	4	2	0
	1,000	1.0	0.5	0.2	0	0	7.5	4	2	0	0
	300	0.5	0.2	0	0	0	4	2	0	0	0
	100	0.2	0	0	0	0	2	0	0	0	0
150 kv[1]	10,000	1.9	1.5	1.0	0.6	0.2	15	11	8	5	2
	3,000	1.5	0.9	0.6	0.2	0	11	7.5	5	2	0
	1,000	1.0	0.6	0.2	0	0	8	5	2	0	0
	300	0.6	0.2	0	0	0	5	2	0	0	0
	100	0.2	0	0	0	0	2	0	0	0	0
200 kv[2]	30,000	5.1	3.9	3.2	2.6	1.6	36	29	25	22	15
	10,000	4.1	3.0	2.4	1.8	0.8	31	24	20	16	9
	3,000	3.2	2.2	1.6	0.9	0.3	26	19	15	10	5
	1,000	2.4	1.4	0.9	0.3	0	21	14	10	5	0
	300	1.6	0.7	0.3	0	0	15	8	5	0	0
250 kv[2]	30,000	7.5	6.0	5.0	4.0	2.5	37	31	27	23	17
	10,000	6.2	4.8	3.8	2.7	1.2	32	26	22	18	10
	3,000	5.0	3.6	2.6	1.4	0.5	27	21	17	11	5
	1,000	3.8	2.4	1.4	0.5	0	22	16	11	5	0
	300	2.5	1.0	0.5	0	0	17	9	5	0	0
300 kv[2]	30,000	14.5	11	9.5	7.5	5	40	32	28	25	19
	10,000	12	8.5	7	5	3	34	27	24	19	14
	3,000	9.5	6	5	3	1	28	22	19	14	8
	1,000	7	4	3	1	0	24	16	14	8	0
	300	5	2.5	1	0	0	19	12	8	0	0

Note: the tabulated values give the shielding required to reduce the exposure dose to 100 mr in a week, the value assumed for design purposes in controlled areas; to compute the shielding required outside controlled areas, it is necessary to add half a tenth-value layer to reduce the weekly exposure dose to about 30 mr, and one tenth-value layer to reduce it to 10 mr.

The figures in the table allow for both scattered and leakage radiation. The scattered radiation at 1 m was assumed to be 0.1 percent of the incident beam. To compute the leakage radiation, it was assumed that the maximum ratings were 180 mA-min in 1 hour and 900 mA-min in 1 hour for the diagnostic-type protective tube housing (1) and the therapeutic-type protective tube housing (2) respectively.

From: Recommendations of the ICRP. *Publication No. 3, Report of Committee 3.* Copyright 1960. Pergamon Press, New York.

Table 54.
^{60}Co SHIELDING REQUIREMENTS FOR 100 mr/week[21]

WUT*			Distance from Source to Occupied Area, meters									
80,000			1.4	2.0	2.8	4.0	5.6	8.0	11.0	16.0		
40,000				1.4	2.0	2.8	4.0	5.6	8.0	11.0	16.0	
20,000				1.0	1.4	2.0	2.8	4.0	5.6	8.0	11.0	16.0
10,000					1.0	1.4	2.0	2.8	4.0	5.6	8.0	11.0	16.0
5,000						1.0	1.4	2.0	2.8	4.0	5.6	8.0	11.0
2,500							1.0	1.4	2.0	2.8	4.0	5.6	8.0
1,250								1.0	1.4	2.0	2.8	4.0	5.6
620									1.0	1.4	2.0	2.8	4.0
310										1.0	1.4	2.0	2.8

	Approximate		Thickness of Lead, cm									
	HVL	TVL										
Primary Barrier	1.20	4.00	22.7	21.5	20.3	19.1	17.9	16.7	15.5	14.3	13.1	11.9	10.7
Leakage† (0.1%) Barrier	1.20	4.00	10.7	9.5	8.4	7.3	6.2	4.9	3.6	2.2	0.8	0.0	0.0
Scatter‡ Barrier													
30°	1.02	3.40	12.4	11.3	10.3	9.2	8.2	7.2	6.2	5.1	4.1	3.1	2.1
45°	0.87	2.90	9.8	8.9	8.1	7.2	6.3	5.4	4.5	3.7	2.8	1.8	1.1
60°	0.75	2.50	7.8	7.0	6.3	5.6	4.8	4.1	3.3	2.6	1.8	1.1	0.4
90°	0.43	1.45	3.7	3.2	2.8	2.4	1.9	1.5	1.1	0.7	0.3	0.1	0.1
120°	0.20	0.65	1.5	1.3	1.1	0.9	0.7	0.5	0.35	0.2	0.1	0.0	0.0
150°	0.14	0.45	0.9	0.8	0.7	0.55	0.45	0.3	0.2	0.1	0.05	0.0	0.0

	Approximate		Thickness of Concrete (2.35 g/cm³), cm									
	HVL	TVL										
Primary Barrier	6.6	21.8	122	116	110	104	97	91	85	79	72	66	60
Leakage† (0.1%) Barrier	6.6	21.8	60	54	48	41	34	27	19.5	12	4.5	0.0	0.0
Scatter‡ Barrier													
30°	6.3	20.8	79	73	67	60	54	48	41	35	29	22	16
45°	6.1	20.3	70	64	58	52	46	39	33	27	21	15	9
60°	5.9	19.2	62	56	50	45	39	33	27	21	15	9	2.5
90°	4.6	15.8	44	39	35	30	25	21	16	11	6	1	0
120°	4.3	14.7	39	35	30	26	21	17	13	8.5	4	0	0
150°	3.8	12.5	32	28	25	21	17	13	10	6	2	0	0

Note: add one tenth-layer (TVL), to reduce radiation to 10 mr/week.
* W = work load in r/week at 1 m, U = use factor, T = occupancy factor.
† Refers to leakage radiation of source housing.
‡ For large field (20 cm diameter) and a source—scatterer distance of 50 cm; these values include scattering from the collimator and from the phantom.
From: Recommendations of the ICRP. *Publication No. 3 Report of Committee 3*. Copyright 1960. Pergamon Press, New York.

Table 55.
^{137}Cs SHIELDING REQUIREMENTS FOR 100 mr/week[21]

WUT*			Distance from Source to Occupied Area, meters									
24,000			1.4	2.0	2.8	4.0	5.6	8.0	11.0	16.0		
12,000			1.0	1.4	2.0	2.8	4.0	5.6	8.0	11.0	16.0	
6,000				1.0	1.4	2.0	2.8	4.0	5.6	8.0	11.0	16.0
3,000					1.0	1.4	2.0	2.8	4.0	5.6	8.0	11.0
1,500						1.0	1.4	2.0	2.8	4.0	5.6	8.0
750							1.0	1.4	2.0	2.8	4.0	5.6
375								1.0	1.4	2.0	2.8	4.0	5.6

	Approximate		Thickness of Lead, cm									
	HVL	TVL										
Primary Barrier	0.65	2.1	10.6	10.0	9.4	8.7	8.1	7.5	6.8	6.2	5.6	4.9	4.2
Leakage† (0.1%) Barrier	0.65	2.1	4.3	3.6	3.0	2.4	1.7	1.1	0.5	0.0	0.0	0.0	0.0
Scatter‡ Barrier													
35°	0.45	1.5	5.4	5.0	4.5	4.0	3.6	3.1	2.7	2.3	1.8	1.4	0.9
56°	0.38	1.3	4.2	3.8	3.4	3.0	2.6	2.2	1.8	1.5	1.1	0.8	0.5
90°	0.22	0.7	2.1	1.9	1.7	1.4	1.2	1.0	0.8	0.6	0.5	0.3	0.2
119°	0.13	0.4	1.0	0.9	0.8	0.7	0.6	0.5	0.3	0.2	0.2	0.1	0.05

	Approximate		Thickness of Concrete (2.35 g/cm³), cm									
	HVL	TVL										
Primary Barrier	4.9	15.8	88	83	78	73	68	63	58	53	49	44	39
Leakage† (0.1%) Barrier	4.9	15.8	39	34	29	24	19	14	6	0	0	0	0
Scatter‡ Barrier													
35°	4.6	15.5	61	56	51	47	42	38	33	28	23	19	14
56°	3.8	12.5	49	45	41	37	33	30	26	22	19	15	11
90°	3.6	12.0	42	39	35	31	28	25	21	17	14	10	6
119°	3.3	11.2	39	36	32	29	25	22	19	15	12	8	4

Note: add one tenth-value layer (TVLL), to reduce radiation to 10 mr/week.
* W = work load in r/week at 1 m, U = use factor, T = occupancy factor.
† Refers to leakage radiation of source housing.
‡ For large field (20 cm diameter) and a source—scatterer distance of 50 cm; these values include only scattering from an obliquely positioned concrete scatterer.

From: Recommendations of the ICRP. *Publication No. 3. Report of Committee 3.* Copyright 1960, Pergamon Press, New York.

Table 56.
THICKNESS OF LEAD REQUIRED TO REDUCE USEFUL BEAM TO 5 PERCENT[26]

Beam Quality		Required Lead Thickness, millimeters
Potential	Half-Value Layer, millimeters	
60 kvp	1.2 Al	0.10
100 kvp	1.0 Al	0.16
100 kvp	2.0 Al	0.25
100 kvp	3.0 Al	0.35
140 kvp	0.5 Cu	0.70
200 kvp	1.0 Cu	1.00
250 kvp	3.0 Cu	1.70
400 kvp	4.0 Cu	2.30
1000 kvp	3.2 Pb	20.50
2000 kvp	6.0 Pb	43.00
2000 kvcp	14.5 Pb	63.00
3000 kvcp	16.2 Pb	70.00
6000 kv	17.0 Pb	74.00
8000 kv	15.5 Pb	67.00
^{60}Co	10.4 Pb	47.00

Note: approximate values for broad beams. Transmission data for brass, steel, and other material for potentials up to 2,000 kvp may be found in Reference 15 of Reference 26. Measurements on 1,000 kvp and 2,000 kvp made with resonant-type therapy units. Data for 6,000 kv taken from Reference 33 for a linear accelerator. Data for 2,000 kvcp, 3,000 kvcp, and 8,000 kv derived by interpolation from graph presented in Reference 34. The third column refers to lead or to the required equivalent lead thickness of lead-containing materials (e.g. lead rubber, lead glass, etc.).

From: Medical X-Ray and Gamma-Ray Protection for Energies up to 110 Mev—Equipment Design and Use. *NCRP Report No. 33*. By permission of the National Council on Radiation Protection and Measurements, Washington, D.C.

Figure 24 presents a graph of the effective energy of an X-ray beam of given constant kilovolt potential as a function of the added g/cm² of copper filtration, illustrating the initial rapid hardening of a beam as filtration is added.[36]

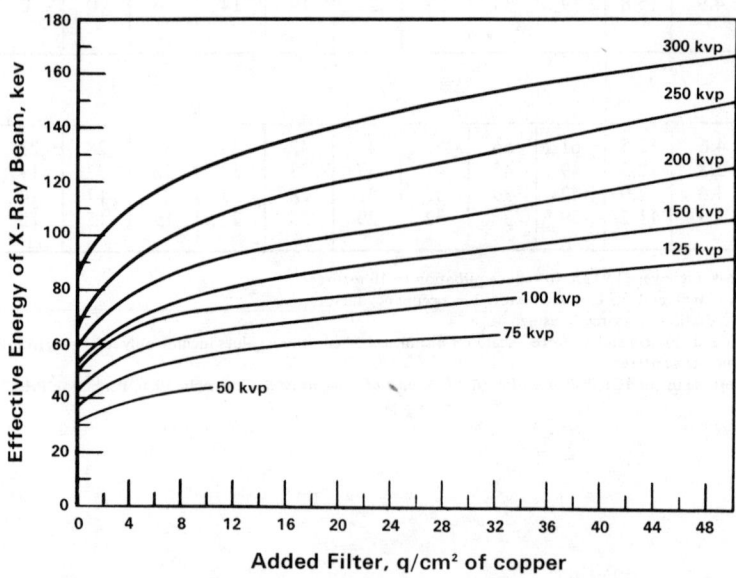

FIGURE 24.
X-Ray Beam Effective Energy vs. Thickness of Copper Filter Added for Various Applied Voltages. (From: Childers, H. M., Brodsky, A., and Nash, A. E., *A Standardized X-Ray Field Range*. Naval Research Laboratory, Washington, D.C., 1958.)

There has been considerable interest in recent years in developing further standards for medical X-ray protection.[37,38] The increased supervision over the safety aspects of the design and manufacture of equipment, the training of medical X-ray technicians in proper radiation-safety practices, and periodic surveys and equipment calibration by certified physicists may be expected to provide adequate radiation safety in medical installations.

REFERENCES

1. Blatz, H., ed., *Radiation Hygiene Handbook*. McGraw-Hill, New York, 1959.
2. International Atomic Energy Agency, Safe Handling of Radioisotopes. *Safety Series No. 1*, p. 99. International Atomic Energy Agency, Vienna, Austria, 1958.
3. Garden, N. B., ed., Report on Glove Boxes and Containment Enclosures. *U.S. Atomic Energy Publications*, *TID-16020*. Office of Technical Services, Department of Commerce, Washington, D.C., 1962.
4. Bradley, F. J., Nuclear-Plant Engineering and Maintenance. *Production Handbook*, 2nd ed., pp. 24-55–24-80, Carson, G. B., ed. Ronald Press, New York, 1958.
5. U.S. Atomic Energy Commission, *High-Efficiency Particulate Air-Filter Units*. Office of Technical Services, Department of Commerce, Washington, D.C., 1961.
6. Brodsky, A., Determining Industrial Hygiene Requirements for Installations Using Radioactive Materials. *Amer. Ind. Hyg. Ass. J.*, 26: 294–310, 1965.
7. International Commission on Radiological Protection, Report of Committee 2. *Health Phys.*, 3: 1–380, 1960.
8. International Commission on Radiological Protection, Report of the ICRP Task Group on Lung Dynamics. *Health Phys.*, 12: 173–207, 1966.
9. Drinker, P. and Hatch, T., *Industrial Dust*. McGraw-Hill, New York, 1959.
10. Hemeon, W. C. L., *Plant and Process Ventilation*, 2nd ed. Industrial Press, New York, 1963.
11. Morgan, G. W., Laboratory Design. *Radiation Hygiene Handbook*, Section 9, Blatz, H., ed. McGraw-Hill, New York, 1959.
12. Silverman, L., Control of Radioactive Air Pollution. *Radiation Hygiene Handbook*, Section 22, Blatz, H., ed. McGraw-Hill, New York, 1959.
13. Tompkins, P. C., Surface Contamination and Decontamination. *Radiation Hygiene Handbook*, Section 18, Blatz, H., ed. McGraw-Hill, New York, 1959.
14. Eliassen, R. and Lauderdale R. A., Liquid and Solid Waste Disposal. *Radiation Hygiene Handbook*, Section 21, Blatz, H., ed. McGraw-Hill, New York, 1959.
15. Hawkins, M. B., Equipment for Handling, Storage, and Transportation of Radioactive Materials. *Radiation Hygiene Handbook*, Section 17, Blatz, H., ed. McGraw-Hill, New York, 1959.
16. Caldwell, R., *Personal Communications*. Nuclear Materials and Equipment Corp., Apollo, Pennsylvania, 1968.
17. Brodsky, A. and Beard, G. V., compilers and eds., *A Compendium of Information for Use in Controlling Radiation Emergencies*, *TID-8206* (Rev.). U.S. Atomic Energy Commission, Washington, D. C., 1960.
18. Gilbert, H. and Palmer, J. H., *High-Efficiency Air-Filter Units*. U.S. Atomic Energy Commission, Washington, D.C., 1961.
19. Brodsky, A., Criteria for Acute Exposure to Mixed Fission-Product Aerosols. *Health Phys.*, 11: 1017–1032, 1965.
20. Wexler, H., et al., Meteorology and Atomic Energy. *AECU 3066, U.S. Weather Bureau-USAEC Document*. U.S. Government Printing Office, Washington, D.C., 1955.
21. Rees, D. J., *Health Physics*. M.I.T. Press, Cambridge, Massachusetts, 1967.
22. International Atomic Energy Agency, Safe Handling of Radioisotopes. *Safety Series, No. 1*. International Atomic Energy Agency, Vienna, Austria, 1960, 1962; also *Safety Series No. 22*, 1967.
23. Cottrell, W., *Compilation of United States Nuclear Standards*. United States of America Standards Institute, Clearing House for Federal Scientific and Technical Information, Springfield, Virginia.
24. Johns, H. E., *The Physics of Radiology*, 2nd ed. Charles C Thomas, Springfield, Illinois, 1964.
25. Glasser, O., Quimby, E. H., Taylor, L. S., Weatherwax, H. L., and Morgan, R. H., *Physical Foundations of Radiology*. Hoeber Medical Division, Harper and Row, New York, 1963.
26. National Committee on Radiation Protection and Measurements, Medical X-Ray and Gamma-Ray Protection for Energies up to 10 Mev—Equipment Design and Use. *NCRP Report No. 33*. NCRP Publications Office, Washington, D.C., 1968.
27. Public Health Service, *Radium and the Physician*. U.S. Government Printing Office, Washington, D.C., 1968.
28. Braestrup, C. B., and Wyckoff, H. O., *Radiation Protection*. Charles C Thomas, Springfield, Illinois, 1958.
29. National Committee on Radiation Protection and Measurements, Medical X-Ray Protection up to 3 Million Volts. *NCRP Report No. 26*. NCRP Publications Office, Washington, D.C., 1961.

30. U.S. Atomic Energy Commission, *USAEC Rules and Regulations, Title 10, Code of Federal Regulations,* Part 20, Standards of Protection Against Radiation, Revised August 9, 1966. U.S. Government Printing Office, Washington, D.C.
31. Trout, E. D. and Kelley, J. P., Leakage Radiation Through Lead-Glass Fluoroscopic-Screen Assemblies. *Radiology,* 82: 977, 1964.
32. Hale, J., The Homogeneity Factor for Pulsating X-Ray Beams in the Diagnostic Energy Region. *Radiology,* 86: 147, 1966.
33. Karzmark, C. J. and Tatiana, C., Measurements of 6-Mev X Rays. I. Primary Radiation Absorption in Lead, Steel, and Concrete. *Brit. J. Radiol.,* 1968.
34. Bly, J. H., and Burrill, E. A., High-Energy Radiography in the 6- to 30-Mev Range. *Symposium on Nondestructive Testing in the Missile Industry, Special Technical Publication No. 278.* American Society for Testing Materials, Philadelphia, Pennsylvania, 1960.
35. International Commission on Radiological Protection, Report of Committee 3. *ICRP Publication No. 3.* Pergamon Press, Oxford, England, 1960.
36. Childers, H. M., Brodsky, A., and Nash, A. E., *A Standardized X-Ray Field Range.* Naval Research Laboratory, Washington, D.C., 1958.
37. U.S. Public Health Service, *Report of the Medical X-Ray Advisory Committee on Public Health Considerations in Medical Diagnostic Radiology (X-Rays).* U.S. Government Printing Office, Washington, D.C., 1967.
38. Morgan, K. Z., Reduction of Unnecessary Medical Exposure. *Testimony Before the Senate Commerce Committee on Senate Bill S-2067.* Oak Ridge National Laboratory, Health Physics Division, Oak Ridge, Tennessee, 1967.

PERSONNEL DOSIMETRY
Allen Brodsky, Sc.D., C.H.P.

The general principles of radiation detection and dosimetry were outlined in the first chapter of this section. In the field of radiation protection various devices have been developed, to be worn by personnel in order to obtain a more direct estimate of their radiation exposures for the following purposes: to limit their cumulative lifetime exposures from routine operations to well within the 5(N-18) rem, as discussed in the second chapter; to detect increasingly hazardous situations or unexpected incidents; and to provide a record of the degree of protection afforded by the employer's facility. These personnel-monitoring devices are designed to be small and wearable, and to measure over the widest range of types and amounts of radiation exposure to which the employee may be exposed. Thus, for control purposes, a wider degree of error in assessing the actual dose equivalent in rem must be accepted. This is a result of the limited number of devices or filters that can be incorporated in a personnel dosimeter, which may be required to detect multiple types of interactions from various types or energies of ionizing radiations. Usually, when accuracy must be compromised for certain types or quantum energies of radiation, the personnel dosimeter is designed with sufficient safety factors in its method of read-out or interpretation, so that radiation exposures at the position where the badge is worn on the body are more likely to be conservatively overestimated, if in error at all.

From the technical complexities indicated by the above considerations, as well as from the principles of dose assessment discussed in the first three chapters, one can understand why the subject of personnel dosimetry is itself a field of specialization within the health physics (radiation-protection) profession. A number of texts are already available to introduce the scientific and technical considerations of personnel dosimetry;[1-13] only the more frequently utilized devices will be briefly described here. There are more than a dozen firms offering film-badge, thermoluminescent-dosimeter, or other personnel-monitoring and record-keeping services on a commercial basis throughout the world. These firms may be contacted through journal advertisements or through the Film Dosimetry Testing Services, National Sanitation Foundation, University of Michigan, Ann Arbor, Michigan.

The first type of personnel-monitoring device routinely utilized in the atomic-energy laboratories after the discovery of nuclear fission was the pocket-type or "pencil" ionization chambers. These pocket chambers consist of a center electrode charged to perhaps 150 volts positive with respect to the outer cylindrical wall; the center electrode has a miniature quartz-fiber electroscope attached to it so that the deflection of the quartz fiber decreases with increasing radiation exposure. The degree of loss of charge may be read out by connecting the chamber to a separate charge reader or electrometer. Some "self-reading" chambers, or "pocket dosimeters", contain a built-in microscope and scale, which allow the position of the quartz fibers to be read by the wearer at any time by simply looking through the chamber toward a light source. Chambers are made with various types and thicknesses of wall material, depending on the type of radiation to be measured. Special chambers coated inside with ^{10}B, which absorbs slow neutrons and emits heavily ionizing alpha particles, may be used in the presence of some gamma radiation to detect neutron exposures. Pocket chambers of various types and ranges are available commercially. They are valuable for obtaining an immediate or daily assessment of individual exposures as well as for area-monitoring. When high accident-exposure potentials are present, or where independent assessment and recording of longer-term cumulative exposures are desired, pocket chambers are supplemented or replaced, as appropriate, by badges containing film, thermoluminescent materials, radiophotoluminescent glass, or other devices that can integrate exposures over a wide dynamic range. An illustration of a typical pocket dosimeter for measuring occupational exposures to gamma radiation is presented in Figure 25.

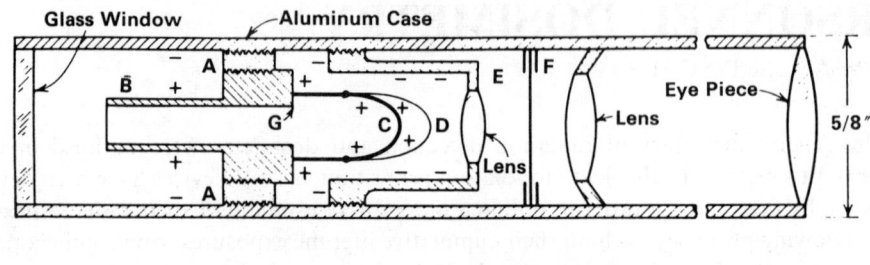

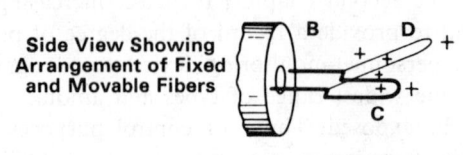

A—Insulating Ring
B—Charging Rod (hollow, to admit light from window)
C—Fixed Heavy-Metal-Coated Quartz Fiber
D—Movable Fine-Metal-Coated Quartz Fiber
E—Metal Cylinder
F—Transparent Scale
G—Metal Support for Fibers

FIGURE 25.
Self-Reading Pocket Dosimeter.
(From: Lapp, R.E. and Andrews, H.L., *Nuclear Radiation Physics*. Copyright 1948, Prentice-Hall, Inc., Englewood Cliffs, New Jersey.)

Probably the most widely used personnel dosimeter at present is the film badge, a badge containing selected films in a packet similar to those used for taking dental X-rays. Usually the badge contains several filters, for use in determining the relative penetration of the type of radiation being measured, or for discriminating between types of radiation in such a way that corrections can be made in estimating the dose at the skin surface and, typically, at an average depth of 5 cm of tissue for blood-forming organs. A summary of the techniques of of film-badge dosimetry is presented in Reference 13, and a more detailed and updated literature survey is presented in Reference 11. Figure 26 shows a series of photographs of a film badge and of films exposed to different types of radiation.

The following description of these photographs is taken from Reference 14.

> The way films are used to monitor different kinds of radiation is really quite a simple process to illustrate. This fact helps make film easily accepted evidence in court.
> Notice in Photograph 1 that the badge holder is composed of plastic front and back sections, which tightly sandwich filters of copper and cadmium on both surfaces of the film. The front filters are circular while the rear filters are rectangular. Darkening of the entire film indicates gamma radiation (Photograph 2). If the cadmium filter is clear while the rest of the film is darkened, it is a sign of X-ray radiation (Photograph 3). If the area under both the cadmium and copper filters remains clear while the rest of the film is darkened, beta radiation is indicated (Photograph 4).
> The subscriber is protected from erroneous results caused by light leaks, since this kind of exposure is easily identified. The exposure of a film packet resulting from a torn package-wrapping is shown in Photograph 5.
> Because of the use of differently shaped filters, exposure patterns can also indicate if the wearer was exposed to radiation from the back rather than the front of the body. Back exposure generally indicates that radiation penetrated the entire body before reaching the film and may have been strongly absorbed (Photograph 6).
> Neutron-monitoring is accomplished with special film, which must be analyzed

by direct observation. The neutron film is interpreted by an experienced technician, who examines the special film with a microscope. The result of neutron bombardment shown in Photograph 7 has been enlarged 800 times.

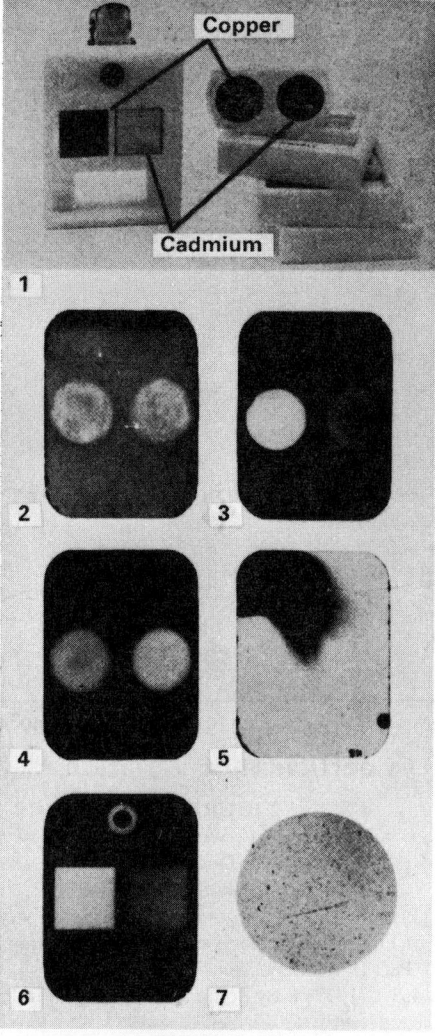

FIGURE 26.
Typical Film Badges.
(From: *Answers to Your Questions About Radiation-Monitoring*, 1968. Courtesy of Nuclear-Chicago Corp., Des Plaines, Illinois.)

Although the development of improved or supplementary types of dosimeters should be continued, the photographic emulsion has many advantages in personnel dosimetry: relatively high sensitivity; wide range of exposure measurement; dose rate independence up to 1000 R/sec or greater; possibilities for multiple-filter interpretations of energy and type of radiation; possibilities for a recheck of density readings or for a professional reexamination of the nature of the photographic image days or years later; low cost and ready availability; adaptability for automatic processing and reading in large quantities; and relative insensitivity to ordinary mechanical shock and vibration.[15] However, films are subject to certain fading and fogging characteristics at high temperature or humidities unless they are protected by suitable wrappers.[16] Also, for proper processing a developing time of approximately 30 minutes is required, which may—in emergency situations, or with a limited number of measurements to be made—be a certain disadvantage.

Figure 27 shows the degrees of precision obtainable for DuPont Type 555 personnel-monitoring film when measuring weekly gamma-ray doses over 12-week and 52-week periods, assuming that density readings are recorded to within the nearest 0.001 density units. Often this degree of precision in density is not practicable under routine conditions, so somewhat less precision would be obtainable, according to the instruments and procedures utilized.

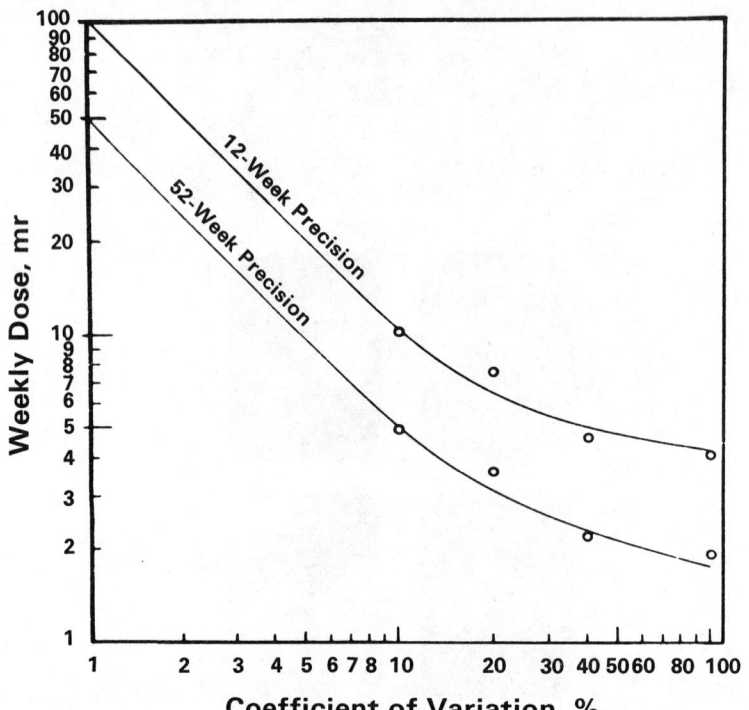

FIGURE 27.
Precision of 12-Week and 52-Week
Cumulative Dose Assignment
from Weekly DuPont Type 555 Films.
(From: Brodsky, A., Accuracy and Sensitivity
of Film Measurements of Gamma Radiation.
Part II. Limits of Sensitivity and Precision.
Health Phys., 9: 463–471, 1963. By permission of Pergamon Press, New York.)

Figure 28 presents the precision obtainable for the more sensitive Kodak Type 3 film for a single film, a 12-week total, and a 52-week total. The confidence bands shown in Figure 29 are indicative of precisions obtainable with ordinary commercial equipment or procedures. The confidence bands on the log-log plots could also be superimposed on the X-ray curves at corresponding densities. The Kodak Type 3 packet also contains a more insensitive film for evaluation of emergency exposures up to several hundred roentgens, but these calibration curves are not shown here. In addition to the problems of protecting films from environmental temperature and humidity variations,[16] care must be taken to insure against the phenomenon of image reversal,[17] in which extremely high doses may bleach out latent images formed at lower exposures.

Within recent years, thermoluminescent materials have been developed that will measure gamma radiation exposure in the range of 10 mr to hundreds of roentgens or more with a fairly linear response. These dosimeters involve the use of a material that emits light when heated after irradiation, and the amount of light emitted is proportional to the amount of radiation exposure when heating conditions are kept constant. Figure 30 shows a schematic diagram of a thermoluminescent dosimetry (TLD) system using a thermoluminescent powder

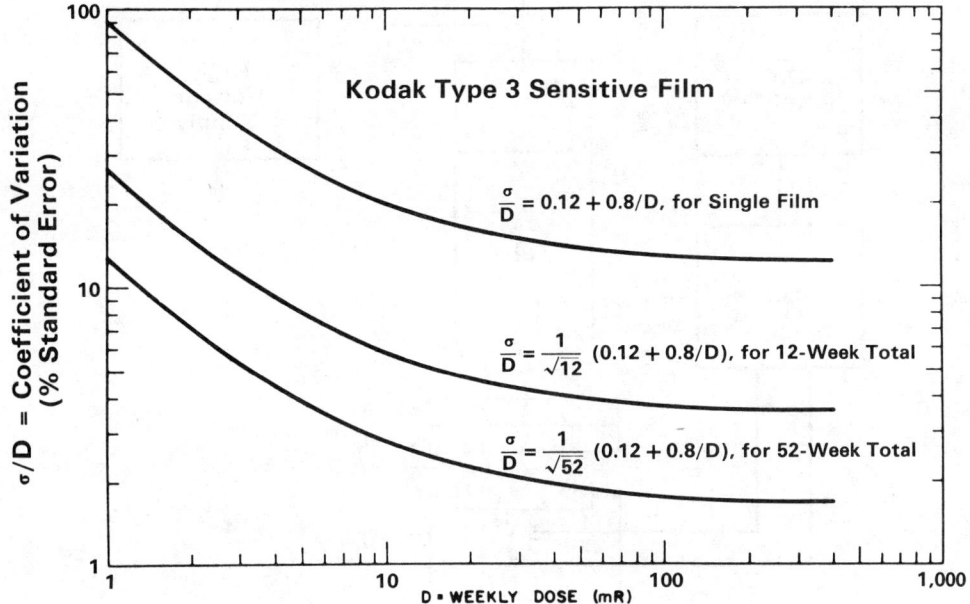

FIGURE 28.
Minimum Percent Error vs. Weekly Dose to Films Evaluated on a Weekly Basis,
for Total Wearing Periods of 1 Week, 12 Weeks, and 52 Weeks.
(From: Brodsky, A., Spritzer, A. A., Feagin, F. E., Bradley, F. J.,
Karches, G. J., and Mandelberg, H. I., Accuracy and Sensitivity
of Film Measurements of Gamma Radiation. Part IV.
Intrinsic and Extrinsic Errors.
Health Phys., 11: 1071–1082, 1965. By permission of Pergamon Press, New York.)

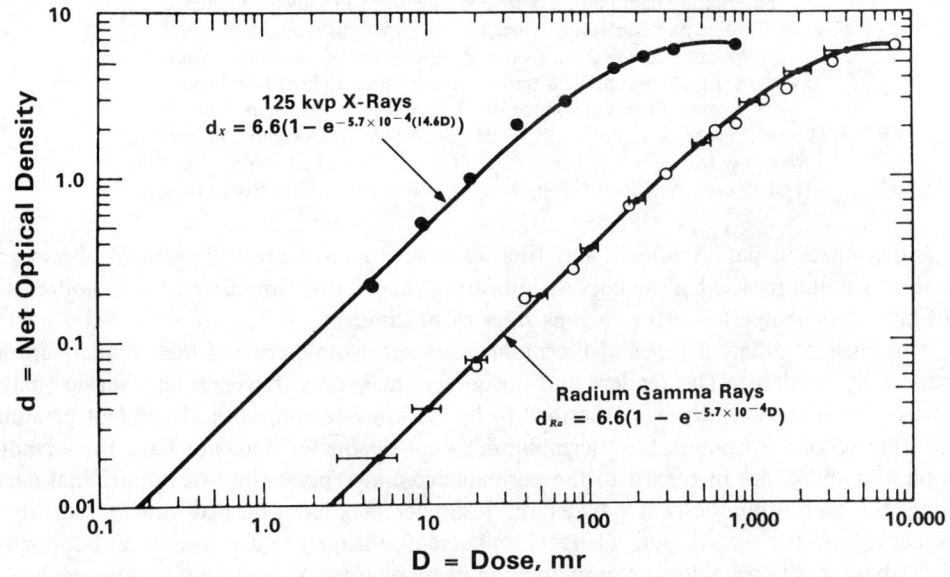

FIGURE 29.
Net Density vs. Dose for Kodak Type 3 Sensitive Film
Developed at 68°F for 3 Minutes in DuPont Liquid X-Ray Developer.
(From: Brodsky, A., Spritzer, A. A., Feagin, F. E., Bradley, F. J.,
Karches, G. J., and Mandelberg, H. I., Accuracy and Sensitivity
of Film Measurements of Gamma Radiation. Part IV.
Intrinsic and Extrinsic Errors.
Health Phys., 11: 1071–1082, 1965. By permission of Pergamon Press, New York.)

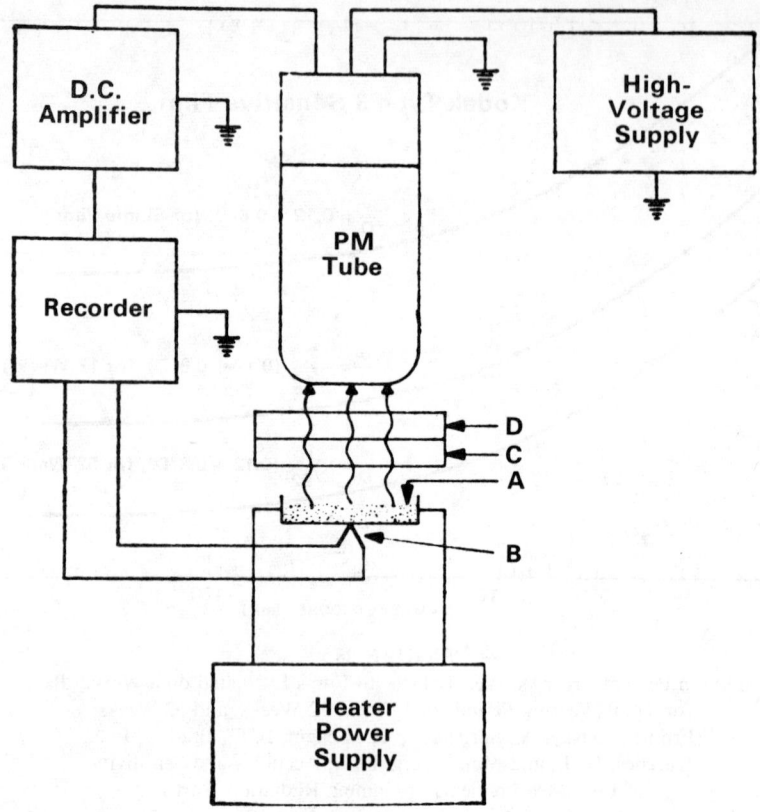

FIGURE 30.
Schematic Diagram of a Thermoluminescent Dosimeter Reader.
A—heater pan containing either loose powdered thermoluminescent phosphor or a sealed dosimeter; B—thermocouple in contact with the pan; C—an optical filter rejecting infrared light; D—a bandpass filter suitable for the thermoluminescent light.
(From: Fowler, J. F. and Attix, F. H., Solid-State Integrating Dosimeters. *Radiation Dosimetry*, 2nd ed., Vol. 2, Ch. 13, Attix, F. H., Tochilin, E., and Roesch, W.C., eds. Copyright Academic Press, New York, 1968.)

that is deposited in pan A after it has been exposed to radiation.[18] Figure 31 shows some typical thermoluminescent glow curves, indicating the relative intensity of the emitted light as a function of temperature for various rates of heating.

A number of different types of thermoluminescent dosimeters and their readers are now commercially available. The readers and dosimeters may be purchased, or a service may be purchased, or a service may be subscribed to by facilities desiring an independent personnel-monitoring service. Although the thermoluminescent dosimeter does not have the advantage of a photographic film in regard to the permanence and repeatability of the original density reading, the thermoluminescent powder or dosimeter may be annealed and reused after its glow curve has been read out. Thermoluminescent dosimeters are also useful for special studies where small dosimeters are implanted in or on phantoms for dose-distribution measurements. A few dozen thermoluminescent dosimeters may be read out more quickly than the same amount of films, since the read-out time for each dosimeter is only about 30 seconds. However, when used in batches of several hundred or more, read-out times for thermoluminescent dosimeters and films are comparable, since films can be developed in batches of 1000 or more, and thus the time to obtain final results is dependent mainly on the time it takes to read the film density or the glow curve—20 to 30 seconds per dosimeter. Thermoluminescent dosimeters may also be utilized in badges containing several dosimeters, placed between

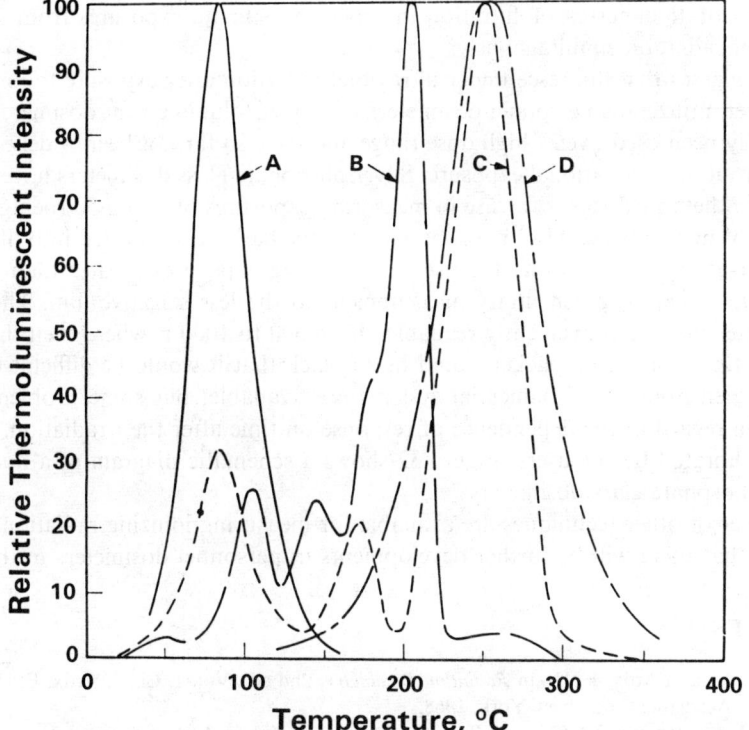

FIGURE 31.
Typical Thermoluminescent Glow Curves, Intensity vs. Temperature.
A—$CaSO_4$: Mn, heating rate 6°C/min; B—LiF (TLD-100), 10^4 r exposure, heating rate 20°C/min; C—CaF_2 (MBLE), about 20 rad from beta rays, heating rate 60°C/min; D—CaF_2 : Mn (NRL), 10^4 r exposure, heating rate 20°C/min.
(From: Fowler, J. F. and Attix, F. H., Solid-State Integrating Dosimeters. *Radiation Dosimetry*, 2nd ed., Vol. 2, Ch. 13, Attix, F. H., Tochilin, E., and Roesch, W.C., eds. Copyright Academic Press, New York, 1968.)

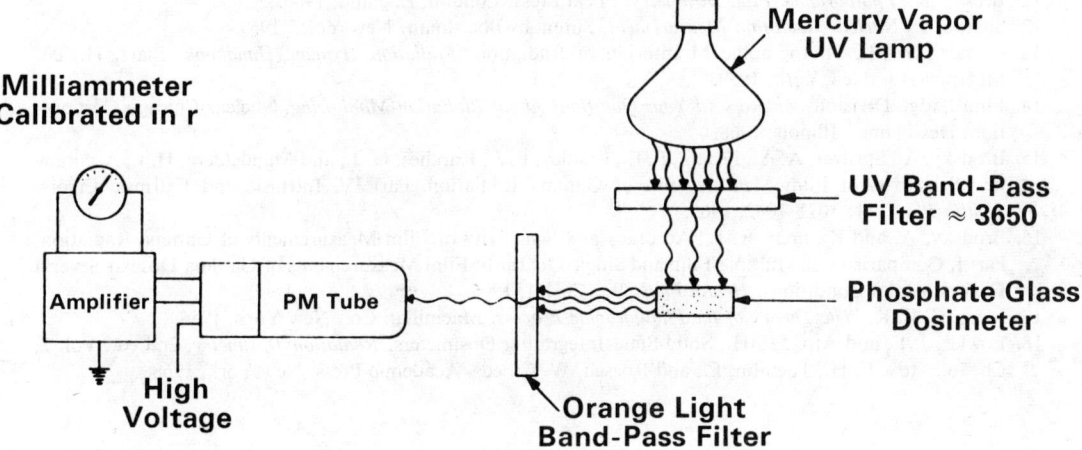

FIGURE 32.
Schematic Diagram of a Fluorimeter for Measuring Radiophotoluminescence in Silver-Activated Phosphate Glass Dosimeters.
(From: Fowler, J. F. and Attix, F. H., Solid-State Integrating Dosimeters. *Radiation Dosimetry*, 2nd ed., Vol. 2, Ch. 13, Attix, F. H., Tochilin, E., and Roesch, W.C., eds. Copyright Academic Press, New York, 1968.)

various types or thicknesses of filtration in order to estimate exposures from several types or energies of radiation simultaneously.

Certain glasses that fluoresce under ultraviolet radiation after exposure to ionizing radiation have been utilized as personnel dosimeters (radiophotoluminescent dosimeters), but they have generally been used over a high dose range and have so far not been widely employed in monitoring routine occupational exposure. Silver phosphate glass dosimeters have been widely distributed to the armed forces for use in measuring exposures of about 50 roentgens or more in the event of nuclear war. Also, phosphate glass rods have been inserted into film badges in many atomic-energy installations for use in measuring exposures greater than about 25 to 50 r as an immediate supplementary measurement to the less sensitive film. The phosphate glass dosimeter may be conveniently readable above 500 to 1000 r, where even the insensitive film in a routine monitoring packet would be so black that it would be difficult to read with an ordinary densitometer. Commercial systems are available, but some problems have still been noted in regard to the dependence of response on time after the irradiation. Any system should be calibrated by the users. Figure 32 shows a schematic diagram of a fluorimeter for measuring phosphate glass dosimeters.

A number of other techniques are available for measuring ionizing radiation, and it may be expected that there will be further developments in personnel dosimeters in the future.

REFERENCES

1. Roesch, W. C. and Attix, F. H., in *Radiation Dosimetry*, 2nd ed., Vol. 1, Ch. 1, Attix, F. H. and Roesch, W. C., eds. Academic Press, New York, 1968.
2. Attix, F. H. and Roesch, W. C., eds., *Radiation Dosimetry*, Vol. 2. Academic Press, New York, 1966.
3. Hine, G. J. and Brownell, G. L., eds., *Radiation Dosimetry*, 1st ed. Academic Press, New York, 1956.
4. Price, W. J., *Nuclear Radiation Detection*, 2nd ed. McGraw-Hill, New York, 1966.
5. Morgan, K. Z. and Turner, J. E., eds., *Principles of Radiation Protection—A Textbook of Health Physics*. John Wiley and Sons, New York, 1967.
6. Johns, H. E., *The Physics of Radiology*, 2nd ed. Charles C Thomas, Springfield, Illinois, 1964.
7. Fitzgerald, J. J., Brownell, G. L., and Mahoney, F. J., *Mathematical Theory of Radiation Dosimetry*. Gordon and Breach, New York, 1967.
8. Rees, D. J., *Health Physics*. M.I.T. Press, Cambridge, Massachusetts, 1967.
9. Braestrup, C. B. and Wyckoff, H. O., *Radiation Protection*. Charles C Thomas, Springfield, Illinois, 1958.
10. Handloser, J. S., *Health-Physics Instrumentation*. Pergamon Press, New York, 1959.
11. Becker, K., *Photographic-Film Dosimetry*. Focal Press, London, England, 1966.
12. Sharpe, J., *Nuclear-Radiation Measurement*. Simmons-Boardman, New York, 1959.
13. Corney, G. M., Photographic Monitoring of Radiation. *Radiation Hygiene Handbook*, Blatz, H., ed. McGraw-Hill, New York, 1959.
14. Film-Badge Division, *Answers to Your Questions About Radiation-Monitoring*. Nuclear-Chicago Corporation, Des Plaines, Illinois, 1968.
15. Brodsky, A., Spritzer, A. A., Feagin, F. E., Bradley, F. J., Karches, G. J., and Mandelberg, H. I., Accuracy and Sensitivity of Film Measurements of Gamma Radiation. Part IV. Intrinsic and Extrinsic Errors. *Health Phys.*, 11: 1071–1082, 1965.
16. Brodsky, A. and Kathren, R. L., Accuracy and Sensitivity of Film Measurements of Gamma Radiation. Part I. Comparison of Multiple-Film and Single-Quarterly-Film Measurements of Gamma Dose at Several Environmental Conditions. *Health Phys.*, 9: 453–461, 1965.
17. Mees, C. E. K., *The Theory of the Photographic Process*. Macmillan Co., New York, 1954.
18. Fowler, J. F. and Attix, F. H., Solid-State Integrating Dosimeters. *Radiation Dosimetry*, 2nd ed., Vol. 2, Ch. 13, Attix, F. H., Tochilin, E., and Roesch, W. C., eds. Academic Press, New York, 1968.

TRANSPORTATION OF RADIOACTIVE MATERIALS

F. J. Bradley, Ph.D., C.H.P.

In 1936, the United States Post Office Department issued the following directive:

RADIUM, THORIUM, OR ANY OTHER RADIOACTIVE SUBSTANCE,
OR ANY MATERIALS CONTAINING RADIOACTIVE SUBSTANCES,
SUCH AS POWDERS CONTAINING RADIUM OR THORIUM,
LIQUIDS CONTAINING RADIUM EMANATION,
RADIUM SALTS, RADIOACTIVE MINERALS,
OR ANY RADIOACTIVE MATERIAL WHATEVER
NOT PERMITTED IN THE MAILS.

Second Assistant Postmaster General,
Washington, July 13, 1936.

The Department has received complaints of damage to mail matter, which upon investigation has been found to be due to the presence of radioactive material in the mails.

Any radioactive material, including radium, thorium or other radioactive substance, or any materials containing radioactive substances, such as powders containing radium or thorium, liquids containing radium emanation, radium salts, or radioactive minerals, is prohibited transmission in the mails under the provisions of paragraph 1 of section 588, Postal Laws and Regulations, which provides "...and all other natural or artificial articles, compositions or material of whatever kind which may kill or in anywise hurt, harm, or injure another or damage, deface, or otherwise injure the mails or other property, whether sealed as first class matter or not, are hereby declared to be nonmailable matter, and shall not be conveyed in the mails or delivered from any post office or station thereof, nor by any letter carrier..."

Postmasters must watch this most carefully and refuse to accept any parcel containing any radioactive material whatever.

J. W. Cole,
Acting Second Assistant Postmaster General

As noted, the immediate complaint was not injury to post-office personnel, but damage to film in transit through the mails. Two exceptions to this order were issued. They were:

RADIUM PAINT IN THE MAILS

Second Assistant Postmaster General
Washington, Sept. 24, 1936.

After careful study it has been ascertained that radium or luminous paint may be transmitted in the mails without damaging other mail matter under the following conditions.

> One gram of luminous paint properly cushioned in a cubic box of wood with all six walls 1 inch thick, and this cubic box enclosed in a strong outside container with a bursting strength of not less than 275 pounds per square inch, measuring at least 7 inches on all sides, so that all outer surfaces will be at least 3 inches from the tube containing the luminous powder or paint. The wooden box must be packed in the outside container so that the tube of luminous paint will be equidistant from the walls of the outside container at all times. Such parcels must be labeled "Radium Paint."
>
> Postmasters may accept radium or luminous paint packed as above described.
>
> Harllee Branch,
> Second Assistant Postmaster General

and a letter exemption dated January 28, 1946, which reads

> The National Bureau of Standards reports that only pure polonium, free from any radioactive members, would be safe to be handled in the mails. Dials painted with pure polonium would be mailable. However, dials that are painted with any material containing radium are not mailable.

In 1939, the Railway Express Agency began to accept parcels containing radium with provisions made for special handling of the shipments. This handling provided separation of the parcels by at least 15 feet from parcels containing undeveloped film. It has been disclosed that, in the nine years in which this procedure was in force, no known case of film darkening attributable to exposure to ionizing radiation during transit was reported.

Due to the half-lives of many radioisotopes used in medicine and research, the ideal mode of transportation is by air. Rule 13-A of Supplement No. 5 to Air Express Division Tariff No. 8, published in 1946, permits shipments of radium and radioactive materials by Air Express, and this was further modified and updated in keeping with the new Interstate Commerce Commission (ICC) Regulations.

The ICC has been the federally designated rule-making body for ground-transportation regulations of dangerous articles. This responsibility was exercised by the Bureau of Explosives of the Association of American Railroads by authorization of the ICC up until 1966, when this responsibility was assumed by the Hazardous Materials Regulations Board of the new Department of Transportation.

The Bureau of Explosives designated regulations as Tariffs, which cover all dangerous articles in transit. The Bureau, anticipating a great expansion of radioisotope shipments following World War II, was instrumental in having delegated to the National Research Council the responsibility of formulating a set of regulations covering the transportation of radioactive material. A technically oriented committee was formed, which devised sensible regulations that were superimposed on the existing regulations. This was one area of commerce in which radioactive materials were handled with the least fanfare, resulting in a most effective and mainly unsung job of assisting in their application. The regulations formulated by the committee were summarized in the USAEC publication *Transportation of Radioactive Materials*, published in 1955 and 1958, with accompanying comments. The one area of increasing importance that is not covered by the original publication is the transportation of fissile material, i.e., material that might become critical in certain mass and geometrical configurations.

A recently proposed rule-making change further updates the original ICC regulations to cover transportation of fissile material and also revised the regulations covering radioactive materials. These regulations are issued by the Hazardous Materials Regulations Board of the Department of Transportation, which issues regulations covering interstate shipments by land and air under Title 49 (Transportation) and Title 14 (Aeronautics and Space); they are listed at the end of this chapter, followed by a bibliography of all current 1968 regulations covering the transportation of radioactive and fissile material.

General Considerations. The basic principle involved in transporting radioactive material is to safely and economically transfer material from one person to another via air, land, or sea transport. There is a tripartite group of responsible persons to insure safety: shipper (consignor), carrier, and receiver (consignee). The U.S. Department of Transportation regulations cover the first two, but not the consignee.

The specifics of transporting radioactive materials can be summarized by a flow diagram as given in Figure 33. The shipper is primarily responsible for the safety and correctness of his packaging and labeling, and for assuring the carrier that all applicable regulations have been fulfilled. The carrier's responsibilities are basically the same as those connected with transporting other dangerous and hazardous materials. The problems that have arisen in transportation of radioactive materials—e.g., incorrect packaging, incorrect labeling, misplaced packages—can be traced in many cases to the shipper rather than the carrier; but transportation is basically a dual responsibility, with a specific job to be done by each of the two parties.

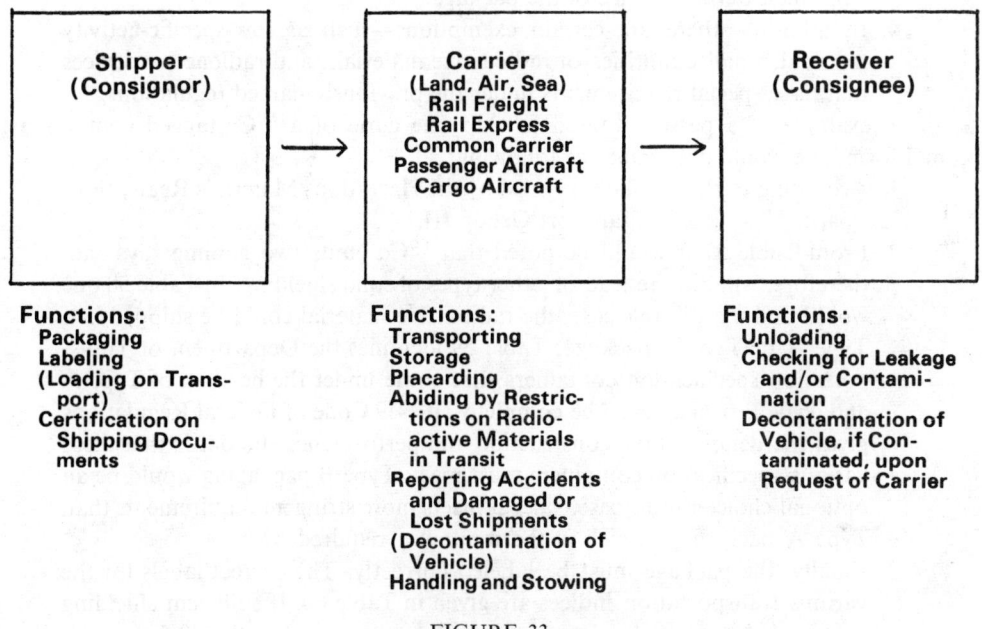

FIGURE 33.
Flow Diagram Indicating Areas of Responsibility in Transporting Radioactive Materials.

Shipper. The shipper of radioactive materials normally holds a license from the United States Atomic Energy Commission or an Agreement State, or he is a prime contractor of the United States Atomic Energy Commission. He is, therefore, familiar with the requirements for transporting radioactive materials.

To summarize, the requirements of the shipper in transporting radioactive materials are: he must package the material correctly; he must label it with the correct label; and, in filling out the shipping documents, he must certify that the package has been properly marked and sealed and must designate the Transport Groups of the radionuclides, name, activity, and label required for radionuclide(s), if they are in normal form. As an example, the shipper must have the following information about the radioactive material that is to be shipped; element and atomic number, e.g., ^{60}Co; activity of the radioactive material, in curie units; radiation dose on the surface of the package; and radiation dose at 3 feet from the surface of the package. Furthermore, if the radioactive material has other hazardous properties—i.e., if it is explosive, gaseous, inflammable, oxidizing, poisonous, infectious, or corrosive—this must be stated as well.

In addition to the above, the following designations must also be determined for the radioactive material.

1. The form of the radioactive material must be known. There are two forms; "special form" is the designation used if the radioactive material is not easily dispersed, as in the case of a radioactive alloy or sealed source; all other radio-active materials are designated as "normal form", e.g., liquids, gases, and dispersible solids.
2. If the radioactive material is in "normal form", one must specify the transport group of the radioactive material as defined in the regulations given under Title 49 and Title 14. There are seven groups, which classify radioactive materials according to their radiotoxicity.
3. For nonfissile radioactive material, the transport index must be known; this is defined as the radiation dose rate in millirem per hour at three feet from the external surface of the package.
4. In addition, there are certain exemptions—such as low-specific-activity material, small quantities of radioactive materials, and radioactive devices that have special requirements under the previously named regulations.

As an example, if a person wanted to ship one curie of a ^{60}Co-tagged compound in "normal form", he would determine the following:

1. According to the regulation issued by the Hazardous Materials Regulations Board, ^{60}Co falls in Transport Group III.
2. From Table 57 it would be noted that ^{60}Co emits two gamma rays and, therefore, will require lead or other types of solid shielding. In Table 58 one would find that, in this case, the radioactive material could be shipped in a Type A or Type B package. Table 59 identifies the Department of Transportation specification containers that come under the heading of Type A and Type B packaging. The complete Title 49 Code of Federal Regulations gives the details on the construction and performance standards that these various specification containers must meet. Type B packaging would be an optional choice in this case, since it meets more stringent requirements than Type A packaging, which is the minimum required.
3. Finally, the package must be labeled correctly. The correct labels for the various transportation indices are given in Table 60. If sufficient shielding is provided so that the dose rate at 3 feet measures less then 0.5 millirem per hour (transport index equal to 0.5), then a "Radioactive Yellow II" label must be used. If the dose rate at 3 feet is greater than 0.5 and less than or equal to 10 millirem (transport index greater than 0.5, but less than or equal to 10), a "Radioactive Yellow III" label must be used. These labels have been changed from the previous Groups I, II, or III Class D Poison labels into the "Radioactive White I", "Radioactive Yellow II", and "Radioactive Yellow III".
4. The shipper must certify on the shipping paper the following: "This is to certify that the above-named requests are properly classified, described, packaged, marked, and labeled, and are in proper containers for transportation according to the applicable regulations of the Department of Transportation."

Table 57.
SOME COMMON GAMMA-EMITTING RADIONUCLIDES

Isotope	Half-Life	Energy, Mev	Yield per Disintegration
^{24}Na	14.97 hours	2.76	100.0%
		1.37	100.0%
^{72}Ga	14.30 hours	2.51	17.0%
		2.49	10.0%
		2.20	30.0%
		0.89	10.0%
		0.83	83.0%
		0.63	21.0%
^{140}La	40.40 hours	2.57	4.0%
		1.60	96.0%
		0.89	11.0%
		0.82	26.0%
		0.49	46.0%
		0.33	24.0%
^{144}Pr	17.50 min.	2.20	~0.8%
		1.50	~0.3%
		0.69	~2.0%
^{76}As	27.60 hours	2.05	1.6%
		1.40	0.89%
		1.20	0.9%
		0.64	8.0%
		0.55	41.0%
^{124}Sb	60.60 days	2.11	10.0%
		1.71	46.0%
		1.38	6.2%
		0.99	5.4%
		0.60	100.0%
^{194}Ir	19.00 hours	2.05	weak
		1.80	weak
		1.66	weak
		1.62	weak
		1.51	weak
		1.48	weak
		1.34	weak
		1.22	weak
		1.18	
		1.15	
		0.94	
		0.64	
		0.62	
		0.47	
		0.33	
		0.29	
^{166}Ho	27.00 hours	1.61	weak
		1.53	weak
		1.36	11.0%
		0.08	6.0%
^{142}Pr	19.00 hours	1.57	7.0%
^{42}K	12.40 hours	1.53	18.0%

Table 57. (*Continued*)
SOME COMMON GAMMA-EMITTING RADIONUCLIDES

Isotope	Half-Life	Energy, Mev	Yield per Disintegration
^{110m}Ag	270.00 days	1.52 1.39* 0.94* 0.89* 0.81 0.76 0.71 0.68 0.66*	
^{82}Br	35.00 hours	1.47 1.31 1.03 0.82 0.77 0.69 0.61 0.55	15.0% 30.0% 30.0% 25.0% 85.0% 30.0% 40.0% 80.0%
^{152}Eu ^{154}Eu	12.40 years	1.40 1.25 1.10 0.97 0.78 0.36 0.26 0.18 0.12 0.09	
^{134}Cs	2.30 years	1.35 0.79 0.60 0.57	5.0% 95.0% 95.0% 25.0%
^{64}Cu	12.80 hours	1.34	0.5%
^{60}Co	5.30 years	1.33 1.17	100.0% 100.0%
^{114}In	50.00 days	1.30 0.72 0.56	~4.0% ~4.0% ~4.0%
^{115}Cd	43.00 days	1.30 0.94 0.49	1.0% 2.3% 0.3%
^{59}Fe	46.30 days	1.29 1.10 0.19	43.0% 57.0% 2.8%
^{122}Sb	2.80 days	1.26 1.137 0.69 0.57	0.66% 0.73% 3.4% 66.0%
^{182}Ta	111.00 days	1.23 1.22	14.0% 28.0%

Table 57. (*Continued*)
SOME COMMON GAMMA-EMITTING RADIONUCLIDES

Isotope	Half-Life	Energy, Mev	Yield per Disintegration
^{182}Ta (*Cont.*)		1.19	14.0%
		1.12	29.0%
		0.22	10.0%
		0.152	10.0%
		0.10	11.0%
		0.085	1.0%
		0.068	24.0%
		0.066	2.0%
^{91}Y	59.50 days	1.22	0.3%
^{106}Ru	1.00 year	1.20	1.0%
		0.73	17.0%
		0.51	17.0%
^{46}Sc	85.00 days	1.12	98.0%
		0.89	100.0%
^{65}Zn	250.00 days	1.11	45.0%
^{193}Os	31.00 hours	1.10	^{192}Ir
^{198}Au	2.70 days	1.087	~1.0%
		0.676	~1.0%
		0.411	96.4%
^{86}Rb	19.50 days	1.08	8.5%
T-^{58}Co	72.00 days	0.81	~100.0%
^{99}Mo	67.00 hours	0.78	20.0%
		0.74	
		0.372	
		0.181	
		0.142	
		0.14	
		0.041	
^{185}W	73.00 days	0.77	very weak
		0.57	very weak
		0.056	2.5%
^{186}Re	3.80 days	0.76	very weak
		0.63	very weak
		0.14	11.0%
		0.12	~2.0%
^{95}Nb	35.00 days	0.75	100.0%
^{105}Ru	4.00 hours	0.73	100.0%
		0.13	25.0%
^{95}Zr	65.00 days	0.72	43.0%
		0.75	54.0%
^{131}I	8.00 days	0.72	3.0%
		0.64	9.0%
		0.36	80.0%
		0.28	5.0%
		0.08	2.0%
^{187}W	24.10 hours	0.69	
		0.62	

Table 57. (*Continued*)
SOME COMMON GAMMA-EMITTING RADIONUCLIDES

Isotope	Half-Life	Energy, Mev	Yield per Disintegration
^{187}W (*Cont.*)		0.55	
		0.48*	
		0.13	
		0.07	
^{137}Cs	33.00 years	0.66 ^{137m}Ba	82.0%
^{125}Sb	2.70 years	0.64	33.0%
		0.60	
		0.465	49.0%
		0.425	
		0.176	
		0.04	
^{192}Ir	74.00 days	0.61	
		0.60	
		0.48	
		0.47	
		0.32	
		0.31	
		0.30	
^{181}Hf	46.00 days	0.61	very weak
		0.48	~84.0%
		0.35	~12.0%
		0.14	~12.0%
		0.13	29.0%
^{103}Ru	40.00 days	0.61	6.0%
		0.50	90.0%
^{97}Ru	2.80 days	0.57	
		0.33	
		0.22	
		0.11	
^{140}Ba	12.80 days	0.54	30.0%
		0.30	10.0%
		0.16	70.0%
		0.13	
		0.03	100.0%
^{153}Sm	47.00 hours	0.54	~9.0%
		0.17	
		0.103	~31.0%
		0.069	~4.0%
^{147}Nd	11.00 days	0.53	34.0%
		0.092	26.0%
^{77}As	38.00 hours	0.53	0.8%
		0.25	2.5%
		0.16	0.26%
		0.09	0.13%
		0.028	
		0.023	
^{131}Ba	13.00 days	0.49	80.0%
		0.37	20.0%
		0.21	16.0%
		0.12	20.0%
^{75}Se	127.00 days	0.40*	
		0.31	

Table 57. (*Continued*)
SOME COMMON GAMMA-EMITTING RADIONUCLIDES

Isotope	Half-Life	Energy, Mev	Yield per Disintegration
^{75}Se (*Cont.*)		0.28* 0.27* 0.20 0.14 0.12 0.10 0.08 0.07 0.02	
^{133}Ba	7.50 years	0.36 0.29 0.08 0.07	74.0% 26.0% 32.0% 6.0%
^{111}Ag	7.60 days	0.34 0.24	~8.0% ~1.0%
^{51}Cr	26.50 days	0.32	~8.0%
^{105}Rh	36.50 hours	0.31	35.0%
^{203}Hg	46.50 days	0.28	83.0%
^{197}Hg	65.00 hours	0.28 0.19 0.16 0.13 0.08	~0.5% 29.0%
^{113}Sn	118.00 days	0.26 0.39 ^{113}In via 104 min. ^{113}I	~70.0%
^{199}Au	3.15 days	0.21 0.16 0.05	11.0% 48.0% 0.4%
^{197m}Hg	23.00 hours	0.28 0.19 0.16 0.13	3.0% 3.0% 97.0% 97.0%
^{114m}In	49.00 days	0.19	20.0%
^{141}Ce	32.50 days	0.32 0.15	 60.0%
^{131}Cs	9.60 days	0.15	
^{144}Ce	285.00 days	0.13 0.10 0.08 0.05 0.04 0.03	8.5% 1.5% 3.0% weak weak weak
^{155}Eu	1.70 years	0.13 0.11 0.09 0.06	
^{191}Os	15.00 days	0.134 0.0	33.0% weak

Table 57. (Continued)
SOME COMMON GAMMA-EMITTING RADIONUCLIDES

Isotope	Half-Life	Energy, Mev	Yield per Disintegration
^{109}Pd	13.60 hours	0.09	7.0%
^{127}Te	115.00 days	0.09	weak
^{170}Tm	127.00 days	0.08	3.0%
^{102}Pb (Ra)	22.00 years	0.05	
^{129}I	1.72×10^7 years	0.04	
^{227}Ac	22.00 years	0.04	

* Most abundant gammas.
Based on material in publication of the Edlow Lead Company, Columbus, Ohio.

Table 58. QUANTITY LIMITATIONS AND PACKAGING REQUIREMENTS

Form of Radioactive Material	Maximum Quantity per Package				
	Exempt	Manufactured Articles	Low-Specific-Activity Material	Less Than	Less Than
Special form	—	20.000 Ci	—	20.000 Ci	5,000 Ci
Normal form, Transport Group					
I	0.00001 Ci	0.001 Ci	0.001 Ci	0.001 Ci	20 Ci
II	0.00010 Ci	0.050 Ci	0.050 Ci	0.050 Ci	20 Ci
III	0.00100 Ci	3.000 Ci	3.000 Ci	3.000 Ci	200 Ci
IV	0.00100 Ci	3.000 Ci	20.000 Ci	20.000 Ci	200 Ci
V	0.00100 Ci	1.000 Ci	20.000 Ci	20.000 Ci	5,000 Ci
VI	0.00100 Ci	1.000 Ci	1,000.000 Ci	1,000.000 Ci	50,000 Ci
VII	—	200.000 Ci	1,000.000 Ci (see Note 1)	1,000.000 Ci	50,000 Ci
Fissile Material, Class I	15 g	—	—	—	—
II	—	—	500 g	—	—
III	see Note 2	see Note 2	see Note 2	see Note 2	see Note 2
Container required	see Note 3	see Note 4	Type C	Type A	Type B

Note 1. Tritium as a gas, as luminous paint, or as absorbed on solid material, and tritium oxide in aqueous solutions of not more than 0.5 mCi per milliliter.

Note 2. Fissile Class III material must be transported under controls providing nuclear-criticality safety in vehicles assigned for sole use by the consigner who is responsible, to protect against loading, transporting, or storing such material with other fissile material, or the material must be transported under an escort having the capability, equipment, authority, and instruction to provide compliance with the above conditions.

Note 3. Radioactive materials exempted from specification packaging must be in an inner container bearing the marking "Radioactive" on its exterior surface and packaged in a strong, tight package, so that there will be no leakage of radioactive materials under conditions normally incident to transportation. The package must be such that there is no detectable exterior-surface contamination and that the dose rate at any point on the exterior of the surface does not exceed 0.5 millirems per hour.

Note 4. Manufactured articles—such as instruments, clocks, electronic tubes, or other similar devices having radioactive materials in a nondispersible form (others have liquids) and having a dose rate not exceeding 10 millirem per hour at 4 inches from the unpackaged item—must be packed in strong, tight packages, so that there will be no leakage of radioactive contents under conditions normally incident to transportation. The outside of the package must be marked "Radioactive" and be such that there is no detectable radioactive surface contamination and that the dose rate at the surface of the package does not exceed 0.5 millirem per hour. For carload or truckload lots the external dose rate may not exceed 0.5 millirem at the surface of the vehicle.

From: Conlon, F. B., Internal Memorandum, *Comments on Proposed Rule Making*, Department of Transportation, Hazardous Materials Regulations. NCRH, 1968.

Table 59.
SPECIFICATION CONTAINERS FOR TYPE A, B, AND C PACKAGING

Container Type	Authorized Specification Containers					
	Metal Drums	Wooden Boxes	Fiberboard Drums	Fiberboard Boxes	Cylinders	Special Containers
B						6M (see Notes 1 and 2)
A	5B 5D 6B 6C 6J 6K 17C 17H 42B 42C	15A 15B 15C 15D 19A 19B	21C	12A 12B 12C 12D 12E 12H 12P 12R		7A
C	5 5A 5C 5F 5H 5K 5L 5M 5P 5X 6A 6D 6L 13A 17E 17F 17X 37A 37B 37C 37K 37M 37P 42D 42E 42F 42G 42H	14 15E 15L 15M 15P 15X 16A 16B 16D 19A 19B	21P	23F 23H	3A 3AA 3B 3BN 3C 3D 3E 3HT 4 4A 4B 4BA 4C 4D 4DA 4DA 4L	

Note 1. Packages designed to meet the environmental and test conditions incident to normal transportation included in Article 173.393 (d) (1) and (2) and the hypothetical accident conditions delineated in Article 173.350-2, which have been specifically authorized by the Department under Article 176 of this chapter, may be used in lieu of specification packages listed above.

Note 2. Quantities of radioactive materials in special form or normal form and fissile materials exceeding quantities authorized for Type A packaging in Article 173 must meet the requirements of Note 1 and the criteria in the regulation of the U.S. Atomic Energy Commission delineated in Title 10, Code of Federal Regulations, Part 71, and must be specifically approved by the Department under Article 170 of this chapter.

From: Conlon, F. B., Internal Memorandum, *Comments on Proposed Rule Making*, Department of Transportation, Hazardous Materials Regulations. NCRH, 1968.

Table 60.
LABELING REQUIREMENTS
FOR PACKAGES CONTAINING RADIOACTIVE MATERIALS

Labels	Radioactive Materials		
	Normal or Special Form, Having		Fissile
	Surface Dose Rates	Transport Index	
Radioactive White I	under 0.5 mrem/hr	0	Class I
Radioactive Yellow II	under 10 mrem/hr	under 0.5	Class II
Radioactive Yellow III	over 10 mrem/hr	over 0.5	Class III

From: Conlon, F. B., Internal Memorandum, *Comments on Proposed Rule Making, Department of Transportation, Hazardous Materials Regulations.* NCRH, 1968.

These new regulations considerably simplify the shipment of fissile materials, which previously were considered almost on a case-by-case basis. The standards that such shipments must meet are covered in the regulations at the end of this chapter.

Carrier. Transportation of radioactive material may be via land, sea, or air, or by various combinations of these common modes of transport. Basic responsibilities of the carrier are transporting, placarding, safe storage in transit, and immediate reporting of accidents and of damaged or lost shipments; in addition, he is responsible for determining whether his transport vehicle is contaminated. To discharge these responsibilities, he may ask the consignee to assist him. Maximum permissible surface contaminations for packages and transport vehicles are given in §173.397 of the regulations.

In transporting via land, one can transport by rail freight or by rail express. Parts 174 and 175 of the regulations cover requirements on shipping papers, loading of low-specific-activity material commonly shipped by this mode, placarding of the rail car, and cleaning of the car in case of contamination. The handling of radioactive material during transit is restricted according to the transport index on the package; in any rail car or storage location the total transport index number of all the packages must not exceed 50. Furthermore, packages bearing radioactive-material labels " Radioactive Yellow II " or " Radioactive Yellow III " must not be placed in rail cars, depots, or other places closer than three feet to an area that may be continuously occupied by persons, nor closer than fifteen feet to any undeveloped film. This requirement is for a transport index of 10 or less; the various safe distances for other transport indices are given in the regulations. There are, in addition, restrictions on train position, which prohibit placing a rail car with a " Radioactive-Material " placard next to a rail car with an " Explosives " placard or next to rail-car shipments of undeveloped film.

A more common means of transport via land is by common, contract, or private carrier. The carrier will demand certification similar to that given earlier from a shipper of packages bearing " Radioactive Yellow II " or " Radioactive Yellow III " labels. Parallel to the rail car restrictions, packages with a total transport index of no more than 50 can be transported in a single vehicle, and there are restrictions on the distance between continuously occupied areas and the storage locations of packages with various transport indices. These would be different for shipments in vehicles that are used solely for transporting radioactive materials. Regulations governing this type of shipment and the certification required are given at the end of this chapter.

Shipments via passenger-carrying and cargo-only aircraft are given in Title 14, CFR, Part 103, which is also included in the regulations presented at the end of this chapter. There must be certification by the shipper that the materials are packaged in a prescribed manner, as given above; one signed copy must accompany the package, and a duplicate copy is retained by the originating air carrier. In addition, radioactive liquids within the restrictions of the regulations are permitted.

Consignee. The Department of Transportation regulations given below do not cover the consignee; however, since most consignees are licensed recipients of radioactive materials, they are responsible—at least indirectly—for determining, upon receipt, that the radioactive material has not leaked or contaminated the transporting vehicle. If leaking, damaged, or contaminated shipments are received by the consignee, or if a shipment is lost, he must immediately notify the carrier and the shipper, or consignor. If he has the required expertise, he may be asked by the carrier to decontaminate his transport vehicle in the event of contamination. In like manner, he is required to immediately notify responsible authorities about lost shipments or expected shipments that may have gone astray. The responsible authority may be the United States Atomic Energy Commission, United States Public Health Service, or state and local government radiation-control groups who are responsible at the local level for public health and safety and will provide the necessary expertise in personnel and equipment for locating lost or astray shipments of radioactive materials.

U.S. Postal Regulations. Packages containing radioactive materials that conform to the paragraph cited below will be accepted by the U.S. Postal Service.

.25 Radioactive Materials.

In addition to special packaging precautions, as prescribed in this part, a package containing radioactive materials must not emit from its exterior any significant alpha, beta, or neutron radiations, and the gamma radiation at any surface of the package must be less than 10 milliroentgens for 24 hours. The package must not contain more than 0.1 millicurie of radium or polonium, or that amount of strontium-89, strontium-90, or barium-140 which disintegrates at a rate of more than 5 million atoms per second, or that amount of any other radioactive substance which disintegrates at a rate of more than 50 million atoms per second. See 124.24 regarding disposition of dangerous radioactive materials that are nonmailable.

Post Office Service TL-169, 11-19-65—Issue 906

RULES AND REGULATIONS

Because of the complex nature of these amendments and the impact that they will have on the transportation of radioactive materials, and to allow a reasonable time for compliance with the changes made herein, the effective date of the amendments is December 31, 1968. However, compliance with these amendments is authorized on and after the date of publication in the Federal Register.

In consideration of the foregoing, the Hazardous Materials Regulations of the Department of Transportation (14 CFR, Part 103, and 49 CFR, Parts 170–190) are amended effective December 31, 1968, as set forth below.

> (Title 18, U.S.C., secs. 831–835; sec. 9, Department of Transportation Act (49 U.S.C. 1657); Title VI and sec. 902(h), Federal Aviation Act of 1958 (49 U.S.C. 1421-1430, 1472(h)))

Issued in Washington, D.C., on September 26, 1968.
> W. J. Smith,
> Commandant,
> United States Coast Guard

Issued in Washington, D.C., on September 26, 1968.
> Sam Schneider,
> Board Member for the Federal Aviation Administration

Issued in Washington, D.C., on September 26, 1968.
Lowell K. Bridwell,
Administrator,
Federal Highway Administration
Issued in Washington, D.C., on September 26, 1968.
A. Scheffer Lang,
Administrator,
Federal Railroad Administration

1. Chapter I of Title 49 is amended as follows:

PART 171—GENERAL INFORMATION AND REGULATIONS

I. Section 171.8 is amended by adding the following new paragraphs at the end thereof:

§ 171.8 Definitions.

* * * * *

(i) "Packaging" means the assembly of the containers and any other components necessary to assure compliance with the prescribed packaging requirements.

(j) "Package" means the packaging plus its content of explosives or other dangerous articles, as presented for transportation.

(k) "Transport vehicle" means the conveyance used for the transportation of explosives or other dangerous articles, and includes any motor vehicle, rail car, or aircraft. Each cargo-carrying body (trailer, van, box car, etc.) is a separate vehicle.

(l) "Department" means the Department of Transportation.

PART 172—COMMODITY LIST OF EXPLOSIVES AND OTHER DANGEROUS ARTICLES CONTAINING THE SHIPPING NAME OR DESCRIPTION OF ALL ARTICLES SUBJECT TO PARTS 171–179 OF THIS CHAPTER

II. Part 172 is amended as follows:

(A) By amending § 172.2 to read as follows:

§ 172.2 Articles Not Described.

(a) An article whose proper shipping name is not shown in the commodity list in § 172.5, and which must be classified as dangerous under the definitions in § 172.53, § 173.88, § 173.100, § 173.115, § 173.150, § 173.151, § 173.240, § 173.300, § 173.326, § 173.343, § 173.381, or §173.389 of this chapter, must be prepared and offered for shipment in compliance with the regulations for the appropriate hazard classification.

§ 172.4 (Amended)

(B) In paragraph (a) of § 172.4, amend the listing as follows:

1. Cancel: "Poison D—Radioactive materials, Class D."
2. Add: "R.A.M.—Radioactive materials."

§ 172.5 (Amended)

(C) In paragraph (a) of § 172.5, amend the commodity list as follows:

Radiation Protection and Regulations

Article	Classed As	Exemptions and Packing, see Section	Label Required, if Not Exempt	Maximum Quantity in One Outside Container by Rail Express
Change				
Fissile radioactive materials	Radioactive material	173.393 173.396	Radioactive	See § 173.396
Radioactive devices	Radioactive material	173.391 (b)	None	See § 173.391
Radioactive materials, low specific activity (LSA)	Radioactive material	173.392 173.393	Radioactive	See § 173.392 (a)
Thorium nitrate, solid	Radioactive material, oxygen material	173.392	Radioactive and yellow	100 pounds
Uranyl nitrate, solid	Radioactive material, oxygen material	173.392 173.396	Radioactive and yellow	100 pounds
Add				
Radioactive materials, small quantities	Radioactive material	173.391(a)	None	See § 173.391
Radioactive materials, n.o.s.	Radioactive material	173.393 173.395	Radioactive	See §§ 173.393, 173.395
Radioactive materials, special form	Radioactive material	173.393 173.394	Radioactive	See §§ 173.393, 173.394
Cancel				
Magnesium-thorium alloys in formed shapes (not powdered), and which shall contain not more than 4% nominal ^{232}Th	Poison D	173.392(e)	Radioactive materials, red	See § 173.393 (L)
Radioactive materials, n.o.s.	Poison D	173.392 173.393	Radioactive materials, blue or red	See § 173.393 (f) and (L)
Uranium, normal or depleted, in solid metal form (not borings, chips, or pieces)	Poison D	173.392(f)	Radioactive materials, red	See § 173.393 (L)

PART 173—SHIPPERS

III. Part 173 is amended as follows:
    (A) The table of contents is amended by adding §§ 173.389, 173.390, and 173.397 through 173.399; amending §§ 173.2, 173.22, 173.23, 173.24, 173.28, 173.325, 173.391 through 173.396, and 173.402; amending Subpart G to read as follows:

Sec.
173.2   Classification; dangerous articles
173.22   Shipper's responsibility
173.23   Previously authorized packaging
173.24   Standard requirements for all packages
173.28   Reuse of containers

Subpart G—Poisonous Materials and Radioactive Materials: Definition and Preparation

173.325 Classes of poisonous materials
173.389 Radioactive materials: definitions
173.390 Transport groups of radionuclides
173.391 Small quantities of radioactive materials and radioactive devices
173.392 Low-specific-activity radioactive material
173.393 General packaging requirements
173.394 Radioactive material in special form
173.395 Radioactive material in normal form
173.396 Fissile radioactive material
173.397 Contamination control
173.398 Special tests
173.399 Labeling of packages of radioactive materials
173.402 Labeling of explosives or other dangerous articles

(B) By amending § 173.2 to read as follows:

§ 173.2 Classification: Dangerous Articles.

(a) Dangerous articles other than explosives having more than one hazardous characteristic, as defined in Parts 170–190 of this chapter, must be classified according to the greatest hazard present. However, such articles which are also Class A poisons or radioactive materials must be classified according to both hazardous characteristics, as defined in this part.

(C) By amending § 173.22 in its entirety to read as follows:

§ 173.22 Shipper's Responsibility.

(a) Where containers are supplied by the shipper, the shipper shall be responsible to determine that shipments of explosives and other dangerous articles are made in containers which, unless otherwise provided in this part (see § 173.9 (c)), have been made assembled, with all parts or fittings in their proper place, and are marked in compliance with applicable specifications prescribed in Parts 178 and 179 of this chapter or with specifications of the Department in effect at date of manufacture of container. The shipper may accept the manufacturer's certification or specification marking to determine that the containers were manufactured in accordance with applicable specifications. Where containers are supplied by the carrier, the shipper shall determine that the containers in which commodities are to be loaded are proper containers for the transportation of such commodities by examining the manufacturer's identification plate, specification marking, or certification by the carrier.

(D) By amending § 173.23 in its entirety to read as follows:

§ 173.23 Previously Authorized Packaging.

(a) Where regulations require Spec. 6D or 37M (§ 178.102 or § 178.134 of this chapter) cylindrical steel overpacks, Spec. 5B, 6J, or 37A (single-trip container) (§ 178.82, § 178.100, or §178.131 of this chapter) metal drums manufactured before March 18, 1964, having inside Spec. 2S, 2SL, 2T, or 2TL (§ 178.21, §178.27, § 178.35, or § 178.35(a) of this chapter) polyethylene container, may be continued in use for the commodities and gross weights for which they were previously authorized.

(b) Reusable molded polyethylene containers for use without overpack complying with Spec. 34 (§ 178.19 of this chapter), manufactured before September 5, 1966, may be continued in use if they are plainly marked "ICC-34" and are embossed with the maker's name or symbol, rated capacity, and month and year of manufacture.

(c) Containers manufactured before January 1, 1967, and approved by the Bureau of Explosives before July 12, 1966
  (1) may be continued in use for the shipment of fissile and other radioactive materials under the approved conditions until that approval is terminated by the Department or the Bureau of Explosives, but in no case after February 28, 1969;
  (2) may not be used for export unless specifically approved by the Department.

Petitions for continuing use of such containers may be filed with the Department under § 170.13 of this chapter.

(E) By amending § 173.24 in its entirety to read as follows:

§ 173.24 Standard Requirements for All Packages.

(a) Each package used for shipping explosives or other dangerous articles under this chapter shall be so designed and constructed, and its contents so limited, that under conditions normally incident to transportation
  (1) there will be no significant release of the explosive or other dangerous article to the environment;
  (2) the effectiveness of the packaging will not be substantially reduced, and
  (3) there will be no mixture of gases or vapors in the package, which could, through any credible spontaneous increase of heat or pressure, or through an explosion, significantly reduce the effectiveness of the packaging.

(b) Materials for which detailed specifications for packaging are not set forth in this part must be securely packaged in strong, tight packages meeting the requirements of this section.

(c) Packaging used for the shipment of explosives or other dangerous articles under this chapter shall, unless otherwise specified or exempted therein, meet all of the following design and construction criteria.
  (1) Each specification container shall be marked in an unobstructed area with letters and numerals identifying that specification (e.g., ICC-6J, DOT-6L, DOT-MC 306, ICC-105A200-F);
    (i) the marking is a certification that the packaging complies with all specification requirements;
    (ii) the name and address or the symbol of the manufacturer or the user who assumes responsibility for compliance with the specification requirements shall be included; symbol letters must be registered with the Bureau of Explosives; duplicate symbols are not authorized;
    (iii) the markings shall be stamped, embossed, burned, printed, or otherwise marked on the packaging to provide adequate accessibility, permanency, and contrast so as to be readily apparent and understood;
    (iv) unless otherwise specified, letters and numerals shall be at least $\frac{1}{2}$ inch high;
    (v) packaging which does not comply with the applicable specification listed in Parts 178 and 179 of this chapter must not be marked to indicate such compliance.
  (2) Steel used shall be low-carbon, commercial-quality steel; stainless, open-hearth, electric, basic-oxygen, or other similar quality steels are acceptable. Steel sheets of specified gauges shall comply wth the following:

Gauge No.	Nominal Thickness, inches	Minimum Thickness, inches
12	0.1046	0.0946
13	0.0897	0.0817
14	0.0747	0.0677
15	0.0673	0.0603
16	0.0598	0.0533
17	0.0538	0.0478
18	0.0478	0.0428
19	0.0418	0.0378
20	0.0359	0.0324
22	0.0299	0.0269
24	0.0239	0.0209
26	0.0179	0.0159
28	0.0149	0.0129
30	0.0120	0.0110

(3) Lumber used shall be well seasoned, commercially dry, and free from decay, loose knots, knots that would interfere with nailing, and other defects that would materially lessen the strength.

(4) Welding and brazing shall be performed in a workman-like manner, using suitable and appropriate techniques, materials, and equipment.

(5) Packaging materials and contents shall be such that there will be no significant chemical or galvanic reaction among any of the materials in the package.

(6) Closures shall be adequate to prevent inadvertent leakage of the contents under normal conditions incident to transportation; gasketed closures shall be fitted with gaskets of efficient material which will not be deteriorated by the contents of the container.

(7) Nails, staples, and other metallic devices shall not protrude into the interior of the outer packaging in such a manner as to be likely to cause failures.

(8) The nature and thickness of the packaging shall be such that friction during transport does not generate any heating likely to decrease the chemical stability of the contents.

(d) For specification containers, compliance with the applicable specifications in parts 178 and 179 of this chapter shall be required in all details, except as otherwise provided in this chapter.

(F) By amending the heading, the introductory text of paragraph (a) and paragraph (h) in § 173.28 to read as follows.

§ 173.28 Reuse of Containers.

(a) Containers used more than once (refilled and reshipped after having been previously emptied) must be in such condition, including closure devices and cushioning materials, that they comply in all respects with the prescribed requirements for those containers. Repairs must be made in an efficient manner in accordance with requirements for materials and construction as prescribed in Parts 178 and 179 of this chapter for new containers, or as otherwise prescribed. Parts that are weak, broken, or otherwise deteriorated must be replaced.

* * * * *

(h) Except as provided in this section, single-trip containers made under specifications prescribed in Part 178 of this chapter, from which contents have once been removed following use for shipment of any commodity, shall not be again used as shipping containers for explosives or other dangerous articles. Single-trip containers may be reused, if retested before each reuse in accordance with methods approved by the Bureau of Explosives, for service for specific commodities or classes of commodities. Applications for permission for reuse should be made to the Bureau of Explosives, 2 Pennsylvania Plaza, New York, N.Y. 10001.

* * * * *

(G) By amending § 173.29(e) to read as follows:

§ 173.29 Empty Containers.

(e) All packagings and accessories which have been used for shipments of radioactive materials, when shipped as empty, must be securely closed; the external surface must be free of significant removable radioactive contamination, as provided in § 173.397(a), and the radiation at the external surface of the packaging must not exceed 0.5 millirem per hour. The "Empty" label, described in § 173.413, must be affixed to the packaging.

(H) By amending the title of Subpart G to read as follows:

SUBPART G—POISONOUS MATERIALS AND RADIOACTIVE MATERIALS: DEFINITION AND PREPARATION

(I) By amending the section heading and the introductory text of paragraph (a); cancel paragraph (a)(4) of § 173.325 as follows:

§ 173.325 Classes of Poisonous Materials.

(a) Poisonous materials for the purpose of Parts 170–190 of this chapter are divided into three classes, according to degree of hazard in transportation.

* * * * *

(4) (Canceled)

(J) By amending the introductory text of paragraph (a) of § 173.343 as follows:

§ 173.343 Less Dangerous Poisons, Class B, Liquid or Solid, (Poison Label; Definition).

(a) For the purposes of Parts 170–190 of this chapter, and except as otherwise provided in this part, Class B poisons are those substances, liquid or solid (including pastes and semisolids), other than Class A or Class C poisons, which are known to be so toxic to man as to afford a hazard to health during transportation, or which, in the absence of adequate data on human toxicity, are presumed to be toxic to man because they fall within any one of the following categories when tested on laboratory animals:

* * * * *

(K) By adding new §§ 173.389 and 173.390 as follows:

§ 173.389 Radioactive Materials; Definitions.

For the purpose of Parts 170–190 of this chapter:

(a) "Fissile radioactive material" means the following material: plutonium-238, plutonium-239, plutonium-241, uranium-233, or uranium-235, or any material containing any of the foregoing materials. See § 173.396(a) for exclusions. Fissile radioactive-material packages are classified according to the controls needed to provide nuclear-criticality safety during transportation, as follows.

(1) Fissile Class I: packages which may be transported in unlimited numbers and in any arrangement, and which require no nuclear-criticality safety controls during transportation. For purposes of nuclear-criticality safety control, a transport index is not assigned to Fissile Class I packages. However, the external radiation levels may require a transport index number.

(2) Fissile Class II: packages which may be transported together in any arrangement, but in numbers which do not exceed an aggregate transport index of 50. For purposes of nuclear-criticality safety control, individual packages may have a transport index of not less than 0.1 and not more than 10. However, the external radiation levels may require a higher transport index number, but this must not exceed 10. Such shipments require no nuclear-criticality safety control by the shipper during transportation.

(3) Fissile Class III: shipments of packages which do not meet the requirements of Fissile Classes I and II, and which are controlled in transportation by special arrangements between the shipper and the carrier.

>Note 1. Uranium-235 exists only in combination with various percentages of uranium-234 and uranium-238. "Fissile radioactive material," as applied to uranium-235, refers to the amount of uranium-235 actually contained in the total quantity of uranium being transported.

>Note 2. Radioactive material may consist of mixtures of fissile and nonfissile radionuclides. "Fissile radioactive material" refers to the amount of plutonium-238, plutonium-239, plutonium-241, uranium-233, uranium-235, or any combination thereof actually contained in the mixture. The "radioactivity" of the mixture consists of the total activity of both the fissile and nonfissile radionuclides. All mixtures containing "fissile material" shall be subject to §173.396.

(b) "Large-quantity radioactive materials" means a quantity the aggregate radioactivity of which exceeds that specified as follows:
  (1) Groups I or II (see paragraph (h) of this section) radionuclides: 20 curies;
  (2) Groups III or IV radionuclides: 200 curies;
  (3) Groups V radionuclides: 5,000 curies;
  (4) Groups VI or VII radionuclides: 50,000 curies;
  (5) special-form material: 5,000 curies;

(c) "Low-specific-activity material" means any of the following:
  (1) uranium or thorium ores, and physical or chemical concentrates of those ores;
  (2) unirradiated natural or depleted uranium, or unirradiated natural thorium;
  (3) tritium oxide in aqueous solutions, provided the concentration does not exceed 5 millicuries per milliliter;
  (4) material in which the activity is essentially uniformly distributed, and in which the estimated average concentration per gram of contents does not exceed:
    (i) 0.0001 millicuries of Group I (see § 173.389(h)) radionuclides, or
    (ii) 0.005 millicuries of Group II radionuclides, or
    (iii) 0.3 millicuries of Groups III or IV radionuclides.
    >Note: this includes, but is not limited to, materials

of low radioactivity concentration, such as residues or solutions from chemical processing; wastes, such as building rubble, metal, wood, and fabric scrap, glassware, paper and cardboard; solid or liquid plant waste, sludges, and ashes.

(d) "Normal-form radioactive materials" means those which are not special-form radioactive materials. Normal-form radioactive materials are grouped into transport groups (see paragraph (h) of this section).

(e) "Radioactive material" means any material or combination of materials which spontaneously emits ionizing radiation. Materials in which the estimated specific activity is not greater than 0.002 microcuries per gram of material and in which the radioactivity is essentially uniformly distributed are not considered to be radioactive materials.

(f) "Removable radioactive contamination" means radioactive contamination which can be readily removed in measurable quantities by wiping the contaminated surface with an absorbent material. The measurable quantities shall be considered as being not significant if they do not exceed the limits specified in § 173.397.

(g) "Special-form radioactive materials" means those which, if released from a package, might present some direct radiation hazard, but would present little hazard due to radiotoxicity and little possibility of contamination. This may be the result of inherent properties of the material (such as metals or alloys) or acquired characteristics (as through encapsulation). The criteria for determining whether a material meets the definition of special form are prescribed in § 173.398(a).

(h) "Transport group" means any one of seven groups into which normal-form radionuclides are classified according to their radiotoxicity and their relative potential hazard in transportation, and as listed in § 173.390.

(i) "Transport index" means the number placed on a package to designate the degree of control to be exercised by the carrier during transportation. The transport index to be assigned to a package of radioactive materials shall be determined by either subparagraph (1) or (2) of this paragraph, whichever is larger. The number expressing the transport index shall be rounded up to the next highest tenth (e.g., 1.01 becomes 1.1);

  (1) the highest radiation dose rate, in millirem per hour, at three feet from any accessible external surface of the package; or
  (2) for Fissile Class II packages only, the transport index number calculated by dividing the number 50 by the number of similar packages which may be transported together (see § 173.396), as determined by the procedures prescribed in the regulations of the U.S. Atomic Energy Commission, Title 10, Code of Federal Regulations, Part 71.

(j) "Type A packaging" means packaging which is designed in accordance with the general packaging requirements of §§ 173.24 and 173.393, and which is adequate to prevent the loss or dispersal of the radioactive contents and to retain the efficiency of its radiation shielding properties if the package is subject to the tests prescribed in § 173.398(b).

(k) "Type B packaging" means packaging which meets the standards for Type A packaging and, in addition, meets the standards for hypothetical accident conditions of transportation as prescribed in § 173.398(c).

(l) "Type A quantity" and "Type B quantity" radioactive materials means a quantity the aggregate radioactivity of which does not exceed that specified as follows:

Transport Group (see § 173.389(h))	Type A Quantity, curies	Type B Quantity, curies
I	0.001	20
II	0.05	20
III	3	200
IV	20	200
V	20	5,000
VI and VII	1,000	50,000
Special form	20	5,000

§ 173.390 Transport Groups of Radionuclides.
(a) List of radionuclides:

Element	Atomic Number	Radionuclide	Transport Group						
			I	II	III	IV	V	VI	VII
Actinium	89	^{227}Ac	x						
		^{228}Ac	x						
Americium	95	^{241}Am	x						
		^{243}Am	x						
Antimony	51	^{122}Sb				x			
		^{124}Sb			x				
		^{125}Sb			x				
Argon	18	^{37}Ar						x	
		^{41}Ar		x					
		^{41}Ar (uncompressed*)					x		
Arsenic	33	^{73}As				x			
		^{74}As				x			
		^{76}As				x			
		^{77}As				x			
Astatine	85	^{211}At				x			
Barium	56	^{131}Ba				x			
		^{133}Ba		x					
		^{140}Ba				x			
Berkelium	97	^{249}Bk	x						
Beryllium	4	^{7}Be				x			
Bismuth	83	^{206}Bi				x			
		^{207}Bi			x				
		^{210}Bi			x				
		^{212}Bi				x			
Bromine	35	^{82}Br				x			
Cadmium	48	^{109}Cd				x			
		^{115m}Cd			x				
		^{115}Cd				x			
Calcium	20	^{45}Ca				x			
		^{47}Ca				x			

Element	Atomic Number	Radionuclide	Transport Group						
			I	II	III	IV	V	VI	VII
Californium	98	^{249}Cf	x						
		^{250}Cf	x						
		^{252}Cf	x						
Carbon	6	^{14}C				x			
Cerium	58	^{141}Ce				x			
		^{143}Ce				x			
		^{144}Ce			x				
Cesium	55	^{131}Cs				x			
		^{134m}Cs			x				
		^{134}Cs			x				
		^{135}Cs				x			
		^{136}Cs				x			
		^{137}Cs			x				
Chlorine	17	^{36}Cl			x				
		^{38}Cl				x			
Chromium	24	^{51}Cr				x			
Cobalt	27	^{56}Co			x				
		^{57}Co				x			
		^{58m}Co				x			
		^{58}Co				x			
		^{60}Co			x				
Copper	29	^{64}Cu				x			
Curium	96	^{242}Cm	x						
		^{243}Cm	x						
		^{244}Cm	x						
		^{245}Cm	x						
		^{246}Cm	x						
Dysprosium	66	^{154}Dy			x				
		^{165}Dy				x			
		^{166}Dy				x			
Erbium	68	^{169}Er				x			
		^{171}Er				x			
Europium	63	^{150}Eu			x				
		^{152m}Eu				x			
		^{152}Eu			x				
		^{154}Eu		x					
		^{155}Eu				x			
Fluorine	9	^{18}F				x			
Gadolinium	64	^{153}Gd				x			
		^{159}Gd				x			
Gallium	31	^{67}Ga			x				
		^{72}Ga				x			
Germanium	32	^{71}Ge				x			

Element	Atomic Number	Radionuclide	Transport Group						
			I	II	III	IV	V	VI	VII
Gold	79	^{193}Au			x				
		^{194}Au			x				
		^{195}Au			x				
		^{196}Au				x			
		^{198}Au				x			
		^{199}Au				x			
Hafnium	72	^{181}Hf				x			
Holmium	67	^{166}Ho				x			
Hydrogen	1	see tritium							
Indium	49	^{113m}In				x			
		^{114m}In			x				
		^{115m}In				x			
		^{115}In				x			
Iodine	53	^{124}I			x				
		^{125}I			x				
		^{126}I			x				
		^{129}I			x				
		^{131}I			x				
		^{132}I				x			
		^{133}I			x				
		^{134}I				x			
		^{135}I				x			
Iridium	77	^{190}Ir				x			
		^{192}Ir			x				
		^{194}Ir				x			
Iron	26	^{55}Fe				x			
		^{59}Fe				x			
Krypton	36	^{85m}Kr			x				
		^{85m}Kr (uncompressed)					x		
		^{85}Kr			x				
		^{85}Kr (uncompressed)						x	
		^{87}Kr		x					
		^{87}Kr (uncompressed)					x		
Lanthanum	57	^{140}La				x			
Lead	82	^{203}Pb				x			
		^{210}Pb		x					
		^{212}Pb		x					
Lutecium	71	^{172}Lu			x				
		^{177}Lu				x			
Magnesium	12	^{28}Mg			x				
Manganese	25	^{52}Mn				x			
		^{54}Mn				x			
		^{56}Mn				x			
Mercury	80	^{197m}Hg				x			
		^{197}Hg				x			
		^{203}Hg				x			
Mixed Fission Products	—	MFP		x					
Molybdenum	42	^{99}Mo				x			

Radiation Protection and Regulations

Element	Atomic Number	Radionuclide	Transport Group						
			I	II	III	IV	V	VI	VII
Neodymium	60	^{147}Nd				x			
		^{149}Nd				x			
Neptunium	93	^{237}Np	x						
		^{239}Np	x						
Nickel	28	^{56}Ni			x				
		^{59}Ni				x			
		^{63}Ni				x			
		^{65}Ni				x			
Niobium	41	^{93m}Nb				x			
		^{95}Nb				x			
		^{97}Nb				x			
Osmium	76	^{185}Os				x			
		^{191m}Os				x			
		^{191}Os				x			
		^{193}Os				x			
Palladium	46	^{103}Pd				x			
		^{109}Pd				x			
Phosphorus	15	^{32}P				x			
Platinum	78	^{191}Pt				x			
		^{193}Pt				x			
		^{193m}Pt				x			
		^{197m}Pt				x			
		^{197}Pt				x			
Plutonium	94	^{238}Pu†	x						
		^{239}Pu†	x						
		^{240}Pu	x						
		^{241}Pu†	x						
		^{242}Pu	x						
Polonium	84	^{210}Po	x						
Potassium	19	^{42}K				x			
		^{43}K			x				
Praseodymium	59	^{142}Pr				x			
		^{143}Pr				x			
Promethium	61	^{147}Pm				x			
		^{149}Pm				x			
Protactinium	91	^{230}Pa	x						
		^{231}Pa	x						
		^{233}Pa			x				
Radium	88	^{223}Ra		x					
		^{224}Ra		x					
		^{226}Ra	x						
		^{228}Ra	x						
Radon	86	^{220}Rn				x			
		^{222}Rn		x					
Rhenium	75	^{183}Re				x			
		^{186}Re				x			
		^{187}Re				x			
		^{188}Re				x			
		^{natural}Re				x			
Rhodium	45	^{103m}Rh				x			
		^{105}Rh				x			
Rubidium	37	^{86}Rb				x			
		^{87}Rb				x			
		^{natural}Rb				x			

Element	Atomic Number	Radionuclide	Transport Group							
			I	II	III	IV	V	VI	VII
Ruthenium	44	^{97}Ru				x			
		^{103}Ru				x			
		^{105}Ru				x			
		^{106}Ru			x				
Samarium	62	^{145}Sm			x				
		^{147}Sm			x				
		^{151}Sm				x			
		^{153}Sm				x			
Scandium	21	^{46}Sc			x				
		^{47}Sc				x			
		^{48}Sc				x			
Selenium	34	^{75}Se				x			
Silicon	14	^{31}Si				x			
Silver	47	^{105}Ag				x			
		^{110m}Ag			x				
		^{111}Ag				x			
Sodium	11	^{22}Na			x				
		^{24}Na				x			
Strontium	38	^{85m}Sr				x			
		^{85}Sr				x			
		^{89}Sr			x				
		^{90}Sr		x					
		^{91}Sr			x				
		^{92}Sr				x			
Sulfur	16	^{35}S				x			
Tantalum	73	^{182}Ta			x				
Technetium	43	^{96m}Tc				x			
		^{96}Tc				x			
		^{97m}Tc				x			
		^{97}Tc				x			
		^{99m}Tc				x			
		^{99}Tc				x			
Tellurium	52	^{125m}Te				x			
		^{127m}Te				x			
		^{127}Te				x			
		^{129m}Te			x				
		^{129}Te				x			
		^{131m}Te			x				
		^{132}Te				x			
Terbium	65	^{160}Tb			x				
Thallium	81	^{200}Tl				x			
		^{201}Tl				x			
		^{202}Tl				x			
		^{204}Tl			x				
Thorium	90	^{227}Th		x					
		^{228}Th	x						
		^{230}Th	x						
		^{231}Th	x						
		^{232}Th				x			
		^{234}Th		x					
		^{natural}Th				x			

Element	Atomic Number	Radionuclide	Transport Group						
			I	II	III	IV	V	VI	VII
Thullium	69	^{168}Tm			x				
		^{170}Tm			x				
		^{171}Tm				x			
Tin	50	^{113}Sn				x			
		^{117m}Sn			x				
		^{121}Sn			x				
		^{125}Sn				x			
Tritium	1	^{3}H				x			
		^{3}H (as a gas, as luminous paint, or adsorbed on solid material)							
Tungsten	74	^{181}W				x			
		^{185}W				x			
		^{187}W				x			
Uranium	92	^{230}U		x					
		^{232}U	x						
		^{233}U†		x					
		^{234}U		x					
		^{235}U†			x				
		^{236}U		x					
		^{238}U			x				
		^{natural}U			x				
		^{enriched}U†			x				
		^{depleted}U			x				
Vanadium	23	^{48}V				x			
		^{49}V			x				
Xenon	54	^{125}Xe			x				
		^{131m}Xe			x				
		^{131m}Xe (uncompressed)					x		
		^{133}Xe			x				
		^{133}Xe (uncompressed)						x	
		^{135}Xe		x					
		^{135}Xe (uncompressed)					x		
Ytterbium	70	^{175}Yb				x			
Yttrium	39	^{88}Y			x				
		^{90}Y				x			
		^{91m}Y			x				
		^{91}Y			x				
		^{92}Y				x			
		^{93}Y				x			
Zinc	30	^{65}Zn				x			
		^{69m}Zn				x			
		^{69}Zn				x			
Zirconium	40	^{93}Zr				x			
		^{95}Zr			x				
		^{97}Zr				x			

* "Uncompressed" means at a pressure not exceeding 14.7 psi (absolute).
† Fissile radioactive material.

(b) Any radionuclide not listed in the above table shall be assigned to one of the groups in accordance with the following table:

Radionuclide	Radioactive Half-Life		
	0 to 1,000 Days	1,000 Days to 10^6 Years	Over 10^6 Years
Atomic number 1 to 81	Group III	Group II	Group III
Atomic number 82 and over	Group I	Group I	Group III

Note: no unlisted radionuclides shall be assigned to Groups IV, V, VI, or VII.

(c) For mixtures of radionuclides the following shall apply:
   (1) if the identity and respective activity of each radionuclide are known, the permissible activity of each radionuclide shall be such that the sum, for all groups present, of the ratio between the total activity for each group and the permissible activity for each group will not be greater than unity;
   (2) if the groups of the radionuclides are known, but the amount in each group cannot be reasonably determined, the mixture shall be assigned to the most restrictive group present;
   (3) if the identity of all or some of the radionuclides cannot be reasonably determined, each of those unidentified radionuclides shall be considered as belonging to the most restrictive group which cannot be positively excluded;
   (4) mixtures consisting of a single radioactive decay chain where the radionuclides are in the naturally occurring proportions shall be considered as consisting of a single radionuclide; the group and activity shall be that of the first member present in the chain, except if a radionuclide "x" has a half-life longer than that of that first member and an activity greater than that of any other member, including the first, at any time during transportation; in that case, the transport group of the nuclide "x" and the activity of the mixture shall be the maximum activity of that nuclide "x" during transportation.

(L) By amending § 173.391 in its entirety to read as follows:

§ 173.391 Small Quantities of Radioactive Materials, and Radioactive Devices.

(a) Radioactive materials in normal form not exceeding 0.01 millicurie of Group I radionuclides, 0.1 millicurie of Group II radionuclides, 1 millicurie of Groups III, IV, V, or VI radionuclides, 24 curies of Group VII radionuclides, tritium oxide in aqueous solution with a concentration not exceeding 0.5 millicuries per milliliter and with a total activity per package of not more than 3 curies, or 1 millicurie of radioactive material in special form, and fissile radioactive materials not containing more than 15 g of uranium-235 are exempt from specification packaging, marking and labeling, and are exempt from the provisions of § 173.393, if the following conditions are met:
   (1) the materials are packaged in strong, tight packages, so that there will be no leakage of radioactive materials under conditions normally incident to transportation;
   (2) the packages must be such that the radiation dose rate at any point on the external surface of the package does not exceed 0.5 millirem per hour;

(3) there must be no significant removable radioactive surface contamination on the exterior of the package (see § 173.397);
(4) the outside of the inner container must bear the marking "Radioactive."
(b) Manufactured articles, such as instruments, clocks, electronic tubes or apparatus, or other similar devices, having radioactive materials (other than liquids) in a nondispersible form as a component part are exempt from specification packaging, marking and labeling, and are exempt from the provisions of § 173.393, if the following conditions are met:

> Note: for radioactive gases, the requirement for the radioactive material to be in a nondispersible form does not apply.

(1) radioactive materials are securely contained within the device, or are securely packaged in strong, tight packages, so that there will be no leakage of radioactive materials under conditions normally incident to transportation;
(2) the radiation dose rate at four inches from any unpackaged device does not exceed 10 millirem per hour;
(3) the radiation dose rate at any point on the external surface of the outside container does not exceed 0.5 millirem per hour; however, for carload or truckload lots only, the radiation at the external surface of the package or the item may exceed 0.5 millirem per hour, but must not exceed 2 millirem per hour;
(4) there must be no significant removable radioactive surface contamination on the exterior of the package (see § 173.397);
(5) the total radioactivity content of a package containing radioactive devices must not exceed the quantities shown in the following table:

Transport Group	Quantity, curies	
	Per Device	Per Package
I	0.0001	0.001
II	0.001	0.05
III	0.01	3
IV	0.05	3
V or VI	1	1
VII	25	200
Special form	0.05	20

(6) no package may contain more than 15 grams of fissile material.
(c) Manufactured articles, other than reactor fuel elements, in which the sole radioactive material is natural or depleted uranium are exempt from specification packaging, marking and labeling, and are exempt from the provisions of § 173.393, if the following conditions are met:
(1) the radiation dose rate at any point on the external surface of the outside container does not exceed 0.5 millirem per hour;
(2) there must be no detectable radioactive surface contamination on the exterior of the package (see § 173. 397);
(3) the total radioactivity content of each article must not exceed 3 curies;
(4) the outer surface of the uranium is enclosed in an inactive metallic sheet.

> Note: such articles may be packagings for the transportation of radioactive materials.

(d) Shipments made under this section for transportation by motor carriers are exempt from Part 177, except § 177.817, of this chapter.

(M) By amending § 173.392 in its entirety to read as follows:

§ 173.392 Low-Specific-Activity Materials.

(a) Low-specific-activity materials, when transported in other than transport vehicles assigned for the sole use of the consignor, are exempt from the provisions of § 173.393(a) through (g); they must be packaged in accordance with the requirements of § 173.395, and must be marked and labeled as required in §§ 173.401 and 173.402.

(b) Low-specific-activity materials which are transported in transport vehicles (except aircraft) assigned for the sole use of that consignor are exempt from specification packaging, marking and labeling, provided the shipments meet the requirements of paragraph (c) or (d) of this section.

(c) Packaged shipments of low-specific-activity materials transported in transport vehicles (except aircraft) assigned for the sole use of that consignor must comply with the following:

  (1) materials must be packaged in strong, tight packages, so that there will be no leakage of radioactive material under conditions normally incident to transportation;

  (2) packages must not have any significant removable surface contamination (see § 173.397);

  (3) external radiation levels must comply with § 173.393(j);

  (4) shipments must be loaded by consignor, and unloaded by consignee from the transport vehicle in which originally loaded;

  (5) there must be no loose radioactive material in the car or vehicle;

  (6) shipments must be braced so as to prevent leakage or shift of lading under conditions normally incident to transportation;

  (7) except for shipments of uranium or thorium ores, unconcentrated, the transport vehicle must be placarded with the placards prescribed in accordance with § 174.541(b) or § 177.823 of this chapter, as appropriate;

  (8) the outside of each outside package must be stenciled or otherwise marked "Radioactive—LSA".

(d) Unpackaged (bulk) shipments of low-specific-activity materials transported in closed transport vehicles (except aircraft) assigned for the sole use of that consignor must comply with the following:

  (1) authorized materials are limited to the following:

    (i) uranium or thorium ores, and physical or chemical concentrates of those ores;

    (ii) uranium metal or natural thorium metal, or alloys of these materials; or

    (iii) materials of low radioactive concentration, if the average estimated radioactivity concentration does not exceed 0.001 millicurie per gram and the contribution from Group I material does not exceed one percent of the total radioactivity:

    (iv) objects of nonradioactive material externally contaminated with radioactive material, if the radioactive material is not readily dispersible and the surface contamination, when averaged over one square meter, does not exceed 0.0001 millicurie per square centimeter of Group I radionuclides or 0.001 millicurie per square centimeter of other radionuclides; such objects must be suitably wrapped or enclosed;

(2) bulk liquids must be transported in the following:
- (i) Spec. 103C-W (§§ 179.200, 179.201, and 179.202 of this chapter) tank cars; bottom fittings and valves are not authorized;
- (ii) Spec. MC 310, MC 311, MC 312, or MC 331 (§ 178.330, § 178.331, § 178.337, or § 178.343 of this chapter) cargo tanks, authorized only where the radioactivity concentration does not exceed 10 percent of the specified low-specific-activity levels (see § 173.389(c)); the requirements of § 173.393(g) do not apply to these cargo tanks; bottom fittings and valves are not authorized; trailer-on-flat-car service is not authorized;

(3) external radiation levels must comply with subparagraphs (2), (3), and (4) of § 173.393(j);

(4) shipments must be loaded by the consignor, and unloaded by the consignee from the transport vehicles in which originally loaded;

(5) except for shipments of uranium or thorium ores, unconcentrated, the transport vehicle must be placarded with the placards prescribed in accordance with § 174.541(b) or § 177.823 of this chapter, as appropriate;

(6) there must be no leakage or radioactive materials from the vehicle.

(N) By amending § 173.393 in its entirety to read as follows:

§ 173.393 General Packaging Requirements.

(a) Unless otherwise specified, all shipments of radioactive materials must meet all requirements of this section, and must be packaged as prescribed in §§ 173.391 through 173.396.

(b) The outside of each package must incorporate a feature such as a seal which is not readily breakable and which, while intact, will be evidence that the package has not been illicitly opened.

(c) The smallest outside dimension of any package must be 4 inches or greater.

(d) Radioactive materials must be packaged in packagings which have been designed to maintain shielding efficiency and leak tightness, so that, under conditions normally incident to transportation, there will be no release of radioactive material. If necessary, additional suitable inside packaging must be used. Each package must be capable of meeting the standards in § 173.389 (b) (see also § 173.24). Specification containers listed as authorized for radioactive-materials shipments may be assumed to meet those standards, provided the packages do not exceed the gross weight limits prescribed for those containers in Part 178 of this chapter.

(1) Internal bracing or cushioning, where used, must be adequate to assure that, under the conditions normally incident to transportation, the distance from the inner container or radioactive material to the outside wall of the package remain within the limits on which the package design was based, and that the radiation dose rate external to the package does not exceed the transport index number shown on the label. Inner shield closures must be positively secured, to prevent loss of the contents.

(e) The packaging must be so designed, constructed and loaded that, when transporting large quantities of radioactive material,

(1) the heat generated within the package because of the radioactive materials present will not, at any time during transportation, affect the efficiency of the package under the conditions normally incident to transportation; and

(2) the temperature of the accessible external surfaces of the package will not exceed 122°F in the shade when fully loaded, assuming still air at ambient temperature; if the package is transported in a transport vehicle consigned for the sole use of the consignor, the maximum accessible external surface temperature shall be 180°F.

(f) Pyrophoric materials, in addition to the packaging prescribed in this subpart, must also meet the packaging requirements of § 173.134 or § 173.154. Pyrophoric radioactive liquids may not be shipped by air.

(g) Liquid radioactive material must be packaged in or within a leak-resistant and corrosion-resistant inner container. In addition,

    (1) the packaging must be adequate to prevent loss or dispersal of the radioactive contents from the inner container, if the package were subjected to the 30-foot drop test prescribed in § 172.298(c)(2)(i); or

    (2) enough absorbent material must be provided to absorb at least twice the volume of the radioactive liquid contents; the absorbent material may be located outside the radiation shield only if it can be shown that, if the radioactive liquid contents were taken up by the absorbent material, the resultant dose rate at the surface of the package would not exceed 1,000 millirem per hour.

(h) There must be no significant removable radioactive surface contamination on the exterior of the package (see § 173.397).

(i) Except for shipments described in paragraph (j) of this section, all radioactive materials must be packaged in suitable packaging (shielded, if necessary), so that at any time during the normal conditions incident to transportation the radiation dose rate does not exceed 200 millirem per hour at any point on the external surface of the package and the transport index does not exceed 10.

(j) Packages for which the radiation dose rate exceeds the limits specified in paragraph (i) of this section, but does not exceed at any time during transportation any of the limits specified in subparagraphs (1) through (4) of this paragraph, may be transported in a transport vehicle (except aircraft) assigned for the sole use of that consignor, and unloaded by the consignee from the transport vehicle in which originally loaded.

    (1) 1,000 millirem per hour at 3 feet from the external surface of the package (closed transport vehicle only);

    (2) 200 millirem per hour at any point on the external surface of the car or vehicle (closed transport vehicle only);

    (3) 10 millirem per hour at 6 feet from the external surface of the car or vehicle; and

    (4) 2 millirem per hour in any normally occupied position in the car or vehicle, except that this provision does not apply to private motor carriers.

(k) When radioactive materials are loaded by the shipper into a transport vehicle assigned for the sole use of that shipper, the shipper must observe all applicable requirements of Part 174, 175, or 177 of this chapter, as appropriate.

(l) Packages consigned for export are also subject to the regulations of the foreign governments involved in the shipment (see §§ 183.8 and 173.9).

(O) By amending § 173.394 in its entirety to read as follows:

§ 173.394 Radioactive Material in Special Form.

(a) Type A quantities of special-form radioactive materials must be packaged as follows:

    (1) Spec. 5B, 5D, 6A, 6B, 6C, 6J, 6K, 6L, 6M, 17C, 17H, 42B, or 42C (§§ 178.82, 178.84, 178.97, 178.98, 178.99, 178.100, 178.101, 178.103,

178.104, 178.107, 178.108, 178.115, and 178.118 of this chapter) metal drums;
   (2) Spec. 21C (§ 178.224 of this chapter) fiber drums;
   (3) Spec. 14, 15A, 15B, 15C, 15D, 19A, or 19B (§§ 178.168, 178.169, 178.170, 178.171, 178.190, and 178.191 of this chapter) wooden boxes;
   (4) any Spec. 12 series (§§ 178.205 through 178.212 of this chapter) fiberboard boxes, 200-pound test minimum, or Spec. 23E or 23H (§ 178.214 or § 178.219 of this chapter) fiberboard boxes;
   (5) Spec. 55 (§ 178.250 of this chapter) metal-encased shielded container (additionally authorized for not more than 300 curies per package, for domestic shipments only);
   (6) Spec. 7A (§ 178.350 of this chapter) Type A general package;
   (7) foreign-made packagings which bear the symbol "Type A" may be used for transportation of radioactive materials from the point of entry in the United States to their destination in the United States or through the United States en route to a point of destination outside of the United States.
(b) Type B quantities of special-form radioactive materials must be packaged as follows:
   (1) Spec. 55 (§ 178.250 of this chapter) metal-encased shielded container (authorized only for not more than 300 curies per package; authorized for domestic shipments only; see also § 178.394(a)(5) of this chapter);
   (2) Spec. 6M (§ 178.304 of this chapter) metal packaging;
   (3) any Type B packaging specifically approved for such use by the Department.
(c) Large quantities of radioactive materials in special form must be packaged as follows:
   (1) Spec. 6M (§ 178.104 of this chapter) metal packaging (radioactive thermal-decay energy must not exceed 10 watts);
   (2) any Type B packaging which meets the standards in the regulations of the U.S. Atomic Energy Commission (Title 10, Code of Federal Regulations, Part 71) or the 1967 regulations of the International Atomic Energy Agency, and which has been specifically authorized for such use by the Department under Part 170 of this chapter; in applying for Departmental authorization of packages for large quantities of radioactive materials to be used in shipments by the U.S. Atomic Energy Commission or one of its contractors or licensees, a copy of the license amendment or other approval issued by that Commission will be accepted in place of the package structural-integrity evaluation.
(P) By amending § 173.395 in its entirety to read as follows:
§ 173.395 Radioactive Material in Normal Form.
(a) Type A quantities of normal-form radioactive materials must be packaged as follows:
   (1) Spec. 5B, 5D, 6A, 6B, 6C, 6J, 6K, 6L, 6M, 17C, 17H, 42B, or 42C (§§ 178.82, 178.84, 178.97, 178.98, 178.99, 178.100, 178.101, 178.103, 178.104, 178.107, 178.108, 178.115, and 178.118 of this chapter) metal drums;
   (2) Spec. 21C (§ 178.224 of this chapter) fiber drums;
   (3) Spec. 14, 15A, 15B, 15C, 15D, 19A, or 19B, (§§ 178.165, 178.168, 178.169, 178.170, 178.171, 178.190, and 178.191 of this chapter) wooden boxes;

(4) any Spec. 12 series (§§ 178.205 through 178.212 of this chapter) fiberboard boxes, 200-pound test minimum, or Spec. 23F or 23H (§ 178.214 or § 178.219 of this chapter) fiberboard boxes;

(5) any Spec. 3 or 4 series (§§ 178.36 through 178.44 or §§ 178.47 through 178.58 of this chapter) cylinders;

(6) Spec. 55 (§ 178.250 of this chapter) metal-encased shielded container;

(7) Spec. 7A (§ 178.350 of this chapter) Type A general package;

(8) foreign-made packagings which bear the symbol "Type A" may be used for transportation of radioactive materials from their point of entry in the United States to their destination in the United States or through the United States en route to a point of destination outside of the United States.

(b) Type B quantities of radioactive materials in normal form must be packaged as follows:

(1) Spec. 6M (§ 178.104 of this chapter) metal packaging (authorized only for solid or gaseous radioactive materials which will not decompose at temperatures up to 250°F);

(2) any Type B packaging specifically approved for such use by the Department.

(c) Large quantities of radioactive materials in normal form must be packaged as follows:

(1) Spec. 6M (§ 178.104 of this chapter) metal packaging (authorized only for solid or gaseous radioactive materials which will not decompose at temperatures up to 250°F; radioactive thermal-decay energy must not exceed 10 watts);

(2) any Type B packaging which meets the standards prescribed in the regulations of the U.S. Atomic Energy Commission (Title 10, Code of Federal Regulations, Part 71) or the 1967 regulations of the International Atomic Energy Agency, and which has been specifically authorized for such use by the Department under Part 170 of this chapter; in applying for Departmental authorization of package for large quantities of radioactive materials to be used in shipments by the U.S. Atomic Energy Commission or one of its contractors or licensees, a copy of the license amendment or other approval issued by that Commission will be accepted in place of the package structural-integrity evaluation.

(Q) By amending § 173.396 in its entirety to read as follows:

§ 173.396 Fissile Radioactive Material.

(a) The following materials are not classified as fissile radioactive materials, are exempted from this section, and must instead be packaged in accordance with the other provisions of this subpart, as appropriate:

(1) not more than 15 grams of fissile material;

(2) thorium or uranium containing not more than 0.72 percent, by weight, of fissile material;

(3) uranium compounds other than metal (e.g., UF_4, UF_6, or uranium oxide in bulk form, not pelleted or fabricated into shapes) and aqueous solutions of uranium, in which the total amount of uranium-233 and plutonium present does not exceed 1 percent, by weight, of the uranium-235 content and the total fissile content does not exceed 1 percent, by weight, of the total uranium content;

(4) homogeneous hydrogenous solutions or mixtures containing not more than:

(i) 500 grams of any fissile material, provided the atomic ratio of hydrogen to fissile material is greater-than 7,600; or
(ii) 800 grams of uranium-235, if the atomic ratio of hydrogen to fissile material is greater than 5,200 and the content of other fissile material is not more than 1 percent, by weight, of the total uranium-235 content; or
(iii) 500 grams of uranium-233 and uranium-235, if the atomic ratio of hydrogen to fissile material is greater than 5,200 and the content of plutonium is not more than 1 per cent, by weight, of the total uranium-233 and uranium-235 content;
(5) a package containing less than 350 grams of fissile material, if there is not more than 5 grams of fissile material in any cubic foot within the package.
(b) Fissile radioactive materials containing not more than Type A quantities of radionuclides in either normal form or special form must be packaged as follows:
(1) Spec. 6L (§ 178.103 of this chapter) metal packaging (authorized only for not more than 14 kilograms of uranium-235 as metal or oxide, or as compounds or alloys which will not decompose at temperatures up to 750°F); each package shipped as Fissile Class II shall be assigned a transport index of 1.3 (unless external radiation levels require a higher assignment); the atomic ratio of hydrogen to uranium-235 shall not exceed 3, all sources of hydrogen within the inner packaging being considered; the gross weight of the loaded package shall not exceed 350 pounds for the 55-gallon size, or 480 pounds for sizes up through 110 gallons;
(2) Spec. 6M (§ 178.104 of this chapter) metal packaging (see paragraph (c)(2) of this section for authorized contents);
(3) any packaging listed in § 173.395(a); authorized only for not more than the following:
(i) 500 grams of uranium-235 as Fissile Class III, or not more than 40 grams of uranium-235 as Fissile Class II; for Fissile Class II shipments, the transport index to be assigned to each package shall be 0.4 for each gram of uranium-235 above 15 grams, up to a maximum of 40 grams (transport index of 10);
(ii) 320 grams of plutonium-239 as plutonium-beryllium neutron sources in special form; total radioactivity content must not exceed 20 curies; the transport index to be assigned to each package shall be 0.5 for each 20 grams, or fraction thereof, of fissile plutonium;
(4) any other packaging which meets the standards in the regulations of the U.S. Atomic Energy Commission (Title 10, Code of Federal Regulations, Part 71) or the 1967 regulations of the International Atomic Energy Agency, and which has been specifically authorized for such use by the Department under Part 170 of this chapter.
(c) Fissile radioactive materials containing Type B quantities of radionuclides in either normal form or special form must be packaged as follows:
(1) Spec. 6L (§ 178.103 of this chapter) metal packaging (authorized only for enriched uranium, the fissile content not to exceed 14 kilograms uranium-235 as metal or oxide, or as compounds or alloys which will not decompose at temperatures up to 750°F); each package shipped as Fissile Class II shall be assigned a transport index of 1.3 (unless

external radiation levels require a higher assignment); the atomic ratio of hydrogen to uranium-235 shall not exceed 3, all sources of hydrogen within the inner packaging being considered; the gross weight of the loaded package shall not exceed 350 pounds for the 55-gallon size, or 480 pounds for sizes up through 110 gallons;

(2) Spec. 6M (§ 178.104 of this chapter) metal packaging (authorized only for solid radioactive materials which will not decompose at temperatures up to 250°F; radioactive thermal-decay energy output shall not exceed 10 watts); large-quantity radioactive materials in normal form must be packaged in one or more sealed and leaktight metal cans or polyethylene bottles within the Spec. 2R containment vessel;

   (i) *Fissile Class I Packages*—the following quantities of fissile radioactive material are authorized for Fissile Class I packages: 1.6 kilograms of uranium-235; 0.9 kilograms of plutonium (see note); 0.5 kilograms of uranium-233; the maximum ratio of hydrogen to fissile material must not exceed 3, all sources of hydrogen within the Spec. 2R containment vessel being considered;

      Note: because of the 10-watt thermal-decay heat limitation, the limit for plutonium-238 is only 0.02 kilograms.

   (ii) *Fissile Class II and III Packages*—quantities of fissile radioactive material as shown in the following table are authorized for Fissile Class II and Fissile Class III packages; where a maximum ratio of hydrogen to fissile material is specified in the table, only the hydrogen interspersed with the fissile material need be considered; for Fissile Class II packages, the minimum transport index to be assigned is shown in the table; for Fissile Class III packages, the maximum number of similar packages per transport vehicle is shown; Fissile Class III shipments are also subject to paragraph (g) of this section;

TABLE OF AUTHORIZED CONTENTS*

Uranium-235[†]			Plutonium [‡††]			Fissile Class II Transport Index	Fissile Class III Maximum Number of Packages per Transport Vehicle
Metal or Alloy $H/X = 0$	Compounds		Metal or Alloy $H/X = 0$	Compounds			
	$H/X = 0$	$H/X \leq 3$		$H/X = 0$	$H/X \leq 3$		
7.2	7.6	5.2	3.1	4.1	3.4	0.1	1,250
8.7	9.6	6.4	3.4	4.5	4.1	0.2	625
11.2	13.9	8.3	4.2	—	4.5	0.5	250
13.5	16.0	10.1	4.5	—	—	1.0	125
—	26.0	16.1	—	—	—	5.0	25
—	32.0	19.5	—	—	—	10.0	12

* Quantity in kilograms.
† Maximum ^{235}U enrichment is 93 weight percent.
‡ Minimum percentage of ^{240}Pu is 5 weight percent.
†† 4.5-kilogram limitation on plutonium, due to 10-watt decay heat limitation.

  (3) any other packaging which meets the standards prescribed in the regulations of the U.S. Atomic Energy Commission (Title 10, Code of Federal Regulations, Part 71) or the 1967 regulations of the International Atomic Energy Agency, and which has been specifically

authorized for such use by the Department under Part 170 of this chapter.

(d) Petitions for authorization of nonspecification packagings for fissile radioactive materials must be submitted as prescribed in Part 170 of this chapter, and must also include the following:

    (1) type and amount of fissile radioactive materials which are to be carried in each package, including

        (i) the transport index to be assigned to the package for the proposed package loadings when shipped as Fissile Class II; and

        (ii) the maximum number of packages proposed when shipped as Fissile Class III;

    (2) a nuclear-criticality safety evaluation demonstrating that the packaging design and limitation on its contents are adequate to assure nuclear-criticality safety; any tests performed in this respect should be described.

> Note: In applying for Departmental authorization of packages for fissile radioactive materials to be used in shipments by the U.S. Atomic Energy Commission or one of its contractors or licensees, a copy of the license amendment or other approval issued by that Commission will be accepted in place of the nuclear-criticality safety evaluation and the package structural-integrity evaluation.

(e) Mixing of packages of other types of radioactive materials, including Fissile Class I, with Fissile Class II packages is permitted, if the total transport index in any one transport vehicle or storage location does not exceed 50.

(f) For Fissile Class II packages shipped under the exclusive-use provisions of § 173.393(j) to provide for packages with high radiation dose rates, the transport index number which is calculated for nuclear-criticality control purposes must not exceed 10 for any single package, or a total of 50 for the full load, unless specifically authorized by the Department for Fissile Class III shipments.

(g) Fissile Class III shipments may be made only in accordance with subparagraph (1) or (2) of this paragraph, or in accordance with other procedures authorized by the Department; the transport controls must provide nuclear-criticality safety, and shall be carried out by the shipper or carrier, as appropriate, to protect against loading, transporting, or storing of that shipment together with other fissile material;

    (1) transportation in a transport vehicle assigned for the sole use of that consignor, with a specific restriction for such sole use to be provided in the special arrangements, and with instructions to that effect issued with the shipping papers;

    (2) transportation under escort by a person in a separate vehicle, with the escort having the capability, equipment, authority and instructions to provide administrative controls adequate to assure compliance with this paragraph.

(R) By adding new §§ 173.397, 173.398, and 173.399 to read as follows:

§ 173.397 Contamination Control.

(a) Removable radioactive contamination is not significant, if the average amount of radioactive contamination which can be removed by wiping the external surface of the package with an absorbent material, as measured on the wiping material, does not exceed

(1) 10^{-11} curies per square centimeter beta-gamma (2,200 disintegrations/minute per 100 square centimeters) and 10^{-12} curies per square centimeter alpha (220 disintegrations/minute per 100 square centimeters) for all contaminants except natural or depleted uranium and natural thorium; or

(2) 10^{-10} curies per square centimeter beta-gamma (22,000 disintegrations/minute per 100 square centimeters) and 10^{-11} curies per square centimeter alpha (2,200 disintegrations/minute per 100 square centimeters) where the only contaminant is known to be natural or depleted uranium or natural thorium.

(b) Each transport vehicle used for transporting low-specific-activity radioactive materials in carload or truckload lots under § 173.392(d) must be surveyed with appropriate radiation detection instruments after each use; vehicles must not again be placed in service until the radiation dose rate at any accessible surface is not more than 0.5 millirem per hour and there is no significant removable radioactive surface contamination (see § 173.399).

(1) This paragraph does not apply to any closed transport vehicle (except aircraft) used solely for the transportation of radioactive materials, if a survey of its interior surface shows that the radiation dose rate does not exceed 10 millirem per hour at the interior surface or 2 millirem per hour at 3 feet from any interior surface. These vehicles must be stenciled with the words "For Radioactive-Materials Use Only" in lettering at least 3 inches high, in a conspicuous place or places on both sides of the exterior of the vehicle; these vehicles must be kept closed at all times other than loading and unloading.

§ 173.398 Special Tests.

(a) *Special-form material.* To qualify as special form material, the radioactive material must either be in massive solid form or encapsulated. Each item in massive solid form, or each capsule, must either have no overall dimension less than 0.5 millimeters or must have at least one dimension greater than 5 millimeters. Each item, or the capsule material, must not dissolve or convert into dispersible form to the extent of more than 0.005 percent, by weight, by immersion for 1 week in water at pH 6 to pH 8 and 68°F and with a maximum conductivity of 10 micromhos/cm, and by immersion in air at 86°F. If in massive solid form, the radioactive material must not break, crumble, or shatter when subjected to the percussion test prescribed in this section, and must not melt, sublime, or ignite at temperatures below 1,000°F. If encapsulated, the capsule must retain its contents when subjected to all of the performance tests prescribed in this section, and must not melt, sublime, or ignite at temperatures below 1,475°F.

(1) *Free Drop.* A free drop through a distance of 30 feet onto a flat, essentially unyielding horizontal surface, striking the surface in such a position as to suffer maximum damage.

(2) *Percussion.* Impact of the flat circular end of a one-inch-diameter steel rod weighing three pounds, dropped through a distance of 40 inches. The capsule or material shall be placed on a sheet of lead, of hardness number 3.5 to 4.5 on the Vickers scale and not more than one inch thick, supported by a smooth, essentially unyielding surface.

(3) *Heating.* Heating in air to a temperature of 1,475°F, and remaining at that temperature for a period of 10 minutes.

(4) *Immersion.* Immersion for 24 hours in water at room temperature. The water shall be at pH 6 to pH 8, with a maximum conductivity of 10 micromhos/cm.

(b) Standards for Type A packaging.
- (1) Type A packaging must be so designed and constructed that, if it were subject to the environmental and test conditions prescribed in this section,
  - (i) there would be no release of radioactive material from the package;
  - (ii) the effectiveness of the packaging would not be substantially reduced; and
  - (iii) there would be no mixture of gases or vapors in the package, which could, through any credible increase of pressure or an explosion, significantly reduce the effectiveness of the package.
- (2) Environmental conditions.
  - (i) *Heat*. Direct sunlight at an ambient temperature of 130°F in still air.
  - (ii) *Cold*. An ambient temperature of $^-40°F$ in still air and shade.
  - (iii) *Reduced Pressure*. Ambient atmospheric pressure of 0.5 atmosphere (absolute) (7.3 psi).
  - (iv) *Vibration*. Vibration normally incident to transportation.
- (3) Test conditions. The packaging shall be subject to all of the following tests, unless specifically exempted therefrom, and also to the consecutive application of at least two of the following tests from which it is not specifically exempted.
  - (i) *Water Spray*. A water spray heavy enough to keep the entire exposed surface of the package, except the bottom, continuously wet during a period of 30 minutes. Packages for which the outer layer consists entirely of metal, wood, ceramic, or plastic, or combinations thereof, are exempt from the water spray test.
  - (ii) *Free Drop*. Between $1\frac{1}{2}$ to $2\frac{1}{2}$ hours after the conclusion of the water spray test, a free drop through a distance of 4 feet onto a flat, essentially unyielding horizontal surface, striking the surface in a position for which maximum damage is expected.
  - (iii) *Corner Drop*. A free drop onto each corner of the package in succession, or, in the case of a cylindrical package onto each quarter of each rim, from a height of 1 foot onto a flat, essentially unyielding horizontal surface. This test applies only to packages which are constructed primarily of wood or fiberboard and do not exceed 110 pounds gross weight, and to all Fissile Class II packagings.
  - (iv) *Penetration*. Impact of the hemispherical end of a vertical steel cylinder $1\frac{1}{4}$ inches in diameter and weighing 13 pounds, dropped from a height of 40 inches onto the exposed surface of the package which is expected to be most vulnerable to puncture. The long axis of the cylinder shall be perpendicular to the package surface.
  - (v) *Compression*. For packages not more than 10,000 pounds in weight, a compressive load equal to either five times the weight of the package or to 2 pounds per square inch multiplied by the maximum horizontal cross section of the package, whichever is greater. The load shall be applied during a period of 24 hours, uniformly against the top and bottom of the package, in the position in which the package would normally be transported.

(c) Standards for hypothetical accident conditions of transportation for Type B packagings.
   (1) Type B packaging must meet the applicable Type A packaging standards, and must be designed and constructed, and its contents so limited, that, if subjected to the hypothetical accident conditions prescribed in this paragraph, it will meet the following conditions:
      (i) the reduction of shielding would not be enough to increase the radiation dose rate at 3 feet from the external surface of the package to more than 1,000 millirem per hour;
      (ii) no radioactive material would be released from packages containing Type B quantities of radioactive material; the allowable release of radioactivity from packages containing large quantities of radioactive material is limited to gases and contaminated coolant containing total radioactivity exceeding neither 0.1 percent of the total radioactivity of the package contents, nor 0.01 curie of Group I radionuclides, 0.5 curie of Group II radionuclides, and 10 curies of Groups III and IV radionuclides, except that for inert gases the limit is 1,000 curies.
   (2) Test conditions. The conditions which the package must be capable of withstanding must be applied sequentially, to determine their cumulative effect on a package, in the following order.
      (i) *Free Drop*. A free drop through a distance of 30 feet onto a flat, essentially unyielding horizontal target surface, striking the surface in a position for which maximum damage is expected.
      (ii) *Puncture*. A free drop through a distance of 40 inches, striking, in a position for which maximum damage is expected, the top end of a vertical cylindrical mild-steel bar mounted on an essentially unyielding horizontal surface; the bar shall be 6 inches in diameter, with the top horizontal and its edge rounded to a radius of not more than $\frac{1}{4}$ inch, and of such a length as to cause maximum damage to the package, but not less than 8 inches long. The long axis of the bar shall be perpendicular to the unyielding horizontal surface.
      (iii) *Thermal*. Exposure to a thermal test in which the heat input to the package is no less than that which would result from exposure of the whole package for 30 minutes to a radiation environment of 1,475°F with an emissivity coefficient of 0.9, assuming the surfaces of the package have an absorption coefficient of 0.8. The package shall not be cooled artificially until 3 hours after the test period, unless it can be shown that the temperature on the inside of the package has begun to fall in less than 3 hours.
      (iv) *Water Immersion* (fissile radioactive-materials packages only). Immersion in water to the extent that all portions of the package to be tested are under at least 3 feet of water for a period of not less than 8 hours.
(d) It is not necessary to actually conduct the tests prescribed in this section, if it can be clearly shown, through engineering evaluations or comparative data, that the material or item would be capable of performing satisfactorily under the prescribed test conditions.

§ 173.399 Labeling of Packages of Radioactive Materials.

(a) Each package of radioactive materials, unless exempted by § 173.391 or § 173.392, shall be labeled as provided in this section (see § 173.414 for description of labels). The label to be used shall be determined by the transport index or other considerations, as follows.

   (1) "Radioactive White I" label:
      (i) each package not exceeding 0.5 millirem per hour at any point on the external surface of the package, and which does not come within the provisions of subparagraph (2) or (3) of this paragraph (not authorized for Fissile Class II packages).
   (2) "Radioactive Yellow II" label: when the limit in subparagraph (1) of this paragraph is exceeded, but the provisions of subparagraph (3) of this paragraph are not met; and
      (i) each package not exceeding 10 millirem per hour at any point on the external surface of the package and not exceeding 0.5 millirem per hour at 3 feet from the external surface of the package; or
      (ii) each package for which the transport index does not exceed 0.5 at any time during transportation.
   (3) "Radioactive Yellow III" label: when either of the limits in subparagraph (2) of this paragraph is exceeded; in addition, the following types of packages must also bear this label:
      (i) each Fissile Class III package;
      (ii) each package containing a large quantity of radioactive material as defined in § 173.389; or
      (iii) each package being transported under a permit issued as authorized in § 173.23(c).

(b) Radioactive materials having other hazardous characteristics, as defined elsewhere in this part, must also be labeled with other labels, as required by this part, according to the hazards of the commodity (see §§ 173.2 and 173.402). For example:

   (1) packages containing the solid nitrates of uranium or thorium must bear both a "Radioactive" label and a yellow "Oxidizing Materials" label;
   (2) packages containing nitric acid solutions of radioactive materials must bear both a "Radioactive" label and a white "Corrosive Acid" label.

(S) By amending the heading and paragraphs (a)(1) through (8), (11), and (14), (b), introductory text of (c), and (c)(1); by adding paragraphs (a)(15) and (c)(2); canceling paragraphs (a)(9) and (10), (b)(1), and (d) in § 173.402 to read as follows:

§ 173.402 Labeling of Explosives or Other Dangerous Articles.

(a) Each package containing explosives or other dangerous articles, as defined in this part, must be conspicuously labeled by the shipper as follows, except as otherwise provided:

   (1) red label, as described in § 173.405(a) or (b), on packages of flammable liquids, except when exempted from labeling requirements in Subpart C of this part;
   (2) yellow label, as described in § 173.406(a) or (b), on packages of flammable solids or oxidizing materials, except when exempted from the labeling requirements in Subpart D of this part;
   (3) white label, as described in § 173.407(a) or (b), on packages of acids,

alkaline caustic liquids, or other corrosive liquids, except when exempted from the labeling requirements in Subpart E of this part;
(4) red "Gas" label, as described in § 173.408(a)(1) or (b), on packages of flammable compressed gases, except when exempted from the labeling requirements in Subpart F of this part;
(5) green label, as described in § 173.408(a)(2) or (b)(1), on packages of nonflammable compressed gases, except when exempted from the labeling requirements in Subpart F of this part;
(6) "Poison Gas" label, as described in § 173.409(a)(1), on packages of Class A poisons;
(7) "Poison" label, as described in § 173.409(a)(2) or (b), on packages of Class B poisons, except when exempted from the labeling requirements in Subpart G of this part;
(8) "Radioactive" (White I, Yellow II, or Yellow III) label, as described in § 173.414, on packages of radioactive materials, except when exempted from the labeling requirements in Subpart G of this part; each package must be labeled with two such labels, affixed to opposite sides of the package; the method of determination of which label to use is given in § 173.399;
   (i) labels which conform to the model prescribed in the regulations of the International Atomic Energy Agency, and which are similar in appearance to the labels prescribed herein (although the inscriptions on the labels may be in a foreign language), are authorized in place of the labels prescribed herein for import or export shipments only;
(9) (canceled);
(10) (canceled);
(11) "Tear Gas" label, as described in § 173.409(a)(3) or (b), on packages of Class C poisons;

* * * * *

(14) labels authorized for shipment of explosives or other dangerous articles by air, as shown in §§ 173.405 through 173.412, may be used in place of the labels otherwise prescribed therein;
(15) packages containing materials which are either Class A poisons or radioactive materials, and which are also flammable gases, liquids or solids corrosive liquids, Class B poisons, oxidizing materials, or compressed gases, must also bear additional labels showing those other hazardous characteristics (see also § 173.2).
(b) Labels required for shipments of explosives or other dangerous articles by air are shown in §§ 173.405(b), 173.406(b), 173.407(b), 173.408(b), 173.409(b), 173.410(b), 173.411(b), 173.412(b), and 173.414.
   (1) (canceled)
(c) Labels are not required on carload or truckload lots of dangerous articles, except for the commodities listed in this paragraph, when the shipments are loaded by the shipper and are unloaded by the consignee from the transport vehicle in which originally loaded. The commodities for which this exemption does not apply include Class A or C poisons, etiological agents, and radioactive materials.
   (1) labels are not required on carload or truckload lots of Class A or C poisons, etiological agents, or radioactive materials on shipments made by, for, or to the Department of Defense, if the shipments are loaded by the shipper, and unloaded by the consignee from the transport

vehicle in which originally loaded, and if the shipments are accompanied by qualified personnel supplied with equipment to repair leaks or other container failures which would permit escape of contents.

(2) The proper shipping name of the contents (see § 172.5 of this chapter) must be marked on each package shipped under the exemption in this paragraph.

(d) (canceled)

* * * * *

(T) By amending § 173.414 to read as follows:

§ **173.414 Radioactive-Materials Labels.**

(a) Labels for packages of radioactive materials must be of diamond shape, in colors specified in this section, with each side at least 4 inches long. Printing must be in black inside of a black line border, measuring at least $3\frac{1}{2}$ inches on each side and as shown in this section.

(b) "Radioactive White I" label for radioactive materials. Label must be white in color. The single vertical bar on the lower half of the label must be bright red in color.

(c) "Radioactive Yellow II" label for radioactive materials. The upper half of the label must be bright yellow, and the bottom half must be white. The two vertical bars on the lower half of the label must be bright red in color.

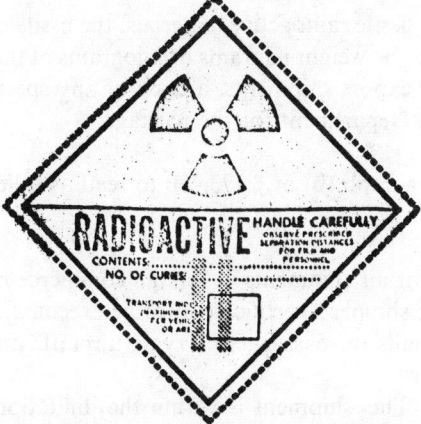

(d) "Radioactive Yellow III" label for radioactive materials. The upper half of the label must be bright yellow, and the bottom half must be white. The three vertical bars on the lower half of the label must be bright red in color.

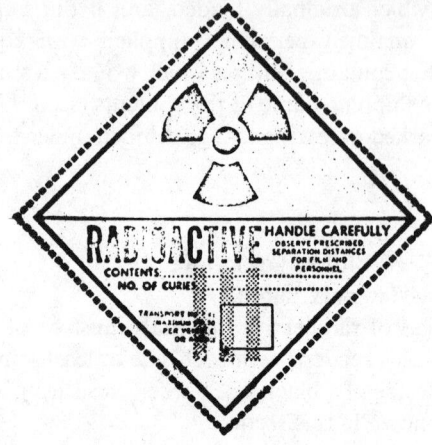

(U) By amending paragraph (a)(2) and adding paragraph (a)(5) in § 173.427 to read as follows:

§ 173.427 Shipping Papers.

(a) ******

    (2) Where the regulations (except § 173.402) exempt the packages from labeling, the exemption must be indicated by the words "No Label Required" immediately following the description on the shipping paper.

* * * * *

    (5) For shipments of radioactive materials, the shipping paper description must include:

        (i) the transport group or groups of the radionuclides in the radioactive material, if the material is in normal form;

        (ii) the name of the radionuclides in the radioactive material, and a description of its physical and chemical form, if the material is in normal form;

        (iii) the activity of the radioactive material, in curies;

        (iv) the type of label applied to the package, i.e., "Radioactive White I", "Radioactive Yellow II", or "Radioactive Yellow III";

        (v) for fissile radioactive materials, the fissile class of the package, and the weight in grams or kilograms of the fissile isotope; and

        (vi) for export shipments, a copy of any special permit issued by the Department for the package.

* * * * *

(V) By amending paragraph (b) of § 173.430 to read as follows:

§ 173.430 Certificate.

* * * * *

(b) Shipping papers for air shipments in foreign commerce must be made out in duplicate, and the shipper's certificate must be executed on both copies.

    (1) For shipments on passenger-carrying aircraft, the shipper must also add the words:

        "This shipment is within the limitations prescribed for passenger-carrying aircraft."

    (2) The shipper may also add the words: "* * * and to the IATA Restricted Articles Regulations."

* * * * *

PART 174—CARRIERS BY RAIL FREIGHT

IV. Part 174 is amended as follows:
   (A) By amending paragraph (b) of § 174.510 to read as follows:

§ 174.510 Shipping Papers.

* * * * *

(b) Where the regulations (except § 173.402(c) of this chapter) exempt the packages from labeling, the exemption must be indicated by the words "No Label Required" immediately following the description on the shipping paper.

* * * * *

(B) By amending paragraphs (j)(1) and (2) of § 174.532 to read as follows:

§ 174.532 Loading Other Dangerous Articles.

(j) * * *

   (1) Shipments of low-specific-activity materials, as defined in § 173.389(c) of this chapter, must be loaded so as to avoid spillage and scattering of loose material. Loading restrictions are prescribed in § 137.392 of this chapter.
   (2) Storage and loading restrictions are prescribed in § 174.586(h).

* * * * *

§ 174.538 (Amended)

(C) By deleting the phrase "Class D poisons" from item 15 in vertical and horizontal columns in § 174.538(a) chart; item 15 will then read: "Radioactive materials."

(D) By amending § 174.541(b) to read as follows:

§ 174.541 "Dangerous" Placards;
   "Dangerous—Radioactive Material" Placards;
   or "Caution—Residual Phosphorus" Placards.

(b) "Dangerous—Radioactive Material" placards, as prescribed in § 174.553, must be applied to cars containing packages bearing a "Radioactive Yellow III" label (three vertical red stripes) as prescribed in § 173.414(d) of this chapter, and to carload lots as under §§ 173.392 and 173.393(j) and (k) of this chapter.

* * * * *

(E) By amending § 174.544(a)(6) to read as follows:

§ 174.544 Placards Not Required.

(a) * * *

   (6) Cars containing packages or radioactive material which are exempted from labeling under § 173.391 of this chapter; which bear only the labels prescribed in § 173.414(b) and (c) of this chapter; or which are exempted from placarding under § 173.392(c)(7) of this chapter.

(F) By amending the introductory text of paragraph (a) of § 174.553 to read as follows:

§ 174.443 "Dangerous—Radioactive Material" Placard.

(a) The "Dangerous—Radioactive Material" placard for radioactive materials must be of diamond shape, measuring $10\frac{3}{4}$ inches on each side, and must bear the wording in red letters as shown in the following cut:

* * * * *

(G) By amending paragraph (d) and adding paragraph (e) in § 174.566 to read as follows:

§ 174.566 Cleaning Cars.

* * * * *

(d) Cars contaminated with radioactive materials.
   (1) Each car used for transporting low-specific-activity radioactive

materials in carload lots under the provisions of § 173.392(d) of this chapter must be surveyed with appropriate radiation detection instruments after each use. Carriers must not return such cars to service until the radiation dose rate at any accessible surface is not more than 0.5 millirem per hour and there is no significant removable radioactive surface contamination (see § 173.399 of this chapter).

(2) This paragraph does not apply to any car used solely for transporting radioactive materials, if a survey of the interior surface shows that the radiation dose rate does not exceed 10 millirem per hour at the interior surface, or 2 millirem per hour at 3 feet from any interior surface. These cars must be stenciled with the words " For Radioactive-Materials Use Only" in lettering at least three inches high in a conspicuous place on both sides of the exterior of the car. These cars must be kept closed at all times other than loading and unloading.

(e) In case of fire, wreck, breakage, or unusual delay involving shipments of radioactive material, see § 174.588.

§ 174.584 (Amended)

(H) By amending the table in paragraph (a) of § 174.584 as follows and canceling footnote 1 to the table:

Commodity	Label Notation to Follow Entry of the Article on the Billing	Placard Notation to Follow Entry of the Article on the Billing	Placard Endorsement*
Add			
For radioactive materials with " Radioactive White I " or " Radioactive Yellow II " labels	" Radioactive White I " or " Radioactive Yellow II "	None	None
For radioactive materials with " Radioactive Yellow III " label	" Radioactive Yellow III "	" Dangerous—Radioactive Material " placard	" Radioactive Material "
Cancel			
For radioactive materials, Class D poison ***	"Radioactive Material" label ***	" Dangerous—Radioactive Material " placard ***	" Radioactive Material " ***

* Must be 3/8 inches high and appear on the billing near the space provided for the car number.
¹ Canceled.

(I) By amending the heading and paragraph (h) in § 174.586 to read as follows:

§ 174.586 **Handling Hazardous Materials.**

* * * * *

(h) Radioactive Materials.

(1) The number of packages of radioactive materials, as provided in §§ 173.393 through 173.396 of this chapter, in any rail car or storage location must be limited so that the total transport index number, as defined in § 173.389(i) of this chapter and determined by adding together the transport index numbers on the labels of the individual packages, does not exceed 50. This provision does not apply to sole-use shipments described in § 173.393(j) or (k) or in § 173.392 of this chapter.

(2) Packages of radioactive material bearing "Radioactive Yellow II" or "Radioactive Yellow III" labels must not be placed in cars, depots, or other places closer than 3 feet to an area (or dividing partition between areas) which may be continuously occupied by passengers, employees or shipments of animals, nor closer than 15 feet to any packages containing undeveloped film (if so marked). If more than one of these packages is present, the distance must be computed from the table below on the basis of the total transport index number (determined by adding together the transport index numbers on the labels of the individual packages) of packages in the car or storeroom.

Total Transport Index	Minimum Separation Distance to Nearest Undeveloped Film, feet	Minimum Distance to Area of Persons, or from Dividing Partition of a Combination Car, feet
None	0	0
0.1—10	15	3
10.1—20	22	4
20.1—30	29	5
30.1—40	33	6
40.1—50	36	7

Note: the distance in the table must be measured from the nearest point on the packages of radioactive materials.

(J) By amending § 174.588(c)(1) to read as follows:
§ Disposition of Damaged or Astray Shipments.

* * * * *

(c) * * *

(1) Radioactive materials. In case of fire, accident, breakage, or unusual delay involving shipments of radioactive materials, the carrier shall immediately notify the shipper and the Department. Cars, buildings, areas, or equipment in which radioactive materials have been spilled may not be again placed in service or routinely occupied until the radiation dose rate at any accessible surface is less than 0.5 millirem per hour and there is no significant removable radioactive surface contamination (see § 173.399 of this chapter).

Note 1. In these instances, the package or materials should be segregated as far as practicable from personnel contact. If radiological advice or assistance is needed, the U.S. Atomic Energy Commission should also be notified. In case of obvious leakage, or if it appears likely that the inside container may have been damaged, care should be taken to avoid inhalation, ingestion, or contact with the radioactive material. Any loose radioactive materials should be left in a segregated area and held pending disposal instructions from qualified persons.

Note 2. Details involving the handling of radioactive materials in the event of an accident can be found in Bureau of Explosives Pamphlet No. 22, *Recommended Practices for Handling Collisions and Derailments Involving Explosives, Gasoline, and*

Other Dangerous Articles, available from the Bureau of Explosives, Association of American Railroads, 2 Pennsylvania Plaza, New York, N.Y. 10001.

* * * * *

(K) By amending paragraph (n) of § 174.589 to read as follows:
§ **174.589 Handling Cars.**

* * * * *

(n) Position of cars containing radioactive materials in train: In a freight train or mixed train, either standing or during transportation thereof, a car placarded "Dangerous—Radioactive Material" must not be handled next to cars placarded "Explosives" or next to carload shipments of undeveloped film.

PART 175—CARRIERS BY RAIL EXPRESS

V. Part 175 is amended as follows:
 (A) By amending paragraph (b) of § 175.652a to read as follows:
§ **175.652a Shipping Papers.**

* * * * *

(b) Where the regulations (except § 173.402(c) of this chapter) exempt the packages from labeling, the exemption must be indicated by the words "No Label Required" immediately following the description on the shipping paper.

* * * * *

(B) By amending paragraph (j) of § 175.655 to read as follows:
§ **175.655 Protection of Packages.**

* * * * *

(j) Radioactive materials.
 (1) The number of packages of radioactive materials, as provided in §§ 173.393 through 173.396 of this chapter, in any rail car or storage location must be limited so that the total transport index number, as defined in § 173.389(h) of this chapter and determined by adding together the transport index numbers shown on the labels of the individual packages, does not exceed 50. This provision does not apply to sole-use shipments described in § 173.393(j) or (k) or in § 173.397 of this chapter.
 (2) Packages of radioactive material bearing "Radioactive Yellow II" or "Radioactive Yellow III" labels shall not be placed in cars, depots, or other places closer than 3 feet to an area (or dividing partition between areas) which may be continuously occupied by passengers, exployees or shipments of animals, nor closer than 15 feet to any package containing undeveloped film (if so marked). If more than one of these packages is present, the distance shall be computed from the following table on the basis of the total transport index number (determined by adding together the transport index numbers on the labels of the individual packages) of packages in the car or storeroom.

Total Transport Index	Minimum Separation Distance to Nearest Undeveloped Film, feet	Minimum Distance to Area of Persons, or from Dividing Partition of a Combination Car, feet
None	0	0
0.1—10	15	3
10.1—20	22	4
20.1—30	29	5
30.1—40	33	6
40.1—50	36	7

Note: the distance in the table must be measured from the nearest point on the packages of radioactive materials.

(3) In case of fire, accident, breakage, or unusual delay involving shipments of radioactive materials, the carrier shall immediately notify the shipper and the Department. Cars, buildings, areas, or equipment in which radioactive materials have been spilled may not be again placed in service or routinely occupied until the radiation dose rate at any accessible surface is less than 0.5 millirem per hour and there is no significant removable radioactive surface contamination (see § 173.399 of this chapter).

Note 1. In these instances, the package or materials should be segregated as far as practicable from personnel contact. If radiological advice or assistance is needed, the U.S. Atomic Energy Commission should also be notified. In case of obvious leakage, or if it appears likely that the inside container may have been damaged, care should be taken to avoid inhalation, ingestion, or contact with the radioactive material. Any loose radioactive material should be left in a segregated area and held pending disposal instructions from qualified persons.

Note 2. Details involving the handling of radioactive materials in the event of an accident can be found in Bureau of Explosives Pamphlet No. 22, *Recommended Practices for Handling Collisions and Derailments Involving Explosives, Gasoline, and Other Dangerous Articles*, available from the Bureau of Explosives, Association of American Railroads, 2 Pennsylvania Plaza, New York, N.Y. 10001.

* * * * *

PART 177—SHIPMENTS MADE BY WAY OF COMMON, CONTRACT, OR PRIVATE CARRIERS BY PUBLIC HIGHWAY

VI. Part 177 is amended as follows:
  (A) By adding the following new sections to the table of contents:
      Sec.
      177.842 Radioactive material
      177.843 Contamination of vehicles
      177.861 Accidents; radioactive materials

(B) By amending paragraph (b) and introductory text of paragraph (c) in § 177.815 to read as follows:

§ 177.815 Labels.

* * * * *

(b) Labels are not required on truckload lots of dangerous articles, except for the commodities listed in this paragraph, when the shipments are loaded by the shipper and are unloaded by the consignee from the transport vehicle in which originally loaded. The commodities for which this exemption does not apply include Class A poisons, etiological agents, and radioactive materials.

    (1) Labels are not required on truckload lots of Class A or C poisons, etiologic agents, or radioactive materials on shipments made by, for, or to the Department of Defense, if the shipments are loaded by the shipper and unloaded by the consignee from the transport vehicle in which originally loaded, and if the shipments are accompanied by qualified personnel supplied with equipment to repair leaks or other container failures which would permit escape of contents.

    (2) The proper shipping name of the contents must be marked on each package shipped under the exemption in this paragraph.

(c) Except on packages of Class A or C poisons, etiologic agents, or radioactive materials labels are not required on less-than-truckload shipments by motor vehicle by public highway, when the articles are readily identifiable by reason of type of container, or when the container is plainly marked to indicate its contents; and

* * * * *

(C) By amending paragraph (b) of § 177.817 to read as follows:

§ 177.817 Shipping Papers.

(b) Where the regulations (except § 173.402) exempt the packages from labeling, the exemption must be indicated by the words "No Label Required" immediately following the description on the shipping paper.

* * * * *

§ 177.823 (Amended)

(D) By amending the ninth listing in § 177.823(a)(1) to read as follows:

Commodity	Type of Marking or Placard
Radioactive material requiring "Radioactive Yellow III" label, any quantity (see §173.414(d))	RADIOACTIVE (black letters on yellow background)

* * * * *

§ 177.841 (Amended)

(E) By canceling § 177.841(d)

(F) By adding new §§ 177.842 and 177.843 as follows:

§ 177.842 Radioactive Material.

(a) The number of packages of radioactive materials, as provided for in §§ 173.393 through 173.396 of this chapter, in any motor vehicle, trailer, or storage location must be limited so that the total transport index number, as defined in § 173.389(h) of this chapter and determined by adding together the transport index numbers shown on the labels of the individual packages, does not exceed 50. This provision does not apply to sole-use shipments described into § 173.393(j) or (k) or in § 173.397 of this chapter.

(b) Packages of radioactive material bearing "Radioactive Yellow II" or "Radioactive Yellow III" labels shall not be placed in motor vehicles or other places closer than the distance shown in the following table to any area which may be continuously occupied by passengers, employees or shipments of animals, nor closer than the distances shown in the table below to any package containing undeveloped film (if so marked). If more than one of these packages is present, the distance shall be computed from the following table on the basis of the total transport index number (determined by adding together the transport index numbers on the labels of the individual packages) of packages in the vehicle or storeroom.

Total Transport Index	Minimum Separation Distance, in feet, to Nearest Undeveloped Film, for Various Times of Transit					Minimum Distance, in feet, to Area of Persons, or from Dividing Partition of Cargo Compartments
	Up to 2 Hours	2 to 4 Hours	4 to 8 Hours	8 to 12 Hours	Over 12 Hours	
None	0	0	0	0	0	0
0.1—1	1	2	3	4	5	1
1.1—5	3	4	6	8	11	2
5.1—10	4	6	9	11	15	3
10.1—20	5	8	12	16	22	4
20.1—30	7	10	15	20	29	5
30.1—40	8	11	17	22	33	6
40.1—50	9	12	19	24	36	7

Note: the distance in the table must be measured from the nearest point on the packages of radioactive materials.

(c) Shipments of low-specific-activity materials, as defined in § 173.391 of this chapter, must be loaded so as to avoid spillage and scattering of loose materials. Loading restrictions are set forth in § 173.397 of this chapter.

(d) Packages must be so blocked and braced that they cannot change position during conditions normally incident to transportation.

(e) Persons should not remain unnecessarily in a vehicle containing radioactive materials.

§ 177.843 Contamination of Vehicles.

(a) Each motor vehicle used for transporting low-specific-activity radioactive materials in truckload lots under the provisions of § 173.392(d) of this chapter must be surveyed with appropriate radiation detection instruments after each use. Carriers must not return such vehicles to service until the radiation dose rate at any accessible surface is not more than 0.5 millirem per hour and there is no significant removable radioactive surface contamination (see § 173.399 of this chapter).

(b) This section does not apply to any vehicle used solely for transporting radioactive material, if a survey of the interior surface shows that the radiation dose rate does not exceed 10 millirem per hour at the interior surface, or 2 millirem per hour at 3 feet from any interior surface. These vehicles must be stenciled with the words "For Radioactive Materials Use Only" in lettering at least 3 inches high in a conspicuous place on both sides of the exterior of the vehicle. These vehicles must be kept closed at all times other than loading and unloading.

(c) In case of fire, accident, breakage, or unusual delay involving shipments of radioactive material, see § 177.861.

§ 177.848 (Amended)

(G) By deleting the phrase "Class D poisons" from item 15 in vertical and horizontal columns in § 177.848(a) chart. Item 15 will then read: "Radioactive materials."

§ **177.860** (Amended)

(H) By canceling paragraphs (c) and (d) in § 177.860.

(I) By adding § 177.861 to read as follows:

§ **177.861 Accidents; Radioactive Materials.**

(a) Radioactive materials. In case of fire, accident, breakage, or unusual delay involving shipments of radioactive materials, the carrier shall immediately notify the shipper and the Department. Vehicles, buildings, areas, or equipment in which radioactive materials have been spilled may not be again placed in service or routinely occupied until the radiation dose rate at any accessible surface is less than 0.5 millirem per hour and there is not significant removable radioactive surface contamination (see § 173.399 of this chapter).

> Note 1. In these instances, the package or materials should be segregated as far as practicable from personnel contact. If radiological advice or assistance is needed, the U.S. Atomic Energy Commission should also be notified. In case of obvious leakage, or if it appears likely that the inside container may have been damaged, care should be taken to avoid inhalation, ingestion, or contact with the radioactive material. Any loose radioactive material should be left in a segregated area and held pending disposal instruction from qualified persons.
>
> Note 2. Details involving the handling of radioactive materials in the event of an accident can be found in Bureau of Explosives Pamphlet No. 22, *Recommended Practices for Handling Collisions and Derailments Involving Explosives, Gasoline and Other Dangerous Articles*, available from the Bureau of Explosives, Association of American Railroads, 2 Pennsylvania Plaza, New York, N.Y. 10001.

(b) Cleaning Vehicles. See § 177.843.

(J) By amending § 177.870(g) to read as follows:

§ **177.870 Regulations for Passenger-Carrying Vehicles.**

* * * * *

(g) Radioactive materials. In addition to the limitations prescribed in paragraphs (b) and (e) of this section, no person may transport any radioactive material requiring labels under § 173.402 of this chapter in or on any motor vehicle carrying passengers for hire, except where no other practicable means of transportation is available. Packages of radioactive materials must be stored only in the trunk or baggage compartment of the vehicle, and must not be stored in any compartment occupied by persons. Packages of radioactive materials must be handled and placed in the vehicle as prescribed in §177.841 (d).

PART 178—SHIPPING CONTAINER SPECIFICATIONS

VII. Part 178 is amended as follows:

(A) By amending the title of § 178.103, and by adding §§ 178.104 and 178.350 and Subpart K to the table of contents to read as follows:

Sec.

178.103 Specification 6L; metal packaging

178.104 Specification 6M; metal packaging
Subpart K—Specifications for General Packagings
178.350 Specification 7A; general packaging, Type A

(B) By amending § 178.34-5 to read as follows:

§ 178.34-5 Marking.

(a) Each container shall be marked with the words "Radioactive Material" in letters at least ¼ inch in height, either by embossing or diestamping directly onto the container, or by securely affixing by welding or brazing a metal plate bearing this notation to the container.

§§ 178.38-13, 178.39-13, 178.40-13, 178.41-13, 178.42-10, 178.43-13, 178.44-13, 178.48-13, 178.49-13, 178.50.13, 178.51-13, 178.52-13, 178.53-12, 178.54-13, 178.55-13, 178.56-13, 178.59-11, 178.60-13, 178.63-12, 178.66-12, 178.67-12, 178.68-12 (Amended)

(C) In the following sections change the reference "§ 173.34(f)" to read "§ 173.34(d)", and change the reference to "§ 173.301(i)" to read "§ 173.301(g)":
§§ 178.38-13, 178.39-13, 178.40-13, 178.41-13, 178.42-10, 178.43-13, 178.44-13, 178.48-13, 178.49-13, 178.50-13, 178.51-13, 178.52-13, 178.53-12, 178.54-13, 178.55-13, 178.56-13, 178.59-11, 178.60-13, 178.63-12, 178.66-12, 178.67-12, and 178.68-12.

§§ 178.47-13, 178.58-13 (Amended)

(D) In §§ 178.47-13 and 178.58-13 change the reference to "§ 178.34(f)" to read "§ 178.34(d)".

§ 178.57-13 (Amended)

(E) In § 178.57-13 change the reference to "§ 178.34(f)" to read "§ 173.34(d)", and change the reference to "§ 173.304(f)" to read "§ 173.304(b)(2)".

(F) By amending the title of § 178.103, and by amending § 178.103-2(a) to read as follows:

§ 178.103 Specification 6L; Metal Packaging.

§ 178.103-2 Rated Capacity.

(a) Rated capacity as marked (see § 178.103-6). Not less than 55 gallons, nor more than 110 gallons, for the outer steel drum. Not more than 17.74 liters for the inner vessel.

(G) By amending § 178.103-3 in its entirety to read as follows:

§ 178.103-3 General Requirements.

(a) Outside drum must conform to specifications 6J (§ 178.100) steel drum or equivalent, except as otherwise specified herein. The drum wall must be at least 18-guage steel, and may be either a single sheet of steel or may be produced by welding together two appropriate lengths of such drums. The removable head must be constructed of at least 16-gauge steel, with one or more corrugations in the cover near the periphery.

(b) Inner vessel must conform to specification 2R (§ 178.340) or equivalent (except that cast iron is not authorized), with maximum usable inside diameter of $5\frac{1}{4}$ inches, maximum height of 50 inches (with cap in place), and minimum wall thickness of $\frac{1}{4}$ inch. Flanged closures are not authorized. Pipe threads must be luted with appropriate nonhardening compound, to prevent inleakage of water or loosening of the cap due to vibration or heat.

(c) Inner vessel must be fixed within the outer drum with appropriate centering devices of adequate physical strength and fire resistance to be able to withstand the accident test conditions of § 173.398 of this chapter without a displacement of the inner container of more than 2 inches in any direction. The following types of centering mechanisms meet this requirement without need for performing the accident test; and other type of centering device must be specifically approved by the Department.

(1) Not less than four steel-rod spacers of at least $\frac{1}{4}$ inch (for packages of 55-gallon capacity) or $\frac{3}{8}$ inch (for packages with greater than 55-gallon capacity) cold-rolled steel, welded to the pipe at each end by minimum 2-inch continuous weld. Rods must be welded to the pipe at radial positions not exceeding 90°, and so as not to interfere with closure of inner vessel. Each spacer rod must extend at least $2\frac{1}{4}$ inches beyond the inner vessel at each end, then radially to the wall of the outer drum (to provide a springlike snug fit), and along the entire length of the wall of the outer drum. For packages of more than 55-gallon capacity each spacer rod shall be braced by welding a $\frac{1}{4}$-inch × 2 inch steel plate to the spacer rod and the pipe with a continuous weld at each joint, the joints being located approximately halfway along the length of the drum.

(2) At least three steel "spiders", not more than 24 inches apart, with each spider having at least four legs. Each leg must be constructed of materials having dimensions not less than those listed in this subparagraph, welded by continuous weld at each joint to inner and outer steel bands of at least $\frac{1}{4}$-inch × 1-inch steel. The inner steel band must be welded to the inner vessel by at least six 2-inch welds on both edges of the band. The outer steel band must be welded to the outer drum by at least six 2-inch welds on both edges of the top outer band, so that the inner vessel is at least $2\frac{1}{4}$ inches from the top and bottom of the drum. Authorized construction materials are:

(i) $\frac{1}{4}$-inch × $\frac{1}{4}$-inch × 1-inch steel angle iron;
(ii) $\frac{3}{16}$-inch × $\frac{3}{16}$-inch × $1\frac{1}{4}$-inch steel angle iron;
(iii) $\frac{1}{4}$-inch thick, 1-inch outer diameter schedule 40 steel pipe;
(iv) $1\frac{1}{2}$-inch diameter solid-steel rods, with only two such spiders required instead of three.

(d) The void between the inner vessel and the outer drum shall be filled with either vermiculite (expanded mica) with a density of at least 4.5 pounds per cubic foot or with other material having an equivalent thermal and shock-absorbing effect.

(H) By amending § 178.103-5 in its entirety to read as follows:

§ 178.103-5 Closure.

(a) The outer drum closure shall be at least a 12-gauge bolted ring with drop-forged lugs, one of which is threaded, and having at least a steel bolt (at least $\frac{3}{8}$-inch for 55- gallon size, and at least $\frac{5}{8}$-inch for larger than 55-gallon size) and a lock nut or equivalent device.

(b) The closure device must have a means for the attachment of a tamper-proof lock wire and seal or equivalent.

(I) By adding subparagraph (4) to § 178.103-6(a) to read as follows:

§ 178.103-6 Markings.

(a) * * *

(4) Gauge of metal of the outer steel drum in the thinnest part, rated capacity of the outer steel drum, in gallons, and the year of manufacture of the assembled package (e.g., 18-110-68). When the gauge of the metal in the drum wall differs from that in the head, both must be indicated with a slanting line between, and with the gauge of the body indicated first (e.g., 18/16-110-68) for 18-gauge body and 16-gauge head).

(J) By adding the following new § 178.104:

§ 178.104 Specification 6M; Metal Packaging.

§ 178.104-1 General Requirements.
(a) Each package must meet the applicable requirements of § 173.24 of this chapter.

§ 178.104-2 Rated Capacity.
(a) Rated capacity as marked (see § 178.104-5); not less than 10 gallons, nor more than 110 gallons, for the outer steel drum; not less than 1.24 liters for the inner containment vessel.

§ 178.104-3 General Construction Requirements.
(a) Outside drum must conform to specification 6C or 17C (§§ 178.99 and 178.115) steel drum or equivalent, except as otherwise specified herein. The drum wall may be either a single sheet of steel, or may be produced by welding together two appropriate lengths of such drums. Removable head for drums of 55-gallon or larger size must have one or more corrugations in the cover near the periphery. Maximum gross weight, metal thickness, and minimum end insulation for the marked capacity is as follows:

Marked Capacity, gallons	Authorized Gross Weight, pounds	Minimum Thickness of Uncoated Sheets and Heads, gauge	Minimum Thickness of End Insulation, inches
10	160	20	1-7/8
15	160	20	1-7/8
30	480	18	3-3/4
55	880	16	3-3/4
110	880	16	3-3/4

(b) Inner containment vessel must conform to specification 2R (§ 178.34) or equivalent, with maximum usable inside diameter of 5.25 inches, minimum usable inside diameter of 4 inches, and minimum height of 6 inches. Material of construction must be steel with a minimum wall thickness of 0.125 inch for vessels up to 12 inches in height, and 0.25 inch for vessels over 12 inches in height. Pipe threads must be luted with an appropriate nonhardening compound, to prevent inleakage of water or loosening of the cap due to vibration or heat.

(c) Inner containment vessel must be fixed within the outer drum with appropriate centering devices of adequate physical strength and fire resistance to be able to withstand the accident test conditions prescribed in § 173.398 of this chapter without a displacement of the inner vessel of more than 2 inches in any direction. The following types of centering mechanisms meet this requirement without need for performing the accident tests; any other type of centering device must be specifically approved by the Department.

   (1) Machined discs and rings made either of solid industrial-cane fiberboard insulation having a density of at least 15 pounds per cubic foot; or of hardwood or plywood, at least $\frac{1}{2}$-inch thick, having a density of at least 28 pounds per cubic foot; or of other material having an equivalent thermal, neutron-absorbing and shock-absorbing effect. The sides of the inner vessel shall be protected by at least 3.75 inches of such material, and the ends by at least the thickness of such material prescribed in § 173.104-3(a) of this chapter. There must be no gap or direct heat path to the inner containment vessel.

(d) Any radiation shielding material used must be placed within the inner containment vessel, or must be protected in all directions by at least the thickness of the thermal insulating material prescribed in paragraph (a) of this section.

§ **178.104-4 Closure.**

(a) The outer drum closure must be at least a 16-gauge bolt-type locking ring, having at least a $5/16$-inch steel bolt for drum sizes not over 15 gallons, or a 12-gauge bolted ring with drop-forged lugs, one of which is threaded, and a $5/8$-inch steel bolt for drum sizes over 15 gallons. Each bolt must be provided with a lock nut or equivalent device.

(b) The closure device must have means for the attachment of a tamper-proof lock wire and seal or equivalent.

§ **178.104-5 Marking.**

(a) Marking must be as prescribed in § 173.24 of this chapter.

(b) Marking on the outside of each package must be as follows: "DOT-6M, Type B", "Radioactive Materials", or "Fissile Radioactive Materials", as appropriate; and the gauge of metal of the outer drum in the thinnest part, rated capacity of the outer drum, in gallons, and year of manufacture (for example, 18-30-69). When the gauge of the metal in the drum wall differs from that in the head, both must be indicated with a slanting line between, and with a gauge of the body indicated first (e.g., 18/16-55-69 for 18-gauge body and 16-gauge head).

§ **178.205-38** (Canceled)

(J) By canceling § 178.205-38.

(K) By adding a new Subpart K to read as follows:

SUBPART K—SPECIFICATIONS FOR GENERAL PACKAGING

(L) By adding a new § 178.350 to read as follows:

§ **178.350 Specification 7A; General Packaging, Type A.**

§ **718.350-1 General Requirements.**

(a) Each packaging must meet all applicable requirements of § 173.24 of this chapter.

§ **178.350-2 Specific Requirements.**

(a) Each packaging must be so designed and constructed that it meets the standards for Type A packaging (see §§ 173.389(j) and 173.398(b) of this chapter).

§ **178.350-3 Marking.**

(a) Marking on the outside of each packaging as follows: "USA DOT-7A, Type A" and "Radioactive Material".

(b) Marking to conform with §173.24 of this chapter.

2. In Title 14, Code of Federal Regulations, Part 103 is amended as follows:

(A) By amending § 103.1(b) and (c)(3) to read as follows, and by canceling paragraph (c)(4):

§ **103.1 Applicability.**

* * * * *

(b) For the purposes of this part, "dangerous article" means the material defined and regulated in the applicable regulations of the Department of Transportation (49 CFR, Parts 170–190), and includes:

    (1) explosives;

    (2) flammable liquids and solids;

    (3) oxidizing materials;

    (4) corrosive liquids;

    (5) compressed gases;

    (6) poisons;

    (7) etiologic agents;

    (8) radioactive materials.

(c) * * *
   (3) Shipments of radioactive materials via cargo-only aircraft made by or under the direction or supervision of the U.S. Atomic Energy Commission or the Department of Defense, which are escorted by personnel especially designated by or under the authority of that Commission or Department for the purposes of national security.
   (4) (Canceled)

(B) By amending § 103.3(b) to read as follows:

§ **103.3 Certification Requirements.**

* * * * *

(b) The shipper shall execute the required certificates in duplicate. One signed copy accompanies the shipment, and the originating air carrier retains the other signed copy.

* * * * *

(C) By amending § 103.7 to read as follows:

§ **103.7 Passenger-Carrying Aircraft.**

No person may carry any dangerous article in a passenger-carrying aircraft, except

(a) articles specified by 49 CFR, Part 173, as exempted from the specification packaging, marking and labeling requirements of 49 CFR, Part 173, when these articles are shipped as required for the exemption; and

(b) the following articles, when packaged, marked and labeled as specifically provided in 49 CFR, Parts 171 through 173, for shipment by rail express:
   (1) small-arms ammunition and practice-cartridge ammunition;
   (2) Class C explosives, other than those permitted under subparagraph (1) of this paragraph, with a net weight of not more than 50 pounds in each outside container;
   (3) subject to § 103.19(a), nonflammable compressed gases, except anhydrous ammonia, boron trifluoride, chlorine, hydrogen bromide, hydrogen chloride, nitrosyl chloride, and sulfur dioxide;
   (4) X-ray film or motion picture film with a nitrocellulose base, either exposed or unexposed;
   (5) pyroxylin plastics containing nitrocellulose, in sheets, rolls, rods, or tubes;
   (6) subject to § 103.19(b), radioactive materials.

(D) By amending § 103.9 to read as follows:

§ **103.9 Cargo-Only Aircraft.**

(a) No person may carry any dangerous article in a cargo-only aircraft, except those articles permitted on passenger-carrying aircraft under § 103.7, and except articles that
   (1) are specified in 49 CFR, 172.5, as acceptable for shipment by rail express;
   (2) do not exceed the maximum quantity for each outside container specified in 49 CFR, 172.5, for rail express; and
   (3) are packaged, marked and labeled as specified in 49 CFR, Part 173, for shipment by rail express.

(b) For the purposes of this part, a cargo-only aircraft is any aircraft that is not a passenger aircraft.

(E) By amending § 103.19(b) to read as follows:

§ **103.19 Quantity Limitations.**

* * * * *

(b) No person may carry aboard an aircraft a number of packages of radioactive materials that make the total transport index number (determined by adding together the transport index numbers shown on the labels of the individual packages) more than 50.

* * * * *

§ 103.21 (Canceled)

(F) By canceling § 103.21

(G) By amending § 103.23 to read as follows:

§ **103.23 Special Requirements For Radioactive Materials.**

(a) No person may place packages of radioactive materials bearing "Radioactive Yellow II" or "Radioactive Yellow III" labels in aircraft closer than the distance shown in the following table to a space (or dividing partition between spaces) which may be continuously occupied by people or shipments of animals, nor closer than the distances shown in the following table to any package containing undeveloped film (if so marked). If more than one of these packages is present, the distance shall be computed from the following table on the basis of the total transport index numbers shown on the labels of the individual packages in the aircraft:

Total Transport Index	Minimum Separation Distance, in feet, to Nearest Undeveloped Film, for Various Times of Transit					Minimum Distance, in feet, to Area of Persons, or from Dividing Partition of Cargo Compartments
	Up to 2 Hours	2 to 4 Hours	4 to 8 Hours	8 to 12 Hours	Over 12 Hours	
Non	0	0	0	0	0	0
0.1—1	1	2	3	4	5	1
1.1—5	3	4	6	8	11	2
5.1—10	4	6	9	11	15	3
10.1—20	5	8	12	16	22	4
20.1—30	7	10	15	20	29	5
30.1—40	8	11	17	22	33	6
40.1—50	9	12	19	24	36	7

Note: the distance in the table must be measured from the nearest point on the packages of radioactive materials.

(b) In case of fire, accident, breakage, or unusual delay involving shipments of radioactive materials, the operator of the aircraft shall immediately notify the shipper and the Department of Transportation. Aircraft in which radioactive materials have been spilled may not be again placed in service or routinely occupied until the radiation dose rate at any accesible surface is less than 0.5 millirem per hour and there is no significant removable radioactive surface contamination (see 49 CFR, 173.399). In these instances, the package or materials should be segregated as practicable from personnel contact. If radiological advice or assistance is needed, the U.S. Atomic Energy Commission should also be notified. In case of obvious leakage, or if it appears likely that the inside container may have been damaged, care should be taken to avoid inhalation, ingestion, or contact with the radioactive materials. Any loose radioactive materials should be left in a segregated area pending disposal instructions from qualified persons.

(H) By amending paragraph (c) of § 103.31 to read as follows:

§ **103.31 Cargo Location.**

* * * * *

(c) No person may place a package of yellow-label material (flammable solids or oxidizing materials) next to, or in a position to allow contact with, a package of white-label material (poisons) in any aircraft.

* * * * *

(F. R. Doc. 68-11880; filed, Oct. 3, 1968, 8:45 a.m.)

Regulations Issued by the U.S. Department of Transportation, Hazardous Materials Regulations Board, Effective December 31, 1968.

(d) Each carrier offering or delivering for rail transportation any loaded motor vehicle, trailer, semitrailer, or container containing dangerous articles must show on the shipping paper the information required in paragraphs (a) and (b) of this section, the description of the vehicle or container, and the kind of placards applied.

Cancel § 177.818, page 156 of Tariff.

Cancels and supersedes § 177.819, page 156 of Tariff and page 10 of Supplement No. 2 to read as follows:

§ 177.819 Certificate.

(a) Carriers must not accept for transportation, nor transport, any dangerous article subject to the regulations in this chapter, unless it has been certified by the shipper, using the following certificate, which must be signed by the shipper:

"This is to certify that the above named articles are properly classified, described, packaged, marked and labeled, and are in proper condition for transportation, according to the applicable regulations of the Department of Transportation."

However, preprinted shipping papers bearing previously authorized certificates may continue to be used until existing stocks are depleted, but not after June 1, 1968.

(b) Shipper certification is not required for shipments to be transported by the shipper as a private carrier, except for shipments which are to be reshipped or transferred from one carrier to another, or for bulk shipments in cargo tanks supplied by the carrier.

Cancels and supersedes § 177.823, page 156 of Tariff, to read as follows:

§ 177.823 Required Exterior Marking on Motor Vehicles and Combinations.

(a) Every carrier operating, hauling, or in any manner using a motor vehicle or trailer containing any explosive or other dangerous article as specified in subparagraph (a)(1) and paragraph (b) of this section shall cause every motor vehicle, trailer, or combination vehicle—at all times while containing such explosive or other dangerous article, or combination of such articles—to display markings or placards in accordance with the following requirements.

(1) The marking or placards required to be displayed on each motor vehicle or trailer shall be as follows:

Commodity	Type of Marking or Placard
Explosives, Class A, any quantity	EXPLOSIVES A (red letters on white background)
Explosives, Class B, any quantity	EXPLOSIVES B (red letters on white background)
Combination of Class A and Class B explosives	EXPLOSIVES A (red letters on white background)
Poison, Class A, any quantity	POISON (blue letters on white background)

Commodity	Type of Marking or Placard
Poison, Class B, 1,000 lbs or more gross weight	POISON (blue letters on white background)
Flammable liquid, 1,000 lbs or more gross weight	FLAMMABLE (red letters on white background)
Flammable solid, 1,000 lbs or more gross weight	FLAMMABLE (red letters on white background)
Oxidizing material, 1,000 lbs or more gross weight	OXIDIZERS (yellow letters on black background)
Nonflammable compressed gas, 1,000 lbs or more gross weight	COMPRESSED GAS (green letters on white background)
Corrosive liquid, 1,000 lbs or more gross weight	CORROSIVES (blue letters on white background)
Flammable compressed gas, 1,000 lbs or more gross weight	FLAMMABLE GAS (red letters on white background)
Radioactive material requiring red label as prescribed in § 173.414(a) and (c)	RADIOACTIVE (black letters on yellow background)
Mixed ladings (see subparagraph (a) (4) of this section)	DANGEROUS (red letters on white background)

(2) Each marking or placard shall consist of letters not less than 4 inches high, in the color specified, using approximately a $\frac{5}{8}$-inch stroke. The placard must be larger than the lettering required thereon by at least 1 inch at the top and bottom. Such marking or placard described in subparagraph (a)(1) shall be contained in an area on the vehicle which has no other marking, lettering, or other graphic display for at least 3 inches in each direction, except as specified in subparagraph (a)(4) and paragraph (c) of this section.

(3) Such markings or placards shall be displayed at the front, rear, and on each side of the motor vehicle, trailer, or other cargo-carrying body while it contains explosives or other dangerous articles of such type and in such quantity as specified in subparagraph (a)(1) and in paragraphs (b) and (c) of this section. The front marking or placard may be displayed on the front of either the truck, truck body, truck tractor, or the trailer.

(4) Any motor vehicle, trailer, or other cargo-carrying body containing more than one kind of explosives or other dangerous articles requiring different placards under the provisions of subparagraph (a)(1) of this section, the aggregate gross weight of which totals 1,000 pounds or more, shall be marked or placarded "DANGEROUS" instead of being marked or placarded as required by that subparagraph. Any such vehicle which contains any quantity of Class A explosives, Class B explosives, Class A poison, or radioactive materials requiring red label, as prescribed in § 173.414(a) and (c), shall display the marking or placard "EXPLOSIVES A", "EXPLOSIVES B", "POISON", or "RADIOACTIVE", as appropriate, in addition to the marking or

placard "DANGEROUS". If Class A explosives and Class B explosives are loaded on the same vehicle, the "EXPLOSIVES B" marking need not be displayed.

    (5) In any combination of two or more vehicles containing explosives or other dangerous articles, each vehicle shall be marked or placarded as to its contents and in accordance with paragraphs (a)(1) and (a)(4) of this section.

(b) Tank motor vehicles.

    (1) Every tank motor vehicle or tank trailer used for the transportation of any explosive or other dangerous article, regardless of quantity or whether loaded or empty, shall be marked or placarded in accordance with the requirements of paragraph (a) of this section, except as otherwise provided in subparagraphs (2), (3) and (4) of this paragraph, provided, however, that no such marking or placard shall be displayed during such time as such vehicle or trailer is laden only with a commodity not classified as a dangerous article.

> Note: permanent markings on tank motor vehicles in compliance with the regulations prior to January 1, 1967, may be displayed until such vehicles are repainted or remarked, but not later than July 1, 1967.

    (2) Tank motor vehicles transporting gasoline may be marked or placarded "GASOLINE" in lieu of the required "FLAMMABLE" marking of the placard, in the same size and color as required for the "FLAMMABLE" marking or placard.

    (3) Tank motor vehicles transporting any flammable compressed gas shall be marked in letters at least 4 inches high with the words "FLAMMABLE GAS" or "FLAMMABLE COMPRESSED GAS". In addition, the common name of the contents shall be marked on the tank in letters at least 2 inches high, using approximately a $\frac{1}{4}$-inch stroke, in the colors specified in subparagraph (a)(1) of this paragraph.

    (4) Tank motor vehicles transporting any nonflammable compressed gas shall be marked in letters not less than 4 inches high with the words "COMPRESSED GAS". In addition, the common name of the contents shall be marked on the tank in letters at least 2 inches high, using approximately a $\frac{1}{4}$-inch stroke, in the colors specified in subparagraphs (a)(1) of this paragraph.

(c) In addition to displaying the marking or placards required by paragraphs (a) and (b) of this section, a carrier shall display markings or placards reading "CARGO FIRE—AVOID WATER", or words of similar meaning that denote the incompatibility of water with the lading, in letters at least 2 inches high, when such wording is specified or requested by the shipper on the shipping papers, or when the carrier knows that such warning is appropriate. Such wording shall be displayed immediately adjacent to all required markings or placards on the truck, trailer, or trailers containing the commodity involved.

(d) The marking or placarding required by this section shall be removed from or covered on any motor vehicle to which it is attached when such vehicle does not contain the article for which the marking is required, except in the case of tank motor vehicles used exclusively for transportation of the article for which such marking is required.

> Regulation of Department of Transportation Issued as Supplement No. 4 to Agent T. C. George's Tarriff No. 19, Effective December 1, 1967.

BIBLIOGRAPHY OF REGULATIONS COVERING THE TRANSPORTATION OF RADIOACTIVE AND FISSILE MATERIAL, PREPARED BY THE STEERING COMMITTEE OF USASI N 14 COMMITTEE ON TRANSPORTATION OF RADIOACTIVE AND FISSILE MATERIAL

1. *Regulation for the Safe Transport of Radioactive Materials*, 1967 edition. International Atomic Energy Agency, Vienna, Austria. Distributed by International Publications, Inc., 317 East 34th Street, New York, N.Y.
2. *International Air Transport Association Regulations Relating to Carriage of Restricted Articles by Air*. AITA Terminal Center Building, Montreal, Canada.
3. *Code of Federal Regulations, Title 14, Part 103* (Air Transport) and *Title 49, Parts 170–190* (Land Transport). Department of Transportation, Hazardous Materials Regulations Board (see *Federal Register*, 33(194): 14918—14936, 1968).
4. *U.S. Postal Guide (Postal Laws and Regulations)*, §§124.24—124.244 and 125.25.
5. *Code of Federal Regulations, Title 46, Parts 146—149* (U.S. Coast Guard Regulations). The references to the ICC are being withdrawn, and the regulations will be amended to reflect DOT requirements.
6. *Official Air Transport Restricted Articles Tarriff*. Air Traffic Conference of America, Washington, D.C.
7. *Supplements to Tarriff No. 19*. Bureau of Explosives, Association of American Railroads. To be published.
8. *Code of Federal Regulations, Title 10, Part 71*.
9. *AEC Manual, Chapter 0529 for Contractors*. Engineering requirements the same as for *Code of Federal Regulations, Title 10, Part 71*, but administrative details are different.

RADIOACTIVE-WASTE DISPOSAL
F. J. Bradley, Ph.D, C.H.P.

Operations involving work with radioactive materials generate radioactive waste. The physical form may be solid, liquid, or gaseous, and chemically there is seldom any limitation, except at high specific activities. Since an underlying philosophy in work with radionuclides is to keep the potential problem within bound at all times, a conservative approach has been adopted with respect to waste disposal. Everyone is familiar today with problems of air pollution in the cities arising from high car densities on the highways and industrial air contaminants during periods of atmospheric inversion. We blithely went ahead and continued to push industrial expansion and car population, giving little thought to the consequences, since the problem increased rather slowly in severity rather than becoming a major problem overnight. With such a background for a developing field like the application of radionuclides it, has been possible to regulate their safe use; hopefully, we will then not be presented one day with an intolerable problem.

Radioactive wastes have been divided in three rather arbitrary categories, based on specific activities; these are, according to Straub*, for liquids as follows:

low-level wastes—in the range of microcuries per liter, or 10^{-3} μCi/cm^3;
medium-level wastes—in the range of millicuries per liter, or 10^{-3} mCi/cm^3;
high-level wastes—in the range of curies per liter, or 10^{-3} Ci/cm^3.

Obviously this classification has many drawbacks; it does not take into account the different radiotoxicity of the various radionuclides, and is based on an assumption that the state of the material is liquid. But despite these limitations the classifications have been found useful; by transforming a liter into 10^3 grams, the system can be extended to solids:

low-level wastes—in the range of 10^{-3} μCi/g;
medium-level wastes—in the range of 10^{-3} mCi/g;
high-level wastes—in the range of 10^{-3} Ci/g.

Most of the wastes that are encountered in university, hospital, and industrial laboratories are usually classified as low-level radioactive wastes; therefore, most of the discussion will cover this type of waste.

There are two basic approaches to radioactive-waste disposal: concentration and containment, and dilution and dispersal into the environment. Today most high-level, highly toxic radioactive waste is concentrated and stored in large underground tanks and vaults. Dilution and dispersal into the environment is practiced, on a limited scale, for most low- and intermediate-level wastes. The increase in radioactivity caused by this dispersal into the environment is continuously monitored, and if the activity increases above certain limits, the dispersal is stopped or modified. One might compare this situation to air and water pollution, which were permitted to go on unabated until even the basic ecology of a region was altered, to emphasize the importance of control over dispersal.

The standards of the Federal Radiation Council, called Radiation Protection Guide, are given in Table 61; on the basis of these values one can determine the Radioactivity Concentration Guide, which is defined as "the concentration of radioactivity in the environment that is determined to result in organ doses equal to the Radiation Protection Guide". Actually only two organs are specified for population exposures: the whole body for individual members of population, and the gonads for an average population dose over a 30-year period. One

---

* Straub, C. P., *Low-Level Radioactive Wastes*. U.S. Atomic Energy Commission, Division of Technical Information, U.S Government Printing Office, Washington, D.C., 1964.

normally takes $\frac{1}{10}$ of the radiation worker dose for the other organs as the value for nonradiation workers.

Table 61.
RADIATION PROTECTION GUIDE
(Published by the Federal Radiation Council)

Type of Exposure	Condition	Dose, rem
Radiation Worker		
Whole body, head and trunk, active blood-forming organs, gonads, or lens of eye*	Accumulated dose 13 weeks	5 times the number of years beyond age 18, or 5 (N—18) 3†
Skin of whole body and thyroid	Year 13 weeks	30 10
Hands and forearms, feet and ankles	Year 13 weeks	75 25
Bone	Body burden	0.1 microgram of ^{226}Ra or its biological equivalent
Other organs	Year 13 weeks	15 5
Population		
Individual	Year	0.5 (whole body)
Average	30 years	5 (gonads)

* New recommendations by ICRP take the lens of eye and place it with other organs.
† New recommendations by ICRP limit 3 rem in 13 weeks to men and 1.3 rem in 13 weeks to women.

When radionuclides are dispersed into the environment at concentrations that might be deemed safe by themselves, reconcentrations by flora or fauna in the environment must be studied, to insure that the basic guides are not exceeded through these processes.

SOURCES OF RADIOACTIVE WASTES, AND METHODS OF DISPOSAL

In biomedical research, there has been a continuing increase in the use of tagged compounds; in many institutions the work with radioactive isotopes has doubled every 3 to 4 years. This expansion obviously will not continue, but it has strained the capabilities of some institutions in the initial stages, and this is particularly true for the handling and disposal of "hot" wastes.

In laboratories, solid wastes include contaminated paper, syringes, labware, and even laboratory equipment—such as centrifuges, pH meters, and the like—as well as carcasses of animals containing radioisotopes. The concentration of the activity is usually in the range of 1 to 100 µCi/kg for intact animals. If the animal is sacrificed and certain organs are removed, these might have a higher specific activity, but the remainder of the animal could certainly be classified as low-level waste. Because many of the radionuclides used in biochemical research—such as ^{3}H, ^{14}C, ^{35}S, and ^{125}I—are pure beta emitters or weak gamma emitters, external radiation is normally not a major problem.

Disposal of Animal Carcasses

The disposal of "hot" animals has been particularly troublesome over the past few years. The following are suggested means of disposal: storage with and without refrigeration, burial, commercial waste disposal service, pulverization, and incineration. These methods will be detailed in the following paragraphs.

Storage With and Without Refrigeration. If the animal contains a radionuclide with a half-life of about one week, it is practical to store the animal in a deep freeze for a time period of ten times the half-life. At the end of that time period the activity present in the animal will be 0.1 percent of the initial activity, and (depending on the initial activity) the total activity is normally exempt from regulation. The generally licensed quantities and exempt concentrations are given in Tables 62 and 63, which are taken from Title 10, Code of Federal Regulations.

Dead animals in containers that are not refrigerated will decay and give off gases, unless preserved in formaldehyde. These gases can cause rupture of the container, with subsequent release of noxious fumes. It is possible to preserve animals in concrete sealed with paraffin for 2 to 3 months without refrigeration. The animals can also be preserved and stored in polyethylene bags, as described in the following procedure*.

A procedure has been worked out for the preservation of whole carcasses or portions of tissue of animals not greater than 3 kg in weight by sealing them in polyethylene bags together with a mixture of fresh bleaching powder and Vermiculite (a commercial expanded mica).

The following details are essential to the success of the process.
1. It is desirable to split the carcass ventrally along the midline, if this has not already been done, before placing it in the bag.
2. If the carcass is soiled with tissue fluid or blood, it should be dusted with bleaching powder and allowed to cool before being placed in the bag. This avoids the development inside the bag of heat due to the rapid interaction of the bleaching powder and organic matter. The heating may also be avoided by applying Vermiculite to the opened carcass. The finer grades of this material absorb about four times their weight of fluid, and thus the interaction with the bleaching powder is slowed down.
3. The open end of the bag (which should be made of the heavier gauges, 250 to 500, of polyethylene) should have been previously folded back in a double cuff, fold outwards, to protect the surfaces to be sealed from dust or organic matter.
4. The weights of bleaching powder and vermiculite should each be not less than one fifth of the weight of the carcass. This represents, in practice, a volume of about 300 ml and 1,500 ml respectively per kg weight of carcass.
5. The animal's claws should be cut before the start of the radioactive experiment. The feet of the carcass should be bound with adhesive tape, lint or bandage before it is placed in the bag, to avoid the risk of puncturing the bag; for the same reason it is unwise to sever the foot of the animal with bone forceps, lest sharp splinters are left.
6. The bag should be sealed with a suitable type of electrically heated soldering gun so that only a small amount of air is included, but care should be taken in expelling the air not to blow dust on the surfaces to be sealed.
7. The "cocooned" carcasses should be stored in open containers, to allow for free circulation of air outside the bags, until the radioactive isotope has decayed sufficiently for incineration.

---

* Boursnell, J. C. and Gleeson-White, M. H. *Nature*, 179:54, 1959.

Table 62.
GENERALLY LICENSED QUANTITIES FROM 10 CFR 31.100

By-Product Material	Not as a Sealed Source, μCi	As a Sealed Source, μCi
^{124}Sb	1	10
^{76}As	10	10
^{77}As	10	10
^{140}Ba + ^{140}La	1	10
^{7}Be	50	50
^{109}Cd + ^{109}Ag	10	10
^{45}Ca	10	10
^{14}C	50	50
^{144}Ce + ^{144}Pr	1	10
^{137}Cs + ^{137}Ba	1	10
^{36}Cl	1	10
^{51}Cr	50	50
^{60}Co	1	10
^{64}Cu	50	50
^{154}Eu	1	10
^{18}F	50	50
^{72}Ga	10	10
^{71}Ge	50	50
^{198}Au	10	10
^{199}Au	10	10
^{3}H	10	10
^{114}In	1	10
^{131}I	10	10
^{192}Ir	10	10
^{55}Fe	50	50
^{59}Fe	1	10
^{140}La	10	10
^{52}Mn	1	10
^{56}Mn	50	50
^{99}Mo	10	10
^{59}Ni	1	10
^{63}Ni	1	10
^{95}Nb	10	10

Table 62. (Continued)
GENERALLY LICENSED QUANTITIES
FROM 10 CFR 31.100

By-Product Material	Not as a Sealed Source, μCi	As a Sealed Source, μCi
^{109}Pd	10	10
^{103}Pd + ^{103}Rh	50	50
^{32}P	10	10
^{210}Po	0.1	1
^{42}K	10	10
^{143}Pr	10	10
^{147}Pm	10	10
^{186}Re	10	10
^{105}Rh	10	10
^{86}Rb	10	10
^{106}Ru + ^{106}Rb	1	10
^{135}Sm	10	10
^{46}Sc	1	10
^{105}Ag	1	10
^{111}Ag	10	10
^{22}Na	10	10
^{24}Na	10	10
^{89}Sr	1	10
^{90}Sr + ^{90}Y	0.1	1
^{35}S	50	50
^{182}Ta	10	10
^{96}Tc	1	10
^{99}Tc	1	10
^{127}Te	10	10
^{129}Te	1	10
^{204}Tl	50	50
^{113}Sn	10	10
^{185}W	10	10
^{48}V	1	10
^{90}Y	1	10
^{91}Y	1	10
^{65}Zn	10	10
Beta and/or gamma-emitting by-product material not listed above	1	10

Table 63.
EXEMPT CONCENTRATIONS FROM 10 CFR 30.70

Element	Atomic Number	Isotope	Gas Concentration, µCi/ml*	Liquid or Solid Concentration, µCi/ml†
Antimony	51	^{122}Sb ^{124}Sb ^{125}Sb		3×10^{-4} 2×10^{-4} 1×10^{-3}
Argon	18	^{37}Ar ^{41}Ar	1×10^{-3} 4×10^{-7}	
Arsenic	33	^{73}As ^{74}As ^{76}As ^{77}As		5×10^{-3} 5×10^{-4} 2×10^{-4} 8×10^{-4}
Barium	56	^{131}Ba ^{140}Ba		2×10^{-3} 3×10^{-4}
Beryllium	4	^{7}Be		2×10^{-2}
Bismuth	83	^{206}Bi		4×10^{-4}
Bromine	35	^{82}Br	4×10^{-7}	3×10^{-3}
Cadmium	48	^{109}Cd ^{115m}Cd ^{115}Cd		2×10^{-3} 3×10^{-4} 3×10^{-4}
Calcium	20	^{45}Ca ^{47}Ca		9×10^{-5} 5×10^{-4}
Carbon	6	^{14}C	1×10^{-6}	8×10^{-3}
Cerium	58	^{141}Ce ^{143}Ce ^{144}Ce		9×10^{-4} 4×10^{-4} 1×10^{-4}
Cesium	55	^{131}Cs ^{134m}Cs ^{134}Cs		2×10^{-2} 6×10^{-2} 9×10^{-5}
Chlorine	17	^{38}Cl	9×10^{-7}	4×10^{-3}
Chromium	34	^{51}Cr		2×10^{-2}
Cobalt	27	^{57}Co ^{58}Co ^{60}Co		5×10^{-3} 1×10^{-3} 5×10^{-4}
Copper	29	^{64}Cu		3×10^{-3}
Dysprosium	66	^{165}Dy ^{166}Dy		4×10^{-3} 4×10^{-4}
Erbium	68	^{169}Er ^{171}Er		9×10^{-4} 1×10^{-3}
Europium	63	^{152}Eu (T/2=9.2 hrs) ^{155}Eu		6×10^{-4} 2×10^{-3}
Fluorine	9	^{18}F	2×10^{-6}	8×10^{-3}

Table 63. (*Continued*)
EXEMPT CONCENTRATIONS FROM 10 CFR 30.70

Element	Atomic Number	Isotope	Gas Concentration, μCi/ml*	Liquid or Solid Concentration, μCi/ml†
Gadolinium	64	^{153}Gd		2×10^{-3}
		^{159}Gd		8×10^{-4}
Gallium	31	^{72}Ga		4×10^{-4}
Germanium	32	^{71}Ge		2×10^{-2}
Gold	79	^{196}Au		2×10^{-3}
		^{198}Au		5×10^{-4}
		^{199}Au		2×10^{-3}
Hafnium	72	^{181}Hf		7×10^{-4}
Hydrogen	1	^{3}H	5×10^{-6}	3×10^{-2}
Indium	49	^{113m}In		1×10^{-2}
		^{114m}In		2×10^{-4}
Iodine	53	^{126}I	3×10^{-9}	2×10^{-5}
		^{131}I	3×10^{-9}	2×10^{-5}
		^{132}I	8×10^{-8}	6×10^{-4}
		^{133}I	1×10^{-8}	7×10^{-5}
		^{134}I	2×10^{-7}	1×10^{-3}
Iridium	77	^{190}Ir		2×10^{-3}
		^{192}Ir		4×10^{-4}
		^{194}Ir		3×10^{-4}
Iron	26	^{55}Fe		8×10^{-3}
		^{59}Fe		6×10^{-4}
Krypton	36	^{85m}Kr	1×10^{-6}	
		^{85}Kr	3×10^{-6}	
Lanthanum	57	^{140}La		2×10^{-4}
Lead	82	^{203}Pb		4×10^{-3}
Lutetium	71	^{177}Lu		1×10^{-3}
Manganese	25	^{52}Mn		3×10^{-4}
		^{54}Mn		1×10^{-3}
		^{56}Mn		1×10^{-3}
Mercury	80	^{197m}Hg		2×10^{-3}
		^{197}Hg		3×10^{-3}
		^{203}Hg		2×10^{-4}
Molybdenum	42	^{99}Mo		2×10^{-3}
Neodymium	60	^{147}Nd		6×10^{-4}
		^{149}Nd		3×10^{-3}
Nickel	28	^{65}Ni		1×10^{-3}
Niobium	41	^{95}Nb		1×10^{-3}
		^{97}Nb		9×10^{-3}
Osmium	76	^{185}Os		7×10^{-4}
		^{191m}Os		3×10^{-2}
		^{191}Os		2×10^{-3}
		^{193}Os		6×10^{-4}

Table 63. (*Continued*)
EXEMPT CONCENTRATIONS FROM 10 CFR 30.70

Element	Atomic Number	Isotope	Gas Concentration, µCi/ml*	Liquid or Solid Concentration, µCi/ml†
Palladium	46	^{103}Pd		3×10^{-3}
		^{109}Pd		9×10^{-4}
Phosphorus	15	^{32}P		2×10^{-4}
Platinum	78	^{191}Pt		1×10^{-3}
		^{193m}Pt		1×10^{-2}
		^{197m}Pt		1×10^{-2}
		^{197}Pt		1×10^{-3}
Potassium	19	^{42}K		3×10^{-3}
Praseodymium	59	^{142}Pr		3×10^{-4}
		^{143}Pr		5×10^{-4}
Promethium	61	^{147}Pm		2×10^{-3}
		^{149}Pm		4×10^{-4}
Rhenium	75	^{183}Re		6×10^{-3}
		^{186}Re		9×10^{-4}
		^{188}Re		6×10^{-4}
Rhodium	45	^{103m}Rh		1×10^{-1}
		^{105}Rh		1×10^{-3}
Rubidium	37	^{86}Rb		7×10^{-4}
Ruthenium	44	^{97}Ru		4×10^{-3}
		^{103}Ru		8×10^{-4}
		^{105}Ru		1×10^{-3}
		^{106}Ru		1×10^{-4}
Samarium	62	^{153}Sm		8×10^{-4}
Scandium	21	^{46}Sc		4×10^{-4}
		^{47}Sc		9×10^{-4}
		^{48}Sc		3×10^{-4}
Selenium	34	^{75}Se		3×10^{-3}
Silicon	14	^{31}Si		9×10^{-3}
Silver	47	^{105}Ag		1×10^{-3}
		^{110m}Ag		3×10^{-4}
		^{111}Ag		4×10^{-4}
Sodium	11	^{24}Na		2×10^{-3}
Strontium	38	^{89}Sr		1×10^{-4}
		^{91}Sr		7×10^{-4}
		^{92}Sr		7×10^{-4}
Sulfur	16	^{35}S	9×10^{-8}	6×10^{-4}
Tantalum	73	^{182}Ta		4×10^{-4}
Technetium	43	^{96m}Tc		1×10^{-1}
		^{96}Tc		1×10^{-3}

Table 63. (*Continued*)
EXEMPT CONCENTRATIONS FROM 10 CFR 30.70

Element	Atomic Number	Isotope	Gas Concentration, µCi/ml*	Liquid or Solid Concentration, µCi/ml†
Tellurium	52	^{125m}Te		2×10^{-3}
		^{127m}Te		6×10^{-4}
		^{127}Te		3×10^{-3}
		^{129m}Te		3×10^{-4}
		^{131m}Te		6×10^{-4}
		^{132}Te		3×10^{-4}
Terbium	65	^{160}Tb		4×10^{-4}
Thallium	81	^{200}Tl		4×10^{-3}
		^{201}Tl		3×10^{-3}
		^{202}Tl		1×10^{-3}
		^{204}Tl		1×10^{-3}
Thulium	69	^{170}Tm		5×10^{-4}
		^{171}Tm		5×10^{-3}
Tin	50	^{113}Sn		9×10^{-4}
		^{125}Sn		2×10^{-4}
Tungsten	74	^{181}W		4×10^{-3}
		^{187}W		7×10^{-4}
Vanadium	23	^{48}V		3×10^{-4}
Xenon	54	^{131m}Xe	4×10^{-6}	
		^{133}Xe	3×10^{-6}	
		^{135}Xe	1×10^{-6}	
Ytterbium	70	^{175}Yb		1×10^{-3}
Yttrium	39	^{90}Y		2×10^{-4}
		^{91m}Y		3×10^{-2}
		^{91}Y		3×10^{-4}
		^{92}Y		6×10^{-4}
		^{93}Y		3×10^{-4}
Zinc	30	^{65}Zn		1×10^{-3}
		^{69m}Zn		7×10^{-4}
		^{69}Zn		2×10^{-2}
Zirconium	40	^{95}Zr		6×10^{-4}
		^{97}Zr		2×10^{-4}
Beta- and/or gamma-emitting by-product material not listed above with half-life less than 3 years			1×10^{-10}	1×10^{-6}

* Values are given only for those materials normally used as gases.
† µCi/g for solids.

Note 1. Many radioisotopes disintegrate into isotopes that are also radioactive. In expressing the concentration in Schedule A, the activity stated is that of the parent isotope and takes into account the daughters.

Note 2. For purposes of §30.14. where a combination of isotopes is involved, the limit for the combination should be derived as follows: determine for each isotope in the product the ratio between the concentration present in the product and the exempt concentration established in Schedule A for the specific isotope when not in combination. The sum of such ratios may not exceed 1 (i.e., unity).
Example:

$$\frac{\text{concentration of isotope A in product}}{\text{exempt concentration of isotope A}} + \frac{\text{concentration of isotope B in product}}{\text{exempt concentration of isotope B}} = 1$$

Another means of animal storage that has been tried is the vacuum-canning of carcasses.

Burial. On-site burial is ideal, and if the proper approval is obtained, this is the recommended practice. USAEC regulations for burial in soil, as published in Title 10, Code of Federal Regulations, Part 20, are as follows:
- (a) the total quantity of licensed and other radioactive materials buried at any one location and time does not exceed, at the time of burial, 1,000 times the amount specified in Appendix B, Column 1 (see Table 19, pp. 616 to 633);
- (b) burial is at a minimum depth of 4 feet; and
- (c) successive burials are separated by distances of at least 6 feet, and not more than 12 burials are made in any year.

Commercial Waste Disposal Service. As shown in Table 64, a number of commercial waste disposal firms will provide the service of picking up properly packaged animals and transporting them to licensed nuclear-waste burial sites. The wastes must be packaged in a manner that insures against decomposition of the animals for upwards of 24 months; in many cases this is not done.

Pulverization. Many small animals given tagged radiochemicals have a total quantity of activity that permits disposal through the sanitary sewer. Animals can be pulverized in a commercial blender, and disposed in a liquid mixture. Alternately, animals can be disposed through drains with built-in "dispose-all" units. The criteria for disposal through a sanitary sewer, as given in Title 10, Code of Federal Regulations, Part 20, are as follows:
- (a) it is readily soluble or dispersible in water;
- (b) the quantity of any licensed or other radioactive material released into the system by the licensee in any one day does not exceed the larger of subparagraph (1) or (2) of this paragraph:
  - (1) the quantity which, if diluted by the average daily quantity of sewage released into the sewer by the licensee, will result in an average concentration equal to the limits specified in Appendix B, Table I, Column 2 of this part (see Table 19, pp. 616 to 633); or
  - (2) ten times the quantity of such material specified in Appendix C of this part;
- (c) the quantity of any licensed or other radioactive material released in any one month, if diluted by the average monthly quantity of water released by the licensee, will not result in an average concentration exceeding the limits specified in Appendix B, Table I, Column 2 of this part (see Table 19, pp. 616 to 633); and
- (d) the gross quantity of licensed and other radioactive material released into the sewage system by the licensee does not exceed one curie per year.

Excreta from individuals undergoing medical diagnosis or therapy with radioactive materials shall be exempt from any limitations contained in this section.

Incineration. The use of incineration is, at first glance, an ideal method of animal disposal, but when coupled with radioactive materials, it has created more problems than it solved. First of all, let it be stated that incineration of animals with a low specific activity or with amounts of activity equal to or less than the exempt concentrations given in Table 63 is an acceptable means of disposal. Some of the problems that have lessened the importance of incineration for animal disposal are: 1) many radionuclides commonly encountered—such as ^{14}C, ^{3}H, ^{35}S, ^{125}I, and ^{131}I—are volatilized when heated or oxidized, and consequently one must ensure that the maximum permissible concentration in the exhaust is not exceeded even with filtration; 2) the ash remaining after incineration is normally radioactive, and its concentration and containment by personnel who are not trained radiation workers presents further problems.

Table 64.
COMMERCIAL FIRMS LICENSED TO RECEIVE RADIOACTIVE WASTES

Commercial Firms Licensed by the U.S. Atomic Energy Commission to Receive Radioactive Wastes (as of April 12, 1967)	Commercial Firms Licensed by Agreement States to Receive Radioactive Wastes (as of June 6, 1967)
ATCOR, Inc. 360 Bradhurst Avenue Hawthorne, New York 10532	Atomic Energy Industrial Laboratories of the Southwest 6413 South Main Street Houston, Texas 77025
Atomic Disposal Company, Inc. 14532 South Kedzie Avenue Midlothian, Illinois 60445	California Nuclear, Inc. P.O. Box 59 Cornell, California 91315
California Nuclear, Inc. 2323 South 9th Street Lafayette, Indiana 47905	General Nuclear, Inc. 6740 Telean Houston, Texas 77034
California Salvage Company 700–745 North Pacific Avenue San Pedro, California 90731	Laboratory for Electronics Tracerlab Division 2030 Wright Avenue Richmond, California 94804
Laboratory for Electronics, Inc. Tracerlab Division 1601 Trapelo Road Waltham, Massachusetts 02154	McCormack's Highway Transportation, Inc. 151 Erie Boulevard Schenectady, New York 12305
Long Island Nuclear Service Corp. Station Road Bellport, New York 11713	Nuclear Engineering Company, Inc. Box 146 Barns Building Morehead, Kentucky 40351
Nuclear Diagnostic Laboratories, Inc. 1000 Lower South Street Peekskill, New York 10566	Nuclear Fuel Services, Inc. P.O. Box 124 West Valley, New York 14171
Nuclear Engineering Company, Inc. Box 594 Walnut Creek, California 93002	U.S. Nuclear Corporation 801 North Lake Street Burbank, California 91502
Radiological Service Company, Inc. 35 Urban Avenue Westbury, New York, 11590	William Wayne Electronics, Inc. Hastings Radiochemical Works P.O. Box 60448 Houston, Texas 77060
The Walker Trucking Company 1283–1285 East Street New Britain, Connecticut 06053	

These Firms Should Be Contacted Directly For Information Regarding Services They Can Provide

From: U.S. Atomic Energy Commission, Division of Industrial Participation, *The Nuclear Industry, 1967*. U.S. Government Printing Office, Washington, D.C.

Disposal of Radioactive Solid Waste

In a manner similar to animal disposal, any solid radioactive waste of low specific activity can be disposed of by storage until the specific activity is exempt, by burial, commercial waste disposal service, pulverization and sewage disposal, or by incineration.

The storage of low-specific-activity waste should be in metal specification containers, as given in Title 49, Code of Federal Regulations, Part 178, and appended to the Transportation section.

Incineration of radioactive materials above exempt concentrations must be done with approval of proper regulatory agencies, which normally include the U.S. Atomic Energy Commission and State (City) Radiological Health Agency.

For burial, radioactive waste can be shipped to the commercial nuclear-waste burial sites shown in Figure 34. The waste can be shipped by rail express or by common, contract, or private carrier, with the laboratory assuming the responsibility of shipper and certifying to the carrier that the waste is packaged and labeled in accordance with U.S. Department of Transportation regulations. Alternatively, the waste can be turned over to the licensed commercial waste disposal firms given in Table 64, who normally assume the responsibilities of shipper in regard to storage and transfer to nuclear-waste burial sites.

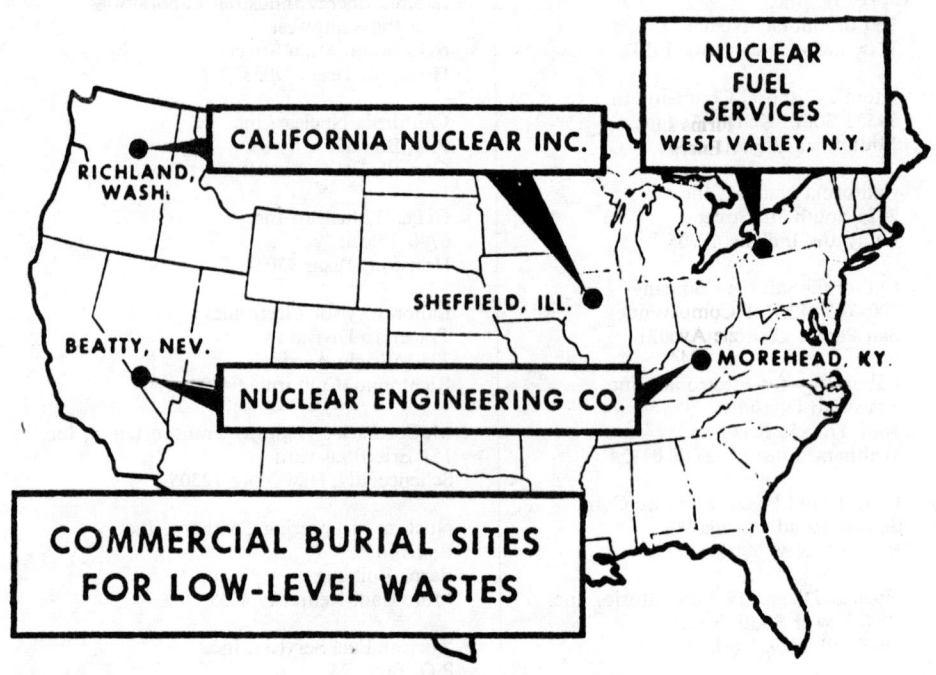

FIGURE 34.
Low-Level-Waste Burial Sites.

On May 14, 1963, the Commission approved withdrawal of the AEC from its interim low-level-waste burial operations provided to licensees, based upon the availability of commercial service from Nuclear Engineering Company (NECO) burial sites at Beatty, Nevada, and Morehead, Kentucky. In November of 1963, Nuclear Fuel Services, Inc. (NFS) established a low-level-waste disposal facility near West Valley, New York. A new waste burial site was recently established by California Nuclear, Inc. (CNI) on state-owned land near Sheffield, Illinois. The AEC issued an amendment to an existing CNI waste-handling license, permitting them to begin operations on August 1, 1967 (see AEC public release K-176). CNI also operates a waste disposal site at Richland, Washington.
(From: U.S. Atomic Energy Commission, Division of Industrial Participation, *The Nuclear Industry, 1967*. U.S. Government Printing Office, Washington, D.C.)

Disposal of Radioactive Liquid Waste

Low-level liquid wastes can be disposed of by the following methods: storage until activity decays to exempt quantity, solidification and disposal as solid waste, commercial waste disposal, dispersal, and evaporation and/or chemical treatment.

In the case of short-lived radioisotopes and small volumes, the liquid can be stored, if space is available. Precautions must be taken to reduce external exposure by shielding of containers, if necessary, and by reducing the chance for breakage to a minimum, if glassware is used. Glass containers must be stored in fireproof metal containers, to avoid spread of contamination in case of fire.

Solidification of liquid can be accomplished by various means. An aqueous liquid can be incorporated into concrete, if it has a pH greater than or equal to 7.0; if its pH is less than 7.0, then the liquid should be made basic before mixing with concrete. Two other commonly used solidification agents are Vermiculite and Speedi-Dry; the former is a commercial expanded mica, and the latter is a commercial designation for a floor liquid absorbent. Very little research has been done on the absorbent properties of these two agents, but it can be stated that they are practically useless for tritiated water solutions. It is best to attempt to match the liquid with the solidifying agent by test prior to actual use; the proportion of liquid to solid is an important parameter often forgotten in practice. If dealing with organic liquid, then one must rely on the chemistry of the compound for conversion to the solid phase.

Liquids may be accepted by commercial radioactive-waste disposal firms, and each of these should be consulted for their requirements. Certain waste burial sites will not accept radioactive liquids, and they also must be consulted before shipment, if direct shipments are made. Radioactive liquids are accepted by rail express and common carriers, if they are packaged in DOT specification containers. A problem is presented by shipments that claim to be solid, but are in fact liquid; extra care must, therefore, be exercised at the shipping end in order to ensure correct packaging and proper notification of carrier and consignee.

Evaporation of liquid wastes is normally not practiced for low- and medium-activity wastes, but if the liquid volume is sufficiently great, it is a convenient and proven technique for reducing the volume of such wastes. The residue can then be incorporated into concrete, and disposed as solid wastes.

A summary of various water treatment processes that can be used to remove radionuclides from the liquid before discharge if given in Table 65.

Table 65.
DECONTAMINATION FACTORS
FOR VARIOUS WATER TREATMENT PROCESSES

Process	Decontamination Factors	
	Individual Radionuclides	Mixed Fission Products
Conventional Treatments		
Coagulation and settling	0.0–100 +	2.0–9.1
Clay addition, coagulation, and settling	0.0–100	1.1–6.2
Coagulation and settling		2.6–9.1
Sand filtration	1.0–100	
Coagulation, settling, and filtration	1.0–50	
Coagulation, settling, and filtration	1.8–14	3.3–3.7
Coagulation, settling and filtration		1.4–13.3
Lime-soda ash softening	2.0–100	
Ion exchange, cation	1.1–500	
Ion exchange, anion	0.0–125	
Ion exchange, mixed bed	11.0–3,300	
Solids-contact clarifier	1.9–15	2.0–6.1
Ion exchange, cation		3.0–9.1
Ion exchange, mixed bed		50.0–100.0
Nonconventional Treatments		
Ion exchange		2.3–19.0
Phosphate	1.2–1,000	125.0–250.0
Metallic dusts	1.1–1,000	1.1–8.6
Clay treatment	0.0–100 +	2.0–6.0
Diatomaceous earth	1.1–	

From: Straub, C. P., *Low-Level Radioactive Wastes,* Table 8.31, p. 199. Prepared under the auspices of the U.S. Atomic Energy Commission, Division of Technical Information. U.S. Government Printing Office, Washington, D.C., 1964.

One would assume the disposal of radioactive gases to be straightforward, but it raises some potential problems. Controlling the release of the radioactive waste gases to the environment so that concentrations remain within the maximum permissible limits (see Table 19, pp. 616 to 633) is the only feasible procedure for the radioactive isotopes of neon, argon, krypton, xenon, and radon. Some of these gases, or their radioactive daughters, adhere to dust particles, which can be filtered out of an exhaust duct. Tritium and ^{14}C, as H_2 and CO_2 respectively, can also be released to the environment under controlled conditions, as these are two relatively stable chemical compounds. Elements such as iodine and mercury, which sublime, must be used with care in order to limit the need for air disposal. Special filters specific for such nuclides must be used, if large quantities of these nuclides are handled.

ADMINISTRATION OF A RADIATION-PROTECTION PROGRAM

F. J. Bradley, Ph.D., C.H.P.

RADIATION-SAFETY OFFICER

Whether the facility utilizing ionizing radiation is a university, research institute, government lab, or industrial installation, it is recommended that a single person be designated as the Radiation-Safety Officer (RSO), whose responsibility it is to maintain a radiation-safe environment. Some of the functions that may be assigned to the RSO are enumerated in the *N.B.S. Handbook* 76 as follows:
1. Establishing and maintaining operational procedures so that the radiation exposure of each worker is kept as far below the maximum permissible exposure as is practicable.
2. Instructing personnel in safe working practices and in the nature of injuries resulting from overexposure to radiation.
3. Assuring that personnel-monitoring devices are used where indicated, and that records are kept of the results of such monitoring.
4. Conducting periodic radiation surveys where indicated, and keeping records of such surveys, including descriptions of corrective measures.
5. Investigating each case of excessive or abnormal exposure to determine the cause, and taking steps to prevent its recurrence.

In addition, the Radiation-Safety Officer will normally maintain liaison with federal, state, county, and city radiation-control groups; he is usually listed as the RSO (or under a similar title) on license applications for radioactive materials to the U.S. Atomic Energy Commission or Agreement States (states that have assumed licensing and inspection responsibility under an agreement with the USAEC). Figure 35 indicates the Agreement States as of 1967.

In many organizations one can draw lines from one function to another, with personnel in charge of each function responsible to the person above him; these persons are in the so-called "line organization". There are other functions in a facility, which are of a service nature and usually assist the line organization in various advisory capacities; such functions are designated as "staff functions", and may include safety, plant engineering, security, quality control, and the like. It is not recommended that the functions of the RSO be put into this category, but rather that a single person be designated RSO by top management and have facility-wide responsibility. Diffusion of this responsibility is certain approach to failure in a radiation-protection program, as was originally noted in government labs at the beginning of the atomic-energy programs, and later in universities.

If the use of ionizing radiation is limited and on a small scale, the function of the RSO will be a part-time function, and the user may also be the RSO; for large users, however, these functions should be separate. If the use of ionizing radiation is sufficiently diverse, then a committee should function as advisor to the RSO, and possibly as an arbitrator for disagreements between RSO and user, if so directed by top management. In addition it is becoming more prevalent for large-scale users to periodically have independent outside audits of their programs by consultants.

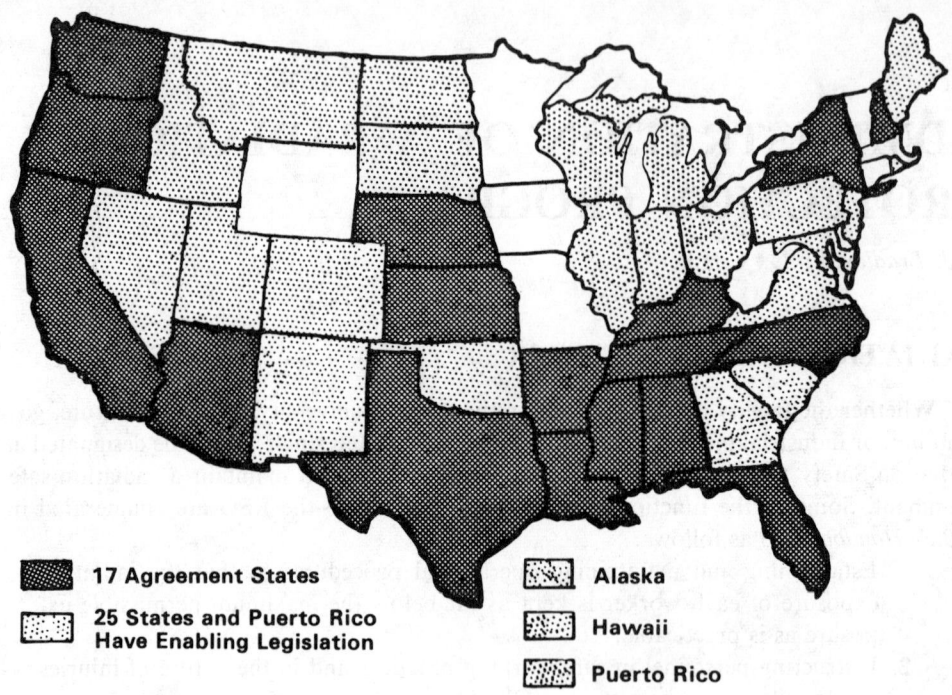

FIGURE 35.
Transfer of Materials-Licensing Functions to States,
Current to December 1968.
Shaded outline of the United States indicates the states that have entered into an agreement with the U.S. Atomic Energy Commission under subsection 274b of the Atomic Energy Act of 1954 for assuming certain of the Commission's regulatory responsibilities. Two additional states, Colorado and Idaho, have become Agreement States since this chart was published.
(From: U.S. Atomic Energy Commission, Division of Industrial Participation, *The Nuclear Industry, 1967.* U.S. Government Printing Office, Washington, D.C.)

SERVICES

As the use of ionizing radiation and of radioactive materials expands, the need for services such as film badge suppliers, radioactive-waste disposal firms, and laundries for contaminated clothing increase. Tables 66 and 67 give a part of this information; commercial waste disposal sites and firms are listed in the preceding chapter. The film badge processor should be contacted to determine if he complies with an objective standard, such as National Sanitary Foundation Standard No. 16, or if he is approved by the National Sanitation Foundation Testing Laboratory, Inc., Ann Arbor, Michigan.

At the present time (1968), two firms provide whole-body counting services by the use of portable whole-body counters; these are:

> Helgeson Nuclear Services
> 872 Abbie Street
> Pleasanton, California 94566
>
> Nuclear Electronics Counter Corporation
> McCulloch Building, Box 604
> Okmulgee, Oklahoma 74447

Table 68 indicates firms that provide decontamination services after spilling radium and other radionuclides.

Table 66. FILM BADGE PROCESSORS

1. Applied Health Physics, Inc.
   Bethel Park, Pennsylvania
2. Atomic Energy Industrial Laboratories
   Houston, Texas
3. Controls for Radiation, Inc.
   Cambridge, Massachusetts
4. Eberline Instrument Corporation
   Santa Fe, New Mexico
5. T. M. Gaines Company
   Berkeley, California
6. Gard-Ray Film Badge Processors
   Burlington, Massachusetts
7. Health Physics Services, Inc.
   Baltimore, Maryland
8. R. S. Landauer, Jr. & Company
   Matteson, Illinois
9. Nuclear-Chicago Corporation
   Des Plaines, Illinois
10. Nuclear Consultants Corporation
    St. Louis, Missouri
11. Nuclear Services Laboratory
    Knoxville, Tennessee
12. Nucleonic Corporation of America
    Brooklyn, New York
13. Radiation Detection Company
    Mountain View, California
14. St. John X-Ray Laboratory
    Califon, New Jersey
15. Technical Associates
    Burbank, California
16. Tracerlab
    Waltham, Massachusetts
17. United States Testing Company
    Hoboken, New Jersey

From: U.S. Atomic Energy Commission, Division of Industrial Participation, *The Nuclear Industry, 1967.* U.S. Government Printing Office, Washington, D.C.

Table 67. COMPANIES ACTIVELY PROCESSING CONTAMINATED LAUNDRY

1. Interstate Industrial Laundry Service, Inc.
   Pleasanton, California
2. Interstate Industrial Laundry Service, Inc.
   Santa Fe, New Mexico
3. Interstate Industrial Uniform Rental Service, Inc.
   North Charleston, South Carolina
4. Interstate Industrial Uniform Service, Inc.
   Indian Orchard, Massachusetts
5. Nuclear Laundry Rental Services, Inc.
   Jeanette, Pennsylvania
6. Nuclear Laundry Service, Inc.
   Waterbury, Connecticut
7. Nuclear Materials and Equipment Corporation
   Apollo, Pennsylvania
8. Penn Overall Supply Company
   Pittsburgh, Pennsylvania
9. Southern Space, Inc.
   Macon, Georgia

From: U.S. Atomic Energy Commission, Division of Industrial Participation, *The Nuclear Industry, 1967.* U.S. Government Printing Office, Washington, D.C.

Table 68.
COMPANIES OFFERING
DECONTAMINATION SERVICES

1. Applied Health Physics, Inc.
   2986 Industrial Boulevard
   P.O. Box 197
   Bethel Park, Pennsylvania 15102
2. ATCOR, Inc.
   360 Bradhurst Avenue
   Hawthorne, New York 10523
3. Isotopes, Division of Teledyne Company
   50 Van Buren Avenue
   Westwood, New Jersey 07675
4. Laboratory for Electronics
   230 Wright Avenue
   Richmond, California 94802
5. Long Island Nuclear Service Corporation
   Station Road
   Bellport, New York 11713
6. Nuclear Diagnostic Labs, Inc.
   755 Park Avenue
   New York, New York 10021
7. Nuclear Engineering Company, Inc.
   Box 594
   Walnut Creek, California 93002
8. Radiological Service Company, Inc.
   Division of Isotopes
   50 Van Buren Avenue
   Westwood, New Jersey 07675

From: U.S. Atomic Energy Commission, Division of Industrial Participation, *The Nuclear Industry, 1967*. U.S. Government Printing Office, Washington, D.C.

RECORDS

Because most of the diseases that result from overexposure to ionizing radiation are not peculiar to radiation causation, there is a certain probability of their natural occurrence in the general population. Furthermore, in the minds of some people radiation has been the causative agent for many strange maladies and occurrences. Consequently the preservation of certain records is wise, and is, in fact, mandatory in certain states and under the regulations of the USAEC.

The records that are of fundamental importance are the following:
1) a radiation worker's preemployment physical;
2) all dosimetry records, such as film badge and pocket dosimeter records;
3) all bioassay records;
4) all medical records generated during the course of a radiation worker's employment, especially blood examination records, ophthalmological examination records of those persons working with fast neutrons, and radiographic lung examination records;
5) area-monitoring results and radiation-equipment checks and tests.

The length of time that these records should be preserved has been a matter of controversy. At present this time interval depends on the record, location of the installation, and the control body having jurisdiction, and may be for 0, 3, 5, or 30 years, or forever. In practical terms this will probably be determined by the economics of preserving the record versus the probability of a claim for delayed injury arising from radiation exposure.

EMERGENCY PLANNING AND PROCEDURES

Robert G. Gallaghar
Applied Health Physics, Inc., Bethel Park, Pennsylvania.

DOCUMENTATION AND REPORTING OF RADIATION ACCIDENTS

During the first hectic hours following a radiation emergency, a steady stream of facts is sifted as efficiently as possible from personal witnesses, from different viewpoints (possibly involving guesses and personal emotions), and from surveyors' observations. Based upon these facts, evaluations are made as to the magnitude of the accident, and decisions arrived at to prevent further damage and initiate remedial action. This is not a time conducive to documentation or scholarly reporting of information. Yet it is during this period, immediately after the discovery of a radiation emergency, that certain documentation is vitally needed as a basis for specific reports, which must be made to official agencies, top management, insurance carriers, and the news media. Certain regulations [1-3] require that official agencies be notified within periods ranging from immediately after discovery—in the case of the most severe accidents—to 30 days following discovery of a minor accident.

The opportunity to obtain valid documentation of certain facts concerning an incident fades rapidly with time. The ability to recollect specific details about an accident suffers from the frailties of the human mind. Unfortunately, the tasks of notifying top management and official agencies and of filling out the required accident reports usually fall on management personnel at a time of maximum stress, when they can least afford the difficulty of extracting accurate data from feeble recollections. Furthermore, it is highly probable that management may not be familiar with specific requirements on reporting of radiation emergencies. Considerable savings in time, money, and reputation can be achieved by obtaining data and complying with incident-reporting procedures by the use of some of the methods outlined below.

Initial Accident Notification

Invariably, the first notification of a radiation accident and the calls for help will be made by telephone. Hopefully, these calls will be made by someone who has received prior instructions concerning *whom* to notify and *what* information must be given. It is impossible in an emergency to expect an individual to remember and adhere to more than a few very simple rules. Thus, as a guide for reporting fires, accidents and injuries, the use of a simple emergency notification procedure, similar to the one shown in Appendix 1, is suggested. The procedure should be printed on a bright color (e.g., yellow) and posted in strategic areas—naturally, near a telephone. This form contains a few essential facts that have been reviewed and approved by several major municipal fire departments, physicians, and insurance carriers. Copies of this form, printed on yellow, are available from the writer.

The person receiving an initial report of an accident involving radiation must be able to obtain essential information in order to evaluate the magnitude of the accident and institute remedial action, as well as to report the episode as required by specific regulations. To obtain these facts, the use of a form similar to that shown in Appendix 2 is recommended. This form contains a list of questions that have been found helpful in establishing the urgency of the situation and in determining the most effective methods of bringing the situation under control.

The use of a portable tape recorder or the aid of shorthand transcription to obtain a verbatim record of telephone and radio messages can be invaluable. An inexpensive ($20 to $30) battery-operated tape recorder can be used for this purpose. Most telephone companies also have special recording equipment, which can be installed on any telephone, and which has the warning "beep" tone at fifteen-second intervals required by telephone company policy. When recording radiation emergency communiques, it is essential to indicate the *time, date,* and *identity* of the speakers; in some instances the location and origin of the call may also be useful. Care must be taken to identify the date and contents of tape recordings of important events, so that these tapes can easily be used in the future without being lost or accidentally erased.

Hazard Evaluation and Requests for Assistance

Following the receipt of an initial report of a radiation incident, certain key management personnel will review the information obtained in response to the questions listed in Appendix 2. These persons need to refer to recordings or transcriptions of the incoming emergency communiques that provide the facts on which to base a preliminary evaluation of the accident. Appendix 3 contains a list of some of the questions that must be answered as part of this hazard evaluation. Obviously, management's primary objective is to bring the emergency under control as effectively as possible, and then to institute remedial action, so that routine operations can be resumed without delay. Radiological risks and potential costs of the accident will be primary considerations.

Immediate action is taken to obtain outside help—such as fire, medical, police, and radiological assistance, if such is needed. An effective communications system is a prime necessity during this phase, when efforts are being made to control the accident.

The severity of a radiation accident should never be underestimated. Effective communications systems should have been established within the organization prior to the accident —augmented by the necessary telephone, public-address, and perhaps radio equipment—to permit an uninterrupted flow of information to a communication center for interpretation, and for the issuance of instructions for radiation-risk control. An unlisted telephone number can be very valuable for such use. All personnel should clearly understand and observe instructions concerning communications relative to the accident. All information should be channeled to one or two key personnel, and released—especially to outsiders—through one designated official.

Certain officials, key employees, consultants, and trusted associates will be contacted in order to assist the organization to recover from the accident as safely, quickly, and economically as possible. Obviously, top management is informed, along with the organization's legal counsel, physician, and health-physics consultant. Each will be required to perform his services as needed in order to aid in recovery from the accident, and to coordinate and document such work with a responsible official within the organization. Insurance carriers must be notified relative to possible claims under existing workmen's compensation, public-liability, and physical-damage coverage.

Obviously, the occurrence of a radiation accident will soon become known to a substantial number of people. Employees and their families may become concerned, and soon rumors may exaggerate the magnitude of the radiation risks. This is why the preparation and issuance of a concise factual statement concerning the accident is recommended, and why one responsible individual should be officially designated to release information. The employees and news media should be told what has happened, that steps are being taken by competent specialists to control any radiation risk, and that additional information will be released by the organization as it becomes available. All requests for information also should be directed to the individual designated to issue such information.

One of the most significant facts that should be established concerning a radiation accident is the length of time that has elapsed since the accident occurred—especially if the accident involves the release of unsealed radioactive materials, a number of people, or personnel other than employees (e.g., public, personal property, subcontractor, or client personnel). The identity of such personnel, and the description and location of property or equipment potentially involved in the accident, must be obtained promptly, and appropriate action must be taken to evaluate and control any radiological risks and to avoid adverse public reactions resulting from the emergency. It has been found that the duration of time between the occurrence and the beginning of organized efforts to control the radiation emergency and abate undue alarm has a profound effect upon the total cost of the episode and its impact on reputations.

Experienced assistance and technical advice on bringing any type of radiation emergency under control can be obtained by telephone 24 hours a day by contacting a private health-physics service firm whose professional assistance is offered as a public service to radiation users in the United States and Canada. Organizations licensed by the U.S. Atomic Energy Commission can obtain radiological assistance by contacting the appropriate Atomic Energy Commission offices shown in Appendix 4.

There is no substitute for qualified personnel experienced in the health-physics and medical aspects of radiation emergencies. Such personnel must be technically competent and certified by appropriate professional boards, so that their work and reports will satisfy future medicolegal requirements. Furthermore, they must be able to assume the technical responsibility and possible financial risks associated with the professional services offered. Prompt notification of such personnel by telephone can bring advice on the method for controlling the emergency, and assistance on some of the post-emergency problems (e.g., public relations, insurance, regulatory, and medicolegal), through methods and contacts established and proven effective on prior radiation emergencies.

The initial phase of the radiation emergency has passed when effective action has been taken to control the emergency by quarantining the source of the radiation hazard along with all potentially affected areas and equipment. As a part of this procedure, prompt and effective medical attention must be obtained. Reference 4 contains information and advice that can be of considerable help to the physician.

Simultaneously with the detection and notification of the incident, radiological-safety surveys are made, to determine the magnitude of the radiation risks. Essential health-physics data must be gathered and interpreted. References 5 and 6 provide useful information for obtaining and utilizing such data.

Health-Physics Data, Results of Radiation Surveys, and Medical Reports

Health-physics data, results of radiation surveys, and medical reports on personnel involved in a radiation incident are essential forms of documentation. However, in a number of accidents the long-term value of these data has been overlooked in the process of concentrating upon the immediate use of the information to evaluate the situation and to restore the installation to normal operations quickly. Technical personnel who provide such data, as well as those who review and use them, should question themselves as to whether the data a) will provide valid and pertinent information several years hence, b) can easily be related to the episode and the individuals involved in the analysis and interpretation of the data, and c) will be retained, identified, and easily accessible for review in the future.

At the present time we are without universally accepted standards for the preparation of health-physics records. To describe and discuss the documentation of medical and health-physics data completely is beyond the scope of this chapter. However, a few important items, which are the result of experience gained during the management of several radiation emergencies and during subsequent litigation, will be included.

Personnel-Monitoring Data and Bioassay Results

Personnel-monitoring data and bioassay results can be of much greater value if some additional facts concerning their use during and subsequent to a radiation emergency are

documented. For example, immediately following the discovery of an accident, personnel involved in the episode submit their film badges and dosimeters for processing. Film badges must be processed without delay, once the dosimeter has gone "off scale". Some of the important facts that must be added to the reports of the personnel-monitoring results are:
1) the location of the body where the dosimeter and/or film badge was worn at the time of the accident;
2) time and date when these devices were zeroed and/or issued;
3) the location of the individual during the accident, and the route by which he left the accident scene;
4) the time and date when the personnel-monitoring equipment was turned in;
5) the names of those who received and processed it.

The circumstances under which the personnel-monitoring devices were exposed should be documented by the use of a form letter similar to that shown in Appendix 5.

It is just as important to document the fact that no radiation risks were present as it is to determine accurately the magnitude of such risks whenever and wherever they exist. This is especially true in the case of a radiation emergency involving potential exposure of personnel who are not employees of the radiation user, or who are not considered radiation workers, but may have been in the area at the time of the accident.

Sometimes samples of clothing, metal buttons, jewelry, hair or tissue may be collected for analyses, as in the case of neutron exposure. The location on the individual's body and his position relative to the source of exposure are essential pieces of information, which must be adequately documented. The same applies to bioassay samples (e.g., urine, blood, breath, excised tissue, etc.).

The result of radiological analyses of samples of air, water, surface smears, etc., must be transmitted without delay to the health physicist. It becomes his responsibility to receive, plot, interpret, and utilize these data to control radiation risks. His responsibility for risk control does not end there; he must consider future interpretations, reviews, and possible litigation, which may necessitate the use of such data as evidence to establish—possibly years later, in a court of law—certain facts concerning the accident. It would be desirable for such data to be reviewed by an individual experienced in the possible insurance and legal use of such documents, before they are filed away following the radiation accident.

Eyewitness Accounts

Eyewitness accounts of a radiation accident provide essential documentation that can be extremely valuable, especially if they are obtained shortly after the incident, when facts are fresh in the person's mind. One of the best times to obtain such information is immediately after the required health-physics and medical procedures have been completed and the individual has had a chance to regain his composure. Written statements should be obtained from any individual potentially exposed as a result of the radiation accident, or who claims to have been affected by it, as well as from those persons who may have pertinent information concerning the accident. These statements can be in the form of tape recordings or handwritten narrative reports of what the individual saw, did, saw others doing, and where or when certain events relative to the accident took place. Such an individual should also describe how and why he believes the accident occurred. These accounts should not be prepared as a cooperative or "team" effort, nor should they be written under the "helpful" eye of a supervisor. They need not be in scholarly prose, but should be clear and accurate accounts in the individual's own words. Each should be signed and dated by the individual. Again, a tape recording of this report, which is later typed, signed by the individual, and dated, may be used. It is recommended that reports of this type be witnessed or notarized.

When such statements are reviewed by an experienced investigator, they sometimes produce clues that may be extremely important in reconstructing pertinent facts at the time of the accident and in determining its cause and practical methods of avoiding recurrences. Such statements have uncovered other undetected or unreported accidents that had occurred in the past. For example, in investigating the statement made by a radiographer during one radiographic accident, we learned about two others involving the same piece of equipment, but of lesser severity.

Photographic Reenactments

Photographic reenactments of the radiation accident are invaluable in establishing the location and identity of the individuals with respect to the radiation source. Eyewitness accounts are used to reconstruct the scene. A nonradioactive "dummy" radioisotope source is set up to duplicate the actual conditions during the radiation accident. Either the individuals involved in the incident or substitutes are asked to attempt to duplicate as closely as possible the actions that took place. We have used at least one motion picture camera, and sometimes several, as well as still cameras, to photograph such reenactments.

We have found it valuable to have a narrative account of the reenactment recorded on tape. The tapes also contain periodic time checks along with announcements of the running times; that is, while the narrator describes the actions being filmed, a timekeeper can be heard in the background counting off the seconds that have elapsed during each important phase of the reenactment. Thus, when the recording is played back along with the movies or matched with sequence photographs, determinations can be made relative to the approximate times and distances where personnel may have received significant radiation exposures. Another very valuable technique is to include in the background of the picture at least one large electric clock that has been synchronized to start at an appropriate time during the reenactment in order to document the running time of the various stages of the reenactment.

Documentary photographs should also be taken of the equipment present in the incident. It is our practice to use a shielded periscope for photographing a source involved in a radiation accident. In some cases such photographs reveal wear or defects in the source or in the equipment. Radiation injuries should be photographed in color and properly labeled as to date and identity of the individual. The documentary photographs, together with the statements by personnel involved in the accident, enable the health physicist to reconstruct the scene for the purpose of obtaining radiation dosimetry data. Polaroid photographs can be very helpful during the dosimetry experiments as well as during the reenactment.

Reporting of Radiation Accidents

Every organization has a policy according to which good or bad news concerning that organization is released to its own officials and to outside firms, to official agencies, and eventually to the public. Unfortunately, many organizations fail to realize that they can have a radiation accident and may not be prepared administratively, technically, or financially to cope with such a situation; top management may not have recognized the need for an official of the organization to become aware of the legal responsibility for reporting a radiation accident within a prescribed period of time to certain official agencies, as required by law. A responsible individual within the organization must be clearly designated and briefed as to *whom* he is to notify, *when, where,* and *how* such reports must be transmitted, and *what* should be reported.

Naturally, a radiation accident is not the type of occurrence that an organization would want anyone to know about. In fact, it may be of interest that in our work on radiation accidents we have detected two, and possibly three, cases of what psychologists refer to as "subconscious denial" on the part of the radiation workers that a radiation accident had occurred; that is, an individual refused to accept the fact that he had received a potentially serious exposure.

However, top management must make clear to each radiation worker that failure to notify certain designated personnel (e.g., a Radiation-Safety Officer or a corporate official) of an accident involving radioactive materials or radiation-producing equipment can be reason for dismissal. Furthermore, all radiation workers must be made aware, of their responsibilities as specified by federal regulations[1, 2] and other appropriate state and local regulations relative to their assigned duties, which involve radiation safety and the preparation of radiation reports.

Usually these responsibilities are clearly defined in the operating-procedures manual, and therefore the worker involved in a radiation accident should have received prior instructions as contained in the operating-procedures manual, augmented by instructions on radiation emergencies. He should know what to do to protect himself and his fellow workers in the event of an emergency, and whom to notify by using the procedure shown in Appendix 1. The periodic use of the check list (Appendix 6) will facilitate understanding and adherence to the emergency plan.

Once the notification of the radiation emergency has been received and initial action has been taken to get the situation under control, the magnitude of the accident must be determined in order to establish which of the various reporting procedures is to be followed in accordance with federal regulations[1-3] as well as appropriate state and local laws.

In order to aid key management personnel in determining the magnitude of the accident, we have prepared a table (see Appendix 7) for the evaluation of incidents involving radiation. The classification of the radiation accident can be established by reviewing the classification criteria and marking an "X" in the table following any statement that, at the time of the review, describes the magnitude of the situation. Depending upon the accident classification, as determined by the table, appropriate notification is required in accordance with the "Radiation-Incident Reporting Procedures" shown in Appendix 8.

It is important to note that there may be several official agencies (e.g., U.S. Atomic Energy Commission, Walsh-Healy Section of the U.S. Labor Department, state or local radiation-regulatory agencies) that must be notified in accordance with their specific regulations. The content of such notification should state the facts as briefly and concisely as possible. If possible, mention of the names of individuals involved in the incident should be avoided. Since such notification can become a public record, open to use by the news media, care should be taken in its preparation in order to comply with the existing regulations, yet not produce a severe public-relations problem. Naturally, copies of any notification are a vital form of documentation and should be carefully kept as a permanent record.

Summary

A radiation emergency can occur where radioactive materials or radiation equipment are used, stored, or transported. Plans for the prompt notification and documentation of essential facts should be established as part of the radiation emergency procedure. The value of such preparedness for the possibility of a radiation accident—however remote it may seem—will provide the radiation user with protection against unnecessary risks to personnel and reputations. The value of proper documentation of a radiation accident should not be underestimated, and the legal requirements of reporting such accidents within the prescribed reporting period must not be overlooked. In the case of individuals potentially exposed as a result of an accident, it is just as important to document the absence of such exposure as it is to measure it accurately. The methods of obtaining documentation and preparation of the necessary reports are described. Certain recommendations for reducing the possible risks resulting from radiation accidents are given.

RADIATION EMERGENCY PROCEDURES FOR USE BY RADIOGRAPHERS*

There may be an emergency situation arising from the use or transport of radioactive material, which will require immediate action. Emergency measures that must be taken to effectively cope with the emergency are so broad in scope that we could not list every step that must be followed to remedy the situation. Instead, we must place our confidence in the intelligence of mature, well-trained personnel who can apply their knowledge of radiological safety to deal effectively with emergencies.

* Presented as a part of a lecture during Picker X-Ray Corporation's Radiation-Safety Training Course in Industrial Radiography with Radioisotopes at the Picker Research Center, Cleveland, Ohio. Copyrighted 1964 by Applied Health Physics, Inc.

"*Think before you act!*" is the most important order to give personnel in attempting to cope with a radiation accident. There have been numerous instances of a minor radiation incident becoming a major accident through misguided and impulsive actions. Indecision by supervisors and lack of initiative to thoroughly evaluate and control a suspected radiation risk is another principal cause of additional losses following a radiation mishap. Unfortunately, to every list of causes of severe radiation accidents—such as equipment failure, human error, etc. —must be added "lack of emergency planning". Such planning can only start with those of us who have prepared the manual and investigated and reported the tragic results from radiation accidents. The actual establishment of practical emergency plans rests with responsible supervisors and radiographers who take the time to consider the following questions.

1. What can go wrong?
2. What action must be taken to avoid injury?
3. What remedial steps must be taken?
4. Who must be called for help?
5. What action must be taken to prevent a recurrence, not only here, but anywhere?

The time to think about what to do about a radiation accident is *before* it occurs. Take the time to consider the following list of common radiation emergency situations, and start your personnel planning to handle this along the lines described here.

1. Loss or theft of radioisotope source.
2. Malfunction of equipment.
3. Fire.
4. Accidental damage to equipment, survey meter.
5. Highway accident.
6. Human error.
7. Unauthorized or accidental entry into a radiation area.

Responsibility

It is the responsibility of each manager to see that all personnel—including the Radiation Safety Officer—have practical and well-understood plans that can be effective in controlling a radiation emergency.

The Radiation Safety Officer (RSO) has the responsibility to see that each radiation worker knows how to

a) limit or confine the radiation accident,
b) exclude all personnel from possible risk of exposure,
c) immediately contact his supervisor and/or the RSO for assistance.

Precautions must be taken, when evaluating the emergency, to minimize the potential hazards to individuals near the emergency and to personnel performing the recovery or corrective measures.

It is the responsibility of the Radiation Safety Officer (RSO) to take the measures necessary to cope with the emergency, and to report by telephone to the management immediately when the emergency exists or immediately after it has been remedied, depending on the seriousness of the emergency. Any abnormality must be reported, so that the necessary corrective action can be taken and described in the report considered for the AEC general information.

Notification Requirements

The seriousness of the emergency will determine what notification requirements will be followed. Included as Appendix 10 are the Notification Requirements as specified in AEC Regulation Title 10, Code of Federal Regulations, Part 20. Immediately after evaluating the seriousness of the emergency, the Radiation Safety Officer will report by telephone to his management any incident that caused, or threatened to cause, an emergency such as those listed in Appendix 10 (Notification Requirements 10–CFR–20).

The company official whose name appears on the AEC license, or his designated representative of top management, will aid the Radiation Safety Officer (RSO) in determining the seriousness of the emergency. They should then notify the proper agencies by telephone and telegraph of the conditions that exist and of the corrective steps that will be followed. Final analysis of the episode will be made by the RSO, who will report the facts to the proper agencies —such as the AEC or State Health Department—within 30 days, as prescribed by regulations.

When additional aid is needed, the RSO should call experienced health physicists, available through the sources given earlier.

Control of Radiation Accidents

Loss or Theft.

The Radiation Safety Officer (RSO) will notify the authorized personnel at job location and local authorities of the nature of the hazards that exist, and aid them in recovery. He must evaluate all facts reported to him, and attempt to locate the radioisotope equipment with the aid of a survey meter.

Malfunction of Equipment.

The Radiation Safety Officer (RSO) will retrieve the source or repair the equipment, depending on the circumstances involved. He must retrieve the source or repair the equipment in such a manner that it will not jeopardize his own safety, nor that of any other individual. The radiographer must restrict the area until the RSO arrives.

The following steps must be used when retrieving a loose or free source; they are performed only by the RSO.

Remote-Exposure Radiographic Devices.

1. Locate the position of the source in the tube with the aid of an operable survey meter by triangulation.
2. Evaluate the condition of the source tube.
3. Develop a recovery plan and calculate exposures for approximate recovery time.
4. Zero dosimeter pencil and position dosimeter and film badge on the body most likely to receive the maximum exposure.
5. Do not permit radiographer or authorized aides to receive more radiation than limited by the regulations.
6. Provided the source tube is not blocked or damaged, to retrieve the source by reverse method, do the following:
    a) lock exposure device;
    b) remove source tube tip from end of source tube, *provided source is not located at that point*; if so, elevate source tube, using a remote-handling tool or rod, and shake source down the tube, as far as possible from end of source tube, to a shielded tunnel; if source tube tip will not screw off, cut off with saw;
    c) remove control cables from exposure device;
    d) dosimeter pencils should be checked periodically at this point; zero pencil, if necessary, and record radiation received;
    e) position control cables in source tube tip, which is now open, and hold control cables against source tube tip;
    f) have radiographer or helper crank control cable out slowly to push source back to shielded position within exposure device;
    (*Caution:* crank slowly—excessive speed might override source and jam.)
    g) check dosimeter pencils and record readings; zero, if necessary;
    h) unlock exposure device;

i) have radiographer crank control cables slowly and lock source into position when in proper location within exposure device;
(*Caution:* make certain that source does not jump out of lock box.)
j) secure and lock the source;
k) check dosimeter pencils and record readings;
l) notify the RSO, and inform others in authority who were not aware of the emergency and remedial actions.
7. If nobody's dosimeter went beyond range, it is not considered a "reportable incident"; however, paragraph 8, following below, *must be observed.*
8. Immediately complete a detailed report in writing. The Radiation Safety Officer (RSO), will evaluate the incident thoroughly and will notify the AEC within 30 days, giving them details of the accident, provided it falls within their classification of a reportable accident.

Accidents Involving a "Free" Source (Open-Air Radiography Technique).
1. Check dosimeter pencil.
2. Using survey meter, locate source.
3. Develop a recovery plan.
4. Calculate exposure for approximate recovery time before attempting to make recovery.
5. Using remote-handling rod or improvised handling device (magnet on a pole, chewing gum on a stick, etc.), pick up source with rod and secure in exposure device.
6. Check dosimeter pencil.

Note: free sources may contain as much as 1 curie of ^{60}Co; radiation intensities will be quite high; dosimeter pencil must be read frequently and carefully, and personnel exposure limits must be watched.

7. If recovery cannot be made by the above method, then more consideration must be given to emergency; a recovery plan must be developed, which must be approved by the Radiation Safety Officer (RSO). Many complications may hinder safe recovery, but take the time to insure safety.
8. If assistance is needed, the Radiation Safety Officer will notify proper authorities. *Do not minimize the seriousness of the incident.* All facts must be considered for the maximum when evaluating the accident.

Fire.

Notify the fire company nearest your area, and give them the address of the location of the fire. Upon their arrival, caution firemen where radioactive materials are stored or where radioisotopes were being used; give them the present location of the radioactive material, and advise them of the best entrance route to the radiation area.

Notify the Radiation Safety Officer immediately, giving him complete details.

Do not permit firemen to enter the radiation area, after fire has been extinguished, until a thorough examination can evaluate the extent of the damage to the radioactive material.

Accidental Damage of Radiological Equipment.

The Radiation Safety Officer (RSO) will determine the extent of the damage, and analyze the recovery plan to repair damaged equipment. If it is obvious that the damaged equipment can be repaired without any risk to personnel, repair the equipment. A report must be prepared in writing.

Note: *under no conditions* can any repair be made to the actual radioisotope source or actual source housing—only to the accessories (e.g., cables, tubes, etc.).

Remember that the restricted area must be posted and kept under constant surveillance until the condition is remedied. The damaged device may be in such condition that special equipment may have to be constructed in order to retrieve the source safely.

Highway Accident.

Situations arising from this type of accident could vary tremendously, depending on the seriousness of the accident. However, maximum security must be maintained around the vehicle, and all bystanders must be excluded, until the situation can be fully evaluated. Top priority must always be given to care of injured and to saving of human life.

Road Accidents Involving Vehicles Containing Radioactive Materials.

See p. 813, "Recommendations for Local Authorities in Dealing with Incidents Involving Radioactive Materials". If a vehicle accident involves radioactive material and there are no injuries, a restricted area must be established, as specified in the following procedure.

If the survey meter is operable, use it to establish the perimeter of the restricted area.

If the survey meter is inoperable, use calculations to establish the perimeter of the restricted area, assuming that the source is in an exposed position inside of the vehicle. In case of a minor accident, where it can be visually determined that the source is safely stored in its container, no restricted area is required.

If the survey meter is operable, if no radiation hazard exists, and if the vehicle is movable, continue on to destination.

In any case, immediately after establishing the restricted area, have a responsible person notify your supervisor.

The Radiation Safety Officer shall notify the AEC (when required by 20.402 and 20.403 of 10–CFR–20).

If the survey meter has been damaged, and if a visual examination detects any indication of damage to the source container, the vehicle shall be isolated, and arrangements shall be made to secure an operable survey instrument.

Operable survey meters are available from a number of companies, from various colleges and universities, and from State Health Departments. Survey instruments can be borrowed or rented in case of an emergency by calling 412–563–2242.

Radiation warning signs must be used in accordance with AEC and state regulations.

Unauthorized or Accidental Entry to Restricted Area.

Gather all the facts and report to the Radiation Safety Officer as indicated below.
 1. Immediately by telephone, if possible overexposure is evident.
 2. In writing within 15 days, if no overexposure is evident.

The report must contain name and address, employer, telephone number, and social-security number of unauthorized person or persons involved, as well as the kind and quantity of by-product material, distance, time, and portion of body that was exposed.

Accidents Involving Possible Overexposure.

If radiation exposure has occurred, or is suspected to have occurred, proceed as follows.
 1. Remove personnel involved from radiographic operations immediately.
 2. Send film badge to supplier for immediate processing. Complete report in quadruplicate (see Applied Health Physics, Inc.'s Form RR–20); send two (2) copies with badge(s), one (1) to your consultant, and keep one (1) on file.
 3. Have the person or persons complete a statement of facts on how exposure occurred and send it to the Radiation Safety Officer.
 4. Upon receipt of all facts and exposure results, your consultant and the Radiation Safety Officer will determine what action must be taken before the person or persons can return to normal radiographic operations.

PROCEDURES FOR GENERAL EMERGENCY PLANNING AND ACTION

Orientation Information for Fire and Police Officials

It is difficult for fire and police officials to accurately evaluate the magnitude of a radiation risk at the time of an emergency without having had prior information about the user's operations, location of radiation sources, and the potential risks that may be involved during a fire emergency. It is suggested that the radiation user use a form similar to the one below for discussion with these officials. They can fill in the information on the items below from the information given to them by the licensee's health physicist or Radiation Safety Officer (RSO).

ORIENTATION SESSION ON RADIATION ACCIDENT PREPAREDNESS

Date: _____ Location: _____

Names of Persons Attending:

_____ _____ Title
_____ _____ Title
_____ _____ Title
_____ _____ Title
_____ _____ Title
_____ _____ Title

Session Conducted by:

_____ _____ Title

A. Description of Normal Operations Involving Radiation

1. Radiation is used for: _____
2. Types of radiation emitted: _____
3. Description of radiation sources: _____
4. Location of radiation sources: _____
5. Radiation used licensed by: _____
6. Radiological safety under direction of:

    Name: _____

    Title: _____

7. Radiation-safety inspections made by: _____
8. Frequency of surveys: _____
9. Date of last inspection by a health physicist: _____
10. Security measures maintained on radiation sources: _____
11. Types of radiation warnings (alarm systems, interlock systems, warning signs, etc.): _____

B. Radiation-Hazard Evaluation During an Emergency
1. Are the containers used to shield radioactive material fire resistant? _____
2. Are all sources shielded? _____
3. Are the radioisotope containers apt to be affected by fire releasing radioactive air contamination? _____
4. Should the firemen wear self-contained breathing apparatus? _____
5. Should the firemen practice contamination control? _____
6. Is the radioactive material itself combustible or flammable material? _____
7. Will the radioactive material be affected chemically by smoke, heat, or water? _____
8. Should the firefighters use radiation instruments before entering, or not? _____
9. Types and locations of radiation instruments (at site): _____
10. Are spare instruments located outside radiation areas? _____
11. Auxiliary monitoring instrumentation can be obtained from: _____
12. If personnel-monitoring instruments (dosimeters and film badges) are required, they can be obtained from: _____

C. The Following Personnel Are Qualified to Advise Fire and Police Officials:

Name:	Telephone Numbers	
	Day	Night
_____	_____	_____
_____	_____	_____
_____	_____	_____
_____	_____	_____
_____	_____	_____
_____	_____	_____

Procedures for Radioactive Spills

Emergencies will generally be in the nature of spills, fires, or explosions, as a result of which radioactive materials are spread around the installation. Emergency procedures adopted from *NBS Handbook 48* are given here as a guide. It must be recognized that these procedures are general, and any specific emergency will call for further adaptations and changes in procedure.

Major Spills, Involving Radiation Hazards to Personnel.
1. Notify all persons not involved in the spills to vacate the room at once. Limit the movement of displaced persons to confine the spread of contamination.
2. If the spill is liquid and the hands are protected, right the container; otherwise, use a stick or lever.
3. If the spill is on the skin, flush thoroughly.
4. If the spill is on clothing, discard outer or protective clothing at once.
5. Switch off all fans.
6. Vacate the rooms.

Radiation Protection and Regulations

7. Notify the Radiation Safety Officer as soon as possible.
8. Take immediate steps to decontaminate personnel involved as necessary.
9. Decontaminate the area (personnel involved in decontamination must be adequately protected). The Radiation Safety Officer will direct the decontamination.
10. Monitor all persons involved in the spill and cleaning.
11. Permit no person to resume work in the area without the approval of the Radiation Safety Officer.
12. A complete history of the accident and subsequent activity must be submitted to the Radiation Safety Officer.

Minor Spills, Involving No Radiation Hazard to Personnel.
1. Notify all other persons in the room and area at once.
2. Survey people before they become dispersed, and change clothes as necessary.
3. Permit only the minimum number of persons necessary to deal with the spill into the area.
4. Confine the spill immediately.
   A. Liquid Spills.
      Don protective gloves.
      Drop absorbent paper on spill.
   B. Dry Spills.
      Don protective gloves.
      Dampen thoroughly, taking care not to spread the contamination. Water may generally be used, except where chemical reaction with the water would generate an air contaminant; oil should be used instead.
5. Decontaminate; make a plan first.
6. A complete history of the accident and subsequent remedial or protective measures must be submitted to the Radiation Safety Officer.

Fires or Other Major Emergencies

1. Notify all other persons in the room and building at once.
2. Notify the fire department and other local plant safety personnel as well as the Radiation Safety Officer.
3. Attempt to put our fires by approved means if radiation hazard is not immediately present.
4. Govern fire fighting or other emergency activities by the restrictions of the Radiation Safety Officer. Avoid, if possible, the tracking of contamination or passing of contaminated equipment into clean areas by emergency workers.
5. Monitor all persons involved in combating the emergency.
6. Following the emergency, monitor the area and determine the protective devices necessary for safe decontamination.
7. Decontaminate; follow a plan.
8. Permit no person to return to work without the approval of the Radiation Safety Officer.
9. Prepare a complete history of the emergency and subsequent activity related thereto for the Radiation Safety Officer.

Decontamination Techniques

Decontamination techniques are needed for personnel and areas. In cleaning objects and areas, the initial step depends on whether the contaminant is in powder or liquid form. If the material is dry or powdered, vacuuming is a most valuable technique, since much adherent material can be removed, lessening the chance for penetration into the surfaces when wetting agents are applied. A suitable method of filtration of effluent air from the vacuum cleaner must be provided, so that there is no further spread of radioactivity. Damp wiping and mopping with water and detergent are the next steps. If the chemical characteristics of the contaminant are not known, detergents of neutral pH are preferable to soaps, which—in some instances—may cause fixation of certain nuclides rather than removal. Complexing agents, e.g., citric acid or chelating agents (EDTA or DTPA) in combination with detergent or soap increase the cleaning efficiency; the action of chelating agents is accelerated by warming. Occasionally, weak hydrochloric or nitric acid may be of value. The procedure for decontamination is given below*.

Preoperational.
1. Plan the decontamination operation thoroughly, and obtain supplies.
2. Provide adequate protection for all decontamination personnel, and allow for replacements.
3. Provide safe storage of all radioactive wastes and decontamination supplies.

Operational.
1. Always work towards the center of contamination.
2. Take care not to spread or track contamination to cleaner (lower-activity) areas.
3. Monitor frequently and thoroughly.
4. Cover clean areas with plastic sheets, kraft paper, or its equivalent.
5. Monitor all personnel and materials before permitting their movement to clean areas.

Postoperational.
1. Quarantine all used cleaning solutions and decontamination equipment until they can be monitored.

Some Recommendations for Disaster Preparedness at Hospitals†

1. An interhospital-communication program should be organized to aid in orderly transportation and assignment of patients in order to distribute the community load as equally as possible.
2. Control of traffic in the involved area is essential in order to implement care of the injured and minimize confusion of volunteers and spectators.
3. Hospital receiving areas should be designed to permit ease of access and egress of ambulances, so that the injured can be unloaded without requiring the vehicle to turn around.
4. Emergency patient identification and treatment tags must be available; recording of therapy should be provided.
5. Physicians trained in triage should be continually on duty.
6. Standard treatment routines for common injuries must be utilized.
7. Relatively well and convalescent patients should be sent home or to hotels or dormitories, where simpler care facilities can be employed.
8. X-rays may often be postponed for 24 to 48 hours, until the emergency has been stabilized.
9. Blood bank facilities should be provided sufficiently far from the hospital to minimize the flow of people through the institution.

---

* Radiation protection guides for skin contamination are not rigidly established; suitable levels will depend on the types of radionuclides present.
† From: Saenger, E. L., ed., *Medical Aspects of Radiation Accidents*. U.S. Government Printing Office, Washington, D.C., 1963

10. Public information via press and radio should be planned so that no excessive quantities of blood are obtained.
11. Hospitals should be provided with unlisted telephone lines, so that important emergency calls can be made and received.
12. Emergency morgue and embalming procedures should be prepared.
13. A careful public-relations program for the hospital is required, with attention to the following aspects:
    a) transmission of information to the public by press, radio, and television;
    b) identification of dead and injured;
    c) tactful handling of volunteers and crowds of curious onlookers;
    d) community reaction towards the hospital on a long-term basis; public relations are most important in minimizing resentment by individuals and the community at large.

Recommendations for Local Authorities in Dealing with Incidents Involving Radioactive Materials

The handling of such incidents, with maximum protection to emergency personnel, will be aided if the following rules are observed.

1. The state's Director of Radiological Health and/or the nearest Atomic Energy Commission office should be notified as soon as possible (see Appendix 4).
2. If the incident involves wreckage and a person is believed to be alive and entrapped, every possible effort should be made to rescue him.
3. The area of the accident should be restricted. The public should be kept as far from the scene as is practical. Local authorities should make only necessary entries and investigation in the accident area. No attempt should be made to clean up any debris or material involved in the accident.
4. Any persons who have had possible contact with the radioactive material should be segregated and confined until they can be examined further. The names and addresses of those involved should be obtained.
5. The injured should be removed from the area of the accident with as little contact as possible and held at a transfer point. All lifesaving measures should be performed promptly, but elective first aid and surgical procedures should be delayed until advice or help can be obtained from a physician familiar with radiation medicine. Except in extreme emergency, patients should not be moved to a local hospital or doctor's office before a radiological survey has been made.
6. If the incident involves fire, attempts to extinguish it should be made from as great a distance as possible, avoiding smoke, fumes, or dust as much as possible. The fire should be treated as one involving toxic chemicals. Suspected material should not be handled until it has been monitored and released by monitoring personnel. Clothing and tools used at the fire should be segregated until they can be checked by emergency monitoring teams.
7. In the event of a radiological incident involving a vehicle accident, all traffic should be detoured around the scene of the accident. If this is not possible, vehicles should be moved the shortest distance necessary to clear right of way. If radioactive material is spilled, passage through the area should be prevented unless absolutely necessary. If right of way must be cleared before radiological assistance has arrived, the spillage should be covered with dirt or washed to the shoulders of right of way with minimum dispersal of wash water.
8. Eating, drinking, or smoking in the area of the accident should be prohibited.

Food or drinking water that may have been in contact with material from the accident should not be used.
9. Ambulances or other evacuation vehicles should not be permitted to travel faster than prevailing speed limits. They should not be returned to normal use until a radiological survey has been made of them.

Emergency Action for X-Ray Accidents

1. Shut off power to the unit.
2. Have exposed individuals examined by a physician.
3. Obtain and record all details of the incident.
4. Consult appropriate experts to determine the extent of the hazards. Except for obvious first aid, do not treat the patient until the dose received has been determined.
5. There will not be any induced radioactivity in the patient in such an accident.

Emergency Procedures for Incidents Involving Teletherapy Equipment

The emergency procedures to be used in case of shutter malfunction of a teletherapy unit depend on the individual installation. The following is an example.

If the warning signals indicate that the beam-control system has failed to terminate irradiation at the preset time, the following steps are to be carried out in a calm manner.

1. Turn off the main switch at the control (unless this interferes with the removal of the patient).
2. Open the door to the treatment room.
3. If the patient is ambulatory, direct him to leave the room. If the patient is not ambulatory, enter treatment room, avoiding exposure to the useful beam as much as possible, and try to close the manual emergency beam control; if not immediately successful, remove the patient from the treatment room.
4. Close and lock the door.
5. Notify the Radiation Safety Officer.
6. No attempt to repair equipment should be made by inexperienced persons.
7. Repairs should be made only by experienced technicians under the control and/or supervision of the Radiation Safety Officer.

Accidents Involving Radioactive Dusts, Mists, Fumes, Organic Vapors, and Gases

1. Notify all other persons to vacate the room immediately.
2. Hold breath and close air vents.
3. Vacate the room. Seal off area, if possible.
4. Notify the Radiation Safety Officer at once.
5. Ascertain that all doors giving access to the room are closed. Post conspicuous warnings or guards to prevent accidental opening of the doors.
6. Monitor all persons suspected of contamination. Proceed with decontamination of personnel.
7. Report at once to the Radiation Safety Officer all known or suspected inhalations of radioactive materials.
8. Evaluate the hazard and the safety devices necessary for safe reentry.
9. Determine the cause of contamination and rectify the condition.
10. Decontaminate the area only upon the advice of the Radiation Safety Officer.
11. Perform an air survey of the area before permitting work to be resumed.
12. Submit a complete history of the accident and subsequent activities to the Radiation Safety Officer.

Injuries to Personnel, Involving Radiation Hazard

1. Wash minor wounds immediately under running water, spreading the edges of the gash.
2. Report all radiation accidents involving personnel (wounds, overexposure, ingestion, inhalation) to the Radiation Safety Officer as soon as possible.
3. Call at once a physician qualified to treat radiation injuries.
4. Permit no person involved in a radiation injury to return to work without the approval of the attendant physician and the Radiation Safety Officer.
5. Prepare a complete history of the accident and subsequent activity related thereto for the Radiation Safety Officer.

REFERENCES

1. Code of Federal Regulations, Title 10, Atomic Energy; Part 20, Standards for Protection Against Radiation. *Federal Register*. U.S. Government Printing Office, Washington, D.C., 1960.
2. Code of Federal Regulations, Title 41–CFR–Part 50–204, Safety and Health Standards for Federal Supply Contracts, The Walsh-Healy Act. *Federal Register*. U.S. Government Printing Office, Washington, D.C., 1964.
3. U.S. Atomic Energy Commission, Division of Operational Safety, Reporting and Investigating Accidents and Radiation Exposures, *U.S. Atomic Energy Commission Manual*, Appendix 0502. U.S. Government Printing Office, Washington, D.C., 1962.
4. Saenger, E. L., ed., *Medical Aspects of Radiation Accidents*. U.S. Government Printing Office, Washington, D.C., 1963.
5. Brodsky, A. and Beard, G. V., *A Compendium of Information for Use in Controlling Radiation Emergencies* U.S. Atomic Energy Commission, Washington, D.C., 1960.
6. National Committee on Radiation Protection and Measurement, *Exposure to Radiation in an Emergency*, Report No. 29, Section of Nuclear Medicine. The University of Chicago, Chicago, Illinois, 1962.
7. Morgan, K. Z. and Turner, J. E., eds., *Principles of Radiation Protection*. John Wiley and Sons, New York. 1967.

GENERAL REFERENCES ON RADIATION EMERGENCIES

Bahler, K. W., et al., *Prevention and Handling of Radiation Emergencies, K-1436*. Office of Technical Information, U.S. Atomic Energy Commission, Washington, D.C.

Gallaghar, R. G. and Saenger, E. L., Radium Capsules and Their Associated Hazards. *Amer. J. Roentgenol. Radium Therapy, Nucl. Med.* 77: 511–513, 1957.

Lanzl, L. H., Pingel, J. H., and Rust, J. H., eds., *Radiation Accidents and Emergencies in Medicine, Research, and Industry*. Charles C Thomas, Springfield Illinois, 1965.

Oerlein, K. F., *Radiological Emergency Procedures for the Non-Specialist*. U.S. Atomic Energy Commission, Washington, D.C., 1964.

Saenger, E. L., Planning for a Radiation Accident. *Amer. Ind. Hyg. Ass. J.*, 20: 482–487, 1959.

Saenger, E. L., Radiation Accidents. *Amer. J. Roentgenol., Radium Therapy, Nucl. Med.*, 84: 715–728, 1960.

U.S. Atomic Energy Commission Clearing House for Federal Scientific and Technical Information, *Radiological Emergency Operations* (Instructor's Manual *TID-24918*, Student's Manual *TID-24918*). National Bureau of Standards, U.S. Department of Commerce, Springfield, Virginia, 1969.

U.S. Atomic Energy Commission and HEW, Peacetime Radiation Hazards in the Fire Service. *Basic Course Resource Manual OE-84019*. U.S. Government Printing Office, Washington, D.C., 1961.

APPENDIX 1
EMERGENCY NOTIFICATION INSTRUCTIONS

These instructions are to be read and followed by all employees in the event of an emergency.

Immediate notification of an emergency is to be given by telephoning in numerical order:

Organization or Individual   **Telephone No.**

Fire.

1. Fire Department
_____ _____
                (Location)

2. RSO
_____ _____
                (Name)   (Home)

3. Other
_____ _____

Information Given by Caller
1. Your name
2. Company name
3. Company address
4. Location of fire
5. Entrance firemen should use
6. Caution, if radioisotopes are involved
7. Warn, if airborne contamination, to use masks

Injury.

1. Regular Physician
Dr._____ _____
                                                                  (Office)

                                                                  (Office)

                                                                  (Home)

2. Alternate Physician
Dr._____ _____
                                                                  (Office)

                                                                  (Office)

                                                                  (Home)

or
Emergency room of _____ Hospital

3. Company Official
_____ _____

Information Given by Caller
1. Extent of injury
2. Present condition
3. First-aid measures taken

Accidents.

1. No Injuries, No Radiation

   _____   _____
              (Company Official)

2. Radiation Involved

   _____   _____
                 (RSO)   (Home)

3. Off-Site Accidents

If accident occurred off-site (highway, etc.), have someone maintain security over radiation source and keep personnel away while you call "collect" the individuals named above *plus* highway patrol and/or police.

Information Given by Caller
1. What happened
2. When
3. Where
4. Who was involved
5. What has been done to control loss
6. Current status
7. Where you can be reached.

APPENDIX 2
RADIATION-INCIDENT
EVALUATION RECORD

In the event of a radiation accident, certain essential facts must be obtained as promptly and accurately as possible. These facts are needed to estimate the magnitude of the incident, limit the extent of the damage, and institute remedial measures. Try to obtain these facts calmly; use a recorder, if possible. Be sure to note specific details on this sheet, including the time the information was received and the name of the person who furnished it.

Initial Evaluation Record

Question	Response	By	Time
1. What happened?			
2. When did it occur? (time, date)			
3. Where did it happen? (building, floor, area)			
4. Who was involved? (names, employer)			
5. Who was exposed or injured? (name, extent of exposure or injury)			
6. Where are the injured or exposed now?			
7. How much damage to facilities?			
A. Was damage confined to company property?			
B. What damage was done to property of others?			
8. Is radioactive contamination a problem?			
A. If so, how extensive is contamination? (on-site, off-site)			
B. What is being done to control the contamination?			
9. Who, other than I, has been notified?			
10. Where can you be reached, if you are needed?			

APPENDIX 3
RADIATION-INCIDENT
EVALUATION RECORD

Preliminary Evaluation

1. Is outside help (fire, police, AEC) required? _____
   (If so, follow Emergency Notification Instructions)
2. Is medical assistance required? _____
3. Should personnel be evacuated? _____
   - A. From the incident area or building? _____
   - B. From the site? _____
   - C. From the locations off-site? _____
4. Has the Radiation Emergency Plan been activated? _____
   - A. Was the health-physics consultant notified? _____
   - B. If so, is additional help required? _____
   - C. Were decontamination procedures mobilized? _____
5. What is the accident classification? _____
6. Have official agencies (AEC, state) been notified as required? _____
7. What are the immediate problems? _____
   - A. Off-site? _____
   - B. On-site? _____
8. What are the long-range problems? _____
   - A. Off-site? _____
   - B. On-site? _____
9. Estimate the duration of these problems. _____
   - A. Immediate _____
   - B. Long-range _____
10. Have the families of the injured been contacted? _____
11. What information should be given to the employees? By whom? _____
12. Has a newspaper release been prepared by company officials authorized to disseminate information to news media? _____

APPENDIX 4
U.S. ATOMIC ENERGY COMMISSION REGIONAL OFFICE AREAS OF RESPONSIBILITY FOR RADIOLOGICAL ASSISTANCE

Region No.	Operations Office	Post Office Address	Telephone for Assistance	
			Area Code	Number (Extension)
1	New York	476 Hudson Street New York New York 10014	212	989–1000
2	Oak Ridge	P.O. Box E Oak Ridge Tennessee 37830	615	433–8611 (3–4510)
3	Savannah River	North Augusta South Carolina 29841	803	824–6311 (3333)
4	Albuquerque	P.O. Box 5400 Albuquerque New Mexico 87115	505	264–4667
5	Chicago	9800 South Cass Ave. Argonne Illinois 60439	312	739–7711 (2111) nights and holidays (4401)
6	Idaho	P.O. Box 2108 Idaho Falls Idaho 83401	208	526–0111 (1515)
7	San Francisco	2111 Bancroft Way Berkeley California 94704	415	841–5121 nights and holidays 841–9244
8	Hanford	P.O. Box 550 Richland Washington 99352	509	942–1111 (6–5441)

Radiation Protection and Regulations

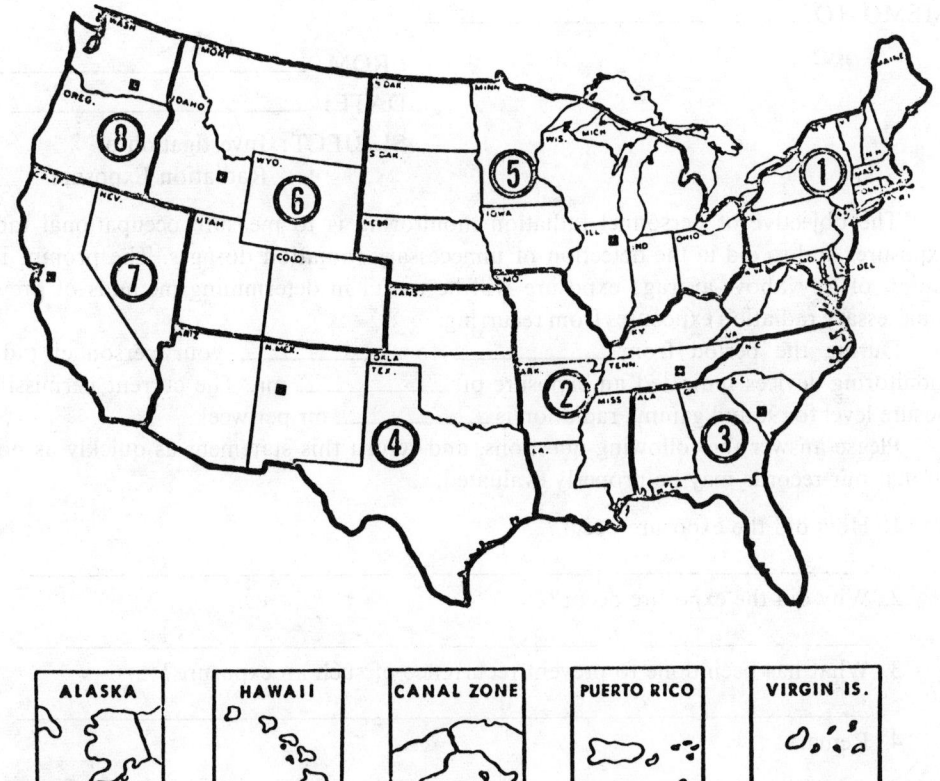

APPENDIX 5
EVALUATION OF
PERSONNEL-MONITORING RESULTS

MEMO TO: _____

                                        FROM: _____
                                        DATE: _____
                                        SUBJECT: Investigation of
                                                             Radiation Exposure

    The objective of personnel radiation monitoring is to measure occupational radiation exposures and to aid in the detection of unnecessary radiation dosages. The prompt investigation of any above-average exposure will be useful in determining methods of preventing unnecessary radiation exposures from recurring.

    During the period from _____ to _____, your personnel radiation-monitoring devices indicated an exposure of _____ mr. The current permissible exposure level for X and gamma radiation is _____ mr per week.

    Please answer the following questions, and return this statement as quickly as possible, so that our records may be properly evaluated.

    1. How did the exposure occur?
        _____

    2. Why did the exposure occur?
        _____

    3. What has been done to prevent recurrence of such an exposure?
        _____

    4. Remarks:
        _____
        _____
        _____
        _____

Signed _____   Date _____
                        (Employee)

Forwarded _____   Date _____
                        (Supervisor)

APPENDIX 6
A SUGGESTED CHECK LIST ON RADIATION ACCIDENT PREPAREDNESS

The following items are some, but not necessarily all, of the essential points that should be considered when formulating an emergency plan. These items should be included in periodic evaluations of radiation users and their readiness for an emergency that—either directly or indirectly—may involve radiation. The current status of each item should be designated as "Satisfactory" (S), "Marginal" (M), or "Unsatisfactory" (U), and appropriate action should be taken to achieve the desired degree of protection.

Item	Status (S,M,U)
Management	
1. Channel of command clearly established:	
2. Emergency Procedures well defined and accepted:	
3. Copy of Emergency Procedures on hand, for key management personnel to use	
a) at office:	
b) at home:	
4. Communication channel established and posted:	
A. Plan designates *who* to call, and gives office and home telephone numbers:	
B. Plan tells *what* information must be given:	
C. Plan specifies a key employee as Information Officer authorized to cope with all official communications (e.g., news media) other than emergency calls:	
5. Incident-Evaluation Plan formulated and ready:	
6. Damage-Control Plan formulated and ready:	
7. Medical-Assistance Plan formulated and ready:	
8. Incident-Notification Plan formulated and ready:	
9. Recovery Plan formulated and ready:	
Health Physics; Safety and Security	
1. Supervisor and radiation workers familiar with emergency procedures:	
2. Copies of Emergency Procedures on hand at office and at home:	
3. Emergency equipment and instrumentation on "stand-by":	
A. Rescue equipment available and accessible:	
B. Portable instruments strategically stored:	
C. Extra personnel-monitoring equipment and protective clothing on "stand-by":	

Item	Status (S, M, U)

4. Channel of command clearly understood: _____

5. Emergency communications network planned and in readiness: _____

6. Radiation inventory current, and radiation locations well defined: _____

7. Radiation-Risk Evaluation Plan and data sheets formulated and ready: _____

8. Orientation on possible radiation emergencies given to
    a) supervisors and all radiation workers: _____
    b) medical personnel to be called upon in an emergency: _____
    c) local fire officials: _____
    d) local police officials: _____
    e) local public-health officials: _____

9. In the event of an emergency (e.g., after hours), the above (Item 8, a to e) know *what* to do, *whom* to contact, and *how* to cope with the incident as it pertains to their responsibilities: _____

10. Auxiliary personnel available to assist:
    A. Health physics: _____
    B. Medical: _____
    C. Security: _____
    D. Services (temporary power, procurement, communications): _____

Radiation Workers

1. Simplified Radiation-Emergency Plan understood: _____
    A. Radiation warnings and alarms known: _____
    B. Evacuation signal and route known: _____
    C. Reassembly points (in safe area) specified: _____

2. Channels of command clearly understood: _____

3. Damage-Control Measures clearly stated: _____

4. Emergency Plan practiced at least once a year: _____

5. Each worker has received safety training relative to his radiological-safety responsibilities, and is familiar with the written operating procedures and regulations governing the routine and emergency conditions affecting his work: _____

Medical Services

1. Availability of medical assistance for the type of potential injuries from this installation: _____
    A. Specific M.D.'s have reviewed potential injuries, including those which may involve radiation: _____
    B. Plans are formulated and in readiness for
        a) removal of injured to specific hospitals: _____
        b) notification of physician, emergency medical facilities: _____

2. Specific measures are to be taken to
    a) provide prompt and effective medical aid for injured: _____

Item	Status (S, M, U)
b) control radiation risks to medical team:	
c) furnish radiation exposure evaluation data to physicians:	
3. Key management and health physicists have established communication network (e.g., unlisted telephone numbers for emergency calls to hospital officials, M.D., etc.):	
4. Adequate health-physics assistance (certified health physicist or radiation physicist) is available, should the M.D. require a consultation:	

APPENDIX 7
EVALUATION OF INCIDENTS OR OCCURRENCES INVOLVING RADIATION

Accident Description	Accident Classification					
	For AEC Licensees, per 10–CFR–20			For AEC Operations, per AEC Manual		
	a	b	c	A	B	C
Type A: Serious Incident, Immediate Notification Required						
For AEC licensees and/or AEC contractors:						
1. Loss or theft of licensed radioactive material	x			x		
2. Exposure of individual's whole body to 25 rems or more	x			x		
3. Exposure of individual's skin of whole body to 150 rems or more	x			x		
4. Exposure of individual's hands and forearms or feet and ankles to 375 rems or more	x			x		
5. Release of radioactive materials in concentrations which, if averaged over 24 hours, would exceed 5,000 times the limits specified in Appendix B, Table II of Title 10–CFR–20 (see Table 19, pp. 616 to 633)	x			x		
6. Loss of one workweek or more of the operation of any facility	x			x		
7. Damage to property exceeding $100,000.00	x			x		
In addition to the above, the following applies *only* to AEC installations and prime contractors:						
8. Any incident involving an atomic weapon that results in injury or damage to private property				x		
9. Any notice that an individual may have received 25 rems or more in a calendar year				x		
10. Any injury or illness diagnosed by an M.D. competent in nuclear medicine as having conceivably resulted from cumulative or massive radiation exposure				x		
11. Allegations the ex-employees of the AEC or its contractors are disabled as a result of exposure to toxic materials or radiation related to atomic-energy operations				x		
12. Any injury or industrial illness of 5 or more persons in AEC operation				x		
13. Any accident or radiation exposure that creates a significant public-relations problem				x		
14. Any off-site accident that involves vehicles carrying AEC shipments of radioactive materials				x		
Type B: Reportable Accident, 24-hour Notification Required						
1. Exposure of individual's whole body to 5 rems or more		x				
2. Exposure of individual's skin of whole body to 30 rems or more		x				

	Accident Classification					
Accident Description	For AEC Licensees, per 10–CFR–20			For AEC Operations, per AEC Manual		
	a	b	c	A	B	C
3. Exposure of individual's hands and forearms or feet and ankles to 75 rems or more		x				
4. Release of radioactive materials in concentrations which, if averaged over 24 hours, would exceed 500 times the limits specified in Appendix B, Table II, Title 10–CFR–20 (see Table 19, pp. 616 to 633)		x				
5. Loss of one workday or more of the operation of any facility		x				
6. Property damage exceeding $1,000.00		x				
7. Incident may not result in any of the above, but may create serious public-relations problems		x*				

Type C: Minor Reportable Occurrences, Notification Required Within 30 Days

	a	b	c	A	B	C
1. Radiation exposure of any individual in excess of limits specified in Title 10–CFR–20 or in AEC license conditions			x			
2. Radiation levels or radioisotope released in unrestricted areas, which have not resulted in excessive exposures, but have been in excess of 10 times any applicable limit set forth in Title 10–CFR–20 or in the conditions specified in an AEC license			x			
3. Bodily injury, property damage, or claim connected with the possession or use of radioactive material at a facility or in the course of transportation which would pertain to an indemnity agreement between the licensee and the AEC			x			

* AEC does not *require* notification, but it is suggested that they be notified.

APPENDIX 8
RADIATION-INCIDENT
REPORTING PROCEDURES

REPORTING OF INCIDENTS

Type A incidents require IMMEDIATE reports to the U.S. Atomic Energy Commission, the U.S. Department of Labor, and—in certain states—to appropriate officials in the Health Department.

Type A-1

1. Immediately telephone the Director of the appropriate Atomic Energy Commission Regional Compliance Office listed in Appendix D of 10–CFR–20.
2. Confirm the telephone report with a telegram.
3. Within 30 days, the licensee must make a formal written report of the incident to the Director, Division of Compliance, U.S. Atomic Energy Commission, Washington, D.C. 20545, with a copy to the U.S. Atomic Energy Commission Regional Compliance Office listed in Appendix D of 10–CFR–20.
4. It is also recommended that the licensee maintain a close liaison by telephone with the nearest U.S. Atomic Energy Commission Compliance Office during the first few days following discovery of the incident.

Type A-2

Refer to those requiring Atomic Energy Commission contractors and Atomic Energy Commission installations to report IMMEDIATELY to the Atomic Energy Commission headquarters responsible for these operations. Specific instructions are contained in the Atomic Energy Commission Appendix 0502 of the U.S. Atomic Energy Commission's Operational Safety Handbook. These procedures do not apply to the average radioisotope licensee.

REPORTING OF INCIDENTS
REQUIRING REPORTS WITHIN 24 HOURS

1. The licensee must notify the Director of the appropriate U.S. Atomic Energy Commission Division of Regional Compliance Office listed in Appendix D of 10–CFR–20. Within 24 hours after learning of an accident that may have caused, or threatens to cause, those effects listed in Part B, shall report as follows.
    A. Telephone the U.S. Atomic Energy Commission Director of Compliance within 24 hours and report the incident.
    B. Confirm the telephone call by telegram within 24 hours.
    C. Provide a written report of the incident within 30 days to the Director of the U.S. Atomic Energy Commission Regional Compliance Office, Washington, D.C. 20545, with a copy to the appropriate Atomic Energy Commission Compliance Office. This written report must

describe the extent of the exposures, levels of concentrations, cause of exposure, and corrective steps taken to avoid a recurrence.

MINOR INCIDENTS

Part C of the classification outline requires reports within 30 days. A report should be sent to the Director, Division of Compliance, U.S. Atomic Energy Commission, Washington, D.C. 20545, with a copy to the appropriate Atomic Energy Commission Regional Compliance Office. This report must describe the extent of the exposures, levels of concentrations, cause of exposure, and the corrective measures taken to avoid a recurrence.

REPORTING TYPE B AND TYPE C INCIDENTS

The reporting of Type B and Type C incidents by Atomic Energy Commission contractors and Atomic Energy Commission operations is specifically covered in the Atomic Energy Commission Manual, Appendix 0502, Parts II and III, and should be followed as necessary.

OTHER ACCIDENTS AND OCCURRENCES THAT MUST BE INCLUDED IN ROUTINE REPORTS BY ATOMIC ENERGY COMMISSION CONTRACTORS (ALSO KNOWN AS TYPE C)

1. All accidents normally reported quarterly and annually.
2. Disabling occupational injuries.
3. Property damage or loss resulting from fires or other accidental causes.
4. All motor vehicle accidents (government vehicles only).

APPENDIX 9
U.S. ATOMIC ENERGY COMMISSION REGIONAL COMPLIANCE OFFICES

	Address	Telephone	
		Daytime	Nights and Holidays
Region I			
Connecticut, Delaware, District of Columbia, Maine, Maryland, Massachusetts, New Hampshire, New Jersey, New York, Pennsylvania, Rhode Island, Vermont	Division of Compliance, USAEC 970 Broad Street Newark, New Jersey 07102	201–645–3960	212–989–1000
Region II			
Alabama, Arkansas, Florida, Georgia, Kentucky, Louisiana, Mississippi, North Carolina, Panama Canal Zone, Puerto Rico, South Carolina, Tennessee, Virginia, Virgin Islands, West Virginia	Division of Compliance, USAEC Suite 818 230 Peachtree Street, N.W. Atlanta, Georgia 30303	404–526–4537	404–526–4537
Region III			
Illinois, Indiana, Iowa, Michigan, Minnesota, Missouri, Ohio, Wisconsin	Division of Compliance, USAEC 799 Roosevelt Road Glen Ellyn, Illinois 60137	312–858–2660	312–858–2660
Region IV			
Colorado, Idaho, Kansas, Montana, Nebraska, New Mexico, North Dakota, Oklahoma, South Dakota, Texas, Utah, Wyoming	Division of Compliance, USAEC 10395 West Colfax Avenue Denver, Colorado 80215	303–297–4211	303–237–5095
Region V			
Alaska, Arizona, California, Hawaii, Nevada, Oregon, Washington, U.S. territories and possessions in the Pacific	Division of Compliance, USAEC 211 Bancroft Way Berkeley, California 94704	415–841–5121 Ext. 651	415–841–9244

APPENDIX 10
RECORDS, REPORTS
AND NOTIFICATIONS PERTINENT
TO RADIATION INCIDENTS

The following excerpts were taken from Title 10, Code of Federal Regulations, Part 20, "Standards for Protection Against Radiation".

§20.401 Records of Surveys, Radiation Monitoring, and Disposal.

(a) Each licensee shall maintain records showing radiation exposures of all individuals for whom personnel monitoring is required under §20.202 of the regulations in this part. Such records shall be kept on Form AEC-5, in accordance with the instructions contained in that form, or on clear and legible records containing all the information required by Form AEC-5. The doses entered on the forms or records shall be for periods of time not exceeding one calendar quarter.

(b) Each licensee shall maintain records in the same units used in the appendices to this part, showing the results of surveys required by §20.201(b), and disposals made under §§20.302, 20.303, and 20.304.

(c) Records of individual radiation exposure which must be maintained pursuant to the provisions of this subsection shall be preserved until December 31, 1970, or until a data five years after termination of the individual's employment, whichever is later. Records which must be maintained pursuant to this part may be maintained in the form of microfilms.

> Note: prior to December 31, 1970, the Commission may (and probably) will amend this paragraph to assure the further preservation of records which it determines should not be destroyed.

§20.402 Reports of Theft or Loss of Licensed Material.

Each licensee shall report by telephone and telegraph to the Director of the appropriate Atomic Energy Commission Regional Compliance Office listed in Appendix D, immediately after its occurrence becomes known to the licensee, any loss or theft of licensed material in such quantities and under such circumstances that it appears to the licensee that a substantial hazard may result to persons in unrestricted areas.

§20.403 Notifications of Incidents.

(a) Immediate notification. Each licensee shall immediately notify the Director of the appropriate Atomic Energy Commission Regional Compliance Office shown in Appendix D by telephone and telegraph of any incident involving by-product, source, or special nuclear material possessed by him, and which may have caused, or threatens to cause,

(1) exposure of the whole body of any individual to 25 rems or more of radiation; exposure of the skin of the whole body of any individual to 150 rems or more of radiation; or exposure of the feet, ankles, hands, or forearms of any individual to 375 rems or more of radiation; or

(2) the release of radioactive material in concentrations which, if averaged over a period of 24 hours, would exceed 5,000 times the limit specified for such materials in Appendix B, Table II (see Table 19, pp. 616 to 633); or

(3) a loss of one workweek or more of the operation of any facility affected; or

(4) damage to property in excess of $100,000.00.

(b) Twenty-four-hour notification. Each licensee shall within 24 hours notify the Director of the appropriate Atomic Energy Commission Regional Compliance Office listed in Appendix D by telephone and telegraph of any incident involving licensed material possessed by him, and which may have caused, or threatens to cause,

(1) exposure of the whole body of any individual to 5 rems or more of radiation; exposure of the skin of the whole body of any individual to 30 rems or more of radiation; or exposure of the feet, ankles, hands, or forearms to 75 rems or more of radiation; or

(2) the release of radioactive material in concentrations which, if averaged over a period of 24 hours, would exceed 500 times the limits specified for such materials in Appendix B, Table II (see pp. 616 to 633); or

(3) a loss of one day or more of the operation of any facilities affected; or

(4) damage to property in excess of $1,000.00.

§20.404 Report of Exposure to Radiation to Former Employees.

At the request of a former employee, each licensee shall furnish to the former employee a report of the former employee's exposure to radiation as shown in records maintained by the licensee pursuant to §20.401.

(a) Such report shall be furnished within 30 days from the time the request is made; it shall cover each calendar quarter of the individual's employment involving exposure to radiation, or such lesser periods as may be requested by the employee. The report shall also include the results of any calculations and analyses of radioactive material deposited in the body of the employee made pursuant to the provisions of §20.108. The report shall be in writing, and contain the following statement:

> This report is furnished to you under the provisions of the Atomic Energy Commission regulation entitled "Standards for Protection Against Radiation" (Title 10, CFR, Part 20). You should preserve this report for future reference.

(b) The former employee's request should include appropriate identifying data, such as social-security number and dates and locations of employment.

§20.405 Reports of Overexposures and Excessive Levels and Concentrations.

(a) In addition to any notification required by §20.403, each licensee shall make a report in writing within 30 days to the Director, Division of Compliance, U.S. Atomic Energy Commission, Washington, D.C. 20545, with a copy to the Director of the appropriate Atomic Energy Commission Regional Compliance Office listed in Appendix D, of

(1) each exposure of an individual to radiation or concentrations of radioactive material in excess of any applicable limit in this part or in the licensee's license;

(2) any incident for which notification is required by §20.403; and

(3) levels of radiation or concentrations of radioactive material (not involving excessive exposure of any individual) in an unrestricted area in excess of 10 times any applicable limit set forth in this part or in the licensee's license.

Each report required under this paragraph shall describe the extent of exposure of persons to radiation or to radioactive material; levels of radiation and concentrations of radioactive material involved; the cause of the exposure, levels, or concentrations; and corrective steps taken or planned to assure against a recurrence.

(b) In any case where a licensee is required pursuant to the provisions of this section to report to the Commission any exposure of an individual to radiation or to concentrations of radioactive material, the licensee shall also notify such individual of the nature and extent of exposure. Such notice shall be in writing, and shall contain the following statement:

> This report is furnished to you under the provisions of the Atomic Energy Commission regulations entitled "Standards for Protection Against Radiation" (Title 10, CFR, Part 20). You should preserve this report for future reference.

§20.406 Notice of Exposure to Radiation to Employees.

At the request of any employee, each licensee shall advise such employee annually of the employee's exposure to radiation as shown in records maintained by the licensee pursuant to §20.401(a).

ENFORCEMENT

§20.601 Violations.

An injunction or other court order may be obtained, prohibiting any violation of any provision of the act or of any regulation or order issued thereunder. Any person who willfully violates any provision of the act or any regulation or order issued thereunder may be guilty of a crime, and upon conviction may be punished by fine or imprisonment, or both, as provided by law.

PART IX
RADIATION INJURY AND ITS MANAGEMENT

Niel Wald, M.D.
University of Pittsburgh
Department of Radiation Health

PART IX

RADIATION INJURY AND ITS MANAGEMENT

STANDARD MAN

The purpose of this section is to make readily available to users of radioactive nuclides those data concerning the biological effect of radiation that are of practical value in the need to understand, recognize, and deal with the deleterious biomedical effects of radiation in man. It is, therefore, limited to considerations relevant to radiation safety and to the management of radiation injury. A more complete treatment of this large subject is available in the following recent references:

> Morgan, K. Z. and Turner, J. E., eds., *Principles of Radiation Protection*. John Wiley and Sons, New York, 1967.
>
> Saenger, E. L., ed., *Medical Aspects of Radiation Accidents*. United States Atomic Energy Commission, Washington, D.C., 1963.
>
> Cronkite, E. P. and Bond, V. P., *Radiation Injury in Man*. Charles C Thomas, Springfield, Illinois, 1960.
>
> Blakely, J., *The Care of Radiation Casualties*. Charles C Thomas, Springfield, Illinois, 1968.

ORGANS

The hypothetical standard man serves as the basis for many of the calculations upon which the setting of radiation exposure limits depends. The diagnosis and prognosis for individuals accidentally contaminated externally and/or internally with radionuclides also rests upon quantitative information developed with the aid of computations utilizing these standardized biologic dimensions. Table 1 lists organ weights and related values for the standard man.

PHYSIOLOGICAL FUNCTIONS

Certain physiological functions of man are utilized in the determination of estimated internal radiation exposure and its probable duration.

Fluid Balance

Mean values for intake and excretion are given in Table 2.

Respiration

Normal Respiratory Values.

Respiratory values directly related to the current theoretical models for deposition and clearance of particles from the respiratory tract are presented in Table 3. They include the more important respiratory variables in individuals representative of different ages, sizes and sexes. In addition, the more useful and reliable parameters of ventilatory performance under conditions of normal and moderate activity states are listed. All the measurements shown are subject to considerable individual variation. For lung volume and ventilation, it is approximately true that 95 percent of a normal population falls within plus or minus 20 percent of the value given in the table The measurements of ventilation during resting are subject to greater variation than this, and the values given in this table are designed as an approximate guide rather than as invariable values.

Table 1. ORGANS OF THE STANDARD MAN

Organ	Mass, g	Percent of Total Body	Effective Radius, cm
Total body*	70,000	100	30
Muscle	30,000	43	30
Skin and subcutaneous tissue†	6,100	8.7	0.1
Fat	10,000	14	20
Skeleton			
without bone marrow	7,000	10	5
red marrow	1,500	2.1	
yellow marrow	1,500	2.1	
Blood	5,400	7.7	
Gastrointestinal tract*	2,000	2.9	30
Contents of gastrointestinal tract			
lower large intestine	150		5
stomach	250		10
small intestine	1,100		30
upper large intestine	135		5
Liver	1,700	2.4	10
Brain	1,500	2.1	15
Lungs (2)	1,000	1.4	10
Lymphoid tissue	700	1.0	
Kidneys (2)	300	0.43	7
Heart	300	0.43	7
Spleen	150	0.21	7
Urinary bladder	150	0.21	
Pancreas	70	0.10	5
Salivary glands (6)	50	0.071	
Testes (2)	40	0.057	3
Spinal cord	30	0.043	1
Eyes (2)	30	0.043	0.25
Thyroid gland	20	0.029	3
Teeth	20	0.029	
Prostate gland	20	0.029	3
Adrenal glands or suprarenal (2)	20	0.029	3
Thymus	10	0.014	
Ovaries (2)	8	0.011	3
Hypophysis (pituitary)	0.6	8.6×10^{-6}	0.5
Pineal gland	0.2	2.9×10^{-6}	0.04
Parathyroids (4)	0.15	2.1×10^{-6}	0.06
Miscellaneous (blood vessels, cartilage, nerves, etc.)	390	0.56	

* Does not include contents of the gastrointestinal tract.
† The mass of the skin alone is taken to be 2,000 grams.
From: Shilling, C. W., ed., *Atomic-Energy Encyclopedia in the Life Sciences*, p. 404. W. B. Saunders, London, England, and Philadelphia, 1964.

Table 2.
INTAKE AND EXCRETION OF THE STANDARD MAN (WATER BALANCE)

Intake, cm³/day		Excretion, cm³/day	
Food	1,000	Urine	1,400
Fluids	1,200	Sweat	600
Oxidation	300	From lungs	300
		Feces	200
	2,500		2,500

From: Shilling, C. W., ed., *Atomic-Energy Encyclopedia in the Life Sciences*, p. 405. W. B. Saunders, London, England, and Philadelphia, 1964.

Table 3. APPROXIMATE NORMAL RESPIRATORY VALUES

Individual	Age, years	Height, cm	FRC,[a] liters	TLC,[b] liters	V_D,[c] cm³	Weight of Two Lungs,[d] grams	MPV,[e] l/min	ISA,[f] m²	Resting						Light Exercise					
									V_E,[g] l/min	V_T,[h] cm³	f,[i] min	V_C,[j] cm³	V_E,[k] l/min	V_{O_2},[l] l/min	f,[m] min	V_T,[n] liters	Q_C,[o] l/min	V_C,[p] cm³
Child (male or female)	10	140	1.4	3.0	60	400	70	40	4.8	300	16	44	14.4	0.55	24	0.60	—	—
Woman	30	160	2.7	4.8	130	885	91	65	6.0	400	15	61	18.8	0.75	20	0.94	10.0	120
Man	30	170	3.5	6.4	160	1,169	135	80	7.5	500	15	76	20.0	0.82	16	1.25	10.0	150

Footnotes on the Use of the Tabulations.

It must be recognized that all the measurements shown in this tabulation are subject to considerable individual variation. In the case of measurements of lung volume and ventilation, it is approximately true that 95 percent of a normal population fall within ±20 percent of the value given in the table. The measurements of ventilation during rest and during exercise are subject to greater variation than this, and the values given in this table are designed as approximate guides rather than as invariable values. Explanations of the individual measurements are given below.

[a] Functional Residual Capacity—the volume of gas in the lungs at the end of a normal quiet expiration.
[b] Total Lung Capacity—the volume of gas in the lungs at the end of a maximal inspiration.
[c] Inert-gas deadspace representing approximately the volume of the airways down as far as the gas-exchanging surface; this value is similar when measured with gases of different densities, and may be taken to represent upper and lower airway volume; in a normal subject, approximately half of this volume is above the larynx, but the fraction which this represents is greatly dependent on the position of the neck.
[d] These data include the blood contained in the lungs, and it is thought that the lung tissue *per se* is probably just over half of the weight given in this column.
[e] Maximal Possible Volume—this represents the greatest volumetric rate a subject can breathe, but he cannot sustain the maximal figure given in this column for longer than a few minutes; normally, when the subject is exercising maximally, he can sustain about 75 percent of this value.
[f] Internal Surface Area—this has been computed from knowledge of the total lung capacity.
[g] This is the minute volume of ventilation, i.e., the expired gas per minute at rest, as measured at the mouth.
[h] This is the tidal volume of each breath.
[i] This represents average respiratory frequency.
[j] Pulmonary Capillary Blood Volume—the volume of blood believed to be in contact with alveolar gas at any one instant; the blood in the lung is predominantly contained in vessels other than the capillaries.
[k] This is the minute volume of ventilation during flat walking at about 3 miles per hour.
[l] This is the oxygen uptake during the performance of this task.
[m] This is the average respiratory frequency seen at this level of exercise.
[n] This is the tidal volume during performance of this level of exercise.
[o] This is the pulmonary blood flow or cardiac output; no reliable data exist for children at this level of exercise; the data for adults are to be regarded as approximate only.
[p] Pulmonary capillary blood volume increases approximately to twice its resting value at this level of exercise.

Values given under footnotes, [g], [h] and [i] are not basal values; they were different when measured after sedation, but represent average resting values for normal subjects studied in a pulmonary-function laboratory. Basal values are somewhat lower than those given in this table, but are more representative of the resting subject as he is normally encountered.

From: ICRP Task Group on Lung Dynamics, Deposition and Retention Models for Internal Dosimetry of the Human Respiratory Tract. *Health Phys.*, 12: 204–205, 1966.

The Lung Model.

In 1965, the Task Group on Lung Dynamics of the International Radiological Protection Commission proposed a new and improved lung model, i.e., a scheme for computing dust deposition in and clearance from the human respiratory tract, to provide a basis for lung dosimetry and the setting of exposure limits.[1] In this model, the respiratory tract was divided into three clearly defined compartments.

1. The nasopharynx (N–P). This consists of the anterior nares, and extends through the anterior pharynx, back, and down through the posterior pharynx to the level of the larynx or epiglottis.
2. Trachea and broncheal tree (P–B). Extends to include the terminal bronchioles.
3. Pulmonary (P). This includes respiratory bronchioles, alveolar ducts, atria, alveoli, and alveolar sacs.

Dust Deposition Model. In developing a model for dust deposition, certain assumptions are made. Dusts cannot be regarded as having uniform physical characteristics, but rather must be treated as distributions. The log-normal distribution has been deemed most applicable for this purpose. Similarly, human ventilation cannot be regarded as constant, since it is influenced by the activity state of man, which in turn affects deposition. Finally, it appears valid to consider that the aerodynamic properties of the dust, the physiology of respiration, and the anatomy of the respiratory tract govern dust deposition together.

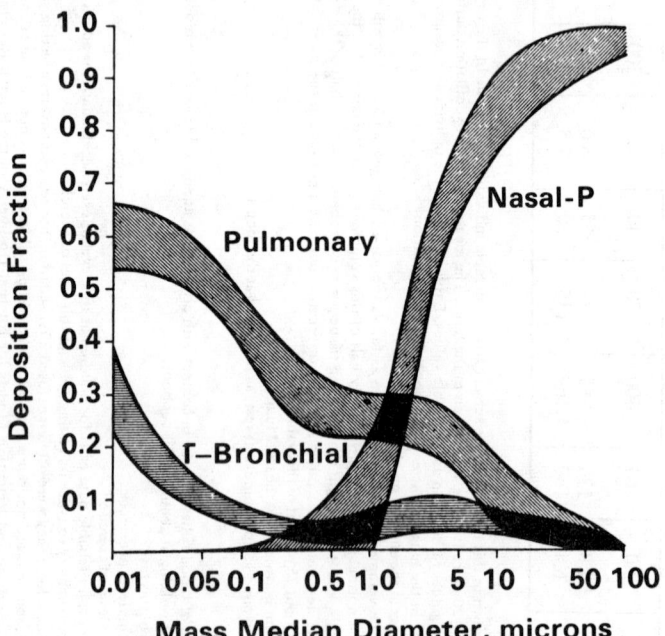

FIGURE 1. Compartmental Deposition Curves.
Each of the shaded areas indicates the variability of deposition for a given mass median (aerodynamic) diameter in each compartment when the distribution parameter σ_g varies from 1.2 to 4.5 and the tidal volume is 1,450 ml.
(From: Task Group on Lung Dynamics, Deposition and Retention Models for Internal Dosimetry of the Human Respiratory Tract. *Health Phys.*, 12: 181, 1966.)

The evaluation of inhalation hazards can be approached initially by calculating the deposition pattern for any given aerosol with the aid of curves that relate deposition in the respiratory tract to some parameter of the aerosol distributions, such as a median aerodynamic diameter (diameter of a unit density sphere with the same settling velocity as the particle in question). Figure 1 can be used for this purpose; the curves indicate a relationship between the mass aerodynamic diameter and the gravimetric fraction of the inhaled dust that will be deposited in each anatomical compartment. Within this limit (0.01 to 100 microns), this diameter serves to characterize the deposition probabilities of the entire distribution from which it emerges. One may generalize the relationship shown in this figure by interpreting the abscissa as the activity median aerodynamic diameter; the ordinate then represents the correspondent fraction of inhaled activity that will be deposited in a given compartment. The concepts used with respect to the mass distribution are equally valid for a log-normal activity distribution, regardless of the relationship between particle size and activity.

Dust Clearance Model. Dust clearance is a nonuniform process. Many variables are involved, such as the aerodynamic quality of the dust, the "solubility", and the specific mechanisms whereby the clearance processes operate. Adequate information concerning many of these factors is lacking; nevertheless, the proposed clearance model is a composite of cellular or endocytotic activity, the ciliary-mucus transport mechanism, and the physicochemical properties of the dust—especially those affecting the tendency or ability of dust, intact or dissipated, to cross the pulmonary membranes.

A model that relies upon as much quantitative information as possible and includes all known clearance mechanisms and pathways is presented schematically in Figure 2.

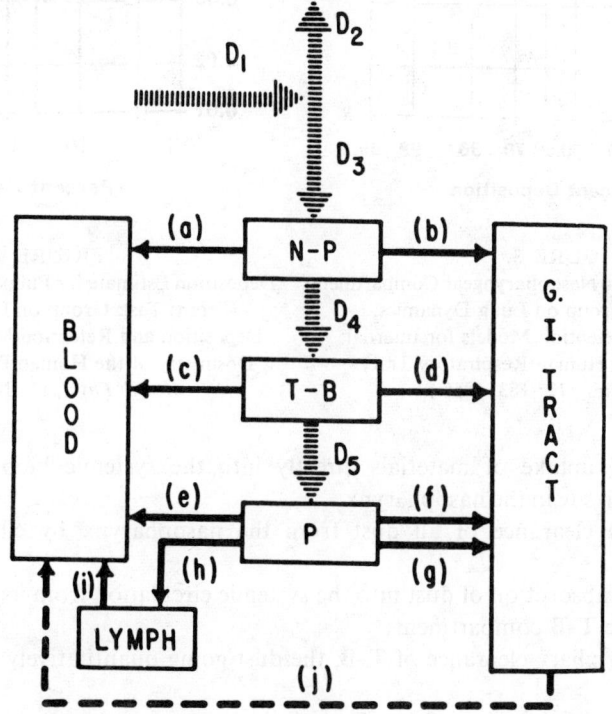

FIGURE 2. Clearance Model.
(From: Task Group on Lung Dynamics, Deposition and Retention Models for Internal Dosimetry of the Human Respiratory Tract. *Health Phys.*, 12: 190, 1966.)

D_1 through D_5 are the amounts of dust in various respiratory volumes or areas; D_1 is the total dust inhaled (global air concentration), D_2 is the total dust in the exhaled air, D_3 is the amount of dust deposited in the N–P (obtainable from Figure 3), D_4 is the dust deposited in the T–B zone, and D_5 is the pulmonary dust deposition (obtainable from Figure 4). Ordinarily, D_3, D_4, and D_5 are expressed as percentages of D_1 and are determinable from the deposition models.

In addition to the major respiratory regions, three other closely allied compartments are listed: the gastrointestinal tract, systemic blood, and pulmonary lymph. The letters (a) through (j) indicate the different absorption and translocation processes that are associated with the clearance of various compartments. The letters and the processes they represent are as follows:

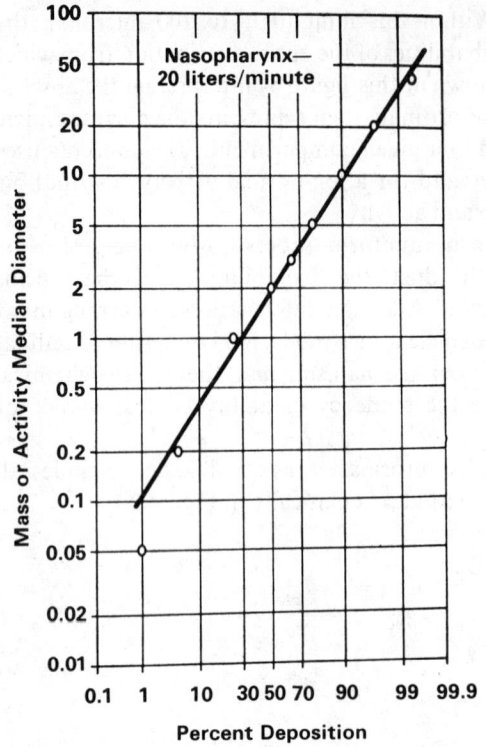
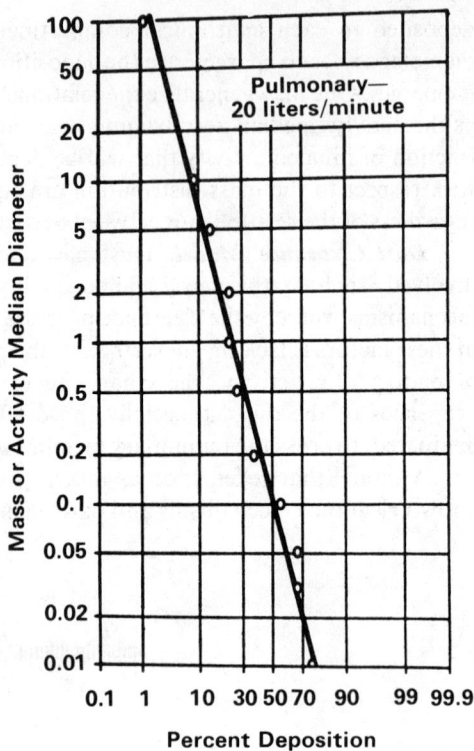

FIGURE 3.
Deposition Estimate for Nasopharyngeal Compartment.
(From: Task Group on Lung Dynamics,
Deposition and Retention Models for Internal
Dosimetry of the Human Respiratory Tract.
Health Phys., 12: 183, 1966.)

FIGURE 4.
Deposition Estimate for Pulmonary Compartment.
(From: Task Group on Lung Dynamics,
Deposition and Retention Models for Internal
Dosimetry of the Human Respiratory Tract.
Health Phys., 12: 184, 1966.)

(a) the rapid uptake of materials directly into the systemic blood from its deposition site in the nasopharynx;
(b) the rapid clearance of all dust from the nasopharynx by ciliary-mucus transport;
(c) the rapid absorption of dust into the systemic circulation from its deposition site in the T–B compartment;
(d) the rapid ciliary clearance of T–B, the dust going quantitatively to the G.I. tract;
(e) the direct translocation of dust from the pulmonary region to the blood;
(f) the relatively rapid clearance phase of the pulmonary regions (this presumably depends on recruitable macrophages, and is in turn coupled to the ciliary-mucus transport process; the dust cleared goes to the G.I. tract via the tracheobronchial tree);
(g) the second pulmonary clearance process, which is typically much slower than (f), but still depends upon endocytosis and ciliary-mucus transport; the clear dust goes via T–B to the G.I. tract; it is apparently rate-limited in the pulmonary region by the nature of the deposited dust;
(h) the slow removal of dust from the pulmonary compartment via the lymphatic system;
(i) the secondary pathway, in which dust is cleared by the lymphatic system (h), introduces it into the systemic blood; the material must penetrate lymph

tissue, which implies either partial or complete dissolution of the dust particles, although turnover of lymphocytes may contribute;

(j) the collective absorption of clear dust from the gastrointestinal tract by direct and indirect pathways.

To facilitate employment of this model, all the quantitative clearance information has been utilized to provide a grouping of nine major classes of inorganic compounds into three categories (Table 4). On the basis of the three clearance classifications, the values of (a) to (i) can now be assigned in the clearance model. Suggested values for the various clearance processes are given in Table 5.

Table 4.
PULMONARY-CLEARANCE CLASSIFICATION OF INORGANIC COMPOUNDS

Class Y: Avid Retention, Cleared Slowly (years)

Carbides	Actinides, lanthanides, Zr, Y, Mn
Sulfides	None
Sulfates	None
Carbonates	None
Phosphates	None
Oxides and hydroxides	Lanthanides, actinides, Groups 8 (V and VI), 1b, 2b (IV and V), 3b except Sc^{3+}, and 6b
Halides	Lanthanide fluorides
Nitrates	None

Class W: Moderate Retention, Intermediate Clearance Rates (weeks)

Carbides	Cations of all Class W hydroxides, except those listed as Class Y carbides
Sulfides	Groups 2a (V and VI), 5a (IV to VI), 1b, 2b, and 6b (V and VI)
Sulfates	Groups 2a (IV to VII) and 5a (IV to VI)
Carbonates	Lanthanides, Bi^{3+}, and Groups 2a (IV to VII)
Phosphates	Zn^{2+}, Sn^{3+}, Mg^{2+}, Fe^{3+}, Bi^{3+}, and lanthanides
Oxides and hydroxides	Groups 2a (II to VII), 3a (III to VI), 4a (III to VI), 5a (IV to VI), 6a (IV to VI), 8, 2b (VI), 4b, 6b, and 7b, and Sc^{3+}
Halides	Lanthanides (except fluorides), Groups 2a, 3a (III to VI), 4a (IV to VI), 5a (IV to VI), 8, 1b, 2b, 3b (IV and V), 4b, 5b, 6b, and 7b
Nitrates	All cations whose hydroxides are Class Y or Class W

Class D: Minimal Retention, Rapid Clearance (days)

Carbides	See hydroxides
Sulfides	All except those listed as Class W
Sulfates	All except those listed as Class W
Carbonates	All except those listed as Class W
Phosphates	All except those listed as Class W
Oxides and hydroxides	Groups 1a, 3a (II), 4a (II), 5a (II and III), 6a (III)
Halides	Groups 1a and 7a
Nitrates	All except those listed as Class W
Noble gases	Group 0

Note: where reference is made from one chemical form to another, it implies that an *in-vivo* conversion occurs, e.g., hydrolysis reaction.

The following periodic table of the elements is used with the foregoing classification:

Period	1a	2a	3b	4b	5b	6b	7b	8			1b	2b	3a	4a	5a	6a	7a	0
I	H																	He
II	Li	Be											B	C	N	O	F	Ne
III	Na	Mg											Al	Si	P	S	Cl	Ar
IV	K	Ca	Sc	Ti	V	Cr	Mn	Fe	Co	Ni	Cu	Zn	Ga	Ge	As	Se	Br	Kr
V	Rb	Sr	Y	Zr	Nb	Mo	Tc	Ru	Rh	Pd	Ag	Cd	In	Sn	Sb	Te	I	Xe
VI	Cs	Ba	La*	Hf	Ta	W	Rc	Os	Ir	Pt	Au	Hg	Tl	Pb	Bi	Po	At	Rn
VII	Fr	Ra	Ac†															

* Lanthanides	Ce	Pr	Nd	Pm	Sm	Eu	Gd	Tb	Dy	Ho	Er	Tm	Yb	Lu
† Actinides	Th	Pa	U	Np	Pu	Am	Cm	Bk	Cf	Es	Fm	Md	No	Lw

From: ICRP Task Group on Lung Dynamics, Deposition and Retention Models for Internal Dosimetry of the Human Respiratory Tract. *Health Phys.*, 12: 192, 1966.

Table 5. CONSTANTS FOR USE WITH CLEARANCE MODEL

Compartment	Absorption or Translocation Process	Class D (Clearance in Days) Soluble (biological $T_{1/2}$ = <1 to 10 days)	Class W (Clearance in Weeks) Moderately Soluble (biological $T_{1/2}$ = >10 to 100 days)	Class Y (Clearance in Years) Insoluble (biological $T_{1/2}$ = >100 days)
N–P	(a)	4 min./0.50	4 min./0.10	4 min./0.01
	(b)	4 min./0.50	4 min./0.90	4 min./0.99
T–B	(c)	10 min./0.50	10 min./0.10	10 min./0.01
	(d)	10 min./0.50	10 min./0.90	10 min./0.99
P	(e)	30 min./0.80	90 days/0.15	360 days/0.05
	(f)	n.a.	24 hrs. /0.40	24 hrs. /0.40
	(g)	n.a.	90 days/0.40	360 days/0.40
	(h)	30 min./0.20	90 days/0.05	360 days/0.15
Lymph	(i)	30 min./1.00	90 days/1.00	360 days/0.10

Note: the first value is the biological half-time, the second value is the regional fraction.
From: ICRP Task Group on Lung Dynamics, *Report to Committee 2.* 1965.

The Gastrointestinal Tract

Some quantitative characteristics of the gastrointestinal tract are given in Table 6.

Table 6. GASTROINTESTINAL TRACT OF THE STANDARD MAN

Portion of the G.I. Tract That Is the Critical Tissue	Mass of Contents, grams	Time Food Remains, hours	Fraction from Lung to G.I. Tract	
			Soluble	Insoluble
Stomach (S)	250	1	0.50	0.625
Small intestine (SI)	1,100	4	0.50	0.625
Upper large intestine (ULI)	135	8	0.50	0.625
Lower large intestine (LLI)	150	18	0.50	0.625

From: Shilling, C. W., ed., *Atomic-Energy Encyclopedia in the Life Sciences*, p. 404. W. B. Saunders, London, England, and Philadelphia, 1964.

REFERENCE

1. Task Group on Lung Dynamics, Deposition and Retention Models for Internal Dosimetry of the Human Respiratory Tract. *Health Phys.*, 12: 173–205, 1966.

RADIATION INJURY

The sequence of events produced by the absorption of ionizing radiation extends from the atomic level through the population level. The time development of these changes ranges from 10^{-16} seconds to years and decades. Figure 5 provides a concise view of radiobiology, useful for orientation and perspective. For the purposes of our concern with radiation safety, this section will focus on certain limited aspects of the effects of radiation on cells, tissues, organs, and systems of particular relevance, as well as on whole-body radiation injury.

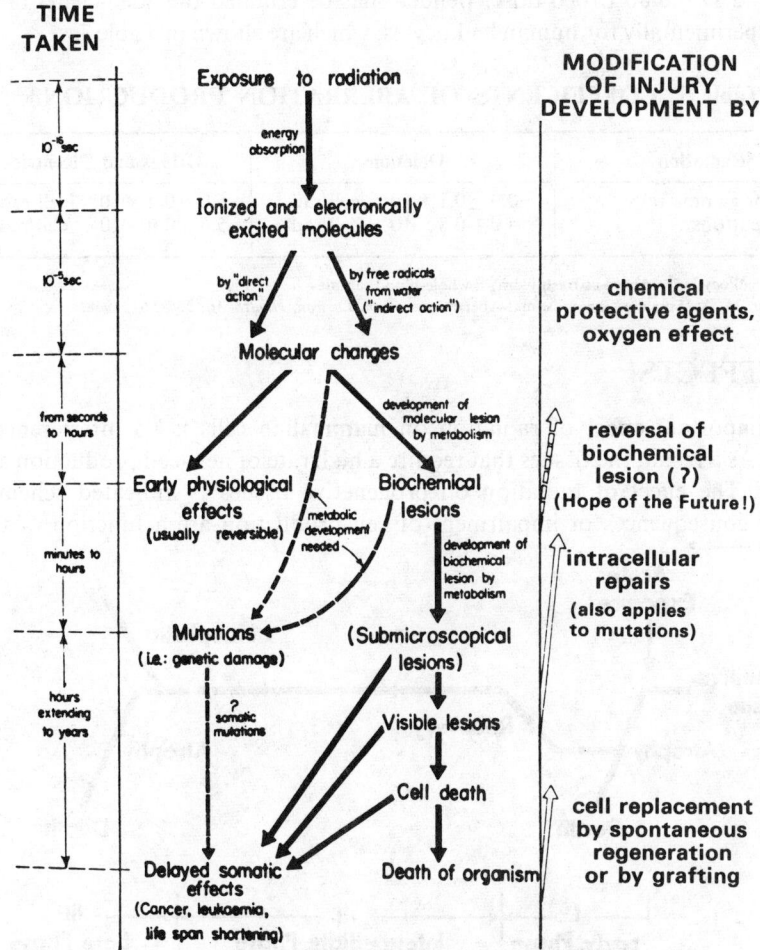

FIGURE 5. Schematic View of Radiobiology.
(From: Alexander, P. and Bacq, Z. M., *Fundamentals of Radiobiology*.
Copyright 1961, Pergamon Press, New York.)

CELL EFFECTS

The main effects of radiation at the cell level are lethality and reproductive damage to cells; these lead to clinical sequelae at the tissue and organ level, which are described below.

One of the visible manifestations of the latter type of damage is the appearance of chromosomal aberrations in somatic cells. Recent developments in cytogenetic techniques have made it possible to study dividing cells from the circulating blood of man. Because of certain quantitative relationships, it is possible to use the frequency of appearance of certain chromosomal aberrations in the first two weeks after an accidental exposure as an indicator of the type and magnitude of radiation exposure. One type of aberration scored for this purpose is deletion of chromosomal material, which increases in a linear relationship to dose. The other type of aberration is the ring or dicentric chromosome, which occurs with a frequency that increases with the square of the dose. The relationship may be expressed as follows:

$$Y_1 = a + b_1 D \text{ (for deletions)},$$
$$Y_2 = b_2 D^2 \text{ (for rings and dicentrics)},$$

where Y is the aberration yield, a the spontaneous deletion frequency, b_1 and b_2 the aberration coefficients, and D the absorbed dose. Bender[1] has determined the coefficients of aberration production experimentally for human leukocytes, which are shown in Table 7.

Table 7. COEFFICENTS OF ABERRATION PRODUCTION*

Radiation	Deletions	Rings and Dicentrics
Hard X or gamma rays	$0.9 \pm 0.1 \times 10^{-3}$/cell/rad	$6.0 \pm 0.4 \times 10^{-6}$/cell/rad sq
Fission neutrons	$4.0 \pm 0.8 \times 10^{-3}$/cell/rad	$5.6 \pm 0.6 \times 10^{-3}$/cell/rad

*For human leukocytes irradiated as freshly drawn whole-blood samples.
From: Bender, M. A., Somatic Chromosomal Aberrations. *Arch. Environ. Health*, 16: 556–564, 1968.

TISSUE EFFECTS

A most important effect of radiation on mammalian cells is its interference with cell reproduction. As a result, the tissues that require a high rate of new-cell production are affected most severely. The effect of radiation on progenetive tissues is indicated schematically in Figure 6. The consequences of impairment of cell production are a function of the normal

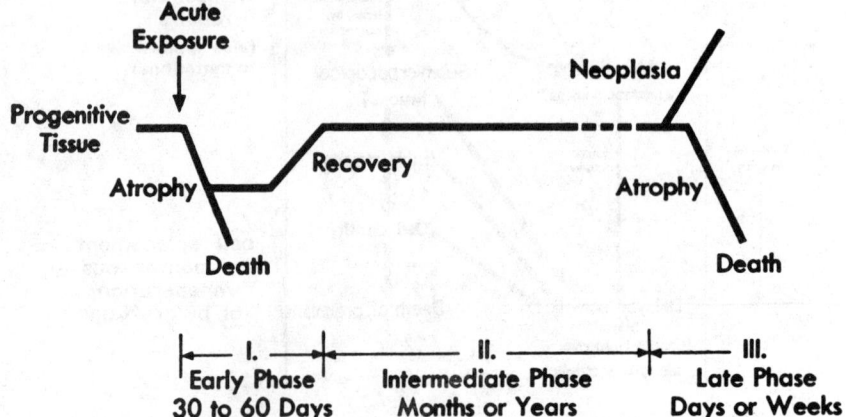

FIGURE 6. Pattern of Irradiation Injury
(From: Claus, W. D., ed., *Radiation Biology and Medicine*, p. 331.
Copyright Addison-Wesley, Reading, Massachusetts, 1958.)

maturation time and the survival time of mature cells thereafter. The sequelae of irradiation are expressed by specific organs in manners appropriate to their particular functions. The effect on progenitive tissues of some key organs is portrayed in Figure 7.

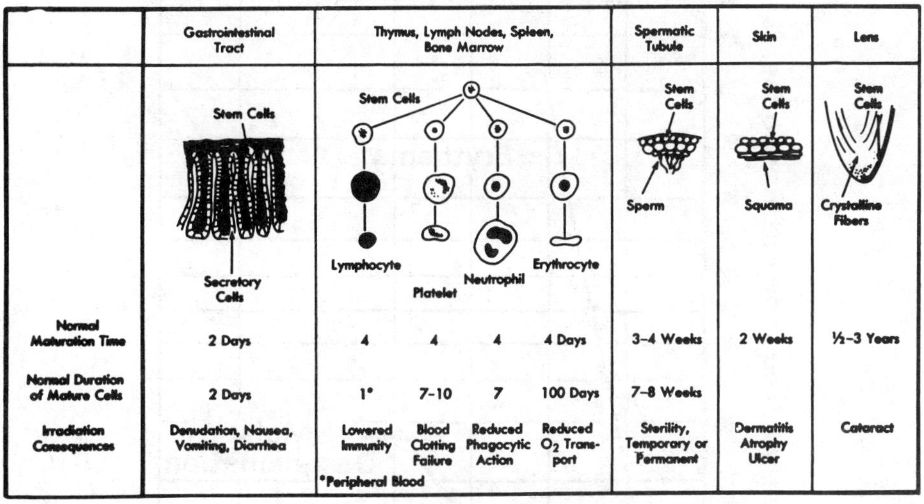

	Gastrointestinal Tract	Thymus, Lymph Nodes, Spleen, Bone Marrow				Spermatic Tubule	Skin	Lens
	Stem Cells → Secretory Cells	Stem Cells → Lymphocyte, Platelet, Neutrophil, Erythrocyte				Stem Cells → Sperm	Stem Cells → Squama	Stem Cells → Crystalline Fibers
Normal Maturation Time	2 Days	4	4	4	4 Days	3–4 Weeks	2 Weeks	½–3 Years
Normal Duration of Mature Cells	2 Days	1*	7–10	7	100 Days	7–8 Weeks		
Irradiation Consequences	Denudation, Nausea, Vomiting, Diarrhea	Lowered Immunity	Blood Clotting Failure	Reduced Phagocytic Action	Reduced O₂ Transport	Sterility, Temporary or Permanent	Dermatitis Atrophy Ulcer	Cataract

*Peripheral Blood

FIGURE 7.
Progenitive Tissues: Maturation,
Duration of End Product, and Irradiation Consequences.
(From: Claus, W. D., ed., *Radiation Biology and Medicine*, p. 326.
Copyright Addison-Wesley, Reading, Massachusetts, 1958.)

ORGAN AND SYSTEM EFFECTS

Skin

The immediate effects of external radiation exposure are seen most often as changes in the skin. The rate of appearance, extent, and severity of these changes is a function of the energy and quantity of radiation received. The general pattern is shown in Table 8.

Table 8. RADIATION DAMAGE TO THE SKIN

Epilation (Loss of Hair)	Erythema (First-Degree Burns)	Wet Dermatitis and Blistering (Second-Degree Burns)	Ulceration (Third-Degree Burns)
Rare at less than 200 r			
Partial epilation at 350 to 450 r	Response is dependent on energy, dose rate, area exposed, and complexion of the individual; full effect in 1 to 3 weeks after: 200 to 400 r (<150 kev),		
Complete epilation in 16 to 18 days at >450 r	500 to 600 r (200 to 400 kev), 800 to 1000 r (>400 kev);		
Permanent epilation at >700 r	response in first hours at 1000 r	Effect in 1 to 2 weeks at >1000 r	
			Rapidly progressive effect at >2000 r

Note: 1 r = 1 rad, since these statements are based on air doses.
From: Webb, P., *Bioastronautic Data Book*, p. 153. National Aeronautics and Space Administration, Scientific and Technical Information Division, Washington, D.C., 1964.

A dose–frequency relationship of erythema and moist desquamation is presented in Figure 8.

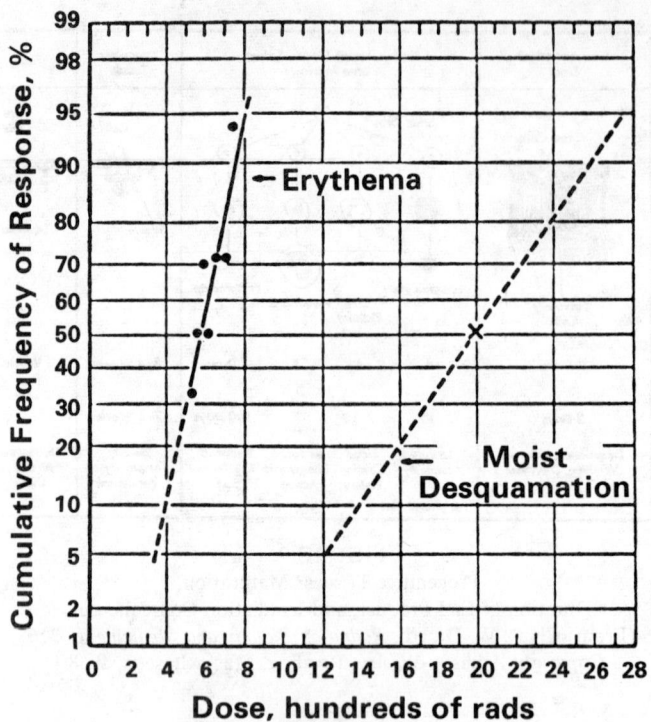

FIGURE 8.
Dose–Frequency Relationship of Minimal Erythema (Duffy et al., 1934)
and Moist Desquamation (Clinical Tolerance Response)
for Acute Exposure to 200-kvp X Rays.
Site of interest for dose estimates was 0.1 mm depth; area exposed was 35 to 100 cm².
(From: Langham, W. H., ed., *Radiobiological Factors in Manned Space Flight* (*Publication No. 1487*), p.63. National Academy of Sciences, National Research Council, Washington, D.C., 1967.)

Since transepidermal injury from beta radiation is strongly dependent on the sources and energies involved, a tabulation of the surface dose required for injuries from various isotopes is presented in Table 9. Fractionation of exposure produces a different biological response to the same total dose. This has been estimated for the production of late skin necrosis in Table 10.

Table 9.
AMOUNT OF TRANSEPIDERMAL BETA RADIATION (TO A DEPTH OF 0.09 mm) REQUIRED FOR PRODUCTION OF RECOGNIZABLE TRANSEPIDERMAL INJURY*

Isotope	Energy, Mev	Surface Dose Required to Produce Recognizable Transepidermal Injury, rads	Estimated Amount of Radiation That Penetrated Skin to a Depth of 0.09 mm, rads
^{35}S	0.17	20,000	1,200
^{60}Co	0.31	4,000	1,600
^{137}Cs	0.55	2,000	1,700
^{91}Y	1.53	1,500	1,200
^{90}Sr	0.61	1,500	1,400
^{90}Y	2.20		
Average			1,460

* Porcine skin (Moritz and Henriques, 1952).
From: Langham, W. H., ed., *Radiobiological Factors in Manned Space Flight* (*Publication No. 1487*), p. 71. National Academy of Sciences, National Research Council, Washington, D.C., 1967.

Table 10.
SUGGESTED ABSORBED DOSES* OF REFERENCE RADIATION FOR PRODUCTION OF LATE SKIN NECROSIS†

Probability of Response, percent	High-Intensity Single (1-Day) Dose, rads	Fractionated or Protracted Dose‡ rads
10	2,000	4,600
50	2,800	6,400
90	3,600	8,200

* Point of interest for dose estimation, 0.1–mm depth: area exposed < 150 cm².
† See Permanent or Late Skin Effects, p. 147 of reference cited below.
‡ Dose protracted over 7 weeks or longer.
From: Langham, W. H., ed., *Radiobiological Factors in Manned Space Flight* (Publication No. 1487), p. 262. National Academy of Sciences, National Research Council, Washington, D.C., 1967.

The clinical phenomena associated with radiation injury to the skin have been described[2] as follows:

> Erythema appears on exposure to more than 200 or 300 rads. It is equivalent to a mild sunburn or a thermal burn of the first degree. A transient first wave of erythema may be present within hours after exposure, associated with a sensation of warmth or itching. The major redness appears two or three weeks later, the interval depending on the dose. In the lower dose range no further changes may occur, and medical care is not necessary.
>
> A counterpart to this reaction has been described in the conjunctiva, where inflammatory changes have been observed promptly after exposure. An inflammatory response has also been reported in the anterior chamber inside the eye within 30 minutes after irradiation, reaching a peak in one day, and ending in one week.
>
> Epilation, or loss of hair, may occur after exposure to any form of radiation in doses of about 200 rads or more. Regardless of the dosage, it generally becomes apparent only two or three weeks after exposure. Associated skin or scalp tenderness may occur one or two days preceding the actual hair loss. With doses greater than about 500 rads, epilation may be complete. If the exposure is much greater than 600 rads, hair may not regrow.
>
> Transepidermal injury (dry or wet dermatitis) is equivalent to a thermal burn of the second degree. Blisters form and break open, leaving raw, painful wounds that are vulnerable to infection. These may occur within one to two weeks after exposure, depending on the dose. They are preceded or accompanied by erythema or epilation. Recognizable injuries of this grade require skin doses in excess of 1,000 rads, or 200 μCi/hr/cm². The need for medical care depends on the size and severity of the burn.
>
> With higher doses, probably on the order of 5,000 rads, a more serious version of transepidermal injury occurs, in which the lesion resembles a scalding or chemical burn. Pain occurs promptly and is intense. The raw areas may be very slow in healing, or may not heal until surgical resection and skin grafting is performed. Epilation is permanent.

Blood

Measurement of the levels of circulating blood cells is one of the simplest and most useful indices of radiation effect. The fluctuations in the count are the result of changes in production and destruction rates as well as in the life span of the particular cell line. These levels are associated with various degrees of radiation injury. The changes themselves are nonspecific; however, when combined with other historical or clinical information, they are useful indicators of radiation overexposure. Figure 9 presents the changes typical of patients undergoing mild to moderately severe radiation injury in whom survival is possible. In Figure 10 the hematologic changes associated with more severe degrees of radiation injury are presented; the first example is associated with major hematologic damage, in which survival is possible; the second is associated with gastrointestinal damage, in which survival is improbable; the third is associated with central-nervous-system and cardiovascular injury, in which survival cannot be expected.

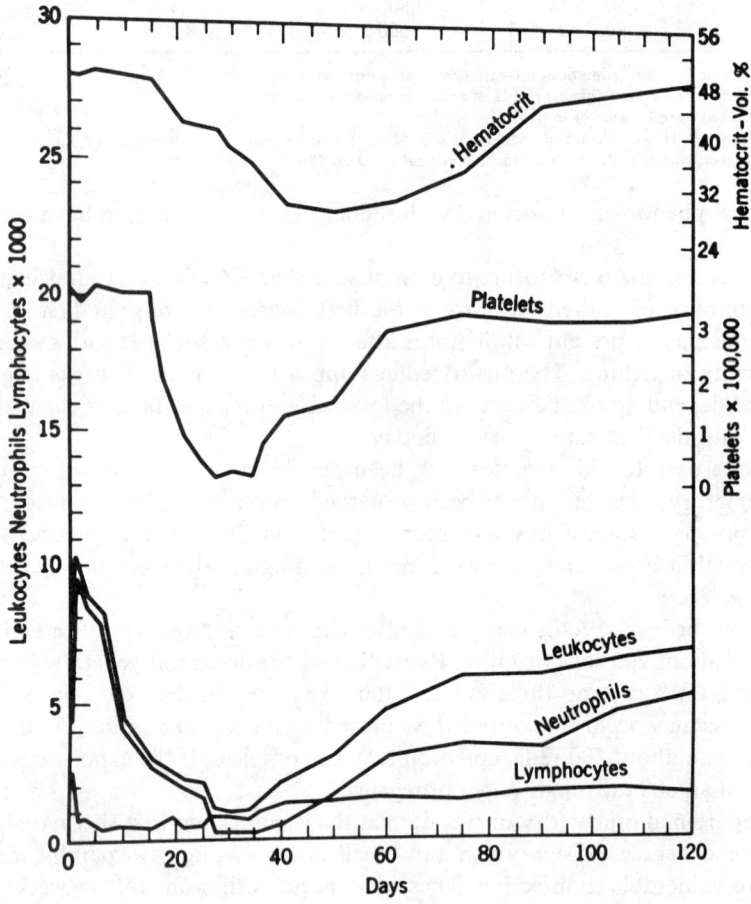

FIGURE 9.
Blood Counts—Injury Group II.
Moderate Injury—Survival Possible.
(From: Thoma, G. E., Jr. and Wald N., The Diagnosis and Management of Accidental Radiation Injury. *J. Occup. Med.*, 1: 421—447, 1959.)

Hematologic change as a function of exposure dose is of some interest, although the correlation with the severity of radiation injury is probably of greater importance. Figure 11 depicts the changes in the average values for the various blood elements in five individuals exposed to between 250 and 350 rads of neutron and gamma radiation in a criticality accident.

Smooth average time-course graphs for changes in the various blood elements as a function of dose have been derived from human accidental radiation exposure cases, and are presented in Figures 12, 13, and 14; the nadir, or low point, in the counts of these blood elements is plotted against dose in an idealized graphic presentation in Figure 15. The absorbed dose estimated to produce 45, 50, and 75 percent depression of three blood cell lines is given in Table 11.

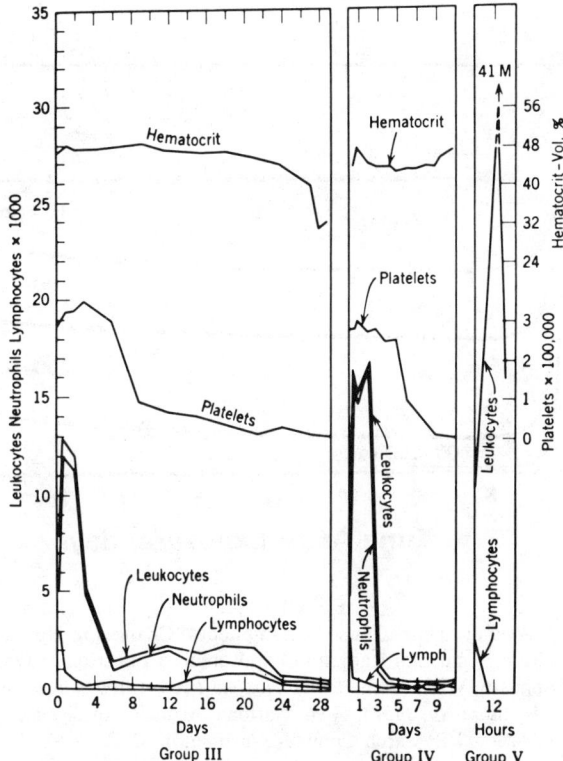

FIGURE 10.
Blood Counts—Injury Groups III, IV, and V.
III = Severe Injury—Survival Possible.
IV and V = Mortal Injury—Survival Unlikely.
(From: Thoma, G. E., Jr. and Wald, N., The Diagnosis and Management of Accidental Radiation Injury. *J. Occup. Med.*, 1: 421—447, 1959.)

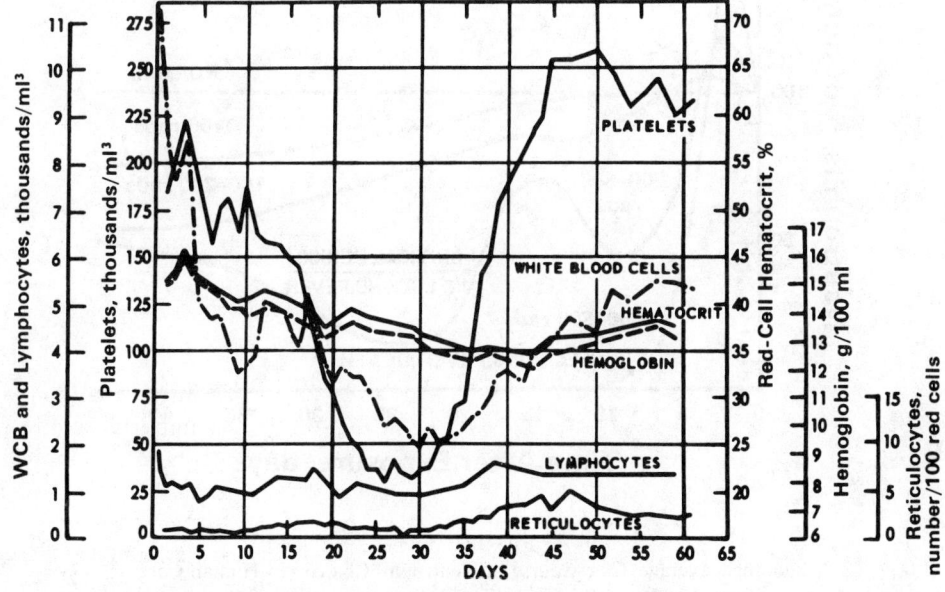

FIGURE 11.
Average Values of Blood Elements of Five Individuals Exposed to Estimated Doses of 250 to 350 rads at the Oak Ridge Criticality Accident of June 16, 1958.
(From: Webb, P., ed., *Bioastronautic Data Book*, p. 149.
National Aeronautics and Space Administration, Scientific and Technical Information Division, Washington, D.C., 1964.)

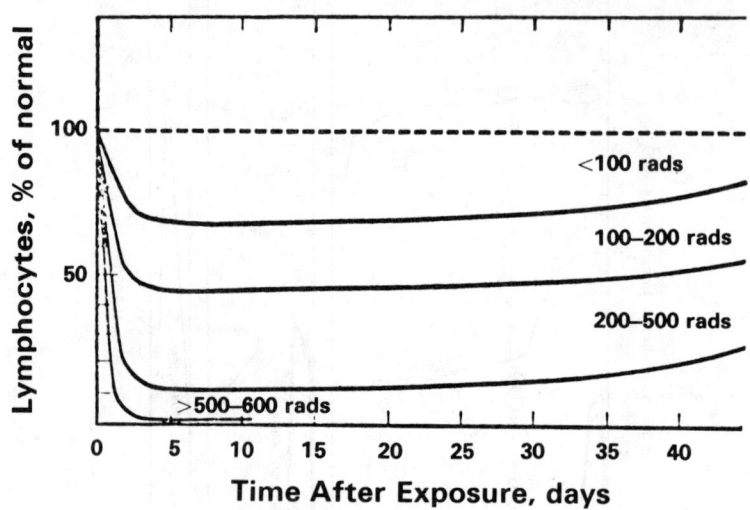

FIGURE 12.
Smoothed Average Time-Course of Lymphocyte Changes in Human Cases
from Accidental Radiation Exposure as a Function of Dose.
(From: Langham, W. H., ed., *Radiobiological Factors in Manned Space Flight*
(*Publication No. 1487*), p. 94. National Academy of Sciences,
National Research Council, Washington, D.C., 1967.)

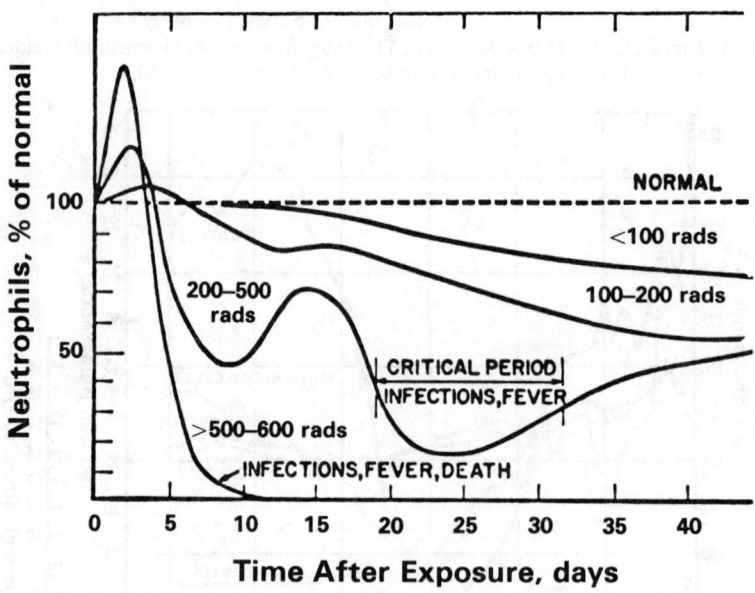

FIGURE 13.
Smoothed Average Time-Course of Neutrophil Changes in Human Cases
from Accidental Radiation Exposure as a Function of Dose.
(From: Langham, W. H., ed., *Radiobiological Factors in Manned Space Flight*
(*Publication No. 1487*), p. 94. National Academy of Sciences,
National Research Council, Washington, D.C., 1967.)

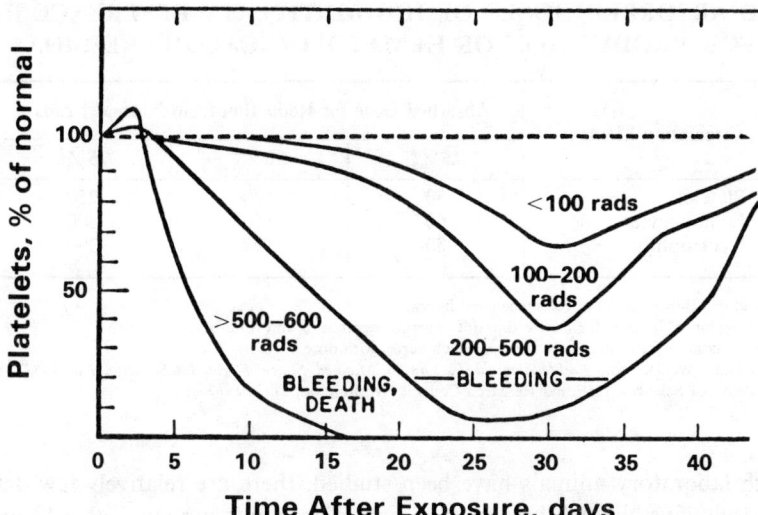

FIGURE 14.
Smoothed Average Time-Course of Platelet Changes in Human Cases
from Accidental Radiation Exposure as a Function of Dose.
(From: Langham, W. H., ed., *Radiobiological Factors in Manned Space Flight*
(*Publication No. 1487*), p. 95. National Academy of Sciences,
National Research Council, Washington, D.C., 1967.)

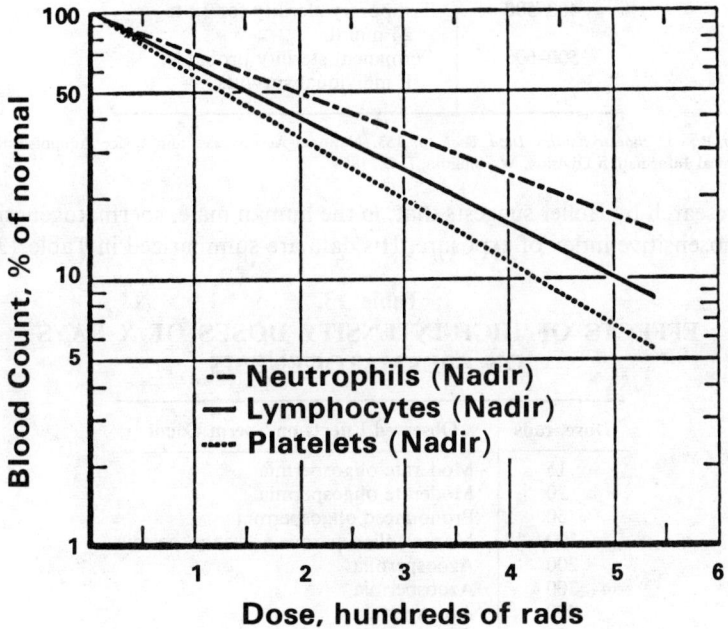

FIGURE 15.
Idealized Dose–Response Relationship for Lymphocytes, Neutrophils, and Platelets,
In Which the Nadir of Each Blood Element Is Plotted Against Dose.
(From: Langham, W. H., ed., *Radiobiological Factors in Manned Space Flight*
(*Publication No. 1487*), p. 95. National Academy of Sciences,
National Research Council, Washington, D.C., 1967.)

Table 11.
ESTIMATED ABSORBED DOSES OF HIGH-INTENSITY REFERENCE RADIATION FOR PRODUCTION OF HEMATOLOGICAL DEPRESSION*

Circulating Element	Absorbed Dose for Reduction from Normal,† rads		
	25%‡	50%‡	75%‡
Platelets	50	120	250
Lymphocytes	60	150	300
Neutrophils	80	190	390

* See Hematologic Effects, p. 90 of reference cited below.
† Anatomical region of interest for dose estimation, average depth of 5 cm, total-body exposure.
‡ At point of maximum depression, the time of which varies with dose.
From: Langham, W. H., ed., *Radiobiological Factors in Manned Space Flight* (Publication No. 1487), p. 249. National Academy of Sciences, National Research Council, Washington, D.C., 1967.

Gonads

Although laboratory animals have been studied, there are relatively few data based on observations following clinical therapy or radiation accidents in man. Table 12 presents some generalized dose–response relationships.

Table 12. RADIATION DAMAGE TO GONADS

Dose, rads	Response
< 100	Reduced fertility
200–300	Temporary sterility for approximately 12 to 15 months
400–500	Temporary sterility for 18 to 24 months
500–600	Permanent sterility probable, if individual survives

From: Webb, P., ed., *Bioastronautics Data Book*, p. 153. National Aeronautics and Space Administration, Scientific and Technical Information Division, Washington, D.C., 1964.

Current research by Heller suggests that, in the human male, spermatogenesis provides an extremely radiosensitive index of exposure. His data are summarized in Table 13.

Table 13.
EFFECTS OF HIGH-INTENSITY DOSES OF X RAYS ON SPERMATOGENESIS

Dose, rads	Observed Effects on Sperm Count
15	Moderate oligospermia
20	Moderate oligospermia
50	Pronounced oligospermia
100	Marked oligospermia and azoospermia
200	Azoospermia
300	Azoospermia
400	Azoospermia
600	Azoospermia

From: Langham, W. H., ed., *Radiobiological Factors in Manned Space Flight* (Publication No. 1487), p. 128. National Academy of Sciences, National Research Council, Washington, D.C., 1967.

The time-course of the effect of radiation on spermatogenesis is shown graphically in Figure 16.

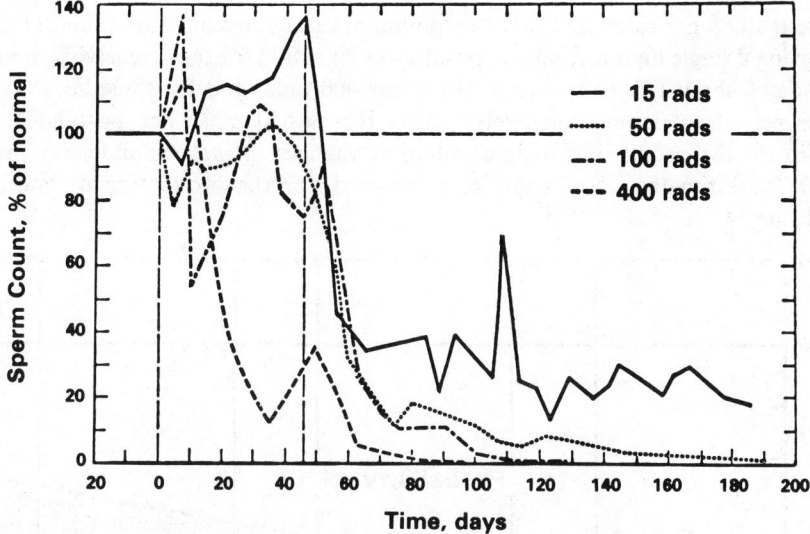

FIGURE 16.
Time-Course of Sperm Counts of Normal Man
Following High-Intensity Exposure to Various Doses
of 190-kvp X Rays (Heller, 1966).
(From: Langham, W. H., ed., *Radiobiological Factors in Manned Space Flight*
(*Publication No. 1487*), p. 127. National Academy of Sciences,
National Research Council, Washington, D.C., 1967.)

Eye

The lens of the eye is a radiosensitive, but slowly responding, tissue. The dose–response relationship for lens opacities following protracted exposure ranging from two to twelve weeks is shown in Figure 17.

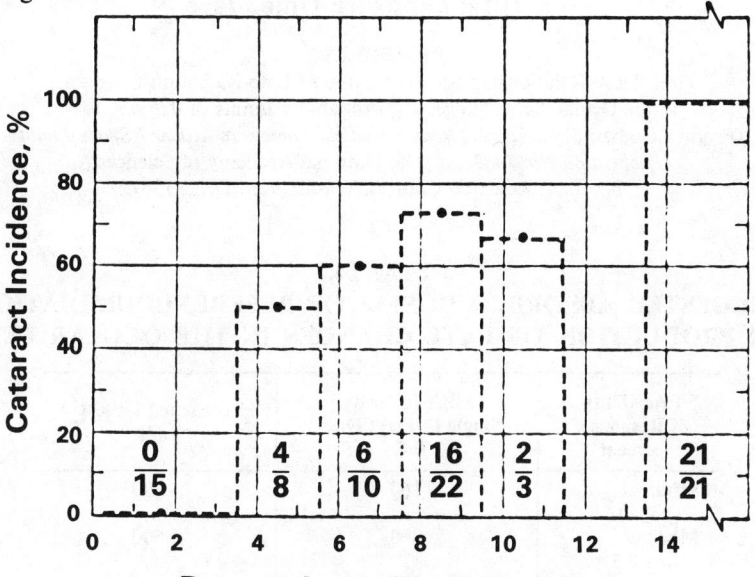

FIGURE 17.
Incidence of Late Lenticular Changes in Patients
Who Received Photon Exposures Fractionated over 2 to 12 Weeks.
(From: Langham, W. H., ed., *Radiobiological Factors in Manned Space Flight*
(*Publication No. 1487*), p. 142. National Academy of Sciences,
National Research Council, Washington, D.C., 1967.)

A recent study has estimated that the minimum cataractogenic dose found at the lens for those receiving a single high-intensity exposure was 200 rads; for those receiving multiple doses over periods of about two to twelve weeks it was 400 rads; and for those receiving multiple doses over periods of more than twelve weeks it was 550 rads. The postulated time–dose relationship for the production of late radiation changes in the ocular lens is presented in Figure 18. The estimated dose, single or protracted, for the production of late changes is given in Table 14.

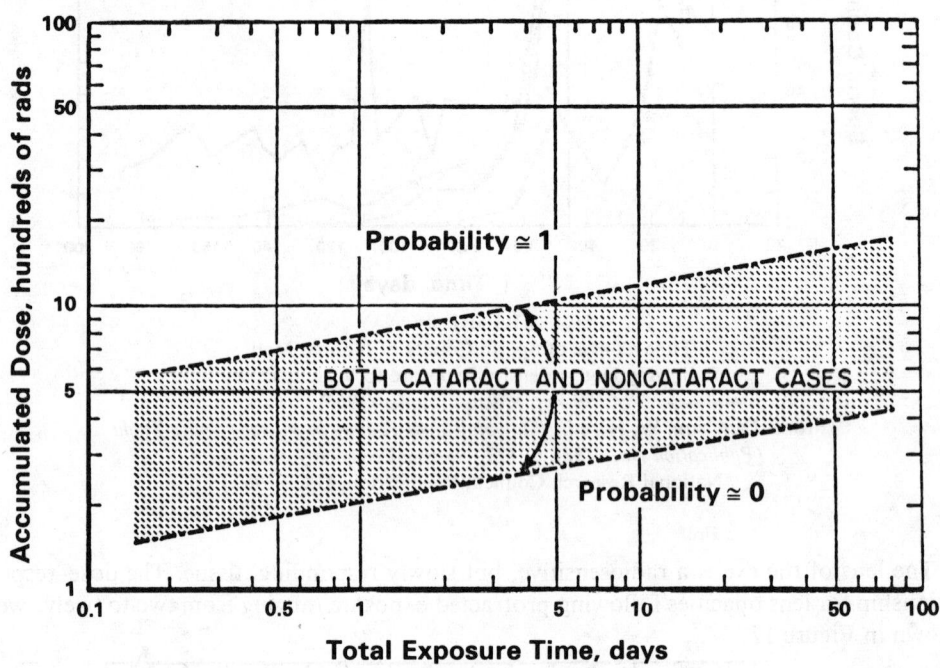

FIGURE 18.
Time–Dose Relationship for Production of Late Radiation Changes
in the Ocular Lens, Suggesting Probability Limits of $0 \leq p \leq 1$.
(From: Langham, W. H., ed., *Radiobiological Factors in Manned Space Flight*
(*Publication No. 1487*), p. 139. National Academy of Sciences,
National Research Council, Washington, D.C., 1967.)

Table 14.
SUGGESTED ABSORBED DOSES* OF REFERENCE RADIATION
FOR PRODUCTION OF LATE CHANGES IN THE OCULAR LENS[†]

Probability of Response, percent	High-Intensity Single (1-Day) Dose, rads	Protracted Dose,[‡] rads
Minimal ($p \gtreqqless 0$)	150	300
Median ($p \simeq 0.5$)[††]	300	600
Maximal ($p = 1$)	650	1,300

* Point of interest for dose estimation, 3 mm depth.
[†] See Effects on the Ocular Lens, p. 135 of reference cited below.
[‡] Dose protracted over 7 weeks or longer.
[††] Assuming log-normal distribution of response.
From: Langham, W. H., ed., *Radiobiological Factors in Manned Space Flight* (*Publication No. 1487*), p. 261. National Academy of Sciences, National Research Council, Washington, D.C., 1967.

Lethality

Lethality is usually expressed in terms of that dose which would prove fatal for 50 percent of a group of humans so exposed within a time period of either 30 or 60 days. Table 15 reviews estimates from various sources for the LD_{50} of man.

Table 15.
ESTIMATES OF LD_{50} OF MAN FOR HIGH-INTENSITY RADIATION EXPOSURE

Source	Exposure, roentgens	Dose, rads*	Reference
Patients, pathology of atomic-bomb casualties	~450	~300†	Warren and Bowers, 1950
Committee evaluation	400–600	260–400†	NAS–NRC, 1960
Marshallese observations, large-animal studies	~350	~300	Bond and Robertson, 1957 Cronkite and Bond, 1960
Committee evaluation	—	300–500‡	United Nations, 1962
Whole-body exposure of patients	370†	243 ± 22	Lushbaugh et al., 1966
Whole-body exposure of patients	380†	250 ± 28	NASA, p. 76, 1967
Whole-body exposure of patients and accident cases	430†	285 ± 25	NASA, p. 106, 1967

* Average absorbed dose near midline of the body.
† Assuming radiation of the quality of ^{137}Cs gamma rays.
‡ Nature of radiation and point of dose assessment not specified.
From: Langham, W. H., ed., *Radiobiological Factors in Manned Space Flight (Publication No. 1487)*, p. 106. National Academy of Sciences, National Research Council, Washington, D.C., 1967.

The dose–response relationship derived from dog, monkey, and human patients, together with a postulated relationship for normal man, are given in Figure 19.

The doses estimated for various probability levels for the production of early lethality are tabulated in Table 16.

Table 16.
ESTIMATED ABSORBED DOSES OF HIGH-INTENSITY REFERENCE RADIATION FOR PRODUCTION OF EARLY LETHALITY*

Response Probability Level, percent	Absorbed Dose, rads†
10	220
50	285
90	350

* See section on Early Lethality, p. 105 of reference cited below.
† Anatomical site of interest for dose estimation, 11-cm depth, total-body exposure.
From: Langham, W. H., ed., *Radiobiological Factors in Manned Space Flight (Publication No. 1487)*, p. 250. National Academy of Sciences, National Research Council, Washington, D.C., 1967.

Life Expectancy

There are no human data from which to draw definite information concerning a dose–response relationship relating radiation exposure and decreased life expectancy. Some extrapolations have been carried out on the basis of animal experimental studies. The estimated life expectancy of a twenty-year-old population, exposed to fixed daily doses of whole-body radiation until the time of death, is depicted in Figure 20. Deaths would be from natural causes, with an apparent acceleration of the normal processes of aging being invoked as the mechanism.

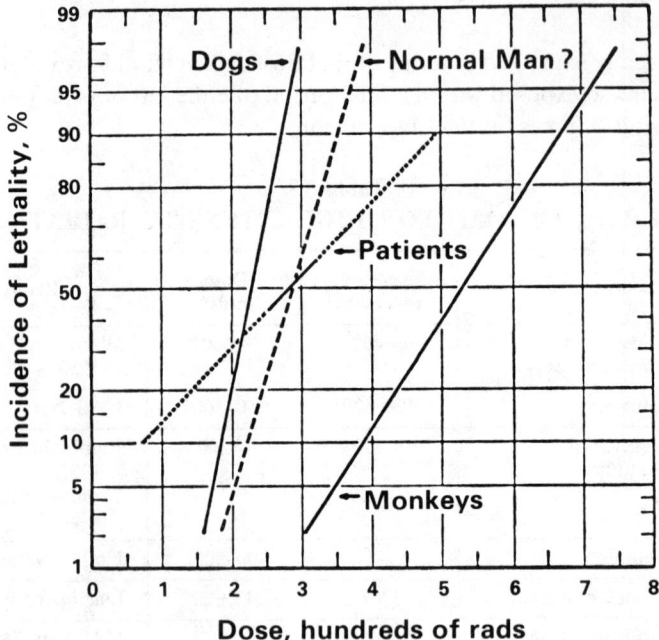

FIGURE 19.
Derived Dose–Lethality Relationships for Dogs, Monkeys,
and Human Patients, and a Postulated Relationship for Normal Man.
(From: Langham, W. H., ed., *Radiobiological Factors in Manned Space Flight*
(*Publication No. 1487*), p. 114. National Academy of Sciences,
National Research Council, Washington, D. C., 1967.)

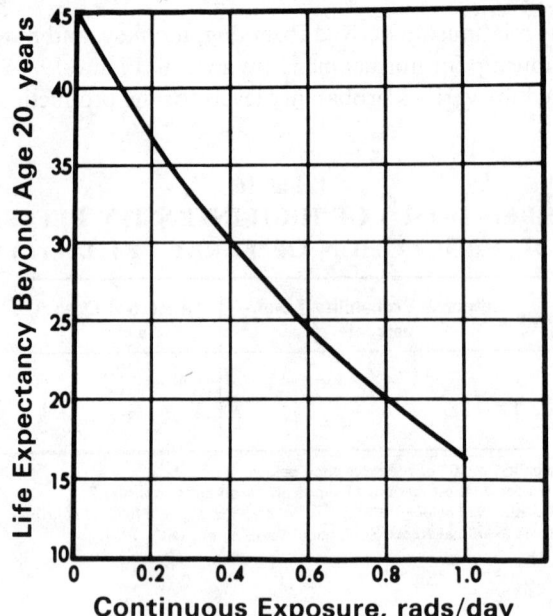

FIGURE 20.
Estimated Life Expectancy of a Twenty-Year-Old Population
Exposed to Fixed Daily Doses of Whole-Body Radiation
Until Time of Death (Based on Animal Studies).
(From: Webb, P., ed., *Bioastronautic Data Book*, p. 154.
National Aeronautics and Space Administration, Scientific and
Technical Information Division, Washington, D.C., 1964.)

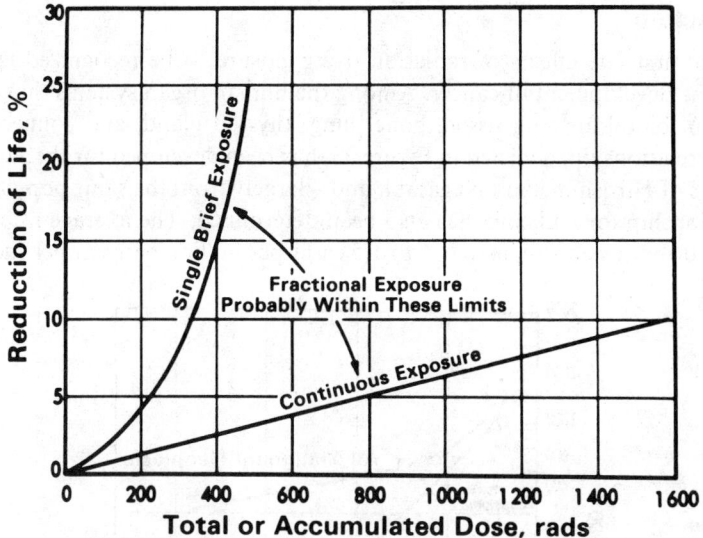

FIGURE 21.
Comparison of Effects of Brief Single Doses of Whole-Body Radiation
and Those of Long-Term Continuous Exposure on Life Expectancy
(Based on Animal Studies).
(From: Webb, P., ed., *Bioastronautic Data Book*, p. 154.
National Aeronautics and Space Administration, Scientific and
Technical Information Division, Washington, D.C., 1964.)

Figure 21 compares the effects of brief single doses of whole-body radiation on life expectancy with those of long-term continuous exposure. Multiple brief exposures separated by days or weeks—termed "fractional exposures"—will be expected to have an effect somewhere between the limits shown for single and continuous exposure.

A comparison between the effects of accumulated low-intensity and high-intensity radiation are graphically presented in Figure 22.

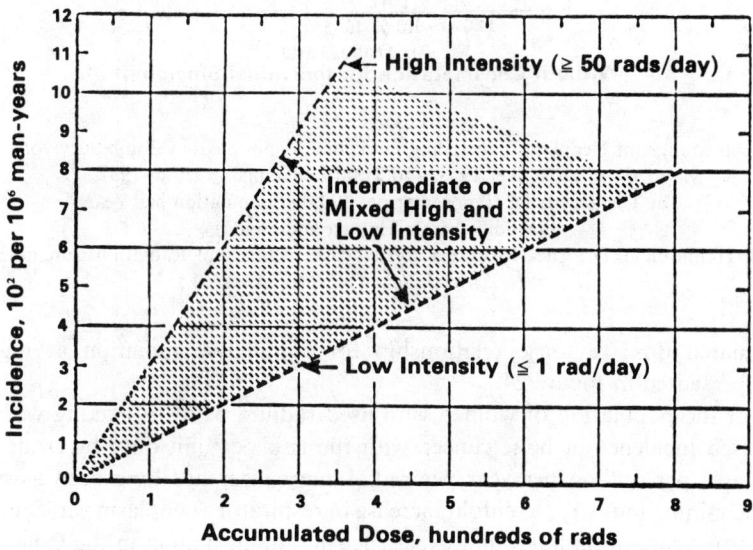

FIGURE 22.
Relationship of Accumulated Dose and Intensity of Reference-Quality
Whole-Body Radiation to Life-Shortening Probability.
(From: Langham, W. H., ed., *Radiobiological Factors in Manned Space Flight*
(*Publication No. 1487*), p. 264. National Academy of Sciences,
National Research Council, Washington, D.C.. 1967.)

Cancer Induction

One of the first late effects of radiation overexposure to be recognized is the increased likelihood of the development of cancer. Among the human organ systems that show evidence of this are skin, blood-forming tissue, bone, lung, thyroid gland, and connective tissues. A dose–response relationship, depicted in Figure 23, has been developed for the Japanese atomic-bomb survivors of Hiroshima and Nagasaki, and—largely from the same population—a dose–response relationship for leukemia has also been determined. The average rate of increase in incidence with dose was approximately 1 to 1.5 cases per million per year per rad.

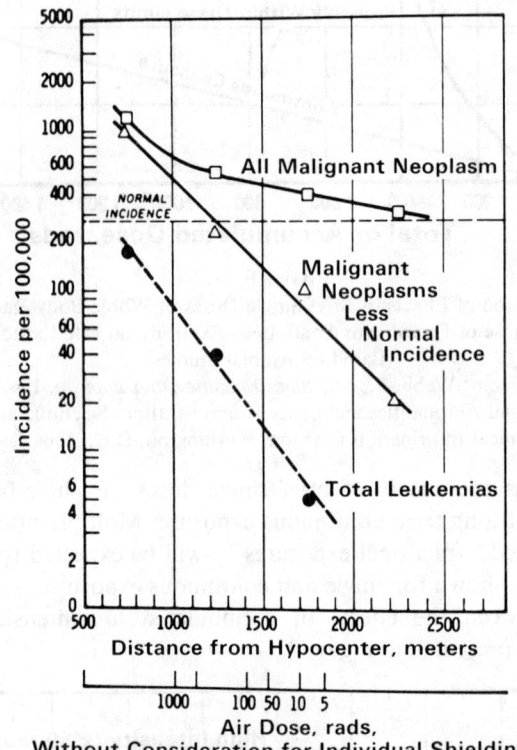

FIGURE 23.
All Malignant Neoplasms, Including Leukemia, Among Atom Bomb Survivors,
May 1957 to December 1958, and Total Leukemias, 1950 to 1957,
by Distance from Hypocenter per 100,000 Population per Year
(Modified from Harada and Ishida).
(From: Hamilton, L.D., The Hiroshima and Nagasaki Data and Radiation Carcinogenesis.
Ann. N.Y. Acad. Sci., 114: 241–247, 1964.)

An estimated dose–response relationship for cumulative radiation at high and low intensities is presented in Figure 24.

Studies of the population of women who used radium paint in making watch dials has shown increased incidence of bone cancer, with the risk per unit dose by crude estimate at about four cases per million per year per rad. Lung cancer has long been associated with workers in the mining industry; a tenfold increase in respiratory neoplasm was found in miners with five or more years of underground experience in radium mining in the Colorado plateau area. An increase in benign and malignant thyroid neoplasms has been observed in several populations receiving radiation as a form of therapy or accidentally following nuclear testing; the risk estimate appears to be between 0.5 and 1.5 cases per million per year per rad. Malignant tumors in reticuloendothelial systems are increased in populations receiving thorium dioxide

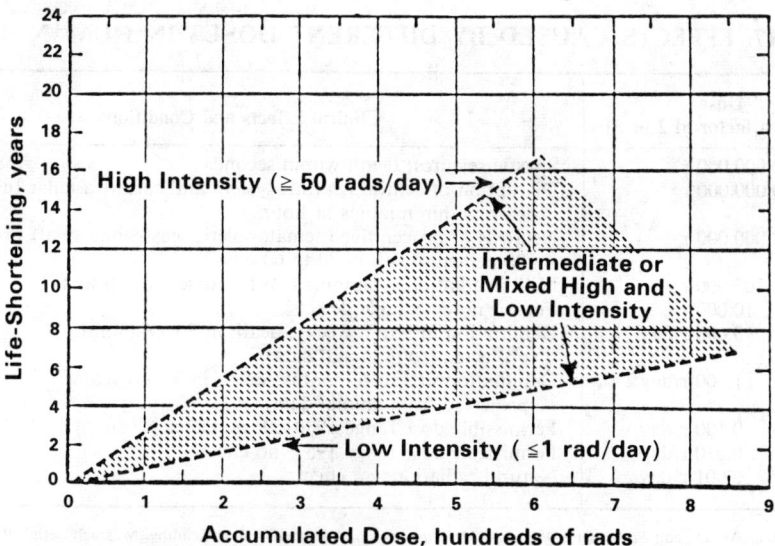

FIGURE 24.
Relationship of Accumulated Dose and Intensity of Reference-Quality
Whole-Body Radiation to Increased Probability of Leukemia.
(From: Langham, W. H., ed., *Radiobiological Factors in Manned Space Flight*
(*Publication No. 1487*), p. 265. National Academy of Sciences,
National Research Council, Washington, D.C., 1967.)

internally as a tool in diagnostic radiology for visualization of various organs and cavities of the body.

Genetic Changes

Large-scale observations of the progeny of atomic-bomb survivors in Hiroshima and Nagasaki, Japan, have not demonstrated any significant increase in genetic abnormalities. However, a large body of plant and animal experimentation suggests that the many difficulties in making observations in man may preclude the recognition of very subtle changes. Based on such animal experimental data, knowledge of the radiation dose received by the gonad permits an estimate of the probability of mutation in the spermatogonial cells and of its expression in the offspring. Such a calculation assumes reasonable knowledge of the following parameters (the values given in parentheses are based largely on animal data).

1. The mutation rate per rad per gene ($\mu = 5 \times 10^{-8}$).
2. The number of genes per haploid set ($n = 10^4$).
3. The probability of expression of the new mutation in the first-generation heterozygote ($s = 5 \times 10^{-2}$).
4. The gonad dose, D.

Assuming a dose of 100 rads, for example, the probability of a new mutation in the immediate offspring would be

$$(5 \times 10^{-8})(10^4)(5 \times 10^{-2})(10^2) = 2.5 \times 10^{-3}.$$

Such probability statements must obviously be accepted with considerable reserve.[3]

WHOLE-BODY EFFECTS

General Dose–Effect Relationships

In order to place a whole-body radiation injury in perspective, it is useful to consider the dose range from natural radiation backgrounds to 100,000 r. The clinical effects of exposures within this range are presented in Table 17.

Table 17. EFFECTS CAUSED BY DIFFERENT DOSES IN HUMAN BEINGS

Dose (within a factor of 2 or 3)	Clinical Effects and Conditions
100,000.000 r	Spastic seizures; death within seconds
10,000.000 r	Disruption of central nervous system and cardiovascular function; death within minutes or hours
1,000.000 r	Necrosis of progenitive (hematopoietic, gastrointestinal) tissues; 100 percent death in 30 to 60 days
100.000 r	Mild irradiation symptoms in a few cases; no deaths
10.000 r	Few or no detectable defects
10.000 r/day	Debilitation in 3 to 6 weeks; death in 3 to 6 months (projected from animal data)
1.000 r/day	Debilitation in 3 to 6 months; death in 3 to 6 years (projected from animal data)
0.100 r/day	Permissible dose range 1930–50; no effect
0.010 r/day	Permissible dose range 1957; no effect
0.001 r/day	Natural radiation; no effect

From: Claus, W. D., ed., *Radiation Biology and Medicine*, p. 336. Addison-Wesley, Reading, Massachusetts, 1958.

A graphic presentation of the relationship between radiation dose and possible biological effects of whole-body radiation exposure is given in Figure 25.

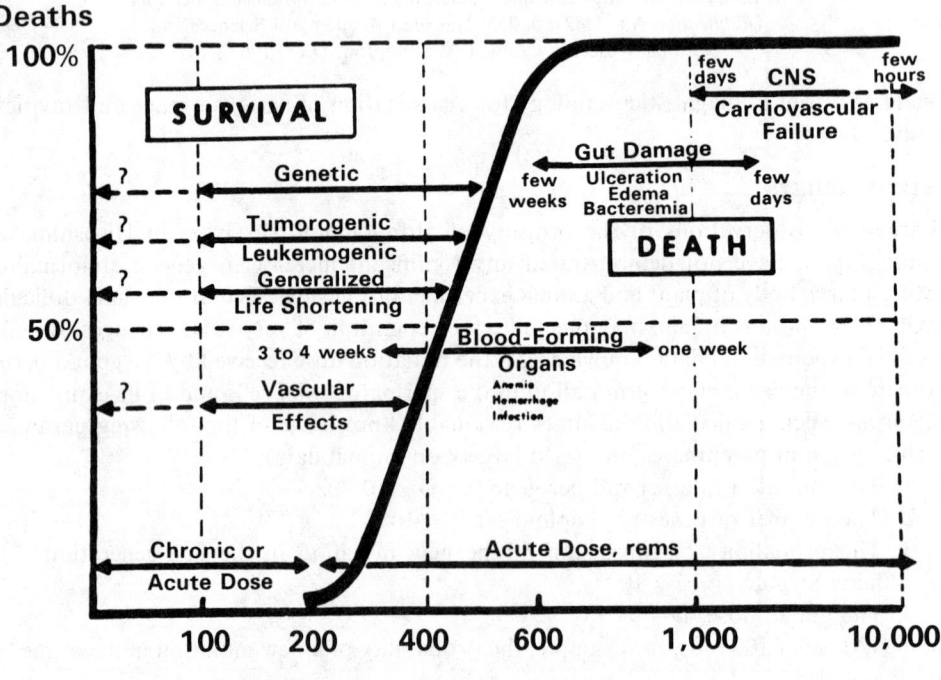

FIGURE 25.
Possible Biological Effects of Whole-Body Radiation Exposure.
(From: Shilling, C. W., ed., *Atomic Energy Encyclopedia in the Life Sciences*; p. 468.
W. B. Saunders, London, England, and Philadelphia, 1964.)

Acute Radiation Syndrome

When a man is exposed to a single large short-term whole-body dose of ionizing radiation, the resultant injury is expressed as a complex of clinical symptoms, signs, and laboratory findings, which are collectively termed the "acute radiation syndrome". This syndrome represents the clinical expression of the damage to many important organs simultaneously by depletion of the radiosensitive cells, which are unable to reproduce themselves because of radiation-induced disturbances at the molecular level. The various clinical signs and symptoms resulting from short-term whole-body exposure are tabulated in Table 18. A graphic version of the time sequence of the symptoms is given in Figure 26.

The typical chronological sequence of events following a large whole-body radiation exposure can be divided into four clinical stages.

1. The initial or prodromal stage—0 to 48 hours post exposure.
2. The latent stage—48 hours to 2 or 3 weeks post exposure.
3. The manifest-illness stage—2 or 3 weeks to 6 or 8 weeks post exposure.
4. The recovery stage—6 or 8 weeks to several months post exposure.

The signs and symptoms of the prodromal stage are shown in Table 19, and the estimated absorbed doses required for varying probabilities of some of the key prodromal symptoms are given in Table 20. The signs and symptoms of the manifest-illness stage are presented in Table 21.

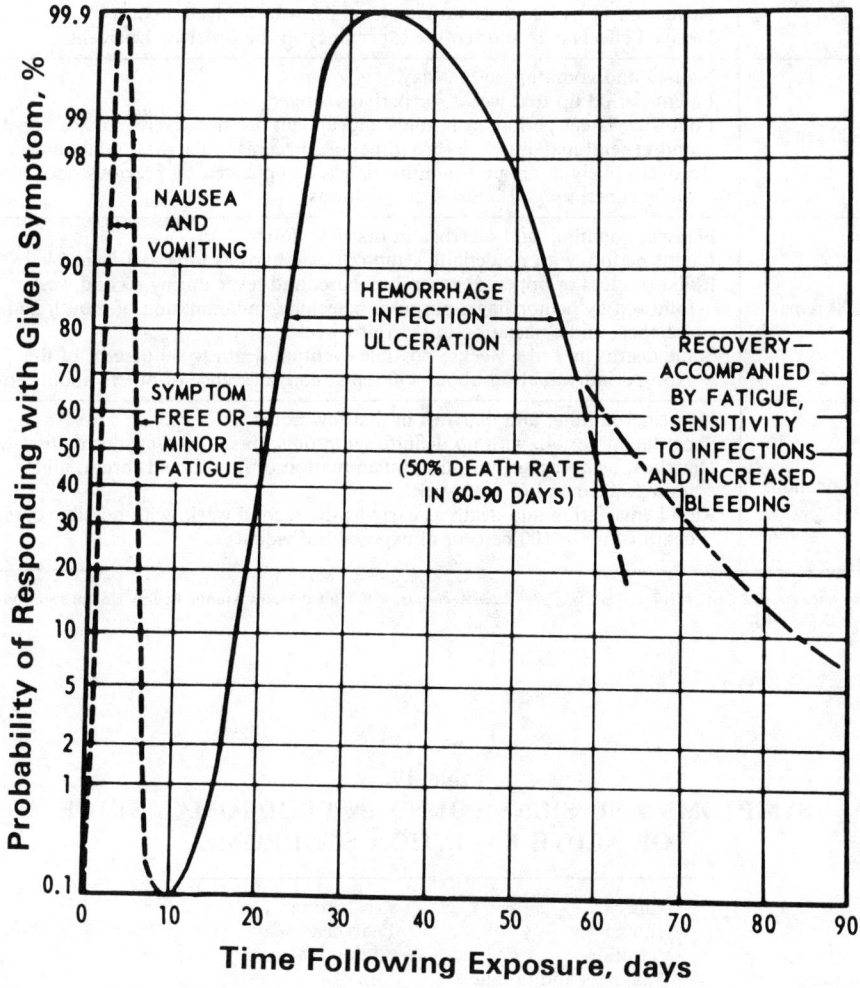

FIGURE 26.
Idealized Description of the Time-Course of Symptoms
of Acute Radiation Illness Following Midlethal Acute Exposure
to Whole-Body Radiation of 250 to 500 rads.
(From: Webb, P., ed., *Bioastronautic Data Book*, p. 146.
National Aeronautics and Space Administration, Scientific and
Technical Information Division, Washington, D.C., 1964.)

Table 18.
SUMMARY OF EFFECTS ON MAN AS A RESULT OF SHORT-TERM WHOLE-BODY EXTERNAL EXPOSURE TO RADIATION

Dose Equivalent	Effects on Man
0 to 25 rems	No detectable clinical effects Delayed effects may occur
25 to 100 rems	Slight transient reductions in lymphocytes and neutrophils Disabling sickness not common; exposed individuals should be able to proceed with usual tasks Delayed effects possible, but serious effects on average individual very improbable
100 to 200 rems	Nausea and fatigue, with possible vomiting above 125 rems in about 20 to 25 percent of people Reduction in lymphocytes and neutrophils, with delayed recovery Delayed effects may shorten life expectancy in the order of 1 percent
200 to 300 rems	Nausea and vomiting on first day Latent period up to 2 weeks, or perhaps longer Following latent period, symptoms appear, but are not severe: loss of appetite, and general malaise, sore throat, pallor, petechiae, diarrhea, moderate emaciation Recovery likely in about 3 months, unless complicated by previous poor health or by superimposed injuries or infections
300 to 600 rems	Nausea, vomiting, and diarrhea in first few hours Latent period, with no definite symptoms, perhaps as long as 1 week Epilation, loss of appetite, general malaise, and fever during second week, followed by hemorrhage, purpura, petechiae, inflammation of mouth and throat, diarrhea, and emaciation in the third week Some deaths in 2 to 6 weeks; possible eventual death to 50 percent of the exposed individuals at about 450 rems; convalescence of others about 6 months
600 rems or more	Nausea, vomiting, and diarrhea in first few hours Short latent period, with no definite symptoms, in some cases, during first week Diarrhea, hemorrhage, purpura, inflammation of mouth and throat, and fever toward end of first week Rapid emaciation and death as early as the second week, with possible eventual death of up to 100 percent of exposed individuals

From: Saenger, E. L., ed., *Medical Aspects of Radiation Accidents*, p. 9. United States Atomic Energy Commission, Washington, D.C., 1963.

Table 19.
SYMPTOMS AND SIGNS FOUND IN PRODROMAL STAGE OF ACUTE RADIATION SYNDROME

Anorexia	Prostration
Nausea	Diarrhea
Vomiting	Abdominal pain
Weakness and fatigue	
Conjunctivitis	Sweating
Erythema	Oliguria
Fever	
Hyperesthesia	Paresthesia
Ataxia	Coma
Disorientation	Death
Shock	

Table 20.
ESTIMATED ABSORBED DOSES OF HIGH-INTENSITY REFERENCE RADIATION FOR PRODUCTION OF EARLY PRODROMAL SEQUELAE

Clinical Sign	Absorbed Dose for Probability of Response, rads*		
	10%	50%	90%
Anorexia	40	100	240
Nausea	50	170	320
Vomiting	60	215	380
Diarrhea	90	240	390

* Anatomical region of interest for dose estimation: a sphere, 26 cm in diameter, in the midepigastric region.
From: Langham, W. H., ed., *Radiobiological Factors in Manned Space Flight* (*Publication No. 1487*), p. 248. National Academy of Sciences, National Research Council, Washington, D.C., 1968.

Table 21.
SYMPTOMS AND SIGNS FOUND IN MANIFEST-ILLNESS STAGE OF ACUTE RADIATION SYNDROME

Anorexia	Sweating
Nausea	Oliguria
Vomiting	Weakness and fatigue
Diarrhea	Prostration
Abdominal pain	Weight loss
Abdominal distention	Hyperesthesia
Conjunctivitis	Paresthesia
Erythema	Ataxia
Jaundice	Disorientation
Fever	Shock
Infection	Epilation
Purpura	Coma
Hemorrhage	Death
Scalp pain	

From: Saenger, E. L., ed., *Medical Aspects of Radiation Accidents*, p. 42. United States Atomic Energy Commission, Washington, D.C., 1963.

A general summary of five well-defined clinical injury patterns into which radiation-overexposure patients can be grouped has been developed as follows.[4]

Group I.

Most of these patients are asymptomatic; a few may have minimal prodromal symptoms, such as nausea and anorexia; little therapy is required.

Group II.

These patients develop the acute radiation syndrome in a mild form; after transient prodromal nausea and vomiting, laboratory and mild clinical evidence of hematologic derangement dominates the picture. Careful observation of symptoms, but conservative therapy, is indicated.

Group III.

Severe manifestations occur in these patients; complications of hematologic malfunction, including hemorrhage and infections, are pronounced; in the sicker members of this group, some evidence of gastrointestinal damage may also be present; survival of these patients may depend on wise clinical management.

Group IV.

An accelerated version of the acute radiation syndrome takes place; complications of gastrointestinal injury, such as diarrhea and ileus, dominate the clinical picture; the severity of hematopoietic disturbances is related to the length of survival time following exposure; recovery is not likely.

Group V.

A fulminating course occurs, with marked central-nervous-system impairment, including paresthesia, excitement and coma, as well as cardiovascular complications, such as severe hypotension; a fatal outcome may be expected.

Internal Contamination

Internal contamination with radionuclides in most instances does not present immediate evidence of acute radiation injury. Inadvertent overdosage with ^{32}P, ^{131}I, or ^{198}Au are exceptions to this generalization. The usual problem, however, is one of long-term low-level exposure by the radionuclides deposited in specific organs of concentration. The factors influencing the entry and distribution of radionuclides in living organisms are listed in Table 22. The organs of preferential concentration for particular radionuclides are tabulated in Table 23.

Table 22.
FACTORS INFLUENCING THE ENTRY AND DISTRIBUTION OF RADIONUCLIDES IN LIVING ORGANISMS AND CELLS

1. Physicochemical properties of the radionuclide
   a) physical state
   b) solubility
   c) particle size
   d) tendency to form colloids
   e) formation of unabsorbable complexes or chelates
2. Radiochemical properties of the radionuclide
   a) physical half-life
3. Biological considerations
   a) transfer of the radionuclide in food chain
   b) metabolic pathway of the radionuclide *in vivo*
   c) degree and route of absorption of the radionuclide
   d) nutritional and physiological status of the organism

From: Morgan, K. Z. and Turner, J. E., eds., *Principles of Radiation Protection*, p. 438. John Wiley and Sons, New York, 1967.

Table 23.
RADIOISOTOPES, AND THE ORGANS AND TISSUES IN WHICH THEY MAY BE CONCENTRATED

Organs	Radioisotopes
Lungs	^{63}Ni, ^{222}Rn, ^{210}Po, ^{238}U, ^{239}Pu
Kidneys	^{51}Cr, ^{56}Mn, ^{71}Ge, ^{76}As, ^{105}Rh, ^{106}Ru + ^{106}Rh, ^{127}Tc, ^{129}Te, ^{190}Ir, ^{192}Ir, ^{198}Au, ^{238}U
Liver	^{56}Mn, ^{59}Ni, ^{60}Co, ^{64}Cu, ^{105}Ag, ^{109}Cd + ^{109}Ag, ^{111}Ag
Bone	^{7}Be, ^{14}C, ^{18}F, ^{32}P, ^{45}Ca, ^{48}V, ^{65}Zn, ^{72}Ga, ^{89}Sr, ^{90}Sr + ^{90}Y, ^{91}Y, ^{95}Nb, ^{99}Mo, ^{113}Sn, ^{140}Ba + ^{140}La, ^{143}Pr, ^{144}Ce + ^{144}Pr, ^{147}Pm, ^{151}Sm, ^{154}Eu, ^{166}Ho, ^{170}Tm, ^{177}Lu, ^{185}W, ^{203}Pb, ^{226}Ra, ^{233}U, ^{234}Th, ^{239}Pu, ^{241}Am, ^{242}Cm

From: RCA Service Company, *Atomic Radiation*, p. 53. RCA Service Co., Inc., Camden, New Jersey, 1957.

REFERENCES

1. Bender, M. A., Somatic Chromosomal Aberrations. *Arch. Environ. Health,* 16: 556–564, 1968.
2. Morgan, K. Z. and Turner, J. E., *Principles of Radiation Protection,* pp. 461, 462. John Wiley and Sons, New York, 1967.
3. Langham, W. H., ed., *Radiobiological Factors in Manned Space Flight (Publication No. 1487)*, p. 182. National Academy of Sciences, National Research Council, Washington, D.C., 1967.
4. Thoma, G. E., Jr. and Wald, N., The Diagnosis and Management of Accidental Radiation Injury, *J. Occup. Med.,* 1: 421–447, 1959.

MEDICAL MANAGEMENT OF RADIATION EMERGENCIES

IMMEDIATE EMERGENCY PROCEDURES

The immediate goals in a radiation emergency involving people are removal of the individuals from any further sources of radiation exposure and carrying out any necessary first aid for life-threatening conventional injury. Table 24 provides an operational plan in a logical sequence of priorities for this purpose.

Table 24. EMERGENCY PROCEDURES FOR RADIATION ACCIDENTS

1. Survey and then evacuate possibly exposed personnel from accident area, and give urgent first aid.
2. Notify: Medical Authority
      Phone _____
      Health Physics (Industrial Hygiene)
      Phone _____
      Management
      Phone _____
3. Close off radiation area. Turn off air supply. Seal area if contamination is likely.
4. Confine and survey all contaminated people, then give first aid for traumatic injury and burns.
5. Evaluate situation in regard to:
      contamination by radionuclides
      neutron exposure
      level of radiation exposure.
6. If contamination is present, perform simple decontamination and resurvey patient.
7. Put patient to bed. Do brief physical examination.
8. Save all samples of clothes, jewelry, blood, urine, stool, and vomitus. Label with name, time, and date.
9. Do routine blood counts.
10. Obtain careful history of accident.
11. Send patient to hospital if exposure of 100 r or more is suspected.

From: Saenger, E. L., ed., *Medical Aspects of Radiation Accidents,* form provided. United States Atomic Energy Commission, Washington, D.C., 1963.

PATIENT CARE

Initial Care

The following eleven rules have been proposed for the use of physicians and other appropriate personnel after the immediate emergency procedures have been carried out. They are presented in Table 25. It is evident that expert advice can and should be sought, if needed, following the emergency period of medical care.

Emergency Care for Possibly Contaminated Persons

The following procedures have been proposed to facilitate care of the patients while minimizing further radiation exposure to the patients themselves as well as to those in attendance of them. In addition, these procedures will facilitate the necessary examination for diagnostic and prognostic purposes.

1. All suspected persons should be surveyed for radioactive contamination.
2. If no monitoring instrument is available, all possibly exposed persons should be regarded as contaminated. Wipes from various parts of the bodies of these persons and their clothing should be made with some type of disposable tissue,

filter paper, or blotting paper, and the samples placed in separate labeled envelopes for future study.
3. Contaminated clothing should be carefully removed and placed in some type of disposable container or bag; if this is not available, the clothing should be put on sheets of paper, to prevent contamination of floor and furniture. The clothing and paper can later be monitored to determine the possibility of contamination or the need for disposal.
4. If necessary, contaminated persons should be taken to a shower area for bathing.
5. Bathing should be done under showers; commercially available detergents and soaps can be used. Several separate washings should be performed. Highly alkaline soaps, abrasives, organic solvents, or cleaners that tend to increase permeability of the skin should not be used. Special emphasis should be given to cleaning the fingernails, toenails, nostrils, scalp, ears, and body folds.

Table 25. RULES PRIMARILY, BUT NOT EXCLUSIVELY, FOR PHYSICIANS

Rule 1.
If there is radionuclide contamination, all exposed individuals shall be surveyed and decontaminated before being examined and treated.
    If this step is neglected, personnel whose duty is to care for patients may become so contaminated themselves as to be rendered ineffective. Proper protective clothing shall be available.

Rule 2.
Do simple decontamination of patients if radionuclides are present.
    The operator should wear coveralls or scrub suit, shoe covers, gloves, cap, and respiratory mask as indicated by initial survey before working with patients.
If patient is ambulatory:
Spread sheet or paper for patient to stand on. Patient disrobes, putting clothing in suitable container (bag, waste basket, etc.) for later survey. Nose, mouth, and ear wipes are done. Patient takes shower and is resurveyed until clean.
If patient is not ambulatory:
Patient is placed on sheet, and clothing is cut off. Clothing is saved as above for later surveying. Patient is washed by operator and repeatedly surveyed.

Rule 3.
If there is evidence of massive exposure to external radiation, and if large numbers of individuals are involved, triage will be needed.
    It may be necessary to limit drastically any extensive medical care to moribund individuals in order to utilize available facilities for the care of those potentially salvageable. Dose estimates of over 2,000 r for external radiation, or over 2,000 rad/hr (rem, r) for radionuclide contamination, would indicate that little could be offered to such patients. In the latter case, contamination of treatment personnel may become a major problem.

Rule 4.
Put patients at bed rest. Obtain careful history. Do brief physical examination.

Rule 5.
Obtain routine blood count. Obtain additional 20-cc sample of venous blood, if there is neutron exposure.

Rule 6.
Patients with severe nausea and vomiting should be hospitalized.

Rule 7.
No specific therapy is indicated for acute radiation injury within the first few days after exposure.

Rule 8.
If a dose estimate of less than 100 rad (r) can be estimated for certain individuals, they do not require emergency care of a major degree.

Rule 9.
Individuals receiving only external exposure from alpha, beta, gamma, and X radiation are not radioactive.

Rule 10.
Individuals receiving neutron radiation are slightly radioactive, but produce no hazard to personnel caring for them.

Rule 11.
Individuals contaminated with radionuclides present a hazard to personnel caring for them. The degree of hazard depends upon the level and type of contamination.

From: Saenger, E. L., ed., *Medical Aspects of Radiation Accidents*, p. 5. United States Atomic Energy Commission, Washington, D.C., 1963.

6. Scrub brushes should be used, but care should be taken that the skin surfaces do not become abraded.
7. After the body is well washed, the person should be surveyed with a suitable monitoring instrument, and additional smears should be taken, using disposable tissues, cotton-tipped applicators, or filter paper. The ear canals and nostrils should be swabbed for contamination. Smear tests are especially important if alpha survey instruments are not available. Clothing known not to be contaminated should be put on.
8. Small cuts and other breaks in the skin surface should be carefully searched for, since absorption of radionuclides can occur by this route. Such lesions should be decontaminated by repeated 5-minute scrubs after removal of scabs and crusts, following the above-mentioned washes.
9. Suitable syringes, curved basins, and appropriate irrigating solutions should be readily available for conjunctival irrigation. The used solutions should be saved for counting, preferably in separate labeled bottles marked consecutively in the order of collection.
10. If there is a possibility of alpha contamination, it will be necessary to use circular filter paper discs or other suitable type of filter paper that can be counted in the available type of alpha chamber. This detail should be prepared in advance by consultation with the laboratory that will carry out the counting of these wipes.
11. A physician should be called immediately for recommendations and advice on taking further action. Any and all of the following steps may assist in the formulation of his recommendations:
    a) complete medical history and physical examination, with special emphasis on previous occupational history and possible exposure to radiation, and a chest roentgenogram;
    b) complete blood count, including hematocrit reading, and routine urinalysis;
    c) quantitative collection of urine for the first 72 hours for assay of the radionuclide; each day's specimens may be collected in bottles containing 10 ml of dilute nitric acid (approximately 10 ml of concentrated nitric acid should be added to the specimens after the collection is complete, since nitric acid prevents absorption of radium on vessel walls);
    d) feces collected for the first 72 hours for determination of radioactivity; each day's specimens should be put in a separate container; these can be collected in round 1-quart (1-liter) ice cream containers.
    e) breath samples for radon, if the accident involves radium;
    f) arrangement for surveys of the total-body gamma radiation with a sensitive measuring device;
    g) samples of blood within 72 hours for determination of radioactivity;
    h) the specimens of urine, feces and blood should be refrigerated, and kept until arrangement can be made for analysis at a qualified laboratory; proper collection and storage of these samples will be of great value to the contaminated persons, and also in obtaining further data concerning the metabolism of the nuclide involved.[1]

Skin Decontamination Procedures

A suggested plan of decontamination is presented in Table 26.

The safe removal of radionuclide contamination from the skin requires a systematic approach based on both technical information and an awareness that the objectives are not to merely remove the contamination from the individual, but also to contain it and dispose of it,

Table 26. SUGGESTED PLAN OF DECONTAMINATION

A. Gross Decontamination
    1. Removal of contaminated clothing
    2. Washing and removal of contaminated hair
    3. Removal of gross wound contamination
B. Intermediate Stage (at clean location, if necessary)
    1. Removal of contaminated clothing
    2. Further local decontamination, swabs of body orifices
    3. Supportive measures, first aid
C. Final Stage
    1. Patient discharged with fresh clothing
    2. More definitive decontamination (surgical) and other therapy at dispensary or hospital

From: Saenger, E. L., ed., *Medical Aspects of Radiation Accidents*, p. 78. United States Atomic Energy Commission, Washington, D.C., 1963.

so that no other individuals or facilities are unnecessarily contaminated. The recommended procedure for washing contaminated skin and hands is as follows.

1. The skin area should be monitored by a suitable survey meter, if immediately available. If a survey instrument is not immediately available and a person is suspected of being contaminated, wash immediately; useful detergent preparations, in order of increasing strength, are listed in Table 27.

Table 27.
USEFUL DETERGENT PREPARATIONS FOR DECONTAMINATION OF SKIN AND WOUNDS

A. Aqueous Preparations
    1. Soap and water
    2. Abrasive soap and water
    3. Commercial detergents (10%): Tide, Dreft, Alconox, HemoSol
    4. Complexing agent (1% citric-acid solution)
    5. Chelating agent (1% versene solution) with or without detergent
B. Waterless Preparations
    1. Corn meal and commercial powdered detergent in equal parts made into a water paste and used without additional water; scrubbing with brush, removal with cotton or soft tissues
    2. Waterless mechanics' hand cream, used without additional water; scrubbing with brush, removal with cotton or soft tissues
    3. Homogenized cream of 8% carboxymethylcellulose, 3% commercial powdered detergent, 1% versene, and 88% distilled water, used without additional water; scrubbing with brush, removal with cotton or soft tissues

From: Saenger, E. L., ed., *Medical Aspects of Radiation Accidents*, p. 81. United States Atomic Energy Commission, Washington, D.C., 1963.

    Do not delay for want of a monitoring instrument. All readings should be recorded. Wipes of contaminated areas may be useful; these can be made with filter paper of appropriate size to fit in available gas-flow counters for alpha or beta activity; usually a disc of 1-inch diameter is adequate. These wipes can also be counted by beta end-window counters, or can be placed in test tubes and counted in gamma scintillation well counters. Cotton-tipped swabs should be used for detection of activity in nares, ear canals, mouth, and other orifices, as needed. If possible, clean areas should be covered with plastic drapes or towels, to prevent spread of radioactive materials.

2. Wash thoroughly for 2 to 3 minutes by the clock, using tepid (not hot) water and a mild detergent or soap; cover the entire contaminated surface with a

good lather; rinse off completely with running water. Repeat at least three times. Do not use abrasive or highly alkaline soaps or powders. The skin area should be surveyed between washes.
3. If the above procedures do not remove all dirt and contamination, scrub the hands for a period of 8 to 10 minutes by the clock, using liquid or cake soap, hand brush, and tepid water. Light pressure should be exerted on the brush; do not press so hard that the bristles are bent out of shape. There should be at least three complete changes of soap and water. All surfaces should be covered with a minimum of four brush strokes. A convenient routine is to start by scrubbing one thumb, being certain to brush all surfaces, and then proceed to the web space between thumb and index finger, and similarly over the other fingers; next scrub the palm and dorsal surface; each nail and cuticle must be similarly scrubbed. The hand should then be thoroughly rinsed, and checked by a survey meter. An emollient cream should then be applied. The brush and towels should be discarded into a radioactive-waste can or bag suitably labeled. If contamination is localized to a specific part, masking above the contaminated area with plastic material and cellophane tape will prevent the spread of the contaminant.
4. Where water is limited in quantity, waterless cleaners can be used (see Table 27), although they tend to defat the skin. Thus a protective barrier may be removed, and percutaneous absorption may be increased.
5. Contamination of the eyes, nose, and mouth can be handled only by copious amounts of water at first. Isotonic irrigants should be obtained as soon as possible. Eye cups should not be used.
6. Fission-product removal, using titanium dioxide: titanium dioxide may be used as a paste or slurry, made by shaking the powder into the wet palm of the hand until a good paste is formed; run tap water over the hands continually, so that the paste is kept wet, and apply this lather thoroughly to all hand surfaces—especially around the fingernails—for a minimum time of 2 minutes; rinse off thoroughly with lukewarm water; follow with a thorough washing, using soap and water and a handbrush. If any of the paste is left under the nails after washing, it will form a rather hard cake that is difficult to remove.
7. Sweating techniques will sometimes be useful in decontamination. The affected area is covered with a rubber or plastic dam or glove. After local application of heat to promote sweating, the affected area can be washed gently, and then resurveyed.

Ambulatory individuals can carry out their own decontamination, if there are sufficient washing facilities; however, they must be instructed to wash without splashing contaminated water on the floor and walls in order not to spread contamination. A trained attendant with suitable monitoring instruments should be assigned to them. An experienced team of four decontaminators working in a good physical layout with two sinks and two showers can probably process about four to six moderately contaminated individuals per hour. The most heavily contaminated individuals should be processed first.

If wounds or other injuries are present, decontamination will take much longer. An autopsy table provides an excellent area for such work in behalf of nonambulatory patients. If such facilities are not available at the accident site, suitable arrangements should be made with one or more community hospitals. In such situations, gross decontamination by removal of clothing should be carried out at the accident site, and the individual should be given necessary first aid. The patient should be transported to the hospital wrapped in a sheet or blanket, and definitive decontamination should then be done.

Care of Contaminated Wounds

At the accident site the following first-aid procedures are suggested.
1. Wash wounded area with large volumes of water, if available; spread cut edges so as to flush the wound well and stimulate bleeding.
2. A light tourniquet may be applied, to restrict venous return without restricting arterial flow. Suitable precautions in tourniquet use must be followed.
3. Report immediately thereafter for definitive medical care.

The general surgical principles of asepsis, debridement of necrotic and severely damaged tissues, restoration of tissue continuity to preserve normal function and appearance, and treatment of shock and infection apply in these situations. The only modifications are in the aseptic procedures and debridement.

Decontamination can be combined with asepsis by scrubbing the skin with detergents or mild soaps and a brush, and by more frequent washing with water for periods of 10 to 30 minutes. Antiseptics should not be used. In some situations, shaving may be the only method for rapid removal of the bulk of the contaminant. It is important that the contaminants on the skin not be introduced into the wound. Preliminary isolation of the contaminated area by application of barriers or cellophane drapes will provide some protection. Further contamination of the wound during decontamination of skin may be prevented by preliminary suture of the wound. Repeated monitoring must be used during the entire procedure. After the skin is decontaminated, the wound must be thoroughly irrigated to remove foreign substances, which may not necessarily be radioactive. Isotonic sodium chloride is the only solution that can be used safely for this purpose; complexing or chelating agents may enhance absorption, or they may be absorbed in toxic amounts.

Obvious necrotic and devitalized tissue must be removed surgically; tissue so injured may be an efficient trap for contaminants. All such material should be saved for analysis by conventional counting or by radiochemical or radioautograph techniques; labeled containers have to be provided for this purpose. As uncontaminated tissue is encountered during debridement, contaminated instruments, linen and drapes should be discarded. Frequent monitoring of the wound, debrided tissues and the instruments should be employed. The extent of debridement is determined by the level of activity, the radionuclide present, its metabolism and toxic effects, and the age of the patient. In treating penetrating wounds such as punctures, splinters, or slivers in the skin, a small block dissection is frequently indicated. If such excision is done promptly, absorption within the body can be significantly decreased. Obviously such therapy should be more radical for alpha emitters, other bone seekers, such as ^{90}Sr, and for millicurie or curie amounts. For microcurie quantities of shorter-lived radionuclides—such as ^{131}I, ^{32}P, or ^{198}Au—more conservative therapy can be employed. In such situations all possible methods leading to accurate bioassay must be employed. All debrided material, surgical sponges, collected irrigating solutions, and instruments must be monitored. A venous blood sample should be drawn as soon as possible after injury under conditions that will assure that no surface contamination is inadvertently introduced into the body by venipuncture. The venipuncture site must be surveyed, and decontaminated, if necessary. The blood sample should be assayed promptly for radioactivity. All urine and fecal specimens should be collected daily, and assayed as long as necessary. If the absorbed material is a gamma emitter, external scintillation counting with probe counters and total-body counters will be of value.[2]

Bioassay Collection Methods

The measurement of samples drawn or excreted from various parts of the body provides a basis for quantitative assessment of internal radionuclide deposition and translocation. The collection techniques for these so-called "bioassay" examinations are presented in Table 28. The performance and interpretation of the results of these examinations are complex, and appropriate specialized professional advice and assistance are necessary.

Table 28. COLLECTION TECHNIQUES FOR RADIONUCLIDES

Radionuclide	Radiation Protection Guide, μCi	Sample of Choice	Other Samples	Method of Preservation	Notes
^{3}H as ^{3}H$_2$O or HTO	10^3	Urine		None	Determination in blood, feces, or tissue not useful; urine concentration is in equilibrium with other body tissues after several hours.
^{89}Sr	40	Urine	Feces		Feces constitute bulk of excretion.
^{90}Sr	20	Urine	Feces		Feces constitute bulk of excretion.
^{103}Ru	50				Gamma spectrometry would be preferred method.
^{106}Ru	10				Gamma spectrometry would be preferred method.
^{131}I	50	Thyroid	Urine	Add NaOH pellet to urine	Thyroid ^{131}I can be measured with external scintillation counter; urine ^{131}I useful only as index of exposure, unless all postexposure urine can be obtained.
^{137}Cs + ^{137}Ba	30	Urine			Total-body gamma-ray spectrometer very sensitive.
^{140}Ba + ^{140}La	9 10	Urine			
^{226}Ra + daughters	0.2	Urine	Feces, breath	Urine—dilute HNO$_3$	Breath of no value, if radium is in sulfate form.
^{239}Pu	0.04	Urine	Feces	Thymol crystals in urine	24-hour sample most desirable.
^{210}Po, soluble	0.4	Urine	Blood	Preserve urine with sulfamic acid 1 mg/ml, and store at 3°C	At least 100 ml urine needed.
Th	0.07	Urine	Feces	Collect in polyethylene container; add HCl	24-hour urine specimen desirable.
U	0.2	Urine			
Gross beta activity		Urine			Useful as first indication of unspecified activity.
Gross alpha activity		Urine			Useful as first indication of unspecified activity.

From: Saenger, E. L., ed., *Medical Aspects of Radiation Accidents*, p. 47. United States Atomic Energy Commission, Washington, D.C., 1963.

DIAGNOSTIC PROCEDURES

The first objective of the diagnostic process is to determine whether biologically significant radiation overexposure has taken place. If so, the next objective is to determine the nature and extent of the damage either immediately present or to be anticipated. Four possible forms of damage are the acute radiation syndrome, local radiation injury, local traumatic injury with radionuclide contamination, and internal contamination without evident injury. The careful history of the events leading to the suspicion of overexposure to radiation, a detailed physical examination, and the performance of some readily available laboratory procedures are generally sufficient to distinguish among these four possibilities.

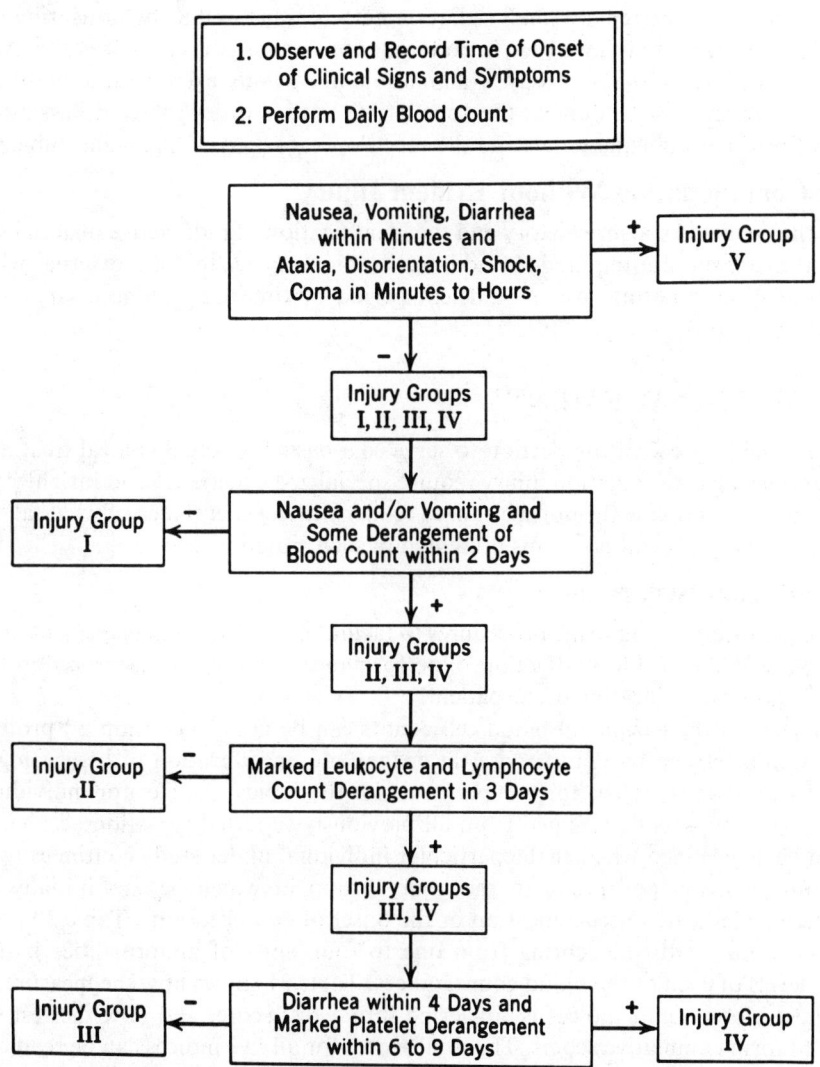

FIGURE 27. Preliminary Evaluation of Radiation Injury.
(From: Thoma, G. E., Jr. and Wald, N., The Diagnosis and Management of Radiation Injury. *J. Occup. Med.*, 1: 421—447, 1959.)

Acute Radiation Syndrome

This syndrome has already been defined, and the classification of patients into one of the five clinical injury groups has already been described in the preceding chapter. It may be identified by the history of possible exposure and the observation of the time of occurrence of various signs and symptoms, as well as by recognition of certain abnormalities developing in the daily peripheral-blood cell count. Figure 27 presents a full chart that facilitates classification of the individual patient into the appropriate injury group; the classification then serves as a basis for proper clinical management and estimation of prognosis in the individual case.

Local Radiation Injury

Careful exposure history and good surveillance of the evolving clinical phenomenon are essential in order to associate local injury with radiation exposure.

Local Traumatic Injury With Radionuclide Contamination

The diagnosis of contamination in traumatic injury is made by the use of appropriate "wound probes", i.e., extremely small radiation detectors, as well as by measuring the radioactivity of cotton swabs that have been introduced into various parts of the wound. An indirect diagnostic procedure is the bioassay examination of various excreta and body fluids for radioactivity, which must be done at the appropriate time in order to avoid missing the rapid excretion of readily soluble materials and the very slow excretion of highly insoluble materials.

Internal Contamination Without Evident Injury

Here the detailed exposure history and the identification of radioactive material within the body or in its excreta are required. In addition to bioassay techniques, external whole-body low-level radiation counting over a sufficient period of time may be necessary in order to establish the diagnosis.

CLINICAL MANAGEMENT

It is not the purpose of this section to serve as a basis for actual clinical treatment, since the problems peculiar to radiation injury require specialized experience and insight. It is hoped that the section will provide the nonspecialized reader with a general idea of what takes place in dealing with the biomedical problems presented in such situations.

Acute Radiation Syndrome

The recommended diagnostic procedures to facilitate the clinical management of radiation injury are presented in Table 29; the time of performance of these various procedures is related to the injury group classification of the patient.

The results of the peripheral-blood cell counts can be used to develop a "profile score", which uses an arbitrary system to score the magnitude and duration of hematologic abnormalities. This provides a basis for comparison with the typical profiles for individuals in the various injury groups as determined from all previously reported radiation accidents. In this way it can be determined whether the particular individual under study continues to show the blood count changes appropriate to the injury group in which he was initially classified. Deviations may indicate misclassification or the onset of complications. Table 30 presents the hematologic values, with the scoring from one to four units of abnormalities based on the particular levels of each of the blood count indices. Table 31 shows how the measured result of each of these blood count indices is converted into a test score, and how the test scores are added up to form cumulative scores. The total scores for all five indices can be read off for any individual or for the cumulative time period of interest. The cumulative score of the individual patient of interest may then be plotted and superimposed on the graph in Figure 28, which shows the group mean cumulative scores for the previously reviewed accident cases. Deviation from the group mean and range should alert the physician to the possibility of previous misclassification or the onset of complications.

The actual treatment of acute radiation syndrome is conservative, no means being presently available for reversing the changes set in motion by the absorption of ionizing radiation at the molecular level. Therapy is primarily supportive, utilizing rest, superior nursing care, strict asepsis to avoid infection, transfusion of appropriate cells, antibiotics chosen because of

Table 29. RECOMMENDED DIAGNOSTIC PROCEDURES FOR CLINICAL MANAGEMENT OF RADIATION INJURY

	Group										
	1-2-3-4			1		2		3		4	5
Time, days	1	2	3	STT	STT	18-48	STT	4-48	STT	4n	1n
Type A Procedures (Diagnostic and Prognostic)											
History											
Symptoms and Signs											
Onset	x	x	x	x	x	D	x	D	x	D	D
Duration	x	x	x	x	x	D	x	D	x	D	D
Past Medical	x										
Physical Examination											
General	x			x	x	d 21	3 mo+	d 15-30	6 mo+	d 6	D
Body weight	x	x	x	x	x	D	x	D	x	D	D
Urinary output						D		D		D	6 hrs
Laboratory Tests											
Hematology											
Hematocrit	x	x	x	x	x	D	x	STT	x	D	6 hrs
Leukocytes	x	x	x	x	x	D	x	D	x	D	6 hrs
Differential count	x	x	x	x	x	D	x	D	x	D	6 hrs
Calculation of total neutrophils and lymphocytes	x	x	x	x	x	D	x	D	x	D	6 hrs
Platelets		x	x	x	x	D	x	STT	x	D	6 hrs
Bone marrow aspiration				d 30	14 d	14 d	mo 6	14 d	mo 6	7 d	d 1
Radioassay											
Blood ^{24}Na	x	x									
Whole-body counting	x	x									

Table 29. RECOMMENDED DIAGNOSTIC PROCEDURES FOR CLINICAL MANAGEMENT OF RADIATION INJURY (*Continued*)

	1-2-3-4			Group 1	Group 2		Group 3		Group 4	Group 5
Time, days	1	2	3	STT	18–48	STT	4–48	STT	4n	1n

Type B Procedures (Complication Detection)

Laboratory Tests

	1	2	3	STT	18–48	STT	4–48	STT	4n	1n
Hematology										
Sedimentation rate	x	x		x	D	x	D	x	D	6 hrs
Reticulocytes	x	x		x	D	x	STT	x	D	6 hrs
Bleeding and Clotting times	x				STT		STT	–75 d	3 d	6 hrs
Biochemistry										
Blood										
NPN	x	prn	prn				prn		STT	6 hrs
Sodium	x	prn	prn				prn		prn	6 hrs
Chloride	x	prn	prn				prn		prn	d 1
Potassium	x	prn	prn				prn		prn	d 1
pH or CO_2	x	prn	prn				prn		prn	d 1
Urine										
Routine analysis	x	x	x	x	D	x	D	x	D	6 hrs
Stool										
Occult Blood	x				D		d 12+		D	all
Opthalmology										
Slit lamp			x	6 mo+		6 mo+		6 mo+		

Type C Procedures (Possibly Valuable)

	1	2	3	STT	18–48	STT	4–48	STT	4n	1n
Biochemistry										
Serum Bilirubin	x	x		–30 d		STT		–30 d	D	6 hrs
Urine BAIBA	x	x		–30 d		D			D	D

Table 29. RECOMMENDED DIAGNOSTIC PROCEDURES FOR CLINICAL MANAGEMENT OF RADIATION INJURY (Continued)

Recommended Frequency of Time of Performance:
STT ...... Standard Testing Times: 6, 9, 12, 15, 18, 21, 24, 27, 30, 33, 36, 40, 44, 48, 60, 75, 90, 105, and 120 days; 6 months, 1 year, and annually
x .... at times indicated in column heading
d .... day(s)
D .... daily during time indicated in column heading
nd........ frequency in days
−nd...... up to and including day at times indicated in column heading
dn†...... on and after day at times indicated in column heading
dn........ specific day recommended
n .... all time after day specified
prn ...... as indicated by clinical course

From: Thoma, G. E. Jr. and Wald, N., The Diagnosis and Management of Accidental Radiation Injury. *J. Occup. Med.*, 1:421–447, 1959

Table 30. PROFILE VALUES ASSIGNED FOR VARIOUS RANGES OF ABNORMALITY HEMATOLOGY: PERIPHERAL COUNTS

Test	Units	Normal*	Increase (above normal)				Decrease (below normal)			
			1	2	3	4	1	2	3	4
Hemoglobin	g%									
Male		15.8	18.0	19.0	20.0	21.0	14.0	12.0	10.0	8.0
Female		13.9	16.0	17.0	18.0	19.0	11.5	10.0	8.5	7.0
Erythrocytes	mill/mm^3									
Male		5.4	6.0	7.0	8.0	9.0	4.5	3.5	2.5	1.5
Female		4.7	5.5	6.5	7.5	8.5	4.0	3.0	2.0	1.0
Hematocrit	vol. %									
Male		47.0	54.0	56.0	58.0	60.0	40.0	35.0	30.0	25.0
Female		42.0	47.0	49.0	51.0	53.0	37.0	32.0	27.0	22.0
Leukocytes	1,000/mm^3	7.4	12.0	18.0	24.0	30.0	4.0	3.0	2.0	1.0
Neutrophils	1,000/mm^3	4.4	7.7	14.0	21.0	28.0	1.8	1.3	0.9	0.5
Lymphocytes	1,000/mm^3	2.5	4.8	7.0	10.0	12.0	1.0	0.75	0.5	0.3
Platelets	1,000/mm^3									
Rees-Ecker		405.0	545.0	700.0	850.0	1,000.0	273.0	200.0	100.0	30.0
Brecher-Cronkite		257.0	440.0	600.0	750.0	900.0	140.0	100.0	50.0	30.0
Fonio		234.0	350.0	500.0	650.0	800.0	130.0	100.0	50.0	30.0
Dameshek		716.0	900.0	1,000.0	1,500.0	2,000.0	500.0	350.0	100.0	30.0
Reticulocytes	% RBC	1.5	4.0	8.0	15.0	25.0	0.5	0.0	0.0	0.0
Erythrocyte sedimentation rate	mm/hr									
Male		5.0	10.0	20.0	30.0	40.0				
Female		10.0	20.0	30.0	40.0	50.0				

* Expressed as "universal mean" (taken from Albritton).

From: Thoma, G. E., Jr. and Wald, N., The Diagnosis and Management of Accidental Radiation Injury. *J. Occup. Med.*, 1:421-447, 1959.

Table 31.
BLOOD COUNTS AND PROFILE SCORING:
HYPOTHETICAL CASE ILLUSTRATING USE OF PROFILE-SCORING METHOD

T = Test Score, C = Cumulative Score

Day	Hematocrit Vol. %	Hematocrit Score T	Hematocrit Score C	Leukocytes Count	Leukocytes Score T	Leukocytes Score C	Neutrophils Count	Neutrophils Score T	Neutrophils Score C	Lymphocytes Count	Lymphocytes Score T	Lymphocytes Score C	Platelets Count ×1,000	Platelets Score T	Platelets Score C	Total Score T	Total Score C
1	48	0	0	13,000	1	1	13,000	1	1	800	1	1	320	0	0	3	3
2	47	0	0	12,000	1	2	11,200	1	2	290	4	5	330	0	0	6	9
3	47	0	0	5,300	0	2	5,000	0	2	100	4	9	350	0	0	4	13
4				3,900			3,700			150							
5	48	0	0	2,900	3	5	2,550	2	4	280	4	13	300	0	0	9	22
6				1,400			1,160			260							
7				1,550			1,300			200							
8	48	0	0	1,600	3	8	1,350	1	5	180	4	17	90	2	2	10	32
9				1,700			1,500			150							
10				2,000			1,800			100							
11				2,050			1,950			90							
12	47	0	0	2,100	2	10	2,000	0	5	90	4	21	60	2	4	8	40
13				2,050			1,800			300							
14				1,950			1,500			400							
15	47	0	0	1,950	3	13	1,250	2	7	650	3	24	50	2	6	10	50
16				2,050			1,280			760							
17				2,150			1,300			820							
18	47	0	0	2,250	2	15	1,350	2	9	880	1	25	40	3	9	8	58
19				2,250			1,300			900			30				
20				2,150			1,200			920			20				
21	45	0	0	2,100	2	17	1,150	2	11	950	1	26	10	4	13	9	67
22				1,750			1,000			750			20				
23				1,300			700			600			30				
24	44	0	0	900	4	21	550	3	14	300	3	29	50	4	17	14	81
25				850			500			250			30				
26				800			550			200			20				
27	39	1	1	750	4	25	600	3	17	100	4	33	10	4	21	16	97
28	31			600			500			84			10				
29	32	2	3	500	4	29	400	4	21	75	4	37	10	4	25	18	115

From: Thoma, G. E., Jr. and Wald, N., The Diagnosis and Management of Accidental Radiation Injury. *J. Occup. Med.*, 1:421–447, 1959.

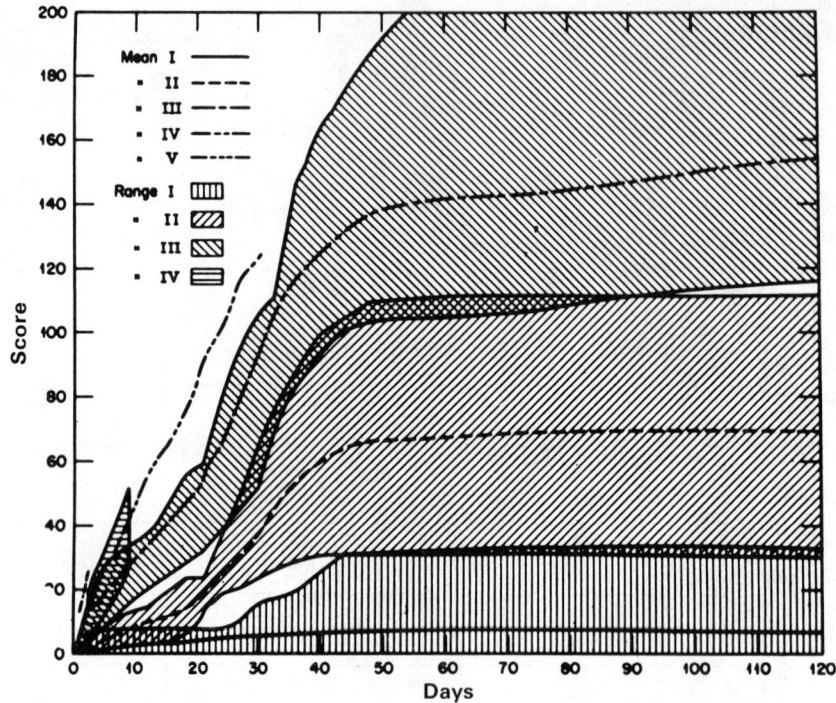

FIGURE 28.
Group Mean Cumulative Profile Scores, Hematology, Total Blood Count.
(From: Wald, N. and Thoma, G.E., Jr., Radiation Accidents: Medical Aspects of Neutron and Gamma-Ray Exposure. *ORNL-2748*, Part B, p. 43. Technical Information Service, Oak Ridge, Tennessee, 1959.)

the sensitivity of specific organisms, maintenance of proper fluid and electrolyte balance, and maintenance of nutrition. Group I patients can be released after 72 hours of hospital observation, when they are asymptomatic. For the remainder, the main hazards are infection and hemorrhage. Reverse isolation procedures and careful screening of attendant personnel should limit the introduction of potentially infectious agents. In more severely injured patients whose granulocyte levels fall below 1,000/mm³, antibiotic sterilization of the gastrointestinal tract may be useful. Control of environmental bacteria with "clean-room" or "life-island" techniques may also be warranted. Transfusions of red cells and platelets should be carried out when indicated; granulocyte transfusions are not readily available as yet. For severely injured patients in Group III, bone marrow transplantation may be considered; this should be carried out promptly, if an identical twin is available; a best-matching relative by red- and white-cell testing techniques may be used in the event that the prognosis is sufficiently poor to warrant the risks involved in foreign-graft incompatibility reactions.

Local Radiation Injury

The treatment of local radiation injury is somewhat similar to that of thermal burns. Surgical removal of necrotic tissue, temporary covering of the denuded area, and, if necessary, the use of skin grafts are the general procedures employed.

Local Traumatic Injury With Radionuclide Contamination

The removal of deposited radionuclides is necessary, using the procedures already outlined above. This is followed by the routine treatment of the traumatic wounds. It is necessary to have adequate information from radiation detection equipment and bioassay measurements in order to assure that sufficient, but not excessive, tissue debridement has been carried out.

Internal Contamination Without Evident Injury

If evidence from such procedures as external whole-body radiation measurements and the radiation bioassays of urine and fecal specimens suggests that a radionuclide deposition within the body is present in larger quantities than seems advisable on the basis of medical and regulatory agency guides, various methods may be used to attempt the acceleration of excretion of the material. Radionuclides in the gastrointestinal tract may be removed more rapidly by the use of laxatives, such as magnesium sulfate. If the radionuclides are in a more soluble form and have circulated within the body, chelating agents, such as EDTA, zirconium citrate, and—more recently—DTPA and BAETA, may enhance the excretion of such nuclides as plutonium, yttrium, lanthanum, uranium, and mixed fission products. It should be noted that the last two chelates mentioned are currently investigational drugs, and the Atomic Energy Commission should be consulted concerning their availability. For ^{90}Sr, high calcium intake and acidification of the blood may reduce bone absorption of the material as well as increase its excretion.

REFERENCES

1. Saenger, E. L., ed., *Medical Aspects of Radiation Accidents,* pp. 21–22. United States Atomic Energy Commission, Washington, D.C., 1963.
2. Saenger, E. L., ed., *Medical Aspects of Radiation Accidents,* pp. 81–84. United States Atomic Energy Commission, Washington, D.C., 1963.

PART X
REFERENCE DATA

GREEK ALPHABET

Alpha	A	α	Iota	I	ι	Rho	P	ρ
Beta	B	β	Kappa	K	κ	Sigma	Σ	σ
Gamma	Γ	γ	Lambda	Λ	λ	Tau	T	τ
Delta	Δ	δ	Mu	M	μ	Upsilon	Υ	υ
Epsilon	E	ε	Nu	N	ν	Phi	Φ	φ
Zeta	Z	ζ	Xi	Ξ	ξ	Chi	X	χ
Eta	H	η	Omicron	O	o	Psi	Ψ	ψ
Theta	Θ	θ	Pi	Π	π	Omega	Ω	ω

SIGNS AND SYMBOLS OF PARTICULAR INTEREST IN RADIATION AND RADIOACTIVITY

A	activity (radio)	l	length
	area	$\bar{l}$	mean free path
	atomic mass number	LD_{50}	median lethal dose
Å	Angstrom unit	m	mass
A_0	activity, original	m_e	rest mass of electron
A_t	activity at time t	m_H	mass of the hydrogen atom
a	acceleration, linear	m_n	mass of the neutron
b	build-up factor	m_p	mass of the proton
C	capacitance, concentration µCi/g	mCi	millicurie
c	velocity of light in vacuum	mµ	millimicron
Ci	curie	mu	mass unit
D	density, film	N	Avogadro's number
	deuterium		number
	dose, dosage	N_0	Avogadro's number
d	density, general		number of radioactive atoms at time zero
	deuteron		
	distance, linear		number, original
E	energy	N_t	number of counts at time t
e	base of natural logarithm	n	neutron
$\mathscr{E}$	electric-field intensity		number, any
e, e$^-$	charge of the electron		number, original
$-1e^0$	electron	$_0p^1$	neutron
	beta particle	p	momentum
$+1e^0$	positron		pressure
e$^+$	charge of the positron		proton
F	Faraday's constant	P.E.	potential energy
	force	Q	electric charge
f	frequency		reaction energy in Mev
G	gravitational constant	q	maximum permissible radionuclide body burden (µCi)
$_1H^1$	proton		
h	Planck's constant	R	range (radiation)
hv	photon energy		rate, count
	quantum		resistance
I	radiation intensity		universal gas constant
I_0	initial intensity	R_α	range of alpha
	exposure	r	radial distance
I_x	radiation intensity, transmitted		radius
°K	degrees Kelvin		roentgen
	absolute temperature	rd	Rutherford
k	Stefan-Boltzmann constant	S	observed standard deviation
K.E.	kinetic energy	s	distance, linear

$T_{1/2}$	half-life, physical	λ	decay constant
T_b	half-life, biological		wave length
T_{eff}	half-life, effective	$\bar{\lambda}$	mean free path
T	temperature, absolute	λ_b	biological elimination constant
t	temperature, general	μ	absorption coefficient
	time		effective or apparent, linear
	triton		micro, micron
t_c	temperature, degrees centigrade	μ_a	$\tau + \kappa + \sigma_a =$ energy absorption coefficient for air
t_f	temperature, degrees Fahrenheit		
v	potential	μ_0	total absorption coefficient
	potential drop	μCi	microcurie
	velocity, linear or particle	$\mu\mu$	micromicron
	volt	$\mu\mu Ci$	micro-microcurie
v	volume	μs	microsecond
W	work	υ	frequency (wave motion quantum theory)
wt	weight		
x	absorber thickness		neutrino
Z	atomic number	ρ	density, general or vapor
		σ	area
			barn (cross section)
α	alpha particle		theoretical standard deviation
β	beta ray (particle)		Compton collision coefficient
β^-	beta ray (particle)	σ_a	absorption cross section in barns
$-1^{\beta 0}$	beta ray (particle)		Compton absorption coefficient
β^+	positron	σ_{ac}	activation cross section in barns
$+1^{\beta 0}$	positron	σ_{eff}	effective cross section in barns
γ	gamma ray	σ_s	Compton scatter coefficient
Δ	finite increment		scattering cross section in barns
ε	dielectric constant	σ_t	total cross section in barns
	electron capture	τ	resolving time
θ	angle between incident and scattered radiation		photoelectric coefficient
		Φ	work function
κ	pair production coefficient		time increment

SIGNS AND SYMBOLS IN MATHEMATICS

$+$	plus, addition, positive	$\therefore$	therefore
$-$	minus, subtraction, negative	$\sqrt{}$	square root
$\pm$	plus or minus, positive or negative	$\sqrt[n]{}$	nth root
$\mp$	minus or plus, negative or positive	a^n	nth power of a
		log	common logarithm
$\div$	division	$\log_{10}$	common logarithm
$/$	division	ln	natural logarithm
——	division	$\log_e$	natural logarithm
$\times$	multiplication	e or ε	base of natural logs, 2.718
$\cdot$	multiplication	π	pi, 3.1416
()()	multiplication	$\angle$	angle
()[]	collection	$\perp$	perpendicular to
$=$	is equal to	$\parallel$	parallel to
$\neq$	is not equal to	n	any number
$\equiv$	is identical to	$\lvert n \rvert$	absolute value of n
$\cong$	equals approximately, congruent	$\bar{n}$	average value of n
$>$	greater than	a^{-n}	reciprocal of nth power of $a = \dfrac{1}{a^n}$
$\not>$	not greater than		
$\geq$	greater than or equal to	$n°$	n degrees
$<$	less than	n'	n minutes, n feet
$\not<$	not less than	n''	n seconds, n inches
$\leq$	less than or equal to	f(x)	function of x
$::$	proportional to	Δx	increment of x
$:$	ratio	dx	differential of x
$\sim$	similar to	$\sum$	summation of
$\propto$	varies as, proportional to	sin	sine
$\rightarrow$	approaches	cos	cosine
∞	infinity	tan	tangent

ABBREVIATIONS

Å	Angstrom unit	colog	cologarithm
abs.	absolute	conc.	concentrated
a.c.	alternating current	const.	constant
alk.	alkali	cos	cosine
amal.	amalgam	cu.	cubic
	amalgamated	cu. cm.	cubic centimeter
amp	ampere	cu. ft.	cubic foot
amu	atomic mass unit	cu. in.	cubic inch
antilog	antilogarithm	cu. m.	cubic meter
appr.	approximately	cu. yd.	cubic yard
aq.	aqua	cwt.	hundredweight
	aqueous	cyl.	cylinder
	water	d	deuteron
asym.	asymmetrical		mass density
atm.	atmosphere	d.	day
	atmospheric	D	deuterium
atmos.	atmosphere		weight density
	atmospheric	d.c.	direct current
at. no.	atomic number	deci-	prefix meaning 1/10
at. wt.	atomic weight	def.	definition(s)
av.	average	deg.	degree (thermometric), generally absolute C
avg.	average		
bar.	barometer	deka-	prefix meaning 10
Bev	billon electron volts	dens.	density
b.p.	boiling point	D.F.	decontamination factor
Btu	British thermal unit	diam.	diameter
C	centigrade	dil.	dilute
ca.	about	dis.	disintegrations
	approximately	dm	decimeter
cal.	calorie (gram)	EC	orbital electron capture
cc	cubic centimeter	emf	electromotive force
c. cm.	cubic centimeter	emu	electromagnetic unit
cf.	confer	equiv.	equivalent
cfm	cubic foot per minute	esu	electrostatic unit
cgs	centimeter-gram-second system of units		cgs unit
		etc.	et cetera (and so forth)
chem.	chemical	ev	electron volt
	chemistry	exp	exponential function
cir.	circular	F	Fahrenheit
circum.	circumference	fahr.	Fahrenheit
cm	centimeter	fps	foot-pound-second system of units
c.m.	circular mill		
cm^2	square centimeter	F.S.	factor of safety
cm^3	cubic centimeter	ft.	foot
coef.	coefficient	$ft.^2$	square foot

ft.³	cubic foot	Mev	million electron volts
ft.-lb.	foot-pound	mg	milligram
G.	gravitation constant	micro-	prefix meaning 1/1,000,000 or 10^{-6}
g	acceleration of gravity		
	gram	micro-micro-	prefix meaning 10^{-12}
gm.	gram		
gal	gallon	milli-	prefix meaning 1/1,000
h	hot	milli-micro-	prefix meaning 10^{-9}
	hour		
hr.	hour	min.	minute
ibid.	ibidem (in the same place)	ml	milliliter
i.e.	id est (that is)	mm	millimeter
in.	inch	mm²	square millimeter
in.²	square inch	mm³	cubic millimeter
in.³	cubic inch	mol.	molecule
insol.	insoluble	mol.wt.	molecular weight
iso.	isotropic	MPC	maximum permissible concentration
isoth.	isothermal		
IT	isomeric transition	MPD	maximum permissible dose
I.T.	isomeric transition	mµ	millimicro-
K	electron capture from K shell		millimicron
k	kilo-	N	number (in mathematical tables)
kev	1,000 electron volts		numeric
kg	kilogram	Obs.	observer
kg-cal.	kilogram-calorie	oz.	ounce
kilo-	prefix meaning 1,000	p.f.	power factor
kv	kilovolt	p'p't'd	precipitated
kva	kilovolt-ampere	precip.	precipitated
kw	kilowatt	pt.	pint
kw.-hr.	kilowatt-hour		point
l	liter	Q	quantity
$\bar{l}$	mean free path		energy
lb	pound	q	maximum permissible radionuclide body burden (µCi)
lin.	linear		
liq.	liquid	qt.	quart
ln	natural logarithm	rad	radian, measure of angle
log	logarithm		radiation absorbed dose
log.	logarithm	rad.	radius
$\log_e$	logarithm to the base e	RBE	relative biological effectiveness
	natural, hyperbolic, or Napierian logarithm	rem	roentgen equivalent man
		R.F.	reliability factor
$\log_{10}$	common logarithm	rpm	revolutions per minute
	logarithm to the base 10	s	second
m	meter		soluble
	milli-	sin	sine
m²	square meter	sol.	soluble
m³	cubic meter		solution
m.	minute	soln.	solution
max.	maximum	sp.	specific
mb	millibarn	SpA	specific activity
med.	medium	sp. gr.	specific gravity
mega-	prefix meaning 1,000,000 or 10^6	sq.	square

sq. ft.	square foot	vert.	vertical
sq. in.	square inch	visc.	viscous
sq. yd.	square yard	vol.	volume
std	standard	wt.	weight
sym	symmetrical	yd	yard
t	triton	yr	year
tan	tangent	μ	micro- micron
temp.	temperature		
th.	thermal	μμ	micromicro- micromicron
tn.	ton		
vel.	velocity	μsec	microsecond
veloc.	velocity		

COMMONLY USED UNITS

UNITS OF RADIOACTIVITY

μμCi	micromicrocurie	10^{-12} curie
μCi	microcurie	10^{-6} curie
Ci	curie	
kCi	kilocurie	10^{3} curies
MCi	megacurie	10^{6} curies

UNITS OF RADIATION

μr	microroentgen	10^{-6} roentgen
mr	milliroentgen	10^{-3} roentgen
r	roentgen	

UNITS OF ABSORBED DOSE

μrad	10^{-6} rad
mrad	10^{-3} rad
rad	

UNITS OF BIOLOGICAL EFFECTIVENESS

μrem	10^{-6} rem
mrem	10^{-3} rem
rem	

UNITS OF ENERGY

ev	electron volt	
kev	kiloelectron volt	10^{3} electron volts
Mev	million electron volts	10^{6} electron volts
Bev	billion electron volts	10^{9} electron volts

FUNDAMENTAL CONSTANTS

Name	Value
Avogadro's number	$N_0 = 6.025 \times 10^{23}$ molecules/g mole
Base of natural logarithm	$e = 2.7183\ldots$
Curie	$Ci = 3.7 \times 10^{10}$ disintegrations/sec
Electron charge	$e = 4.8 \times 10^{-10}$ statcoulomb $= 1.6 \times 10^{-19}$ coulomb
Energy equivalent of electron mass	$mc^2 = 0.51$ Mev
Faraday's constant	$F = 96{,}514$ coulombs/g equivalent (physical scale)
Gravitational acceleration	$g = 980.665$ cm/sec^2
Mass, alpha particle	$m_\alpha = 6.64 \times 10^{-24}$ g $= 4.002777\ mu$
Mass, electron	$m_e = 9.1066 \times 10^{-28}$ g $= 0.000548\ mu$
Mass, H atom	$m_H = 1.67339 \times 10^{-24}$ g $= 1.008142\ mu$
Mass, neutron	$m_n = 1.6751 \times 10^{-24}$ g $= 1.008982\ mu$
Mass, proton	$m_p = 1.67248 \times 10^{-24}$ g $= 1.007594\ mu$
Mass unit	$mu = 1.66035 \times 10^{-24}$ g $= 1.000\ mu$
Microcurie	$\mu Ci = 10^{-6}$ curie $= 3.7 \times 10^4$ disintegrations/sec
Micromicrocurie	$\mu\mu Ci = 10^{-12}$ curie $= 3.7 \times 10^{-2}$ disintegrations/sec
Millicurie	$mCi = 10^{-3}$ curie $= 3.7 \times 10^7$ disintegrations/sec
Pi	$\pi = 3.1416$
Planck's constant	$h = 6.624 \times 10^{27}$ erg-sec
Rad	$rad = 100$ ergs/g of tissue
Roentgen	$r = 1$ esu/0.001293 g of air
Rem	$rem = rads \times RBE$
Rutherford	$rd = 10^6$ disintegrations/sec

Name	Value
Stefan-Boltzmann constant	$k = 5.67 \times 10^{-5}$ erg/cm^2-deg^4-sec $= 1.380 \times 10^{-16}$ erg/degree
Universal gas constant	$R = 0.08206$ liter-atm/g-mole/°K
Velocity of light	$c = 2.99776 \times 10^{10}$ cm/sec
Wavelength associated with 1 ev	$\lambda_0 = 12394.8$ Å

THE STANDARD MAN (CONVENTIONAL)

MASSES OF ORGANS IN THE HUMAN BODY

Organ	Mass, g
Total body	70,000
Muscle	30,000
Skin and subcutaneous tissue	6,100
Skin only	2,000
Fat	10,000
Skeleton	
without bone marrow	7,000
red marrow	1,500
yellow marrow	1,500
Blood	5,400
Gastrointestinal tract	2,000
Contents of gastrointestinal tract	
stomach	250
small intestine	1,000
upper large intestine	135
lower large intestine	150
Liver	1,700
Brain	1,500
Lungs (2)	1,000
Lymphoid tissue	700
Kidneys (2)	200
Heart	300
Spleen	150
Urinary bladder	150
Pancreas	70
Salivary glands (6)	50
Testes (2)	40
Spinal cord	30
Eyes (2)	30
Thyroid gland	20
Teeth	20
Prostate gland	20
Adrenal glands or suprarenal (2)	20
Thymus	10
Miscellaneous (blood vessels, cartilage, nerves, etc.)	390

DAILY RATES OF INGESTION AND INHALATION

Water intake in food	700 cm^3
Water intake in fluids	1,500 cm^3
Water of oxidation	300 cm^3
Total water consumption	2,500 cm^3

Air inhaled during 8-hour working day	10^7 cm^3
Air inhaled during 16 hours not at work	10^7 cm^3
Total inhaled	2×10^7 cm^3

DENSITIES OF COMMON METALS AT 20 TO 25°C

Metal	Symbol	Density	Metal	Symbol	Density
Aluminum	Al	2.700	Nickel	Ni	8.900
Antimony	Sb	6.685	Osmium	Os	22.480
Arsenic	As	5.700	Palladium	Pd	12.100
Barium	Ba	3.500	Platinum	Pt	21.400
Beryllium	Be	1.800	Potassium	K	0.860
Bismuth	Bi	9.800	Praseodymium	Pr	6.500
Cadmium	Cd	8.600	Rhenium	Re	20.530
Calcium	Ca	1.550	Rhodium	Rh	12.500
Cerium	Ce	6.900	Rubidium	Rb	1.530
Cesium	Cs	1.900	Ruthenium	Ru	12.200
Chromium	Cr	7.100	Selenium	Se	4.800
Cobalt	Co	8.900	Silver	Ag	10.500
Copper	Cu	8.930	Sodium	Na	0.970
Gallium	Ga	5.910	Strontium	Sr	2.600
Germanium	Ge	5.360	Tantalum	Ta	16.600
Gold	Au	19.300	Tellurium	Te	6.240
Indium	In	7.300	Thallium	Tl	11.850
Iridium	Ir	22.400	Thorium	Th	11.400
Iron	Fe	7.860	Tin, white	Sn	7.310
Lanthanum	La	6.150	Tin, gray	Sn	5.750
Lead	Pb	11.340	Titanium	Ti	4.500
Lithium	Li	0.530	Uranium	U	18.700
Magnesium	Mg	1.740	Vanadium	V	5.960
Manganese	Mn	7.200	Wolfram	W	19.300
Mercury	Hg	13.546 (liq.)	Yttrium	Y	5.510
Molybdenum	Mo	10.200	Zinc	Zn	7.140
Neodymium	Nd	6.900	Zirconium	Zr	6.400

RELATION BETWEEN THICKNESSES OF ORDINARY CONCRETE AND OF LEAD FOR RADIUM AND ^{60}Co GAMMA RAYS

Thickness of Concrete, in.		Thickness of Lead, mm	
A*	B†	Radium Gamma Rays	^{60}Co Gamma Rays
2	$1\frac{7}{8}$	5	7
4	$3\frac{3}{4}$	12	15
6	$5\frac{5}{8}$	19	24
8	$7\frac{1}{2}$	27	32
10	$9\frac{3}{8}$	35	41
12	$11\frac{1}{4}$	43	50
14	$13\frac{1}{8}$	52	59
16	15	60	69
18	$16\frac{7}{8}$	69	79
20	$18\frac{3}{4}$	77	90
22	$20\frac{5}{8}$	86	100
24	$22\frac{1}{2}$	94	110

* A. Concrete density of 2.2 g/cm^3, as used in England.
† B. Concrete density of 2.35 g/cm^3, as used in the U.S.

NOMOGRAM OF ABSORPTION OF BETA PARTICLES

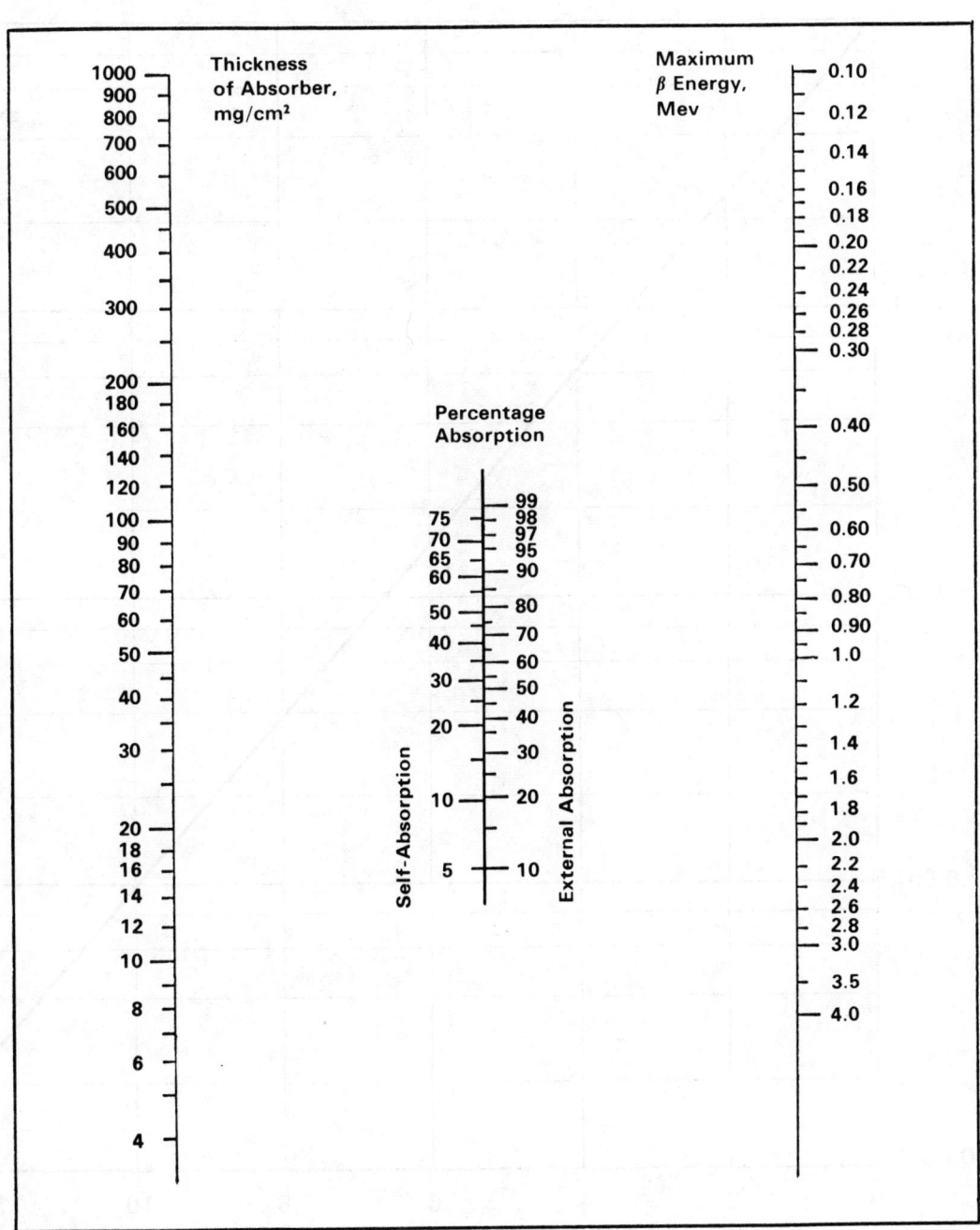

DECAY CURVE FOR RADIOACTIVE MATERIALS

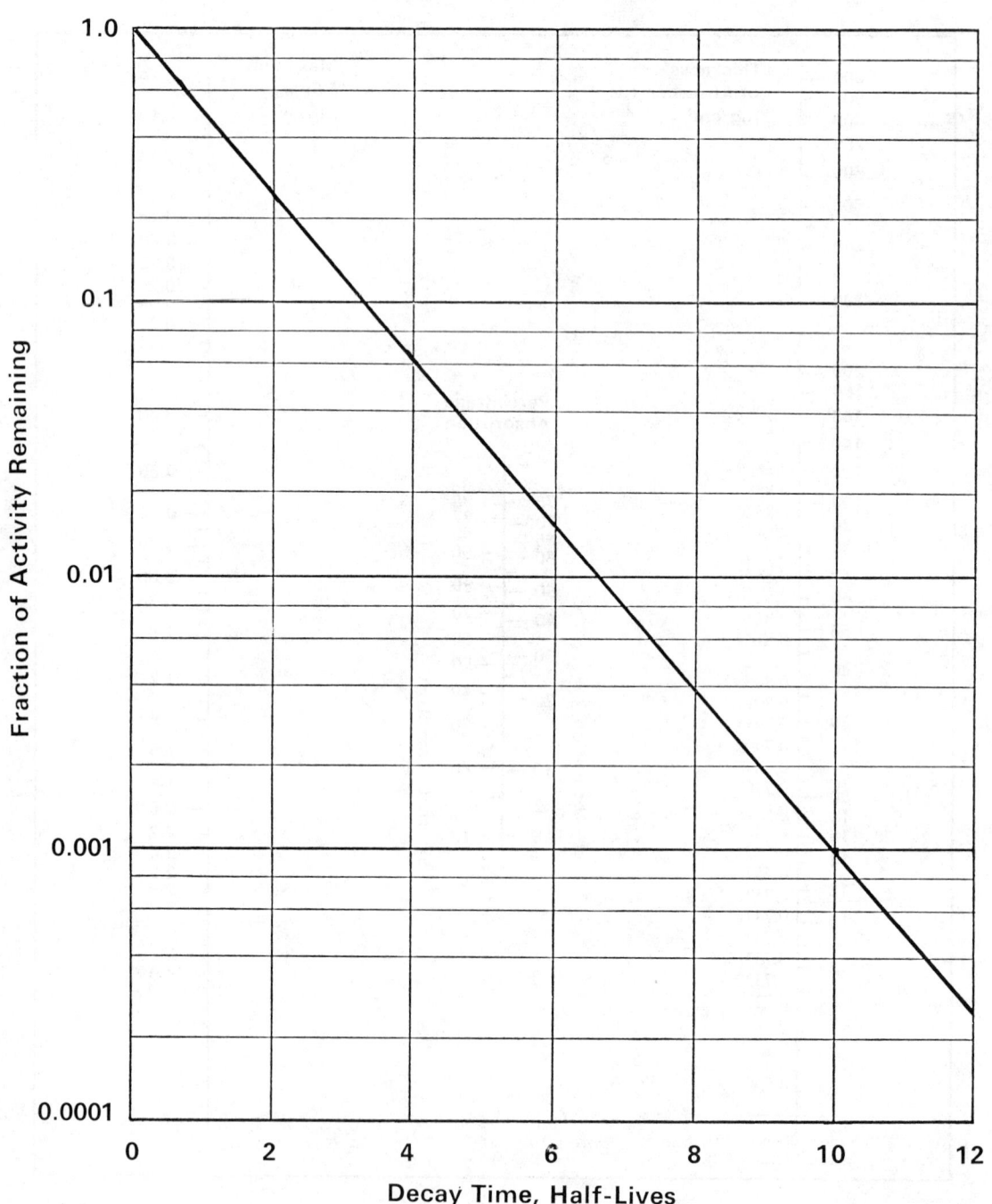

RADIOACTIVE DECAY

t/T	$e^{-\lambda t}$	t/T	$e^{-\lambda t}$	t/T	$e^{-\lambda t}$	t/T	$e^{-\lambda t}$
0	1.0000	0.52	0.6974	1.54	0.3439	3.80	0.0718
0.01	0.9931	0.54	0.6878	1.56	0.3391	3.85	0.0693
0.02	0.9862	0.56	0.6783	1.58	0.3345	3.90	0.0670
0.03	0.9794	0.58	0.6690	1.60	0.3299	3.95	0.0647
0.04	0.9726	0.60	0.6597	1.62	0.3253	4.00	0.0625
0.05	0.9659	0.62	0.6507	1.64	0.3209	4.10	0.0583
0.06	0.9593	0.64	0.6417	1.66	0.3164	4.20	0.0544
0.07	0.9526	0.66	0.6329	1.68	0.3121	4.30	0.0508
0.08	0.9461	0.68	0.6242	1.70	0.3078	4.40	0.0474
0.09	0.9395	0.70	0.6156	1.75	0.2973	4.50	0.0442
0.10	0.9330	0.72	0.6071	1.80	0.2872	4.60	0.0412
0.11	0.9266	0.74	0.5987	1.85	0.2774	4.70	0.0385
0.12	0.9202	0.76	0.5905	1.90	0.2679	4.80	0.0359
0.13	0.9138	0.78	0.5824	1.95	0.2588	4.90	0.0335
0.14	0.9075	0.80	0.5744	2.00	0.2500	5.00	0.0312
0.15	0.9013	0.82	0.5664	2.05	0.2415	5.10	0.0292
0.16	0.8950	0.84	0.5586	2.10	0.2333	5.20	0.0272
0.17	0.8888	0.86	0.5509	2.15	0.2253	5.30	0.0254
0.18	0.8827	0.88	0.5434	2.20	0.2176	5.40	0.0237
0.19	0.8766	0.90	0.5359	2.25	0.2102	5.50	0.0221
0.20	0.8705	0.92	0.5285	2.30	0.2031	5.60	0.0206
0.21	0.8645	0.94	0.5212	2.35	0.1961	5.70	0.0192
0.22	0.8586	0.96	0.5141	2.40	0.1895	5.80	0.0179
0.23	0.8526	0.98	0.5070	2.45	0.1830	5.90	0.0167
0.24	0.8467	1.00	0.5000	2.50	0.1768	6.00	0.0156
0.25	0.8409	1.02	0.4931	2.55	0.1708	6.20	0.0136
0.26	0.8351	1.04	0.4863	2.60	0.1649	6.40	0.0118
0.27	0.8293	1.06	0.4796	2.65	0.1593	6.60	0.0103
0.28	0.8236	1.08	0.4730	2.70	0.1539	6.80	0.0090
0.29	0.8179	1.10	0.4665	2.75	0.1487	7.00	0.0078
0.30	0.8122	1.12	0.4601	2.80	0.1436	7.20	0.0068
0.31	0.8066	1.14	0.4538	2.85	0.1387	7.40	0.0059
0.32	0.8011	1.16	0.4475	2.90	0.1340	7.60	0.0052
0.33	0.7955	1.18	0.4413	2.95	0.1294	7.80	0.0045
0.34	0.7900	1.20	0.4353	3.00	0.1250	8.00	0.0039
0.35	0.7846	1.22	0.4293	3.05	0.1207	8.20	0.0034
0.36	0.7792	1.24	0.4234	3.10	0.1166	8.40	0.0030
0.37	0.7738	1.26	0.4175	3.15	0.1127	8.60	0.0026
0.38	0.7684	1.28	0.4118	3.20	0.1088	8.80	0.0022
0.39	0.7631	1.30	0.4061	3.25	0.1051	9.00	0.0020
0.40	0.7579	1.32	0.4005	3.30	0.1015	9.20	0.0017
0.41	0.7526	1.34	0.3950	3.35	0.0981	9.40	0.0015
0.42	0.7474	1.36	0.3896	3.40	0.0948	9.60	0.0013
0.43	0.7423	1.38	0.3842	3.45	0.0915	9.80	0.0011
0.44	0.7371	1.40	0.3789	3.50	0.0884	10.00	0.0010
0.45	0.7320	1.42	0.3737	3.55	0.0854	10.50	0.0007
0.46	0.7270	1.44	0.3685	3.60	0.0825	11.00	0.0005
0.47	0.7220	1.46	0.3635	3.65	0.0797	11.50	0.0004
0.48	0.7170	1.48	0.3585	3.70	0.0770	12.00	0.0002
0.49	0.7120	1.50	0.3536	3.75	0.0743	13.00	0.0001
0.50	0.7071	1.52	0.3487				

With acknowledgements to D. E. Hull and the American Chemical Society; originally published in *J. Phys. Chem.*, 45: 1310, 1941 (By permission from Williams & Wilkins, copyright owners).

REFERENTIAL CONVERSION FACTORS

TIME

Time	Year	Day	Hour	Minute	Second
1 Year	1	365.24	8.766×10^3	5.259×10^5	3.156×10^7
1 Day	2.738×10^{-3}	1	24	1.440×10^3	8.640×10^4
1 Hour	1.141×10^{-4}	0.04167	1	60	3.600×10^3
1 Minute	1.901×10^{-6}	6.944×10^{-4}	0.01667	1	60
1 Second	3.169×10^{-8}	1.157×10^{-5}	2.778×10^{-4}	0.01667	1

ENERGY

Energy	kw-hr	Btu	ft-lb	cal	Mev
1 kw-hr	1	3412	2.66×10^6	8.60×10^5	2.24×10^{19}
1 Btu	2.93×10^{-4}	1	778.1	252	6.58×10^{15}
1 ft-lb	3.77×10^7	1.29×10^{-3}	1	0.324	8.46×10^{12}
1 cal	1.16×10^{-6}	3.97×10^{-3}	3.088	1	2.61×10^{13}
1 Mev	4.45×10^{-20}	1.52×10^{-16}	1.18×10^{-13}	3.83×10^{-14}	1

MASS ENERGY

Mass Energy	Mass Unit	Mev	erg	cal
1 Mass Unit (*mu*)	1	931	1.49×10^{-3}	3.51×10^{-11}
1 Mev	1.07×10^{-3}	1	1.60×10^{-6}	3.82×10^{-14}
1 erg	670	6.24×10^5	1	2.39×10^{-8}
1 cal	2.81×10^{10}	2.62×10^{13}	4.186×10^7	1

POWER

Power	hp	kw	Btu/hr	cal/sec	Mev/sec
1 hp	1	0.7457	2544	178.1	4.65×10^{15}
1 kw	1.341	1	3412	239	6.24×10^{15}
1 Btu/hr	3.93×10^{-4}	2.93×10^{-4}	1	0.070	1.82×10^{12}
1 cal/sec	5.61×10^{-3}	4.18×10^{-3}	14.29	1	2.61×10^{13}
1 Mev/sec	2.15×10^{-16}	1.60×10^{-16}	5.47×10^{-13}	3.83×10^{-14}	1

CONVENIENT CONVERSION FACTORS

Multiply	By	To Obtain
atmospheres	14.7	pounds/square inch
	76	centimeters Hg
barns	10^{-24}	square centimeters
British thermal units	1.055×10^3	joules
	0.252	kilogram-calories
British thermal units/pound	0.556	gram-calories/gram
centimeters	0.3937	inches
	3.28×10^{-2}	feet
centimeters Hg	0.1934	pounds/square inch
	1.316×10^{-2}	atmospheres
	0.4465	feet of water
centimeters/second	0.6	meters/minute
	2.237×10^{-2}	miles/hour
chemical scale	1.000272	physical scale
coulombs	6.28×10^{18}	electronic charges
	2.997×10^9	statcoulombs
cubic centimeters	6.102×10^{-2}	cubic inches
	3.531×10^{-5}	cubic feet
	2.642×10^{-4}	gallons*
	10^{-3}	liters
cubic centimeters/second	2.119×10^{-3}	cubic feet/minute
	8.64×10^{-2}	cubic meters/day
	1.585×10^{-2}	gallons/minute
cubic centimeters/gram	1.602×10^{-2}	cubic feet/pound
cubic feet	2.832×10^{-2}	cubic meters
	7.481	gallons
	28.32	liters
cubic feet/pound	62.43	cubic centimeters/gram
cubic feet/minute	4.72×10^2	cubic centimeters/second
cubic feet/second	4.488×10^2	gallons/minute
cubic inches	16.39	cubic centimeters
	5.787×10^{-4}	cubic feet
	1.639×10^{-2}	liters
	4.329×10^{-3}	gallons
cubic meters	35.31	cubic feet
	2.642×10^2	gallons
cubic meters/day	11.57	cubic centimeters/second
curies	2.22×10^{12}	disintegrations/minute
	3.7×10^{10}	disintegrations/second
	10^3	millicuries
	10^6	microcuries
	10^{-3}	kilocuries

* U.S. gallon used throughout the table, except where stated otherwise.

Multiply	By	To Obtain
disintegrations/minute	4.55×10^{-10}	millicuries
	4.55×10^{-7}	microcuries
disintegrations/second	2.7×10^{-8}	millicuries
	2.7×10^{-5}	microcuries
dynes	1.02×10^{-3}	grams
	2.248×10^{-6}	pounds
electron volts (ev)	10^{-6}	million electron volts (Mev)
	1.6×10^{-12}	ergs
	1.6×10^{-19}	joules
ergs	10^{-7}	joules
	6.24×10^{5}	million electron volts (Mev)
	6.24×10^{11}	electron volts (ev)
feet	30.48	centimeters
	0.3048	meters
feet of water	2.230	centimeters Hg
feet/minute	0.508	centimeters/second
	1.667×10^{-2}	feet/second
	1.136×10^{-2}	miles/hour
feet/second	18.29	meters/minute
	0.6818	miles/hour
foot-pounds	1.356	joules
	3.238×10^{-4}	kilogram-calories
gallons	231	cubic inches
	0.1337	cubic feet
	3.785×10^{3}	cubic centimeters
	3.785	liters
	0.83	gallons, Imperial
gallons/minute	2.228×10^{-3}	cubic feet/second
grams	980.7	dynes
	2.205×10^{-3}	pounds
gram-calories	3.968×10^{-3}	British thermal units
gram-calories/gram	1.8	British thermal units/pound
gram mole gas	22.4	liters (STP*)
gram/cubic centimeter	62.43	pounds/cubic foot
horsepower	10.7	kilogram-calories/minute
	0.7457	kilowatts
inches	2.54	centimeters
	10^{3}	mils
inches Hg	0.4912	pounds/square inch
joules	10^{7}	ergs
	0.7376	foot-pounds
	9.48×10^{-4}	British thermal units
kilocuries	10^{3}	curies
kilograms	9.807×10^{5}	dynes
	2.205	pounds
kilogram-calories	3.968	British thermal units
	3.09×10^{3}	foot-pounds
kilogram-calories/minute	9.35×10^{-2}	horsepower
kilowatts	14.34	kilogram-calories/minute
	1.341	horsepower

* Standard temperature and pressure (0°C, 760 mm Hg).

Reference Data

Multiply	By	To Obtain
kilowatt-hours	2.247×10^{19}	million electron volts (Mev)
	3.6×10^{13}	ergs
liters	3.53×10^{-2}	cubic feet
	61.02	cubic inches
	0.2642	gallons
	10^3	cubic centimeters
liters/hour	0.278	cubic centimeters/second
liters/minute	15.851	gallons/hour
meters	3.281	feet
	39.37	inches
meters/minute	1.667	centimeters/second
	5.468×10^{-2}	feet/second
	3.728×10^{-2}	miles/hour
microcuries	3.7×10^4	disintegrations/second
	2.22×10^6	disintegrations/minute
miles	5280	feet
miles/hour	44.7	centimeters/second
	88	feet/minute
	1.467	feet/second
	26.82	meters/minute
millicuries	3.7×10^7	disintegrations/second
	2.22×10^9	disintegrations/minute
million electron volts (Mev)	1.6×10^{-6}	ergs
mils	0.001	inches
ounces	28.35	grams
	6.25×10^{-2}	pounds
physical scale	0.999728	chemical scale
pounds	4.448×10^5	dynes
	453.6	grams
pounds/cubic foot	1.602×10^{-2}	grams/cubic centimeter
pounds/cubic inch	27.68	grams/cubic centimeter
pounds/gallon	0.1198	grams/cubic centimeter
pounds/square inch	6.805×10^{-2}	atmospheres
	2.036	inches Hg
	5.17	centimeters Hg
radians	57.3	degrees
roentgens	1	electrostatic units/cubic centimeter air (STP*)
	2.082×10^9	ion pairs/cubic centimeter air (STP*)
	1.61×10^{12}	ion pairs/gram air
	7.03×10^4	million electron volts (Mev)/cubic centimeter air (STP*†)
	5.44×10^7	million electron volts (Mev)/gram air†
	87	ergs/gram air†
	2.0×10^{-6}	calories/gram air
square centimeters	1.076×10^{-3}	square feet
	0.155	square inches
	10^{24}	barns

* Standard temperature and pressure (0°C, 760 mm Hg).
† These values are assuming that the average energy expended per ion pair formed is 5.4×10^{-11} ergs (34 ev).

Multiply	By	To Obtain
square feet	929	square centimeters
	144	square inches
	9.29×10^{-2}	square meters
square inches	6.452	square centimeters
	6.944×10^{-3}	square feet
	6.452×10^{-4}	square meters
square meters	10.76	square feet
statcoulombs	3.34×10^{-10}	coulombs
	2.095×10^{9}	electronic charge
temperature degrees C + 273	1	absolute degrees Kelvin
temperature degrees C	1.8	temperature degrees F − 32
temperature degrees F + 459	1	absolute degrees Rankine
temperature degrees F − 32	0.5555	temperature degrees C
watts	10^{7}	ergs/second
	0.7376	foot-pounds/second

EQUATIONS

LOGARITHMIC RELATIONS

$\log N$ = the exponent or power to which the base 10 must be raised to obtain a value of N (the common logarithm of N).

$\ln N$ = the power to which the base 2.718... (e) must be raised to obtain a value N (the natural logarithm of N).

(1) $\qquad \log N = 0.4343 \ln N$

(2) $\qquad \ln N = 2.3026 \log N$

(3) $\qquad \log MN = \log M + \log N$

(4) $\qquad \log M/N = \log M - \log N$

(5) $\qquad \log N^a = a \log N$

(6) $\qquad \log \sqrt[a]{N} = (\log N)/a$

CLASSICAL PHYSICS

Unless otherwise noted, symbols and dimensions in this section are used consistently as follows:

m = mass (g) $\qquad$ F = force (g-cm/sec², dynes)

v = velocity (cm/sec) $\qquad$ r = radius of action (cm)

a = acceleration (cm/sec²) $\qquad$ s = distance (cm)

(1) Linear Force.

$$F = ma = (g)(cm/sec^2) = g\text{-}cm/sec^2 = dynes$$

(2) Momentum.

$$p = mv = (g)(cm/sec)$$

(3) Conservation of Momentum (any impact between Body A and Body B).

$$i = \text{initial} \qquad f = \text{final}$$

$$m_A v_{A_i} + m_B v_{B_i} = m_A v_{A_f} + m_B v_{B_f}.$$

(4) Work.

$$W = Fs = m\,a\,s = (g)(cm/sec^2)(cm) = g\text{-}cm^2/sec^2 = dyne\text{-}cm = erg.$$

(5) Energy.

$$E = (work) = Fs = (g\text{-}cm/sec^2)(cm) = g\text{-}cm^2/sec^2 = erg.$$

(6) Kinetic Energy.

$$K.E. = \tfrac{1}{2}mv^2 = (g)(cm/sec)^2 = g\text{-}cm^2/sec^2 = erg.$$

(7) **Conservation of Kinetic Energy** (elastic impact between Body A and Body B).

$$\tfrac{1}{2}m_A v_{A_i}^2 + \tfrac{1}{2}m_B v_{B_i}^2 = \tfrac{1}{2}m_A v_{A_f}^2 + \tfrac{1}{2}m_B v_{B_f}^2.$$

(8) **Power.**

$$P = (\text{work/time}) = F\,s/t = (\text{g-cm/sec}^2)(\text{cm})/\text{sec} = \text{ergs/sec}.$$

WAVE AND QUANTUM RELATIONS

Unless otherwise noted, symbols and dimensions in this section are used consistently as follows:

v = velocity of wave or particle (cm/sec)

h = Planck's constant (6.6×10^{-27} erg-sec)

ν = frequency of wave or quanta (cycles/sec)

λ = wavelength (cm)

λ_0 = wavelength of incident radiation (Å)

λ_θ = wavelength of scattered radiation at angle θ (Å)

E = energy (ergs)

θ = angle between incident and scattered radiation

c = velocity of light (3×10^{10} cm/sec)

m = mass of particle (g)

Φ = work function (ergs)

(1) **Wave Equation.**

$$\text{Wave velocity } (v \text{ or } c) = \lambda \nu$$

(2) **Associated Wavelength of Particle.**

$$\text{Wavelength} = \lambda = \frac{h}{mv}$$

(3) **Photoelectric Equation.**

$$E = \Phi + \tfrac{1}{2}mv^2$$

(4) **Photon Energy.**

$$E = h\nu$$

$$E = \frac{hc}{\lambda}$$

$$\text{Energy (ev)} = \frac{1.242 \times 10^4}{\text{Wavelength (Å)}}$$

(5) **Mass–Energy Relation.**

$$E = mc^2$$

(6) **Momentum of Photon.**

$$mv = \frac{h}{\lambda}$$

(7) **Compton Scattering of Gamma and X Rays.**

$$\lambda_0 = \lambda_0 + 0.0242(1 - \cos \theta)$$

ELECTROSTATICS

The following units apply in this section:

$$F = \text{force (dynes)}$$
$$Q = \text{electrostatic charge (statcoulombs)}$$
$$s = \text{distance (cm)}$$
$$V = \text{potential (statvolts)}$$
$$C = \text{capacitance (statfarads)}$$
$$W = \text{work (ergs)}$$
$$\varepsilon = \text{dielectric constant}$$

(1) **Force Between Two Charges, a and b, Coulomb's Law.**

$$F = \frac{Q_a Q_b}{\varepsilon s^2}$$

(2) **Work.**

$$W = QV$$

(3) **Capacitance.**

$$C = Q/V$$

(4) **Potential.**

$$V = Q/s$$

RADIOACTIVE DECAY

The following symbols will be used in this section:

N_0 = number of unstable nuclei at some original time

N = number of unstable nuclei remaining after a time interval t

I_0 = intensity of radiation at some original time

I = intensity of radiation after a time interval t

A_0 = activity of sample at some original time

A = activity remaining after a time interval t

λ = decay constant for particular radioactive element

e = base of natural logarithms, 2.718.

$T_{\frac{1}{2}}$ = half-life

$n = t/T_{\frac{1}{2}}$ = number of half-lives

(1) $\qquad N = N_0 e^{-\lambda t} \quad$ or $\quad N = N_0 e^{-0.693 t/T_{1/2}}$

(2) $\qquad A = A_0 e^{-\lambda t} \quad$ or $\quad A = A_0 e^{-0.693 t/T_{1/2}}$

(3) $$I = I_0 e^{-\lambda t} \quad \text{or} \quad I = I_0 e^{-0.693 t/T_{1/2}}$$
$$N = N_0 2^{-n} \quad \text{or} \quad N/N_0 = 1/2^n$$

Decay Constant

(5) $$\lambda = 0.693/T_{1/2}$$

Fission Product Decay

(6) $$I_1 t_1^{1.2} = I_2 t_2^{1.2},$$

where I_1 = radiation intensity at time t_1 after fission,
I_2 = radiation intensity at time t_2 after fission.

(7) $$A = k_1 t^{-0.89},$$

where fission product age is 1 minute $\leq t <$ 30 minutes.

(8) $$A = k_2 t^{-1.11},$$

where fission product age is 30 minutes $\leq t <$ 1 day.

(9) $$A = k_3 t^{-1.25},$$

where fission product age is 1 day $\leq t <$ 4 days.

(10) $$A = k_4 t^{-1.03},$$

where fission product age is 4 days $\leq t <$ 100 days.

(11) $$A = k_5 t^{-1.60},$$

where fission product age is 100 days $\leq t <$ 3 years.

$$k_2 = 2.036 \, k_1$$
$$k_3 = 5.571 \, k_1$$
$$k_4 = 0.857 \, k_1$$
$$k_5 = 771.4 \, k_1$$

The values of k_1, k_2, k_3, k_4, or k_5 will depend upon the number of atoms fissioned; hence, for the general condition, these constants cannot have assigned values. The ratios between k_1 and k_2, k_3, k_4, or k_5 will remain the same.

SPECIFIC ACTIVITY

$$\text{Specific activity} = \lambda N = 0.693 \, N/T_{1/2} = \text{dis/sec/g},$$

where $T_{1/2}$ = half-life in seconds,
N = number of atoms per gram.

$$\text{Specific activity} = \lambda N/(3.7 \times 10^{10}) = \frac{N \times 1.873 \times 10^{-11}}{T_{1/2}} = \text{Ci/g}$$

RADIATION ABSORPTION

(1) **Alpha Particle Range.**

For $E < 4$ Mev
$$R_\alpha = 0.56 E,$$

for $4 < E < 8$ Mev
$$R_\alpha = 1.24 E - 2.62,$$

where R_α = range in cm of air at 1 atm amd 15°C,
    E = energy in Mev.

(2) Beta Particle Range.

For $0.01 \leqq E \leqq 2.5$ Mev

$$R = 412E^{1.265 - 0.0954 \ln E},$$

$$\ln E = 6.63 - 3.2376(10.2146 - \ln R)^{1/2},$$

where R = range in mg/cm^2,
    E = maximum energy in Mev.

For $E \geqq 2.5$ Mev

$$R = 530E - 106,$$

where R and E are the same as above.

Sargent's Rule ($E > 0.8$ Mev)

$$R = 0.0526E - 0.094,$$

where R = range in g/cm^2,
    E = maximum energy in Mev.

Feather's Rule ($E > 0.6$ Mev)

$$R = 0.542 - 0.133,$$

where R and E are the same as for Sargent's Rule.

(3) Gamma Ray Absorption.

The following symbols will be used in this section:

I_0 = original radiation exposure rate

I = attenuated radiation exposure rate

μ = linear absorption coefficient (cm^{-1}) = $\dfrac{0.693}{x_{1/2}}$

μ/ρ = mass absorption coefficient (cm^2/g)

ρ = absorber density (g/cm^3)

x = absorber thickness (cm)

$x_{1/2}$ = half-value layer of absorber (cm)

e = base of natural logarithms (2.718...)

b = "build-up" factor (see Part IV for values)

For monoenergetic or monochromatic narrow-beam radiation,

$$I = I_0 e^{-\mu x}$$

or

$$I = I_0 e^{-(\mu/\rho)(\rho)(x)};$$

for monoenergetic or monochromatic wide-beam radiation,

$$I = bI_0 e^{-\mu x}.$$

(4) **Neutron Absorption** (for a collimated beam of monoenergetic neutrons).

$$I = I_0 e^{-\sigma N x},$$

where I_0 = initial neutron intensities,
I = final neutron intensities,
N = number of atoms per cc in the absorber,
σ = cross section (cm^2),
x = thickness of absorber (cm),
e = base of the natural logarithm (2.718...).

Since this equation is only an approximation of neutron attenuation, average neutron energies can be used for determining the value of o. The equation is not accurate enough to justify the use of neutron buildup factors.

(5)* **Approximate Range–Energy Relation for Protons.**

$$R = (E/9.3)^{1.8},$$

where E = energy in Mev (few Mev to 200 Mev),
R = range in meters in air.

BETA PARTICLE COUNTING

(1) **Self-Absorption.**

$$\frac{R_0}{R} = \frac{1}{mx}(-e^{-mx}),$$

where R_0 = measured counting rate,
R = true counting rate,
x = sample thickness (mg/cm^2)
m = absorption coefficient (cm^2/mg)

(2) **Resolving-Time Determination.**

$$\tau = \frac{R_1 + R_2 - R_{12}}{2(R_1 R_2)},$$

where τ = resolving time in seconds,
R_1 = counting rate, source 1 (c/s)
R_2 = counting rate, source 2 (c/s)
R_{12} = counting rate, combined sources 1 and 2 (c/s)

(3) **Resolving-Time Correction.**

$$R = \frac{R_0}{1 - R_0 \tau},$$

where R = true counting rate (c/s),
R_0 = observed counting rate (c/s),
τ = resolving time in seconds.

---

* Segré, E., *Experimental Physics*, Vol. I

STATISTICS OF COUNTING

See Part II, p. 80.

CALIBRATION PROCEDURE

The equations in this section are applicable to dose in air from gamma emitters.

(1) Exposure Rate from a Point Source.

$$I_\gamma = 0.156 nE(10^5 \mu_a),$$

where I_γ = mr/hr at 1 meter per mCi,

n = gamma quanta per disintegration,

E = energy of gamma quanta in Mev,

μ_a = energy absorption coefficient for gamma in air (STP) in cm^{-1}.

The equation assumes that one ion pair in air causes an average energy expenditure of 32.7 ev.

(2) Exposure Rate from a Point Source of Radium Gamma Radiation.

$$\text{mr/hr} = \frac{\text{number of mg of Ra}}{s^2},$$

where s = distance to source in yards;

$$\text{mr/hr} = \frac{8400 \text{ (number of mg of Ra)}}{s^2},$$

where s = distance in cm.

(3) Approximate Exposure Rate from Any Gamma Point Source.

$$\text{r/hr at 1 foot} \cong 6CEn,$$

or

$$\text{mr/hr/mCi at 1 meter} \cong 0.5 \, nE,$$

where C = number of curies,

E = gamma-ray energy in Mev,

n = gamma quanta per disintegration.

(4) Exposure Rate from Any Gamma Point Source.

$$\text{mr/hr} = nI_\gamma/s^2,$$

where n = number of millicuries,

I_γ = mr/hr at 1 meter per mCi,

s = distance in meters.

(5) Exposure Rate from Gamma Emitters Other than Point Sources.

For the examples which follow, the following terminology will be used:

S = source activity in photons per second (for a point source, this will be the total activity; for a line source, it will be the activity per unit length; for a plane source, it will be the activity per unit area)

Φ = flux at point of interest in photons per square centimeter per second

r = distance from source to point of interest, P

Linear Source

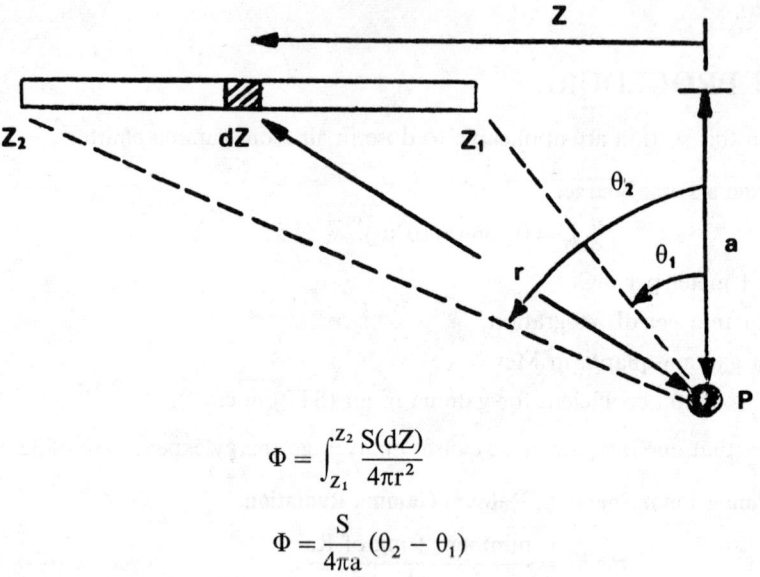

$$\Phi = \int_{Z_1}^{Z_2} \frac{S(dZ)}{4\pi r^2}$$

$$\Phi = \frac{S}{4\pi a}(\theta_2 - \theta_1)$$

Plane Circular Source

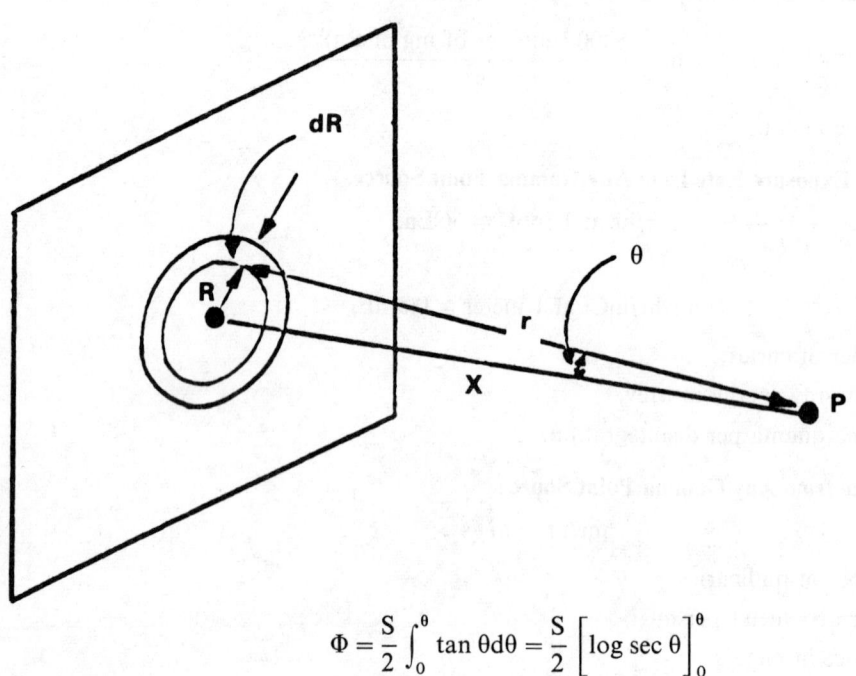

$$\Phi = \frac{S}{2}\int_0^\theta \tan\theta\, d\theta = \frac{S}{2}\left[\log \sec\theta\right]_0^\theta$$

INTERNAL RADIATION DOSAGE

(1) **Biological Half-life.**

$$T_b = \frac{0.693}{\lambda_b},$$

where λ_b = biological rate of elimination constant,
T_b = biological half-life

(2) **Effective Half-Life.**

$$T_{eff} = \frac{(T_{1/2})(T_b)}{T_{1/2} + T_b},$$

where T_{eff} = effective half-life,
$T_{1/2}$ = radioactive (physical) half-life
T_b = biological half-life.

(3) **Beta-Emitter Dose.**

$$D = 88ET_{eff}C(1 - e^{-\lambda_{eff}t}),$$

where D = dose (reps),
E = average energy of beta particle (Mev),
C = Ci/g of radioisotope in tissue
λ_{eff} = effective decay constant (days^{-1}),
t = time (days).

DECONTAMINATION FACTOR

$$D.F. = \frac{\text{Initial Activity}}{\text{Final Activity}}$$

ISOTOPIC DILUTIONS

Single-Addition Method

$$w = w'\left(\frac{SpA'}{SpA} - 1\right),$$

where w = total weight of diluent material (weight of stable material),
w' = total weight of labeled material (weight of radioactive material),
SpA' = specific activity of labeled material,
SpA = specific activity of mixture.

Double Dilution

The following symbols will be used in this section:

S_0 = initial specific activity
S_1 = specific activity of first dilution
S_2 = specific activity of second dilution
G_1 = weight of carrier added for first dilution
G_2 = weight of carrier added for second dilution
Z = weight of original radioactive material

(1) $$S_0 = \frac{S_1 S_2 (G_1 - G_2)}{S_2 G_2 - S_1 G_1}$$

(2) $$Z = \frac{S_2 G_2 - S_1 G_1}{S_1 - S_2}$$

NEUTRON ACTIVATION METHODS

Thin Target*.

$$A_\phi = k\sigma_{ac}fn(1 - e^{-\lambda t})e^{-\lambda \phi},$$

where A_ϕ = measured activity, in net counts per second, at time ϕ,

   k = efficiency of the counter for measuring the induced radioactivity,

   σ_{ac} = activation across section for neutron capture by the target material, cm^2 per atom per neutron,

   f = flux of neutrons, neutrons per cm^2 per second,

   n = total number of target nuclei,

   λ = disintegration constant of radioactive material,

   t = time duration of exposure to neutron flux,

   ϕ = time increment between end of irradiation and the time at which the target is counted,

   e = base of natural logarithm (2.718...).

GEOMETRY OF A COUNTER

Point Source.

$$G = 0.5(1 - \cos \alpha) = \sin^2 \tfrac{1}{2}\alpha,$$

where $\alpha = \tan^{-1} \frac{r}{d}$,

   r = radius of counter window or phosphor,

   d = distance between counter and source,

   G = geometry factor.

* A thin target is one that will not reduce the neutron flux by more than the error involved for the experiment.

FOUR-PLACE MANTISSAS FOR COMMON LOGARITHMS OF DECIMAL FRACTIONS

This table gives the logarithms of the decimal fractions, which are negative numbers; for example, $\log 0.61 = -0.2147 = 9.7853 - 10$.

It should be noted that the entries as given can be used conveniently to find cologarithms of positive numbers. Every positive number N equals $P \cdot (10)^k$, where $0 < P \leq 1$. Since colog $N = -\log N$, it follows that colog $N = -\log P - k$; for example, colog $0.61 = 0.2147$, colog $61 = 0.2147 - 2$, and colog $0.00061 = 3.2147$.

N	0	1	2	3	4	5	6	7	8	9
.10	−1.000	−.9957	−.9914	−.9872	−.9830	−.9788	−.9747	−.9706	−.9666	−.9626
.11	−.9586	−.9547	−.9508	−.9469	−.9431	−.9393	−.9355	−.9318	−.9281	−.9245
.12	−.9208	−.9172	−.9136	−.9101	−.9066	−.9031	−.8996	−.8662	−.8928	−.8894
.13	−.8861	−.8827	−.8794	−.8761	−.8729	−.8697	−.8665	−.8633	−.8601	−.8570
.14	−.8539	−.8508	−.8477	−.8447	−.8416	−.8386	−.8356	−.8327	−.8297	−.8268
.15	−.8239	−.8210	−.8182	−.8153	−.8125	−.8097	−.8069	−.8041	−.8013	−.7986
.16	−.7959	−.7932	−.7905	−.7878	−.7852	−.7825	−.7799	−.7773	−.7747	−.7721
.17	−.7696	−.7670	−.7645	−.7620	−.7595	−.7570	−.7545	−.7520	−.7496	−.7471
.18	−.7447	−.7423	−.7399	−.7375	−.7352	−.7328	−.7305	−.7282	−.7258	−.7235
.19	−.7212	−.7190	−.7167	−.7144	−.7122	−.7100	−.7077	−.7055	−.7033	−.7011
.20	−.6990	−.6968	−.6946	−.6925	−.6904	−.6882	−.6861	−.6840	−.6819	−.6799
.21	−.6778	−.6757	−.6737	−.6716	−.6696	−.6676	−.6655	−.6635	−.6615	−.6596
.22	−.6576	−.6556	−.6536	−.6517	−.6498	−.6478	−.6459	−.6440	−.6421	−.6402
.23	−.6383	−.6364	−.6345	−.6326	−.6308	−.6289	−.6271	−.6253	−.6234	−.6216
.24	−.6198	−.6180	−.6162	−.6144	−.6126	−.6108	−.6091	−.6073	−.6055	−.6038
.25	−.6021	−.6003	−.5986	−.5969	−.5952	−.5935	−.5918	−.5901	−.5884	−.5867
.26	−.5850	−.5834	−.5817	−.5800	−.5784	−.5768	−.5751	−.5735	−.5719	−.5702
.27	−.5686	−.5670	−.5654	−.5638	−.5622	−.5607	−.5591	−.5575	−.5560	−.5544
.28	−.5528	−.5513	−.5498	−.5482	−.5467	−.5452	−.5436	−.5421	−.5406	−.5391
.29	−.5376	−.5361	−.5346	−.5331	−.5317	−.5302	−.5287	−.5272	−.5258	−.5243
.30	−.5229	−.5214	−.5200	−.5186	−.5171	−.5157	−.5143	−.5129	−.5114	−.5100
.31	−.5086	−.5072	−.5058	−.5045	−.5031	−.5017	−.5003	−.4989	−.4976	−.4962
.32	−.4949	−.4935	−.4921	−.4908	−.4895	−.4881	−.4868	−.4855	−.4841	−.4828
.33	−.4815	−.4802	−.4789	−.4776	−.4763	−.4750	−.4737	−.4724	−.4711	−.4698
.34	−.4685	−.4672	−.4660	−.4647	−.4634	−.4622	−.4609	−.4597	−.4584	−.4572
.35	−.4559	−.4547	−.4535	−.4522	−.4510	−.4498	−.4486	−.4473	−.4461	−.4449
.36	−.4437	−.4425	−.4413	−.4401	−.4389	−.4377	−.4365	−.4353	−.4342	−.4330
.37	−.4318	−.4306	−.4295	−.4283	−.4271	−.4260	−.4248	−.4237	−.4225	−.4214
.38	−.4202	−.4191	−.4179	−.4168	−.4157	−.4145	−.4134	−.4123	−.4112	−.4101
.39	−.4089	−.4078	−.4067	−.4056	−.4045	−.4034	−.4023	−.4012	−.4001	−.3990
.40	−.3979	−.3969	−.3958	−.3947	−.3936	−.3925	−.3915	−.3904	−.3893	−.3883
.41	−.3872	−.3862	−.3851	−.3840	−.3830	−.3820	−.3809	−.3799	−.3788	−.3778
.42	−.3768	−.3757	−.3747	−.3737	−.3726	−.3716	−.3706	−.3696	−.3686	−.3675
.43	−.3665	−.3655	−.3645	−.3635	−.3625	−.3615	−.3605	−.3595	−.3585	−.3575
.44	−.3565	−.3556	−.3546	−.3536	−.3526	−.3516	−.3507	−.3497	−.3487	−.3478

N	0	1	2	3	4	5	6	7	8	9
.45	−.3468	−.3458	−.3449	−.3439	−.3429	−.3420	−.3410	−.3401	−.3391	−.3382
.46	−.3372	−.3363	−.3354	−.3344	−.3335	−.3325	−.3316	−.3307	−.3298	−.3288
.47	−.3279	−.3270	−.3261	−.3251	−.3242	−.3233	−.3224	−.3215	−.3206	−.3197
.48	−.3188	−.3179	−.3170	−.3161	−.3152	−.3143	−.3134	−.3125	−.3116	−.3107
.49	−.3098	−.3089	−.3080	−.3072	−.3063	−.3054	−.3045	−.3036	−.3028	−.3019
.50	−.3010	−.3002	−.2993	−.2984	−.2976	−.2967	−.2958	−.2950	−.2941	−.2933
.51	−.2924	−.2916	−.2907	−.2899	−.2890	−.2882	−.2874	−.2865	−.2857	−.2848
.52	−.2840	−.2832	−.2823	−.2815	−.2807	−.2798	−.2790	−.2782	−.2774	−.2765
.53	−.2757	−.2749	−.2741	−.2733	−.2725	−.2716	−.2708	−.2700	−.2692	−.2684
.54	−.2676	−.2668	−.2660	−.2652	−.2644	−.2636	−.2628	−.2620	−.2612	−.2604
.55	−.2596	−.2588	−.2581	−.2573	−.2565	−.2557	−.2549	−.2541	−.2534	−.2526
.56	−.2518	−.2510	−.2503	−.2495	−.2487	−.2480	−.2472	−.2464	−.2457	−.2449
.57	−.2441	−.2434	−.2426	−.2418	−.2411	−.2403	−.2396	−.2388	−.2381	−.2373
.58	−.2366	−.2358	−.2351	−.2343	−.2336	−.2328	−.2321	−.2314	−.2306	−.2299
.59	−.2291	−.2284	−.2277	−.2269	−.2262	−.2255	−.2248	−.2240	−.2233	−.2226
.60	−.2218	−.2211	−.2204	−.2197	−.2190	−.2182	−.2175	−.2168	−.2161	−.2154
.61	−.2147	−.2140	−.2132	−.2125	−.2118	−.2111	−.2104	−.2097	−.2090	−.2083
.62	−.2076	−.2069	−.2062	−.2055	−.2048	−.2041	−.2034	−.2027	−.2020	−.2013
.63	−.2007	−.2000	−.1993	−.1986	−.1979	−.1972	−.1965	−.1959	−.1952	−.1945
.64	−.1938	−.1931	−.1925	−.1918	−.1911	−.1904	−.1898	−.1891	−.1884	−.1878
.65	−.1871	−.1864	−.1858	−.1851	−.1844	−.1838	−.1831	−.1824	−.1818	−.1811
.66	−.1805	−.1798	−.1791	−.1785	−.1778	−.1772	−.1765	−.1759	−.1752	−.1746
.67	−.1739	−.1733	−.1726	−.1720	−.1713	−.1707	−.1701	−.1694	−.1688	−.1681
.68	−.1675	−.1669	−.1662	−.1656	−.1649	−.1643	−.1637	−.1630	−.1624	−.1618
.69	−.1612	−.1605	−.1599	−.1593	−.1586	−.1580	−.1574	−.1568	−.1561	−.1555
.70	−.1549	−.1543	−.1537	−.1530	−.1524	−.1518	−.1512	−.1506	−.1500	−.1494
.71	−.1487	−.1481	−.1475	−.1469	−.1463	−.1457	−.1451	−.1445	−.1439	−.1433
.72	−.1427	−.1421	−.1415	−.1409	−.1403	−.1397	−.1391	−.1385	−.1379	−.1373
.73	−.1367	−.1361	−.1355	−.1349	−.1343	−.1337	−.1331	−.1325	−.1319	−.1314
.74	−.1308	−.1302	−.1296	−.1290	−.1284	−.1278	−.1273	−.1267	−.1261	−.1255
.75	−.1249	−.1244	−.1238	−.1232	−.1226	−.1221	−.1215	−.1209	−.1203	−.1198
.76	−.1192	−.1186	−.1180	−.1175	−.1169	−.1163	−.1158	−.1152	−.1146	−.1141
.77	−.1135	−.1129	−.1124	−.1118	−.1113	−.1107	−.1101	−.1096	−.1090	−.1085
.78	−.1079	−.1073	−.1068	−.1062	−.1057	−.1051	−.1046	−.1040	−.1035	−.1029
.79	−.1024	−.1018	−.1013	−.1007	−.1002	−.0996	−.0991	−.0985	−.0980	−.0975
.80	−.0969	−.0964	−.0958	−.0953	−.0947	−.0942	−.0937	−.0931	−.0926	−.0921
.81	−.0915	−.0910	−.0904	−.0899	−.0894	−.0888	−.0883	−.0878	−.0872	−.0867
.82	−.0862	−.0857	−.0851	−.0846	−.0841	−.0835	−.0830	−.0825	−.0820	−.0814
.83	−.0809	−.0804	−.0799	−.0794	−.0788	−.0783	−.0778	−.0773	−.0768	−.0762
.84	−.0757	−.0752	−.0747	−.0742	−.0737	−.0731	−.0726	−.0721	−.0716	−.0711
.85	−.0706	−.0701	−.0696	−.0691	−.0685	−.0680	−.0675	−.0670	−.0665	−.0660
.86	−.0655	−.0650	−.0645	−.0640	−.0635	−.0630	−.0625	−.0620	−.0615	−.0610
.87	−.0605	−.0600	−.0595	−.0590	−.0585	−.0580	−.0575	−.0570	−.0565	−.0560
.88	−.0555	−.0550	−.0545	−.0540	−.0535	−.0531	−.0526	−.0521	−.0516	−.0511
.89	−.0506	−.0501	−.0496	−.0491	−.0487	−.0482	−.0477	−.0472	−.0467	−.0462
.90	−.0458	−.0453	−.0448	−.0443	−.0438	−.0434	−.0429	−.0424	−.0419	−.0414
.91	−.0410	−.0405	−.0400	−.0395	−.0391	−.0386	−.0381	−.0376	−.0372	−.0367
.92	−.0362	−.0357	−.0353	−.0348	−.0343	−.0339	−.0334	−.0329	−.0325	−.0320
.93	−.0315	−.0311	−.0306	−.0301	−.0297	−.0292	−.0287	−.0283	−.0278	−.0273
.94	−.0269	−.0264	−.0259	−.0255	−.0250	−.0246	−.0241	−.0237	−.0232	−.0227
.95	−.0223	−.0218	−.0214	−.0209	−.0205	−.0200	−.0195	−.0191	−.0186	−.0182
.96	−.0177	−.0173	−.0168	−.0164	−.0159	−.0155	−.0150	−.0146	−.0141	−.0137
.97	−.0132	−.0128	−.0123	−.0119	−.0114	−.0110	−.0106	−.0101	−.0097	−.0092
.98	−.0088	−.0083	−.0079	−.0074	−.0070	−.0066	−.0061	−.0057	−.0052	−.0048
.99	−.0044	−.0039	−.0035	−.0031	−.0026	−.0022	−.0017	−.0013	−.0009	−.0004

FOUR-PLACE MANTISSAS FOR COMMON LOGARITHMS

N	0	1	2	3	4	5	6	7	8	9	1	2	3	4	5	6	7	8	9
											Proportional Parts								
10	0000	0043	0086	0128	0170	0212	0253	0294	0334	0374	*4	8	12	17	21	25	29	33	37
11	0414	0453	0492	0531	0569	0607	0645	0682	0719	0755	4	8	11	15	19	23	26	30	34
12	0792	0828	0864	0899	0934	0969	1004	1038	1072	1106	3	7	10	14	17	21	24	28	31
13	1139	1173	1206	1239	1271	1303	1335	1367	1399	1430	3	6	10	13	16	19	23	26	29
14	1461	1492	1523	1553	1584	1614	1644	1673	1703	1732	3	6	9	12	15	18	21	24	27
15	1761	1790	1818	1847	1875	1903	1931	1959	1987	2014	*3	6	8	11	14	17	20	22	25
16	2041	2068	2095	2122	2148	2175	2201	2227	2253	2279	3	5	8	11	13	16	18	21	24
17	2304	2330	2355	2380	2405	2430	2455	2480	2504	2529	2	5	7	10	12	15	17	20	22
18	2553	2577	2601	2625	2648	2672	2695	2718	2742	2765	2	5	7	9	12	14	16	19	21
19	2788	2810	2833	2856	2878	2900	2923	2945	2967	2989	2	4	7	9	11	13	16	18	20
20	3010	3032	3054	3075	3096	3118	3139	3160	3181	3201	2	4	6	8	11	13	15	17	19
21	3222	3243	3263	3284	3304	3324	3345	3365	3385	3404	2	4	6	8	10	12	14	16	18
22	3424	3444	3464	3483	3502	3522	3541	3560	3579	3598	2	4	6	8	10	12	14	15	17
23	3617	3636	3655	3674	3692	3711	3729	3747	3766	3784	2	4	6	7	9	11	13	15	17
24	3802	3820	3838	3856	3874	3892	3909	3927	3945	3962	2	4	5	7	9	11	12	14	16
25	3979	3997	4014	4031	4048	4065	4082	4099	4116	4133	2	3	5	7	9	10	12	14	15
26	4150	4166	4183	4200	4216	4232	4249	4265	4281	4298	2	3	5	7	8	10	11	13	15
27	4314	4330	4346	4362	4378	4393	4409	4425	4440	4456	2	3	5	6	8	9	11	13	14
28	4472	4487	4502	4518	4533	4548	4564	4579	4594	4609	2	3	5	6	8	9	11	12	14
29	4624	4639	4654	4669	4683	4698	4713	4728	4742	4757	1	3	4	6	7	9	10	12	13
30	4771	4786	4800	4814	4829	4843	4857	4871	4886	4900	1	3	4	6	7	9	10	11	13
31	4914	4928	4942	4955	4969	4983	4997	5011	5024	5038	1	3	4	6	7	8	10	11	12
32	5051	5065	5079	5092	5105	5119	5132	5145	5159	5172	1	3	4	5	7	8	9	11	12
33	5185	5198	5211	5224	5237	5250	5263	5276	5289	5302	1	3	4	5	6	8	9	10	12
34	5315	5328	5340	5353	5366	5378	5391	5403	5416	5428	1	3	4	5	6	8	9	10	11
35	5441	5453	5465	5478	5490	5502	5514	5527	5539	5551	1	2	4	5	6	7	9	10	11
36	5563	5575	5587	5599	5611	5623	5635	5647	5658	5670	1	2	4	5	6	7	8	10	11
37	5682	5694	5705	5717	5729	5740	5752	5763	5775	5786	1	2	3	5	6	7	8	9	10
38	5798	5809	5821	5832	5843	5855	5866	5877	5888	5899	1	2	3	5	6	7	8	9	10
39	5911	5922	5933	5944	5955	5966	5977	5988	5999	6010	1	2	3	4	5	7	8	9	10
40	6021	6031	6042	6053	6064	6075	6085	6096	6107	6117	1	2	3	4	5	6	8	9	10
41	6128	6138	6149	6160	6170	6180	6191	6201	6212	6222	1	2	3	4	5	6	7	8	9
42	6232	6243	6253	6263	6274	6284	6294	6304	6314	6325	1	2	3	4	5	6	7	8	9
43	6335	6345	6355	6365	6375	6385	6395	6405	6415	6425	1	2	3	4	5	6	7	8	9
44	6435	6444	6454	6464	6474	6484	6493	6503	6513	6522	1	2	3	4	5	6	7	8	9
45	6532	6542	6551	6561	6571	6580	6590	6599	6609	6618	1	2	3	4	5	6	7	8	9
46	6628	6637	6646	6656	6665	6675	6684	6693	6702	6712	1	2	3	4	5	6	7	7	8
47	6721	6730	6739	6749	6758	6767	6776	6785	6794	6803	1	2	3	4	5	5	6	7	8
48	6812	6821	6830	6839	6848	6857	6866	6875	6884	6893	1	2	3	4	4	5	6	7	8
49	6902	6911	6920	6928	6937	6946	6955	6964	6972	6981	1	2	3	4	4	5	6	7	8
N	0	1	2	3	4	5	6	7	8	9	1	2	3	4	5	6	7	8	9

* Interpolation in this section of the table is inaccurate.

N	0	1	2	3	4	5	6	7	8	9	Proportional Parts								
											1	2	3	4	5	6	7	8	9
50	6990	6998	7007	7016	7024	7033	7042	7050	7059	7067	1	2	3	3	4	5	6	7	8
51	7076	7084	7093	7101	7110	7118	7126	7135	7143	7152	1	2	3	3	4	5	6	7	8
52	7160	7168	7177	7185	7193	7202	7210	7218	7226	7235	1	2	2	3	4	5	6	7	7
53	7243	7251	7259	7267	7275	7284	7292	7300	7308	7316	1	2	2	3	4	5	6	6	7
54	7324	7332	7340	7348	7356	7364	7372	7380	7388	7396	1	2	2	3	4	5	6	6	7
55	7404	7412	7419	7427	7435	7443	7451	7459	7466	7474	1	2	2	3	4	5	5	6	7
56	7482	7490	7497	7505	7513	7520	7528	7536	7543	7551	1	2	2	3	4	5	5	6	7
57	7559	7566	7574	7582	7589	7597	7604	7612	7619	7627	1	2	2	3	4	5	5	6	7
58	7634	7642	7649	7657	7664	7672	7679	7686	7694	7701	1	1	2	3	4	4	5	6	7
59	7709	7716	7723	7731	7738	7745	7752	7760	7767	7774	1	1	2	3	4	4	5	6	7
60	7782	7789	7796	7803	7810	7818	7825	7832	7839	7846	1	1	2	3	4	4	5	6	6
61	7853	7860	7868	7875	7882	7889	7896	7903	7910	7917	1	1	2	3	4	4	5	6	6
62	7924	7931	7938	7945	7952	7959	7966	7973	7980	7987	1	1	2	3	3	4	5	6	6
63	7993	8000	8007	8014	8021	8028	8035	8041	8048	8055	1	1	2	3	3	4	5	5	6
64	8062	8069	8075	8082	8089	8096	8102	8109	8116	8122	1	1	2	3	3	4	5	5	6
65	8129	8136	8142	8149	8156	8162	8169	8176	8182	8189	1	1	2	3	3	4	5	5	6
66	8195	8202	8209	8215	8222	8228	8235	8241	8248	8254	1	1	2	3	3	4	5	5	6
67	8261	8267	8274	8280	8287	8293	8299	8306	8312	8319	1	1	2	3	3	4	5	5	6
68	8325	8331	8338	8344	8351	8357	8363	8370	8376	8382	1	1	2	3	3	4	4	5	6
69	8388	8395	8401	8407	8414	8420	8426	8432	8439	8445	1	1	2	2	3	4	4	5	6
70	8451	8457	8463	8470	8476	8482	8488	8494	8500	8506	1	1	2	2	3	4	4	5	6
71	8513	8519	8525	8531	8537	8543	8549	8555	8561	8567	1	1	2	2	3	4	4	5	5
72	8573	8579	8585	8591	8597	8603	8609	8615	8621	8627	1	1	2	2	3	4	4	5	5
73	8633	8639	8645	8651	8657	8663	8669	8675	8681	8686	1	1	2	2	3	4	4	5	5
74	8692	8698	8704	8710	8716	8722	8727	8733	8739	8745	1	1	2	2	3	4	4	5	5
75	8751	8756	8762	8768	8774	8779	8785	8791	8797	8802	1	1	2	2	3	3	4	5	5
76	8808	8814	8820	8825	8831	8837	8842	8848	8854	8859	1	1	2	2	3	3	4	5	5
77	8865	8871	8876	8882	8887	8893	8899	8904	8910	8915	1	1	2	2	3	3	4	4	5
78	8921	8927	8932	8938	8943	8949	8954	8960	8965	8971	1	1	2	2	3	3	4	4	5
79	8976	8982	8987	8993	8998	9004	9009	9015	9020	9025	1	1	2	2	3	3	4	4	5
80	9031	9036	9042	9047	9053	9058	9063	9069	9074	9079	1	1	2	2	3	3	4	4	5
81	9085	9090	9096	9101	9106	9112	9117	9122	9128	9133	1	1	2	2	3	3	4	4	5
82	9138	9143	9149	9154	9159	9165	9170	9175	9180	9186	1	1	2	2	3	3	4	4	5
83	9191	9196	9201	9206	9212	9217	9222	9227	9232	9238	1	1	2	2	3	3	4	4	5
84	9243	9248	9253	9258	9263	9269	9274	9279	9284	9289	1	1	2	2	3	3	4	4	5
85	9294	9299	9304	9309	9315	9320	9325	9330	9335	9340	1	1	2	2	3	3	4	4	5
86	9345	9350	9355	9360	9365	9370	9375	9380	9385	9390	1	1	2	2	3	3	4	4	5
87	9395	9400	9405	9410	9415	9420	9425	9430	9435	9440	0	1	1	2	2	3	3	4	4
88	9445	9450	9455	9460	9465	9469	9474	9479	9484	9489	0	1	1	2	2	3	3	4	4
89	9494	9499	9504	9509	9513	9518	9523	9528	9533	9538	0	1	1	2	2	3	3	4	4
90	9542	9547	9552	9557	9562	9566	9571	9576	9581	9586	0	1	1	2	2	3	3	4	4
91	9590	9595	9600	9605	9609	9614	9619	9624	9628	9633	0	1	1	2	2	3	3	4	4
92	9638	9643	9647	9652	9657	9661	9666	9671	9675	9680	0	1	1	2	2	3	3	4	4
93	9685	9689	9694	9699	9703	9708	9713	9717	9722	9727	0	1	1	2	2	3	3	4	4
94	9731	9736	9741	9745	9750	9754	9759	9763	9768	9773	0	1	1	2	2	3	3	4	4
95	9777	9782	9786	9791	9795	9800	9805	9809	9814	9818	0	1	1	2	2	3	3	4	4
96	9823	9827	9832	9836	9841	9845	9850	9854	9859	9863	0	1	1	2	2	3	3	4	4
97	9868	9872	9877	9881	9886	9890	9894	9899	9903	9908	0	1	1	2	2	3	3	4	4
98	9912	9917	9921	9926	9930	9934	9939	9943	9948	9952	0	1	1	2	2	3	3	4	4
99	9956	9961	9965	9969	9974	9978	9983	9987	9991	9996	0	1	1	2	2	3	3	3	4
N	0	1	2	3	4	5	6	7	8	9	1	2	3	4	5	6	7	8	9

NATURAL TRIGONOMETRIC FUNCTIONS, SINE, COSINE, TANGENT, COTANGENT, FOR ANGLES, IN DEGREES AND IN DECIMALS

Degrees	Sin	*Tan	*Cot	Cos		Degrees	Sin	Tan	Cot	Cos	
0.0	0.00000	0.00000	∞	1.0000	**90.0**	**4.0**	0.06976	0.06993	14.301	0.9976	**86.0**
.1	.00175	.00175	573.0	1.0000	89.9	.1	.07150	.07168	13.951	.9974	85.9
.2	.00349	.00349	286.5	1.0000	.8	.2	.07324	.07344	13.617	.9973	.8
.3	.00524	.00524	191.0	1.0000	.7	.3	.07498	.07519	13.300	.9972	.7
.4	.00698	.00698	143.24	1.0000	.6	.4	.07672	.07695	12.996	.9971	.6
.5	.00873	.00873	114.59	1.0000	.5	.5	.07846	.07870	12.706	.9969	.5
.6	.01047	.01047	95.49	0.9999	.4	.6	.08020	.08046	12.429	.9968	.4
.7	.01222	.01222	81.85	.9999	.3	.7	.08194	.08221	12.163	.9966	.3
.8	.01396	.01396	71.62	.9999	.2	.8	.08368	.08397	11.909	.9965	.2
.9	.01571	.01571	63.66	.9999	89.1	.9	.08542	.08573	11.664	.9963	85.1
1.0	0.01745	0.01746	57.29	0.9998	**89.0**	**5.0**	0.08716	0.08749	11.430	0.9962	**85.0**
.1	.01920	.01920	52.08	.9998	88.9	.1	.08889	.08925	11.205	.9960	84.9
.2	.02094	.02095	47.74	.9998	.8	.2	.09063	.09101	10.988	.9959	.8
.3	.02269	.02269	44.07	.9997	.7	.3	.09237	.09277	10.780	.9957	.7
.4	.02443	.02444	40.92	.9997	.6	.4	.09411	.09453	10.579	.9956	.6
.5	.02618	.02619	38.19	.9997	.5	.5	.09585	.09629	10.385	.9954	.5
.6	.02792	.02793	35.80	.9996	.4	.6	.09758	.09805	10.199	.9952	.4
.7	.02967	.02968	33.69	.9996	.3	.7	.09932	.09981	10.019	.9951	.3
.8	.03141	.03143	31.82	.9995	.2	.8	.10106	.10158	9.845	.9949	.2
.9	.03316	.03317	30.14	.9995	88.1	.9	.10279	.10334	9.677	.9947	84.1
2.0	0.03490	0.03492	28.64	0.9994	**88.0**	**6.0**	0.10453	0.10510	9.514	0.9945	**84.0**
.1	.03664	.03667	27.27	.9993	87.9	.1	.10626	.10687	9.357	.9943	83.9
.2	.03839	.03842	26.03	.9993	.8	.2	.10800	.10863	9.205	.9942	.8
.3	.04013	.04016	24.90	.9992	.7	.3	.10973	.11040	9.058	.9940	.7
.4	.04188	.04191	23.86	.9991	.6	.4	.11147	.11217	8.915	.9938	.6
.5	.04362	.04366	22.90	.9990	.5	.5	.11320	.11394	8.777	.9936	.5
.6	.04536	.04541	22.02	.9990	.4	.6	.11494	.11570	8.643	.9934	.4
.7	.04711	.04716	21.20	.9989	.3	.7	.11667	.11747	8.513	.9932	.3
.8	.04885	.04891	20.45	.9988	.2	.8	.11840	.11924	8.386	.9930	.2
.9	.05059	.05066	19.74	.9987	87.1	.9	.12014	.12101	8.264	.9928	83.1
3.0	0.05234	0.05241	19.081	0.9986	**87.0**	**7.0**	0.12187	0.12278	8.144	0.9925	**83.0**
.1	.05408	.05416	18.464	.9985	86.9	.1	.12360	.12456	8.028	.9923	82.9
.2	.05582	.05591	17.886	.9984	.8	.2	.12533	.12633	7.916	.9921	.8
.3	.05756	.05766	17.343	.9983	.7	.3	.12706	.12810	7.806	.9919	.7
.4	.05931	.05941	16.832	.9982	.6	.4	.12880	.12988	7.700	.9917	.6
.5	.06105	.06116	16.350	.9981	.5	.5	.13053	.13165	7.596	.9914	.5
.6	.06279	.06291	15.895	.9980	.4	.6	.13226	.13343	7.495	.9912	.4
.7	.06453	.06467	15.464	.9979	.3	.7	.13399	.13521	7.396	.9910	.3
.8	.06627	.06642	15.056	.9978	.2	.8	.13572	.13698	7.300	.9907	.2
.9	.06802	.06817	14.669	.9977	86.1	.9	.13744	.13876	7.207	.9905	82.1
4.0	0.06976	0.06993	14.301	0.9976	**86.0**	**8.0**	0.13917	0.14054	7.115	0.9903	**82.0**
	Cos	*Cot	*Tan	Sin	Degrees		Cos	Cot	Tan	Sin	Degrees

* Interpolation in this section of the table is inaccurate.

Degrees	Sin	Tan	Cot	Cos		Degrees	Sin	Tan	Cot	Cos	
8.0	0.13917	0.14054	7.115	0.9903	**82.0**	13.5	.2334	.2401	4.165	.9724	76.5
.1	.14090	.14232	7.026	.9900	81.9	.6	.2351	.2419	4.134	.9720	.4
.2	.14263	.14410	6.940	.9898	.8	.7	.2368	.2438	4.102	.9715	.3
.3	.14436	.14588	6.855	.9895	.7	.8	.2385	.2456	4.071	.9711	.2
.4	.14608	.14767	6.772	.9893	.6	.9	.2402	.2475	4.041	.9707	76.1
.5	.14781	.14945	6.691	.9890	.5	**14.0**	0.2419	0.2493	4.011	0.9703	**76.0**
.6	.14954	.15124	6.612	.9888	.4	.1	.2436	.2512	3.981	.9699	75.9
.7	.15126	.15302	6.535	.9885	.3	.2	.2453	.2530	3.952	.9694	.8
.8	.15299	.15481	6.460	.9882	.2	.3	.2470	.2549	3.923	.9690	.7
.9	.15471	.15660	6.386	.9880	81.1	.4	.2487	.2568	3.895	.9686	.6
9.0	0.15643	0.15838	6.314	0.9877	**81.0**	.5	.2504	.2586	3.867	.9681	.5
.1	.15816	.16017	6.243	.9874	80.9	.6	.2521	.2605	3.839	.9677	.4
.2	.15988	.16196	6.174	.9871	.8	.7	.2538	.2623	3.812	.9673	.3
.3	.16160	.16376	6.107	.9869	.7	.8	.2554	.2642	3.785	.9668	.2
.4	.16333	.16555	6.041	.9866	.6	.9	.2571	.2661	3.758	.9664	75.1
.5	.16505	.16734	5.976	.9863	.5	**15.0**	0.2588	0.2679	3.732	0.9659	**75.0**
.6	.16677	.16914	5.912	.9860	.4	.1	.2605	.2698	3.706	.9655	74.9
.7	.16849	.17093	5.850	.9857	.3	.2	.2622	.2717	3.681	.9650	.8
.8	.17021	.17273	5.789	.9854	.2	.3	.2639	.2736	3.655	.9646	.7
.9	.17193	.17453	5.730	.9851	80.1	.4	.2656	.2754	3.630	.9641	.6
10.0	0.1736	0.1763	5.671	0.9848	**80.0**	.5	.2672	.2773	3.606	.9636	.5
.1	.1754	.1781	5.614	.9845	79.9	.6	.2689	.2792	3.582	.9632	.4
.2	.1771	.1799	5.558	.9842	.8	.7	.2706	.2811	3.558	.9627	.3
.3	.1788	.1817	5.503	.9839	.7	.8	.2723	.2830	3.534	.9622	.2
.4	.1805	.1835	5.449	.9836	.6	.9	.2740	.2849	3.511	.9617	74.1
.5	.1822	.1853	5.396	.9833	.5	**16.0**	0.2756	0.2867	3.487	0.9613	**74.0**
.6	.1840	.1871	5.343	.9829	.4	.1	.2773	.2886	3.465	.9608	73.9
.7	.1857	.1890	5.292	.9826	.3	.2	.2790	.2905	3.442	.9603	.8
.8	.1874	.1908	5.242	.9823	.2	.3	.2807	.2924	3.420	.9598	.7
.9	.1891	.1926	5.193	.9820	79.1	.4	.2823	.2943	3.398	.9593	.6
11.0	0.1908	0.1944	5.145	0.9816	**79.0**	.5	.2840	.2962	3.376	.9588	.5
.1	.1925	.1962	5.097	.9813	78.9	.6	.2857	.2981	3.354	.9583	.4
.2	.1942	.1980	5.050	.9810	.8	.7	.2874	.3000	3.333	.9578	.3
.3	.1959	.1998	5.005	.9806	.7	.8	.2890	.3019	3.312	.9573	.2
.4	.1977	.2016	4.959	.9803	.6	.9	.2907	.3038	3.291	.9568	73.1
.5	.1994	.2035	4.915	.9799	.5	**17.0**	0.2924	0.3057	3.271	0.9563	**73.0**
.6	.2011	.2053	4.872	.9796	.4	.1	.2940	.3076	3.251	.9558	72.9
.7	.2028	.2071	4.829	.9792	.3	.2	.2957	.3096	3.230	.9553	.8
.8	.2045	.2089	4.787	.9789	.2	.3	.2974	.3115	3.211	.9548	.7
.9	.2062	.2107	4.745	.9785	78.1	.4	.2990	.3134	3.191	.9542	.6
12.0	0.2079	0.2126	4.705	0.9781	**78.0**	.5	.3007	.3153	3.172	.9537	.5
.1	.2096	.2144	4.665	.9778	77.9	.6	.3024	.3172	3.152	.9532	.4
.2	.2113	.2162	4.625	.9774	.8	.7	.3040	.3191	3.133	.9527	.3
.3	.2130	.2180	4.586	.9770	.7	.8	.3057	.3211	3.115	.9521	.2
.4	.2147	.2199	4.548	.9767	.6	.9	.3074	.3230	3.096	.9516	72.1
.5	.2164	.2217	4.511	.9763	.5	**18.0**	0.3090	0.3249	3.078	0.9511	**72.0**
.6	.2181	.2235	4.474	.9759	.4	.1	.3107	.3269	3.060	.9505	71.9
.7	.2198	.2254	4.437	.9755	.3	.2	.3123	.3288	3.042	.9500	.8
.8	.2215	.2272	4.402	.9751	.2	.3	.3140	.3307	3.024	.9494	.7
.9	.2233	.2290	4.366	.9748	77.1	.4	.3156	.3327	3.006	.9489	.6
13.0	0.2250	0.2309	4.331	0.9744	**77.0**	.5	.3173	.3346	2.989	.9483	.5
.1	.2267	.2327	4.297	.9740	76.9	.6	.3190	.3365	2.971	.9478	.4
.2	.2284	.2345	4.264	.9736	.8	.7	.3206	.3385	2.954	.9472	.3
.3	.2300	.2364	4.230	.9732	.7	.8	.3223	.3404	2.937	.9466	.2
.4	.2317	.2382	4.198	.9728	.6	.9	.3239	.3424	2.921	.9461	71.1
13.5	.2334	.2401	4.165	.9724	76.5	**19.0**	0.3256	0.3443	2.904	0.9455	**71.0**
	Cos	Cot	Tan	Sin	Degrees		Cos	Cot	Tan	Sin	Degrees

Reference Data

Degrees	Sin	Tan	Cot	Cos		Degrees	Sin	Tan	Cot	Cos	
19.0	0.3256	0.3443	2.904	0.9455	**71.0**	24.5	.4147	.4557	2.194	.9100	65.5
.1	.3272	.3463	2.888	.9449	70.9	.6	.4163	.4578	2.184	.9092	.4
.2	.3289	.3482	2.872	.9444	.8	.7	.4179	.4599	2.174	.9085	.3
.3	.3305	.3502	2.856	.9438	.7	.8	.4195	.4621	2.164	.9078	.2
.4	.3322	.3522	2.840	.9432	.6	.9	.4210	.4642	2.154	.9070	65.1
.5	.3338	.3541	2.824	.9426	.5	**25.0**	0.4226	0.4663	2.145	0.9063	**65.0**
.6	.3355	.3561	2.808	.9421	.4	.1	.4242	.4684	2.135	.9056	64.9
.7	.3371	.3581	2.793	.9415	.3	.2	.4258	.4706	2.125	.9048	.8
.8	.3387	.3600	2.778	.9409	.2	.3	.4274	.4727	2.116	.9041	.7
.9	.3404	.3620	2.762	.9403	70.1	.4	.4289	.4748	2.106	.9033	.6
20.0	0.3420	0.3640	2.747	0.9397	**70.0**	.5	.4305	.4770	2.097	.9026	.5
.1	.3437	.3659	2.733	.9391	69.9	.6	.4321	.4791	2.087	.9018	.4
.2	.3453	.3679	2.718	.9385	.8	.7	.4337	.4813	2.078	.9011	.3
.3	.3469	.3699	2.703	.9379	.7	.8	.4352	.4834	2.069	.9003	.2
.4	.3486	.3719	2.689	.9373	.6	.9	.4368	.4856	2.059	.8996	64.1
.5	.3502	.3739	2.675	.9367	.5	**26.0**	0.4384	0.4877	2.050	0.8988	**64.0**
.6	.3518	.3759	2.660	.9361	.4	.1	.4399	.4899	2.041	.8980	63.9
.7	.3535	.3779	2.646	.9354	.3	.2	.4415	.4921	2.032	.8973	.8
.8	.3551	.3799	2.633	.9348	.2	.3	.4431	.4942	2.023	.8965	.7
.9	.3567	.3819	2.619	.9342	69.1	.4	.4446	.4964	2.014	.8957	.6
21.0	0.3584	0.3839	2.605	0.9336	**69.0**	.5	.4462	.4986	2.006	.8949	.5
.1	.3600	.3859	2.592	.9330	68.9	.6	.4478	.5008	1.997	.8942	.4
.2	.3616	.3879	2.578	.9323	.8	.7	.4493	.5029	1.988	.8934	.3
.3	.3633	.3899	2.565	.9317	.7	.8	.4509	.5051	1.980	.8926	.2
.4	.3649	.3919	2.552	.9311	.6	.9	.4524	.5073	1.971	.8918	63.1
.5	.3665	.3939	2.539	.9304	.5	**27.0**	0.4540	0.5095	1.963	0.8910	**63.0**
.6	.3681	.3959	2.526	.9298	.4	.1	.4555	.5117	1.954	.8902	62.9
.7	.3697	.3979	2.513	.9291	.3	.2	.4571	.5139	1.946	.8894	.8
.8	.3714	.4000	2.500	.9285	.2	.3	.4586	.5161	1.937	.8886	.7
.9	.3730	.4020	2.488	.9278	68.1	.4	.4602	.5184	1.929	.8878	.6
22.0	0.3746	0.4040	2.475	0.9272	**68.0**	.5	.4617	.5206	1.921	.8870	.5
.1	.3762	.4061	2.463	.9265	67.9	.6	.4633	.5228	1.913	.8862	.4
.2	.3778	.4081	2.450	.9259	.8	.7	.4648	.5250	1.905	.8854	.3
.3	.3795	.4101	2.438	.9252	.7	.8	.4664	.5272	1.897	.8846	.2
.4	.3811	.4122	2.426	.9245	.6	.9	.4679	.5295	1.889	.8838	62.1
.5	.3827	.4142	2.414	.9239	.5	**28.0**	0.4695	0.5317	1.881	0.8829	**62.0**
.6	.3843	.4163	2.402	.9232	.4	.1	.4710	.5340	1.873	.8821	61.9
.7	.3859	.4183	2.391	.9225	.3	.2	.4726	.5362	1.865	.8813	.8
.8	.3875	.4204	2.379	.9219	.2	.3	.4741	.5384	1.857	.8805	.7
.9	.3891	.4224	2.367	.9212	67.1	.4	.4756	.5407	1.849	.8796	.6
23.0	0.3907	0.4245	2.356	0.9205	**67.0**	.5	.4772	.5430	1.842	.8788	.5
.1	.3923	.4265	2.344	.9198	66.9	.6	.4787	.5452	1.834	.8780	.4
.2	.3939	.4286	2.333	.9191	.8	.7	.4802	.5475	1.827	.8771	.3
.3	.3955	.4307	2.322	.9184	.7	.8	.4818	.5498	1.819	.8763	.2
.4	.3971	.4327	2.311	.9178	.6	.9	.4833	.5520	1.811	.8755	61.1
.5	.3987	.4348	2.300	.9171	.5	**29.0**	0.4848	0.5543	1.804	0.8746	**61.0**
.6	.4003	.4369	2.289	.9164	.4	.1	.4863	.5566	1.797	.8738	60.9
.7	.4019	.4390	2.278	.9157	.3	.2	.4879	.5589	1.789	.8729	.8
.8	.4035	.4411	2.267	.9150	.2	.3	.4894	.5612	1.782	.8721	.7
.9	.4051	.4431	2.257	.9143	66.1	.4	.4909	.5635	1.775	.8712	.6
24.0	0.4067	0.4452	2.246	0.9135	**66.0**	.5	.4924	.5658	1.767	.8704	.5
.1	.4083	.4473	2.236	.9128	65.9	.6	.4939	.5681	1.760	.8695	.4
.2	.4099	.4494	2.225	.9121	.8	.7	.4955	.5704	1.753	.8686	.3
.3	.4115	.4515	2.215	.9114	.7	.8	.4970	.5727	1.746	.8678	.2
.4	.4131	.4536	2.204	.9107	.6	.9	.4985	.5750	1.739	.8669	60.1
24.5	.4147	.4557	2.194	.9100	65.5	**30.0**	0.5000	0.5774	1.732	0.8660	**60.0**
	Cos	Cot	Tan	Sin	Degrees		Cos	Cot	Tan	Sin	Degrees

Degrees	Sin	Tan	Cot	Cos		Degrees	Sin	Tan	Cot	Cos	
30.0	0.5000	0.5774	1.7321	0.8660	**60.0**	35.5	.5807	.7133	1.4019	.8141	54.5
.1	.5015	.5797	1.7251	.8652	59.9	.6	.5821	.7159	1.3968	.8131	.4
.2	.5030	.5820	1.7182	.8643	.8	.7	.5835	.7186	1.3916	.8121	.3
.3	.5045	.5844	1.7113	.8634	.7	.8	.5850	.7212	1.3865	.8111	.2
.4	.5060	.5867	1.7045	.8625	.6	.9	.5864	.7239	1.3814	.8100	54.1
.5	.5075	.5890	1.6977	.8616	.5	**36.0**	0.5878	0.7265	1.3764	0.8090	**54.0**
.6	.5090	.5914	1.6909	.8607	.4	.1	.5892	.7292	1.3713	.8080	53.9
.7	.5105	.5938	1.6842	.8599	.3	.2	.5906	.7319	1.3663	.8070	.8
.8	.5120	.5961	1.6775	.8590	.2	.3	.5920	.7346	1.3613	.8059	.7
.9	.5135	.5985	1.6709	.8581	59.1	.4	.5934	.7373	1.3564	.8049	.6
31.0	0.5150	0.6009	1.6643	0.8572	**59.0**	.5	.5948	.7400	1.3514	.8039	.5
.1	.5165	.6032	1.6577	.8563	58.9	.6	.5962	.7427	1.3465	.8028	.4
.2	.5180	.6056	1.6512	.8554	.8	.7	.5976	.7454	1.3416	.8018	.3
.3	.5195	.6080	1.6447	.8545	.7	.8	.5990	.7481	1.3367	.8007	.2
.4	.5210	.6104	1.6383	.8536	.6	.9	.6004	.7508	1.3319	.7997	53.1
.5	.5225	.6128	1.6319	.8526	.5	**37.0**	0.6018	0.7536	1.3270	0.7986	**53.0**
.6	.5240	.6152	1.6255	.8517	.4	.1	.6032	.7563	1.3222	.7976	52.9
.7	.5255	.6176	1.6191	.8508	.3	.2	.6046	.7590	1.3175	.7965	.8
.8	.5270	.6200	1.6128	.8499	.2	.3	.6060	.7618	1.3127	.7955	.7
.9	.5284	.6224	1.6066	.8490	58.1	.4	.6074	.7646	1.3079	.7944	.6
32.0	0.5299	0.6249	1.6003	0.8480	**58.0**	.5	.6088	.7673	1.3032	.7934	.5
.1	.5314	.6273	1.5941	.8471	57.9	.6	.6101	.7701	1.2985	.7923	.4
.2	.5329	.6297	1.5880	.8462	.8	.7	.6115	.7729	1.2938	.7912	.3
.3	.5344	.6322	1.5818	.8453	.7	.8	.6129	.7757	1.2892	.7902	.2
.4	.5358	.6346	1.5757	.8443	.6	.9	.6143	.7785	1.2846	.7891	52.1
.5	.5373	.6371	1.5697	.8434	.5	**38.0**	0.6157	0.7813	1.2799	0.7880	**52.0**
.6	.5388	.6395	1.5637	.8425	.4	.1	.6170	.7841	1.2753	.7869	51.9
.7	.5402	.6420	1.5577	.8415	.3	.2	.6184	.7869	1.2708	.7859	.8
.8	.5417	.6445	1.5517	.8406	.2	.3	.6198	.7898	1.2662	.7848	.7
.9	.5432	.6469	1.5458	.8396	57.1	.4	.6211	.7926	1.2617	.7837	.6
33.0	0.5446	0.6494	1.5399	0.8387	**57.0**	.5	.6225	.7954	1.2572	.7826	.5
.1	.5461	.6519	1.5340	.8377	56.9	.6	.6239	.7983	1.2527	.7815	.4
.2	.5476	.6544	1.5282	.8368	.8	.7	.6252	.8012	1.2482	.7804	.3
.3	.5490	.6569	1.5224	.8358	.7	.8	.6266	.8040	1.2437	.7793	.2
.4	.5505	.6594	1.5166	.8348	.6	.9	.6280	.8069	1.2393	.7782	51.1
.5	.5519	.6619	1.5108	.8339	.5	**39.0**	0.6293	0.8098	1.2349	0.7771	**51.0**
.6	.5534	.6644	1.5051	.8329	.4	.1	.6307	.8127	1.2305	.7760	50.9
.7	.5548	.6669	1.4994	.8320	.3	.2	.6320	.8156	1.2261	.7749	.8
.8	.5563	.6694	1.4938	.8310	.2	.3	.6334	.8185	1.2218	.7738	.7
.9	.5577	.6720	1.4882	.8300	56.1	.4	.6347	.8214	1.2174	.7727	.6
34.0	0.5592	0.6745	1.4826	0.8290	**56.0**	.5	.6361	.8243	1.2131	.7716	.5
.1	.5606	.6771	1.4770	.8281	55.9	.6	.6374	.8273	1.2088	.7705	.4
.2	.5621	.6796	1.4715	.8271	.8	.7	.6388	.8302	1.2045	.7694	.3
.3	.5635	.6822	1.4659	.8261	.7	.8	.6401	.8332	1.2002	.7683	.2
.4	.5650	.6847	1.4605	.8251	.6	.9	.6414	.8361	1.1960	.7672	50.1
.5	.5664	.6873	1.4550	.8241	.5	**40.0**	0.6428	0.8391	1.1918	0.7660	**50.0**
.6	.5678	.6899	1.4496	.8231	.4	.1	.6441	.8421	1.1875	.7649	49.9
.7	.5693	.6924	1.4442	.8221	.3	.2	.6455	.8451	1.1833	.7638	.8
.8	.5707	.6950	1.4388	.8211	.2	.3	.6468	.8481	1.1792	.7627	.7
.9	.5721	.6976	1.4335	.8202	55.1	.4	.6481	.8511	1.1750	.7615	.6
35.0	0.5736	0.7002	1.4281	0.8192	**55.0**	40.5	0.6494	0.8541	1.1708	0.7604	49.5
.1	.5750	.7028	1.4229	.8181	54.9	.6	.6508	.8571	1.1667	.7593	.4
.2	.5764	.7054	1.4176	.8171	.8	.7	.6521	.8601	1.1626	.7581	.3
.3	.5779	.7080	1.4124	.8161	.7	.8	.6534	.8632	1.1585	.7570	.2
.4	.5793	.7107	1.4071	.8151	.6	.9	.6547	.8662	1.1544	.7559	49.1
35.5	.5807	.7133	1.4019	.8141	54.5	**41.0**	0.6561	0.8693	1.1504	0.7547	**49.0**
	Cos	Cot	Tan	Sin	Degrees		Cos	Cot	Tan	Sin	Degrees

Degrees	Sin	Tan	Cot	Cos		Degrees	Sin	Tan	Cot	Cos	
41.0	0.6561	0.8693	1.1504	0.7547	**49.0**	**43.0**	0.6820	0.9325	1.0724	0.7314	**47.0**
.1	.6574	.8724	1.1463	.7536	48.9	.1	.6833	.9358	1.0686	.7302	46.9
.2	.6587	.8754	1.1423	.7524	.8	.2	.6845	.9391	1.0649	.7290	.8
.3	.6600	.8785	1.1383	.7513	.7	.3	.6858	.9424	1.0612	.7278	.7
.4	.6613	.8816	1.1343	.7501	.6	.4	.6871	.9457	1.0575	.7266	.6
.5	.6626	.8847	1.1303	.7490	.5	.5	.6884	.9490	1.0538	.7254	.5
.6	.6639	.8878	1.1263	.7478	.4	.6	.6896	.9523	1.0501	.7242	.4
.7	.6652	.8910	1.1224	.7466	.3	.7	.6909	.9556	1.0464	.7230	.3
.8	.6665	.8941	1.1184	.7455	.2	.8	.6921	.9590	1.0428	.7218	.2
.9	.6678	.8972	1.1145	.7443	48.1	.9	.6934	.9623	1.0392	.7206	46.1
42.0	0.6691	0.9004	1.1106	0.7431	**48.0**	**44.0**	0.6947	0.9657	1.0355	0.7193	**46.0**
.1	.6704	.9036	1.1067	.7420	47.9	.1	.6959	.9691	1.0319	.7181	45.9
.2	.6717	.9067	1.1028	.7408	.8	.2	.6972	.9725	1.0283	.7169	.8
.3	.6730	.9099	1.0990	.7396	.7	.3	.6984	.9759	1.0247	.7157	.7
.4	.6743	.9131	1.0951	.7385	.6	.4	.6997	.9793	1.0212	.7145	.6
.5	.6756	.9163	1.0913	.7373	.5	.5	.7009	.9827	1.0176	.7133	.5
.6	.6769	.9195	1.0875	.7361	.4	.6	.7022	.9861	1.0141	.7120	.4
.7	.6782	.9228	1.0837	.7349	.3	.7	.7034	.9896	1.0105	.7108	.3
.8	.6794	.9260	1.0799	.7337	.2	.8	.7046	.9930	1.0070	.7096	.2
.9	.6807	.9293	1.0761	.7325	47.1	.9	.7059	.9965	1.0035	.7083	45.1
43.0	0.6820	0.9325	1.0724	0.7314	**47.0**	**45.0**	0.7071	1.0000	1.0000	0.7071	**45.0**
	Cos	Cot	Tan	Sin	Degrees		Cos	Cot	Tan	Sin	Degrees

EXPONENTIAL FUNCTIONS

These tables give the value of e^x and e^{-x}, where e is the base of the natural system of logarithms, 2.71828..., and x has values from 0 to 10. Facilitating the solution of exponential equations, these tables also serve as a table of natural, or Naperian, antilogarithms; for instance, if the logarithm or exponent $x = 3.25$, the corresponding number or value of e^x is 25.790, and its reciprocal, e^{-x}, is 0.38774.

x	e^x	e^{-x}	x	e^x	e^{-x}
0.00	1.0000	1.000000	0.40	1.4918	0.670320
0.01	1.0101	0.990050	0.41	1.5068	.663650
0.02	1.0202	.980199	0.42	1.5220	.657047
0.03	1.0305	.970446	0.43	1.5373	.650509
0.04	1.0408	.960789	0.44	1.5527	.644036
0.05	1.0513	0.951229	0.45	1.5683	0.637628
0.06	1.0618	.941765	0.46	1.5841	.631284
0.07	1.0725	.932394	0.47	1.6000	.625002
0.08	1.0833	.923116	0.48	1.6161	.618783
0.09	1.0942	.913931	0.49	1.6323	612626
0.10	1.1052	0.904837	0.50	1.6487	0.606531
0.11	1.1163	.895834	0.51	1.6653	.600496
0.12	1.1275	.886920	0.52	1.6820	.594521
0.13	1.1388	.878095	0.53	1.6989	.588605
0.14	1.1503	.869358	0.54	1.7160	.582748
0.15	1.1618	0.860708	0.55	1.7333	0.576950
0.16	1.1735	.852144	0.56	1.7507	.571209
0.17	1.1853	.843665	0.57	1.7683	.565525
0.18	1.1972	.835270	0.58	1.7860	.559898
0.19	1.2092	.826959	0.59	1.8040	.554327
0.20	1.2214	0.818731	0.60	1.8221	0.548812
0.21	1.2337	.810584	0.61	1.8404	.543351
0.22	1.2461	.802519	0.62	1.8589	.537944
0.23	1.2586	.794534	0.63	1.8776	.532592
0.24	1.2712	.786628	0.64	1.8965	.527292
0.25	1.2840	0.778801	0.65	1.9155	0.522046
0.26	1.2969	.771052	0.66	1.9348	.516851
0.27	1.3100	.763379	0.67	1.9542	.511709
0.28	1.3231	.755784	0.68	1.9739	.506617
0.29	1.3364	.748264	0.69	1.9937	.501576
0.30	1.3499	0.740818	0.70	2.0138	0.496585
0.31	1.3634	.733447	0.71	2.0340	.491644
0.32	1.3771	.726149	0.72	2.0544	.486752
0.33	1.3910	.718924	0.73	2.0751	.481909
0.34	1.4049	.711770	0.74	2.0959	.477114
0.35	1.4191	0.704688	0.75	2.1170	0.472367
0.36	1.4333	.697676	0.76	2.1383	.467666
0.37	1.4477	.690734	0.77	2.1598	.463013
0.38	1.4623	.683861	0.78	2.1815	.458406
0.39	1.4770	.677057	0.79	2.2034	.453845

Reference Data

x	e^x	e^{-x}	x	e^x	e^{-x}
0.80	2.2255	0.449329	1.70	5.4739	0.182684
0.81	2.2479	.444858	1.72	5.5845	.179066
0.82	2.2705	.440432	1.74	5.6973	.175520
0.83	2.2933	.436049	1.76	5.8124	.172045
0.84	2.3164	.431711	1.78	5.9299	.168638
0.85	2.3396	0.427415	1.80	6.0496	0.165299
0.86	2.3632	.423162	1.82	6.1719	.162026
0.87	2.3869	.418952	1.84	6.2965	.158817
0.88	2.4109	.414783	1.86	6.4237	.155673
0.89	2.4351	.410656	1.88	6.5535	.152590
0.90	2.4596	0.406570	1.90	6.6859	0.149569
0.91	2.4843	.402524	1.92	6.8210	.146607
0.92	2.5093	.398519	1.94	6.9588	.143704
0.93	2.5345	.394554	1.96	7.0993	.140858
0.94	2.5600	.390628	1.98	7.2427	.138069
0.95	2.5857	0.386741	2.00	7.3891	0.135335
0.96	2.6117	.382893	2.05	7.7679	.128735
0.97	2.6379	.379083	2.10	8.1662	.122456
0.98	2.6645	.375311	2.15	8.5849	.116484
0.99	2.6912	.371577	2.20	9.0250	.110803
1.00	2.7183	0.367879	2.25	9.4877	0.105399
1.02	2.7732	.360595	2.30	9.9742	.100259
1.04	2.8292	.353455	2.35	10.486	.095369
1.06	2.8846	.346456	2.40	11.023	.090718
1.08	2.9447	.339596	2.45	11.588	.086294
1.10	3.0042	0.332871	2.50	12.182	0.082085
1.12	3.0649	.326280	2.55	12.807	.078082
1.14	3.1268	.319819	2.60	13.464	.074274
1.16	3.1899	.313486	2.65	14.154	.070651
1.18	3.2544	.307279	2.70	14.880	.067206
1.20	3.3201	0.301194	2.75	15.643	0.063928
1.22	3.3872	.295230	2.80	16.445	.060810
1.24	3.4556	.289384	2.85	17.288	.057844
1.26	3.5254	.283645	2.90	18.174	.055023
1.28	3.5966	.278037	2.95	19.106	.052340
1.30	3.6693	0.272532	3.00	20.086	0.049787
1.32	3.7434	.267135	3.05	21.115	0.47359
1.34	3.8190	.261846	3.10	22.198	.045049
1.36	3.8962	.256661	3.15	23.336	.042852
1.38	3.9749	.251579	3.20	24.533	.040762
1.40	4.0552	0.246597	3.25	25.790	0.038774
1.42	4.1371	.241714	3.30	27.113	.036883
1.44	4.2207	.236928	3.35	28.503	.035084
1.46	4.3060	.232236	3.40	29.964	.033373
1.48	4.3929	.227638	3.45	31.500	.031746
1.50	4.4817	0.223130	3.50	33.115	0.030197
1.52	4.5722	.218712	3.55	34.813	.028725
1.54	4.6646	.214381	3.60	36.598	.027324
1.56	4.7588	.210136	3.65	38.475	.025991
1.58	4.8550	.205975	3.70	40.447	.024724
1.60	4.9530	0.201897	3.75	42.521	0.023518
1.62	5.0531	.197899	3.80	44.701	.022371
1.64	5.1552	.193980	3.85	46.993	.021280
1.66	5.2593	.190139	3.90	49.402	.020242
1.68	5.3656	.186374	3.95	51.935	.019255

x	e^x	e^{-x}	x	e^x	e^{-x}
4.00	54.598	0.018316	6.75	854.06	0.001171
4.05	57.397	.017422	6.80	897.85	.001114
4.10	60.340	.016573	6.85	943.88	.001060
4.15	63.434	.015764	6.90	992.27	.001008
4.20	66.686	.014996	6.95	1,043.1	.000959
4.25	70.105	0.014264	7.00	1,096.6	0.000912
4.30	73.700	.013569	7.05	1,152.9	.000868
4.35	77.478	.012907	7.10	1,212.0	.000825
4.40	81.451	.012277	7.15	1,274.1	.000785
4.45	85.627	.011679	7.20	1,339.4	.000747
4.50	90.017	0.011109	7.25	1,408.1	0.000710
4.55	94.632	.010567	7.30	1,480.3	.000676
4.60	99.484	.010052	7.35	1,556.2	.000643
4.65	104.58	.009562	7.40	1,636.0	.000611
4.70	109.95	.009095	7.45	1,719.9	.000581
4.75	115.58	0.008652	7.50	1,808.0	0.000553
4.80	121.51	.008230	7.55	1,900.7	.000526
4.85	127.74	.007828	7.60	1,998.2	.000501
4.90	134.29	.007447	7.65	2,100.6	.000476
4.95	141.17	.007083	7.70	2,208.3	.000453
5.00	148.41	0.006738	7.75	2,321.6	0.000431
5.05	156.02	.006409	7.80	2,440.6	.000410
5.10	164.02	.006097	7.85	2,565.7	.000390
5.15	172.43	.005799	7.90	2,697.3	.000371
5.20	181.27	.005517	7.95	2,835.6	.000353
5.25	190.57	0.005248	8.00	2,981.0	0.000336
5.30	200.34	.004992	8.05	3,133.8	.000319
5.35	210.61	.004748	8.10	3,394.5	.000304
5.40	221.41	.004517	8.15	3,463.4	.000289
5.45	232.76	.004296	8.20	3,641.0	.000275
5.50	244.69	0.004087	8.25	3,827.6	0.000261
5.55	257.24	.003888	8.30	4,023.9	.000249
5.60	270.43	.003698	8.35	4,230.2	.000236
5.65	284.29	.003518	8.40	4,447.1	.000225
5.70	298.87	.003346	8.45	4,675.1	.000214
5.75	315.19	0.003183	8.50	4,914.8	0.000204
5.80	330.30	.003028	8.55	5,166.8	.000194
5.85	347.23	.002880	8.60	5,431.7	.000184
5.90	365.04	.002739	8.65	5,710.1	.000175
5.95	383.75	.002606	8.70	6,002.9	.000167
6.00	403.43	0.002479	8.75	6,310.7	0.000159
6.05	424.11	.002379	8.80	6,634.2	.000151
6.10	445.86	.002243	8.85	6,974.4	.000143
6.15	468.72	.002134	8.90	7,332.0	.000136
6.20	492.75	.002030	8.95	7,707.9	.000130
6.25	518.01	0.001931	9.00	8,103.1	0.000123
6.30	544.57	.001836	9.05	8,518.5	.000117
6.35	572.49	.001747	9.10	8,955.3	.000112
6.40	601.85	.001662	9.15	9,414.4	.000106
6.45	632.70	.001581	9.20	9,897.1	.000101
6.50	665.14	0.001503	9.25	10,405	0.000096
6.55	699.24	.001430	9.30	10,938	.000091
6.60	735.10	.001360	9.35	11,499	.000087
6.65	772.78	.001294	9.40	12,088	.000083
6.70	812.41	.001231	9.45	12,708	.000079

x	e^x	e^{-x}	x	e^x	e^{-x}
9.50	13,360	0.000075	9.75	17,154	0.000058
9.55	14,045	.000071	9.80	18,034	.000056
9.60	14,765	.000068	9.85	18,958	.000053
9.65	15,522	.000064	9.90	19,930	.000050
9.70	16,318	.000061	9.95	20,952	.000048
			10.00	22,026	0.000045

Condensed from: Selby, S. H., *Handbook of Tables for Mathematics*, 3rd ed. The Chemical Rubber Co., Cleveland, Ohio, 1967.

SQUARES, SQUARE ROOTS, CUBES, AND CUBE ROOTS

n	n^2	$\sqrt{n}$	n^3	$\sqrt[3]{n}$	n	n^2	$\sqrt{n}$	n^3	$\sqrt[3]{n}$
1	1	1.000 000	1	1.000 000	46	2,116	6.782 330	97,336	3.583 048
2	4	1.414 214	8	1.259 921	47	2,209	6.855 655	103,823	3.608 826
3	9	1.732 051	27	1.442 250	48	2,304	6.928 203	110,592	3.634 241
4	16	2.000 000	64	1.587 401	49	2,401	7.000 000	117,649	3.659 306
5	25	2.236 068	125	1.709 976	50	2,500	7.071 068	125,000	3.684 031
6	36	2.449 490	216	1.817 121	51	2,601	7.141 428	132,651	3.708 430
7	49	2.645 751	343	1.912 931	52	2,704	7.211 103	140,608	3.732 511
8	64	2.828 427	512	2.000 000	53	2,809	7.280 110	148,877	3.756 286
9	81	3.000 000	729	2.080 084	54	2,916	7.348 469	157,464	3.779 763
10	100	3.162 278	1,000	2.154 435	55	3,025	7.416 198	166,375	3.802 952
11	121	3.316 625	1,331	2.223 980	56	3,136	7.483 315	175,616	3.825 862
12	144	3.464 102	1,728	2.289 428	57	3,249	7.549 834	185,193	3.848 501
13	169	3.605 551	2,197	2.351 335	58	3,364	7.615 773	195,112	3.870 877
14	196	3.741 657	2,744	2.410 142	59	3,481	7.681 146	205,379	3.892 996
15	225	3.872 983	3,375	2.466 212	60	3,600	7.745 967	216,000	3.914 868
16	256	4.000 000	4,096	2.519 842	61	3,721	7.810 250	226,981	3.936 497
17	289	4.123 106	4,913	2.571 282	62	3,844	7.874 008	238,328	3.957 892
18	324	4.242 641	5,832	2.620 741	63	3,969	7.937 254	250,047	3.979 057
19	361	4.358 899	6,859	2.668 402	64	4,096	8.000 000	262,144	4.000 000
20	400	4.472 136	8,000	2.714 418	65	4,225	8.062 258	274,625	4.020 726
21	441	4.582 576	9,261	2.758 924	66	4,356	8.124 038	287,496	4.041 240
22	484	4.690 416	10,648	2.802 039	67	4,489	8.185 353	300,763	4.061 548
23	529	4.795 832	12,167	2.843 867	68	4,624	8.246 211	314,432	4.081 655
24	576	4.898 979	13,824	2.884 499	69	4,761	8.306 624	328,509	4.101 566
25	625	5.000 000	15,625	2.924 018	70	4,900	8.366 600	343,000	4.121 285
26	676	5.099 020	17,576	2.962 496	71	5,041	8.426 150	357,911	4.140 818
27	729	5.099 020	19,683	3.000 000	72	5,184	8.485 281	373,248	4.160 168
28	784	5.291 503	21,952	3.036 589	73	5,329	8.544 004	389,017	4.179 339
29	841	5.385 165	24,389	3.072 317	74	5,476	8.602 325	405,224	4.198 336
30	900	5.477 226	27,000	3.107 233	75	5,625	8.660 254	421,875	4.217 163
31	961	5.567 764	29,791	3.141 381	76	5,776	8.717 798	438,976	4.235 824
32	1,024	5.656 854	32,768	3.174 802	77	5,929	8.774 964	456,533	4.254 321
33	1,089	5.744 563	35,937	3.207 534	78	6,084	8.831 761	474,552	4.272 659
34	1,156	5.830 952	39,304	3.239 612	79	6,241	8.888 194	493,039	4.290 840
35	1,225	5.916 080	42,875	3.271 066	80	6,400	8.944 272	512,000	4.308 869
36	1,296	6.000 000	46,656	3.301 927	81	6,561	9.000 000	531,441	4.326 749
37	1,369	6.082 763	50,653	3.332 222	82	6,724	9.055 385	551,368	4.344 481
38	1,444	6.164 414	54,872	3.361 975	83	6,889	9.110 434	571,787	4.362 071
39	1,521	6.244 998	59,319	3.391 211	84	7,056	9.165 151	592,704	4.379 519
40	1,600	6.324 555	64,000	3.419 952	85	7,225	9.219 544	614,125	4.396 830
41	1,681	6.403 124	68,921	3.448 217	86	7,396	9.273 618	636,056	4.414 005
42	1,764	6.480 741	74,088	3.476 027	87	7,569	9.327 379	658,503	4.431 048
43	1,849	6.557 439	79,507	3.503 398	88	7,744	9.380 832	681,472	4.447 960
44	1,936	6.633 250	85,184	3.530 348	89	7,921	9.433 981	704,969	4.464 745
45	2,025	6.708 204	91,125	3.556 893	90	8,100	9.486 833	729,000	4.481 405

Reference Data

n	n^2	$\sqrt{n}$	n^3	$\sqrt[3]{n}$	n	n^2	$\sqrt{n}$	n^3	$\sqrt[3]{n}$
91	8,281	9.539 392	753,571	4.497 941	146	21,316	12.083 05	3,112,136	5.265 637
92	8,464	9.591 663	778,688	4.514 357	147	21,609	12.124 36	3,176,523	5.277 632
93	8,649	9.643 651	804,357	4.530 655	148	21,904	12.165 53	3,241,792	5.289 572
94	8,836	9.695 360	830,584	4.546 836	149	22,201	12.206 56	3,307,949	5.301 459
95	9,025	9.746 794	857,375	4.562 903	150	22,500	12.247 45	3,375,000	5.313 293
96	9,216	9.797 959	884,736	4.578 857	151	22,801	12.288 21	3,442,951	5.325 074
97	9,409	9.848 858	912,673	4.594 701	152	23,104	12.328 83	3,511,808	5.336 803
98	9,604	9.899 495	941,192	4.610 436	153	23,409	12.369 32	3,581,577	5.348 481
99	9,801	9.949 874	970,299	4.626 065	154	23,716	12.409 67	3,652,264	5.360 108
100	10,000	10.000 000	1,000,000	4.641 589	155	24,025	12.449 90	3,723,875	5.371 685
101	10,201	10.049 88	1,030,010	4.657 010	156	24,336	12.490 00	3,796,416	5.383 213
102	10,404	10.099 50	1,061,208	4.672 329	157	24,649	12.529 96	3,869,893	5.394 691
103	10,609	10.148 89	1,092,727	4.687 548	158	24,964	12.569 81	3,944,312	5.406 120
104	10,816	10.198 04	1,124,864	4.702 669	159	25,281	12.609 52	4,019,679	5.417 502
105	11,025	10.246 95	1,157,625	4.717 694	160	25,600	12.649 11	4,096,000	5.428 835
106	11,236	10.295 63	1,191,016	4.732 623	161	25,921	12.688 58	4,173,281	5.440 122
107	11,449	10.344 08	1,225,043	4.747 459	162	26,244	12.727 92	4,251,528	5.451 362
108	11,664	10.392 30	1,259,712	4.762 203	163	26,569	12.767 15	4,330,747	5.462 556
109	11,881	10.440 31	1,295,029	4.776 856	164	26,896	12.806 25	4,410,944	5.473 704
110	12,100	10.488 09	1,331,000	4.791 420	165	27,225	12.845 23	4,492,125	5.484 807
111	12,321	10.535 65	1,367,631	4.805 896	166	27,556	12.885 10	4,574,296	5.495 865
112	12,544	10.583 01	1,404,928	4.820 285	167	27,889	12.922 85	4,657,463	5.506 878
113	12,769	10.630 15	1,442,897	4.834 588	168	28,224	12.961 48	4,741,632	5.517 848
114	12,996	10.677 08	1,481,544	4.848 808	169	28,561	13.000 00	4,826,809	5.528 775
115	13,225	10.723 81	1,520,875	4.862 944	170	28,900	13.038 40	4,913,000	5.539 658
116	13,456	10.770 33	1,560,896	4.876 999	171	29,241	13.076 70	5,000,211	5.550 499
117	13,689	10.816 65	1,601,613	4.890 973	172	29,584	13.114 88	5,088,448	5.561 298
118	13,924	10.862 78	1,643,032	4.904 868	173	29,929	13.152 95	5,177,717	5.572 055
119	14,161	10.908 71	1,685,159	4.918 685	174	30,276	13.190 91	5,268,024	5.582 770
120	14,400	10.954 45	1,728,000	4.932 424	175	30,625	13.228 76	5,359,375	5.593 445
121	14,641	11.000 00	1,771,561	4.946 087	176	30,976	13.266 50	5,451,776	5.604 079
122	14,884	11.045 36	1,815,848	4.959 676	177	31,329	13.304 13	5,545,233	5.614 672
123	15,129	11.090 54	1,860,867	4.973 190	178	31,684	13.341 66	5,639,752	5.625 226
124	15,376	11.135 53	1,906,624	4.986 631	179	32,041	13.379 09	5,735,339	5.635 741
125	15,625	11.180 34	1,953,125	5.000 000	180	32,400	13.416 41	5,832,000	5.646 216
126	15,876	11.224 97	2,000,376	5.013 298	181	32,761	13.453 62	5,929,741	5.656 653
127	16,129	11.269 43	2,048,383	5.026 526	182	33,124	13.490 74	6,028,568	5.667 051
128	16,384	11.313 71	2,097,152	5.039 684	183	33,489	13.527 75	6,128,487	5.677 411
129	16,641	11.357 82	2,146,689	5.052 774	184	33,856	13.564 66	6,229,504	5.687 734
130	16,900	11.401 75	2,197,000	5.065 797	185	34,225	13.601 47	6,331,625	5.698 019
131	17,161	11.445 52	2,248,091	5.078 753	186	34,596	13.638 18	6,434,856	5.708 267
132	17,424	11.489 13	2,299,968	5.091 643	187	34,969	13.674 79	6,539,203	5.718 479
133	17,689	11.532 56	2,352,637	5.104 469	188	35,344	13.711 31	6,644,672	5.728 654
134	17,956	11.575 84	2,406,104	5.117 230	189	35,721	13.747 73	6,751,269	5.738 794
135	18,225	11.618 95	2,460,375	5.129 928	190	36,100	13.784 05	6,859,000	5.784 897
136	18,496	11.661 90	2,515,456	5.142 563	191	36,481	13.820 27	6,967,871	5.758 965
137	18,769	11.704 70	2,571,353	5.155 137	192	36,864	13.856 41	7,077,888	5.768 998
138	19,044	11.747 34	2,628,072	5.167 649	193	37,249	13.892 44	7,189,057	5.778 997
139	19,321	11.789 83	2,685,619	5.180 101	194	37,636	13.928 39	7,301,384	5.788 960
140	19,600	11.832 16	2,744,000	5.192 494	195	38,025	13.964 24	7,414,875	5.798 890
141	19,881	11.874 24	2,803,221	5.204 828	196	38,416	14.000 00	7,529,536	5.808 786
142	20,164	11.916 38	2,863,288	5.217 103	197	38,809	14.035 67	7,645,373	5.818 648
143	20,449	11.958 26	2,924,207	5.229 322	198	39,204	14.071 25	7,762,392	5.828 477
144	20,736	12.000 00	2,985,984	5.241 483	199	39,601	14.106 74	7,880,599	5.838 272
145	21,025	12.041 59	3,048,625	5.253 588	200	40,000	14.142 14	8,000,000	5.848 035

n	n^2	$\sqrt{n}$	n^3	$\sqrt[3]{n}$	n	n^2	$\sqrt{n}$	n^3	$\sqrt[3]{n}$
201	40,401	14.177 45	8,120,601	5.587 766	256	65,536	16.000 00	16,777,216	6.349 604
202	40,804	14.212 67	8,242,408	5.867 464	257	66,049	16.031 22	16,974,593	6.357 861
203	41,209	14.247 81	8,365,427	5.877 131	258	66,564	16.062 38	17,173,512	6.366 097
204	41,616	14.282 86	8,489,664	5.886 765	259	67,081	16.093 48	17,373,979	6.374 311
205	42,025	14.317 82	8,615,125	5.896 369	260	67,600	16.124 52	17,576,000	6.382 504
206	42,436	14.352 70	8,741,816	5.905 941	261	68,121	16.155 49	17,779,581	6.390 677
207	42,849	14.387 49	8,869,743	5.915 482	262	68,644	16.186 41	17,984,728	6.398 828
208	43,264	14.422 21	8,998,912	5.924 992	263	69,169	16.217 27	18,191,447	6.406 959
209	43,681	14.456 83	9,129,329	5.934 472	264	69,696	16.248 08	18,399,744	6.415 069
210	44,100	14.491 38	9,261,000	5.943 922	265	70,225	16.278 82	18,609,625	6.423 158
211	44,521	14.525 84	9,393,931	5.953 342	266	70,756	16.309 51	18,821,096	6.431 228
212	44,944	14.560 22	9,528,128	5.962 732	267	71,289	16.340 13	19,034,163	6.439 277
213	45,369	14.594 52	9,663,597	5.972 093	268	71,824	16.370 71	19,248,832	6.447 306
214	45,796	14.628 74	9,800,344	5.981 424	269	72,361	16.401 22	19,465,109	6.455 315
215	46,225	14.662 88	9,938,375	5.990 726	270	72,900	16.431 68	19,683,000	6.463 304
216	46,656	14.696 94	10,077,696	6.000 000	271	73,441	16.462 08	19,902,511	6.471 274
217	47,089	14.730 92	10,218,313	6.009 245	272	73,984	16.492 42	20,123,648	6.479 224
218	47,524	14.764 82	10,360,232	6.018 462	273	74,529	16.522 71	20,346,417	6.487 154
219	47,961	14.798 65	10,503,459	6.027 650	274	75,076	16.552 95	20,570,824	6.495 065
220	48,400	14.832 40	10,648,000	6.036 811	275	75,625	16.583 12	20,796,875	6.502 957
221	48,841	14.866 07	10,793,861	6.045 944	276	76,176	16.613 25	21,024,576	6.510 830
222	49,284	14.899 66	10,941,048	6.055 049	277	76,729	16.643 32	21,253,933	6.518 684
223	49,729	14.933 18	11,089,567	6.064 127	278	77,284	16.673 33	21,484,952	6.526 519
224	50,176	14.966 63	11,239,424	6.073 178	279	77,841	16.703 29	21,717,639	6.534 335
225	50,625	15.000 00	11,390,625	6.082 202	280	78,400	16.733 20	21,952,000	6.542 133
226	51,076	15.033 30	11,543,176	6.091 199	281	78,961	16.763 05	22,188,041	6.549 912
227	51,529	15.066 52	11,697,083	6.100 170	282	79,524	16.792 86	22,425,768	6.557 672
228	51,984	15.099 67	11,852,352	6.109 115	283	80,089	16.822 60	22,662,187	6.565 414
229	52,441	15.132 75	12,008,989	6.118 033	284	80,656	16.852 30	22,906,304	6.573 138
230	52,900	15.165 75	12,167,000	6.126 926	285	81,225	16.881 94	23,149,125	6.580 844
231	53,361	15.198 68	12,326,391	6.135 792	286	81,796	16.911 53	23,393,656	6.588 532
232	53,824	15.231 55	12,487,168	6.144 634	287	82,369	16.941 07	23,639,903	6.596 202
233	54,289	15.264 34	12,649,337	6.153 449	288	82,944	16.970 56	23,887,872	6.603 854
234	54,756	15.297 06	12,812,904	6.162 240	289	83,521	17.000 00	24,137,569	6.611 489
235	55,225	15.329 71	12,977,875	6.171 006	290	84,100	17.029 39	24,389,000	6.619 106
236	55,696	15.362 29	13,144,256	6.179 747	291	84,681	17.058 72	24,642,171	6.626 705
237	56,169	15.394 80	13,312,053	6.188 463	292	85,264	17.088 01	24,897,088	6.634 287
238	56,644	15.427 25	13,481,272	6.197 154	293	85,849	17.117 24	25,153,757	6.641 852
239	57,121	15.459 62	13,651,919	6.205 822	294	86,436	17.146 43	25,412,184	6.649 400
240	57,600	15.491 93	13,824,000	6.214 465	295	87,025	17.175 56	25,672,375	6.656 930
241	58,081	15.524 17	13,997,521	6.223 084	296	87,616	17.204 65	25,934,336	6.664 444
242	58,564	15.556 35	14,172,488	6.231 680	297	88,209	17.233 69	26,198,073	6.671 073
243	59,040	15.588 46	14,348,907	6.240 251	298	88,804	17.262 68	26,463,592	6.679 420
244	59,536	15.620 50	14,526,784	6.248 800	299	89,401	17.291 62	26,730,899	6.686 883
245	60,025	15.652 48	14,706,125	6.257 325	300	90,000	17.320 51	27,000,000	6.694 330
246	60,516	15.684 39	14,886,936	6.265 827	301	90,601	17.349 35	27,270,901	6.701 759
247	61,009	15.712 23	15,069,223	6.274 305	302	91,204	17.378 15	27,543,608	6.709 173
248	61,504	15.748 02	15,252,992	6.282 761	303	91,809	17.406 90	27,818,127	6.716 570
249	62,001	15.779 73	15,438,249	6.291 195	304	92,416	17.435 60	28,094,464	6.723 951
250	62,500	15.811 39	15,625,000	6.299 605	305	93,025	17.464 24	28,372,625	6.731 315
251	63,001	15.842 98	15,813,251	6.307 994	306	93,636	17.492 86	28,652,616	6.738 664
252	63,504	15.874 51	16,003,008	6.316 360	307	94,249	17.521 42	28,934,443	6.745 997
253	64,009	15.905 97	16,194,277	6.324 704	308	94,864	17.549 93	29,218,112	6.753 313
254	64,516	15.937 38	16,387,064	6.333 064	309	95,481	17.578 40	29,503,629	6.760 614
255	65,025	15.968 72	16,581,375	6.341 326	310	96,100	17.606 82	29,791,000	6.767 899

Reference Data

n	n^2	$\sqrt{n}$	n^3	$\sqrt[3]{n}$	n	n^2	$\sqrt{n}$	n^3	$\sqrt[3]{n}$
311	96,721	17.635 19	30,080,231	6.775 169	366	133,956	19.131 13	49,027,896	7.153 090
312	97,344	17.663 52	30,371,328	6.782 423	367	134,689	19.157 24	49,430,863	7.159 599
313	97,969	17.691 81	30,664,297	6.789 661	368	135,424	19.183 33	49,836,032	7.166 096
314	98,596	17.720 05	30,959,144	6.796 884	369	136,161	19.209 37	50,243,409	7.172 581
315	99,225	17.748 24	31.255,875	6.804 092	370	136,900	19.235 38	50,653,000	7.179 054
316	99,856	17.776 39	31,554,496	6.811 285	371	137,641	19.261 36	51,064,811	7.185 516
317	100,489	17.804 49	31,855,013	6.818 462	372	138,384	19.287 30	51,478,848	7.191 966
318	101,124	17.832 55	32,157,432	6.825 624	373	139,129	19.313 21	51,895,117	7.198 405
319	101,761	17.860 57	32,461,759	6.832 771	374	139,876	19.339 08	52,313,624	7.204 832
320	102,400	17.888 54	32,768,000	6.839 904	375	140,625	19.364 92	52,734,375	7.211 248
321	103,041	17.916 47	33,076,161	6.847 021	376	141,376	19.390 72	53,157,376	7.217 652
322	103,684	17.944 36	33,386,248	6.854 124	377	142,129	19.416 49	53,582,633	7.224 045
323	104,329	17.972 20	33,698,267	6.861 212	378	142,884	19.442 22	54,010,152	7.230 427
324	104,976	18.000 00	34,012,224	6.868 285	379	143,641	19.467 92	54,439,939	7.236 797
325	105,625	18.027 76	34,328,125	6.875 344	380	144,400	19.493 59	54,872,000	7.243 156
326	106,276	18.055 47	34,645,976	6.882 389	381	145,161	19.519 22	55,306,341	7.249 505
327	106,929	18.083 14	34,965,783	6.889 419	382	145,924	19.544 82	55,742,968	7.255 842
328	107,584	18.110 77	35,287,552	6.896 434	383	146,689	19.570 39	56,181,887	7.262 167
329	108,241	18.138 36	35,611,289	6.903 436	384	147,456	19.595 92	56,623,104	7.268 482
330	108,900	18.165 90	35,937,000	6.910 423	385	148,225	19.621 42	57,066,625	7.274 786
331	109,561	18.193 41	36,264,691	6.917 396	386	148,996	19.646 88	57,512,456	7.281 079
332	110,224	18.220 87	36,594,368	6.924 356	387	149,769	19.672 32	57,960,603	7.287 362
333	110,889	18.248 29	36,926,037	6.931 301	388	150,544	19.697 72	58,411,072	7.293 633
334	111,556	18.275 67	37,259,704	6.938 232	389	151,321	19.723 08	58,863,869	7.299 894
335	112,225	18.303 01	37,595,375	6.945 150	390	152,100	19.748 42	59,319,000	7.306 144
336	112,896	18.330 30	37,933,056	6.952 053	391	152,881	19.773 72	59,776,471	7.312 383
337	113,569	18.357 56	38,272,753	6.958 943	392	153,664	19.798 99	60,236,288	7.318 611
338	114,244	18.384 78	38,614,472	6.965 820	393	154,449	19.824 23	60,698,457	7.324 829
339	114,921	18.411 95	38,958,219	6.972 683	394	155,236	19.849 43	61,162,984	7.331 037
340	115,600	18.439 09	39,304,000	6.979 532	395	156,025	19.874 61	61,629,875	7.337 234
341	116,281	18.466 19	39,651,821	6.986 368	396	156,816	19.899 75	62,099,136	7.343 420
342	116,964	18.493 24	40,001,688	6.993 191	397	157,609	19.924 86	62,570,773	7.349 597
343	117,649	18.520 26	40,353,607	7.000 000	398	158,404	19.949 94	63,044,792	7.355 762
344	118,336	18.547 24	40,707,584	7.006 796	399	159,201	19.974 98	63,521,199	7.361 918
345	119,025	18.574 18	41,063,625	7.013 579	400	160,000	20.000 00	64,000,000	7.368 063
346	119,716	18.601 08	41,421,736	7.020 349	401	160,801	20.024 98	64,481,201	7.374 198
347	120,409	18.627 94	41,781,923	7.027 106	402	161,604	20.049 94	64,964,808	7.380 323
348	121,104	18.654 76	42,144,192	7.033 850	403	162,409	20.074 86	65,450,827	7.386 437
349	121,801	18.681 54	42,508,549	7.040 581	404	163,216	20.099 75	65,939,264	7.392 542
350	122,500	18.708 29	42,875,000	7.047 299	405	164,025	20.124 61	66,430,125	7.398 636
351	123,201	18.734 99	43,243,551	7.054 004	406	164,836	20.149 44	66,923,416	7.404 721
352	123,904	18.761 66	43,614,208	7.060 697	407	165,649	20.174 24	67,419,143	7.410 795
353	124,609	18.788 29	43,986,977	7.067 377	408	166,464	20.199 01	67,917,312	7.416 860
354	125,316	18.814 89	44,361,864	7.074 044	409	167,281	20.223 75	68,417,929	7.422 914
355	126,025	18.841 44	44,738,875	7.080 699	410	168,100	20.248 46	68,921,000	7.428 959
356	126,736	18.867 96	45,118,016	7.087 341	411	168,921	20.273 13	69,426,531	7.434 994
357	127,449	18.894 44	45,499,203	7.093 971	412	169,744	20.297 78	69,934,528	7.441 019
358	128,164	18.920 89	45,882,712	7.100 588	413	170,569	20.322 40	70,444,997	7.447 034
359	128,881	18.947 30	46,268,279	7.107 194	414	171,396	20.346 99	70,957,944	7.453 040
360	129,600	18.973 67	46,656,000	7.113 787	415	172,225	20.371 55	71,473,375	7.459 036
361	130,321	19.000 00	47,045 881	7.120 367	416	173,056	20.396 08	71,991,296	7.465 022
362	131,044	19.026 30	47,437,928	7.126 936	417	173,889	20.420 58	72,511,713	7.470 999
363	131,769	19.052 56	47,832,147	7.133 492	418	174,724	20.445 05	73,034,632	7.476 966
364	132,496	19.078 78	48,228,544	7.140 037	419	175,561	20.469 49	73,560,059	7.482 924
365	133,225	19.104 97	48,627,125	7.146 569	420	176,400	20.493 90	74,088,000	7.488 872

n	n²	√n	n³	∛n	n	n²	√n	n³	∛n
421	177,241	20.518 28	74,618,461	7.494 811	476	226,576	21.817 42	107,850,176	7.807 925
422	178,084	20.542 64	75,151,448	7.500 741	477	227,529	21.840 33	108,531,333	7.813 389
423	178,929	20.566 96	75,686,967	7.506 661	478	228,484	21.863 21	109,215,352	7.818 846
424	179,776	20.591 26	76,225,024	7.512 572	479	229,441	21.886 07	109,902,239	7.824 294
425	180,625	20.615 53	76,765,625	7.518 473	480	230,400	21.908 90	110,592,000	7,829 735
426	181,476	20.639 77	77,308,776	7.524 365	481	231,361	21.931 71	111,284,641	7.835 169
427	182,329	20.663 98	77,854,483	7.530 248	482	232,324	21.954 50	111,980,168	7.840 595
428	183,184	20.688 16	78,402,752	7.536 122	483	233,289	21.977 26	112,678,587	7.846 013
429	184,041	20.712 32	78,953,589	7.541 987	484	234,256	22.000 00	113,379,904	7.851 424
430	184,900	20.736 44	79,507,000	7.547 842	485	235,225	22.022 72	114,084,125	7.856 828
431	185,761	20.760 54	80,062,991	7.553 689	486	236,196	32.045 41	114,791,256	7.862 224
432	186,624	20.784 61	80,621,568	7.559 526	487	237,169	22.068 08	115,501,303	7.867 613
433	187,489	20.808 65	81,182,737	7.565 355	488	238,144	22.090 72	116,214,272	7.872 994
434	188,356	20.832 67	81,746,504	7.571 174	489	239,121	23.113 34	116,930,169	7.878 368
435	189,225	20.856 65	82,312,875	7.576 985	490	240,100	22.135 94	117,649,000	7.883 735
436	190,096	20.880 61	82,881,856	7.582 787	491	241,081	22.158 52	118,370,771	7.889 095
437	190,969	20.904 54	83,453,453	7.588 579	492	242,064	22.181 07	119,095,488	7.894 447
438	191,844	20.928 45	84,027,672	7.594 363	493	243,049	22.203 60	119,823,157	7.899 792
439	192,721	20.952 33	84,604,519	7.600 139	494	244,036	22.226 11	120,553,784	7.905 129
440	193,600	20.976 18	85,184,000	7.605 905	495	245,025	22.248 60	121,287,373	7.910 460
441	194,481	21.000 00	85,766,121	7.611 663	496	246,016	22.271 06	122,023,936	7.915 783
442	195,364	21.023 80	86,450,888	7.617 412	497	247,009	22.293 50	122,763,473	7.921 099
443	196,249	21.047 57	86,938,307	7.623 152	498	248,004	22.315 91	123,505,992	7.926 408
444	197,136	21.071 31	87,528,384	7.628 884	499	249,001	22.338 31	124,251,499	7.931 710
445	198,025	21.095 02	88,121,125	7.634 607	500	250,000	22.360 68	125,000,000	7.937 005
446	198,916	21.118 71	88,716,536	7.640 321	501	251,001	22.383 03	125,751,501	7.942 293
447	199,809	21.142 37	89,314,623	7.646 027	502	252,004	22.405 36	126,506,008	7.947 574
448	200,704	21.166 01	89,915,392	7.651 725	503	253,009	22.427 66	127,263,527	7.952 848
449	201,601	21.189 62	90,518,849	7.657 414	504	254,016	22.449 94	128,024,064	7.958 114
450	202,600	21.213 20	91,125,000	7.663 094	505	255,025	22.472 21	128,787,625	7.963 374
451	203,401	21.236 76	91,733,851	7.668 766	506	256,036	22.494 44	129,554,216	7.968 627
452	204,304	21.260 29	92,345,408	7.674 430	507	257,049	22.516 66	130,323,843	7.973 873
453	205,209	21.283 80	92,959,677	7.680 086	508	258,064	22.538 86	131,096,512	7.979 112
454	206,116	21.307 28	93,576,664	7.685 733	509	259,081	22.561 03	131,872,229	7.984 344
455	207,025	21.330 73	94,196,375	7.691 372	510	260,100	22.583 18	132,651,000	7.989 570
456	207,936	21.345 16	94,818,816	7.697 002	511	261,121	22.605 31	133,432,831	7.994 788
457	208,849	31.377 56	95,443,993	7.702 625	512	262,144	22.627 42	134,217,728	8.000 000
458	209,764	21.400 93	96,071,912	7.708 239	513	263,169	22.649 50	135,005,697	8.005 205
459	210,681	21.424 29	96,702,579	7.713 845	514	264,196	22.671 57	135,796,744	8.010 403
460	211,600	21.447 61	97,336,000	7.719 443	515	265,225	22.693 61	136,590,875	8.015 595
461	212,521	21.470 91	97,972,181	7.725 032	516	266,256	22.715 63	137,388,096	8.020 779
462	213,444	21.494 19	98,611,128	7.730 614	517	267,289	22.737 63	138,188,413	8.025 957
463	214,369	21.517 43	99,252,847	7.736 188	518	268,324	22.759 61	138,991,832	8.031 129
464	215,296	21.540 66	99,897,344	7.741 753	519	269,361	22.781 57	139,798,359	8.036 293
465	216,225	21.563 86	100,544,625	7.747 311	520	270,400	22.803 51	140,608,000	8.041 452
466	217,156	21.587 03	101,194,696	7.752 861	521	271,441	22.825 42	141,420,761	8.046 603
467	218,089	21.610 18	101,847,563	7.758 402	522	272,484	22.847 32	142,236,648	8.051 748
468	219,024	21.633 31	102,503,232	7.763 936	523	273,529	22.869 19	143,055,667	8.056 886
469	219,961	21.656 41	103,161,709	7.769 462	524	274,576	22.891 05	143,877,824	8.062 018
470	220,900	21.679 48	103,823,000	7.774 980	525	275,625	22.912 88	144,703,125	8.067 143
471	221,841	21.702 53	104,487,111	7.780 490	526	276,676	22.934 69	145,531,576	8.072 262
472	222,784	21.725 56	105,154,048	7.785 993	527	277,729	22.956 48	146,363,183	8.077 374
473	223,729	21.748 56	105,823,817	7.791 488	528	278,784	22.978 25	147,197,952	8.082 480
474	224,676	21.771 54	106,496,424	7.796 975	529	279,841	23.000 00	148,035,889	8.087 579
475	225,625	21.794 49	107,171,875	7.802 454	530	280,900	23.021 73	148,877,000	8.092 672

Reference Data

n	n²	√n	n³	³√n	n	n²	√n	n³	³√n
531	281,961	23.043 44	149,721,291	8.097 759	586	343,396	24.207 44	201,230,056	8.368 209
532	283,024	23.065 13	150,568,768	8.102 839	587	344,569	24.228 08	202,262,003	8.372 967
533	284,089	23.086 79	151,419,437	8.107 913	588	345,744	24.248 71	203,297,472	8.377 719
534	285,156	23.108 44	152,273,304	8.112 980	589	346,921	24.269 32	204,336,469	8.382 465
535	286,225	23.130 07	153,130,375	8.118 041	590	348,100	24.289 92	205,379,000	8.387 207
536	287,296	23.151 67	153,990,656	8.123 096	591	349,281	24.310 49	206,425,071	8.392 942
537	288,369	23.173 26	154,854,153	8.128 145	592	350,464	24.331 05	207,474,688	8.396 673
538	289,444	23.194 83	155,720,872	8.133 187	593	351,649	24.351 59	208,527,857	8.401 398
539	290,521	23.216 37	156,590,819	8.138 223	594	352,836	24.372 12	209,584,584	8.406 118
540	291,600	23.237 90	157,464,000	8.143 253	595	354,025	24.392 62	210,644,875	8.410 833
541	292,681	23.259 41	158,340,421	8.148 276	596	355,216	24.413 11	211,708,736	8.415 542
542	293,764	23.280 89	159,220,088	8.153 294	597	356,409	24.433 58	212,776,173	8.420 246
543	294,849	23.302 36	160,103,007	8.158 305	598	357,604	24.454 04	213,847,192	8.424 945
544	295,936	23.323 81	160,989,184	8.163 310	599	358,801	24.474 48	214,921,799	8.429 638
545	297,025	23.345 24	161,878,625	8.168 309	600	360,000	24.494 90	216,000,000	8.434 327
546	298,116	23.366 64	162,771,336	8.173 302	601	361,201	24.515 30	217,081,801	8.439 010
547	299,209	23.388 03	163,667,323	8.178 289	602	362,404	24.535 69	218,167,208	8.443 688
548	300,304	23.409 40	164,566,592	8.183 269	603	363,609	24.556 06	219,256,227	8.448 361
549	301,401	23.430 75	165,469,149	8.188 244	604	364,816	24.576 41	220,348,864	8.453 028
550	302,500	23.452 08	166,375,000	8.193 213	605	366,025	24.596 75	221,445,125	8.457 691
551	303,601	23.473 39	167,284,151	8.198 175	606	367,236	24.617 07	222,545,016	8.462 348
552	304,704	23.494 68	168,196,608	8.203 132	607	368,449	24.637 37	223,648,543	8.467 000
553	305,809	23.515 95	169,112,377	8.208 082	608	369,664	24.657 66	224,755,712	8.471 647
554	306,916	23.537 20	170,031,464	8.213 027	609	370,881	24.677 93	225,866,529	8.476 289
555	308,025	23.558 44	170,953,875	8.217 966	610	372,100	24.698 18	226,981,000	8.480 926
556	309,136	23.579 65	171,879,616	8.222 899	611	373,321	24.718 41	228,099,131	8.485 558
557	310,249	23.600 85	172,808,693	8.227 825	612	374,544	24.738 63	229,220,928	8.490 185
558	311,364	23.622 02	173,741,112	8.232 746	613	375,769	24.758 84	230,346,397	8.494 807
559	312,481	23.643 18	174,676,879	8.237 661	614	376,996	24.779 02	231,475,544	8.499 423
560	313,600	23.664 32	175,616,000	8.242 571	615	378,225	24.799 19	232,608,375	8.504 035
561	314,721	23.685 44	176,558,481	8.247 474	616	379,456	24.819 35	233,744,896	8.508 642
562	315,844	23.706 54	177,504,328	8.252 372	617	380,689	24.839 48	234,885,113	8.513 243
563	316,969	23.727 62	178,453,547	8.257 263	618	381,924	24.859 61	236,029,032	8.517 840
564	318,096	23.748 68	179,406,144	8.262 149	619	383,161	24.879 71	237,176,659	8.522 432
565	319,225	23.769 73	180,362,029	8.267 029	620	384,400	24.899 80	238,328,000	8.527 019
566	320,356	23.790 75	181,321,496	8.271 904	621	385,641	24.919 87	239,483,061	8.531 601
567	321,489	23.811 76	182,284,263	8.276 773	622	386,884	24.939 93	240,641,848	8.536 178
568	322,624	23.832 75	183,250,432	8.281 635	623	388,129	24.959 97	241,804,367	8.540 750
569	323,761	23.853 72	184,220,009	8.286 493	624	389,376	24.979 99	242,970,624	8.545 317
570	324,900	23.874 67	185,193,000	8.291 344	625	390,625	25.000 00	274,140,625	8.549 880
571	326,041	23.895 61	186,169,411	8.296 190	626	391,876	25.019 99	245,314,376	8.554 437
572	327,184	23.916 52	187,149,248	8.301 031	627	393,129	25.039 97	246,491,883	8.558 990
573	328,329	23.937 42	188,132,517	8.305 865	628	394,384	25.059 93	247,673,152	8.563 538
574	329,476	23.958 30	189,119,224	8.310 694	629	395,641	25.079 87	248,858,189	8.568 081
575	330,625	23.979 16	190,109,375	8.315 517	630	396,900	25.099 80	250,047,000	8.572 619
576	331,776	24.000 00	191,102,976	8.320 335	631	398,161	25.119 71	251,239,591	8.577 152
577	332,929	24.020 82	192,100,033	8.325 148	632	399,424	25.139 61	252,435,968	8.581 681
578	334,084	24.041 63	193,100,552	8.329 954	633	400,689	25.159 49	235,636,137	8.586 205
579	335,241	24.062 42	194,104,539	8.334 755	634	401,956	25.179 36	254,840,104	8.590 724
580	336,400	24.083 19	195,112,000	8.339 551	635	403,225	25.199 21	256,047,875	8.595 238
581	337,561	24.103 94	196,122,941	8.344 341	636	404,496	25.219 04	257,259,456	8.599 748
582	338,724	24.124 68	197,137,368	8.349 126	637	405,769	25.238 86	258,474,853	8.604 252
583	339,889	24.145 39	198,155,287	8.353 905	638	407,044	25.258 66	259,694,072	8.608 753
584	341,056	24.166 09	199,176,704	8.358 678	639	408,321	25.278 45	260,917,119	8.613 248
585	342,225	24.186 77	200,201,625	8.363 447	640	409,600	25.298 22	262,144,000	8.617 739

n	n^2	$\sqrt{n}$	n^3	$\sqrt[3]{n}$	n	n^2	$\sqrt{n}$	n^3	$\sqrt[3]{n}$
641	410,881	25.317 98	263,374,721	8.622 225	696	484,416	26.381 81	337,153,536	8.862 095
642	412,164	25.337 72	264,609,288	8.626 706	697	485,809	26.400 76	338,608,873	8.866 338
643	413,449	25.357 44	265,847,707	8.631 183	698	487,204	26.419 69	340,068,392	8.870 576
644	414,736	25.377 16	267,089,984	8.635 655	699	488,601	26.438 61	341,532,099	8.874 810
645	416,025	25.396 85	268,336,125	8.640 123	700	490,000	26.457 51	343,000,000	8.879 040
646	417,316	25.416 53	269,586,136	8.644 585	701	491,401	26.476 40	344,472,101	8.883 266
647	418,609	25.436 19	270,840,023	8.649 004	702	492,804	26.495 28	345,948,408	8.887 488
648	419,904	25.455 84	272,097,792	8.653 497	703	494,209	26.514 15	347,428,927	8.891 706
649	421,201	25.475 48	273,359,449	8.657 947	704	495,616	26.533 00	348,913,664	8.895 920
650	422,500	25.495 10	274,625,000	8.662 391	705	497,025	26.551 84	350,402,625	8.900 130
651	423,801	25.517 70	275,894,451	8.666 831	706	498,436	26.570 60	351,895,816	8.904 337
652	425,104	25.534 29	277,167,808	8.671 266	707	499,849	26.589 47	353,393,243	8.908 539
653	426,409	25.553 86	278,445,077	8.675 697	708	501,264	26.608 27	354,894,912	8.912 737
654	427,716	25.573 42	279,726,264	8.680 124	709	502,681	26.627 05	356,400,829	8.916 931
655	429,025	25.592 97	281,011,375	8.684 586	710	504,100	26.645 83	357,911,000	8.921 121
656	430,336	25.612 50	282,300,416	8.688 963	711	505,521	26.664 58	395,425,431	8.925 308
657	431,649	25.632 01	283,593,393	8.693 376	712	506,944	26.683 33	360,944,128	8.929 490
658	432,964	25.651 51	284,890,312	8.697 784	713	508,369	26.702 06	362,467,097	8.933 669
659	434,281	25.671 00	286,191,179	8.702 188	714	509,796	26.720 78	363,994,344	8.937 843
660	435,600	25.690 47	287,496,000	8.706 588	715	511,225	26.739 48	365,525,875	8.942 014
661	436,921	25.709 92	288,804,781	8.710 983	716	512,656	26.758 18	367,061,696	8.946 181
662	438,244	25.729 36	290,117,528	8.715 373	717	514,089	26.776 86	368,601,813	8.950 344
663	439,569	25.748 79	291,434,247	8.719 760	718	515,524	26.795 52	370,146,232	8.954 503
664	440,896	25.768 20	292,754,944	8.724 141	719	516,961	26.814 18	371,694,959	8.958 658
665	442,225	25.787 59	294,097,625	8.728 519	720	518,400	26.832 82	373,248,000	8.962 809
666	443,556	25.806 98	295,408,296	8.732 892	721	519,841	26.851 44	374,805,361	8.966 957
667	444,889	25.826 34	296,740,963	8.737 260	722	521,284	26.870 06	376,367,048	8.971 101
668	446,224	25.845 70	298,077,632	8.741 625	723	522,729	26.888 66	377,933,067	8.975 241
669	447,561	25.865 03	299,418,309	8.745 985	724	524,176	26.907 25	379,503,424	8.979 377
670	448,900	25.884 36	300,763,000	8.750 340	725	525,625	26.925 82	381,078,125	8.983 509
671	450,241	25.903 67	302,111,711	8.754 691	726	527,076	26.944 39	382,657,176	8.987 637
672	451,584	25.922 96	303,464,448	8.759 038	727	528,529	26.962 94	384,240,583	8.991 762
673	452,929	25.942 24	304,821,217	8.763 381	728	529,984	36.981 48	385,828,352	8.995 883
674	454,276	25.961 51	306,182,024	8.767 719	729	531,441	27.000 00	387,420,489	9.000 000
675	455,625	25.980 76	307,546,875	8.772 053	730	532,900	27.018 51	389,017,000	9.004 113
676	456,976	26.000 00	308,915,776	8.776 383	731	534,361	27.037 01	390,617,891	9.008 223
677	458,329	26.019 22	310,288,733	8.780 708	732	535,824	27.055 50	392,223,168	9.012 329
678	459,684	26.038 43	311,665,752	8.785 030	733	537,289	27.073 97	393,832,837	9.016 431
679	461,041	26.057 63	313,046,839	8.789 347	734	538,756	27.092 43	395,446,904	9.020 529
680	462,400	26.076 81	314,432,000	8.793 659	735	540,225	27.110 88	397,065,375	9.024 624
681	463,761	26.095 98	315,821,241	8.797 968	736	541,696	27.129 32	398,688,256	9.028 715
682	465,124	26.115 13	317,214,568	8.802 272	737	543,169	27.147 74	400,315,553	9.032 802
683	466,489	26.134 27	318,611,987	8.806 572	738	544,644	27.166 16	401,947,272	9.036 886
684	467,856	26.153 39	320,013,504	8.810 868	739	546,121	27.184 55	403,583,419	9.040 966
685	469,225	26.172 50	321,419,125	8.815 160	740	547,600	27.202 94	405,224,000	9.045 042
686	470,596	26.191 60	322,828,856	8.819 447	741	549,081	27.221 32	406,869,021	9.049 114
687	471,969	26.210 68	324,242,703	8.823 731	742	550,564	27.239 68	408,518,488	9.053 183
688	473,344	26.229 75	325,660,672	8.828 010	743	552,049	27.258 03	410,172,407	9.057 248
689	474,721	26.248 81	327,082,769	8.832 285	744	553,536	27.276 36	411,830,784	9.061 310
690	476,100	26.267 85	328,509,000	8.836 556	745	555,025	27.294 69	413,493,625	9.065 368
691	477,481	26.286 88	329,939,371	8.840 823	746	556,516	27.313 00	415,160,936	9.069 422
692	478,864	26.305 89	331,373,888	8.845 085	747	558,009	27.331 30	416,832,723	9.073 473
693	480,249	26.324 89	332,812,557	8.849 344	748	559,504	27.349 59	418,508,992	9.077 520
694	481,636	26.343 88	334,255,384	8.853 599	749	561,001	27.367 86	420,189,749	9.081 563
695	483,024	26.362 85	335,702,375	8.857 849	750	562,500	27.386 13	421,875,000	9.085 603

Reference Data

n	n^2	$\sqrt{n}$	n^3	$\sqrt[3]{n}$	n	n^2	$\sqrt{n}$	n^3	$\sqrt[3]{n}$
751	564,001	27.404 38	423,564,751	9.089 639	806	649,636	28.390 14	523,606,616	9.306 328
752	565,504	27.422 62	425,259,008	9.093 672	807	651,249	28.407 75	525,557,943	9.310 175
753	567,009	27.440 85	426,957,777	9.097 701	808	652,864	28.425 34	527,514,112	9.314 019
754	568,516	27.459 06	428,661,064	9.101 727	809	654,481	28.442 93	529,475,129	9.317 860
755	570,025	27.477 26	430,368,875	9.105 748	810	656,100	28.460 50	531,441,000	9.321 698
756	571,536	27.495 45	432,081,216	9.109 767	811	657,721	28.478 06	533,411,731	9.325 532
757	573,049	27.513 63	433,798,093	9.113 782	812	659,344	28.495 61	535,387,328	9.329 363
758	574,564	27.531 80	435,519,512	9.117 793	813	660,969	28.513 15	537,367,797	9.333 192
759	576,081	27.549 95	437,245,479	9.121 801	814	662,596	28.530 69	539,353,144	9.337 017
760	577,600	27.568 10	438,976,000	9.125 805	815	664,225	28.548 20	541,343,375	9.349 839
761	579,121	27.586 23	440,711,081	9.129 806	816	665,856	28.565 71	543,338,496	9.344 657
762	580,644	27.604 35	442,450,728	9.133 803	817	667,489	28.583 21	545,338,513	9.348 473
763	582,169	27.622 45	444,194,947	9.137 797	818	669,124	28.600 70	547,343,432	9.352 286
764	583,696	27.640 55	445,943,744	9.141 787	819	670,761	28.618 18	549,353,259	9.356 095
765	585,225	27.658 63	447,697,125	9.145 774	820	672,400	28.635 64	551,368,000	9.359 902
766	586,756	27.676 71	449,455,096	9.149 758	821	674,041	28.653 10	553,387,661	9.363 705
767	588,289	27.694 76	451,217,663	9.153 738	822	675,684	28.670 54	555,412,248	9.367 505
768	589,824	27.712 81	452,984,832	9.157 714	823	677,329	28.687 98	557,441,767	9.371 302
769	591,361	27.730 85	454,756,609	9.161 687	824	678,976	28.705 40	559,476,224	9.375 096
770	592,900	27.748 87	456,533,000	9.165 656	825	680,625	28.722 81	561,515,625	9.378 887
771	594,441	27.766 89	458,314,011	9.169 623	826	682,276	28.740 22	563,559,976	9.382 675
772	595,984	27.784 89	460,099,648	9.173 585	827	683,929	28.757 61	565,609,283	9.386 460
773	597,529	27.802 88	461,889,917	9.177 544	828	685,584	28.774 99	567,663,552	9.390 242
774	599,076	27.820 86	463,684,824	9.181 500	829	687,241	28.792 36	569,722,789	9.394 021
775	622,625	27.838 82	465,484,375	9.185 453	830	688,900	28.809 72	571,787,000	9.397 796
776	602,176	27.856 78	467,288,576	9.189 402	831	690,561	28.827 07	573,856,191	9.401 569
777	603,729	27.874 72	469,097,433	9.193 347	832	692,224	28.844 41	575,930,368	9.405 339
778	605,284	27.892 65	470,910,952	9.197 290	833	693,889	28.861 74	578,009,537	9.409 105
779	606,841	27.910 57	472,729,139	9.201 229	834	695,556	28.879 06	280,093,704	9.412 869
780	608,400	27.928 48	474,552,000	9.205 164	835	697,225	28.896 37	582,182,875	9.416 630
781	609,961	27.946 38	476,379,541	9.209 096	836	698,896	28.913 66	584,277,056	9.420 387
782	611,524	27.964 26	478,211,768	9.213 025	837	700,569	28.930 95	586,376,253	9.424 142
783	613,089	27.982 14	480,048,687	9.216 950	838	702,244	28.948 23	588,480,472	9.427 894
784	614,656	28.000 00	481,890,304	9.220,873	839	703,921	28.965 50	590,589,719	9.431 642
785	616,225	28.017 85	483,736,625	9.224 791	840	705,600	28.982 75	592,704,000	9.435 388
786	617,796	28.035 69	485,587,656	9.228 707	841	707,281	29.000 00	594,823,321	9.439 131
787	619,369	28.053 52	487,443,403	9.232 619	842	708,964	29.017 24	596,947,688	9.442 870
788	620,944	28.071 34	489,303,872	9.236 528	843	710,649	29.034 46	599,077,107	9.446 607
789	622,521	28.089 14	491,169,069	9.240 433	844	712,336	29.051 68	601,211,584	9.450 341
790	624,100	28.106 94	493,039,000	9.244 335	845	714,025	29.068 88	603,351,125	9.454 072
791	625,681	28.124 72	494,913,671	9.248 234	846	715,716	29.086 08	605,495,736	9.457 800
792	627,264	28.142 49	496,793,088	9.252 130	847	717,409	29.103 26	607,645,423	9.461 525
793	628,849	28.160 26	498,677,257	9.256 022	848	719,104	29.120 44	609,800,192	9.465 247
794	630,436	28.178 01	500,566,184	9.259 911	849	720,801	29.137 60	611,960,049	9.468 966
795	632,025	28.195 74	502,459,875	9.263 797	850	722,500	29.154 76	614,125,000	9.472 682
796	633,616	28.213 47	504,358,336	9.267 680	851	724,201	29.171 90	616,295,051	9.476 396
797	635,209	28.231 19	506,261,573	9.271 559	852	725,904	29.189 04	618,470,208	9.480 106
798	636,804	28.248 89	508,169,592	9.275 435	853	727,609	29.206 16	620,650,477	9.483 814
799	638,401	28.266 59	510,082,399	9.279 308	854	729,316	29.223 28	622,835,864	9.487 518
800	640,000	28.284 27	512,000,000	9.283 178	855	731,025	29.240 38	625,026,375	9.491 220
801	641,601	28.301 94	513,922,401	9.287 044	856	732,736	29.257 48	627,222,016	9.494 191
802	643,204	28.319 60	515,849,608	9.290 907	857	734,449	29.274 56	629,422,793	9.498 615
803	644,809	28.337 25	517,781,627	9.294 767	858	736,164	29.291 64	631,628,712	9.502 308
804	646,416	28.354 89	519,718,464	9.298 624	859	737,881	29.308 70	633,839,779	9.505 998
805	648,025	28.372 52	521,660,125	9.302 477	860	739,600	29.325 76	636,056,000	9.509 685

n	n^2	$\sqrt{n}$	n^3	$\sqrt[3]{n}$	n	n^2	$\sqrt{n}$	n^3	$\sqrt[3]{n}$
861	741,321	29.342 80	638,277,381	9.513 370	916	839,056	30.265 49	768,575,296	9.711 772
862	743,044	29.359 84	640,503,928	9.517 052	917	840,889	30.282 01	771,095,213	9.715 305
863	744,769	29.376 86	642,735,647	9.520 730	918	842,724	30.298 51	774,620,632	9.718 835
864	746,496	29.393 88	644,972,544	9.524 406	919	844,561	30.315 01	776,151,559	9.722 363
865	748,225	29.410 88	647,214,625	9.528 079	920	846,400	30.331 50	778,688,000	9.725 888
866	749,956	29.427 88	649,461,896	9.531 750	921	848,241	30.347 98	781,229,961	9.729 411
867	751,689	29.444 86	651,714,363	9.535 417	922	850,084	30.364 45	783,777,448	9.732 931
868	753,424	29.461 84	653,972,032	9.539 082	923	851,929	30.380 92	786,330,467	9.736 448
869	755,161	29.478 81	656,234,909	9.542 744	924	853,776	30.397 37	788,889,024	9.739 963
870	756,900	29.495 76	658,503,000	9.546 403	925	855,625	30.413 81	791,453,125	9.743 476
871	758,641	29.512 71	660,776,311	9.550 059	926	857,476	30.430 25	794,022,776	9.746 986
872	760,384	29.529 65	663,054,848	9.553 712	927	859,329	30.446 67	796,597,983	9.750 493
873	762,129	29.546 57	665,338,617	9.557 363	928	861,184	30.463 09	799,178,752	9.753 998
874	763,876	29.563 49	667,627,624	9.561 011	929	863,041	30.479 50	801,765,089	9.757 500
875	765,625	29.580 40	669,921,875	9.564 656	930	864,900	30.495 90	804,357,000	9.761 000
876	767,376	29.597 30	672,221,376	9.568 298	931	866,761	30.512 29	806,954,491	9.764 497
877	769,129	29.614 19	674,526,133	9.571 938	932	868,624	30.528 68	809,557,568	9.767 992
878	770,884	29.631 06	676,836,152	9.575 574	933	870,489	30.545 05	812,166,237	9.771 485
879	772,641	29.647 93	679,151,439	9.579 208	934	872,356	30.561 41	814,780,504	9.774 974
880	774,400	29.664 79	681,472,000	9.582 840	935	874,225	30.577 77	817,400,375	9.778 462
881	776,161	29.681 64	683,797,841	9.586 468	936	876,096	30.594 12	820,025,856	9.781 946
882	777,924	29.698 48	686,128,968	9.590 094	937	877,969	30.610 46	822,656,953	9.785 429
883	779,689	29.715 32	688,465,387	9.593 717	938	879,844	30.626 79	825,293,672	9.788 909
884	781,456	29.732 14	690,807,104	9.597 337	939	881,721	30.643 11	827,936,019	9.792 386
885	783,225	29.748 95	693,154,125	9.600 955	940	883,600	30.659 42	830,584,000	9.795 861
886	784,996	29.765 75	695,506,456	9.604 570	941	885,481	30.675 72	833,237,621	9.799 334
887	786,769	29.782 55	697,864,103	9.608 182	942	887,364	30.692 02	835,896,888	9.802 804
888	788,544	29.799 33	700,227,072	9.611 791	943	889,249	30.708 31	838,561,807	9.806 271
889	790,321	29.816 10	702,595,369	9.615 398	944	891,136	30.724 58	841,232,384	9.809 736
890	792,100	29.832 87	704,969,000	9.619 002	945	893,025	30.740 85	843,908,625	9.813 199
891	793,881	29.849 62	707,347,971	9.622 603	946	894,916	30.757 11	846,590,536	9.816 659
892	795,644	29.866 37	709,732,288	9.626 202	947	896,809	30.773 37	849,278,123	9.820 117
893	797,449	29.883 11	712,121,957	9.629 797	948	898,704	30.789 61	851,971,392	9.823 572
894	799,236	29.899 83	714,516,984	9.633 391	949	900,601	30.805 84	854,670,349	9.827 025
895	801,025	29.916 55	716,917,375	9.636 981	950	902,500	30.822 07	857,375,000	9.830 476
896	802,816	29.933 26	719,323,136	9.640 569	951	904,401	30.838 29	860,085,351	9.833 924
897	804,609	29.949 96	721,734,273	9.644 154	952	906,304	30.854 50	862,801,408	9.837 369
898	806,404	29.966 65	724,150,792	9.647 737	953	908,209	30.870 70	865,523,177	9.840 813
899	808,201	29.983 33	726,572,699	9.651 317	954	910,116	30.886 89	868,250,664	9.844 254
900	810,000	30.000 00	729,000,000	9.654 894	955	912,025	30.903 07	870,983,875	9.847 692
901	811,801	30.016 66	731,432,701	9.658 468	956	913,936	30.919 25	873,722,816	9.851 128
902	813,604	30.033 31	733,870,808	9.662 040	957	915,849	30.935 42	876,467,492	9.854 562
903	815,409	30.049 96	736,314,327	9.665 610	958	917,764	30.951 58	879,217,912	9.857 993
904	817,216	30.066 59	738,763,264	9.669 176	959	919,681	30.967 73	881,974,079	9.861 422
905	819,025	30.083 22	741,217,625	9.672 740	960	921,600	30.983 87	884,736,000	9.864 848
906	820,836	30.099 83	743,677,416	9.676 302	961	923,521	31.000 00	887,503,681	9.868 272
907	822,649	30.116 44	746,142,643	9.679 860	962	925,444	31.016 12	890,277,128	9.871 694
908	824,464	30.133 04	748,613,312	9.683 417	963	927,369	31.032 24	893,056,347	9.875 113
909	826,281	30.149 63	751,089,429	9.686 970	964	929,296	31.048 35	895,841,344	9.878 530
910	828,100	30.166 21	753,571,000	9.690 521	965	931,225	31.064 45	898,632,125	9.881 945
911	829,921	30.182 78	756,058,031	9.694 069	966	933,156	31.080 54	901,428,696	9.885 357
912	831,744	30.199 34	758,550,528	9.697 615	967	935,089	31.096 62	904,231,063	9.888 767
913	833,569	30.215 89	761,048,497	9.701 158	968	937,024	31.112 70	907,039,232	9.892 175
914	835,396	30.232 43	763,551,944	9.704 699	969	938,961	31.128 76	909,853,209	9.895 580
915	837,225	30.248 97	766,060,875	9.708 237	970	940,900	31.144 82	912,673,000	9.898 983

Reference Data

n	n^2	$\sqrt{n}$	n^3	$\sqrt[3]{n}$	n	n^2	$\sqrt{n}$	n^3	$\sqrt[3]{n}$
971	942,841	31.160 87	915,498,611	9.902 384	986	972,196	31.400 64	958,585,256	9.953 114
972	944,784	31.176 91	918,330,048	9.905 782	987	974,169	31.416 56	961,504,803	9.956 478
973	946,729	31.192 95	921,167,317	9.909 178	988	976,144	31.432 47	964,430,272	9.959 839
974	948,676	31.208 97	924,010,424	9.912 571	989	978,121	31.448 37	967,361,669	9.963 198
975	950,625	31.224 99	926,859,375	9.915 962	990	980,100	31.464 27	970,299,000	9.966 555
976	952,576	31.241 00	929,714,176	9.919 351	991	982,081	31.480 15	973,242,271	9.969 910
977	954,529	31.257 00	932,574,833	9.922 738	992	984,064	31.496 03	976,191,488	9.973 262
978	956,484	31.272 99	935,441,352	9.926 122	993	986,049	31.511 90	979,146,657	9.976 612
979	958,441	31.288 98	938,313,739	9.929 504	994	988,036	31.527 77	982,107,784	9.979 960
980	960,400	31.304 95	941,192,000	9.932 884	995	990,025	31.543 62	985,074,875	9.983 305
981	962,361	31.320 92	944,076,141	9.936 261	996	992,016	31.559 47	988,047,936	9.986 649
982	964,324	31.336 88	946,966,168	9.939 636	997	994,009	31.575 31	991,026,973	9.989 990
983	966,289	31.352 83	949,862,087	9.943 009	998	996,004	31.591 14	994,011,992	9.993 329
984	968,256	31.368 77	952,763,904	9.946 380	999	998,001	31.606 96	997,002,999	9.996 666
985	970,225	31.384 71	955,671,625	9.949 748	1,000	1,000,000	31.622 78	1,000,000,000	10.000 000

INDEX

A

Abbreviations	891–893
Absorbed-dose calculations	
anatomical and metabolic data	197–199
body weights, organ weights, and calculated geometrical factors for various ages	198
gastrointestinal tract of standard man	198
intake and excretion of standard man	199
organs of standard man	197
concentration of activity as a function of time	194–197
biological, physical, and effective half-life	194
equations for cumulated activity and cumulated concentration	194–197
tissue distribution data	194
definition of units	167, 168
dose–reciprocity theorem	171
equations	169, 170
equilibrium absorbed-dose constant	171
beta-minus and beta-plus particles	172
characteristic X rays	178
electron-binding energy and characteristic X-ray energy	174–177
electron capture	172, 173
gamma rays	178
internal conversion	173, 178
K-shell fluorescent yield	174
maximum permissible dose	168
nonpenetrating radiation	179
penetrating photon radiations	179–194
absorbed fraction	179, 180, 181–189
energy-absorption build-up factor	180, 190, 191–193
geometrical factors	190, 193, 194
quantities, symbols, and notations	168, 169
Accelerators, *see* Charged-particle activations	
Accidents	
documentation and reporting	799–804
eyewitness accounts	802, 803
hazard evaluation and requests for assistance	800, 801
health-physics data, results of surveys, and medical data	801
initial notification	799, 800
personnel-monitoring data and bioassay results	801, 802
photographic reenactments	803
emergency procedures	804–808
control	806–809
notification requirements	805, 806
general planning and action	809–815
accidents involving radioactive dusts, mists, fumes, organic vapors and gases	814, 815
decontamination techniques	811, 812
disaster preparedness at hospitals	812, 813
fires or other major emergencies	811
incidents involving teletherapy equipment	814
orientation information for fire and police officials	809, 810
radioactive spills	810, 811
recommendations for local authorities	813, 814
X-ray accidents	814
preparedness, check list	823–825
shipping	770
Acute radiation syndrome	861–866
clinical management	876–882
diagnosis	875, 876
Aerospace, application of radionuclides	557–568
gaging	558
heat and power	559–568
ionizing radiation	558
radiography	558
tracing	557

Agriculture, application of radionuclides	550–556
gaging	553
ionizing radiation	553–555
radiography	553
tracing	550–553
Arc geometry	139
Assistance, radiological, USAEC	820, 821

B

Beta particles	
equations, counting	914
nomogram of absorption	901
Beta radiation, penetration ability	73
Biological half-life	
absorbed-dose calculations	194
elements, various	25–33
radioactive substances in the lung	650–653
Biosynthesis of labeled compounds	343
carbon-14 compounds	346–359
phosphorus-32 compounds	370
sulfur-35 compounds	365
tritium compounds	363
Body burdens	
maximum permissible, various radioisotopes	655–658
organs other than bone	612
skeletal	612
Bone scanning	468, 469
Bragg-Gray principle	577
Brain scanning	388–394
comparison of neurological diagnostic tests	393
instruments	390, 391
interpretation	391, 392
abnormal brain scan	392
normal brain scan	391, 392
mechanism	388
methodology	388, 390
Burial sites, low-level wastes	792

C

Calculations	
absorbed dose for biologically distributed radionuclides	167, 168
dose from administered radionuclides	201–216
dose from ingestion or inhalation of soluble radionuclides	220–225
dose to the lung, per μCi inhaled	649
exposures	647–663
beta surface dose in air	648
gamma dose	647
resolving time	84
statistical error	77–84
Calibration procedure, equations	915, 916
Cameras, radioisotope	148–163
collimators	149–150
detector efficiency and figure of merit	150, 151
digital autofluoroscope	154–156
extrinsic and intrinsic resolution	148, 149
image-intensifier camera (Magnacamera)	158–160
scintillation camera	156–158
spark-imaging (spark chamber) camera	152–154
Cardiac output	423–426
Cardiovascular disease, diagnosis	423–433
cardiac output	423–426
circulation time	429
congenital heart disease	429–431
left-to-right shunts	429, 430
right-to-left shunts	431
localization of placenta	431–432
radionuclide angiography	427–429
Cerebral blood-flow measurement	383–387
inhalation techniques	387
intracarotid injection techniques	383, 384, 385
diffusable tracer	385
nondiffusable tracer	384
intravenous techniques	383, 386

Index

Certification	
shipper's	762, 775, 777
standards of activity	232
Charged-particle activations	13
Chemical processing, application of radionuclides	519–530
gaging	523–526
ionizing radiation	527–530
radiography	526
tracing	519–523
Chemical synthesis of labeled compounds	343
carbon-14 compounds	347–351, 352–356
chlorine-36 compounds	375
phosphorus-32 compounds	370, 371, 372
radioiodine compounds	373
sulfur-35 compounds	365
tritium compounds	360–362
addition (hydrogenation)	360, 362
exchange catalyzed by radiation	360, 361, 362
exchange with heterogeneous catalysis	360
exchange with homogeneous catalysis	360, 361
substitution by chemical reduction	360, 362
Classification of radioactive materials	737–739
Collimators for radioisotope cameras	149, 150
Commercially available isotopes, nuclear data	34–63
radioactive isotopes	34–59
stable isotopes	60–63
Commercial waste-disposal services	791
Compounds, labeled	
isomerisms	341, 342
stability of organic compounds	274–284
storage and stability	285–338
synthesis	339–379
biosynthesis	343
chemical synthesis	343
radiation synthesis	343, 344
Concentrations, permissible, in air and water	611–614
Congenital heart disease	429–431
left-to-right shunts	429, 430
external pulmonary monitoring	430
inhalation technique	429
injection technique	430
right-to-left shunts	431
Consumer-products industries, application of radionuclides	531–539
gaging	532, 534, 535
ionizing radiation	535–529
tracing	531, 532
Contamination	
control, shipping	755, 756
radioactive, removable, definition	739
Conversion factors	904–908
referential	904
standard measurements	905–908
Cows, *see* Radioisotope generators	15
Crude-petroleum and natural-gas industries, application of radionuclides	540–543
gaging	543
heat and power	543
tracing	540–543
Cyclotron, *see* Charged-particle activations	

D

Dead time, radiation detectors	102, 103
Decay, radioactive	
curve	902
equations	911, 912
nomograms for correction	76
semi-log plots	74, 75
values	903
Decomposition	
chemical	281
radioactive	275
Decontamination	
factor	917
services	798
techniques	811, 812

Definitions
- acute radiation syndrome ... 861
- fissile materials ... 737, 738
- isotopes ... 3
- K-capture ... 5
- low-specific-activity materials ... 738, 739
- normal-form radioactive materials ... 739
- quantities and units, radiation measurement ... 573–576
- quenching ... 98
- radioactive material ... 739
- removable radioactive contamination ... 739
- special-form radioactive materials ... 739
- transport group ... 739
- transport index ... 739
- units of absorbed-dose calculations ... 167, 168

Densities, common metals ... 899
Depth dose percentages, various ... 582–597
Derivative dilution ... 239–242
Detectors, radiation
- comparisons ... 101
- dead time ... 102, 103
- gas-filled ... 89
- Geiger-Mueller counters ... 90
- liquid scintillation counters ... 98–101
- nonparalyzing systems ... 102
- paralyzing systems ... 102
- personnel-monitoring devices ... 130–132
  - photographic film (film badges) ... 130, 712–714
  - pocket ionization chambers ... 130, 711
  - radiophotoluminescent materials ... 130, 718
  - thermoluminescent materials ... 131, 714, 716
- portable, for personnel monitoring ... 132
- proportional counters ... 90
- scintillation ... 90
- semiconductor ... 101
- whole-body counter systems ... 134–140

Digital autofluoroscope ... 154–156
Dilution, isotopic
- analysis ... 237–245
  - derivative dilution ... 239–242
  - direct dilution ... 237–239
  - saturation analysis ... 242
- equations ... 917

Dose–effect relationships ... 581, 600–606
Dose measurement, physical ... 576–581
Dose-reciprocity theorem ... 171
Dosimetry
- absorbed dose for biologically distributed radionuclides ... 167–200
  - anatomical and metabolic data ... 197–199
  - concentration of activity as a function of time ... 194–197
  - definition of units ... 167, 168
  - dose–reciprocity theorem ... 171
  - equations ... 169, 170
  - equilibrium absorbed-dose constant ... 171
  - nonpenetrating radiation ... 179
  - penetrating photon radiations ... 179–194
  - quantities, symbols and notations ... 168, 169
- dose from administered radionuclides ... 201–219
  - calculation of dose ... 201–216
  - dose to the gastrointestinal tract ... 216, 217
- dose from ingestion or inhalation of soluble radionuclides ... 220–225
  - calculation of dose ... 220–225
  - calculation of iodine conversion factors applying to the thyroid ... 222
  - dose conversion factors for the thyroid
  - nomogram for infinity dose calculations ... 224
- personnel ... 711–718
  - specific dose rates, various, internal and external exposure ... 659

E

Electrical industry, application of radionuclides ... 515, 516
- gaging ... 515
- heat and power ... 516
- ionizing radiation ... 516
- radiography ... 515
- tracing ... 515

Index

Elements	
commercially available isotopes	34–63
general data	25–33
nuclear data	34–63
Emergencies	
acceptable doses from inhalation	637, 638
food and water levels	644
incident evaluation record	818, 819
maximum permissible levels for surface contamination	643
medical management of radiation injury	868–883
notification instructions	816–817
planning for	799–833
procedures	699, 700, 799–833
radiological assistance, USAEC	820, 821
recommendations for limiting exposures	634–644
Endocrine system, use of radionuclides	395–422
parathyroid localization	420
pituitary irradiation	412–415
radioimmunoassay of hormones	415–419
therapy with iodine-131	407–412
thyroid metabolism and function tests	398–406
Energy absorption versus photon energy	71
Equations	
absorbed-dose calculations	169–170
cumulated activity and cumulated concentration	194–197
beta-particle counting	914
calibration procedure	915, 916
classical physics	909, 910
counter geometry	918
decontamination factor	917
dose to the lung, per μCi inhaled	649
electrostatics	911
internal radiation dosage	916, 917
isotopic dilutions	917
logarithmic relations	909
maximum permissible concentrations, normal occupational exposure	612–614
neutron-activation methods	918
radiation absorption	912–914
radioactive decay	911, 912
specific activity	912
wave and quantum relations	910, 911
Equilibrium absorbed-dose constant	171
Estimation of radiation exposures	647–663
beta surface dose in air	648
gamma dose	647
Estimation of risks to humans, low-level exposure	606, 607
Exponential functions	928–931
Exposures	
effects expected from inhalation	635, 636
emergencies, limiting of	634–644
estimation of, from internal and external sources	647–663
betasurface dose in air	648
gamma dose	647
limits following reactor incidents	641
recommended limits	609–644
normal occupational	612–614
regulatory limits	614–634
maximum permissible concentrations in air and water	616–633

F

Fat-absorption test, gastroenterologic	459, 460
Film badges	
for personnel monitoring	130, 712–714
processing services	797
Fissile materials	
definition	737, 738
packaging requirements	
Fission products	
Fixed radiographic equipment	695–697
design recommendations	695, 696
guidelines for use	696, 697
Fluoroscopic equipment	693–695
design recommendations	693, 694
guidelines for the fluoroscopist	694, 695
Fundamental constants	895, 896

G

Gaging, application in
    aerospace and other environmental uses   558
    agriculture   553
    chemical processing   523–526
        density gaging   524, 526
        level gaging   526
        thickness gaging   523, 524
    consumer-products industries   532, 534, 535
    crude-petroleum and natural-gas industries   543
    electrical industry   515
    metals industries   511–513
        density gaging   513
        level gaging   513
        thickness gaging   511
    mining   544, 545
    transportation-equipment industries   517
    utilities   548
Gamma-emitter dose, exposure rates   70, 71
Gamma factor versus photon energy   71
Gas-filled radiation detectors   89
Gastroenterologic disease, diagnosis   443–465
    in vitro tests   459–462
        clearance of plasma albumin and of circulating red blood cells   461, 462
        fat absorption   459, 460
        liver-function evaluation with radiopharmaceuticals   462
        vitamin B_{12} absorption   460, 461
    liver imaging   444–456
        imaging techniques   451
        interpretation   451–456
        liver anatomy by gamma imaging techniques   445–451
        liver labeling   444, 445
    pancreas imaging   456–458
    salivary-gland imaging   458, 459
Geiger-Mueller counters   90
Genitourinary-system studies   492–500
    kidney function   492–494
        radiochlormerodrin renal uptake   494
        radioiodinated orthoiodohippurate renogram   492, 493
    renal-clearance methods   498, 499
    renal scanning   494–498
        radiochlormerodrin renal imaging   494–496
        radiohippurate renal imaging   496–498
    residual-urine determination   498

H

Half-life
    biological, of radioactive substances in the lung   650–653
    biological, physical, and effective, in absorbed-dose calculations   194
    decay in relation to time   4
    definition   3
    typical decay curve   3
Half-thickness versus photon energy
    concrete   69
    lead   69
    water   69
Hazard evaluation, incident report   800, 801
Hazardous materials regulations amendments   731–780
Hazard summary report   664, 675, 676
Heat and power, application in
    aerospace and other environmental uses   559–568
        mechanical-energy devices   563, 567, 568
        thermal applications   559, 562
        thermoelectric devices   563
    crude-petroleum and natural-gas industries   543
    electrical industry
Hematopoietic-system studies   477–491
    deficiency and malabsorption of vitamin B_{12} 488, 489
    hemolytic disease and hypersplenism   489–491
    iron metabolism and red-cell formation   481, 486–488
    volume measurement, total blood, plasma, and red-cell mass   477–481, 482–485
Hemolytic disease and hypersplenism   489–491

Index

I

Image-intensifier camera	158–160
Incident evaluation	826, 827
Industrial applications	501–568
aerospace and other environmental uses	557–568
agriculture	559–556
chemical processing	519–530
consumer-products industries	531–539
crude-petroleum and natural-gas industries	540–543
electrical industry	515, 516
metals industries	509–514
mining	544–546
transportation-equipment industries	517, 518
utilities	547–549
Information storage and retrieval in radioisotope imaging systems	141–147
data-handling operations	141
practical considerations	144
quantitation and image manipulation	145
theoretic considerations	142
total-information storage	141
Injury, radiation	845–867
cell effects	845, 846
medical management of	868–883
clinical management	876–883
diagnostic procedures	875, 876
immediate emergency procedures	868
patient care	868–874
organ and system effects	847–861
blood	849–854
cancer induction	860, 861
eye	855, 856
gonads	854, 855
genetic changes	861
lethality	857, 858
life expectancy	857, 858, 859
skin	847–849
tissue effects	846, 847
whole-body effects	861–866
acute radiation syndrome	862–866
dose–effect relationships	861, 862
internal contamination	866
Instrument calibration	233, 234
Ionization chambers	
personnel monitoring	130, 711
radiation detection	112, 113
Ionizing radiation, application in	
aerospace and other environmental uses	558
agriculture	553–555
plant mutations	554, 555
sex sterilization of insects	554, 555
sterilization of goat hair	554
chemical processing	527–530
consumer-products industries	535–539
electrical industry	516
mining	546
transportation-equipment industries	518
utilities	548
Iron metabolism and red-cell formation	481, 486–488
Isotopes	
anisotopic (stable) elements	3
availability	505, 506
half-life	4
decay in relation to time	4
definition	3
typical decay curve	3
physical data	15
reactor production	7
Isotopes, radioactive	
definition	3
dilution analysis	237–245
nuclear data	34–59
Isotopes, stable	
enriched targets, various	11
calcium	12
iron	12
nuclear data	60–63

K

K-capture, definition	5
K-shell fluorescent yield	
Kidney-function studies	492–494
radiochlormerodrin renal uptake	494
radioiodinated orthoiodohippurate renogram	492, 492

L

Labeling problems	344, 345
Labeling processes, primary	360–362
Labels, radioactive materials	
descriptions	761, 762
White-I	759, 761
Yellow-II	759, 761
Yellow-III	759, 761, 762
Laundry services, contaminated clothing	797
Limits of exposure	
reactor incidents	641
regulatory	614–634
Line spread function, determination of MTF(ν)	124, 125
Liquid scintillation counters	98–101
channels-ratio method	99
evaluation of solutes	114–122
external-standardization method	99
improved solubilization procedures for counting of biological materials	246–255
internal-standardization method	98
primary-solute concentration	118–120
primary-solute efficiencies	120, 121
sample preparation	256–273
counting dry samples	264, 265
counting emulsions	265, 270, 271
counting flowing aqueous solutions	266, 267
counting gases and vapors	265, 266
counting solutions	258–264, 268
counting suspensions	263, 264, 268–270
determination of counting efficiency	267
solvents and solutes	258
secondary solutes	115–118
Liver-function evaluation	462
Liver imaging	444–456
imaging techniques	451
interpretation	451–456
liver anatomy by gamma imaging techniques	445–451
liver labeling	444, 445
Logarithms, common, four-place mantissas	919–922
of decimal fractions	919, 920
Low-specific-activity materials	
definition	738, 739
packaging requirements	748, 749
Lung-scan characteristics, differential	439–441
diseases of the pleura	441
obstructive pulmonary disease	439, 440
pulmonary embolism	440
pulmonary hypertension	441

M

Magnacamera, *see* Image-intensifier camera	
Mass absorption coefficients versus photon energy	67, 68
air	67
aluminum	68
lead	68
sodium iodide	68
water	67
Mathematical signs and symbols	890
Mathematical tables	
exponential functions	928–931
logarithms, common, four-place mantissas	919–922
of decimal fractions	919, 920
natural trigonometric functions	923–927
squares, square roots, cubes, and cube roots	923–941
Maximum permissible concentrations	
emergencies	637, 638, 643, 644

Index

food and water	644
inhalation	637, 638
surface contamination	643
in air and water	616–633
normal occupational exposure	612–614
body burden to critical organ	612
concentration based on body burden	613
concentration based on gastrointestinal tract	613
concentration of insoluble compounds in lung	614
concentration of noble gases	613, 614
skeletal body burden	612
values for various radioisotopes	655–658
Medical applications	381–500
cardiovascular	423–433
cardiac output	423–426
circulation time	429
congenital heart disease	429–431
placenta localization	431, 432
radionuclide angiography	427–429
cerebronervous	383–394
brain scanning	388–394
measurement of cerebral blood flow	383–387
endocrine	395–422
parathyroid localization	420
pituitary irradiation	412–415
radioimmunoassay of hormones	415–419
therapy with iodine-131	407–412
thyroid metabolism and function tests	398–406
gastroenterologic	443–465
in-vitro tests	459–462
liver imaging	444–456
pancreas imaging	456–458
salivary-gland imaging	458, 459
genitourinary	492–500
kidney-function studies	492–494
renal-clearance methods	498, 499
renal scanning	494–498
residual-urine determination	498
hematopoietic	477–491
deficiency and malabsorption of vintamin B_{12}	488, 489
hemolytic disease and hypersplenism	489–491
iron metabolism and red-cell formation	481, 486–488
volume of plasma, red-cell mass, and total blood	477–481
osseous and cartilaginous	466–476
circulatory studies	474–475
dilution studies	469, 470
kinetic studies	470–474
quantitative-distribution studies	468, 469
tracer-distribution studies	467, 468
respiratory	434–442
artifacts and frequent normal variants	438
differential lung-scan characteristics	439–441
external detection	437–438
pulmonary arterial blood flow, normal	438
radioactive gases	434, 435
radioactive particles	435, 436
Medical management of radiation emergencies	868–883
clinical management	876–883
acute radiation syndrome	876–882
internal contamination without evident injury	883
local radiation injury	882
local traumatic injury with radionuclide contamination	882
diagnostic procedures	875, 876
acute radiation syndrome	875, 876
internal contamination without evident injury	876
local radiation injury	876
local traumatic injury with radionuclide contamination	876
immediate emergency procedures	868
patient care	868–874
bioassay collection methods	873, 874
care of contaminated wounds	873
emergency care for possibly contaminated persons	868–870
initial care	868
skin-decontamination procedures	870–672
Medical practice, diagnosis	
cardiovascular disease	423–433
gastroenterologic	443–465

Medical practice, protective measures	692–709
emergency procedures	699, 700
for the radiation protection supervisor	700
for the technician	699, 700
fixed radiographic equipment	695–697
design recommendations	695, 696
guidelines for use	696, 697
fluoroscopic equipment	693–695
design recommendations	693, 694
guidelines for the fluoroscopist	694, 695
general guidelines in the clinical use of radiation	692, 693
general working conditions	698–700
medical examination	699
personnel monitoring	689, 699
responsibilities of owner	698
responsibilities of radiation protection supervisor	698
vacations	699
Metals industries, application of radionuclides	509–514
gaging	511–513
radiography	513, 514
tracing	509, 510
Mining, application of radionuclides	544–546
gaging	544, 545
ionizing radiation	546
tracing	544
Modulation transfer function, MTF(ν)	123–129
applications	123
components	126
determination from the line spread function	124, 125
Multidetector array	139

N

Neutron-activation methods, equations	918
Neutron reactions	
common types	9
energetics	11
secondary capture	11
Nonpenetrating radiation, absorbed-dose calculations	179
Normal-form radioactive materials	
definition	739
packaging requirements	751, 752

P

Packaging requirements	
contamination control	755
definitions	739
empty containers	737
fissile materials	752–755
general	749, 750
labeling	759
low-specific-activity material	748, 749
normal-form materials	751, 752
previously authorized packaging	734, 735
quantity limitations	728
reuse of containers	736, 737
shipping container specifications	729, 770–774
small quantities	746–748
special-form materials	750, 751
special tests	757, 758
standard specifications	735, 736
Pancreas imaging	456–458
Parathyroid localization	420
Particle-distribution method	435, 436
use of indium-113m-labeled particles	436
validity of the method	436
Penetrating photon radiations, absorbed-dose calculations	179–194
Personnel monitoring	
activity levels above which required	671–674
comparison of various techniques	131
data, radiation incident report	801, 802
evaluation of results	822
methods	130
occupational exposure	698, 699

Index

radiation detectors	130, 132, 717, 718
regulations, AEC or state	130
Photographic film	
damage in transit	719
film badges for personnel monitoring	130, 712–714
minimum separation distance from radioactive material in transit	765, 767, 769
Photomultipliers, characteristics	93–97
Pituitary irradiation	412–415
Placenta localization	431, 432
Plasma albumin clearance, gastrointestinal	461, 462
Primary labeling processes	360
addition (hydrogenation)	360, 362
exchange catalyzed by radiation	360, 361, 362
exchange with heterogeneous catalysis	360
exchange with homogeneous catalysis	360, 361
substitution by chemical reaction	360, 362
Proportional counters	90
Protection guides and regulatory limits of exposure	609–646
determination of permissible concentrations in air and water	611–614
maximum permissible concentration, occupational exposure	612–614
guides for limiting organ and whole-body doses	609–611
recommendations for limiting exposures in emergencies	633–644
Protective measures	
facilities and equipment	667–670
air samplers	667–670
clothing	691
exhaust-stack monitors	667–670
filters	667–670, 687, 688
glove boxes	667–670, 677, 679
radiosotope hoods	667–670, 677, 678
medical practice	692–709
fixed radiographic equipment	695–697
fluoroscopic equipment	693–695
general guidelines in the clinical use of radiation	692, 693
general working conditions	698, 699
personnel dosimetry	711–718
photographic film	712–714
pocket ionization chambers	711
radiophotoluminescent materials	718
thermoluminescent materials	714, 716
procedures	671–674
environmental monitoring	671–674
measurement of high external doses	671–674
personnel monitoring	671–674
preoperational analysis of maximum credible accidents	671–674
preplanned emergency procedures and drills	671–674
radioassays	671–674
shielding	678, 680–687
standard precautions	691, 692
Pulmonary arterial blood flow, normal	438

Q

Quantities and units, radiation measurement	573–575
Quantities, symbols, and notations for absorbed-dose calculations	168, 169
Quenching, definition	98

R

Radiation absorption, equations	912–914
Radiation dosimetry	165–225
internal dosage, equations	916, 917
Radiation effects	
biological	845–867
cells	845, 846
organic and systemic	847–861
tissue	846, 847
whole body	861–866
characteristic	503, 504
emergency exposures	635, 636
Radiation measurement, basic units	573–608
Radiation protection and regulations	571–833
Radiation-protection program	
administration of	795–798
Radiation Safety Officer (RSO)	795, 796

records 798
services 796–798
    decontamination 798
    film-badge processing 797
    laundry 797
    whole-body counting 796

Radiation-protection supervisor, see Radiation Safety Officer

Radiations
  alpha particles 5
  beta particles 5
  electromagnetic 5
  electrons 5
  gamma rays 5
  nonpenetrating 179
    absorbed-dose calculations 179
  photons, penetrating 179–194
    absorbed-dose calculations 179–194
  types, various 5
  X rays 5

Radiation Safety Officer
  emergency procedures 700
    in medical practice 700
  responsibilities 698, 795, 805

Radiation synthesis 343, 344
  carbon-14 compounds 357
  phosphorus-32 compounds 370
  radioiodine compounds 370
  sulfur-35 compounds 365

Radioactive-isotope dilution analysis 237–245
  developments 239–242
    derivative dilution 239–242
    saturation analysis 242
  laboratory and material requirements 237
  principle 237–239

Radiography, application in
  aerospace and other environmental uses 558
  agriculture 535
  chemical processing 526
  electrical industry 515
  metals industries 513, 514
  transportation-equipment industries 517, 518
  utilities 548

Radioimmunoassay of hormones 415–419

Radioisotope generators 15

Radioisotope imaging systems
  cameras 148–163
    digital autofluoroscope 154–156
    image-intensifier camera (Magnacamera) 158–160
    scintillation camera 156–158
    spark-imaging (spark chamber) camera 152–154
  information storage and retrieval 141–147
    data-handling operations 141
    practical considerations 144
    quantitation and image manipulation 145
    theoretic considerations 142
    total-information storage 141
  modulation transfer function 123–129

Radioisotopes
  definition, see Isotopes, radioactive
  maximum permissible body burdens, various 655–658
  maximum permissible concentrations, various 655–658
  production and processing 7–15
    charged-particle activations 13
    fission products 15
    radioisotope generators (cows) 15
    reactor production 7
  reasonable expectancy of routine production 17–23
  utilization 507, 508
    gaging 508
    heat and power 508
    ionizing radiation 508
    radiography 508
    tracing 508

Radionuclides
  administered, calculation of dose 201–216
  applications 381–568
    industrial 501–568

Index

medical	381–500
definition, *see* Isotopes, radioactive	
ingested or inhaled, calculation of dose	220–225
long-lived, production	12
relative toxicity	665, 666
short-lived, production	12
Radiophotoluminescent materials	130, 718
Radiotoxicity	
levels above which safeguards are required	665–674
relative	665, 666
Range–energy relation for electrons	72
Reactor production	7
neutron reactions	8
product yield and characteristics	11
Records	
incident evaluation	818, 819
radiation incidents	799–804
radiation-protection program	798
regulations, federal	831
Reference data	885–941
abbreviations	891–893
commonly used units	894
conversion factors	904–908
referential	904
standard measurements	905–908
decay curve for radioactive materials	902
densities of common metals	899
equations	909–918
beta-particle counting	914
calibration procedure	915, 916
classical physics	909, 910
counter geometry	918
decontamination factor	917
electrostatics	911
internal radiation dosage	916, 917
isotopic dilutions	917
logarithmic relations	909
neutron-activation methods	918
radiation absorption	912–914
radioactive decay	911, 912
specific activity	912
wave and quantum relations	910, 911
exponential functions	928–931
fundamental constants	895, 896
Greek alphabet	887
logarithms, common, four-place mantissas	919–922
natural trigonometric functions	923–927
nomogram of absorption of beta particles	901
radioactive decay	903
relation between thickness of ordinary concrete and of lead for radium and ^{60}Co gamma rays	900
signs and symbols	888–980
mathematics	890
radiation and radioactivity	888, 889
squares, square roots, cubes, and cube roots	932–941
standard man	897, 898
Regulations	
exposure limits	614–634
incident-reporting	831–833
personnel-monitoring, AEC or state	130
transportation	731–780
Renal-clearance methods	498, 499
Renal scanning	494–498
radiochlormerodrin renal imaging	494–496
radiohippurate renal imaging	496–498
Reports	
incident	799–804, 828, 829
bioassay results	801, 802
eyewitness accounts	802, 803
hazard evaluation	800, 801
health-physics data	801
medical	801
radiation surveys	801
regulations, federal	831–833
Resolving-time calculations	84
Respiratory-system studies	434–442
artifacts and frequent normal variants	438, 439

differential lung-scan characteristics	439–441
diseases of the pleura	441
obstructive pulmonary disease	439, 440
pulmonary embolism	440
pulmonary hypertension	441
external detection	437, 438
pulmonary arterial blood flow, normal	438
radioactive gases	434, 435
radioactive particles	435, 436
use of indium-113m-labeled particles	436
validity of the particle-distribution method	436

S

Safety facilities and equipment	667–670
air samplers	667–670
exhaust-stack monitors	667–670
filters	667–670, 687, 688
glove boxes	667–670, 677, 679
protective clothing	691
radioisotope hoods	667–670, 677, 678
Safety procedures	671–674
environmental monitoring	671–674
measurement of high external doses	671–674
personnel monitoring	671–674
preoperational analysis of maximum credible accidents	671–674
preplanned emergency procedures and drills	671–674
radioassays	671–674
shielding	678, 680–687
standard precautions	691, 692
Salivary-gland imaging	458, 459
Sample counting and calibration	106–113
ionization chambers	112, 113
sodium iodide well counters	106–112
Sample preparation, liquid scintillation counting	256–273
counting dry samples	264, 265
counting emulsions	265, 270, 271
counting flowing aqueous solutions	266, 267
counting gases and vapors	265, 266
counting solutions	258–263, 268
biological materials	259, 261, 262
inorganic salts	259, 260
solution counting with toluene/Triton X-100 scintillant	262, 263
water and aqueous solutions	258, 259
counting suspensions	263, 264, 268–270
determination of counting efficiency	267
solvents and solutes	258
Saturation analysis	242
Scan geometry	138
Scintillation camera	156–158
Scintillation detectors	90
Scintillator geometry, large-volume	138
Semiconductor radiation detectors	101
Shielding	678, 680–687
Shipping regulations, see Transportation of radioactive materials	
Signs and symbols	
mathematics	888–889
	890
radiation and radioactivity	888, 889
Skeletal-system studies	466–476
circulatory	474, 475
dilution	469, 470
kinetic	470–474
quantitative distribution	468, 469
radioisotope scanning of bone	468, 469
whole-body counting	469
tracer distribution	467, 468
Sodium iodide well counters	106–112
components	106
count rate	107, 110, 111
pulse-height spectra	107
volume effects	107
Solubilization procedures, improved, for liquid scintillation counting of biological materials	246–255
calculated specific activities in various digests	252
materials and instrumentation	247
method	247, 248

Index

optimal sample composition for counting ^{3}H in dry materials	251
optimal sample composition for counting ^{3}H and ^{14}C in wet tissue	253
Spark-imaging (spark chamber) camera	152–154
Spatial resolution, measure of	123
Special-form radioactive materials	
definition	739
packaging requirements	750, 751
Specific activity, equations	912
Stability of labeled organic compounds	274–284
decomposition during experiments	283
definition of terms	274, 275
precautions	282, 283
radiation decomposition	275
special cases	282
underlying factors	275–282
decomposition independent of radiation	281
primary (external) effect	276
primary (internal) effect	275
secondary effects	276–281
Standard man	
body weight, organ weight and calculated geometrical factors	198
characteristics	653
gastrointestinal tract	198
ingestion and inhalation, daily rates	897, 898
intake and excretion	199, 838
organs	197, 654, 838, 897
particulate retention, respiratory tract	654
physiological functions	837–844
fluid balance	838
gastrointestinal tract	844
respiration	837, 839–844
Standards of activity	229–236
accuracy	229, 230
definition	229
statement of	229, 230
available forms	230–232
bulk materials	232
mock standards	231, 232
solid sources	231
solutions	230, 231
certification	232, 233
uses	233–235
calibration of laboratory instruments	233, 234
precautions in the use of reference solutions	234, 235
precautions in the use of standardized solutions	234
Statistical-error calculations	77–84
Storage of labeled compounds	
analysis of labeled compounds	288, 289
control of self-irradiation decomposition	311, 312
decomposition rates, various	324–338
carbon-14 compounds	325–335
phosphorus-32 compounds	338
selenium-75 compounds	335
sulfur-35 compounds	335, 336, 337
tritium compounds	324, 325
effect of impurities on tracer use	323, 324
experimental observations	289–311
carbon-14 compounds	293, 295–305
chlorine-36 compounds	289
cobalt-57 and cobalt-58 compounds	310, 311
iodine-125 and iodine-131 compounds	309, 310
phosphorus-32 compounds	393, 294
selenium-75 compounds	306
sulfur-35 compounds	289, 291, 292
tritium compounds	306, 307, 308
percentage decomposition in relation to the G(–M) value	287, 288
reasons for and modes of decomposition	286, 287
self-radiolysis of compounds in aqueous solution	316–320
self-radiolysis of compounds in nonaqueous solvents	320–323
Subject–detector geometry	134–139
arc geometry	139
large-volume liquid or plastic scintillator geometry	138
multidetector array	139
scan geometry	138
tilting-chair geometry	137
Synthesis of labeled compounds	339–379
chemical and biological methods	343

isomerism 341, 342
isotopic and nonisotopic labeling 340, 341
labeling problems 344, 345
radiation synthesis 343, 344
"shot-gun" methods 344
synthesis of carbon-14 compounds 346–359
    biosynthesis 351, 357
    chemical synthesis 347–351, 352–356
    radiation synthesis 357
    requirements and basic factors 346
    specific activities 346, 347
    syntheses via acetic acid 350
    syntheses via acetylene 354
    syntheses via carboxyl-labeled acids 348, 349
    syntheses via cyanamide 356
    syntheses via cyanide 353
    syntheses via formic acid 355
    syntheses via methanol 352
synthesis of chlorine-36 compounds 374, 375
    basic routes 375
synthesis of phosphorus-32 compounds 369, 370
    approach to synthesis 370
    direct irradiation 370
    pattern of demand 369
    routes to intermediates 371
    syntheses from phosphorus trichloride 372
synthesis of radioiodine compounds 370, 373, 374
    approach to synthesis 373
    choice of isotopes 374
    pattern of demand 370, 373
    stability 373, 374
synthesis of sulfur-35 compounds 364–369
    approach to synthesis 365–368
    direct neutron irradiation 365
    pattern of demand 364, 365
    routes from benzyl mercaptan to DL-methionine 368
    routes to intermediates 366
    specific activity 369
    stability 369
    syntheses via sulfur dioxide 361
synthesis of tritium compounds 359–364
    addition (hydrogenation) 360, 362
    biosynthesis 363
    essential features 359, 360
    exchange catalyzed by radiation 360, 361, 362
    exchange with heterogeneous catalysis 360
    exchange with homogeneous catalysis 360, 361
    specific activities 362, 363
    stability 363, 364
    substitution by chemical reaction 360, 362

T

Targets, enriched
    calcium 12
    iron 12
    various 11
Thermoluminescent materials 130, 714, 716
Thyroid metabolism and function tests 398–406
Tilting-chair geometry 137
Total-information storage 141
Tracing, application in
    aerospace and other environmental uses 557
    agriculture 550–553
        dairy industry 552, 553
        fertilizers 550
        insects and pesticides 552
        metabolism and photosynthesis 550
        plant diseases 550, 552
        weeds 552
    chemical processing 519–523
    consumer-products industries 531, 532
    crude-petroleum and natural-gas industries 540–543
    electrical industry 515
    metals industries 509, 510
    mining 544

Index

transportation-equipment industries	517
utilities	547, 548
Transportation-equipment industries, application of radionuclides	517, 518
gaging	517
ionizing radiation	518
radiography	517, 518
tracing	517
Transportation of radioactive nuclides	719–780
accidents	770
carriers	730, 731, 763–770
aircraft	775, 776
public highway	767–770
rail express	766, 767
rail freight	763–766
consignee	731
contamination control	755, 756, 769
fissile materials	752–755
general considerations	721
labels	759, 761, 764, 768
normal-form materials	751, 752
packaging requirements, general	729, 749, 750
packaging standards	757, 758
placards	763, 764, 768, 777–779
rules and regulations, hazardous materials	731–780
carriers	763–770
shippers	733–762
shipper's responsibilities	721, 722
shipping container specifications	770–774
shipping papers	762, 763, 766, 768
special-form materials	750, 751
transport groups	739, 740–746
definition	739
list of radionuclides	740–746
transport index, definition	739
U.S. postal regulations	719, 720, 731
Transport groups	
assignment of unlisted radionuclides	746
definition	739
list of radionuclides	740–745
Transport index, definition	739
Trigonometric functions, natural	923–927

U

Uniform isotropic model	171
Urine, residual, determination	498
U.S. Atomic Energy Commission	
regional compliance offices	830
regional office areas of responsibility for radiological assistance	820, 821
Utilities, application of radionuclides	547–549
gaging	548
ionizing radiation	548
radiography	548
tracing	547, 548

V

Vitamin B_{12}	
absorption test, gastrointestinal	460, 461
deficiency and malabsorption	488, 489
Volume measurement, total blood, plasma, and red-cell mass	477–481, 482–455
clinical interpretation	481

W

Waste disposal	781–794
animal carcasses	783–791
liquids	792–794
methods	783, 790, 792–794
air dispersal	794
burial	790, 792, 793
chemical treatment	793
commercial services	790, 791, 792, 793

evaporation	793
incineration	790, 791
pulverization	790, 791
solidification	792, 793
storage	783, 790, 791, 792
solids	791, 792
Wavelength shifters	114, 117
Whole-body counter systems	134–140
clinical diagnostic and research applications	134, 469
general characteristics	137
sensitivity requirements for individual units	134
subject-detector geometry	134–139
arc geometry	139
large-volume liquid or plastic scintillator geometry	138
multidetector array	139
scan geometry	138
tilting-chair geometry	137
Whole-body counting services	796
Work area, design of	690, 691

X

X-ray accidents, emergency action	814